Medicinal Chemistry

Medicinal Chemistry

Ashutosh Kar

ANSHAN LTD
6 Newlands Road
Tunbridge Wells
Kent.
TN4 9AT. UK

Co-published in the U. K. by

ANSHAN LTD, 6 Newlands Road, Tunbridge Wells, Kent TN4 9AT
In 2006

Tel/Fax: +44(0)1892 557767
e-mail: info@anshan.co.uk
Web Site: www.anshan.co.uk

ISBN 1 904798 764
ISBN 978 1904798 767

British Library Cataloguing in Publication Data
A Catalogue record for this book is available from the British Library

Not for sale in India, Pakistan, Sri Lanka, Bangladesh and Nepal.

Note - every effort has been made to ensure that the drug dosage schedules in this book are accurate and in accord with the standards accepted at the time of publication. However the reader is urged to consult drug manufacturer's printed instructions, particularly regarding the recommended dose, indications and contra-indications for administration and adverse reactions, before administering any of the drugs.

CONTENTS

CHAPTER 14: Antihistamines — 414

CHAPTER 15: Non-Steroidal Anti-Inflammatory Drugs (NSAIDs) — 450

CHAPTER 19: Antimalarials 532

CHAPTER 20: Anthelmintics 571

CHAPTER 25: Antipsychotics (Tranquilizers)   712

CHAPTER 26: Antiviral Drugs   729

1 Drug Design—A Rational Approach

1. INTRODUCTION

In the past few decades there has been a hiatus in the momentum of research and discovery of 'novel' medicinal compounds. This particular trend in drug development perhaps is augmented due to **two** vital factors, namely : *first*, strict empirical and rational approach to drug design ; and *secondly*, high standards of safety and therapeutic efficacy together with tremendous increased costs of research and development and finally the clinical trials.

'**Drug design**' or '*tailor-made compound*' aims at developing a drug with high degree of chemotherapeutic index and specific action. It is a logical effort to design a drug on as much a rational basis as possible thus reducing to the minimum the trial and error approach. It essentially involves the study of biodynamics of a drug besides the interaction between drug molecules and molecules composing the biological objects.

Drug design seeks to explain :

(a) Effects of biological compounds on the basis of molecular interaction in terms of molecular structures or precisely the physico-chemical properties of the molecules involved.

(b) Various processes by which the drugs usually produce their pharmacological effects.

(c) How the drugs specifically react with the protoplasm to elicit a particular pharmacological response.

(d) How the drugs usually get modified or detoxicated, metabolized or eliminated by the organism.

(e) Probable relationship between biological activity with chemical structure.

In short, drug design may be considered as an integrated whole approach which essentially involves various steps, namely : chemical synthesis, evaluation for activity-spectrum, toxicological studies, metabolism of the drug, *i.e.;* biotransformation and the study of the various metabolites formed, assay procedures, and lastly galenical formulation and biopharmaceutics.

The 'drug design' in a broader sense implies random evaluation of synthetic as well as natural products in bioassay systems, creation of newer drug molecules based on biologically-active-prototypes derived from either plant or animal kingdom, synthesis of congeners displaying interesting biological actions, the basic concept of isosterism and bioisosterism, and finally precise design of a drug to enable it to interact with a receptor site efficaciously.

In the recent past, another terminology 'prodrugs' has been introduced to make a clear distinction from the widely used term 'analogues'. Prodrugs are frequently used to improve pharmacological or biological properties. Analogues are primarily employed to increase potency and to achieve specificity of action.

2.   ANALOGUES AND PRODRUGS

In the course of drug design the *two* major types of chemical modifications are achieved through the formation of **analogues** and **prodrugs.**

An analogue is normally accepted as being that modification which brings about a carbon-skeletal transformation or substituent synthesis. *Examples : oxytetracycline, demclocycline, chlortetracycline, trans-diethylstilbesterol* with regard to oestradiol.

The term **prodrug** is applied to either an appropriate derivative of a drug that undergoes *in vivo* hydrolysis to the parent drug, *e.g.,* testosterone propionate, chloramphenicol palmitate and the like ; or an analogue which is metabolically transformed to a biologically active drug, for instance : phenylbutazone undergoes *in vivo* hydroxylation to oxyphenbutazone.

3.   CONCEPT OF 'LEAD'

Another school of thought views **'drug design'** as the vital process of envisioning and preparing specific new molecules that can lead more efficiently to useful drug discovery. This may be considered broadly in terms of two types of investigational activities. These include :

(*a*) *Exploration of Leads*, which involves the search for a new lead ; and

(*b*) *Exploitation of Leads,* that requires the assessment, improvement and extension of the lead.

From the practical view-point it is the latter area wherein rational approaches to drug design have been mostly productive with fruitful results.

3.1   Examples

It is worthwhile to look into the right perspective of a few typical and classical examples of drug design as detailed below :

(i) Narcotic Analgesics

In the year 1939, Schaumann first identified and recognized the presence of a quaternary-carbon-atom in the morphine molecule, which eventually formed an altogether new basis and opened up a new horizon in the field of drug design of narcotic analgesics. Intensive research further led to the evolution of pethidine (meperidine) which incidentally combines both the properties of morphine and atropine. It possesses a quaternary carbon-atom and quite astonishingly a much simpler chemical structure to that of morphine.

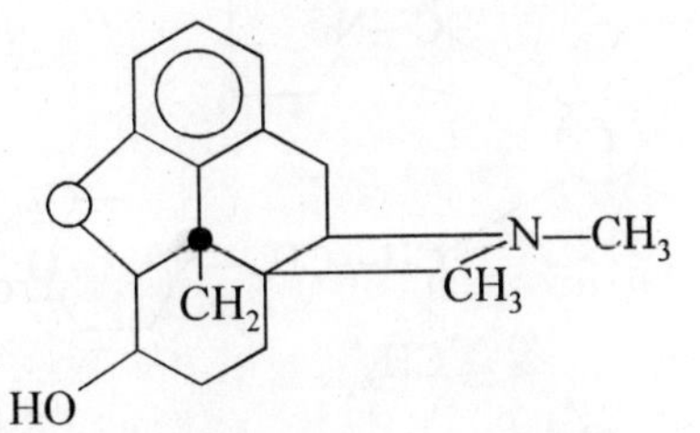

Morphine

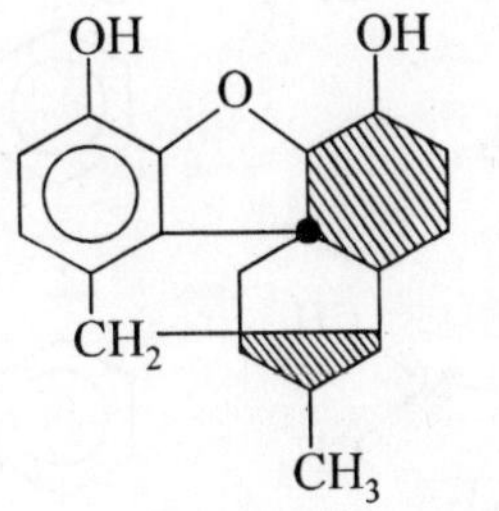

Morphine
(Schaumann's Model)

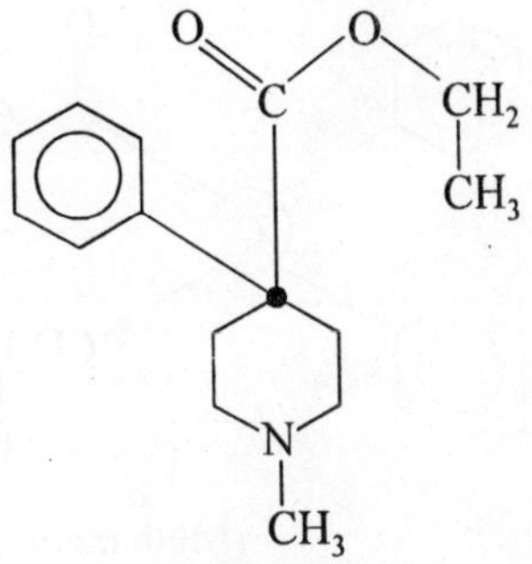

Pethidine
(Meperidine)

Ehrhardt suggested a general formula relevant to the analgesic activity in 1949 as stated below :

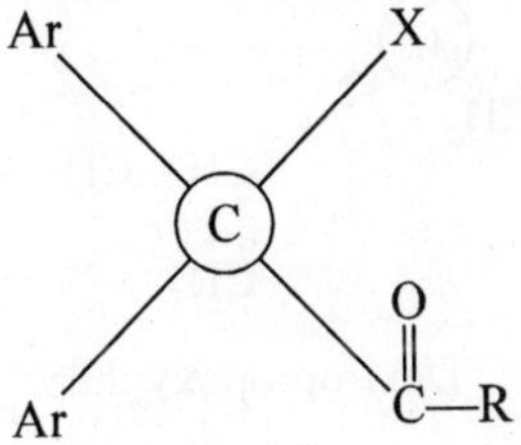

where, Ar is the aromatic ring, X the basic side chain and (—C—) carbonyl function in the form of an ester, ketone or an amide.

Later on, the above general formula was modified slightly as follows :

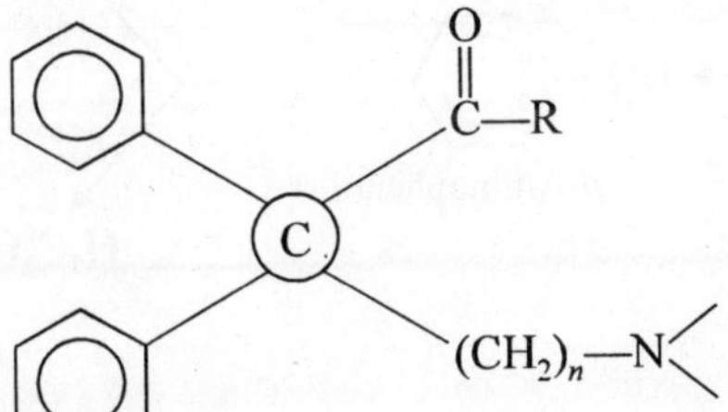

which successfully led to the development of the following *three* narcotic analgesics, namely : methadone, dextromoramid and dextropropoxyphen.

(ii) Antipyretic Analgesics

Another fruitful approach in drug design is the meticulous screening of the metabolite for probable pharmacological activity. The most interesting example is the bio-oxidation of acetanilide into *para*-aminophenol which subsequently on chemical manipulation has yielded better tolerated antipyretic-analgesics like paracetamol and phenacetine.

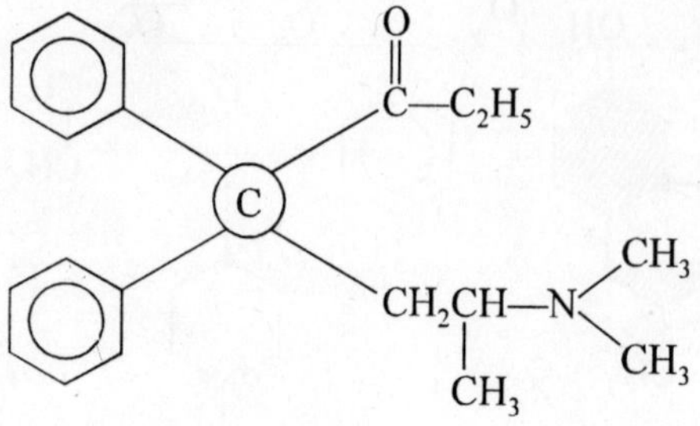

Methadone

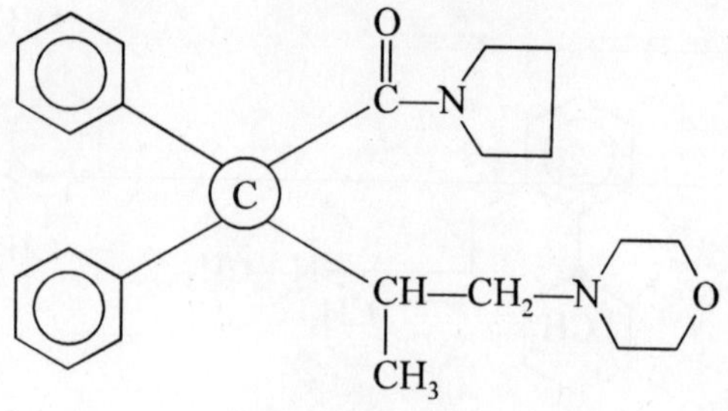

Dextromoramid

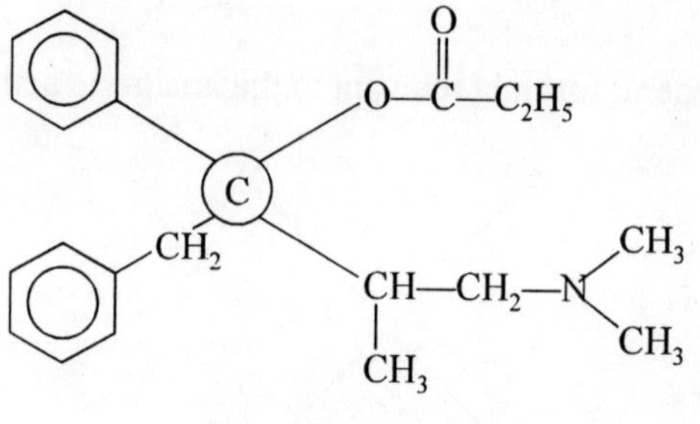

Dextropropoxyphen

Quite recently phenacetine has been withdrawn completely because of its toxic after effects, though it dominated the therapeutic field for over 30 years as a potent antipyretic analgesics.

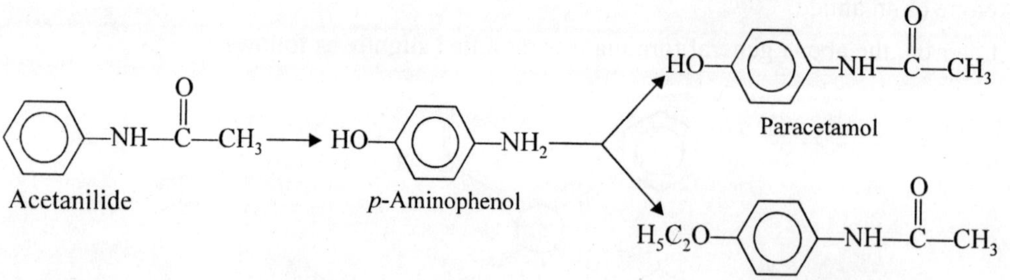

Acetanilide

p-Aminophenol

Paracetamol

Phenacetine

(iii) *Antirheumatic Drugs*

The study of the metabolite conversion of the antirheumatic drug phenylbutazone resulted in the introduction of a better tolerated drug oxyphenylbutazone as an antirheumatic drug and phenylbutazone alcohol as an uricosuric agent.

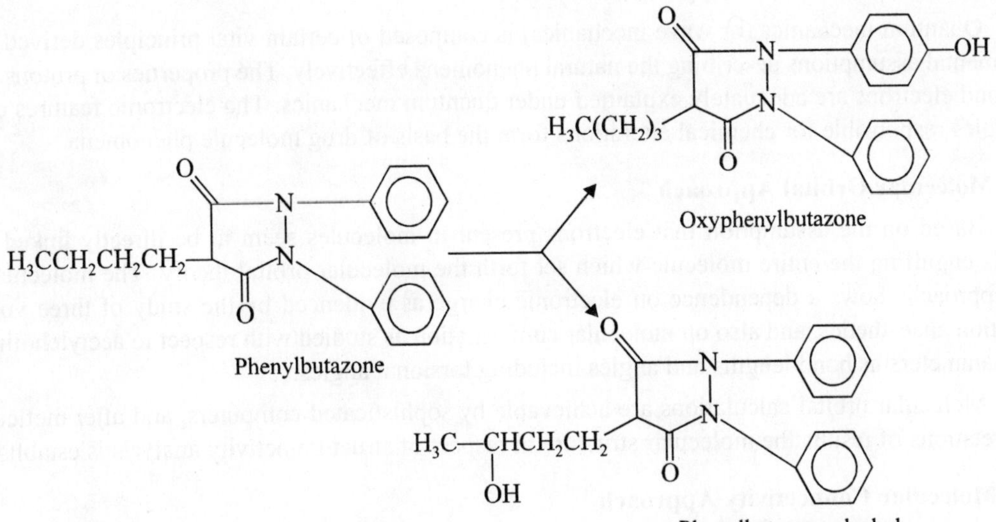

Oxyphenylbutazone

Phenylbutazone

Phenylbutazone alcohol

4. FACTORS GOVERNING DRUG-DESIGN

A few cardinal factors governing the efficacy towards the evaluation of drug design include :

(a) The smaller the expenditure of human and material resources involved to evolve a new drug of a particular value, the more viable is the design of the programme.

(b) Experimental animal and clinical screening operations of the new drugs.

(c) Relationships between chemical features and biolgoical properties need to be established retrospectively.

(d) Quantitative structure-activity relationships (QSARs) vary to an appreciable extent in depth and sophistication based on the nature of evaluation of structure or activity. A purposeful relation of structural variables must include steric factors, electronic features of component functional groups and, in general, the molecule as a whole.

(e) The trend to synthesize a huge number of newer medicinal compounds indiscriminately for exploratory evaluation still prevails which exclusively reflects the creative genuineness and conceptual functions of a highly individualized expression of novelty by a medicinal chemist.

(f) Introduction of functional groups in a molecule that need not essentially resemble metabolites, but are capable of undergoing bonding interactions with important functional groups of biochemical components of living organisms affords an important basis for exploration.

(g) Disease etiologies and various biochemical processes involved prove useful.

5. RATIONAL APPROACH TO DRUG DESIGN

A rational approach to drug design may be viewed from different angles, namely :

5.1. Quantum Mechanical Approach

Quantum mechanics (or wave mechanics) is composed of certain vital principles derived from fundamental assumptions describing the natural phenomena effectively. The properties of protons, neutrons and electrons are adequately explained under quantum mechanics. The electronic features of the molecules responsible for chemical alterations form the basis of drug molecule phenomena.

5.2. Molecular Orbital Approach

Based on the assumption that electrons present in molecules seem to be directly linked with orbitals engulfing the entire molecule which set forth the molecular orbital theory. The molecular orbital approach shows a dependence on electronic charge as evidenced by the study of three volatile inhalation anaesthetics, and also on molecular conformation as studied with respect to acetylcholine by such parameters as bond lengths and angles including torsional angles.

Molecular orbital calculations are achievable by sophisticated computers, and after meticulous interpretations of results the molecular structure in respect of structure-activity analysis is established.

5.3. Molecular Connectivity Approach

This approach establishes the presence of structural features like cyclization, unsaturation, skeletal branching, and the position and presence of heteroatom in molecules with the aid of a series of numerical indices. For example : an index was determined to possess a correlative factor in the SAR study of amphetamine-type hallucinogenic drugs.

Molecular connectivity approach has some definite limitations, such as, electronegativity variance between atoms, non-distinguishable entity of *cis-trans* isomerism.

5.4. Linear Free-Energy Approaches

This method establishes the vital link between the proper selection of physicochemical parameters with a specific biological phenomenon. However, such a correlation may not guarantee and allow a direct interpretation with regard to molecular structure, but may positively offer a possible clue towards the selection of candidate molecules for synthesis.

6. DRUG-DESIGN : THE METHOD OF VARIATION

Under this method a new drug molecule is developed from a biologically active prototype. The various advantages are as follows :

(a) At least one new compound of known activity is found.

(b) The new structural analogues even if not superior may be more economical.

(c) Identical chemical procedure is adopted and hence, considerable economy of time, library and laboratory facilities.

(d) Screening of a series of congener (*i.e.,* member of the same gene) gives basic information with regard to pharmacological activity.

(e) Similar pharmacological technique for specific screening may be used effectively.

The cardinal objectives of the method of variation are :

• To improve potency

• To modify specificity of action

- To improve duration of action
- To reduce toxicity
- To effect ease of application or administration or handling
- To improve stability
- To reduce cost of production

In order to obtain a therapeutically potent and better-tolerated drug there exists invariably an apparent conflict of pure scientific objectives and practical objectives. *This may be expatiated by citing the instance of an exceedingly toxic congener (say an anti-neoplastic agent) that possesses a very high degree of specificity and the researcher may have in mind to prepare still more toxic compounds so as to develop the highest possible specificity of action.* On the contrary, absolutely from the practical aspect, the proposed clue may not be pursued solely depending on the policy of the organization and not the individual or group of researchers.

In fact, there are a few generalized approaches utilizing the method of variation. In this particular context, the familiarity with the molecular structure is of the prime importance. The various possible approaches in designing newer drugs by applying variation of a prototype are quite numerous. Once the molecular structure of the compound in question is drawn on the drawing board, one takes into consideration such information as the following :

(*a*) study of the core nucleus of the hydro-carbon skeleton ;

(*b*) variation of functional groups and their proximity to one another ;

(*c*) various probable rotational and spatial configurations ;

(*d*) possibility of steric hindrance between various portions of the molecule in different configurations in space ; and

(*e*) probability of electronic interactions between various portions of the molecule including such matters as inductive and mesomeric effects, hyper-conjugation, ionizability, polarity, possibility of chelation, asymmetric centres and zwitterion formation.

The application of the method of variation, depending on the considerations enumerated above, is exploited in two different manners to evolve a better drug. The two main approaches for this goal can be indicated as :

(*a*) drug design through disjunction ; and

(*b*) drug design through conjunction.

6.1. Drug Design through Disjunction

Disjunction comes in where there is the systematic formulation of analogues of a prototype agent, in general, toward structurally simpler products, which may be viewed as partial or quasi-replicas of the prototype agent.

The method of disjunction is usually employed in three *different* manners, namely :

(*i*) unjoining of certain bonds ,

(*ii*) substitution of aromatic cyclic system for saturated bonds ; and

(*iii*) diminution of the size of the hydrocarbon portion of the parent molecule.

Example :

    The extensive study on the estrogenic activity of oestradiol *via* drug design through disjunction ultimately rewarded in the crowning success of the synthesis and evaluation of *trans*-diethylstilbesterol. The flow-sheet of estrogen design is stated below :

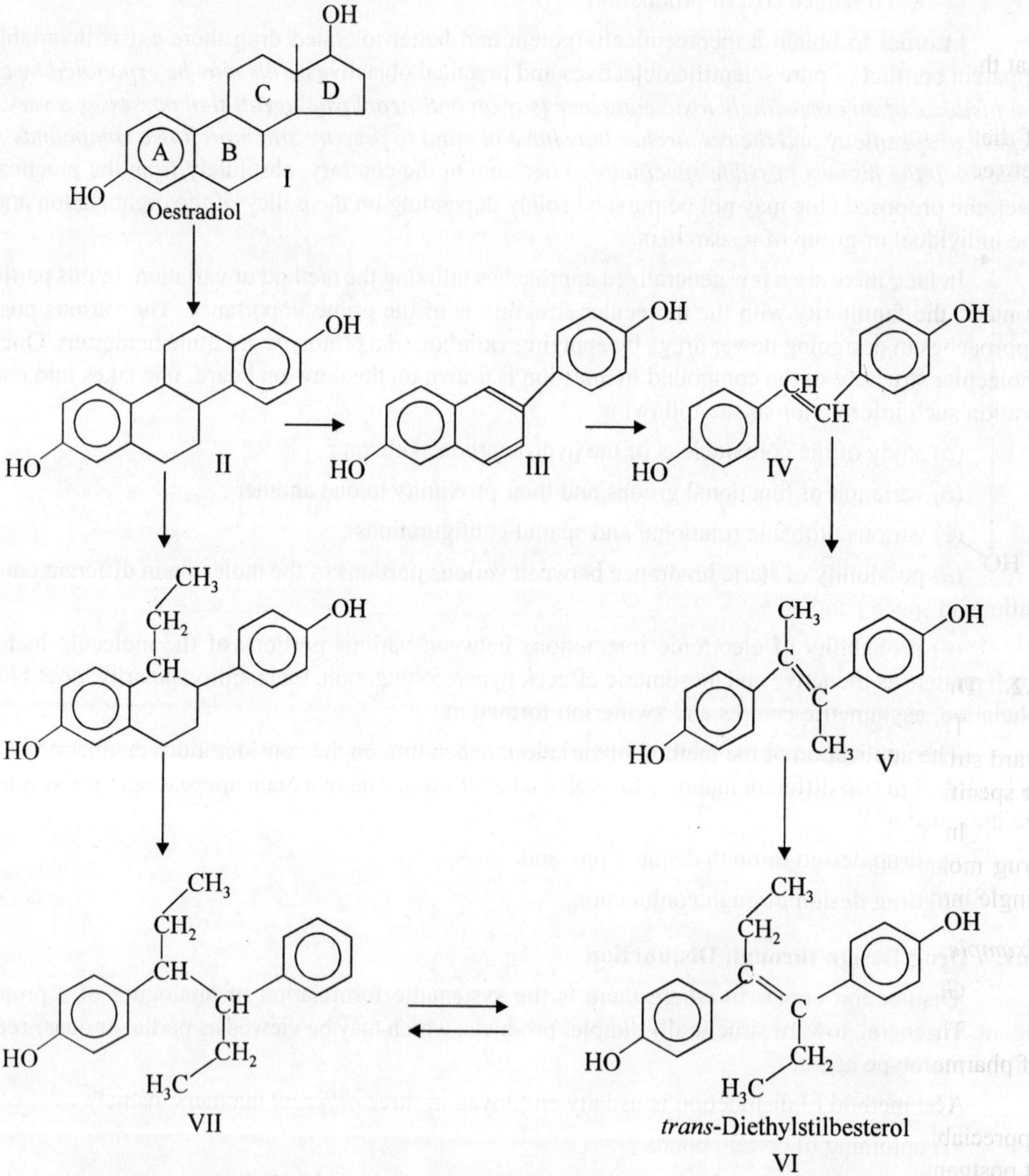

Flow-sheet of Estrogen Design

From the above the following *three* observations may be made. They include :

(*i*) Various steps in design of II to III to IV designate nothing but successive simplification through total elimination of the rings *B* and *C* in oestradiol (I).

(*ii*) The above manner of drug design finally led to successively less active products (*i.e.,* II, III, IV).

(*iii*) Upon plotting oestrogenic activity against various structures (I to VII) it was quite evident that the maximal activity in this series was attributed to *trans*-diethylstilbesterol.

It is, however, pertinent to mention here that in the following **three** different possible structures of diethylstilbesterol analogues, the oestrogenic potency decreases substantially as the distance '*D*' between the two hydroxyl groups decreases.

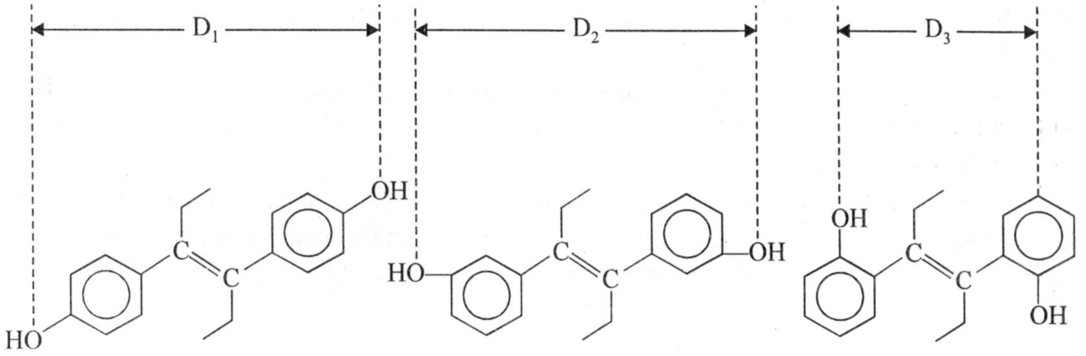

$$D_1 > D_2 > D_3 > [D_1 = 14.5°A]$$

6.2. Drug Design through Conjunction

This is known as the systematic formulation of analogues of a prototype agent, in general, toward structurally more complex products, which may be viewed as structures embodying, in a general or specific way, certain or all of the features of the prototype.

In this type of drug-design, the main principle involved is the **'principle of mixed moieties'**. A drug molecule is essentially made up with two or more pharmacophoric moieties embedded into a single molecule.

Example :

Ganglionic blocking agent—its development based on the principle of mixed moieties.

The principle of mixed moieties actually involve the conjunction of two or more different types of pharmacophoric moieties within a single molecule.

Acetylcholine is an effective postganglionic parasympathetic stimulant in doses that afford no appreciable changes in the ganglionic function ; whereas hexamethonium possesses only a slight action at postganglionic parasympathetic endings in doses that produce a high degree of ganglionic blockade.

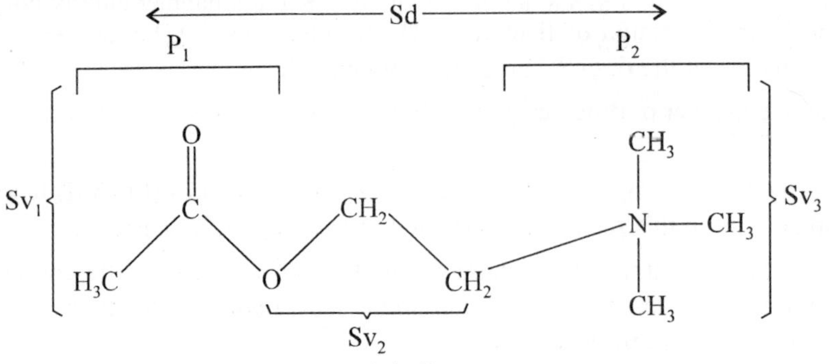

Acetylcholine

(where Sv_1 = steric factor 1 ; Sv_2 = steric factor 2 ; Sv_3 = steric factor 3 ; Sd = steric distance factor ; P_1 = polarity factor 1 and P_2 = polarity factor 2).

The moiety requirements for postganglionic parasympathetic stimulant action (muscarinic moiety) have been duly summarized for convenience to the above structure of acetylcholine wherein the various operating factors have been highlighted.

The foregoing generalization of the muscarinic moiety on being studied in relation to the particular bisquaternary type of structure, *e.g.* hexamethonium, promptly suggests the following proposed design, thus embodying the ganglionic moiety and the muscarinic moiety into a single molecule.

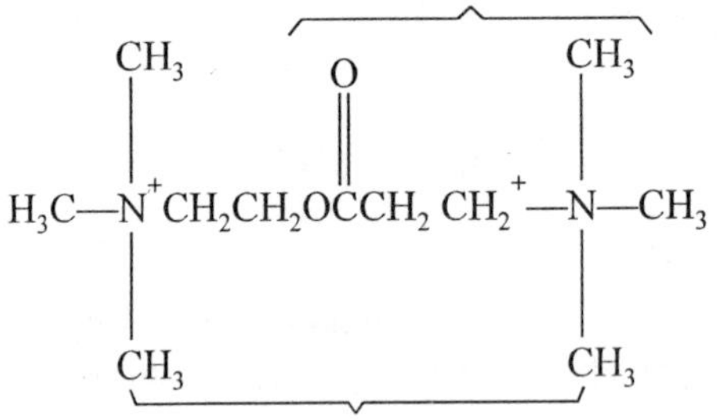

Ganglionic Active Moiety

It is, however, pertinent to mention here that the 'internitrogen distance' essentially constitute an important factor in many series of bisquaternary salts that possess ganglionic blocking activity. It is worthwhile to note that this distance is almost similar to that present in hexamethonium in its most extended configuration.

However, the actual synthesis and pharmacological evaluation of the above hexamethyl analogue reveal the presence of both muscarinic stimulant and ganglionic blocking actions. Interestingly, the corresponding hexaethyl analogue possesses a ganglionic blocking effect and a weak muscarinic stimulant action.

7.   DRUG DESIGN AND DEVELOPMENT : AN OVERVIEW

7.1.   Preamble

The overwhelming qualified success in the evolution of **'ethical pharmaceutical industry'** in the twentieth century have not only registered an unquestionable growth in improving the fabric of society to combat dreadful diseases across the globe but also made a significant legitimate cognizance of an individual's quality of life and above all the life expectancy.

The twentyfirst century may obviously record and witness an apparent positive tilt in population demographics ultimately leading to a much healthier, stronger and happier elderly population.

However, in the 21st century, the **'ethical pharmaceutical industry'** has been fully geared towards the production of relatively safer, less toxic, more effective, higher therapeutic index, novel, innovative medicaments that will evidently help the mankind to afford a disease-free society ; besides, the elder ones with a glaring hope to live a still longer life span.

Following is the brief description in a chromological order for the development of **'ethical pharmaceutical industry'** in the world :

Year	Country	Historical Development
1600s	Japan	—Takeda in 1637*.
1800s	Europe and USA	—Fine chemical industries**.
1880s	Germany and UK	—Hoechst (Germany) and Wellcome (UK) for immunological drugs.
1889	UK	—Aspirin (as NSAID)
1990	France	—Rhone Poulenc
1914	Europe	—Engaged in US-operations
1929	USA	—Aureomycin (Lederle) ; Chloromycetin (Parke-Davis) ; Teramycin (Pfizer) ;
1950	France and Belgian	—Chlorpromazine [Rhone-Poulenc (France)] ; Haloperidol [Janssen (Belgium)]—both psychotropic drugs
1950s to 1970s	USA	—Pharmaceutical Industry showed a steady growth***
1970s	USA	—Greater advancement on molecular focus in the regimen of *'drug discovery'* picked up substantial momentum with the strategic induction of noted scientists in the US National Academy of Sciences, namely : Needleman P (Monsanto) ; Cuatrecasas P (Burroughs Wellcome) ; and Vagelos PR (Merck).

*Sneader WJ, **'Drug Discovery : The Evolution of Modern Medicines,'** John Wiley, Chichester, UK, 1997.

** Di Masi, d J *et al*. **Research and Development costs for new drugs for therapeutic category'**, *Pharmaco.Econ.,* **7** : *52, 169, 1995.*

***Drayer JI and Burns JP. From discovery to market : the development of pharmaceuticals. In : Wolff ME, ed, **'Burgers Medicinal Chemistry and Drug Discovery,** 5th edn, Vol I, Wiley, New York, 1995, pp 251-300.

The various phases of transformations in **'ethical pharmaceutical industry'** between 1600 to 1970s brought about a sea-change with a significant shift from the core techniques of molecular pharmacology and biochemistry to those of molecular biology and genomics (biotechnology). Based upon these fundamental newer concepts amalgamated with various paradigm shifts resulted into the evolution of an exclusive progressive change in the scenario of both culture and the environment of the **'ethical pharmaceutical industry'** in developed as well as developing countries in the world.

7.2. Revolutions in Drug Discovery

A tremendous noticeable change in the *'process of drug discovery'* in the past three decades has been focused solely on the *'biotechnology revolution'*. In short, the techniques employed invariably in 'molecular biology' and 'biotechnology' opened up an altogether **'new trend in biomedical research'**.

In 1997, a staggering 1150 companies were established based on **'biotechnology'**, engaging three lacs research scientists working round-the-clock, and generated USD 12 billion. The six major biotech companies in USA, established in mid 1980s, now proudly enjoys the number one status not only in US but also in rest of the world, namely :

(*a*) **Genentech**—Presently subsidiaries of *Roche Biosciences ;*

(*b*) **Genetics Institute**—Presently subsidiaries of *American Home Products,*

(*c*) **Amgen ; Genzyme ; Chiron** and **Biogen**—Presently emerged as *major pharmaceutical companies.*

In the light of the huge accelerated costs for drug development, touching USD 359 million in 1991, to almost USD 627 million in 1995 and a projected USD 1.36 billion in 2000, have virtually pumped in lots of force geared towards superb efficacies and efficiencies in the pharmaceutical industry.* And this could only be accomplished through appreciable consolidation amalgamated with continued efforts of outsourcing of higher risk, early drug discovery to venture *capital-aided-biotech units ;* besides, clinical trials to the *clinical-research organizations* exclusively.

In order to significantly cut down the overhead expenses, and encash on sizable profitability various giants in the pharmaceutical industry have more or less adopted the following stringent measures to face the cut-throat competition in the global market and also survive gainfully, such as :

(*a*) To enhance the required productivity in the R and D activities of major pharmaceutical companies to sustain and maintain profitability,

(*b*) Increased productivity without enhancing R and D resources,

(*c*) Focusing on new research activities/strategies thereby creating a possible balance between internal research and external alliances,

(*d*) Merger and alliances in Pharmaceutical Industries dates back to 1970s with the formation of **Ciba-Geigy*** ; and till 2000 more than 20 such acquisitions/mergers have already been materialized across the globe.

---

*Carr G : **The Alchemists : A survey of the Pharmaceutical Industry, Economist**, February 21, 1998, pp 3-18.

** de Stevens G., **Conflicts and Resolutions.**, *Med. Res Rev.,* 1995, **15**, pp 261-275.

7.3. Research and Development Strategies

It has been proved beyond any reasonable doubt that the *'rate of success'* in **drug discovery** is exclusively dependent on the ability to identify, characterize novel, patentable newer *'target-drug-molecules'* usually termed as **New Chemical Entities** (NCEs), which essentially possess the inherent capability and potential in the management and control of a specific disease/ailment ; besides, being efficacious and safer in character. With the advent of latest technological advancements in the specialized areas related to **genomics and combinatorial chemistry** an appreciable advancement has been accomplished in the R & D strategies. It is, however, pertinent to mention here that a proprietary NCE status, position and recognition is an absolute must not only to ensure marketing exclusively but also to aptly justify the huge investment in the ensuing R & D process thereby making *medicinal chemitry* a more or less core element of the entire *'drug discovery process'.*

Interestingly, the *'drug discovery process'* may be categorized into **four** distinct heads, namely :

(*i*) Target identification and selection,

(*ii*) Target optimization,

(*iii*) Lead identification, and

(*iv*) Lead optimization.

The concerted efforts encompassing various intangible and critical methodologies that ultimately relate to the activities, expertise, wisdom and integration of the individual scientist directly or indirectly involved in *'drug discovery process'* virtually leads to advance drug discovery profiles.*

In short, the qualified success in the *'drug discovery process'* predominantly revolves around the following cardinal factors, namely :

- Articulated project management processses
- Prioritization
- Well-defined aims and objectives
- Company organization(s) and culture
- Resourcing *modus operandi*
- Prompt decision making factors.

8.   MOLECULAR HYBRIDISATION

The molecular hybridisation essentially embodies the synthesis of strategically designed of altogether newer breeds of **'bioactive agents'** either from two or even more compounds having different characteristic features by the aid of **covalent-bond synthesis.**

Necki (1886) first conceived the interesting *'salol principle',* whereby he exploited the beneficial properties of phenols and carboxylic acids possessing potent antibacterial characteristic features into the *'design'* of newer drug molecules with better and improved pharmacological activities by means of simple esterification.

---

*Sapienza A.M., **Managing Scientists**, New York, Wiley, 1995.

A few typical examples wherein the hyberdisation was accomplished commencing from two **'bioactive entities'** *i.e.,* implementation of the **full-salol principle** occurred, as stated under :

Examples :

(*a*) **Antibacterial Agent :** *Streptoniazid ;*

A molecule of streptomycin and a molecule of isoniazid by means of a strong double bond between C and N with the elimination of a mole of water. **The 'hyberdised molecule'** exhibits a significant potentiated antibacterial and tuberculosstatic agent.

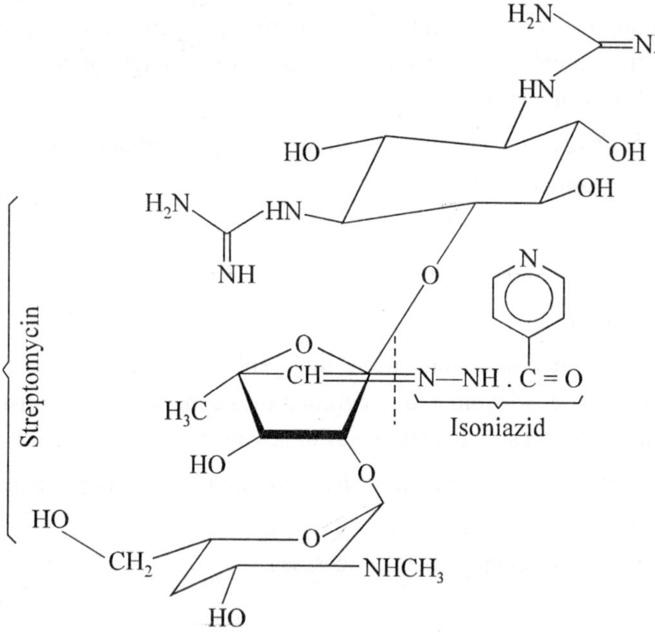

(*b*) **Antitussive Expectorant Drug :** *Guaicyl phenyl cinchoninate ;*

A mole each of cincophen and guaiacol gets hyberdised by forming an ester-linkage and losing a mole of water. The new product shows an improved antitussive and expectorant activity.

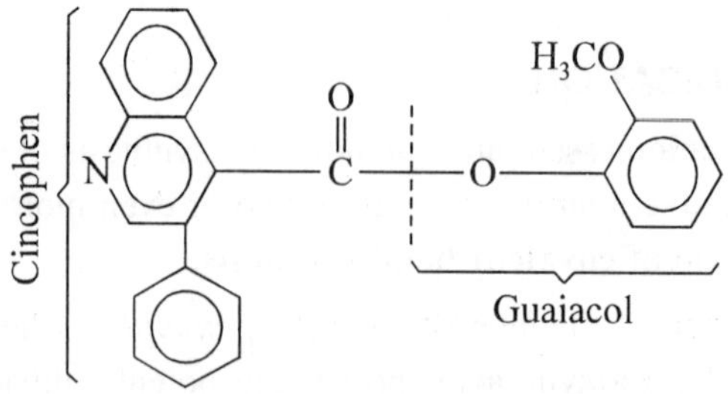

(*c*) **Antipyretic-Analgesic Agent :** *Quinine acetylsalicylate ;*

Hybridisation takes place between a mole of acetylsalicylic acid (*i.e.,* aspirin) and quinine (*i.e.,* a potent antimalarial agent) to lose a mole of water ; and the resulting hyberdised product potentiates the antimalarial activity along with substantial antipyretic—analgesic activity.

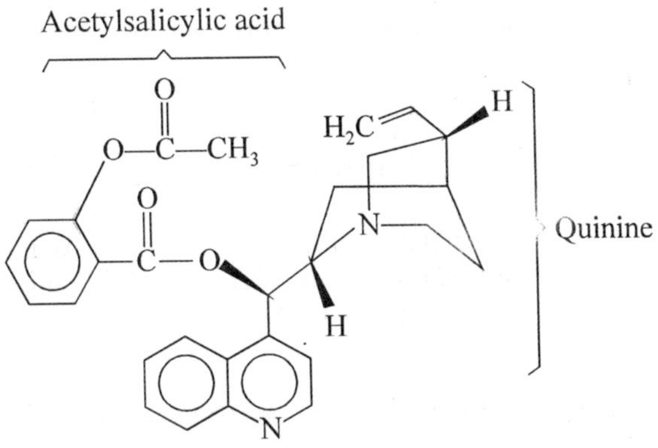

Acetylsalicylic acid

Quinine

9. RIGIDITY AND FLEXIBILITY VS DRUG DESIGN :

It has been observed beyond any reasonable doubt whatsoever that the structure-activity relationship invariably affords certainly a molecular complementary prevailing evidently between the bioactive compound and the probable receptor site. At this point in time *two* different situations may usually crop up, namely :

(*a*) *increased rigidity* — that may ultimately lead to improved potencies ; and

(*b*) *increased flexibility*—that may give rise to better and improved activity.

These two aforesaid situations shall now be discussed with typical examples so that one may have a better understanding of these aspects *vis-a-vis* drug design of **newer targetted** drug molecules.

9.1. Increased Rigidity

There are a plethora of *'drug molecules'* which are inherently flexible in nature *i.e.,* they can assume a wide-range of shapes (*spatial arrangements*). Of these structural variants quite a few are absolutely not so favourably acceptable for reaction at a specific **'receptor site'**. Therefore, the *'design'* or *'search'* for a relatively more rigid structural analogue essentially having the required, correct and desired **'dimensions'** must be looked into in order to obtain a more potent *drug substance.*

Besides, the actual distance existing between two vital functional moieties may be almost fixed arbitrarily in rigid molecular structural variants. These restructured and strategically positioned newer targetted-drug molecules may be subjected to vigorous and critical examinations by the aid of several sophisticated latest physicochemical analytical devices, such as : X-Ray diffraction analysis ; Optical Rotary Dispersion (ORD) ; NMR-spectroscopy ; Mass Spectroscopy ; FTIR-Spectrophotometry and the like.

Examples : Structural analogues of *acetylcholine* (ACh) *i.e.,* a short-acting cholinergic drug, with **'increased rigidity'** having 5- or 6- membered saturated rings were synthesized ; and their activities were compared using ACh as the reference drug :

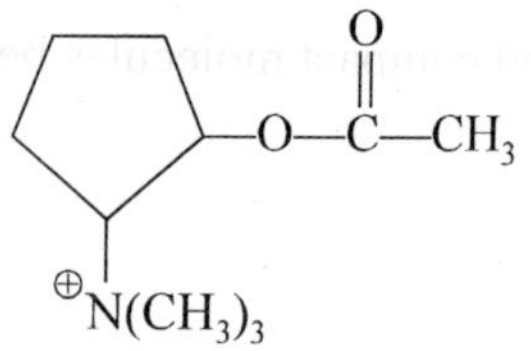

MODIFIED ACh
WITH 5-MEMBERED
CYCLIC RING
(A)

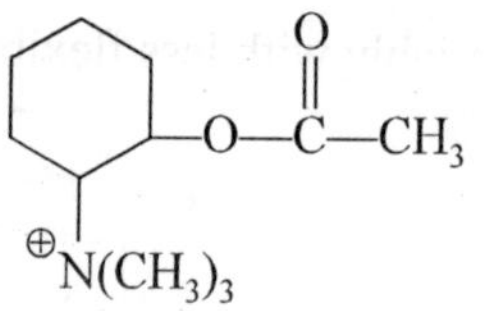

MODIFIED ACh
WITH 6-MEMBERED
CYCLIC RING
(B)

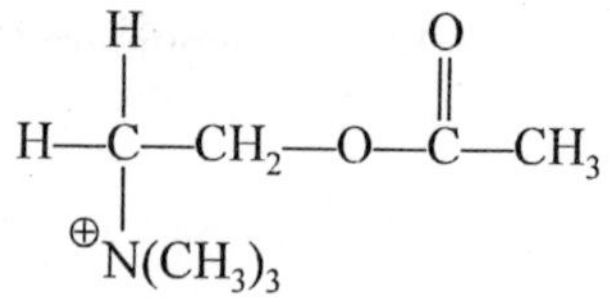

ACETYLCHOLINE ACh
(AN OPEN-CHAIN COMPOUND)

Interestingly, either of the *two* structural analogues (A) and (B) can be further resolved into their respective *trans*- and *cis*-isomers *i.e.,* spatially rearranged structures, as given below :

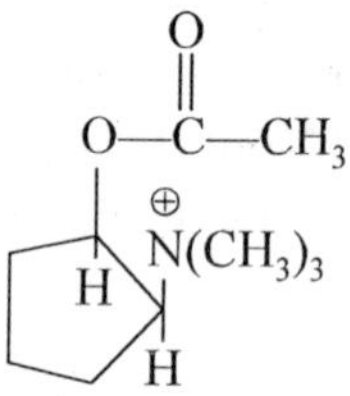

A-*cis*—

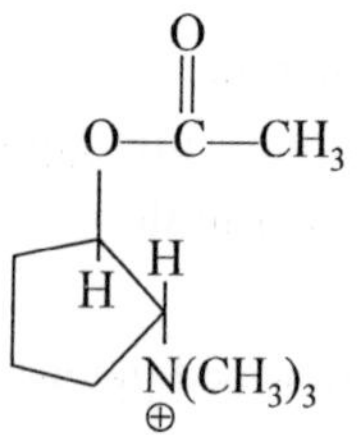

A-*trans*—

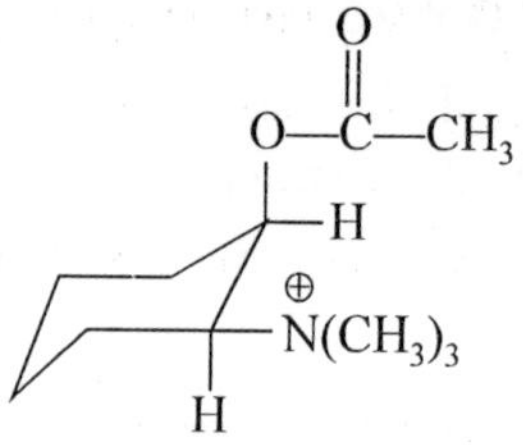

B-*cis*—

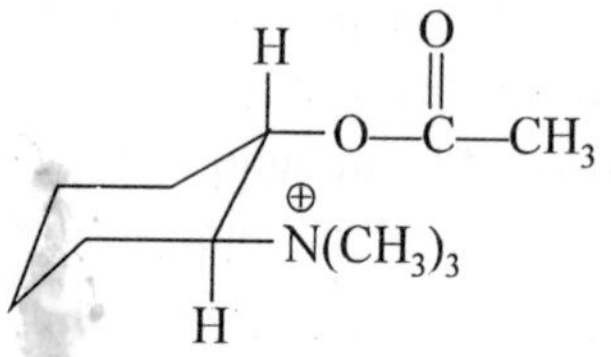

B-*trans*—

It has been observed that the *'intraatomic distance' between* 'O' and 'N' atoms for the *cis*-isomers (A & B) ranged between 2.5—2.9 Å ; whereas, between the corresponding *trans*-isomers (A & B) varied between 2.9—3.7 Å. Furthermore, the relative cholinergic activities of the *cis*-isomers were found to be greater than the corresponding *trans*-isomers using ACh as the reference drug.

The results of these findings have been summarized in the following table, wherefrom certain important clues may be derived with regard to some important functional group(s) located on the enzymes and the existing distances between such moieties.

S. No.	Drugs	Intra-Atomic Distances between 'O' and 'N' (Å)	Relative cholinergic Activity
1.	ACh	—	1.00
2.	A-*cis*-	2.51	1.43
3.	A-*trans*-	3.45	1.07
4.	B-*cis*-	2.5—2.9	1.14
5.	B-*trans*-	2.9—3.7	1.06

Thus, A-*cis*— is found to be almost 50% more active than ACh, and B-*cis*-only upto 15% than ACh. However, the corresponding *trans*-isomers of A and B did not show any improvement in their cholinergic activities.

9.2. Increased Flexibility

The problems encountered invariably with less flexible, rigid and compact molecules being that their manoeuvrability are comparatively much less. In other words, they either possess little or practically negligible capacity to have them rearranged to a more favoured conformation that may ultimately give rise to enhanced bioactivity.

Example :

Propoxyphene (I) is an open-chain structural analogue having narcotic analgesic activity ; whereas, its corresponding cyclic analogue (II) is almost found to be devoid of the pharmacological activity.

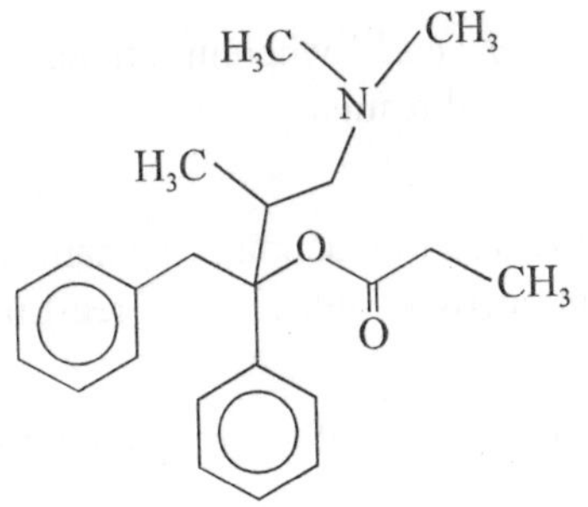

PROPOXYPHENE
(I)
(ACTIVE)

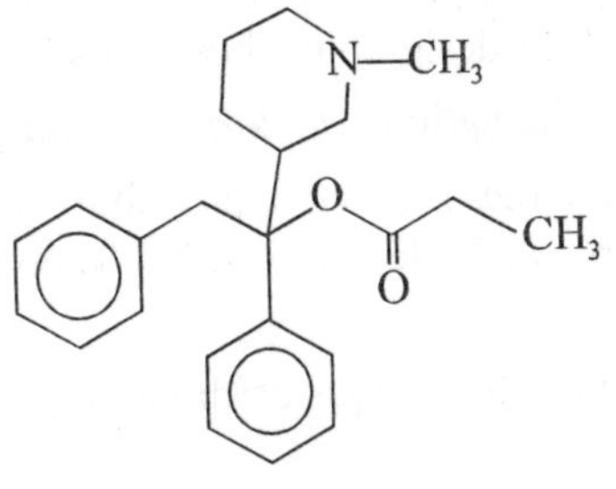

PROPOXYPHENE CYCLIC ANALOGUE
(II)
(INACTIVE)

10. 'TAILORING' OF DRUGS

With the advent of enormous in-depth knowledge of 'modern chemistry', the 'tailoring' of drugs has become a skilful art that may result fruitful results through specific modes of attack on a drug molecule.

Various configurational and stereochemical changes afford flexibility and overall dimension of a drug molecule. Such alterations may be conveniently achieved through different means and ways, namely : ring fission or fusion, formation of lower or higher homologues, introduction of optically active centres, formation of double bonds towards geometrical isomerism, and lastly introduction of bulky groups towards restricted rotation or the removal and replacement of such groups.

Alterations of various physical and chemical characteristics through the insertion of newer functional moieties or by the replacement of such groups already present by others that essentially differ in degree or in type. These types of changes may be effectively brought about by : isosteric replacement, changes of orientation or position of given moieties, introduction of polar character of given functional groups or replacement of other groups with different electrical features, and finally such changes which either promote or inhibit the presence of different electronic conditions achieved through inductive effects, mesomeric effects, tautomerism, chelation, hyperconjugation, etc.

11. GENERAL CONSIDERATIONS

Molecules, in general, may be viewed as dynamic electric entities. Hence, even the slightest alteration made in a relatively remote section of a molecule may cause either through spatial or through the overall matrix of the molecule, additional changes in some or all of its inherent characteristics.

An effective drug design from a biologically active prototype, whether approached through disjunction or conjunction or both, normally aims at modifying collectively all the moiety attributes that are absolutely essential capacities of a drug eliminated to a great extent which otherwise would have reduced its specificity of action, or interference with the primary type of action sought.

Probable Questions for B. Pharm. Examinations

1. Jutify the following statements :

    (*a*) Drug design aims at developing a drug with high degree of chemotherapeutic index and specific action.

    (*b*) From the practical view-point it is the '*Exploitation of Leads*' wherein rational approaches to drug-design have been mostly productive with fruitful results.

2. Discuss the variuos '*factors governing drug-design*'.

3. Eloborate the 'rational approach to drug design' weith regard to Quantum Mechanics (or Wave Mechanics), Molecular Orbital Theory, Molecular Connectivity and Linear Free-Energy Concepts.

4. Enumerate the various cardinal objectives of '*the Methods of Variation*' giving appropriate examples.

5. The first synthetic oestrogen *trans*-diethylstilbesterol came into existence by applying the principle of 'drug-design through disjunction' from 'oestradiol'. Explain.

6. The development of 'ganglionic block agent' is exclusively based on the '*pricniple of mixed molecular*' as drug design through conjunction.

7. '*Tailoring of Drugs*' is the outcome of an unique blend of skillful an involving various configurational and stereochemical changes attributing its flexibility and overall dimension. Explain.

8. Discuss the various possible approaches in designing newer drugs by applying variation of a '*biologically active prototype*'.

9. Bio-oxidation and acetanilide and metabolic conversion of phenylbutazone gave rise to two better tolerated drug molecule used frequently and profusely in the therapeutic armamentarium. Explain.

10. Differentiate the basic concepts of '*analogues*' and '*prodrugs*' with the help of suitable examples of parent drug molecule(s).

RECOMMENDED READINGS

1. M.E. Wolff (Ed) **Medicinal Chemistry and Drug Discovery,** John Wiley & Sons, New York, 5th edn., 1995.

2. **Foye's Principles of Medicinal Chemistry,** Williams DA and Lemke T.L. (Eds), Lippincot Willams & Wilkins, New York, 5th edn., 2002.

3. Kenny BA *et al.* **The Application of high throughput Screening to novel lead discovery,** *Prog. Drug Res.,* **41** : 246-269, 1998.

4. Blundell T., **'Structure-based Drug Design',** *Nature,* **384,** 23-26, 1996.

5. Williams M., **Strategies for Drug Discovery,** *NIDA Research Monograph,* **132** : 1-22, 1993.

2 Physical-Chemical Factors and Biological Activities

1. INTRODUCTION

The quest for knowledge to establish how the drugs act in a living system has been a thought-provoking topic to scientists belonging to various disciplines such as medicinal chemistry, molecular pharmacology and biochemistry. Since the turn of the twentieth century these researches have more or less established the basis of drug action on a more scientific and logically acceptable hypothesis. With the advent of such newer fields of study, for instance : tracer techniques, genetic engineering, bio-technology, electron microscopy and computer-aided physico-chemical methods, a new direction has been achieved towards more vivid explanation of the intricacies of drug interaction sequel to drug design.

In the recent past, receptors and drug-receptor interactions theories have highlighted the importance of physical and chemical characteristics with regard to drug action. Such salient features may include : partition coefficients, solubility, degree of ionization, isosterism and bio-isosterism, surface activity, thermodynamic activity, intramolecular and intermolecular forces, redox potentials, stereochemistry and interatomic distances between various functional groups.

Medicinal chemistry undoubtedly rests its main focus on the broad based variations embracing the influence of numerous possible manipulations with regard to the chemical structure on the biological activity. In the light of the above statement of facts supported by copious volumes of scientific evidences reported in literatures, it is almost important and necessary for the 'medicinal chemist' to decepher and logically understand not only the *'mechanism of drug action' in vivo* by which a **drug substance** exerts its effect, but also the overall physicochemical properties of the molecule. In a rather most recent conceptualized theoretical basis the specific terminology *'physicochemical characteristics'* invariably refers to the cognizable influence of the plethora of organic functional moieties strategically positioned within a **drug substance,** namely : acid/base characteristics, partition coefficient, water solubility, lipoidal solubility, crystal structure, stereochemistry, chirality to name a few. It is, however, pertinent to mention here that most of the aforesaid properties covertly and overtly exert a significant influence upon the various biological phenomenon *in vivo,* such as : **absorption, distribution, metabolism** and **excretion** (ADME) of the newer *'target-drug molecule'.*

Therefore, a creative **'medicinal chemist'** should ponder over the intricacies, complexities and legitimate presence of each functional moiety to the overall physical chemical properties of the *'target-drug molecule'* with a view to arrive at or design safer, better and efficacious medicinal agents. Nevertheless, such critical studies have to be carried out in a rather methodical and systematic manner *vis-a-vis* their affect upon biological activities. Generally, such elaborated studies are commonly referred to as *'structure-activity relationship'* (SAR) ; and more recently as *'quantitative-structure-activity relationship'* (QSAR).

It would be worthwhile to look into the physical and chemical aspects of the drug separately and an attempt made to establish the relation of such properties to biological activities.

2. PHYSICAL PROPERTIES

A plethora of physical properties play an important role in modifying the biological activities of a good number of medicinal compounds. A few such properties are *features governing drug action at active site, factors governing ability of drugs to reach active site, dissociation constants, isosterism* and *bio-isosterism.*

2.1. Features Governing Drug Action at Active Site

The various factors that govern the action of drugs at the active site may be due to structurally specific and non-specific drugs.

2.2. Structurally Specific Drugs

A number of compounds that possess remarkable pharmacological actions are essentially the structurally specific drugs. Though the physical characteristics of the drug play an important role in the biological activity, yet the chemical properties do exert their justified influence on the activity.

Effects of Minor Structural Modifications on Biological Acitivity

Structure	Biological Activity	Pharmacological Classification
(a) NaRC structure, $R = O$, (pentobarbitone sodium) / $R = S$, (thiopental sodium)	Short-acting / Ultra-shortacting	Hypnotic
(b) R—⬡—S(O)(O)—NH—C(O)—NH—R′, $R = CH_3$, $R' = C_4H_9$ (tolbutamide) / $R = Cl$, $R' = C_3H_7$ (chloropropamide)	Short-acting / Long-acting	Hypoglycemic
(c) R—C(O)—O—CH$_2$—CH$_2$—$\overset{\oplus}{N}(CH_3)_3$, $R = CH_3$ (acetylcholine) / $R = NH_2$ (carbamylcholine)	Short-acting / Long-acting	Cholinergic

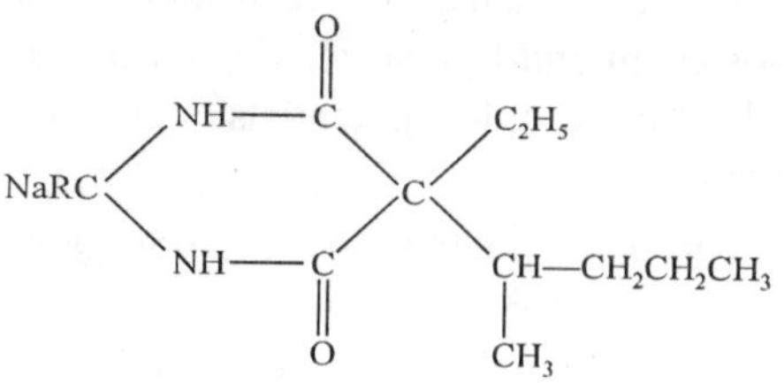

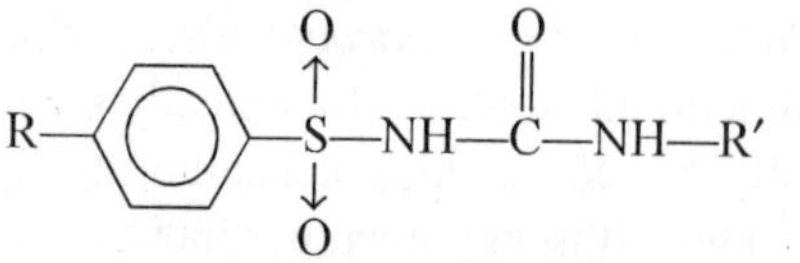

The apparent effect of structure on biological activity may be observed in the following examples where such a change is mainly due to minor group alterations. For instance, alterations in groups in parent structures bring about appreciable difference in the hynotic, hypoglycemic and cholinergic activities. It is pertinent to mention here that such changes only affect the duration of action without any influence on the biological response.

2.3. Structurally Non-specific Drugs

The structurally non-specific drugs include general anaesthetics, hypnotics together with a few bactericidal compounds and insecticides. However, it is important to note here that the biological characteristic of such drugs is solely linked with the physical properties of the molecules rather than the chemical feature.

It has been reported that the toxic depressant concentration of such a drug bears a close resemblance with the physical features, namely : partition coefficient, solubility, vapour pressure and surface activity. As most of these characteristics are entirely based on an equilibrium phenomenon, the ultimate effects of the drug on the biological system are directly linked to an equilibrium model.

2.4. Thermodynamic Activity

Structurally non-specific action is usually due to the accumulation of a drug in an important part of a cell which possesses dominant lipid characteristics. Substances like alkanes, alkenes, alkynes, ketones, amides, chlorinated hydro-carbons, ethers and alcohols display narcotic activity which is directly proportional to the partition coefficient of each individual substances.

2.5. Meyer-Overton and Meyer-Hemmi Theory

Meyer and Overton in 1899* observed that the narcotic efficacy of drugs was directly related to their partition coefficients between oil and water. In other words, the degree of narcosis produced by a chemically indifferent substance solely depends on its ability to attain a certain molar concentration specifically in cell lipids. Such a concentration of a drug in cell lipids is usually governed by two major factors, namely : *first*, the partition coefficient of the drug ; and *secondly*, the least molar concentration of the drug need to be present in the extracellular fluids to cause narcosis in a test animal. Thus the lipid concentration may now be attained by the multiplication of the molar concentration of the narcotic present in the extracellular fluids by its partition coefficient.

For instance, phenobarbital, has narcotic concentration for tadpoles (moles/litre of water) 0.008 and partition coefficient (oil/water) 5.9 gives rise to the value of lipid concentration for narcosis (moles/litre) ($0.008 \times 5.9 = 0.048$).

The results achieved by means of the above experimental approach fully coincides with the theory of Meyer and Hemmi.

Meyer and Overton further expanded their theory and suggested that the correlation may be established and observed between lipid solubility and the central nervous system (CNS) depressant activity profile. The CNS-depressant activity is found to be directly proportional to the partition coefficient of the *'drug substance'*.

Example : *Three* widely different drug substances having altogether divergent chemical structures, such as : *thymol, valeramide-OM* and *2-nitroaniline* exhibited partition coefficient ranging between 0.030 to 950.0 ; and the calculated depressant concentration in cellular lipids varying between 0.021 to 0.045 as summarized below :

---

*H. Meyer. *Arch Exptl. Pathol. Pharmakol,* **42,** 109 (1899).

E. Overton. *Viertljahrsschr. Naturforsch. Ges. Ziirich,* **44, 88** (1899).

CH$_3$

OH

H$_3$C   CH$_3$

Thymol

CH$_3$ CH$_3$

C—CH$_2$—CH—N

O=C—NH$_2$ CH$_3$

Valeramide-OM

NH$_2$

NO$_2$

2-Nitroaniline

S.No.	Compounds	Partition coefficient (n-Octanol/Water)	Required conc. to immobilise Tadpole	Calcd. Depressant conc. in cellular lipids
1.	Thymol	950.0	4.7×10^{-5} moles	0.045 [*i.e.,* $950 \times 4.7 \times 10^{-5}$]
2.	Valeramide-OM	0.030	0.07 moles	0.021 [*i.e.,* 0.30×0.07]
3.	2-Nitroaniline	14.0	0.0025 moles	0.035 [*i.e.,* 14×0.0025]

From the above *three* structural variants one may infer that the value of 0.03 mole in the cellular lipids strongly indicates the required concentration for *'immobilization of tadpoles'*. In short, it adequately confers the bioactivity of a drug substance to its inherent hydrophobicity and hydrophilicity.

2.6.   Ferguson's Theory

Ferguson observed that a number of physical characteristics, namely : partition coefficients, vapour pressure, solubility in water, effect on the surface tension of water, and capillary activity, are usually altered in accordance with the geometric progression in stepping up a homologous series. A plot between the logs of geometric progression values and the number of carbon atoms will form a straight line. Thus, each of the physical characteristics stated earlier, designates a heterogenous phase distribution at equilibrium. According to Ferguson— *"the molar toxic concentrations in a homologous series change on ascending the series not by equal steps but that instead their logarithms decrease by equal steps, it is to be concluded that they are largely determined by a distribution equilibrium between heterogenous phase — the external circumambient phase where the concentration is measured and a biophase (i.e., the phase at the site of action) which is the primary seat of toxic action"*.

The thermodynamic activity of a non-volatile drug may be calculated from the expression S/S_o, where S is the molar concentration of the drug and S_o its solubility. Likewise, the thermodynamic activity values of volatile substances may be calculated by using the expression P/P_o, where P is the partial pressure of the substances in solution and P_o is the saturated vapour pressure of the substance. These findings coined the Ferguson's principle which states that — *"substances which are present at the same proportional saturation in a given medium have the same degree of biological action."*

2.7.   Van der Waal's Constants

Van der Waal's equation may be expressed as :

$$(P + a/V^2)(V - b) = RT,$$

where P = Pressure,

V = Volume,

T = Temperature,

R = Gas constant,

a = Constant for attractive forces between molecules and

b = Constant for volume occupied by molecules of gas.

It has been proved that in the case of narcotic agents the activity enhanced proportionately with the increased volume and size of the molecules. This may further explain their activity in terms of their ability to fit into chemical structures inside the cell membranes thereby checking the passage of essential elements.

2.8. The Cut-off Point

It has been observed at several instances that the biological activity of a homologous series of synthesized analogous may not increase endlessly. Normally one comes across a point at which the biological activity starts falling very rapidly as one ascends a homologous series. This particular situation is termed as the 'cut-off point'. Such a peculiar behaviour may be caused due to either the solubility of the drug in water or the minimum concentration required in water to exhibit the biological response.

Examples : The various examples expatiating the importance and utility of **'cut-off point'** may be observed in the following typical instances :

(1) In a series of *alkylated resorcinols* the length of the *alkyl chain* is directly proportional to the **antibacterial activity,**

(2) Likewise, in a series of *para*-aminobenzoic acid esters the length of the alkyl chain is directly proportional to the **local anaesthetic activity,** and

(3) Similarly, in a homologous series of *n-alkanols* correlation between the length of chain and **antibacterial** activity could be observed.

However, it has been reported that a positive enhancement in the physical characteristics usually take place with the ascending homologous series, such as : viscosity, surface activity, boiling point and above all the partition coefficient ; but the water-solubility decreases appreciably.

Salient Features : The various salient features of **'cut-off point'** are as started under :

(*a*) In ascending homologous series of **'alkylated resorcinols'** the maximum antibacterial activity, in terms of the **phenol coefficient,** is found to be with *6-carbon atoms* located in the side-chain. Hence, the **cut-off point** in this specific instance stands at when $n = 5$.

(*b*) A logical explanation may be given by virtue of the fact that increase in number of C-atoms in the side-chain increases the antibacterial activity, which is due to the enhanced partition coefficient ; and like ultimately affords an enhancement in **penetration of the cell wall.**

(*c*) A situation, when $n = 6$, the antibacterial activity falls sharply due to the drastic poor water solubility of the resultant compounds.

(*d*) Ferguson plotted 'log of aqueous solubility' *Vs* 'number of carbon-atoms in the side-chain' and observed that as the number of C-atoms in the alkyl group increases, a stage is reached when an *'interaction'* between the **"saturation line S"** and **"log water solubility"** takes place, as shown in Fig. 2.1.

(*e*) Fig. 2.2 represents a plot between **"Log conc. (Mole $\times 10^{-6}$ L^{-1}) for bactericidal action"** *Vs* **"Number of C-atoms"**. At 'X', when there are 6-carbon atoms in the side chain, the water solubility decreases than the previous compound having 5-C-atom. Similarly, at 'Y', when there are 9-carbon atoms, the water-solubility dips down considerably. Therefore, the **'cut-off-point'** for the microorganisms are essentially located between (X + Y) when there are 6 and 9 C-atoms present in the alkyl group respectively.

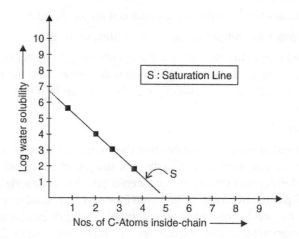

Fig. 2.1. Interaction between 'Saturation-Line S' and 'Log water solubility'.

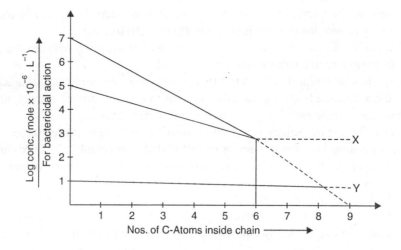

Fig. 2.2. Plot between Log conc. and Nos. of C-Atoms in side-chain.

2.9.   Steric Factors

Interestingly, it is absolutely necessary for a *'drug molecule'* to engage into a viable and plausible interaction either with a drug receptor or with an enzyme, it has got to *first* approach ; and *secondly*, attach to a binding site. Obviously, this essentially demands certain specific criteria that a *'drug molecule'* must fulfil, for instance : bulk, size and shape of the **'drug'**. Precisely, the *'bulky substituent'* more or less serve as a shield that eventually hinders the possible and feasible interaction taking place between a **'drug'** and a **'receptor'**.

Meticulous and intensive in-depth studies in this particular aspect has practically failed to *justify* and *quantify* steric characteristics in comparison to quantifying either electronic or hydrophobic characteristics. In fact, a plethora of methodologies have been tried and tested to ascertain the steric factor(s). In the present context only *three* such methods shall be discussed briefly, namely :

   (*a*) Taft's Steric Factor (Es),

   (*b*) Molar Refractivity (MR), and

   (*c*) Verloop Steric Parameter.

2.9.1.   Taft's Steric Factor (Es) :

   An attempt has been mode to quantify the steric features of various substitutents (*i.e.,* functional moieties) by the help of Taft's steric factor (Es).

   In fact, there are *three* predominant constants, namely :

   (*i*) Hammett substitution constant (σ),

   (*ii*) Resonance effect (R), and

   (*iii*) Inductive effect (F).

can only be employed for aromatic substituents ; and are hence suitable exclusively for such **'drugs'** that contain *aromatic rings.*

   A. Hammett Substitution Constant (σ). It is a measure of either the *electron-withdrawing* or *electron-donating* capability of a substituent (*i.e.,* the functional moiety). Hammet substitution constant may be determined conveniently by actual measurement of the dissociation of a series of benzoic acid substituted derivatives *vis-a-vis* the dissociation of pure benzoic acid itself.

   However, benzoic acid being a **'weak-acid'** gets partially ionized in an aqueous medium (H_2O) as depicted under :

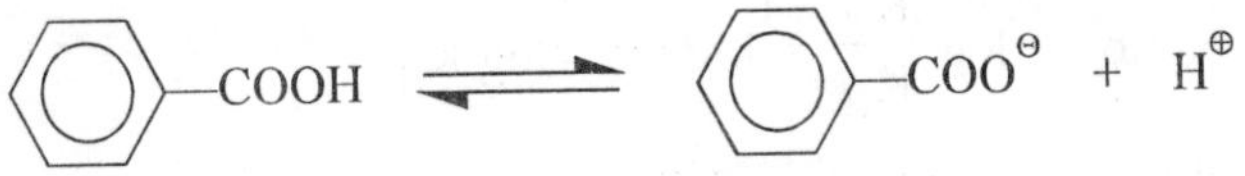

$$\langle\bigcirc\rangle\!-\!COOH \rightleftharpoons \langle\bigcirc\rangle\!-\!COO^{\ominus} + H^{\oplus}$$

   Explanation. An equilibrium is established between the two distinct species *i.e.,* the *ionized* and *non-ionized* forms. Thus, the relative proportion of the said two species is usually termed as the **'dissociation'** or **'equilibrium'** constant ; and invariable designated by K_H (wherein the *'subscript H'* represents/signifies that there is no substituents normally attached to the aromatic nucleus *i.e.,* the phenyl ring).

$$\therefore \quad K_H = \frac{[PhCOO^{\ominus}]}{[PhCOOH]}$$

   As soon as a substituent is strategically positioned on the aromatic (phenyl) ring, this *'equilibrium'* gets imbalanced. At this juncture *two* situations may crop up distinctly by virtue of the fact that :

   (*i*) An electron-withdrawing moiety, and

   (*ii*) An electron-releasing (donating) moiety

could be present in the aromatic ring thereby giving rise to altogether different electronic status to the **'Aryl Nucleus'.**

   (*a*) **Electron-Withdrawing Moiety.** A host of electron-withdrawing groups, such as : NO_2, CN, COOH, COOR, $CONH_2$, CONHR, $CONR_2$, CHO, COR, SO_2R, SO_2OR, NO ; cause and result in the aromatic ring (with a π electron cloud both on its top and bottom) having a marked and stronger electron withdrawing and stabilizing influence on the carboxylate anion as illustrated below. Hence, the overall equilibrium shall influence and shift more to the ionized form thereby rendering the '*substituted benzoic acid*' into a **much stronger acid** (benzoic acid as such is a weak acid). The resulting substituted benzoic acid exhibits a larger K_X value (where, X designates the substituent on the aromatic nucleus) (see Fig. 2.3.).

(*b*) **Electron-Donating Moiety :** A plethora of electron-donating groups, for instance : R, Ar, F, Cl, I, Br, SH, SR, O^-, S^-, NR_2, NHR, NH_2, NHCOR, OR, OH, OCOR, influence and render the ensuing aromatic ring into a distinctly much less stable to stabilize the *carboxylate ion*. Thus, the equilibrium gets shifted to the left overwhelmingly ; thereby ultimately forming a relatively **much weaker acid** having a smaller K_X value (see Fig. 2.3).

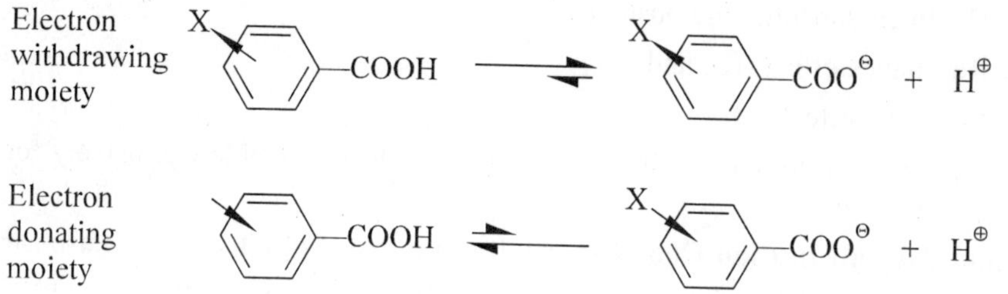

Fig. 2.3 : Influence of Substituent Moiety X on the Status of Equilibrium in Reaction.

Now, the Hammett substitution constant $[\sigma_X]$ with reference to a specific substituent X is usually defined by the following expression :

$$\sigma_X = \log \frac{K_X}{K_H} = \log K_X - \log K_H$$

Therefore, for all benzoic acids essentially possessing electron-withdrawing substituents shall have larger K_X values than the parent benzoic acid itself (K_H) ; thereby the value of Hammett substitution constant σ_X for an electron-withdrawing substituent shall be always *positive*.

Similarly, for most benzoic acid variants essentially having electron-donating substituents shall have comparatively smaller K_X values than benzoic acid itself ; and, therefore, the value of Hammett substitution constant σ_X for an electron-donating substituent will always be *negative*.

Furthermore, the Hammett substitution constant essentially and importantly takes cognizance of *two* vital and critical supportive effects, such as : *resonance effect,* and *inductive effect.* Consequently, the value of σ with respect to a specific substituent may exclusively depend upon whether the attached '*substituent*' is located either at *meta*-or at *para*-position. Conventionally, such particular substituent is invariably indicated by the subscript *m* or *p* first after the symbol σ.

Example : The nitro ($-NO_2$) substituent on the benzene nucleus has two distinct σ values, namely : $\sigma_m = 0.71$ and $\sigma_p = 0.78$.

Explanation. From the σ values, one may evidently observe that the electron-withdrawing strength at the *para*-position is solely contributed by both '*inductive*' and '*resonance*' effects combinedly which justifies the greater value of σ_p, as shown in Fig. 2.4(*a*). Likewise, the *meta*-position, only affords the electron-withdrawing power by virtue of the '*inductive*' influence of the substituent ($-NO_2$ group), as shown in Fig. 2.4(*b*).

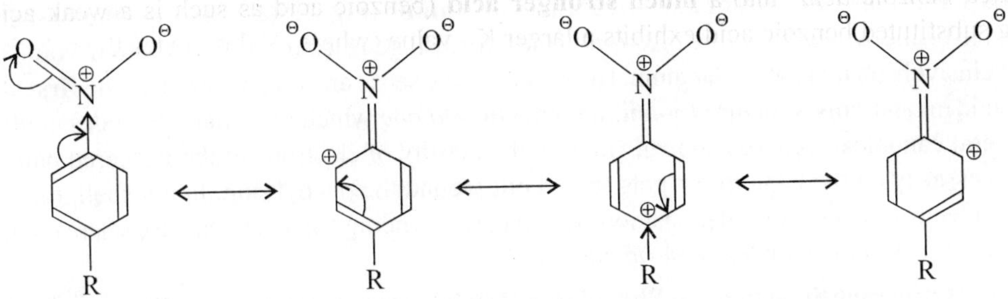

Fig. 2.4(*a*) : Electronic Influence on R caused due to Resonance
and Inductive Effects of *p*-nitro Function.

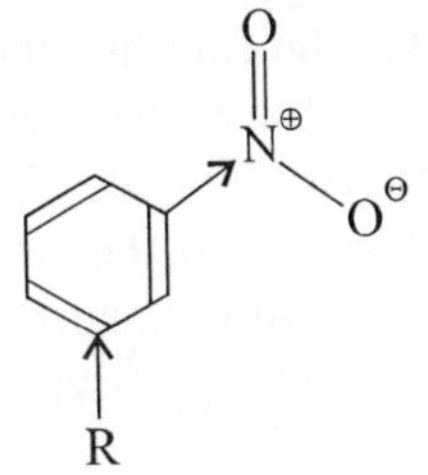

Fig. 2.4(*b*) : Electronic Influence on R caused due to
Inductive Effect alone of *m*-nitro Function.

B. Resonance Effect (R) : It has been observed that '**resonance**' mostly gives rise to an altogether *different distribution of electron density* than would be the situation if there existed absolutely no resonance.

Examples : The resonance effects, as observed in *two* electron donating functional moieties, such as : –NH$_2$ (amino) ; and –OH (hydroxyl), attached to an aromatic nucleus, are depicted in Fig. 2.5(*a*) and (*b*) as under :

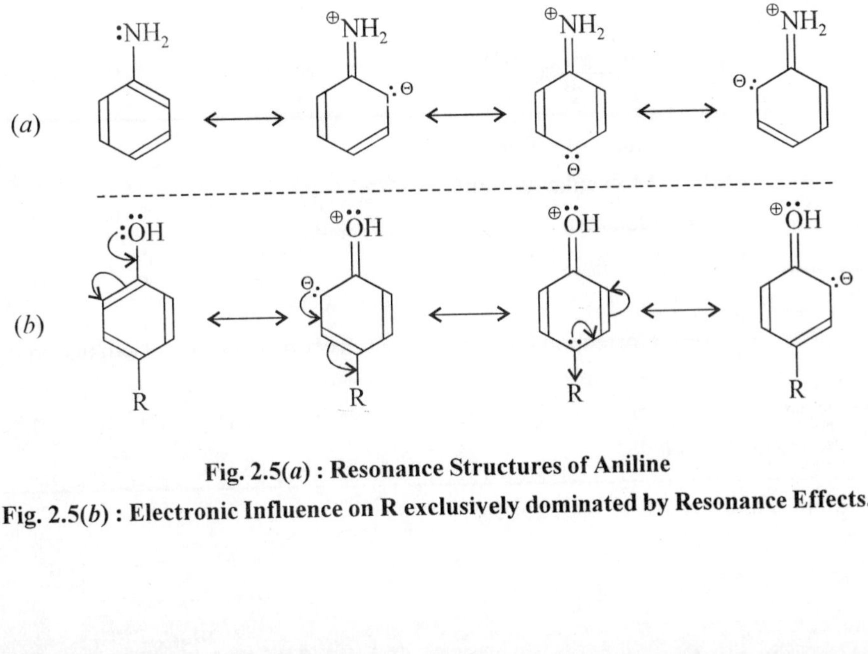

Fig. 2.5(*a*) : Resonance Structures of Aniline

Fig. 2.5(*b*) : Electronic Influence on R exclusively dominated by Resonance Effects.

Explanations : For Resonance structures of Aniline [Fig. 2:5(*a*)] : In case, the **first structure** happened to be the *'actual structure of aniline'*, the two unshared electrons of the N-atom would certainly reside exclusively on that particular atom. However, in true sense and real perspective the **first structure** is **not** the ideal and only structure for aniline but a *hybrid one* which essentially includes contributions from several canonical forms as shown, wherein the density of electrons of the unshared pair does not reside necessarily on the N-atom but gets spread out around the phenyl ring. In nut shell, this observed density of electron at one particular position (with a corresponding enhancement elsewhere) is invariably known as the *'resonance'* or *'mesomeric effect'*.

For Resonance Structures of Phenol [Fig. 2:5(*b*)] : Here, the influence of R at the *para* position, and the electron-donating effect caused due to resonance is more marked, pronounced and significant as compared to the electron-withdrawing influence due to induction.

C. Inductive Effect (F) : The C—C single bond present in *'ethane'* has practically no polarity as it simply connects two equivalent atoms. On the contrary, the C—C single bond in *'chloroethane'* gets solemnly polarized by the critical presence of the electronegative *chlorine-atom*. In fact, the prevailing polarization is actually the sum of *two* **separate effects.** *First*, being the C-1 atom that has been duly deprived of a part of its electron density evidently by the greater electronegativity of Cl. It is, however, compensated partially by drawing the C—C electrons located closer to itself, thereby causing polarization of this bond and consequently rendering a slightly positive charge on the C-2 atom as shown below :

$$\underset{2}{H_3 C} \overset{\delta\delta^{\oplus}}{} \longrightarrow \underset{1}{CH_2} \overset{\delta^{\oplus}}{} \longrightarrow \overset{\delta^{\ominus}}{Cl}$$

Chloroethane

Secondly, the effect is caused not through bonds, but directly either through *space* or *solvent molecules*, and is usually termed as the **field effect.**[*]

2.9.2. Molar Refractivity (MR)

Another vital and equally important criterion to measure the *'steric factor'* is adequately provided by a parameter called as **molar refractivity** (MR). It is usually designated as a simple measure of the volume occupied either by an individual atom or a cluster (group) of atoms. However, the MR may be obtained by the help of the following expression :

$$MR = \frac{(n^2 - 1)}{(n^2 + 2)} \times \frac{MW}{d}$$

where, n = Index of refraction,

MW = Molecular Weight,

d = Density,

MW/d = Volume and

$\dfrac{n^2 - 1}{n^2 + 2}$ = Correction factor (*i.e.,* how easily the substituent can undergo polarization)

Molar refractivity is specifically significant in a situation when the substituent possesses either π *electron* or *lone pairs of electrons*.

---

[*]Roberts ; Moreland., *J.Am. Chem. Soc.,* **75,** 2167, 1953.

2.9.3.   Verloop Steric Parameter

The unique revelation and wisdom of a latest computer researched programme termed as **sterimol** has indeed helped a long way in measuring the *steric factor* to a reasonably correct extent. It essentially aids in the calculation of desired steric substituent values (otherwise known as Verloop steric parameters) based on various standard physical parameters, such as : Van der Waals radii, bond lengths, bond angles, and ultimately the proposed most likely conformations for the substituent under examination. It is, however, pertinent to mention here that unlike the Taft's steric factor (E_S) (see section 2.9.1) the Verloop steric parameters may be measured conveniently and accurately for any substituent.

Example : *Carboxylic acid (say, Benzoic Acid) :* The ensuing Verloop steric parameters for a carboxylic acid moiety are duly measured as shown in Fig. 2.5 below, where L represents the length of the substituent, and $B_1 - B_4$ designate the radii (*i.e.,* **longitudinal** and **horizontal**) of the *two* functional groups *viz., carboxyl and hydroxyl* (—O—H).

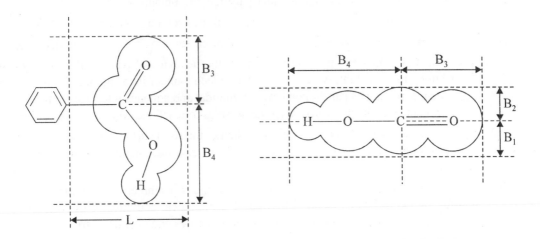

Fig. 2.5 : Verloop Steric Parameters for a Carboxylic acid (—COOH) moiety.

Interestingly, most quantitative structural activity relationship (QSAR) studies usually commence by considering σ (Hammett substitution constant) and, in case there exists more than one substituent, the σ values are represented in a summed up manner as $\Sigma\sigma$. Keeping in view the enormous quantum of *synthetic newer target drug molecules*, it has now become almost necessary and possible either to modify/refine or fine tune-up the QSAR equation. In fact, a substituent's resonance effect (R) and inductive effect (F) may be quantified as far as possible with the help of available *'tables of constants'*. In certain instances one may evidently observe that :

- a substituent's effect on biological activity is solely on account of F rather than R, and *vice versa.*

- a substituent exerts a more prominent and appreciable activity when strategically located at a specific position on the aromatic nucleus ; and moreover it may also be embedded in the 'equation' appropriately.

2.10. Hansch Equation

Integrating various factors, namely : Taft's steric factor, resonance, inductive, Verloop steric parameters with the partition behaviour of *'drug molecules'* Hansch* and Fujita** exploited these principles in determining the establishing quantitative structure-activity relationship (QSAR) of **drugs,** which has undergone a sea change both in expansion and improvement with the help of computer researched soft-wares.

The *hydrophobic characteristic,* designated by π_x, may be correlated to a drug's distribution pattern, within which a given substituent 'x' affects molecular behaviour and conduct with regard to its :

➤   distribution and transport, and

➤   drug-receptor activities.

The hydrophobic characteristic π_x of a drug substance may be expressed as :

$$\pi_x = \log P_x - \log P_{yH}$$

where,   log P = logarithm of 1-octanol-water partition coefficient

y = A parent compound (*i.e.,* an unsubstituted reference compound/drug).

Salient Features : The various salient features of Hansch equation are as enumerated under :

(1) Value of π is indicative, to a certain extent, the behavioural pattern of a *'substituent'* contributing to the solubility behaviour of a molecule under investigation. It also reflects upon the manner it gets partitioned between lipoidal and aqueous interfaces in the reputed compartments it happens to cross as a *'drug'* so as to reach the *'site of action'* ultimately.

(2) It is, however, not very clear and definite whether the solid surface of a *'drug'* undergoes adsorption on colloidally suspended plasma proteins while establishing the hydrophobic characteristisc π.

(3) Interestingly, the concurrent considerations of π and σ (Hammett's constant) has evolved gainful vital correlations existing between the biological activities of quite a few drug substances with their corresponding physical properties and chemical structures.

Therefore, Hansch's correlations piece together valuable information(s) of a newly designed *'drug molecule'* in a more plausible, predictive and quantifiable manner than before — and apply it to a biological system more logistically and judiciously. This particular concept and idea was further substantiated and expanded by assuming that all the *three* substituents *viz.,* π, σ and Es, exert a significant effect on the efficacy and hence the potency of a *'drug substance'* ; and are found to be additive in nature independently. Therefore, it has given rise to the underlying linear Hansch equation :

$$\log \left(\frac{1}{C}\right) = a \log P + b\, E_S + \rho(\sigma) + d$$

where ,   C = Concentration of drug producing the biological response being measured,

log P = Substituent constant for solubility (*i.e.,* π),

E_S = Taft constant (for steric effects),

ρ = (rho) Proportionality constant designating the sensitivity of the reaction to electron density.

---

* Hansch *et al. J.Am. Chem. Soc.,* **85,** 2817, 1963, *ibid,* **86,** 1616, 1964 ;

** Fujita *et al. J.Am. Chem. Soc.,* **86,** 5175, 1964.

σ = Hammett substitution constant

a, b, d = Constants of the system (which are determined by computer to obtain the *'best fitting line'*).

It is pertinent to state at this juncture that *not* all the parameters shall necessarily be significant.

Example : β-*Halo-arylamines :* The *adrenergic blocking profile* of β-halo-arylamines was observed to be solely related to the two constants, π and σ ; and specifically excluded the steric factor altogether.

i.e., $\log\left(\dfrac{1}{C}\right) = 1.22\ \pi - 1.59\ \sigma + 7.89$

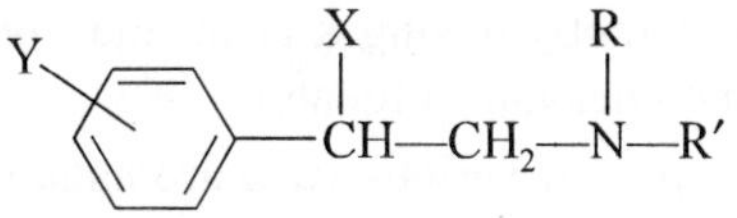

The aforesaid equation offers a dictum that the *'biological response'* gets enhanced if the substituents possess a positive π value and a negative σ value ; or more explicitly the substituents must preferentially be both hydrophobic in nature and electron donating in character.

It has been established beyond any reasonable doubt that there exists no correlation between the **π factor** and the **P value** ; therefore, it is quite feasible to have Hansch equations essentially comprising of these two stated components :

Example : *Phenanthrene aminocarbinols :* An analogous series of more than one hundred phenanthrene aminocarbinols were successfully synthesized and subsequently screened for their *antimalarial profile.* Interestingly, the analogous series fitted appropriately into the following version of Hansch equation :

$$\log\left(\frac{1}{C}\right) = -0.015\,(\log P)^2 + 0.14\log P + 0.27\,\Sigma\pi_x + 0.40\,\Sigma\pi_y + 0.65\,\Sigma\sigma_x + 0.88\,\Sigma\sigma_y + 2.34$$

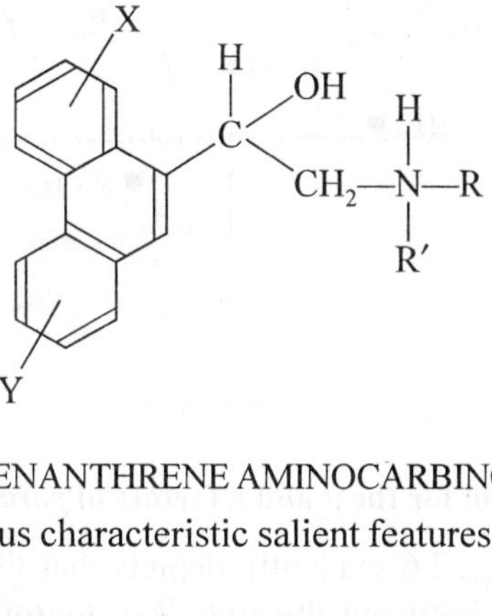

PHENANTHRENE AMINOCARBINOL

Salient Features : The various characteristic salient features that may be derived from the above equation are, namely :

(1) As the hydrophobicity of the molecule (P) enhances there exists a very nominal increase in the antimalarial activity.

(2) The corresponding constant is low (0.14) which reflects that the increase in antimalarial activity is also low.

(3) The value of $(\log P)^2$ evidently reveals that there prevails a maximum P value for activity.

(4) Further the above equation suggests that the antimalarial activity gets enhanced appreciably when the hydrophobic moieties are strategically located either on ring 'X' or more specifically on ring 'Y'. It further ascertains that the hydrophobic interaction(s) are virtually taking place at these sites.

(5) The electron-withdrawing substituents on rings 'X' and 'Y' contribute enormously to the antimalarial activity ; however, the effect is more on ring 'Y' than in ring 'X'.

2.11.   The Craig Plot

The Craig plot is nothing but an actual plot between the 'π *factor*' taken along the **X-axis** and the 'σ *factor*' taken along the **Y-axis**, thereby having a clear and vivid idea with regard to the relative properties of different functional moieties (substituents).

Fig. 2.6 illustrates the Craig plot of various *para* aromatic substituents for the σ and π factors respectively.

Salient Features : The various advantageous salient features of a Craig plot are enumerated as under :

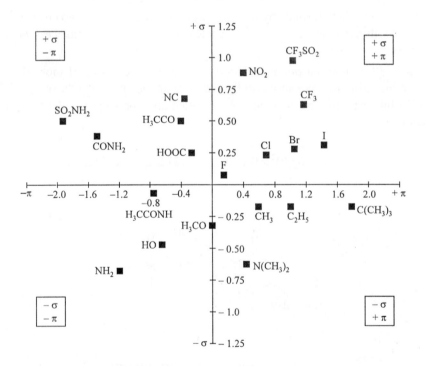

Fig. 2.6 : The Craig Plot for the σ and π factors of *para*-Aromatic Substituents.

(1) The Craig plot in Fig. 2.6 evidently depicts that there is absolutely no clearly defined overall relationship between the two '*key factors*' σ and π. However, the various functional moieties (*i.e.*, substituents) are strategically positioned around all the four quadrants of the plot based on their inherent physicochemical status and integrity.

(2) From the above Craig plot one may obviously identify the substituents that are particularly responsible for +ve π and σ parameters, $-$ve π and σ parameters, and lastly one +ve and one $-$ve parameter.

(3) Further it is quite convenient and easy to observe which substituents have nearly identical π values, such as : *dimethyl-amino, fluoro* and *nitro* on one hand ; whereas, *ethyl, bromo, trifluoromethyl,* and *trifluoromethyl sulphonyl* moieties on the other are found to be almost located on the same **'vertical line'** on the Craig plot. Thus, theoretically all these functional groups are legitamately interchangeable on a newly designed drug molecule wherein the major critical and principal factor that significantly affects the **'biological characteristics'** is essentially the π factor.

Interestingly, in the same vein, the various functional moieties that are located on the **'horizontal line'**, for instance : methyl, ethyl *tert-butyl* on one hand ; whereas, carboxy, chloro, bromo, and iodo moieties on the other can be regarded and identified as being *iso-electric* in nature or possessing identical σ values.

(4) QSAR studies are exclusively and predominantly governed and guided by the Craig plot with regard to the various substituents in a new drug molecule. Therefore, in order to arrive at the most preferred **'accurate equation'** essentially consisting of π and σ, —the various structural analogues must be synthesized having appropriate substituents pertaining to each of the four quadrants.

Examples :

(*i*) **Alkyl Moieties :** These substituents contribute exclusively +ve p values and $-$ve s values.

(*ii*) **Acetyl Moieties :** These are responsible for attributing $-$ve π values and +ve σ values.

(*iii*) **Halide Groups :** These functional moieties essentially enhance both electron-withdrawing characteristics and hydrophobicity in the *'drug molecule'* by virtue of their +ve σ and +ve π effects.

(*iv*) **Hydroxy Groups :** These functional moieties exert progressively more hydrophilic and electron-donating characteristics on account of the $-$ve π and $-$ve σ effects.

(5) Importantly, the very establishment and derivation of Hansch equation will certainly give a better reliable and meaningful clue with regard to attaining a reasonably good biological property based on the fact whether π and σ must be $-$ve or +ve in character. However, further improvements in the *'drug molecule'* could be accomplished by exploring various other possible substituents picked up judiciously from the relevant quadrant (see Fig. 2.6).

Example : In case, the Hansch equation rightfully demands that +ve σ and $-\pi$ values are an absolute necessity, additional relevant substituents must be picked up from the top-left quadrant.

(6) The Craig plot may also be exploited to compare the *MR* and *hydrophobicity*.

2.12. The Topliss Scheme

Keeping in view the enormous cost incurred with regard to the synthesis of a large range of structural analogues necessarily required for a *Hansch equation,* it has become almost necessary to restrict the synthesis of a relatively lesser number of drug molecules that may be produced in a limited span of time having viable biological activity. Based on the actual outcome of the biological activity *vis-a-vis* the actual structure of the 'drug' ultimately helps to determine the next analogue to be synthesized.

The Topliss scheme is nothing but an organized **'flow diagram'** which categorically permits such a procedure to be adopted with a commendable success rate.

In actual practice, however, there are *two* distinct Topliss Schemes, namely : (*a*) For *aromatic substituents* ; and (*b*) For *aliphatic side-chain substituents.* It is pertinent to mention here that the said *two schemes* were so meticulously designed by taking into consideration both **electronic** and **hydrophobicity** features (*i.e.,* substituents) with a common objective to arrive at the **'optimum biological active substituents'.**

It may be made abundantly clear and explicit that the Topliss Schemes are not a replacement for the Hansch analysis. Hence, the former may be made useful and effective only when a good number of tailor-made structures have been designed and synthesized.

[A]. Fig. 2.6. represents the *Topliss Scheme for Aromatic substituents* ; and has been based on the assumption that the *'lead compound'* essentially possesses a single monosubstituted aromatic ring and that it has already been screened for its desired biological activity.

Salient Features : The various salient features with respect to the Topliss scheme for aromatic substituents are as described below :

(1) 4-chloro derivative happens to be the *'first structural analogue'* in this particular scheme perhaps because it is easy to synthesize.

(2) The π and σ values are both positive by virtue of the fact that the chloro substituent is much more hydrophobic and electron-withdrawing than hydrogen-atom.

(3) The synthesized chloro-analogue is subjected to the biological activity measurements accordingly.

(4) Three situations may arise, namely : (*a*) analogue possessing less activity (L) ; (*b*) equal activity (E) ; and (*c*) more activity (M). Thus, the type of observed activity is solely the determining factor as to which *'branch'* of the Topliss scheme is to adopted next.

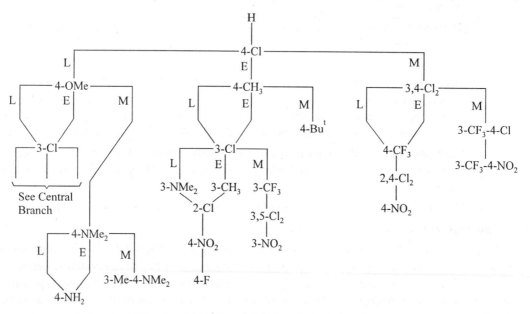

Fig. 2.6 : Topliss Scheme for Aromatic Substituents.

(5) Further line of action towards the synthesis of structural analogues of 4-chloro aromatic substituents are entirely guided and based on the following *three* options, namely :

Biological Activity	Series Followed	Next Analogues Synthesized
(*a*) Increases	M-series	3, 4-Dichloro substitued derivatives
(*b*) Same profile	E-series	4-Methyl derivatives
(*c*) Decreases	L-series	4-Methoxy derivatives

(6) Let us consider the second analogous series which shows the same biological activity. The various situations that may arise are as follows :

(*i*) *4-Chloro derivative enhances the desired biological property* :

As the Cl-substituent exerts both positive π and σ values it evidently shows that either one or both of three characteristic features are quite critical and important to biological property. In case, both characteristic features are important, addition of the second Cl-moiety shall enhance the biological activity to the positive side further more. If it fails, there may exist either an excess hydrophobic character or an obstructive steric hindrance is exhibited. Thus, the situation demands further modification based on subsequent biological screening *vis-a-vis* the comparative importance and status of π as well as the steric features.

(*ii*) *4-Chloro derivative lowers the desired biological activity* :

It gives a clue that either the location of the *para*-substituent is absolutely unsuitable sterically or the $-$ve π and/or σ values are prominently important with regard to the biological activity. It has been established that the reduced activity is solely attributed due to an unfavourable σ effect ; and hence, one may assign the *para*-methoxy moiety as the next probable substituent having a $-$ve σ factor. If by doing so there is an apparent improvement in activity, further alterations are accomplished with a view to ascertain the prevailing relative importance of the σ and π values. Now, if the above modifications *i.e., para*-methoxy moiety fails to make any improvement in the activity, one may draw an inference that an undesired steric factor is playing the havoc, and the next possible entrance is of the *meta*-chloro group. However, further modifications of this functional group shall then be persued as depicted in the middle series of Fig. 2.6.

(*iii*) *4-Methyl derivative equals the desired biological activity* :

In this specific instance, the overall biological activity of the 4-chloro structural analogue exerts practically no change as compared to the **'lead compound'**. It might have emanated from the *'drug substance'* essentially looking for a negative π value and a positive σ value. As it is quite evident that the two said values attributed by the chloro moiety are apparently positive, the useful effect of the positive π value should have been nullified due to the detrimental influence of a positive σ value. Therefore, the most preferred substituent would be the *para*-methyl group, which adequately possesses positive π value and negative σ value. In case, it still exhibits no useful effect, one may draw a conclusion that there exists an unfavourable steric interaction prevailing at the *para*-position. Hence, the next preferable line of action would be the introduction of chloro group at the *meta*-position. However, any additional changes shall affect the values attributed by both π and σ factors.

The **Topliss scheme** has been thoroughly investigated, tested and above all validated by various researcher after evaluating their structure-activity relationships (SARs) for a host of *'drug substances'*.

Example : *Substituted phenyltetrazolylalkanoic acid :* A total of 28 structural analogues of substituted phenyltetrazolylalkanoic acids were synthesized in the laboratory and screened duly for their anti-inflammatory activities. Nevertheless, if the whole exercise would have been based on the **Topliss Scheme** only the first eight compounds (out of 28) should have yielded *three* most active compounds as given below :

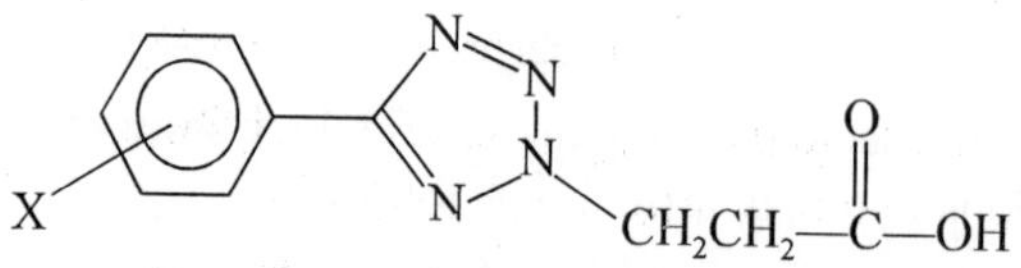

Substituted phenyltetrazolylalkanoic acids

Sequence of Synthesis	X	Biological Activity (Observed)	Maximum Potency (Observed)
1	H	—	
2	*para*-Cl	L	
3	*para*-OCH$_3$	L	
4	*meta*-Cl	M	++++
5	*meta*-CF$_3$	L	++++
6	*meta*-Br	M	
7	*meta*-I	L	
8	3, 5—Cl$_2$	M	++++

L = Less Activity
M = More Activity
E = Equal Activity

[B] Fig. 2.7 designates the **Topliss Scheme** for the **Aliphatic** side-chains was adopted in the same vein and rationale as the aforementioned *'aromatic scheme'* (section 'A'). The present scheme is expanded exactly in the same fashion for the side functional moieties strategically linked to a variety of such functional groups as : amine, amide or carbonyl.

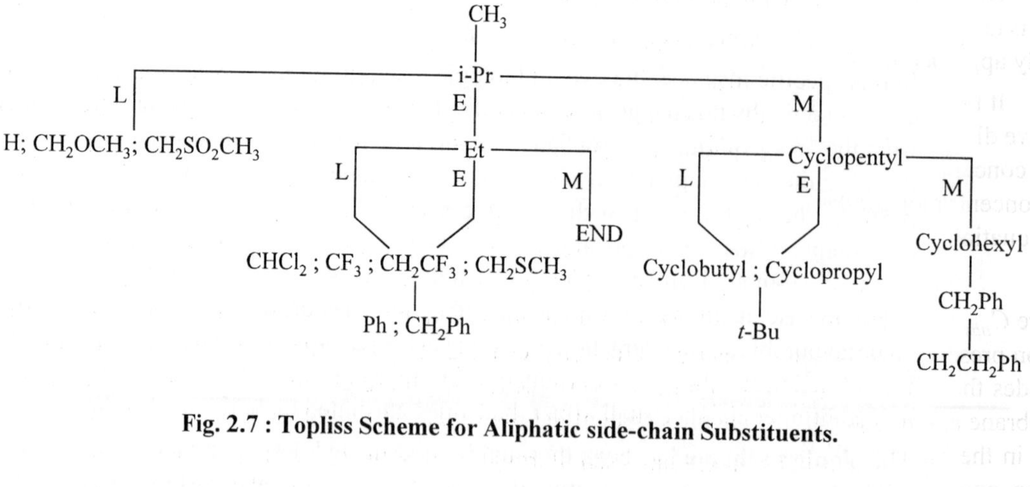

Fig. 2.7 : Topliss Scheme for Aliphatic side-chain Substituents.

Interestingly, the **Topliss Scheme** helps to make a clear cut distinction between the two pronounced physical characteristic features, namely : electronic effect, and hydrophobic effect, caused due to the various substituents ; and not the steric characteristic features. Perhaps that could be the possible line of thought judiciously utilized in the selection of appropriate substituents so as to reduce any steric differences. Let us have an assumption that the 'lead compound' possesses a—CH_3 functional moiety.

A variety of typical situations may crop up during the said studies :

(a) *Rise in anti-inflammatory activity* : A cyclopentyl moiety is now utilized which provides a much larger π value, and simultaneously holds the steric influence to a bear minimum level. In case, a further rise in activity is observed one may institute more hydrophobic substituents. On the contrary if the activity fails to rise, there could be two possible reasons, namely : (*i*) optimum hydrophobicity has superseded ; and (*ii*) electronic effect (σ_1) has triggered action. Of course, an elaborated further study would reveal and ascertain the exact substituents to substantiate which of the two explanations stands valid.

(b) *Static anti-inflammatory activity* : The activity exerted by the isopropyl structural analogue almost remains the same as that of the methyl one. It could be explained most logically that both methyl and isopropyl moieties are actually located on either side of the 'hydrophobic optimum'. Hence, an intermediate functional group *i.e.*, an *ethyl group*, possessing an intermediate π value, is employed as the next substituent (see Fig. 2.7). In case, it still registers practically no plausible or appreciable improvement in the activity profile, one may switch over to an electron-withdrawing moiety instead of an electron-donating moiety, having identical π values, as futuristic suggestive approach.

3. FACTORS GOVERNING ABILITY OF DRUGS TO REACH ACTIVE SITE

There are certain vital factors that govern the ability of a drug to reach the active site soon after its administration through various modes known to us. These factors essentially include absorption, distribution, biotransformation (metabolism) and elimination. However, in all these instances, the drug molecule has to cross a few biological membrane in one form or the other. These factors shall now be treated briefly with appropriate typical examples wherever necessary.

3.1. Absorption

Biological membranes play a vital role towards the absorption of a drug molecule. Soon after a drug is taken orally, it makes its way through the gastrointestinal tract, cross the various membranes and finally approach the site or cell where it exerts its desired pharmacological action.

It has been observed that a plethora of drug molecules normally cross biological membranes by passive diffusion from a region of high drug concentration (*viz* : gastrointestinal tract) to a region of low drug concentration (*viz* : blood). However, the rate of diffusion solely depends upon the magnitude of the concentration gradient (ΔC) across the biological membrane and may be represented by the following equation :

$$\text{Rate} = -K\Delta C = -K(C_{abs} - C_{bl}) \qquad \qquad ...(1)$$

where C_{abs} represents the concentration of drug at the absorption site and C_{bl} is the respective concentration present in the blood. The constant of proportionality K, is a complex constant which essentially includes the area and thickness of membrane, partition of drug molecule between aqueous phase and membrane and finally the diffusion coefficient of the drug. It may be assumed that the concentration of drug in the blood is fairly negligible as compared to the concentration in the gastrointestinal lumen. Hence, equation (1) simplifies to

$$\text{Rate} = -KC_{abs} \qquad \qquad \qquad ...(2)$$

As one may observe from equation (2) that absorption by passive diffusion is nothing but a first-order process, hence the rate of drug absorption is directly proportional to the concentration of drug at the absorption site. In other words, the larger the concentration of drug, the faster is the rate of absorption. At any time after the administration of the drug, the percentage of the dose absorbed remains the same irrespective of the dose administered.

Lipid solubility of the drug is the determining factor for the penetration of cell membranes. Therefore, the passage of many drug molecules across the membranes of the skin, oral cavity, bile, tissue cells, kidneys, central nervous system and the gastrointestinal epithelium is very much related to the lipid solubility of the drug molecule.

3.2. Distribution

As soon as a drug finds its way into the blood stream, it tries to approach the site of biological action. Hence, the distribution of a drug is markedly influenced by such vital factors as tissue distribution and membrane penetration, which largely depends on the physico-chemical characteristics of the drug. For instance, the effect of the ultra-short acting barbiturate thiopental may be explained on its dissociation constant and lipid solubility. It is worthwhile to observe here that the duration of thiopental is not influenced by its rate of excretion or metabolism, but by its rate of distribution.

3.3. Metabolism (Biotransformation)

When a drug molecule gets converted into the body to an altogether different form, which may be either less or more active than the parent drug, the phenomenon is termed as biotransformation. Mostly the drug metabolism occurs in liver. In fact, a number of pathways are genuinely responsible for carrying out various diverse metabolism reactions in the body.

It may be pertinent to observe here that most of the metabolised products are usually more polar in character than the parent drug molecule. This increased polarity renders the metabolism less absorbable through the renal tubules and also makes it transient in the body.

A large number of barbiturates are metabolized by liver microsomes. Isoniazid is quickly metabolized in *Japanese race* to the extent of 86.7%, whereas approximately half of it (44.9%) in *American and Canadian whites*. This disparity is due to the genetic differences in the said races.

In a broader sense a plethora of metabolic processes which usually **detoxify** the foreign substances *in vivo*, such as : oxidation, reduction, hydrolysis, esterification or conjugation ; thereby rendering the *'drug substance'* normally more water-soluble, so as to enhance its excretion from the body. It has been duly observed that in a good number of cases a 'drug metabolite' actually may serve as the *active compound,* almost showing identical biological activity to the original compound. Interestingly, after having undergone several biotransformations, the ultimate modified form of the drug is excreted finally.

Though liver is considered to be the **primary site** of *'detoxification'* ; however, many enzymatic degradation processes may also take place in the stomach, intestine, pancreas, and other locations in the body. Generally, the metabolic processes occurring in the liver may be conveniently categorized under the following **two** heads, namely :

> (*a*) *Functional Group Changes :* Here, the *'drug substance'* undergoes functional group changes, for instance : side-chain or ring hydroxylation, reduction of nitrogroup, reduction, aldehyde oxidation, deamination or dealkylation, and

(b) *Conjugation :* In this instance, the *'drug substance'* undergoes conjugation whereby the metabolized product subsequently combines with various solubilizing groups, such as : **glycine** (an amino acid) or glucuronic acid (glucuronides) to result into the formation of excretable conjugates ultimately.

Hence, during the course of designing/development of a **'new target-drug-molecule'** the *'medicinal chemist'* must take into cognizance of such metabolic phenomena and modify the structure of the drug substance in question so as to alter the course in which it should have been metabolized otherwise.

3.4. Excretion

Excretion of drugs from their sites of action is of paramount importance and may be effectively carried out with the help of a number of processes, namely : renal excretion, biliary excretion, excretion through lungs and above all by drug metabolism (biotransformation).

Drugs which are either water-soluble or get metabolized gradually are mostly eliminated through the kidneys by the aid of these three essential phenomena, *viz* : secretion, glomerular filtration and tubular reabsorption. For instance, probenecid considerably retards tubular secretion of penicillin thereby enhancing its duration of action appreciably.

Another aspect of excretion is the biliary excretion of drugs or its metabolites which essentially affects excretion of drugs by liver cells into the bile and subsequently into intestine. Invariably, a drug undergoes 'enterohepatic cycling', *i.e.,* instead of its elimination through the faeces it gains entry into the system through the intestines, *eg.,* penicillin, fluorescein, etc.

3.5. Intramolecular Distances and Biological Activity

The intramolecular distance is regarded as a structural feature of the drug molecule which falls within the regimen of physical property. In can be effectively measured eitehr by X-ray or by electron diffraction measurements.

The intramolecular distance present in the grouping $-X$ CH_2CH_2N- between the nitrogen atom and X (where $X = N$, O etc.) that could be seen in a variety of medicinal compounds (*i – iii*) as stated below falls in the vicinity of 5A :

| (*i*)   Procaine | (*ii*)   Acetylcholine |
| (local anaesthetic) | (parasympathomimetic) |

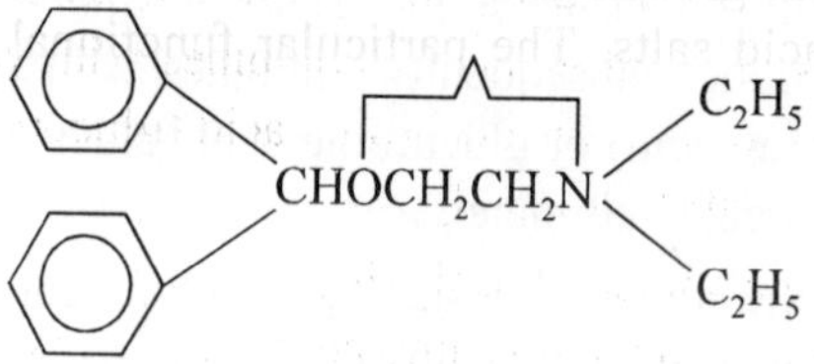

Diphenhydramine
(antihistaminic)

However, the distance between the two centres may alter based on the shape of the drug molecule.

Another interesting example can be observed in the two geometrical isomers of the artificial oestrogenic hormone, namely : *trans*-diethylstilbesterol and *cis*-diethylstilbesterol.

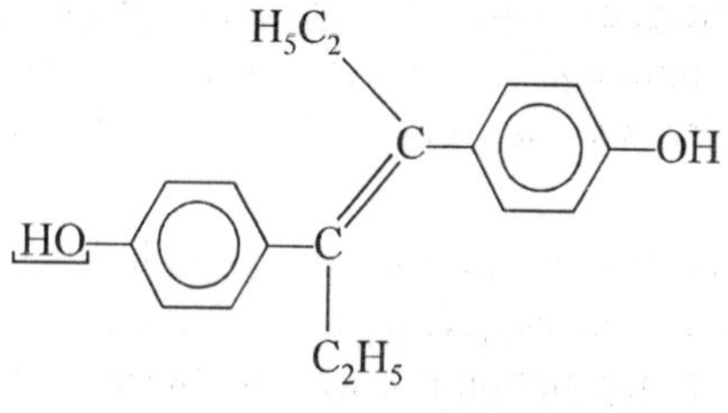

trans-Diethylstilbesterol 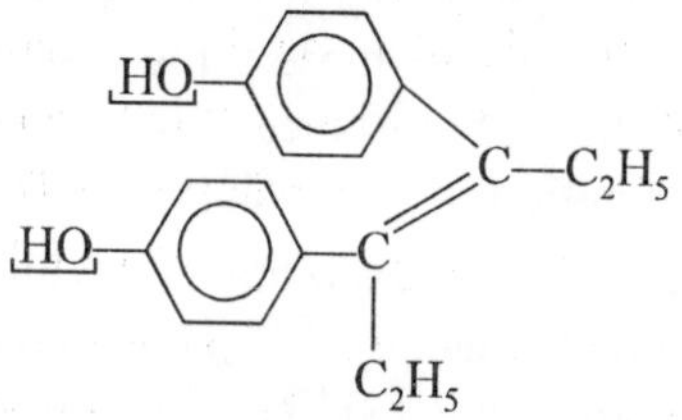   *cis*-Diethylstilbesterol

The distance between the two hydroxyl groups present in the *trans*-diethylstilbesterol is 14.5A which is being higher than the *cis*-diethylstilbesterol and this is actually responsible for the more potent oestrogenic activity of the former.

4.   DISSOCIATION CONSTANTS

Usually the dissociation constant of week electrolytes is expressed by the Henderson-Hasselbalch equations as stated below :

$$pKa = pH + \log [acid]/[conjugate\ base]$$

For acids :

$$pKa = pH + \log [undissociated\ acid]/[ionized\ acid]$$

For bases :

$$pKa = pH + \log [ionized\ base]/[undissociated\ base]$$

However, the dissociation constant of a weak base or acid may be conveniently determined by one of these several established methods, namely : ultraviolet or visible absorption spectroscopy, conductivity measurements and finally the potentiometric pH measurement.

It has been observed that most drugs exert their pharmacodynamic action either as undissociated molecules or as ionized molecules. These *two* different aspects shall be discussed briefly.

4.1.   Drugs Exerting Action as Undissociated Molecules

In a large number of potent medicinal compounds the dissociation constants play a vital role for their respective biological characteristics.

The unusual structural groupings in the tetracyclines results *three* distinct acidity constants in aqueous solutions of the acid salts. The particular functional groups responsible for each of the thermodynamic pKa values has been determined by Lessen and co-workers as described below :

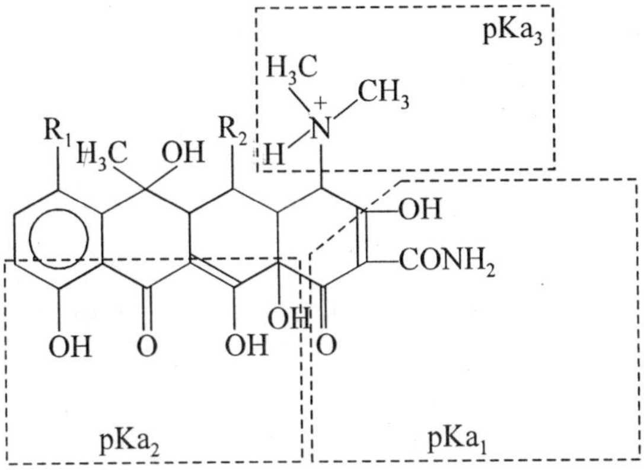

The approximate *pKa* values for each of these groups in *four* commonly used tetracyclines are shown on the next page.

pKa Values (of Hydrochlorides) in Aqueous Solutions at 25°C.

S.No.	Names	pKa₁	pKa₂	pKa₃
1.	Tetracycline	3.3	7.7	9.5
2.	Chlorotetracycline	3.3	7.4	9.3
3.	Demeclocycline	3.3	7.2	9.3
4.	Oxytetracycline	3.3	7.3	9.1

Besides, the activity of several local anaesthetics, *d*-tubucurarine and phenol has also been proved to be related to their degree of ionization.

4.2. Drugs Exerting Action as Ionized Molecules

A plethora of medicinal compounds exert their pharmacodynamic action exclusively as their ionized molecules, *viz :* acetylcholine, quaternary salts as ganglionic blocking agents and muscle relaxants (discussed elsewhere in this book), and antiseptics.

5. ISOSTERISM AND BIO-ISOSTERISM

The constant endeavour toward newer and more potent biologically active compound has paved the way for research into more specific, more effective, structurally similar compounds either possessing same or opposite activity.

Langmuir* suggested that any two ions or molecules possessing essentially an identical number and arrangement of electrons must exhibit similar characteristics ; and all such pairs he named as 'isosteres' *e.g.,* CO and N ; CO_2 and N_2O ; and N_3^- and NCO^-. However, it is quite evident that such isosteres which are isoelectric in nature must show good similarity in properties.

---

* I. Langmuir, *J. Am. Chem. Soc.,* **41**, 868, 1543 (1919).

Isosterism is of vital importance to a medicinal chemist because the biological characteristics of isosteres appear to be similar more frequently than their physical or chemical characteristics.

Keeping in view the numerous advantageous applications of isosterism in resolving biological problems effectively, Friedman* proposed the following definition of 'bio-isosterism' — *the phenomenon by which compounds usually fit the broadest definition of isosteres and possess the same type of biological activity.*

For instance, among antihistaminics it is always preferable to have small compact substituents on the terminal nitrogen.

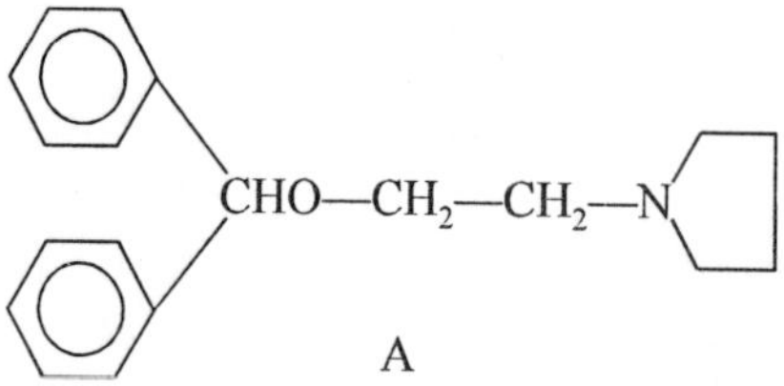

Pyrrolidine analogue

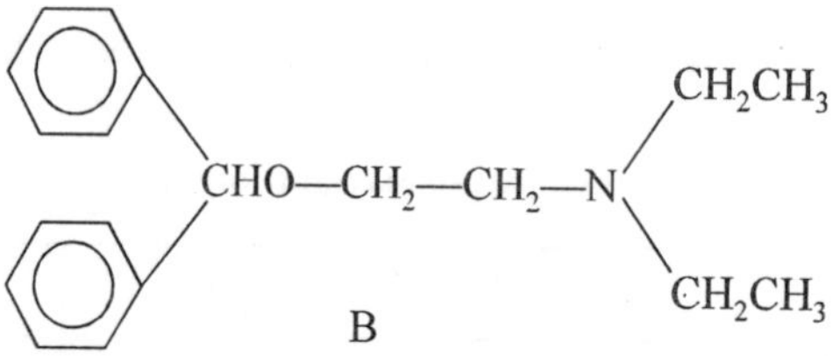

A

Diethyl analogue

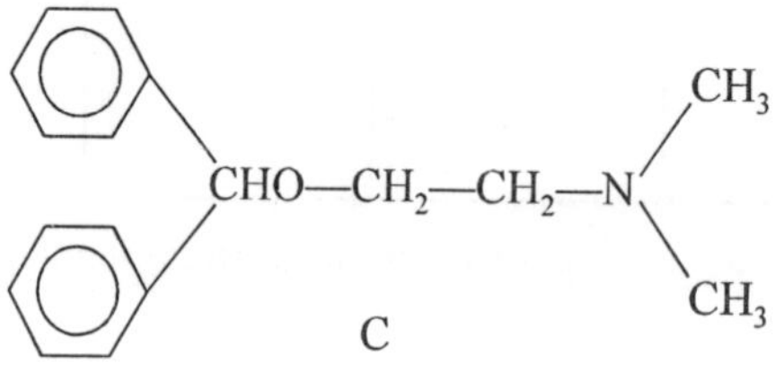

B

Dimethyl analogue

C

In the above *three* structural analogues it has been observed that *A* possesses twice the activity of *C*, whereas it showed an activity many times greater than that of the open-chain diethylamino analogue.

It has been duly observed that it is more or less difficult to correlate the **biological properties** *vis-a-vis* **physico-chemical** properties inherited by specific individual atoms, functional groups or entire molecules by virtue of the glaring and established fact that a host of physical and chemical parameters are invovled simultaneously and are, therefore, extremely difficult to quantitate them justifiably. Besides, simpler relationships *e.g., 'isosterism'* invariably do not delay across the several varieties of biological systems that are often encountered with medicinal agents (*i.e., drug substances*). In other words, a specific isoelectric replacement in one particular *biological system* (or a given *drug receptor*) may either work or fail to response in another.

---

* H.L. Friedman, *Influence of Isosteric Replacements upon Biological Activity,* National Academy of Sciences-National Research Council Publication No. 206, Washington, D.C., 1951, p. 295.

In order to expatiate further the terminology **'bioisosters'**. Burger* expanded the definition of Friedman to take into consideration the *biochemical views of the biological activity* :

*"Bioisosteres are compounds or groups that essentially possess near equal molecular shapes and volumes, approximately the same distribution of electrons, and that exhibit similar physical characteristics, such as : hydro-phobicity. Bioisosteric compounds affect the same biochemically associated systems as agonists or antagonists and thereby produce biological properties that are more or less related to each other**''.*

Bioisosteres may be classified under **two** categories, namely :

(*a*) Classical Bioisosteres, and

(*b*) Nonclassical Bioisosteres.

5.1. Classical Bioisosteres

Functional moieties which either fulfil or satisfy the original conditionalities put forward by Langmuir** and Grimm*** are termed as *'classical bioisosteres'*. More explicitely, in animals the occurrence of several hormones, neurotransmitters etc., having almost idential structural features and above all similar biological activities may be classified as bioisosteres.

Example : Insulins isolated from various mammalian species are found to differ by a substantial quantum of *'aminoacid residues'* but surprisingly they do exert the same biological effects (*i.e.,* lowering of blood-sugar). However, if this did not occur ; the actual usage of *'insulin'* to treat, control and manage **diabetes** might had to wait for another half-a-century for the development and recognition of recombinant DNA technology to allow production of human insulin****.

Actual applications of **'bioisosteres'** in the successful design of a specific given molecule interacting with a particular **'receptor'** in one glaring example, very often either fails or negates the biological characteristics in another environment (system). Therefore, it is pertinent to state at this juncture that the logical use of *biological replacement* (classical or nonclassical) in the design of a **'new target-drug molecule'** is solely and significantly dependent on the specific biological system under critical investigation. Hence, there are no predetermined, well-established, predictable hard and fast guidelines or laid-out generalized rules that may be useful to a *'medicinal chemist'* to affect biosteric replacement gainfully towards improved biological activity. The wisdom, inituition, skill, experience and creative imagination of a *'medicinal chemist'* contribute a major role to zero-down or pin-point or hit the bull's-eye to obtain the best possible results towards **'new target-drug'** molecules.

Table 2.1. Evidently shows the various *'classical bioisosteres'* with their appropriate examples :

---

* Burger, A., **'Isosterism and Bioisosterism in Drug Design'**, *Progress in Drug Research,* 1991, **37,** 288–371.

** Langmuir, I. J., *Amer. Chem. Soc.,* 1919 ; **41** : 868 ; *ibid,* **41,** 1543.

*** Grimm, H.G., *Z. Elekrochemie.,* 1925 ; **31** : 474.

****A Danish recombinant-DNA-technology based firm has produced **'insulin'** from the yeast cells that almost meets all the stringent requirements of human insulin [**Humulin**[(R)]].

Table 2.1 : Classical Bioisosteres*

S.No.	Types of Classical Bioisosteres	Various suitable examples
1.	Monovalent atoms and groups.	F, H ; OH, NH ; F, OH, NH or CH_3 for H ; SH, OH ; Cl, Br, CF_3 ;
2.	Divalent bioisosteres	$-C=S, -C=O, -C=NH, -C=C-,$
3.	Trivalent atoms and groups	$-\overset{\mid}{\underset{H}{C}}=, -N=; -P=, -AS=;$
4.	Tetrasubstituted atoms	$-\overset{\mid}{\underset{\mid}{N}}\overset{\oplus}{} \quad -\overset{\mid}{\underset{\mid}{C}}- \quad -\overset{\mid}{\underset{\mid}{P}}\overset{\oplus}{} \quad -\overset{\mid}{\underset{\mid}{As}}\overset{\oplus}{}-$
5.	Ring equivalents	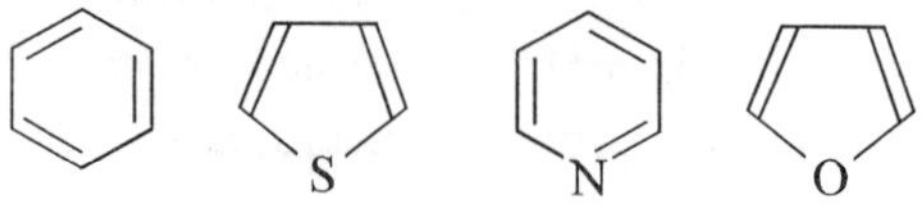

* Groups within the row can replace each other conveniently.

Salient Features : Following are the various salient features of *'classical bioisosteres'* :

(1) **Hydrogen replaced by Fluorine :** It is regarded as one of the commonest monovalent isosteric replacements. Both H and F are fairly identical with their Van der Waal's radii being 1.2 Å and 1.35 Å respectively. Fluorine being the *most electronegative element* in the periodic table ; therefore, the augmentation in the biological profile of drugs containing F may be attributed to this specific characteristic.

Example : [1] 5-Fluorouracil from uracil, obtained by replacement of H with F gives rise to the formation of an extremely therapeutically potent antineoplastic drug :

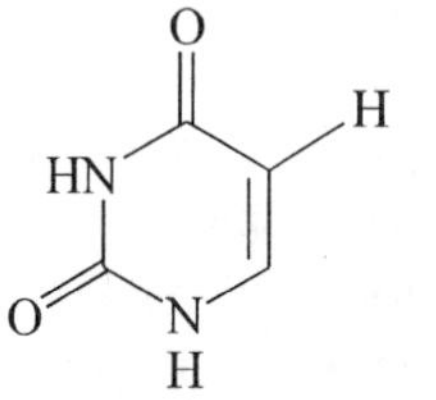

Uracil 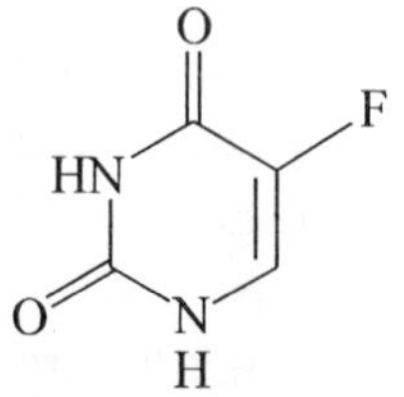   5-Fluorouracil

[2] Aminopterin (I) mimics the tautomeric forms of folic acid (II), thereby giving rise to the formation of suitable H-bondings to the corresponding **enzyme active site,** as illustrated below :

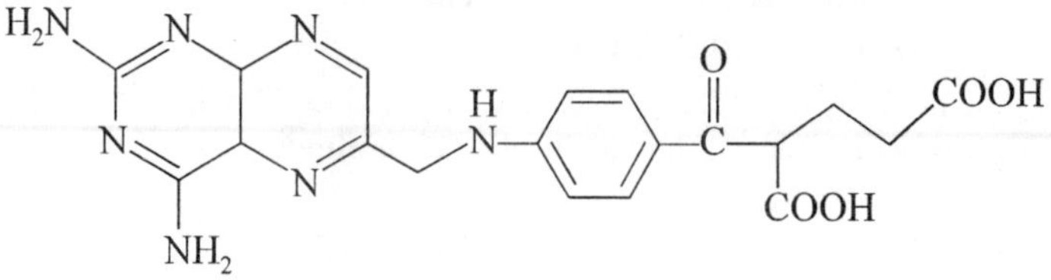

Aminopterin (I)

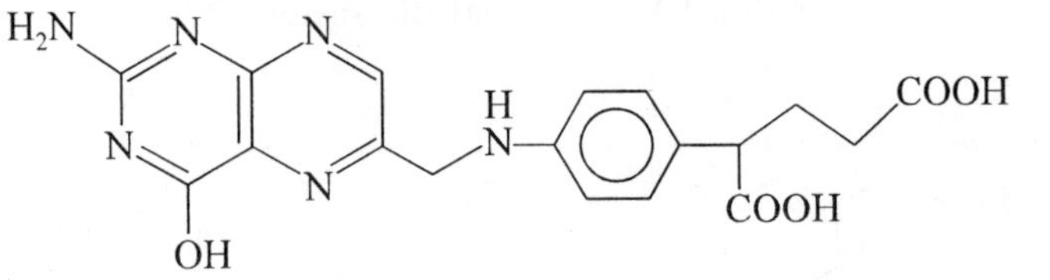

Folic acid (*enol*-form) (II)

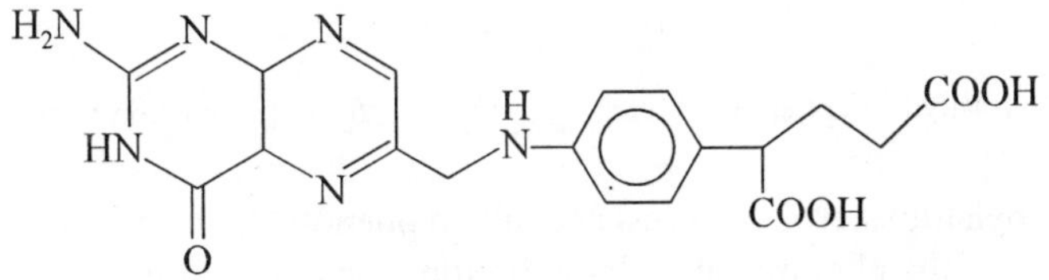

Folic acid (*keto*-form)

5.2. Nonclassical Bio-isosteres

Importantly, the nonclassical bioisosteres are precisely the replacements of functional groups not falling within the regimen by classical definitions. Although, several of these functional moieties practically just behave as one of the following characteristic specific features, such as :

- Electronic proporties,
- Physicochemical property of the molecule,
- Spatial arrangements,
- Functional moiety critical for biological activity.

Examples : [1] Non-cyclic analogues of oestradiol :

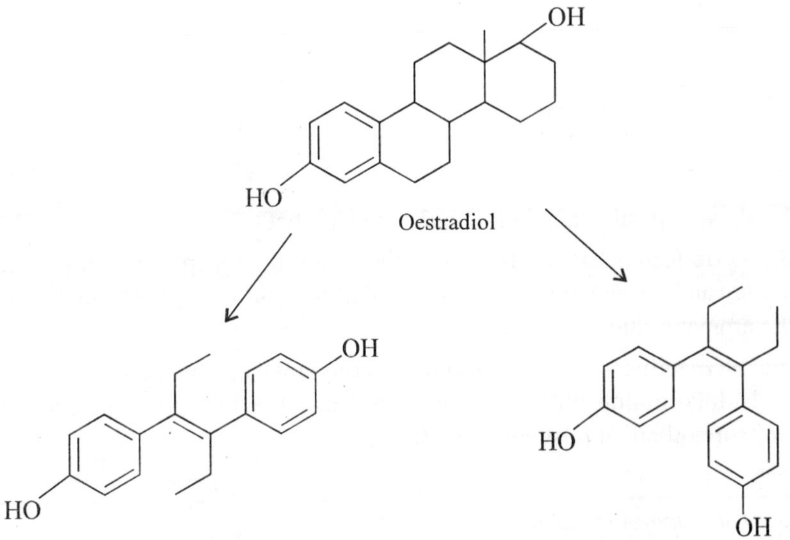

Oestradiol

trans-Diethylstilbesterol
(Active)

cis-Diethylstilbesterol
(Inactive)

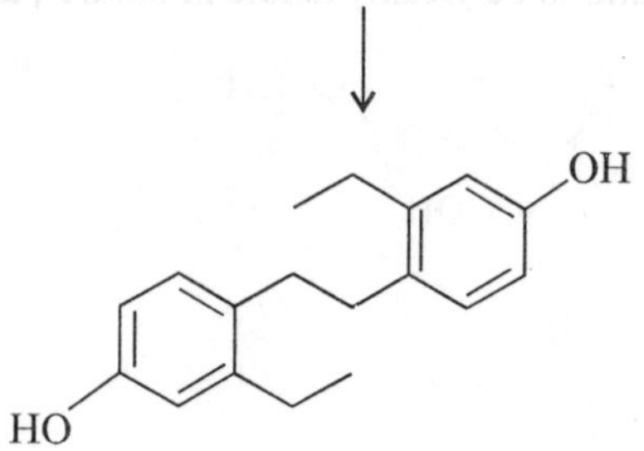

1, 2-*bis*-(2-Ethyl-4-hydroxyphenyl) ethane 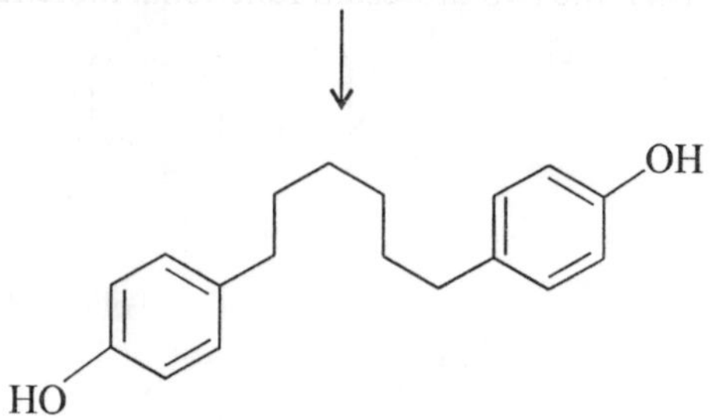   1, 6-*bis*-(*p*-Hydroxyphenyl) hexane

The *trans*-diethylstilbesterol, the presence of two phenolic hydroxy functions very closely mimic the correct orientation of the phenolic and alcoholic functions present in the natural oestradiol*. However, it is *not* accomplished by the corresponding *cis*-diethylstilbesterol, besides some other more flexible analogues that practically show no activity at all**.

[2] Nonclassical replacement of a sulphonamide function for a phenol in catecholamines :

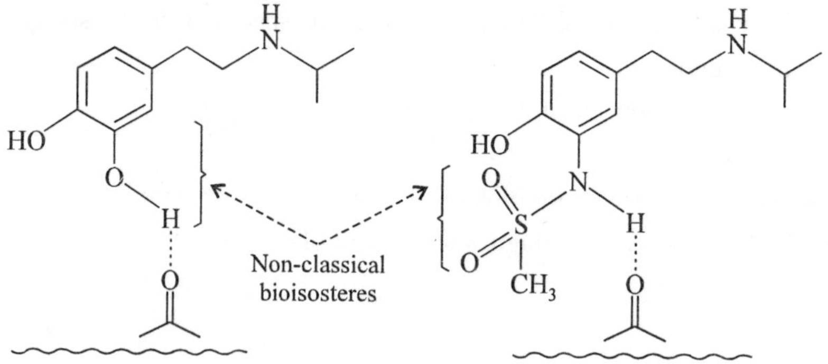

Isoproterenol
[A prototypic, non-
selective β-agonist]

Soterenol

From the above typical example it is quite evident that —

(*i*) the steric factors exert a little influence upon the receptor binding in comparison to the *acidity* and *hydrogen bonding* potential of the in-coming functional moiety positioned on the aromatic ring,

(*ii*) the *p*K*a* values (*i.e.*, dissociation constant) of the acidic proton present in the arylsulphonimide moiety of **soterenol** and the two phenolic hydroxyl functions in **isoprotenol** are almost equal to 10***.

---

*Dodds E.C *et al. Nature,* **141** : 247, 1938

** Blanchard EW. *et al. Endocrinology,* **32** : 307, 1943.

*** Baker BR. *J. Amer. Chem. Soc.,* **65** : 1572, 1943.

(*iii*) the *two* aforesaid functional moieties are found to be weakly acidic in nature ; and hence, capable of :

- losing a proton

- interacting with the receptor as anions

- participating as H-bond donors at the receptor site

- as the above cited 'replacement' is NOT influenced to metabolism by *catechol O-methyl transferase* enzyme, thereby enhancing not only the span of action *in vivo* but also rendering the compound active orally.

6. STEREOCHEMISTRY AND DRUG ACTION

The potential biological activity (*i.e.,* drug action) of a *'targetted-drug molecule'* is solely dependent on its physicochemical characteristics essentially comprise of the nature and type of functional moieties ; and also the spatial arrangement of such groups in the molecule. Interestingly, the human body itself represents an asymmetric environment wherein the drug molecules interact with proteins and biological macromolecules (receptors). An elaborated study of the 3D-orientation of the organic functional moieties present *in vivo* provides a substantial evidence about the most probable mechanism in the interaction existing between a specific *'drug substance'* and biological macromolecules. Hence, it is virtually important and necessary that the **decisive functional moieties** must be strategically located with respect to the exact spatial region encircling the *'targetted-drug molecule'* so as to enable the crucial and productive bonding interaction(s) particularly with the receptor (biological molecule), thereby potentially accomplishing the desired pharmacologic effect. It is, however, pertinent to state here that the right fitment of correct 3D-orientation of the functional moieties in a *'drug substance'* may ultimately result into the formation of an extremely viable and reasonably strong interaction with its receptor.

Stereochemistry is that branch of chemistry which deals with atoms in their space relationship, and the effect of such a relationship on the action and effects of the molecule.

Stereoisomers are compounds having the same number and kinds of atoms, the same configuration (arrangement) of bonds, but altogether different 3D-structures *i.e.,* they specifically differ in the 3D-arrangements of atoms in space.

Stereoisomers may be further sub-divided into *two* types, namely : (*a*) enantiomers ; and (*b*) diastereoisomers.

6.1. Enantiomers

Enantiomers are isomers whose 3D-configuration (arrangement) of atoms gives rise to the formation of nonsuperimposable mirror images.

These are also invariably termed as *chiral compounds, enantiomorphs* or *antipodes*. Furthermore, these compounds essentially possess identical physical as well as chemical characteristic features except for their inherent ability to rotate the plane of polarized light in just opposite directions with almost equal magnitude, quantum and extent.

Predominantly, when enantiomeric features are introduced strategically right into either a chiral environment or an asymmetric one, for instance : *the human body*, enantiomers shall evidently show marked and pronounced variant physical chemical properties thereby exhibiting appreciable and significant differences in their respective **'pharmacokinetic'** and **'pharmacodynamic'** behaviour.

Thus, the presence of variant biological activities based on their diverse enantiomeric features in a *'drug substance'* may lead to :

- adverse side effects,
- toxicity caused due to one of the isomers,
- exhibit appreciable differences in absorption *i.e.,* active transport,
- show significant variations in serum protein binding,
- extent/degree of metabolism,
- conversion into a toxic substance (impaired metabolism), and
- influence the metabolism of an altogether another drug.

6.2. Diastereoisomers

Diastereoisomers are all stereoisomeric compounds which are *not* enantiomers. In other words, the terminology *'distereoisomer'* essentially includes compounds containing both ring systems and double-bonds simultaneously. In apparent contrast to *'enantiomers'* diastereoisomers invariably display different physical and chemical characteristics, namely :

- chromatographic behaviour
- solubility
- melting point
- boiling point

Based on these glaring differences prevailing in their physical chemical properties one may effectively cause the separation of a mixture of diastereoisomers by the aid of established and standard chemical separation techniques, for instance :

- Crystallization
- Column chromatography.

Note : It may be pointed out at this juncture that 'enantiomers' cannot be separated by using any one of such methods unless either these are either converted to *diastereoisomers* or a chiral-environment is provided.

A few typical examples of stereoisomers *viz.,* enantiomers and diastereoisomers are illustrated below :

ENANTIOMERS

(*i*)

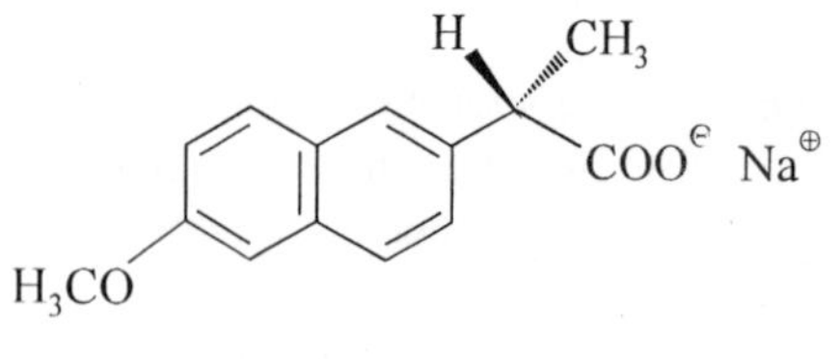

(S)–(+) Naproxen sodium
[2-Naphthaleneacetic acid, (+)-6-
methoxy-α-methyl-, sodium salt]
*Active as an analgesic, antipyretic
and anti-inflammatory*

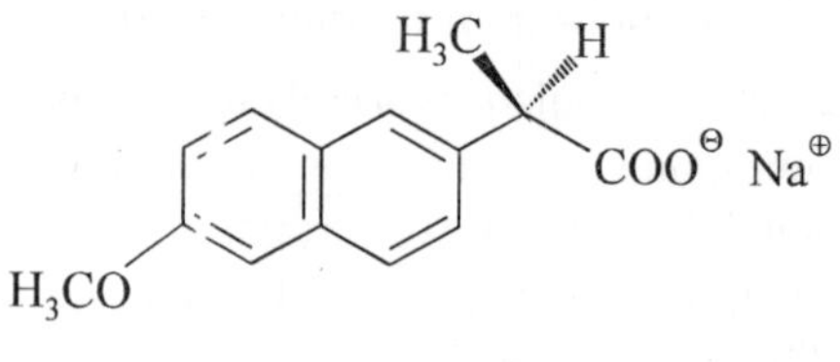

(R)–(–) Naproxen sodium
[Inactive]

(*ii*)

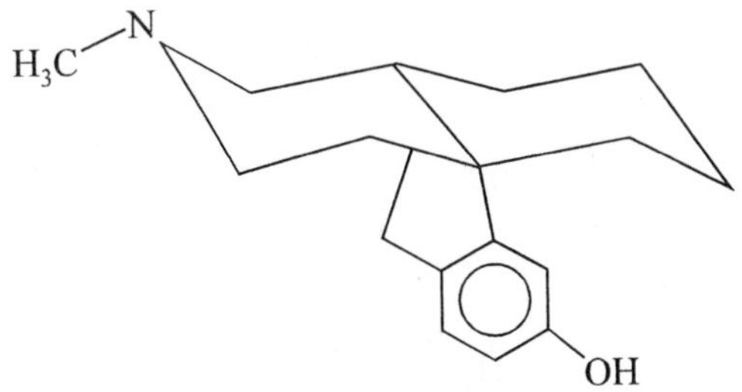

Levorphanol
[17-Methylmorphinan-3-ol]
**A potent synthetic analgesic
related to morphine**

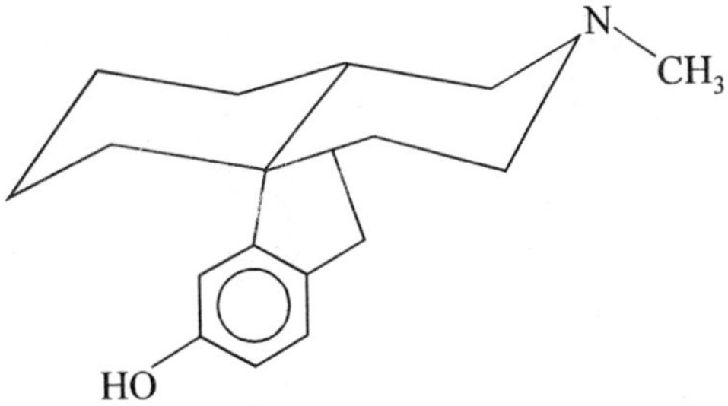

Dextrorphan
(Antitussive)

Example (*i*) shows that the priority sequence in (S) – (+) naproxen sodium is to the left ; and it exhibits activity as an antipyretic, analgesic and anti-inflammatory drug. In contrast, the R – (–) naproxen sodium is **inactive.**

Example (*ii*) illustrates that levorphanol exhibits a potent analgesic activity, whereas the counterpart *i.e.,* dextrorphan exclusively shows an antitussive activity.

DIASTEREOISOMERS

(*i*)

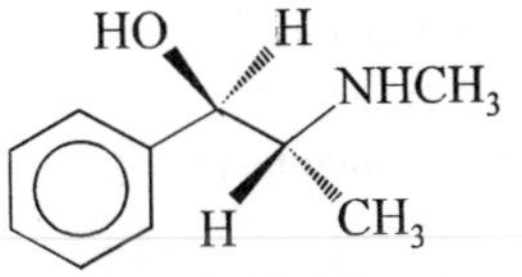

(–)–Pseudoephedrine
(Inactive)

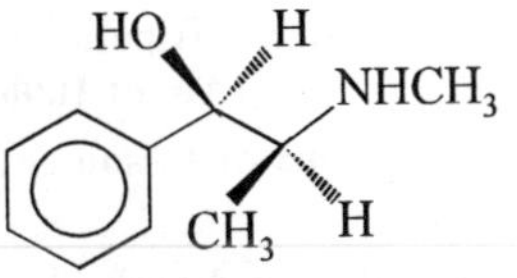

(–)–Ephedrine
**[CNS-Stimulatory actions ; OTC-cold,
allergy, and asthma remedies]**

---

* **Stereoisomers :** The nomenclature of stereoisomers was postulated by Cahn, Ingold and Prelog (1956) and is commonly known as the *Sequence Rule System* (or CIP-system). Here, the atoms attached to a chiral centre (*i.e.,* asymmetric C-atom) are RANKED as per their atomic number according to the following laid-down norms : (*a*) Maximum (highest) priority is given to the atom with highest atomic number and subsequent atoms are ranked accordingly from highest to lowest ; (*b*) In a situation when a decision cannot be reached with respect to '*priority*' *i.e.,* 2-atoms having the same atomic number attached to the chiral centre, the process continues to the next atom until a decision could be arrived. The molecule is then viewed from the side opposite the lowest priority atom ; and the sequence of priority from highest to lowest is determined ; (*c*) In case, the sequence is to the right, or clockwise, the **chiral centre** is designated as the R absolute configuration ; when the priority sequence is to the left, or anticlockwise, the designation is S.

(*ii*)

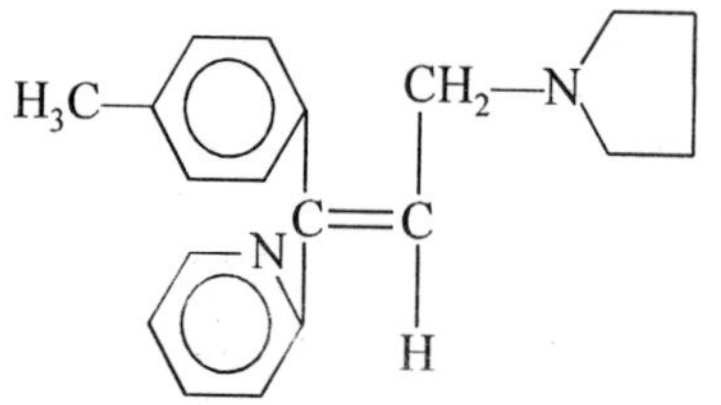

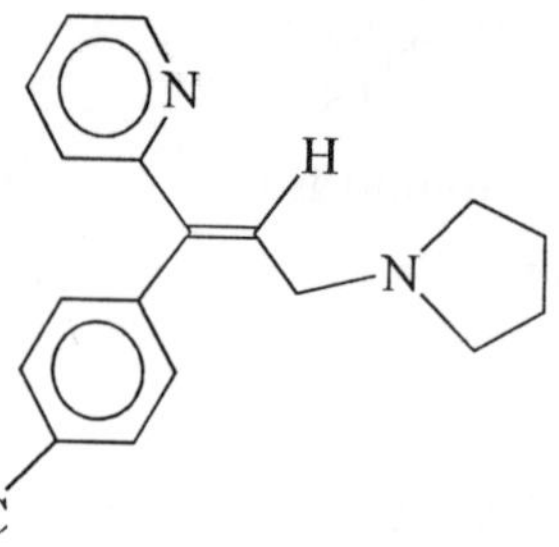

(E)*–Tripolidine or *trans*-Tripolidine **[Active]**
(Antihistaminic activity, anticholinergic activity,
and as in OTC cold and sinus preparations)

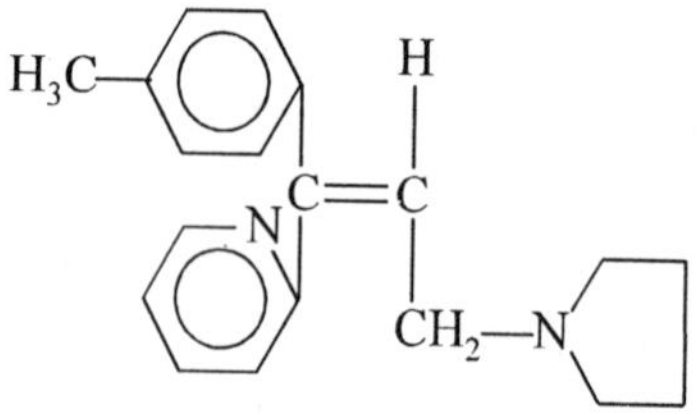

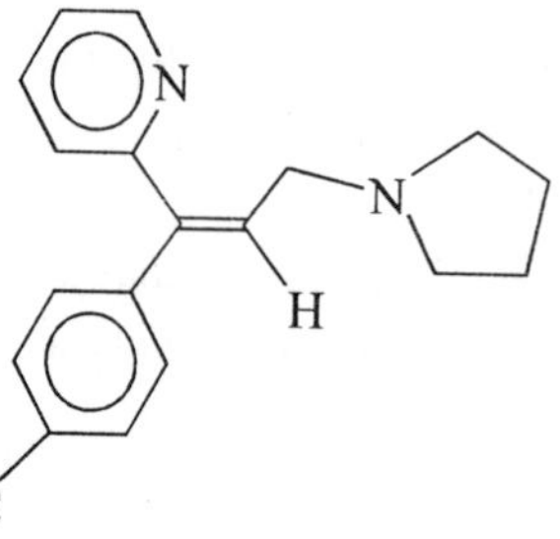

(Z)*–Tripolidine or *cis*-Tripolidine **[Inactive]**
(About 1000 times less potent as the
trans-or E-isomer as a H1-histamine antagonist)

In **Example (*i*)** *i.e.,* (–)-pseudoephedrine the two H-atoms are on the *opposite side of the plane of the ring i.e.,* one withbroken line is viewed as projecting beneath the ring plane (α-substituent) ; and the other with solid line is viewed as projecting above the ring plane. It does not exhibit any biological activity. However, in (–)-Ephedrine the said two H-atoms are located on the same side *i.e.,* beneath the ring plane ; and hence, it shows biological activities as stated above.

In **Example (*ii*)** *i.e.,* (Z)-Tripolidine the two heterocyclic aromatic rings are strategically located on the same side of the double bond (*cis*-configuration/Z-configuration) ; and is found to be inactive. Interestingly, simply by swapping the said two rings on either sides of the double bond (*trans*-configuration/E-configuration) the new compound (E)-tripolidine shows potent pharmacological actions as mentioned above.

6.3. Stereochemistry and Biologic Activity

An intensive and extensive research carried out till date on *'drug-design'* has not only established but also paved the way in the specialized aspect of **'stereochemistry'** of the *'targetted-drug molecules'*. This particular approach has inspired the *'medicinal chemist'* to tailor-made such newer drug substance(s) in which the proper strategical positioning of various functional moieties are introduced (or inducted) so that they are capable of interacting optimally with either an **enzyme** or a **receptor**.

---

*E and Z-isomer : When the two aromatic heterocyclic rings are on the same side of the molecule, having the double-bond, it is known as the *cis*-or Z-isomer (from the German zusamer or ''together'') ; when these are located on opposite sides the designation is *trans*- or E-(from the German entagegen or ''opposite'').

Interestingly, the following *five* different aspects of stereochemistry (*i.e.*, types of isomeric drugs) shall now be discussed in the sections that follows :

6.3.1. Positional Isomers (or Constitutional Isomers) :

In this specific instance the compounds essentially possess the same *emperical formula* but the atoms of the molecule are rearranged in an altogether different order.

Examples :

(1) **Pentobarbital and Amobarbital :**

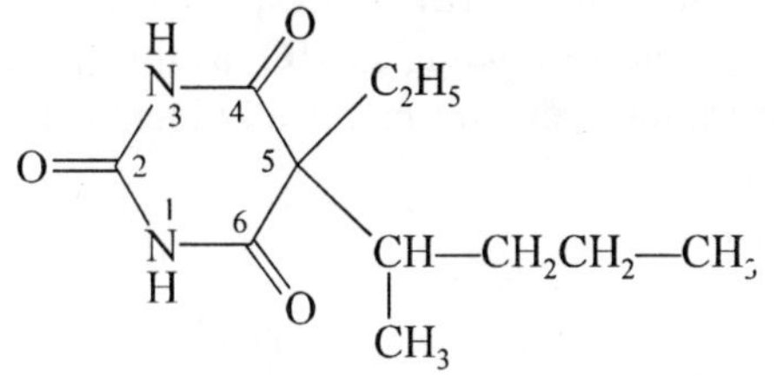

Pentobarbital (I) 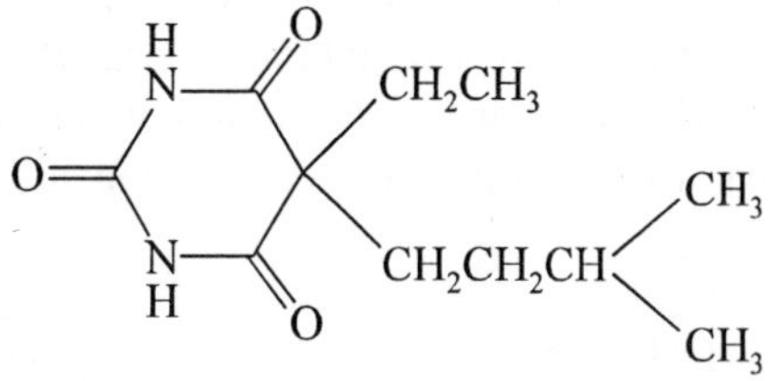   Amobarbital (II)

These positional isomers (I) and (II) belong to the barbiturate family. However, these *positional isomers* specifically differ only in the formation of the 5-carbon side chain attached to the C-5 position to the barbiturate ring system. Thus, compound (I) is a *short-acting barbiturate* ; whereas, compound (II) is an *intermediate-acting barbiturate.*

(2) **Terbutaline and N-*tert*-Butyl norepinephrine :**

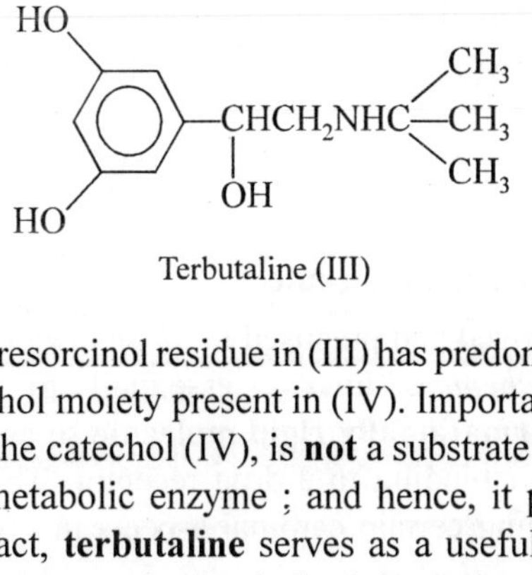

Terbutaline (III) 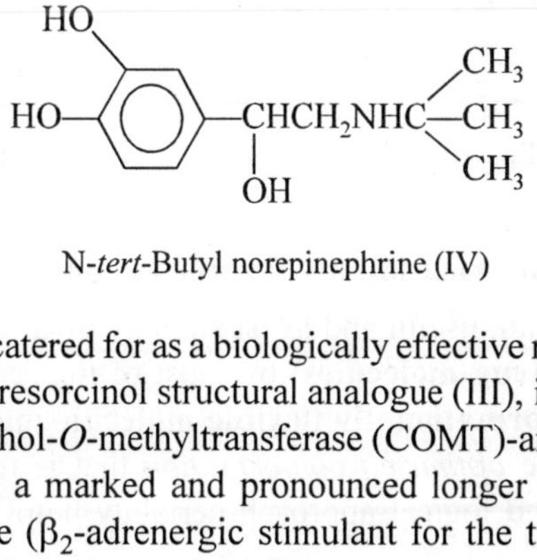   N-*tert*-Butyl norepinephrine (IV)

The resorcinol residue in (III) has predominantly catered for as a biologically effective replacement of the catechol moiety present in (IV). Importantly, the resorcinol structural analogue (III), in a striking contrast to the catechol (IV), is **not** a substrate for catechol-*O*-methyltransferase (COMT)-an extremely important metabolic enzyme ; and hence, it possesses a marked and pronounced longer duration of action. In fact, **terbutaline** serves as a useful selective (β_2-adrenergic stimulant for the treatment of bronchial asthma and related physiological conditions (*administered orally*).

6.3.2. Geometrical Isomers

In geometrical isomers there exists a spatial arrangement of either atoms of functional groups in the carbon-carbon double bond locations, which has been duly expatiated earlier as under :

(*a*) section 5.2 *i.e.*, non-cyclic analogues of oestradiol, and

(*b*) section 6.2 *i.e.*, diastereoisomers example (*ii*).

6.3.3.   Absolute Configuration

The terminogloy *'absolute configuration'* particularly refers to the arrangement of atoms in space of a chiral compound. It has been observed that there is a stark and distinct difference in specific biologic activity of the **optical isomers** (*enantiomers*) having the (R) and (S) configuration. A typical example of *Levorphanol* and *Dextrorphan* has already been discussed under section 6.2 in this chapter.

6.3.4.   Easson-Stedman Theory*

According to this theory put forward in 1974—the relative order of activity of the *'isomers' viz.,* R(–) isomer epinephrine, S(+) isomer epinephrine and epinine on the *adrenergic receptors* are in the order of R > S ~ deoxy. Besides, the **R isomer** can bind to all the **three sites,** namely : (*i*) *catechol binding site* **'A'** ; (*ii*) *hydroxy binding site* **'B'** ; and (*iii*) *anionic binding site* **'C'** as illustrated below ; whereas ; the **S isomer** and the **deoxy isomer,** that essentially exhibit practically identical biological activity, can exclusively bind to *two* of the sites.

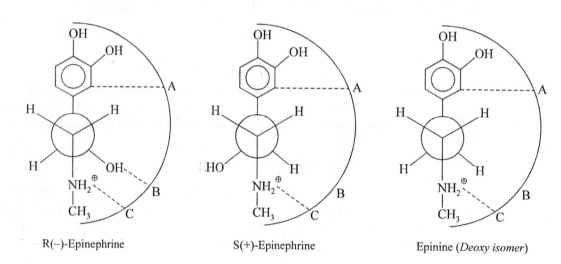

R(–)-Epinephrine   S(+)-Epinephrine   Epinine (*Deoxy isomer*)

6.3.5.   Conformationally flexible to conformationally rigid molecule

A vital, useful and latest strategy invariably employed and practised in *'drug design'* of **'newer targetted drug molecules'** by most of the *'medicinal chemists'* involves essentially of converting a rather **conformationally flexible molecule** into a **conformationally rigid molecule** so as to establish and find the *optimized conformation* that is required for binding to a drug receptor. This particular scientific and logical approach certainly helps in revealing certain cardinal aspects in *'drug design'*, such as :

■   in incorporating selectivity for receptors

■   to minimise and eliminate undesired side effects

■   to learn more with regard to spatial relationships of functional moieties for receptors.

Example : The aforesaid critical and important aspect may be expatiated with the aid of the following example of **DOPAMINE**, which enhances cardiac output by stimulating β-receptors.

---

*Patil, PN. *et al. Pharmacol. Rev.* **26** : 232, 1974

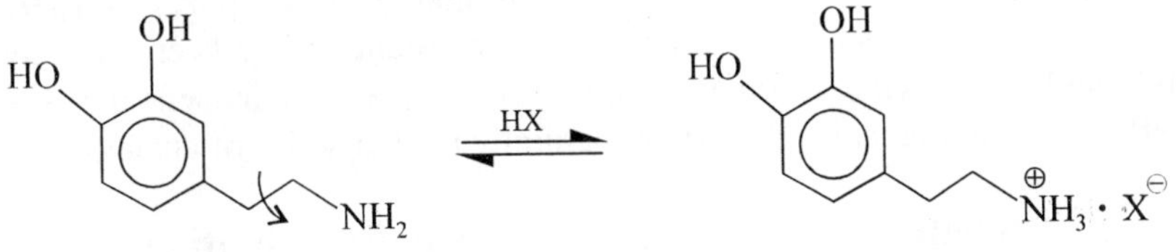

(I)   (IV)

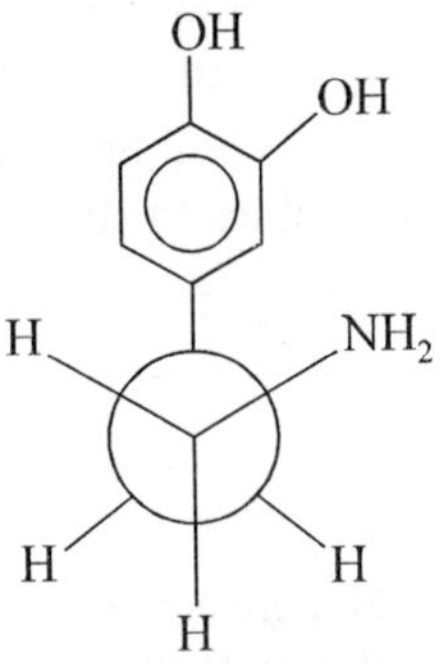

(II)   (III)

$\theta = 60°$ *gauche* $\theta = 180°$ *gauche*

trans-conformation   *trans*-conformation

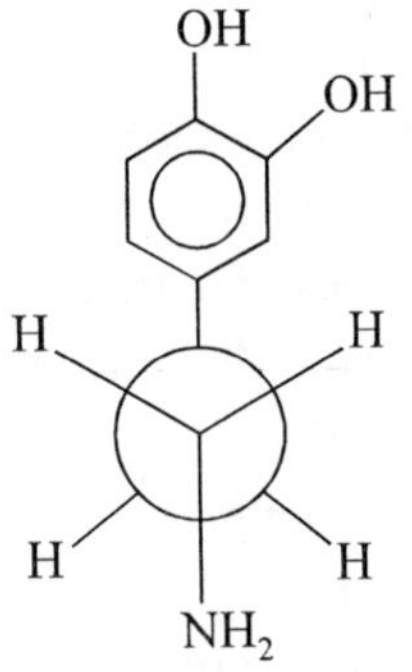

In reality, **dopamine** (I) may exist in an **infinite number** of conformation about the **single** side-chain C—C bond. However, *two* such conformations, namely : (II) [$\theta = 60°$ *gauche*] and (III) [$\theta = 180°$ *gauche*], both having *trans*-conformation may show the maximum biological activity.

7. CHEMICAL PROPERTIES

Modern approaches to the design of bioactive molecules are usually based on the quantification of bioactivity as a function of a molecular structure. According to the concept of receptor theory, biological activity solely depends on the recognition of bioactive substrate by a receptor site followed by binding of the bioactive substrate to the receptor site. Realizing the ultimate dependence on configuration of a drug molecule one takes cognizance of the fact that steric effects of one type or another served as a major determining factor toward the potency of bioactive drug.

Following are some of the chemical parameters that have been put forward to buttress the above-mentioned facts. They are :

7.1. Molecules Negentropy

Molecule negentropy is a summation of the negative information entropy calculated from the multiplicity and probability of equivalent sets of atoms in any selected "pharmacea".

7.2. Cammarata Correlation

It essentially establishes and also determines the prevailing relationship amongst electronic, hydrophobic and steric effects of a substituent and a change in biological effect.

Cammarate correlation may be expressed as :

$$\sigma \, An = \bar{a}\,\sigma + \bar{b}\,\pi + \bar{c}\,E_s + \bar{d} \, ,$$

where $\bar{a}$, $\bar{b}$, $\bar{c}$ and $\bar{d}$, are constants and σ, π and E_S represent the Hammett, the Hansch and the Taft constants, respectively.

It is, however, pertinent to mention here that the biological transport characteristics of a drug molecule *i.e.,* the ability of diffusion of a drug across membranes, have been extensively studied by Hansch, Leo and Smith. These vital informations obviously help a medicinal chemist to a great extent in predicting the efficacy of biological activity in structurally related series of analogues.

Probable Questions for B. Pharm. Examinations

1. How do the newer disciplines like : Computer-aided physico-chemical methods, tracer techniques, genetic engineering, biotechnology and electron microscopy have evolved a new direction to expatiate the intricacies of drug interaction sequel to drug design ? Explain.

2. Effects on minor structural modifications of the hypoglycemic and cholinergic activities fo the parent compound alter their duration of action. Give suitable examples to support answer.

3. The biological activities of a drug molecule can be modified by—

   (*a*) Meyer-Overton and Meyer-Hemmi Theory, and

   (*b*) Ferguson's Theory.

   Explain with typical examples.

4. The intramolecular distance present in the grouping—XCH_2CH_2N- between N-atom and X (where X = N, O etc.), which are present in procalne, acetylcholine and diphenylhydramine govern the biological activities. Explain.

5. Why the geometrical isomer *trans*-diethylstibesterol exhibit higher oestrogenic activity than the *cis*-isoner ?

6. Discuss the specific role of absorption, distribution, excertion and biotransformation (*i.e.,* metabolism) to enable a 'drug' to reach the '*active site*'.

7. Explain Henderson-Hasselbalch equations with regard to dissociation constant of weak electrolytes.

8. Lessen *et at* described the unusual structural grouping in tetracyclines which attribute three distinct acidity constants in aqueous medium of the acid salts. Explain.

9. Drugs exerting action as '*ionized molecules*' viz., muscle relaxants, ganglionic blocking agents vis-a-vis their pharmacodynamic action(s).

10. Give a comprehensive account of the imporatnce of 'Isosterism' and 'Bio-isosterism' in drug design.

RECOMMENDED READINGS

1. Foye's *Principles of Medicinal Chemistry* (5th edn.), New York, Lippincott Williams and Wilkins, 2002.

2. ME. Wolff (*Ed.*), *Burger's Medicinal Chemistry and Drug Discovery* (5th edn.) New York, John Wiley and Sons (1995).

3. A Cammarata and R L Stein *J Med. Chem-829* (1968).

4. C Hansch *Qualitative, Structure-Activity Relationship in Drug Design, in Drug-Design*, Vol. I, Ch. 2 (Ed. E J Ariens) Academic Press, New York (1971) pp. 271-342.

5. A. Leo, C Hansch and D Elkins *Chem. Rev. 71* (1971) 525.

6. RN. Smith, C Hansch and M M Ames *J Pharm Sci 64* (1975) 599.

7. Gennaro AR., Remington ; The Science and Practice of Pharmacy, Vol. I and II., Lippincott Williams and Wilkins, New York, 20th edn., 2000.

It is, however, pertinent to mention here that the biological response characteristics of a drug molecule are aptly...

General Anaesthetics

1. INTRODUCTION

General anaesthetics are a group of drugs that produce loss of consciousness ; and, therefore, loss of all sensation. The absolute loss of sensation is termed as *anaesthesia* (derived from the Greek word meaning insensitivity or lack of feeling). General anaesthetics bring about descending depression of the central nervous system ; starting with the cerebral cortex, the basal ganglia, the cerebellum and finally the spinal cord.

These drugs are used in surgical operations to induce unconsciousness ; and, therefore, abolish the sensation of pain.

Horace Wells, a Hartford dentist first and foremost demonstrated the usage of nitrous oxide ('laughing gas') as an effective surgical anaesthetic in 1844. However, its application as a *'general anaesthetic'* resurfaced in mid 1860's when it was dispensed in steel cylinders as an admixture with oxygen. Interestingly, nitrous oxide finds its application even today, particularly in combination with other anaesthetic and analgesic agents.

Later on, William Morton — a Boston dentist demonstrated the anaesthetic actions of diethyl ether in 1846 at the historical *"Ether Dome"* located at the Massachusetts General Hospital. In actual practice, the usage of diethyl ether followed by cyclopropane were withdrawn completely for being highly toxic amalgamated with equally dangerous physical properties, such as : flammable and explosive.

The anaesthetic agents that have gained cognizance today are invariably hydrocarbons and ethers with halogen (F, Br, Cl) substitution.

2. CLASSIFICATION

The general anaesthetics may be divided into *three* groups based solely on the method of administration. These are : inhalation anaesthetics ; intravenous anaesthetics ; and basal anaesthetics.

The above *three* categories of general anaesthetics shall be discussed here with appropriate examples.

2.1. Inhalation Anaesthetics

Inhalation anaesthetics could be either volatile liquids or gases and they are administered through inhalation process. As few typical examples are discussed below :

A. *Ether* USAN, *Anaesthetic Ether* BAN,

$$CH_3CH_2—O—CH_2CH_3$$

Ethane, 1, 1'-oxybis- ; Ethyl ether ; Diethyl ether ; Sulphuric ether ;

U.S.P., B.P., Eur. P., Int. P., Ind. P.,

Synthesis :

Method-I (From Alcohol) :

$$C_2H_5OH + H_2SO_4 \longrightarrow C_2H_5HSO_4 + H_2O$$
Ethanol $\qquad\qquad\qquad$ Ethyl sulphuric acid

$$C_2H_5HSO_4 + C_2H_2OH \longrightarrow (C_2H_5)_2O + H_2SO_4$$
$\qquad\qquad\qquad\qquad\qquad$ Ether

It may be prepared by the interaction of alcohol with sulphuric acid between 130—137°C commonly termed as the *etherifying temperature.*

Method-II (From Ethylene) :

$$H_2C = CH_2 + H_2SO_4 \longrightarrow C_2H_5HSO_4$$
Ethylene $\qquad\qquad\qquad$ Ethyl sulphuric acid

$$C_2H_5HSO_4 + C_2H_5OH \longrightarrow (C_2H_5)_2O + H_2SO_4$$
$\qquad\qquad\qquad\qquad\qquad$ Ether

Ethylene reacts with sulphuric acid to form ethyl sulphuric acid which on subsequent treatment with ethanol results into the formation of ether.

It still continues to be employed as an anaesthetic for producing insensitivity to pain in surgical trauma. It has a broad spectrum of usefulness besides its acclaimed potency for very painful surgical conditions and for being relatively benign with regard to the metabolic processes of the body.

Dose : By inhalation as required.

B. *Ethyl Chloride* BAN, USAN,

$$C_2H_5—Cl$$

Ethane, chloro- ; Chloroethane ; Monochloroethane ; Kelene ;

B.P., U.S.P., Int. P., Ind. P.,

Ethyl Chloride[R] (Bengue' U.K.) ;

Synthesis :

$$C_2H_5OH + NaCl + H_2SO_4 \longrightarrow C_2H_5Cl + NaHSO_4 + H_2O$$
Ethanol $\qquad\qquad\qquad\qquad\qquad$ Ethyl chloride

Ethyl chloride is prepared conveniently by distilling together a mixture of alcohol, sodium chloride and sulphuric acid.

Used in the past as a general anaesthetic by inhalation, particularly for minor operations, it is also employed as a local anaesthetic by 'freezing'. *It is no longer used as a general anaesthetic because of its damage to the liver and serious disturbances of the cardiac rhythm.*

Dose : Topical, as spray on intact skin.

C. *Vinyl Ether* BAN, USAN,

$$CH_2 = CO - O - CH = CH_2$$

Ethene, 1, 1'-oxybis- ; Vinyl ether ; Divinyl oxide ;

B.P., U.S.P., Int. P., Ind. P.,

Vinesthene[R] (May and Baker) ; Vinydan[R] (Byk Gulden) ;

Synthesis :

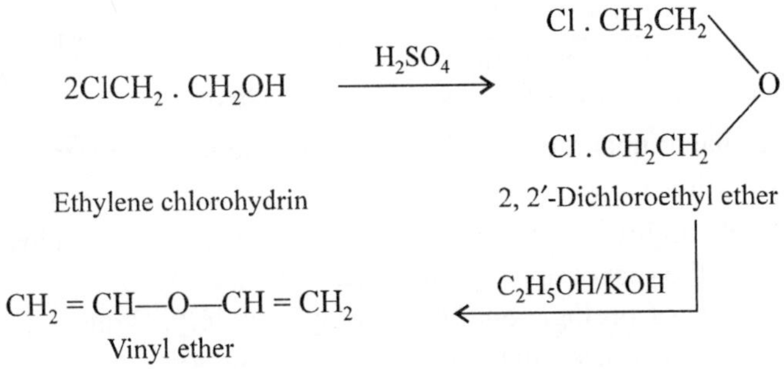

$$2ClCH_2 . CH_2OH \xrightarrow{H_2SO_4} \begin{array}{c} Cl . CH_2CH_2 \\ \\ Cl . CH_2CH_2 \end{array} O$$

Ethylene chlorohydrin   2, 2′-Dichloroethyl ether

$$CH_2 = CH—O—CH = CH_2 \xleftarrow{C_2H_5OH/KOH}$$
Vinyl ether

It is prepared by treating ethylene chlorohydrin with sulphuric acid to obtain 2, 2′-dichloroethyl ether, which on subsequent treatment with alkaline ethanol yields vinyl ether.

A volatile anaesthetic administered by inhalation, it is 4 times as potent as ether. Its prolonged inhalation for long operations is rather harmful because of the risk of liver necrosis.

Dose : By inhalation as required.

D. *Cyclopropane* INN, BAN, USAN,

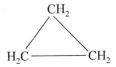

Trimethylene ;

B.P., U.S.P., Ind. P.,

Cyclopropane[R] (CI Pharmaceuticals U.K.) ;

Synthesis :

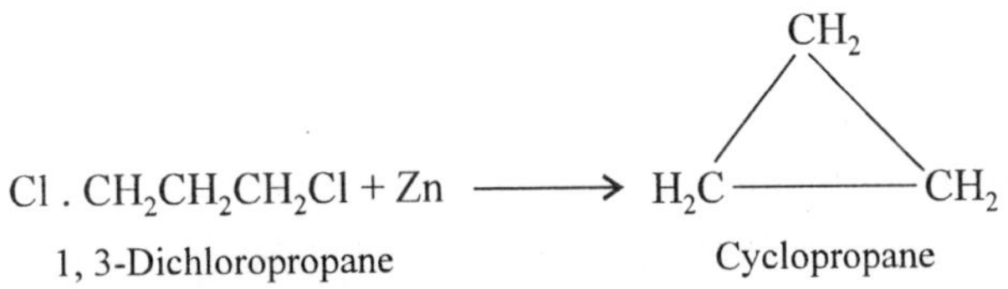

$$Cl . CH_2CH_2CH_2Cl + Zn \longrightarrow \begin{array}{c} CH_2 \\ \diagup \diagdown \\ H_2C——CH_2 \end{array}$$
1, 3-Dichloropropane   Cyclopropane

It may be prepared by the action of zinc metal (or sodium or magnesium) on 1, 3-dichloropropane.

Cyclopropane is one of the most potent anaesthetics administered by inhalation with the help of a brass regulator. The chief merits of this anaesthetics over others are due to its non-irritant nature and rapid recovery from anaesthesia. Its demerits include : *depressant effects on respiration, tendency to induce cardiac arrhythmias and to enhance haemorrhage.*

Dose : By inhalation as required.

E : *Fluroxene* INN, USAN,

$$CF_3CH_2 — O — CH = CH_2$$

2, 2, 2-Trifluoroethyl vinyl ether ; Ethene, (2, 2, 2-trifluoroethoxy)- ;

N.F. XIV ;

Fluoromar[R] (Anaquest) ;

Synthesis :

$$CF_3—CH_2—OH \xrightarrow[\text{Under pressure}]{\text{Basic catalyst}} CF_3CH_2—O—CH=CH_2$$

2, 2, 2-Trifluoroethanol   + C_2H_2 (Acetylene)   Fluroxene

2, 2, 2-Trifluoroethanol undergoes addition to acetylene in the presence of basic catalyst under moderate pressure to yield fluroxene.

It is a comparatively highly volatile pleasant-smelling anaesthetic employed most frequently for procedures requiring either the first or upper second plane of anaesthesia, *e.g.*, cardiac, dental, obstetric, orthopedic and certain types of urologic and gynaecologic surgery. Being a good analgesic it brings recovery rapidly. Postoperative nausea and vomiting are uncommon.

Dose : By inhalation as required.

F. *Halothane* INN, BAN, USAN,

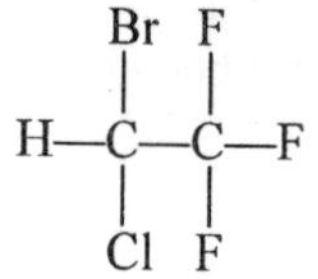

$$\begin{array}{ccc} & Br & F \\ & | & | \\ H—&C—&C—F \\ & | & | \\ & Cl & F \end{array}$$

2-Brome-2-chloro-1, 1, 1-trifluoroethane ; Ethane, 2-bromo-2-chloro-1, 1, 1-trifluoro-; B.P., U.S.P., Eur. P.,

Fluothane[R] (Ayerst) ; Halothane[R] (May and Baker) ;

Synthesis :

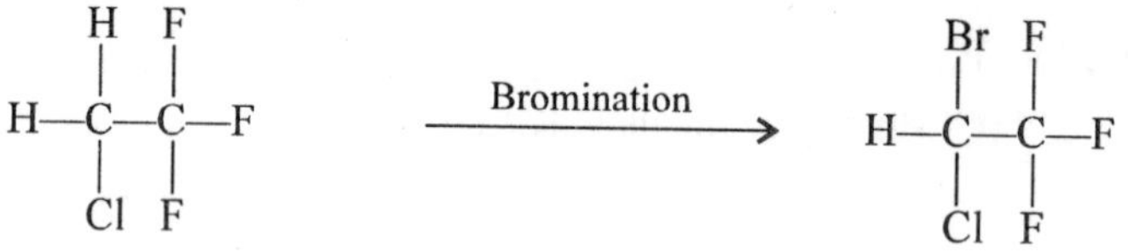

$$\begin{array}{ccc} H & F \\ | & | \\ H—C—&C—F \\ | & | \\ Cl & F \end{array} \xrightarrow{\text{Bromination}} \begin{array}{ccc} Br & F \\ | & | \\ H—C—&C—F \\ | & | \\ Cl & F \end{array}$$

2-Chloro-1, 1, 1-trifluoroethane   Halothane

Bromination of 2-chloro-1, 1, 1-trifluoroethane yields halothane which is isolated from the reaction product by fractional distillation.

It is a relatively safe potent volatile anaesthetic administered by inhalation. It is twice as potent as chloroform and 4 times that of ether. It may produce any depth of anaesthesia without causing hypoxia. Being a non-irritant, its inherent hypotensive effect retards capillary bleeding and renders a comparatively bloodless field.

Dose : By inhalation as required.

G. *Methoxyflurane* INN, BAN, USAN,

$$CHCl_2CF_2—O—CH_3$$

2, 2-Dichloro-1, 1-difluoroethyl methyl ether ; Ethane, 2, 2-dichloro-1, 1-difluoro-1-methoxy-; B.P., B.P.C., U.S.P., N.F. ;

Penthrane[R] (Abbott) ;

It is one of the most potent anaesthetic agents frequently used in practice today. In fact, it is employed to cause comparatively light anaesthesia with deep analgesic and muscle relaxation, features which make it convenient for short surgical operations, *e.g.*, obstetrics.

Dose : By inhalation as required.

H. *Trichloroethylene* INN, BAN,

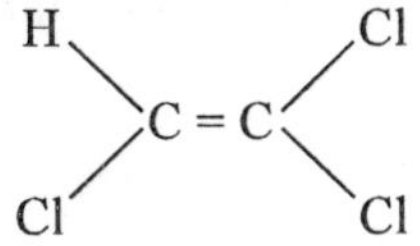

$$\underset{Cl}{\overset{H}{\diagdown}}C = C\underset{Cl}{\overset{Cl}{\diagup}}$$

Ethene, trichloro-;

B.P., U.S.P. 1970 ; Eur. P., N.F. XIV, Int. P., Ind. P., Trilene[R] (Ayerst) ; Trimar[R] (Ohio Medical)

Synthesis

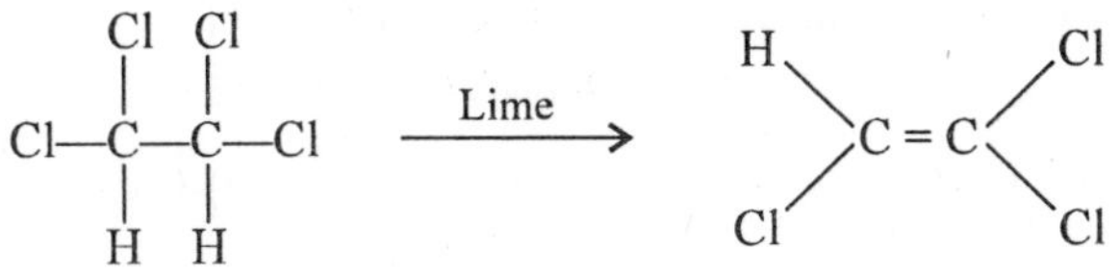

$$\underset{\underset{H\ \ H}{|\ \ \ \ |}}{\overset{\overset{Cl\ \ Cl}{|\ \ \ \ |}}{Cl-C-C-Cl}} \quad \xrightarrow{\text{Lime}} \quad \underset{Cl}{\overset{H}{\diagdown}}C = C\underset{Cl}{\overset{Cl}{\diagup}}$$

Sym-tetrachloro ethane   Trichloroethylene

It may be prepared by the careful abstraction of the elements of hydrogen chloride from *sym*-trichloroethane with the aid of lime.

It may be used sporadically as a weak volatile anaesthetic administered by inhalation. It possesses an excellent analgesic property but is wanting miserably as a muscle relaxant. It is frequently employed in short surgical operations where a mild anaesthesia having a potent analgesia is desired, as in obstetrics.

Dose : By inhalation as required.

I. *Nitrous Oxide* BAN, USAN,

$$N_2O$$

Nitrogen oxide ; Dinitrogen monoxide ; Laughing gas ;

B.P., U.S.P., Eur. P. Int. P., Ind. P.,

Entomox[R] (BOC Medishield U.K.) ($N_2O : O_2 :: 1 : 1$)

Synthesis :

$$NH_4NO_3 \xrightarrow{200°C} N_2O + 2H_2O$$

Ammonium nitrate   Nitrous
                                    oxide

It may be prepared by heating ammonium nitrate up to 200°C.

It is the weakest but the safest inhalation general anaesthetic. It is usually administered in conjunction with other potent inhalation anaesthetics, such as methoxyflurane and halothane. However, its anaesthetic regimen may be further broadened by the incorporation of neuromuscular blocking agents whereby the muscle relaxant characteristics are increased to a considerable extent. Some patients often get an attack of hysteria and for this reason it is invariably termed as '*laughing gas*'. It is an inhalation anaesthesia of choice in dental surgery by dint of its ability of the rapid recovery.

Dose : By inhalation as required.

J. *Chloroform* BAN, USAN,

$$CHCl_3$$

Trichloromethane ; Methane, trichloro-; Chloroformum pro Narcosi ;

B.P., N.F., Int. P., Ind. P.,

Synthesis :

$$CaOCl_2 + H_2O \longrightarrow Ca(OH)_2 + Cl_2$$
Bleaching powder   Slaked lime

$$C_2H_5 - OH + Cl_2 \longrightarrow CH_3CHO + 2HCl$$
Ethanol   Acetaldehyde

$$CH_3CHO + Cl_2 \longrightarrow Cl_3C.CHO + 3HCl$$
                                     Tri-chloroacetaldehyde

$$2Cl_3C.CHO + Ca(OH)_2 \longrightarrow CHCl_3 + (HCOO)_2Ca$$
                                     Chloroform

It may be prepared from bleaching powder and ethanol after a series of chemical reactions as shown above.

It is a potent anaesthetic administered by inhalation. It has both reasonably good muscle relaxant and analgesic properties. It is no longer used because of its liver and kidney toxicity.

Dose : By inhalation as required.

2.2. Intravenous Anaesthetics

Intravenous anaesthetics usually cause unconsciousness when administered parenterally. However, the duration of action can be safely monitored depending on the amount of drug administered.

A few such potent intravenous anaesthetics shall be discussed here.

A. *Thiopental Sodium* INN, USAN, *Thiopentone Sodium* BAN,

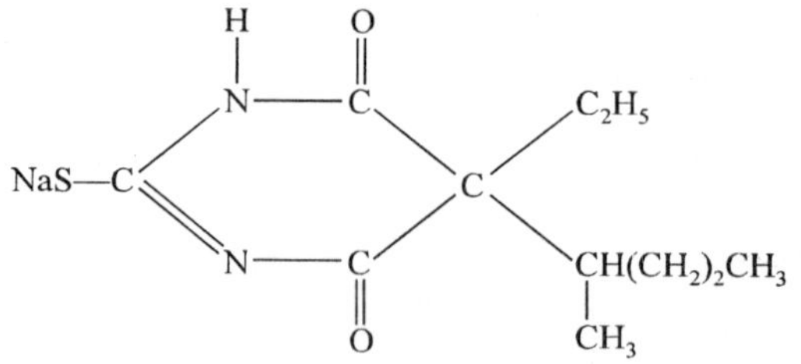

Sodium 5-ethyl-5-(1-methylbutyl)-2-thiobarbiturate ;

U.S.P., B.P., Eur. P., Int. P., Ind. P.,

Pentothal Sodium[R] (Abbott) ;

Synthesis :

It has been described in the chapter on sedatives and hypnotics.

It belongs to the category of ultra-short-acting barbiturates which are usually administered intravenously for the production of complete anaesthesia of a short duration. It is also used as a basal anaesthesia.

Dose : 100 to 500 mg intravenously.

B. *Ketamine Hydrochloride* INN, BAN, USAN,

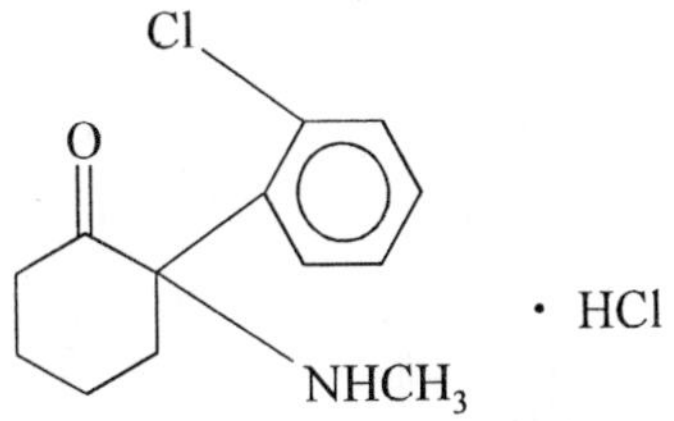

(±)-2-(*o*-Chlorophenyl)-2-(methylamino) cyclohexanone hydrochloride ; Cyclohexanone, 2-(2-chlorophenyl)-2-(methylamino)-, hydrochloride ;

U.S.P., N.F.,

Ketalar[R] (Parke-Davis) ; Ketaject[R] (Bristol) ;

Synthesis

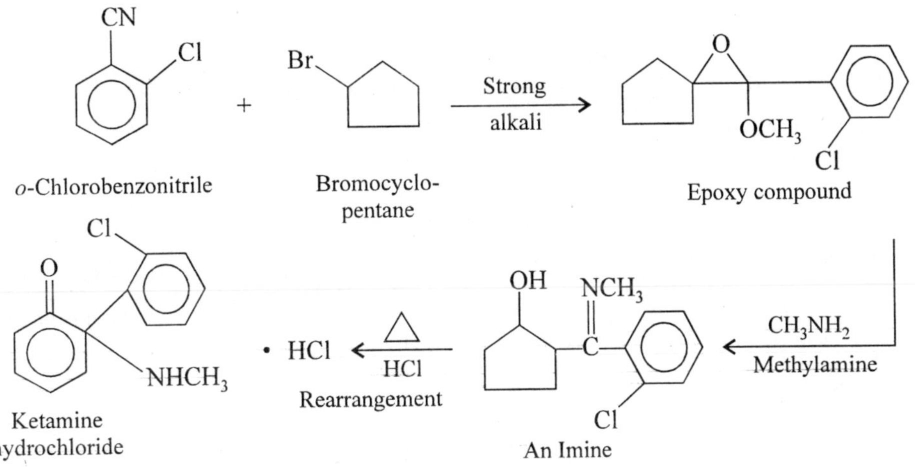

It is prepared first by the interaction of *o*-chlorobenzonitrile and bromo-cyclopentane in the presence of strong alkali to yield an epoxy compound.

Secondly, the resulting epoxy compound on treatment with methylamine forms an imine which undergoes molecular rearrangement upon heating in the presence of hydrochloric acid to yield ketamine hydrochloride.

It is a rapid-acting general anaesthetic drug which causes anaesthesia accompanied by deep analgesia, slightly modified skeletal muscle tone and appreciable cardiovascular and respiratory stimulation. Of course, it is an intravenous anaesthetic agent of choice for surgical operations of short duration, but *with additional doses it may effect anaesthesia for a span of 6 hours or even longer.*

Dose : Induction, intravenous 1 to 4.5 mg/kg.

C. *Methohexital Sodium* USAN, *Methohexitone Sodium* BAN,

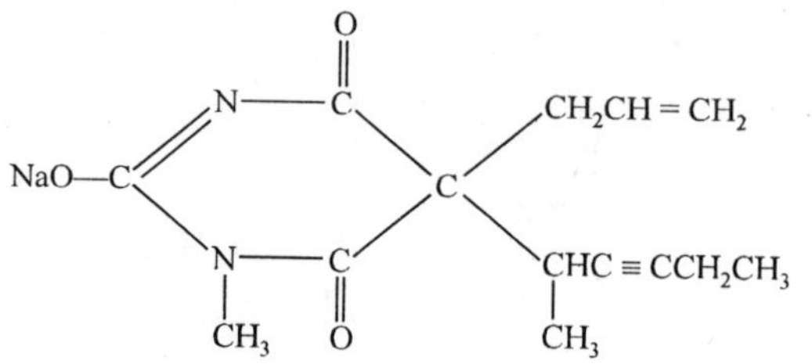

Sodium 5 allyl-1-methyl-5-(1-methyl-2-pentynyl) barbiturate ; 2, 4, 6 (1H, 3H, 5H)-Pyrimidinetrione, 1-methyl-5-(1-methyl-2-pentynyl)-5-(2-propenyl)-, (±)- , monosodium salt ; U.S.P., Brevital Sodium[R] (Lilly) ;

Synthesis :

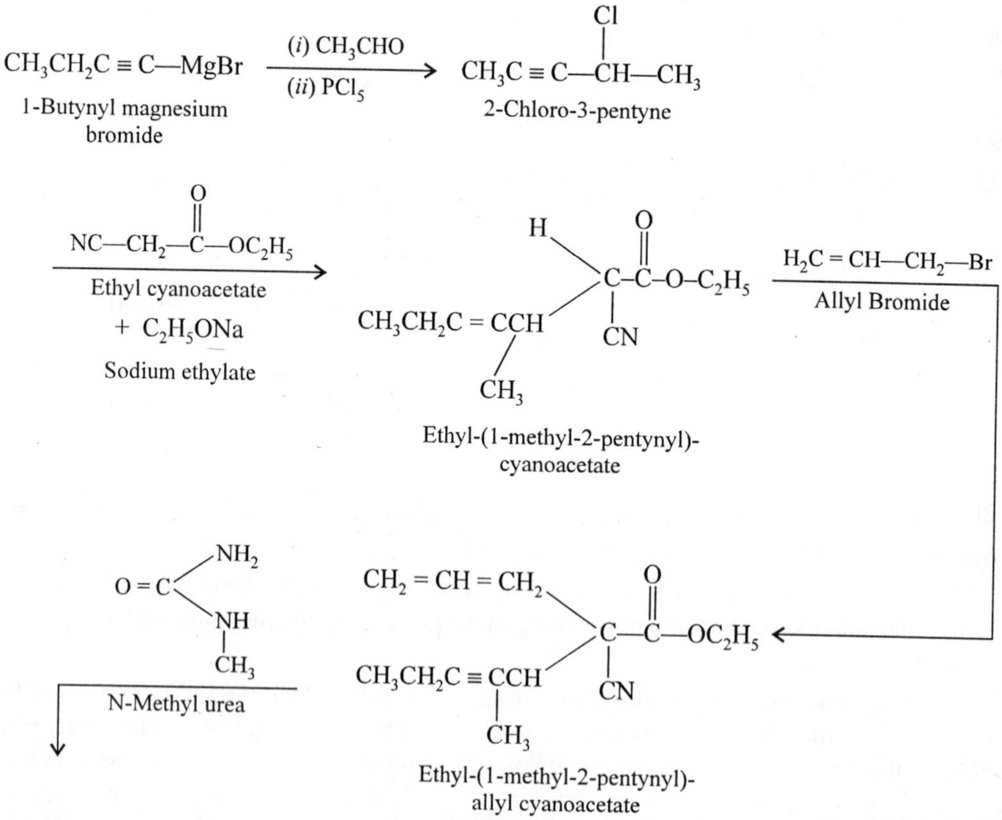

$CH_3CH_2C \equiv C—MgBr$
1-Butynyl magnesium bromide

$\xrightarrow[\text{(ii) PCl}_5]{\text{(i) CH}_3\text{CHO}}$

$CH_3C \equiv C—\overset{\overset{\displaystyle Cl}{|}}{CH}—CH_3$
2-Chloro-3-pentyne

Ethyl-(1-methyl-2-pentynyl)-cyanoacetate

Ethyl-(1-methyl-2-pentynyl)-allyl cyanoacetate

N-Methyl urea

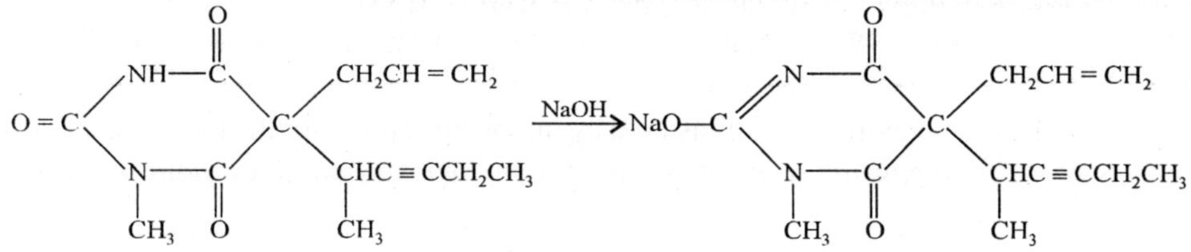

Methohexital   Methohexital Sodium

Grignardization of 1-butynyl magnesium bromide with acetaldehyde and subsequent treatment of the resulting alcohol with PCl_5 yields 2-chloro-3-pentyne. Now, ethyl (1-methyl-2-pentynyl)-cyanoacetate is obtained therefrom by its condensation with ethyl cyanoacetate in the presence of sodium ethylate. Further condensation of the resulting product with allyl bromide gives rise to ethyl-(1-methyl-2-pentynyl) allylcyanoacetate. Condensation with N-methyl urea and subsequent neutralization with NaOH produces methohexital sodium.

It is used for the induction of anaesthesia through the intravenous administration. It has two advantages over thiopental sodium ; *first,* being its less affinity towards fatty tissues and *secondly*, its greater potency. Its onset of action is quite rapid comparable to thiopental sodium while its recovery is more rapid. For these reasons this intravenous anaesthetic is specifically useful for short surgical operation, such as : oral surgery, gynaecologic investigation, genitourinary procedures and eletroconvulsive therapy.

Dose : 5 to 12ml of 1% solution, at the rate of 1ml every 5 seconds ; usual intravenous administration ; maintenance 2 to 4 ml every 4 to 7 min.

D. *Hydroxydione Sodium Succinate* INN, BAN,

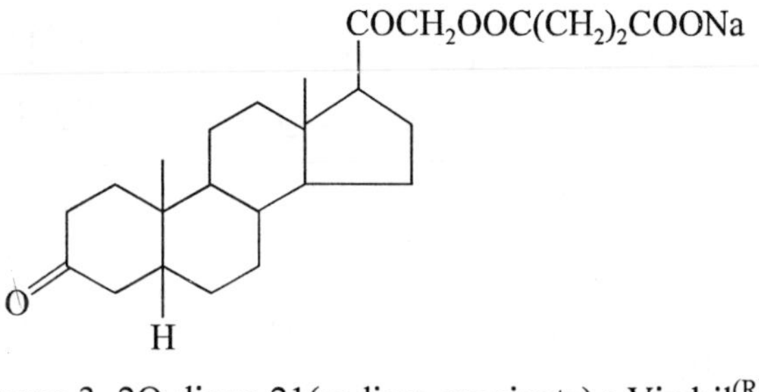

$$COCH_2OOC(CH_2)_2COONa$$

21-Hydroxy-5-pregnane-3, 2O-dione-21(sodium succinate) ; Viadril[(R)] (Pfizer)

Synthesis :

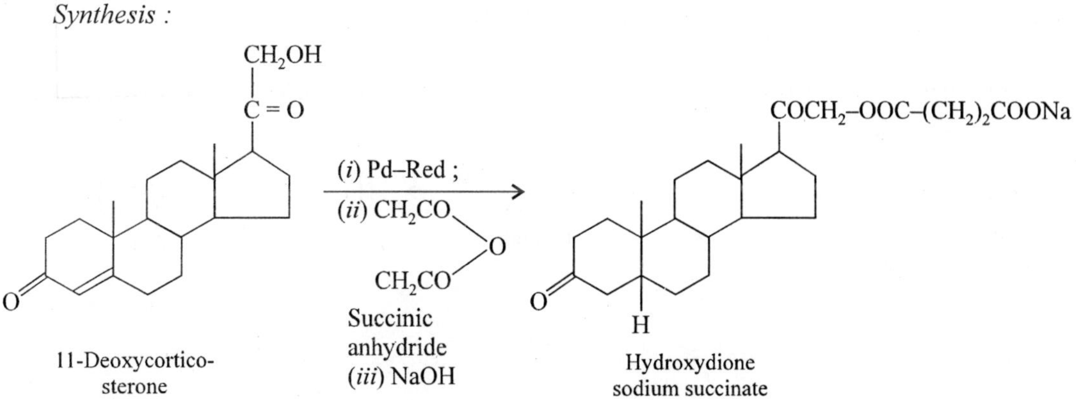

11-Deoxycortico-
sterone

Succinic
anhydride
(*iii*) NaOH

Hydroxydione
sodium succinate

It may be prepared *first* by the reduction of 11-deoxycorticosterone in the presence of palladium, *secondly* by treatment with succinic anhydride and *thirdly* the formation of its sodium salt from sodium hydroxide.

It is a steroidal drug previously administered by intravenous route for the induction of anaesthesia. *Its toxic effects range from causing respiratory depression, hypotension and venous irritation.*

Dose : 0.5 to 1.5g.

E. *Thiamylal Sodium* USAN, *Sodium Thiamylal* BAN,

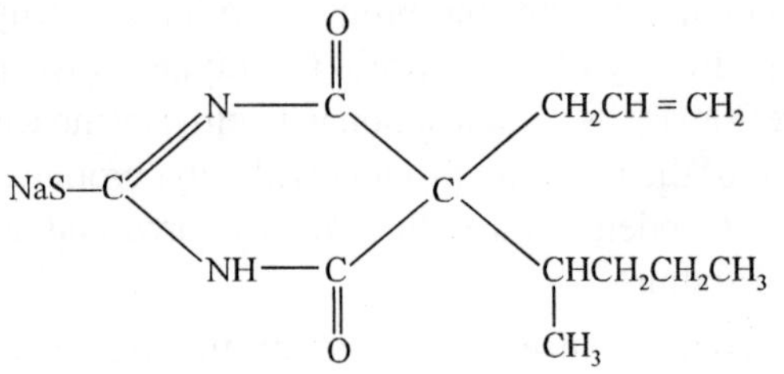

Sodium 5-allyl-5-(1-methylbutyl)-2-thiobarbiturate ; 4, 6-(1H, 5H)-Pyrimidine-dione, dihydro-5-(1-methylbutyl)-5-(2-propenyl)-2-thioxo-, monosodium salt ; U.S.P.,

Surital[R] (Parke-Davis) ;

Synthesis :

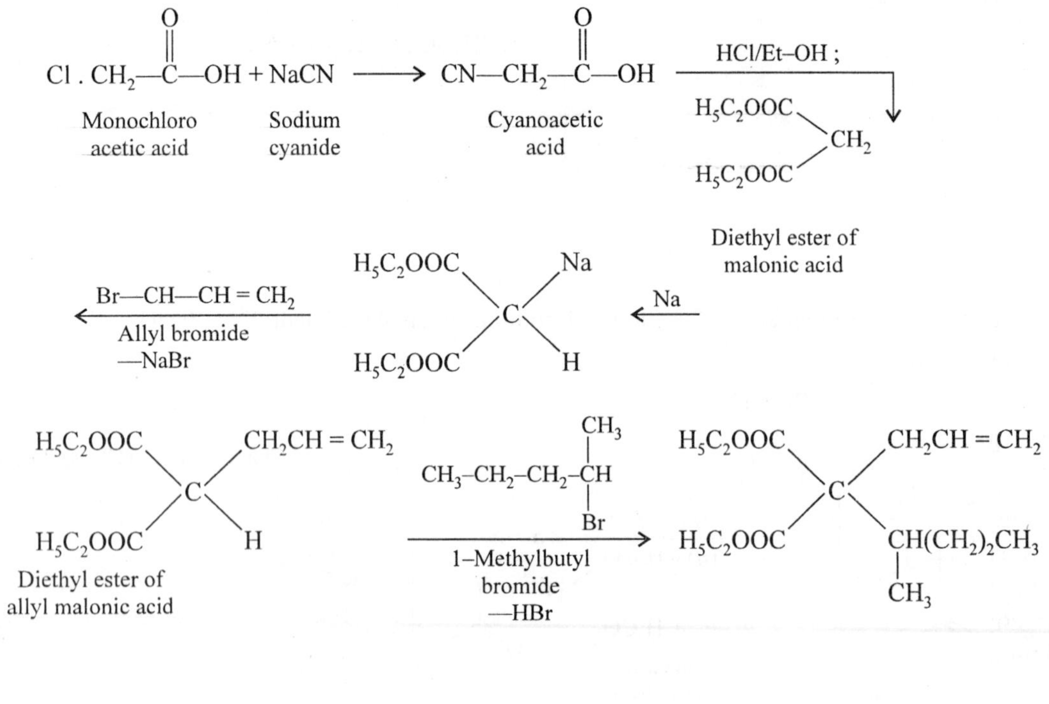

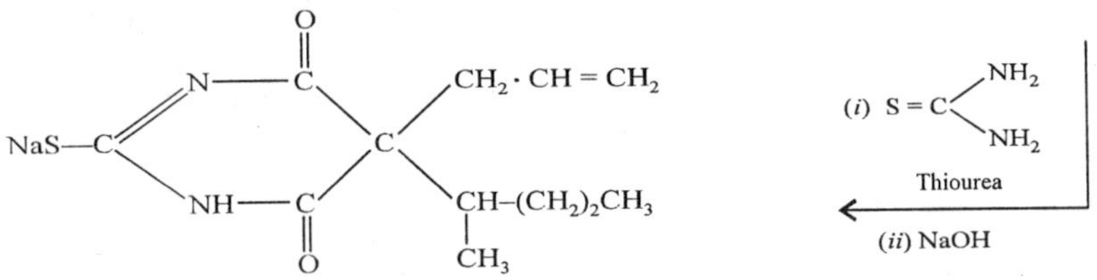

Thiamylal Sodium

The cyanoacetic acid obtained from monochloroacetic acid and sodium cyanide, is treated with hydrochloric acid and ethanol to yield the diethyl ester of malonic acid. The ester, in absolute ethanol, is reacted with the stoichiometric proportion of metallic sodium so as to replace only one active hydrogen of the methylene (CH_2) group. Thereupon a slight excess of the calculated amount of allyl bromide is added. The second replaceable hydrogen is abstracted with 1-methyl butyl bormide and the resulting product is made to react with a theoretical amount of thiourea to yield thiamylal. The free acid thus obtained is conveniently transformed into the official sodium salt by neutralization with a stoichiometric proportion of sodium hydroxide (1 : 1).

It is an ultra-short acting barbiturate mainly used for intravenous anaesthesia in conditions of comparatively short-duration. It is also effective for the termination of convulsions of unknown origin.

Dose : Usual, intravenous, 3 to 6 ml of 2.5% solution at the rate of 1 ml every 5 seconds ; maintenance dose being 0.5 to 1 ml as per requirement.

F. *Propanidid* INN, BAN, USAN,

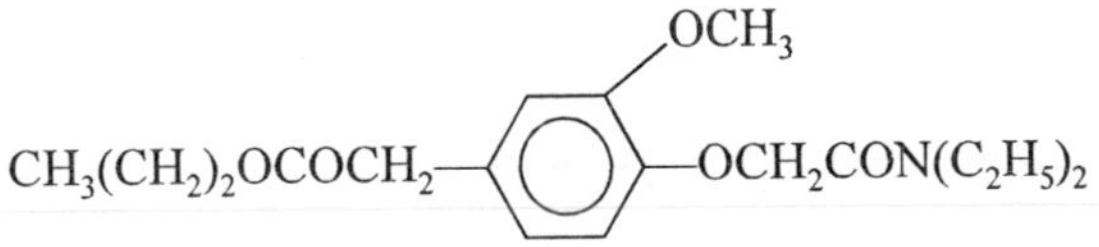

Propyl {4-[(diethylcarbamoyl)-methoxy]-3 methoxyphenyl} acetate ; Benzene-acetic acid, 4-[2-(diethylamino)-2-oxoethyl]-3-methoxy-, propyl ester ;

B.P.,

Epontol[R] (Bayer) :

It is an ultra-short acting anaesthetic which is administered intravenously either for producing complete anaesthesia of short duration or for induction of general anaesthesia. It is relatively less potent than thiopentone. It possesses neither analgesic nor muscle relaxant activities.

Dose : 5 to 10 mg per kg body weight, normally administered as a 5% solution.

2.3. Basal Anaesthetics

Basal anaesthetics are agents which induce a state of unconsciousness but *the depth of unconsciousness is not enough for surgical procedures.* They are often used to induce basal anaesthesia before the administration of inhalation anaesthetics. They are also used for repeated short procedures in children like the changing of painful dressings. *Basal anaesthetics offer three cardinal merit points, namely : devoid of mental distress, pleasant induction and lesser respiratory irritation.* They are often administered through the rectum. Few deserve mention.

A. *Fentanyl Citrate* INN, BAN, USAN,

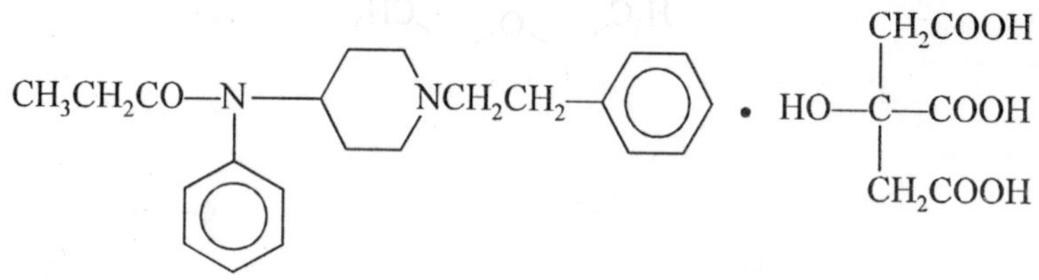

$$CH_3CH_2CO-N \underset{\displaystyle \bigcirc}{\quad} \overline{} NCH_2CH_2 - \bigcirc \quad \cdot \quad HO-\underset{\displaystyle CH_2COOH}{\overset{\displaystyle CH_2COOH}{\underset{|}{\overset{|}{C}}}} -COOH$$

N-(1-Phenethyl-4-piperidyl) propionanilide citrate (1 : 1) ; Propanamide, N-phenyl-N-[1-(2-phenylethyl)-4-piperidinyl]-, 2-hydroxy-1, 2, 3-propanetricarboxylate (1 : 1) ;

B.P., U.S.P.,

Sublimaze[R] (Janssen) ; Innovar[R] (McNeil) ;

It is employed basically as an analgesic for the control of pain associated with all kinds of surgery. It may also be used an an adjunct to all drugs commonly employed for regional and general anaesthesia. It is one of the components in '*Fentanyl citrate and Droperidol Injection*' which is used as premedication for anaesthesia and also as an supplement for induction and maintenance of anaesthesia.

Dose : Usual, intramuscular, 0.05 to 0.1 mg 30 to 60 minutes before operation.

B. *Tribromoethanol* USAN, *Tribromoethyl Alcohol* BAN,

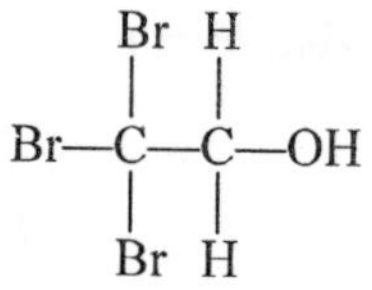

$$Br-\underset{\underset{\displaystyle Br}{|}}{\overset{\overset{\displaystyle Br}{|}}{C}}-\underset{\underset{\displaystyle H}{|}}{\overset{\overset{\displaystyle H}{|}}{C}}-OH$$

2, 2, 2-Tribromoethanol ; Tribromoethyl alcohol ;

B.P. 1953, Int. P., N.F. XIII ;

Avertin[R] (Winthrop) ;

Synthesis :

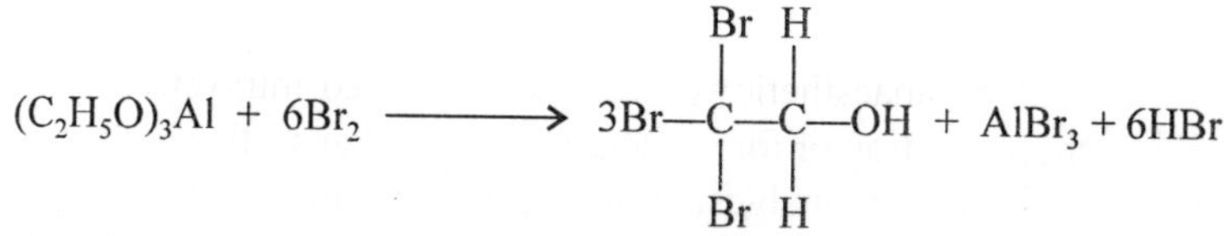

$$(C_2H_5O)_3Al + 6Br_2 \longrightarrow 3Br-\underset{\underset{\displaystyle Br}{|}}{\overset{\overset{\displaystyle Br}{|}}{C}}-\underset{\underset{\displaystyle H}{|}}{\overset{\overset{\displaystyle H}{|}}{C}}-OH + AlBr_3 + 6HBr$$

Aluminium ethoxide   Tribromoethanol

It is prepared by the interaction of a solution of bromine with aluminium ethoxide or preferably aluminium isopropoxide.

It is a basal anaesthetic agent of choice which is administered through rectum in the form of its solution. The main advantage of such an anaesthesia being its pleasant induction amalgamated with lack of irritating vapours.

Dose : Usual, rectal, 60 to 80 mg/kg body weight, not more than 8 g for woman and 10 g for man.

C. *Paraldehyde* BAN, USAN,

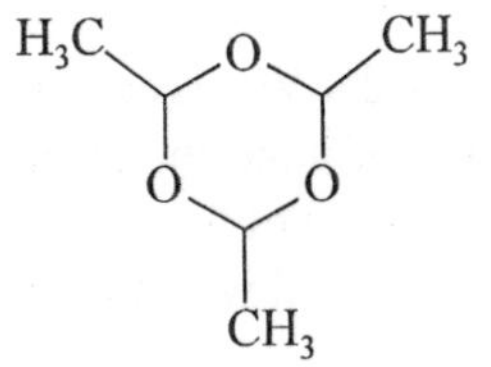

2, 4, 6-Trimethyl-s-trioxane ; 1, 3, 5-Trioxane, 2, 4, 6-trimethyl-; Paracetaldehyde ;

The trimer of acetaldehyde ;

B.P., U.S.P., Eur. P., Ind. P.,

Paral[R] (O'Neal, Jones and Feldman) ;

Synthesis :

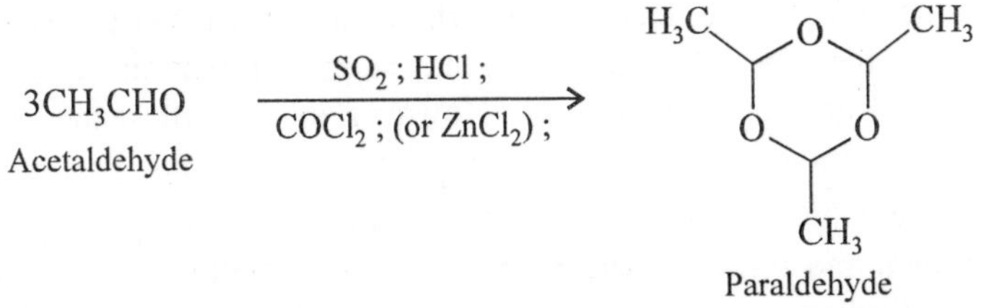

$$3CH_3CHO \xrightarrow[\text{COCl}_2 ; \text{(or ZnCl}_2) ;]{\text{SO}_2 ; \text{HCl} ;}$$

Acetaldehyde

Paraldehyde

It may be prepared by treating acetaldehyde with small amounts of sulphur dioxide, hydrochloric acid, zinc chloride or carbonyl chloride, when almost complete conversion is caused. The resulting liquid is freezed and then distilling the crystallized substance under reduced pressure yields the pure paraldehyde.

It is one of the oldest and best hypnotics having anti-convulsant effects. It is sometimes used as an obstetrical analgesic, in which situation large doses are administered, usually through rectum.

Dose : Adult, oral, sedative 5 to 10 ml ; as hypnotic 10 to 30 ml ; Intra-muscular sedative, 5 ml ; hypnotic 10 ml.

3. MODE OF ACTION OF GENERAL ANAESTHETICS

General anaesthetics have been in use for more than a century, but unfortunately so far no exact mechanism of action has been put forward. Of course, a few theories, namely ; lipid, physical, biochemical, miscellaneous, *Meyer-Overton, minimum alveolar concentration (MAC), stereochemical effects and ion-channel and protein receptor theories have been advocated from time to time in support of the mode of action of the general anaesthetics. These will be discussed briefly in this context.

3.1. Lipid Theory

There exists a direct relationship between the anaesthetic activity of an agent and its lipid solubility. It offers a reasonably acceptable correlation between anaesthetic activity (pharmacologic) of a compound and its oil/water or oil/gas partition coefficient (physical). This hypothesis rightly advocates that the site of action of anaesthetics is usually hydrophobic in nature. It may be anticipated that *the greater the lipid solubility of an anaesthetic agent the higher would be its potency.*

3.2. Physical Theory

An increase in the strength of the Van der Waals correlation factors exerts a positive improvement of anaesthetic potency. Likewise, the size of anaesthetic molecules is a determining factor for their ability to reach the site of action. Another school of thought suggests that anaesthesia commences when the critical space within the membrane is charged with the anaesthetic molecules.

3.3. Biochemical Theory

It has been observed that barbiturates normally interfere with the creation of high-energy products by decoupling oxidative phosphorylation. There also exists ample evidence to prove that the cerebral oxygen consumption gets decreased considerably in a human being treated with a variety of anaesthetic agents. This may be expatiated by the fact that *anaesthesia invariably retards the central nervous system activity thereby resulting in a diminished oxygen intake.*

3.4. Miscellaneous Theory

According to one concept the decrease in surface tension caused by an anaesthetic is directly related to its potency. Another possible explanation may suggest a relationship between the anaesthetic action and the function and structure of membrane. However, many of these theories converge to a point indicating that the lipid portion of membranes is the site of anaesthesia.

A good deal of factual information has been accumulated in connection with physical properties together with biochemical and physiological processes of anaesthetic agents, but unfortunately not a single theory proved and substantiated by experimental facts of anaesthesia is known.

3.5. Meyer-Overton Theory

Meyer and Overton put forward that the potency of a drug substance as an anaesthetic exhibited a direct relationship to its ability to attain lipid solubility, or oil-gas-partition coefficient*. They have studied various *membrane-like lipids*, octanol and olive oil to establish and determine the lipid-soluble properties of the *'volatile anaesthetics'* that existed at that time. However, it has been observed critically the drug substances having reasonably high lipid solubility essentially needed appreciably lower concentrations [*i.e.,* lower Minimum Alveolar Concentration (MAC)] to cause anaesthesia. It was suggested further that the possible interaction between the hydrophobic section of the membrane and the anaesthetic molecules afforted an apparent distortion of the former very close to the channels that particularly conducted Na^+ ions. Thus, the membrane has been subjected to **squeeze** and **bloat** in onto the corresponding channel to bring about *two* distinct biological actions, namely :

(*a*) interference with *sodium conductance*, and

(*b*) interference with usual *neuronal depolarization.*

In short, the recent development in the field of *protein : drug interactions* has more or less challenged this theory squarely.

3.6. Minimum Alveolar Concentration (MAC)

MAC may be defined as — *"the concentration at 1 atmosphere of anaesthetic in the alveoli required to produce immobility in 50% of adult patients being subjected to a surgical procedure.* It has been duly observed that a further increase to 30% MAC (*i.e.,* 1.3 MAC) invariably affords apparent *immobility in 99% subjects.*

---

* Meyer HH., *J. Am. Med. Assoc.,* 1906, **26** : 1499—1502 ; Overton E., *Jena : Gustav Fischer,* 1901.

Mechanism : Probably at equilibrium, the prevailing concentration of a volatile anaesthetic in the alveoli is equal to that in the brain ; and consequently this particular concentration in the brain that very intimately exhibits the concentration at the site responsible for the anaesthetic activities. Therefore, the MAC of a volatile anaesthetic is most frequently employed as a reliable **'yardstick'** to ascertain the exact potency of an individual general anaesthetic agent. Table-1 depicts the MAC values of several gaseous and volatile anaesthetics commonly put into practice nowadays.

Table 1. MACs Partition Coefficients and Metabolism of Volatile Anaesthetics

Volatile Anaesthetics	MAC (% of 1 ATM)*		Partition Coefficients(At 37°C**)		Metabolism (%)
	Without N_2O	with N_2O(%)	Oil/Gas	Blood/Gas	
Halothane	0.77	0.29 (66)	224	2.3	20
Isoflurane	1.15	0.50 (70)	90.8	1.4	0.17
Sevoflurane	1.71	0.66 (64)	53.4	0.60	4–6
Nitrous oxide	104	—	1.4	0.47	None

* MAC is the minimum alveolar concentration, expressed as volume %, which is essentially needed to cause immobility in 50% of middle-aged human beings.

** Stoelting RK., **'Pharmacology and Physiology of Anaesthetic Practice'**, 3rd, edn., Lippincott Williams and Wilkins, Philadelphia, 1999.

Interestingly, when these general (volatile) anaesthetics are employed in *combination*, the MACs for the **Inhaled Anaesthetics are Additive.**

Example : The intensity as well as depth accomplished with 0.5 MAC of **enflurane** together with 0.5 MAC of **nitrous oxide** is almost equivalent to what is produced by 1 MAC of either of the said two agents employed alone. The major advantage of such a combination being that a patient is not exposed to excessive quantum of any one of the individual agents, which in other words drastically minimises the probable risk of **adverse reactions,** if any.

3.7.   Stereochemical Effects

It is pertinent to observe here that a number of volatile anaesthetics *viz.*, halothane, isoflurane, enflurane and the like essentially contain in each of them an **asymmetric carbon atom** (*i.e.,* a *chiral centre*) ; therefore, may invariably occur both as (+)-or (–)-enantiomers. It has been a common practice to make use of these volatile anaesthetics as their racemates commercially ; however, another school of thought devised a mean to establish and determine the anaesthetic characteristics of individual enantiomers.

Salient Features : The following are some of the salient features of such investigations, namely :

(1) Lysco *et al**. (1994) reported that (+)-isoflurane (MAC 1.06%) is approximately 50% more potent as an anaesthetic in the rat than its corresponding (–)-isoflurane compound (MAC 1.62%).

(2) However, in another study by Graf *et al.*** (1994) it was revealed beyond any reasonable doubt that the potency of the individual enantiomers to cause depression of mycardial activity was determined to be almost identical.

---

* Lysco GS *et al. Eur. J. Pharmacol.,* **263,** 25-29 (1994).

Graf BM *et al. Anaesthisiology.,* **81, 129-136 (1994).

In short, the above findings in (1) and (2) evidently drive out attention to a rather more intricate and complex mechanism responsible for such distinct and apparent variations in their activities ; and that is the **protein-anaesthetic interactions.***

3.8. Ion Channel and Protein Receptor Hypotheses

Importantly, a relatively more recent intensive studies have not only established but also helped in determining the cardinal effects of a host of volatile anaesthetics on a good number of protein receptors very much within the realm of central nervous system (CNS). The various characteristic features which critically and overwhelmingly support the possibility of an important and a vital interaction with a protein essentially include are, namely :

(*a*) Steep dose-response curves observed,

(*b*) Stereochemical requirements of different volatile anaesthetics,

(*c*) Observations with regard to enhanced molecular weight *vis-a-vis* lipid solubility profile of a general anaesthetic may eventually either decrease or negate absolutely the desired anaesthetic activity, and

(*d*) Revelations that the presence of particular *ion channels* and *nerotransmitter receptor* systems are a vital need and basic requirements for a plethora of the noticeable activities of the volatile anaesthetics.

Mechanism : The most *pivotal theme* to explain the actual mechanism of action of volatile (general) anaesthetics logically and legitimately involves the interaction of the anaesthetics with the receptors which critically regulate the performance of the ion-channels, such as : K^+, Cl^- ; or with the ion-channel in a direct fashion (*e.g.,* Na^+).

4. MECHANISM OF ACTION OF GENERAL ANAESTHETICS

The probable mechanism of action of certain general anaesthetics dealt with in this chapter are enumerated as under :

4.1. Ethyl Chloride

An extremely volatile liquid with an agreeable pleasant odour. When sprayed on the skin, it evaporates so rapidly that the tissue is *cooled* immediately. By virtue of this characteristic property, the skin gets anaesthetized ; and hence, used in minor surgery for very short durations.

4.2. Vinyl ether

An anaesthetic agent which has become virtually obsolete because it is explosive or highly inflammable in the concentration needed to cause anaesthesia.

4.3. Cyclopropane

It also enjoyed some popularity earlier, but has to be abandoned because of its highly explosive nature just like diethyl ether.

4.4. Fluroxene

It is a fluorinated unsaturated ethereal compound showing a rapid on-set of action by inhalation.

---

*Sidebotham DA, and Schug SA., *Clin. Exp. Pharmacol. Physiol.,* **24,** 126-130, 1997.

4.5. Halothane

It is a noninflammable, nonexplosive fluorinated volatile anaesthetic which is invariably mixed with either air or oxygen. Importantly, the presence of the C and halogen bonds generously contributes to its noninflammability nature. It was so designed to accomplish certain characteristic properties, namely : (*a*) chemical stability ; (*b*) exert an intermediate blood solubility ; and (*c*) appreciable anaesthetic potency. Nevertheless, it is the only useful general anaesthetic having a bromine atom, which is exclusively responsible for its enhanced potency. Likewise, the presence of three strategically positioned in halothane is believed to increase its inherent potency, volatility and chemical stability of the hydrocarbon structure to a considerable extent.

Salient Features :

(*i*) It exhibits a rapid onset of action followed by rapid recovery from the induced anaesthetic effect in *two* different situations ; *first*, when used alone with high potency ; and *secondly,* when used in combination along with nitrous oxide.

(*ii*) A large number of metals, with the exception of chromium (Cr), nickel (Ni) and titanium (Ti), are very quickly tarnished (to lose lustre) by it.

(*iii*) Though it is comparatively stable ; however, it undergoes *spontaneous oxidative decomposition* resulting into the formation of hydrobromic acid (HBr), hydrochloric acid (HCl), and phosgene ($COCl_2$). Therefore, it is specifically dispensed in dark-amber coloured glass containers with the addition of **'thymol'** as a preservative to reduce the chances of oxidation.

(*iv*) It has observed duly that nearly 20% of an administered dosage of *'halothane'* gets metabolized that ultimately is responsible for the enhanced observed hepatotoxicity.

4.6. Methoxyflurane

A general anaesthetic usually administered by inhalation for surgical procedures of relatively short duration. Its renal toxicity prevents its being used for prolonged anaesthesia.

4.7. Trichloroethylene

It is mostly used as an analgesic and anaesthetic agent to supplement the action of nitrous oxide. It should not be used with epinephrine.

4.8. Nitrous Oxide

The tasteless, odourless and colourless and sweet smelling N_2O (*'laughing gas'*) having minimum alveolar concentration (MAC) value almost exceeding 105% is observed to be **incapable** of inducing surgical anaesthesia if administered alone. It has already been established for N_2O to have an MAC value ranging between 105–140% ; and, therefore, is **not** able to succeed in accomplishing in **'surgical anaesthesia'** under conditions prevailing at *standard barometric pressure.* Interestingly, Bert in 1879 demonstrated that an MAC more than 100% could be achieved by employing an admixture of 85% N_2O with O_2 at 1.2 atmospheres in a pressurized vessel, which would provide an MAC fairly sufficient for causing *'surgical anaesthesia'*.

In fact, N_2O is usually employed alone as an anaesthetic agent during some special localized *'dental procedures'* only. However, most frequently N_2O is utilized with other volatile anaesthetics to cause a sufficient desirable depth of anaesthesia essentially required for various **'surgical procedures'.**

Mechanisms :

Importantly, to date no definite mechanism(s) have been put forward to explain how nitrous oxide exerts its anaesthetic activity.

While some theories have been advocated which probably suggest the *'irreversible oxidation'* of the cobalt atom in vitamin B_{12} by the help of N_2O that may ultimately render the inactivation of certain specific enzymes* dependent on Vit B_{12} having resultant deviations from its normal course.

4.9. Chloroform :

It has been established that the addition of halogens to the hydrocarbon backbone not only enhances potency but also retards flammability to a great extent. Chloroform ($CHCl_3$) is a very potent anaesthetic agent having appreciable analgesic and neuromuscular relaxing activity. It is a known **'carcinogen'** and proved to be both *hepatotoxic* and *nephrotoxic* ; besides, causing severe adverse circulatory effects, for instance : *arrythmias* and *hypotension.* Hence, by virtue of its absolutely unacceptable therapeutic index it is no longer used as a volatile anaesthesia.

4.10. Thiopental Sodium

Thiopental sodium belonging to the class of ultrashort-acting barbiturates (*e.g.,* thiopental) are mostly employed IV to cause a rapid on set of unconsciousness for both surgical and basal anaesthesia. Importantly, it may be used first and foremost to cause anaesthesia, which subsequently should be adequately subtained as well as maintained in the course of a surgical operative procedure with the aid of a general anaesthetic.

Mechanism : Barbiturates, in general, afford a marked decrease in the specific functional activities in the brain. They are found to increase considerably the GABAergic inhibitory response, by categorically influencing conductance at the **chloride channel** (just like the benzodiazepines). It has been observed that at a relatively higher doses, these may cause potentiation of the existing **$GABA_A$–mediated chloride ion conductance,** thereby strengthening the bondage between *GABA* and *benzodiazepine.*

Another school of thought suggests the following mechanisms of barbiturates, such as : (*a*) uncoupling of oxidative phosphorylation ; (*b*) prevention of the electron-transport system ; and (*c*) inhibition of the prevailing cerebral carbonic-anhydrase activity ; occurring all at relatively higher concentrations. Barbiturates are also found to induce liver microsomal enzymes which may invariably lead to an enhanced rate of biotransformation of a host of other commonly employed drugs. They also exert an appreciable affect on the transport of sugars *in vivo.*

4.11. Ketamine hydrochloride

It is an extremely potent fast-acting anaesthetic agent and having comparatively short duration of action (10–25 minutes).

Mechanism :

   (*i*) It does not relax skeletal muscles ; and hence, it may be employed safely in such cases of short-duration wherein muscle-relaxation is not needed at all.

  (*ii*) Cessation of the acute action is caused mostly due to its redistribution from the brain into other tissues of the body.

 (*iii*) It has been observed that a plethora of **'metabolites'** invariably occur on account of the formation of the *glucoronide conjugate* and *metabolism in the* liver.

 (*iv*) *Norketamine,* a metabolite is generated *via* the action of **cytochrome P450.** Interestingly, this particular *demethylated structural analogue* does retain an appreciable activity at the site of the N-methyl-D-aspartate (NDMA) receptor, which may eventually be responsible

---

*Methionine synthetase and **thymidylate synthetase** are necessary in the synthetic pathways thereby leading to the production of *myelin* and *thymidine* respectively.

for attributing towards the longer duration of action of this anaesthetic drug. Norketamine gets converted to its corresponding hydroxylated metabolites which upon further conjugation form certain metabolites that gets eliminated through the kidney.

(v) Ketamine is believed to act very much alike the *phencyclidine* **(PCP)** that essentially serves as an antagonist very much within the *cationic channel* of the **NDMA-receptor complex***. By preventing the flow of cations through the cationic channel, it evidently checks neuronal activation that is usually needed for holding the conscious state.

(vi) It is largely able to sustain and produce a **'dissociative'** anaesthesia, that is particularly characterized by electraencephalogram (EEG) alterations thereby showing a marked and pronounced dissociation occurring between the **thermocortical** and **limbic systems****.

(vii) Ketamine's analgesic activity could be due to an interaction either with an *opioid receptor* or a *sigma receptor* (which is relatively not-so-well understood).

4.12. Hydroxydione sodium succinate

It is basically a steroidal drug used intravenously as an anaesthetic agent. It has, however, no hormonal activity.

4.13. Thiamylal Sodium

Thiamylal is a highly hydrophobic thiobarbiturate having its structural features very much related to thiopental. Besides, its biological activities are almost identical to *thiopental.* After IV administration, unconsciousness is induced within a span of a few seconds only, while complete recovery of consciousness occurs within 30 minutes. Therefore, it is mostly used effectively in short surgical procedures.

4.14. Fentanyl citrate

Fentanyl citrate is a potent narcotic analgesic with rapid onset and short duration of action when administered parenterally. It shows a profile of pharmacological action quite similar to *morphine,* but with two glaring exceptions, namely : (a) does not cause emesis ; and (b) releases histamine. After IV administration, the peak analgesia seems to occur within a span of 3–5 minutes and lasts for 30–60 minutes. Interestingly, it is employed primarily as an analgesic for the acute control and management of pain associated with all types of surgery. It also finds its enormous application as a supplement to all such agents that are invariably used either for general and regional anaesthesia.

Advantage : The administration of *'fentanyl'* (*i.e.,* the base) *via* a **transdermal patch** exhibits a much slower onset (8 to 12 hours) and significantly longer duration of action (more than 72 hours) ; and, therefore , quite frequently is employed to manage chronic pain that essentially requires an *'opiate analgesic'*.

4.15. Paraldehyde

It is one of the 'oldest *sedatives and hypnotic'* which gets absorbed very quickly after oral administration and helps to induce sleep within 10–15 minutes after a 4-to 8-mL dose. Its application has been resticted in patients with a history of asthma or other pulmonary diseases because it gets partially excreted through the lungs thereby imparting an odour to the exhaled air that produces undesirable irritation.

---

*Yamamura T *et al. Anesthesiology, 72* : 704—710 (1990)

Reich DL and Silvay G., *Can. J. Anaesth,* **36 : 186—197 (1989).

Mechanism : Paraldehyde mostly gets detoxified by the liver (70 to 80%) and 11 to 28% is excreted by the lungs. However, only a negligible small extent is excreted through the urine.

Probable Questions for B. Pharm. Examinations

1. What is the importance of 'Inhalation Anaesthetics' over 'Intravenous' and 'Basal' Anaesthetics ? How would you synthesize the following :

    (*a*) Ethyl Chloride   (*b*) Cyclopropane   (*c*) Fluroxene   (*d*) Halothane   (*e*) Trichloroethylene

2. What are intravinous anaesthetics ? Discuss the synthesis of the following :

    (*i*) Thiopental Sodium   (*ii*) Ketamine Hydrochloride   (*iii*) Methahexital Sodium

    (*iv*) Hydroxydione sodium succinate   (*v*) Thiamylal Sodium.

3. What are the merits and demerits of '*Basal Anaesthetics*' ? Describe the synthesis of the following :

    (*i*) Fentanyl citrate   (*ii*) Tribromoethanol   (*iii*) Paraldehyde.

4. Give a brief account of the following theories put forward to explain the '*mode of action*' of general anaesthetics :

    (*a*) Lipid Theory   (*b*) Physical Theory   (*c*) Biochemical Theory   (*d*) Miscellaneous Theory.

5. How would you classify the '*General Anaesthetics*' ? Give the structure, chemical name and uses of one potent drug from each class.

6. Give the synthesis of a '*General Anaesthetic*' having a steroidal nucleus.

7. Name the **three** modified varsions of '*barbiturates*' that are used abundantly as intravenous anaesthetics.

8. Discuss three fluorinated compounds employed mostly as inhalation anaesthetics.

RECOMMENDED READINGS

1. RR Macintosh and FB Bannister *Essentials of General Anaesthesia*, London, Blackwell Scientific Publication (1947).

2. EI Eger II *Anaesthetic Uptake and Action* Baltimore, Williams and Wilkins Co. (1974).

3. VJ Collins *Principles of Anesthesiology,* Philadelphia, Lea and Febiger (1966).

4. TC Daniels and EC Jorgensen General Anaesthetics in *Textbook of Organic Medicinal and Pharmaceutical Chemistry (Eds)* (CO Wilson), O Gisvold and RF Doerge) Philadelphia, JB Lippincott (1977).

5. LS Goodman and A Gilman *The Pharmacological Basis of Therapeutics* (9th edn) London, Macmillan Co. (1995).

6. HG Mautner and HC Clemson Hypnotics and Sedatives, in ; M.E. Wolff (*Ed.*) *Burger's Medicinal Chemistry and Drug Discovery* (5th edn), New York, Wiley & Sons. Inc., (1995).

7. DLednicer and LA Mitscher *The Organic Chemistry of Drug Synthesis* New York, John Wiley and Sons (1995).

8. Alex Gringauz, *'Introduction to Medicinal Chemistry'*, New York, Wiley-VCH, 1997.

9. Graham L. Patrick, *'An Introduction to Medicinal Chemistry'*, Oxford University Press, New York, 2nd, edn, 2002.

4   Local Anaesthetics

1. INTRODUCTION

Local anaesthetics are drugs which reversibly prevent the generation and the propagation of active potentials in all excitable membranes including nerve fibres by stabilizing the membrane. They achieve this by preventing the transient increase in sodium permeability of the excitable membrane. Generally, small diameter cells are more sensitive to their action than larger diameter cells. Thus conduction in nerve fibres is more readily blocked than conduction in muscle fibres. Also usually *the smaller diameter nerve fibres are more susceptible to the action of local anaesthetics than the larger diameter ones.* For example, on a mixed sensory fibre, local anaesthetic will abolish conduction on the fibre conveying the sensation of pain first. This will be followed by block of the fibre responsible for the sensations fo cold, warmth, touch and deep pressure in that order. This is in accordance with the fibre diameter. It is important to mention here that *fibre diameter though very important is not always the only factor that determines the relative sensitivity of the nerve fibres to the action of the local anaesthetics.*

It is, however, pertinent to mention here that both *'local anaesthetics'* and *'general anaesthetics'* essentially afford anaesthesia by blocking nerve conductance in motor neurons as well as sensory neurons. The ultimate blockade of nerve conduction causes not only a *loss of pain sensation* but also affects *impairment of motor functions.* Interestingly, the anaesthesia produced by the *'local anaesthetics'* does not necessarily cause complete loss of either consciousness or significant impairment of important central functions. It is, however, believed that *'local anaesthetics'* normally exert their action by blocking nerve conductance after having bondage to selective site(s) of the Na-channels in the particular excitable membranes. In this manner, the passage of Na through the pores gets reduced significantly and hence cause direct interference with the action potentials. Evidently, a local anesthetic helps in drastically minimising the excitability of the nerve membranes with touching the resting potentials. In short, local anaesthetics neither *interact with the pain receptors* nor *affect the biosynthesis of pain mediators.*

Local anaesthetics are used to abolish the sensation of pain in a restricted area of the body and for minor surgical operations when loss of consciousness is not desirable. The area is determined by the site and the technique of administration of the anaesthetic agent. The main uses are as follows :

(*a*) *Surface or Topical Anaesthesia*

The local anaesthetic is applied to the mucous membrane, *.e.g.,* conjunctiva, larynx, throat, damaged skin surface, etc.

(b) *Infiltration Anaesthesia*

The drug is injected subcutaneously to paralyse the sensory nerve endings around the area to be rendered insensitive, *e.g.,* an area to be incised or for tooth extraction.

(c) *Nerve Block Anaesthesia*

The local anaesthetic is injected as close as possible to the nerve trunk supplying the specific area to be anaesthetised. This blocks conduction in both sensory and motor fibres and minor operations on the limb are possible.

(d) *Spinal Anaesthesia*

The drug is injected into the subarachnoid space, *i.e.,* into the cerebrospinal fluid, to paralyse the roots of the spinal nerves. This method is used to induce anaesthesia for abdominal or pelvic surgical operations.

Saddle block is a variation of spinal anaesthesia where the injection is made into the lower part of the subarachnoid space. The drug normally settles in the lower part of the dural space. It is used in obstetrics and for surgery in the perineal region.

(e) *Epidural Anaesthesia*

This is a special type of nerve block anaesthesia in which the drug is injected into the epidural space. It is technically a more difficult procedure. The roots of the spinal nerves are anaesthetized.

(f) *Caudal Anaesthesia*

This is smaller to epidural anaesthesia where the injection is made through *sacral hiatus* into the vertebral canal which contains the *cauda equina.* It is used for operations on the pelvic viscera.

The first local anaesthetic to be used was cocaine as alkaloid isolated from the leaves of *Erythroxylon coca (Erythroxylum coca).* Carl Koller,[1] an Australian ophthalmologist in 1884, made an epoch making observation that cocaine hydrochloride causes anaesthesia in the eye. A follow-up by Willstätter and Müller[2] towards an elaborated elucidation of the structure of cocaine ultimately paved the way to the synthesis of a large number of compounds exhibiting local anaesthetic characteristics.

The scheme shown on next page, represents the formation of active benzoyl ecgonine and tropocaine from cocaine and inactive ecognine and pseudo tropane therefrom.

The portion of the cocaine molecule enclosed by a dotted line in the above scheme represents the **'anaesthesiophoric moiety'** and may be designed by the general structure : $Ar—COO—(CH_2)_n NR_1R_2$. The next step in the advance of the knowledge of anaesthesia was the recognition that the presence of a basic nitrogen atom in the esterified alcohol was highly desirable. It permitted the formation of the neutral salt, solutions of which could be injected conveniently and it might also influence the anaesthetic activity effectively.

---

[1]C. Koller, *Wien. Med. Bl. 7,* 1224 (1884) ; C. Koller, *Lancet, 2,* 990 (1884).

[2]R. Willstatter and W. Muller, *Chem. Ber. 31,* 2655 (1898).

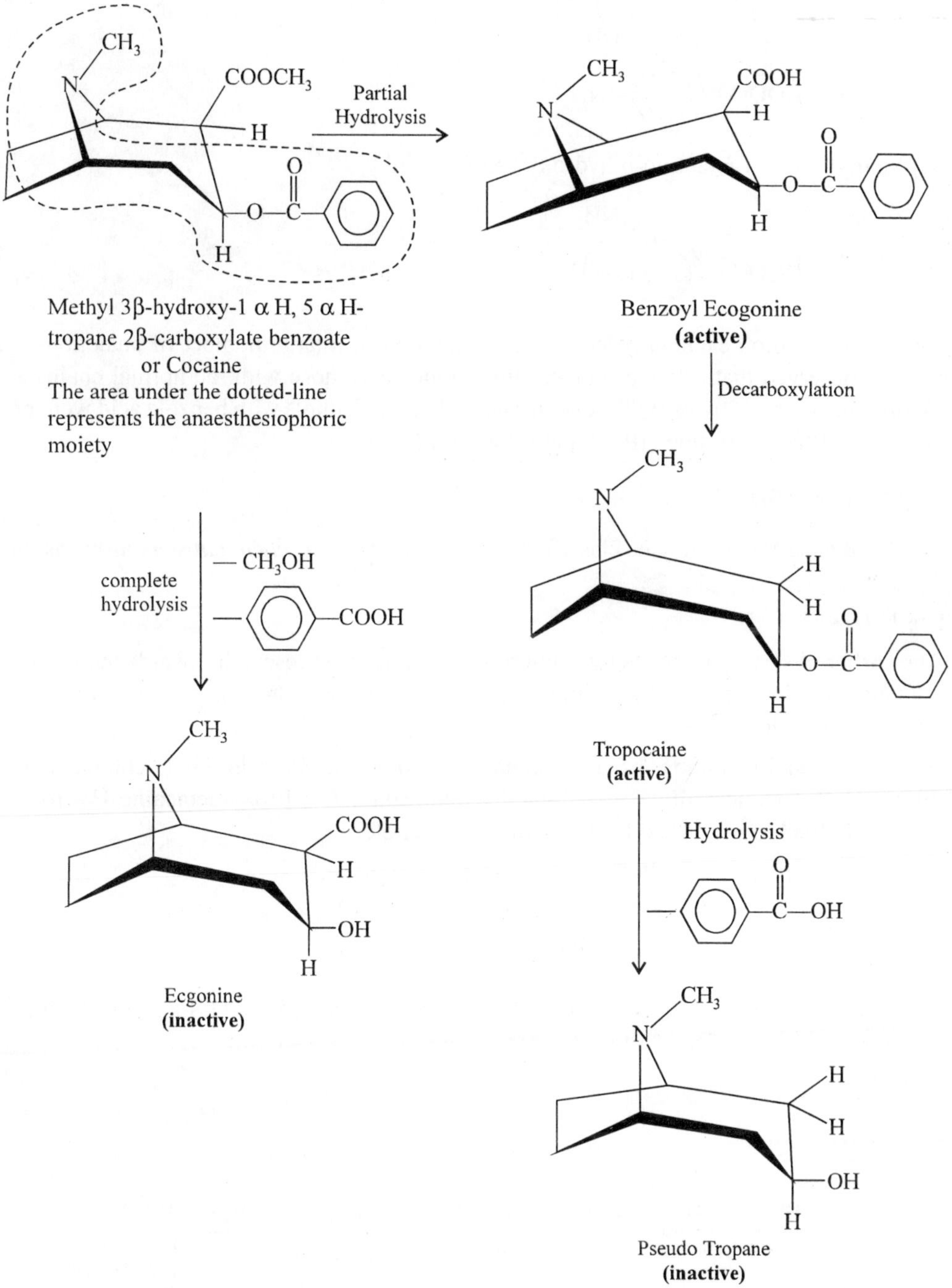
CH3
COOCH3
H
O
H
Partial
Hydrolysis
Methyl 3β-hydroxy-1 α H, 5 α H-
tropane 2β-carboxylate benzoate
or Cocaine
The area under the dotted-line
represents the anaesthesiophoric
moiety
CH3
COOH
H
O
H
Benzoyl Ecogonine
(active)
Decarboxylation
complete
hydrolysis
— CH3OH
COOH
CH3
N
COOH
H
OH
H
Ecgonine
(inactive)
CH3
N
H
H
O
H
Tropocaine
(active)
Hydrolysis
O
C—OH
CH3
N
H
H
OH
H
Pseudo Tropane
(inactive)

Einhorn generalised that all aromatic esters possess the ability to cause anaesthesia and hence he prepared *two* types of esters, namely :

(*a*) Methyl ester of *p*-amino-*m*-hydroxy benzoic acid

i.e.,   HOOC—⟨benzene ring with OH and NH$_2$⟩

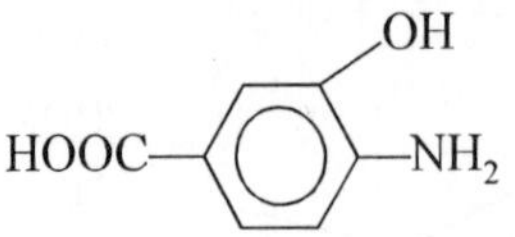

(*b*) Methyl ester of *m*-amino-*p*-hydroxy benzoic acid

i.e.,   HOOC—⟨benzene ring with NH$_2$ and OH⟩

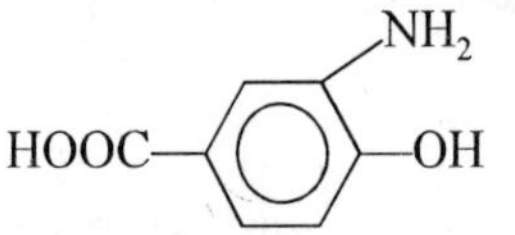

This generalization eventually led to the preparation of a large number of analogous esters. It was, however, observed that esters of higher alkyl groups and those with the normal chains are most active. During the period 1904 to 1909, several esters of basic alcohols with benzoic acid were prepared, *viz.,* amylocaine (1904), procaine (1906) and orthoform (1909).

2.   CLASSIFICATION

The local anaesthetics may be classified on the basis of their *'chemical structures'* as described below :

2.1.   The Esters

The earlier observations made by Einhorn really stimulated research towards the synthesis of a number of benzoic acid esters which exhibited significant local anaesthetic properties. A few important esters are described below :

Examples : Ethyl-*p*-amino benzoate ; Butamben ; Orthocaine ; Procaine Hydrochloride ; Tetracaine Hydrochloride ; Butacaine Sulfate ; Cyclomethycaine Sulphate ; Proxymetacaine Hydrochloride ; Propoxycaine Hydrochloride ; Hexylcaine Hydrochloride, etc.

A. *Ethyl p-aminobenzoate* INN, *Benzocaine* BAN, USAN,

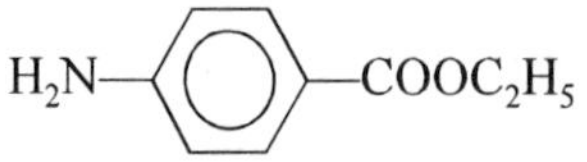

$$H_2N—⟨benzene ring⟩—COOC_2H_5$$

Benzoic acid, 4-amino-, ethyl ester ; Ethyl Aminobenzoate ; B.P., U.S.P., Eur. P., Int. P., N.F., Ind. P., Americaine$^{(R)}$ (American Critical Care) :

Synthesis :

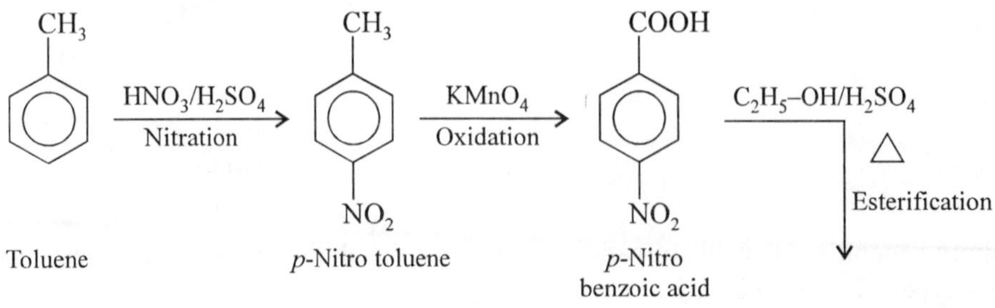

Toluene → HNO$_3$/H$_2$SO$_4$ (Nitration) → *p*-Nitro toluene → KMnO$_4$ (Oxidation) → *p*-Nitro benzoic acid → C$_2$H$_5$–OH/H$_2$SO$_4$ (Esterification)

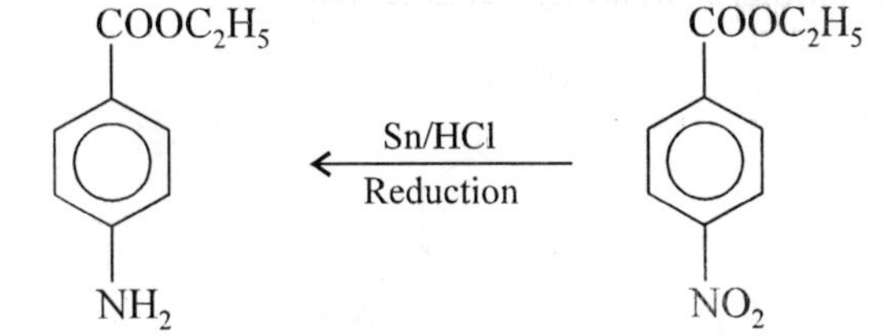

Ethyl-*p*-amino benzoate — Sn/HCl Reduction — Ethyl-*p*-nitro benzoate

p-Nitrobenzoic acid, may be prepared by nitration of toluene and oxidation of the resulting *p*-nitrotoluene, is esterified to the corresponding ethyl ester by heating with absolute ethanol and a few drops of sulphuric acid. The resulting ethyl *p*-nitrobenzoate is reduced with tin and hydrochloric acid to give the official compound.

Being insoluble it is generally used as an ointment to get rid of the pain caused due to wounds, ulcers and mucous surfaces. Its anaesthetic action is usually displayed for the period it remains in contact with the skin or mucosal surface. Hence, it forms an important ingredient in various types of creams, ointments, powders, lozenges and aerosol sprays so as to relieve the pain of naked surfaces and severely inflamed mucous membranes.

Dose : Topical, 1 to 20% in ointment, cream, aerosol for skin.

B. *Butamben* USAN, *Butyl Aminobenzoate* BAN,

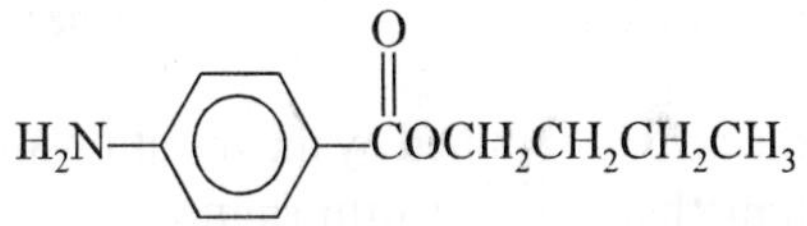

$$H_2N-\langle\bigcirc\rangle-\overset{\overset{O}{\|}}{C}OCH_2CH_2CH_2CH_3$$

Butyl *p*-aminobenzoate ; Benzoic acid, 4-amino-, butyl ester ; U.S.P., N.F. XIII ; Butesin[R] (Abbott).

Synthesis :

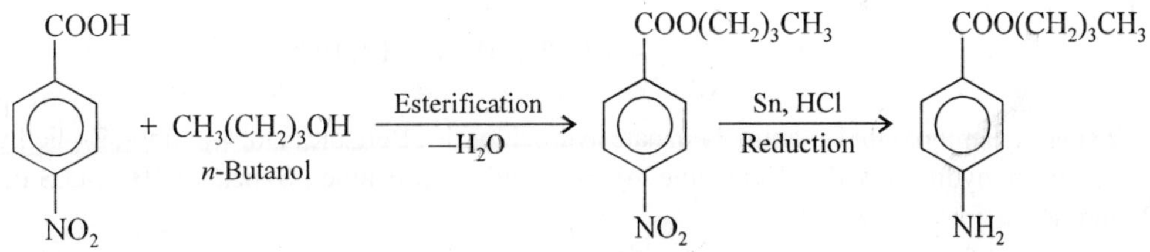

p-Nitrobenzoic acid + CH₃(CH₂)₃OH *n*-Butanol — Esterification —H₂O → *p*-Nitrobutyl benzoate — Sn, HCl Reduction → Butamben

It may be prepared by the esterification of *p*-nitrobenzoic acid with *n*-butanol, then reducing the nitro group with tin and hydrochloric acid.

It is a local anaesthetic of relatively low solubility and used in a similar manner to benzocaine. It has been reported to be more efficacious than its corresponding ethyl ester when applied to intact mucous membranes.

Dose : Topical, 1 to 2% in conjunction with other local anaesthetics in creams, ointments, sprays and suppositories.

C. *Orthocaine* BAN,

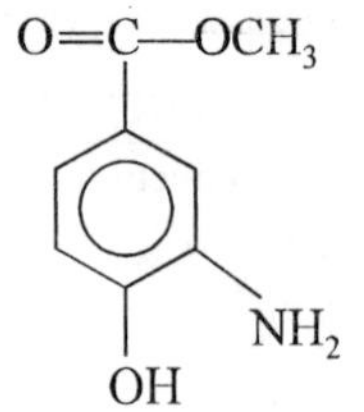

$$O=C-OCH_3$$

Methyl 3-amino-4-hydroxybenzoate ;

B.P. 1953

Synthesis :

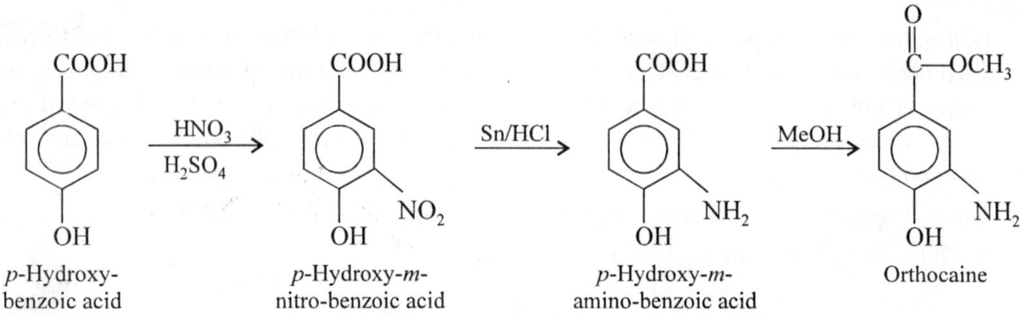

COOH	COOH	COOH	$\overset{O}{\overset{\|}{C}}-OCH_3$

$\xrightarrow[H_2SO_4]{HNO_3}$ $\xrightarrow{Sn/HCl}$ $\xrightarrow{MeOH}$

OH   OH NO_2 OH NH_2 OH NH_2

p-Hydroxy-   *p*-Hydroxy-*m*-   *p*-Hydroxy-*m*-   Orthocaine
benzoic acid   nitro-benzoic acid   amino-benzoic acid

p-Hydroxy-*m*-aminobenzoic acid is obtained by the nitration and reduction of *p*-Hydroxy benzoic acid, which on esterification with methanol yields orthocaine.

It is used for surface anaesthesia, but now it is obsolete due to its *irritation and necrosis effects.*

D. *Procaine Hydrochloride* BAN, USAN,

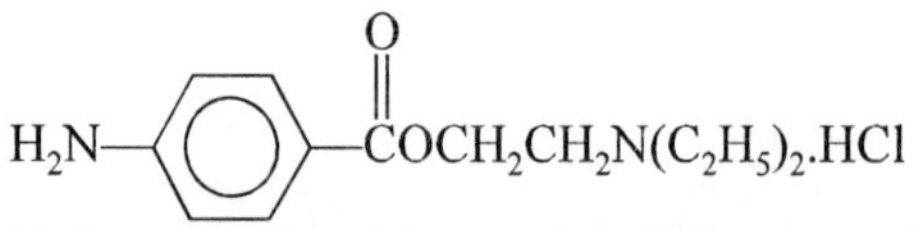

$$H_2N-\text{<benzene ring>}-\overset{O}{\overset{\|}{C}}OCH_2CH_2N(C_2H_5)_2.HCl$$

2-(Diethylamino) ethyl-*p*-aminobenzoate hydrochloride ; Benzoic acid, 4-amino-, 2-(diethylamino) ethyl ester, monohydrochloride ; Ethocaine hydrochloride ; Allocaine ; Syncaine ; B.P. U.S.P., Eur. P. Int. P., Ind. P.,

Novocaine[R] (Sterling) ;

Synthesis :

It may be prepared by any one of the following *three* methods :

Method-1 : From 2-Chloroethyl p-amino benzoate :

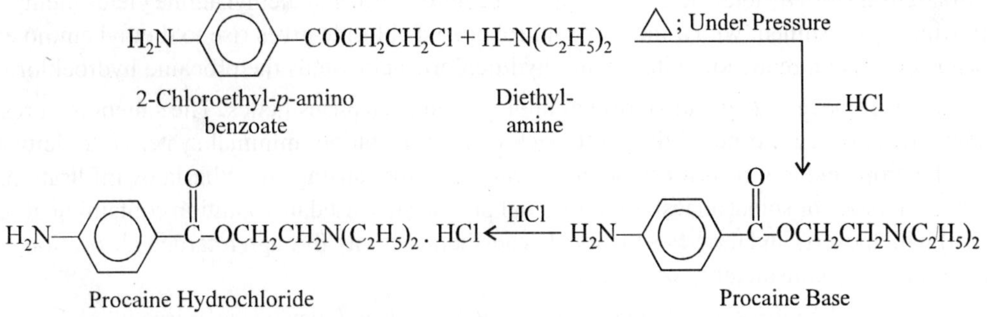

H_2N—⟨○⟩—$COCH_2CH_2Cl$ + H—$N(C_2H_5)_2$ △ ; Under Pressure

2-Chloroethyl-*p*-amino   Diethyl-
benzoate   amine   — HCl

H_2N—⟨○⟩—$\overset{O}{\overset{\|}{C}}$—$OCH_2CH_2N(C_2H_5)_2$. HCl ←—$\overset{HCl}{}$— H_2N—⟨○⟩—$\overset{O}{\overset{\|}{C}}$—$OCH_2CH_2N(C_2H_5)_2$

Procaine Hydrochloride   Procaine Base

Procaine base may be prepared by the interaction of 2-chloro ethyl-*p*-amino benzoate and diethylamine at an elevated temperature under pressure. The base is converted into its hydrochloride subsequently.

Method-II : From p-Aminobenzoic Acid :

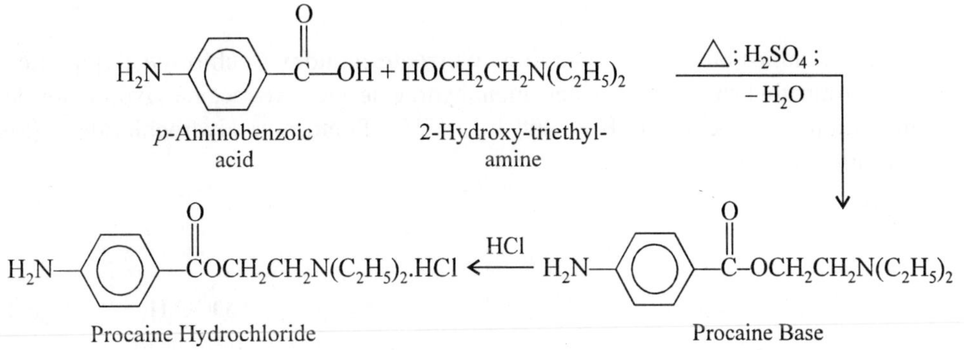

H_2N—⟨○⟩—$\overset{O}{\overset{\|}{C}}$—$OH$ + $HOCH_2CH_2N(C_2H_5)_2$ △ ; H_2SO_4 ;
                                                                       — H_2O

p-Aminobenzoic   2-Hydroxy-triethyl-
acid   amine

H_2N—⟨○⟩—$\overset{O}{\overset{\|}{C}}OCH_2CH_2N(C_2H_5)_2$.HCl ←—$\overset{HCl}{}$— H_2N—⟨○⟩—$\overset{O}{\overset{\|}{C}}$—$OCH_2CH_2N(C_2H_5)_2$

Procaine Hydrochloride   Procaine Base

The procaine base is obtained by the dehydration of molecule of *p*-amino benzoic acid and 2-hydroxy triethyl amine, which on treatment with hydrochloric acid yields the official compound.

Method-III : From Ethylene Chlorohydrin :

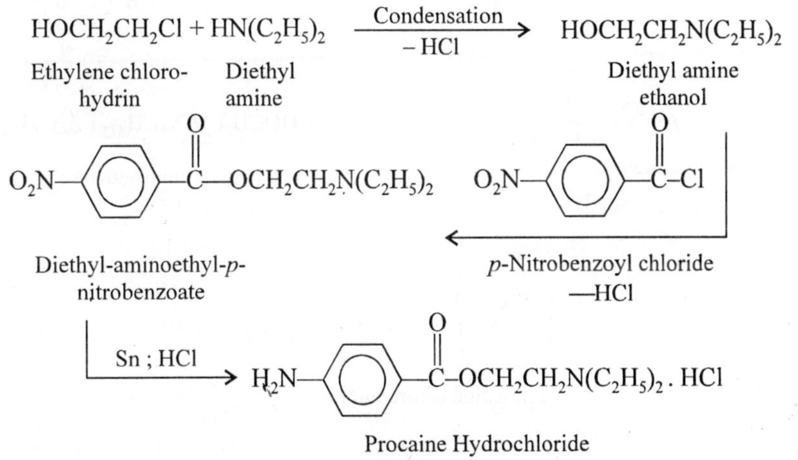

$HOCH_2CH_2Cl$ + $HN(C_2H_5)_2$ $\overset{Condensation}{\underset{- HCl}{\longrightarrow}}$ $HOCH_2CH_2N(C_2H_5)_2$

Ethylene chloro-   Diethyl   Diethyl amine
hydrin   amine   ethanol

O_2N—⟨○⟩—$\overset{O}{\overset{\|}{C}}$—$OCH_2CH_2N(C_2H_5)_2$ O_2N—⟨○⟩—$\overset{O}{\overset{\|}{C}}$—$Cl$

Diethyl-aminoethyl-*p*-   *p*-Nitrobenzoyl chloride
nitrobenzoate   —HCl

$\overset{Sn\ ;\ HCl}{\longrightarrow}$ H_2N—⟨○⟩—$\overset{O}{\overset{\|}{C}}$—$OCH_2CH_2N(C_2H_5)_2$. HCl

Procaine Hydrochloride

Condensation of a molecule each of ethylene chlorohydrin and diethyl amine yields diethyl amino ethanol, which on treatment with a mole of *p*-nitrobenzoyl chloride gives rise to diethyl amino ethyl-*p*-nitrobenzoate. This on reduction with tin and hydrochloric acid yields the procaine hydrochloride.

It is one of the least toxic and most commonly used local anaesthetics. The salient features for its wide popularity may be attributed due to its lack of local irritation, minimal systemic toxicity, longer duration of action, and low cost. It can be effectively used for causing anaesthesia by infiltration, nerve block, epidural block or spinal anaesthesia. In usual practice it is used in a solution containing adrenaline (1:50,000) which exerts and modifies the local anaesthetic activity through retarded absorption, and the duration of action is considerably prolonged.

Dose : Usual, infiltration, 50 ml of a 0.5% solution ; usual, peripheral nerve block, 25 ml of a 1 or 2% solution ; usual, epidural, 25 ml of a 1.5% solution.

E. *Tetracaine Hydrochloride* INN, USAN, *Amethocaine Hydrochloride* BAN,

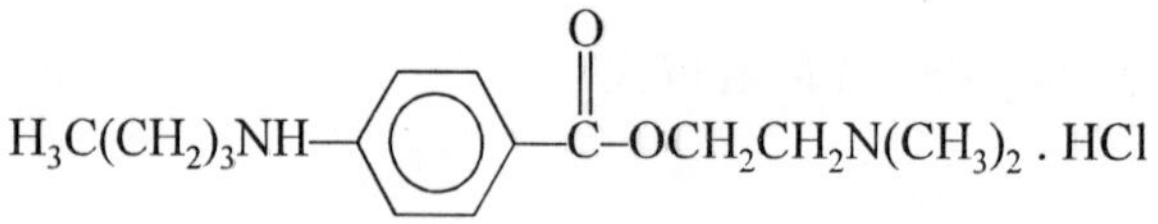

$$H_3C(CH_2)_3NH-\underset{}{\bigcirc}-\overset{\overset{O}{\|}}{C}-OCH_2CH_2N(CH_3)_2 . HCl$$

2-(Dimethylamino)-ethyl-*p*-(butylamino) benzoate monohydrochloride ; Benzoic acid, 4-(butylamino)-, 2-(dimethylamino) ethyl ester, monohydrochloride ; Tetracaine Hydrochloride U.S.P., Eur. P., Amethocaine Hydrochloride B.P., Int. P., Ind. P., Pontocaine Hydrochloride[R] (Sterling) ; Anethaine[R] (Farley, U.K.) ;

Synthesis :

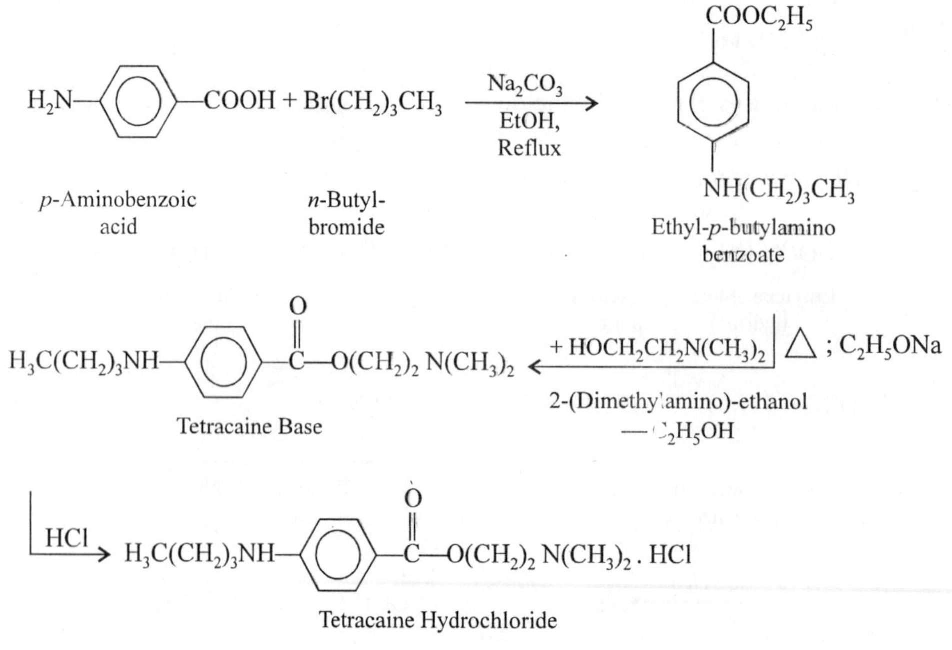

p-Aminobenzoic acid   *n*-Butyl-bromide   Ethyl-*p*-butylamino benzoate

Tetracaine Base

2-(Dimethylamino)-ethanol

Tetracaine Hydrochloride

Butylation of *p*-aminobenzoic acid with *n*-butyl bromide under reflux in ethanolic solution and in the presence of sodium carbonate yields ethyl-*p*-butylamino benzoate. This is then caused to undergo transesterification by heating a solution of it in 2-(dimethylamino) ethanol in the presence of sodium ethoxide in such a manner that the liberated ethanol is continuously removed from the reaction mixture by distillation. The tetracaine base is dissolved in benzene and hydrogen chloride is passed through the solution to obtain the corresponding monohydrochloride salt.

It is an all-purpose local anaesthetic drug used frequently in surface infiltration, block, caudal and spinal anaesthesia. It is reported to be 10 times more toxic and potent than procaine, whereas its duration of action is twice than that of procaine.

Dose : Usual, subarachnoid 0.5 to 2 ml as a 0.5% solution ; topically, 0.1 ml of a 0.5% solution to the conjunctiva.

F. *Butacaine Sulfate* USAN, *Butacaine Sulphate* BAN, Butacaine INN,

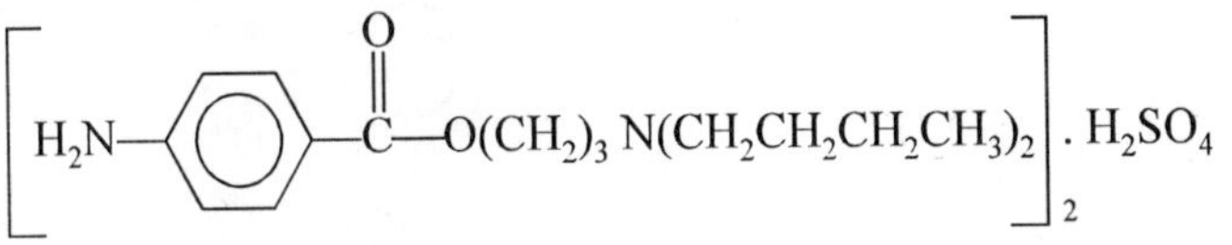

$$\left[H_2N{-}\bigcirc{-}\overset{\overset{O}{\|}}{C}{-}O(CH_2)_3\ N(CH_2CH_2CH_2CH_3)_2 \right]_2 \cdot H_2SO_4$$

3-(Dibutylamino)-1-propanol-*p*-aminobenzoate (ester) sulfate (2 : 1) ; 1-Propanol, 3-(dibutylamino)-, 4-amino benzoate (ester) sulfate (salt) (2 : 1) ; U.S.P., B.P.C. 1968, Ind. P.,

Butyn Sulphate[R] (Abbott)

Synthesis :

It may be prepared from either of the following *two* methods :

Method-I : From 3-Dibutylaminopropyl chloride :

	Sodium-*p*-amino benzoate	3-Dibutylamino- propyl chloride		Butacaine Base

Butacaine Sulfate

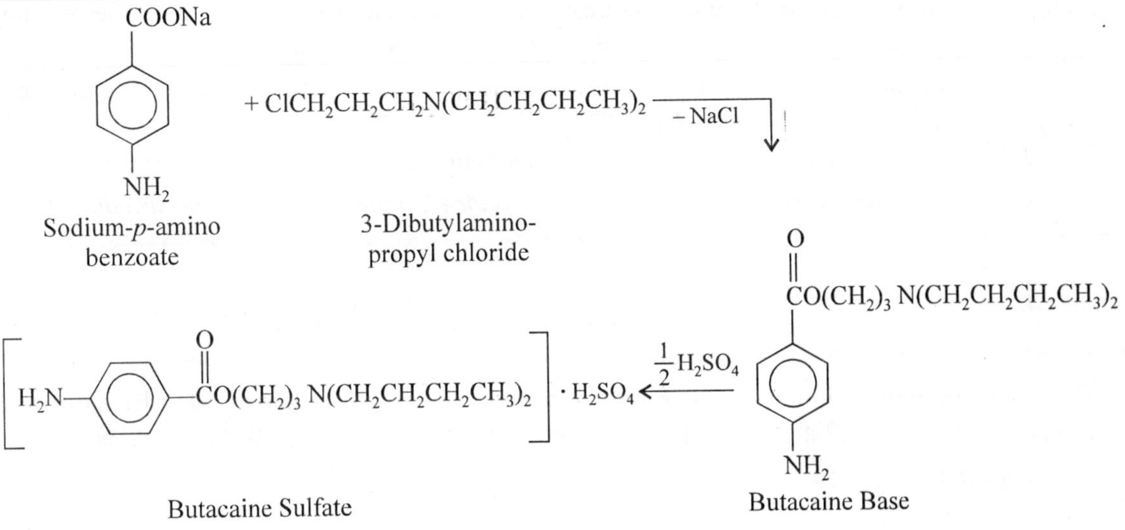

Butacaine base may be prepared by the interaction of a mole each of 3-dibutylaminopropyl chloride with sodium *p*-amino benzoate, which is then treated with a half-molar quantity of sulphuric acid to obtain the official compound.

Method-II : From 3-(Dibutylamino)-1-Propanol :

COOCl

+ HO—CH$_2$CH$_2$CH$_2$ N(CH$_2$CH$_2$CH$_2$CH$_3$)$_2$ ⟶

NO$_2$

p-Nitrobenzoyl   3-Dibutylamino-*l*-
chloride   propanol

$$\underset{\text{NO}_2}{\bigcirc}\overset{O}{\underset{\|}{C}}—O(CH_2)_3\ N(CH_2CH_2CH_2CH_3)_2$$

$$\left[H_2N—\bigcirc—\overset{O}{\underset{\|}{C}}—O(CH_2)_3\ N(CH_2CH_2CH_2CH_3)_2 \right] . H_2SO_4 \xleftarrow[\text{(ii) 1/2 H}_2\text{SO}_4]{\text{(i) Sn/HCl}}$$

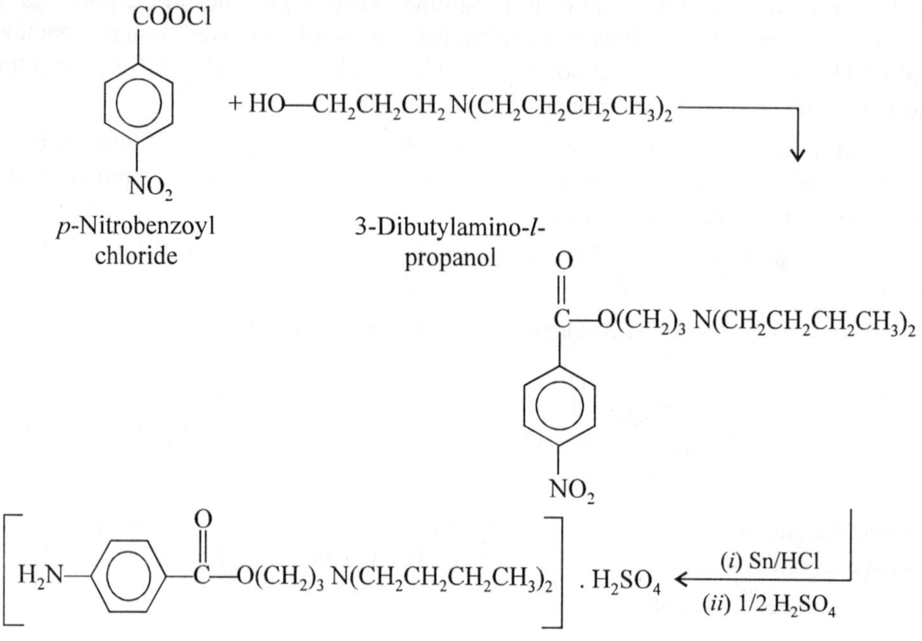

Butacaine Sulfate

3-(Dibutylamino)-1-propanol is coupled with *p*-nitrobenzoyl chloride and the corresponding nitro group is subsequently reduced to amino by treatment with tin and hydrochloric acid. The resulting butacaine base is made to react with a half-molar quantity of sulphuric acid.

It is a surface anaesthetic having effects similar to those of cocaine, but it exhibits more *rapid onset of action followed by a prolonged action.*

Dose : Several instillations of a 2% solution about 3 minutes apart allow most surgical procedures.

G. *Cyclomethycaine Sulphate* BAN, *Cyclomethycaine Sulfate* USAN, *Cyclomethycaine* INN,

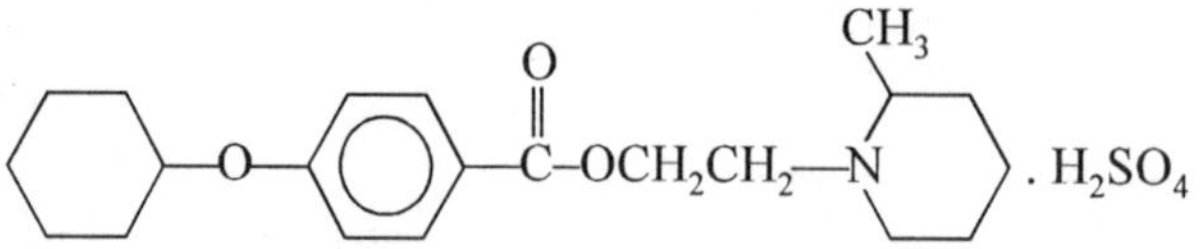

$$\bigcirc—O—\bigcirc—\overset{O}{\underset{\|}{C}}—OCH_2CH_2—\underset{}{N}\overset{CH_3}{\bigcirc} . H_2SO_4$$

3-(2-Methylpiperidino)-propyl *p*-(cyclohexyloxy) benzoate sulfate (1 : 1) ; Benzoic acid, 4-(cyclohexyloxy)- , 3-(2-methyl-1-piperidinyl) propyl ester sulfate (1 : 1) ; B.P., U.S.P., N.F.,

Surfacaine[R] (Lilly)

Synthesis :

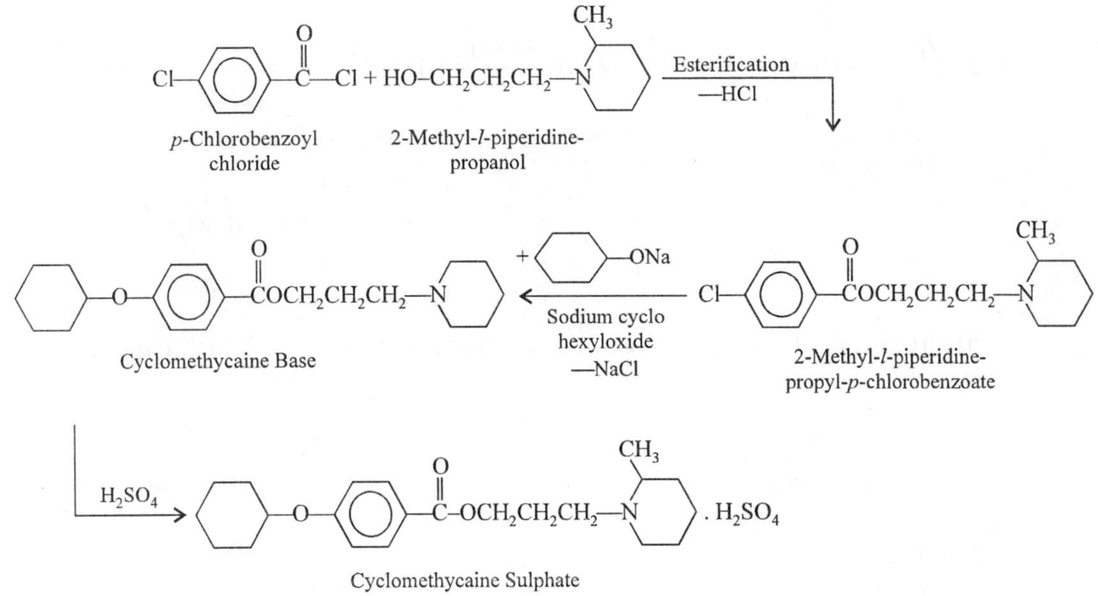

p-Chlorobenzoyl chloride undergoes esterification with 2-methyl-1-piperidinepropanol with the elimination of a mole of hydrogen chloride. This on treatment with sodium cyclohexyl oxide yields the cyclomethycaine base which on further treatment with sulphuric acid gives the official compound.

It is extensively used as an effective topical anaesthetic in thermal and chemical burns ; in dermatological lesions, sunburn and skin abrasions ; in urology, gynaecology, obstetrics and anaesthetic procedures.

Dose : Topical, 0.25 to 1.0% in suitable form.

H. *Proxymetacaine Hydrochloride* BAN, *Proxymetacaine* INN, *Proparacaine Hydrochloride* USAN,

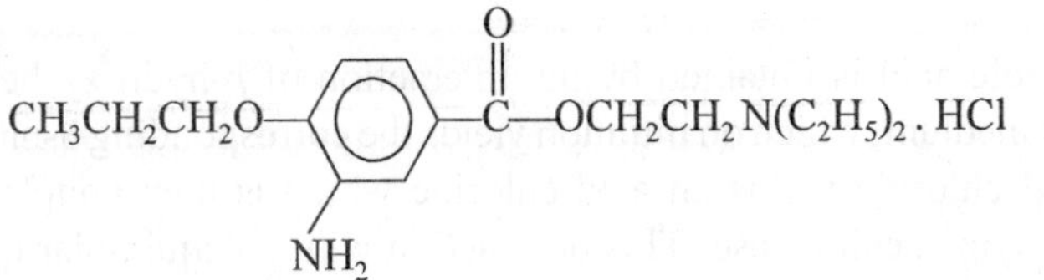

2-(Diethylamino) ethyl 3-amino-4-propoxybenzoate monohydrochloride ; Benzoic acid, 3-amino-4-propoxy-, 2-(diethylamino)-ethyl ester, monohydrochloride ;

Proxymetacaine Hydrochloride B.P.C. (1973) ; Proparacaine Hydrochloride U.S.P.

Alcaine[(R)] (Alcon) ; Ophthaine[(R)] (Squibb) ; Ophthetic[(R)] (Allergan)

Synthesis :

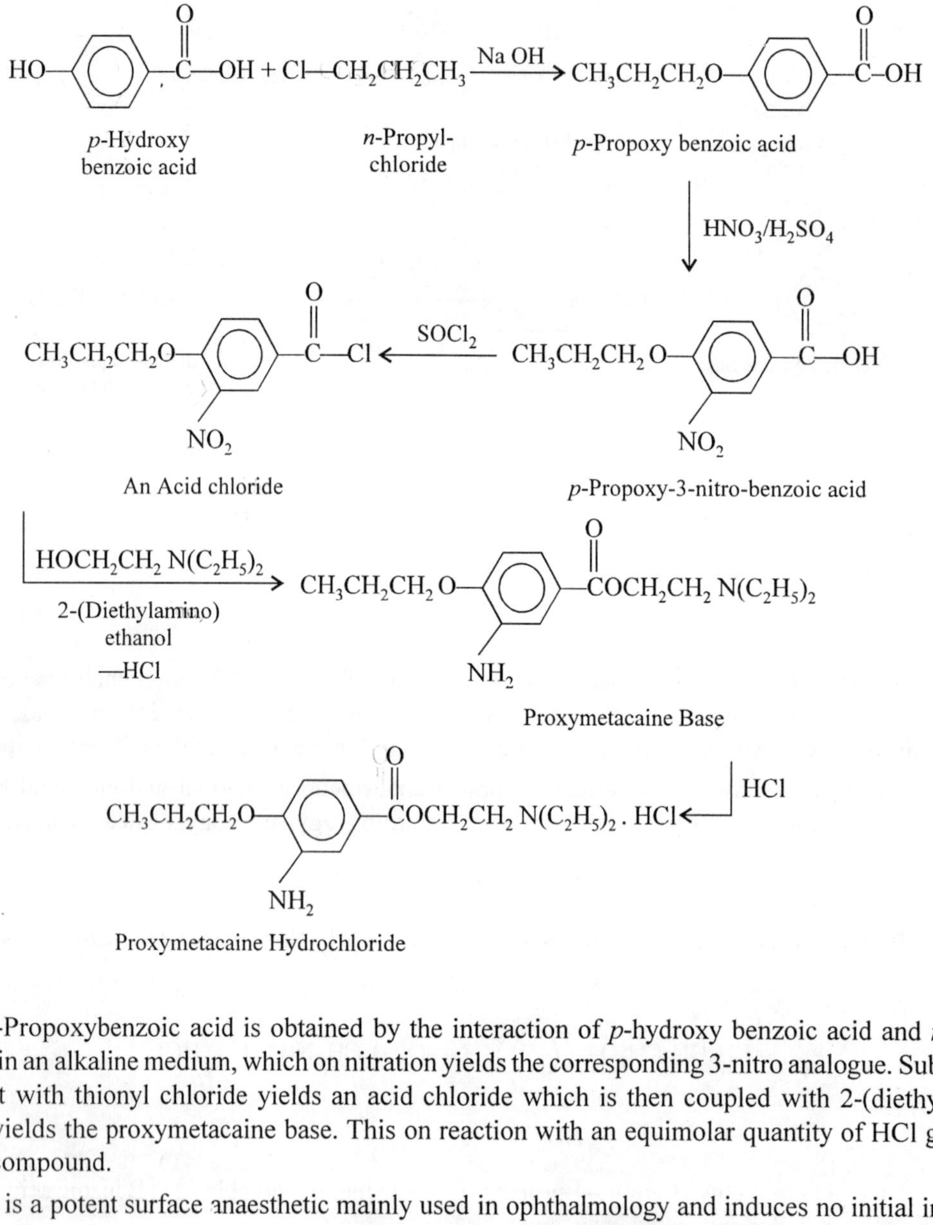

p-Propoxybenzoic acid is obtained by the interaction of *p*-hydroxy benzoic acid and *n*-propyl chloride in an alkaline medium, which on nitration yields the corresponding 3-nitro analogue. Subsequent treatment with thionyl chloride yields an acid chloride which is then coupled with 2-(diethylamino) ethanol yields the proxymetacaine base. This on reaction with an equimolar quantity of HCl gives the official compound.

It is a potent surface anaesthetic mainly used in ophthalmology and induces no initial irritation. Because of its rapid onset of action it is useful for most occular procedures that require topical anaesthesia such as tonometry, removal of foreign particles, gonioscopy and various short operative procedures which may involve the conjunctiva and cornea. It has also been reported to be employed frequently as a surface anaesthesia in glaucoma surgery and in cataract operations.

Dose : Topical, 0.05 ml of a 0.5% solution to the conjunctiva.

I. *Propoxycaine Hydrochloride* BAN, USAN, *Propoxycaine* INN,

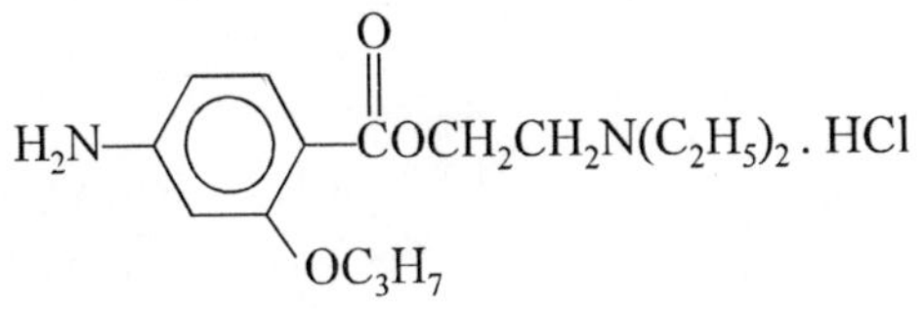

$$H_2N-\underset{OC_3H_7}{\overset{O}{\underset{\parallel}{\overset{\parallel}{C}}OCH_2CH_2N(C_2H_5)_2}} . HCl$$

2-(Diethylamino) ethyl 4-amino-2-propoxybenzoate monohydrochloride ; Benzoic acid, 4-amino-2-propoxy-, 2-(diethylamino) ethyl ester, monohydrochloride ;

U.S.P., N.F.,

Blockain[R] (Breon) ; Ravocaine Hydrochloride[R] (Cook Waite)

Its local anaesthetic potency is reported to be 7 or 8 times more than that of procaine. It is mainly used for infiltration and nerve block anaesthesia.

Dose : Usual, 2 to 5 ml of a 0.5% solution.

J. *Hexylcaine Hydrochloride* USAN, *Hexylcaine* INN,

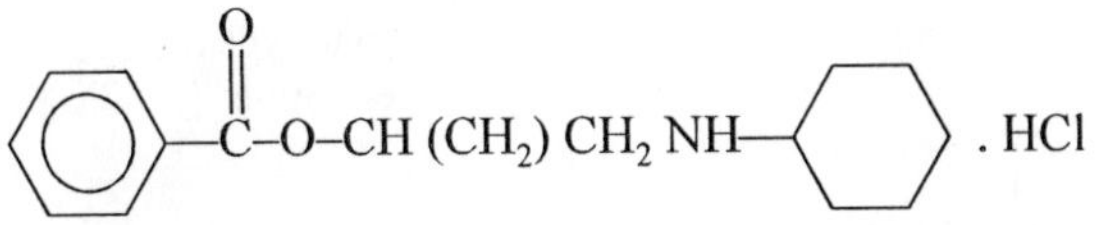

$$\underset{}{\overset{O}{\underset{\parallel}{C}}-O-CH (CH_2) CH_2 NH-\bigcirc} . HCl$$

1-(Cyclohexylamino)-2-propanol benzoate (ester) hydrochloride ; 2-Propanol, 1-(cyclohexylamino)-, benzoate (ester), hydrochloride ;

U.S.P., N.F.,

Cyclaine[R] (MSD)

It is regarded as an all-purpose soluble local anaesthetic agent. The onset and duration of action is almost similar to that of lignocaine.

Dose : For infiltration anaesthesia 1% ; for nerve block anaesthesia 1 and 2% solution ; and for topical application to skin and mucous membranes 1 to 5%.

2.2. Piperidine or Tropane Derivatives

The members of this particular group of compounds essentially contain the piperidine nucleus

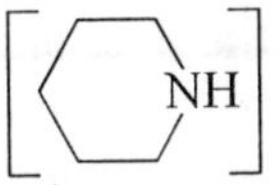

$$\left[\bigcirc NH \right] . \text{ *Examples* : } \alpha\text{-Eucaine ; Benzamine Hydrochloride ; Euphthalmin}$$

A. α-Eucaine

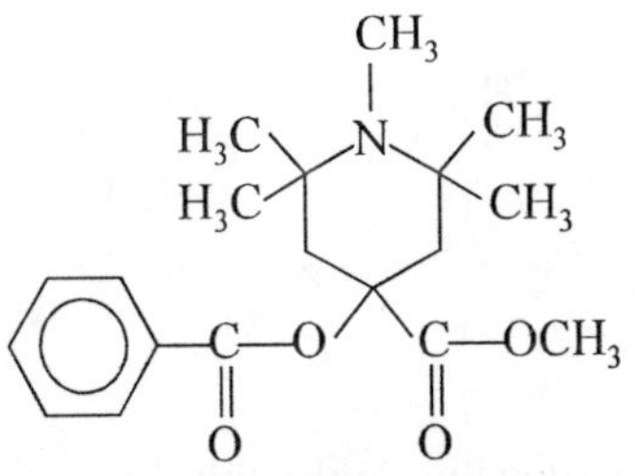

$$\underset{O \qquad O}{\overset{CH_3}{\underset{}{}}}$$

2, 2, 6, 6-Tetramethyl-4-benzoxy-4-methyl carboxylate-N-methyl piperidine.

Synthesis :

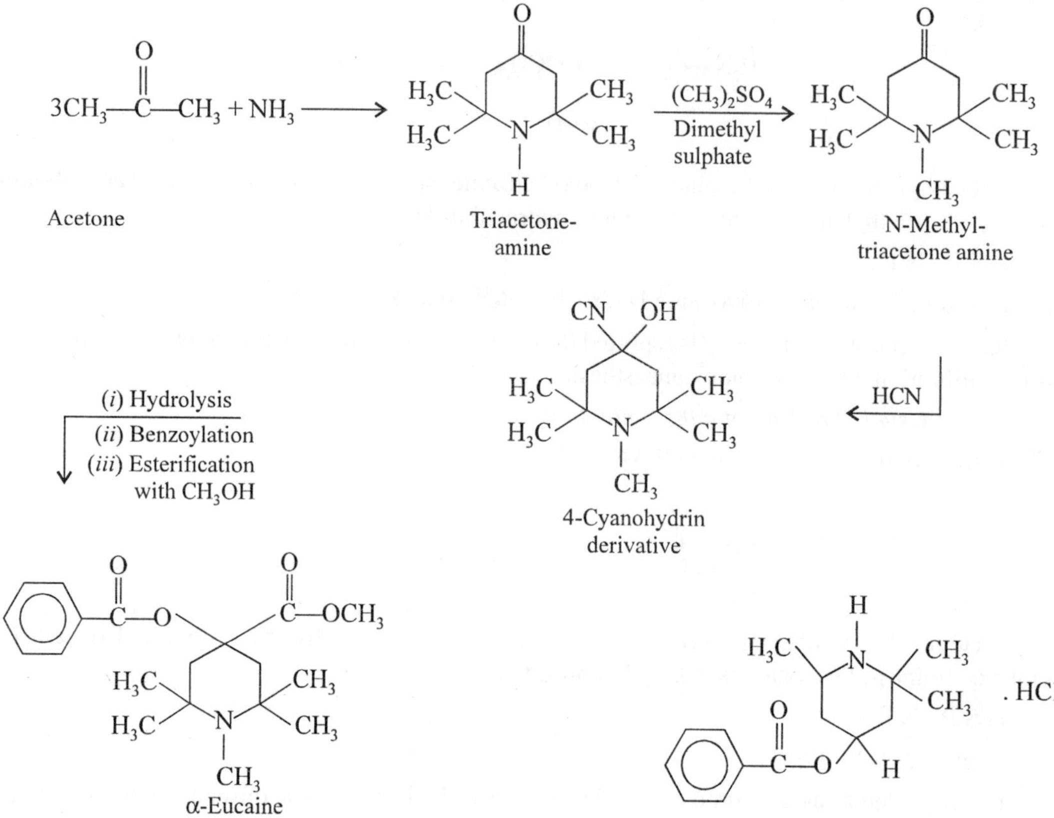

Acetone   Triacetone-amine   N-Methyl-triacetone amine

4-Cyanohydrin derivative

(i) Hydrolysis
(ii) Benzoylation
(iii) Esterification with CH₃OH

α-Eucaine

Triacetoneamine is first prepared by the condensation of three moles of acetone with one mole of ammonia. This on methylation with dimethyl sulphate yields the corresponding N-methyl triacetoneamine which on treatment with hydrocyanic acid gives cyanohydrin analogue. Finally, when the resulting product is subjected to hydrolysis, followed by benzoylation and esterification with methanol yields α-eucaine.

It is comparatively less toxic than cocaine, but is more painful and irritant than the later and hence ; it has been replaced by β-eucaine.

B. *Benzamine Hydrochloride* BAN

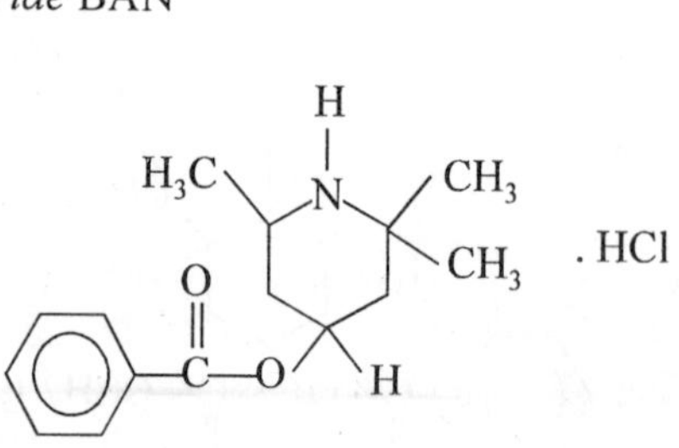

2, 2, 6-Trimethyl-4-piperidyl benzoate hydrochloride ; 2, 2, 6-Trimethyl-4-benzoxy piperidine hydrochloride ; β-Eucaine ; B.P.C. 1954 ;

Synthesis :

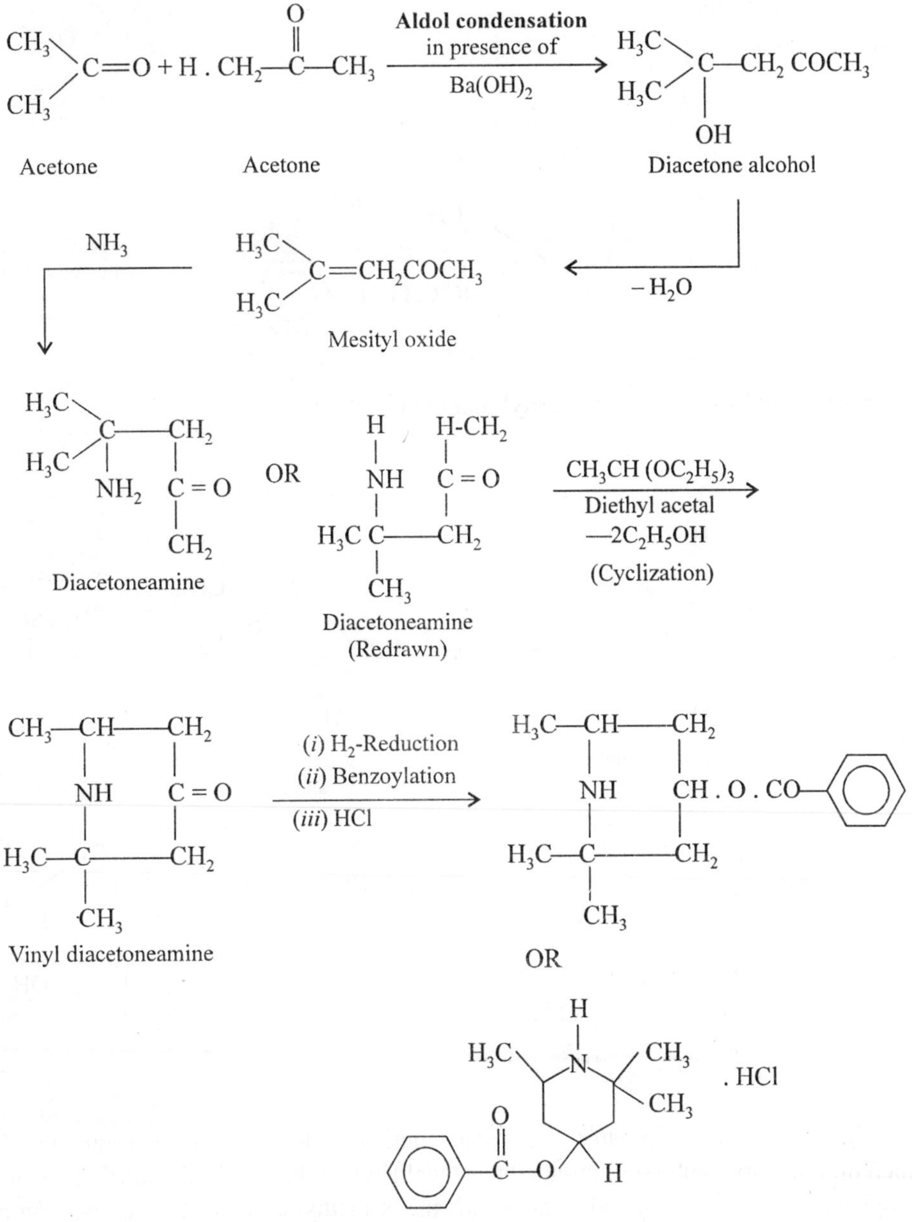

Aldol condensation in presence of Ba(OH)$_2$

Acetone + Acetone → Diacetone alcohol

Mesityl oxide ← $-H_2O$

NH_3

Diacetoneamine   OR   Diacetoneamine (Redrawn)

$CH_3CH(OC_2H_5)_3$
Diethyl acetal
$-2C_2H_5OH$
(Cyclization)

Vinyl diacetoneamine

(i) H$_2$-Reduction
(ii) Benzoylation
(iii) HCl

OR

Benzamine Hydrochloride

Diacetone alcohol is obtained by the *Aldol condensation* of two moles of acetone in the presence of barium hydroxide. On dehydration diacetone alcohol yields mesityl oxide, which upon amination gives diacetoneamine. This on treatment with diethyl acetal undergoes cyclization with the elimination of two moles of ethanol to give vinyl diacetoneamine. The cyclized product on reduction followed by benzoylation and treatment with hydrogen chloride yields benzamine hydrochloride.

It is a local anaesthetic formerly used for surface anaesthesia. Its anaesthetic property is fairly comparable to that of cocaine.

C. *Euphthalmin*

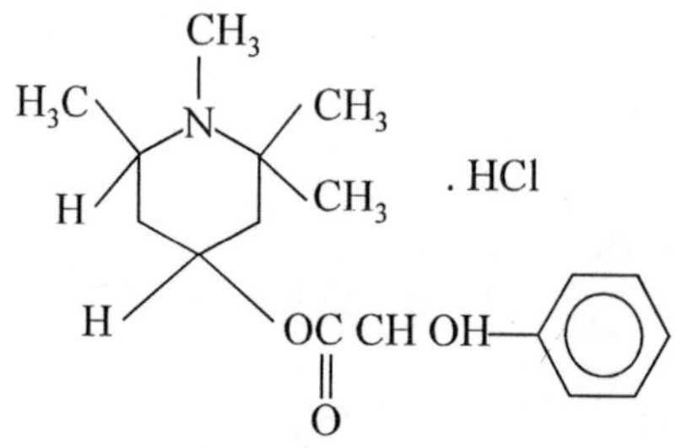

2, 2, 6-Trimethyl-4-N-methyl piperidyl mandeloate ;

Synthesis :

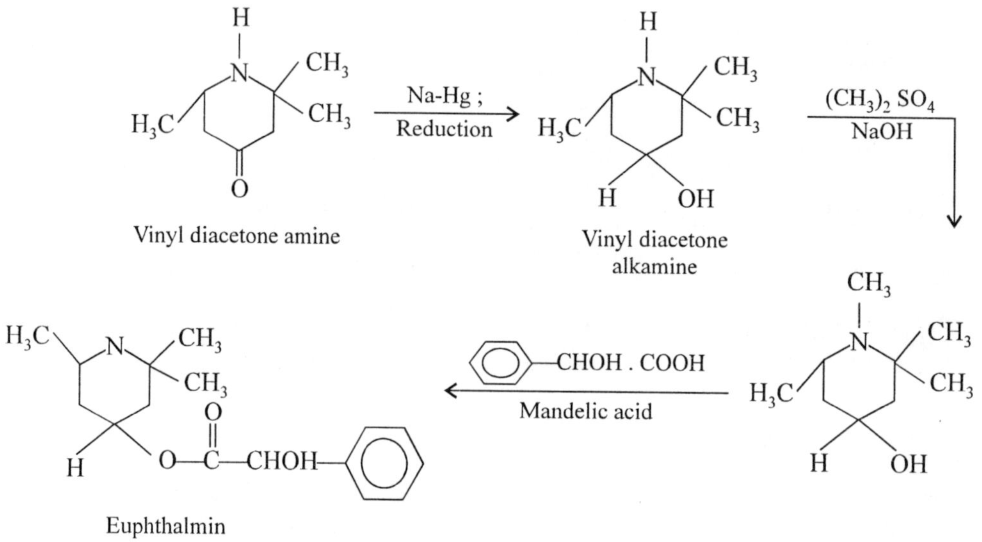

Vinyl diacetoneamine is prepared by the same method as described for benzamine hydrochloride above, which on reduction with (sodium-amalgam) yields vinyl diacetone alkamine. This on treatment with dimethylsulphate and sodium hydroxide yields the N-methyl derivative. The resulting product on treatment with mandelic acid undergoes esterification to yield euphthalmin.

2.3. The Amides

The earlier usage of acetanilide and methylacetanilide (exalgin) in the therapeutic armamentarium as antipyretic and analgesic drugs strongly advocate the idea and belief that when the $COCH_3$ moiety of these simple amides is further extended to the $COCH_2NR_2NR_2$, the resulting compounds offer remarkable local anaesthetic properties. This, in fact, led to synthesis of a large number of compounds of the amide type and a few classical examples are described below.

Examples : Lignocaine Hydrochloride ; Prilocaine Hydrochloride ; Mepivacaine Hydrochloride ; Bupivacaine Hydrochloride ; Pyrrocaine Hydrochloride ; Diperodon.

A. *Lignocaine Hydrochloride* BAN, *Lidocaine* INN, *Lidocaine Hydrochloride* USAN.

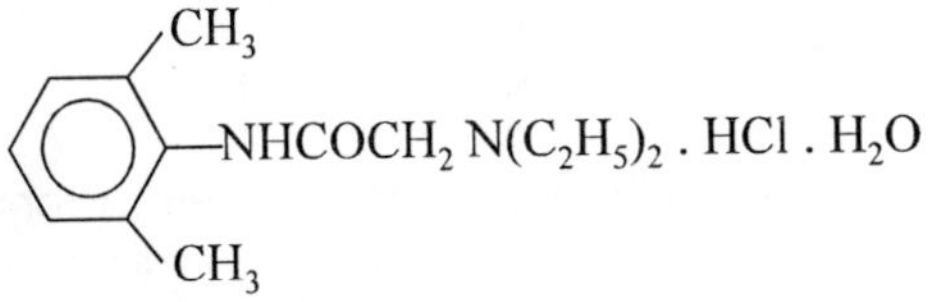

$$\text{Ar}-NHCOCH_2\,N(C_2H_5)_2\,.\,HCl\,.\,H_2O$$

(with CH$_3$ groups at 2,6 positions on the ring)

2-(Diethylamino)-2', 6'-acetoxylidide monohydrochloride monohydrate ; Acetamide, 2-(diethylamino)-N-(2, 6-dimethyl-phenyl-, monohydrochloride, monohydrate ;

Lidocaine Hydrochloride U.S.P., Eur. P., Lignocaine Hydrochloride B.P., Int. P., Ind. P.

Xylocaine[R] (Astra) ; Dolicaine[R] (Reid-Provident) ;

Synthesis :

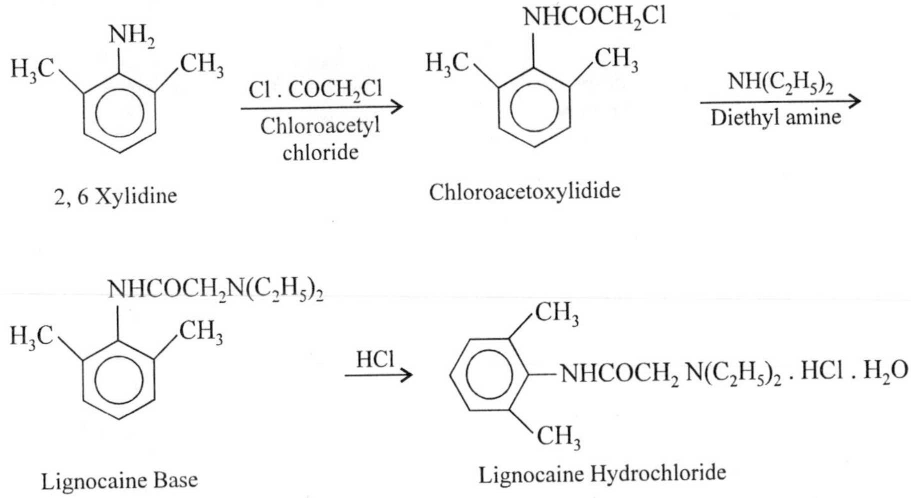

2, 6 Xylidine   Chloroacetoxylidide

Lignocaine Base   Lignocaine Hydrochloride

Chloroacetoxylidide is prepared by the interaction of 2, 6-xylidine with chloroacetyl chloride which on treatment with diethylamine yields the lignocaine base. This when treated with an equimolar quantity of hydrochloric acid gives the respective lignocaine hydrochloride.

It is a potent local anaesthetic and is reported to be twice as active as procaine hyrdrochloride in the same concentration. A 0.5% solution is toxic, *but at 2% its toxicity is increased to 50% than that of procaine hydrochloride.*

Dose : Usual, infiltration, 50 ml of a 0.5% solution ; Usual, peripheral nerve block, 25 ml of a 1.5% solution, usual epidural 15 to 25 ml of a 1.5% solution ; Topical, up to 250 mg as a 2-4% solution or as a 2% jelly to mucous membranes.

B. *Prilocaine Hydrochloride* BAN, USAN, *Prilocaine* INN,

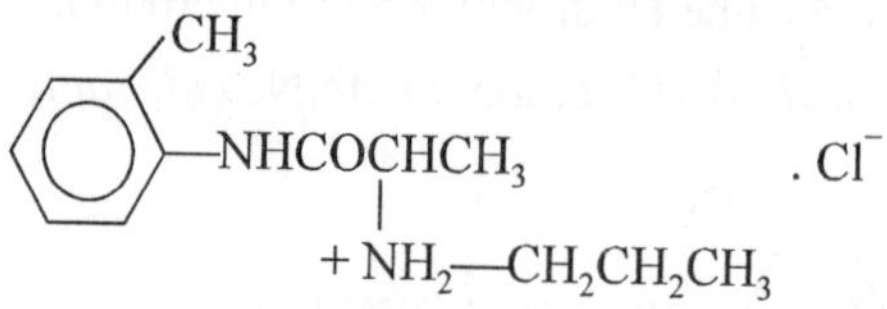

$$
\begin{array}{c}
\text{CH}_3 \\
\bigcirc\text{—NHCOCHCH}_3 \qquad \text{. Cl}^- \\
\overset{|}{+\,\text{NH}_2\text{—CH}_2\text{CH}_2\text{CH}_3}
\end{array}
$$

2-(Propylamino)-*o*-propionotoluidine monohydrochloride ; Propanamide, N-(2-methylphenyl)-2-propylamino-, monohydrochloride ;

B.P., U.S.P., N.F.,

Citanest Hydrochloride[R] (Astra)

Synthesis :

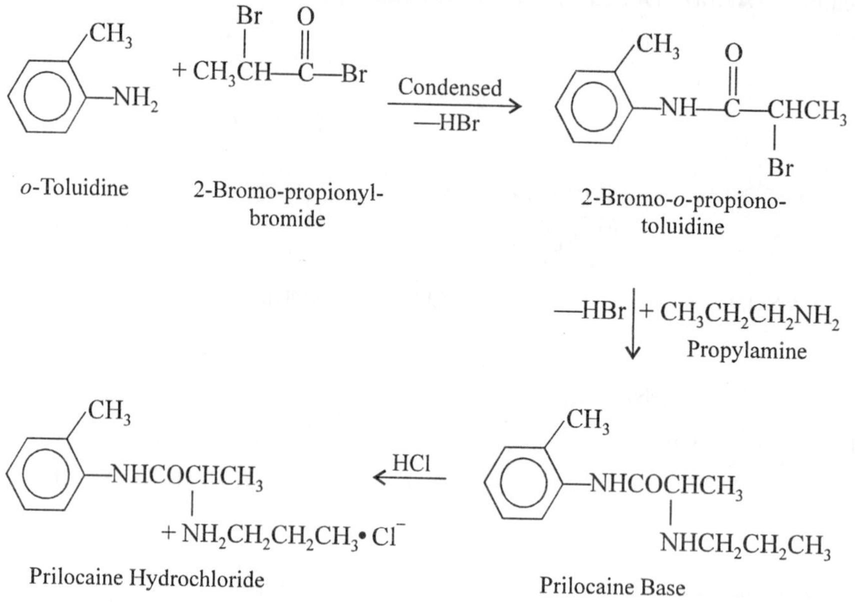

o-Toluidine	2-Bromo-propionyl-bromide	2-Bromo-*o*-propiono-toluidine

—HBr ⎪ + CH₃CH₂CH₂NH₂
        Propylamine

Prilocaine Hydrochloride   Prilocaine Base

2-Bromo-*o*-propiono toluidine may be prepared by the condensation of a mole each of *o*-toluidine and 2-bromopropionyl bromide with the elimination of a mole of hydrogen bromide. This on further condensation with one mole of propyl amine yields the prilocaine base which on reaction with an equimolar amount of hydrochloric acid gives the official compound.

It is a local anaesthetic of the amide type which is employed for surface, infiltration and nerve block anaesthesia. Its duration of action is in between the shorter-acting lidocaine and longer-acting mepivacaine. It possesses less vaso-dilator activity than lidocaine and hence may be used without adrenaline. Therefore, solutions of prilocaine hydrochloride are specifically beneficial for such patients who cannot tolerate vasopressor agents ; patients having cardiovascular disorders, diabetes, hypertension and thyrotoxicosis.

Dose : Usual, therapeutic nerve block, 3 to 5 ml of a 1 or 2% solution ; infiltration, 20 to 30 ml of a 1 or 2% solution ; peridural, caudal, regional, 15 to 20 ml of a 3% solution ; infiltration and nerve block, 0.5 to 5 ml of a 4% solution.

C. *Mepivacaine Hydrochloride* BAN, USAN, *Mepivacaine* INN,

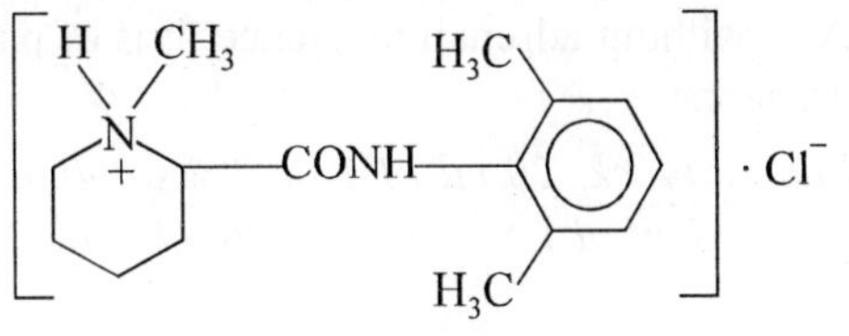

1-Methyl-2′, 6′-pipecoloxylidide monohydrochloride ; 2-Piperidinecarboxamide, N-(2, 6-dimethylphenyl)-1-methyl-, monohydrochloride ;

U.S.P.

Carbocaine Hydrochloride[(R)] (Winthrop) ; Polocaine[(R)] (Astra) ;

Synthesis :

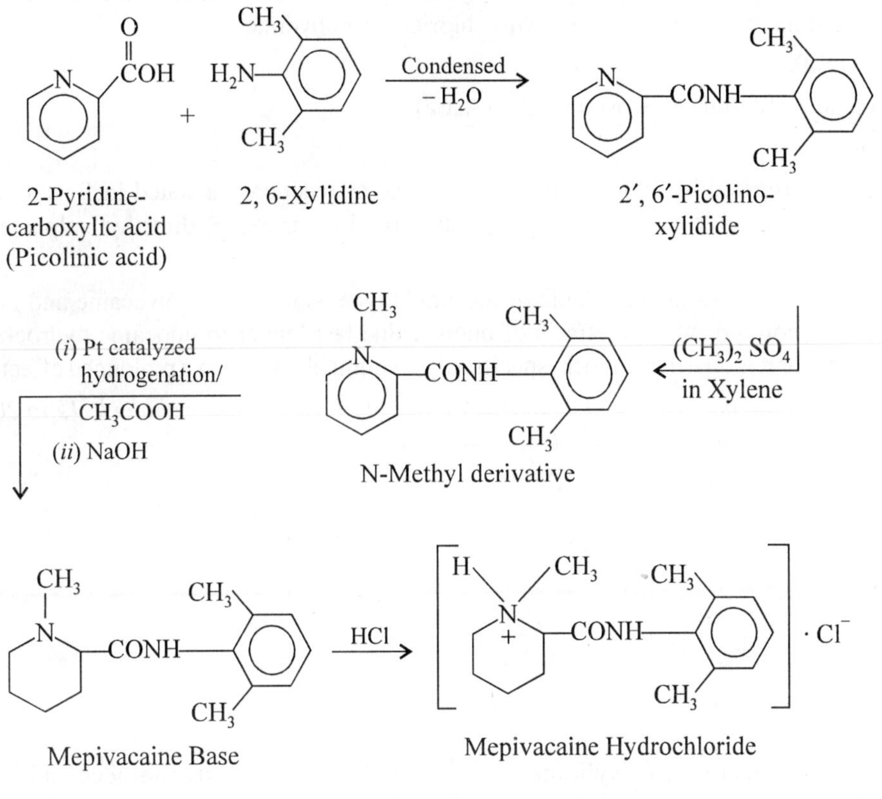

2′, 6′-Picolinoxylidide is prepared by the condensation of picolinic acid (2-pyridinecarboxylic acid) with 2, 6-xylidine which on methylation with dimethyl sulphate in xylene yields the corresponding N-methyl derivative. This when subjected to platinum-catalysed hydrogenation in active acetic acid followed by alkalinization yields mepivacaine base, which is then dissolved in an inert solvent and made to react with an equimolar amount of hydrochloric acid.

It is a local anaesthetic used for infiltration, peridural, nerve block, and caudal anaesthesia. It is found to be twice as potent as procaine. It has been reported that its duration of action is significantly longer than that of lidocaine, even without adrenaline. Hence, it is of particular importance in subjects showing contraindication to adrenaline.

Dose : Infiltration and nerve block, 20 ml of 1 or 2% solution is sterile saline ; Caudal and peridural, 15 to 30 ml of 1%, 10 to 25 ml of 1.5% or 10 to 20 ml of a 2% solution in modified Ringer's solution.

D. *Bupivacaine Hydrochloride* BAN, USAN, *Bupivacaine* INN,

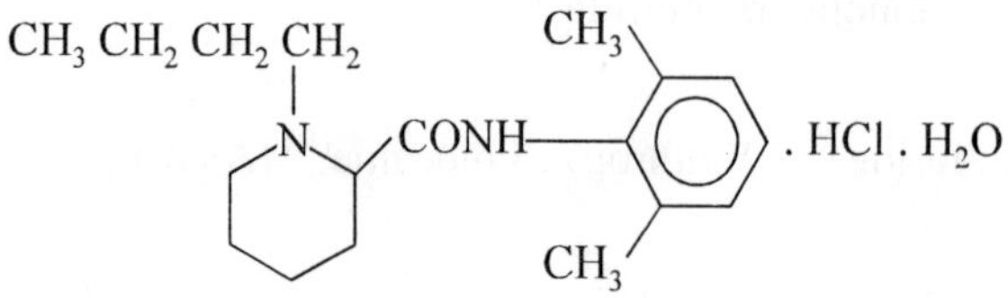

(±)-1-Butyl-2', 6'-pipecoloxylidide monohydrochloride monohydrate ; 2-Piperidinecarbonxamide, 1-butyl-N (2, 6-dimethyl-phenyl)-, monohydrochloride monohydrate ;

B.P., U.S.P.

Marcaine[R] (Sterling) ; Sensorcaine[R] (Astra)

Synthesis :

It may be prepared by adopting the same course of reactions as stated in for the synthesis of mepivacaine, hydrochloride except employing butyl bromide instead of dimethyl sulphate for the N-alkylation.

It is a long-acting local anaesthetic of the amide type, similar to mepivacaine and lidocaine but about four times more potent. The effects of bupivacaine last longer to lidocaine hydrochloride. It is mainly employed for regional nerve block, specifically epidural block, when a prolonged effect is required.

Dose : Regional nerve block, 0.25 to 0.5% solution ; Lumbar epidural block, 15 to 20 ml of 0.25 to 0.5% solution ; Caudal block, 15 to 40 ml of 0.2% solution.

E. *Pyrrocaine Hydrochloride* BAN, *Pyrrocaine* INN, USAN,

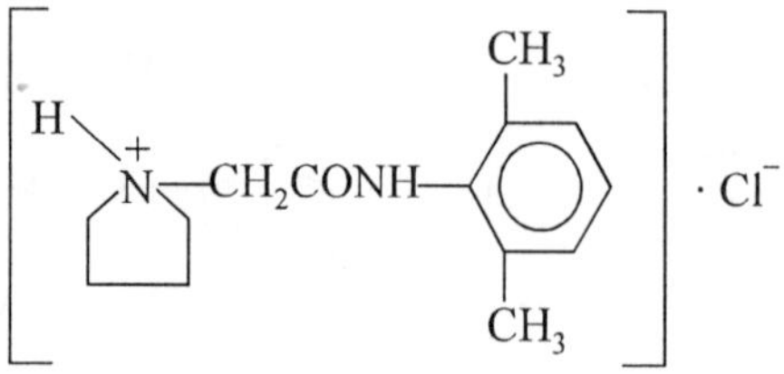

1-Pyrrolidinoaceto-2', 6'-xylidide monohydrochloride ; 1-Pyrrolidineacetamide, N-(2-6-dimethylphenyl)- ; N.F.,

Endocaine Hydrochloride[R] (Endo)

Synthesis :

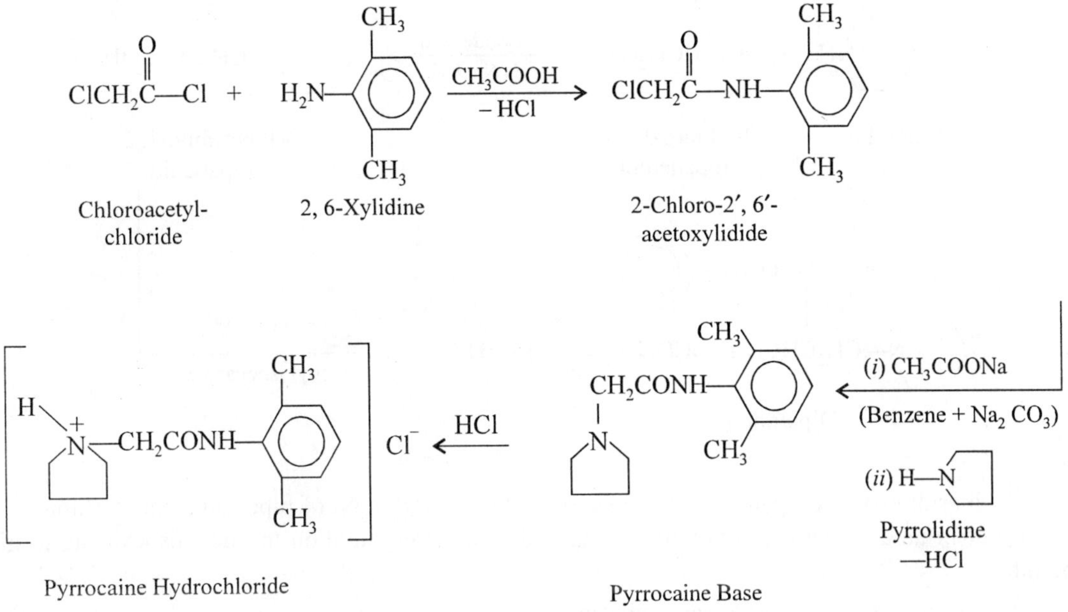

Chloroacetyl-
chloride   2, 6-Xylidine   2-Chloro-2', 6'-
acetoxylidide

Pyrrocaine Hydrochloride   Pyrrocaine Base

Pyrrolidine
—HCl

2-Chloro-2', 6'-acetoxylidide is prepared by treating a solution of 2, 6-xylidine in glacial acetic acid with chloroacetyl chloride which is precipitated with sodium acetate, dissolved in benzene containing suspended sodium carbonate. This on treatment with pyrrolidine yields the pyrrocaine base which on reacting with an equimolar quantity of hydrochloric acid produces the pyrrocaine hydrochloride.

It is used in dentistry for infiltration and block anaesthesia. Its duration of action as well as potency is almost similar to that of lidocaine.

Dose : Usual, infiltration, 1 ml of a 2% solution ; Nerve block, 1.5 to 2 ml of a 2% solution.

F. *Diperodon* INN, BAN, USAN,

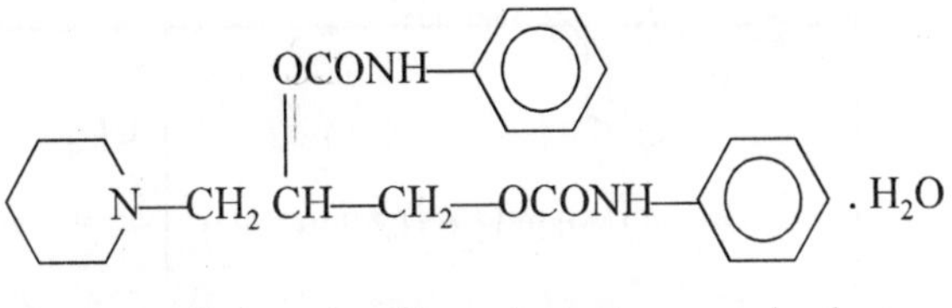

3-Piperidino-1, 2-propanediol carbanilate (ester), monohydrate ; 1, 2-Propanediol, 3-(1-piperidinyl)-, *bis*-(phenyl-carbamate) (ester), monohydrate ;

U.S.P., N.F.,

Diothane[(R)] (Merrell)

Synthesis :

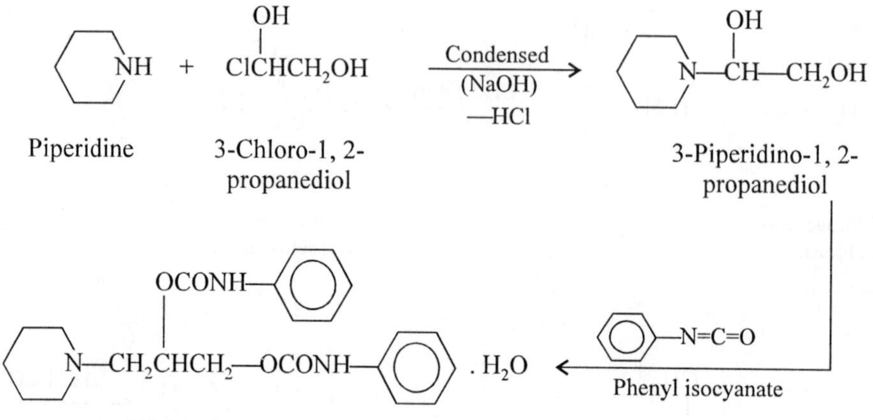

Piperidine   3-Chloro-1, 2-propanediol   3-Piperidino-1, 2-propanediol

Diperodon

3-Piperidino-1, 2-propanediol is prepared by the condensation of piperidine and 3-chloro-1, 2-propanediol in an alkaline medium which is caused to undergo addition to phenylisocyanate to give diperodon.

It is used as a potent surface anaesthesia.

Dose : Topical, 0.5 to 1% solution, to the mucous membranes.

2.4. The Quinoline and Iso-quinoline Analogues

In an attempt to search for more potent and better tolerated local anaesthetics, it has been observed that the amides of quinoline derivatives are more potent than common local anaesthetics and the most important member in this class is **dibucaine hydrochloride.** Besides, another member which essentially contains an iso-quinoline nucleus but without an amide moiety, namely, **dimethisoquine hydrochloride** is reported to be one of the most potent local anaesthetics. These *two* members will be described below :

A. *Dibucaine Hydrochloride* USAN, *Cinchocaine* INN, *Cinchocaine Hydrochloride* BAN,

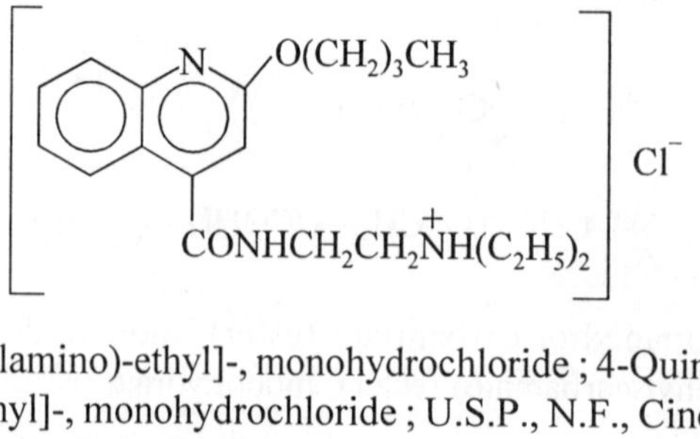

2-Butoxy-N-[2-(diethylamino)-ethyl]-, monohydrochloride ; 4-Quinolinecarboxamide, 2-butoxy-N-[2-diethylamino)-ethyl]-, monohydrochloride ; U.S.P., N.F., Cinchocaine Hydrochloride B.P., Nupercaine Hydrochloride[(R)] (Ciba-Geigy)

Synthesis :

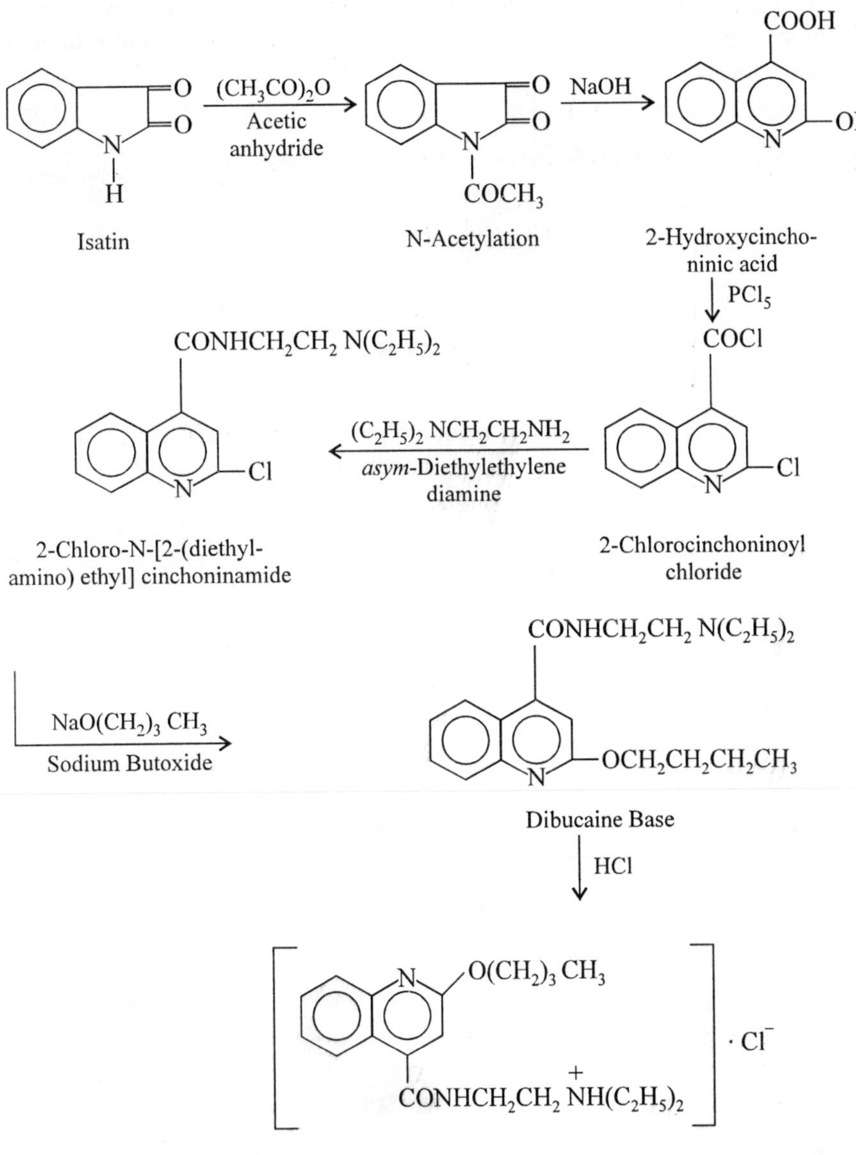

Isatin — N-Acetylation — 2-Hydroxycincho-ninic acid

$(CH_3CO)_2O$ / Acetic anhydride

NaOH

PCl_5

2-Chloro-N-[2-(diethyl-amino) ethyl] cinchoninamide ← $(C_2H_5)_2 NCH_2CH_2NH_2$ / *asym*-Diethylethylene diamine — 2-Chlorocinchoninoyl chloride

$NaO(CH_2)_3 CH_3$
Sodium Butoxide

Dibucaine Base

HCl

Dibucaine Hydrochloride

N-Acetylation of isatin with acetic anhydride yields N-acetylation which undergoes intra molecular rearrangement with alkali to the quinoline compound 2-hydroxycinchoninic acid under the name *Pfitzinger Reaction.* The diamino group is now introduced under controlled condition at room temperature so as to avoid reaction with relatively less reactive 2-chloro substituent.

The subsequent treatment with sodium butoxide yields the dibucaine base which is then dissolved in a suitable organic solvent and precipitated by bubbling hydrogen chloride through the liquid.

It is one of the most toxic, most potent and longest acting of the frequently used local anaesthetics. It may be used as infiltration, surface, epidural and spinal anaesthesia. It finds its use in dentistry. Its anaesthetic activity is similar to those of procaine or cocaine when injected. However, it is several times more potent than procaine when injected subcutaneously and about 5 times more toxic than cocaine when injected intravenously.

Dose : Subarachnoid, 0.5 to 2 ml of 0.5% solution ; usual, 1.5 ml of a 0.5% solution.

B. *Dimethisoquin Hydrochloride* BAN, USAN, *Quinisocaine* INN,

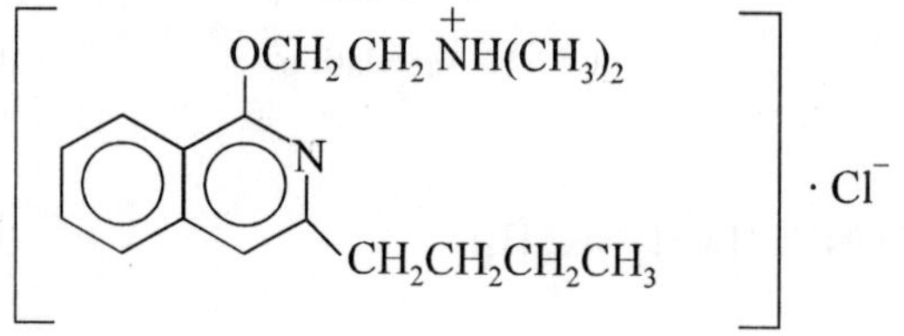

$$\left[\underset{\underset{CH_2CH_2CH_2CH_3}{\overset{OCH_2\,CH_2\,\overset{+}{N}H(CH_3)_2}{\bigg|}}}{\text{isoquinoline}} \right] \cdot Cl^-$$

3-Butyl-1-[2-(dimethylamino)-ethoxy] isoquinoline monohydrochloride ; Ethanamine, 2-[(3-butyl-1-isoquinolinyl) oxy]-N, N-dimethyl-, monohydrochloride ; U.S.P., N.F.,

Quotane[R] (SK & F)

Synthesis :

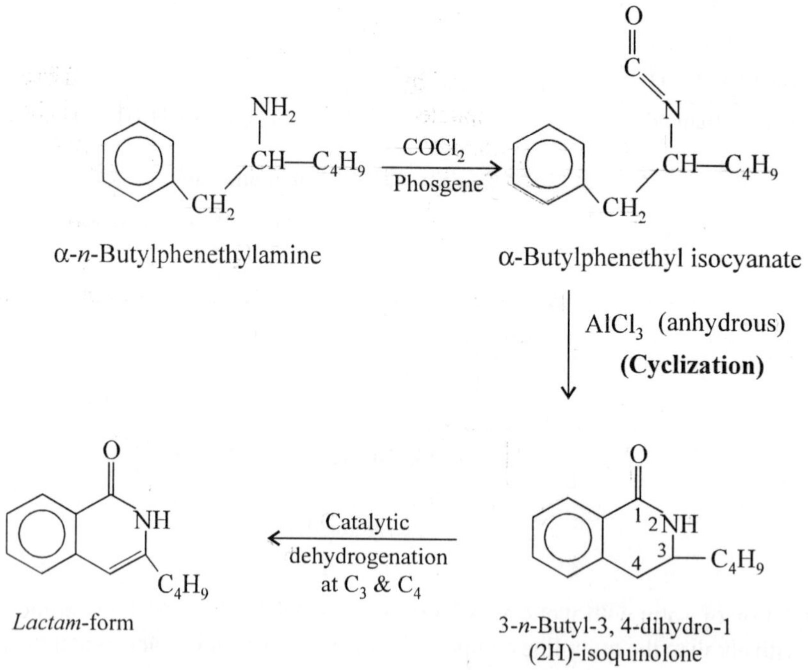

α-*n*-Butylphenethylamine

$\xrightarrow[\text{Phosgene}]{COCl_2}$

α-Butylphenethyl isocyanate

AlCl₃ (anhydrous)
(Cyclization)

$\xleftarrow[\substack{\text{dehydrogenation} \\ \text{at } C_3 \& C_4}]{\text{Catalytic}}$

Lactam-form

3-*n*-Butyl-3, 4-dihydro-1
(2H)-isoquinolone

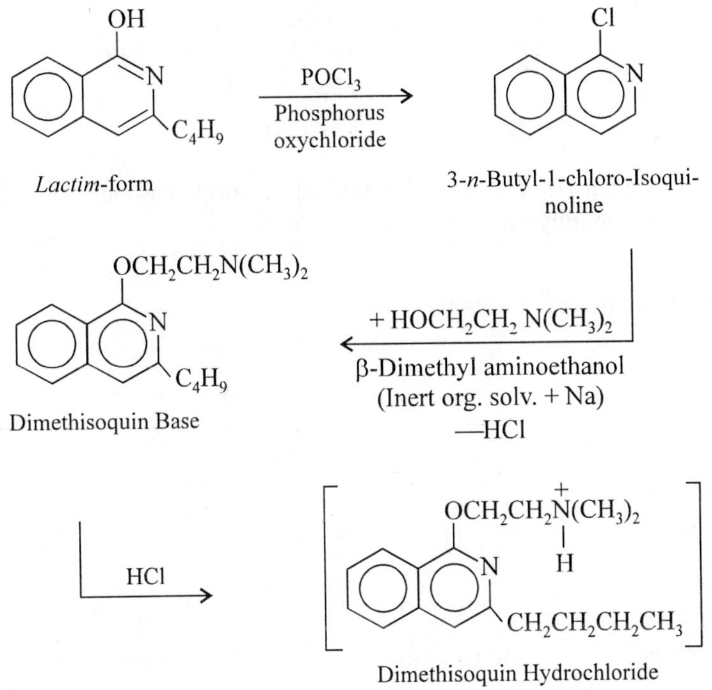

Lactim-form → POCl₃ / Phosphorus oxychloride → 3-n-Butyl-1-chloro-Isoqui-noline

Dimethisoquin Base ← + HOCH₂CH₂ N(CH₃)₂ / β-Dimethyl aminoethanol (Inert org. solv. + Na) / —HCl

HCl → Dimethisoquin Hydrochloride

α-Butylphenethyl isocyanate is prepared by treating α-*n*-butylphenethylamine with phosgene in an appropriate organic solvent which upon treatment with anhydrous aluminium chloride undergoes cyclization to yield 3-*n*-butyl-3, 4-dihydro-1 (2H)-isoquinolone. This on catalytic dehydrogenation at C_3 and C_4 and the subsequent conversion to the *Lactim*-form, which is then reacted with phosphorus oxychloride to yield 3-*n*-butyl-1-chloroisoquinoline. The resulting product when dissolved in an inert organic solvent in the presence of sodium metal and is reacted with β-dimethylaminoethanol it produces the dimethisoquin base. The crude base may be purified by distillation under reduced pressure, dissolved in an appropriate organic solvent, and treated with an equimolar amount of hydrogen chloride to yield the official compound.

It is a surface anaesthetic and has been used as a lotion or ointment in a concentration of 0.5% for the relief of irritation, itching, burning or pain in dermatoses, including mild sunburn and nonspecific pruritus. It is reported to be less toxic than dibucaine but more toxic than procaine.

Dose : Topical, to the skin, as a 0.5% ointment or lotion 2 to 4 times daily.

2.5. Miscellaneous Type

There are a few medicinal compounds which have proved to be potent local anaesthetics and could not be accommodated conveniently into any one of the previous categories discussed, are grouped together under this heading.

Examples : Phenacaine Hydrochloride ; Pramoxine Hydrochloride ; Eugenol etc.

A. *Phenacaine Hydrochloride* BAN, USAN, *Phenacaine* INN,

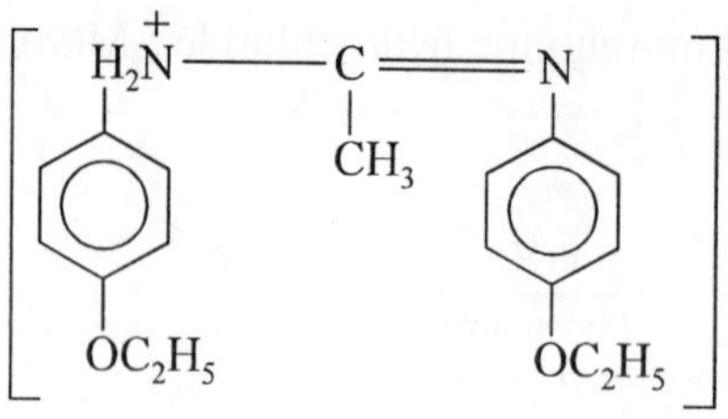

$$\left[\begin{array}{c} \overset{+}{H_2N}\!\!-\!\!\underset{\underset{CH_3}{|}}{C}\!\!=\!\!N \\ \end{array}\right]. Cl^-. H_2O$$

N, N-Bis(*p*-ethoxyphenyl) acetamidine monohydrochloride monohydrate ; Ethanimidamide N, N'-*bis* (4-ethoxyphenol)-, monohydrochloride, monohydrate ;

U.S.P., N.F.,

Holocaine Hydrochloride[R] (Abbott)

Synthesis :

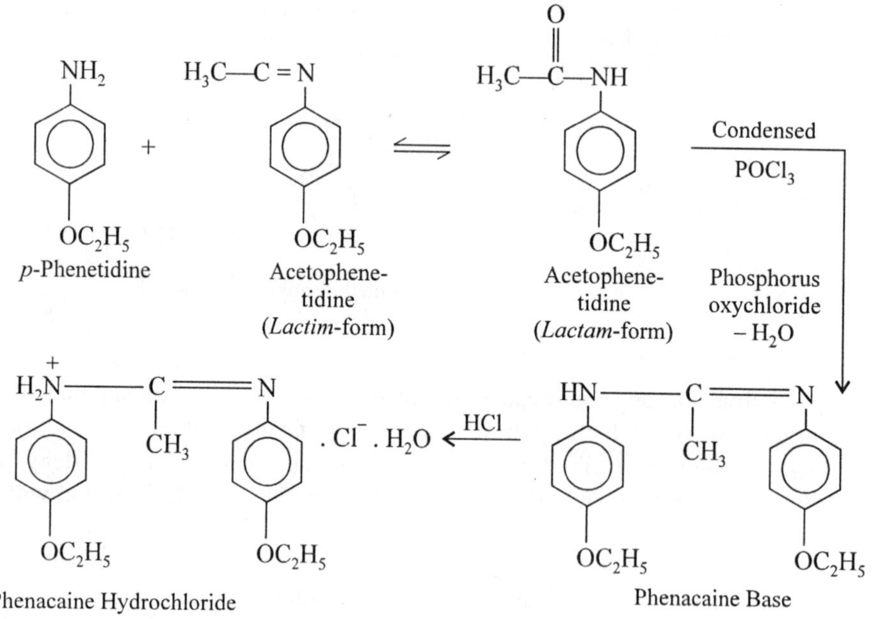

Condensation of *para*-phenetidine and acetophenetidine in *Lactim*-form in the presence of phosphorus oxychloride yields phenacaine base with the elimination of a molecule of water, which on treatment with an equimolar quantity of hydrochloric acid gives the official compound.

It is one of the oldest synthetic local anaesthetics. It is chiefly employed as a 1% solution for effecting local anaesthesia of the eye.

Dose : To the conjuctiva as 1-2% ointment or as a 1% solution.

B. *Pramoxine Hydrochloride* BAN, USAN, *Pramocaine* INN,

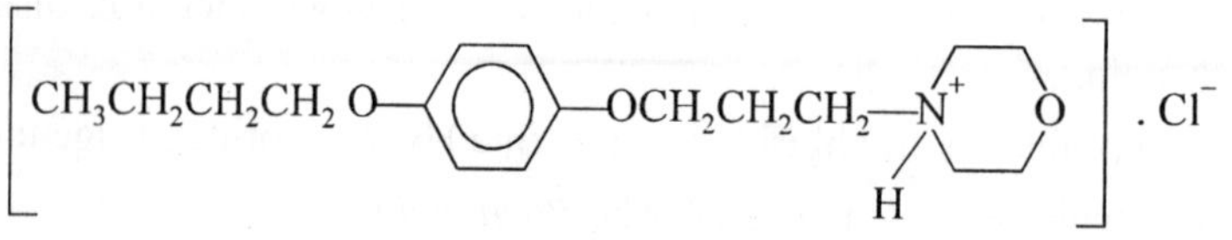

$$\left[CH_3CH_2CH_2CH_2\,O\!-\!\!\left\langle\!\!\bigcirc\!\!\right\rangle\!\!-\!OCH_2CH_2CH_2\!-\!\overset{+}{\underset{H}{N}}\!\!\bigcirc\!O \right]. Cl^-$$

4-[3-(*p*-Butoxyphenoxy) propyl] morpholine hydrochloride ; Morpholine,

4-[3-(4-butoxyphenoxy) propyl]-, hydrochloride ;

U.S.P., N.F.,

Tronothane[R] (Abbott)

Synthesis :

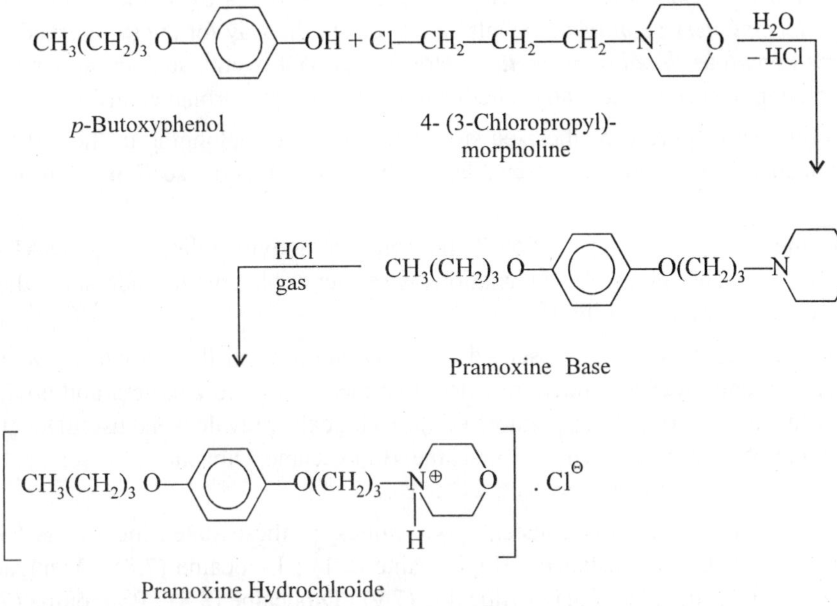

p-Butoxyphenol   4- (3-Chloropropyl)-
morpholine

Pramoxine Base

Pramoxine Hydrochlroide

    The condensation of *p*-buoxyphenol and 4-(3-chloro-propyl)-morpholine is carried out by refluxing them together in an aqueous medium. On cooling the reaction mixture the pramoxine base is *first* extracted with benzene, *secondly* purified by distillation under reduced pressure, *thirdly* dissolved in an appropriate organic solvent, and *fourthly* converted to the corresponding hydrochloride by means of a stream of hydrogen chloride.

    It is a surface anaesthetic which possesses very low degree of toxicity and sensitization. It may be applied locally in a 1% strength to soothen pain in hemorrhoids and rectal surgery, itching dermatoses, some intubation procedures, anogenital pruritus and moderate burns and sunburn.

    Dose : Topical as a 1% jelly or cream every 3 to 4 hours.

    C. *Eugenol* BAN, USAN,

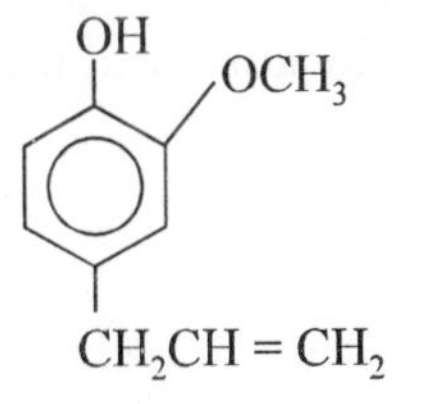

4-Allyl-2-methoxyphenol ; Phenol, 2-methoxy-4-(2-propenyl)- ; Synthetic Clove Oil

B.P., U.S.P.

It is used frequently in dentistry as an obtundent for hypersensitive dentine, caries or exposed pulp, as mild rubefacient in dentifrices and as a temporary anodyne dental filling.

Dose : Topical, in dental protectives.

3. CHEMICAL CONSIDERATIONS OF LOCAL ANAESTHETIC DRUG SUBSTANCES

The presence of the **'anaesthesiophoric moiety'** in *cocaine,* discovered in the year 1880, employed profusely as a vital local anaesthetic in various surgical procedures by virtue of its ability to cause anaesthesia by *blocking nerve conductance,* ultimately paved the way for the synthesis of thousands of new compounds known as *'local anaesthetics'.* However, about twenty such compounds have gained recognition and congnizance as local anaesthetics in the therapeutic armamentarium.

A good number of purely synthesized structural analogues belonging to the different chemical classifications of such compounds have been adequately dealt with under section 2.1 through 2.5 in this chapter.

It is pertinent to state under section 3 the structure activity relationships (SARs) of certain compounds with reasonably plausible explanations wherever applicable to understand the mechanism of action more explicitely and vividly.

Broadly speaking, it has been observed critically amongst all the known *'local anaesthetics',* which are invariably employed in clinical practice, that there exists no clear-cut and obvious structure activity relationship in them. Besides, a plethora of these clinically prevalent and useful local anaesthetics are importantly **'tertiary amines',** such as : Butacaine, Bupivacaine, Dibucaine, Lidocaine, Mepivacaine, Prilocaine, Pramoxine, Proxymetacaine, Tetracaine etc.

Interestingly, the dissociation constant, pKa values, of these stated local anaesthetics usually range between 7.0 and 9.0, for instance : Bupivacaine (8.1) ; Lidocaine (7.8) ; Mepivacaine (7.6) ; Proxymetacaine (9.1) ; Tetracaine (8.4) ; Prilocaine (7.9) ; Dibucaine (8.8) ; Pramoxine (7.1).

It is worthwhile to mention at this juncture that the above stated purely synthetic **'local anaesthetics'** evidently display their biological activities on account of their ability of the binding between the corresponding **onium ions** and a **selective site** very much existing within the *sodium channels.* Therefore, it is absolutely necessary to acquire the following *two* vital and characteristic features in the 'design' of a *local anaesthetic* so as to retain maximum therapeutic activity :

(*a*) Alteration in the lipoidal solubility (*i.e., pKa*) of the 'drug' ;

(*b*) Effect on the ability of a *'drug substance'* to first reach and then get bound to the **hypothetical receptor site(s).**

3.1.   Löfgren's Classification

Bean *et al.** assumed it to be true as a logical basis for reasoning the structure activity relationships amongst the known *'local anaesthetics'* as per the following specific structural characteristic features put forward under Löfgren's classification as illustrated below in Fig. 4.1.

---

*Bean BP *et al. J. Gen. Physiol.,* **81** : 613–642, 1983.

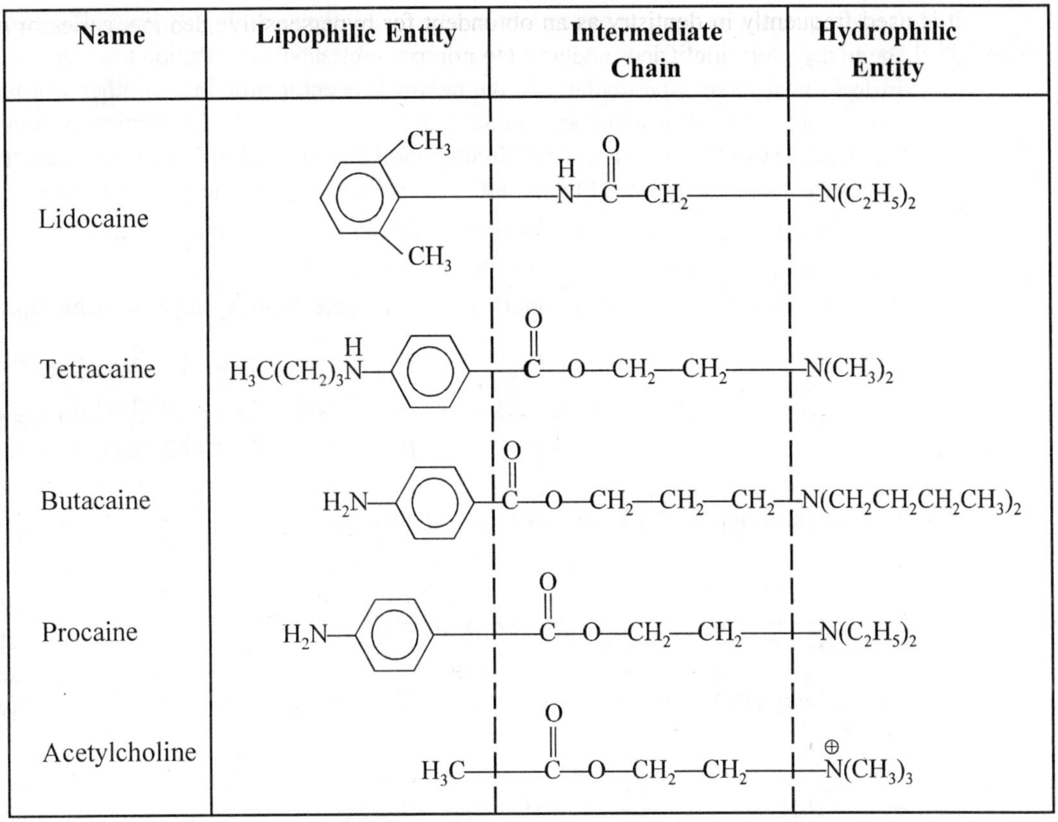

Name	Lipophilic Entity	Intermediate Chain	Hydrophilic Entity
Lidocaine			
Tetracaine			
Butacaine			
Procaine			
Acetylcholine			

Fig. 4.1 Local Anaesthetics and Cholinergic Agent : A Comparison.

From Fig. 4.1 one may evidently observe that a entire molecule of a *'local anaesthetic'* has been judiciously divided into *three* distinct compartments/zones, otherwise termed as : **lipophilic entity, intermediate chain** and **hydrophilic entity.** However these *three* zones have been clearly illustrated in a few typical examples *e.g.,* Lidocaine, Tetracaine, Butacaine, Procaine and a cholinergic agent Acetylcholine.

It will be now worthwhile to elaborate and discuss the structure-activity relationship of certain specific exampels of *'local anaesthetics' vis-a-vis* the aforesaid *three* separate zones imbeded into the drug molecule.

3.1.1. Lipophilic Entity

The various **salient features** essentially associated with the lipophilic portion (entity) and the anaesthetic activity are as given under :

(*a*) the presence of an *aryl function* attached directly to a carbonyl $\left(\begin{smallmatrix} O \\ \| \\ -C- \end{smallmatrix}\right)$ moiety, such as : **amino-ester series ;**

   Examples : Tetracaine ; Butacaine ; Procaine ; etc.,

(*b*) the presence of a *2, 6-dimethylphenyl function* usually linked to a carbonyl moiety by means of an imino (—NH—) group, for instance : **amino-amide series ;**

   Examples : Lidocaine ; Pyrocaine ; etc.,

Thus, the amino-ester and the amino-amide series attribute a highly lipophilic property to the **'drug molecule'** ; and is believed to afford a substantial contribution towards the binding of local anaesthetics particularly to the channel-receptor proteins. In other words, whatever structural modifications are intended to be carried out in this particular zone of the molecule, it would certainly reflect directly upon the physical and chemical characteristics thereby causing an appreciable alteration in its local anaesthetic profile ultimately.

(*c*) the effect of *electron-donating moieties* in the *amino-ester series* are quite vital and significant *i.e.,* present in both or *para-* or *ortho-*positions.

     Examples : (*i*) **Amino (—NH$_2$) Function.** *e.g.,* Procaine, Propoxycaine, and Chloroprocaine :

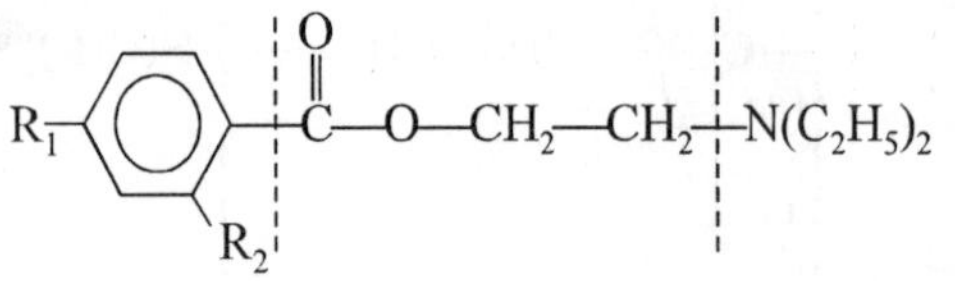

$R_1 = - NH_2$; $R_2 = - H$: Procaine ;
$R_1 = - NH_2$; $R_2 = - OC_3H_7$ Propoxycaine ;
$R_1 = - NH_2$; $R_2 = - Cl$; Chloroprocaine ;

(*ii*) **Alkylamino (RHN) Function :** *e.g.,* Tetracaine ;

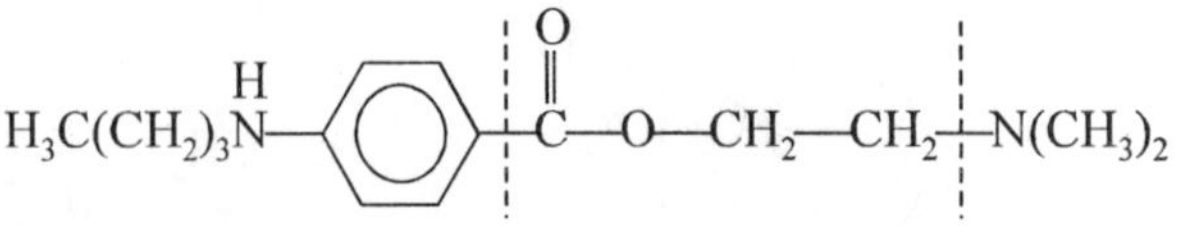

(*iii*) **Alkoxy (RO) Function :** *e.g.,* Propoxycaine ; Proparacaine ;

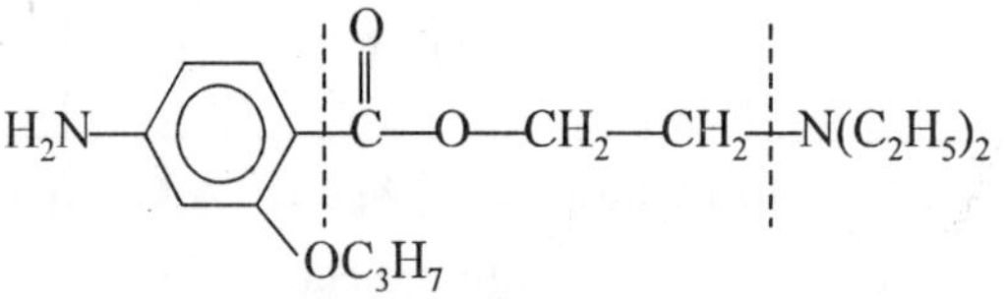

Propoxycaine

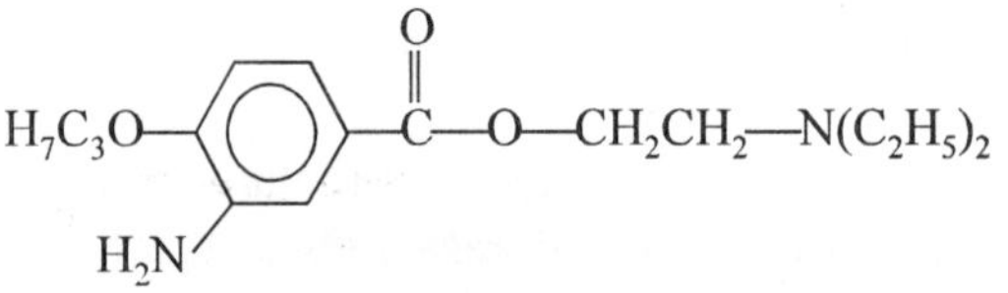

Proparacaine

Note : All these *three* functional moieties being electron-donating in nature do help in profusely contributing electron density to the π-clouds of electrons present in the aromatic ring by such effects as — **'resonance'** and **'inductive',** thereby increasing ultimately the local anaesthetic potency in comparison to the non-substituted structural analogues, namely : Hexylcaine ; Meprylcaine ;

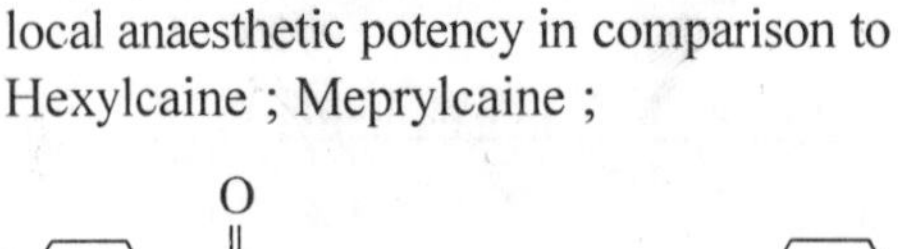

Hexylcaine 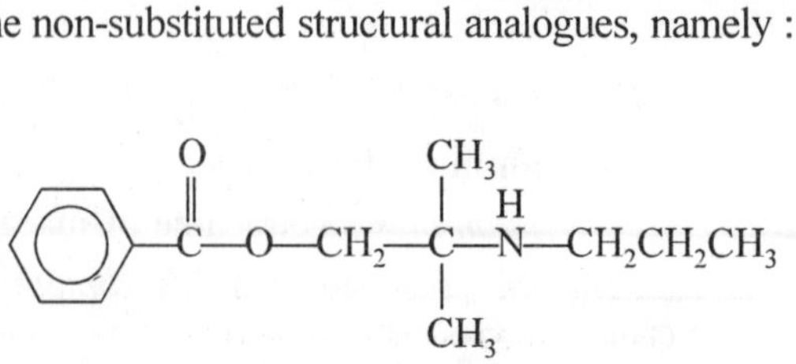   Meprylcaine

(d) the **'Resonance Effects'** do play a predominant role in explaining specifically the influence of various substituents of the aromatic portion of the molecule upon the **'local anaesthetic'** actions.

Examples : Possible resonance structures of procaine, lidocaine and tetracaine are as describe below :

(*i*) **Procaine :**

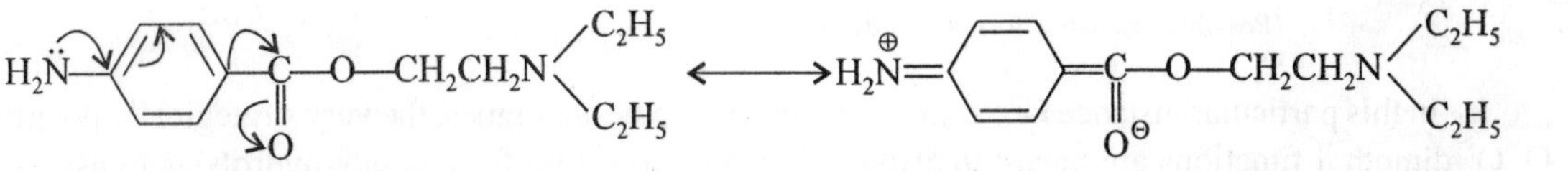

(Possible resonance form of procaine) I   (Protonated onium ion of procaine) II

The above two structures, I and II, are two resonance forms of procaine, neither of which actually represents the 'drug'. In fact, a *'hybrid structure'*, wherein the carbonyl moiety attains a partially ionic (not totally as in II), would perhaps be a more probable and correct representation. The aforesaid hypothesis may be further substantiated by varying the anaesthetic potency of structrural variants *vis-a-vis* the altered nature of the respective *para* substituents on the benzene ring ; and relate it to the **bond order** of the *'ester carbonyl'* by adequately measuring the corresponding observed **IR-stretching frequency,** as summarized in Table : 4.1 below :

Table 4.1. Relationship of Local Anaesthetic Profile and Bond order

S.No.	R-Substituent	IR (C = 0) cm^{-1}	ED_{50} (m mol/100 mL)*
1.	H_3CO —	1.708	0.060
2.	H_5C_2O —	1.708	0.012
3.	H_2N —	1.711	0.075
4.	HO —	1.714	0.125
5.	O_2N —	1.731	0.740

From Table 4.1, it may be observed that both *amino* and *alkoxy* functions are regarded as *'electron donors'* by virtue of their resonance characteristic ; and, therefore, increase the dipolar (*i.e.,* ionic) nature of the carbonyl (C = 0) moiety. Further, *para*-substituents do exert a marked and pronounced electron-withdrawing effect in the carbonyl function having more double-bond character and ultimately resulting in *less intense* local anaesthetic activity.

* Galinsky AM *et al. J. Med. Chem.,* **6** : 320, 1963.

(*ii*) **Lidocaine :**

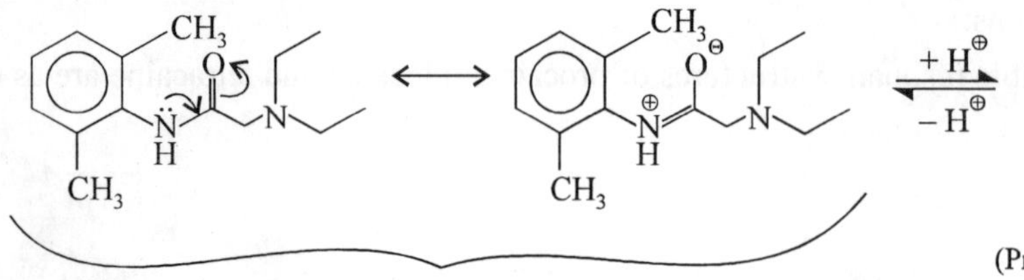

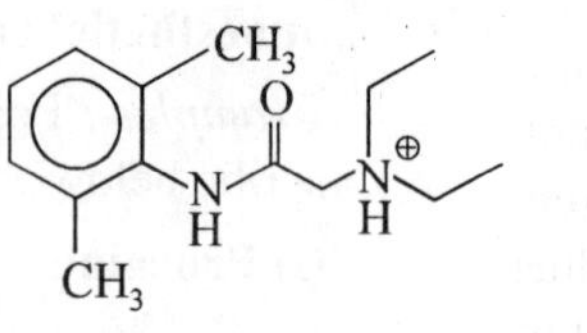

(Protonated onium ion of Lidocaine)

(Possible resonance forms of Lidocaine)

In this particular instance *i.e.,* amino-amides, lidocaine analogues, the very strategically positioned *O, O'*-dimethyl functions are meant to afford adequate protection from amide hydrolysis to ascertain a predicted and desirable duration of action.

In the same vein, one may derive logical conclusions to only justify but also rationalize the possible enhancement in the duration of action of *propoxycaine* by the presence of *ortho*-propoxy moiety.

However, in another instance *i.e., chloroprocaine*, the observed relatively shorter duration of action, in comparison to procaine, may be evidently expatiated by the *inductive effect* due to the presence of *ortho*-chloro function, which might help in pulling out the density of electron away from the carbonyl group, thereby rendering it more prone to *nucleophilic attack* by the *plasma esterases.*

(*iii*) **Tetracaine :**

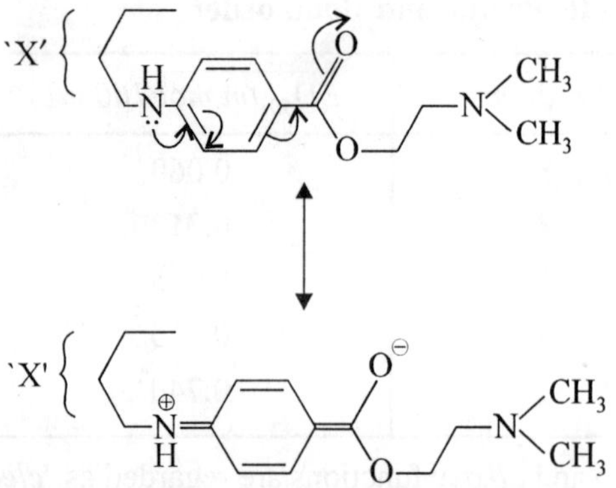

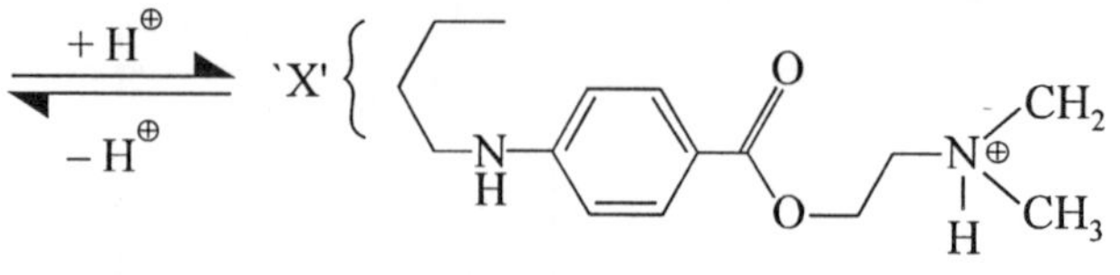

(Protonated resonance form of tetracaine)

(Resonance forms of unionized tetracaine)

Tetracaine is found to be 50-times more active than *procaine*. However, this phenomenal enhancement in potency may not be explained and proved experimentally by virtue of the presence of the overwhelming surge in lipid-solubility attributed by the *n*-butyl moiety marked 'X'. Logistically, the marked and pronounced potentiation of local anaesthetic profile offered by tetracaine may be attributed to the distinct electron-releasing activity of the said *n*-butyl function *via* the inductive effect, that preferentially increases the prevailing electron density of the *para*-amino moiety, which subsequently enhances the creation of the *'resonance form'* readily available for enabling to get bound to the viable receptor proteins.

3.1.2. Intermediate Chain

The intermediate chain essentially comprise of a *'distinct short alkylene chain'*, made up of 1-3 C-atoms hooked on to the aromatic ring *via* a plethora of organic functional moieties. It has been established that the *'intermediate chain'* categorically serve as a *'determinant factor'* to decide the **'chemical stability of the drug molecule'**. In addition to this, it also exerts its marked and pronounced influence on the duration of action and relative toxicity of the drug substance. It has been duly observed that invariably both the **amino carbamates** and the **amino amides** exert a much higher resistance to the **'metabolic inactivation'** particularly in comparison to the corresponding *'amino esters'*, which property ascertains importantly a longer-acting local anaesthetic criterion.

A few other equally vital characteristic features are, namely :

(a) induction of small alkyl moieties in the viccinity of the ester function *e.g.*, **hexylcaine** and **meprylcaine.** Both these compounds check and prevent amide and ester hydrolysis thereby increasing the duration of action appreciably.

Hexylcaine (pKa = 9.3)

Meprylcaine (pKa = 7.8)

(b) induction of amide function in the close range of simple alkyl amino or branched alkyl amino functions located in the viccinity of an amide moiety *e.g.*, **prilocaine** and **etidocaine.** Interestingly, these two compounds hinder ester hydrolysis thereby enhancing the duration of action significantly.

Prilocaine (pKa = 7.9)

Etidocaine (pKa = 7.9)

Salient Features : The salient features are as follows :

(i) Increasing the length of the alkylene chain in the lidocaine structural analogues from one to two (increases pKa from 7.7. to 9.9) ; and from two to three (increases pKa from 9.9 to 9.5)

(ii) Enhancement of the **'intermediate chain'** (Fig. 4.1) drastically lowers the potency of local anaesthetics due to the reduction of **onium ions** specifically under the prevailing physiologic conditions. Nevertheless, the onium ions are essentially utilized in promoting the binding to the **'channel receptor'** efficiently and effectively.

3.1.3. Hydrophilic Entity

If one examines the most potent and commonly used *'local anaesthetics'* in the therapeutic armamentarium it may be abundantly clear that they essentially possess a **tertiary alkylamine** moiety which rapidly converts the *'base'* into the corresponding water-soluble salts with the various mineral acids. Obviously, the **'basic entity'** is frequently regarded as the *hydrophilic entity* of the drug molecule.

However, it has been adequately proved both logically and scientifically, that the **voltage-activated sodium channel** along with the **most probable mechanism of action discussed earlier,** one may safely infer and suggest that the onium ions generated by protonation of the tertiary amine function are urgently required in carrying out the binding phenomenon with the *'receptors'*.

4. BENZOIC ACID AND ANILINE ANALOGUES WITH POTENTIAL LOCAL ANAESTHETIC PROFILE

It fact, the *'esters'* of benzoic acid derivatives are derived from **cocaine,** while the *'amides'* of aniline derivatives are obtained from **isogramine.**

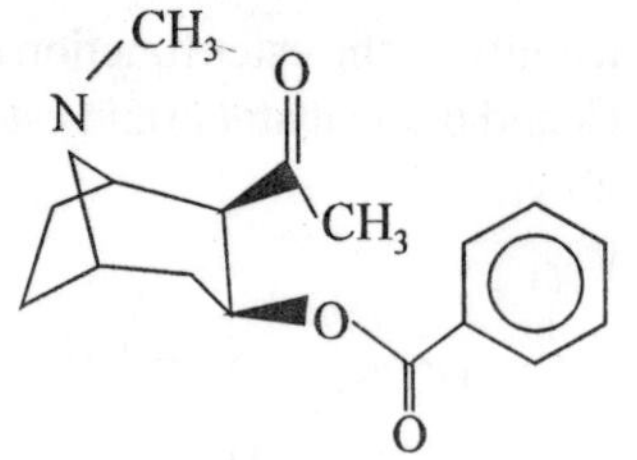

Cocaine 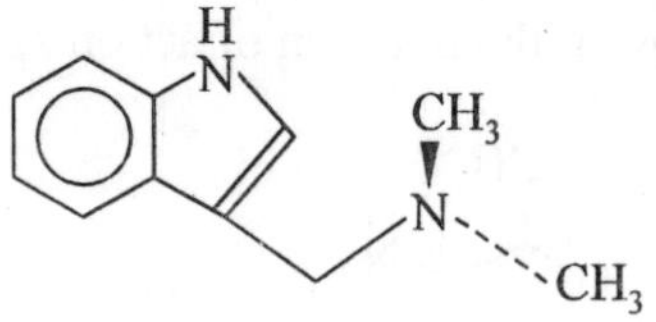   Isogramine

The following general arrangement holds good with regard to the chemical structures of *'esters'* and *'amides'* :

| LIPOPHILIC CENTRE | ESTER OR AMIDE FUNCTION | CONNECTING BRIDGE | HYROPHILIC CENTRE |

It is, however, pertinent to mention here that the *'ester'* as well as the N-substituted functional moieties are nothing but **bioisosteres** as illustrated below ; and it further expatiates and justifies the strategical presence of such groups in almost identical locations in the various tailor-made structurally designed local anaesthetics.

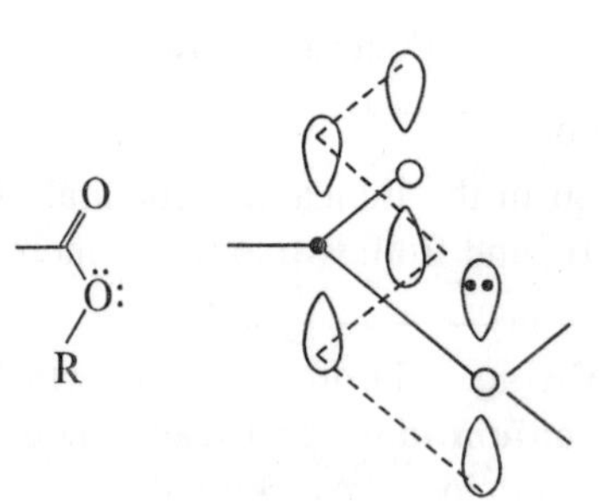

An 'Ester' 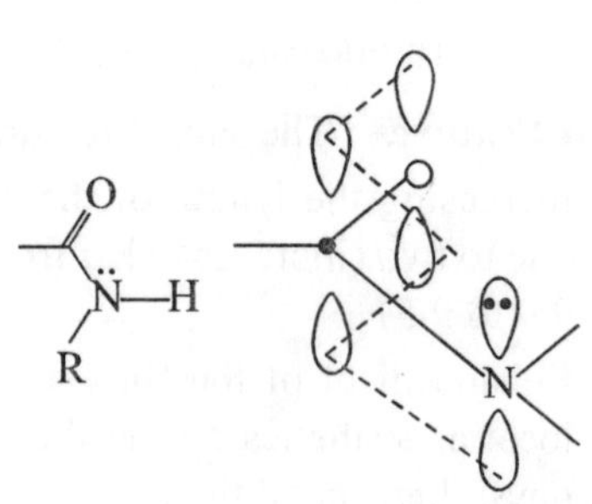   An 'Amide'

Salient Features : The various salient features influencing the functional moieties *vis-a-vis* the lipophilic and hydrophilic characterstics of the **'local anaesthetics'** are enumerated as under :

    (1) A heterocyclic-ring system or a carboxylic moiety invariably gives rise to a distinct and prominent *'lipophilic centre'*,

(2) A secondary or tertiary amine that could be either cyclic or an open-chain analogue usually affords a marked and pronounced *'hydrophilic centre'*,

(3) A *'hydrophilic centre'* may be suitably linked to either an amide function or an ester moiety either by S, N, O-atoms or apporpriately by a short hydrocarbon unit. However, it has been observed that the latter (*i.e.,* short hydrocarbon unit) happens to be the most preferred choice in majority of synthesized local anaesthetics,

(4) Evidently, the **lipophilicity** of a local anaesthetic is exclusively attributed by the embedded **'lipophilic centre'**,

(5) The inducted lipophilicity of a local anaesthetic solely depends on its capability to penetrate right into the cell membrane of the axon,

(6) The water-solubility characteristics are predominantly provided by the corresponding hydrophilic centre of the molecule. Undoubtedly, this constitutes a cardinal factor in the transportation of the *'drug substance'* (*i.e.,* local anaesthetic) to the membrane and once slipping inside the cell, subsequently moves on to the desired receptor site. Besides, hydrophilicity also aids towards the binding of the 'drug molecule' ultimately to the receptor.

(7) An ideal *'local anaesthetic'* characteristic is best accomplished by balancing the lipophlilic and hydrophilic centres. In case the lipophilic centre is prevailing as the dominant structure, the ensuing anaesthetic action of the *'drug substance'* is poor by virtue of the fact that it is able to penetrate the lipoidal membrane of the axon, while its expected solubility in both the intracellular and extracellular fluids remains equally poor. In another situation, when the hydrophilic centre is found to be predominant in the structure, the ensuing anaesthetic action of the drug is weak in nature on account of its poor membrane penetration ability.

(8) The dissociation constant, pKa, values, of various 'local anaesthetic' drugs have been employed as an important and critical measure of their extent of **'ionization'** ; and, therefore, serves as a means of evaluating their lipophilic/hydrophilic ratio. Broadly speaking, the pKa values of quite a few therapeutically potent 'local anaesthetic' drugs essentially fall within the range 7.5 to 9.5.

(9) Based on the above findigns the pKa values essentially give rise to the following *two* distinct situations, such as :

(*a*) *Local anaesthetics having pKa values less than 8.0* — are observed to be not so adequately ionizable at the prevailing physiological pH in order to exert their effective influence in causing anaesthesia even though they are capable of penetrating the axon*, and

(*b*) *Local anaesthetics having pKa values more than 9.5* — are found to be practically fully ionized at the prevailing physiological pH ; as a result these 'local anaesthetic' drugs exert a significant less effective activity by virtue of the fact that they experience an obvious difficulty in penetrating the cell membrane.

10. The partition coefficient characteristic of 'local anaesthetic' drugs, having identical structural features, exhibit an enhancement in activity corresponding to an enhancement in the partition coefficient values, unless and until a maximum activity is accomplished. Once the peak activity is reached, the activity starts decreasing progressively even though the partition coefficient gets enhanced appreciably.

---

*A process of a neuron that conducts impulses away from the cell body.

11. A comprehensive study* of the homologous series obtained meticulously by substituting the **aromatic ring** of 'local anaesthetics' by such substituents as : alkyl, alkyloxy ; and alkyl amino moieties evidently displayed that the partition coefficients of the members of a series enhanced according to an enhancement in the actual number of methylene ($-CH_2-$) functions present in the substituent of that specific series. However, the maximum activity in a homologous series was observed with a C_4 to C_6 methylene chain. Likewise, the increase in the number of C-atoms of the substituents in the *'hydrophilic centre'* exhibited not only an enhancement of the partition coefficient but also an increase in activity.

Example : The introduction of such functional moieties as : **diethyl amino ; piperidino ;** and **pyrrolidino** etc., — gave rise to newer products having almost identical degree of activity. Importantly, the **morpholino** function resulted in lowering the activity considerably.**

12. It is vehemently assumed that the *'local anaesthetics'* get bound to various tissue and plasma proteins by means of *three* vital physical forces, namely : **van der Waals forces ; dipole-dipole attractions ;** and **electrostatic forces** as illustrated below in Fig. 4.2.

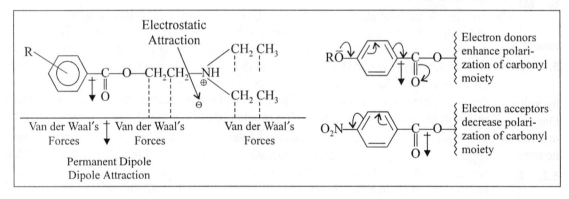

Fig. 4.2 : A diagramatic sketch of the binding phenomenon of an ester-type local anaesthetic drug to a receptor site by various physical forces.

[Adapted from Buchi J and Perlia, X : In Ariens EJ (ed.) **Drug Design,** Academic Press, New York, Volume III, p-243, 1972]

Observations : The various important observations derived from Fig. 4.2 are as follows :

(*i*) The activity of benzoic acid-based 'local anaesthetic' drug gets improved when the aromatic lipophilic centre essentially possesses electron-donor substituents (*e.g.,* O^-, COO^-, CHR_2, CH_2R, CH_3) ; and consequently gets reduced with electron-acceptor substituents (*e.g.,* NO_2 ; COR, CN, NH_3^+, COOR, Cl, Br, I, F, SH).

(*ii*) From (*i*) above one may safely conclude that the electron donor functional moieties enhances the binding capability of the 'local anaesthetic' agent to the receptor site, whereas the electron acceptor functional moieties help in lowering this binding phenomenon significantly.

---

*Buchi J. and Perlia X., **'Local Anaesthetics Encyclopedia of Pharmacology and Therapeutics',** Section 8, Vol. I., Pergamon, New York, 39, 1971.

Ariens E.J. (ed.) **'Drug Design', Academic Press, New York, Vol. III, p-243, 1972.

Buchi and Perlia put forward a plausible explanation for lowering the binding phenomenon on account of an electron acceptor engaged in the withdrawl of electrons from the carbon of carbonyl moiety that eventually retards the polarization ability of this moiety. Furthermore, it also consequently reduces the actual strength of the dipole moment of the carbonyl group, which ultimately weakens its dipole-dipole attraction with the receptor.

(13) It has been proved with substantial evidence that *'local anaesthetics'* essentially containing **'amide'** functional moities show a tendency to afford a reasonably stronger bondage with the receptor site.

Examples : (*a*) Tucker *et. al.** proved taht approximately 95% of **'bupivacaine'** (see section 4.2.3.D) usually get bound to plasma and tissue proteins in comparison to 55% for **'prilocaine'** (see section 4.2.3.B.).

(*b*) Tucker and Mather** further demonstrated that the relatively longer-acting and distinctly more potent *'local anaesthetics'* are more intimately and extensively get bound to the plasma proteins. Nevertheless, it may not be the only critical factor that controls the potency.

5.   MODE OF ACTION OF SOME SELECTED LOCAL ANAESTHETICS

The specific mode of action of certain local anaesthetics, already discussed under section 4.2, are enumerated below :

5.1.   Amethocaine (or Tetracaine)

The onset of action is found to be rather slow but it is of relatively longer duration. It is normally hydrolyzed by plasma esterases to *para*-aminobenzoic acid (PABA) together with some other metabolites. Due to its inherent high level of toxicity it is invariably being employed for several restricted *'topical applications'*, for instance : bronchoscopy, ENT-realted surgical procedures, and local treatment of hemorrhoids.

5.2.   Benzocaine

It is found to possess both low potency and low systemic toxicity. It is mostly employed as a local topical anaesthetic in conjunction with other similar agents ; though some of these mixture may give rise to undesired allergic manifestations. Besides, benzocaine is also employed as a possible **sulphonamide antagonist**.

5.3.   Bupivacaine

It is generally employed in solution form either alone or in combination with adrenaline (a vasodilator). It exhibits upto 95% ability to cause plasma protein binding. Because it exerts a minimal nerve motor block, hence it is specifically suitable for some surgical operations. The drug is largely metabolized in the liver ; and its metabolites are chiefly excreted in the urine.

5.4.   Butacaine Sulphate

Its solution finds usage as a topical local anaesthesia in dentistry. It has also been employed in several ear and nose drops for the relief of pain along with other drugs.

---

*Wilson and Gisvold's : **Text book of Organic Medicinal and Parmaceutical Chemistry,** Delgado J.N. and Remers W.A., Lippincott-Raven, New York, 10th edn, p-647, 1998.

Tucker G.T. and Mather L.E., *Br. J. Anaesthesia,* **47, 213, 1975.

5.5. Dibucaine Hydrochloride (or Cinchocaine)

Its onset of action is found to be rapid, and action is of long duration. The overall extent of anaesthetic action is almost one fourth to that of cocaine. It has an apparent half-life of 11 hr, and gets metabolized mostly in the liver. It has been observed that the *'amide portion'* of the drug does not get hydrolyzed even to a small extent in the serum. Therefore, the prevailing sluggish rate of metabolism is perhaps responsible for giving rise to a relatively high degree of systemic toxicity.

5.6. Diperodon

Diperodon hydrochloride is usually used which is found to be slightly soluble in water ; however, its solubility may be enhanced by adding NaCl-solution. Importantly, even the traces of alkali will precipitate the free-base, and, therefore, it is always preferred that its aqueous solutions must be utilized as soon as possible.

5.7. Lidocaine (or Lignocaine)

It is one of the most widely used local anaesthetics having a plasma-protein binding ability of approximately 64% at therapeutic drug concentrations. Lidocaine is reported to penetrate the placenta ; however, fetal-plasma binding is only about half that found in maternal plasma. It is also found to cause depressive action on the cardiovascular system ; and perhaps based on this characteristic feature it is invariably employed IV for the control and management of cardiac arrythmias. Lidocaine is largely metabolized in the liver by the help of a plethora of *'metabolic pathways'* particularly using oxidases of the mixed function type. It has been duly observed that careful modification of the pH of lignocaine solutions by means of $NaHCO_3$ appreciably retards the noted discomfort caused in patients when the local anaesthetic is administered by infiltration anaesthesia.

5.8. Mepivacaine

Tullar resolved mepivacaine and demonstrated that the (+)-isomer had an S configuration which displayed a significant long-acting profile. It is found to get bound to the plasma proteins to a considerable extent, upto 78%, and exerts identical activities comparable to lidocaine. Its onset of action is rather sluggish and slow ; however, the *'local anaesthetic'* activity lasts much longer as compared to lidocaine. Admixtures with known vasoconstrictors usually prolong the therapeutic action. Mepivacaine is invariably metabolized in the liver, and less than 10% gets excreted unchanged in the urine. However, some other metabolites are normally excreted by the kidney and the bile, whereas relatively smaller amounts of the latter metabolites are found to be excreted through the faecal wastes.

In obstetrics, the maternal plasma concentration varies between 2.9–6.9 mcg.mL^{-1}, whereas the umblical vein concentration ranges between 1.9–4.9 mcg.mL^{-1}. Thus, the faetus is only exposed to 60–70% of that available in maternal plasma. It is found to have a $t_{1/2}$ of 1.9 hour, aVd of 1.2 Lkg^{-1}, and a partition coefficient of 12.1.

5.9. Pramoxine

It is mostly employed as a topical anaesthetic for the relief of insect bites, hemmoroids, and minor wounds. As the drug shows a stinging and burning sensation, hence it must not be used for the eyes, nose and throat at all. The local anaesthetic action commences in 3–5 minutes ; its potency is fairly comparable to that of benzocaine and is not adequate to abolish the gag reflex. It is also indicated in a 1% (*w/v*) solution for the rapid relief of pain in rectal surgery, episiotomies, anogenital pruritus, itching dermatoses and minor burns.

5.10.   Prilocaine Hydrochloride

It has a plasma-protein binding of 55% and also having anaesthetic activity very much identical to that of cocaine. It has an onset of action rather slow, but the duration of action is almost comparable to that of lidocaine. It has been observed that prilocaine hydrochloride is slightly less toxic than lidocaine ; however, large doses of approximately 800 mg or more may cause severe methemoglobinemia*. It may pass across the placenta in due course and thus can cause methemoglobinemia in the *faetus*. It gets mostly metabolized in the liver, but also a portion in the kidney. Interestingly, one of the metabolites happens to be **2-methylaniline,** which seems to get subsequently metabolized to such compounds that are responsible for causing methemoglobinemia. Thus, 2-methyl aniline and other metabolites are duly exereted in the urine. A combination of prilocaine and lignocaine gives rise to an eutectic mixture (eutectic point) having *mp* below-either compounds, and is used exclusively for the preparation of topical-dosage forms.

Approximately 55% of prilocaine is bound to plasma protein. After 600 mg of the drug, peak-plasma levels are accomplished in 20 minutes, at which time span plasma levels average 4 mcg mL^{-1}, the same dose with epinephrine also attains peak at 20 minute, at which time plasma levels average 2 mcg mL^{-1}. As a result of this, prilocaine is mostly used without epinephrine. Therefore, it is specifically useful for such patients who cannot tolerate vasopressor agents, such as : patients having diabetes, hypertension, thyrotoxicosis, or other cardiovascular disorders.

5.11.   Procaine Hydrochloride

It is largely hydrolyzed in the plasma by plasma cholinesterase to produce PABA (this particu-larly prevents the action of sulphonamides) and diethylaminoethanol. Both these metabolites are exereted in the urine, the former being partially in the form of conjugates. Diethylaminoethanol is normally metabolized in the liver upto 70% approximately. It has been duly observed that procaine specifically prolongs the action of certain drugs by the subsequent formation of their corresponding salts which ultimately gets decomposed slowly to release the drug.

Example : IM injection of procaine retards the absorption of **Pencillin-G,** thereby prolonging its action significantly.

As stated earlier, the drug is affected by esterases ; and since the spinal-fluid virtually contains little or no esterase ; therefore, when given by this route of administration it remains active till such time it gradually gets absorbed into the general circulation.

Procaine   *para*-Aminobenzoic acid (PABA)   Diethylaminoethanol

5.12.   Propoxycaine Hydrochloride

It is nothing but a structural isomer of *proparacaine* ; and also being less toxic but slightly having lower potency than proparacaine.

---

*The clinical condition in which more than 1% of haemoglobin in blood has been oxidized to the ferric (Fe^{3+}) form. The principal system is cyanosis because the oxidized haemoglobin is incapable of transporting oxygen.

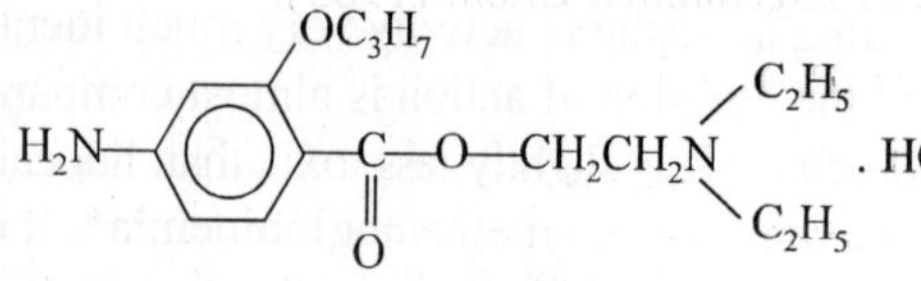

Propoxycaine Hydrochloride
**[Less toxic ; Less potent than
proparacaine]**

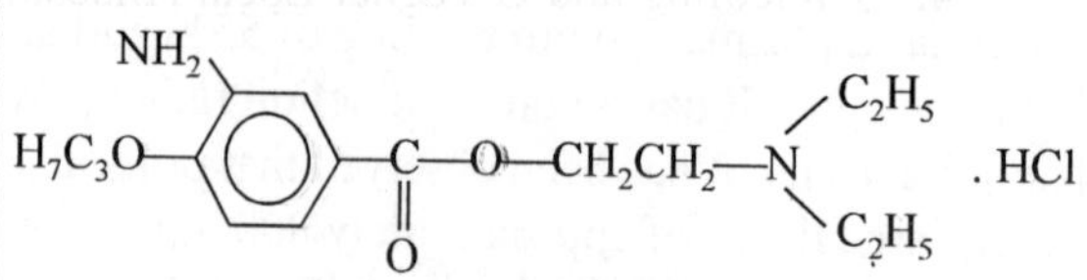

Proparacaine Hydrochloride
(An *isomer* of Propoxycaine)
[More toxic ; More potent than propoxycaine]

Probable Questions for B. Pharm. Examinations

1. Differentiate between the '*Local anaesthetics*' and the '*General Anaesthetics*'. Is it necessary to include local anaesthetics as adjuncts in antiseptic creams used in severe burns and painful skin abrasions ? Explain with typical examples.

2. The '*anaesthesiophoric moiety*' present in local anaesthetics is essentially derived from CO-CAINE. Explain.

3. Einhorn's generalization of aromatic esters gave rise to :

   (*a*) Methyl ester of *p*-amino-*m*-hydroxy benzoic acid,

   (*b*) Methyl ester of *m*-amino-*p*-hydroxy benzoic acid

   This ultimately led to the synthesis of certain potent local anaesthetics. Explain with suitable examples.

4. Explain how *PROCAINE* can be synthesized from :

   (*a*) 2-Chloroethyl *p*-amino benzoate

   (*b*) *p*-Aminobenzoic acid

   (*c*) Ethylene chlorohydrin.

5. Justify why propoxycaine hydrochloride is eight times more potent than procaine hydrochloride.

6. Discuss '*tropane derivatives*' as potent surface anaesthetic agents. Give examples and synthesis of any one compound selected by you.

7. '*Amides*' constitute an important category of local anaesthetic. Describe mepvacaine hydrochloride.

8. Discuss the synthesis of a quinoline analogue *i.e.*, dibucaine hydrochloride from isatin.

9. Dimethisoquin hydrochloride [Quotane® (SK & F)] an iso-quinoline analogue may be synthesized from alpha-*n*-butylphenethyl amine. Explain.

10. How would you synthesize pramoxine hdyrochloride ? Explain its applications.

RECOMMENDED READINGS

1. C L Hewer and J A Lee *Recent Advances in Anaesthesia and Analgesia*, (1958).

2. N Lofgren *Studies on Local Anaesthetics Stockholm,* University of Stockholm (1948).

3. T P Carney Benzoates and Substituted Benzoates as Local Anaesthetics, *Med. Chem* (1951).

4. S Wiedling and C Tegner Local Anaesthetics *Programmed Chem* (1963).

5. P Lechat (*Ed*) Local Anaesthetics in *Int Encycl Pharm Therapeut* I Oxford Pergramon Press (1971).

6. J M Ritchie and P J Cohen in : *The Pharmacological Basis of Therapeutics (Eds)* (L S Goodman and A Gilman) (9th edn) New York Macmillan (1995).

7. I C Geddes Chemical Structure of Local Anaesthetics *Br J Anaesth 34* (1962).

8. B H Takman and H J Adams Local Anaesthetics in : *Medicinal Chemistry (Ed)* M E Wolff (4th edn) New York, Wiley interscience (1980).

9. D Lednicer and L A Mitscher *(Eds)* Local Anaesthetics in *The Organic Chemistry of Drug Synthesis* New York, John Wiley and Sons (1995).

10. Williams DA and Lemke TL, *'Foye's Principles of Medicinal Chemistry'*, Lippincott Williams & Wilkins, New York, 5th edn., 2002.

11. Covino BG, *Local Anaesthetics.* In : *Drugs in Anaesthesia : Mechanisms of Action,* Feldman SA *et al.* eds., Edward Arnold, London, pp.261-291, 1987.

12. Smith J and Williams H : *Introduction to the Principles of Drug Design,* Wright, PSG, Bristol, 1988.

5 Sedatives and Hypnotics

1. INTRODUCTION

Sedatives are drugs which exert a quietening effect accompanied by relaxation and rest but do not necessarily induce sleep. **Hypnotics** are drugs which induce sleep and are synonymous with somnifacient and saporific. Both hypnotic and sedative properties usually reside in the same drug ; a large dose of a drug may act as a hypnotic, whereas a small dose of the same drug would act as a sedative. It is pertinent to mention here that in a few exceptional cases a compound exerts only one specific effect, *e.g.,* potassium bromide is a good sedative and exhibits no hypnotic action ; likewise certain powerful hypnotics, *e.g.,* thiopentone sodium cannot be used as a sedative.

Dier and sudden unforeseen *'emotional crisis'* invariably create such clinical situations that exceptionally demand the essential use of sedatives and hypnotics. A plethora of insomnia (sleeplessness) inadvertently caused due to short-term *'situational stress'* is usually regarded as the most suitable and ideal circumstantial condition(s) which necessiates drug-treatment not only to facilitate but also to augment sleep. After a vigorous and intensive research the introduction of rather newer yet safer and definitely more efficacious sedative and hypnotic drug has always been greeted with great fervour and optimism.

In a broader perspective, therefore, an *'ideal sedative and hypnotic'* must fulfil the following essential pre-requisites and requirements, namely :

(*a*) Must exert transient decrease in the degree of consciousness for the purpose of inducing sleep without any lingering effects,

(*b*) Should possess no potential for either lowering or arresting respirations (even at relatively higher dose regimens), and

(*c*) Essentially cause absolutely little abuse, dependence, addiction, or tolerance.

Interestingly, an honest and dedicated attempt to circumvent the already discussed undesirable characteristic features of *'barbiturates'*, have ultimately resulted to the synthesis of a host of *'nonbarbiturate'* structural analogue as potential sedatives-hypnotics in early 1950s, namely : methaqualone, nitrazepam, glutethimide, methyprylon, that are specifically noteworthy.

2. CLASSIFICATION

The sedatives and hypnotics are broadly classified into the following *two* categories, namely :

(*a*) Barbiturates ; and

(*b*) Non-barbiturates.

2.1 Barbiturates

In general, the barbiturates exert a significant deperssant action on the cerebrospinal axis. The relative degrees of depression, sedation, hypnosis or anaesthesia are exclusively dependent on the nature of the barbiturate, its dose and route of administration.

Barbiturates are *cyclic ureides* and are formed when a dicarboxylic acid reacts with urea. The acids used are generally in the form of ester and are condensed in the presence of sodium ethoxide (*i.e.*, C_2H_5—ONa) *e.g.*,

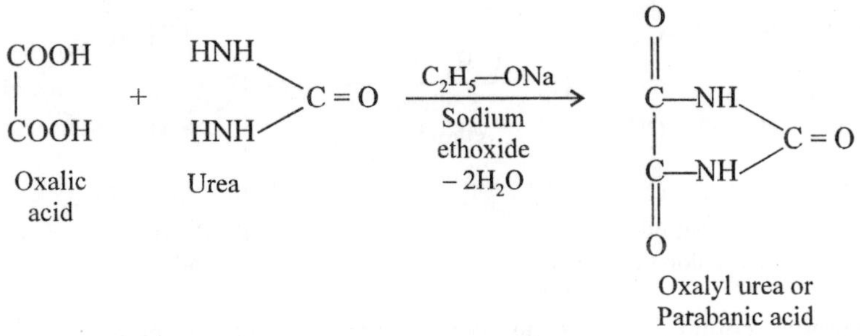

Oxalyl urea or
Parabanic acid

Parabanic acid is a cyclic ureide containing a five membered ring, which on hydrolysis by alkali may regenerate the corresponding acid and urea. The cyclic ureides are acidic owing to '*enolization*' and hence, they may form metallic salts by replacing the H atom of the –OH group as shown below :

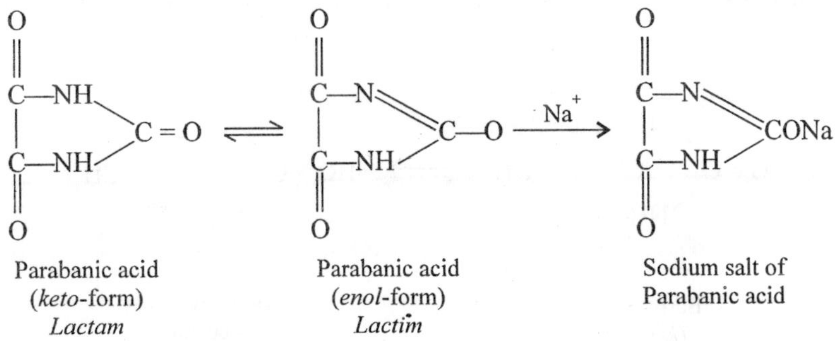

Parabanic acid
(*keto*-form)
Lactam

Parabanic acid
(*enol*-form)
Lactim

Sodium salt of
Parabanic acid

Many cyclic ureides are derived from malonic acid or malonic esters. They are collectively known as '**barbiturates**' because of their relationship of malonyl urea or barbituric acid.

Barbituric acid is prepared by the following *two* methods :

(a) *By the interaction of urea and malonyl dichloride,*

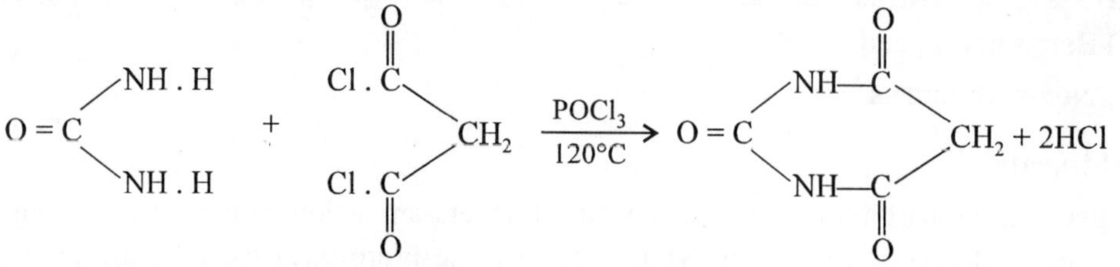

$$O = C \begin{matrix} NH.H \\ \\ NH.H \end{matrix} \quad + \quad \begin{matrix} Cl.C \\ \\ Cl.C \end{matrix} CH_2 \quad \xrightarrow[120°C]{POCl_3} \quad O = C \begin{matrix} NH-C \\ \\ NH-C \end{matrix} CH_2 + 2HCl$$

Urea   Malonyl dichloride   Malonyl urea or
                                                   Barbituric acid

(b) *By the interaction of urea and diethyl malonate,*

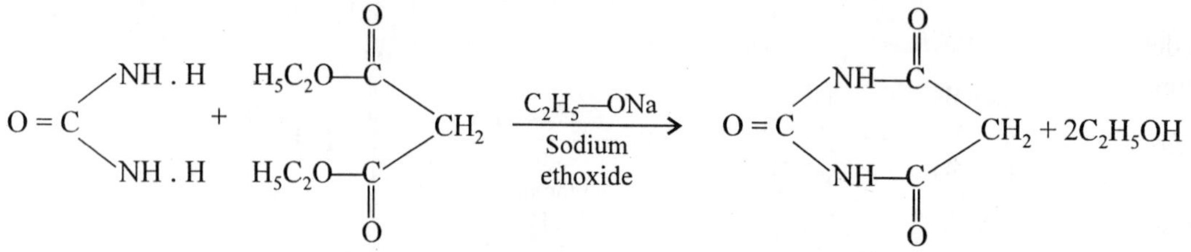

$$O = C \begin{matrix} NH.H \\ \\ NH.H \end{matrix} \quad + \quad \begin{matrix} H_5C_2O-C \\ \\ H_5C_2O-C \end{matrix} CH_2 \quad \xrightarrow[\text{Sodium ethoxide}]{C_2H_5-ONa} \quad O = C \begin{matrix} NH-C \\ \\ NH-C \end{matrix} CH_2 + 2C_2H_5OH$$

Urea   Diethyl   Barbituric   Ethanol
                  malonate   acid

The cyclic ureides containing a six membered ring, are also regarded as derivatives of the funda-

mental type pyrimidine or 1 : 3-diazine $\left[i.e., {}_1N \underset{2 \quad 3}{\overset{6 \quad 5}{\bigcirc}} N \right]$

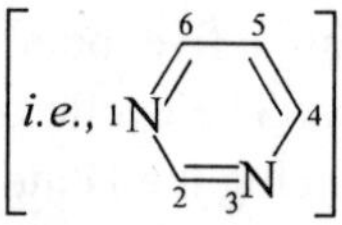

Barbituric acid like parabanic acid exhibits '*keto-enol tautomerism*' as illustrated below :

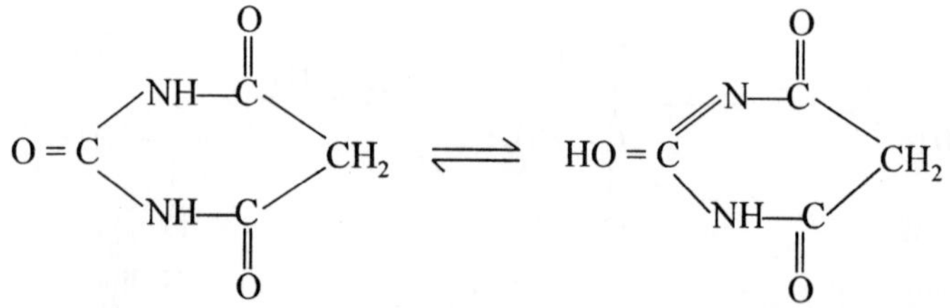

$$O = C \begin{matrix} NH-C \\ \\ NH-C \end{matrix} CH_2 \quad \rightleftharpoons \quad HO = C \begin{matrix} N=C \\ \\ NH-C \end{matrix} CH_2$$

Barbituric acid   Barbituric acid
(*keto*-form)   (*enol*-form)
Lactam   **Lactim**

In thiobarbiturate the oxygen of the $(-\overset{O}{\underset{\|}{C}}-)$ group of the urea residue is replaced by a sulphur
atom *i.e.,*

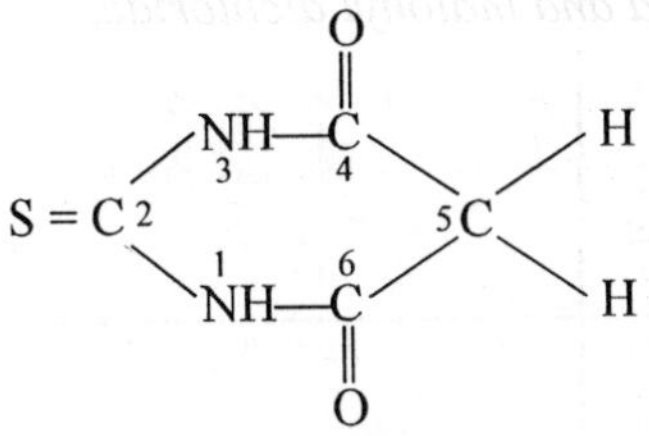

Thiobarbituric acid

However, it is interesting to observe that the barbituric acid itself does not possess any hypnotic properties, but such a characteristic is conferred only when the hydrogen atoms at C-5 are replaced by organic groups (alkyl or aryl).

Classification of Barbiturates

Barbiturates are classified, rather arbitrarily, by the duration of their clinical effects. More than 50 derivatives are used in therapeutics. The more commonly used analogues of barbituric acid are recorded in Table 5.1.

A. Long Acting Barbiturates

The onset of action for long acting barbiturates is visible after an hour or so, and the duration of action lasts for 6–10 hours. They are largely excreted by the kidney. *Examples* : Barbitone ; Barbital Sodium ; Phenobarbital ; Methyl-phenobarbital.

A1. *Barbital* INN, *Barbitone* BAN,

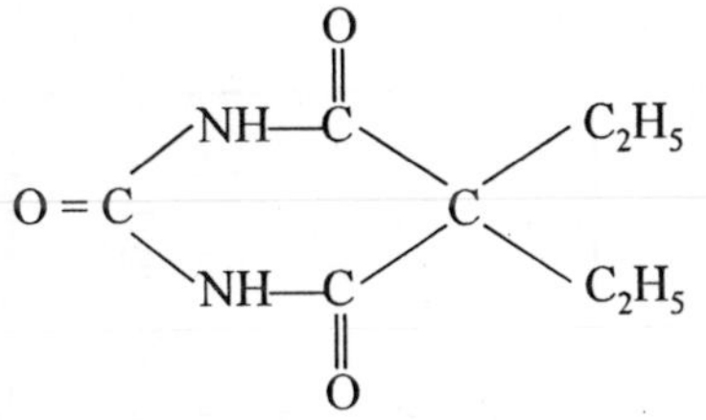

5, 5-Diethyl barbituric acid ; Diethylmalonyl urea ; Malonal ;

B.P.C. 1963 ; Eur. P. ; N.F. XI

Veronal[R] (Winthrop)

Synthesis :

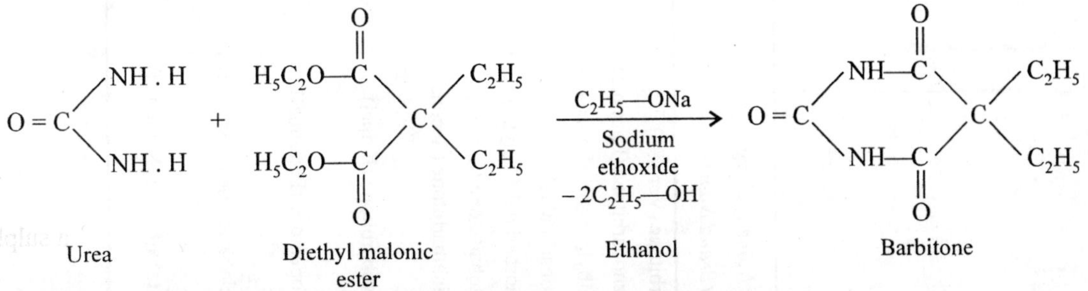

Barbitone is prepared by the condensation of urea with diethyl malonic ester in the presence of sodium ethoxide with the elimination of two molecules of ethanol.

Table 5.1. Barbituric acid and its analogues

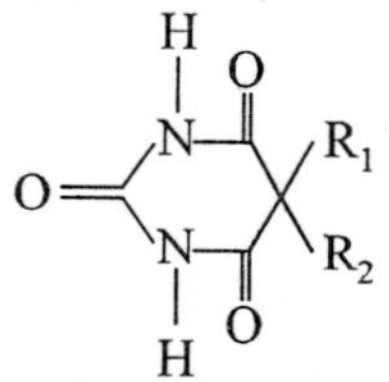

Name of Drug (Brand Name)	Chemical Name (-barbituric acid)	Substitution at R_1	R_2	Other Attachments	Duration of Action (hr.)
Barbitone (Veronal)	5, 5-Diethyl-	C_2H_5	C_2H_5		4—12
Phenobarbitone (Luminal)	5-Ethyl-5-phenyl	C_2H_5	C_6H_5		4—12
Methylphenobarbitone (Prominal)	5-Ethyl-1-methyl-5-phenyl	C_2H_5	C_6H_5	$1\text{-}CH_3$	1—4
Allobarbital (Dial)	5, 5-Diallyl	$CH_2 = CH—CH_2$	$CH_2 = CH—CH_2$		2—8
Amorbarbital (Amytal)	5-Ethyl-5-isopentyl	C_2H_5	$(CH_3)_2CHCH_2CH_2$		2—8
Butobarbitone (Soneryl)	5-Ethyl-5-n-butyl	C_2H_5	$CH_3CH_2CH_2CH_2$		2—6
Quinalbarbitone (Seconal)	5-Allyl-5 (1-methyl-butyl)	C_2H_5	$CH_3—CH—CH_2—CH_2$ CH_3		1—4
Hexobarbitone (Evipal)	5-Δ'-Cyelohexenyl-1-methyl N-methyl	CH_3	$(CH_2)_4CH = C—$	$1\text{-}CH_3$	1—4
Thiopentone (Pentothal)	5-Ethyl-5-(1-methyl butyl)-2-thiobarbituric acid	C_2H_5	$CH_3(CH_2)_2CH(CH_3)$	2-S	1—4
Cyclobarbitone (Phanodorn)	5-Ethyl-5 (1-cyclohexen-1-yl)	C_2H_5	$(CH_2)_4CH = C—$		2—8
Pentobarbitone (Nembutal)	5-Ethyl-5 (1-methyl butyl)	C_2H_5	$CH_3(CH_2)_2CH(CH_3)$		2—4

It is a powerful hypnotic drug and generally used in the treatment of epileptic seizures. It has the main drawback of having a *low therapeutic index.*

Dose : 0.3 to 0.6 g.

The solutions are incompatible with ammonium salts and acidic substances.

A2. *Barbital Sodium* INN, *Barbitone Sodium* BAN :

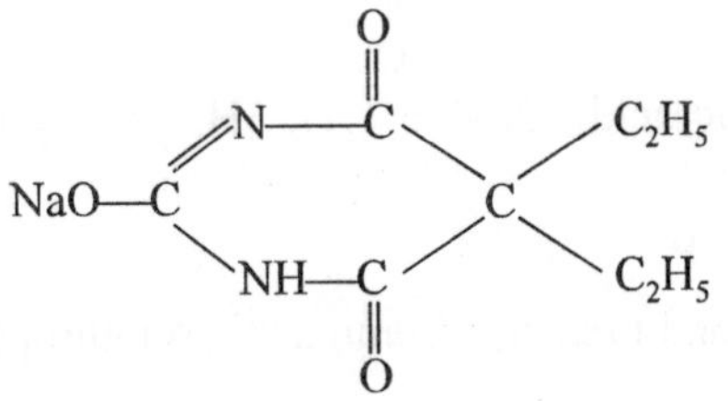

Sodium 5, 5-diethylbarbiturate ; Sodium derivative of 5, 5-diethylbarbituric acid ;

Soluble Barbitone,

B.P. 1973 ; NF XI ; Int. P. ; Ind. P.

Somnylic Tablets[R] (Philip Harris)

Synthesis :

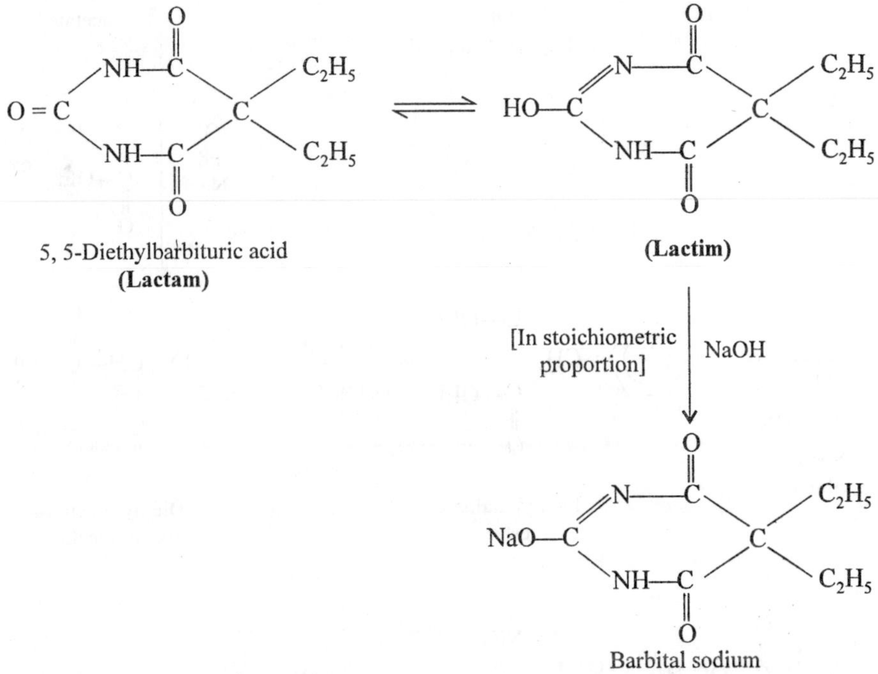

5, 5-Diethylbarbituric acid
(Lactam)

(Lactim)

[In stoichiometric proportion] NaOH

Barbital sodium

It is prepared by the neutralization of an aqueous solution of barbital with sodium hydroxide and then precipitating the salt by the addition of alcohol.

Being water-soluble, barbital sodium is more readily absorbed than its parent compound barbital. *Owing to its slow rate of excretion there exists an element of risk of a cumulative action.*

Dose : 0.34 to 0.6 g.

A3. *Phenobarbital* INN, USAN, *Phenobarbitone* BAN,

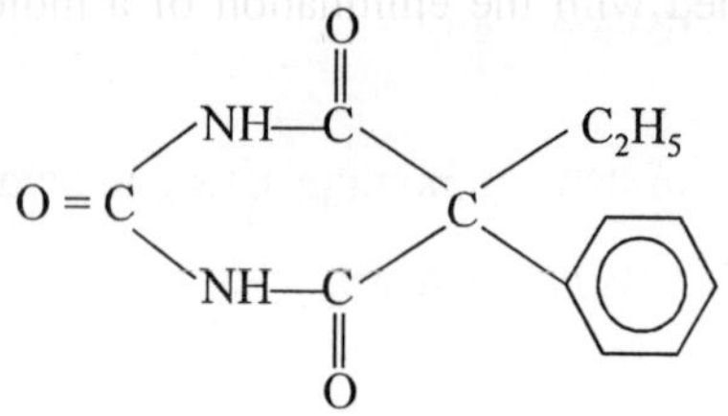

5-Ethyl-5-phenyl barbituric acid ; 2, 4, 6 (1H, 3H, 5H)-pyrimidinetrione, 5-ethyl-5-phenyl ; Phenylethylmalonylurea ;

U.S.P. ; B.P. ; Eur. P. ; Int. P.

Eskabarb(R) (Smith Kline and French) ; Luminal(R) (Winthrop) ; Gardinal(R) (May and Baker) ; Stental(R) (Robins)

Synthesis

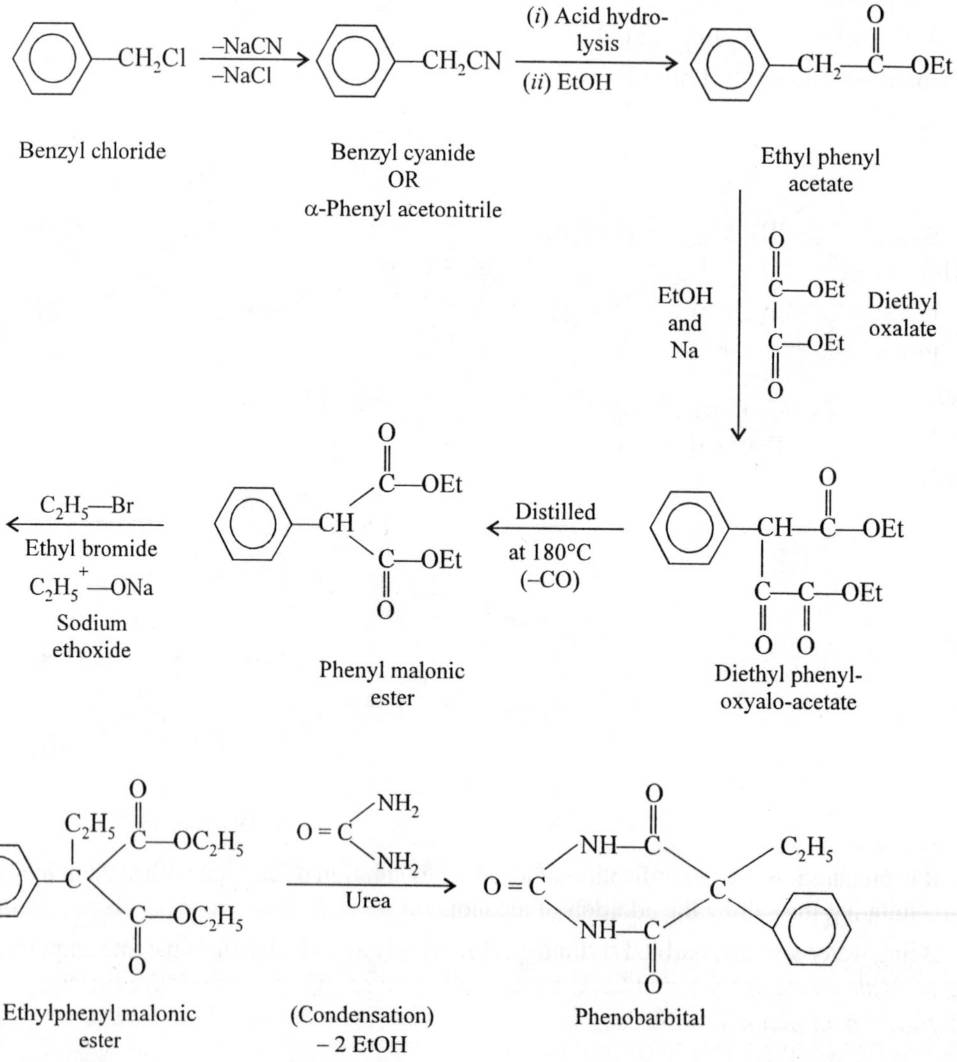

–NaCN
–NaCl
(i) Acid hydrolysis
(ii) EtOH
Benzyl chloride
Benzyl cyanide OR α-Phenyl acetonitrile
Ethyl phenyl acetate
EtOH and Na
Diethyl oxalate
Distilled at 180°C (–CO)
C2H5—Br Ethyl bromide + C2H5—ONa Sodium ethoxide
Phenyl malonic ester
Diethyl phenyl-oxyalo-acetate
Urea
(Condensation) – 2 EtOH
Ethylphenyl malonic ester
Phenobarbital

Phenobarbital is prepared by treating benzyl chloride with sodium cyanide when benzyl cyanide (or α-phenyl acetonitrile) is formed with the elimination of a molecule of sodium chloride. Benzyl cyanide, first on hydrolysis yields phenyl acetic acid which on subsequent esterification with ethanol forms the corresponding ester as ethyl phenyl acetate. This on reaction with diethyl oxalate in the persence of absolute ethanol and sodium metal gives diethyl phenyl oxalo acetate which on distillation at 180°C results into phenyl malonic ester. When it is treated with ethyl bromide and sodium ethoxide, the lonely active hydrogen atom gets replaced with an ethyl group thus forming ethyl phenyl malonic ester. Lastly, this on condensation with urea loses two molecules of ethanol and finally forms the desired compound phenobarbital.

It is used both as sedative and hypnotic. It is the drug of choice in the treatment of *grandmal and petitmal epilepsy.*

Dose : 30 to 120 mg.

A4. *Methylphenobarbital* INN, *Methylphenobarbitone* BAN, *Mephobarbital* USAN,

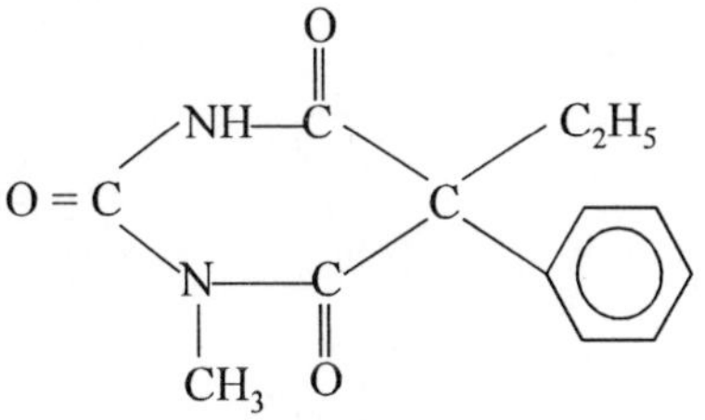

5-Ethyl-1-methyl-5-phenylbarbituric acid ; 2, 4, 6 (1H, 3H, 5H)-pyrimidinetrione, 5-ethyl-1-methyl-5-phenyl- ;

U.S.P., B.P., Eur. P.

Prominal[R] (Winthrop, U.K.)

Synthesis :

Methylphenobarbital can be prepared by the following *two* methods :

Method I : (From Urea)

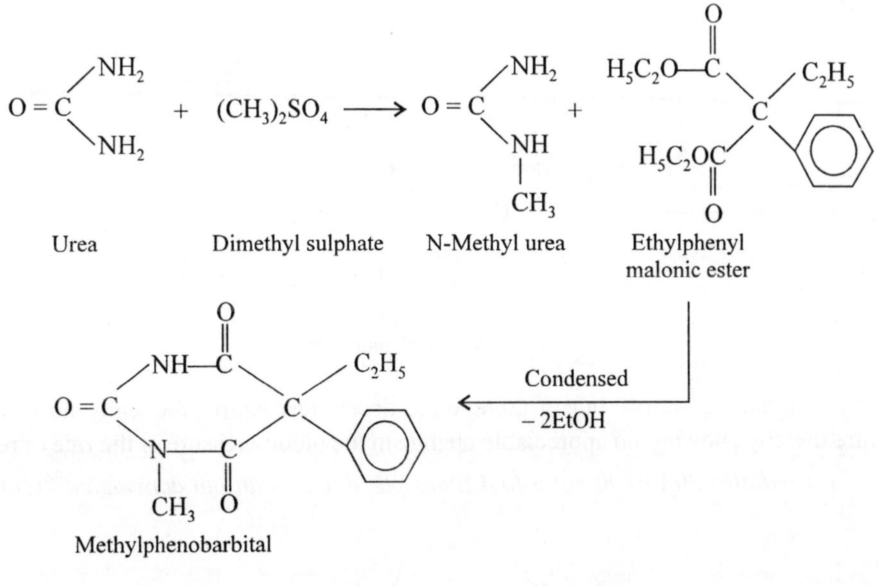

Urea   Dimethyl sulphate   N-Methyl urea   Ethylphenyl malonic ester

Condensed

– 2EtOH

Methylphenobarbital

It may be prepared by the condensation of N-methyl urea, obtained from urea and dimethyl sulphate, and ethylphenyl malonic ester with the elimination of two molecules of ethanol.

Method-II : (From Dicyandiamide)

Methylphenobarbital can also be prepared by the interaction of ethylphenyl malonic ester and dicyandiamide when it results into the formation of 5-ethyl-5-phenyl-3-cyano barbituric acid, which on methylation with dimethyl sulphate yields 5-ethyl-5-phenyl-3-cyano-1-methyl barbituric acid. This on acid hydrolysis converts the cyano moiety at position 3 into a free COOH group which on decarboxylation finally gives methylphenobarbital.

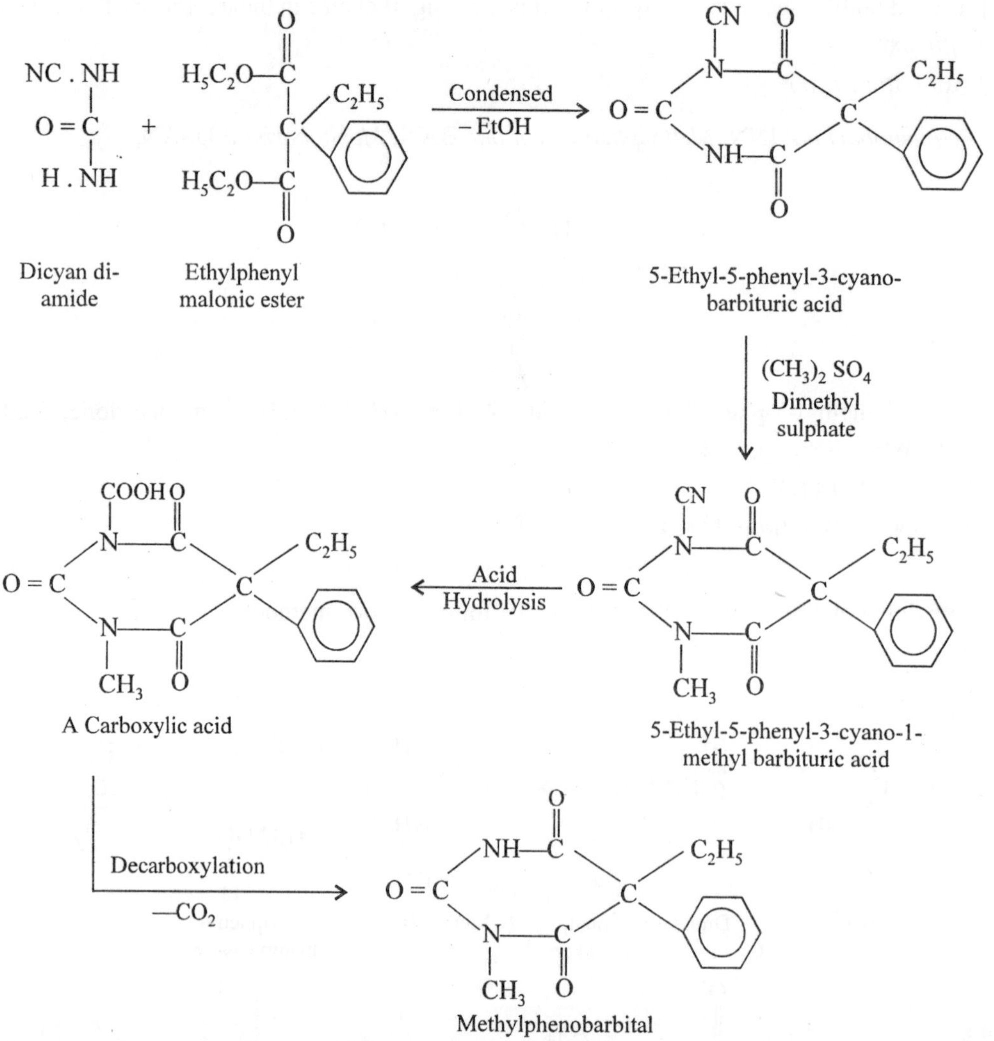

It possesses hypnotic action, but in therapeutic doses they exert practically no effect on the medullary centre thereby allowing no appreciable change in the blood pressure or the rate of respiration.

Dose : As a sedative 30 to 100 mg 3 to 4 times per day ; as an anticonvulsant 400 to 600 mg daily.

B. *Intermediate Acting Barbiturates*

The onset of action for intermediate acting barbiturates is 30 minutes and their hypnotic effect last for 2 to 6 hours. Most of them are first degraded by the liver and the metabolised product subsequently excreted by the kidney. They are generally used in insomnia and also as a pre-operative sedative. They also find their use in the treatment of convulsions when administered intravenously.

Examples : Allobarbital ; Butobarbitone ; Amobarbital

B1. *Allobarbital* INN,

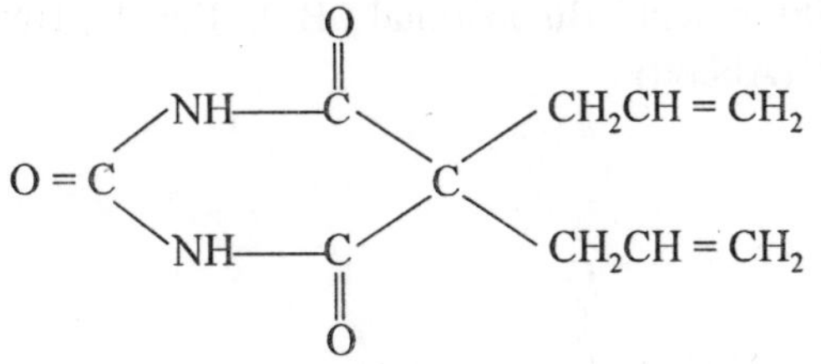

5, 5-Diallylbarbituric acid ; 2, 4, 6 (1H, 3H, 5H)-Pyrimidinetrione, 5, 5-di-2-propenyl-; Allobarbital ; Diallylbarbitone ; Diallylmalonylurea ; Diallymalum ; Allobarbitone (BPC 1959)

Synthesis :

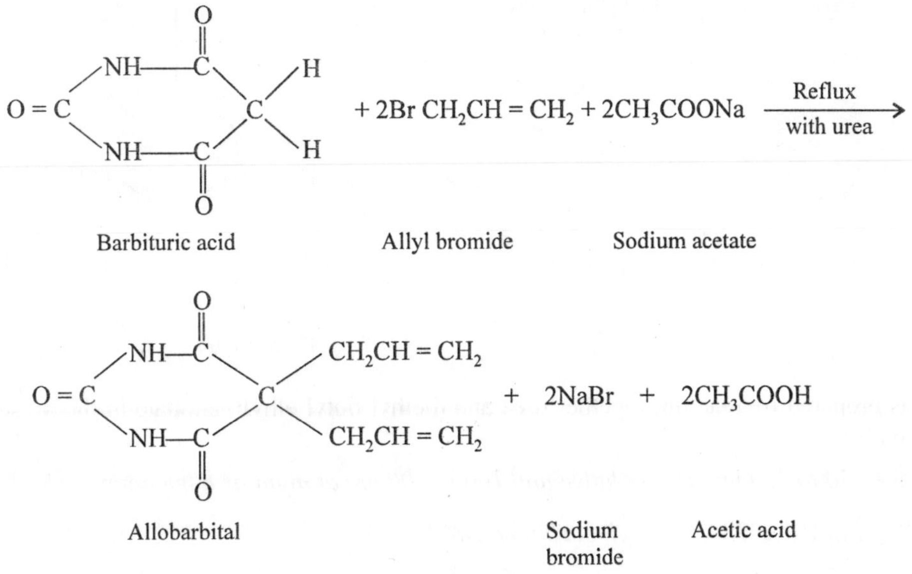

| Barbituric acid | Allyl bromide | Sodium acetate |

| Allobarbital | Sodium bromide | Acetic acid |

It may be prepared by the interaction of barbituric acid with an alcoholic solution of allyl bromide and sodium acetate and on being refluxed with urea.

It can be used both as a sedative and hypnotic at different dose levels.

Dose : As a sedative 30 mg 3 to 4 times a day ; as a hypnotic 100 to 200 mg at night.

B2. *Butobarbitone* BAN,

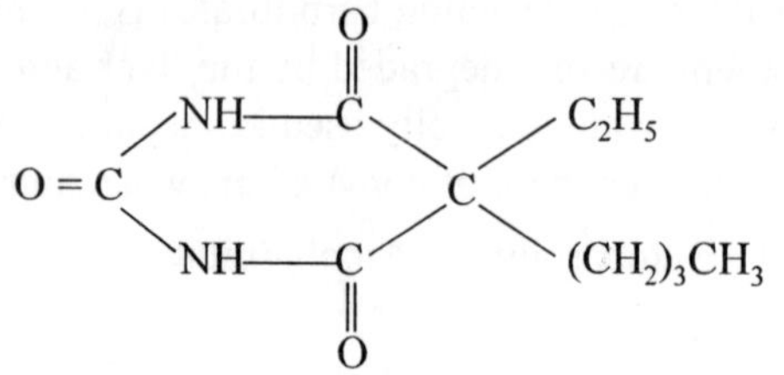

5-Butyl-5-ethyl barbituric acid ; Butobarbital ; B.P., Eur. P., Butethal N.F. X Soneryl[R] (May and Baker, U.K.) ; Neonal[R] (Abbott) ;

Synthesis :

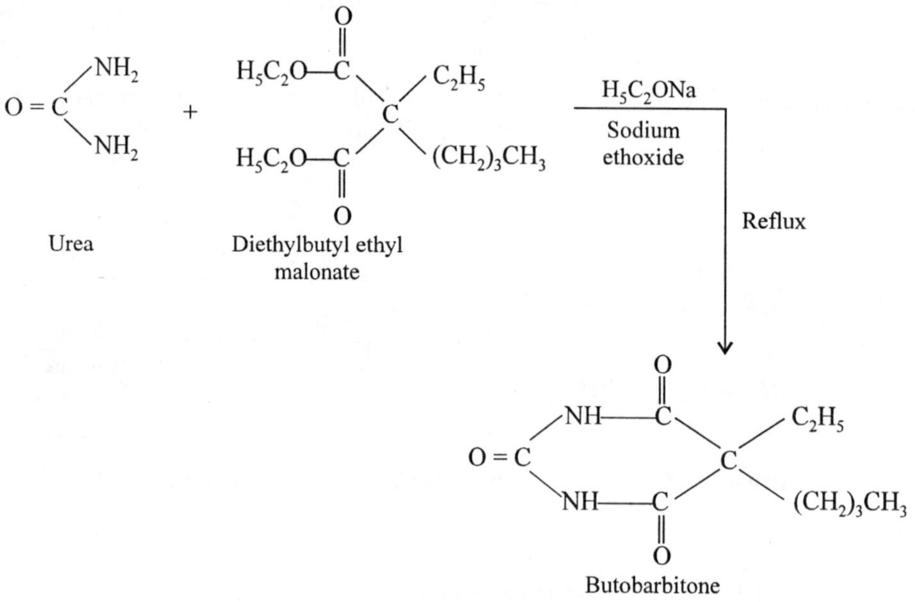

It is prepared by reacting together urea and diethyl butyl ethyl malonate in the presence of sodium ethoxide.

Dose : 30 to 120 mg as a sedative and 100 to 200 mg at night as a hypnotic.

B3. *Amobarbital* INN, USAN, *Amylobarbitone* BAN ;

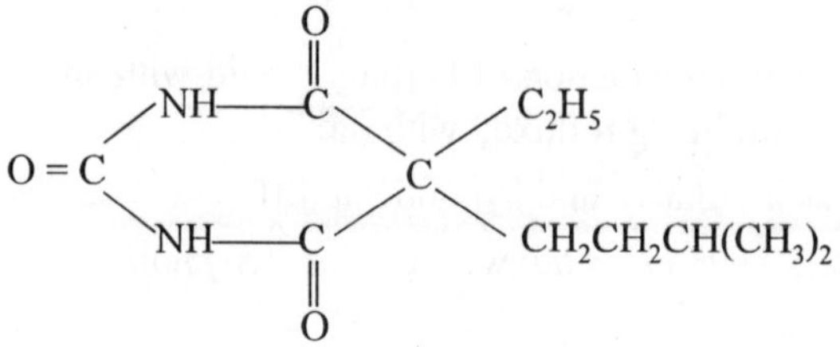

5-Ethyl-5-isopentylbarbituric acid ; 5-Ethyl-5-isoamylbarbituric acid ; 2, 4, 6 (1H, 3H, 5H)-Pyrimidinetrione, 5-ethyl-5-(3-methyl-butyl)- ;

Amobarbital (U.S.P.), Amylobarbitone (B.P., Ind. P., Int. P.),

Amytal[R] (Lilly) ;

Synthesis :

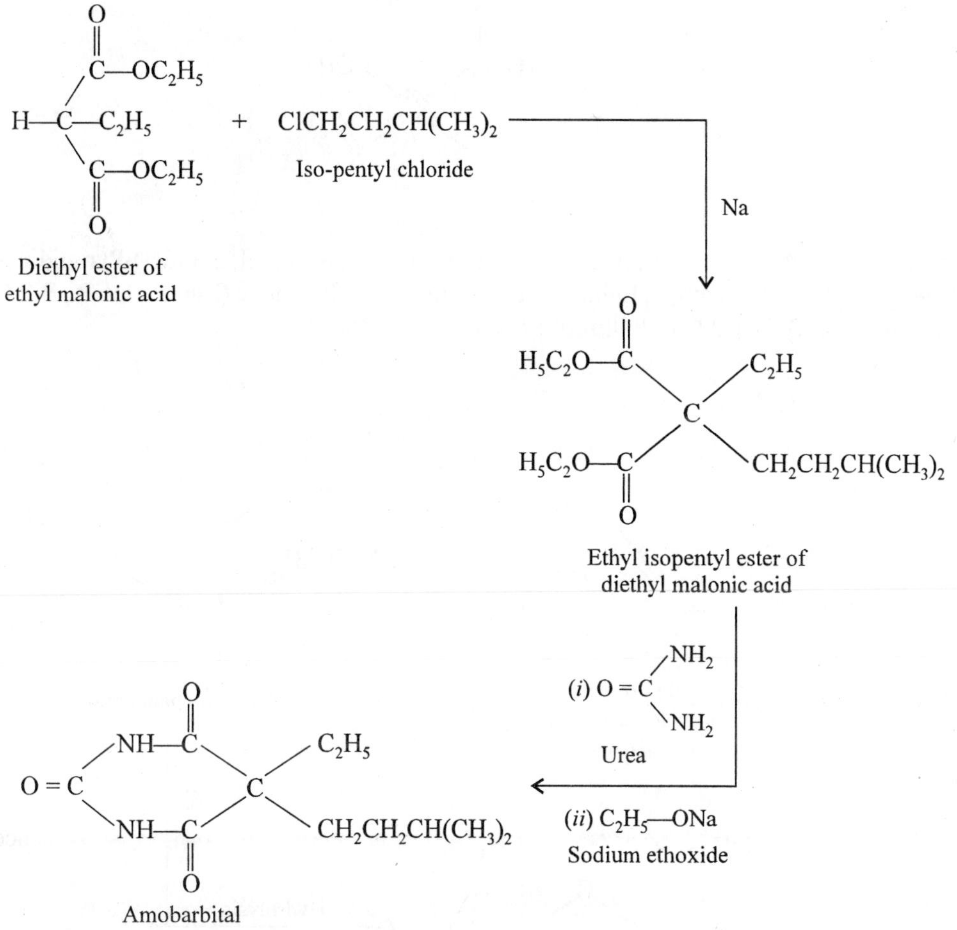

Diethyl ester of
ethyl malonic acid

Ethyl isopentyl ester of
diethyl malonic acid

Amobarbital

It is prepared by the interaction of diethyl ester of ethyl malonic acid and iso-pentyl chloride in the presence of sodium metal, when ethyl is isopentyl ester of diethyl malonic acid is obtained as an intermediate compound. This on condensation with urea in the presence of sodium ethoxide results into the formation of amobarbital.

Dose : 300 mg

C. Short-Acting Barbiturates

The onset of action for short-acting barbiturates falls within 15 minutes and their hypnotic action last for 1 to 2 hours. They are mostly metabolized in the liver. They are invariably used in the treatment of insomnia and pre-operative medication.

Examples : Hexobarbital ; Pentobarbital sodium ; Quinalbarbitone Sodium ; and Cyclobarbital.

C1. *Hexobarbital* INN, USAN,

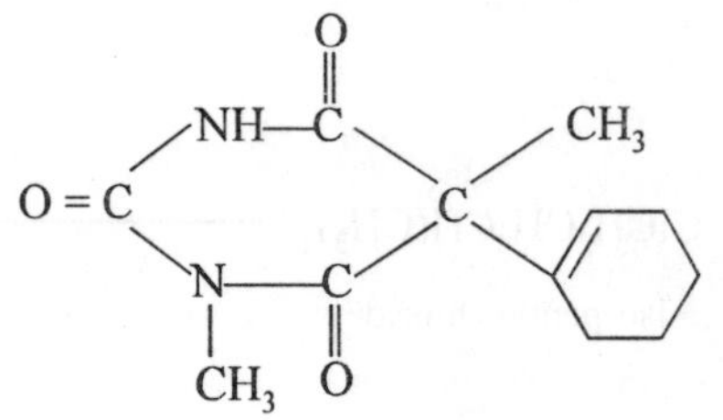

5-(1-Cyclohexen-1-yl)-1, 5-dimethyl barbituric acid ; 2, 4, 6 (1H, 3H, 5H)-Pyrimidinetrione, 5-(1-cyclohexen-1-yl)- 1, 5-dimethyl ; Enimal ; Methexenyl ; Hexobarbitone (Eur. P., Int. P., B.P.C. 1959) ; Hexobarbital (U.S.P.) Evipal[R] (Winthrop) ; Sombulex[R] (Riker)

Synthesis :

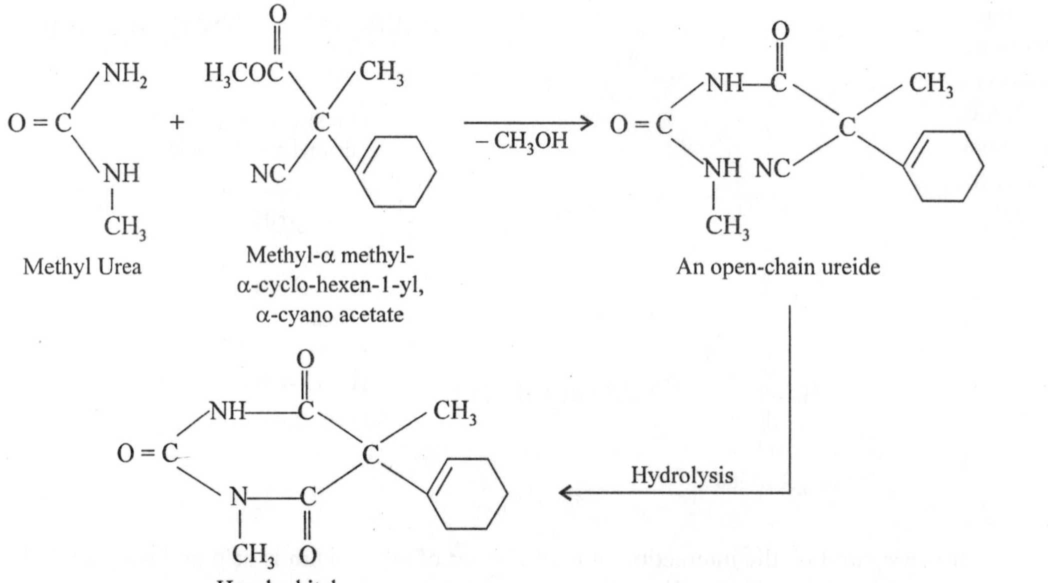

Hexobarbital is prepared by reacting together methyl urea and methyl-α-methyl-α-cyclo-hexen-1-yl-α-cyano acetate when an open-chain ureide is formed as an intermediate with the elimination of a molecule of methanol. This upon hydrolysis affords spontaneous closure of the ring thereby resulting into the formation of hexobarbital.

Dose : Adult, oral, hypnotic, 250 to 500 mg.

C2. *Pentobarbital Sodium* USAN, *Pentobarbitone Sodium* BAN,

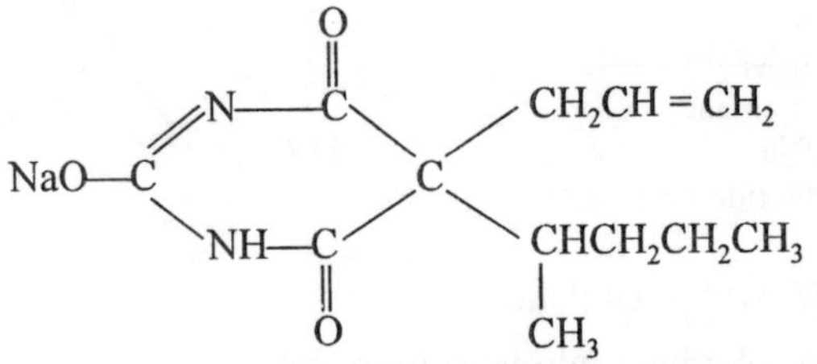

Sodium 5-allyl-5-(1-methylbutyl) barbiturate ; 2, 4, 6 (1H, 3H, 5H)-Pyrimidinetrione, 5-ethyl-5-(1-methylbutyl), monosodium salt ; Ethaminal Sodium ; Soluble Pentobarbitone ; Pentobarbitone Sodium B.P., Eur. P., Ind. P., Nembutal Sodium[R] (Abbott) ; Palapent[R] (Bristol-Myers) ; Sodital[R] (American Critical Care) ;

Synthesis :

It is prepared by :

(*i*) Synthesis of diethylester of ethyl-(1-methyl butyl) malonate ;

(*ii*) Condensation of (*i*) with urea ; and

(*iii*) Conversion of Pentobarbital into its sodium salt.

In the first step the diethyl ester of malonic acid is treated with ethyl bromide in the presence of sodium ethoxide when one of the active hydrogen atoms in the former gets eliminated with bromine atom in the later as a molecule of hydrobromic acid resulting into the formation of the corresponding diethyl ester of ethyl malonic acid. This on subsequent addition of 2-monobromopentane and in the presence of sodium ethoxide gives rise to diethyl ether of ethyl-(1-methyl butyl) malonate with the elimination of one molecule of hydrobromic acid. Urea is made to condense with the product obtained from the previous step when pentobarbital is formed with the elimination of two moles of ethanol. Finally, the pentobarbital is treated with a calculated amount of sodium hydroxide when the required official compound is formed.

It is used mostly in the treatment of insomnia, as a basal anaesthetic and also in strychnine poisoning.

Dose : 100—200 mg

(*i*) Preparation of Diethyl ester of ethyl-(1-methyl butyl) malonate :

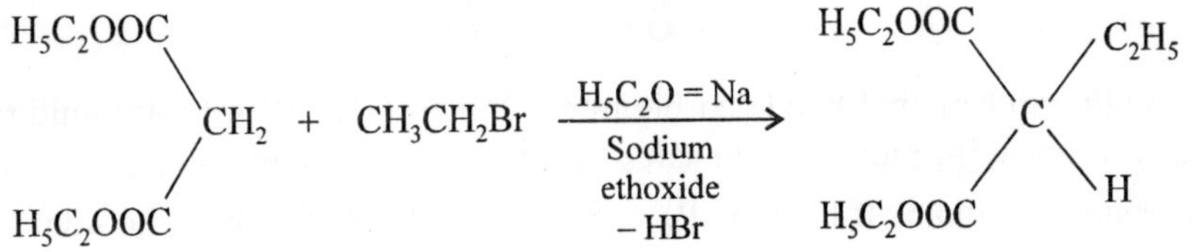

Diethyl ester of malonic acid   Ethyl bromide   Diethyl ester of ethyl malonic acid

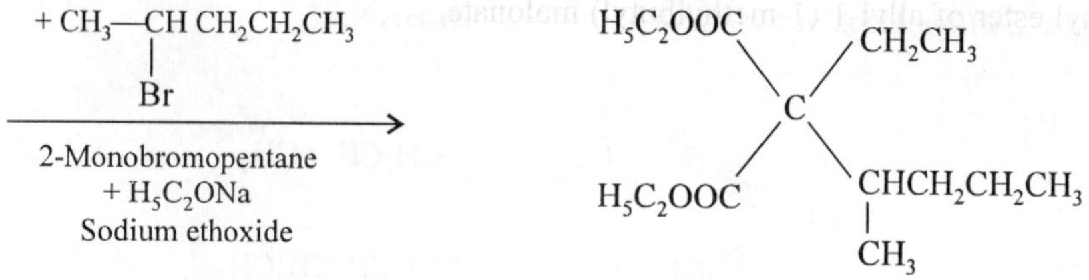

$$+ CH_3{-}CH\ CH_2CH_2CH_3$$
$$\underset{Br}{|}$$

$$\xrightarrow[\substack{\text{2-Monobromopentane} \\ + H_5C_2ONa \\ \text{Sodium ethoxide}}]{}$$

$$\underset{H_5C_2OOC}{\overset{H_5C_2OOC}{>}} C \underset{\underset{CH_3}{|}}{\overset{CH_2CH_3}{<}} CHCH_2CH_2CH_3$$

(*ii*) Condensation of (*i*) with urea ; and

(*iii*) Conversion of Pentobarbital into its sodium salt.

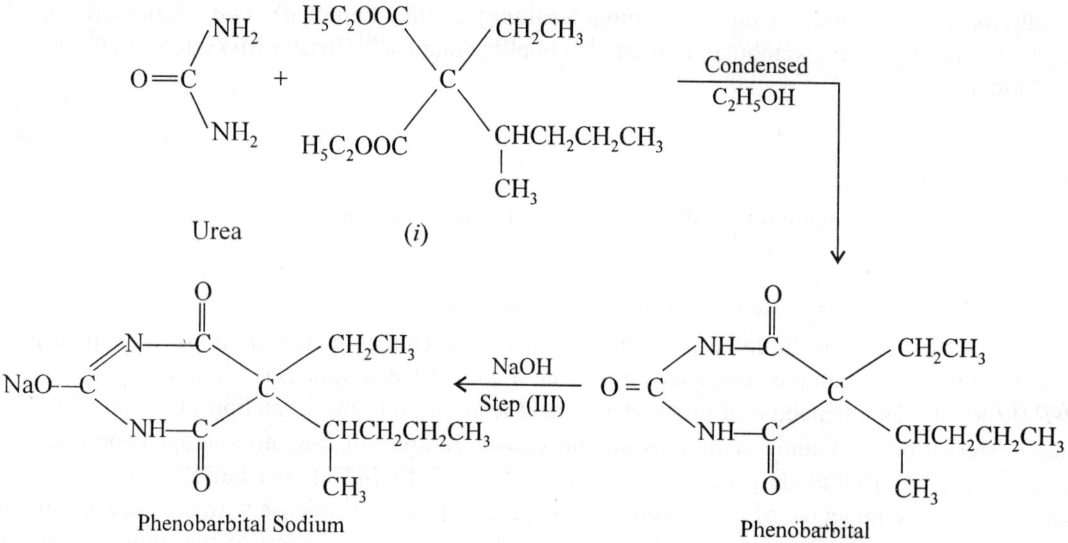

$$O{=}C\underset{NH_2}{\overset{NH_2}{<}} \quad + \quad \underset{H_5C_2OOC}{\overset{H_5C_2OOC}{>}}C\underset{\underset{CH_3}{|}}{\overset{CH_2CH_3}{<}}CHCH_2CH_2CH_3$$

Urea   (*i*)

$$\xrightarrow[\text{C}_2\text{H}_5\text{OH}]{\text{Condensed}}$$

Phenobarbital Sodium

$$\xleftarrow[\text{Step (III)}]{\text{NaOH}}$$

Phenobarbital

C3. *Quinalbarbitone Sodium BAN, Secobarbital Sodium USAN,*

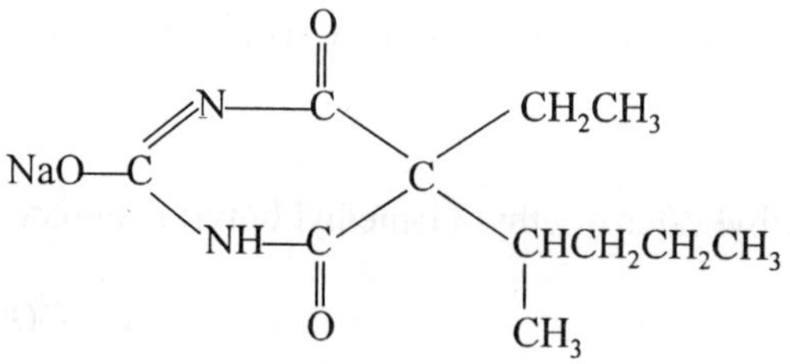

Sodium 5-allyl-5-(1-methyl butyl) barbiturate ; 2, 4, 6 (1H, 3H, 5H)-Pyrimidine-trione, 5-(1-methyl butyl)-5-(2-propenyl)-, monosodium salt ; Secobarbitone sodium.

Quinalbarbitone Sodium, B.P., Eur. P., Ind. P., Int. P., Secobarbital Sodium U.S.P. Seconal Sodium[R] (Lilly)

Synthesis :

Quinalbarbitone sodium can be conveniently prepared by means of the follwoing *three* steps :

(*i*) Preparation of diethyl ester of allyl-(1-methyl butyl) malonate ; (*ii*) Condensation of (*i*) above with urea ; and (*iii*) Preparation of the sodium salt.

Step (i) Diethyl ester of allyl-1 (1-methylbutyl) malonate

Diethyl malonate + 2-Bromopentane → (NaOEt, Sodium ethoxide, —HBr) → Diethyl ester of (1-methyl butyl)-malonate

Diethyl ester of (1-methyl butyl)-malonate → (+ CH₂=CHCH₂Br, Allyl Bromide, −HBr) → Diethyl ester of allyl-(1-methyl butyl)-malonate

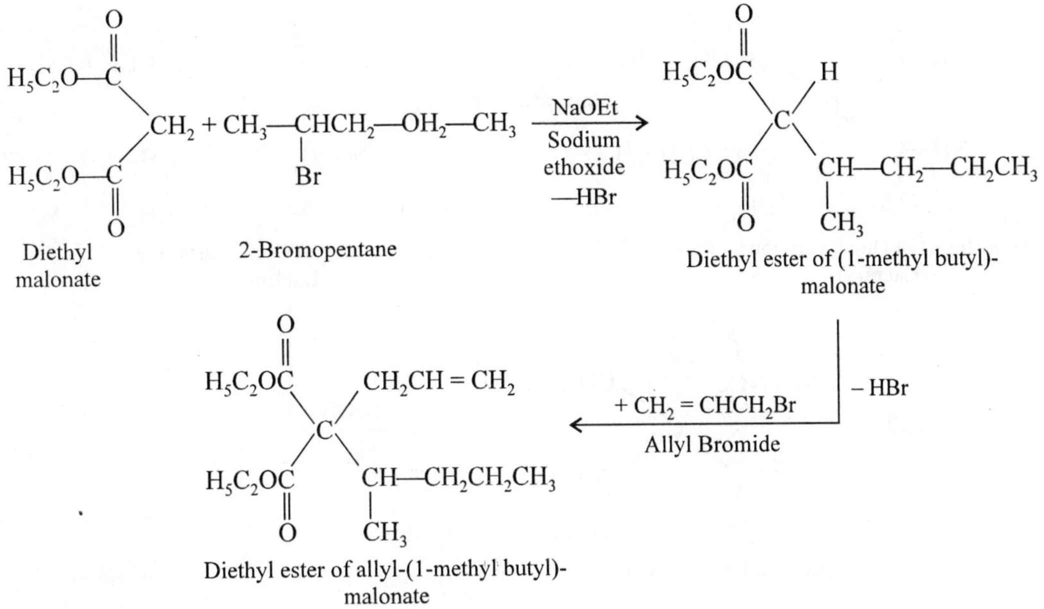

Step (ii) Condensation of *(i)* above with urea

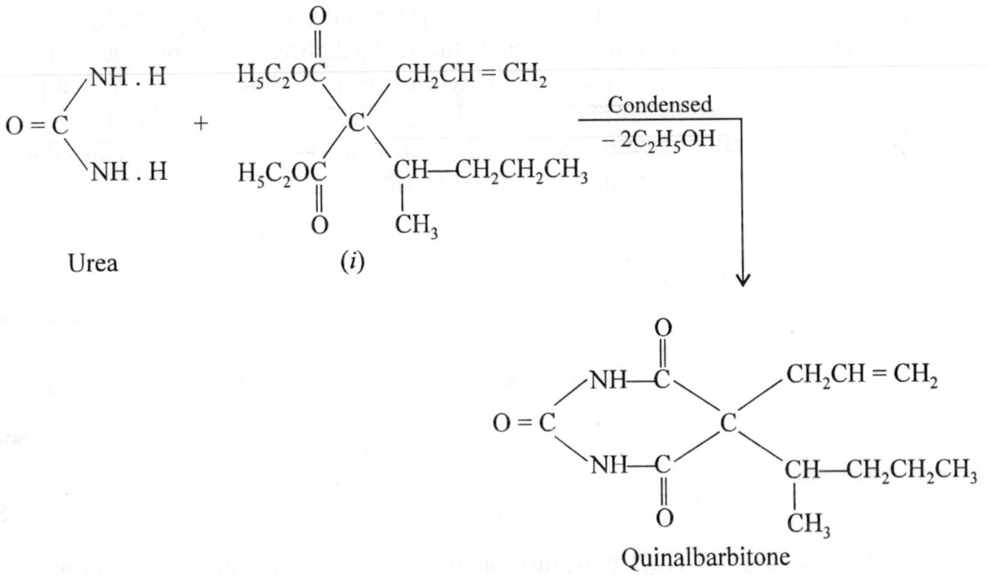

Urea + *(i)* → (Condensed, −2C₂H₅OH) → Quinalbarbitone

Step (*iii*) Preparation of Sodium Salt of Quinalbarbitone

(*keto*-form) of Quinalbarbitone
(Lactam)

(*enol*-form) of Quinalbarbitone
(Lactim)

Quinalbarbitone sodium

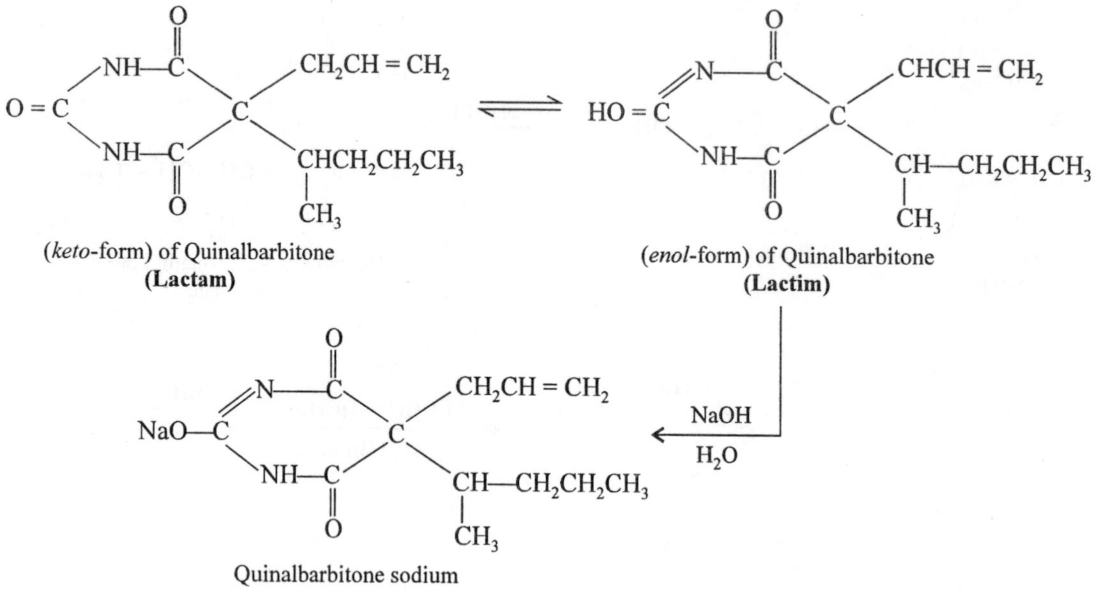

First, diethyl malonate is reacted with 2-bromopentane in the persence of sodium ethoxide when one active hydrogen atom from the former and a bromine atom from the later gets eliminated as a molecule of hydrobromic acid and resulting into the formation of diethyl ester of (1-methyl butyl) malonate. On further treatment of this compound with allyl bromide, the lonely active hydrogen present in the malonate is abstracted with the bromine atom in the allyl halide as hydrobromic acid giving rise to the corresponding diethyl ester of allyl-(1-methyl butyl) malonate. *Secondly*, the resulting product on condensation with urea loses two moles of ethanol yielding quinalbarbitone. *Thirdly*, it undergoes *keto-enol* tautomerism and finally the *enol*-form of quinalbarbitone on reaction with a calculated amount of sodium hydroxide results into the sodium salt of quinalbarbitone.

Dose : 50—200 mg

C4. *Cyclobarbital* INN, *Cyclobarbitone* BAN,

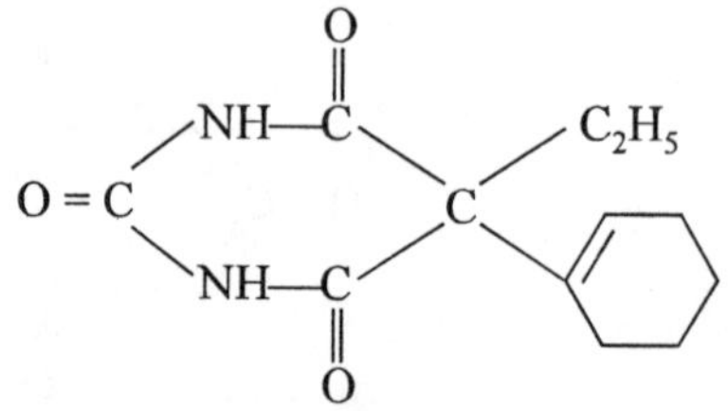

5-(1-Cyclohexen-1-yl)-5-ethyl barbituric acid ; Ethylhexabarbital ; Cyclobarbital N.F. X, Cyclobarbitone B.P., Ind. P., Phanodorn[R] (Winthrop)

Synthesis :

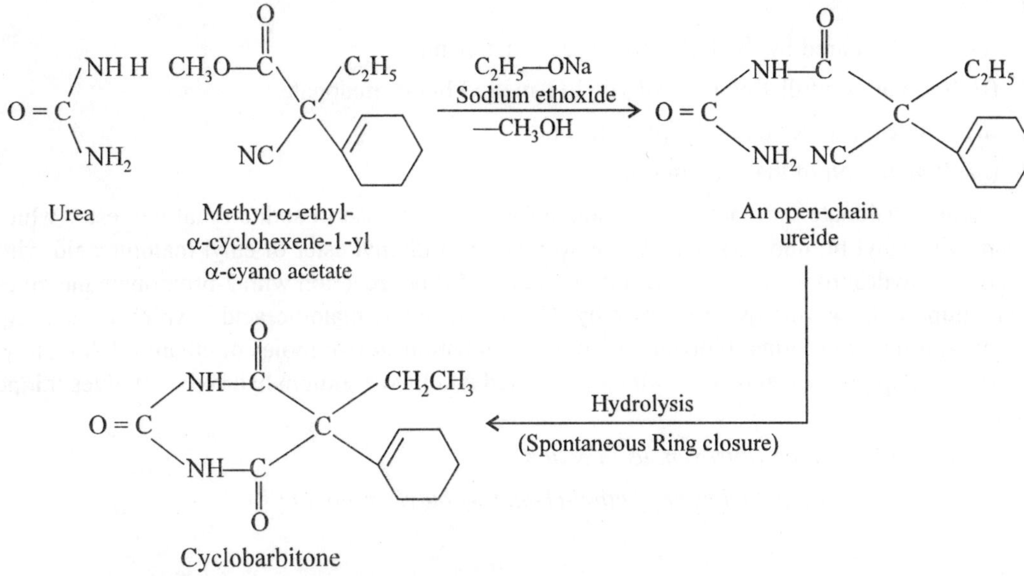

Urea

Methyl-α-ethyl-
α-cyclohexene-1-yl
α-cyano acetate

An open-chain
ureide

Cyclobarbitone

It is prepared by the interaction of urea with methyl-2-ethyl-α-cyclo-hexene-1-yl-α-cyano acetate in the presence of sodium ethoxide when a molecule of methanol is eliminated with the formation of an *open-chain ureide* ; which upon hydrolysis results into spontaneous ring closure and gives rise to cyclobarbital.

Dose : 200 to 400 mg

D. *Ultra-Short-Acting Barbiturates*

These act almost instantaneously, *i.e.,* within a few seconds after administration. Because of this peculiar characteristics they are usually employed to produce general anaesthesia and to control convulsions. They may be used either alone or in conjunction with inhalation anaesthesia. After administration, they are first deposited in adipose tissues but are eventually dependent on the liver and kidney for their ultimate metabolic degradation and elimination.

Examples : Thiopental sodium, methohexital sodium.

D1. *Thiopental Sodium* INN, USAN,

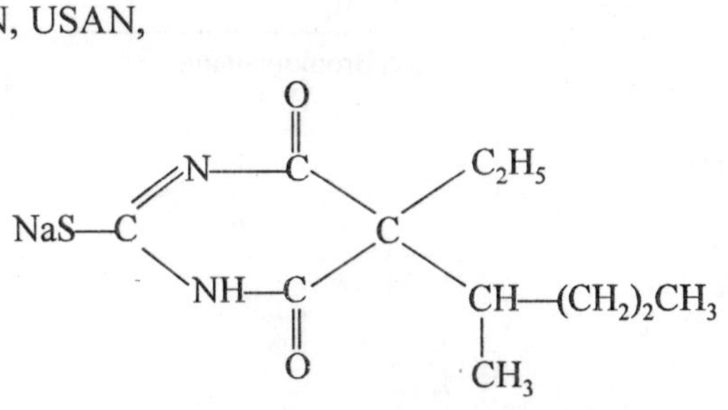

Sodium 5-ethyl-5 (1-methyl butyl)-2-thiobarbiturate ; 4, 6 (1H, 5H)-Pyrimidinedione, 5-ethyldihydro-5-(1-methyl butyl)-2 thioxo-, monosodium salt ; Thiopental sodium U.S.P., Thiopentone sodium B.P., Eur. P., Ind. P., Int. P. Pentothal sodium[R] (Abbott) ;

Intraval sodium[R] (May and Baker)

Synthesis :

It can be prepared by the following *three* steps, namely :

(*i*) Preparation of diethyl ester of ethyl-(1-methyl butyl) malonate ;

(*ii*) Condensation of (*i*) with thiourea ; and

(*iii*) Preparation of the sodium salt.

Diethyl malonate on reaction with sodium metal gives rise to sodium malonic ester which on treatment with ethyl bromide results into the formation of diethyl ester of ethyl malonic acid with the elimination of hydrobromic acid. The resulting ester on further reaction with 2-bromopentane gives the desired compound, *i.e.,* diethyl ester of ethyl (1-methyl butyl) malonic acid ; which on subsequent treatment with thiourea forms thiopental with the elimination of two moles of ethanol. Ultimately, the *enol*-form of thiopental when reacted with a calculated amount of sodium hydroxide, it gives thiopental sodium.

Dose : 100—150 mg (intravenous injection).

(*i*) *Preparation of diethyl ester of ethyl-(1-methyl butyl) malonate :*

Diethyl ester of ethyl-
(1-methyl butyl) malonic acid

Diethyl ester of
ethyl malonic acid

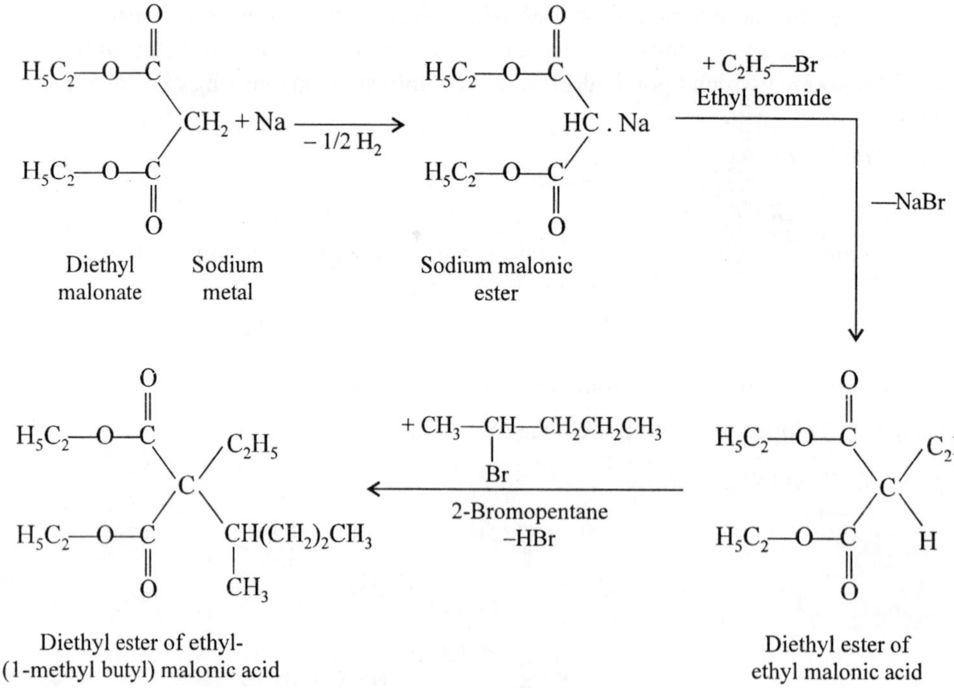

(*ii*) Condensation of (*i*) with thiourea :

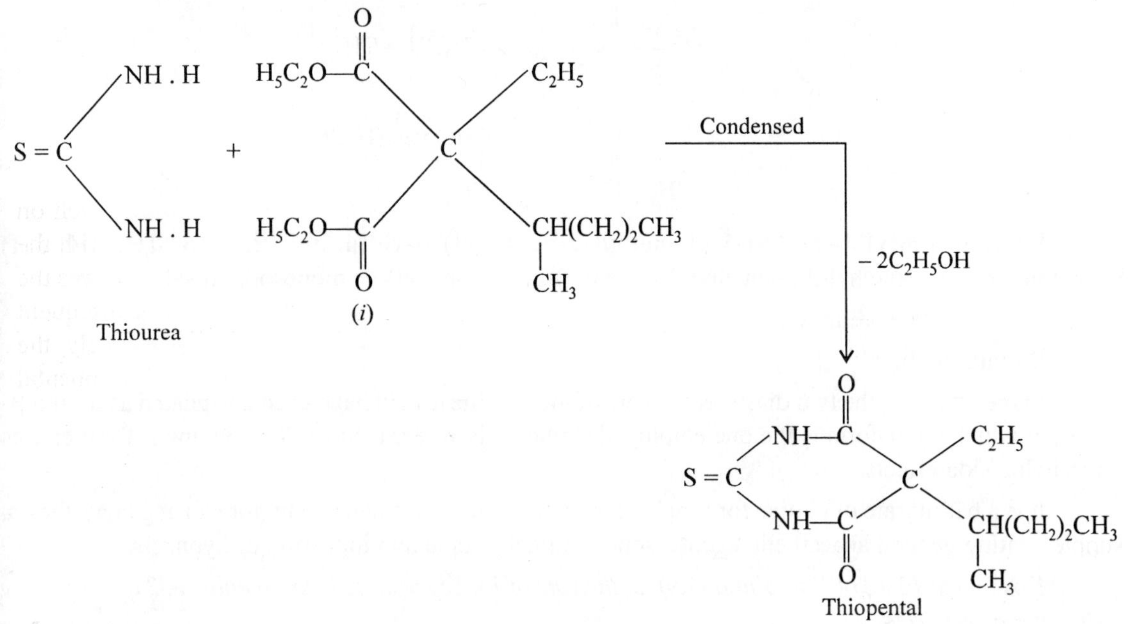

Thiourea
(*i*)

Thiopental

(*iii*) Perparation of the sodium salt :

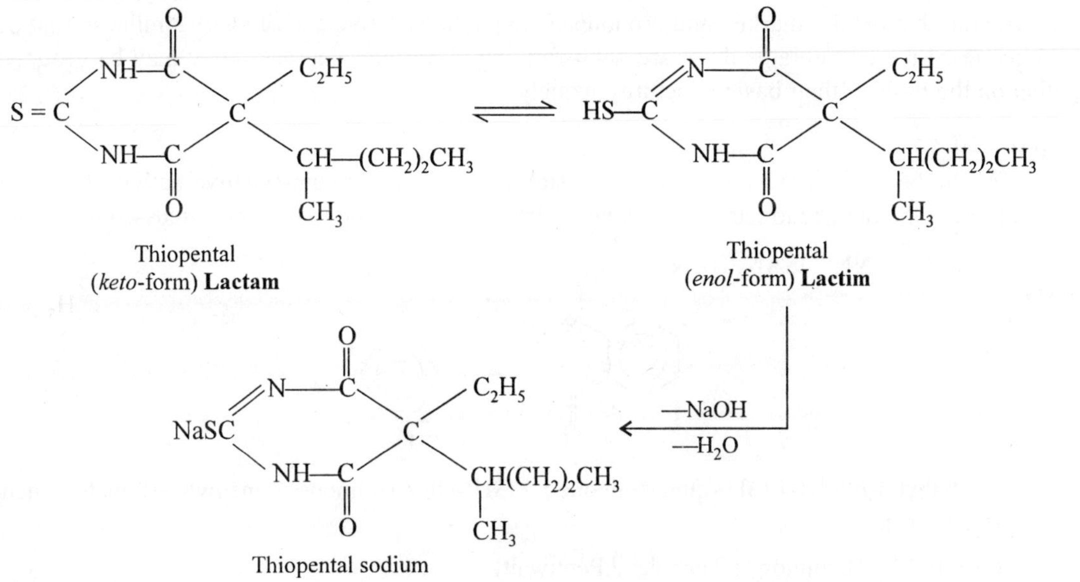

Thiopental
(*keto*-form) **Lactam**

Thiopental
(*enol*-form) **Lactim**

Thiopental sodium

D2. *Methohexital Sodium* USAN,

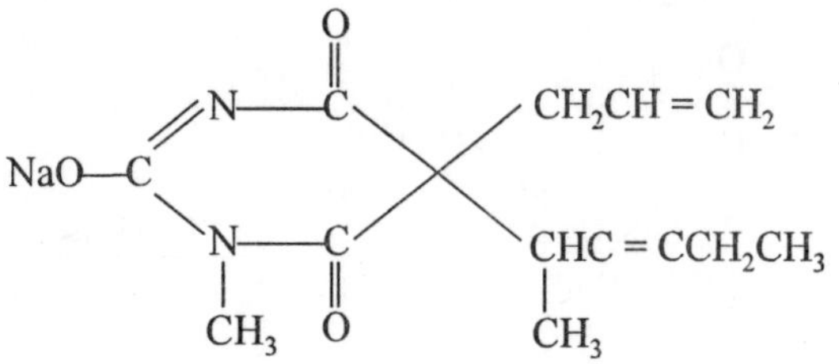

Sodium 5-allyl-1-methyl-5 (1-methyl-2-pentynyl) barbiturate ; 2, 4, 6 (1H, 3H, 5H) Pyrimidinetrione, 1-methyl-5 (1-methyl-2-pentynyl)-5-(2-propenyl)- ± monosodium salt ;

Methohexital sodium U.S.P.

Brevital Sodium[R] (Lilly)

In the literature the two diastereoisomers of the barbituric acid have been designated as α- and β- forms, of which the α-form is the one employed medicinally whereas the corresponding β-form causes undesirable side-effects.

It is a barbiturate of choice for rapid action, administered intravenously, for causing anaesthesia, supplementing general anaesthetic agents, short surgical trauma and induction of hypnosis.

Dose : 5 to 12 ml of 1% solution (iv), at the rate of 1 ml every 5 sec., maintenance, 2 to 4 ml every 4 to 7 min as requried.

2.2   Non-barbiturates

There are a number of compounds which do not essentially possess the malonyl urea or barbitu- rate structure but exhibit marked and pronounced hypnotic-sedative activity very similar to that of the barbiturates. Like barbiturates these are habit-forming to varying degrees. They may be grouped to- gether on the basis of their basic structures, namely :

Heterocyclics

A number of heterocyclics possessing significant hypnotic-cum-sedative activity have gained recognition in the therapeutic armanentarium. A few such compounds shall be discussed here briefly.

A. *Methaqualone* INN, USAN, BAN,

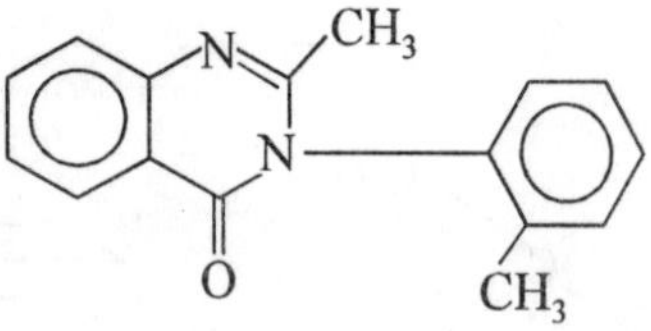

2, Methyl-3-*o*-tolyl-4 (3H)-quinazolinone ; 4 (3H)-Quinazolinone, 2-methyl-3-(2-methylphenyl) ; U.S.P., B.P.

Quaalude[R] (Lemmon) ; Tuazole[R] Pennwalt) ;

Its hypnotic action is similar to the intermediate-acting barbiturate which may be enhanced when administered along with an antihistaminic agent like diphenylhydramine. It may be used as a sedative at lower dose levels.

Dose : 50 to 150 mg

B. *Paraldehyde* USAN, BAN

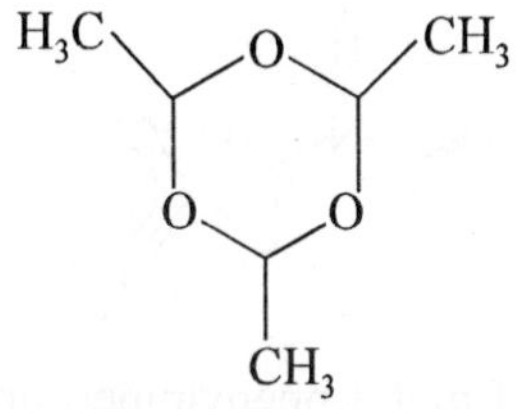

2, 4, 6-Trimethyl-s-trioxane ; 1, 3, 5-Trioxane, 2, 4, 6-trimethyl- ; Paracetaldehyde ; The trimer of acetaldehyde ; U.S.P., B.P., Eur. P., Ind. P.

Paral[(R)] (O'Neal, Jones and Feldman) ;

Synthesis :

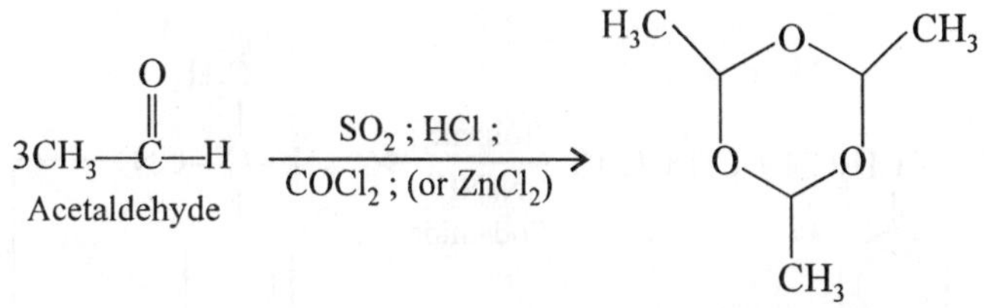

Paraldehyde

It is prepared by treating acetaldehyde with small quantities of sulphur dioxide, hydrochloric acid, carbonyl chloride, or zinc chloride. The resulting liquid is frozen and subsequent distillation of the crystallised material often yields paraldehyde.

It is one of the oldest hypnotics. The drug is usually employed in delirium tremens, status epilepticus and in patients undergoing withdrawal therapy for alcoholism. A certain portion of the administered drug is excreted through the lungs.

Dose : Adult, oral : As sedative 5 to 10 ml ; as hypnotic 10 to 30 ml

C. *Nitrazepam* INN, BAN,

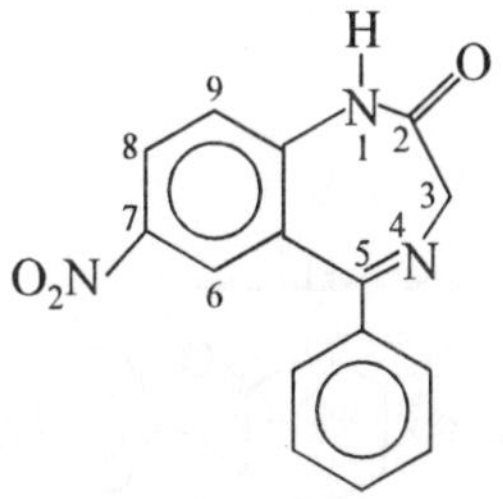

1, 3-Dihydro-7-nitro-5-phenyl-2H-1, 4-benzodiazepin-2-one ; 2H-1, 4-Benzodiazepin-2-one, 1, 3-dihydro-7-nitro-5-phenyl- ; B.P. Eur. P.

Mogadon[(R)] (Roche)

It is one of the members of the group benzodiazepines employed as sedatives and hypnotics. It is widely used in Canada and Europe as a sedative/hypnotic and also in the management of myoclonic seizures. It is extensively metabolized to inactive substances which are ultimately excreted through the urine.

Dose : Usual, oral, adult 2.5 to 10 mg.

D. Glutethimide INN, USAN, BAN,

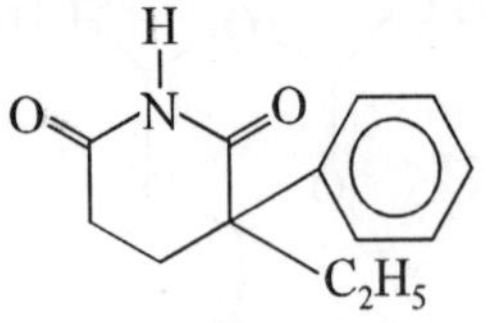

2-Ethyl-2-phenylglutarimide ; 3-Ethyl-3-phenylpiperidine-2, 6-dione ;

U.S.P., B.P., Int. P.

Doriden[R] (Ciba, UK) ; Glutril[R] (Roche)

Synthesis :

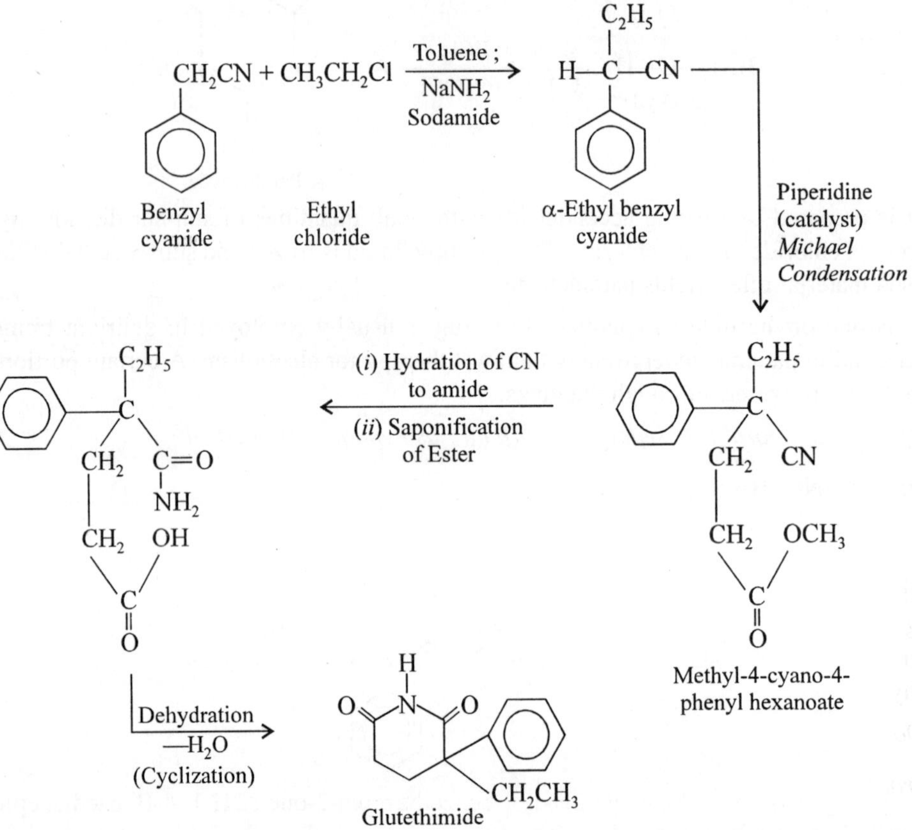

α-Ethylbenzyl cyanide is obtained from the reaction of benzyl cyanide and ethyl chloride in toluene as a medium and sodamide as a catalyst, which is then caused to undergo **Michael condensation** in the persence of piperidine to yield methyl-4-cyano-4-phenyl hexanoate. This on hydration converts the free cyano moiety to an amide, which on saponification gives an intermediate. Cyclization of this intermediate is effected through dehydration between the amide and carbon groups. The pure product may be obtained by recrystallization from ethanol-water mixture (1 : 1).

It is safely used for inducing sleep in all types of insomnia without causing depression of respiration. Its hypnotic action is parallel to intermediate-acting barbiturates.

Dose : 250 to 500 mg

E. Methyprylon INN, USAN, Methyprylone BAN

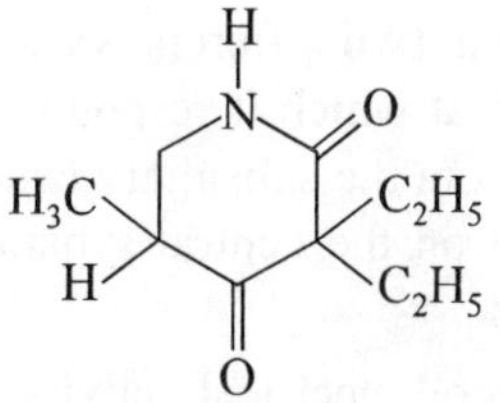

$$\text{H}_3\text{C} \quad \overset{\displaystyle\text{H}}{\underset{\displaystyle\text{O}}{\text{N}}} \quad \overset{\displaystyle\text{O}}{\underset{\displaystyle\text{C}_2\text{H}_5}{\text{C}_2\text{H}_5}}$$

3, 3-Dimethyl-5-methyl-2, 4-piperidinedione ; 2, 4-Piperidine-dione, 3, 3-diethyl-5-methyl ; Methyprylon U.S.P., N.F., Methyprylone B.P.

Nodular[R] (Roche) ;

It is an useful hypnotic agent employed in the management of insomnia of varied etiology. It may induce sleep within a span of 45 minutes and produce sleep ranging between 5 to 8 hours.

Like glutethimide it is also a piperidinedione derivative.

Dose : Usual, sedative, 125 to 250 mg up to 3 times daily ; hypnotic, 500 mg to 1 g.

F. *Chloral Hydrate* USAN, BAN,

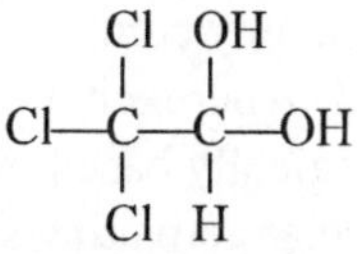

$$\begin{array}{c} \quad\;\; \text{Cl} \;\; \text{OH} \\ \quad\;\; | \quad\;\; | \\ \text{Cl}-\text{C}-\text{C}-\text{OH} \\ \quad\;\; | \quad\;\; | \\ \quad\;\; \text{Cl} \;\; \text{H} \end{array}$$

1, 1-Ethanediol, 2, 2, 2-trichloro- ; Hydrated Chloral ; U.S.P., B.P., Eur. P., Int. P., Ind. P.

Noctec[R] (Squibb) ; SK-Chloral Hydrate[R] (Smith Kline and French) ; Somnos[R] (MSD)

It is mainly used for nocturnal and preoperative sedation. It is frequently administered in combination with barbiturates to allay anxiety in the first stage of labour. However, it finds an additional use as an adjunct to analgesics and opiates in post-operative medication. It must be avoided in subjects suffering from such diseases as : heat, kidney or liver complications. In patients having gastritis, chloral hydrate may be given through rectum in olive oil as a retention enema. Chloral hydrate is being replaced by new hypnotics with fewer side effects.

Dose : As sedative 250 mg 3 times per day ; as hypnotic 500 mg to 1 g before going to bed.

G. *Carbromal* INN, BAN,

$$\text{H}_2\text{NCONHCOC}(\text{C}_2\text{H}_5)_2$$
$$|$$
$$\text{Br}$$

2-Bromo-2-ethylbutyrylurea ; Bromodiethylacetylurea ; Bromadal ; Uradal ; Karbromal ; B.P., N.F. XI.

Adalin[R] (Winthrop)

It is a weak hypnotic drug and is but infrequently used in modern therapeutics. On account of its inherent weak central depressant properties, its overall action is invariably disappointing and unreliable.

Dose : Usual, oral, 500 mg

3. MODE OF ACTION OF BARBITURATES

In general, the hypnotics may act in **two** different ways, namely ; *first* by exerting their action on the sensory cortex raising the threshold at which it responds to afferent stimuli ; *secondly*, they may interfere with the passage of impulses from the subsidiary centre or centres in the hypothalamus to the cortex. In other words, barbiturates act on the central synaptic transmission process of the reticular activating system.

In normal human being the cerebral electrical activity is directly proportional to anxiety, emotional excitement, or administration of a potent central nervous system stimulant (*e.g.,* caffeine, dexamphetamine, lysergic acid diethylamide LSD to name a few). At this juncture the administration of a reasonably overdose of barbiturates would cause a calming effect, which could be measured demonstrably with the help of an *electroencephalogram,* (EEG). Thus, barbiturates depress the reticular activating system by impairing the synaptic transmission.

In a broader perspective, the **excitatory synaptic transmission** is usually depressed appreciably by *barbiturates,* whereas the *inhibitory synaptic transmission* is normally either increased or unaffected absolutely. Interestingly, *barbiturates* are found to exert *antidepolarizing blocking action* that essentially check the causation of excitatory post synaptic potential not only by enhancing the threshold but also extending the refractory span of the post synaptic cell specifically. Paradoxically, however, the overall effect of *barbiturates* takes place whereby comparatively small doses bring about an unexpectedly marked and pronounced agitation and hyperexcitation rather than the expected sedative effect. This sepcific anomaly could be explained logically based on the fact the concentration of barbiturate is not enough to cause a depression in the reticular activating system, but is just capable of impedementing the inhibitory synapses usually present very much within the cortex. Besides, the *barbiturates* are also found to act on the hypothalamic, limbic, and thalamic synaptic systems.*

4. MECHANISM OF ACTION

The pharmacoloical effects of the barbiturates is invariably marked by a decrease with regard to the normal functional activities in the brain. It has been duly observed that at the prevailing therapeutic dose levels *in vivo* the barbiturates cause a distinct marked enhancement of the GABAergic inhibitory response, in a mechanism very much akin to that shown by the *benzodiazepines,* that is, by influencing conductance at the site of chloride channel. It is pertinent to mention here that at comparatively **higher concentration** barbiturates would display *six* marked and pronounced pharmacological actions :

   (*a*) Potentiation of the $GABA_A$-mediated chloride ion conductance.

   (*b*) Enhancement of binding between GABA and benzodiazepine,

   (*c*) Reduction in glutaminergic transmission,

   (*d*) Uncoupling of oxidative phosphorylation,

---

*Richter JA *et al. Prog. Neurobiol.,* **18** : 275-319, 1982.

**GABA* : gamma-aminobutyric acid.

(*e*) Inhibition of cerebral carbonic anhydrase activity, and

(*f*) Inhibition of the electron-transport system.

Barbiturates, also effect the transportation of carbohydrates, and are observed to enhance the activity of liver microsomal enzymes, responsible for the regulatory mode of several drug substances *via* biotransformation.

5. BARBITURATES *VS* BENZODIAZEPINES

The major glaring difference between the two entirely variant chemically structured compounds being that the barbiturate binding site happens to be altogether different from the benzodiazepines. The '*pharmaceutical elegance*' of benzodiazepines is believed to take place at the picrotoxin-binding site located on the chloride channel. These highly specific and high-affinity binding sites for the benzodiazepines* was adequately established by the aid of **radiolabeled benzodiazepines.** It was revealed that the benzodiazepines normally get bound at the specific $GABA_A$ receptors intimately involved in the regulatory function of the *chloride channel.* However, further extensive and intensive studies substantially suggested two additional subclasses of receptors, now known as BZ_1 and BZ_2 receptors. Interestingly, the presence of two distinct structurally benzodiazepine receptor subclasses were duly ascertained by making use of the classical *recombinant techniques* wherein $GABA_A$ receptor subclasses bearing variable subunit composition, very much similar to the BZ_1 and BZ_2 receptors, were coexpressed magnificently.**

In short, it has been advocated, though not yet proven experimentally, that the BZ_2 subclass of the benzodiazepines is solely responsible for attributing the sedative-hypnotic activity ; whereas the BZ_1 subclass specific compounds would be '*non sedative*' in character.***

Barbiturates, in general, exert a distinct marked depresant activity on the cerebrospinal axis and depress the neuronal performance as well. Besides, these are found to retard the activities of smooth muscle, skeletal muscle and cardiac muscle. Perhaps, this could explain the spectrum of CNS-depression ranging from sedatives, hypnotics, anaesthetics, or even anticonvulsants.

6. STRUCTURE-ACTIVITY RELATIONSHIP

Barbituric acid itself does not possess any hypnotic properties. It is only when the two active hydrogen atoms at position 5 : 5 have the appropriate substituent (*e.g.,* alkyl or aryl groups) that the **'hypnotic activity'** is produced by the compound. The following cardinal points must be taken into consideration with respect to the structure-activity relationship amongst the barbiturates. These are :

(*i*) The total number carbon atoms present in the two groups at carbon 5 must not be less than 4 and more than 10 for the optimal therapeutic results.

(*ii*) Only one of the substituent groups at position 5 may be a closed chain.

(*iii*) The branched chain isomer exhibits greater activity and shorter duration. The greater the branching, the more potent is the drug (*e.g.,* pentobarbital > amobarbital).

---

*Mohler H and Okada T, *Science,* **98** : 849-851, 1977.

Luddens H and Wisden W, *Trends Pharmacol-Sci.,* **12 : 49-51, 1991

***Muller WE, In : Kales A, ed, '**The Pharmacology of Sleep**', springer-verlag, Berlin, **116,** 211-242, 1995.

(*iv*) Double bonds in the alkyl substituent groups produce compounds more readily vulnerable to tissue oxidation. Hence, they are short-acting.

(*v*) Stereoisomers have more or less the same potencies.

(*vi*) Aromatic and alicyclic moieties exert greater potency than the corresponding aliphatic moiety having the same number of carbon atoms.

(*vii*) Short chains at carbon 5 resist oxidation and hence are long-acting. Long chains are readily oxidized and thus produce short-acting barbiturates.

(*viii*) Inclusion of a halogen atom in the 5-alkyl moiety enhances activity.

(*ix*) Inclusion of polar groups (*e.g.,* OH, CO, COOH, NH_2, *R*NH, and SO_3H) in the 5-alkyl moiety reduces potency considerably.

(*x*) Methylation of one of the imide hydrogens enhances onset and reduces duration of action (*e.g.,* the transition from 5, 5-disubstituted to 1, 5, 5-trisubstituted barbituric acid).

(*xi*) Replacement of sulphur for the carbonyl oxygen at carbon 2 results in thiobarbiturates which exhibit rapid onset and short duration of action because these are readily detoxified.

(*xii*) Inclusion of more sulphur atoms (*e.g.,* 2, 4-dithio ; 2, 4, 6-trithio) decreases activity. Likewise introduction of imino group(s) into the barbituric acids abolishes activity (*e.g.,* 2-imino ; 4-imino ; 2, 4-diimino and 2, 4, 6-triimino).

(*xiii*) Replacement of the hydrogen atom at carbon atoms 1 and 3 with an alkyl group increases the vulnerability of the molecule to tissue oxidation.

7. BARBITURATES *Vs* DISSOCIATION CONSTANT (pKa)

It has been adequately substantiated that the presence of 5, 5-disubstituted barbituric acid essentially comprises of *three lactam* moieties which can be circumvented conveniently to pH dependent **lactim-lactam tautomerization** as given under :

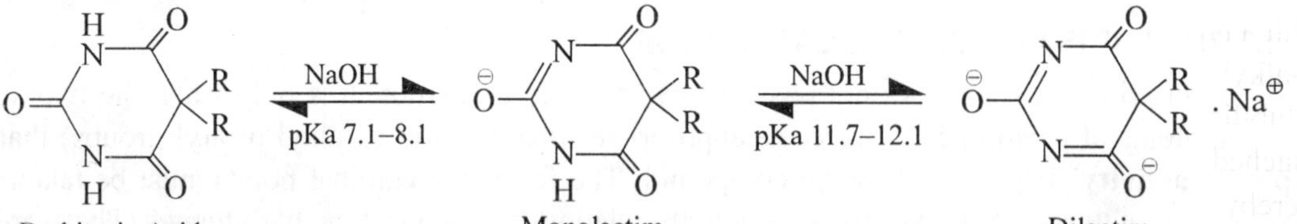

| Barbituric Acid | Monolactim (ionized) | Dilactim (ionized) |

Salient Features

The various salient features are as follows :

(1) 5, 5-Disubstituted barbituric acids* when subjected to UV-spectrophotometric investigations reveal the following facts, namely :

    (*a*) The predominant-forms that usually exist in an aqueous medium are either the *trioxo tautomeric form* (*i.e.,* barbituric acid configuration in an acid medium) or the corresponding *dioxo tautomeric form* (*i.e.,* monolactam in an alkaline medium),

---

*Vida JA. In : Foye WO *et al.* **'Principles of Medicinal Chemistry'**, Williams and Wilkinson, Baltimore, 5th edn, 154-180, 2002.

(*b*) In an aqueous medium the '*acidity*' of the barbiturates is solely guided by the number of substituents attached at C-5 of barbituric acid,

(*c*) The following species, such as : 5, 5-disubstituted barbituric acids ; 5, 5-disubstituted thiobarbituric acids ; and 1, 5, 5-trisubstituted barbituric acids are relatively *weak acids*. Besides, the corresponding salts of these barbiturates are easily produced by interaction with appropriate bases.

(2) The pKa values of 5, 5-disubstituted barbituric acids usually range between 7.1 to 8.1**.

(3) It has been reported*** that 5, 5-disubstituted barbituric acids are capable of undergoing a '*second phase of ionization*' that essentially possess pKa values falling within the range of 11.7 to 12.7.

(4) It has been observed that the '*sodium barbiturates*', in general, exhibit extremely lipophilic property that may cause **distinct and rapid chemical incompatibility reactions,** such as : precipitation, when such compounds are inadvertently brought in contact with the acid salts of relatively weak basic amines.

8. SUBSTITUTIONS ON HETERO ATOMS IN BARBITURATES

The presence of hetero atoms, such as : N and O in the structure of barbiturates play an important role with regard to the wide spectrum of pharmacological activities.

(*a*) **Replacement of Oxygen at C-2.** The replacement of O-atom with an isostere, S-atom, at C-2 position of the barbiturates significantly enhances the lipid solubility profile. The resulting modified versions of the barbiturates thus obtained exert a rapid onset of activity by virtue of the fact that they attain maximal thiobarbiturate-brain levels. Therefore, such drugs as '**thiopental sodium**' find their profuse and abundant application as '*intravenous anaesthetics*'.

(*b*) **Alkyl group substitution on N^1, and /or N^3.** The careful replacement of one *imide-hydrogen atom* either on N^1 or N^3 by alkyl moieties enhances the lipid solubility profile appreciably. The ultimate outcome of such minor structural modification(s) usually give rise to a rapid onset of action but with a reasonable shorter duration of activity. Importantly, an increase in the size and magnitude of the N-alkyl substituent, for instance : *methyl, ethyl* and *propyl*, result into products with increasing lipid solubility and decreasing hydrophilic character. Furthermore, no sooner an alkyl functional moiety is attached to the N^1 or N^3 atom the resulting barbiturate gets converted into a non acidic drug molecule thereby rendering it drastically *inactive*. Therefore, focussed modifications at these two pivotal positions (N^1 and N^3) are normally of cardinal significance and enormous potential in the design of barbiturates intended to be utilized as anaesthetics and anticonvulsants.

9. OH⁻ CATALYZED DEGRADATION OF BARBITURATES

Importantly, barbiturates posed certain serious hydrolytic problems with regard to their incorporation in the liquid dosage formulations, such as : parenterals and elixirs. It has been duly observed that '**acid solvolysis**' is not the root cause of such encountered problems : whereas the major culprit being the *OH⁻ catalyzed degradation* of the *ureide rings* imbedded in the barbiturates.

---

*Butler TC *et al. J. Am. Chem. Soc.,* **77** : 1488-1491, 1955.

Drayton CJ, ed. In : Hansch C *et al.* eds, **Comprehensive Medicinal Chemistry, Pergamon Press, New York, 1990.

Example. *Phenobarbital [Gardinal^(R)].* The following scheme vividly explains the base hydrolysis of phenobarbital wherein the cyclic ureide ring (in barbiturate) undergoes cessation. Besides, it may also be seen that the aforesaid cessation strategically takes place either between C-1/C-2 and/or C-1/C-6 locations in the structure of barbiturate. However, the cleavage between C-1 and C-6 is considered to be the most preferred pathway prevailing in the *'ionized barbiturates'*, such as : aqueous solutions of sodium salts.

Therefore, to pervent the said cleavage between C-1 and C-6 in phenobarbital it is absolutely necessary to *'stabilize'* its liquid dosage forms (*e.g.,* elixir) at pH 6.0. At this specific pH the resulting *'hydroalcoholic solution'* remains fairly stable.*

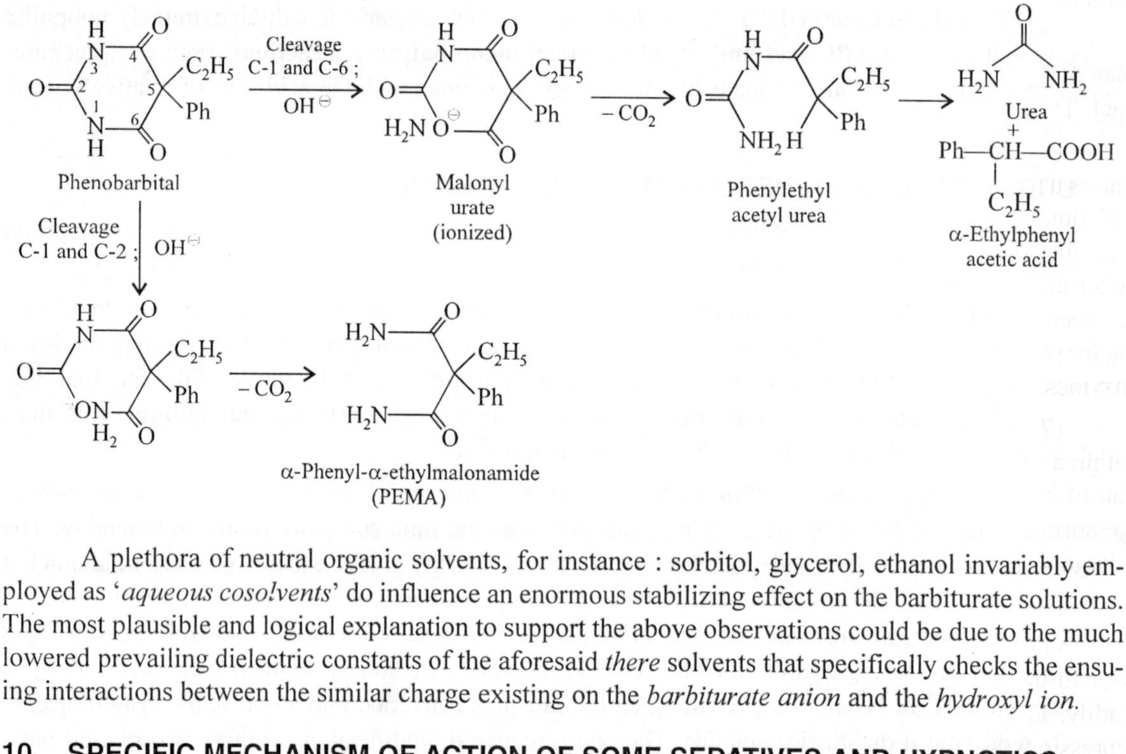

A plethora of neutral organic solvents, for instance : sorbitol, glycerol, ethanol invariably employed as *'aqueous cosolvents'* do influence an enormous stabilizing effect on the barbiturate solutions. The most plausible and logical explanation to support the above observations could be due to the much lowered prevailing dielectric constants of the aforesaid *there* solvents that specifically checks the ensuing interactions between the similar charge existing on the *barbiturate anion* and the *hydroxyl ion.*

10. SPECIFIC MECHANISM OF ACTION OF SOME SEDATIVES AND HYPNOTICS

In this section the *'mechanism of action'* of certain drugs already discussed in this chapter shall be explained in *two* categories, namely : (*a*) Barbiturates ; and (*b*) Non-barbiturates.

[A] **Barbiturates.** The various barbiturates described are as under :

(1) **Phenobarbital.** It gets metabolized upto 65%, mostly to the inactive *para*-hydroxyphenyl derivative ; and upto 35% gets exerted by the kidney totally in unchanged form. However,, the plasma clearance is rather slow and generally approximates to 0.004 L kg^{-1} hr^{-1}. The apparent volume of distribution stands at 0.7 to 1 L kg^{-1} ; and the therapeutic plasma levels range between 10 to 30 mcg. mL^{-1}. Nearly 45 to 50% of phenobarbital is bound to plasma protein. Consequently, the plasma half-life varies from 50 to 120 hr in adults, whereas in children it ranges between 40 to 70 hr.

---

*A 15-20% (*v/v*) ethanol exerts a *'stabilizing effect'*.

(2) **Methylphenobarbital (Mephobarbital).** It is anticonvulsant and sedative of the barbiturate class ; and in therapeutic dose levels it exerts practically no significant effect on the modullary centres thereby causing no apperciable alteration either in the blood pressure or the rate of respiration. Hence, the overall action is strong sedative and anticonvulsant actions but having a comparatively quite mild hypnotic activity.

(3) **Ammobarbital.** It has seven total C-atoms duly substituted on the C-5 position, which particularly enables, its ability to penetrate the liver microsomes. Besides, for such hydrophobic drug substance the legitimate partioning out of the brain to other relevant sites may also be involved.

(4) **Pentobarbital sodium.** This specific barbiturate is considered to effectively lower the cerebral blood flow, and thereby minimise substantially either oedema and/or intracranial perssure.

(5) **Quinalbarbitone sodium (Secobarbital Sodium).** It has been observed that within a short span of 2 hours after the oral administration, approximately 90% gets adequately absorbed from the GI tract. The elimination half-life is nearly 30 hours.

(6) **Methohexital sodium.** In this particular instance an induction dose of 1 mg kg^{-1} reliably causes unconsciousness in just 30 seconds. As a results pharmacological effect gets instantly terminated with quick distribution from the brain to the corresponding peripheral sites. Consequently, the recovery from the brain to the corresponding peripheral sites. Consequently, the recovery from methohexital is rather more rapid and there is less myocardial depression observed than with thiopental. Methohexital has been employed extensively to elicit spiking discharges on the EEG* in patients undergoing screening for seizure activity. It is exclusively metabolized in the liver thereby causing induction of cytochrome enzymes.

(7) **Thiopental sodium.** A single induction dose of 3 to 5 mg kg^{-1} may cause unconsciousness within a short span of 30 to 40 seconds. Its action is, however, terminated by the immediate redistribution of drug away from the brain. It has been observed that there exists a transient decrease in blood pressure (20%), and a simultaneous compensatory enhancement in the heart rate on injection. Thiopental is largely metabolized in the liver, although the kidney and muscle tissue may participate concurrently.

[B] **Non-Barbiturates :** The *five* non-barbiturates discussed are as stated below :

(1) **Methaqualone.** It may cause acroparesthesia (tingling and numbers in the extremities) just prior to the onset of the hypnotic activity, specifically in such situations when sleep does not ensue readily. Large doses of methaqualone do not give rise to significant respiratory and cardiovascular depression. **Mandrax**, a combination of methaqualone and diphenylhydramine , has been profusely abused by addicts in Australia, Canada and United Kingdom. In fact, methaqualone has now been banned altogether in several countries, including the United States (1984), by virtue of its abuse potential figured out to be much larger and dangerous to the '*society*' as compared to the corresponding short-acting barbiturates.

(2) **Glutethimide.** This particular *imide* essentially possess several structural relationships with the '*barbiturates*', and hence resembles the latter in many aspects pharmacologically. Based on its severe hydrophiobic nature the absorption from the GI-tract is found to be erratic to a certain extent. It exhibits an extensive metabolism, and the drug is established to be an '*enzyme inducer*'. Its oral absorption is quite variable, having peak plasma level lasting between 1 to 6 hours. It is observed to induce

---

*EEG : Electroencephalogram ;

liver microsomal enzymes ; therefore, it is absolutely important and necessary for patients on coumarin-anticoagulant therapy for readjustment of the coumarin dose level both during and on discontinuation of such therapy. Its elimination half-life varies from 5 to 22 hours, having an average value of 11.6 hours.

(3) **Methyprylon.** Interestingly, the two **piperidinediones** *viz.,* glutethimide and methyprylon ; and the **quinazolone** *viz.,* methaqualone, are of vital importance within the context of non-barbiturate sedatives and hypnotics. However, *glutethimide* and *methyl prylon* possess a striking resemblance to barbiturates as given below :

| Glutethimide | Methyprylon | Barbiturate |
| [Doriden[(R)]] | [Noludar[(R)]] | [General Structure] |

The metabolism of methyprylon is found to be extensive, and the drug is an enzyme inducer. Its pharmacological effects resemble to those of the barbiturates.

(4) **Paraldehyde.** It is chiefly detoxified by the liver upto 70–80% and is excreted by the lungs (exhalation) upto 11–28%. It has been observed that only a negligible amount is excreted in the urine. Because the drug has a peculiar strong characteristic smell invariably and prominently detectable in the exhaled air from the lungs along with an '*unpleasant palate*' its usage has been almost restricted exclusively to an institutional setting, such as : in the treatment of **delirium tremens.***

(5) **Nitrazepam.** Generally, the various aromatic nitro functional containing drug substances undergo enzymatic reduction to the corresponding *amines*. Therefore, nitrazepam, a 7-nitrodiazepine structural analogue gets metabolized extensively to their respective 7-amino metabolites in humans.**

Probable Questions for B. Pharm. Examinations

1. Explain how the '*cyclic ureide*' barbituric acid may by prepared from :

    (*a*) Urea and malonyl dichloride

    (*b*) Urea and diethyl malonate.

2. Keto-enol tautomerism exists in parabenic acid and barbituric acid. Explain.

3. Why the 'thiobarbiturates' get metabolized *in vivo* faster than the barbiturates ? Explain.

4. How the long-active barbiturate '*mephobarbital*' can be synthesized from :

    (*a*) Urea, and

    (*b*) Dicyandiamide

5. Introduction of two similar allyl functions at C-5 in the barbituric acid yields an intermediaet acting barbiturate. Name the product and give its synthesis.

---

*A dramatic complication of alcoholism.

Rieder J. and Wendt G. : In Garalini S. *et al.* (eds). **The Benzodiazepines, Raven Press, New York, p-99, 1973.

6. A class of barbiturates find their usuage in the treatment of insomnia and preoperative medication. Name two potent drugs and discuss the synthesis of **one** compound.

7. A brand of barbiturates are usually employed to cause general anaesthesia and control convulsions. Discuss one potent member and explain how it is metabolised *in vivo*.

8. Based on Quinozolinone nucleus a potent hypnotic drug was introduced which subsequently was withdrawn because of its 'abuse'. Name the compound and give its structure.

9. Give the structure, chemical name and uses of a potent non-barbiturate drug having a benzodiazepin nucleus.

10. Explain the following with suitable examples :

   (*a*) Mode of Action of Barbiturates

   (*b*) Structure-activity relationship amongst barbiturates.

RECOMMENDED READINGS

1. WJ Doran Barbituric Acid Hypnotics in : *Medicinal Chemistry* Vol. 4 (*Eds.*) FF Blicke and R H Cox New York, John Wiley and Sons (1959).

2. CC Cheng and B Roth in : *Progress in Medicinal Chemistry* (*Eds.*) G P Ellis and G B West, New York, Appleton-Century-Croft (1971).

3. L Cook and J Sepinwall in : *Mechanism of Action of Benzodiazepinse* (*Eds.*) E. Costa and P. Greengard, New York, Raven Press (1975).

4. FH Clarke (*Ed*) *Ann Rep Med Chem* Vol. 12, New Yrok, Academic Press (1977).

5. LH Sternback *et al* in : Drugs Affecting the Central Nervous System, Vol. 2 (*Ed* A Burger) New, York, Marcel Dekker (1968).

6. EW Maynert and HB Van Dyke in : *The Metabolism of Barbiturate Pharmac Rev* (1949).

7. *FM Berger in Spinal Cord Depressants,* Pharmac Rev (1949).

8. Dyson and May *May's Chemistry of Synthetic Drugs* London, Longmans (1959).

9. USAN and the USP Dictionary of Drug Names, Rockville (USA), United States Pharmacopeial Convention, Inc., (1985).

10. Martindale *The Extra Pharmacopoeia* (30th edn.) London, The Pharmaceutical Press (1992).

11. JN Delgado and W A Remers : Wilson and Gisvold's Textbook of Organic and Medicinal Chemistry, (10th edn), Philadelphia, JB Lipincott Company, (1998).

12. D Lednicer and LA Mitscher : *The Organic Chemistry of Drug Synthesis* New York, John Wiley and Sons (1995).

13. Gringauz A : Introduction to Medicinal Chemistry, Wiley-VCH, New York, 1997.

<table>
<tr><td>**6**</td><td></td></tr>
</table>

Anticonvulsants

1. INTRODUCTION

Epilepsy is an age-long disease which often involves convulsive seizures. Hippocrates first coined the word '*epilepsy*' derived from the Greek word *epilambanein* (to seize). The common symptoms of epilepsy were not unfamiliar even in the earliest annals of medical lore. Because of the fact that these seizures are usually accompanied by both change in the rate and force of the electric pulsation of the cerebral cortex, epilepsy is supposed to be a condition of *paroxysmal cerebral dysrhythmia*. The principal types of epilepsy are as follows :

Grand mal is normally characterized by complete loss of consciousness, followed by transient muscular rigidity (tonic phase) and ultimately plunges into violent clonic convulsions embracing all voluntary muscles.

In epilepsy of the *petit mal* type usually momentary loss of consciousness prevails. This particular state is free of convulsions. However, occasionally blinking movements of the eyelids and jerking movements of the head or arms are observed. It may be pertinent to mention here that this kind of epilepsy is more frequently seen in adolescence.

Psychomotor epileptic seizures normally display outbursts of temper, tantrums, mental apathy and sudden irrational and destructive attitude.

Myoclonic seizures is usually characterized by a rapid rhythmic movement of one side of the palate.

The '*anticonvulsants*' are also termed as '*antiepileptic drugs*' and '*antiseizure drugs*' that are used invariably in the adequate and impressive control and management of CNS disorders essentially characterized by recurrent transient attacks of disturbed brain function which ultimately give rise to motor (convulsive), sensory (seizure), and psychic sequence of events.

In fact, there are *two* school of thoughts who opined the *causes of epilepsy**, namely :

(*a*) Epilepsy being a single disease entity ; and, therefore, all variants of it usually possess a common cause, and

(*b*) Different kinds of epilepsy originate from various anatomic, chemical, and functional imbalance (disorders).

---

*Forster FM. ed. **Report on the Panel on Epilepsy,** WI Univ. of Winconsin Press, Madison, p. 91, 1961.

However, a generalized concerted opinion concluded at the **Symposium on Drug Therapy in Neurologic and Sensory Disease,** was that

"epilepsy is a symtom complex characterized by recurrent paroxysmal aberrations of brain functions, usually brief and self limited".

Jackson sometimes in early 19th century legitimately postulated and adequately substantiated the root cause or genesis of the *'seizures'*.* An intense discharge of *'gray matter'* inside the different portions of the brain virtually kicks off the epileptic seizures. Consequently, it is only a normal response of the brain to initiate the phenomenon of convulsive seizures. Precisely, the ensued discharge of an excessive electrical (nervous) energy has virtually been substantiated by an extensive and intensive brain-wave investigations that was made feasible ultimately by the aid of electroencephalography (EEG).

Interestingly, the establishment of the '**Commission on Classification and Terminology of the International League Against Epilepsy**' in the year 1981 came up with an altogether *new proposal.*** In reality this proposal, advocates a classification of *'epileptic seizures'* exclusively based on *two* cardinal clinical seizure variants commonly observed in patients, namely :

(*a*) **ictal** (seizure induced) electroencephalographic (EEG) expression, and

(*b*) **interictal** (those taking place between attacks or paroxyms) EEG expression.

2. CLASSIFICATION

The various **anticonvulsant drugs**, containing essentially the ureide structure (Table 6.1), belong to the chemical categories of barbiturates, hydantoins, oxazolidinediones and succinimides and lastly the miscellaneous types of compounds possessing such characteristics.

2.1 Barbiturates

Barbiturates as a class of durgs mostly possess sedative and hypnotic properties. Surprisingly only a few of them really show anticonvulsant characteristics. Among the most common barbiturates generally employed as anticonvulsants in clinical use are namely : phenobarbital, mephobarbital and methabarbital (discussed in the chapter on 'Sedatives and Hypnotics'); of which phenobarbital is the drug of choice and is used virtually in all the three types of epileptic seizures *viz.*, grand mal, petit mal and psychomotor.

Table 6.1. Anticonvulsant drugs containing the ureide moiety

Ureide Moiety

---

*Jackson JH ; In : **Selected Writings of John Hughlings Jackson**, Vol.1., Taylor IJ, (ed). Hodder and Stoughton, London, 1931.

**Proposal for Revised Clinical and Electrocephalographic Classification of Epileptic Seizures, Epilepsia, 22, 489 (1981).

Group of Compounds	'X'
Barbiturates	＼NH—C〈=O
Hydantoins	＼NH／
Oxazolidinediones	＼O／
Succinimides	＼CH₂／

Phenobarbital gets metabolized *in vivo* in two different stages ; *first,* by hydroxylation to 5-*p*-hydroxyphenyl-5-ethylbarbituric acid ; and *secondly*, this gets conjugated with either glucuronic acid or sulphuric acid. The metabolic product is ultimately excreted through the urine as its corresponding glucuronide or sulphate salts.

By virtue of its inherent liver-enzyme-inducing characteristic phenobarbital helps in enhancing the metabolism of such drugs that are usually metabolized by the microsomal enzymes.

Mephobarbital loses N-methyl group through metabolism and gets readily converted to phenobarbital.

Methabarbital is mostly demethylated to barbital *in vivo*. Also it possesses more sedating property than phenobarbital, it could be safely recommended for grand mal seizures.

2.2 Hydantoin Derivatives

The following five-membered heterocyclic ring, **hydantoin** is present in the following *three* compounds, namely :

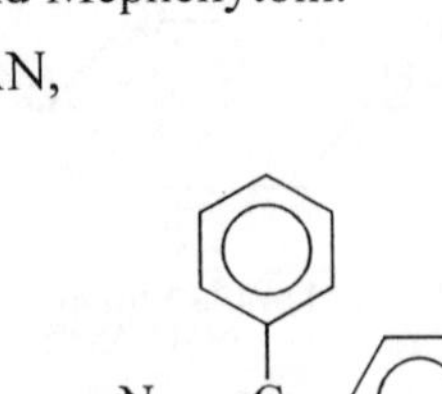

Phenytoin Sodium ; Ethotoin ; and Mephenytoin.

A. *Phenytoin Sodium* USAN, BAN,

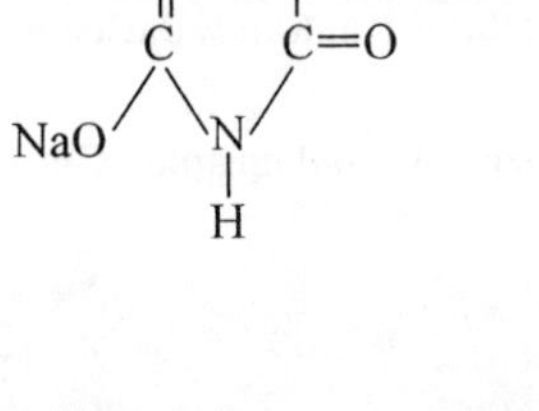

5, 5-Diphenylhydantoin sodium salt ; 2, 4-Imidazolidinedione, 5, 5-diphenyl-monosodium salt ;
Diphenylhydantoin Sodium ; Diphenin ; Soluble Phenytoin U.S.P., B.P., Eur. P., Ind. P.

Diphentoin$^{(R)}$ (Beecham) ; Epanutin Infatabs$^{(R)}$ (Parke-Davis, U.K.) ;

Synthesis :

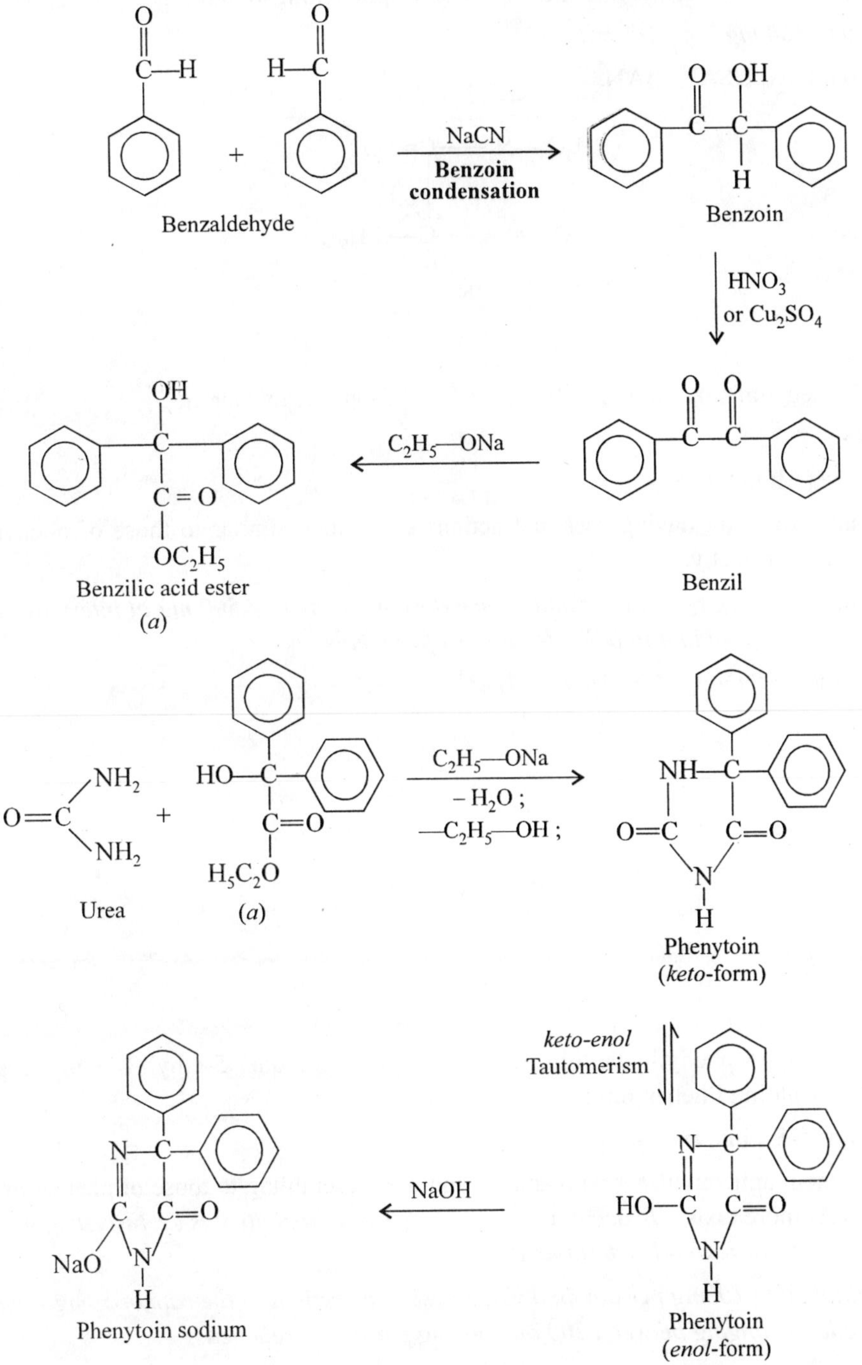

Benzaldehyde   +   NaCN **Benzoin condensation** →   Benzoin

HNO$_3$ or Cu$_2$SO$_4$

Benzil

C$_2$H$_5$—ONa

Benzilic acid ester
(*a*)

Urea   +   (*a*)   C$_2$H$_5$—ONa
− H$_2$O ;
—C$_2$H$_5$—OH ;   →   Phenytoin
(*keto*-form)

keto-enol Tautomerism

Phenytoin
(*enol*-form)

NaOH

Phenytoin sodium

It is prepared by the condensation of two molecules of benzaldehyde with sodium cyanide to get benzoin, which on treatment with nitric acid or cupric sulphate forms benzil. In the presence of sodium ethoxide, benzil in hot condition yields benzilic acid ester. (*a*) The latter on condensation with urea in the presence of sodium ethoxide give rise to phenytoin ; the *enol*-form of which on neutralization with sodium hydroxide ultimately results into the formation of phenytoin sodium.

Dose : 50 to 100 mg

B. *Ethotoin* INN, USAN, BAN,

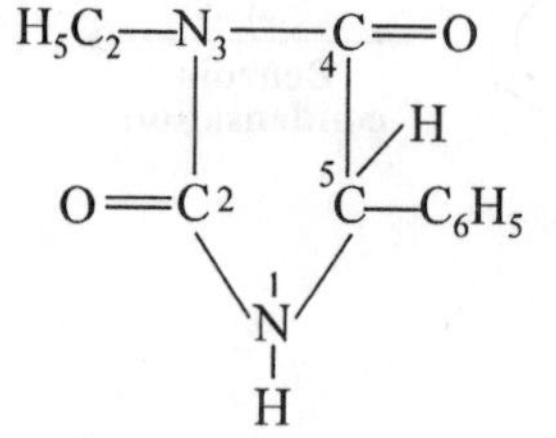

3-Ethyl-5-phenylimidazolidin-2, 4-dione ; 3-Ethyl-5-phenyl hydantoin ;

B.P. 1973 ;

Peganone[R] (Abbott, U.K.)

It is an anticonvulsant having uses and actions very much similar to those of phenytoin, but comparatively it is less effective.

Dose : Initial dose 1g per day ; maintenance dose increased by 500 mg at intervals of several days to 2 to 3g per day, given in 4 to 6 divided doses after meals.

C. *Mephenytoin* INN, USAN, *Methoin* BAN,

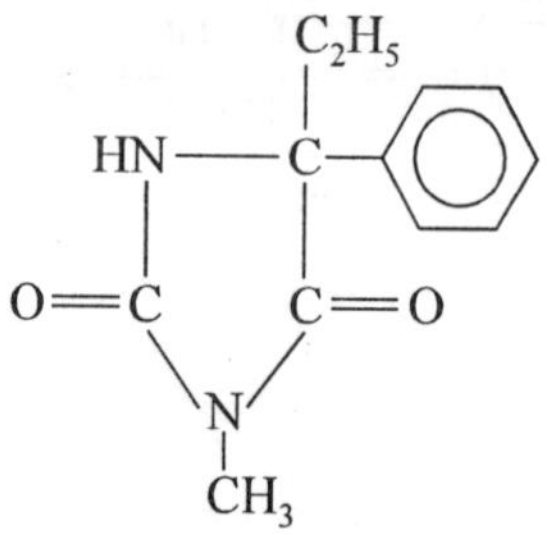

5-Ethyl-3-methyl-5-phenylhydantoin ; 2, 4-Imidazolidinedione, 5-ethyl-3-methyl-5-phenyl ; Mephenetoin ; Methantoin ; Phenantoin ;

B.P. 1973, U.S.P., Ind. P.

It is a hydantoin anticonvulsant with actions and uses resembling to those of phenytoin but it is found to be relatively more toxic. *On account of its high degree of toxicity it is exclusively given to such patients who do not response to other treatments.*

Dose : Initial 50 to 100mg per day in divided doses ; increased at the rate of 50mg weekly until the optimum dose level ranging between 200 and 600mg per day is reached.

2.3. Oxazolidinediones

The ureide function present in **oxazolidine-2, 4-dione** depicts a close resemblance to hydantoin, differing only in the replacement of NH moiety in the latter by an oxygen atom at C_1.

A few important members of this class of compounds are discussed below :

A. *Trimethadione* INN, USAN, *Troxidone* BAN,

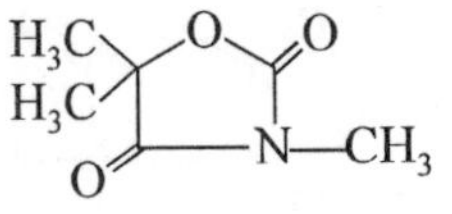

3, 5, 5-Trimethyl-2, 4-oxazolidinedione ; 2, 4-Oxazolidinedione, 3, 5, 5-trimethyl ;

B.P., U.S.P., Int. P., Ind. P.

Tridione[R] (Abbott) ;

Synthesis :

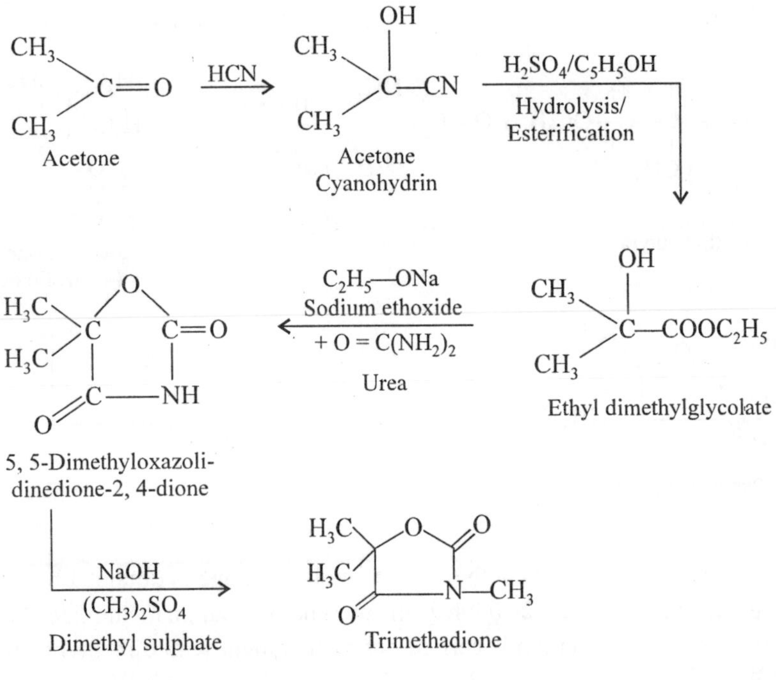

Acetone cyanohydrin is obtained from the interaction of acetone and hydrocyanic acid, which on subsequent hydrolysis followed by esterification with ethanol yields ethyldimethylglycolate. This is condensed with urea in the presence of sodium ethoxide to yield 5, 5-dimethyl-oxazolidinedione-2, 4-dione. Trimethadione is finally obtained by treating the resulting product with dimethyl sulphate in the presence of sodium hydroxide.

It is employed as an anticonvulsant in grand mal epilepsy to arrest status epilepticus and in petit mal epilepsy as a means of resisting control to other treatments. When administered alone trimethadione may not be effective to contain the situation in petit mal epilepsy, but it may be useful when given in conjunction with phenytoin sodium and/or phenobarbital in petit mal seizures. It may be used occasionally in the treatment of behaviour problems encountered in children, status epilepticus and athetoses.

Dose : 900 mg to 2.4g per day ; usual, 300 to 600 mg 2 to 4 times daily.

B. *Paramethadione* INN, BAN, USAN,

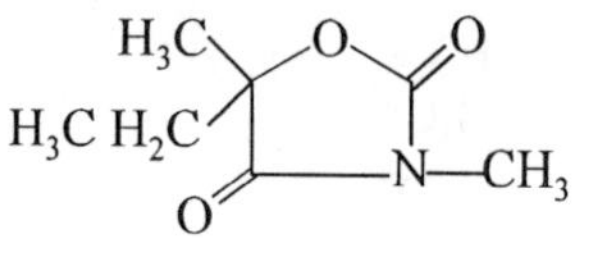

5-Ethyl-3, 5-dimethyl-2, 4-oxazolidinedione ; 2, 4-Oxazolidinedione, 5-ethyl-3, 5-dimethyl- ; Paramethad ;

B.P. 1973, U.S.P., Ind. P.

Paradione[R] (Abbott) ;

Synthesis :

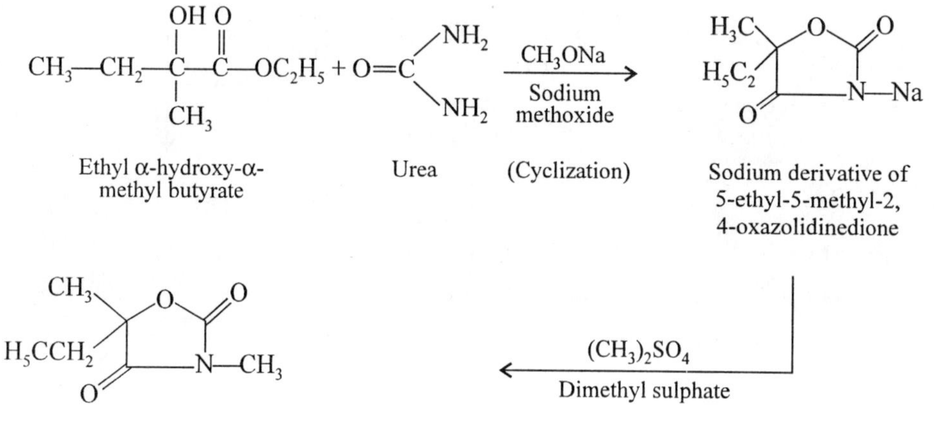

Ethyl α-hydroxy-α-
methyl butyrate   Urea   (Cyclization)   Sodium derivative of
                                                                        5-ethyl-5-methyl-2,
                                                                        4-oxazolidinedione

Paramethadione

Sodium derivative of 5-ethyl-5-methyl-2, 4-oxazolidinedione is obtained by refluxing urea and ethyl-α-hydroxy-α-methylbutyrate for 24 hours in the presence of sodium methoxide, due to condensation followed by cyclization. N-methylation is carried out by treatment with dimethyl sulphate.

It is usually used in the treatment of petit mal epilepsy and possesses similar actions to those of trimethadione. It is relatively less effective and does not exhibit *myasthenia-gravis-like syndrome* in patients.

Dose : 300 mg to 2.4g daily ; usual, 300 mg 3 to 4 times per day.

2.4. Succinimides

Due to the inherent high-level of toxicity attributed by the oxazolidinediones in prolonged therapy as anticonvulsants a vigorous attempt was made to replace them with better effective and less toxic drugs.

Three members of this class of compounds were introduced between early fifties to late fifties, namely ; Phensuximide, Methsuximide and Ethosuximide. All of them gained with acceptance for the treatment of petit mal seizures specifically.

A. *Phensuximide* INN, BAN, USAN,

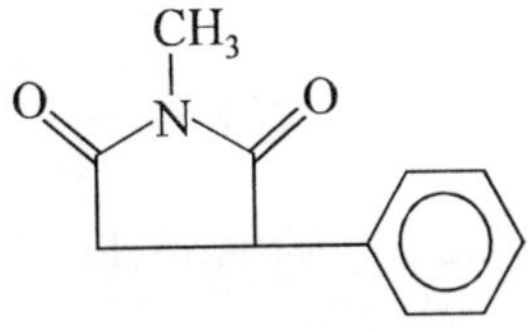

N-Methyl-2-phenylsuccinimide ; 2, 5-Pyrrolidinedione, 1-methyl-3-phenyl ; Fensuximid ;

B.C.P. 1973, U.S.P.

Milontin[R] (Parke-Davis)

Synthesis :

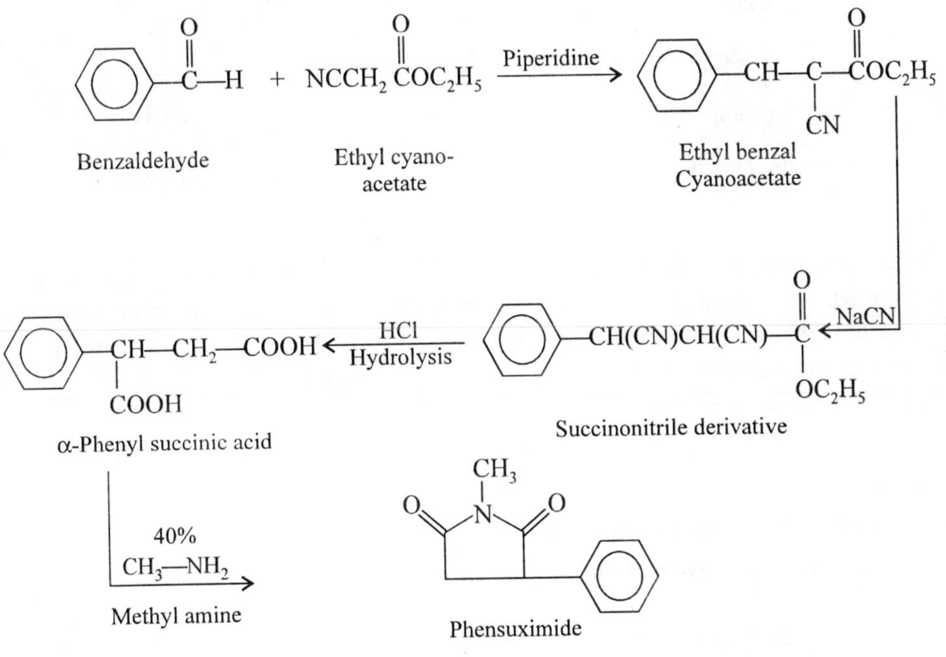

Ethylbenzal cyanoacetate is prepared by reacting together benzaldehyde and ethyl cyanoacetate in the presence of piperidine, which on treatment with sodium cyanide yields the corresponding succinonitrile derivative. This on hydrolysis in an acidic medium (HCl) gives rise to α-phenyl succinic acid which is dissolved in an excess of 40% methylamine to give phensuximide.

It is frequently employed in the treatment of petit mal epilepsy. The arrest of such convulsive seizures is normally caused by depression of the motor cortex together with significant elevation of the threshold of the central nervous system to convulsive stimuli. It is relatively less effectiev than paramethadione and trimethadione.

Dose : 500 mg to 1g 2 to 3 times per day for any age.

B. *Mesuximide* INN, *Methsuximide* BAN, USAN,

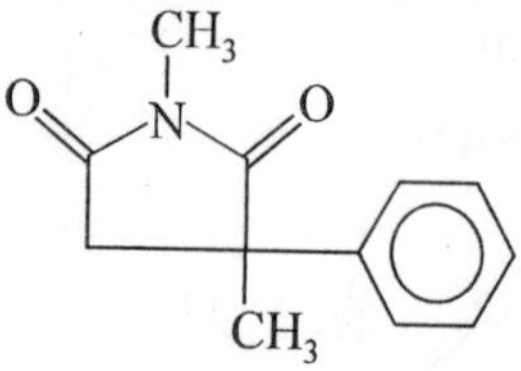

N, 2-Dimethyl-2-phenylsuccinimide ; 2, 5-Pyrrolidinedione, 1, 3-dimethyl-3-phenyl-,
U.S.P.,

Celontin[R] (Parke-Davis)

Synthesis :

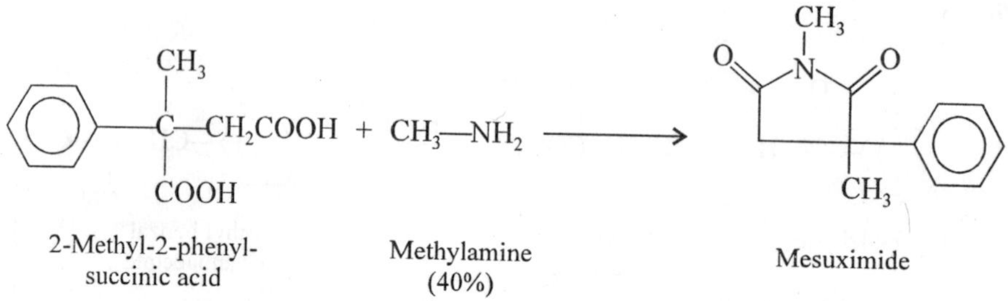

| 2-Methyl-2-phenyl-
succinic acid | Methylamine
(40%) | Mesuximide |

It is conveniently prepared by the interaction of 2-methyl-2-phenyl succinic acid with excess of
40% methylamine. The excess of amine and water are distilled off under reduced pressure. The residue
containing the dimethylamine salt of the acid is pyrolyzed at 250°C until no more distillate is obtained.
The crude product is dissolved in an appropriate solvent, treated with activated charcoal and finally
precipitated by the addition of water.

It is found to be more effective and potent than phensuximide in the cure of petit mal epilepsy
and psychomotor seizures.

Dose : Initial, 300 mg per day ; maintenance 0.3 to 1.2g daily.

C. *Ethosuximide* INN, BAN, USAN,

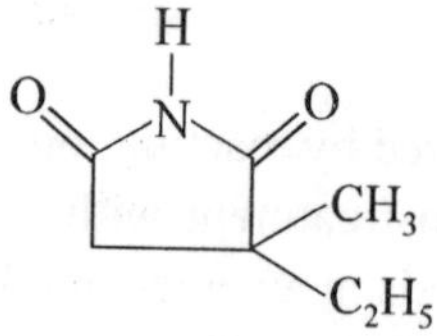

2-Ethyl-2-methylsuccinimide ; 2, 5-Pyrrolidinedione, 3-ethyl-3-methyl ;

B.P., U.S.P.

Zarontin[R] (Parke-Davis)

Synthesis :

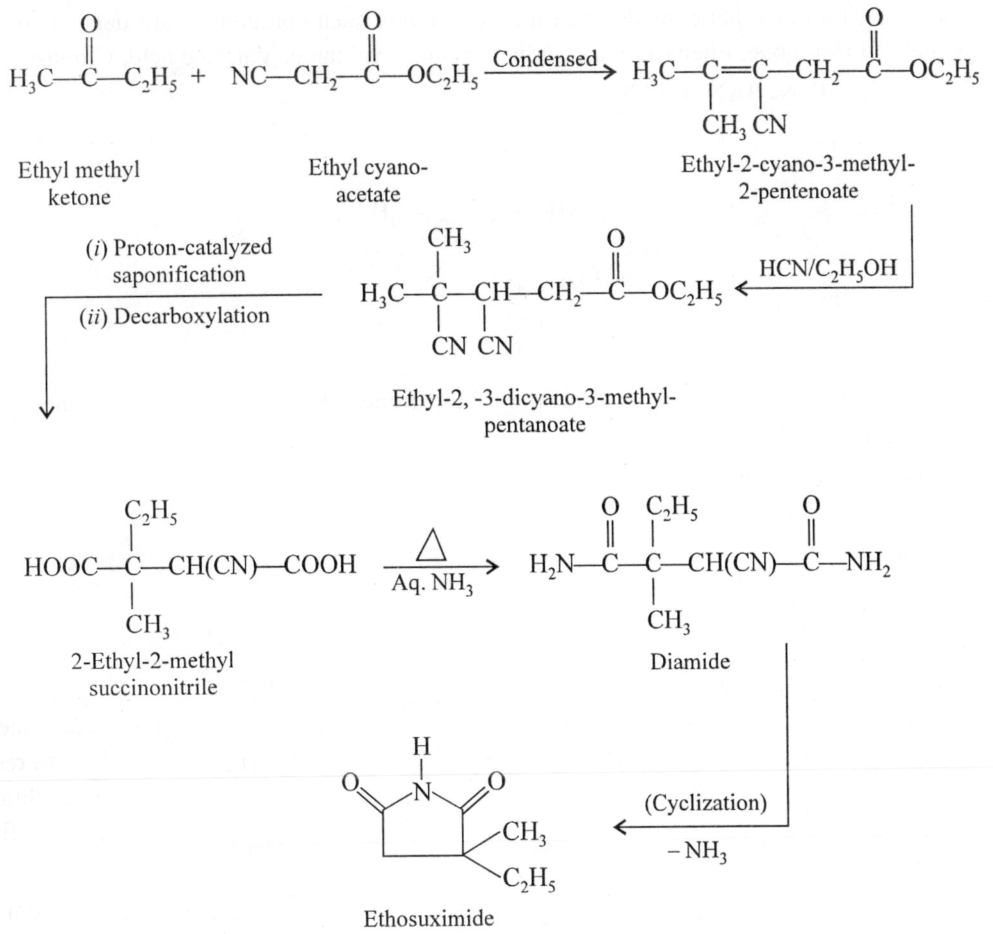

Ethyl 2-cyano-3-methyl-2-pentenoate is first prepared by the condensation of ethyl methyl ketone and ethyl cyanoacetate, which in ethanolic solution takes up a molecule of hydrogen cyanide according to *Markownikoff's rule* to yield ethyl-2, 3-dicyano-3-methyl pentanoate. This upon protoncatalyzed saponification and subsequent decarboxylation gives rise to 2-methyl-2-ethyl succinonitrile, which on heating with aqueous ammonia produces the corresponding diamide. Finally, the diamide undergoes cyclization, through loss of ammonia to give ethosuximide.

It has the reputation for being the most effective succinimide analog in petit mal therapy. It acts by suppressing the EEG pattern of petit mal epilepsy perhaps by depression of the motor cortex and raising the convulsive threshold. Combined drug therapy with either phenobarbital or phenytoin sodium is common in cases where petit mal co-exists with grand mal or other types of epilepsy.

Dose : 500 mg per day, in divided doses.

2.5. Miscellaneous

A number of compounds bearing different miscellaneous basic chemical structures have been found to possess significant anticonvulsant properties. A few such compounds are dealt with in this section, namely ; Primidione, Phenacemide, Carbamazepine, Sultiame, Valproic acid, Clonazepam.

A. *Primidone* INN, BAN, USAN,

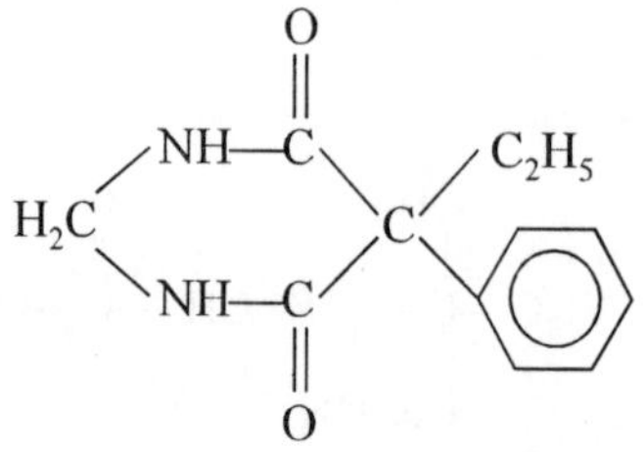

5-Ethyldihydro-5-phenyl-4, 6(1H, 5H)-pyrimidinedione ; 4, 6 (1H, 5H)-Pyrimidinedione, 5-ethyldihydro-5-phenyl- ; Primaclone ;

B.P., U.S.P., Int. P. Ind. P.

Mysoline[R] (Ayerst)

Synthesis :

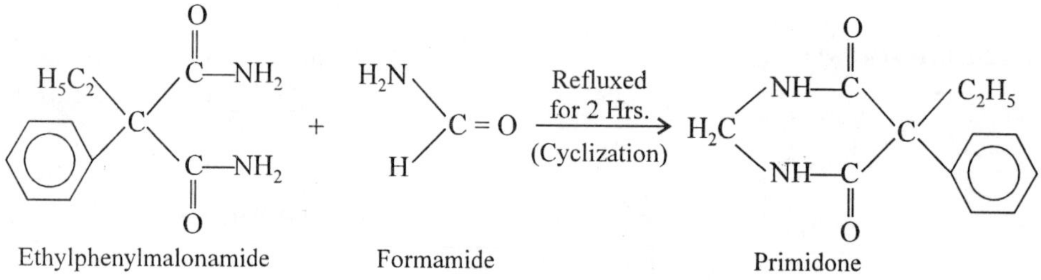

It may be prepared by refluxing together a solution of ethylphenyl-malonamide with a large molar excess of formamide for 2 hours. The probable mechanism of this cyclization may be viewed to have taken place at three different stages, namely : Cannizzaro type of disproportionation of formamide, deammoniation and dehydration between ethylphenylmalonamide and the resulting highly virile methanolamine.

Primidone is a potent anticonvulsant which may be chemically regarded as a 2-deoxy analogue of phenobarbital. It is used either in conjunction with other antiepileptics or alone in the treatment and arrest of psychomotor, grand mal and focal epileptic seizures.

Dose : 500 mg per day, gradually increasing to a maximum of 2g daily.

B. *Phenacemide* INN, BAN, USAN,

Phenylacetyl urea ; Benzeneacetamide, N-(aminocarbonyl)- ;

U.S.P. Phenurone[R] (Abbott)

Synthesis :

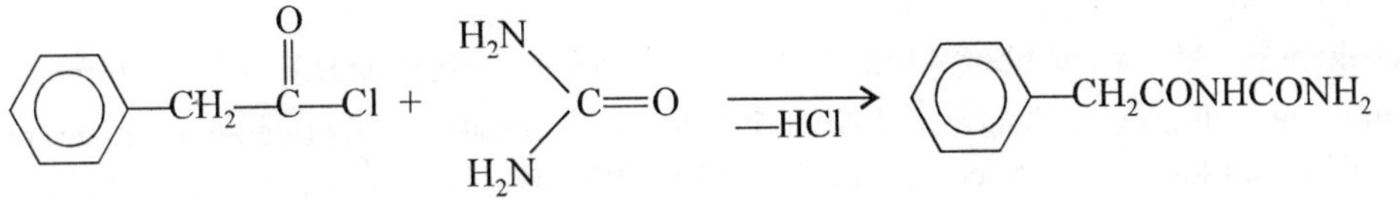
CH2—C—Cl +
O
H2N
C=O
H2N
—HCl
CH2CONHCONH2

Phenylacetyl chloride Urea Phenacemide

It is readily prepared by reacting phenylacetyl chloride and urea with the elimination of a molecule of hydrogen chloride.

It is mainly used in the psychomotor type of seizure ; it is of relatively lesser therapeutic value in petit mal, grand mal and in mixed seizures.

Dose : 0.5g, oral, 3 times daily with meals.

C. *Carbamazepine* INN, BAN, USAN,

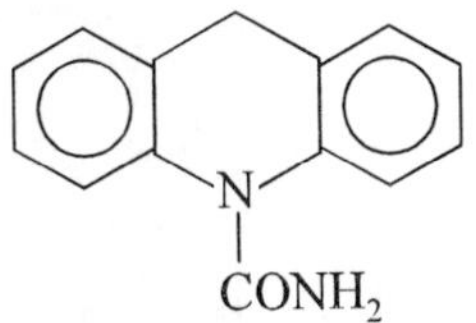
N
CONH2

5H-Dibenz [*b, f*] azepine-5-carboxamide ;

B.P., U.S.P.

Tegretol(R) (Ciba-Geigy)

Carbamazepine essentially contains a dibenzazepine ring system with a carbamoyl moiety hooked on to the nitrogen atom.

It is an anticonvulsant employed in the treatment of grand mal and psychomotor epilepsy. It is considered to be one of the most vital drugs for the relief of pain associated with trigeminal neuralgia.

Dose : 200 mg per day, increasing to 1.2g daily, in divided doses.

D. *Sultiame* INN, *Sulthiame* BAN, USAN,

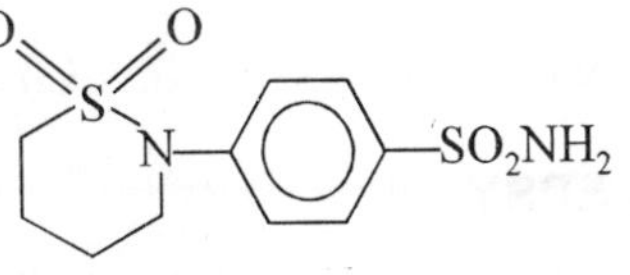
O
O
S
N
SO2NH2

p-(Tetrahydro-2H-1, 2-thiazin-2-yl)-benzenesulphonamide, S, S-dioxide ; Benzenesulfonamide, 4-(tetrahydro-2H- 1, 2-thiazin-2-yl)-, S, S-dioxide

B.P. Conadil(R) (Riker) ; Ospolot(R) (Bayer)

It is a carbonic anhydrase inhibitor which is employed as an anticonvulsant in all types of epilepsy except petit mal seizures. However, it has been shown to elicit a favourable response in the management of myoclonic seizures, hyperkinetic behaviour and focal epilepsy than in controlling grand mal seizures.

Dose : 600mg per day in divided doses.

E. *Valproic Acid* INN, USAN, *Valproate Sodium* USAN, *Sodium Valproate* BAN,

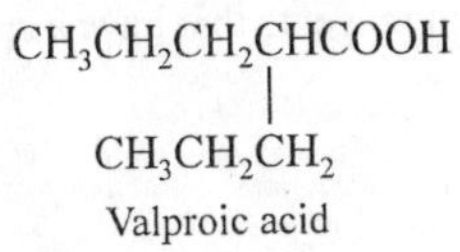
CH3CH2CH2CHCOOH
CH3CH2CH2

Valproic acid

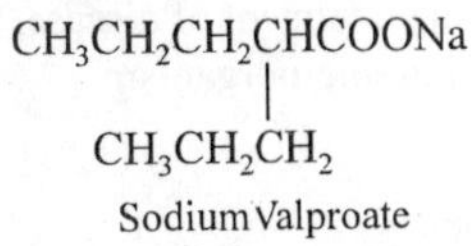
CH3CH2CH2CHCOONa
CH3CH2CH2

Sodium Valproate

Sodium 2-propylvalerate ; Pentanoic acid, 2-propyl-, sodium salt ;

Valproic acid U.S.P.,

Depakene[R] (Abbott) ; Abbott 44089[R] (Abbott) ; Sodium Valproate B.P., Abbott 44090[R] (Abbott)

Valproate sodium was discovered when it showed moderately effective antagonistic property in experimental animal seizures induced chemically and electrically. It also possessed reasonably satisfactory margin of safety.

It is frequently employed as an anticonvulsant in the management and treatment of grand mal, petit mal, mixed and temporal lobe epilepsy.

Its mechanism of action may be related to increased brain levels of the inhibitory neurotransmitter Gamma-aminobutyric acid (GABA). This increase in brain content of GABA is probably due to the inhibition by valproate sodium of the enzymes that metabolize GABA.

Dose : 10 mg per day initially ; 20 to 25 mg daily maintenance dose.

F. *Clonazepam* INN, BAN, USAN,

5-(*o*-Chlorophenyl)-1, 3-dihydro-7-nitro-2H, 1, 4-benzodiazepin-2-one ; 2H-1, 4-Benzodiazepin-2-one, 5-(2-chlorophenyl)-1, 3-dihydro-7-nitro ;

U.S.P. Clonopin[R] (Hoffman-La Roche)

Clonazepam a benzodiazepine is effective in all types of epilepsy *viz.,* grand mal, psychomotor, petit mal, myoclonic and *status epilepticus.* However, diazepam another benzodiazepine is preferred in *status epilepticus.* It possesses minor side effects.

Dose : Usual, adult, oral 4 to 8mg daily in 3 or 4 divided doses.

3.   CHEMOTHERAPY* OF EPILEPSY

The actual chemotherapy of epilepsy dates back to 1850s with the introduction of **'inorganic bromides'**. It is, however, worthwhile to state here that the therapeutic gainful application of *'phenobarbital'* around 1920s virtually made an epoch making meaningful treatment of epilepsy. Almost within a span of two decades the wonderful contributions made by Merritt and Putman were recognized when they discovered that the **'5-substituted hydantoins'** successfully capable of suppressing the electrically induced convulsions in the laboratory animals. This ultimately paved the way towards the synthesis of 5, 5-diphenylhydantoin (or phenytoin) which possessed the best as well as least sedative activity.

Chemotherapy : In the treatment of disease, the application of chemical reagents that have a specific and toxic effect on the disease-causing microorganism.

Interestingly, between 1940 and 1960 a plethora of structurally related chemical agents were synthesized based on the **'common denominator'** structure model as illustrated explicitely below :

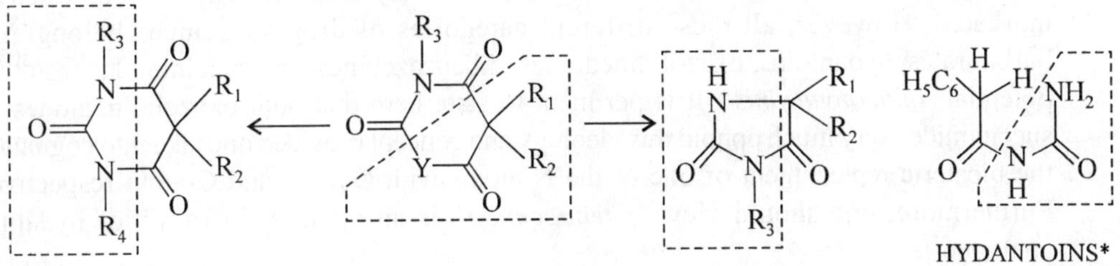

BARBITURATES
[General Structure]

(1) Phenobarbital
$R_1 = C_2H_5$; $R_2 = C_6H_5$
$R_3 = R_4 = H$;

(2) Methabarbital
$R_1 = R_2 = C_2H_5$;
$R_3 = H$; $R_4 = CH_3$

(3) Primidone
$R_1 = C_2H_5$; $R_2 = C_6H_5$;
$R_3 = R_4 = H$;

COMMON DENOMINATOR
[Y = C, N, O]

HYDANTOINS*

(1) Phenytoin
$R_1 = R_2 = C_2H_5$; $R_3 = H$;

(2) Ethotoin
$R_1 = C_2H_5$; $R_2 = H$;
$R_3 = C_2H_5$;

(3) Mephenytoin
$R_1 = C_2H_5$; $R_2 = C_6H_5$;
$R_3 = CH_3$;

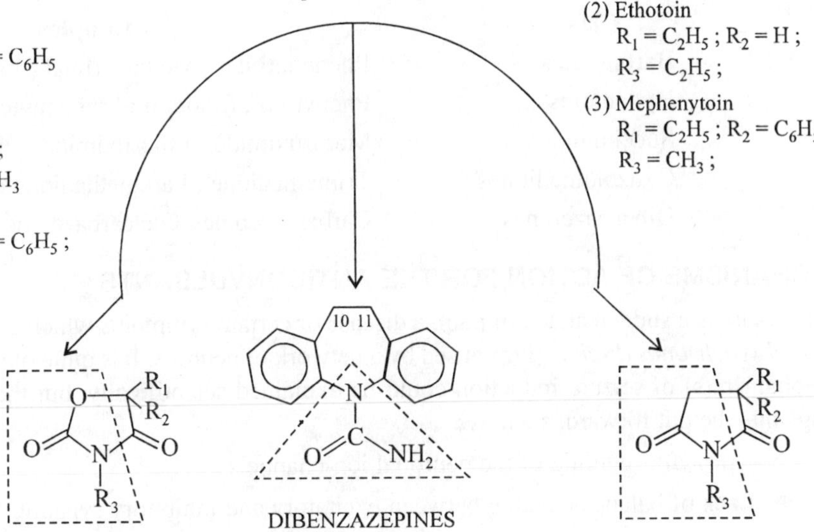

OXAZOLIDINEDIONES

(1) Trimethadione
$R_1 = R_2 = R_3 = CH_3$;

(2) Paramethadione
$R_1 = CH_3$; $R_2 = C_2H_5$;
$R_3 = CH_3$;

DIBENZAZEPINES

(1) Carbamazepine
Double Bond bet. C_{10} and C_{11} ;

(2) Oxacarbazepine

Carbonyl ($-\overset{O}{\underset{\|}{C}}-$) at C_{10} ;

SUCCINIMIDES

(1) Ethasuximide
$R_1 = C_2H_5$; $R_2 = CH_3$;
$R_3 = H$;

(2) Methsuximide
$R_1 = C_6H_5$; $R_2 = R_3 = CH_3$;

(3) Phensuximide
$R_1 = C_6H_5$; $R_2 = H$;
$R_3 = CH_3$;

Salient Features : The salient features with respect to the structural variants in chemical relationships amongst the *'anticonvulsants'* are as given under :

    (1) The ketonic function strategically located in between the nitrogens is reduced to a methylene (—CH_2) function.

---

*Hydantoins : Also known as Imidazole-2, 4-diones ;

(2) Hydantoins essentially have the *'imidazole-2, 4-diones'* moiety.

(3) In dibenzazepines as *'epoxide-ring'* at C—10 and C—11 gives rise to an **active metabolite.**

(4) Based on the **'common denominator'** analogical structures, as shown earlier, quite a few chemically related *'anticonvulsants'* were meticulously designed, synthesized, tested and marketed. However, all these different categories of drug substances belonging to barbiturates, hydantoins, oxazolidinediones, dibenzazepines, and succinimides result into potential *'anticonvulsants'*. It is pertinent to state here that both oxazolidinediones and succinimides very much uphold this ideology and concept provided one take into cognizance the biosteric replacement of one of the N-atoms with O-atom and C-atom respectively. Furthermore, one should view *'phenacemide'* as an open-chain (acyclic) hydantoin analogue.

(5) The various structural variants having an **'uriedo-type moiety'** yield the following major class of *'anticonvulsants'*, namely :

Class	Examples
Barbiturates	Phenobarbital ; Methabarbital ; Primidone etc.,
Hydantoins	Phenytoin ; Ethotoin ; Mephenytoin etc.,
Succinimides	Methsuximide ; Ethasuximide ; Phensuximide etc.,
Oxazoldinediones	Trimethadione ; Paramethadione etc.,
Dibenzazepines	Carbamazepine ; Oxacarbazepine etc.,

4. MECHANISMS OF ACTION FOR THE ANTICONVULSANTS

A seizure is a sudden attack of pain, a disease or certain symptoms which could be due to bursts of *abnormal synchronus discharging* caused by a network of neurons. It is quite obvious that till date no exact mechanism(s) of seizure induction could be explained scientifically, but the following possible reasonings may be put forward, such as :

✦ Unusual triggering of the neuronal ion-channels

✦ Lack of balance existing between excitatory and inhibitory synaptic function

✦ Different antiseizure drugs (AEDs) or anticonvulsants usually display various combined activities on the neuronal function thereby exerting very specific and selective action against the broad range of prevailing seizures*, such as :

(*i*) Ion Channels :

(*ii*) Synaptic Inhibition and Excitation, and

(*iii*) Aberrant Calcium Signalling.

These *three* mechanisms of action for the anticonvulsants shall now be discussed briefly in the sections that follows :

[A] Ion Channels :

It has been observed that both the Na^+ and Cl^- ions are invariably present at much higher concentration *'outside the cell'*, whereas K^+, charged proteins, and organic cations are more abundantly available very much *'inside the cell'*. It is an universal fact that only the smaller ions can permeate through the membrane, whereas the larger ions or proteins fail to do so ; therefore, the neuronal membranes normally maintain and sustain the phenomenon of charge separation. Consequently, a **'resting potential'** having a range of – 50 mV to – 80 mV usually gets established between the inside and outside of the cell.

At this juncture *two* different situations may arise :

(a) *Enhancement in Interior Negativity :* It is also known as *'hyperpolarization'*, which lowers the resting potential down to − 90 mV, thereby rendering it much more difficult for a neuron to accomplish the desired threshold and ultimately fire (*i.e.,* trigger its action), and

(b) *Reduction in Interior Negativity :* It is usually termed as *'depolarization'*, that may essentially cause generation of an *'action potential'*, in case, the extent of depolarization is just enough to accomplish a threshold value nearly − 40 mV. At this point in time the *'neuronal firing'* is commenced by an influx of Na^+ ions.

[B] Synaptic Inhibition and Excitation :

The actual requirement for a neuron, whether or not an *'action potential'* is accomplished exclusively governed by the achievable equilibrium (or balance) between the prevailing excitatory and inhibitory stimulation. It has been established beyond any reasonable doubt that GABA is solely responsible as the predominant inhibitory transmitter in the brain. The *modus operandi* are as stated below :

- synthesized by glutamic acid (an amino acid) and glutamic acid decarboxylase (GAD - an enzyme)
- inactivated by GABA-transaminase (GABA-T)
- GABA binds to two receptor types *viz.,* $GABA_A$ and $GABA_B$
- neuronal *'hyperpolarization'* is caused when $GABA_A$ receptors takes place on Cl^- ion channels, and subsequently the binding of GABA results into chloride influx.
- linkage of $GABA_B$ receptors with G proteins ; besides second messangers to Ca and K channel activity, — also mediating inhibition in the CNS.*
- the oscillation rhythms in certain types of epilepsy is solely caused by $GABA_B$ receptors.**
- plethora of antiseizure drugs (AEDs) are found either to potentiate GABA-mediated inhibition or to affect GABA concentration to a great extent.
- both barbiturates and benzodiazepines augment the activity of GABA precisely on the $GABA_A$ chloride channel.
- activation of such ligand-gated channels *viz.,* excitatory transmission of glutamic acid *via* α-amino-3-hydroxy-5-methyl-4-isoxazole propionic acid (AMPA) ; N-methyl-D-asparate (NMDA) ; and kainate receptors appreciably helps both Ca and Na *influx*, and K *efflux* thereby the phenomenon of depolarization gets facilitated.

[C] Aberrant Calcium Signalling :

It has been amply demonstrated that the T-type calcium currents, at low-threshold levels, play a pivotal role as **'pacemakers'** to maintain and sustain the usual normal brain activity. Furthermore, the thalamic oscillatory currents, believed to be intimately associated with the normal brain activity, help in the generation of absence seizures.***

---

*Jones KA *et. al. Neuropsychopharmacology*, **23**, (4 suppl.) : S41-9, 2000.

Bal T *et al. J. Neuro Sci.,* **20, 7478-88, 2000.

***Browne TR *et. al.,* : In **Hand book of Epilepsy**, Lippincott Williams and Wilkins, New York, 2nd, edn, 1-18, 1999.

Examples : *Oxazolidinediones* and *ethosuximide* that specifically prevent and check T-type currents, are recognized to be very effective against the *'absence seizures'*, whereas absolutely ineffective against either partial or other seizure types.

5.   SPECIFIC MECHANISMS OF SELECTED ANTICONVULSANTS

The specific mechanism of actions by which certain selected anticonvulsants dicussed in this chapter are enumerated below :

5.1.   5, 5-Diphenylhydantoin Sodium (Phenytoin Sodium)

It is found to exert its action on the motor cortex where it stabilizes the neuronal membrane and thereby inhibits the spread of the seizure discharge. Present evidence also suggests that it limits high frequency repititive firing by blocking Na^+-channels in a use — and frequency dependent fashion. Besides, it enhances Ca- binding to the phospholipids present in neuronal membranes.

In fact, these effects collectively give rise to a more stable membrane configuration. Importantly, these critical findings are found to be in perfect harmony with the glaring and supportive fact that its most easily demonstrated characteristic features are by virtue of its ability to limit the development of maximal seizure activity and also to minimize the virtual extension of the seizure phenomenon from the active focus. Interestingly, both of these splendid features in phenytoin are very much related to the clinical usefulness beyond any reasonable doubt.

5.2.   Ethotoin

It is N-dealkylated and *para*-hydroxylated *in vivo*. However, the N-dealkyl metabolite is most presumably the **'active compound'** ; it is similarly metabolized by *para*-hydroxylation, and the resulting hydroxyl function undergoes conjugation subsequently.

This particular *'drug substance'* is used against not-so-specific seizures, but invariably on an adjunctive basis on account of its *low potency*. In a broader perspective, such anticonvulsants which are *not* completely branched on the appropriate C-atom are of definite lower potency than their rather more fully-branched structural analogues.

5.3.   Mephenytoin

It is metabolically N-dealkylated to the corresponding 5-ethyl-5-phenylhydantoin, which is considered to be the *'active agent'*. Interestingly, the said metabolized product, happens to be the *'hydantoin counterpart'* of phenobarbital [Gardinal[R]] as one of the first breed of hydantoins ever introduced into the therapeutic armamentarium. Furthermore, it may be assumed that *'mephenytoin'* is a *'pro-drug'*, that essentially ameliorates a part of its toxicity along with skin and blood disorders of serious nature of the delivered **'active drug'**. The metabolic inactivation of this drug and its corresponding dimethylmetabolite is caused due to the *para*-hydroxylation and subsequent conjugation of the free hydroxyl moiety.

5.4.   Trimethadione

It is believed to get metabolized by N-demethylation to the much reputed *'active metabolite'* **dimethadione***, which is a water-soluble and slowly lipophilic *'drug substance'*. Dimethadione is usually excreted as such without any further metabolic degradation.

5.5.   Paramethadione

Though it is closely related to trimethadione in its structural aspect yet it may be safer. It has been observed that the corresponding N-demethyl metabolite, that subsequently gets excreted very sluggishly and slowly, is actually considered to be the **'active drug'**.

---

*Spinks A and Waring WS. In : Ellis GP and West GB (eds) **Progress in Medicinal Chemistry**, Butterworth, Washington DC. Vol. 3, p. 261, 1963.

5.6. Phensuximide

It has been duly observed that to a certain extent nutrition (trophism) towards antiabsence activity may be caused by the imbedded *'succinimide system'*. Besides, the methylene (– CH₂ –) function could be viewed very much like an *'α-alkyl branch'* condensed into the ring system. Particularly, the inherent phenyl substituent grants certain degree of activity against the generalized tonic-clonic and partial seizures. However, the resulting N-demethylation takes place to give rise to the formation of the **'putative active metabolite'**. Evidently, the parent drug, phensuximide, and the N-dimethyl metabolite are virtually inactivated by either conjugation and/or *para*-hydroxylation.

5.7. Methsuximide

The parent compound and the metabolite are both subjected to N-demethylation and *para*-hydroxylation.

5.8. Ethosuximide

A certain portion of the drug is excreted intact through the kidney. However, the major metabolite is actually produced by oxidation of the ethyl group. It conforms quite closely to the general structural pattern for the *'antiabsence activity'*. It is observed to be much more active and comparatively less toxic than trimethadione. Therefore, it has gained cognizance as a drug of choice for acute absence seizures.

5.9. Primidone

It gets metabolized to phenylethyl malonamide (PEMA) and phenobarbital* (an active long-acting barbiturate). The actual formation to phenobarbital varies between 15-25%. The plasma half-life of PEMA ranges between 24-48 hour, whereas that of phenobarbital is 48-120 hours. However, the two aforesaid metabolites have a tendency to exert a cumulative storage in the body particularly during chronic medication. Primidone is quite often described as a 2-deoxybarbiturate.

5.10. Phenacemide

It gets metabolized by *para*-hydroxylation. It is pertinent to mention here that this *'drug'* must be employed *only* in such instances wherein the patients failed to respond to other anticonvulsant treatment (medication). However, one must exercise extreme caution in the treatment of patients who have a previous history (or record) displaying *liver dysfunction, personality disorder,* and *allergic manifestation.*

5.11. Sultiame

It exerts its action by inhibiting specifically the carbonic anhydrase.

5.12. Valproate Sodium

Importantly, its precise mechanism of its anticonvulsant action is still not fully understood. However, it has been duly advocated that its administration specifically inhibits GABA-transaminase, and thereby enhancing the concentration of cerebral GABA. It has also been observed that a few other straight-chain saturated fatty acids *i.e.,* lower fatty acids, such as : propanoic acid, butyric acid, and pentanoic acid which are devoid of anticonvulsant characteristic features are relatively more potent and efficacious inhibitors of GABA-transaminase than is valproic acid. Furthermore, it has been adequately substantiated that there exists a rather stronger correlation between the anticonvulsant potency of valproate and other branched-chain fatty acids ; besides, their capability to minimise the prevailing concentration of cerebral aspartic acid (an amino acid).

---

*Spinks A and Waring WS. In : Ellis GP and West GB (eds) **Progress in Medicinal Chemistry,** Vol. 3. p. 261, Butterworth, Washington DC., 1963.

Further evidences reveal that valproate sodium may also decrease binding to certain serum proteins or block the hepatic metabolism of phenobarbital. Therefore, administration of the *'drug'* to patients in a steady state, while on phenobarbital concurrently (or primidone, which gets metabolized to phenobarbital) may enhance the plasma levels of phenobarbital from 35-200%, a quantum jump, thereby causing an excessive *somnolence.* * However, the present evidence amply substantiates that this is caused exclusively by an immediate lowering in the prevailing rate of elimination of phenobarbital.

5.13. Clonazepam

It is a partial against at benzodiazepine allosteric binding sites on $GABA_A$ receptors. The metabolism essentially invovles hydroxylation of the 3-position followed by glucuronidation and nitro group reduction, followed by acetylation ultimately. It has been observed that almost 87% of the drug is bound to plasma protein ; volume distribution is 3.2 $L.kg^{-1}$, and its half-life ranges between 19 to 46 hours in adults and from 13 to 33 hours in children. Its concomitant use with valproate may cause absence status.

Probable Questions for B. Pharm. Examinations

1. Classify the drugs used for convulsive seizures. Give the structure, chemical name and uses of one important compound from each class.

2. Explain *'Hydantoins'* as potent anticonvulsants. Give the synthesis of Diphentoin[(R)] (Beecham).

3. Discuss paramethadione as a therapeutic agent used in petitmal epilepsy.

4. 'Succinimides afford better tolerated and less toxic anticonvulsants'. Justify the statement with the help of a detailed account of one of the potent compounds belonging to this category.

5. Name the anticonvulsant drug obtained by replacing 'O' at C-2 of phenobarbital with 2H atom atom. Give its synthesis and uses.

6. Give an account of an anticonvulsant having a dibenzaepine ring system wtih a carbamoyl moiety hooked on to the N-atom.

7. Discuss the relative strucctural differences occuring amongst Phensuximide, Methsuximide and Ethosuximide. Give their chemical names, uses and advantages of one over the other.

8. Describe the mode of action of Hydantoins and Primidone inside the body.

9. How do the mode of action of oxazolidinediones and succinimides differ from hydantoins ?

10. Discuss a **Benzodiazepine based anticonvulsant** which possesses a broad-spectrum activity.

RECOMMENDED READINGS

1. JN Delgado and E I Issacson Anticonvulsants in : *Burger's Medicinal Chemistry and Drug Discovery* M E Wolff (*Ed.*) (5th edn) New york, Wiley-Interscience (1995).

2. JEP Toman and J D Taylor, Mechanism of Action and Metabolism of Anti-convulsants *Epilepsia* I (1952)

3. M Gordon *Psychopharmacological Agents* New York, Academic Press, Vols. 1 and 2 (1964) and (1965).

*Prolonged drowsiness or a condition resembling a trance that may continue for a number of days.

4. DM Woodbury, J K Penry and R P Schmidt (*Eds*) *Antiepileptic Drugs,* New York, Raven Press (1972).

5. JEP Toman and L S Goodman Anti-convulsants *Pharma Rev.* 28 (1948) 409.

6. J Mercier (*Ed.*) Anticonvulsant drug *Int Encycl Pharmac Therapeut* Oxford Pergamon Press, 2 Volumes (1973).

7. E Jucker Some New Developments in the Chemistry of Psychotherapeutic Agents *Angew Chem* (*int Ed*) (1963).

8. JA Vida (*Ed*) Anticonvulsants, *Medicinal Chemistry, A Series of Monographs,* Vol. 15 New York, Academic Press (1977).

9. Gennaro Alfonso R, Remington : *The Science and Practice of Pharmacy,* Lippincott Williams and Wilkins, New York, 20th end., 2000.

10. Williams DA and Lemke TL, *Foye's Principles of Medicinal Chemistry,* Lippincott Williams and Wilkins, New York, 5th edn., 2002.

Muscle Relaxants

1. INTRODUCTION

Drugs which cause depression of motor function leading to relaxation of voluntary muscle are known as **muscle relaxants.**

The skeletal muscle may be relaxed by *two* different groups of drugs, namely : *first,* by those exerting an action on the central nervous system (CNS) and used mainly for the relief of painful muscle spasms of spasticity taking place either in neuromuscular or musculoskeletal disorders ; *secondly,* those affecting neuromuscular transmission that are employed as adjuncts in anaesthesia in order to modify the muscle relaxation ability.

Another school of thought suggests explicitely and with ample evidence that the skeletal muscle in particular could be relaxed by blocking the effect of *somatic motor nerve impulses.* Furthermore, this specific mode of action may be accomplished adequately by affording either substantial depression of the most suitable neurons within the CNS so as to negate the formation of *somatic motor nerve impulses,* or by minimising the availability of Ca^{2+} ions directly to the myofibrillar contractile system. Interestingly, a few local anaesthetics are duly responsible for the interruption of some specific *afferent reflex* pathways that may invariably effect the relaxation of circumscribed muscle groups.

In this chapter an emphasis shall be laid only on those drugs which exert their action at the *myoneural junction i.e.,* the **neuromuscular blocking drugs ;** and such other drug substances which solely act upon the *central neurons i.e.,* the **centrally acting muscle relaxants.**

2. CLASSIFICATION

In general, skeletal muscle relaxants may be classified as (*a*) Neuromuscular Blocking Drugs ; and (*b*) Centrally Acting Muscle Relaxants.

2.1 Neuromuscular Blocking Drugs

Skeletal muscle fibres are innervated by myclinated nerve fibres from the anterior horn cells of the grey matter of the spinal cord. The nerve fibre contains many axons and each axon extends uninterrupted from the spinal cord to the skeletal muscle where it forms many terminal branches. Each branch ends close to a motor end-plate leaving a junctional gap (or cleft) of about 50 nm between nerve ending and the muscle fibre. By means of these branchings, an axon innervates one or more end-plates.

The moror end-plate is a specialized zone of the muscle fibre whose surface is thrown into folds called *junctional folds*. It contains acetylcholine (cholinergic) receptors.

Neuromuscular blocking drugs exert their action by making the motor end-plate membrane of the myoneural junction incapable of reacting to acetylcholine, which functions as the neuro transmitter.

The neuromuscular blocking agents may be further sub-divided into *two* categories, namely :

A. Non-depolarizing Muscle Relaxants

Curare-type drugs containing essentially a bulky structure together with a minimum of one quaternary ammonium group, complete with acetylcholine thereby preventing a free access to the cholinergic receptors.

*Examples :*Tubocurarine chloride ; Metocurine iodide ; Gallamine triethiodide ; Pancuronium bromide ; Hexafluoronium bromide ; Fazadinium bromide ; Alcuronium chloride ; Dacuronium bromide and Stercuronium iodide.

A1. *Tubocurarine chloride* BAN, USAN, *Tubocurarine,* INN,

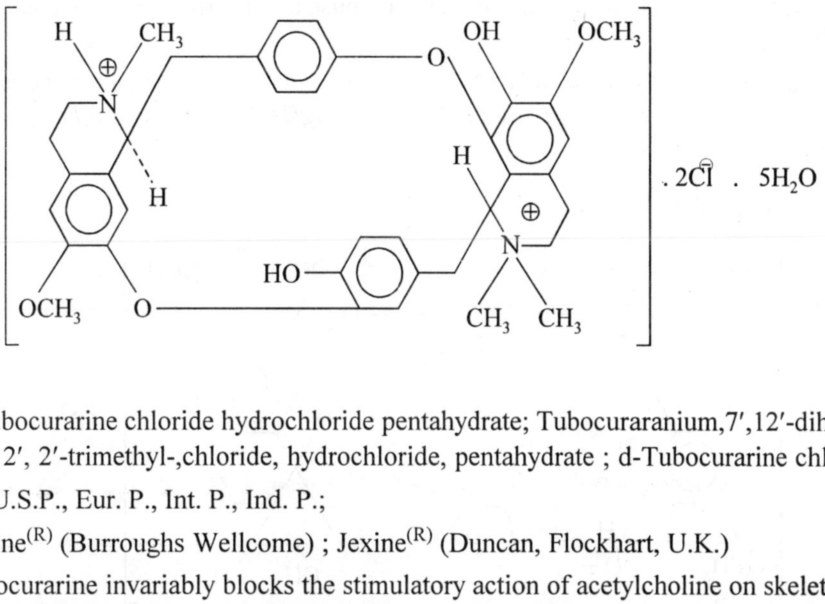

(+)-Tubocurarine chloride hydrochloride pentahydrate; Tubocuraranium,7',12'-dihydroxy-6, 6'-dimethoxy-2, 2', 2'-trimethyl-,chloride, hydrochloride, pentahydrate ; d-Tubocurarine chloride ;

B.P.; U.S.P., Eur. P., Int. P., Ind. P.;

Tubarine[R] (Burroughs Wellcome) ; Jexine[R] (Duncan, Flockhart, U.K.)

d-Tubocurarine invariably blocks the stimulatory action of acetylcholine on skeletal muscles. It fails to produce any effect on involuntary muscles or glands. The fundamental basis of the activity of muscarinic stimulants as related to interatomic distances of vital functional moieties also holds good for this drug. It is interesting to observe that the distance between the three oxygen atoms residing on the same tetrahydroisoquinoline residue and the centre of one N-methyl function falls within a radius of 5-9 Å, which is more or less identical to the linear distances between the functional moieties present in acetylcholine. The average distance between the two quaternary nitrogen atoms is 13-15 Å and that of the two etherial oxygen atoms being 9 Å. Hence, configurationally every 'half-molcule' of tubocurarine is identical to atropine as shown below :

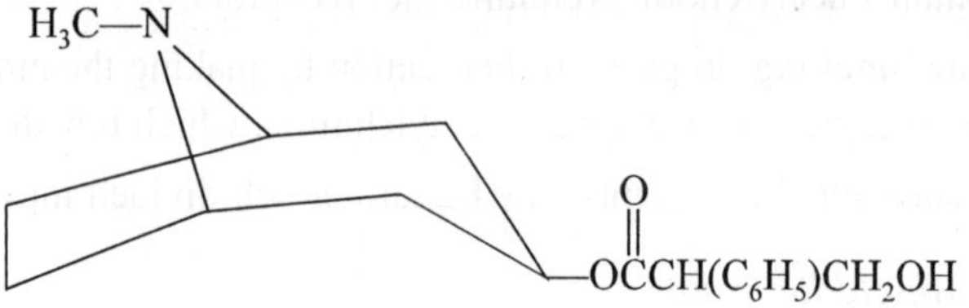

Atropine

However, a striking difference between tubocurarine and atropine exists ; in the former there are double rows of the O-N attachments, while in the latter only a single O-N row is prevalent. This fundamental difference between the two molecules perhaps may be put forward as a possible explanation of the specific action of *d*-tubocurarine on the neuromuscular junction which action atropine is devoid of.

It is a non-depolarizing muscle relaxant and when administered by injection produces paralysis of voluntary muscle by blocking impulses at the neuromuscular junction. It acts primarily by inhibiting the transmission of nervous impulses to skeletal muscle by competing with acetylcholine for cholinergic receptors. It causes muscular relaxation without producing any depression of the nervous system. It is often employed as an adjunct in surgical anaesthesia in order to achieve adequate skeletal muscle relaxation during surgery. Tubocurarine is also used to minimise the severity of muscle contraction during electroshock therapy.

Dose : *For paralysis of limb muscles, 6 to 10 mg in 30 to 90 seconds ; for profound abdominal relaxation and appnea, 15 to 20 mg; for shock therapy, 3 mg/18 kg of body weight ; for diagnosis of myasthenia gravis, 0.3 mg/18 kg.*

A2. *Metocurine Iodide* BAN, USAN,

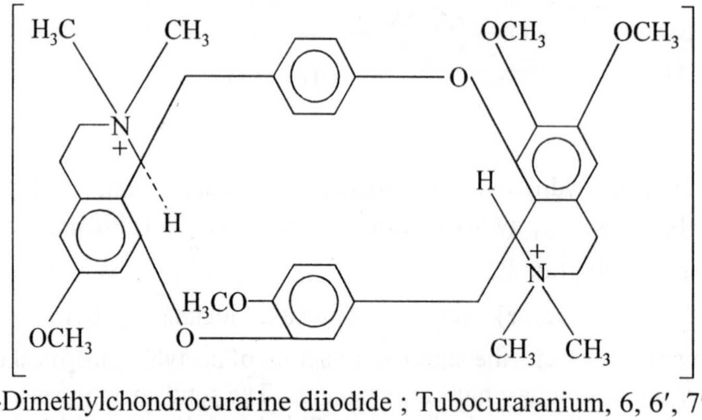

(+)-*O, O'*-Dimethylchondrocurarine diiodide ; Tubocuraranium, 6, 6', 7', 12-tetramethoxy-2, 2, 2', 2'-tetramethyl-, diiodide; Dimethyl Tubocurarine Iodide ;

U.S.P., Dimethyl Tubocurarine Iodide N.F.; Metubine Iodide[(R)] (Lilly)

Synthesis :

$$d\text{-Tubocurarine} \xrightarrow[\substack{(ii)\ HI}]{\substack{(i)\ CH_3-I \\ (Methylation)}} \text{Metocurine Iodide}$$

Metocurine iodide may be prepared by the treatment of *d*-tubocurarine with methyl iodide to effect methylation, followed by reaction with bimolar concentration of hydroiodic acid to form the official compound.

Its actions and uses are almost the same as those of *d*-tubocurarine, however, in man it is found to be *three* times more potent than the latter with longer duration of action.

***Dose :** Initial intravenous, 1.5 to 10 mg stretched over a 60 sec. period;maintenance, 500 μg to 1mg each 25 to 90 min.*

A3. *Gallamine Triethiodide* BAN, USAN, *Gallamine* INN,

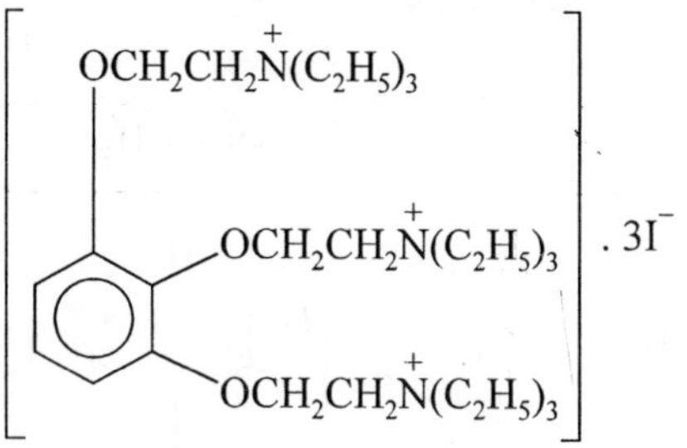

$OCH_2CH_2\overset{+}{N}(C_2H_5)_3$
$OCH_2CH_2\overset{+}{N}(C_2H_5)_3$
$. 3I^-$
$OCH_2CH_2\overset{+}{N}(C_2H_5)_3$

[ν-Phenyl *tris* (oxyethylene)] *tris* [triethylammonium] triiodide ; Ethanaminium, 2, 2′, 2″-[1, 2, 3-benzenetriyltris (oxy) *tris* (N, N, N-triethyl]-, triiodide ; Bencurine Iodide ; B.P., U.S.P., Eur. P., Int. P., Ind. P.

Flaxedil(R) (Davis and Geck) ;

Synthesis :

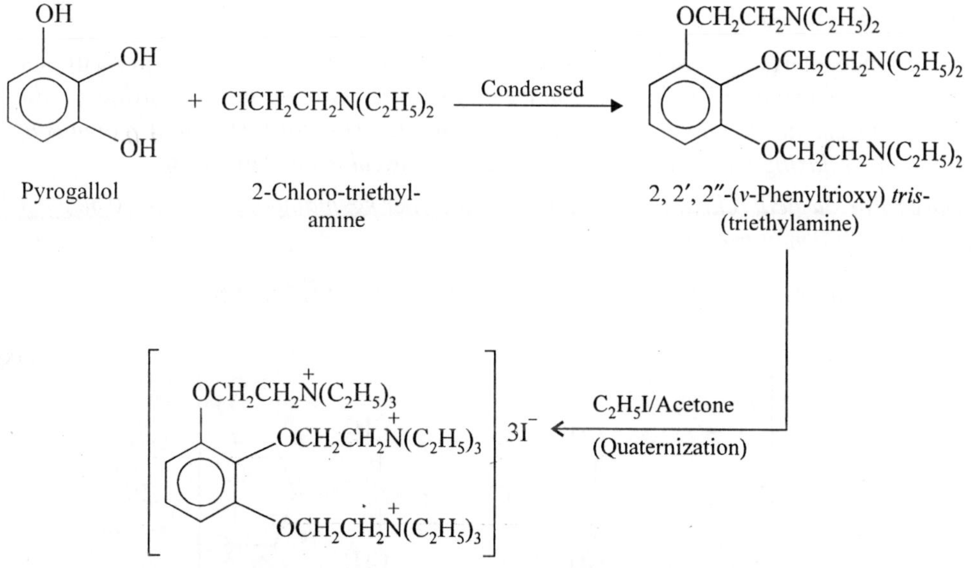

OH
OH
OH
Pyrogallol
+ $ClCH_2CH_2N(C_2H_5)_2$
2-Chloro-triethyl-amine
Condensed
$OCH_2CH_2N(C_2H_5)_2$
$OCH_2CH_2N(C_2H_5)_2$
$OCH_2CH_2N(C_2H_5)_2$
2, 2′, 2″-(ν-Phenyltrioxy) *tris*-(triethylamine)
C_2H_5I/Acetone
(Quaternization)
$OCH_2CH_2\overset{+}{N}(C_2H_5)_3$
$OCH_2CH_2\overset{+}{N}(C_2H_5)_3$
$OCH_2CH_2\overset{+}{N}(C_2H_5)_3$
$3I^-$

Gallamine Triethiodide

The triamine : 2, 2', 2''-(*v*-pheneyltrioxy) *tris* (triethyl-amine) may be prepared by the condensation of pyrogallol and 2-chloro-triethylamine. This is then quaternized with ethyliodide in the presence of boiling acetone to yield the desired official compound.

Its actions are similar to those of tubocurarine choride. It is mostly employed as an adjunct to anaesthesia so as to achieve deeper muscular relaxation to facilitate surgical procedures.

Dose : *For limb muscle paralysis, initial i.v. or i.m. 1 mg/kg of body weight ; for abdominal surgery, 1.5 mg/kg of body weight; Maintenance dose, 500 mcg to 1 mg every 30 to 60 min. intervals if required. The dose must be reduced when used along with anaesthetics, like ether, cyclopropane, etc.*

A4. *Pancuronium Bromide* INN, BAN, USAN,

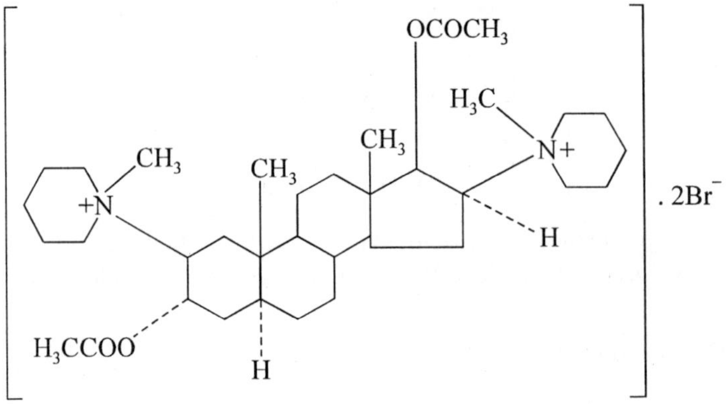

1, 1'-(3α, 17β-Dihydroxy-5α-androstane-2β, 16β-ylene) *bis* [1-methylpiperidinium] dibromide diacetate ; 2β, 16β-Dipiperidino-5α-androstane-3α, 17β-dioldiacetate dimethobromide :

Pavulon[R] (Organon) ;

It is a nondepolarizing muscle relaxant of choice with actions much alike to those of tubocurarine chloride. *It has been employed with greater margin of safety in patients having cardiovascular disease and in the management of status asthamaticus to facilitate artificial respiration (minimising oxygen demand) thereby relaxing the muscles, than other neuromuscular blocking drug.*

Dose : *For surgical relaxation, i.v., 20 to 100 mcg per kg, maintenance dose 10 mcg per kg ; For intubation, 60 to 100 mcg per kg.*

A5. *Hexafluoronium Bromide* INN, *Hexafluorenium Bromide* BAN, USAN,

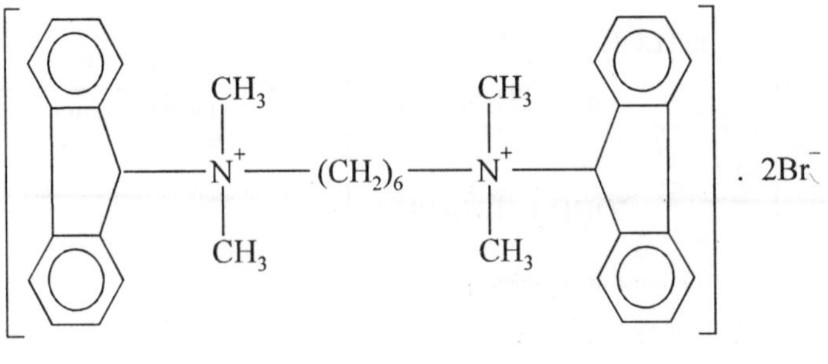

Hexamethylenebis [fluoren-9-yldimethylammonium]-dibromide ; 1, 6-Hexanediaminium, N, N′-di-9H-fluoren-9-yl-N, N, N′, N′-tetramethyl-, dibromide ;

U.S.P.

Mylaxen[R] (Carter-Wallace)

Synthesis :

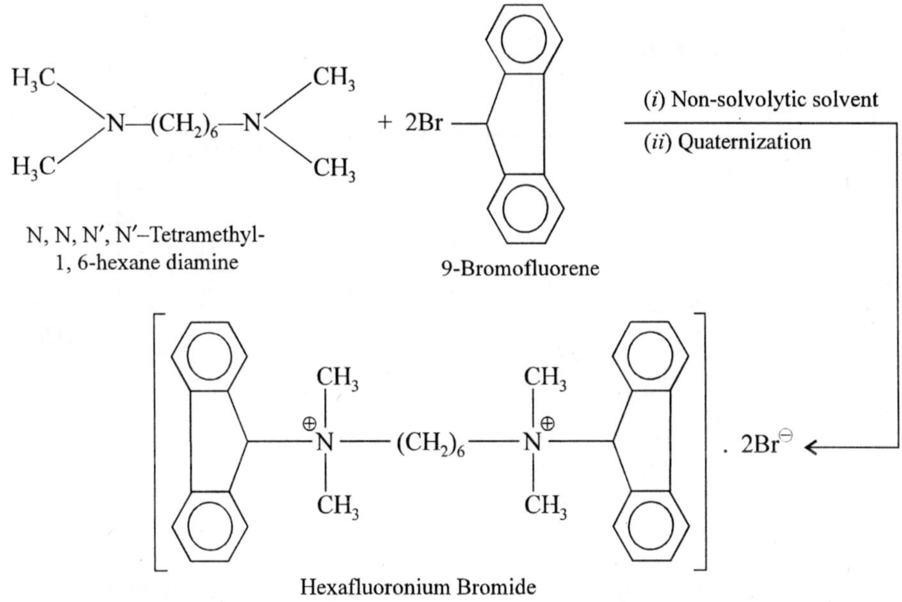

N, N, N′, N′–Tetramethyl-
1, 6-hexane diamine

9-Bromofluorene

(i) Non-solvolytic solvent
(ii) Quaternization

Hexafluoronium Bromide

Hexafluoronium bromide may be prepared by dissolving N, N, N′, N′-tetramethyl-1, 6-hexane diamine in a nonsolvolytic solvent which is quaternized twice with 9-bromofluorene.

It exerts a weak and feeble neuromuscular blocking activity which fails to produce significant muscle relaxation except under deep ether anaesthesia. It has been found to potentiate the neuromuscular blockade caused by tubocurarine and to antagonize the action of decamethonium. Paradoxically, it has been used successfully to prolong and potentiate the relaxant effects of suxamethonium chloride. Besides, it has also been reported to decrease *suxamethonium-induced muscular fasciculations.*

Dose : *Initial, i.v. 300 to 400 mcg per kg ; maintenance dose 100 to 200 mcg at intervals of 80 to 100 min.*

A6. *Fazadinium Bromide* INN, BAN,

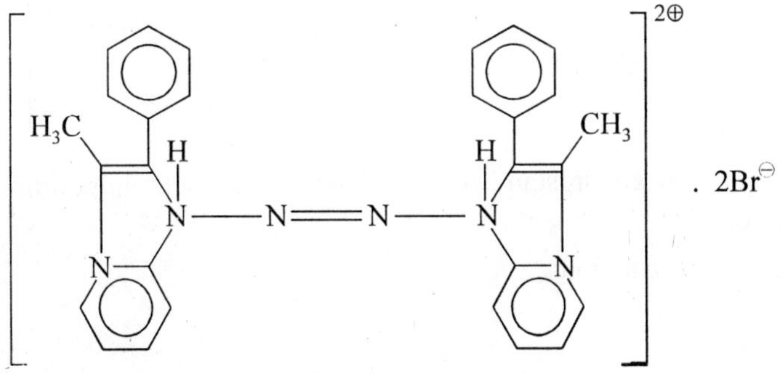

1, 1'-Azobis [(3-methyl-2-phenyl-1H-imidazo [1, 2-*a*] pyridin-4-ium] dibromide ; Fazadon[R] (Duncan, Flockhart, U.K.) ;

It is a muscle relaxant possessing a dose-dependent rapid onset and prolonged duration of action. It is normally employed to aid endotracheal intubation and to produce muscular relaxation during surgi-/ cal procedures.

Dose : *0.5 mg per kg body weight for effects lasting up to 30 minutes ; In surgery, usual, initial, i.v., 0.75 to 1 mg per kg.*

A7. *Alcuronium Chloride* INN, BAN,

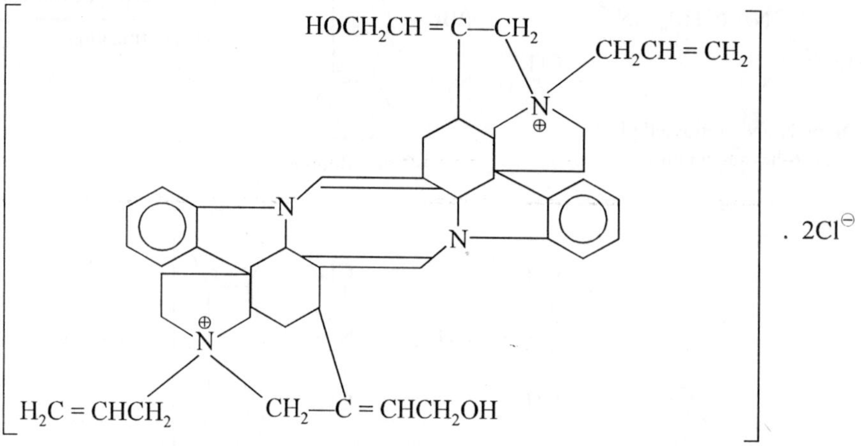

N, N'-Diallylnortoxiferinium dichloride; Toxiferine I, 4, 4'-didemethyl-4, 4'-di-2-propenyl-, dichloride; Alloferin[R] (Hoffman-La Roche-International).

It is employed mostly as an adjuvant to anaesthesia.

Dose : *For neuromuscular blockade, usual, initial, i.v. 250 mcg per kg body weight.*

A8. *Dacuronium Bromide* INN, BAN, USAN,

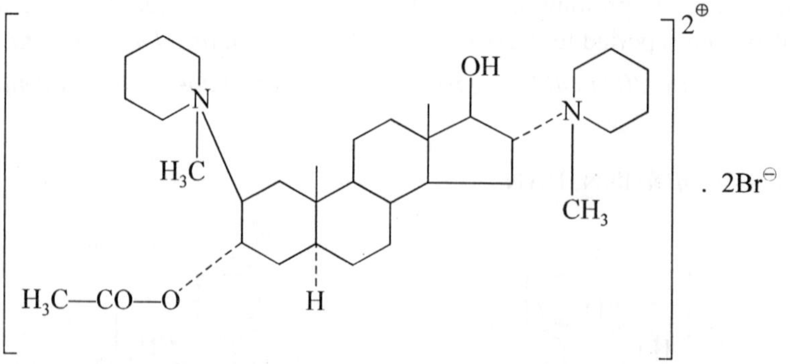

3α, 17β-Dihydroxy-5α-anadrostan-2β, 16β-ylene) *bis*-(1-methylpiperidinium) dibromide-3-acetate ; NB68[R] (Organon) ;

It has anticholinesterase actions.

A9. *Stercuronium Iodide* INN, USAN,

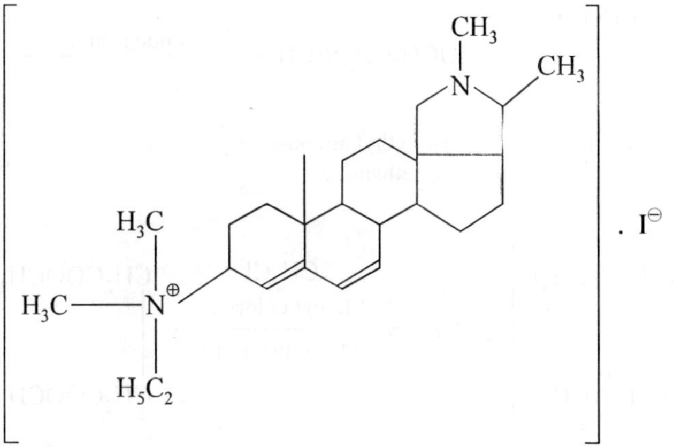

(Cona-4, 6-dienin-3β-yl) dimethylethylammonium iodide; MYSC 1080 (Gist-Brocades) ;
It also possesses anticholinesterase properties.

B. *Depolarizing Neuromuscular Blocking Drugs*

Decamethonium and succinylcholine possessing simple skeleton-like bisquaternary synthetic compounds usually show their presence in the initial phase of action by more or less mimicking acetylcholine and depolarizing the motor end-plate membrane. They produce persistent depolarisation responsible for the initial blockade by preventing the membrane from undergoing repolarisation in order to accept new stimuli. Repetitive doses with depolarizing blocking drugs usually change the initial depolarizing action on the motor end-plate to non-depolarizing type of blockade. Their blocking action is, therefore, rather difficult to antagonize because of the fact that either cholinomimetics or anticholinesterases can cause potentiation or antagonism depending on the blocking phase (depolarizing or non-depolarizing) of the drug at the time.

Examples: Suxamethonium chloride ; Suxethonium bromide ; Decamethonium bromide.

B1. *Suxamethonium Chloride* INN,BAN, *Succinylcholine Chloride* USAN,

$$\left[\begin{array}{l} COOCH_2CH_2\overset{\oplus}{N}(CH_3)_3 \\ | \\ (CH_2)_2 \\ | \\ COOCH_2CH_2\overset{\oplus}{N}(CH_3)_3 \end{array}\right] .\ 2Cl^{\ominus}$$

Choline chloride succinate (2 : 1); Ethanaminium, 2, 2′-[(1, 4-dioxo- 1, 4-butanediyl) bis (oxy)] *bis*-[N, N, N-trimethyl]-dichloride ;

Suxamethonium Chloride B.P., Eur. P., Int. P.,

Succinylcholine Chloride U.S.P.

Acectine(R) (Burroughs Wellcome) ; Quelicin(R) (Abbott) ; Sucostrin Chloride(R) (Squibb)

Synthesis :

$$
\begin{array}{c}
CH_2COCl \\
| \qquad + \; 2HOCH_2CH_2N(CH_3)_2 \quad \xrightarrow{\text{Condensation}} \\
CH_2COCl
\end{array}
$$

$$\downarrow\; -\,2HCl$$

Succinyl Chloride   β-Dimethyl amino-
                                        ethanol

$$
\left[
\begin{array}{c}
COOCH_2CH_2\overset{\oplus}{N}(CH_3)_3 \\
| \\
(CH_2)_2 \\
| \\
COOCH_2CH_2\overset{\oplus}{N}(CH_3)_3
\end{array}
\right] . \; 2Cl^{\ominus}
\quad \xleftarrow[\text{(Quaternization)}]{\overset{CH_3Cl}{\text{Methyl chloride}}}
\quad
\begin{array}{c}
CH_2COOCH_2CH_2N(CH_3)_2 \\
| \\
CH_2COOCH_2CH_2N(CH_3)_2
\end{array}
$$

Suxamethonium Chloride   An Ester

Condensation of succinyl chloride with β-dimethylamino ethanol yields an ester with the elimination of two moles of HCl. Further quaternization with two moles of methyl chloride forms the official compound.

Its depolarizing neuromuscular blocking effect is very transient because of its rapid hydrolysis by cholinesterases. It does not cause histamine liberation and hence it is well tolerated. Single-dose therapy of suxamethonium chloride is generally used to relax the skeletal muscle for orthopedic manipulation, endotracheal intubation, in laryngospasm and also to check the intensity of convulsions in patients receiving electroshock treatment (electroconvulsive therapy).

Dose *: Testing for sensitivity, initial i.v., 10 mg, then 10 to 75 mg ; alternative i.v. or 0.5 to 1 mg per kg ; Usual 20-80 mg i.v. or 0.5 to 10 mg per minute by i.v. infusion as a 0.1 to 0.2% solution.*

B2. *Suxethonium Chloride* INN, USAN, *Suxethonium Bromide* BAN,

$$
\begin{array}{c}
(CH_3)_2\overset{\oplus}{N}(CH_2)_2OOC(CH_2)_2COO(CH_2)_2\overset{\oplus}{N}(CH_3)_2 \\
| \qquad\qquad\qquad\qquad\qquad\qquad\qquad\qquad | \\
C_2H_5 \qquad\qquad\qquad\qquad\qquad\qquad\qquad C_2H_5
\end{array}
\quad . \; 2Cl^{\ominus} \quad\; . \; 2Br^{\ominus}
$$

Suxethonium Chloride   Bromide

Ethyl (2-hydroxyethyl) dimethylammonium chloride (or bromide) succinate ; 2, 2′-Succinyldioxy-bis (diethyldimethylammonium) dichloride (or dibromide) ; Brevidil[R] (May & Baker, U.K.) for Suxethonium Bromide.

Its actions and uses are similar to suxamethonium chloride but it possesses only about half the potency and a relatively shorter duration of action ranging between 2 to 4 minutes.

Dose *: Usual I.V., 1 to 1.25 mg of base per kg body weight (or 1.5 to 1.875 mg bromide).*

B3. *Decamethonium Bromide* BAN, USAN, *Decamethonium* INN,

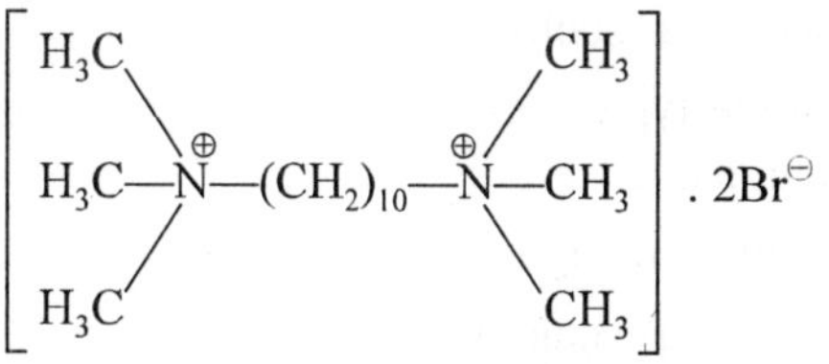

$$\left[\begin{array}{c} H_3C \diagdown \\ H_3C - \overset{\oplus}{N} - (CH_2)_{10} - \overset{\oplus}{N} - CH_3 \\ H_3C \diagup \qquad\qquad\qquad \diagdown CH_3 \end{array}\right] . 2Br^{\ominus}$$

Decamethylenebis [trimethylammonium] dibromide ; 1, 10-Decanediaminium, N, N, N, N', N' N'-hexamethyl-, dibromide ; U.S.P. ;

Syncurine$^{(R)}$ (Burroughs Wellcome)

Synthesis :

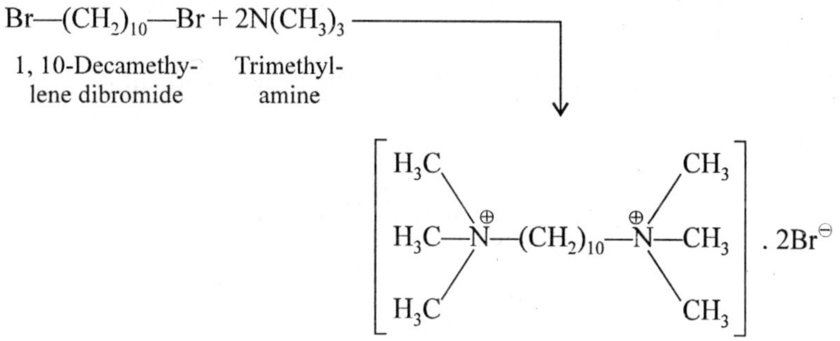

Br—(CH$_2$)$_{10}$—Br + 2N(CH$_3$)$_3$ ——————————

1, 10-Decamethy-   Trimethyl-
lene dibromide   amine

$$\left[\begin{array}{c} H_3C \diagdown \\ H_3C - \overset{\oplus}{N} - (CH_2)_{10} - \overset{\oplus}{N} - CH_3 \\ H_3C \diagup \qquad\qquad\qquad \diagdown CH_3 \end{array}\right] . 2Br^{\ominus}$$

Decamethonium Bromide

It may be prepared by the condensation of one mole of 1, 10-decamethylene dibromide with two moles of trimethylamine.

The two onium groups present in decamethonium bromide are situated at a distance of 15 Å which incidentally compares very closely to that of suxamethonium chloride. Decamethonium cannot be hydrolysed by chloinesterase and hence exerts a much more prolonged effect.

As it exerts an initial action of nicotinic depolarization at the motor end-plate membrane it is regarded as a depolarizing neuromuscular relaxant. It is used as a muscle relaxant especially for comparatively short surgical operations and, also for manipulative procedures.

Dose : *Initial, i.v., 2 to 3 mg administered at the rate of 1 mg per minute ; for maintenance of paralysis, 0.5 to 1 mg at 10-30 min intervals.*

2.2 Centrally Acting Muscle Relaxants

The discovery of centrally acting muscle relaxants dates back to 1910 when antodyne (3-phenoxy-1, 2-propanediol) first gained its entry into the therapeutic armamentarium as an analgesic and antipyretic ad later on as skeletal muscle relaxant. In 1946, Berger and Bradley* observed the muscle relaxant activity present in a large number of glycerol monoethers and analogues.

As of now practically all the centrally acting muscle relaxants seem to depress neuronal activity instead of stimulating the inhibitory nerves.

*F.M. Berger and W. Bradley, *Brit. J. Phannacol.*, **1**, 265 (1946).

As stated earlier they are used to relieve painful muscle spasms and spasticity. They relax the muscle without impairing respiration centrally. They have sedative effects. Some have predominantly tranquilizing effect and are classified as such.

In the following pages some typical examples belonging to different categories of the above class of compounds are described, namely ;

(*a*) Glycerol Monoethers and Analogues

(*b*) Substituted Alkanediols and Analogues

(*c*) Benzoxazole Analogues

(*d*) Imidazoline Analogues

(*e*) Miscellaneous Drugs.

A. *Glycerol Monoethers and Analogues*

A number of α-substituted glycerol ethers possessing potent centrally acting muscle relaxant properties have been used clinically.

Examples : Mephenesin ; Chlorphenesin Carbamate ; Methocarbamol.

A1. *Mephenesin* INN, BAN, USAN,

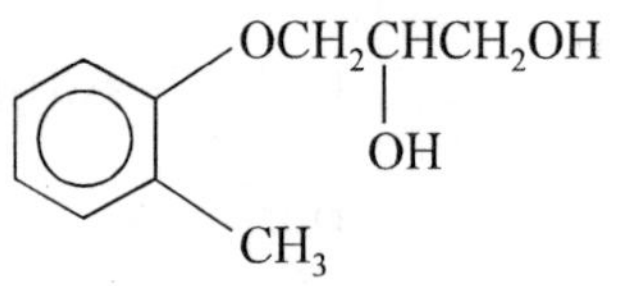

3-(*o*-Methylphenoxy)-1, 2-propanediol ; 3-(*o*-Tolyloxy) propane-1, 2-diol ; Mephenes; Cresioxydiol ; B.P.C. 1973, N.F. XII

Tolserol[R] (Squibb) ; Tolyspaz[R] (Alcon)

Synthesis :

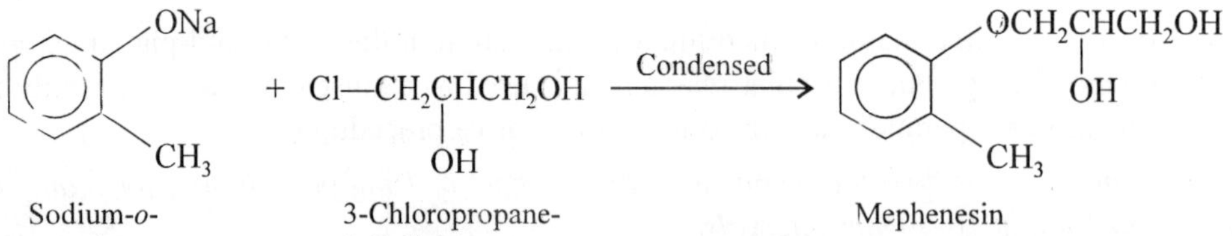

Sodium-*o*-   3-Chloropropane-   Mephenesin
cresolate   1, 2-diol
                     (α-Chlorohydrin)

It may be prepared by the condensation of sodium-*o*-cresolate with 3-chloropropane-1, 2-diol and the resulting product is recrystallized from alcohol.

Mephenesin relaxes hypertonic muscles, decreases response to sensory stimuli, and depresses superficial reflexes. It is used for the symptomatic relief of muscular spasm, and hyperkinetic conditions, such as parkinsonism, athetosis and chorea. It is also used in the treatment of anxiety and tension. Its action lasts up to 3 hours.

Dose : Usual, oral, 0.5 or 1 g 1 to 6 times per day as per requirement.

A2. *Chlorphenesin Carbamate* BAN, USAN, *Chlorphenesin* INN,

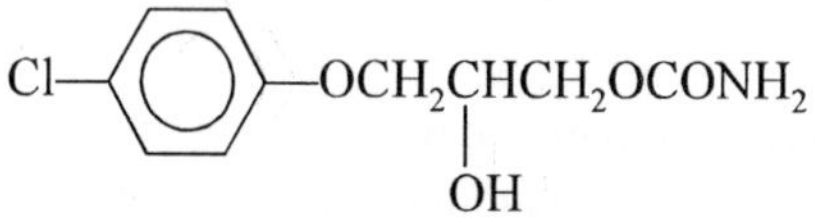

$$Cl—\langle\bigcirc\rangle—OCH_2CHCH_2OCONH_2$$
$$\qquad\qquad\qquad\quad |$$
$$\qquad\qquad\qquad\ OH$$

3-(*p*-Chlorophenoxy)-1, 2-propanediol 1-carbamate ; 1, 2-Propanediol, 3-(4-chlorophenoxy)-, 1-carbamate Maolate[(R)] (Upjohn)

Synthesis :

3-(*p*-Chlorophenoxy) 1, 2-propanediol is first prepared by the alkylation of *p*-chlorophenol with 1-chloropropan-2, 3-diol which on treatment with phosgene selectively forms the terminal carbamoyl chloride, namely, 3-(*p*-chlorophenoxy)-2-hydroxy propanyl oxychloride. The resulting product on amination yields chlorphenesin carbamate.

It is employed for the symptomatic relief of muscular spasm. Chlorphenesin also finds its use as an antifungal agent.

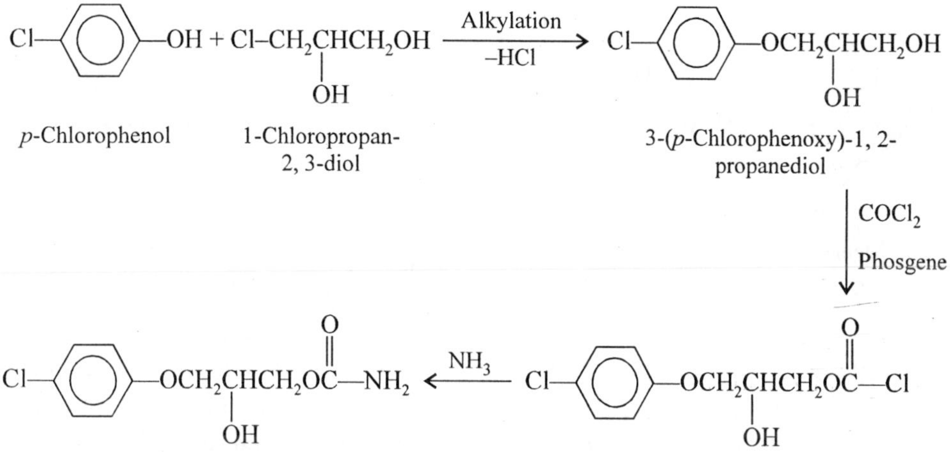

p-Chlorophenol	1-Chloropropan-2, 3-diol

3-(*p*-Chlorophenoxy)-1, 2-propanediol

COCl$_2$
Phosgene

Chlorphenesin Carbamate

3-(*p*-Chlorophenoxy)-2-hydroxy propanyloxy–chloride

Dose : Initial, usual 800 mg 3 times a day reduced to 400 mg 4 times daily or less as required.

A3. *Methocarbamol* INN, BAN, USAN,

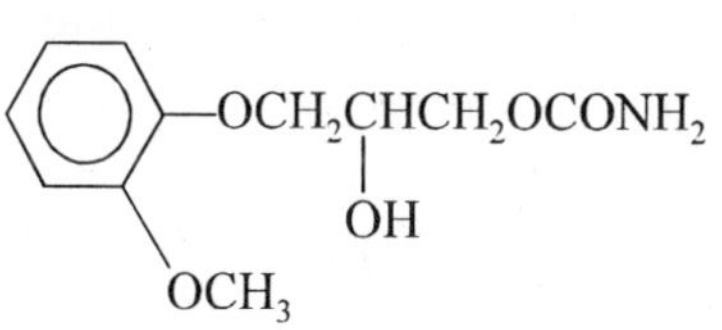

$$\langle\bigcirc\rangle—OCH_2CHCH_2OCONH_2$$
$$\qquad\qquad\qquad |$$
$$\qquad\qquad\quad\ OH$$
$$\quad |$$
$$OCH_3$$

3-(*o*-Methoxyphenoxy)-1, 2-propanediol 1-carbamate ; 1, 2-Propanediol, 3-(2-methoxyphenoxy)-, 1-carbamate ; Guaiphenesin Carbamate ;

U.S.P., N.F.

Robaxin[(R)] (Robins)

Synthesis :

Application of the three-step sequence to catechol mono-methylether affords methocarbamol as shown below.

Owing to its poor solubility, as compared to mephenesin, its absorption through the gastrointestinal tract is rather slow, which is responsible for its longer onset and duration of action. It is employed in the treatment of muscle spasm caused by musculoskeletal disorders, tetanus and injury.

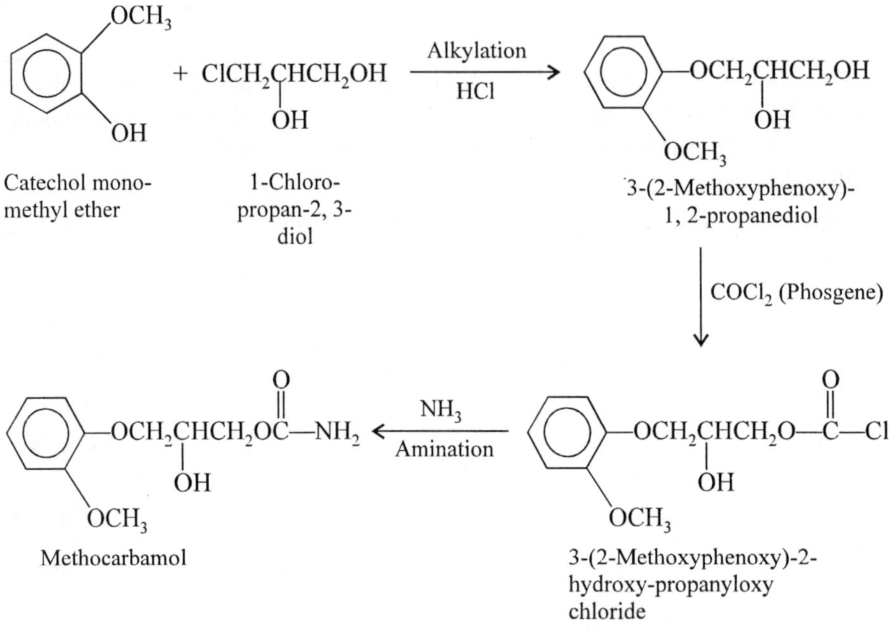

It is also used in the treatment of parkinsonism, cerebrovascular mishaps and cerebral palsy.

Does : *Initial, oral, 1.5 to 2 g 4 times daily for the first 2 or 3 days followed by 2.25 to 4.5 g per day in 2 or 4 divided doses ; i.v., 1 to 3 g per day administered at a rate not exceeding 0.3 g per minute ; i.m., 1 g every 8 hours.*

B. *Substituted Alkanediols and Analogues*

A good number of 1, 3-alkanediols and their structural analogues have been reported to be reasonably potent muscle relaxant drugs. A few examples of this group of compounds are discussed below.

Examples : Meprobamate ; Carisoprodol ; Tybamate ; Metaxalone.

B1. *Meprobamate* INN, BAN, USAN,

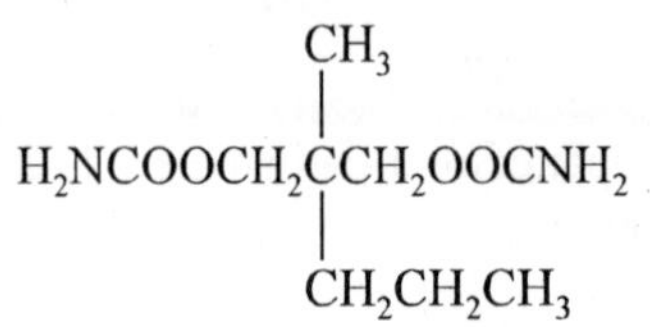

$$\underset{\underset{CH_2CH_2CH_3}{\overset{\overset{CH_3}{|}}{|}}{H_2NCOOCH_2CCH_2OOCNH_2}}$$

2-Methyl-2-propyl-1, 3-propanediol dicarbamate ; 1, 3-Propanediol, 2-methyl-2-propyl-, dicarbamate ; 2-Methyl-2-propyl-trimethylene dicarbamate;

B.P., U.S.P., Eur. P., Int. P., Ind. P.

Equanil[R] (Wyeth) ; SK-Bamate[R] (SK & F) ; Miltown[R] (Wallace)

Synthesis :

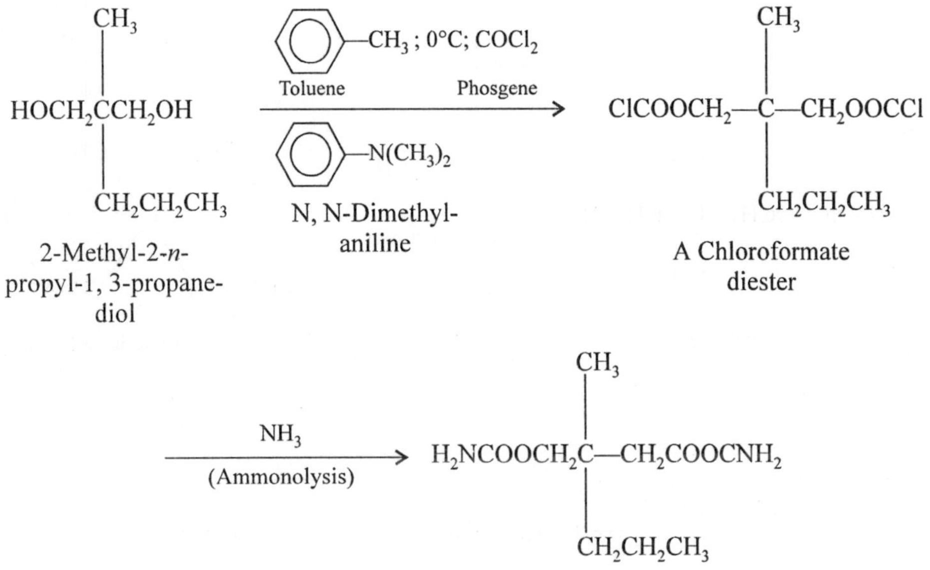

$$
\underset{\substack{\text{2-Methyl-2-}n\text{-}\\\text{propyl-1, 3-propane-}\\\text{diol}}}{\underset{CH_2CH_2CH_3}{\overset{CH_3}{HOCH_2CCH_2OH}}}
\xrightarrow[\substack{\underset{\text{N, N-Dimethyl-}}{\text{aniline}}\\ \ce{C6H4-N(CH3)2}}]{\substack{\ce{C6H4-CH3} ;\ 0°C;\ COCl_2 \\ \text{Toluene} \qquad \text{Phosgene}}}
\underset{\substack{\text{A Chloroformate}\\\text{diester}}}{\underset{CH_2CH_2CH_3}{\overset{CH_3}{ClCOOCH_2-C-CH_2OOCCl}}}
$$

$$
\xrightarrow[\text{(Ammonolysis)}]{NH_3}
\underset{\substack{\text{Meprobamate}}}{\underset{CH_2CH_2CH_3}{\overset{CH_3}{H_2NCOOCH_2C-CH_2COOCNH_2}}}
$$

2-Methyl-2-*n*-propyl-1, 3-propane diol dissolved in toluene is condensed at 0°C with phosgene in the presence of dimethylaniline gives the chloroformate diester, which when subjected to ammonolysis yields meprobamate.

It possesses anticonvulsant and muscle relaxant properties. It has also been used as mild tranquilizer in the treatment of anxiety and tension but has now been more or less replaced by the benzodiazepines, *e.g.,* diazepam.

Dose : *Usual, oral, 400 mg 3 or 4 times per day ; usual, i.m., 400 mg every 3 or 4 hours ; children, 25 mg per kg body weight per day in divided doses.*

B2. *Carisoprodol* INN, BAN, USAN,

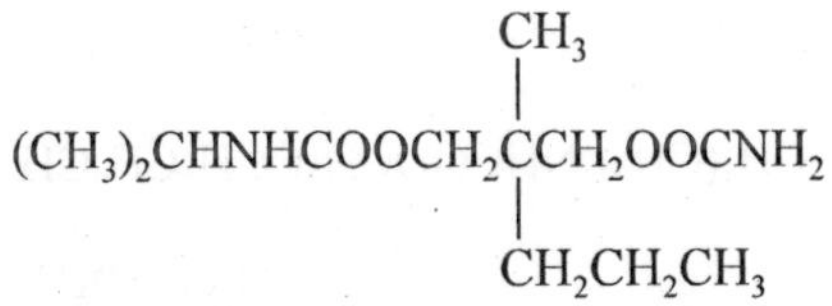

$$
\underset{CH_2CH_2CH_3}{\overset{CH_3}{(CH_3)_2CHNHCOOCH_2CCH_2OOCNH_2}}
$$

2-Methyl-2-propyl-1, 3-propanediol carbamate isopropylcarbamate ; N-Iso-propylmeprobamate.

Carisoma[R] (Pharmax, U.K.) ; Rella[R] (Schering-Plough)

Synthesis :

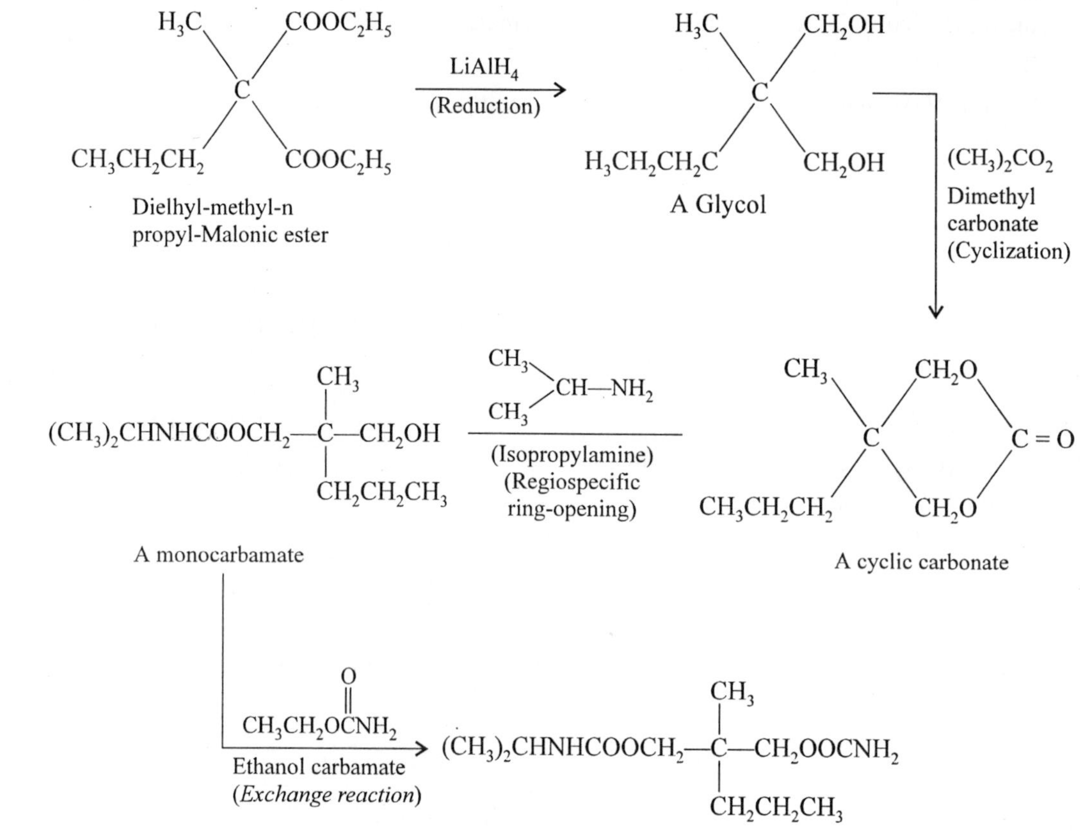

Reduction of diethyl-methyl-*n*-propyl malonic ester with lithium aluminium hydride yields the corresponding glycol, which on treatment with dimethyl carbonate undergoes cyclization and affords the cyclic carbonate. Ring cessation of the resulting product by the aid of isopropyl amine proceeds regiospecifically to afford the corresponding monocarbamate. Finally, the remaining hydroxyl group undergoes an exchange reaction in the presence of ethanol carbamate to yield carisoprodol.

Its actions are similar to those of mephenesin. The duration of action ranges between 4 to 6 hours. It is usually employed for the symptomatic relief of muscular spasm.

Dose : *Usual, adult, 350 mg 4 times a day.*

B3. *Tybamate* INN, BAN, USAN,

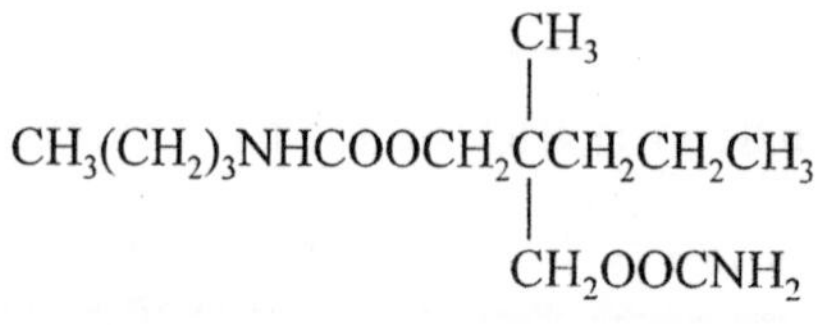

$$CH_3(CH_2)_3NHCOOCH_2\overset{\overset{\displaystyle CH_3}{|}}{\underset{\underset{\displaystyle CH_2OOCNH_2}{|}}{C}}CH_2CH_2CH_3$$

2-(Hydroxymethyl)-2-methylpentyl butylcarbamate ; Carbamic acid, butyl-, 2-[[(aminocarbonyl) oxy] methyl]-2-methylpentyl ester ;

N.F. XIII

Tybatran(R) (Robins)

Synthesis :

$$\underset{\substack{\text{Diethyl methylpropyl-}\\\text{malonate}}}{\text{H}_5\text{C}_2\text{OOC}-\overset{\overset{\displaystyle\text{CH}_3}{|}}{\underset{\underset{\displaystyle\text{COOC}_2\text{H}_5}{|}}{\text{C}}}-\text{CH}_2\text{CH}_2\text{CH}_3} \xrightarrow[\text{H}_2\text{SO}_4 \text{ (dil.)} ;]{\text{Ether; LiAlH}_4 ;} \underset{\substack{\text{2-Methyl-2-propyl-1, 3-}\\\text{propanediol}}}{\text{HOCH}_2-\overset{\overset{\displaystyle\text{CH}_3}{|}}{\underset{\underset{\displaystyle\text{CH}_2\text{OH}}{|}}{\text{C}}}-\text{CH}_2\text{CH}_2\text{CH}_3}$$

COCl$_2$ (Phosgene) ;
Toluene ;
Dimethylaniline ;

$$\underset{\substack{\text{2-Methyl-2-propyl-3-hydroxy-}\\\text{propyl butylcarbamate}}}{\text{CH}_3(\text{CH}_2)_3\text{NHCOOCH}_2-\overset{\overset{\displaystyle\text{CH}_3}{|}}{\underset{\underset{\displaystyle\text{CH}_2\text{OH}}{|}}{\text{C}}}-\text{CH}_2\text{CH}_2\text{CH}_3} \xleftarrow[\text{n-Butylamine} ;]{\text{CH}_3(\text{CH}_2)_3\text{NH}_2} \underset{\substack{\text{2-Methyl-2-propyl-3-hydroxy-}\\\text{propyl chlorocarbonate}}}{\text{ClCOOCH}_2-\overset{\overset{\displaystyle\text{CH}_3}{|}}{\underset{\underset{\displaystyle\text{CH}_2\text{OH}}{|}}{\text{C}}}-\text{CH}_2\text{CH}_2\text{CH}_3}$$

$$\xrightarrow[{[(\text{CH}_3)_2\text{CHO}]_3\text{Al}}]{\text{Ethyl urethane} ;} \underset{\substack{\text{Aluminium Isopropoxide}\\\text{Xylene} ; \Delta;}}{} \underset{\text{Tybamate}}{\text{CH}_3(\text{CH}_2)_3\text{NHCOOCH}_2-\overset{\overset{\displaystyle\text{CH}_3}{|}}{\underset{\underset{\displaystyle\text{CH}_2\text{OOCNH}_2}{|}}{\text{C}}}-\text{CH}_2\text{CH}_2\text{CH}_3}$$

2-Methyl-2-propyl-1, 3-propanediol is prepared by reacting diethyl methylpropyl malonate in ether in the presence of lithium aluminium hydride and then treated with dilute sulphuric acid. This on treatment with phosgene in toluene by means of dimethylaniline yields 2-methyl-2-propyl-3-hydroxypropyl chlorocarbonate, which on reaction with *n*-butylamine forms 2-methyl-2-propyl-3-hydroxypropyl butylcarbamate. The resulting product on treatment with ethyl urethane in the presence of aluminium isopropoxide in boiling xylene yields ethanol during transesterification which is removed from the reaction mixture simultaneously and tybamate is obtained.

Its actions and uses are similar to its congener meprobamate. It has been used in the treatment of anxiety and tension states in patients having psychoneurotic disorders.

Dose : *250 to 500 mg 3 or 4 times per day.*

B4. *Metaxalone* INN, BAN, USAN,

5-[3, 5-Xylyloxy) methyl]-2-oxazolidinone ; 2-Oxazolidinone, 5-[(3, 5-dimethyl-phenoxy) me-thyl]-

Skelaxin[R] (Robins)

Synthesis :

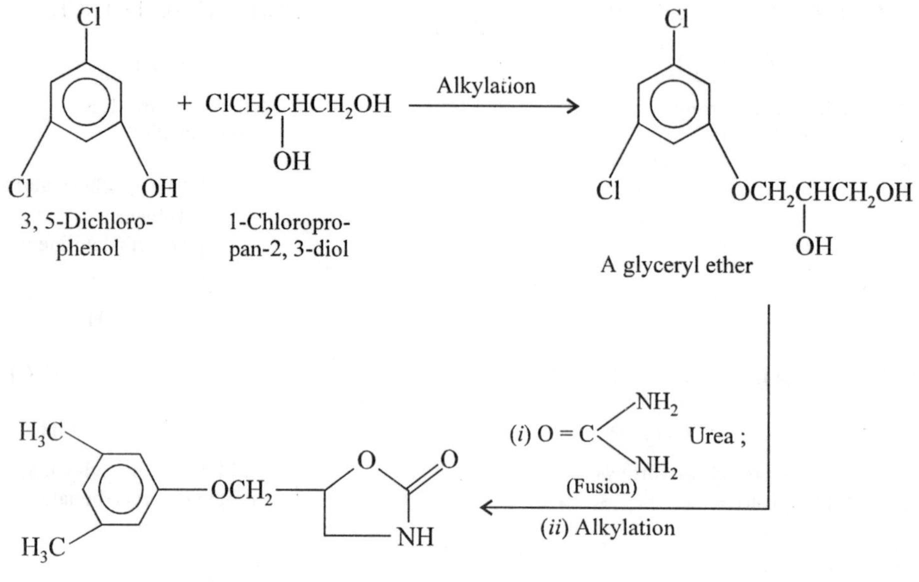

3, 5-Dichloro-
phenol

1-Chloropro-
pan-2, 3-diol

A glyceryl ether

(*i*) Urea ; (Fusion)

(*ii*) Alkylation

Metaxalone

3, 5-Dichlorophenol on alkylation with 1-chloropropan-2, 3-diol affords a glyceryl ether which on treatment with urea and subsequent alkylation yields metaxalone.

Metaxalone is used for the relief of acute muscle spasm resulting from various injuries or strains. Because of its potential toxicity, it has been superseded by other durgs belonging to this class.

Dose : *Usual, 800 mg 3 or 4 times per day.*

C. *Benzoxazole Analogues*

Two structural analogues of benzoxazole have gained prominence as potent muscle relaxants.

Examples : Chlorzoxazone ; Zoxazolamine.

C1. *Chlorzoxazone* INN, BAN, USAN,

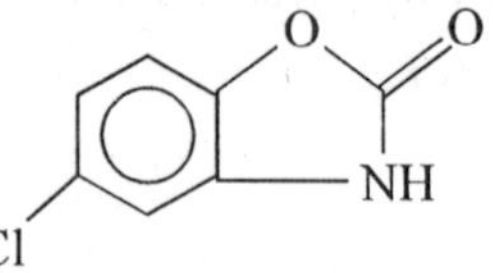

5-Chloro-2-benzoxalinone ; 2(3H)-Benzoxazolone, 5-chloro-; Chlorobenzoxazolinone. U.S.P.

Paraflex[R] (McNeil)

Synthesis :

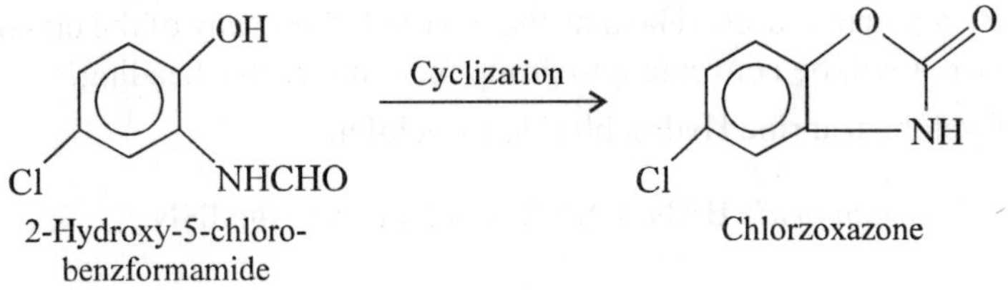

2-Hydroxy-5-chloro-
benzformamide

Chlorzoxazone

Chlorzoxazone may be prepared by the cyclization of the 2-hydroxy-5-chlorobenzformamide.

It is used for the treatment of painful muscle spasm associated with musculoskeletal disorders, such as spondolytis, sprains and muscle strains. It is also recommended sometimes for vertebral disk disorders and cervical root syndrome.

Dose : *Usual, intial, 500 mg 3 or 4 times per day, maintenance dose 250 mg.*

C2. *Zoxazolamine* INN, USAN,

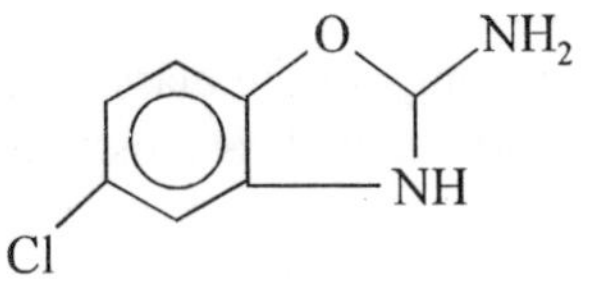

2-Amino-5-chlorobenzoxazole ;

N.F. XI

Due to the significant hepatotoxicity properties of zoxazolamine it is no longer used in clinical therapy.

D. *Imidazoline Analogue*

Dantrolene, an imidazoline analogue has been found to possess muscle relaxant characteristics, *e.g.,* Dantrolene Sodium.

D1. *Dantrolene Sodium* BAN, USAN, *Dantrolene* INN,

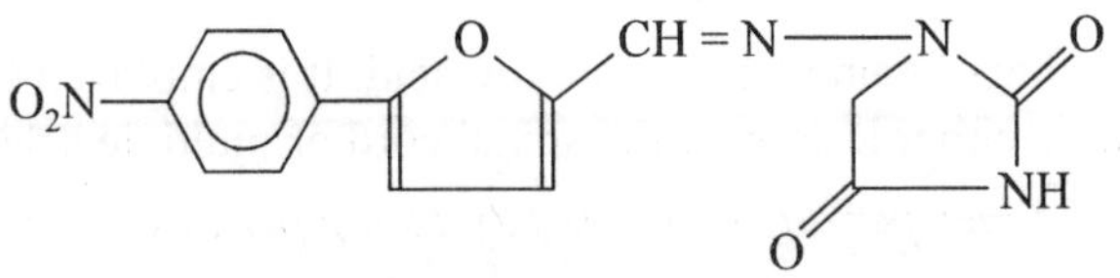

1-[[5-(*p*-Nitrophenyl) furfurylidene] amino] hydantoin sodium salt hydrate ; 2-Imidazolidinedione, 1-[[[5-(4-nitrophenyl)-2-furanyl]-methylene] amino]-, sodium salt, hydrate (2 : 7).

Dantrium [R] (Eaton)

Dantrolene is a skeletal muscle relaxant which may possess either central or peripheral components of action. It is mostly employed for the symptomatic relief of muscular spasm caused by stroke, spinal cord injury and cerebral palsy.

Dose : *Initial, oral, 25 mg per day, slowly increased over a period of 7 weeks to 100 mg 3 to 4 times per day.*

E. *Miscellaneous Drugs*

There exist a few potent muscle relaxants that do not fall into any of the classifications discussed above (A-D) and hence it will be convenient to group them under this heading.

Examples : Cyclobenzaprine Hydrochloride ; Baclofen,

E1. *Cyclobenzaprine Hydrochloride* BAN, USAN, *Cyclobenzaprine* INN,

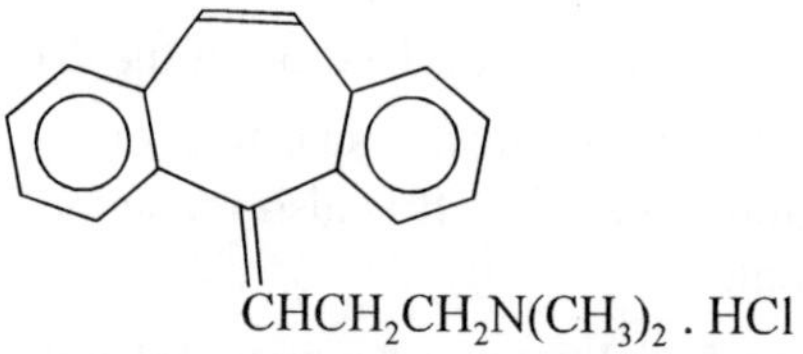

$$CHCH_2CH_2N(CH_3)_2 \text{ . HCl}$$

N, N-Dimethyl-5H-dibenzo [*a, d*] cycloheptane- Δ^5, γ-propylamine hydrochloride; 1-Propanamine, 3-(5H-dibenzo [*a, d*] cyclohepten-5-ylidene)-N, N-dimethyl-, hydrochloride ; Proheptatriene Hydrochloride.

Cyclobenzaprine hydrochloride belongs to the class of centrally acting muscle relaxant and is chemically related to the tricyclic antidepressants. It is mostly employed for the symptomatic relief of muscle spasm.

Dose : *Usual, oral, 10 mg 3 times a day and it must not exceed 60 mg per day.*

E2. *Baclofen* INN, BAN, USAN,

$$H_2NCH_2CHCH_2COOH$$

Cl

β-(Aminomethyl)-*p*-Chlorohydrocinnamic acid ; Butanoic acid, 4-amino-3-(4-chlorophenyl)-; Lioresal[R] (Ciba-Geigy) ;

Baclofen is an analogue of gamma aminobutyric acid. It is employed for the symptomatic relief of muscular spasm caused by either lesions of the spinal cord or multiple sclerosis.

Dose : *Initial 5 mg 3 times per day increased by 15 mg per day every 4th day to 20 mg 3 times a day.*

3. GENERAL MECHANISM OF ACTION OF MUSCLE RELAXANTS

The mechanism of action of the '*muscle relaxants*' shall now be discussed on a broader perspective under the following *two* major categories, namely :

(*a*) Neuromuscular blocking drugs, and

(*b*) Centrally acting muscle relaxants.

[A] Neuromuscular Blocking Drugs

In general, the 'drugs' belonging to this category invariably check the somatic motor nerve impulses from initiating the contractile responses in the effector skeletal (striated) muscles, thereby causing a paralysis of the muscles.

However, this specific category of '*drugs*' may be further sub-divided into *two* heads, namely :

(*i*) Competitive (or stabilizing) paralyzants, and

(*ii*) Depolarizing paralyzants.

(*a*) Competitive Paralyzants

These are also known as the competitive neuromuscular blocking drugs. In a situation, when the impulses in the somatic nerve arrive at the specific region located in the nerve terminals in the motor end-plate, they eventually elicit the release of acetylcholine (ACH), which in turn gets diffused to the post-synaptic motor end-plate membrane. Thus, ACH combines with nicotinic cholinergic receptors to activate them that ultimately leads to the opening of transmembrane ion channels, ion-flow, and as a result affords membrane depolarization. Importantly, end-plate membrane depolarization is usually accompanied by depolarization of the muscle membrane and ultimately leads to '*muscle contraction*'. In short, any plausible and feasible interruption of the aforesaid sequence of events gives rise to the muscular paralysis.

Therefore, the '*competitive paralyzants*' normally found to combine with the nicotinic receptors and occupy them strategically without causing any activation. Furthermore, ACH cannot activate the already preoccupied receptors, consequently the motor nerve impulses are unable to evoke contractions, and, hence, paralysis takes place. A few of them, however, take shelter in the receptor-operated ionophore which subsequently minimise the prevailing activation of the *postsynaptic membrane.*

(*b*) Depolarizing paralyzants

These are also termed as *depolarizing neuromuscular blocking drugs.* They are '*nicotinic agonists*' that essentially interact (just like ACH) with the post synaptic nicotinic receptors to cause a depolarization of the membrane at the motor end-plate specifically. In reality, their 'temporary stay' at the end-plate is a little longer (unlike ACH) and, therefore, the post synaptic membrane virtually remain depolarized. Because, the muscle membrane as well as the resulting contraction can only be excited by a fresh lease of depolarization, the muscle remains paralyzed ultimately. In other words, the virtual initiation for the conducted muscle impulse is due to the short-stayed fall in end-plate membrane potential, and not caused due to the ensued depolarization.

Ultimately, the motor end-plate membrane gets repolarized inspite of the continued presence of the '*drug substance*' by virtue of a shift in receptor conformation. Though the membrane is normally poised for a new lease of depolarization, yet ACH and the motor nerve impulses do not succeed to evoke an appreciable response. Perhaps, this could be due to the fact that the nicotinic receptor is not positioned in its desired configuration. It has also been observed that during this critical phase, the neuromuscular blockade usually occurs specifically on certain characteristic features of competitive blockade ; and this may even get antagonized partially by the aid of *anticholinesterases.*

An Ideal Neuromuscular Blocking Drug. The ultimate objective of an ideal neuromuscular blocking drug should essentially possess the following characteristic features, namely :

(1) A *bis*-quaternary chemical structure having a sufficient separation (gap), between the two N-atoms, so as to cause an effective and significant level of blockade. It has already been established that the distance between the said two N-atoms must be in the range of 10-11 Å (*i.e.,* about 10-12 atoms apart).

(2) Always the choice for a '*competitive antagonist*' is preferred. Experimental evidences have revealed that the presence of a '*bulky and large hydrocarbon environment*' (*i.e.,* causing steric hindrance) in the viccinity of the '*cationic nitrogens*' effectively prevents the access of ACH to the postsynaptic receptor areas located on the motor-end plate.

(3) There should be either minimal or absolutely negative action at the cholinergic receptors except those of the nicotinic subtypes at the neuromuscular junction. In this manner the '*quaternization*' prevents a free access to the centrally located receptors, and, therefore, the proper distance separating the '*cationic zones*' present in the '*drug substance*' decreases the activity on the ganglionic receptors significantly.

(4) It has been well established that the '*molecular drug designs*' which would ultimately pave its way to an efficient metabolic degradation and/or variable pattern of exretions absolutely not found to solely dependent on enzyme catalysts.* Interestingly, a '*drug substance*' having such characteristic qualification would certainly exhibit explicitely both a rapid recovery and a shorter-duration of action from the ensuing blockade whether it could be caused due to either *hepatic dysfunction or renal dysfunction* or *pharmacogentic effects*.

[B] Centrally Acting Muscule Relaxants

It is reasonably proven analogy that the '*cell bodies*' present in the somatic motor nerves invariably lie within the spinal cord and, therefore, very much within the CNS. It has been observed that the prevailing activity of motor neurons is mostly affected by a host of such cardinal factors, such as : facilitatory and inhibitory modulation through feedback from contralateral** and ipsilateral*** stretch, besides other receptors ; various centres of the brain.

Therefore, spasticity**** may arise from particularly the musculoskeletal injury, that invariably give rise to a duration from a standard with regard to the *observed afferent impulse traffic* into the spinal cord. In this manner an inflicted injury slowly leads to the disease related to either motor nerves, or interneurons within the cord, or sensory neurons located in the sensory ganglia ; and ultimately boils down to the '*brain disorders*' thereby changing the regular flow of suprasegmental impulses to the motor neurons. Thus, the virtual cause of impairment to the motor neurons in the brain leads to involuntary movement as could be seen in Parkinsonism, chorea and palsies.

In actual practice it is, however, difficult to make a clear-cut distinction between the disorder caused either within the spinal cord or due to musculoskeletal dysfunction *vis-a-vis* the selectivity of *drug substances*' which evidently remains at a low-ebb. This is perhaps on account of the collective neurons engaged intimately in the reflex arcs that are found to be insufficiently and qualitatively variant from the prevailing motor and sensory neurons with regard to the '*chemical sensitivity*' in order to allow a specific selective depression of the hyperactive influences on the motor neuron.

In short, the centrally acting muscle relaxants find their abundant application in a plethora of conditions, namely : strains and sprains, which may ultimately be responsible for causing acute muscle spasm. Besides, they also particularly possess interneuronal-blocking characteristics at the level of the spinal card, which may give rise to the much desired relaxation of the skeletal muscle*****. Interestingly, most of them exhibit a distinct general CNS-depressant activities.

---

*A drug Albert (1985) suitably describes as '*self-canceling*'.

**Originating in or affecting the opposite side of the body.

***Affecting the same side of the body.

****Increased tone of contractions of muscles causing stiff and awkward movements.

*****Berger FM, In : Usdin E and Forrest IS (eds). Psychotherapeutic Drugs, Pt II, Marcell Dekker, New York, p-1089, 1977.

4. MODE OF ACTION OF SOME SPECIFIC MUSCLE RELAXANTS

The probable and proven mode of action of certain selected muscle relaxants discussed in this chapter shall be dealt with in the section that follows.

4.1. Tubocurarine Chloride

It is found to be not absorbed directly from the gut (intestine). After IV administration the drug simply gets disappeared so readily from the plasma, with a distribution half-life of approximately 12 minutes ; whereas, its terminal plasma half-life ranges between 1 to 3 hours. It is mostly excreted through urine upto 43%; and the remaining gets degraded subsequently either in the liver or in the kidneys. However, in instances where either hepatic failure or renal failure occur it may prolong the half-life of the drug appreciably.

4.2. Metocurine Iodide

It is found to be 2 to 4 times more potent than tubocurarine (+ TB). In man, it is eliminated chiefly by renal and biliary excretions ; the half-life is about 3.5 hour. It is observed that it can safely pass across the placental barrier.

4.3. Gallamine Triethiodide

It is a potent skeletal muscle relaxant which essentially works by blocking the neuromuscular transmission almost identical to that of (+TB). It markedly differs from d-tubo-curarine (+TB) by virtue of its *two* inherent characteristic features, namely ; (*a*) possesses a reasonably strong **vagolytic effect***, and (*b*) it continues to lower the prevailing '*neuromuscular function*' after administration of successive doses which cannot be overcome by *cholinesterase inhibitors*. Besides, it has also been shown to exhibit '*muscarinic antagonistic characteristic features*', and gets bound intimately with greater affinity to the M_2-receptors than the corresponding M_1-receptors. However, the second characteristic feature is solely responsible for attributing its strong and prevalent vagolytic action.**

4.4. Pancuronium Bromide

Its mechanism of action normally is assumed to be almost similar to that of d-tubocurarine (+TB), but the '**dose-response curve**' is rather steeper in nature thereby suggesting a possible difference. The cardinal differences with respect to + TB are, namely : (*a*) it fails to block the autonomic ganglia (side-effect) ; and (*b*) it rarely releases '*histamine*', hence it fails to cause either bronchospasm or hypotension. It is found to have little effect on the circulatory system.

Interestingly, anticholinesterases, ACH and K^+ ion antagonize competitively pancuronium bromide effectively ; however, its activity is virtually enhanced by general anaesthetics, for instance : halothane, ether, enflurane etc. (see Chapter 3). Therefore, the latter substantial potentiation in pharmacological activity is particularly useful to the '**anaesthetist**' due to the fact that it is administered invariably as an '*adjunct*' to the anaesthetic procedure in order to cause simultaneous relaxation of the skeletal muscle.

4.5. Suxamethonium Chloride (Succinylcholine Chloride)

It usually has an extemely transient duration of action by virtue of the fact that it undergoes rapid hydrolysis by the help of serum butyryl (pseudo) cholinesterases. However, a prolonged muscular relaxation may be accomplished by continuous IV infusion, and the degree of muscle paralysis is controlled adequately by fine-tuning the rate of infusion precisely. It has been observed that the '*drug*' does not

---

*An agent, chemical or surgical, that prevents function of the vagus (cranial) nerves.

Benabe JE *et al. AM. J. Hypentens*, **6 : 701, 1993.

cause liberation of histamine, but may give rise to hypersensitivity reactions occassionally. It effectively produces contractions of motor units (*fasciculations*) and axon reflux-conducted impulses on account of its ability in depolarizing the motor end plate. An excessive dose may produce temporary respiratory-depression. Importantly, its action, just contrary to that of +TB, is not antagonized by such drugs as : physostigmine, neostigmine or edrophonium chloride.

4.6. Decamethonium Bromide

It cannot undergo hydrolysis in the presence of cholinesterase ; and, therefore, gives rise to an appreciable prolonged duration of action. It is usually considered to be a depolarizing neuromuscular relaxant by virtue of the fact that it initiates an action of *'nicotinic depolarization'* at the site of motor end-plate membrane.

4.7. Mephenesin

The *'drug'* exerts its action by causing relaxation of the hypertonic (*i.e.,* in a state of greater than normal tension) muscles minimise the response to the sensory stimuli, and also causes a significant depression of the superficial reflexes. Its weak activity and transient effect are on account of the facile metabolism of the primary hydroxyl function. In has been observed that the *'carbamylation'* of the said moiety enhances its activity. Importantly, the *para*-chlorination affords an appreciable increase in the prevailing lipid-water partition coefficient and helps in blocking the *para*-position from undergoing hydroxylation as far as possible.

4.8. Methocarbamol

The most probable site for the metabolic attack are the *secondary hydroxyl function 'a'* and the *two ring positions 'b'* and *'c'* strategically located *opposite the ether moieties,* as shown below :

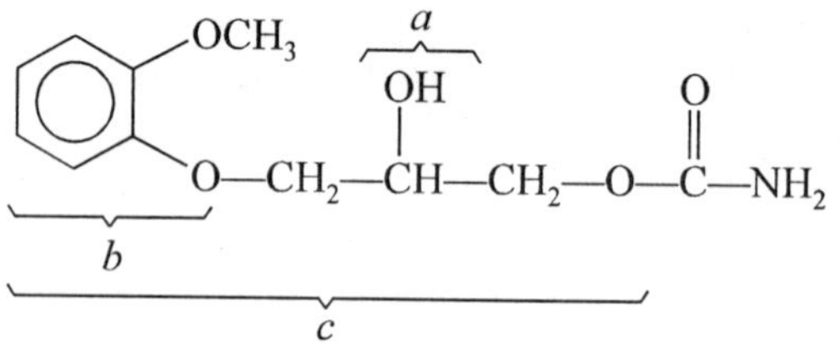

Methocarbamol

Its centrally acting muscle relaxant profile, after due parenteral administration is not only prompt but also intense enough to allow and facilitate orthopaedic procedures.

4.9. Meprobamate

It is also recognized as a potent sedative hypnotic drug ; and exerts a plethora of overall pharmacological characteristic features very much akin to *barbiturates* and *benzodiazepines.* However, the precise mechanism of action causing the anxiolytic effects is still not explicitly understood but it is believed that it may involve effects particularly on conductivity in certain specific areas of the brain.* It has already been shown that it does not seem to act by influencing the prevailing GABAergic systems. Interestingly, it is found to exhibit *interneuronal blocking activities* specifically in the area of the spinal cord ; therefore, it is said to be partially responsible for causing the much desired skeletal muscle relaxation. Besides, the inherent general CNS-depressant activities possessed by it may also mainly attribute towards the skeletal muscle relaxant activity.

---

*Berger FM ; In : Usdin E and Forrest IS (eds.) *Psycliotherapeutic Drugs*, Pt. II, Mercel Dekker, New York, p-1089, 1977

4.10. Chlorphensin Carbamate

The drug undergoes metabolism quite rapidly *via* the '*glucuronidation*' of the secondary hydroxyl function present in it. Its biological half-life in humans is 3.5 hours.

4.11. Carisoprodol

Its sedative and muscle relaxant activities specifically caused due to the reticulospinal depression. It has been established virtually that a certain portion of its muscle relaxant property is contributed due to analgesia, sedation and alleviation of anxiety status. Its on set of action occurs within a span of 30 minutes, while the duration of action lasts between 4 to 6 hours. The drug gets metabolized invariably in the liver ; and its elimination half-life is approximately 8 hours.

4.12. Tybamate

It is closely, related to meprobamate, wherein it has an additional *butyl moiety,* attached to the terminal amino function. It almost possesses the same spectrum of activity as that of meprobamate. Generally, the polyol compounds are not so effective and potent in *spasticity* on account of dyskinesia.

4.13. Metaxalone

This *drug substance* enjoys the reputation to exert muscle relaxant activities having a central nervous system focus of action. It gets metabolised by eliminated *via* hepatic metabolism ; and its half-life is between 2 to 3 hours. It attains peak blood levels within a span of 2 hours, while the duration of action lasts from 4 to 6 hours.

4.14. Chlorzoxazone

It acts by inhibiting the polysynaptic* reflexes both within the spinal cord and subcortical regions of the brain. It has been observed that more than 90% of it gets glucuronidated in the liver. The elimination half-life is about 60 minutes ; and the absorption time is from 3 to 4 hour.

4.15. Dantrolene Sodium

Its mechanism of action essentially differs from the classical neuromuscular blocking drugs, wherein its action is quite distal to either the neuro muscular junction or the nicotinic receptors. Alternatively, it has been amply proven that it suppresses the excitation-contraction coupling sequence by interfering with the corresponding release of calcium from the sarcotubular reticulum. In such a situation, the muscle fibres still very much give response to the nerve impulses ; however, the contractile response is decreased to a significant extent but never abolished completely. Hence, muscle weakness, rather than paralysis is normally accomplished as the ultimate outcome. Consequently, the '*fast muscle fibres*' (white) are affected more significantly in comparison to the '*slow muscle fibres*' (red). As the contractility of the intrafusal fibres present in the muscle spindles gets lowered appreciably, which in turn attenuate the spinal cord-mediated stretch reflexes ; and this provides a plausible explanation of the ability of dantrolene sodium to help in causing tremendous relief in certain types of acute muscular spasm. Furthermore, it is quite possible that perhaps a direct effect on the motor neurom may be involved in this rather narrow spectrum of activity, because the drug seems to exhibit CNS depressant activity to a certain degree.

4.16. Cyclobenzaprine Hydrochloride

The drug exerts its action by causing an appreciable depression in the suprasegmental (*i.e.,* upper) motor neurons in the brainstem. Besides, it also depresses to a certain degree the spinal motor neurons to lower the reflex skeletal muscle activity plus the tonus. Furthermore, it is found to cause inhibition to both the α- and γ-motor systems. It is employed invariably to afford substantial relief in spasm and pain that are linked with *musculoskeletal disorders.* It is intimately bound to plasma albumin. It has been observed to undergo conjugation and biotransformation to the corresponding glucuronides in the liver. It is excreted negligibly in its unchanged form into the urine, but to some extent excreted into milk. The elimination half-life ranges between 1 to 3 days.

4.17. Baclofen

The muscle relaxant actions of this drug invariably result from an action taking place within the *spinal cord,* which being the prime site where both monosynaptic and polysynaptic reflexes are usually prevented by it effectively. Baclofen, being a structural variant of GABA (*i.e.,* γ-aminobutyric acid) which is an inhibitory transmitter within the CNS, the partial activity of it is attributed to its agonist characteristics existing at the site of $GABA_B$ receptor ; and that is subsequently coupled to a *G-protein-activated* K^+ *channel*. Nevertheless, the exact mechanism of its action is still not yet fully understood. Importantly, its inherent ataxis as well as sedative properties are very much consistent with a similar type of action prevailing in the brain. More than 80% of the drug gets excreted in the urine. The elimination half-life ranges between 3 to 4 hours. Its oral absorption period is nearly 2 hours.

Probable Questions for B. Pharm. Examinations

1. Explain with the help of structure that every 'half-molecule' of TUBOCURARINE is identical to ATROPINE. Discuss the striking different between the two drug molecules.

2. How would you classify the neuromuscular blocking drugs ? Give structure, chemical name and uses of **one** potent drug from each category.

3. Explain the following :

   (*i*) The distance between the two onium groups present in decamethonium.

   (*ii*) Why decamethonium bromide exerts a much more prolonged effect.

   (*iii*) A non-depolarizing muscle relaxant having a steroidal moiety.

   (*iv*) Various steps involved in the synthesis of gallamine triethiodide from pyrogallol.

4. Classify the centrally-acting muscle relaxants and give the structure, chemical name and uses of **one** important member of each class.

5. Discuss the synthesis of a potent glycerol monoether analogue prepared from :

   (*i*) *p*-Chlorophenol, and

   (*ii*) 1-Chloropropane-2, 3-diol.

6. How would you synthesize meprobamate ? Discuss the various steps sequentially.

7. Name an important member of benzoxazole analogue employed as muscle relaxant. Give its one-step synthesis.

8. Give the structure, chemical name and uses of Dantrolene Sodium.

9. Discuss the synthesis of the following important muscule relaxants.

   (*i*) Chlorophenesin carbomate, and

   (*ii*) Methocarbamol

10. Enumerate the mode of action of various types of muscule relaxants by giving specific examples.

RECOMMENDED READINGS

1. A Burger (*Ed*) *Drugs Affecting the Central Nervous System, Medicinal Research*, Vol. 2, New York, Dekker (1968).

2. M Gordon *Psychopharmacological Agents* New York, Academic Press, Vols. 1 and 2 (1964) and (1965).

3. C.K. Cain and A P Roszkowski *Psychopharmacological Agents,* Vol. I (*Ed*) M Gordon, New York, Academic Press (1967).

4. M A Lipton, A Dimascio and K. F. Killam (*Eds*) *Psychopharmacology* : *A Generation of Progress* New York, Raven Press (1978).

5. D Lednicer and L A Mitscher *The Organic Chemistry of Drug Synthesis* New York, John Wiley and Sons (1995).

6. J.E.F. Reynolds (*Ed*) *Martindale* : *The Extra Pharmacopoeia* (30th edn) London, The Pharmaceutical Press (1992).

7. M C Griffiths (*Ed*) *USAN and the USP Dictionary of Drug Names-1986* Rockville, United States Pharmacopeial Convention, Inc. (1985).

8. Block JH and Beale JM (eds) : *Wilson and Gisvold's Textbook of Organic Medicinal and Pharmaceutical Chemistry*, Lippincott Williams and Wilkins, New York, 11th edn, 2004.

present in CNs association by exhibiting a distinct depressant action, by virtue of the fact that it has the ability to :

8   Central Nervous System Stimulants

1.   INTRODUCTION

Central nervous system (CNS) stimulants are drugs that produce generalized stimulation of the brain or spinal cord which may lead to convulsion. They are of limited therapeutic value because of their convulsant activities. There are, however, some that are used as respiratory stimulants (*e.g.,* Nikethamide) and others like the xanthine derivatives have many pharmacological actions and uses. Sympathomimetic amines like amphetamines and ephedrine, which are potent CNS stimulants, are discussed elsewhere.

A few central nervous system stimulants exhibit predominant central stimulant action, *e.g.,* strychnine, nikethamide, leptazol, picrotoxine, etc. ; others possess multiple side-effects, *.e.g.,* ephedrine and atropine act on the autonomic nervous system ; and finally a number of drugs do exert temporary stimulation of CNS in toxic doses, *e.g.,* local anaesthetics, santonin, salicylates. In usual practice, the central nervous system stimulants find their use in emergencies for prompt and short-term excitation of CNS, because *a prolonged stimulantion may be followed by depression.*

CNS-stimulants may also be defined as *'drug substances'* that most specifically afford an enhancement in excitability either very much within the different portions of the brain or the spinal cord. The most commonly observed, marked and pronounced stimulatory effects produced by a plethora of CNS-stimulants are essentially comprise of distinct wakening and enhanced motor function which give rise to a host of both covert and overt pharmacological actions, such as : individual feelings of increased mental alertness, lowered feeling of personal fatigue, enhanced concentration, apparent elevation in mood, and above all the increased energy and enthused motivation. It is, however, pertinent to state at this function that an excessive CNS stimulation may ultimately lead to serious and critical dose-dependent adverse effects, namely : mental anxiety, mental agitation, extreme nervousness, and sometimes epileptic seizures as well.

The suggested *modus operandi* of the ensued excitability of the CNS brings about a much so important and an intricate equilibrium between the excitatory and the inhibitory activity inside the brain. It has been established beyond any reasonable doubt that the *'excitatory transmitters'*, such as : glutamic and aspartic acids, are considered to be the most important neurotransmitters prevailing at the excitatory synapses at which strategical location their activities are duly mediated through either NMDA (N-methyl-D-aspartate) or non-NMDA (AMPA/quisqualate or Kainate) receptors. Just contrary to the above the predominant and extremely important *'inhibitory neurotransmitters'* are essentially glycine (α-amino acetic acid) and GABA (γ-aminobutyric acid). Besides, **adenosine-***a* neuromodulator serves as a major player in CNS excitation by exhibiting a distinct depresant action, by virtue of the fact that it has the ability : (*a*) to minimise *impulse-generated transmitter release* ; and (*b*) to check excitation of *post synaptic elements by direct hyperpolarization of the neuronal membrane.* Evidently, a plethora of clinically useful CNS stimulants cause excitation due to their proven antagonism prevailing at **glycine, GABA** and **adenosine** receptors. However, the host of *indirect-acting sympathomimetics* usually produce marked and pronounced CNS stimulation simply by increasing the actions of the *endogenous catecholamines* due to their inherent strong capability to either prevent or release adequately the uptake of endogenous catecholamines.

Salient Features : The important cardinal features of the CNS stimulants are enumerated below :

(1) Amphetamine the well-known *central sympathomimetic agent* and its close structural variants usually exhibit significant *alerting* and *antidepressant* activities but therapeutically find their abundant utility as *anorexiants.**

(2) Because of the immense abuse potential with **amphetamine** and particularly **methamphetamine,** the vigorous and intensive search is already on to suggest better and safer alternative medical treatments for the CNS stimulants to combat such critical disorders efficiently. *Modafinil,* after recent clinical trials, may prove to be a promising alternative candidate drug in treating narcolepsy.**

(3) The relatively older practice of employing CNS stimulants as respiratory stimulants particularly in instances of overdoses with depressant drugs. Recently their therapeutic usage in such critical conditions is no longer recommended on account of the following valid reasons and facts, namely :

(*a*) Antagonism of the *'depressant actions'* are found to be non-selective in nature.

(*b*) Both *epileptic seizures* and *cardiac arrythmias* may be inducted while making an attempt to reverse the process of respiratory depression.

(*c*) In order to provide a relatively safer and equally effective desired treatment of the patient the urgent and dire need of different supportive measures are an absolute necessity.

Interestingly, the proven ability of *caffeine*—a **methylxanthine** to enhance appreciably *'mental alterness'* is perhaps one of the main cardinal factors for the world-wide, high consumption of caffeine-containing beverages and natural *'drug-substances',* such as : tea, coffee, chickory, kola nut, areca nut etc. Caffeine is also found as an integral component in several analgesic drug formulations (dosage forms), including both prescription and non-prescription drugs (OTC-drugs***), although its exact efficaciousness in the control and management of pain is not yet well understood with concrete evidence.

2. CLASSIFICATION

The **central nervous system stimulants** may be classified into the following *three* categories, namelly :

(*i*) Xanthine Derivatives

(*ii*) Analeptics

(*iii*) Miscellaneous Central Nervous System Stilmulants.

---

*Agents that produce loss of appetite.

**A chronic ailment consisting of recurrent attacks of drowsiness and sleep during day time.

***Over-the-counter drugs.

2.1. Xanthine Derivatives

The **xanthine derivatives** as stated earlier have a wide spectrum of therapeutic applications ranging from their stimulation of cardiac muscle, enhanced diuresis, stimulation of CNS and finally a soothing relaxation of bronchi and the coronary arteries.

A few typical examples of the members of this category are : caffeine, theophylline, theobromine, aminophylline, etofylline and proxyphylline.

A. *Caffeine* BAN, USAN,

1, 3, 7-Trimethylxanthine ; 1H-Purine-2, 6-dione ; 7-Methyltheophylline ; B.P. ; U.S.P. ; Eur. P.

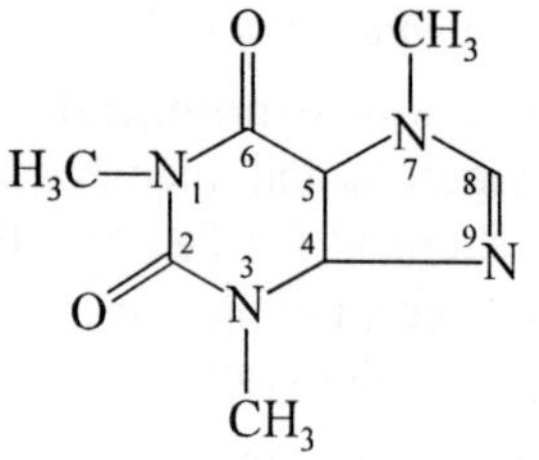

Caffeine is an alkaloid isolated from coffee, tea or the dried leaves of *Camellia sinensis* (*Theaceae*), or prepared synthetically.

Synthesis :

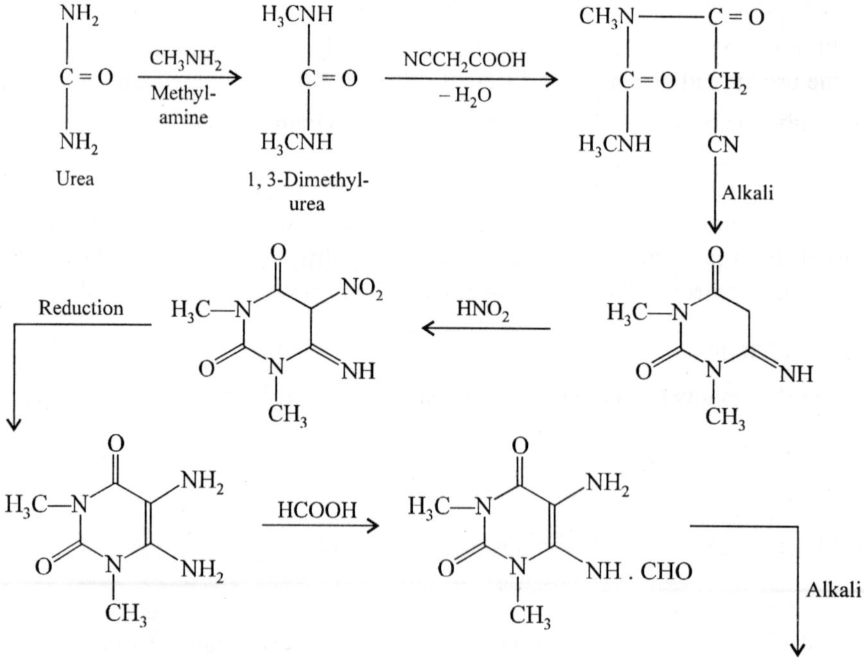

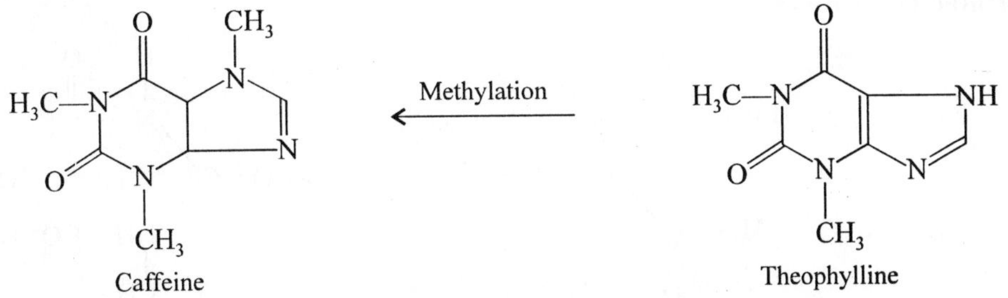

Caffeine   Methylation   Theophylline

1, 3-Dimethyl urea is prepared by the interaction of urea and methylamine, which upon treatment with cyanoacetic acid yields an open-chain nitrite with the elimination of a molecule of water. This resulting compound undergoes cyclization in the presence of alkali. The cyclized compound on treatment with nitrous acid, followed by reduction, reaction with formic acid and subsequently with alkali gives rise to the formation of theophylline, which upon methylation finally yields caffeine.

Caffeine is a potent central stimulant. It also acts on the cardiac muscle and on the kidneys. It stimulates the higher centres of the CNS thereby causing enhanced mental alertness and wakefulness. Caffeine helps in the stimulation of respiratory centres. Its diuretic action is due to enhanced glomerular filtration rate, increased renal blood flow and above all the reduction of the normal tubular reabsorption.

Dose : 100 to 500 mg, usual 200 mg as required.

B. *Theophylline* BAN, USAN,

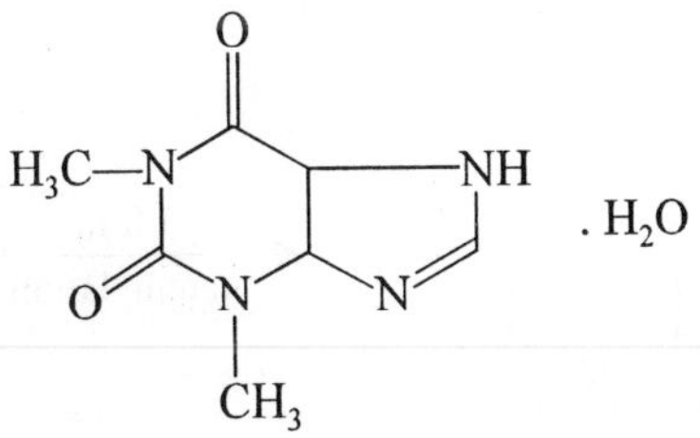

1, 3-Dimethylxanthine ; 1H-Purine-2, 6-dione, 3, 7-dihydro-1, 3-dimethyl monohydrate ; B.P., U.S.P., Eur. P., Int. P., Ind. P.

Constant-T[R] (Ciba-Geigy) ; Elixophyllin[R] (Berlex) ; Theo-24[R] (Searle)

Synthesis :

Theophylline is prepared by the method described under caffeine.

Theophylline is widely used for the treatment and symptomatic relief of acute and chronic bronchial asthma, bronchospasm, cardiac dyspnea and angina pectoris.

Dose : Usual, 200 mg 3 to 4 times per day.

C : *Theobromine* BAN,

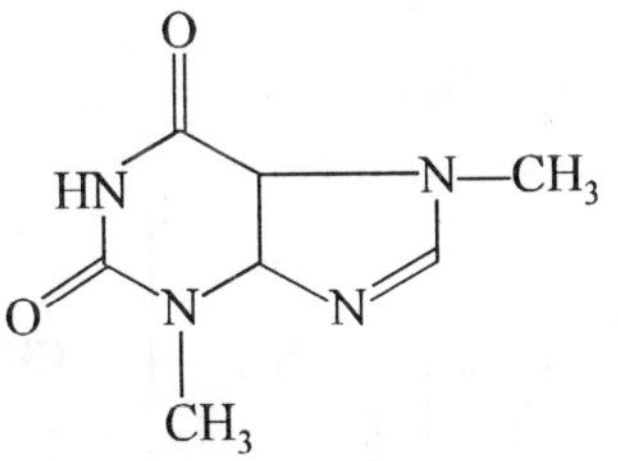

3, 7-Dimethylxanthine ; 3, 7-Dihydro-3, 7-dimethylpurine-2, 6 (1H)-dione ; B.P., Eur. P.

Theosalvose[R] (Techni-Pharma, Mon.)

Synthesis :

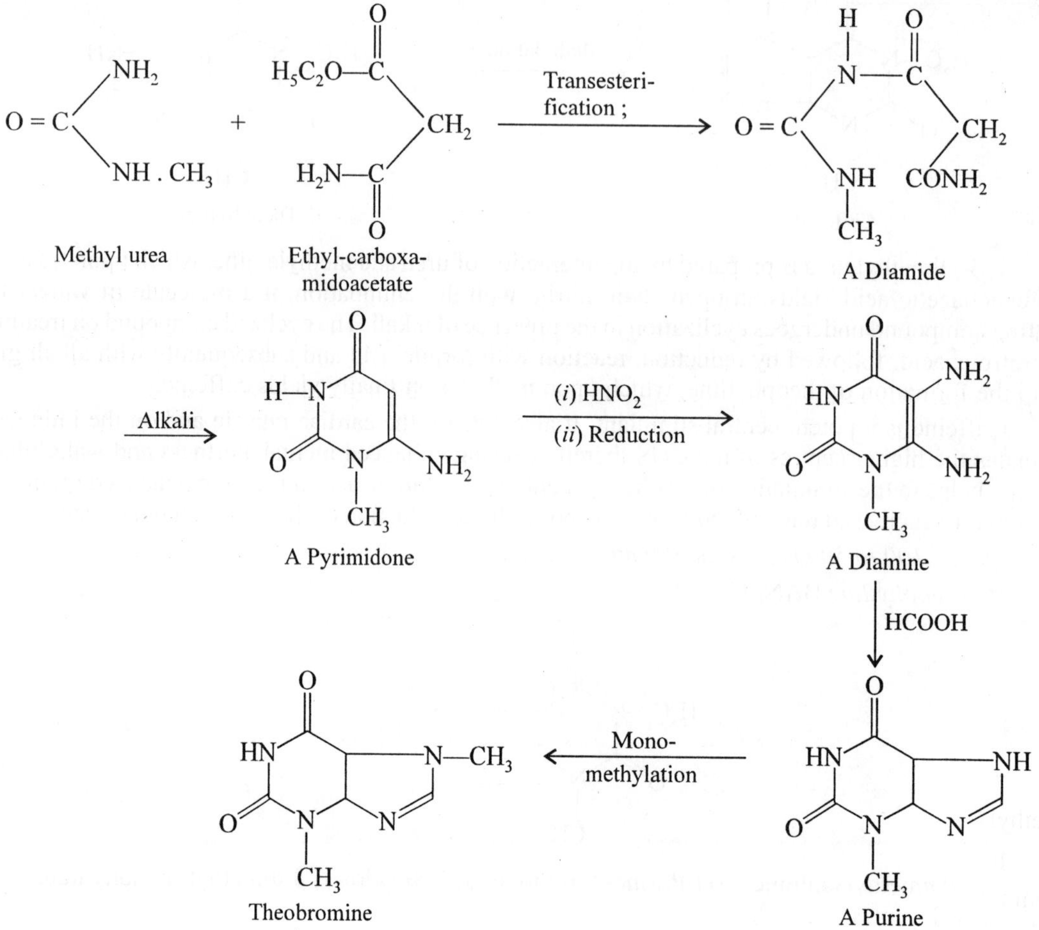

Methyl urea   Ethyl-carboxa-
midoacetate

A Diamide

A Pyrimidone

A Diamine

Theobromine

A Purine

Methyl urea and ethyl carboxamido acetate undergoes transesterification to yield a diamide which on treatment with alkali gives rise to a pyrimidone. The resulting product on reaction with nitrous acid followed by reduction gives a diamine which on being treated with formic acid produces a purine and when subjected to monomethylation yields finally the theobromine.

The action of theobrimine on the CNS is minimal and hence it may be employed for its other effects, namely, diuretic, effect on the coronary arteries, without showing significant side-action of central stimulation.

Dose : Usual, 500 mg.

D. *Aminophylline* INN, BAN, USAN,

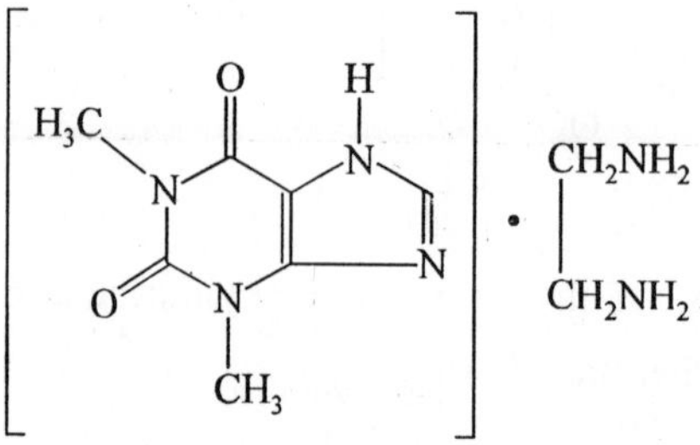

1H-Purine-2, 6-dione, 3, 7-dihydro-1, 3-dimethyl-, compound with 1, 2-ethane-diamine (2 : 1) ; Theophylline compound with ethylene diamine (2 : 1) ;

B.P., U.S.P., Eur. P., Int. P., Ind. P.

Aminophyllin[R] (Searle) ; Phyllocontin[R] (Purdue Frederick) ;

Preparation

Aminophylline is conveniently prepared by the vigorous shaking together of theophyline and ethylenediamine in stoichiometric proportions (2 : 1) in anhydrous ethanol.

Aminophylline is frequently used for the treatment and control of congestive heart failure, bronchial asthma, Cheyne-Stokes respiration, cardiac paroxysmal dyspnea. It is also a vital component in many cough mixtures so as to reduce the cough reflexes and to cause expectoration.

Dose : Oral 300 to 800 mg per day ; usual, 200 mg 3 times daily. Intravenous, slowly, 250 mg to 1.5g per day ; usual, 500 mg slowly 1 to 3 times daily.

E. *Etofylline,* INN, BAN,

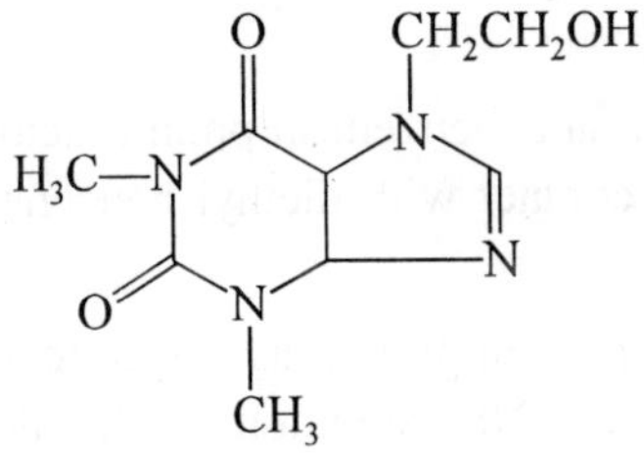

7-(2-Hydroxyethyl)-theophylline ; Hydroxyethyl-theophylline ; 7-(2-Hydroxyethyl)-1, 3-dimethylxanthine ; B.P., Eur. P ; Bio- Phyllin[R] (Bio-Chemical Lab. Canada)

It is claimed to be a *better-tolerated drug than aminophylline* and may be administered orally, intramuscularly and intravenously.

Dose : Up to 1.5g per day.

F. *Proxyphylline* INN, BAN,

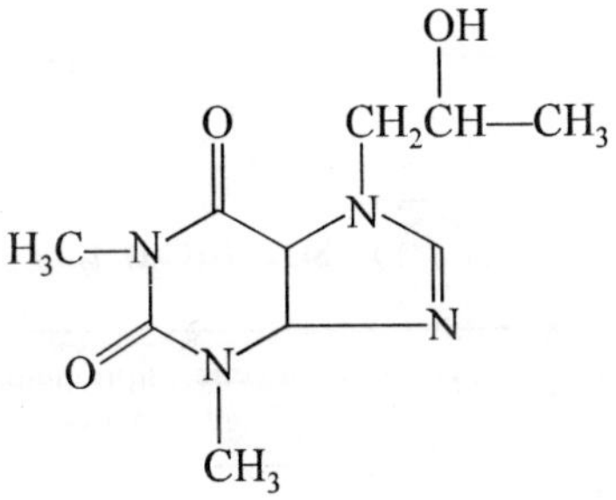

7-(2-Hydroxypropyl)-theophylline ; 7-(2-Hydroxypropyl)-1, 3-dimethylxanthine ; B.P., Eur. P.

It is a theophylline derivative. Its actions and uses are similar to those of aminophylline. It is found to be *better tolerated both orally and intravenously.*

Dose : Usual, oral, 300 mg 3 times per day.

2.2. Analeptics

Analeptics counteract narcosis, with a specific stimulant action on the central nervous system. These are primarily employed to combat the drug-induced respiratory depression. An excessive dose of analeptics may result a wide-spread stimulation of the brain that may ultimately cause convulsions.

A few important drugs under this category are namely : nikethamide, ethamivan, pemoline, pentetrazol, doxapram, and bemegride.

A. *Nikethamide* INN,

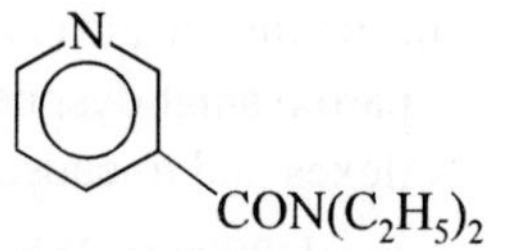

N, N-Diethylnicotinamide ; N, N-Diethylpyridine-3-carboxamide ; Diethylamide nicotinic acid ; B.P., Eur. P., Int. P., Ind. P., N.F. XIII

Coramine[R] (Ciba-Geigy) ; Corazon[R] (Grossmann, Switz)

Synthesis :

A molecule each of nicotinic and benzenesulphonic acid undergoes dehydration to yield a corresponding anhydride, which on contact with diethyl benzenesulphonamide affords an exchange reaction to give nikethamide.

Nikethamide is a weak analeptic employed as respiratory stimulant. It produces respiratory stimulation at doses that have only little CNS excitation. Its duration of action is very transient.

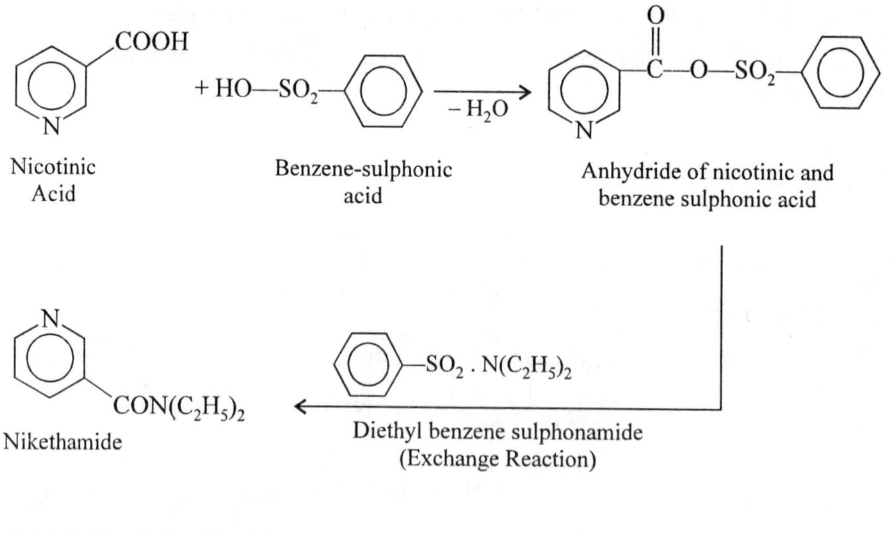

Dose : Usual, 0.5 to 2g intravenously.

B. *Etamivan* INN, *Ethamivan* BAN, USAN,

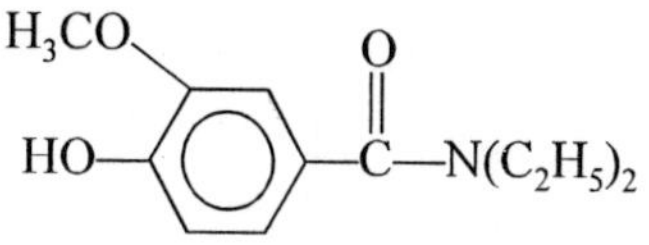

N, N-Diethylvanillamide ; Benzamide, N, N-diethyl-4-hydroxy-3-methoxy- ;
B.P., U.S.P., N.F. ; Emivan[R] (USV)

Synthesis :

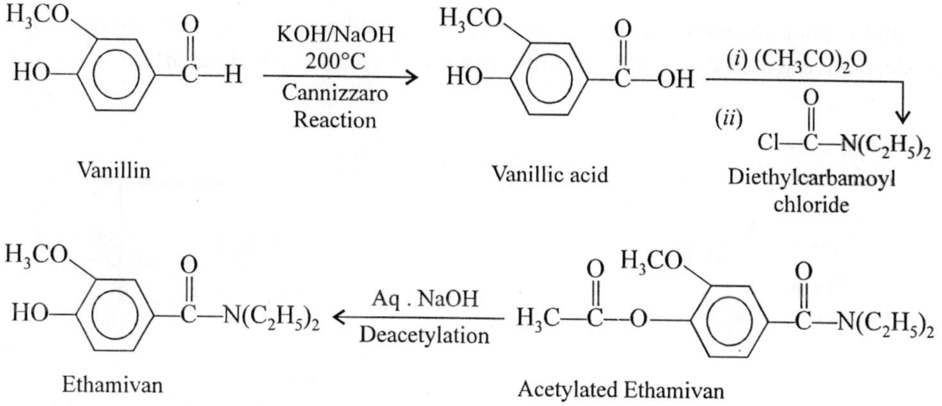

Vanillin subjected to *Cannizzaro reaction* with KOH/NaOH at 200°C yields vanillic acid which upon treatment with acetic anhydride and diethyl carbamoyl chloride gives rise to acetylated ethamivan. This subsequently on deacetylation in aqueous NaOH yields the official compound.

Ethamivan is a respiratory stimulant having actions and uses similar to those of nikethamide. It can cause generalised CNS stimulation, but its action is short-lived.

Dose : Usual, intravenous, 0.5 to 5 mg/kg.

C. *Pemoline* INN, BAN, USAN,

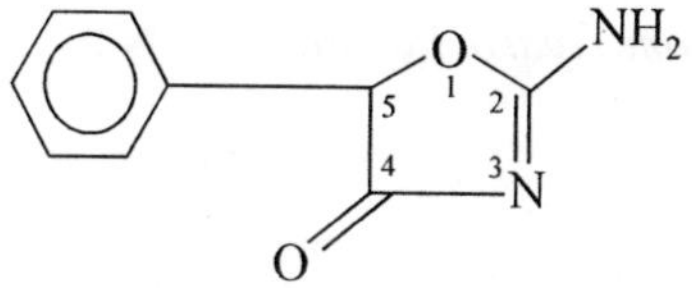

2-Imino-5-phenyloxazolidin-4-one ; 4(5H)-Oxazolone ; 2-amino-5-phenyl ; Phenilone ; Cylert[R] (Abbott) ; Ronyl[R] (Rona, UK)

It is a potent central nervous system stimulant which affects all parts of the nervous system to some extent. It has also been used for the *hyperkinetic states in children.*

Dose : Usual, 20 mg twice daily.

D. *Pentetrazol* INN, BAN,

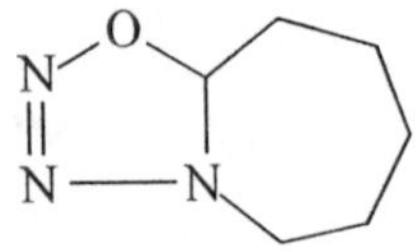

6, 7, 8, 9-Tetrahydro-5H-tetrazoloazepine ; 1, 5-Pentamethylene-tetrazole ; Laptazole ; Pentazol ; Pentylenetetrazol ; Corazol ; B.P., Eur. P., Int. P., Ind. P. ;

Metrazol[R] (Knoll, USA)

Synthesis :

An amino ether obtained from caprolactam when reacted with hydrazine affords the corresponding hydrazino derivative which on treatment with nitrous acid yields the official article.

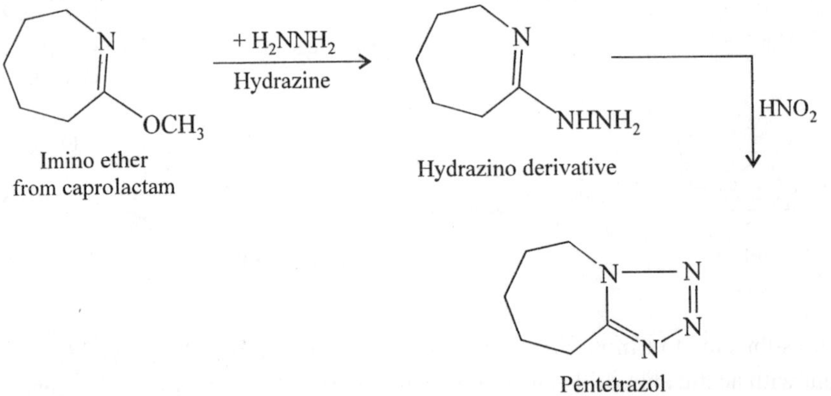

Pentetrazol is a CNS stimulant with actions and uses similar to those of nikethamide. It is used to induce convulsion in animals. It has been employed successfully in the elderly subjects to alleviate the symptoms of mental and physical activity.

Dose : Oral, for treatment of senility, initially 100 or 200 mg 3 or 4 times daily, reduced to half for maintenance.

E. *Doxapram* INN, *Doxapram Hydrochloride* USAN ;

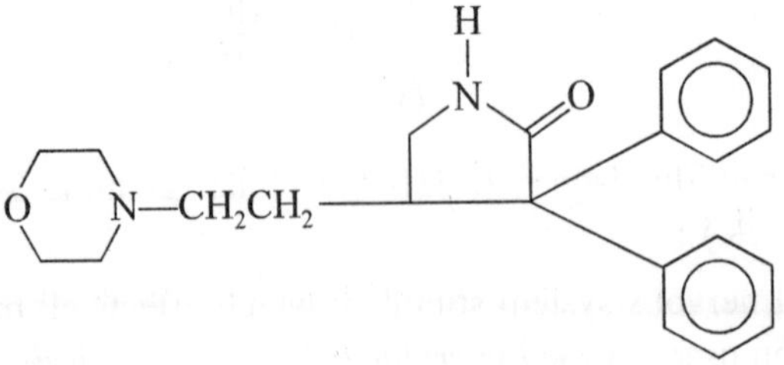

1-Ethyl-4-(2-morpholinoethyl)-3, 3-diphenylpyrrolidin-2-one hydrochloride monohydrate ; U.S.P. ; Dopram[R] (Robins)

It is a respiratory stimulant possessing slight vasopressor characteristics ; frequently employed in the treatment of respiratory depression following anaesthesia.

Dose : Usual, IV., 0.5 to 1 mg/kg.

F. *Bemegride* INN, BAN,

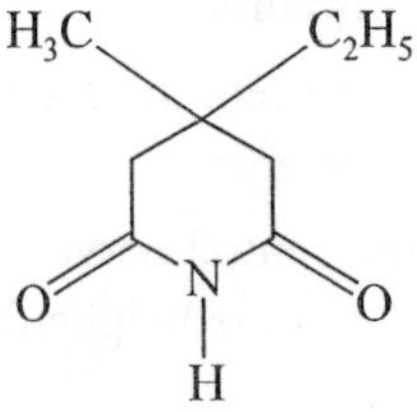

3-Ethyl-3-methyl glutarimide ; 4-Ethyl-4-methyl piperidine-2, 6-dione ; B.P. (1968) ; U.S.P. (XVII) ; Int. P., Ind. P. ; Megimide[R] (Abbott)

Synthesis :

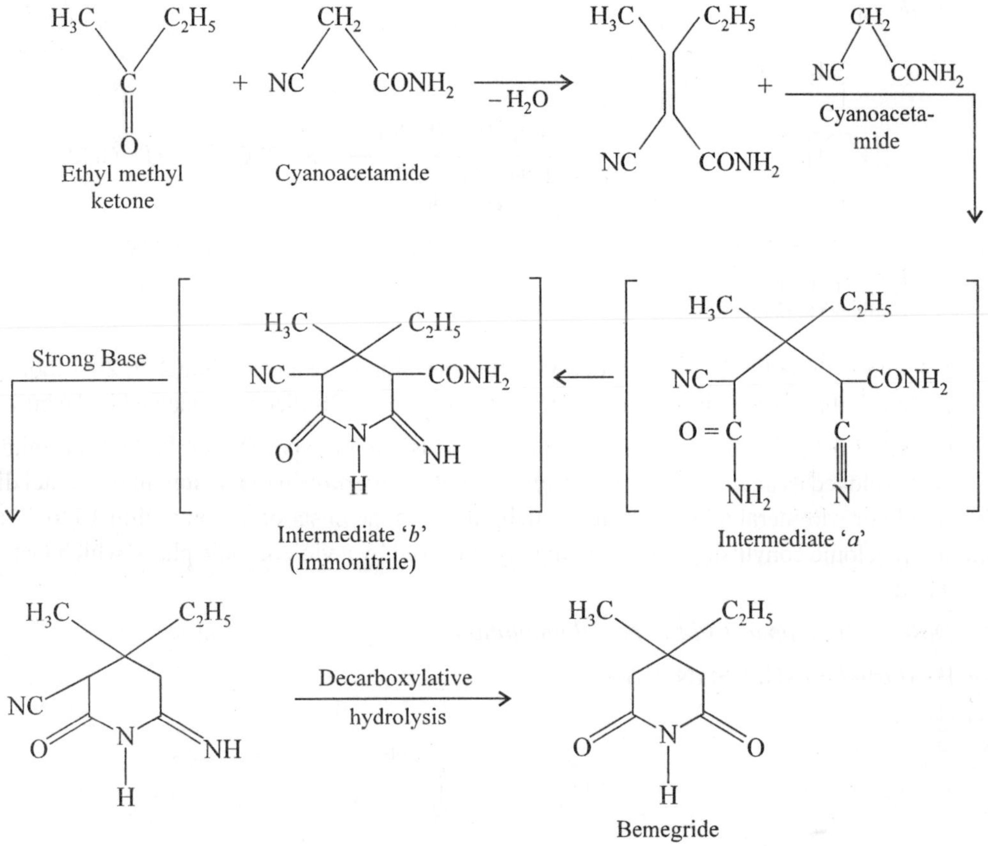

Intermediate '*b*'
(Immonitrile)

Intermediate '*a*'

Bemegride

An *aldol condensation* of ethyl methyl ketone with cyanoacetamide affords a loss of a molecule of water of yield an active methylene compound. Another molecule of cyanoacetamide undergoes congregate addition to give the *intermediate 'a'*. Addition of one of the amide amines to the nitrite will subsequently give rise to the iminonitrile (*intermediate 'b'*), which on treatment with a strong base loses a carboxamide group. The resulting product on decarboxylative hydrolysis yields the official compound.

Bemegride is a respiratory stimulant with actions and uses similar to those of nikethamide.

Dose : Usual, 25 to 50mg i.v.

2.3.   Miscellaneous Central System Stimulants

Several drugs specifically stimulate the central nervous system which are appropriately grouped together as anorexigenic or sympathomimetic agents.

Some drugs that act primarily on the central nervous system are discussed under this category, namely : flurothyl, mazindol, phentermine and methylphenidate hydrochloride.

A. *Flurothyl*

$$F_3CCH_2—O—CH_2CF_3$$

Bis (2, 2, 2-trifluoroethyl) ether ; Ethane, 1, 1'-oxybis 2, 2, 2-trifluoro) ; U.S.P., N.F. ; Indoklon[R] (Ohio Medical)

Synthesis :

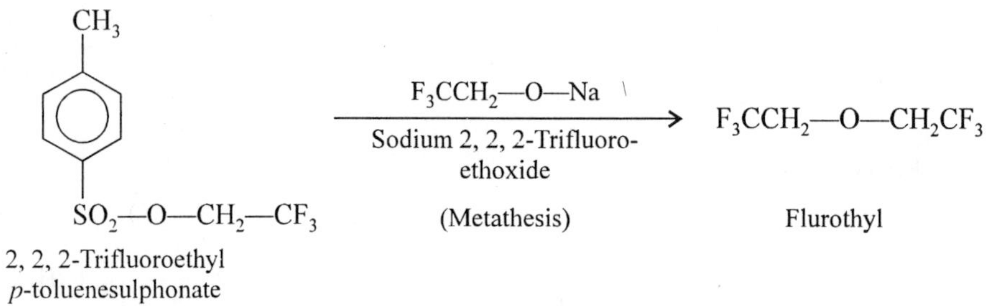

The interaction of 2, 2, 2-trifluoroethyl-*p*-toluenesulphonate and sodium 2, 2, 2-trifluoroethoxide causes metathesis thereby yielding flurothyl which is subsequently distilled and obtained in the purified state.

Flurothyl produces both clonic and tonic convulsions in experimental laboratory animals. It is frequently employed as an *alternative for electroconvulsive therapy* in the treatment of mental disorders. An inhalation or parenteral administration usually helps in the onset of action within 15 to 20 seconds, the initial myoclonic convulsions are immediately followed by a violent tonic phase which lasts from 30 to 90 seconds.

Dose : Usual, up to 1ml by special inhalation.

B. *Mazindol* INN, USAN, BAN,

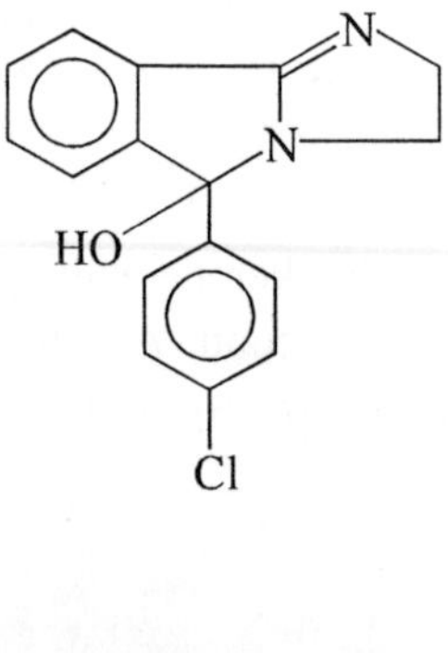

5-(4-Chlorophenyl)-2, 5-dihydro-3H-imidazo 2, 1-a isoindol-5-ol ; U.S.P. ; Mazanor[R] (Wyeth) ; Sanorex[R] (Sandoz)

Synthesis :

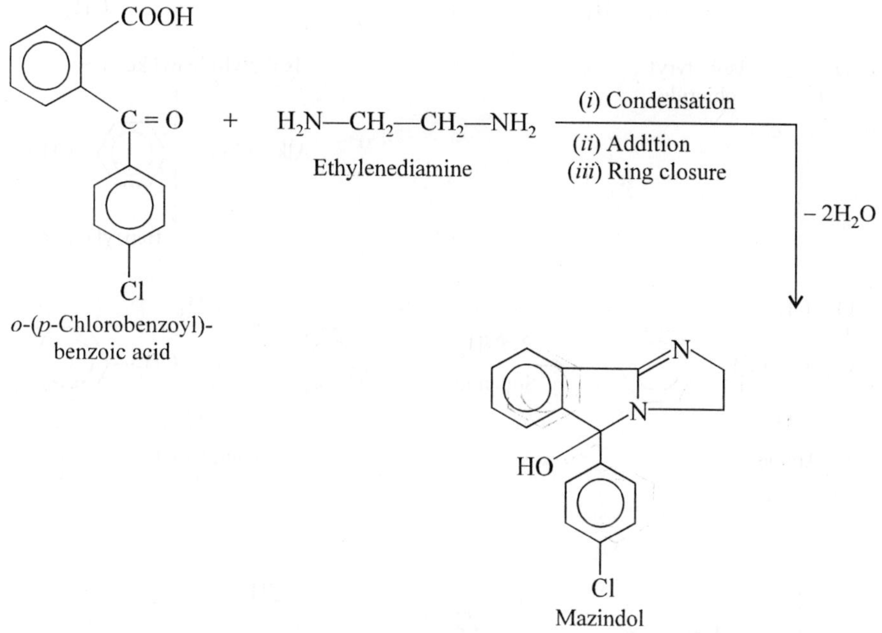

o-(*p*-Chlorobenzoyl)-
benzoic acid

Ethylenediamine

(*i*) Condensation
(*ii*) Addition
(*iii*) Ring closure

$- 2H_2O$

Mazindol

The interaction of *o*-(*p*-chlorobenzoyl)-benzoic acid with ethylenediamine affords condensation, addition and finally cyclization to yield the official product with the elimination of two moles of water.

It is an anorexiant used in the *treatment of obesity*. It also exerts a variable effect on the CNS thereby causing a *mild stimulation* in some subjects and *a mild depression* in others.

Dose : 2 mg once per day 1 hour before lunch.

C. *Phentermine* INN, BAN, USAN,

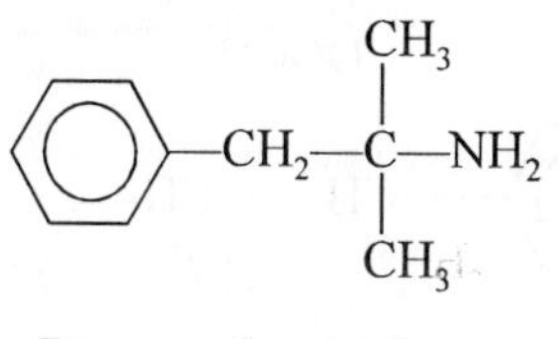

α, α,-Dimethylphenethylamine ; Benzenethanamine, α-α-dimethyl ;

Ionamin[R] (Pennwalt)

Synthesis :

Friedel-Craft's acylation of benzene with isobutyryl chloride yields isobutylphenyl ketone which on alkylation with benzyl chloride gives rise to an intermediate. This intermediate, being a nonenolizable ketone, undergoes cessation at the amide linkage with a strong base like sodamide to yield the corresponding amide which when ultimately subjected to *Hoffmann's degradation* with sodium hypochlorite gives the desired product.

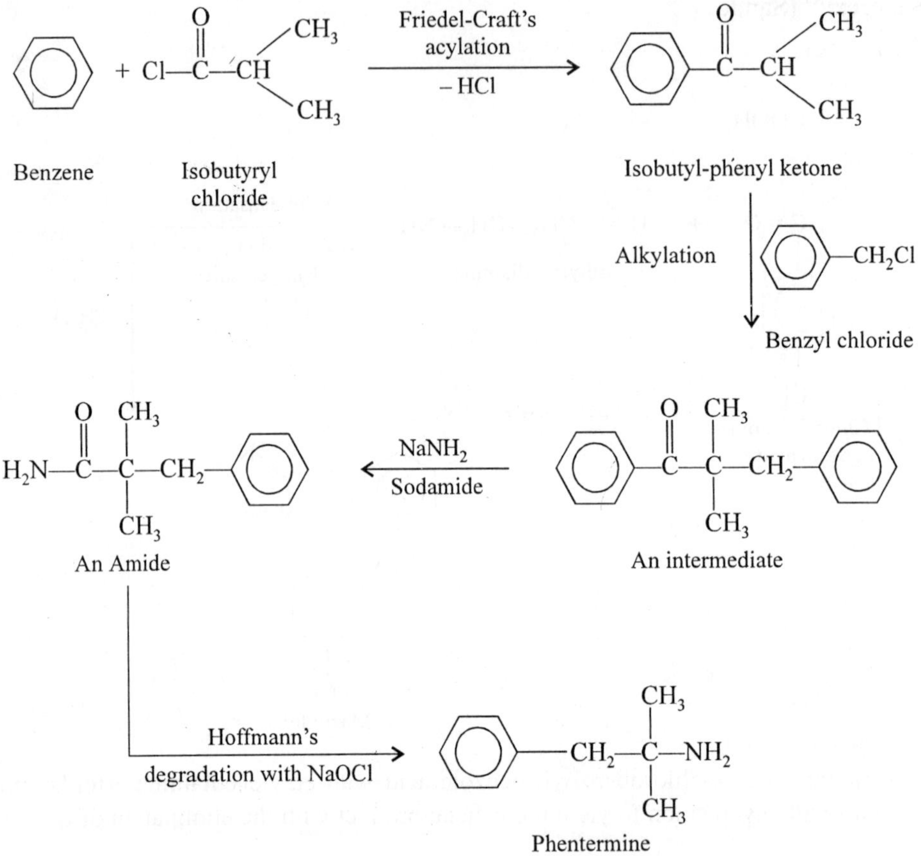

Benzene   Isobutyryl   Isobutyl-phenyl ketone
               chloride

An Amide   An intermediate

Phentermine

Phenetermine is a sympathomimetic agent employed as an anorectic in the *treatment of obesity.*
Dose : Usual, adult 15 to 30 mg at breakfast.

D. *Methylphenidate* INN, BAN, *Methylphenidate Hydrochloride* USAN,

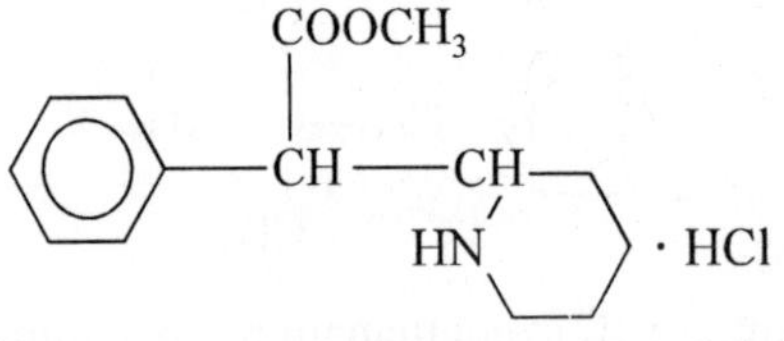

Methylphenidate hydrochloride ; Methyl α-phenyl-α-(2-piperidyl) acetate hydrochloride ; U.S.P. ;
Ritalin Hydrochloride[R] (Ciba-Geigty)

Synthesis :

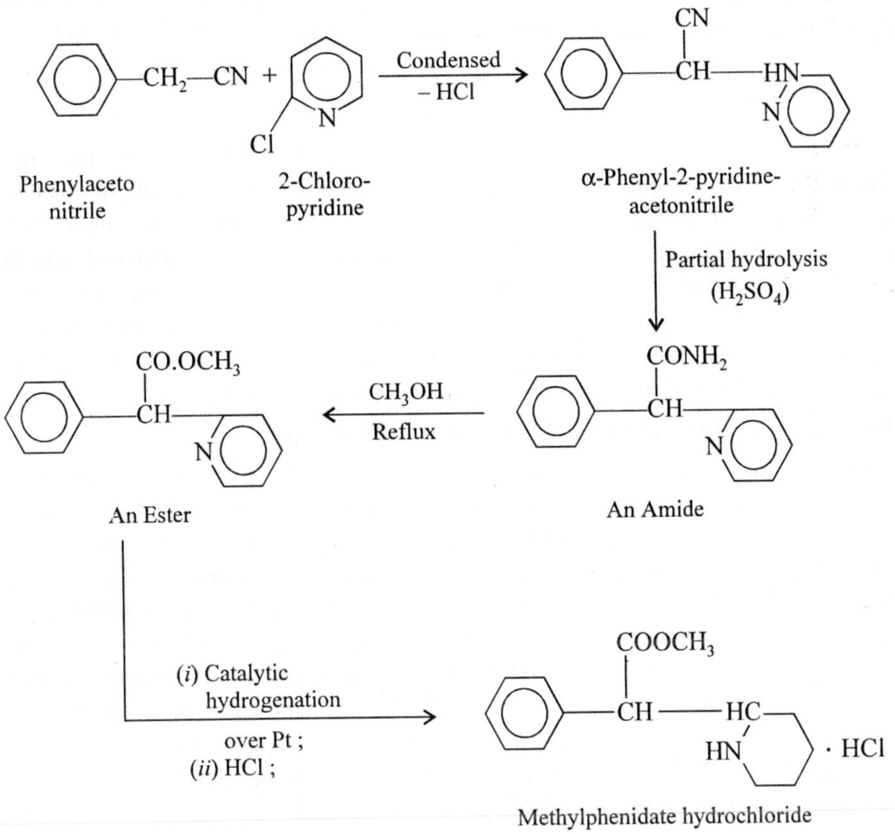

Methylphenidate hydrochloride

Condensation of phenyl acetonitrile and 2-chloro-pyridine yields α-phenyl-2-pyridine acetonitrile which upon partial hydrolysis with sulphuric acid gives an amide. Reflux of this amide with methanol gives the corresponding methyl ester which upon catalytic hydrogenation over platinum and subsequent treatment with a calculated amount of hydrochloric acid yields the official product.

It is a mild CNS stimulant having a therapeutic potency intermediate to caffeine and amphetamine. It is used in the *treatment and management of minimal brain dysfunction in children.*

Dose : Oral or parenteral, 10 to 60 mg per day ; Usual, 10 mg 2 to 3 times per day.

3. CNS-PEPTIDES, S-GLUTAMATE AND BLOCKADE OF NMDA-INDUCED RESPONSES

3.1. CNS-Peptides

It has been duly observed that the endogenous peptide sleep substances in particular seem to regulate the prevailing neuronal activity which are directly linked with the phenomenon of sleep. *Delta-sleep-inducing peptide* (DSIP) is regarded to be the most widely known CNS-peptide that has been proved to be directly associated with the sleep regulatory phenomenon. It has been established experimentally that the *'dialysate'** meticulously derived from the cerebral blood of rabbits, which

---

*The product obtained after carrying out the **'dialysis'**.

were adequately maintained in a state of sleep by careful *'electrical stimulation of the thalamus'*, could induce sleep efficaciously in normal rabbits. Therefore, the exact causative *'factor'* responsible might be the DSIP, which is a **non-apeptide entity.** The amino acid sequence is as given below :

Trp—Ala—Gly—Gly—Asp—Ala—Ser—Gly—Glu

1   2   3   4   5   6   7   8   9

Furthermore, it has also been proved substantially that P-DSIP, *i.e.,* the corresponding phosphorylated structural analogue of DSIP (having Ser at position 7), occurs and in rats it is found to at least five times more active than DSIP. Based scientifically and logically on the above clue it has been observed that prevailing ratio of DSIP/P-DSIP appears to play a vital role in modulating the circadian (*i.e.,* pertaining to events that occur at intervals of approximately 24 hours) time course of sleep and wakefulness in human beings. Besides, the two species of peptides increase non rapid eye movement (NREM) and rapid eye movement (REM) sleep. Interesting, DSIP has also been employed gainfully for the control and management of insomnia and obstructive sleep apnea in patients.

3.2. S-Glutamate and Blockade of NMDA-induced Responses :

It is believed that the acidic amino acid (S)-glutamic acid happens to be the most prevalent and predominant excitatory transmitter in the CNS acting specifically at the *excitatory amino acid* (EAA) receptors.* The first and foremost scientific evidence with regard to the heterogeneity amongst the EAA-receptors was brought to light from the variable activity of a spectrum of agonists entirely based on the glutamic acid structure in various portions of the CNS.** Furthermore, rather concrete evidence surfaced from the wonderful observation suggesting that *(R)-α-aminoadipic acid* (D-αAA) eventually blocked the *N-methyl-(R)-aspartic acid* (NMDA)-induced and synaptic excitation prevailing in the spinal motor neurones ; but not caused due to depolarizations by either of the two natural products, *viz., kainic acid* and *quisqualic acid* as illustrated below. However, the instance for EAA-receptor subtypes was adequately substantiated by the gainful knowledge of the selective blockade of NMDA-induced responses strategically on the spinal neurones by Mg^{2+} ions.***

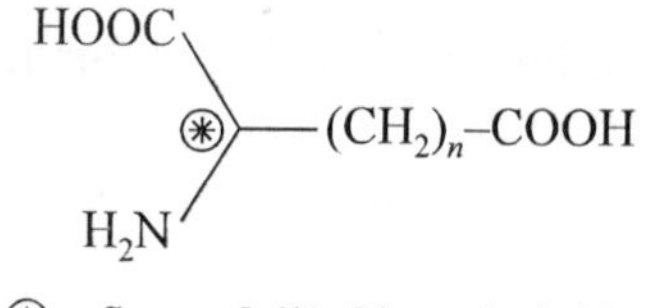

⊛ = S ; n = 2 (S)-Glutamic Acid

⊛ = R ; n = 3 (R)-α-Aminoadipic Acid

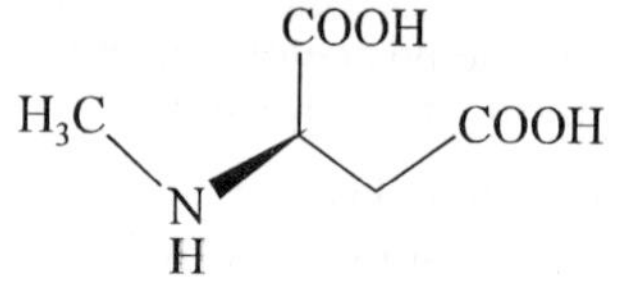

NMDA

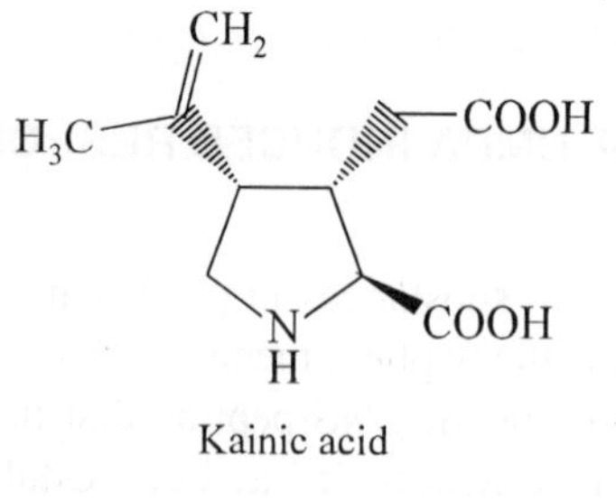

Kainic acid 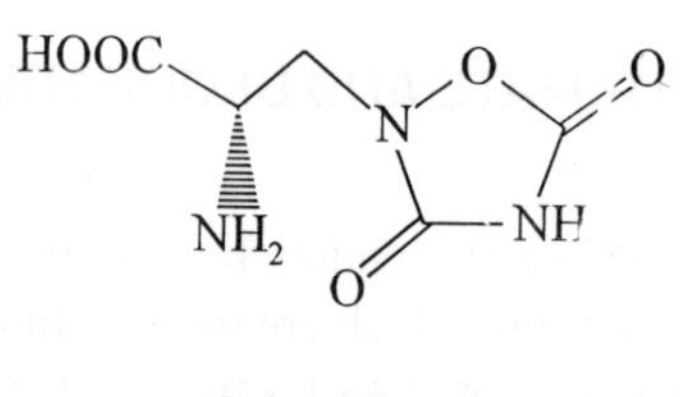   Quisqualic acid 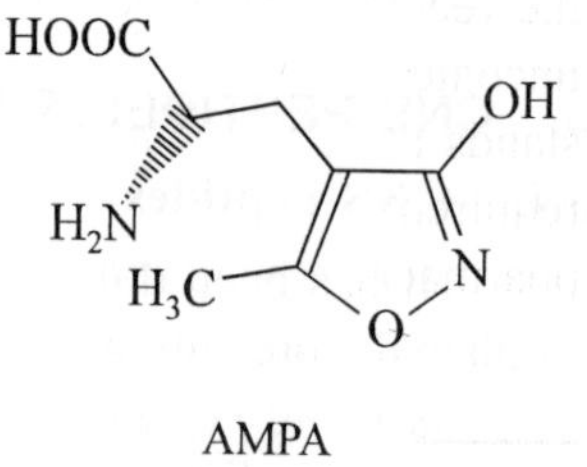   AMPA

*Watkins JC and Evans RH, *Ann. Rev. Pharmacol. and Toxicol.,* **21,** 165, 1981

Duggan AW, *Exp. Brain Res.,* **19, 522, 1974

***Evans RH *et al., Experientia,* **33,** 489, 1977.

4. MECHANISM OF ACTION OF SELECTED CNS-STIMULANTS

The most probable and logical explanations pertaining to the mechanism of action of certain selected CNS-stimulants are described as under :

4.1. Caffeine

It stimulates all levels of the CNS, particularly the cerebral cortex, distinctly producing a more rapid and clarity of thought, improved psychomotor coordination, wakefulness and augmentation of in spirit and feelings in fatigued patients. Its cortical effects are observed to be not only of shorter span but also are of milder nature in comparison to amphetamines. Interestingly, at higher doses it substantially stimulates vasomotor, medullary vagal and respiratory centres, thereby affording an induction of brady cardia, vasoconstriction, and an enhanced rate of respiration.

Caffeine exerts a noticeable inotropic effect* on the myocardium and a positive chronotropic effect** particularly on the sinoatrial mode which ultimately result into a transient observed heart-rate, force of contraction, working of the heart, and above all the cardiac output. However, it is largely belived that the vasoconstriction of the cerebral blood vessels by caffeine remarkably contributes a lot to its exceptional capability to relieve headaches.

It is found to cause stimulation of the voluntary skeletal muscle which is turn enhances the force of muscle contraction ; and, hence, the muscular fatigue.

Caffeine undoubtedly stimulates the parietal cells*** which enhances the excretion of the gastric juice in the stomach (*i.e.,* causes acidity).

It is found to exert a mild diuretic action by increasing the renal blood flow, glomerular filtration rate and by minimising the proximal tubular reabsorption of Na and H_2O.

The metabolism of caffeine resolves around two major phenomena, namely : (*a*) **glycogenolysis** : and (*b*) **lipolysis.** Fortunately, the outcome of these two biochemical processes, leading to enhanced **blood-glucose** and **plasma lipids** do not cause any alarming consequences in relatively healthy human subejcts.

4.2. Aminophylline

It gets converted to almost 79% into its structural analogue *theophylline.* It has been found that the absorption from the GI tract after due oral as well as rectal administration is invariably incomplete, sluggish and variable. The optimal serum therapeutic levels achievable range from 10 to 20 mcg . mL^{-1}.

4.3. Theophylline

It is well absorbed after administration, which may be retarded in the presence of food. However, the rectal suppositories are absorbed slowly and very much erratically. The peak plasma levels for uncoated tablets and liquids are attained within a span of 2 hours ; the average volume of distribution stands at 0.5 L.kg^{-1}. A minimum of 10 to 20 mcg. mL^{-1} in plasma or serum levels is an absolute requirement to cause a maximum bronchodilator response. Generally, it exhibits relatively much more pharmacological response than theobromine in all aspects.

---

*Influencing the force of muscular contractility.

**Influencing the rate of occurence of an event *e.g.,* heart beat.

***A large cell on the margin of the peptic glands of the stomach that secretes HCl and the intrinsic factor.

4.4.   Nipethamide

It has been advocated for the treatment of drug overdosage due to excessive CNS-depressants. It finds its occasional usage having potential *'emergency value'* as a respiratory stimulant prior to other modes of supporting respiratory devices.

4.5.   Pemoline

Though sufficient extensive and intensive laboratory studies have revealed that pemoline may exert its action through the *'dopamenergic mechanisms'* ; however, the exact and precise mechanism and site of action in humans are not yet established and known. It has been observed that about 75% of its oral dosage gets excreted through the urine in 24 hours, another 43% is excreted an *'unchanged'*, and the remaining 22% is excreted as pemoline conjugates.

4.6.   Doxapram

It is found to exert its action by stimulating respiration by an activity directly linked with peripheral carodid chemoreceptors. Therefore, it is normally employed specifically as a respiratory stimulant postanaesthetically in such situations as, namely : (*a*) chronic pulmonary diseases ; (*b*) CNS-depressant drug overdose (inadvertently) ; and (*c*) apneas.

4.7.   Bemegride

Its actions are very much akin to doxapram hydrochloride. It is, however, regarded to be absolutely unsafe in patients having acute *porphyria* because it has been associated with acute attacks.

4.8.   Flurothyl

It causes stimulation of the CNS and also induces convulsions, which is why it was formerly employed as an alternative to *'electro-convulsive therapy'* in the treatment of severe depression either administered through IV injection or inhalation.

4.9.   Mazindol

It is mainly absorbed from the GI-tract and is mostly excreted in the urine, partly unchanged and partly as its corresponding metabolites.

4.10.   Phentermine

The drug is rapidly absorbed from the GI-tract. It is observed that a major portion gets excreted in the urine, partly unchanged and partly as its metabolities. Interestingly, phentermine has a quaternary C-atom with one methyl function exactly oriented like the methyl of (S)-amphetamine and one methyl oriented very much similar to the (R)-amphetamine. Therefore, it exhibits pharmacological characteristic features of both the (R) and (S) isomers of amphetamine. However, it has much less abuse potential in comparison to the not-so-famous *dextroamphetamine*.

4.11.   Methylphenidate Hydrochloride

It is rapidly absorbed from the GI-tract. Approximately, 80% of an orally administered dose gets metabolized to the corresponding ritalinic acid and subsequently excreted in the uine.* In fact, its actions, like cocaine, seem to be regulated by blockade of catecholamine reuptake instead by the release of catecholamines as usually takes place with the amphetamines in general.

---

*The *'urinary excretion'* is not pH dependent.

Probable Questions for B. Pharm. Examinations

1. Name any **eight** potent CNS stimulants, give their structures, chemical names and uses.

2. Why a drug having a short-term excitation of CNS is preferred over the one having a long term effect. Explain with specific examples.

3. Classify the CNS-stimulants and give the synthesis of one potent compound from each category.

4. Give the structures of *Propoxyphylline* and *Aminophylline*. Explain why the former has a better tolerance orally and intravenously than the later.

5. Xanthines represent an important class of CNS stimulants. Give the structure, chemical name and uses of any *Three* potent drugs from this category and discuss the synthesis of any *one* of them.

6. Discuss '*Analeptics*' as an important class of CNS stimulants. Give the synthesis of
   (*i*) Etamivan, and
   (*ii*) Nikethamide

7. Mazindol and Phentermine represents the miscellaneous CNS-stimulants. Give their structure, chemical name and uses.

8. Discuss the synthesis of the following CNS-stimulants :
   (*i*) Phentermine   (*ii*) Bemegride   (*iii*) Fluorothyl   (*iv*) Methylphenidate hydrochloride.

9. A constituent of coca-butter that acts as a CNS-stimulant may be obtained by the interaction of methyl urea and ethyl carboxamidoacetate. Explain.

10. Give a comprehensive account on the various CNS-stimulants used in the therapeutic armamentirium.

RECOMMENDED READINGS

1. Remington's : *The Science and Practice of Pharmacy* Vol. I and II, (20th end.) Lippincott Williams and Wilkins, New York, 2000.

2. H D Fabing The Newer Analeptic Drugs *Med Clin N.A.* 339 (1957).

3. W O Foye *Principles of Medicinal Chemistry* (4th edn.) Philadelphia. Lea and Febiger, Philedelphia, (1996).

4. M E Wolff, *Burger's Medicinal Chemistry and Drug Discovery* (5th edn.) John Wiley and Sons New York (1995).

5. D Lednicer and L A Mitscher *The Organic Chemistry of Drug Synthesis* John Wiley and Sons New York (1995).

6. Patrick GL, '*An Introduction to Medicinal Chemistry*', Oxford University Press, Oxford (U.K.), 2001.

Antipyretic Analgesics

1. INTRODUCTION

Antipyretic analgesics or febrifuges are remedial agents that lower the temperature of the body in pyrexia *i.e.,* in situations when the body temperatures has been raised above normal. In therapeutic doses they do not have any effect on normal body temperature. They exert their action on the heat regulating centre in the hypothalamus. These antipyretic agents also have mild analgesic activity. Amongst the most common group of compounds used as antipyretic analgesics are salicylates, aniline and aminophenol analogues, pyrazolones and quinoline derivatives. Though these heterogenous groups of compounds are analgesics, they have no addictive properties. Their analgesic use is limited to mild aches and pains like headache and backache.

Alternatively, '*antipyretic*' is the terminology quite frequently applied to drugs which essentially help to reduce fever to normal body temperature (*i.e.,* 98.4°F or 37°C). It is, however, worthwhile to mention here that the '*drug substances*' belonging to this particular category usually possess the ability to alleviate the sensation of pain threshold ranging from mild to severe status. These antipyretic agents are also found to be significantly effective in reducing fever to normal levels in humans. The '*drugs*' that are most commonly included here are, namely : acetanilide ; phenacetin (acetophentidin) ; and paracetamol [acetaminophen (known in US), *para*-acetaminophenol]. Interestingly, the aforesaid *three* drug entities are interrelated to one another *metabolically*, as illustrated below :

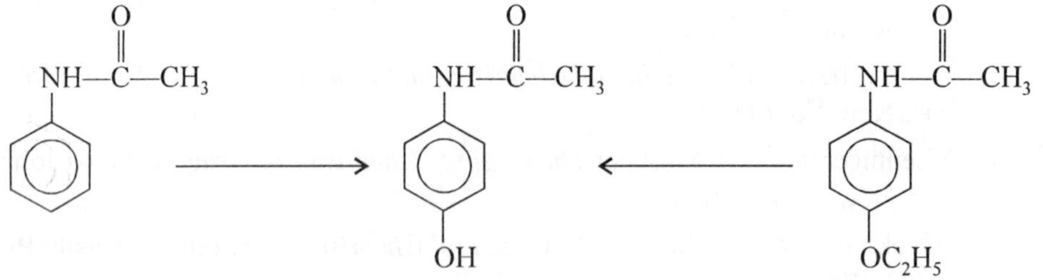

Acetanilide   Paracetamol   Phenacetin
(Acetaminophen)

It is worthwhile to mention here that both acetanilide and phenacetin have has been withdrawn completely from being used because of its numerous toxic and undesirable effects, such as : skin manifestations, jaundice, cardiac irregularities, and a relatively high incidence of methemoglobinemia* ; and quite seldomnly acute blood dyscrasias, for instance : hemolytic anemia. *Phenacetin* has also been dropped as a 'drug' since 1982 in US by virtue of the fact that it earned a bad reputation for causing nephrotoxicity due to its high-dose long-term abuse in several parts of the globle. It was also reported to cause kidney and liver cancer.

Paracetamol (acetaminophen) enjoys still the world-wide recognition as the only '*aniline-based analgetic-antipyretic*' for its abundant utility in controlling fever in most non-inflammatory conditions very much akin to 'aspirin'. It has also been demonstrated adequately that both paracetamol and aspirin are '*equianalgetic*' at a dose of 650 mg.

Analgesics may be defined as–'*agents that relieve pain by elevating the pain threshold without disturbing consciousness or altering other sensory-modalities*'. Besides, '*pain*' may also be defined in psychological perspective as—'*a particular type of sensory experience distinguished by nerve tissue from sensations, such as : touch, heat, pressure and cold*'. In the latest context '*pain*' essentially involves a major chunk of psychological factor which exclusively rests on perception. Therefore, more realistically '*pain*' may be defined introspectively in an exclusive manner.

Broadly speaking, the most probable and logical explanation for the '*mechanism*' by which certain analgesics specifically enhance the pain threshold has been caused solely due to the presence of the '*opiate receptors*' strategically located in selected parts of the CNS overtly and covertly associated with the pain regulation. It has been established that the '*opiate receptors*' are located in the following critical zones, namely :

(*a*) Medial thalamus which processes chronic, deep and burning pain that is usually suppressed by **narcotic analgesics only,**

(*b*) Brainstem's vagus nuclei which triggers the '*cough centres*', and

(*c*) Layers I and II in the spinal cord at the specific zone where the different nerves which solely hold the pain perception first synapse.

Importantly, '*endorphins*'** mostly logistically lower the intensity of pain by modulating particularly the pain threshold the critical material pient at which one may commence to perceive a stimulus as '*painful*' sensation.

2. CLASSIFICATION

Antipyretic analgesics may be classified on the basis of their chemical structures.

2.1. Aniline and p-Aminophenol Analogues

In 1886, Cohn and Hepp first identified the powerful antipyretic activities residing in both aniline and acetanilid. The basic origin of this particular class of compounds from aniline has probably suggested these to be known as '*coal tar analgesics*'. However, the aminophenols (*o, m, p*) are reported to be relatively less toxic than aniline. The *para*-isomer is claimed to be the least toxic of the three isomers of

---

*The clinical condition in which more than 1% of haemoglobin in blood has been oxidized to the (Fe^{3+}) form. The principal symptom is *cyanosis.*

A generic name coined from **endogenous and **morphine** ; and commonly used for all native brain peptides having essentially the opiate-like action.

aminophenols and it also possesses a significant antipyretic action. A few examples belonging to this category of antipyretics are described below.

A. *Paracetamol* INN, BAN, *Acetaminophen* USAN,

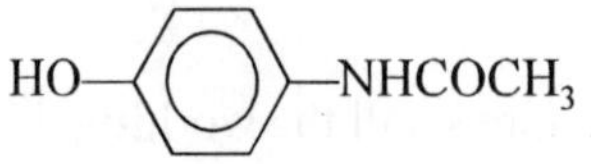

$$HO\text{—}\langle\bigcirc\rangle\text{—}NHCOCH_3$$

4'-Hydroxyacetanilide ; Acetamide, N-(4-hydroxyphenyl)- ; Paracetamol B.P., Eur. P., Acetaminophen U.S.P., Tylenol[R] (McNeil Consumer) ; Tapar[R] (Parke-Davis) ; SK-Apap[R] (Smith Kline & French) ; Valadol[R] (Squibb)

Synthesis

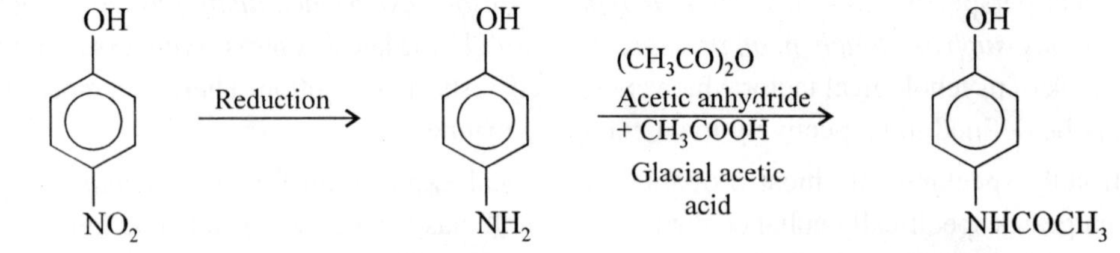

p-Nitrophenol p-Aminophenol   Paracetamol

It may be prepared by the reduction of p-nitrophenol and the resulting p-aminophenol is acetylated by a mixture of acetic anhydride and glacial acetic acid. The crude product can be purified by recrystallization from a water : ethanol mixture (1 : 1) or from other appropriate solvents.

It is a metabolite of acetanilide and phenacetin employed as an anti pyretic and analgesic. It may be used effectively in a broad spectrum of arthritic and rheumatic conditions linked with musculoskeletal pain, headache, neuralgias, myalgias, and dysmenorrhea. It is particularly useful *in aspirin-sensitive patients.*

Dose : Usual oral, adult, 500 mg to 1 g 3 or 4 times per day.

B. *Phenacetin* INN, BAN, USAN,

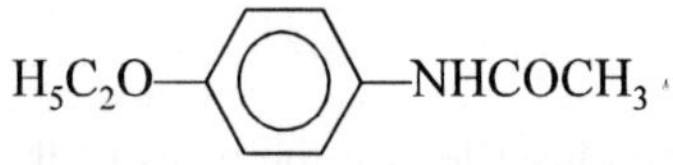

$$H_5C_2O\text{—}\langle\bigcirc\rangle\text{—}NHCOCH_3$$

p-Acetophenetidide ; Acetamide, N-(4-ethoxyphenol)- ; Acetophenetidin ; p-Ethoxyacetanilid ; B.P. (1973), U.S.P., Eur. P., Int. P., Ind. P.

Synthesis

It may be prepared by any one of the following *three* methods, namely :

Method-I : From p-Nitrophenol

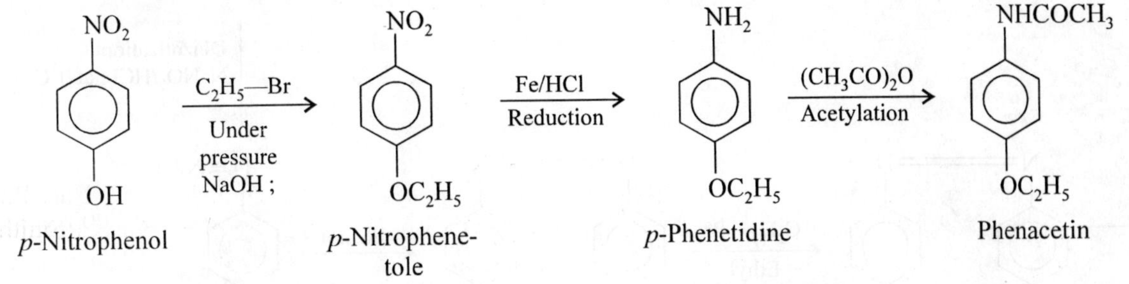

| p-Nitrophenol | p-Nitrophene-tole | p-Phenetidine | Phenacetin |

p-Nitrophenol, dissolved in sodium hydroxide solution, is subjected to condensation with ethyl bromide and the resulting p-nitrophenetole is reduced with suitable reductant. The p-phenetidine thus obtained is acetylated with acetic anhydride to yield the official compound.

Method-II : From Aniline

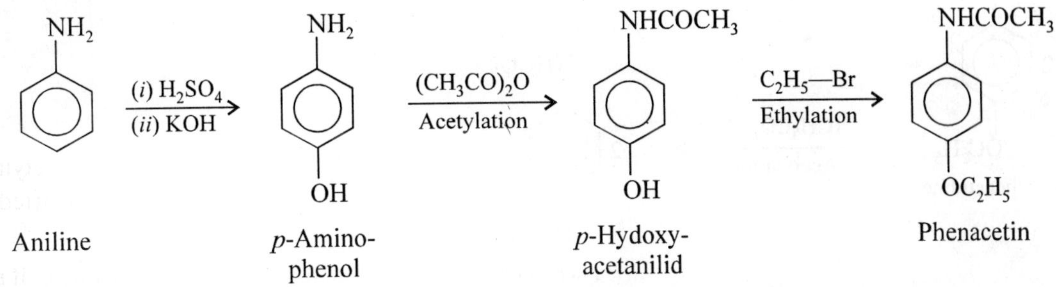

| Aniline | p-Amino-phenol | p-Hydoxy-acetanilid | Phenacetin |

p-Aminophenol is obtained by treating aniline with sulphuric acid and potassium hydroxide, which on acetylation with acetic anhydride yields the p-hydroxy acetanilide. The resulting poduct on ethylation with ethyl bromide forms phenacetin.

Method-III : From Chlorobenzene

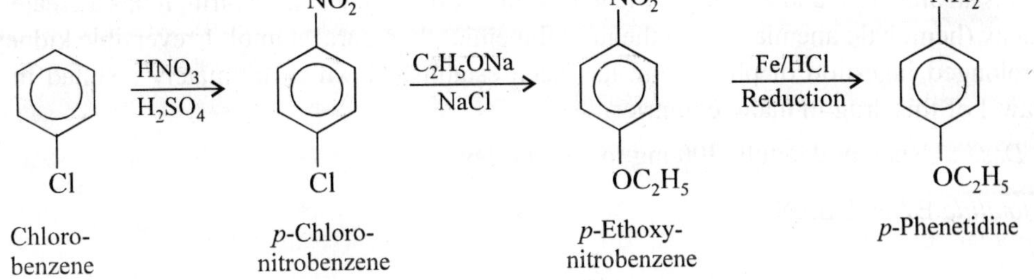

| Chloro-benzene | p-Chloro-nitrobenzene | p-Ethoxy-nitrobenzene | p-Phenetidine |

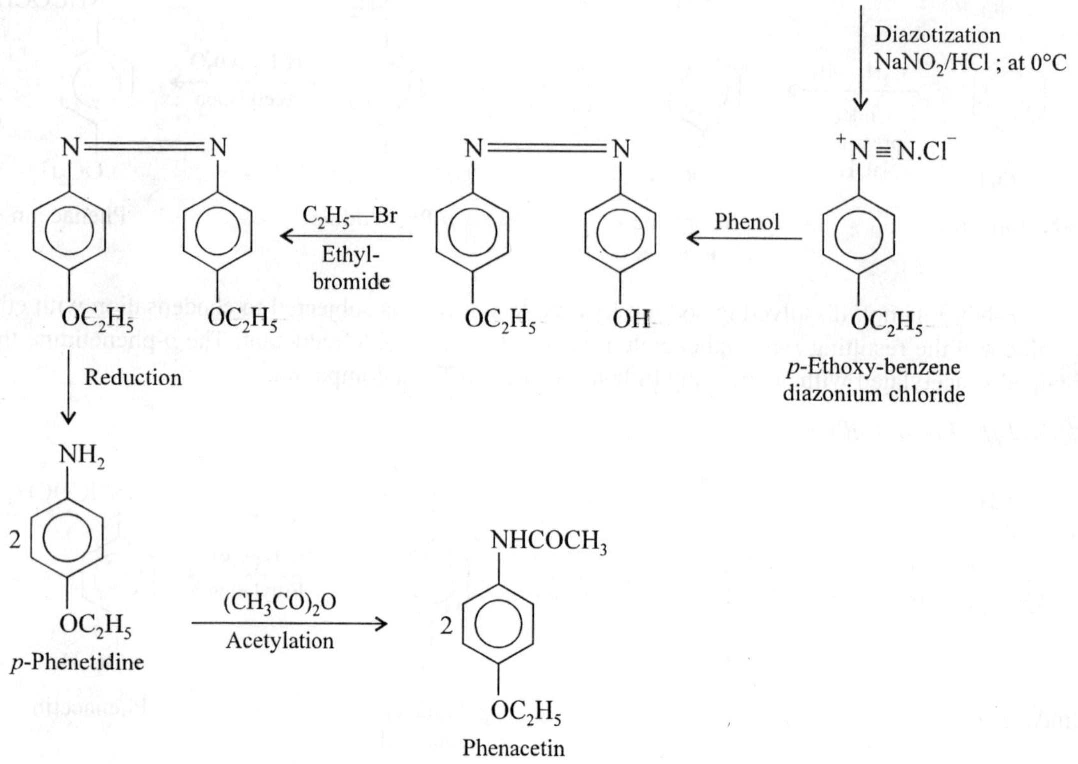

p-Ethoxy nitrobenzene is prepared from chlorobenzene by its nitration followed by treatment with sodium ethoxide, which on reduction yields p-phenetidine. The resulting product is diazotised with nitrous acid at 0°C reacted with phenol, ethyl bromide and reduced to obtain two moles of p-phenetidine which upon acetylation with acetic anhydride yields two moles of phenacetin.

It is an analgesic and an antipyretic with similar effectiveness as aspirin. It has a greater potential for toxicity (hemolytic anemia and methemoglobinemia) than paracetamol. Irreversible kidney damage with prolonged ingestion of phenacetin has been established which ultimately resulted in complete withdrawal of this drug in many countries.

Dose : Usual, oral, adult, 300 mg to 2 g per day.

C. *Acetanilide* BAN, USAN,

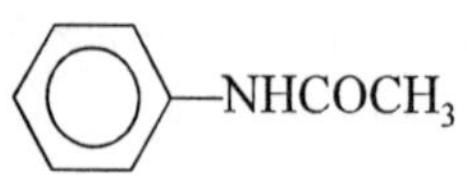

N-Phenylacetamide ; Antifebrin ; B.P.C. 1949, N.F. X ;

Synthesis

It may be prepared from aniline in two different ways, namely :

Method-I : From aniline and acetic anhydride

$$\text{Aniline}-NH_2 + \underset{\substack{CH_3CO \\ CH_3CO}}{\Big\rangle}O \xrightarrow[\substack{\text{Sodium} \\ \text{acetate}}]{CH_3COONa} \text{Acetanilide}-NHCOCH_3 + CH_3COOH$$

| Aniline | Acetic anhydride | Acetanilide | Acetic acid |

It may be prepared by the interaction of aniline and acetic anhydride in the presence of sodium acetate. The crude product may be recrystallized from alcohol.

Method-II : From aniline and acetic acid

$$\text{Aniline}-NH_2 + CH_3COOH \xrightarrow[120-125°C]{ZnCl_2 ;} \text{Acetanilide}-NHCOCH_3 + H_2O$$

| Aniline | Acetic acid | Acetanilide |

It may also be prepared by reacting together redistilled aniline and glacial acetic acid in the presence of zinc chloride at an elevated temperature of 120-125°C.

It is one of the cheapest antipyretic drugs. Owing to its high toxicity caused by liberation of free aniline *in vivo* it has been replaced by much safer antipyretics.

2.2. Salicyclic Acid Analogues

Salicin was the first compound belonging to this category that exhibited medicinal value. It was employed as a substitute for quinine as a febrifuge. In 1838, Paria prepared salicyclic and whose structure was established by Hoffmann. In 1860, Kolbe and Lautermann, introduced the commercial method of preparing salicyclic acid from sodium phenate. Acetylsalicylic acid or aspirin was first synthesized by Gerhardt in 1852, but unfortunately this wonder drug, more or less remained obscure until Felix Hoffmann studied its detailed pharmacodynamic properties in 1899. It gained entry into the world of medicine through Dreser, who coined a new name '*aspirin*' derived from "a" of acetyl and adding to it "spirin", an old name of salicylic or spiric acid, obtained from spirea plants.

A few classical of this series of compounds are discussed here.

A. *Aspirin* BAN, USAN,

$$\text{(structure: benzene ring with } COOH \text{ and } OCOCH_3 \text{ substituents)}$$

Salicylic acid acetate ; Benzoic acid, 2-(acetyloxy)- ; Acetylsalicylic acid ; *o*-Acetylsalicyclic acid ; B.P., U.S.P., Eur. P., Int. P., Ind. P.,

Emipirin[R] (Burroughs Wellcome) ; A.S.A.[R] (Lilly) ; Bufferin[R] (Bristol-Myers)

Synthesis

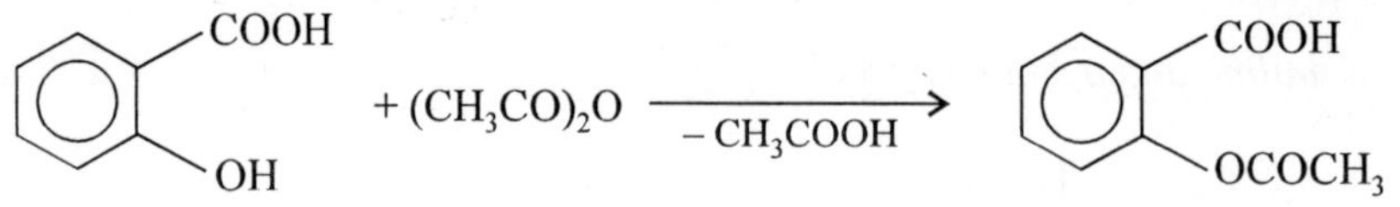

Salicylic acid   Acetic   Aspirin
                    anhydride

Acetylation of salicylic acid with acetic anhydride yields aspirin. The crude product may be recrystallized from benzene, mixture of acetic acid and water (1 : 1) or various other non-aqueous solvents.

It is used as an antipyretic anti-inflammatory and an analgesic in a variety of conditions ranging from headache, discomfort and fever associated with the common cold, and muscular pains and aches. Aspirin is regarded as the drug of choice in the reduction of fever because of its high degree of effectiveness and wide safety margin. As aspirin inhibits platelet function, it has been employed prophylactically to minimise the incidence of myocardial infarction and transient ischemic attacks.

Dose : Usual, adult, oral 300 to 650 mg every 3 or 4 hours ; or 650 mg to 1.3 g as the sustained-release tablet every 8 hours ; Rectal, 200 mg to 1.3 g 3 or 4 times a day.

B. *Salol* BAN, *Phenyl Salicylate* USAN.

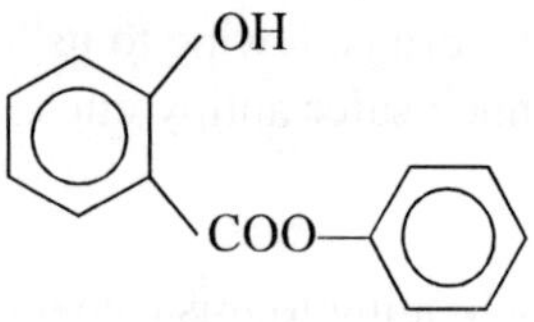

B.P.C. 1954, N.F. XI ; Sola-Stick[R] (Hamilton)

Synthesis

It may be prepared by either of the *two* following methods :

Method-I : By heating salicylic acid alone

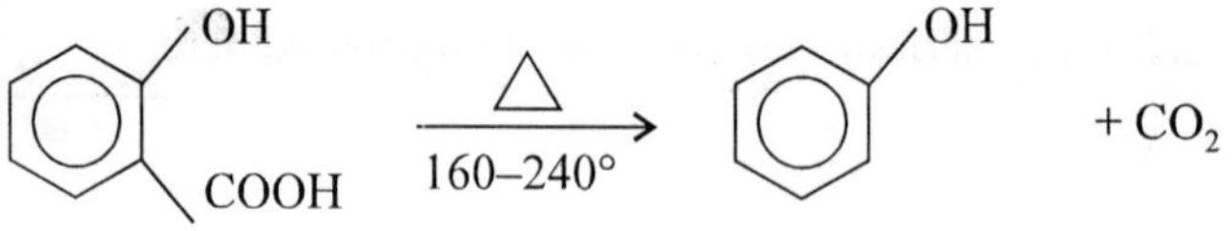

Salicylic acid   Phenol

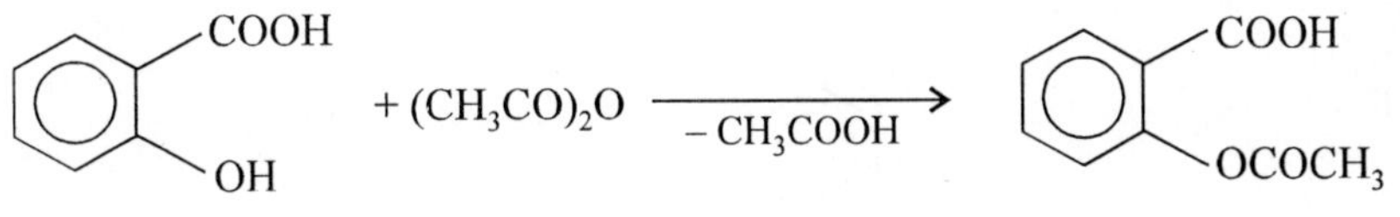

Salicyclic acid   Acetic   Aspirin
                      anhydride

It may be prepared by heating salicylic acid at 160–240°C under vacuum and distilling off the water formed as a by-product.

Method-II : By heating salicylic acid and phenol

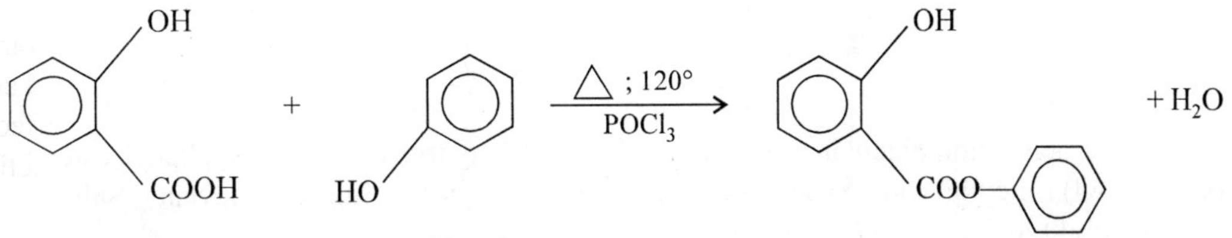

Salicylic acid   Phenol

Phenyl salicylate

It may be prepared by heating together salicylic acid and phenol at 120°C in the presence of phosphorus oxychloride or carbonyl chloride ($COCl_2$).

Salol was first introduced as a drug in 1886 by Nencki. It may be employed as an antipyretic and also as internal antiseptic, but effective doses were toxic owing to the liberation of phenol. It is not usually hydrolysed in the stomach but in the intestine it gradually gets hydrolysed into salicylic acid and phenol respectively. The liberated phenol exerts antiseptic action without any undue toxic effect. Thus the administration of drugs on the above criterion is commonly termed as '*salol principle*' or '*Nencki principle*'. Drugs used on salol principle are generally classified under *two* categories, namely : *true salols* and *partial salols.*

True Salols : are such compounds in which both the compounds *e.g.,* acid and phenol or alcohol are pharmacologically active. *Examples* : Salol and ; betol (β-naphthyl salicylate).

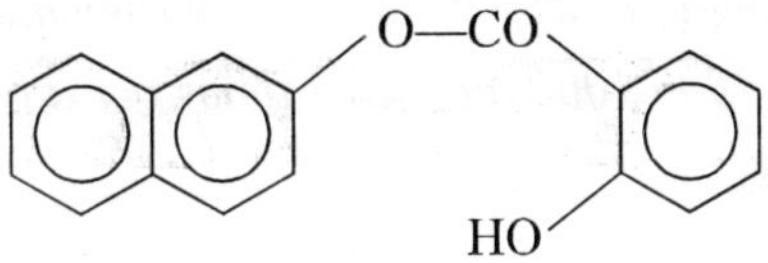

Betol

Partial Salols-are such compounds wherein either the acid or the hydroxylic moiety is active pharmacologically.

Example : Methyl salicylate (oil of wintergreen) in which the salicylic acid constitutes the active component.

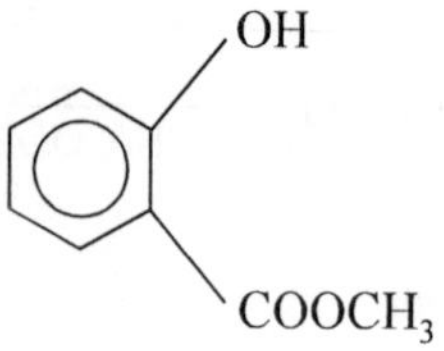

Methyl salicylate

C. *Salsalate* INN, BAN, USAN.

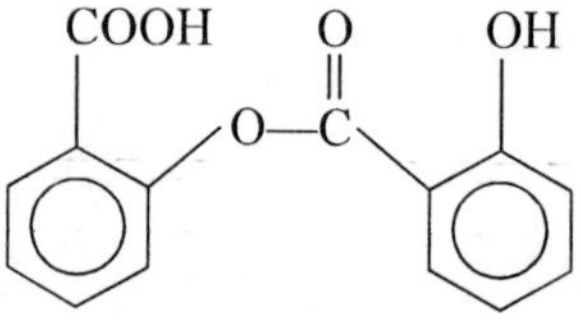

Salicylic acid, biomolecular ester ; Benzoic acid, 2-hydroxy-, 2-carboxyphenyl ester ; *o*-(2-Hydroxybenzoyl) salicylic acid ; Salicylosalicylic acid ; Sasapyrine ; Salicyl Salicylate ; Salysal ;

Disalacid[R] (Riker) ; Saloxium[R] (Whitehall)

Synthesis

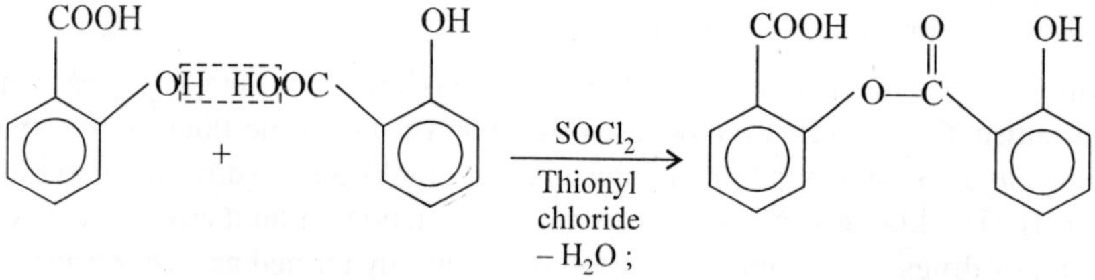

It is prepared by the condensation of two moles of salicylic acid in the presence of thionyl chloride.

It has antipyretic, analgesic and anti-inflammatory properties similar to those of aspirin. It is employed in the *treatment of rheumatoid arthritis and other rheumatic disorders.*

Dose : Usual, adult, oral 325 to 1000 mg 2 to 3 times per day.

D. *Sodium Salicylate* BAN, USAN.

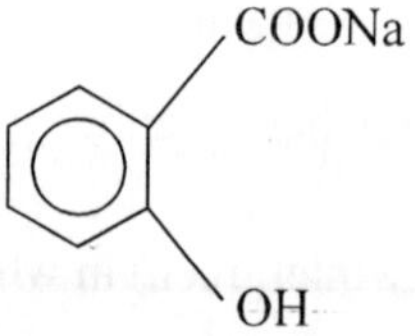

Monosodium salicylate ; Benzoic acid, 2-hydroxy-, monosodium salt ; B.P., U.S.P., Eur. P., Int. P., Ind. P., Entrosalyl (Standard)[R] (Cox Continental, U.K.)

Synthesis

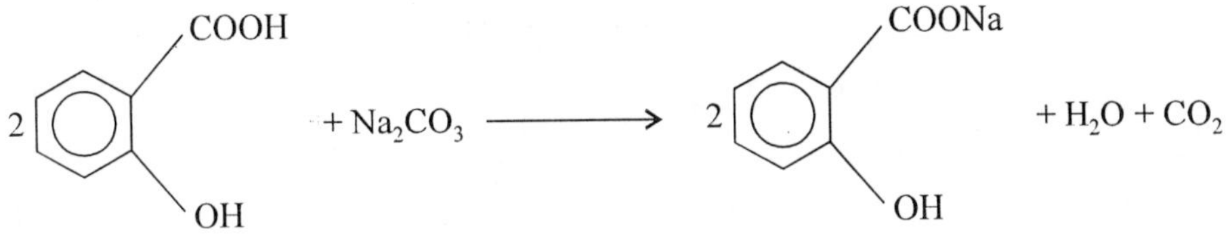

Salicylic acid   Sodium salicylate

It may be prepared by mixing together a paste of salicylic acid in distilled water with sufficient pure sodium carbonate in small lots at intervals. The reaction mixture is filtered through iron-free filter paper and evaporated to dryness under reduced pressure. Caution must be taken to avoid contact with iron which will alter the original white colour of the product.

It is generally used for the reduction of fever and the relief of pain. It also possesses anti-inflammatory actions similar to aspirin. It is recommended in acute rheumatic fever and in the symptomatic therapy of gout.

Dose : In rheumatic fever, 5 to 10 g daily in divided doses.

E. *Salicylamide* INN, BAN, USAN,

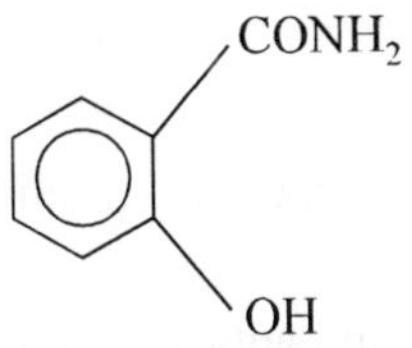

o-Hydroxybenzamide ;

N.F. XIII ; Salined[R] (Medo-Chemicals, U.K.)

Synthesis

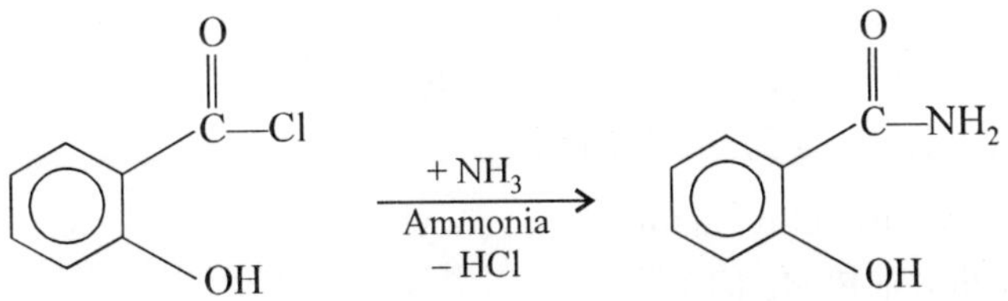

Salicyl chloride   Salicylamide

It is readily prepared from the interaction of salicyl chloride and ammonia.

Its antipyretic and analgesic activity is not more than that of aspirin. It may be used in place of salicylates where apparent sensitivity occurs with the latter.

Dose : 300 mg to 1 g, 3 times per day.

F. *Benorilate* INN, BAN, USAN, *Benorylate* BAN,

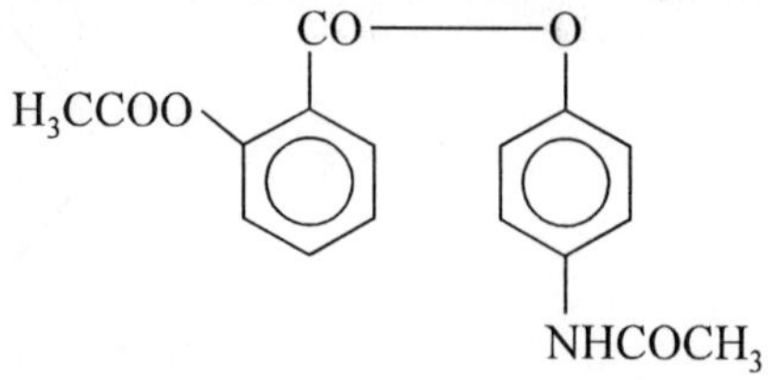

4-Acetamidophenyl salicylate acetate ; 4-Acetamidophenyl-*o*-acetyl-salicylate ; Fenasprate ; Benoral[R] (Winthrop)

Synthesis

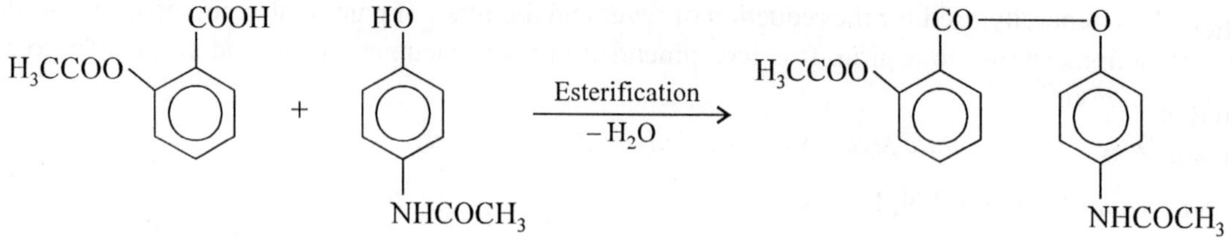

| Acetylsalicylic acid | Paracetamol | Benorilate |

It may be prepared by the esterification of acetyl salicylic acid and paracetamol with the elimination of a mole of water.

It possesses antipyretic, analgesic and anti-inflammatory properties. It is employed in the treatment of rheumatic disorders and in moderate pain, and as an antipyretic.

Dose : Rheumatic conditions 1.5 g, 3 times daily.

G. *Choline Salicylate,* INN, BAN, USAN,

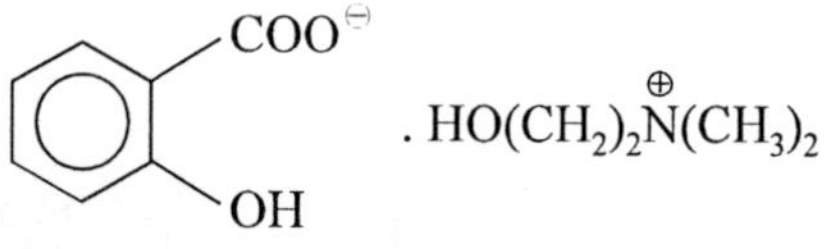

(2-Hydroxymethyl) trimethyl ammonium salicylate ;

Arthropan[R] (Purdue Frederick)

It possesses actions similar to those of aspirin but it is mainly used as a local analgesic by being applied to the painful area by gentle rubbing.

Dose : Adult, usual 0.87 to 1.74 g, 3 or 4 times daily.

H. *Flufenisal* INN, USAN,

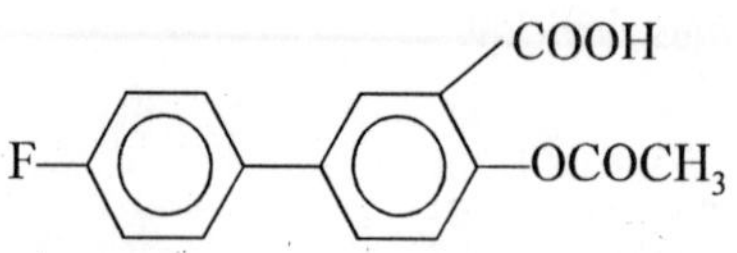

Acetyl-5-(4-fluorophenyl) salicylic acid ; 4'-Fluoro-4-hydroxy-3-biphenyl-carboxylic acid acetate ; [1, 1'-Biphenyl]-3-carboxylic acid, 4-(acetyloxy)-4'-fluoro- ; Flufenisal[(R)] (MSD).

The search for a better drug than aspirin with increased potency, longer duration of action and having less effect on gastric secretion gave birth to flufenisal which essentially has a hydrophobic moiety at C_5. In man, it exhibits a two-fold increase in potency and duration of action than that of aspirin.

Dose : 150 to 300 mg every 3 or 4 hours.

2.3. Quinoline Derivatives

The historical importance and utility of quinine was known in the medical practice for a long time as a potent antipyretic in addition to tis remarkable effect against the malarial fever. The basic quinoline nucleus, present in the quinine molecule, contributes to antipyretic activity to a certain extent. Therefore, an attempt was made to synthesize a number of quinoline derivatives which might exhibit better antipyretic activity.

Two quinoline derivatives frist synthesized though possessed significant antipyretic action, yet could not gain cognizance as a drug because of their high toxic effects on the red blood corpuscles and damaging after-effect on kidneys. These were, thalline and 6-methoxy quinoline.

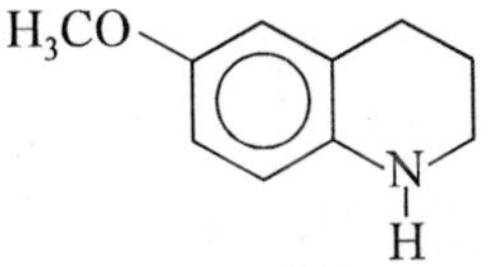

Thalline

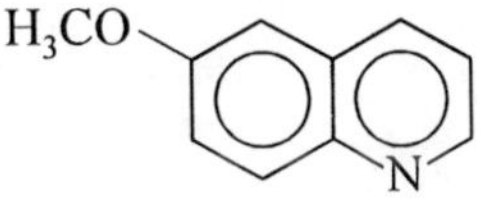

6-Methoxyquinoline

A. *Cinchophen* INN, BAN, USAN,

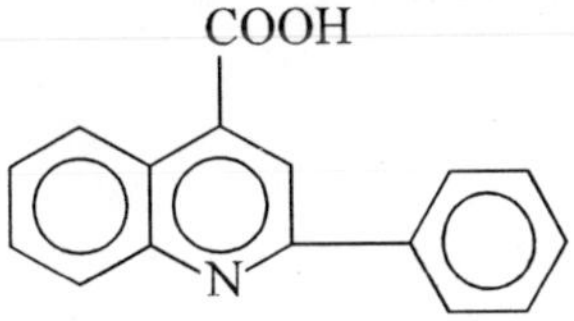

2-Phenyl-cinchoninic acid ; 2-Phenylquinoline-4-carboxylic acid ; Quinophan ; Atophan ; B.P. 1953, N.F. X.

Synthesis

It may be prepared by any one of the following *three* methods :

Method-I : From o-Amino benzaldehyde cyanohydrin

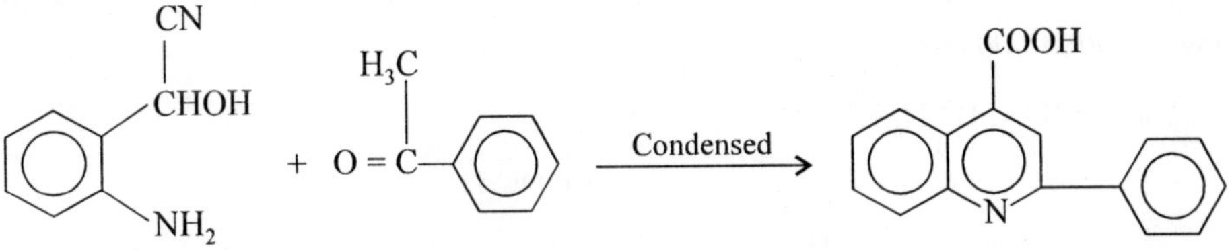

o-Aminobenzalde-
hyde cyanohydrin

Methyl-
phenyl ketone

Cinchophen

Condensation of *o*-aminobenzaldehyde cyanohydrin and methylphenyl ketone yields cinchophen.

Method-II : From Isatin

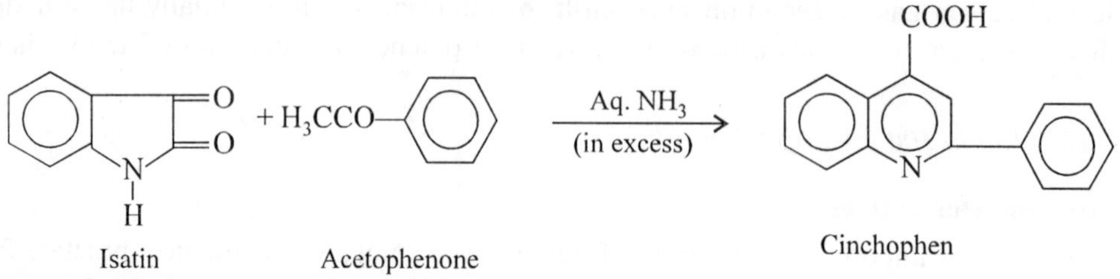

Isatin   Acetophenone   Cinchophen

Cinchophen may be prepared by the interaction of isatin and acetophenone in the presence of excess of aqueous ammonia.

Method-III : From Aniline

The Schiff's base is prepared by the interaction of aniline and benzaldehyde with the elimination of a molecule of water. The resulting base is treated with the *lactim*-form of pyruvic acid thereby resulting into the formation of cinchophen.

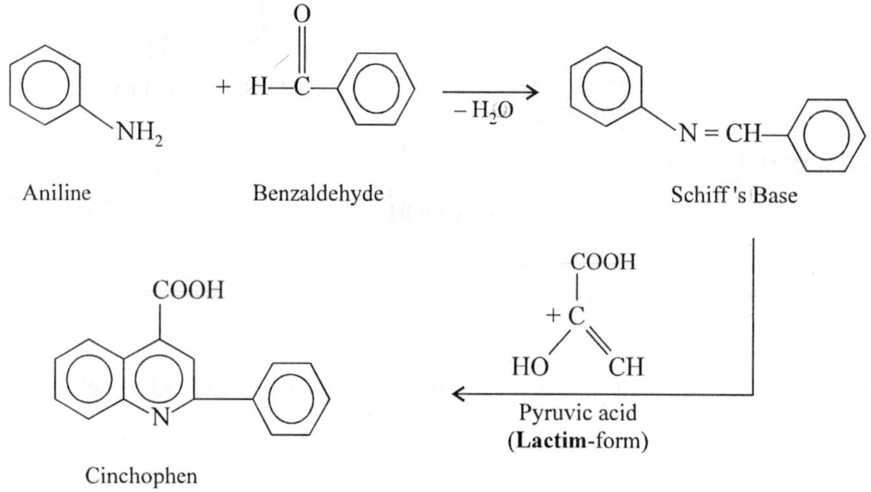

Aniline   Benzaldehyde   Schiff 's Base

Cinchophen

Pyruvic acid
(**Lactim**-form)

Cinchophen possesses antipyretic actions similar to those of the salicylates. It was chiefly used *in the treatment of chronic gout and rheumatic conditions but because of its high toxicity, e.g., liver damage resulting in acute jaundice, it has been completely withdrawn and replaced by safer drugs.*

Dose : 300 to 600 mg.

B. *Neocinchophen* INN, BAN, USAN

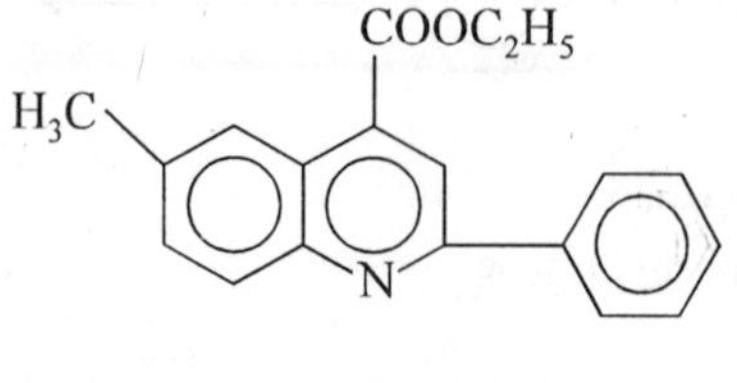

Ethyl-6-methyl-2-phenyl-4-quinolinecarboxylate ; N.F. XI ; Tolysin[R] (Lederle)

Synthesis

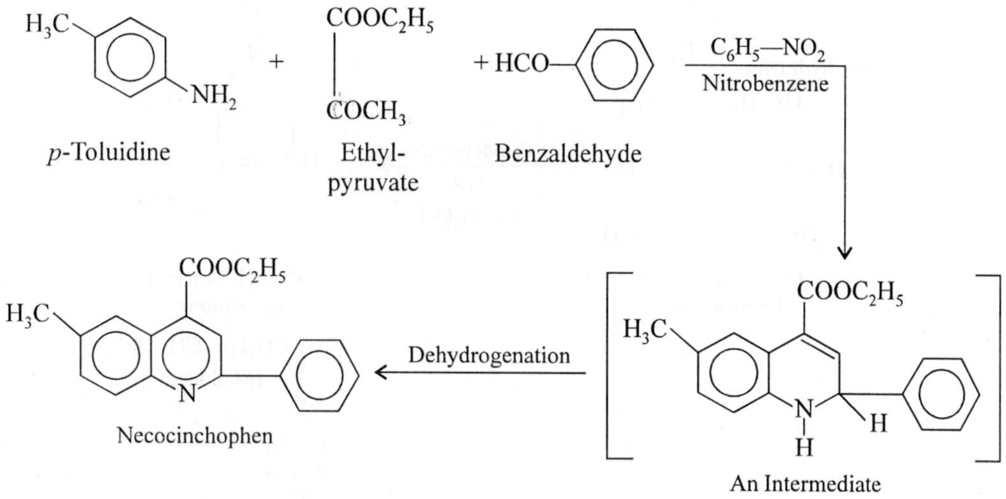

p-Toluidine   Ethyl-   Benzaldehyde
                               pyruvate

Necocinchophen   ← Dehydrogenation   An Intermediate

It occurs through the reaction of *p*-toluidine, ethyl pyruvate and benzaldehyde in the presence of a small amount of nitrobenzene, when the products get condensed to form an intermediate compound. This when subjected to dehydrogenation yields neocinchophen.

It has been used for the same purposes as cinchophen.

Dose : 500 mg.

2.4. Pyrazolones and Pyrazolodiones

One of the first and foremost synthetic organic compounds which were successfully used as drugs was found to be a heterocycle. It is, however, worthwhile to mention here that the pharmacodynamic spectrum of both the above categories of heterocyclic compounds has a close resemblance to that of aspirin. A few such compounds belonging to either of the said classes are discussed here.

A. *Phenazone* INN, BAN, *Antipyrine* USAN,

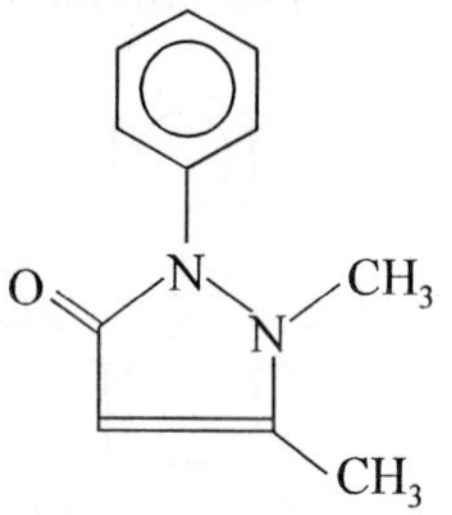

2, 3-Dimethyl-1-phenyl-3-pyrazolin-5-one ; 1, 2-Dihydro-1, 5-dimethyl-2-phenyl-3H-pyrazol-3-one ; Antipyrin ; Phenazone B.P., Eur. P., Int. P., Antipyrine U.S.P. Component of Auralgan[R] (Ayerst) ; Areumal[R] (as Gentisate ; Ecobi. Italy)

Synthesis

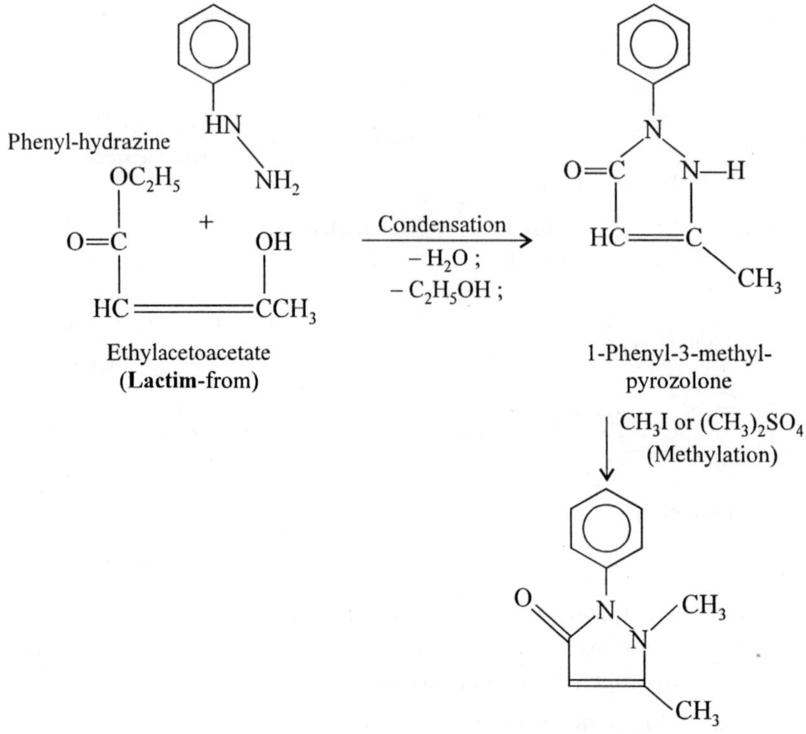

Phenyl-hydrazine

Ethylacetoacetate
(**Lactim**-from)

Condensation
$- H_2O$;
$- C_2H_5OH$;

1-Phenyl-3-methyl-
pyrozolone

CH_3I or $(CH_3)_2SO_4$
(Methylation)

Phenazone

It may be prepared by the condensation of one mole each of phenyl-hydrazine and the *lactim*-form of ethylacetoacetate when 1-phenyl-3-methyl-pyrazolone is obtained by the elimination of a mole each of water and ethanol. The resulting product is subjected to methylation either with methyl iodide or dimethyl sulphate to yield phenazone.

As antipyretic, it possesses local anaesthetic and styptic actions and solutions containing 5% are used locally as ear drops. It has now been replaced by relatively more effective and safer drugs.

Dose : 300 to 600 mg.

B. *Aminophenazone* INN, *Amidopyrine* BAN, *Aminopyrine* USAN,

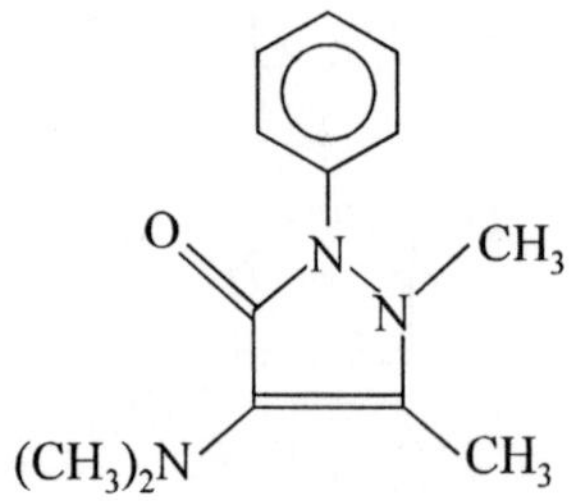

4-Dimethylamino-2, 3-dimethyl-1-phenyl-3-pyrazolin-5-one ; 4-Dimethyl-amino-1, 5-dimethyl-2-phenyl-4-pyrazolin-3-one ; Dimethylaminoantipyrine ; Dimethylaminophenazone ; Amidopyrine B.P.C. 1954, Eur. P., Int. P., Aminopyrine N.F.X. ; Piramidon(R) (Hoechst, Spain)

Synthesis

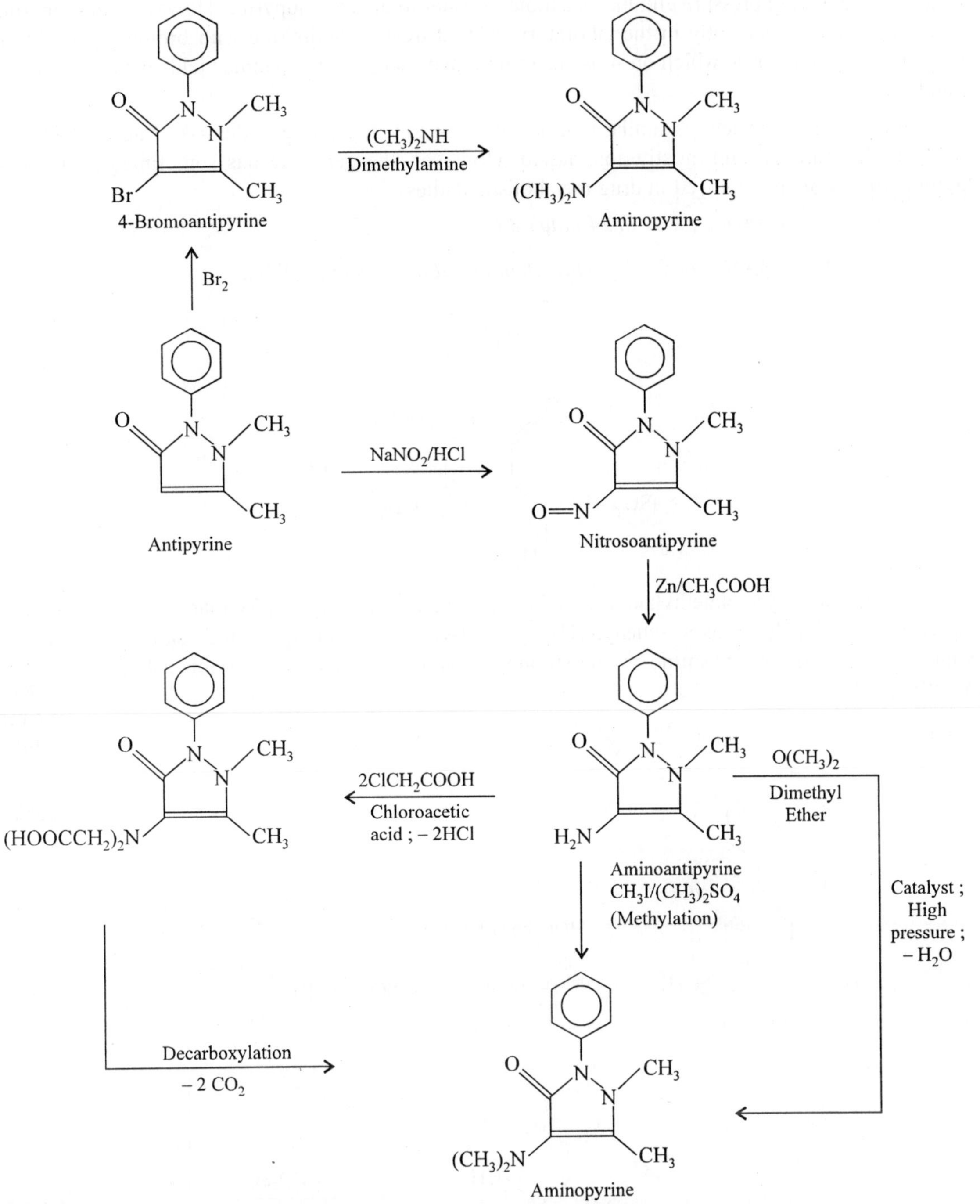

4-Bromoantipyrine → (CH₃)₂NH Dimethylamine → **Aminopyrine**

Br₂

Antipyrine → NaNO₂/HCl → **Nitrosoantipyrine**

Zn/CH₃COOH

Aminoantipyrine ← 2ClCH₂COOH Chloroacetic acid ; – 2HCl

O(CH₃)₂ Dimethyl Ether

CH₃I/(CH₃)₂SO₄ (Methylation)

Catalyst ; High pressure ; – H₂O

Decarboxylation – 2 CO₂

Aminopyrine

Aminopyrine (amidopyrine) may be prepared commercially first by treating antipyrine with nitrous acid to yield nitrosoantipyrine. The resulting product can now be routed through two different course of reactions, namely : (*a*) treatment with two moles of chloroacetic acid followed by decarboxylation producing thereby aminopyrine ; and (b) treatment with dimethyl ether in the presence of catalyst and at high pressure eliminates a mole of water to give aminopyrine. However, aminopyrine can be prepared conveniently in the laboratory by first treating antipyrine with bromine partially to obtain 4-bromo-antipyrine which on subsequent treatment with dimethylamine yields the official compound.

It has antipyretic actions similar to those of phenazone but owing to the risk of agranulocytosis its use is discouraged and mostly abandoned. However, the gentisate has sometimes been used. Aminopyrine is often employed in drug metabolism studies.

Dose : 300 to 500 mg ; max in 24 hours 3 g.

C. *Dipyrone* BAN, USAN, *Noramidopyrine Methanesulfonate Sodium* INN,

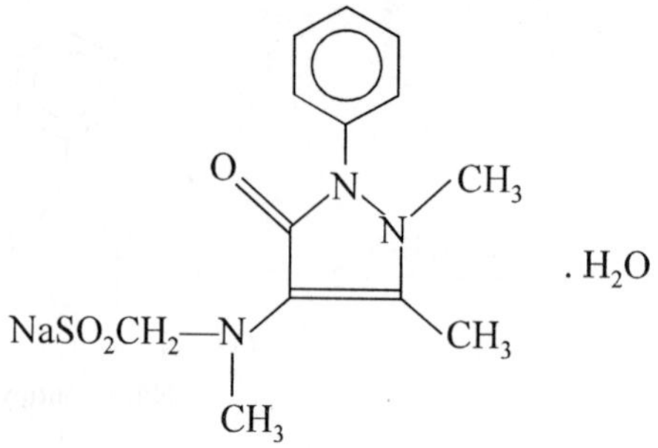

Sodium (antipyrinylmethylamino) methanesulfonate monohydrate ; Methane-sulfonic acid, [(2, 3-dihydro-1, 5-dimethyl-3-oxo-2-phenyl-1H-pyrazol-4-yl) methylamino]-, sodium salt, monohydrate ; Analginum ; Metamizol ; Amino-pyrine-sulphonate sodium ; Novalgin[R] (Hoechst) ; Novaldin[R] (Winthrop)

Synthesis

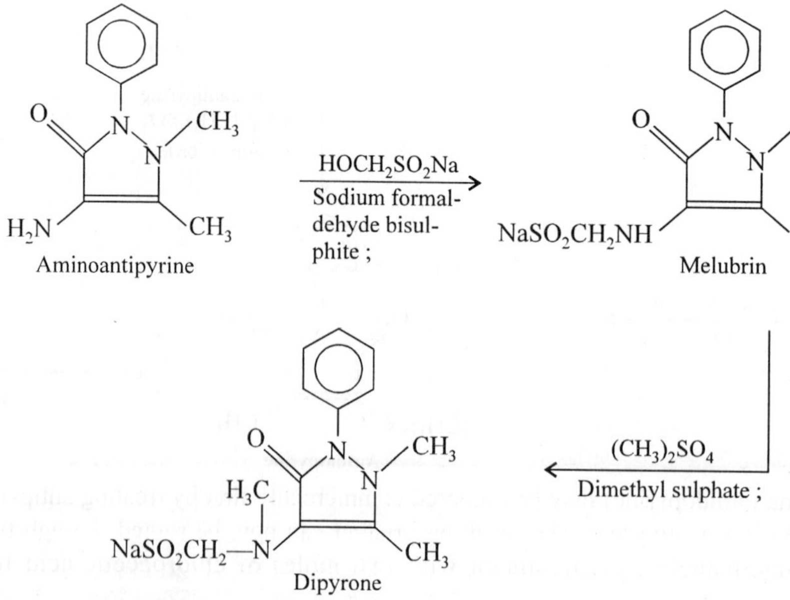

It possesses similar properties to that of amidopyrine. Its use is really justified only in serious or life-threatenting situations where no alternative antipyretic is available or suitable. Its use is restricted in some countries.

Dose : Usual, 0.5 to 1 g, 3 times per day.

D. *Phenylbutazone* INN, BAN, USAN,

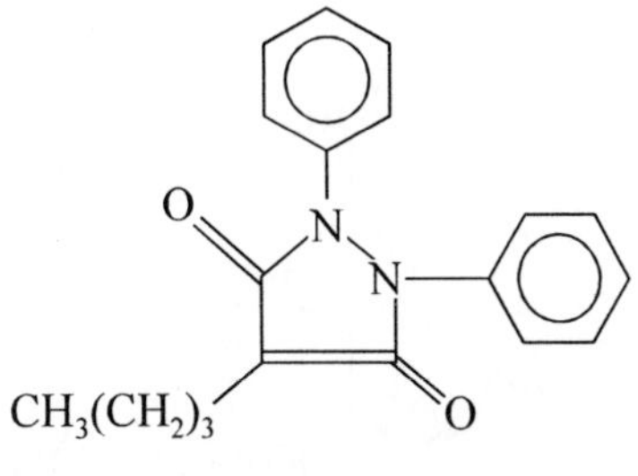

$$CH_3(CH_2)_3$$

4-Butyl-1, 2-diphenyl-3, 5-pyrazolidinedione ; 3, 5-Pyrazolidinedione, 4-butyl-1, 2-diphenyl- ; Butadione ; B.P., U.S.P., Eur. P., Int. P., Butazolidin[R] (Ciba-Geigy) ; Busone[R] (Reid-Provident)

Synthesis :

Phenylbutazone may be prepared by condensation either from diethyl-*n*-butyl malonate or *n*-butyl malonyl chloride with hydrazobenzene in either solution at 0°C with the aid of pyridine. Subsequently, the pyridine is extracted with aqueous hydrochloric acid, the phenylbutazone is extracted with aqueous sodium bicarbonate and finally precipitated by addition of hydrochloric acid.

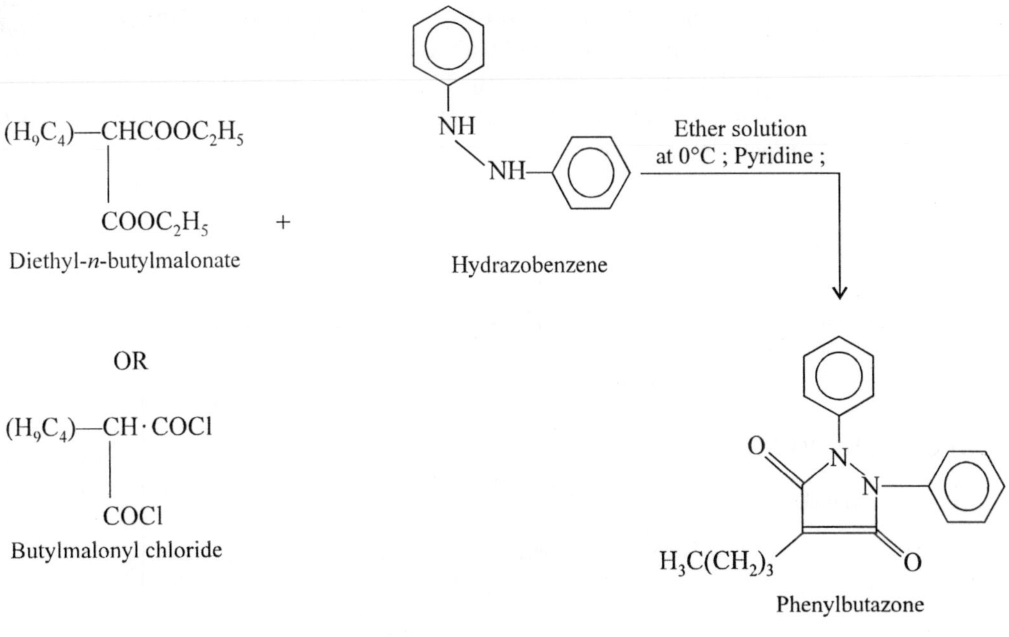

(H_9C_4)—$CHCOOC_2H_5$

$COOC_2H_5$ +

Diethyl-*n*-butylmalonate

NH

NH—

Hydrazobenzene

Ether solution
at 0°C ; Pyridine ;

OR

(H_9C_4)—$CH \cdot COCl$

COCl

Butylmalonyl chloride

$H_3C(CH_2)_3$

Phenylbutazone

It is a pyrazole derivative which has antipyretic, analgesic and anti-inflammatory actions, because of its toxicity it is not used as a general antipyretic or analgesic. It is, an usual practice, reserved for use in the treatment rheumatic disorders, such as : osteoarthrosis, rheumatoid arthritis, ankylosing spondylitis, arthritis, acute superficial thrombophlebitis, painful shoulder and Reiter's disease, where less toxic drugs have failed.

In some countries, its use and that of oxyphenbutazone are now restricted to only ankylosing spondylitis.

Dose : 100 to 600 mg per day.

E. *Oxyphenbutazone* INN, BAN, USAN,

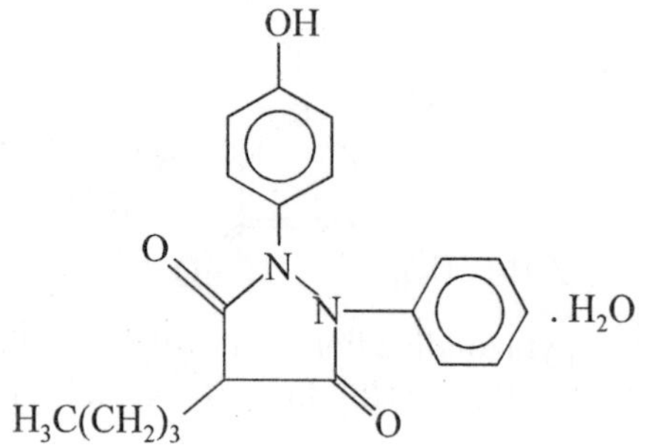

4-Butyl-1-(*p*-hydroxyphenyl)-2-phenyl-3, 5-pyrazolidinedione monohydrate ; 3, 5-Pyrazolidinedione, 4-butyl-1-(4-hydroxyphenyl)-2-phenyl, monohydrate ; B.P., U.S.P., Tandearil[R] (Ciba-Geigy) ; Oxalid[R] (USV).

Synthesis :

Condensation of diethyl butyl malonate and *p*-benzyloxyhydrazo-benzene is done in the presence of sodium ethoxide in anhydrous ethanol to yield 1-(benzyloxy)-2-phenyl-4-butyl-3, 5-pyrazolidinedione. The reaction mixture is heated with xylene to about 140°C for several hours which aids in the removal ethanol eliminated by cyclization. The resulting product is debenzylated by the aid `Raney Nickel hydrogenation at an ambient temperature and pressure. The crude product may be recrystallized from ether/petroleum ether.

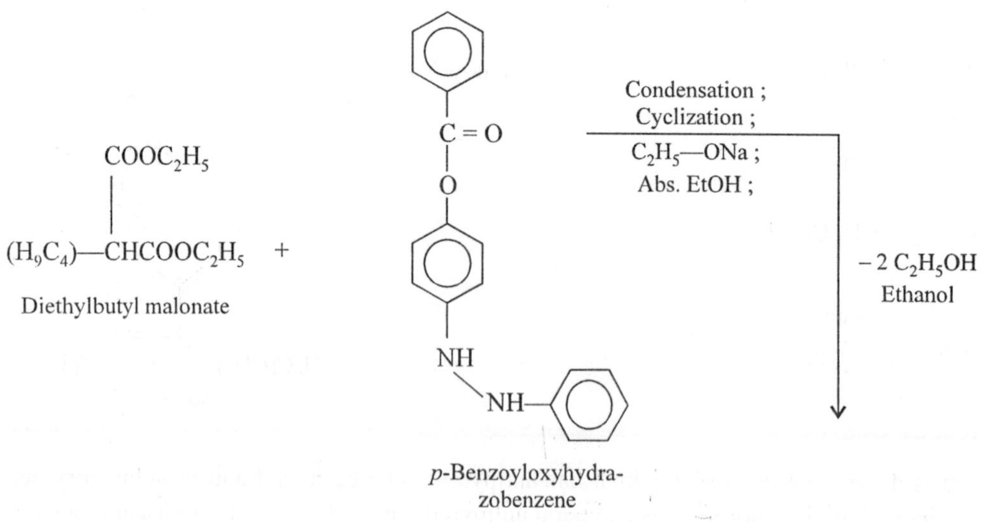

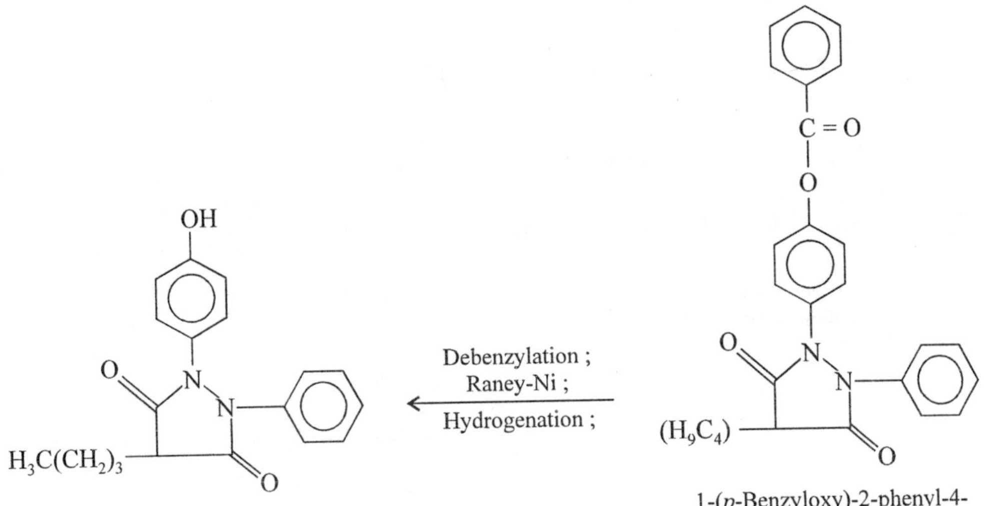

Oxyphenbutazone

Debenzylation ;
Raney-Ni ;
Hydrogenation ;

1-(*p*-Benzyloxy)-2-phenyl-4-
butyl-3, 5-pyrazolidinedione

It is a metabolite of phenylbutazone and possesses the same antipyretic, analgesic, anti-inflammatory and mild uricosuric actions as the parent compound. It also finds its use in rheumatic and other musculo-skeletal disorders. It has also been recommended in the management of thrombophlebits.

Dose : Usual, oral, adult, antirheumatic, 100 or 200 mg 3 times daily ; maintenance 100 mg 1 to 4 times daily ; antigout, 400 mg initially as a loading dose, then 100 mg every 4 hours.

F. *Sulfinpyrazone* INN, USAN, *Sulphinpyrazone* BAN.

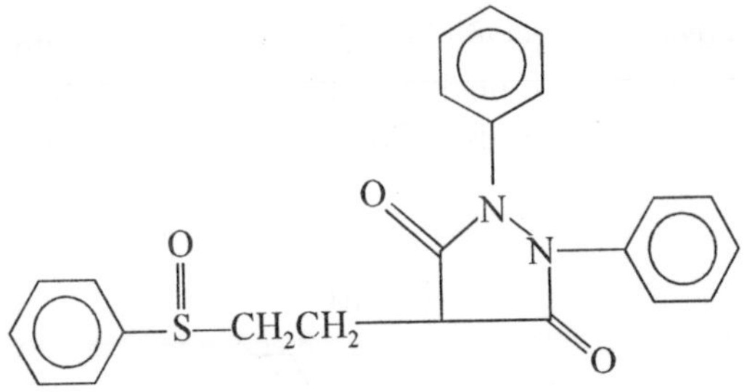

1, 2-Diphenyl-4[2-(phenylsulfinyl) ethyl]-3, 5-pyrazolidinedione ; 3, 5-Pyrazolidinedione, 1, 2-diphenyl-4-[2-(phenylsulfinyl) ethyl]- ; B.P., U.S.P., Anuturane(R) (Ciba-Geigy)

Synthesis :

It may be prepared by the condensation of [2-(phenylsulfinyl) ethyl]- malonic acid diethyl ester with hydrazobenzene in the presence of sodium ethoxide in absolute ethanol. Completion of reaction is achieved by the addition of xylene and subsequent heating at about 130°C whereby the ethanol liberated as a product of condensation is removed completely. The crude product is extracted with a suitable solvent and finally recrystallized from ethanol.

It is a uricosuric agent structurally related to phenylbutazone. It is normally employed for the long term *treatment of chronic gout where it effects slow depletion of urate tophi in the tissues.*

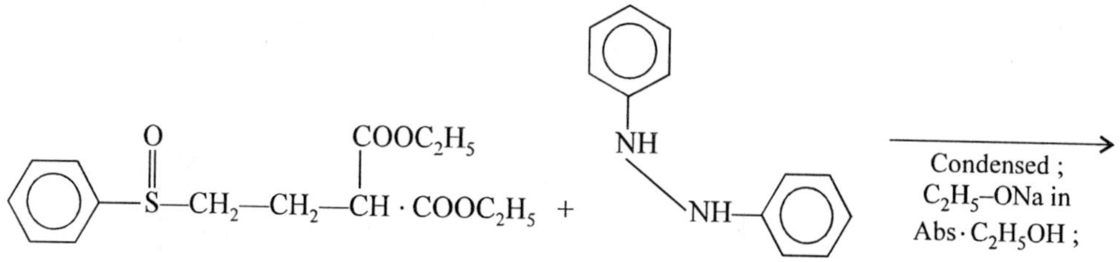

2-(Phenylsulfinyl) ethyl-
malonic acid diethyl ester   Hydrazobenzene

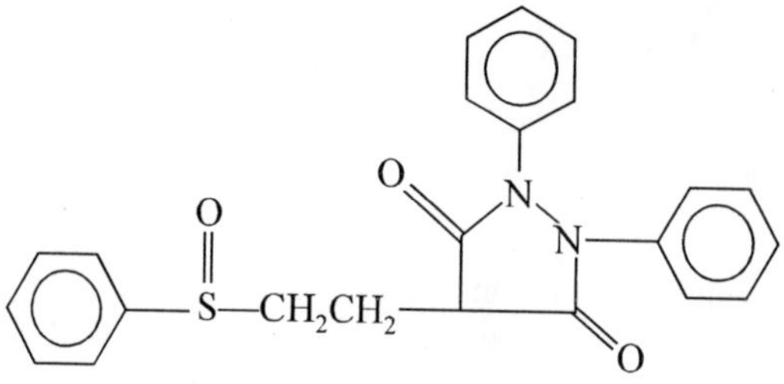

Condensed ;
C_2H_5–ONa in
Abs·C_2H_5OH ;

Sulfinpyrazone

Dose : Initial oral dose 100 to 200 mg per day, taken with meals or milk.

2.5.   The N-Arylanthranilic Acids

The structural analogues of N-arylanthranilic acid opened an altogether new horizon of antipyretic, analgesic and anti-inflammatory compounds which have recently gained recognition in the therapeutic armamentarium. A few compounds belonging to this category are discussed here.

A. *Mefenamic Acid* BAN,

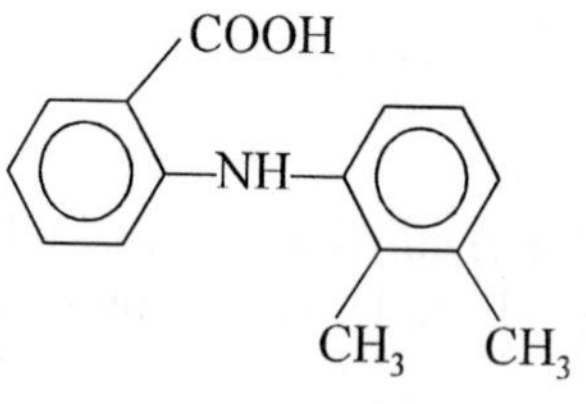

N-(2, 3-Xylyl) anthranilic acid ; Benzoic acid, 2-[(2, 3-dimethylphenyl) amino]- ; B.P., Ponstel[R]
(Parke-Davis).

Synthesis :

It may be prepared by the condensation of *o*-chlorobenzoic acid with 2, 3-xylidine in the presence of potassium carbonate to give the potassium salt of mefenamic acid, which on treatment with
hydrochloric acid yields the official compound.

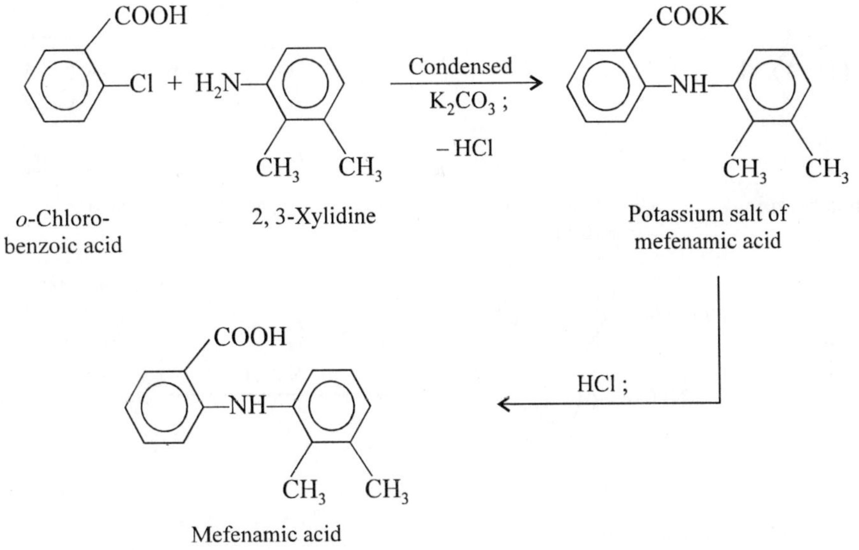

o-Chloro-
benzoic acid

2, 3-Xylidine

Potassium salt of
mefenamic acid

Mefenamic acid

It is an analgesic drug usually indicated for the treatment of primary dysmenorrhea, mild pain and for pain due to dental extractions.

Dose : Usual, adults, children over 14 years of age, oral, 500 mg, followed by 250 mg 4 times daily.

(Caution : Must not be used for more than 7 days).

B. *Meclofenamate Sodium* BAN, USAN, *Meclofenamic Acid* INN.

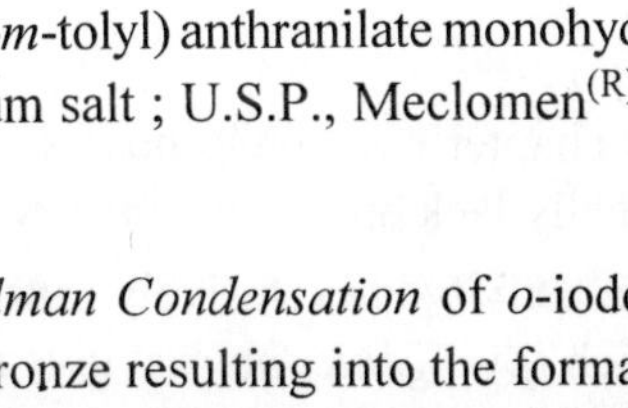

Monosodium N-(2, 6-dichloro-*m*-tolyl) anthranilate monohydrate ; Benzoic acid, 2-[2, 6-(dichloro-3-methylphenyl) amino]-, monosodium salt ; U.S.P., Meclomen[(R)] (Parke-Davis).

Synthesis :

It may be prepared by the *Ulman Condensation* of *o*-iodobenzoic acid with 2, 6-dichloro-*m*-toluidine in the presence of copper-bronze resulting into the formation of meclofenamic acid which on neutralization with equimolar proportion of sodium hydroxide yields meclofenamate sodium.

It possesses analgesic, anti-inflammatory, and antipyretic properties. It is used for the *treatment of acute and chronic rheumatoid arthritis and osteoarthritis.*

Synthesis

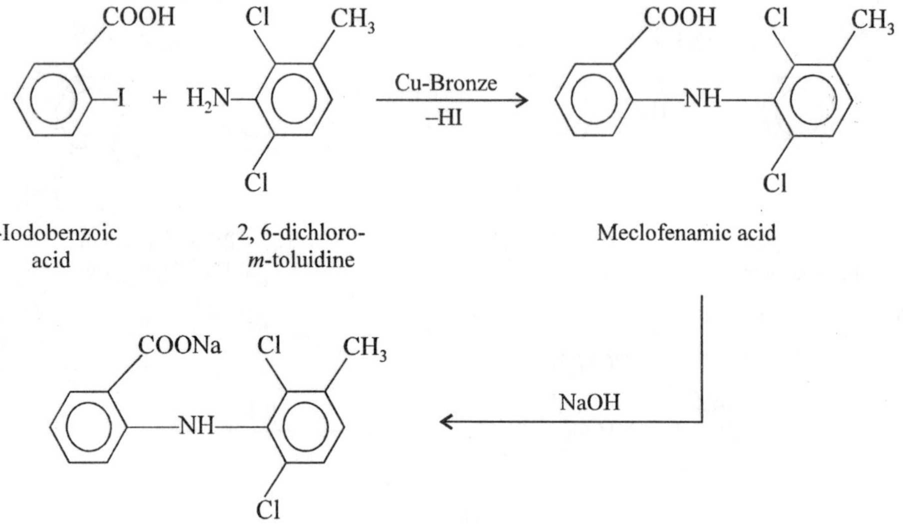

o-Iodobenzoic acid	2, 6-dichloro-*m*-toluidine	Meclofenamic acid

Meclofenamate Sodium

Dose : Usual, oral, 200 to 400 mg daily in 3 or 4 equal doses.

C. *Flufenamic Acid* INN, BAN, USAN.

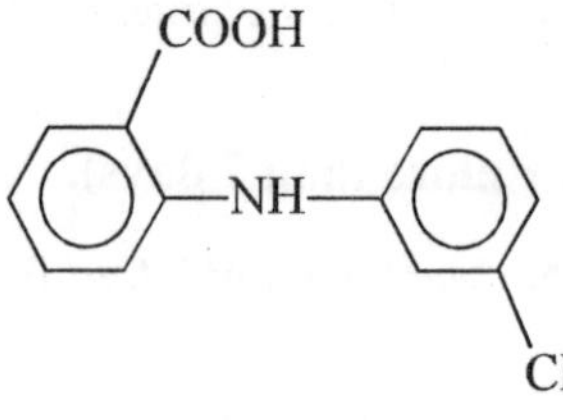

N-(α, α, α-Trifluoro-*m*-tolyl)-anthranilic acid ; Benzoic acid, 2-[[3-(trifluoro-methyl) phenyl] amino]- ; B.P., U.S.P., Meralen[R] (Merrell, U.K.)

It has analgesic, anti-inflammatory and antipyretic actions. It is employed in the *treatment of rheumatic disorders* and dysmenorrhoea.

Dose : 400 to 600 mg per day in divided doses.

3. MECHANISM OF ACTION

The '*drugs*' included in this chapter essentially possess solely the '*antipyretic and analgesic*' pharmacological actions but specifically lack anti-inflammatory effects.

Interestingly, the '*antipyretic*' activity is exclusively caused due to the direct interferences, with such phenomena with the aid of which '*pyrogenic factors*' give rise to **fever** ; however, they are found absolutely unable to bring-down the elevated body temperature ($\gg$ 98.4°F) specifically in afebrile subjects. It has already been established beyond any reasonable doubt that the '*antipyretics*', in general, exert their activities very much within the CNS. These pharmacological actions are significantly located at the *'hypothalamic thermoregulatory centre'* ; however, recent studies categorically advocates the *'peripheral actions'* may also contribute enormously and positively.

In fact, there are *two* different school of thoughts that have been suggested to explain the modalities of antipyretic action :

(*a*) *Endogenous leukocytic pyrogens* are presumably released from the cells that have been duly activated by a host of stimuli, and *'antipyretics'* do exert their action by inhibiting the corresponding activation of these cells by an exogenous pyrogen, and

(*b*) Inhibition of the release of *'endogenous leukocytic pyrogens'* from the cells as soon as these have been adequately activated by the exogenous pyrogen.

Clark* (1979) put forward substantial evidences to prove that there exists a *'central antipyretic mechanism'* which specifically affords an *'antagonism'* that may be caused on account of :

(*i*) A direct competition ensuing of a pyrogen and the antipyretic drug pervailing at the CNS-receptors, and

(*ii*) An inhibition of prostaglandin (PG) synthesis occurring in the CNS.

More logistically, the experimental pharmacologists usually determine the *analgesic activity* in laboratory animals (rat/mice) by measuring the *'pain threshold'* in terms of certain reflex actions essentially produced by noxious stimuli, for instance : pressure, heat and electric, shock. There are several, known methods to determine the exact *'analgesic profile'* in *'synthetic'* as well as *'natural plant products'*, such as : rat tail-flick test ; mouse hot-plate test ; and usage of electricity to tooth-pulp thereby giving rise to almost reproducible results (*i.e.,* end-points) whose appearance may be delayed with respect to *'time'* by *'analgesic drugs'* under examination virtually with a direct relationship to both the **potency** and **efficacy**.

Hughes and Kosterlitz** (1975) isolated (from pig-brain) and identified **'enkephalins'** produced in the body having narcotic-like substances so as to react judiciously with receptors for the narcotic drugs. Thus, the two identified and characterized *'brain-peptides'* essentially differed only in the nature of their N-terminal amino acids, for instance : (*a*) **methionin-enkephalin**-having a tyrosine-glycine-glycine-phenylalamine-methionine sequence ; and (*b*) **leucine-enkephalin**-having a tyrosine-glycine-glycine-phenylalanine-leucine sequence.

4. MECHANISM OF ACTION OF SELECTED ANTIPYRETIC-ANALGESICS

The mechanism of action of certain selected antipyretic-analgesics included in this chapter (section 9.2) are discussed in the sections that follows :

4.1. Paracetamol (Acetaminophen)

It causes antipyresis by exerting its action on the *hypothalamic heat-regulating centre,* and analgesia by enhancing the pain threshold profile appreciably. It is found to lack the anti-inflammatory activity of the salicylates ; therefore, its therapeutic usefulness in inflammatory disorders is very much limited, and hence is not regarded as an NSAID agent. In contrast to the action of *'aspirin'*, paracetamol possesses little affect in antagonizing the actions of uricosuric agents (*i.e.,* increases the urinary excretion of uric acid). It has also been observed that its large doses usually help in potentiating the action of

---

*Clark WG, **Mechanisms of Antipyretic Action**, *Gen. Pharmacol.,* **10** : 71–77, 1979.

Hughes J and Kosterlitz, *Nature,* **258 : 577, 1975.

the anticoagulants, whereas the normal therapeutic dose regimens exert hardly any effect on the *'prothrombin time'* (*i.e.,* coagulation time).

Nearly, 2% of the *'drug'* is excreted almost unchanged in the urine, while approximately 95% is found as its corresponding *glucuronide* and *sulphate-conjugates* that are absolutely devoid of any toxicity. Furthermore, the remaining 3% gets oxidized *via* the *hepatic cytochrome P-450 system* into a respective chemically reactive intermediate which eventually combines specifically with the **liver glutathione** to give rise to the formation of a **'nontoxic'** entity.

4.2. Phenacetin

Its toxic effects are very much comparable to those of *paracetamol* (acetaminophen), the *'active form'* to which it gets converted *in vivo.* The earlier findings revealed that it may cause a damage to the kidneys when used either in excessive dosage or for a longer duration. However, certain interesting recent evidences strongly suggest that phenacetin may *not* be responsible for causing nephritis to any greater extent when compared to *'aspirin'*[*]. Importantly, it has been strongly demonstrated in causing carcinogenesis in rats and associated with the growth of tumours in abuses of phenacetin[**].

4.3. Acetanilide

It is considered to be relatively safer drug in the doses recommended for analgesia. Hence, it may be administered in intermittent periods, not exceeding a few days in any circumstances. The analgesic effect is quite selective for pharmacological action(s) ranging from simple headache to the pain associated with many muscles and joints.

4.4. Aspirin

It has been well established that *'aspirin'* inhibits platelet function ; therefore, it prophylactically minimises the incidence of **myocardial infarction** and **transient ischemic attacks** particularly in men and also postmenopausal women. Interestingly, aspirin is *not* hydrolyzed significantly when it happens to be in contact with the weakly acidic digestive juice present in the stomach (*i.e.,* gastric juice) ; however, as soon as it gains its passage into the intestinal canal it undergoes hydrolysis to a certain extent. A large portion of it usually gets absorbed unchanged. Garrett[***] (1959) put forward a logical explanation with regard to the gastric mucosal irritation of aspirin to the formation of *salicylic acid i.e.,* the natural inherent acidity of aspirin ; besides, the intimate adhesion of undissolved aspirin to the gastric mucosa. Subsequently, Davenport[****] (1967) demonstrated that *aspirin* affords an irreversible modification in the degree of permeability in the mucosal cell, thereby permitting the *'back-diffusion'* of gastric acid (in stomach) that ultimately is responsible in causing permanent damage to the capillaries.

4.5. Sodium Salicylate

It is considered to be one of the *'choicest drug'* specifically for salicylate medication ; and is usually administered with either sodium bicarbonate to minimise effectively the *'gastric distress'* or as enteric-coated dosage forms. However, the usage of $NaHCO_3$ is not advisable as it is found to retard the plasma levels of *'salicylate'* and enhances the elimination of *'free salicylate'* in the urine.

---

[*]Brown DM and Hardy TL., *Brit J. Pharmacol. Chemother.,* **32**, 17, 1968.

[**]Tomatis L. *et. al. Cancer Res.* **38,** 877, 1978.

[***]Garrett ER, *J. Am. Pharm. Assoc. Sci.,* **48**, 676, 1959.

[****]Davenport HW, *N. Engl. J. Med.,* **276**, 1307, 1967.

4.6. Salicylamide

It is believed to exert a moderately faster and deeper analgesic effect in comparison to 'aspirin'. It has also been established that its long term usage in rats no abnormal and untoward physiological and symptomatical reactions observed. Salicylamide gets metaboilized in a manner altogether different from that of other 'salicylates' ; and, importantly it hardly gets hydrolyzed to the corresponding salicyclic acid.*

4.7. Salsalate

The ester is usually hydrolyzed following its immediate systemic absorption. It is believed to afford much less gastric irritation and discomfort in comparison to 'aspirin', by virtue of the fact that the 'drug' is virtually insoluble in the stomach ; and, therefore, never gets absorbed unless and until it happens to gain its access into the small-intestine. It is found to be as effective as 'aspirin' and definitely possess fewer side effects.

4.8. Choline Salicylate

It is observed to be absorbed much more swiftly in comparison to 'aspirin', thereby giving rise to faster peak blood levels.

4.9. Fluferisal

It is found to be more potent, long acting and possesses much less gastric irrilation. All these characteristic features have been duly accomplished, by strategically introducing a hydrophobic functional moiety (4-fluorophenyl) at C-5. Just like other aryl acids the 'drug' is most intimately bound to plasma proteins in the shape of its deacylated metabolite. However, in human beings it seems to be at least twice as effective i.e., having almost twice the duration of activity.

4.10. Cinchophen

Its antipyretic actions are very much akin to those of the salicylates. Its major pharmacological action was in the control and management of chronic gout and rheumatic conditions, but by virtue of its relatively high level of toxic effects, such as : hepatic dysfunction ultimately leading to acute jaundice.

4.11. Phenazone (Antipyrine)

The drug is found to exert an appreciable paralytic action exclusively upon the sensory and the motor nerves which eventually give rise to certain degree of anaesthesia and vasoconstriction. Somatically (i.e., systemically), it is observed to afford pharmacological activities which are very similar to those of acetanilid ; evidently these are normally quite fast and rapid. After due oral administration, it undergoes a free circulation within the system, and finally gets excreted through the kidneys in an 'unchanged form'. It remarkably helps in reducing the abnormally high temperature in an exceptionally rapid manner via an altogether not-so-explicit (unknown) mechanism. Perhaps it is normally caused by a direct effect upon the **serotonin-regulated thermal controlling centre** of the nervous system. Besides, it remarkably minimises certain kinds of perception to pain, without any change in the prevailing central or motor functions, that essentially varies from the effects of morphine.

---

*Smith PK, Ann. N.Y. Acad. Sci., **86**, 38, 1960.

4.12. Aminopyrine

Though its overall antipyretic and analgesic effect is much more powerful and its effect last longer, yet it possesses a major disadvantage because of its ability to produce agranulocytosis* (granulocytopenia). It has been further demonstrated to be caused by drug therapy with a plethora of drug substances *e.g., aminopyrine.*

Note.   Several countries have either banned or adequately restricted its administration.

4.13. Dipyrone

Its pharmacological actions are very much akin to aminopyrine. Because of its high degree of toxicity its usage has been banned or restricted in several countries.

4.14. Phenylbutazone

The 'drug' is absorbed quite rapid after oral administration, and subsequently gets bound to plasma protein very intimately. Its usual time to attain peak serum concentration level is nearly 2.5 hrs. However, the normal span for the overset of antigout activity ranges between 1 to 4 days, and that for antirheumatic activity varies between 3 to 7 days. It has been duly observed that the therapeutic serum concentrations average approximately 43 mg. mL^{-1} ; and the elimination half-life is nearly 84 hours. Interestingly, its major metabolite (oxyphenbutazone, 2%) and the unchanged drug (1%) are both excreted by the kidneys.

Note.   The 'drug' must preferably be taken either with cold milk or with meals to avoid the possible gastric irritation.

4.15. Oxyphenbutazone

It happens to be the '*active metabolite*' of phenylbutazone. It has more or less the same effectiveness, side-effects, indications and contraindications. Undoubtedly, it affords a distinct less frequent incidence of acute gastric irritation.

4.16. Benorylate

The acetaminophen (paracetamol) ester of aspirin, benorylate, is an interestingly novel example of a prodrug where both the individual entities represent active agents. The 'drug' seems to be free from the most undesirable *ulcerogenic characteristic features* ; and, therefore, soonafter the usual absorption, it gets split into its two active components once again by the aid of *serum esterases.* It has been duly reported to serve as an effective analgesic-anti-inflammatory drug.

4.17. Sulfinpyrazone

It belongs to the class of a pyrazone structural analogue having potent uricosuric activity together with some antirhombotic and platelet inhibitory activity. It is invariably employed to minimise **the serum-urate concentration** in the specific instances of chronic and intermittent gouty arthritis. It is observed to get adequately absorbed after the oral administration ; 98 to 99% is bound to plasma protein, plasma half-life is almost nearly 2.2 to 3 hours, and finally 50% of the administered '*drug substance*' is usually gets excreted practically unchanged in the urine.

---

*An acute disease marked by **a deficit or absolute** lack of granulocytic WBC (neutrophils, basophils, and eosinophils).

4.18. Mefenamic Acid

The precise mechanism of action of this '*drug*' is assumed to be related to its ability to block prostaglandin (PG) synthetase almost completely. It has also been observed that it does not bear any relationship whatsoever with respect to partition coefficient, dissociation contant (pKa), and lipid-plasma distribution. Besides, there are several evidences in literature(s) with regard to its anti-UV erythema activities, and antibradykinin activities*. It definitely shows much decreased incidence of *gastrointestinal bleeding*, a prominent drawback of such drugs, when compared to '*aspirin*'. ** Besides, it has been duly approved for the control and management of primary dysmenorrhea, that is believed to be caused by an overwhelming concentrations of endoperoxides as well as prostaglandins (PG).

4.19. Meclofenamate Sodium

This is the 2, 6-dichloro derivative of mefenamic acid, as its sodium salt ; and exerts its most predominant side effects, such as : diarrhea, and gastro intestinal disorders.

4.20. Flufenamic Acid

It is a trifluoromethyl analogue of anthranilic acid, that exerts its three-in-one pharmacological actions *viz.,* antipyretic, analgesic, and anti-inflammatory. It finds its abundant usage in dysmenorrhoea and various types of rheumatic disorders. However, the exact and precise mechanism of antipyretic action of the N-aryl anthranilic acid structural variants has not yet been established. There exists no relationship to lipid plasma distribution, partition coefficient or pKa values of these types of drugs *vis-a-vis* their antipyretic activity.

Probable Questions for B. Pharm. Examinations

1. Classify the **'febrifuges'** and give the structure, chemical name and uses of at least ONE compound from each category.

2. Give the names of **three** drugs belonging to the category of **'aniline and para aminophenol analogues'**. Discuss the synthesis of **one** of them.

3. Discuss **'Salicylic Acid Analogues'** as potent antipyretic analgesics. Give suitable examples of support your answer.

4. What is the structure difference between **Cinchophen** and **Neocinchophen** ? Give the synthesis of any **one** of them.

5. The metabolite of **Phenylbutazone** is a better effecitve drug. Discuss its synthesis and is important uses.

6. Name a Sulphur containing **pyrazolodione drug** used as an antipyretic analgesic and describe its synthesis.

7. 'Structural analogues of N-arylanthranilic acid yielded some potent antipyretic, analgesic and anti-inflammatory compounds'. Justify the statement with **two** important examples along with their synthesis.

---

*Scherrer RA, In : Scherrer RA and Whitehouse MW (eds.) **Antiinflammatory Agents**, Academic Press, New York, *p*-132, 1974.

Lane AZ *et al, J. New Drugs.*, **4, 333, 1964.

8. Discuss the **'mode of action'** of antipyretic analgesics by citing the examples of some typical drugs, which you have studied.

9. What are Salol, Partial Salol and True Salol ? Give the structure, chemical name and uses of **one** typical examples from each type.

10. Give a comprehensive account of antipyretic-analgesics.

RECOMMENDED ERADINGS

1. PAJ Janssen and CAM van der Eycken in : A. Burger (*Ed*) *Drugs Affecting the Central Nervous System Marcell* Dekker, New York (1968).

2. HC Churchill-Davidson Hypothermia *Anaesth* June (1954).

3. D Lednicer and LA Mitscher *The Organic Chemistry of Drug Synthesis* John Wiley and Sons New York (1995).

4. Remington's : The *Science and Practice of Pharmacy* Vol. I and II, (20th edn.), Lippincott Williams & Wilkins, New York, (2000)

5. JEF Reynolds (*Ed*) *Martindale The Extra Pharmacopoeia,* (30th edn) The Pharmaceutical Press London (1992).

6. CO Wilson, O Gisvold and FR Doerge *Textbook of Organic Medicinal and Pharmaceutical Chemistry*, (10th edn) JB Lippincott Company Philadelphia (1998).

7. ME Wolff (*Ed*) *Burger's Medicinal Chemistry and Drug Discovery* (5th edn) John Wiley & Sons, New York (1995).

8. A Gringauz, *Introduction to Medicinal Chemistry*, Wiley-VCH, New York, (1997).

<table>
<tr><td>**10**</td><td></td></tr>
</table>

Narcotic Analgesics
(Opiate Analgesics)

1. INTRODUCTION

Opium was known to man many centuries ago. This is evident from the *Ebers Papyrus* and *Homer's Odyessey* where the use of opium was mentioned. Opium is obtained by making superficial incisions on the immature and unripe capsules of *Papaver somniferum* (or poppy plant). The exudate is air-dried and then powdered to give the official powdered opium. A systematic study of the plant material led to the isolation and identification of the most important alkaloid known as morphine in 1803. Other alkaloids isolated from opium include *codeine, papaverine* and *thebaine*.

The opium class of narcotic drugs are considered not only as the most potent and clinically useful agents causing depression of central nervous system, but also as very strong analgesics. Morphine and morphine-like drugs are referred to as **opioids** or **opiates**. They are also known as **narcotic analgesics** ('*narcotic*' is derived from the Greek word 'narcotic' meaning drowsiness. The term narcotic is now used to refer to dependence producing drugs.

Morphine possesses a host of diverse pharmacological properties and uses, a few of which are, to check diarrhoea, ease dyspnea, suppress cough and above all, to induce sleep in the presence of pain. Though morphine and morphine-like drugs may not alter the sensation of pain but they modify the emotional reaction to pain. The pain may be present but may not be perceived as painful.

The narcotic analgesics tend to produce euphoria which is an important factor in their addictive property which limits their use. Other limitations include : *respiratory depression, decreased gastrointestinal motility leading to constipation, increase biliary tract pressure and pruritus due to histamine release.* Because of these setbacks in the use of morphine there has been a constant effort to develop analgesics with fewer side-effects and minimal addictive actions.

As on date a plethora of CNS-depressants, such as : anti-psychotics, barbiturates and ethanol have been shown to afford effectively a substantial lowering in the '*pain perception*'. It has been already demonstrated beyond any reasonable doubt that two vital phenomenon taking place *in vivo*, namely : (*a*) **norepinephrine** re-uptake (*viz.*, antidepressant drugs) ; and (*b*) preventors of **serotonin*** are extremely beneficial therapeutically when administered either in conjunction (adjuvant) with '*opiates*' or alone in the control and management of certain typical incidences of chronic pain.

---

*A chemical, 5-hydroxy tryptamine (5-HT), present in platelets, gastro intestinal mucosa, mast cells, and carcinoid tumours. It is a potent vasoconstrictor ; and also a neurotransmitter in the CNS, and is important in sleep-walking cycles.

With the advent of '*new mechanisms*' emphatically based on latest trend of research activities geared into the antinociceptive effects of certain centrally acting cannabinoid, α-adrenergic-, and above all the nicotinic-receptor agonists may ultimately give rise to a host of therapeutically potent and efficacious analgesics. Besides, basic fundamental research conducted with inhibitors to tachykinin (neurokinin) receptors evidently shows adequate promising results leading to the discovery of newer breed of analgesic drug substances into the therapeutic armamentarium. It is, however, pertinent to state at this juncture that while the constant efforts are still on with respect to the evolution of '**new-drugs**' the *chronic* as well as *acute pain* is invariably circumvented with the aid of '*opioid analgesics*' most efficaciously.

2. LIMITATIONS OF OPIATE ANALGESICS

There are several disadvantages as well as limitations of '*opiate analgesics*' that are enumerated as under :

(*a*) these are usually *contraindicated* in patients who have essentially a past record of Addison's disease, myxedema, and hepatic cirrhosis.

(*b*) these '*drugs*' exhibit a tendency of minimising ventilation that ultimately give rise to hypercapnia and lead to cerebrovascular dilatation resulting into enhanced intracranial pressure ; therefore, great caution has got to be observed in such situations (conditions) as : cerebral edema, head injuries, and delerium tremens.

(*c*) these are required to be used with utmost caution and restriction in patients having a history of cardiac arrythmias, inpaired kidney function, and chronic ulcerative colitis.

(*d*) these '*drugs*' have a tendency to cross the *placental barrier* ; therefore, newborn infants, whose mothers have been treated with such drugs during labour, must be observed very meticulously for probable symptoms of *respiratory depression*, and should be treated adequately for narcotic overdosage, if so required.

(*e*) an individual who is sensitive either to a specific narcotic agent or a group of agents, must avoid them as far as possible to get into serious complications that may even prove fatal.

(*f*) these '*drugs*' invariably exhibit amalgamated '*analgesic*' and '*depressant*' effects that form the basis for a large number of *drug-drug interactions* with other therapeutic agents.

Examples : (1) A plethora of '*drug substances*', for instance : muscle relaxants, sedatives-hypnotics, tricyclic antidepressants, antipsychotic, antihistaminics, and alcohols are observed to interact with *opiate analgesics* to augment and accelerate their overlapping pharmacological activities, namely : anticholinergic effects and respiratory depression.

(2) Monoamine oxidase inhibitors (MAOIs) must be administered with utmost caution in conjunction with '*narcotic analgesics*' by virtue of their extremely intensified activity, for instance : patients treated with MAOIs when treated with '**meperidine**' give rise to such a severe reaction that may sometimes even prove to be fatal.

In short, one may infer from the aforesaid statement of facts (*a*) through (*f*) that the '*opiate analgesics*' on one hand produce wonderful much needed therapeutic excellence, but on the other extreme care, caution and wisdom need to be applied in their usage in treating specific conditions.

3. CHARACTERISTIC FEATURES OF OPIOIDS

There are several specific characteristic features of **opioids (opiates)** as detailed below which would be treated individually here under :

(*i*) Opioid peptides,

(*ii*) Opioid receptors,

(*iii*) Orphan opioid receptor,

(*iv*) Mu opioid receptors,

(*v*) Kappa opioid receptors,

(*vi*) Delta opioid receptors, and

(*vii*) Opioid receptors : identification and activation.

3.1. Opioid Peptides

Akil *et. al.* (1984) observed that the **endogenous opioid peptides** are invariably synthesized as essential component associated with the structures of specific large precursor proteins. Evidently, each of the major types of opioid peptides does have an altogether different and specific precursor protein.

Examples :	**Opioid Peptide**	**Precursor**
(*i*)	*Proenkephalin A*	— Met- and Leu-Enkephalin
(*ii*)	*Propiomelanocortin* (*PMOC*)	— β-Endorphin
(*iii*)	*Proenkephalin B* (*Prodynorphin*)	— Dynorphin, and α-Neoendorphin

Salient Features. The various salient features of the opioid peptides are stated as under :

(1) Most of the *pro-opioid proteins* are usually synthesized very much within the cell nucleus, and subsequently transported meticulously to the terminals of the nerve cells from where they are being released gradually.

(2) Active peptides are found to be undergoing hydrolysis from the corresponding large proteins by the aid of proteases which particularly, take cognizance of the '*double basic amino acid sequences*' strategically located just prior and immediately after the opioid peptide sequences.

(3) **Endogenous opioid peptides** are found to afford their analgesic activity both at the supraspinal and spinal sites :

(*a*) Cause analgesia alternatively by the help of a peripheral mechanism of action intimately linked with the pervailing inflammatory process, and

(*b*) CNS-happens to be the ideal most preferred site where the opioids are found to exert either a neuromodulator or an inhibitory neurotransmitter action at the following *two* prevalent sites, namely :

(*i*) interconnecting neuronal pathways meant for the exclusive '*pain signals*' within the brain, and

(*ii*) afferent pain signalling neurons located in the dorsal horn of the spinal cord.

3.2. Opioid Receptors

Generally, there exist *three* main categories of the '*opioid receptors*', namely : (*a*) mu ; (*b*) kappa designated by '*k*' ; and (*c*) delta designated by 'δ'.* It is, however, pertinent to state here that all the aforesaid opioid receptors have been adequately characterized and also '*cloned*'.** Based on the most recent universally adopted and recognized '*nomenclature*' classifies the said three opioid receptors in the actual order by which they were eventually cloned.*** According to this classification the various receptors are commonly termed as follows :

OP_1—Receptors : Delta opioid receptors (δ) ;

OP_2—Receptors : Kappa opioid receptors (κ) ;

OP_3—Receptors : mu opioid receptors ;

Interestingly, all the three '*opioid receptor types*' are found to be strategically located either in the human brain or spinal cord tissues ; furthermore, each of them essentially possesses a specific role to play in the control, regulation and management of pain threshold. As on date, however, both **mu** and **kappa** agonists are already in clinical application abundantly across the globe ; whereas, a good number of **delta receptor** selective drug substances are in the regimen of both extensive and intensive '*clinical trial procedures*'.

3.3. Orphan Opioid Receptor

Importantly, an absolutely different 4th receptor, besides mu-delta-kappa opioid receptors, has been duly identified and cloned derived from the homology with the cDNA sequence of the known ones. One of the most predominant feature of the new 4th receptor is that it never got bound to the classical opioid peptide or prevailing antagonists or known non-peptide agonists with high affinity. Therefore, this new receptor has been legitimately termed as the **orphan opioid receptor**. Inspite of the copious volume of research carried out to establish the exact mechanism of this receptor no definite experimental evidence(s) to suggest adequately the importance of this system with respect to the pain transmission and its prevailing association to the classical opioid systems.

3.4. Mu Opioid Receptors

Zadina *et. al.***** (1997) made a pivotal observation that the *two* vital *endogenous opioid peptides*, namely : (*a*) endomorphin-1 ; and (*b*) endomorphin-2. showed an extremely high degree of selectivity for the mu (OP_3) receptors exclusively.

Salient Features.   The salient features of mu opioid receptors are as follows :

1. A plethora of therapeutically potent and useful compounds, such as : morphine, sufetanil, ndomorphin-1, ndomorphin-2, are potent Mu (μ) opioid agonists.

2. A number of other pharmacologically active compounds, for instance : Naloxone, Cyprodime, Naltrexone are Mu (μ) opioid antagonists.

3. Practically all the '*opioid alkaloids*' and most of their synthetic structural analogues are precisely the mu selective agonists.

---

*Lord JAH *et al.* **Endogenous opioid peptides : multiple agonists and receptors,** *Nature,* **267** : 495-499, 1977.

Satoh and Minami, **Molecular pharmacology of the opioid receptors, *Pharmacol. Ther.,* **68**, 343-64, 1995.

***Dhawan BN *et al. Pharmacol. Rev.,* **48**, 567-592, 1996.

****Zadina JE *et. al. Nature,* **386** : 499-502, 1997.

4. Three 'drug substances', namely : morphine, normorphine, and dihydromorphinone are found to have 10–20 times more mu receptor selectivity.

5. Kieffer* (1999) amply demonstrated that almost all the major pharmacologic activities, as studied with mu receptor knockout mice, after having been treated with morphine injection usually take place by interactions with mu receptors. Such observed activities are : decreased gastric motility, emesis, tolerance, analgesia, respiratory depression and withdrawl symptoms.

6. Cyprodime happens to be the most selective non-peptide mu antagonist *i.e.,* showing 100 time more selectivity for mu over delta ; and 30 times more selectivity for mu over kappa.

7. Naltrexone and Naloxone are recognized as opioid antagonists which exhibit only negligible *i.e.,* 5 to 10 fold more selectivity for the mu receptors.

SAR-Mu Antagonists. It has been observed that there are only *two* drug substances which are recognized as '*pure antagonists*' *i.e.,* they behave as antagonists at all opioid receptor sites, such as : *naloxone* (*i.e.,* N-allyl) ; and *naltrexone* (*i.e.,* N-cyclopropylmethyl) structural analogues of **noroxymorphone.** The 14β-OH functional moiety is regarded to be the most important characteristic feature for attributing the pure antagonistic properties of these two aforesaid compounds.

Naloxone   Naltrexone

However, it has not yet been expatiated completely as to why a minor alteration from an N-methyl to an N-allyl moiety can reverse the activity of '*an opioid*' from being a **potent agonist** into a **potent antagonist**. Perhaps a logical explanation may be put forward with regard to the capability of opioid receptor protein to couple with G-proteins** efficaciously in the event when it got bound by an agonist but no such coupling with G-proteins when got bound by an antagonist. Furthermore, one may draw an inference that in the instance of an opioid with an N-substituent of 3-4 carbon number, exerts a distinct conformational change either in the receptor or blocks essential receptor areas that might specifically hinder the possible interaction between the receptor and the G-proteins.

3.5. Kappa Opioid Receptors

The two prominent 6, 7-benzomorphan structural analogues are the racemate of **ethylketazocine** and **bremazocine** which predominantly exhibit kappa opioid receptor selectivity. These two compounds gained prominence as they were used initially to evaluate the kappa receptors ; but later on found to be

---

*Kieffer BL, *Pharmacol, Sci.,* **20**, 19-25, 1999.

**Signal transduction proteins.

possessing not-so-high a selectivity. However, quite recently, a variety *'arylacetamide'* structural ana-logues, which showed a distinctly higher selectivity for kappa in comparison to mu or delta receptors, have seen the light of the day. The *two* new racemic compounds investigated largely are : U50488 and PD117302, whose characteristic features along with the earlier ones are enumerated in Table 10.1.

TABLE 10.1 : CHARACTERISTIC FEATURES OF KAPPA (κ) OPIOID AGONISTS

S. No.	Racemate Compounds	Characteristic Features
1.	(±)-U50488 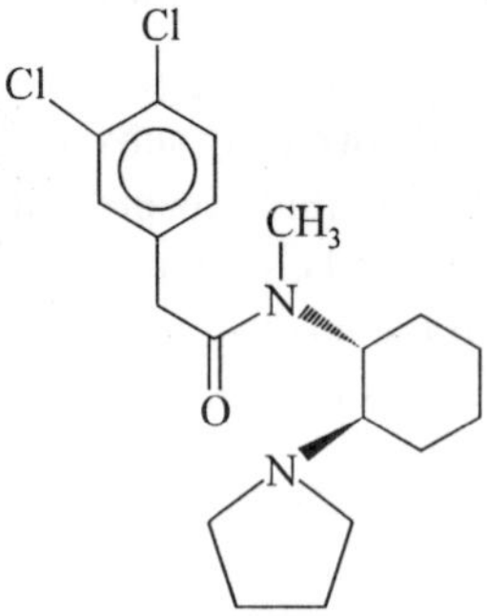	1. It shows 50 times more selectivity for kappa over mu receptors. 2. It enjoys the reputation of being the most-important *'pharmacological tool'* in the characterization of the kappa opioid ac-tivity solely.
2.	(±)-PD117302 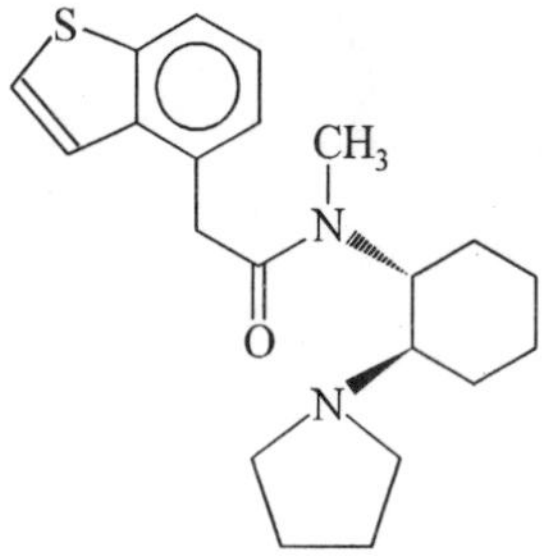	1. It exhibits almost 1000 times selectivity for kappa over mu or delta receptors. 2. Generally, the kappa agonists cause dis-tinct analgesia in humans and animals.
3.	(±)-Ethylketazocine [R_1 = H ; R_2 = H ;] (±)-Bremazocine [R_1 = OH ; R_2 = CH$_3$;] 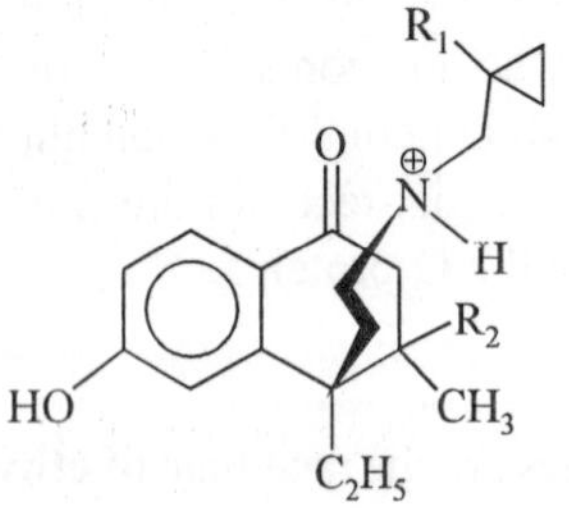	1. Used earlier to help in the investigational studies with regard to kappa receptors. 2. Not found to be possesing enough sensi-tivity and selectivity ; and hence, replaced by U-50488 and PD-117302 (*i.e.,* the *arylacetamides*).

In the search for kappa (κ) opioid antagonists only one drug substance gained cognizance, which is (±) nor-binaltorphimine, and it showed fairly good selectivity for the kappa receptors.*

---

*Choi H *et. al., J Med. Chem.,* **35** : 4638-39, 1992.

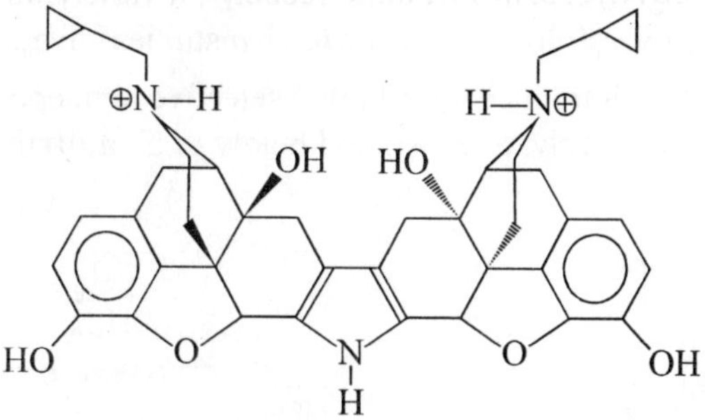

(±) nor-Binaltorphimine

SAR-Kappa Receptor Agonists. Most of the clinically used kappa agonists essentially have their chemical structures very much related to the rather '**rigid opioids**', and those having the additional functional moieties attached to the N-atom, such as : allyl ; cyclopropylmethyl (CPM) ; and cyclobutylmethyl (CBM). Interestingly, all these compounds are observed to be kappa receptor agonists ; besides being mu receptor antagonists. Importantly, the kappa agonist activity may be increased substantially by the following *two* minor structural modifications as stated under :

(a) introduction of the O-atom placed strategically at the 8-position [*e.g.,* ethylketazocine (Table 10.1)], and

(b) introduction of the O-atom right into the N-substituent [*e.g.,* bremazocine (Table 10.1)].

3.6. Delta Opioid Receptors

Adequate modifications and alterations in the amino-acid sequence and composition of the *enkephalins* (pentapeptides produced in the brain) give rise to such compounds that significantly demonstrate both high potency and distinct selectivity for the delta opioid receptors. James *et. al.** (1984) introduced [D-Ala[2], D-Leu[5]] enkephalin, also termed as DADLE ; whereas, Mosberg *et al.*** (1983) introduced the cyclic peptide [D-Pen[2], D-Pen[5]] enkephalin, also known as DPDPE, which enjoyed the reputation of being the peptides invariably and frequently employed as **selective delta receptor ligands.**

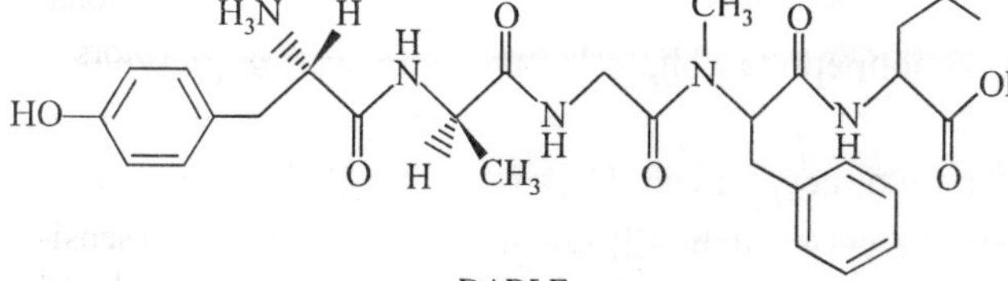

DADLE

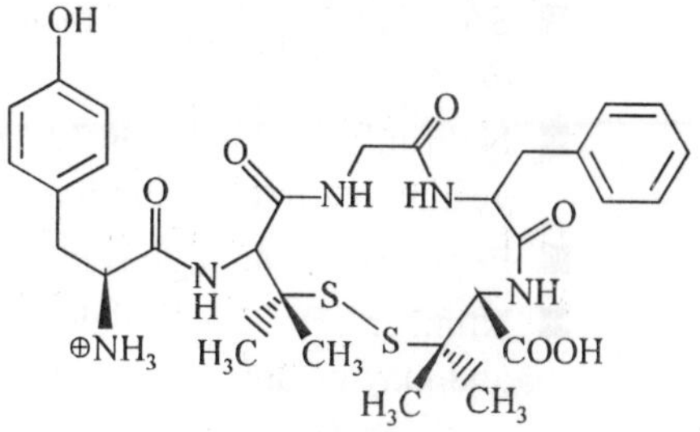

DPDPE

Both DADLE and DPDPE together with some other *delta-receptor-selective* peptides have been employed extensively and intensively for carrying out numerous *in vitro* studies initially. Based on their

---

*James IF *et. al., Mol. Pharmacol.,* **25** : 337-342, 1984.

Moseberg HI *et. al., Proc Natl Acad Sci* USA, **80 : 5871-4, 1983.

two cardinal characteristic negative qualifications, namely : (a) metabolic instability ; and (b) poor distribution properties*, their overall usefulness for *in vivo* studies have been restricted immensely.

Takemori *et. al.*** (1992) introduced *two* highly selective nonpeptide delta (OP$_1$) opioid antagonists for the delta receptors exclusively, as illustrated below : *i.e.*, **naltrindol,** and **naltiben.**

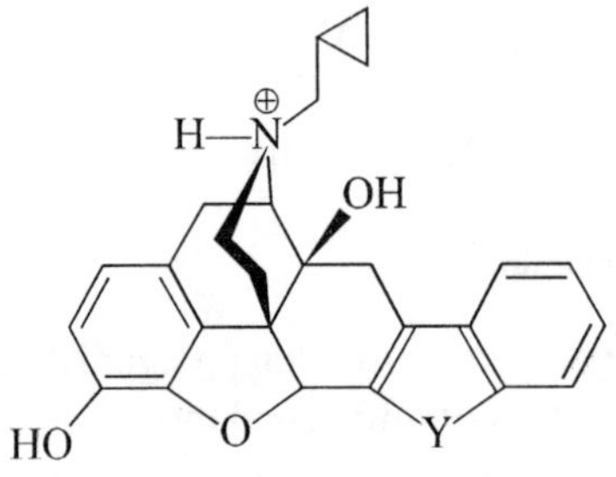

Naltrindol, Y = NH ;
Naltiben, Y = O ;

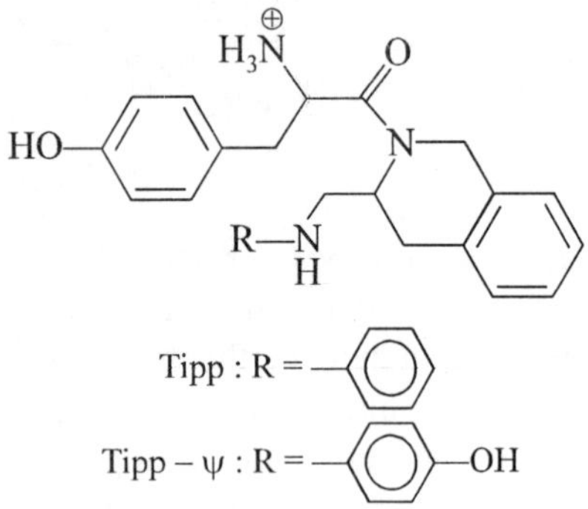

Tipp : R = —⟨O⟩

Tipp – ψ : R = —⟨O⟩—OH

Of the two compounds stated above the former is observed to penetrate the CNS and shows antagonist activity which is particularly selective for the delta (OP$_1$) receptors both *in vitro* and *in vivo* systems.

Schiller *et. al.**** (1993) reported *two* peptidyl antagonists TIPP and TIPP-Ψ, as shown above which are found to be selective for delta receptors. Unfortunately, their clinical usefulness as well as their ability to give fruitful results for *in vivo* studies have been virtually jeopordized and negated by virtue of their absolutely poor pharmacokinetic characteristics.

Interestingly, the opioid receptor antagonists have surprisingly demonstrated appreciable clinical potential both in the treatment of cocaine abuse, and as an immuno suppressive agent.

SAR-Delta Receptor Agonists.   The various cardinal factors with respect to the SAR of delta receptor agonists are enumerated as under :

- ■ SARs for the delta receptor agonists have been least explored/investigated amongst the '*opioid drugs*'.
- ■ Naltrindol and naltiben *i.e.,* the two nonpeptide delta selective agonists, are picking up investigative interest gradually.
- ■ Peptides having high selectivity for delta receptors are already established and documented.
- ■ Number of selective delta agonists are very much still under the required '*clinical trials*', and may be approved as a potential '*drug substance*' of the future.

3.7.   Opioid Receptors : Identification and Activation

The identification of '*multiple opioid receptors*' has been adequately accomplished with the subsequent discovery of certain selective potential agonists as well as antagonists ; and the *two* most sophisticated and reliable '**assay methods**' as given on the next page :

*Penetration into the blood-brain.
Takemori *et. al. Life Sci.,* **149 : 1–5, 1992.
***Schiller PW *et. al. J. Med. Chem.,* **36** : 3182-7, 1993.

(a) **Leslie's Method* (1987).**   For the identification of sensitive assay techniques, and

(b) **Satoh and Minami's Method** (1995).**   For the ultimate cloning of the receptor proteins.

Salient Features.   The various salient features with respect to the identification and activation of the 'opioid receptors' are, namely :

(1) Two techniques are used predominantly, such as : (a) the radioligand binding assays on the brain-tissues ; and (b) the electrically stimulated peripheral muscle preparations.

(2) To differentiate the 'receptor selectivity' of test compounds the following specific assay procedures may be adopted :

(a) computer aided line-filling, and

(b) selective blocking by using either reversible or irreversible binding agents with certain types of receptors.

(3) Signal transduction mechanism for mu, delta, and kappa receptors is via the **Gi/o proteins.** Thus, the activation of the ensuing opioid receptors is directly associated with the G protein to an inhibition of the critical **adenylate cylase activity.** Consequently, the ultimate lowering in cAMP production essentially affords an efflux of k^+ ions, and finally gives rise to hyperpolarization of the nerve cell.***

4. CLASSIFICATION

The narcotic analgesics are usually classified on the basis of their basic chemical structures as discussed below along with a few classical examples from each category.

4.1. Morphine Analogues

Morphine and related drugs possessing potent narcotic analgesic properties, are used in clinical practice. A few examples belonging to this class of compounds are morphine sulphate ; diamorphine hydrochloride ; codeine ; dihydrocodeine phosphate ; hydromorphone hydrochloride ; hydrocodone tartrate ; oxymorphone hydrochloride.

A. *Morphine Sulphate* BAN, *Morphine Sulfate* USAN.

7, 8-Didehydro-4, 5α-epoxy-17-methylmorphinan-3, 6α-diol sulfate (2 : 1) (salt) pentahydrate ; Morphinan-3, 6-diol, 7, 8-didehydro-4, 5-epoxy-17-methyl, (5α, 6α)-, sulfate (2 : 1) (salt), pentahydrate ; B.P., U.S.P., Int. P., Ind. P. ; Duraphine(R) (Elkins-Sinn).

---

*Leslie FM, *Pharmacol. Rev.,* **39,** 197–249, 1987.

Satoh M and Minami M., *Pharmacol. Ther.,* **68, 343-64, 1995.

***Childers SR, *Life Sci.,* **48,** 1991–2003, 1991 ; Georgoussi *et. al. Biochem. J.,* **306,** 71-5, 1995.

Preparation

Morphine can be prepared by total synthesis, but due to the complexity of the molecule renders such an approach not viable commercially. Even today the main bulk of morphine is derived from the natural source and various analogues prepared therefrom.

Method-I. The aqueous extract after concentration is neutralized, a solution of calcium chloride is now added, and the resulting mixture is filtered and further concentrated. The crude morphine hydrochloride separates out and is purified by precipitation with ammonia and recrystallised finally as the sulphate.

Method-II. The concentrated aqueous extract is mixed with ethanol and made sufficiently alkaline with ammonia, when morphine being sparingly soluble in dilute ethanol separates out while the remaining alkaloids are left in solution. The crude morphine thus obtained is usually purified by repeated crystallization as the corresponding sulphate.

It is employed extensively as an analgesic, antitussive, adjunct to anaesthesia and nonspecific antidiarrheal agent. In small doses it helps to alleviate continued dull pain, wheras in large doses to relieve acute pain of traumatic or visceral origin. Morphine is responsible for altering the psychological response to pain and thereby suppresses anxiety and apprehension, and enables the subject to be more tolerant to discomfort and pain. It is specifically used in the *management of postoperative pain and also for alleviating pre-operative apprehension.*

Dose : Usual, adult, oral, 10 to 30 mg 6 times daily.

B. *Diamorphine Hydrochloride* BAN, *Diacetylmorphine Hydrochloride* USAN.

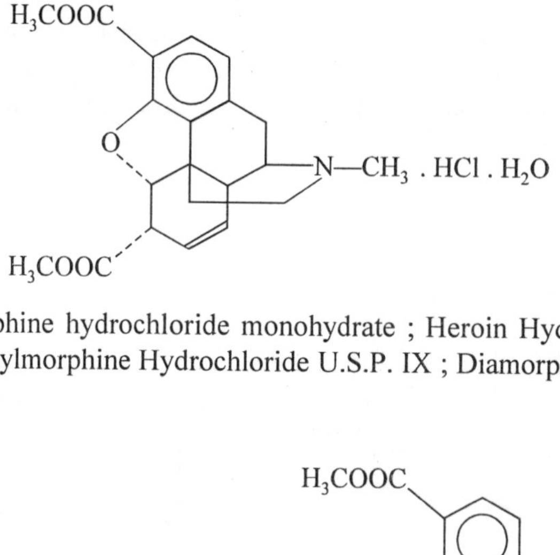

3, 6-*o*-Diacetylmorphine hydrochloride monohydrate ; Heroin Hydrochloride ; Diamorphine Hydrochloride B.P., Diacetylmorphine Hydrochloride U.S.P. IX ; Diamorphine[(R)] (Roche, U.K.)

Synthesis

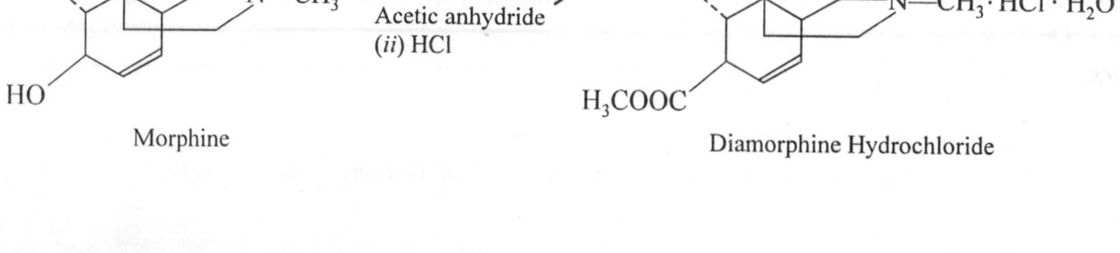

Morphine   Diamorphine Hydrochloride

It may be prepared by the acetylation of morphine and subsequent treatment with hydrochloric acid.

Diamorphine hydrochloride possesses similar actions to that of morphine. It is found to be a more potent analgesic than morphine but it has a shorter duration of action stretching up to 3 hours only. It is generally *employed for the relief of severe pain particularly in terminal illnesses.* Used pre-and post-operatively and being a strong addictive, diamorphine is rigidly controlled and not available in international market.

Dose : Oral, subcutaneous, intramuscular, 5 to 10 mg.

C. *Codeine* BAN, USAN,

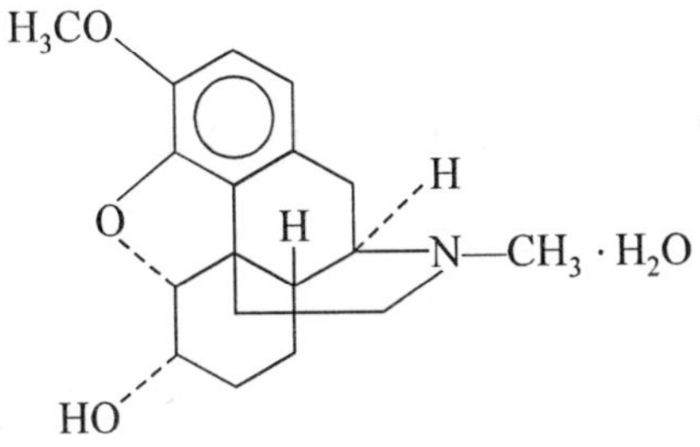

7, 8-Didehydro-5, 5α-epoxy-3-methoxy-17-methyl-morphinan-6α-ol mono-hydrate ; Morphinan-6-ol, 7, 8-didehydro-4, 5-epoxy-3-methoxy-17-methyl-, monohydrate ; B.P., U.S.P., Eur., P., Int. P.

Synthesis

The consumption of codeine is much more than morphine and hence it may be prepared by the partial synthesis of morphine as stated below :

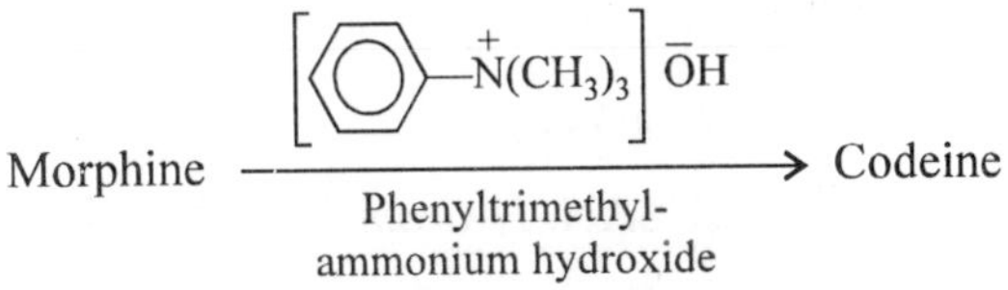

One of the phenolic OH groups in morphine is methylated by phenyltrimethyl ammonium hydroxide. The process involves dissolution of morphine in a solution of KOH in absolute alcohol along with the appropriate quantity of the methylating agent and the resulting solution warmed to 130°C. Afer cooling, water is added and the remaining solution is acidified with sulphuric acid. The generated dimethylaniline is separated and the excess of alcohol is removed by distillation under reduced pressure. The codeine is precipitated by the addition of sodium hydroxide and may be purified by crystallization as the sulphate salt.

It is a narcotic analgesic with utilities similar to those of morphine, but its analgesic activity is relatively much less. It exhibits only mild sedative effects.

Dose : Usual, adult, oral, analgesic, 30 mg 6 times a day ; as antitussive, 5 to 10 mg every 4 hours.

D. *Dihydrocodeine Phosphate* BAN, *Dihydrocodeine* INN, *Drocode* USAN,

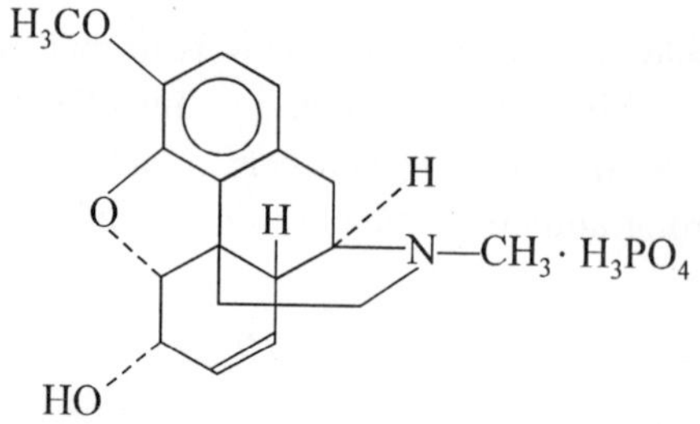

7, 8-Dihydrocodeine phosphate ; Hydrocodeine Phosphate ; Jap. P., Rapocodin[R] (Knoll)

Synthesis

It may be prepared by the catalytic reduction of codeine and treating the resulting product with phosphoric acid.

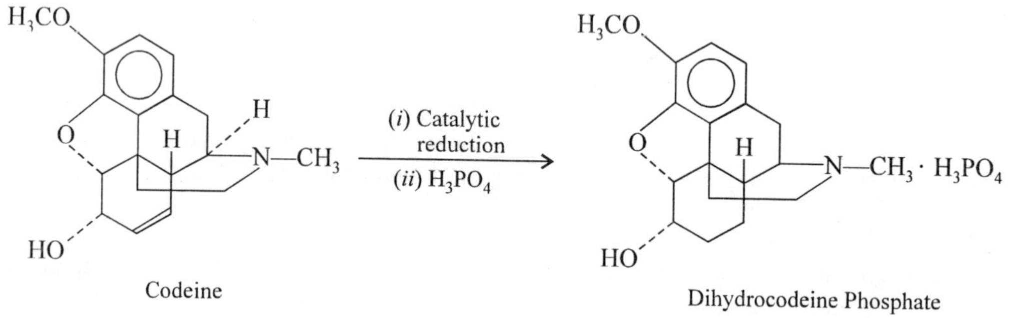

It is used for the relief of mild to moderate pain.

Dose : Usual, oral, 30 mg 4 to 6 times a day.

E. *Hydromorphone Hydrochloride* BAN, USAN, *Hydromorphone* INN.

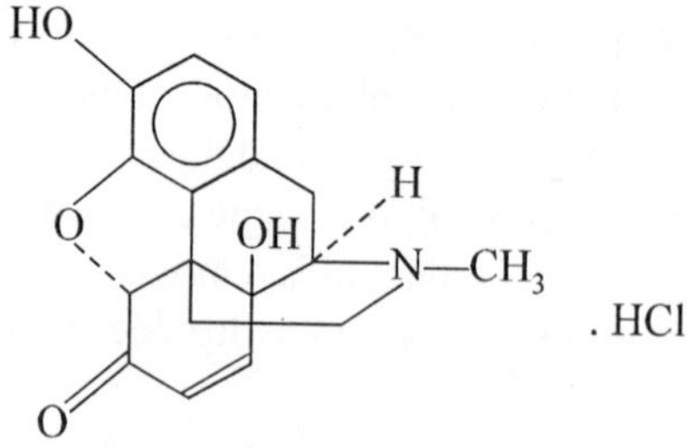

4, 5α-Epoxy-3-hydroxy-17-methylmorphinan-6-one hydrochloride ; Morphinan-6-one, 4, 5-epoxy-3-hydroxy-17-methyl-, hydrochloride, (5α)- ; Dihydromorphinone Hydrochloride ; U.S.P., Int. P., Dilaudid[R] (Knoll) ;

Synthesis

Method-I ; From Morphine

$$\text{Morphine} \xrightarrow[\text{(ii) Reaction with HCl}]{\text{(i) Reduction}} \text{Hydromorphone Hydrochloride}$$

It may be prepared by the reduction of morphine and then treating the resulting product with an equimolar quantity of hyrochloric acid.

Method-II : From Dihydromorphine

$$\text{Dihydromorphine} \xrightarrow[\text{(ii) Reaction with HCl}]{\text{(i) Oxidation}} \text{Hydromorphone Hydrochloride}$$

It is prepared by the oxidation of dihydromorphine and then reacting with an appropriate amount of hydrochloric acid.

It is a semisynthetic opiate analgesic, similar in action to that of morphine, normally used in the *treatment and subsequent relief of mild to severe pain due to cancer, trauma, myocardial infarction, biliary and renal colic, post-operative pain and severe burns.* It is more potent than morphine and the analgesic effect commences within 15 minutes and lasts for about 5 hours.

Dose : Subcutaneous and oral, 1 to 1.5 mg ; Usual, 2 mg every 4 hours.

F. *Hydrocodone Tartrate* BAN, *Hydrocodone Bitartrate* USAN, *Hydrocodone* INN,

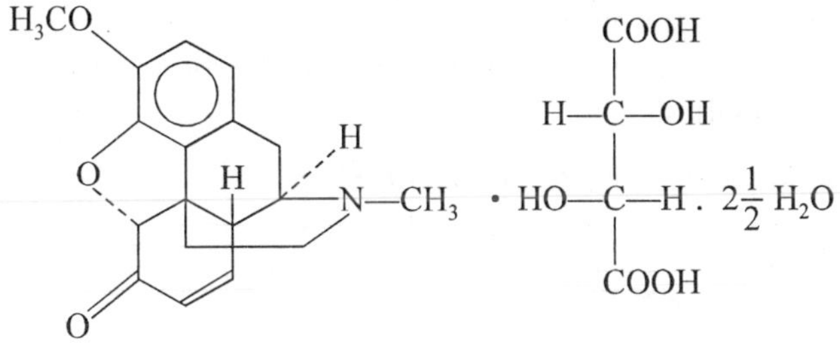

4, 5α-Epoxy-3-methoxy-17-methylmorphinan-6-one, tartrate (1 : 1) hydrate (2 : 5) ; Morphinan-6-one 4, 5-epoxy-3-methoxy-17-methyl-, (5α)-, [R-(R*, R*)]-2, 3-dihydroxybutanedioate (1 : 1), hydrate (2 : 5), U.S.P., Int. P., Dicodid(R) (Knoll) ; Mercodinone(R) (Merrell Dow)

Synthesis

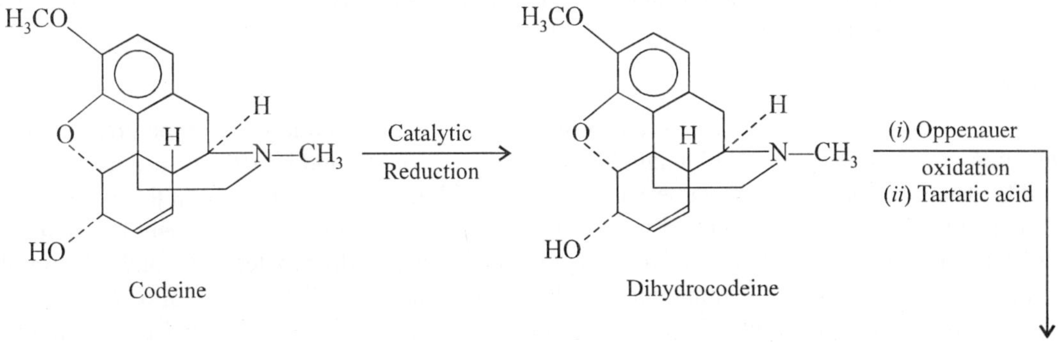

Codeine   Dihydrocodeine

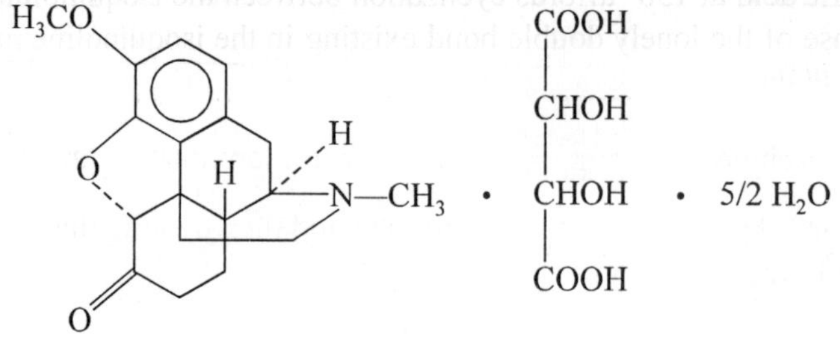

Hydrocodone Tartrate

It may be prepared by the catalytic reduction of codeine to yield dihydrocodeine which on being subjected to Oppenauer oxidation and treatment with equimolar quantity of tartaric acid gives rise to hydrocodone tartrate.

It is mostly used for the *relief of moderate to severe pain and also for the symptomatic treatment of cough.* It is a narcotic analgesic and considered to be more addictive than codeine.

Dose : Usual, adult, oral, 5 to 50 mg per day.

4.2.   Morphinan Analogues

Grewe (1946) introduced a vital alkylation reaction *via* a very specific stereo-selective (*trans*) synthesis followed by acid-catalyzed intramolecular, aromatic substitution, which caused the B/C-*cis* C/D-*trans* ring fusions found to be common in either morphine or its natural congeners. This study has paved the way for an altogether new morphinan analogues known as 'benzomorphans'. A few classical examples of this group of compounds are listed below, *viz.,* levorphanol tartrate ; dextromethorphan hydrobromide ; butorphanol tartrate ;

A. *Levorphanol Tartrate* BAN, USAN, *Levorphanol* INN.

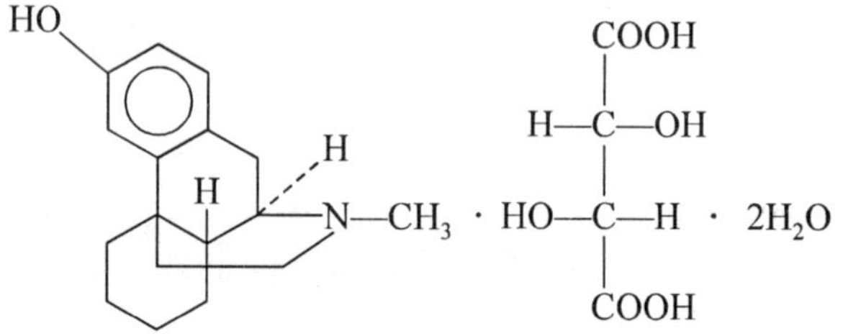

17-Methylmorphinan-3-ol, tartrate (1 : 1) (salt) dihydrate ; Morphinan-3-ol, 17-methyl-, (R-R^*, R^*]-2, 3-dihydroxybutane-diotate (1 : 1) (salt) dihydrate ; B.P., U.S.P., Levo-Dromoran(R) (Roche)

Synthesis

1-(*p*-Methoxybenzyl)-2-methyl-1, 2, 5, 6, 7, 8-hexahydroisoquinoline may be prepared by the interaction of 5, 6, 7, 8-tetrahydro-2-methylisoquinolinium bromide and *p*-methoxy-benzyl magnesium bromide, when the former gets metathesized and the resulting product rearranges at the expense of the 1, 2-double bond. The said compound may be redrawn so as to show the subsequent reactions more vividly. The resulting product is dissolved in hydrochloric acid, hydrogenated at C_3 and C_4 with plati-nized charcoal and treated with ammonia to yield the corresponding *dl*-1, 2, 3, 4, 5, 6, 7, 8-octahydro derivative from which the *1*-enantiomorph is resolved by standard methods. The *1*-enantiomorph on heating with phosphoric acid at 150° affords cyclization between the isoquinoline residue and the benzene ring at the expense of the lonely double bond existing in the isoquinoline nucleus. Conversion of the methoxy group to hydroxy usually takes place during heating with phosphoric acid earlier and the subsequent treatment with aqueous tartaric acid yields the official compound.

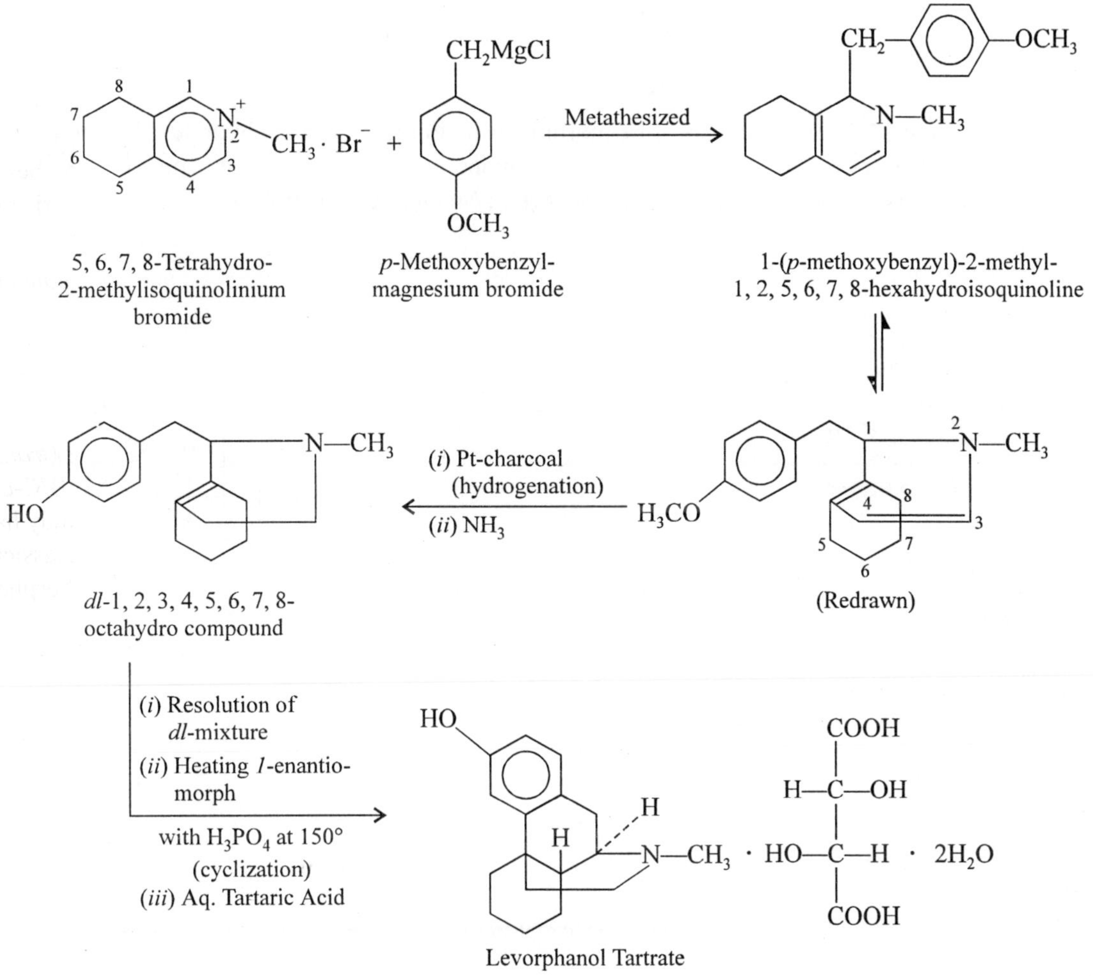

5, 6, 7, 8-Tetrahydro-
2-methylisoquinolinium
bromide

p-Methoxybenzyl-
magnesium bromide

1-(p-methoxybenzyl)-2-methyl-
1, 2, 5, 6, 7, 8-hexahydroisoquinoline

dl-1, 2, 3, 4, 5, 6, 7, 8-
octahydro compound

(i) Pt-charcoal
(hydrogenation)

(ii) NH₃

(Redrawn)

(i) Resolution of
    dl-mixture
(ii) Heating l-enantio-
    morph
with H₃PO₄ at 150°
(cyclization)
(iii) Aq. Tartaric Acid

Levorphanol Tartrate

It is a potent narcotic analgesic having actions and structure similar to that of morphine. It is used effectively for the management of both moderate and severe pain. It produces significant analgesia at a dose level much lower than that of either morphine or meperidine and proves to be longer-acting than these drugs. *It is 2 to 3 times more potent than morphine.*

Dose : Oral, severe pain 1.5 to 4.5 mg 1 or 2 times daily ; Subcutaneous, intramuscular, usual single dose 2 to 4 mg.

B. *Dextromethorphan Hydrobromide* BAN, USAN, *Dextromethorphan* INN,

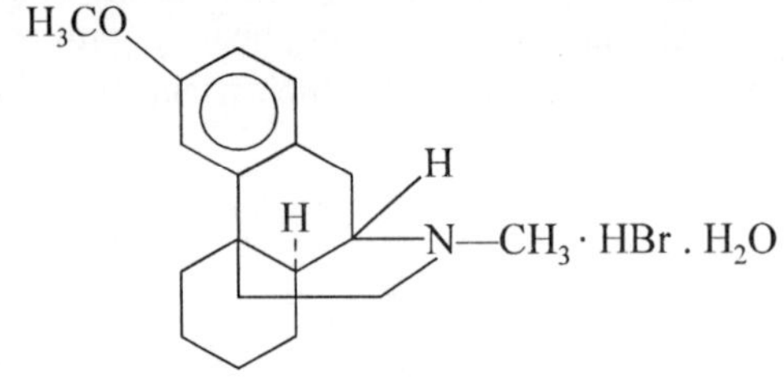

3-Methoxy-17-methyl-9α, 13α, 14α-morphinan hydrobromide monohydrate ; Morphinan, 3-methoxy-17-methyl-, (9α, 13α, 14α)-, hydrobromide, mono-hydrate ; B.P., U.S.P. Romilar[R] (Roche) ; Dormethan[R] (Dorsey) ; Benilyn DM[R] (Parke-Davis) ; Methorate[R] (Upjohn)

Synthesis

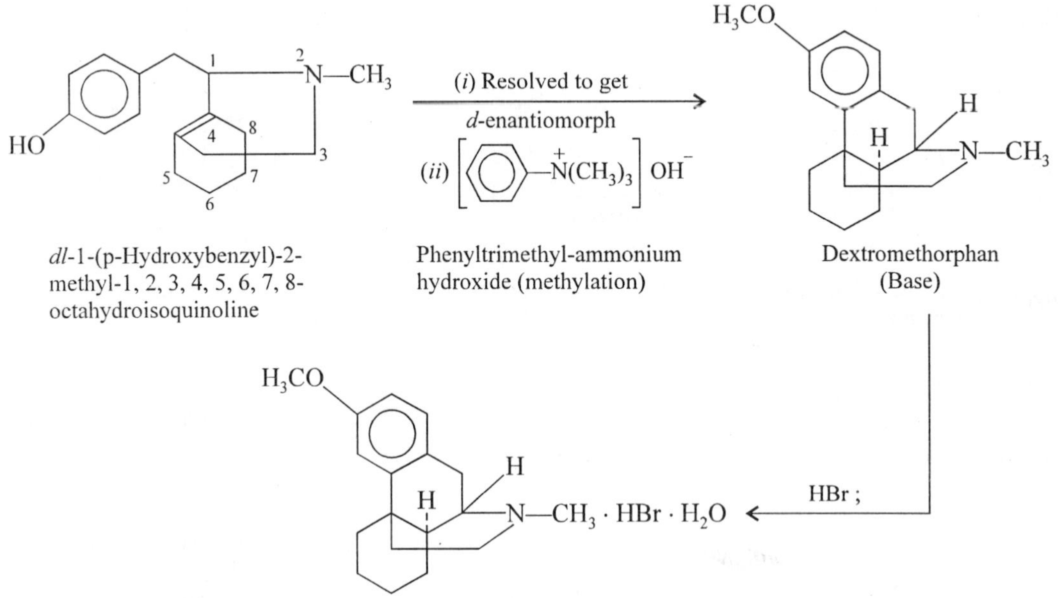

dl-1-(p-Hydroxybenzyl)-2-
methyl-1, 2, 3, 4, 5, 6, 7, 8-
octahydroisoquinoline

Phenyltrimethyl-ammonium
hydroxide (methylation)

Dextromethorphan
(Base)

Dextromethorphan Hydrobromide

The racemic mixture (*dl*) of 1-(*p*-hydroxybenzyl)-2-methyl-1, 2, 3, 4, 5, 6, 7, 8-octahydro-isoquinoline may be obtained exactly in the same manner as described for levorphanol tartrate (in 1 above). This is now resolved to get the *d*-enantiomorph and then treated with phenyltrimethyl ammonium hydroxide to cause methylation and yield the dextromethorphan base. Treatment of the base with appropriate amount of hydrobromide gives the corresponding hydrobromide.

It is a synthetic morphine derivative used as an antitussive agent exclusively. It has been reported to possess a cough suppression potency nearly one-half that of codeine. It exhibits no depression of the central nervous system, lacks analgesic actions and is free from addiction characteristics, which collectively render it possible for its use in cough syrups meant for infants and children.

Dose : Usual, adult, oral, 10 to 30 mg 3 to 6 times a day ; not to exceed 60 to 120 mg in a day ; Children (6 to 12) : 2.5 to 5 mg 6 times a day, not to exceed 40 to 60 mg in a day ; Children (2 to 6) : 1.25 to 2.5 mg 3 to 4 times daily ; not to exceed 30 mg per day.

C. *Butorphanol Tartrate* BAN, USAN, *Butorphanol* INN.

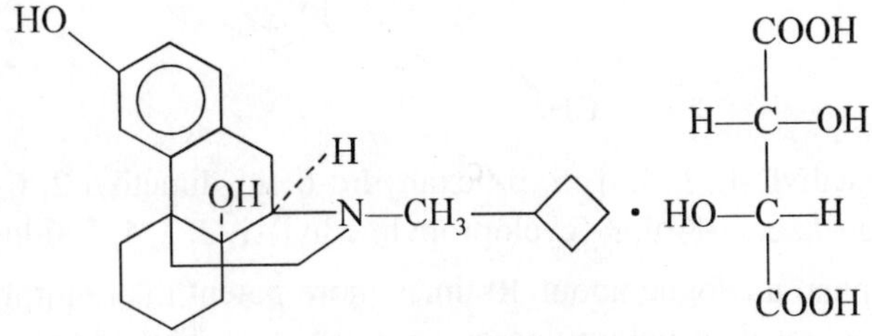

(-)-17-(Cyclobutylmethyl) morphinan-3, 14-diol D-(-)-tartrate (1 : 1) (salt) ; Morphinan-3, 14-diol, 17-(cyclobutylmethyl)-, (-)-, [S-(R^*, R^*)]-2, 3-dihydro-butanedioate (1 : 1) (salt) ; U.S.P., Stadol[R] (Bristol)

It is a synthetic opioid parenteral analgesic with actions and uses similar to those of morphine. It is usually employed *for the relief of moderate to severe post-surgical pain.*

Dose : Usual, adult, intramuscular, 2 mg 6 to 8 times a day ; usual, intravenous, 1 mg every 3 to 4 hours.

4.3. Morphan Analogues

The morphan nucleus is nothing but a bridged perhydroazocine. The numbering pattern of benzomorphan, and the 6, 7-benzomorphan nomenclature has been adopted in the text.

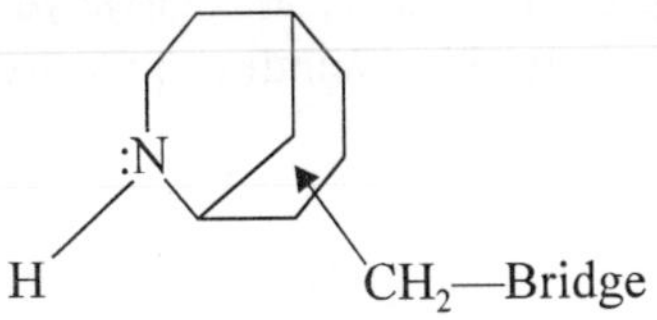

Bridged perhydroazocine 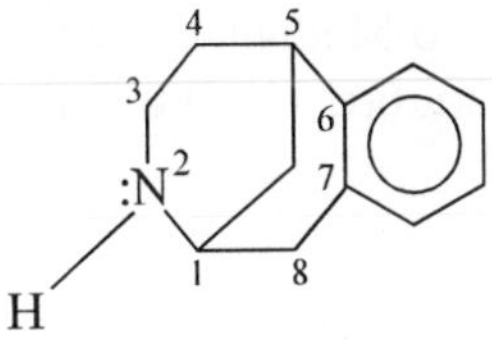   6, 7-Benzomorphan

A few members belonging to the morphan analogues are described here, *e.g.,* metazocine ; cyclazocine ; pentazocine ;

A. *Metazocine* INN, BAN, USAN,

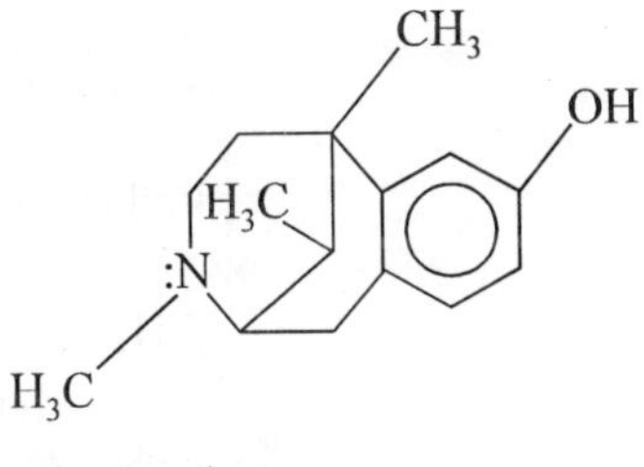

2′-Hydroxy-2, 5, 9-trimethyl-6, 7-benzomorphan

It possesses analgesic activities but owing to its overwhelming psychotomimetic side-effects it is more or less unsuitable for use as an analgesic.

B. *Cyclazocine* INN, BAN, USAN,

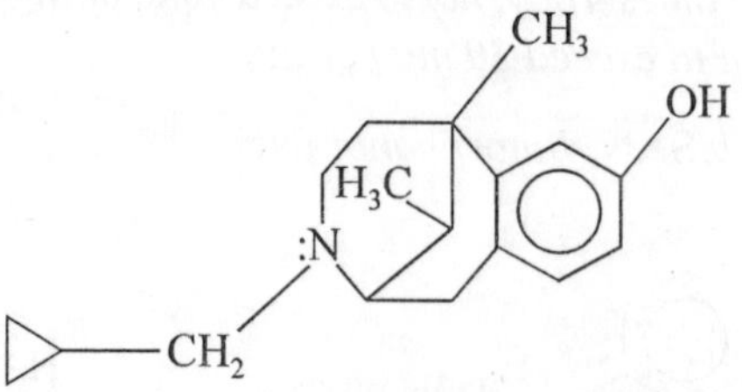

3-(Cyclopropyl-methyl)-1, 2, 3, 4, 5, 6-hexahydro-6, 11-dimethyl-2, 6-methano-3-benzazocin-8-ol ; 2, 6-Methano-3-benzazocin-8-ol, 3-(cyclopropylmethyl)-1, 2, 3, 4, 5, 6-hexahydro-6, 11-dimethyl

It is a benzomorphan analogue about 40 times more potent than morphine as an analgesic and about 100 times more potent than nalorphine as an antagonist. The addiction potential of this drug seems to be much less than that of morphine. It has been used clinically to *treat diamorphine or morphine addicts.*

C. *Pentazocine* INN, BAN, USAN,

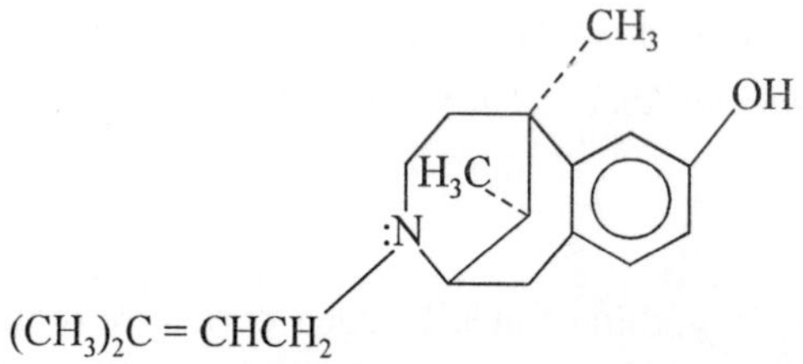

$(CH_3)_2C = CHCH_2$

$(2R^*, 6R^*, 11R^*)$-1, 2, 3, 4, 5, 6-Hexahydro-6, 11-dimethyl-3-(3-methyl-2-butenyl)-2, 6-methano-3-benzazocin-8-ol ; 2, 6-Methano-3-benzazocin-8-ol, 1, 2, 3, 4, 5, 6-hexahydro-6, 11-dimethyl-3-(3-methyl-2-butenyl)-, $(2\alpha, 6\alpha, 11R^*)$- ; B.P., U.S.P., Fortral[(R)] (Winthrop) ; Talwin[(R)] (WInthrop)

Synthesis

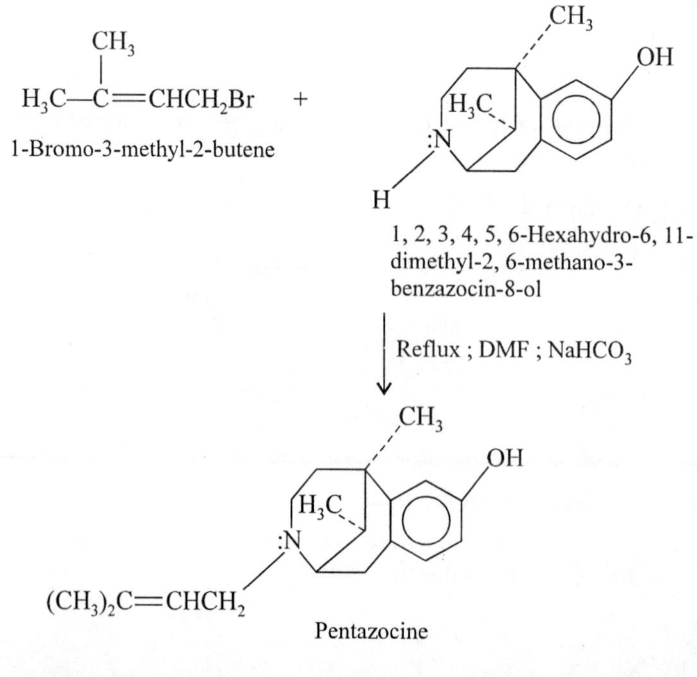

$H_3C-\overset{\underset{|}{CH_3}}{C}=CHCH_2Br$ +

1-Bromo-3-methyl-2-butene

1, 2, 3, 4, 5, 6-Hexahydro-6, 11-
dimethyl-2, 6-methano-3-
benzazocin-8-ol

Reflux ; DMF ; NaHCO$_3$

$(CH_3)_2C = CHCH_2$

Pentazocine

It may be prepared by the condensation of 1, 2, 3, 4, 5, 6-hexahydro-6, 11-dimethyl-2, 6-methano-3-benzazocin-8-ol with 1-bromo-3-methyl-2-butene by refluxing them together in dimethylformamide as a medium and in the presence of sodium bicarbonate. The crude pentazocine is extracted with an appropriate solvent and purified by recrystallization from aqueous methanol.

It is a synthetic analgesic agent commonly used *for the control of moderate to acute pain.* It exerts some sedative actions. It causes incomplete reversal of the respiratory, cardiovascular and behavioural depression produced by either meperidine or morphine.

It behaves both as an agonist and as an antagonist. It is reported to be 3 to 4 times less potent than morphine and about 50 times less potent than nalorphine.

Dose : Parenteral, 20 to 60 mg (as lactate) ; usual, 30 mg 6 to 8 times a day ; daily dose must not exceed 360 mg.

4.4. 4-Phenylpiperidine Analogues

A spectacular accidental discovery of meperidine, in the course of search for structural analogues of atropine with a view to evolve anticholinergic drugs, proved to be a successful attempt towards the synthesis of 4-phenylpiperidine derivatives as narcotic analgesics. This finding has further strengthened the belief that the synthesis of relatively simpler components of the complex molecule of morphine may give rise to a more rational approach towards more efficacious analgesics having lesser nonaddictive liabilities. This ultimately led to the synthesis of a number of the following interesting compounds, namely : pethidine hydrochloride ; diphenoxylate hydrochloride ; fentanyl citrate ; anileridine hydrochloride ; phenoperidine hydrochloride ;

A. *Pethidine Hydrochloride* BAN, *Meperidine Hydrochloride* USAN, *Pethidine* INN,

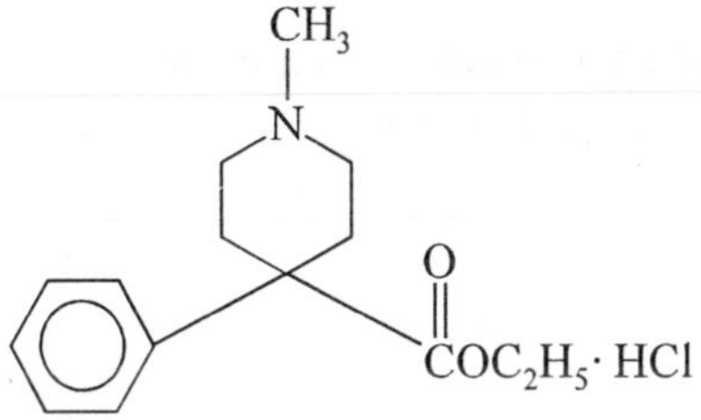

4-Piperidinecarboxylic acid, 1-methyl-4-phenyl-, ethyl ester, hydrochloride ; Ethyl-1-methyl-4-phenyl-isonipecotate hydrochloride ; Pethidine Hydrochloride B.P., Eur. P., Int. P., Ind. P., Meperidine Hydrochloride U.S.P., Denerol[(R)] (Breon) ; Mepadin[(R)] (Merrell Dow)

Synthesis

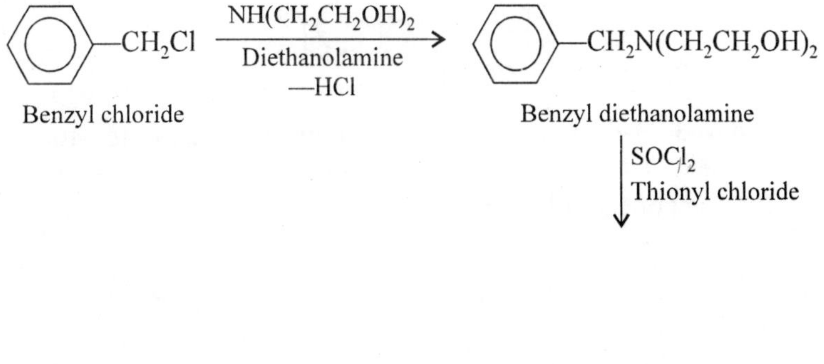

Benzyl chloride   Benzyl diethanolamine

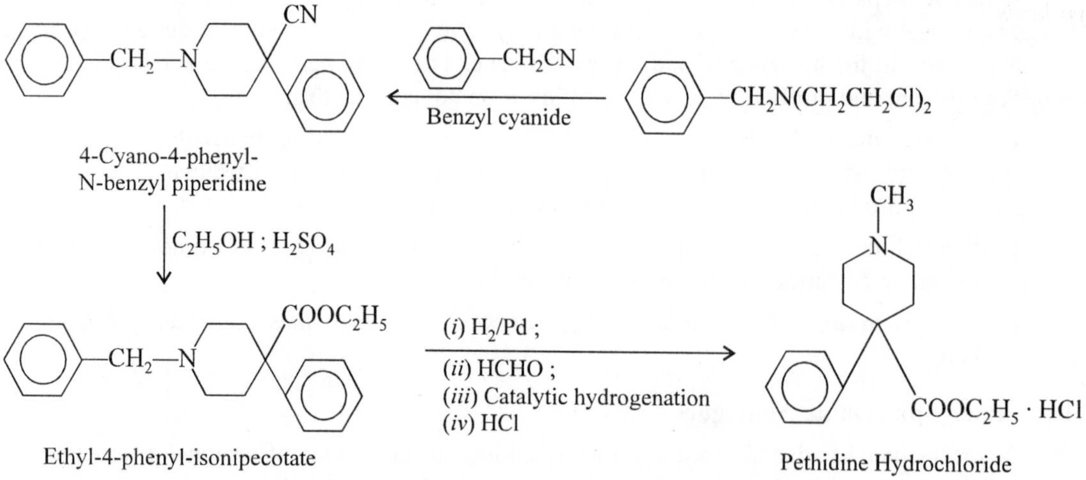

4-Cyano-4-phenyl-
N-benzyl piperidine

Benzyl cyanide

C_2H_5OH ; H_2SO_4

Ethyl-4-phenyl-isonipecotate

(i) H_2/Pd ;
(ii) HCHO ;
(iii) Catalytic hydrogenation
(iv) HCl

Pethidine Hydrochloride

Benzyl diethanolamine is prepared by the interaction of benzyl chloride and diethanolamine with the elimination of a mole of HCl. Chlorination with thionyl chloride gives the corresponding chloride analogue which on treatment with benzyl cyanide yields 4-cyano-4-phenyl-N-benzyl piperidine. Esterification with ethanol in the persence of a small amount of concentrated sulphuric acid yields the ethyl ester. The N-benzyl gruop is removed by means of catalytic hydrogenation in acetic acid solution employing a palladium catalyst. Addition of formaldehyde to the reduction mixture followed by further catalytic hydrogenation yields pethidine which is finally converted to the hydrochloride by neutralization with hydrochloric acid.

It is a synthetic narcotic analgesic which possesses the action and uses of morphine and may be used *for the relief of a variety of moderate to severe pain including the pain of labour and post-operative pain. Pethidine has atropine-like action on smooth muscle.* It is normally used to induce both sedation and analgesia simultaneously.

Dose : Parenteral, usual, adult, oral, 50 to 150 mg 6 to 8 times a day as necessary.

B. *Diphenoxylate Hydrochloride* BAN, USAN, *Diphenoxylate* INN,

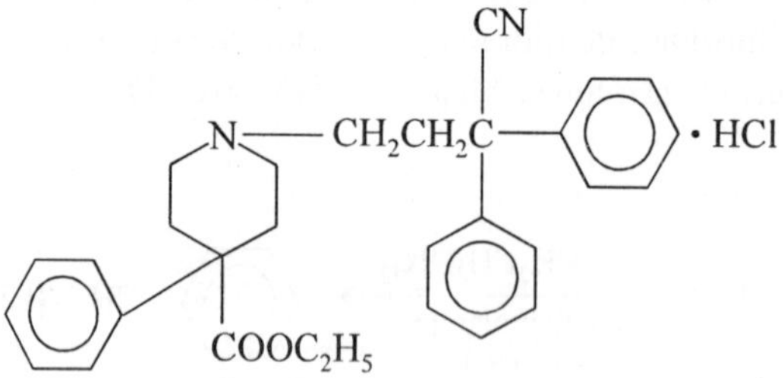

Ethyl, 1-(3-cyano-3, 3-diphenylpropyl)-4-phenylisonipecotate monohydro-chloride ; 4-Piperidinecarboxylic acid, 1-(3-cyano-3, 3-diphenylpropyl)-4-phenyl-, ethyl ester, monohydro-chloride ; B.P., U.S.P., Component of Lomotil[R] (Searle)

Synthesis

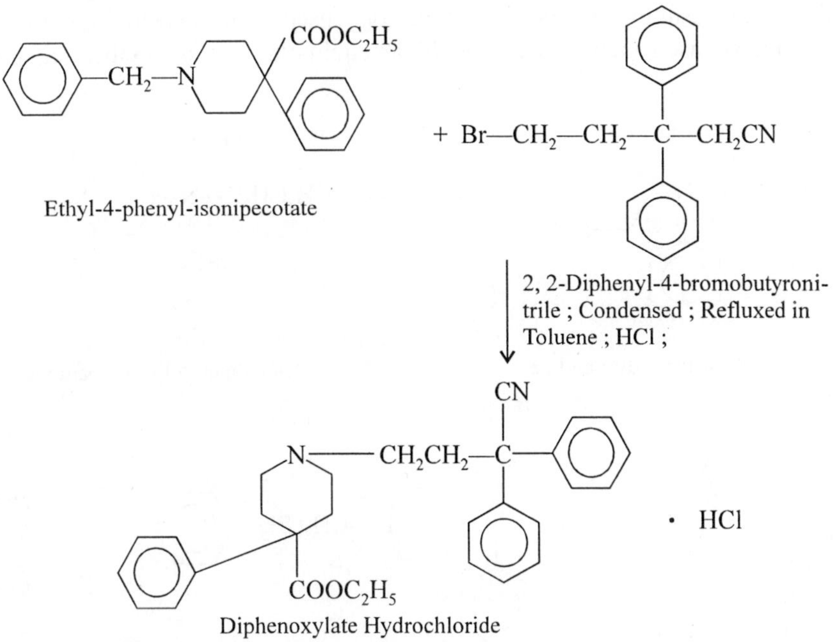

Ethyl-4-phenyl-isonipecotate

$+ Br—CH_2—CH_2—\underset{|}{C}—CH_2CN$

2, 2-Diphenyl-4-bromobutyroni-trile ; Condensed ; Refluxed in Toluene ; HCl ;

Diphenoxylate Hydrochloride

· HCl

Ethyl-4-phenylisonipecotate is perpared as described above in pethidine hydrochloride, which is then condensed with 2, 2-diphenyl-4-bromobutyronitrile by refluxing together in toluene using an excess of the ester.

It is a synthetic analogue of pethidine with some analgesic activity but is mostly used in the *teratment of diarrhoea associated with gastroenteritis, irritable bowel, acute infections, hypermotility, ulcerative colitis and sometimes even food poisoning. It prevents hypergastrointestinal propulsion by reducing intestinal motility.*

Dose : Usual, adult, oral 5, mg 4 times daily.

C. *Fentanyl Citrate* BAN, USAN, *Fentanyl* INN,

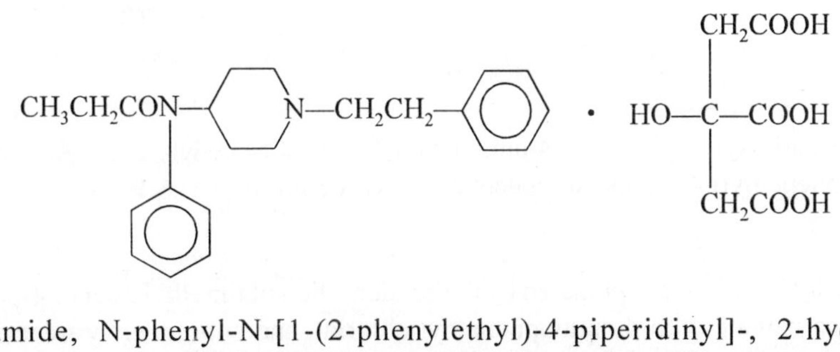

Propanamide, N-phenyl-N-[1-(2-phenylethyl)-4-piperidinyl]-, 2-hydroxy-1, 2, 3-propanetricarboxylate (1 : 1) ; N-(1-Phenethyl-4-piperidyl) propionanilide citrate (1 : 1) ; Phentanyl citrate ; B.P., U.S.P., Sublimaze[R] (Janssen).

Synthesis

N-(4-Piperidyl) propioanilide is prepared by the condensation of propionyl chloride with N-(4-piperidyl)-aniline. The resulting product is further condensed with phenethyl chloride to obtain the corresponding fentanyl base which on reaction with an equimolar portion of citric acid gives rise to the (1 : 1) citrate.

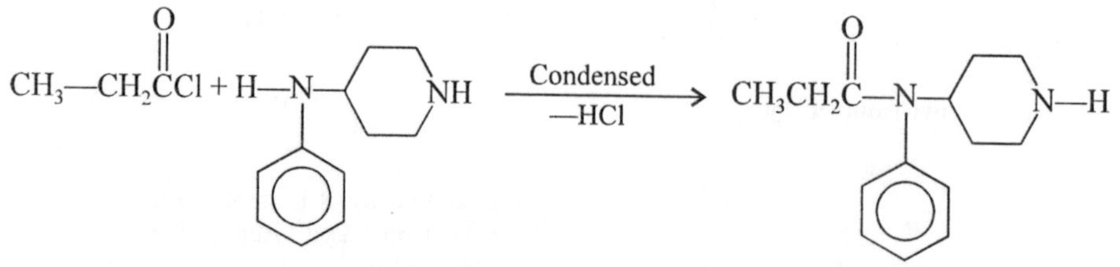

Propionyl   N-(4-Piperidyl)-aniline
chloride

N-(4-Piperidyl)-propioanilide

(i) ClCH$_2$CH$_2$—

Phenethyl chloride
(Condensed)
(ii) Citric acid

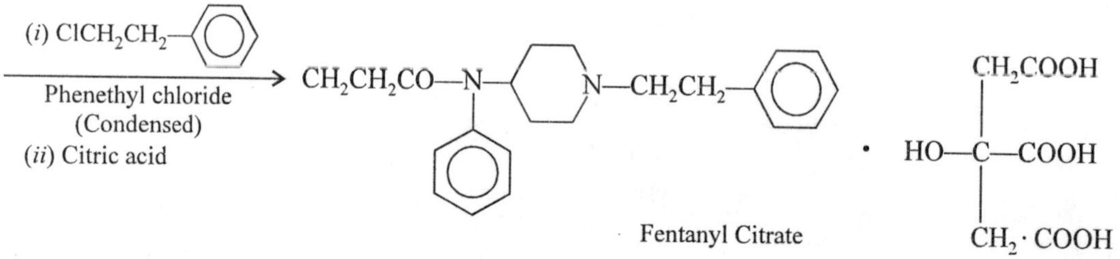

Fentanyl Citrate

It is a potent narcotic analgesic *primarily employed as an analgesic for the arrest of pain after all types of surgical procedures. It possesses an inherent rapid onset and short duration of action. It may be employed also as an adjuvant to all such drugs mostly used for regional and general anaesthesia.*

Dose : Intramuscular, usual, in pre-operative medication 0.05 to to 0.1 mg 30 to 60 minutes before surgical treatment ; for rapid analgesic action, 0.05 to 0.1 mg intravenously.

D. *Anileridine Hydrochloride* BAN, USAN, *Anileridine* INN,

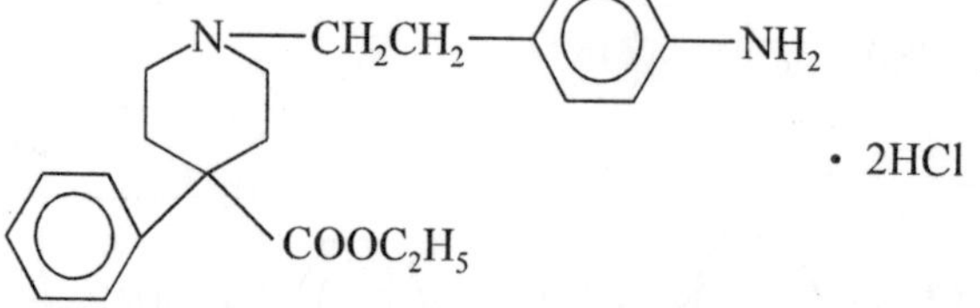

• 2HCl

4-Piperidinecarboxylic acid, 1-[2-(4-aminophenyl) ethyl]-4-phenyl-, ethyl ester, dihydrochloride ; Ethyl-1-(*p*-aminophenethyl)-4-phenylisonipecotate dihydrochloride ; U.S.P.,

Synthesis

Anileridine hydrochloride is prepared by the condensation of the ethyl ester of 4-phenylhexahydro-isonicotinic acid carbamate with 4-aminophenethyl chloride and subsequently treating the base with hydrochloric acid.

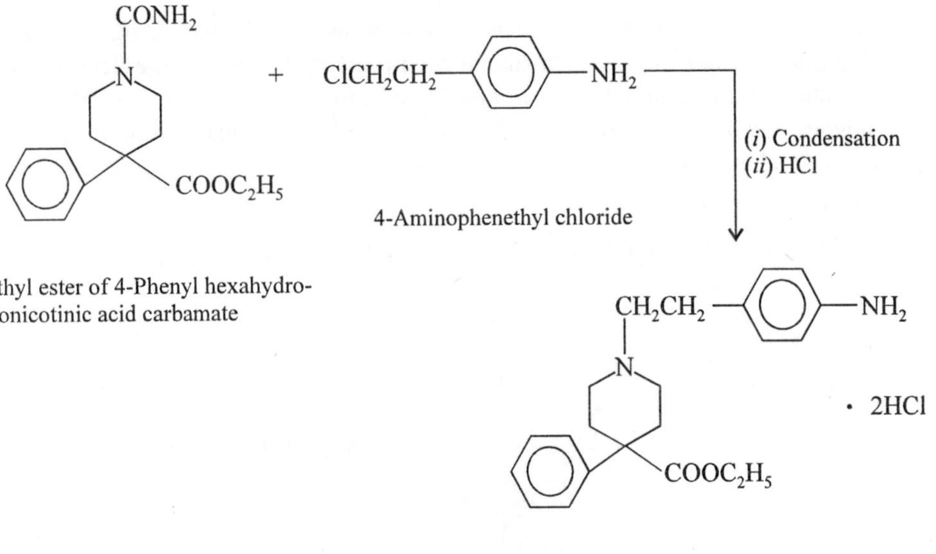

4-Aminophenethyl chloride

Ethyl ester of 4-Phenyl hexahydro-
isonicotinic acid carbamate

(i) Condensation
(ii) HCl

Anileridine Hydrochloride

It is a narcotic analgesic having related chemical structure to that of pethidine and possesses an action similar to it, but with longer duration.

Dose : Usual, oral, 25 mg every 6 hours.

E. *Phenoperidine Hydrochloride* BAN, *Phenoperidine* INN, USAN,

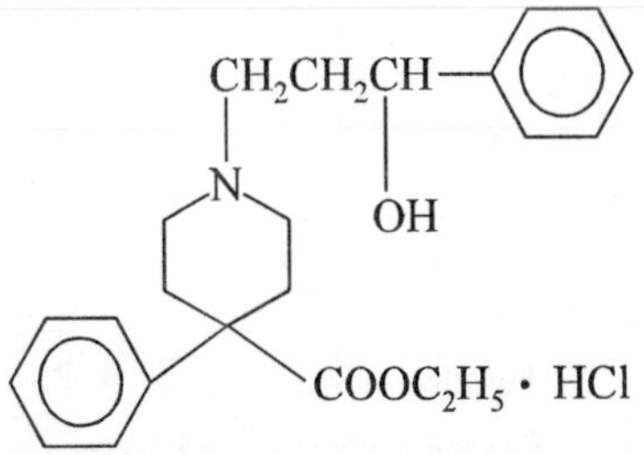

1-(3-Hydroxy-3-phenylpropyl)-4-phenylpiperidine-4-carboxylic acid ethyl ester hydrochloride ; Ethyl 1-(3-hydroxy-3-phenyl-propyl)-4-phenylpiperidine-4-carboxylate hydrochloride ; Operidine[(R)] (Janssen, U.K.)

It is a potent analgesic with actions similar to morphine. It produces neurolepanalgesia, when administered with a major tranquillizer or neuroleptic agent like droperidol, that enables a patient to become calm and indifferent to his environment thereby offering the required co-operation with the surgeon.

Dose : Average, initial, IV, for anaesthesia, 1 mg ; supplemented by 500 mcg every 40 to 60 minutes.

4.5. Phenylpropylamine Analogues

Methadone, a repersentative of this class of compounds may have emerged purely from the molecular design and development of diphenylaminoethyl-propionates ior from the cleavage of piperidine ring present in pethidine molecule. These are considered to be the extremely flexible amongst most analgesic analogues *conformationally*. The following are a few classical examples of this group of analgesics, *viz.*, methadone hydrochloride ; dextro-moramid tartrate ; dextropropoxyphene hydrochloride ; methotrimeprazine

A. *Methadone Hydrochloride* BAN, USAN, *Methadone* INN,

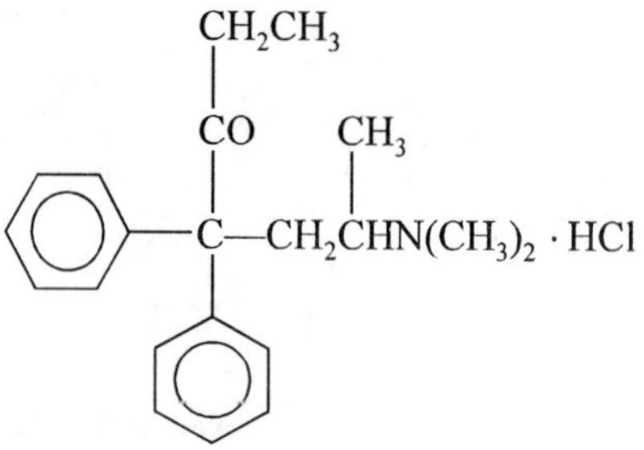

6-(Dimethylamino)-4, 4-diphenyl-3-heptanone hydrochloride ; 3-Heptanone, 6-(dimethylamino)-4, 4-diphenyl-, hydrochloride ; Amidone Hydrochloride ; Phenadone ; B.P., U.S.P., Eur. P., Int. P., Ind. P., Dolophine Hydrochloride[(R)] (Lilly) ; Adanon Hydrochloride[(R)] (Winthrop)

Synthesis

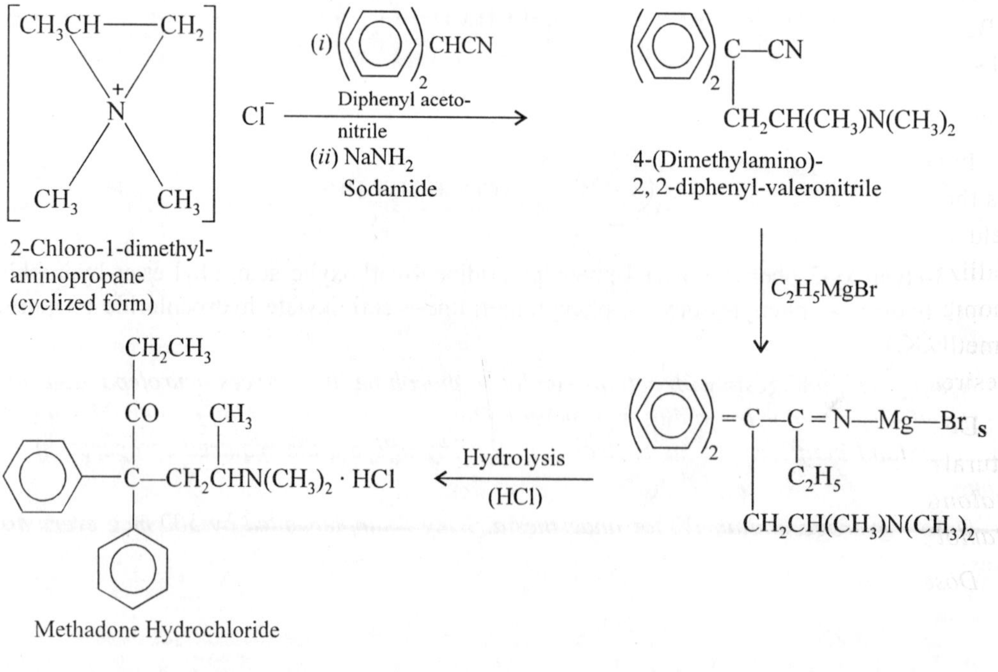

Methadone Hydrochloride

4-(Dimethylamino)-2, 2-diphenylvaleronitrile may be prepared by the condensation of the cyclized form of 2-chloro-1-dimethylaminopropane with diphenyl acetonitrile in the presence of sodamide; together with an undesired equimolar proportion of an isomeric nitrile. The undesired isomer is separated and rejected, while the right isomer is subjected to Grignard reaction with ethyl magnesium bromide to yield an addition compound which on acidic hydrolysis forms the official compound.

It is a potent narcotic analgesics having actions quantitatively comparable to morphine though slightly less potent than morphine as an analgesic. Besides, it exerts sedation and antitussive properties. It also helps in the temporary maintenance and treatment of dependence on narcotic drugs, because its withdrawal syndrome has slow onset and much less intense than mrophine.

Dose : Analgesic, oral, adult, im., 2.5 to 10 mg 6 to 8 times daily.

B. *Dextropropoxyphene Hydrochloride* BAN, *Propoxyphene Hydrochloride* USAN,
   Dextropropoxyphene INN,

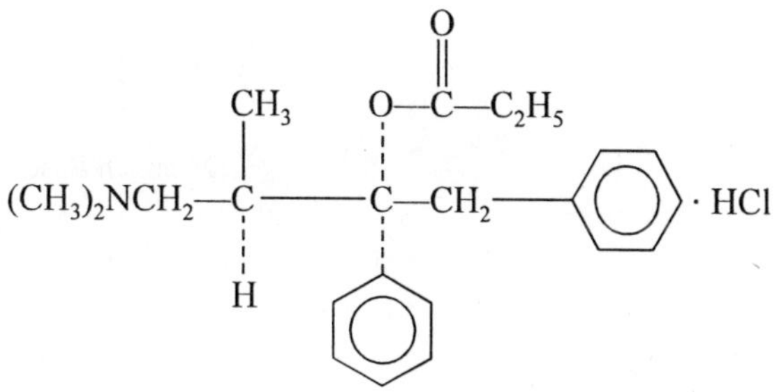

(2S, 3R)-(+)-4-(Dimethylamino)-3-methyl-1, 2-diphenyl-2-butanol propionate (ester) hydrochloride ; Benzeneethanol, α-[2-(dimethylamino)-1-methyl-ethyl]- α-phenyl-, propanoate (ester), hydrochloride, [S- (R*, S*)]- ; B.P., U.S.P., Darvon[R] (Lilly) ; SK 65[R] (SK & F) ; Dolene[R] (Lederle)

Synthesis :

Interaction between propiophenone and dimethylamine in the presence of formaldehyde yields the Mannich base which is subjected to Grignardization with benzyl magnesium chloride to yield a racemic mixture of the two diastereoisomers designated as α- and β-alcohol. Fractional crystallization helps in the separation of α-*dl* form which is subsequently resolved by *d*-camphor-sulphonic acid to obtain (+)-α-form. This is now propionylated with propionic acid in the prsence of trimethylamine tio give dextropropoxyphene which takes up a mole of hydrochloric acid to form the desired official compound.

Dextropropoxyphene is a narcotic analgesic possessing relatively milder actions and bearing structural resemblance to methadone. *It is usually used to control mild to moderate pain and chiefly used along with other analgesics having anti-inflammatory and antipyertic properties like paracetamol and aspirin.*

Dose : Usual, 65 mg, 3 or 4 times per day.

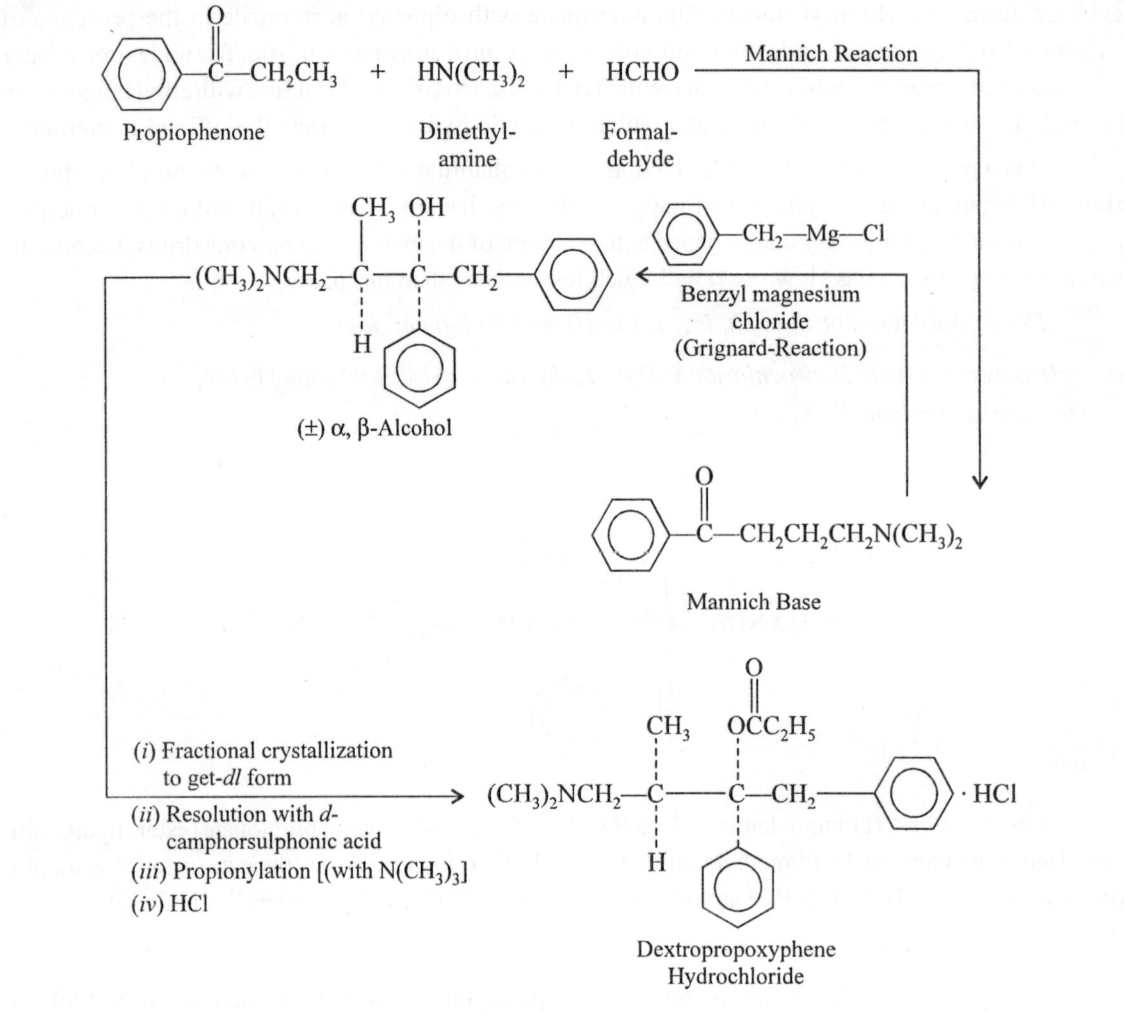

$$\text{Propiophenone} + \text{Dimethylamine} + \text{Formaldehyde} \xrightarrow{\text{Mannich Reaction}}$$

(±) α, β-Alcohol

Benzyl magnesium chloride (Grignard-Reaction)

Mannich Base

(i) Fractional crystallization to get-*dl* form
(ii) Resolution with *d*-camphorsulphonic acid
(iii) Propionylation [(with N(CH$_3$)$_3$]
(iv) HCl

Dextropropoxyphene Hydrochloride

C. *Methotrimeprazine* BAN, USAN, *Levomepromazine* INN,

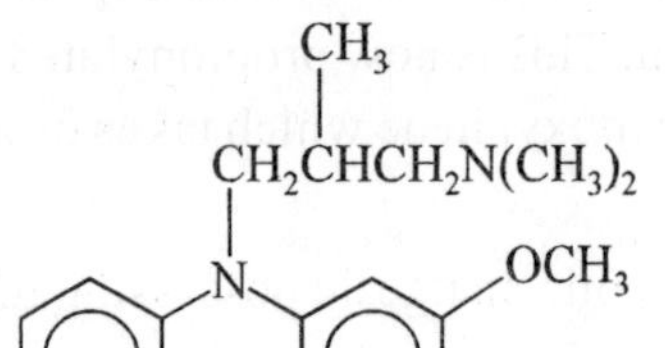

(-)-10-[3-(Dimethylamino)-2-methylpropyl]-2-methoxy-phenothiazine ; 10H-Phenothiazine-10-propanamine, 2-methoxy-N, N, β-trimethyl-, (-)- ; B.P., U.S.P., Levoprome$^{(R)}$ (Lederle)

Methotrimeprazine possesses the histamine-antagonist characteristics of the antihistamines besides CNS effects comparable to those of chloropromazine. It exhibits significant analgesic properties and is *used in the management of severe chronic pain either alone or in conjunction with other analgesics.*

Dose : Usual, adult, oral 25 to 50 mg per day for the treatment of mild psychoses and the severe psychoses 100 to 200 mg with a maximum up to 1 g daily.

4.6. Miscellaneous Analogues

No discourse is usually given a touch of completeness unless and until the miscellaneous structures, which bear essentially the same pharmacological actions are grouped together. There are a few compounds that are analgesic but structurally do not belong to any of the earlier classified groups of compounds (A-E) :

A. *Tilidate Hydrochloride* BAN, *Tilidine Hydrochloride* USAN, *Tilidine* INN,

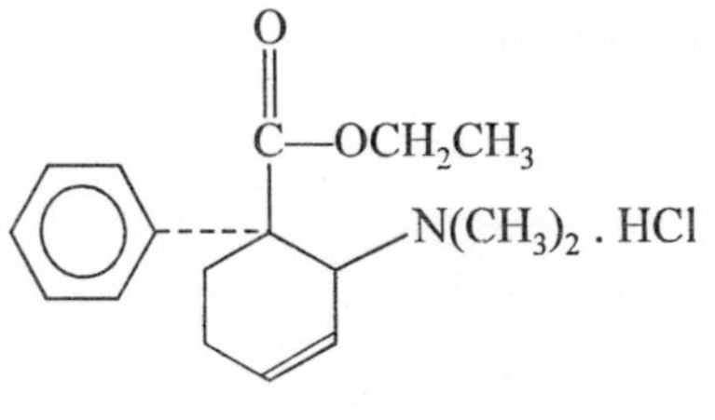

(±)-Ethyl *trans*-2-(dimethylamino)-1-phenyl-3-cyclohexene-1-carboxylate hydrochloride ; 3-Cyclohexene-1-carboxylic acid, 2-(dimethylamino)-1-phenyl-, ethyl ester, hydrochloride (*trans*)-(±)- ; Valoron[R] (Warner) ; Tilidine[R] (Parke-Davis)

It is a narcotic analgesic *mostly employed in the teratment of moderate to severe pain.*

Dose : 50 to 100 mg 4 times a day.

B. *Tramadol Hydrochloride* BAN, USAN, *Tramadol* INN,

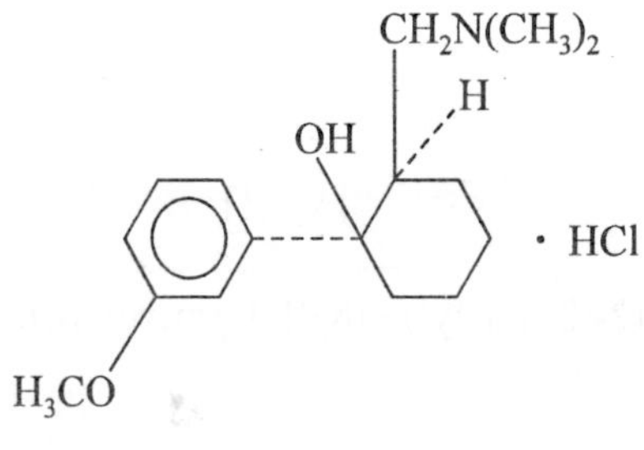

(±)-*trans*-2-[(Dimethylamino) methyl]-1-(*m*-methoxyphenyl) cyclohexanol hydrochloride ; Cyclohexanol, 2-[(dimethylamino) methyl]-1-(3-methoxy-phenyl)-, hydrochloride, *trans*-(±)- ; Melanate[R] (Upjohn) ; Tramal[R] (Grunenthal, W. Ger.)

Tramadol is a potent narcotic analgesic.

Dose : 1 m or iv injection 50 to 100 mg ; as suppository 100 mg.

C. *Dezocine* INN, BAN, USAN,

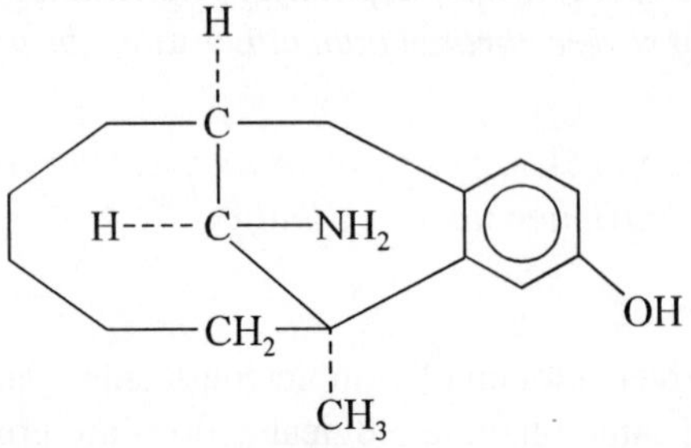

(-)-13 β-Amino-5, 6, 7, 8, 9, 10, 11α, 12-octahydro-5α-methyl-5, 11-methano-benzocyclodecen-3-ol ; 5-11-Methanobenzocyclodecen-3-ol, 13-amino-5, 6, 7, 8, 9, 10, 11, 12-octahydro-5-methyl-, (5α, 11α, 13S*)-, (-)- ; Dalgan[R] (Wyeth).

Dezocine possesses analgesic as well as narcotic antagonist properties and is *usually administered by injection for the relief of severe pain.*

Dose : 10-15 mg.

D. *Sufentanil* INN, BAN, USAN

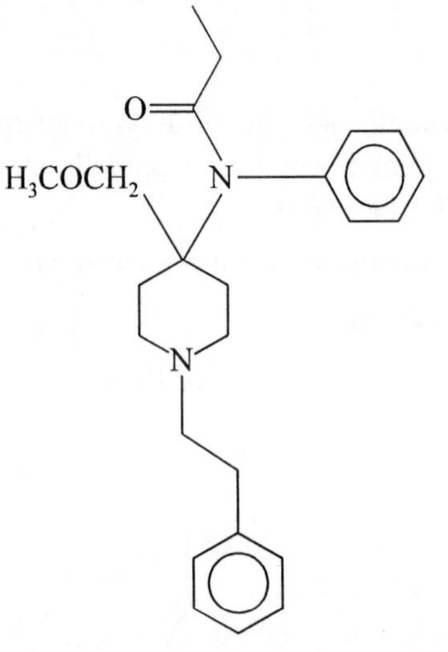

N-[4-(Methoxymethyl)-1-[2-(2-thienyl) ethyl]-4-piperidyl] propionanilide ;

It is a narcotic analgesic.

E. *Nexeridine Hydrochloride* USAN, *Nexeridine* INN,

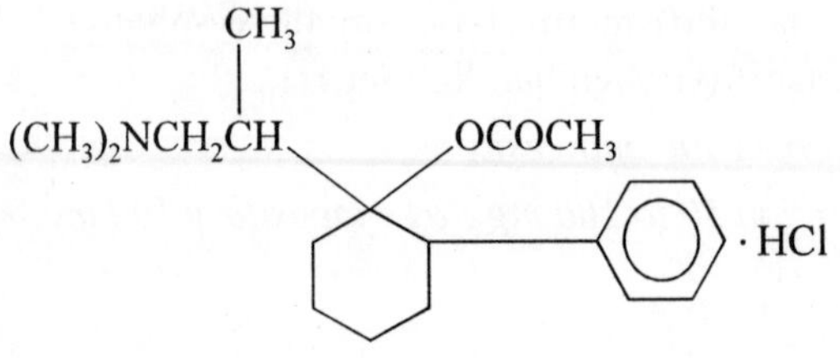

1-[2-(Dimethylamino)-1-methylethyl]-2-phenylcyclohexanol acetate (ester) hydrochloride ; Cyclohexanol, 1-[2-(dimethyl-amino)-1-methylethyl]-2 phenyl-, acetate (ester), hydrochloride.

Nexeridine is a narcotic analgesic.

5. NARCOTIC ANTAGONISTS

In 1915, it was shown that N-allylnorcodeine abolished both heroine- and morphine-induced respiratory depression. Almost 25 years later (1940), it was observed that N-allylnormorphine (commonly known as **nalorphine**) possessed more marked and significant morphine antagonizing actions. Thirteen years later (1953), it was demonstrated that nalorphine had the ability to *arrest severe abstinence syndromes in postaddicts* who were earlier treated briefly with either morphine, methadone or heroine. Examples of narcotic antagonists include : nalorphine hydrochloride ; naloxone hydrochloride ; propiram fumarate and pentazocine.

A. *Nalorphine Hydrochloride* BAN, USAN, *Nalorphine* INN,

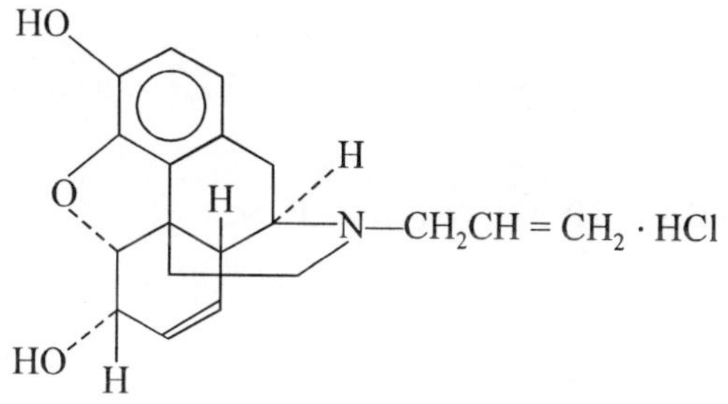

17-Allyl-7, 8-didehydro-4, 5α-epoxymorphinan-3, 6α-diol hydrochloride ; Morphinan-3, 6-diol, 7, 8-didehydro-4, 5-epoxy-17-(2-propenyl)-(5α, 6α)-, hydro-chloride ; U.S.P., Int. P., Ind. P., Nalline[R] (MS & D)

Synthesis

Morphine on treatment with cyanogen bromide gives the corresponding cyano analogue which upon hydrolysis forms the desmethylmorphine. This on reaction with allyl bromide and subsequent treatment with hydrochloric acids yields the official compound.

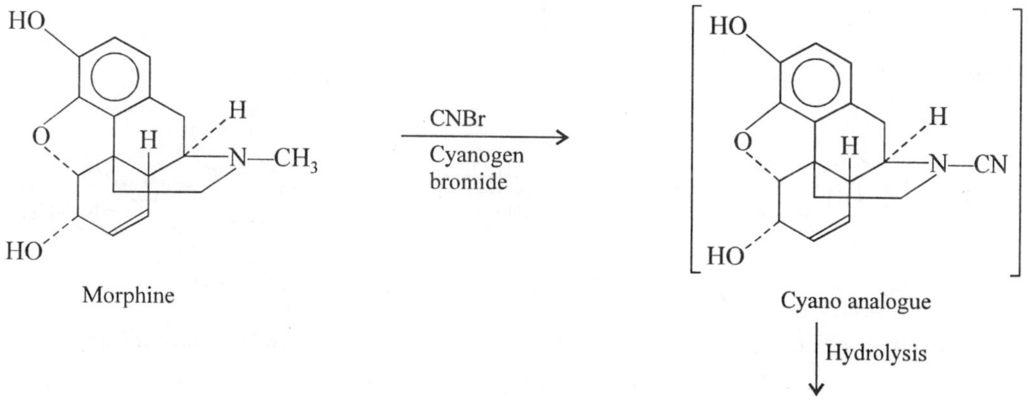

Morphine

Cyano analogue

Hydrolysis

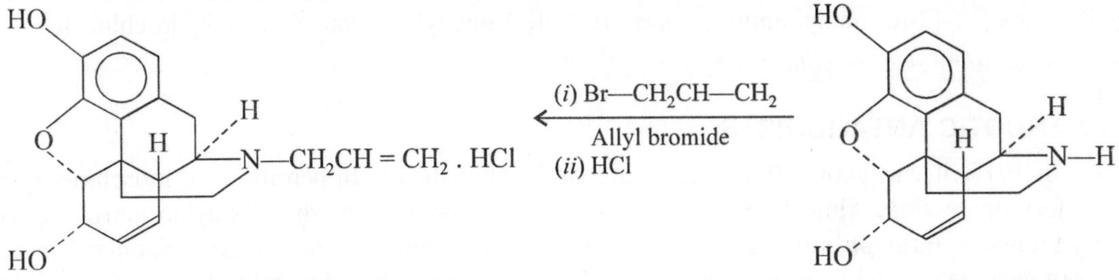
HO
O
H
H
N—CH₂CH = CH₂ . HCl
HO
(i) Br—CH₂CH—CH₂
Allyl bromide
(ii) HCl
HO
O
H
H
N—H
HO

Nalorphine Hydrochloride Desmethylmorphine

It is a narcotic antagonist having certain agonist actions that reduce the depressant actions particularly of morphine together with other narcotic drugs . It is pertinent to observe here that nalorphine does not exert its antagonistic effect caused by either barbiturates or other non-narcotic depressants. It possesses analgesic properties but is not used owing to its undesirable side-effects. *It is effectively employed to reverse narcotic-induced respiratory depression.*

Dose : Intravenous, 2 to 10 mg per dose ; usual, 5 mg repeated twice at 3 minute intervals if required.

B. *Naloxone Hydrochloride* BAN, USAN, *Naloxone* INN,

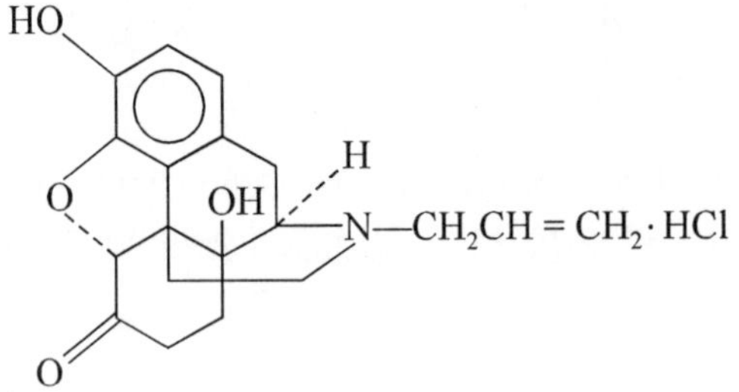
HO
O
OH
H
N—CH₂CH = CH₂ · HCl
O

17-Allyl-4, 5α-3, 14-dihydroxymorphinan-6-one hydrochloride ; Morphinan-6-one, 4, 5-epoxy-3-[4-dihydroxy-17-(2-propenyl)-, hydrochloride, (5α)- ; (-)-N-Allyl-14-hydroxy-nordihydromorphinone hydrochloride ; Allylnoroxy-morphone Hydrochloride ; U.S.P., Narcan(R) (Endo)

Synthesis

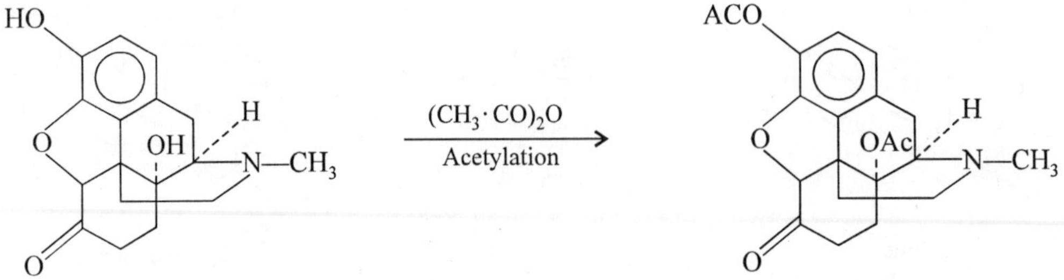
HO
O
OH
H
N—CH₃
O
(CH₃·CO)₂O
Acetylation
ACO
O
OAc
H
N—CH₃
O

Oxymorphone 3, 6-Diacetate derivative

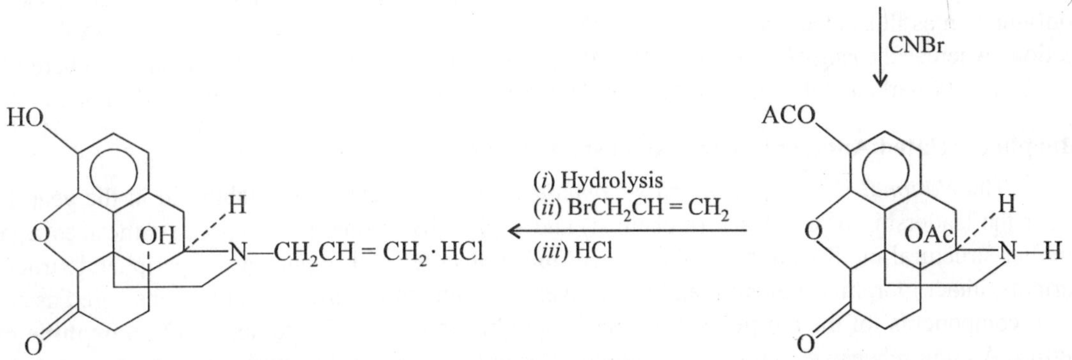

Naloxone Hydrochloride

CNBr

(i) Hydrolysis
(ii) BrCH$_2$CH = CH$_2$

(iii) HCl

Desmethyl derivative

It may be prepared by the acetylation of oxymorphone to give 3, 6-diacetate derivative which on treatment with cyanogen bromide yields the desmethyl derivative. This on hydrolysis, followd by alkylation with allyl bromide and finally treating with hydrochloric acid forms the official compound.

Naloxone is a specific narcotic antagonist which, unlike nalorphine, possesses no morphine-like properties. It is considered to be an effective antagonist for mixed agonist-antagonist like pentazocine. It may also reverse some of the adverse effects of narcotic antagonists having agonist actions. *Owing to its lack of respiratory depressant property, it can be safely administered to patients suspected of narcotic overdosage without having the risk of further increasing respiratory depression. It has been found to reverse narcotic analgesic and possesses little analgesic properties of its own.*

Dose : Usual, parenteral, 0.4 mg (1 ml)

6. MORPHINE : STRUCTURAL REPRESENTATIONS

In fact, the most probable structure of morphine was put forward in the year 1925 ; however, its confirmation by total synthesis was accomplished in 1955. Interestingly, the paucity of the knowledge with regard to the correct structure of morphine, nevertheless subsided the zeal and enthusiasm amongst the medicinal chemists to synthesize several morphine structural analogues by taking advantage of the various known chemical reactions with the peripheral functional moieties present in morphine, such as : C-3 phenolic hydroxyl ; C-6 allylic alcohol ; and C = C between C-7 and C-8 as depicted in the following structure(s). It is, however, pertinent to mention here that several synthesized structural analogues even before 1930 are still constituted as vital and potential 'drugs' in the therapeutic armamentarium, for instance : codeine ; ethyl morphine (*Dionin*[(R)]) ; diacetyl morphine (heroin) ; hydromorphone (*Dilaudid*[(R)]) ; hydrocordone (*Dicodid*[(R)]) ; and methyldihydromorphinone (*Metopon*[(R)]).

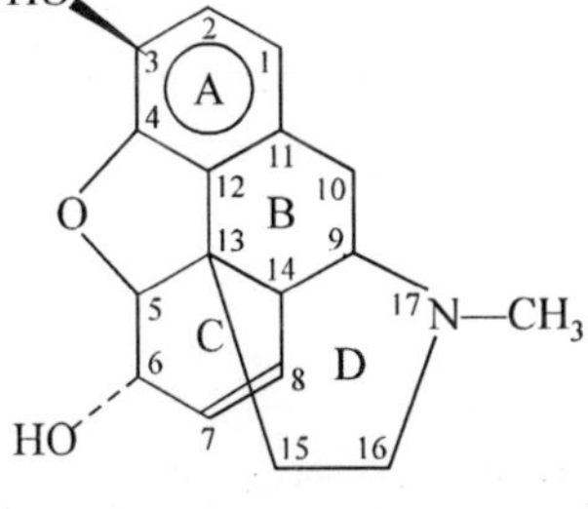

'*Flat*'-Representation
(A)

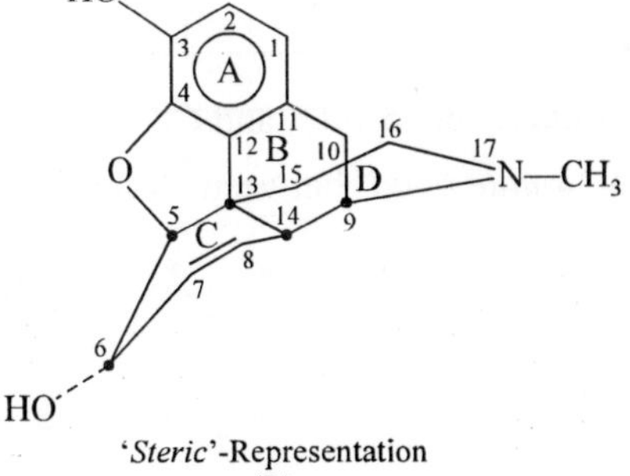

'*Steric*'-Representation
(B)

Morphine may be diagrametically represented as **'flat'** (A) configuration, and also as **'steric'** (B) configuration as illustrated above. Emphatically, in (B) the *ring 'C'* essentially has the **'BOAT'**-conformation ; whereas the *ring-'D'* has the **'CHAIR'** conformation. Besides, the carbon atoms numbered 5, 6, 9, 13, and 14 (marked with a dark spot) are **chiral in nature** (*i.e.,* these are asymmetric C-atoms).

Morphine-related Antagonists and Agonists/Antagonists

The *National Research Council's Committee on Drug Addiction* established in the year 1929 under the leadership of Small LF (a chemist) and Eddie NB (a pharmacologist), synthesized a large number structural modifications of the **'morphine molecule'** with regard to its peripheral structural variants, intact morphine skeleton, and derivatives of compounds which could be considered as structural 'components' of the morphine molecule ; that ultimately gave rise to nearly **125 morphine analogues.** A comprehensive analgestic evaluation certainly helped in the emergence of emperical structure-activity relationships (SARs) as given in Table 10.2.

TABLE. 10.2 : MORPHINE-RELATED ANTAGONISTS and AGONISTS/ANTAGONISTS

General Structure	Name	R	X	Y	Z	Other	Therapeutic Category
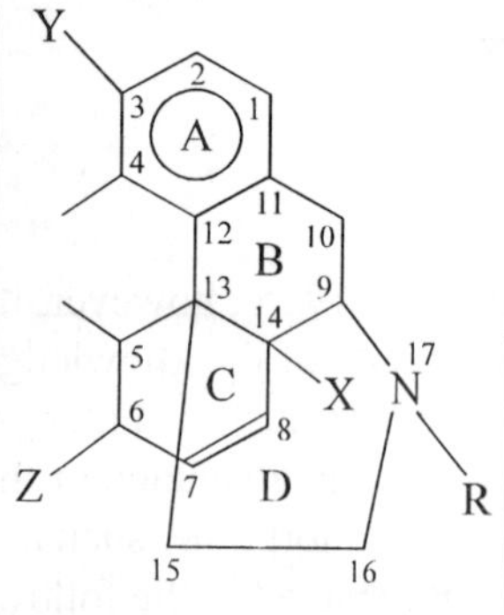	Nalorphine	$-CH_2-CH=CH_2$	H	OH	OH	—	Narcotic antagonist
	Levallorphan	$-CH_2-CH=CH_2$	H	OH	H	a^*	Narcotic antagonist
	Naloxone	$-CH_2-CH=CH_2$	OH	OH	$\overset{\|}{O}$	b^{**}	Narcotic antagonist
	Naltrexone	$-CH_2-\triangleleft$	OH	OH	$\overset{\|}{O}$	b	Narcotic antagonist
	Nalbuphine	$-CH_2-\diamondsuit$	OH	OH	OH	b	Narcotic analgesic
	Butophanol	$-CH_2-\diamondsuit$	OH	OH	H	a, b	Narcotic analgesic

*a = No *o-atom* between C_4 and C_5 ;

**b = No *'double bond'* between C_7 and C_8 ;

7. MECHANISM OF ACTION OF CERTAIN NARCOTIC ANALGESICS

The mechanism of action of certain *'narcotic analgesics'* included in this chapter are discussed below :

7.1. Morphine Sulphate

Its most important action is on the brain more specifically its higher functions. It has been observed that an initial transitory stimulation is usually followed by a distinct depression of the brain, its higher functions, and above all its medullary centres. Besides, the spinal functions and reflexes are normally stimulated. Interestingly, it causes a visible change in perception in such a manner that the patient shows more to tolerance to pain and discomfort perhaps due to possible interference with *'pain conduction'*.

Because of its high addition potential and abuse, the 'drug' is classified as *Schedule II* drug under the *controlled substances Act.*

7.2. Codeine

Codeine is chiefly metabolized in the liver where it undergoes *o*-demethylation, N-demethylation and partial conjugation with glycuronic acid. It is mostly excreted in the urine as *narcodeine* and *morphine* (both as its free and conjugated form). It is found to be less apt than 'morphine' to produce nausea, vomitting, constipation and miosis. It also causes addiction liabilities resulting into enhanced tolerance limits.

Note. **Naloxone is a *'specific antagonist'* in the situations arising from *'acute intoxication'*.**

7.3. Hydromorphone Hydrochloride

It has less tendency to effect sleep than morphine when administered in equivalent analgetic doses. Therefore, the consequent relief from pain may be accomplished either without any sleep or stupefaction. It is a semi-synthetic analgetic, chemically and pharmacologically very much akin to morphine.

7.4. Hydrocodone Bitartrate (Dihydrocodeinone Bitartrate)

The pharmacological action is found to be lying almost midway between those of codeine and morphine. It has been observed that while on one hand it possesses more addition liability than 'codeine', and on the other it displays absolutely very little evidence of its dependence or addiction with long-term administration.

Note. **'Tussionex[R]—is an ion-exchange resin complex with it, that essentially releases the drug gradually in a sustained rate and is said to produce effective cough-suppression over a span of 10-12 hours.**

7.5. Levorphanol Tartrate

It is a potent synthetic analgetic very much related chemically and pharmacologically to *'morphine'* ; and is invariably employed for the relief of acute pair. It is in many aspects closely related to morphine but its action is 6 to 8 times more potent. However, it has been observed that the gastro intestinal effects of this compound are appreciably on the lower range than those experienced with morphine. It is a narcotic with addiction liability quite akin to morphine ; therefore, almost same stringent precautions must be observed when prescribing this *'drug substance'* as for morphine.

7.6. Dextromethorphan Hydrobromide

Dextromethorphan is well absorbed from the GI-tract. It has been observed that the *'drug'* is largely metabolized in the liver ; and consequently, excreted through the urine either as *unchanged dextromethorphan* or as its *demethylated metabolites* including *dextorphan*, that interestingly possesses cough-depressant activity to a certain extent.

7.7. Butorphanol Tartrate

It is a potent synthetic opiate analgesic that gets completely absorbed from the GI tract after oral administration ; and, importantly, it undergoes almost 80% *first-pass metabolism*. It has been duly established that this *'drug'* enhances arterial resistance and the work of the heart (an action very much akin to *'petazocine'*) ; consequently, it is usually contra indicated in such patients who have a history of *acute myocardial infarction*.

7.8. Pentazocine

It happens to be a weak *'antagonist'* (1/30th than 'naloxone') at **mu receptors** ; and also acts as an *'agonist'* at **kappa receptors.** Its half-life after IM administration is 2.1 hour. It is found to exert weakly (nearly 1/50th than *'nalorphine'*) antagonizing effect on the analgesic effect produced by *morphine* and *meperidine*. Besides, it causes incomplete reversal of the cardiovascular, respiratory, and behavioral depression induced by morphine and meperidine. It also possesses certain degree of sedative action. The bioavailability of pentazocine after oral administration is only 20-50% due to the first pass metabolism. The *'drug'* gets metabolized extensively in the liver ; and subsequently, excreted by the urinary tract. It is, however, pertinent to mention here that the *two* major metabolites of **petazocine** are, namely : (*a*) *hydroxylation* of the *two* terminal methyl functional moieties attached to the N-substituent ; and (*b*) 3-*o*-conjugates, which are virtually **inactive**.

7.9. Meperidine Hydrochloride (Pethidine Hydrochloride)

The *'drug'* is largely metabolized in the liver with only a small quantum of it ~ 5% gets excreted unchanged. However, the short duration of action of meperidine is caused on account of its rapid metabolism *in vivo*. Importantly, the *'esterases'* predominantly cause cessation of the *ester linkage* (as ethyl ester at *para*-position) to leave as residue the **inactive**-*carboxylate analogue*. It also undergoes N-demethylation to yield the corresponding product known as **'normeperidine'**—a metabolite which gets accumulated after a prolonged medication with meperidine.

Normeperidine

Note. **Normeperidine has only weak analgesic property, but it gives rise to sufficient CNS stimulation ; and it may end up grand mal seizures. Hence, it must be discontinued immediately in a subject showing the slightest symptoms of CNS stimulation apparently.**

The elimination half-life of the '*drug*' is between 3-4 hours, but it may be simply doubled in patients with the liver malfunction. It has been observed that '*acidification*' of the urine may on one hand *increase the clearance of meperidine*, whereas on the other it may *retard the clearance of the toxic metabolite normeperidine.*

7.10. Diphenoxylate Hydrochloride

The '*drug*' itself possesses relatively low mu opioid agonist activity. It is, however, metabolized rather swiftly by means of ensuing *ester hydrolysis* to the corresponding **'free carboxylate'**, *difenoxin*, that exhibits 5 times more potent activity when administered orally. Interestingly, the inherent excessive higher polarity of difenoxin perhaps restricts its easier penetration into the CNS ; and, therefore, it provides an adequate explanation with regard to the comparatively *low abuse potential* of this narcotic analgesic.

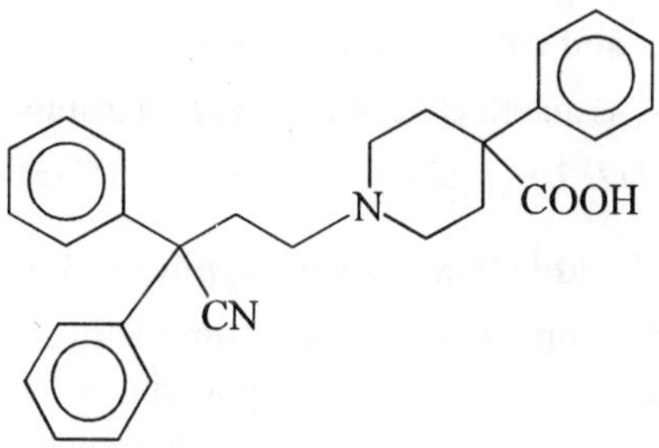

Difenoxin

7.11. Fentanyl Citrate

It exhibits a profile of pharmacological action very much identical to morphine, and differs exceptionally on *two* accounts, namely : *first*-it does not cause emesis ; *secondly,* it does not release histamine. Its safety measure in frequency cases has not yet been fully understood. It is observed to cross the *placental barrier* ; therefore, its usage during labour may ultimately give rise to respiratory depression in the newly born infant. However, Fentanyl's transient action after the parenteral adminis-tration is caused solely on account of redistribution, rather than to '*metabolism*' or '*excretion*'. Hence, longer usage of this '*drug*' may cause in accumulation and toxicities.

Note. Recent advancement in its 'dosage forms' are :

(*a*) **Fentanyl Transdermal Patch : It is used for the treatment of severe chronic pain, and it affords analgesia effitively for a duration ranging between 24—72 hours ; and**

(*b*) **Lollipop Dosage Form. It was introduced in the year 1999 for absorption from the buccal cavity (mouth).**

7.12. Anileridine Hydrochloride

It is found to be more potent as compared to meperidine ; and, hence, possesses the same useful-ness and limitations. Furthermore, its '*dependence capacity*'is significantly much lower ; and, there-fore, it is well accepted as an appropriate and legitimate substitute for meperedine.

7.13. Phenoperidine Hydrochloride

It undergoes absorption from the GI-tract to a certain extent. It has been found that the '*drug*' gets extensively metabolized in the liver to **peltidine** and **norpeltidine**, that are subsequently excreted in the urine.

7.14. Methadone Hydrochloride

The cardinal activities of therapeutic value essentially comprise of : analgesia, sedation and detoxification or temporary maintenance in narcotic addiction. It has been observed that the '*drug*' is most rapidly absorbed (perhaps rather incompletely) after the oral administration, by virtue of the fact that only 52% of a given dosage gets discharged in urine. The mean plasma levels ranging between 182 to 420 mg. mL^{-1} are found in patients administered on a daily oral dose of 40 and 80 mg respectively ; of which 71 to 87% is in the '*bound form*'. Its biological half-life is nearly 25 hour, with a range of 13 to 47 hours.

> Note. (1) It is one of the drugs of choice in the withdrawal management of patients addicted to morphine, heroin, and allied narcotic drugs.
>
> (2) NALOXONE—is an effective '*antagonist*' in instances of acute intoxication.
>
> (3) It is a '*Schedule II Drug*' under the *Controlled Substances Act* in US.

7.15. Propoxyphene Hydrochloride (Dextropropoxyphene Hydrochloride)

It is found, to be absorbed completely after oral administration ; however, *first-pass elimination* ranging between 30-70% reduces its '*bioavailability*' appreciably. The volume of distribution is 700 to 800 L ; oral clearance varies between 1.3 to 3.6 L. min^{-1} ; and the biological half-life is 6 to 12 hours. **Norpropoxyphene** happens to be the '*major metabolite*' having a half-life of 30-36 hours.

7.16. Methotrimeprazine (Levomepromazine)

A phenothiazine structural analogue, very intimately related to chlorpromazine, and exhibits extremely potent analgesic activity. Importantly, it is devoid of any dependence liability, besides it does not produce respiratory depression. It is specifically of **some extent of advantage** in such patients for whom *addiction* as well as *respiratory depression* are serious problems.

7.17. Tramadol Hydrochloride

The '*drug*' exhibits its a analgesic effect by categorically inhibiting the uptake of *norepenephrine* and *serotonin* which is believed to contribute to its analgesic effects. Its major metabolite is about 6 times more potent as an *analgesic* ; besides, it has 200 times greater affinity for the mu receptor.

7.18. Dezocine

It is a synthetic *opioid* '*agonist*' or '*antagonist*' structurally akin to pentazocine, and having analgesic actions almsot identical to morphine. Interestingly, it is a '*primary amine*' whereas the rest of the '*nonpeptide opioids*' are '*tertiary amines*'. Although, its exact receptor selectivity profile has not been reported so far, yet its pharmacological activities are quite similar to that of buprenorphine. It is observed to be a partial agonist at mu receptor sites, practically devoid of any effect at the kappa receptors ; and exhibits agonist effect at delta receptors to a certain extent.

Dezocine gets metabolized largely by glucuranidation of the phenolic hydroxy moiety and also by N-oxidation. Its metabolites are quite inactive, and gets excreted invariably through the renal passage.

7.19. Sufentanil

The introduction of the *para*-methoxymethyl moiety and the subsequent replacement of the bioisosteric phenyl group with a 2-thiophenyl into the *'fentanyl molecule'* give rise to a **10 times enhancement in the prevailing mu opioid activity**.

Hence, the resultant compound *i.e., sufetanil* is found to exhibit higher potency to the extent of 600-800 times in comparison to morphine. When administered IV it gets metabolized rapidly having a biological half-life 2.4 hour. Its volume of distribution is 2.5 L kg^{-1}, 92.5% is bound to plasma protein ; and the plasma clearance is 0.8 L min^{-1}.

7.20. Nalorphine Hydrochloride

It has a *'direct antagonistic effect'* against *morphine, meperidine, methadone* and *levorphanol*. Interestingly, it does not show any antagonistic effect toward barbiturate or general anaesthetic depression. Nalorphine exerts its effect on the circulatory disturbances thereby reversing the effects of morphine. It is found to cause depression in the respiratory activity itself, that may potentiate the pervailing depression produced by morphine.

7.21. Naloxone Hydrochloride

It has been adequately proved based on the available evidence that naloxone specifically antagonizes the opioid effects, such as : respiratory depression, psychotomimetic effects, and pupillary constriction, by means of genuine competition for the receptor sites. The drug disappears from serum in man in a much rapid fashion. After an IV administration it is distributed quite rapidly in the body. It is found that the biological half-life in adults ranges between 30 to 81 minutes ; whereas, the mean half-life in neonates is 3.1 and 0.5 hour.

Naloxone is largely metabolized in the liver, primarily by glucuronide conjugation ; and ultimately excreted in the urine.

> **Note.** Because of its short duration of action it is absolutely necessary to administer a multiple-dosing system that obviously limits its value.

Probable Questions for B. Pharm. Examinations

1. (*a*) What are narcotic analgesics ? Enumerate the **four**-serious limitations of these drugs.

   (*b*) Give the structure, chemical name and uses of MORPHINE.

   (*c*) What are the **three** important alkaloids isolated from *Papaver somniferum* ?

2. Classify narcotic analgesics by giving at least **one** typical example with its structure, chemical name and uses.

3. Discuss the **'morphine analogues'** and give the synthesis of :

   (*a*) Diamorphine Hydrochloride and

   (*b*) Hydrochloride Tartarate

4. Give the structure, chemical name and uses of any **two** important members of Morphinan Analogues. Discuss the synthesis of **one** of them.

5. Based on the **'morphan nucleus'** *i.e.,* a bridged perhyrozocine **three** drugs have been synthesized, namely : Metazocine, Cyclazocine and Pentazocine. Give their structure, chemical names and uses.

6. How would you synthesize **Pentazocine** from 1-bromo-3-methyl-2-butene ? Explain the route of synthesis.

7. The **4-phenylpiperidine analogue** led to the synthesis of much simpler compounds having potent analgesic propertise. Discuss the synthesis of any **one** compound stated below :

(*a*) Meperidine hydrochloride and

(*b*) Fentanyl citrate

8. Give the names of any **two** important drugs based on **phenylpropylamine analogues** and describe the synthesis of **one** of them.

9. Discuss briefly Tilidate hydrochloride, Dazocine, catanlanil and Nexonine as **narcotic analgesics.**

10. Give a brief account of '**Narcotic Antagonists'.** Discuss Nalorphine hydrochlroide in details.

RECOMMENDED READINGS

1. G de Stevens (*Ed*) *Analgesics* Academic Press New York (1965).

2. HW Kosterlitz, H O Collier and J E Vilareal (*Eds*) *Agonist and Antagonist Actions of Narcotic Analgesic Drugs University* Park Press Baltimore (1973).

3. PAJ Janssen *Synthetic Analgetics, Part I, Diphenyl-propylamines,* Pergamon Press, New York (1960).

4. DLednicer and L A Mitscher *The Organic Chemistry of Drug Synthesis* John Wiley and Sons, New York (1995).

5. J Hellerbach *et al. Synthetic Analgetics, Part IIa, Morphinans,* Pergamon Press, New York (1966).

6. NB Eddy and EL May *Synthetic Analgetics (Part II(b)) 6, 7-Benzomorphans* Pergamon Press, New York, (1966).

7. ME Wolff (*Ed*) : *Burger's Medicinal Chemistry and Drug Discovery*, (5th Ed) John Wiley & Sons, Inc., New York (1995).

8. CO Wilson, O Gisvold and R F Doerge *Textbook of Organic Medicinal and Pharmaceutical Chemistry* (10th edn) Philadelphia J B Lippincott Company (1998).

9. JEP Reynolds (*Ed*) *Martindale the Extra Pharmacopoeia,* (30th edn) The Pharmaceutical Press London (1992).

10. *Remington : The Science and Practice of Pharmacy,* Vol. II, Lippincott Williams and Wilkins, New York, Gennaro, A.R., (*Ed.*), 20th edn., 2000.

11 Cardiovascular Drugs

1. INTRODUCTION

Cardiovascular drugs generally exert their action on the heart or blood vessels in a direct or indirect manner thereby affecting the distribution of blood to certain specified portions of the circulatory system. Therefore, they essentially embrace a rather wide spectrum of such drugs which possess cardiovascular actions.

In a broader perspective the term '*cardiovascular drugs*' mostly connotes particularly such drugs that are invariably employed for their cardiovascular actions. In reality, therefore, almost every '*autonomic drug*' normally exerts marked and pronounced cardiovascular activities which are clinically useful in combating the serious human ailments.

Based on several scientific evidences one may observe that there are a plethora of categories of 'drug substances' which may be used effectively as cardiovascular drugs, such as :

(*a*) **Sympathomimetics** are mostly employed to enhance blood pressure, lower the reflexity of heart, stimulate the heart, which are solely dependent on the specific drugs *vis-a-vis* the therapeutic requirements.

(*b*) α-**Adrenergic Blocking Drugs** are commonly used in the *vasospastic** conditions, specifically in the diagnosis, control and management in the malignant and toxemic hypertensive crises ; besides, in *pheochromocytoma***.

(*c*) β-**Adrenergic Blocking Drugs** are invariably made use of in the treatment of essential hypertension, portal hypertension, angina pectoris and certain instances pertaining to dysrhythmias.

(*d*) **Anticholinesterase** (*e.g.* **edrophonium**) is judiciously used in the ensuing diagnosis and treatment of *paroxysmal atrial tachycardia.****

(*e*) **Atropine and other Antimuscarinic Drugs** are beneficially employed to block either the cardiac vagus nerve in the **Adams-Stokes Syndrome** or some other **bradycardias.**

Interestingly, a plethora of '*drug substances*' other than the autonomic agents have been observed to exert powerful as well as useful actions on the cardiovascular system, for instance :

---

*Characterized by *vasospasm i.e.,* spasm of a blood vessel.

**A chromaffin cell tumour of the sympathoadrenal system that produces catecholamines.

***A sudden periodic attack of atrial tachycardia.

(i) Digitalis and its related derivatives, the peripheral and coronary dilators, and above all the antidysrhythmic agents,

(ii) Parenteral fluids that find their application in the treatment of severe shock, and

(iii) Diuretics that are invariably used as an adjunct in the treatment and management of heart failure as well as hypertension.

2. CLASSIFICATION

Cardiovascular drugs may be conveniently classified into the following *four* heads, namely :

(a) Cardiac Glycosides

(b) Antihypertensive and Hypotensive Drugs

(c) Antiarrhythmic Agents, and

(d) Vasopressor Drugs

This chapter mainly deals with the above categories of drugs with specific examples.

3. CARDIAC GLYCOSIDES

Cardiac glycosides or digitalis essentially refer to a group of chemically and pharmacologically related drugs, that act on the heart by causing atrioventricular conduction and vagal tone. They are invariably employed to slow the heart rate in atrial fibrillation and also administered in congestive heart failure.

A large number of '*drug substances*' are able to enhance the force of contraction of the heart. It is, however, pertinent to state here that this ionotropic activity may be specifically of great utility and importance in the ultimate treatment of congestive heart failure*. Evidently, a defective and failing heart is not capable of pumping the requisite quantum of blood supply so as to maintain the bear minimum body needs. Keeping in view the enormous incidence of congestive heart failure across the globe the **'inotropic drugs'** have gained its legitimate importance and cognizance throughout the world.

There is absolutely no reason to believe that the host of '*inotropic drugs*' do help positively in prolonging the life-span of an individual, but nevertheless even with the long-term treatment, the longivity of such patients cannot be improved to an appreciable extent.

3.1.   Designing the Cardiac Glycoside Receptor

In fact, there are *three* most prevalent questions that invariably arise with regard to the ultimate structure-activity relationships (SARs) of any known class of drugs, namely :

(a) How does a '*drug molecule*' fit into the receptor site ?

(b) What are the most probable structures and conformations that may allow the best fit ?

(c) Which parts of the drug molecule are actually responsible for effective binding**, and also for exhibiting the specific pharmacological activity*** ?

An enormous amount of concerted research particularly related to fundamental aspects of drug design, synthesis, conformational energy studies, and computer-aided molecular modelling (CAMM)

---

*Hsu L., *J Am Pharm Assoc.* NS **36**(2) : 93, 1996.

***Binding* : Affinity.

***Intrinsic activity.

have been exploited both extensively and intensively to reveal the intricacies pertaining to the basic structural requirements of the desired cardiac glycoside binding site.

Na$^+$, K$^+$—ATPase is a **'dimer'** which is composed of two catalytic α-*subunits* together with two inert β-*subunits*. It has been well established that the two catalytic α-subunits essentially consist of the required binding sites for the cardiac glycosides, besides ATP, Na$^+$, K$^+$ and phosphorylation. Though the two β-subunits are absolutely necessary for the cardiac activity, yet these are not found to exert any direct catalytic action. Perhaps the β-subunits aid in holding the α-subunits in a strategically active configuration ; however, their exact intention is yet not quite established. It has been observed that the Na$^+$ K$^+$-ATPase enzyme extends across the plasma membrane, having a large quantum of the enzyme specifically onto the extracellular surface.

Interestingly, electron microscopic studies has provided an ample evidence that the α-dimer usually creates a '*deep-cleft*' in the Na$^+$, K$^+$-ATPase ; and that has been suggested as the most preferred binding site for the **cardiac steroidal glycosides** particularly.

Digoxin and digitoxin are the two cardiac glycosides most frequently used nowadays and have obviously replaced digitalis in cardiac therapy.

A. *Digoxin* INN, BAN, USAN,

3β-[0, 2, 6-Dideoxy-β-D-*ribo*-hexopyranosyl-(1 → 4)-0-2, 6-dideoxy-β-D-*ribo*-hexo-pyranosyl-(1 → 4)-2, 6-dideoxy-β-D-*ribo*-hexopyranosyl) oxy]-12β, 14-dihydroxy-5β, 14β-card-20(22)-enolide ; BP ; USP ; Eur. P., Ind. P., Int. P., Lanoxin$^{(R)}$ (Burroughs Wellcome) ; SK-Digoxin$^{(R)}$ (SK & F)

The side chain of digoxin is made up of three molecules of digitoxose in a glycosidic linkage, which upon hydrolysis yields the aglycone, digoxigenin ($C_{23} H_{34} O_5$).

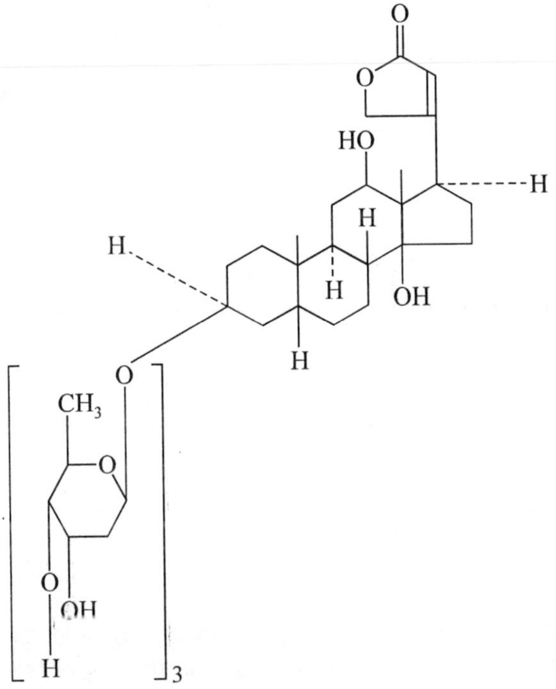

Digoxigenin is obtained from the leaves of *Digitalis lanata* Ehrh. (Family : *Scrophulariaceae*).

Its cardiotonic actions are very much alike to those of digitalis. It is used in the *treatment of congestive heart failure.* It is administered to slow down the ventricular-rate *in the management and treatment of atrial fibrillation.* Digoxin enhances the force of myocardial contraction which in the case of heart-failure ultimately result in an improved cardiac output thereby reducing the size of the dilated heart. Concurrently, venous pressure is lowered as the heart is now in a position to accommodate an increased venous return of blood ; also an improvement in the peripheral circulation modifies renal function thereby causing diuresis and hence an apparent relief of oedema.

Dose : Oral, adults and children more than 10 years of age : For rapid digitalization, 1 to 1.5 mg divided into 2 or more doses after 6-8 hours ; For slow digitalization and maintenance (0.125 to 0.5 mg) once a day.

B. Digitoxin, INN, BAN, USAN,

3β-[0-2, 6-Dideoxy-β-D-*ribo*-hexopyranosyl-(1 $\rightarrow$ 4)-0-2, 6-dideoxy-β-D-*ribo*-hexopyranosyl-(1 $\rightarrow$ 4)-2, 6-dideoxy-β-D-*ribo*-hexopyranosyl) oxyl]-14-hydroxy-5β, 14β-card-20(22)-enolide ; BP ; USP ; Eur. P., Int. P., Ind. P., Crystodigin[R] (Lilly) ; Purodigin, Crystalline[R] (Wyeth)

Digitoxin is obtained from *Digitalis purpurea* Linne, *Digitalis lanata* Ehrh, and other suitable species of Digitalis.

Its side chain is comprised of three molecules of digitoxose in a glycosidic linkage. Hydrolysis affects removal of the side chain to yield the aglycone, digitoxigenin ($C_{23}H_{34}O_4$)... . ...

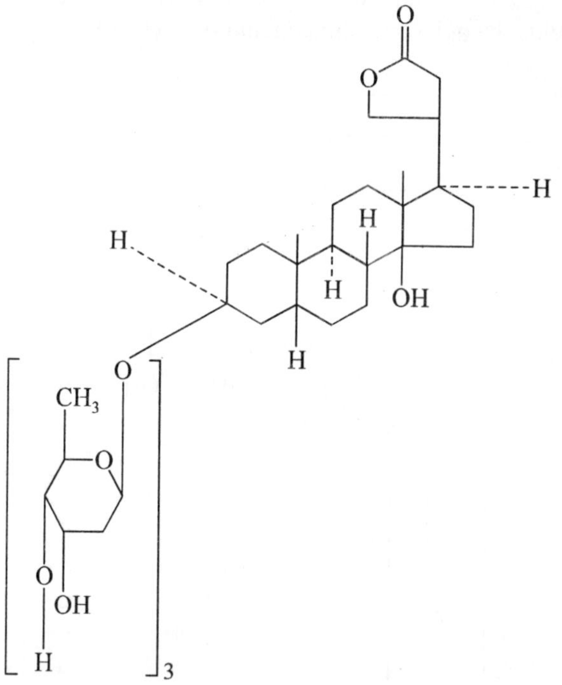

Digitoxin is the most potent of the digitalis glycosides besides possessing its inherent maximum cumulative action. *Though its onset of action is rather slower than that of the other cardiac glycosides, yet its effect persists much longer even extending up to 3 weeks.*

Dose : Oral, intramuscular, or intravenous, adults, for digitalization ; initially 600 mcg folllowed by 200 to 400 mcg every 3 to 6 hours as needed.

3.2. Mechanism of Action

The mechanism of action of digoxin and digitoxin are discussed as under :

3.2.1. Digoxin

The '*drug*' is mostly employed IV for extremely rapid *digitalization*, where upon its action invariably becomes manifest in 15 to 30 minutes, while its effect ultimately attains its peak in 2 to 5 hours. In contrast, its action through the oral administration is usually manifested within a span of 1 to 2 hours, and reaches a peak in 6 to 8 hours. However, after accomplishing complete digitalization, the duration of action is about 6 days. In plasma it is normally protein-bound to the extent of 20 to 30%. It exhibits a high volume of distribution, having a v_d^{ss} of about 5.1 L kg^{-1} in normal healthy adults, neonates and even larger in infants ; whereas, patient with renal failure the value of v_d^{ss} is almost nearly 3.3 L. kg^{-1}. Interestingly, the observed large volume of distribution is exclusively by virtue of its *extensive intracellular binding.*

It has been duly observed that the biliary secretion and the enterohepatic recirculation usually account for almost 7 to 30% of the body burden. However, by the oral administration, approximately 50-80% of the drug gets absorbed from the solid dosage forms, but it may be extended upto 90-100% from the hydroalcoholic solutions in capsules. The overall outcome is the enhanced GI motility which gets diminished ; and hence, the lowered motility enhances the absorption of the drug.

3.2.2. Digitoxin

The '*drug*' after complete digitalization, shows duration of action extended upto 14 days. In fact, nearly 97% of the drug is protein-bound as found in plasma. The therapeutic level of the drug in plasma is optimum at a concentration ranging between 15-25 ng. m L^{-1} ; whereas, at a concentration varying between 35-40 ng. mL^{-1} or even more is regarded to be toxic. It has been observed that the *hepatic metabolism* usually accounts for 52-70% of its elimination. The β-half-life ranges between 2.4 to 9.6 (with an average of 7.6) days.

4. ANTIHYPERTENSIVE AND HYPOTENSIVE DRUGS

A plethora of substances are normally employed to lower the blood pressure, though their effect may be transient. A few of them are used for their hypotensive action. An arbitrary definition of normal adult blood pressure afforded by the World Health Organization (WHO)-'*is a systolic pressure equal to or below 140 mm Hg together with a diastolic pressure equal to or below 90 mm Hg.*'

Antihypertensive drugs are invariably employed in the treatment of hypertension, although a few amongst them, such as : *ganglionic blocking drugs*, do find their scattered applications in a variety of other therapeutic, diagnostic and surgical procedures.

Interestingly, a few of them are used as hypotensive drugs in nonhypertensive subjects. There exist two major categories of '*diastolic hypertension*', namely : (*a*) **primary hypertension** (*e.g.,* essential, idiopathic) ; and (*b*) **secondary hypertension.** However, the *malignant hypertension* is nothing but an acute and rather progressive phase of **primary hypertension.** It has been revealed that there is absolutely no universal therapy for the control and management of primary hypertension ; and, as such, most individual instances do vary widely in response to various therapeutic agents.

In fact, there are several glaring evidences available today that may attribute to certain types of hypertension previously known as *diastolic* or *essential hypertension*, for instances :

(*a*) Renin-angiotensin pathway,

(*b*) Angiotensin II receptor antagonists and

(*c*) Potential-dependent calcium channels.

4.1. Renin-Angiotensin Pathway

It has already been proved adequately that the prevailing *renin-angiotensin pathway* happens to be an extremely complex, highly regulated pathway which is intimately associated with the ultimate regulation of blood volume, electrolyte balance, and above all the arterial blood pressure. It essentially comprises of *two* cardinal enzymes, known as : *renin* and *angiotensin converting enzyme* (ACE). The most predominant and primary objective of these enzymes are to afford adequate release of angiotensin II from its endogenous precusror, usually termed as **angiotensinogen.** Importantly, angiotensinogen is an α_2-**globulin** having a molecular weight ranging between $58,000 - 61,000$. It is essentially made up of 452 amino acids, is available abundantly in the plasma, and is continually replenished by synthesis and secreted by the liver.

In reality, the role of the *renin-angiotensin pathway* in the cardiovascular disorders is extremely vital and critical by virtue of the fact that it exclusively is responsible for the maintenance of blood volume, arterial blood pressure, and the electrolyte balance in the body. Therefore, any slightest abnormalities in this prevailing pathway, such as : excessive release of renin, overproduction of angiotensin II, may ultimately give rise to a plethora of **cardiovascular disorders.**

4.2. Angiotensin II Receptor Antagonists

It is, however, pertinent to state here that *angiotensin II receptor* happened to be the first and foremost '*target approach*' towards the historical developemnt of newer drug substances that could possibly inhibit the renin-angiotensin pathway. In early 1970s, a tremendous effort was geared into action to develop **angiotensin II receptor antagonists** that was solely based on the *peptide-linked structural analogues* of the **natural agonist.** Efforts in this direction gave birth to several drugs of which a few important ones are given below :

(*a*) **Saralasin**

Sar-Arg-Val-Tyr-Val-His-Pro-Ala

1-(N-Methylglycine)-5-*L*-valine-8-*L*-alanineangeotensin II. It is employed as antihypertensive and as a *diagnostic aid* (*i.e.,* renin-dependent hypertension).

(*b*) **Losartan** :

2-Butyl-4-chloro-1-[2'-(1H-tetrazol-5-yl) [1, 1'-biphenyl]-4-yl] methyl]-1H-imidazole-5-methanol.

The wonderful drug 'losartan' was developed in 1982 and since then being used as a potent antihypertensive agent. It specifically blocks the angiotensin II receptor.

4.3. Potential-Dependent Calcium Channels

It has been well established and demonstrated that the *potential-dependent* Ca^{2+} *channels* are solely critical and important in modulating the influx of Ca^{2+} ; and hence, subsequent inhibition of Ca^{2+} flow through these specific channels results in both vasodilation as well as retarded cellular response to the pervailing contractile stimuli. Based on the proven facts that the *arterial smooth muscle* is found to be more sensitive than the *venous smooth muscle* ; besides, the *coronary and cerebral arterial blood vessels* are observed to be more sensitive in comparison to other *arterial beds**.

Consequent to these pharmacological actions, the calcium channel blockers are found to be profusely beneficial in the treatment of hypertension, and ischemic heart disease (IHD).

Examples. Clonidine hydrochloride, hydralazine hydrochloride, methyl-dopa, diaoxide and sodium nitroprusside.

A. *Clonidine* INN, *Clonidine Hydrochloride* BAN, USAN,

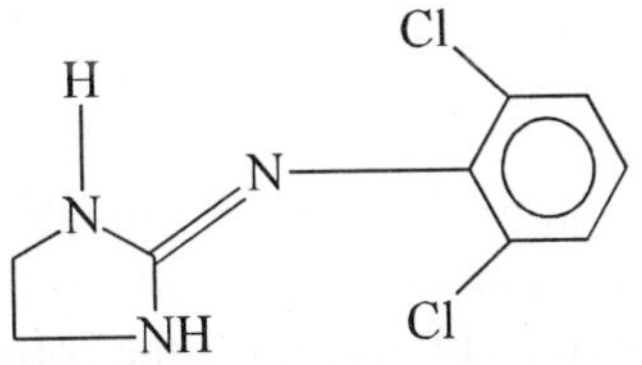

2-(2, 6-Dichloroanilino)-2-imidazoline hydrochloride ; 2-(2, 6-Dichlorophenyl-amino)-2-imidazoline hydrochloride ; 2, 6-Dichloro-N-(imidazolidine-2-ylidene) aniline hydrochloride ; BP ; USP ; Catapres(R) (Boehringer Ingelheim)

The compound was initially investigated as a nasal vasoconstrictor but incidentally has shown to be an effective drug in the *treatment of mild to severe hypertension and prophylaxis of migraine headache.*

Dose : 0.15 to 0.9 mg daily in 2 or 3 divided doses.

B. *Hydralazine* INN, BAN, *Hydralazine Hydrochloride* USAN,

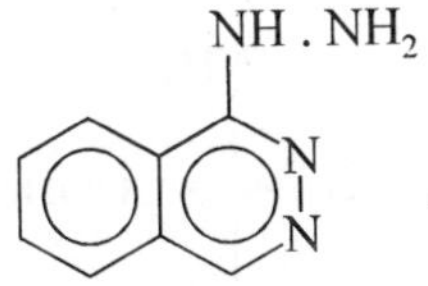

1-Hydrazinophthalazine monohydrochloride ; Phthalazine, 1-hydrazino-, monohydrochloride ; BP ; USP ; Int. P. ; Apresoline Hydrochloride(R) (Ciba-Geigy)

---

*Swamy VC and Triggle DJ. *Calcium channel blockers*, In : Craig GR, Stitzel RE, eds, **Modern Pharmacology with Clinical Applications**, Little Brown, Boston, 5th ed, 1997, 229-34.

Synthesis

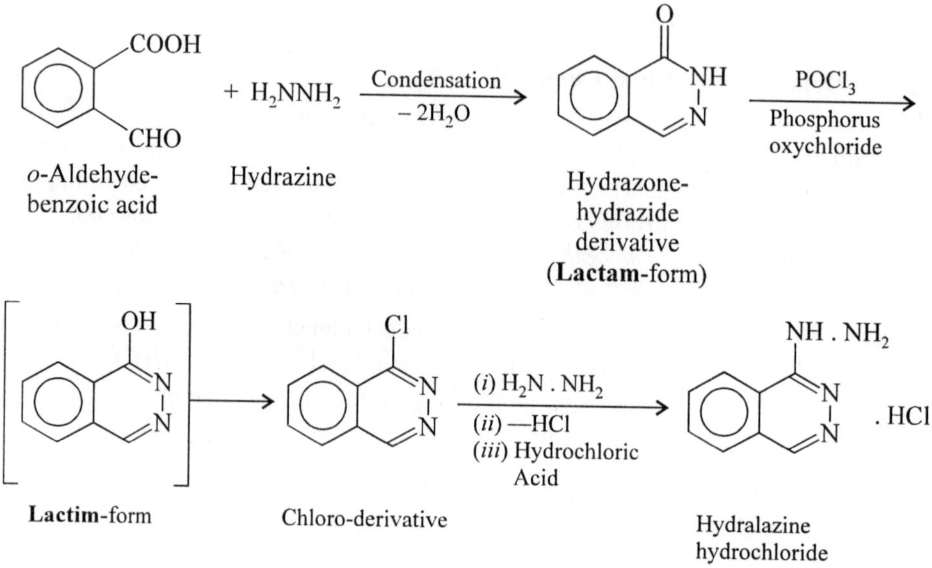

o-Aldehyde benzoic acid, *i.e.,* the half-aldehyde corresponding to phthalic acid, undergoes condensation with hydrazine to yield the hydrazone hydrazide (*lactam*-form). The *lactim*-form of this compound upon chlorination with phosphorus oxychloride gives the corresponding chloro derivative which on *first* treatment with a further mole of hydrazine and *secondly* with a calculated amount of hydrochloric acid affords the official compound.

It is a potent antihypertensive agent which exerts its action mainly by causing direct peripheral vasodilation. It has been observed that its effect on diastolic pressure is more marked and pronounced than on systolic pressure. It is employed in the *treatment of essential and early malignant hypertension usually in conjunction with thiazide diuretics or rauwolfia alkaloids.*

Dose : Oral, initial, 10 mg 4 times daily for 2 to 4 days, then 25 mg 4 times per day for the rest of the week.

C. *Methyldopa* INN, BAN, USAN,

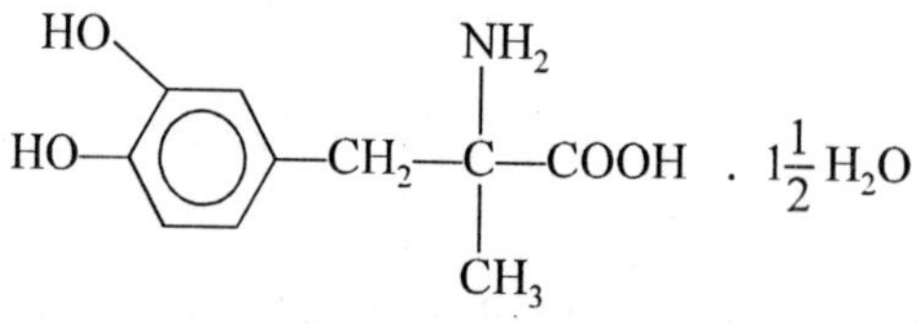

L-3-(3, 4-Dihydroxyphenyl)-2-methylalanine sesquihydrate ; L-Tyrosine, 3-hydroxy-α-methyl-, sesquihydrate ; BP ; USP ; Aldomet$^{(R)}$ (Merck)

Synthesis

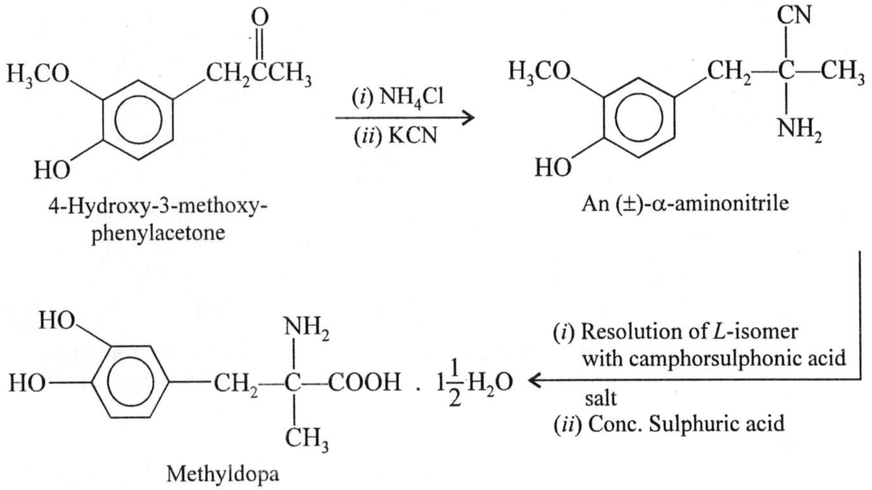

4-Hydroxy-3-methoxy-phenylacetone

(i) NH$_4$Cl
(ii) KCN

An (±)-α-aminonitrile

Methyldopa

(i) Resolution of *L*-isomer with camphorsulphonic acid salt
(ii) Conc. Sulphuric acid

The reaction of 4-hydroxy-3-methoxy phenylacetone with ammonium chloride and potassium cyanide yields the corresponding racemic mixture of α-aminonitrile. The *L*-isomer is separated by means of camphorsulphonic acid salt which on treatment with concentrated sulphuric acid affords two processes simultaneously, namely : hydrolysis of the nitrile function to the acid function ; and cleavage of the methyl ether moiety, to yield the official compound in *its hydrated form.*

Methyldopa is a potent antihypertensive agent that acts centrally by stimulating α-adrenergic receptors. It also helps to minimise the tissue concentrations of adrenaline, noradrenaline and serotonin. It is widely employed *to treat patients having moderate to severe hypertension by reducing the supine blood pressure as well as the standing blood pressure.*

Dose : Usual, initial dose, oral 250 mg of anhydrous methyldopa 2 or 3 times per day for 2 days ; usual maintenance dosage is 0.5 to 2 g of anhydrous methyldopa everyday.

D. *Diazoxide* INN, BAN, USAN,

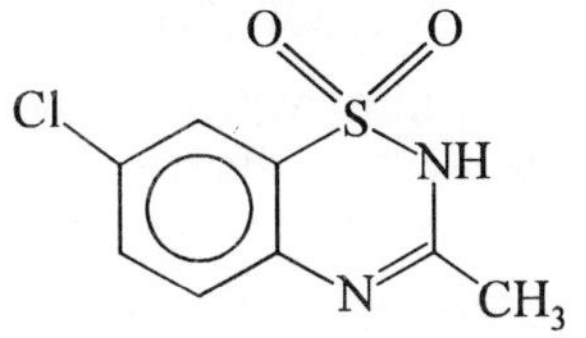

7-Chloro-3-methyl-2H-1, 2, 4-benzothiadiazine 1, 1-dioxide ; 2H-1, 2, 4-Benzothiadiazine, 7-chloro-3-methyl-, 1, 1-dioxide ; BP ; USP ; Hyperstat[R] (Schering-Plough) ; Eudemine[R] (Allen and Hanburys, U.K.)

Synthesis

Interaction between 2, 4-dichloro-nitrobenzene with benzylthiol affords an intermediate thereby eliminating a mole of hydrogen chloride. The resulting product undergoes debenzylation with concomitant oxidation in the presence of aqueous chlorine gives 5-chloro-2-nitrobenzene sulphonyl chloride which on subsequent treatment with ammonia and reduction yields the corresponding sulphonamide. This on condensation with a mole of ethyl *ortho* acetate yields the official product.

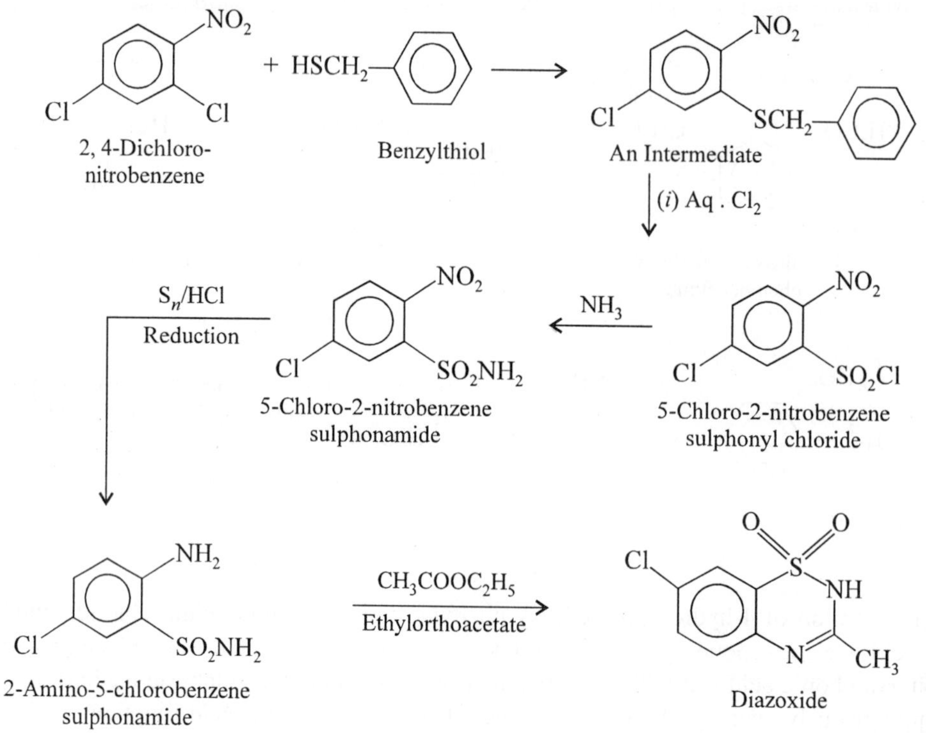

Diazoxide is employed intravenously for the *management and treatment of severe hypertensive crisis thereby lowering the blood pressure by a vasodilator effect on the arterioles.*

Dose : 0.4 to 1 g per day in 2 or 3 divided doses.

E. *Sodium Nitroprusside* USAN,

$$Na_2 [Fe(CN)_5NO] . 2H_2O$$

Sodium nitroferricyanide ; Sodium nitrosylpentacyanoferrate (III) dihydrate ; B.P., U.S.P. ; Nipride[R] (Hoffmann-La Roche) ; Nitropress[R] (Abbott).

Synthesis

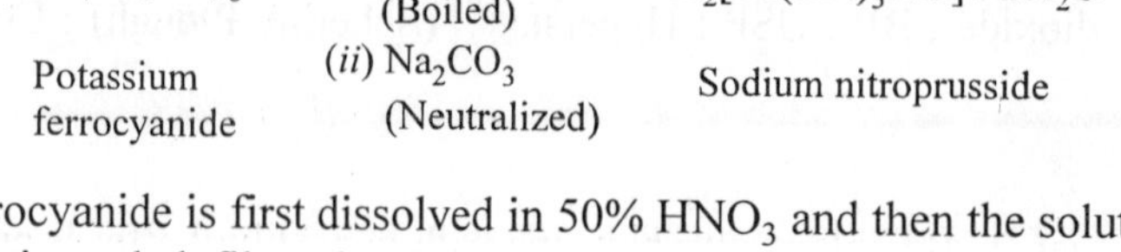

$$K_4Fe(CN)_6 \xrightarrow[\text{(Boiled)}]{\text{(i) 50\% HNO}_3} Na_2[Fe(CN)_5NO] . 2H_2O$$

Potassium ferrocyanide   (ii) Na₂CO₃   Sodium nitroprusside
(Neutralized)

Potassium ferrocyanide is first dissolved in 50% HNO_3 and then the solution boiled for 1 hour. The resulting solution is cooled, filtered and neutralized with sodium carbonate, and finally evaporated to crystallization.

It is a short-acting hypotensive agent. It is mostly *employed as a vasodilator in the emergency treatment of hypertensive crises that normally do not respond to other antihypertensive measures.*

Dose : By continuous infusion of a 0.005 or 0.01% solution in dextrose injection, normally at a rate of 0.5 to 8 mcg per kg body-weight per minute, under physician's observation.

4.4. Mechanism of Action of Selected Antihypertensive and Hypotensive Drugs

The mechanism of action of certain selected antihypertensive and hypotensive drugs shall be discussed in the section that follows :

4.4.1. Clonidine

The antihypertensive actions are, in part, due to a central action. However, an observed retardation in the sympathetic activity gives rise to a variety of pharmacological actions, such as : vasodilation, bradycardia and occasional atrioventicular block, and a decrease in the release of renin from the kidney ; besides, an enhancement in the vagal activity also affords bradycardia.

Interestingly, the central activity, in part, seem to be the outcome of a specific stimulant action on the α_2-*adrenergic receptors* either located in the vasomotor and cardioinhibitory centres, or in the spinal cord on the preganglionic sympathetic neurons. Besides, it may also exert a peripheral action to reduce the release of norepinephrine from the sympathetic nerves. It has been found to cause stimulation of the α_2-*adrenergic receptors* on the sympathetic nerve terminals, which stimulation ultimately affords a feed back almost negatively to put an end to the release of the ensuing mediator.

4.4.2. Hydralazine

The drug acts on the vascular smooth muscle to afford definite relaxation. Its exact mechanism of action is still not quite vivid and clear. It is found to interfere with Ca^{2+} entry and Ca^{2+} release from the prevailing intracellular reserves ; besides, causing a specific activation of *guanylate cyclase* thereby giving rise to an enhanced levels of cGMP. In fact, the concerted effort of all these biochemical events may afford an apparent vasodilation.

It gets excreted rapidly through the kidneys, and within a span of 24 hours nearly 75% of the total quantum administered appears in the urine as its '*metabolites*' or absolutely unchanged form.

The drug invariably undergoes mainly *three* types of chemical transformations, namely : (*a*) benzylic oxidation ; (*b*) glucuronide formation ; and (*c*) N-acetylation by the microsomal enzymes invariably found in the liver and tissues. It has been observed that '*acetylation*' could pose as a main determinant factor of the rate of hepatic removal of the drug from the blood in circulation ; and, hence, of the prevailing systemic availability of the same.* Consequently, the rapid acetylation aids in an extremely high hepatic extraction ratio ensuing from the circulatory blood, which is virtually responsible for the greater portion of the *first-pass elimination.*

4.4.3. Methyldopa

The drug gets converetd to α-methylnorepinephrine that eventually helps in displacing norepinephrine, from the stroage sites ; and thus, release as a '*false transmitter*' by means of the prevailing nervous impulses in the adrenergic nerves. Importantly, the *metabolite* α-*methylnorepinephrine* shows potent α_2-*agonist activity,* and this perhaps acts summararily to lower the blood pressure almost in the same manner as that of clonidine. However, in the spinal cord and the vasomotor centre, the

---

*Zacest R and Koch-Wesres, *J. Clin. Pharmacol.*, 1972, **13**, 4420.

ultimate results is an observed decrease in the vasomotor outflow, that ultimately is responsible for lowering blood pressure besides decreasing the plasma-renin activity.

4.4.4. Diazoxide

The *drug* at therapeutic dose levels, causes **vasodepression** which is primarily the outcome of arteriolar dilatation, in order that the ensuing orthostatic hypotension is normally minimal. However, certain extent of venous dilatation invariably occurs, which occasionaly is responsible to afford orthostatic hypotension. It has been duly observed that the smooth muscle-relaxing effects are usually caused due to the *hyper-polarization* of vascular smooth muscle by activating ATPase-sensitive K-channels. Hence, it is mostly used in IV as a '*hypotensive drug*' in situations arising from acute hypertensive crises.

Diazoxide is found to be 90% protein-bound ; however, fast and rapid IV administration allows quick-distribution to smooth muscle before it gets bound to protein intimately. Therefore, one may attain a greater and longer-lasting drop in blood pressure through faster rates of IV injection. Interestingly, the '*drug*' is found to persist in blood circulation much longer than the corresponding hypotensive effect. The plasma half-life is 20 to 60 hours in subjects having normal renal function, whereas the corresponding hypotensive effects lasts only 2 to 15 hours.

4.4.5. Sodium Nitroprusside

It happens to release nitric oxide (NO), which is also recognised as endogenous, endothelial-derived relaxing factor. Importantly, 'NO' progressively activated *guanylyl cyclase* strategically located in vascular smooth muscle to effect production of *vasodilatation.* However, its specific actions on the ensuing arterioles minimise the total systemic vascular resistance, and that perhaps is the major cause of the fall in blood pressure it evokes eventually. It has been observed to cause milder action on the *capacitance veins* ; and, therefore, with normal doses, venous return is impaired insignificantly in the *recumbent position.* But in the *upright position* there exists an appreciable *orthostatic hypotension.* Evidently, the observed cardiac output gets enhanced in the recumbent status ; whereas, lowered in the upright status distinctly. Besides, there prevails a variable effect particularly on the *renal plasma flow* and the *glomerular filtration rate,* but it is normally found to be enhanced in the recumbent position. One may also observe the **plasma-renin activity** to get enhanced within a range varying between slight to moderate.

5.   ANTIARRHYTHMIC AGENTS

Certain diseases and the effect of some drugs are usually responsible for affecting the rhythm and the normal heart-rate. These cardiac arrhythmias may be caused from disorders in pacemaker function of the sinoatrial node thereby resulting into tachycardia, bradycardia, cardiac arrest, atrial flutter, atrial fibrillation and ventricular fibrillation. Hence, the antiarrhythmic agents are also termed as **'antidysrhythmic drugs'** or **'antifibrillatory drugs'.**

Antiarrythmic drugs may be defined as—'*drugs that are capable of reverting any irregular cardiac rhythm or rate to normal'.* In other words, an arrhythmic situation is that wherein either initiation or propagation of a heart-beat stimulus is found to be invariably abnormal.

At this juncture it is worthwhile to have a little in-depth knowledge with regard to a *normal sinus rhythm* and aspects of the *electrical characteristics* of the heart.

Salient Featrues.   A few salient features are as follows :

(1) *Sinoatrial (SA) node** is situated in the viccinity of the surface at the junction of the right atrium and the superior vena cava, which is solely responsible for the normal orderly maintenance of the sequence of events in the cardiac contraction profile being initiated by a pacemaker. **Automaticity** is one of its main characteristic features. SA-node essentially possesses a normal firing frequency ranging between 60-100 impulses per minute.

(2) Subsequently, the established rhythm is conducted to the atrioventricular (AV) node**, which critically serves to slow down the heart beat so that the atrial contraction may take place prior to the stimulation of the ventricle. The impulse is transmitted from the AV-node ot the '*Bundle of His*' (*i.e.,* to a common bundle of fibers) which typically crosses the right atrium to the left ventricle. The branching of '*Bundle of His*' ultimately leads into the **Purkinje fibers** which essentially innevate the herat musculature of the ventricles.

The electrical activity produced by either the depolarization or the repolarization of myocardial tissue, specifically the nerve fibre cells, may be identified conveniently by the help of suitable electrodes ; and this may be plotted as a graph showing intensity (mV : millivolts) along the Y-axis and time (seconds) along the X-axis, as shown in Fig. 11.1(*a*), also known as the electro-cardiogram (ECG).

Explanations of Fig. 11.1(*a*) are as stated below :

(*i*)   ECG-represents a *'cardiac cycle'* which could be either normal or abnormal.

(*ii*)   P-wave specifically designates the electrical activity which passes over the atrial surface,

(*iii*)   Q, R, S-waves, (*i.e.,* the QRS-complex) are produced by the ventricles.

(*iv*)   T-wave is produced by the repolarization of the ventricular muscle fibres, and

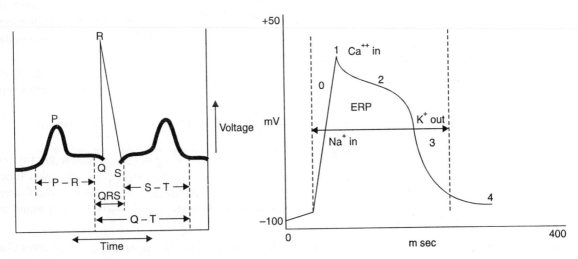

Fig. 11.1(*a*). Tracing of an Electrocardigram (ECG)   **Fig. 11.1(*b*).** Schematic Sketch of Cardiac Electric Profile

[Phases = 0 – 4 ; ERP = Effective Refractory Period]

[Adapted from : Gringauz A : An Introduction to Medicinal Chemistry, Wiley-VCH, New York, 1997]

***SA-node** : Sinoatrial node of the heart.

****AV-node** : Atrioventricular node of the heart.

(v) Irregular distances between the different peaks, such as : a *shortened ST-interval*, may be correlated to the particular rhythmic abnormalities. These vital information(s) could help a physician in the correct diagnosis *vis-a-vis* right choice of medicaments.

In short, any disturbance to the conductance of the electrical impulses in a perfect sequential and orderly manner ultimately forms the basis of an **arrythmia.**

Fig. 11.1(*b*) shows the schematic diagram of the cardiac electrical activity. It is, however, pertinent to state here that the electrophysiology of the heart is overwhelmingly governed by the prevailing *transmembrane resting potential.* Hence, the existing potential inside a **Purkinje fibre cell at rest,** with regard to the outside, is found to be almost equal to 90 mV. Interestingly, the potential difference is maintained by an *active transport system* (*i.e.,* a pumping device) which essentially sustains a higher extracellular Na^+ concentration in comparison to the intracellular K^+ level. It has been observed that upon excitation, the prevailing voltage quite rapidly gets reversed to a positive voltage, most probably spiking around + 30 mV. This situation gives rise to an extremely rapid spontaneous and simultaneous movement of Na^+ into the cell, just like a **gate** had all-of-a-sudden opened a channel (usually known as a '*gating mechanism*'). Therefore, the recovery from excitation status gives rise to the gradual restoration of the ensuing '*resting potential*' in various phases from 1 through 4.

Phase-4 : the resting potential is followed by the rapid depolarization and its reversal.

Phase-0 : gets started with a series of three repolarization phases, namely : Phase-1, 2 and 3.

Importantly, depending on the areas measured*, the distinct separation of phases are not quite feasible ; and the prevailing voltages invariably alter amongst the major cell types of the heart, *viz.,* Purkinje fibres**, AV-node, atrial cells, and the SA-node.

There are *two* types of influx, namely : (*a*) rapid influx ; (*b*) slow influx.

A. Rapid Influx. The rapid influx of Na^+ through the channels (or gates) during phase 0 results in the cell a rapid depolarization, which in turn **"closes"** the gate behind them to enable further influx to occur.

B. Slow Influx. The slow influx of Ca^{2+} gets triggered off to equalize (balance) the K^+ loss besides maintaining a proper relative voltage plateau as shown in Fig. 11.1(*b*). In fact, as the Ca^{2+} entry slows down, the membrane potential becomes low very swiftly to the predepolarization levels (*i.e.,* phase 4). Thus, in the heart muscle the **elcetrical activity** is coupled to a **mechanical activity** by Ca^{2+} as the potential trigger.

An effective refractory period (ERP) [See Fig. 11.1(*b*)]. Comprising of several hundred milliseconds follows during which no further stimulus may propagate an impulse effectively. However, the '*impulse initiation*' happens to be an inherent characteristic features of the cardiac fibres which evidently enables them to modulate action potentials almost spontaneously, and hence, the corresponding impulses.

Anti-arrhythmic agents may be classified on the basis of their different pharmacological actions as follows :

(*a*) membrane-stabilizing agents ;

(*b*) antisympathetic drugs ;

Measured—with the aid of intracellular microelectrode recorders ;

**A cardiac muscle cell beneath the endocardium of the ventricles of the heart. These extend from the bundle branches to the ventricular myocardium and form the last part of the cardiac conduction system.

(c) prolonging cardiac action ; and

(d) interference with calcium conductance.

5.1. Membrane-Stabilizing Agents

These drugs are found to interfere directly with depolarization of the cardiac membrane. Quite often they also exhibit local anaesthetic properties.

A few important members of this category are discussed below :

A. *Quinidine Sulphate* BAN, *Quinidine Sulfate* USAN,

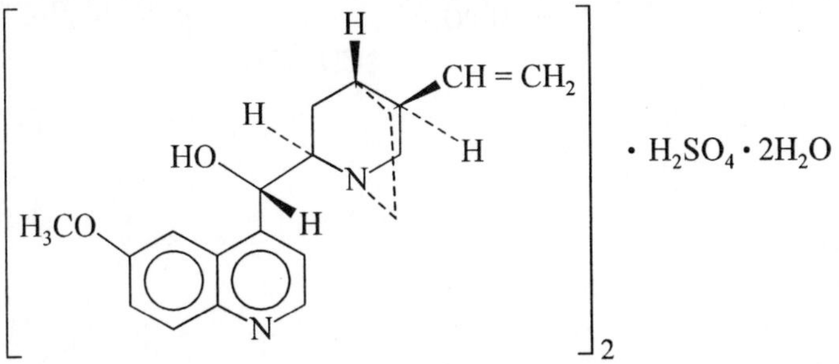

Quinidine sulphate (2 : 1) (salt) dihydrate ; Cinchonan-9-ol, 6'-methoxy-, (9S)-, sulphate (2 : 1) (salt), dihydrate ; BP ; USP ; Eur. P., Int. P., Ind. P. ; SK-Quinidine Sulfate(R) (SK and F) ; Quinidex(R) (Robins)

Preparation

Quinidine sulphate may also be prepared from the mother liquors remaining after separation of quinine from the extracts of *Cinchona* by a known process.

It is very effective in the *suppression of atrial premature contractions and also in shielding the recurrences of atrial fibrillation.* However, it has been found to be effective moderately against ventricular premature systoles and paroxysmal atrial tachycardia.

Dose : Initial, adult, oral, 200 to 800 mg after each 2 or 3 hours up to 5 g on the first day, followed by 100 to 200 mg 3 to 6 times per day.

There are quite a few quinidine-like agents that are used frequently as membrane-stabilizing drugs. For instance

B. *Disopyramide* INN, *Disopyramide Phosphate* BAN, USAN

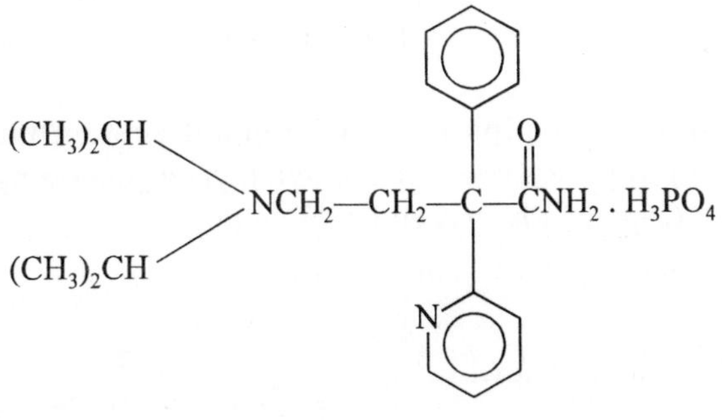

α-[2-(Diisopropylamino)-ethyl]-α-phenyl-2-pyridine-acetamide phosphate (1 : 1) ; 4-Di-isopropylamino-2-phenyl-2-(2-pyridyl) butyramide phosphate (1 : 1) ; BP ; USP ; Norpace(R) (Searle)

Synthesis

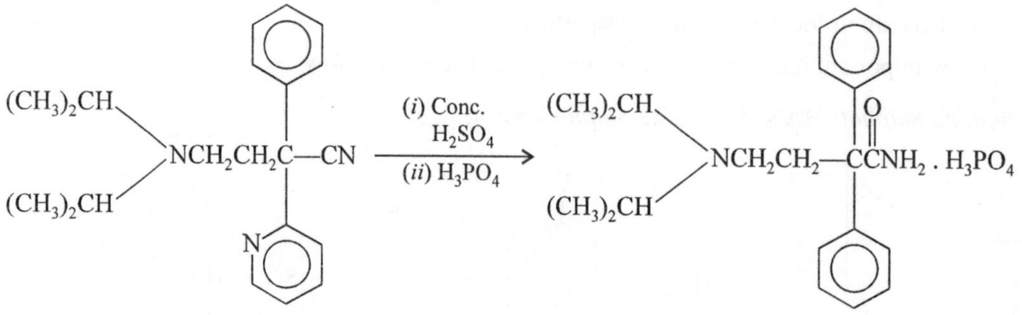

4-Diisopropylamino-
2-phenyl-2-(2-pyridyl)
butyronitrile

Disopyramide phosphate

Conversion of 4-diisopropylamino-2-phenyl-2-(2-pyridyl) butyronitrile to the corresponding amide is caused by heating with concentrated sulphuric acid which on treatment with phosphoric acid yields the official compound. Disopyramide phosphate

Disopyramide phosphate is recommended orally as a prophylaxis of either unifocal or multifocal premature ventricular contractions and ventricular tachycardia. It also exhibits both anticholinergic and local anaesthetic properties.

Dose : Initial, adult, oral, 200 to 300 mg followed by 100 to 200 mg after every 6 hours.

C.   *Lorcainide* INN, *Lorcainide Hydrochloride* BAN, USAN,

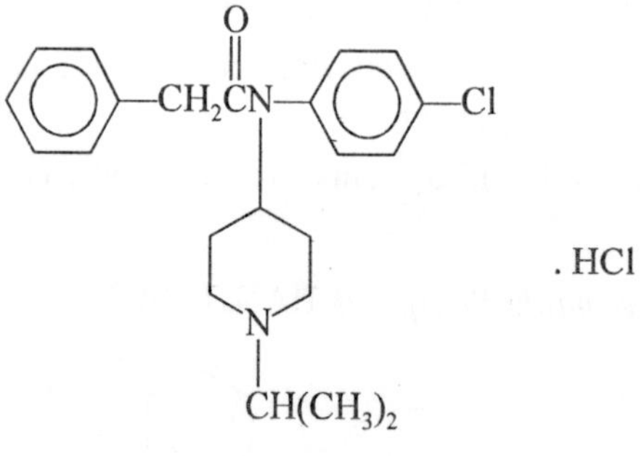

4'-Chloro-N-(1-isopropyl-4-piperidinyl)-2-phenyl-acetanilide monohydrochloride ; Benzeneacetamide, N-(4-chlorophenyl)-N-1-(1-methylethyl)-4-piperidinyl-, monohydrochloride ; Lorcainide Hydrochloride(R) (Janssen Pharmaceutica, Belgium)

It is very effective particularly during *chronic thearpy perhaps due to its inherent high first-pass metabolism orally.*

Dose : 100 mg 2 or 3 times per day.

D. *Procainamide* INN, *Procainamide Hydrochloride* BAN, USAN,

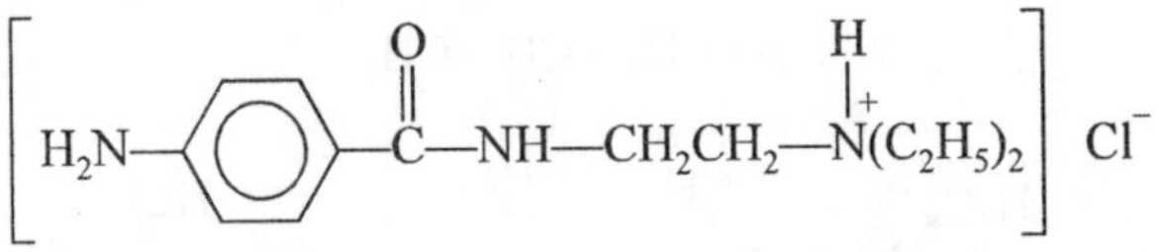

$$\left[H_2N-\underset{}{\bigcirc}-\overset{\overset{O}{\|}}{C}-NH-CH_2CH_2-\overset{\overset{H}{|}}{\underset{+}{N}}(C_2H_5)_2 \right] Cl^-$$

p-Amino-N-[2-diethylamino)-ethyl] benzamide hydrochloride ; Benzamide, 4-amino-N-[2-(diethylamino) ethyl]-, monohydrochloride ; BP ; USP ; Int. P., Ind. P. ; Procan$^{(R)}$ (Parke-Davis) ; Pronestyl$^{(R)}$ (Squibb)

Synthesis

   p-Nitro-N-[2-(diethylamino) ethyl] benzamide is obtained by the interaction of *p*-nitrobenzoyl chloride and β-diethylamino ethylamine. The resulting product on reduction with tin and hydrochloric acid gives the procainamide base which yields the official compound on passing a stream of hydrogen chloride gas.

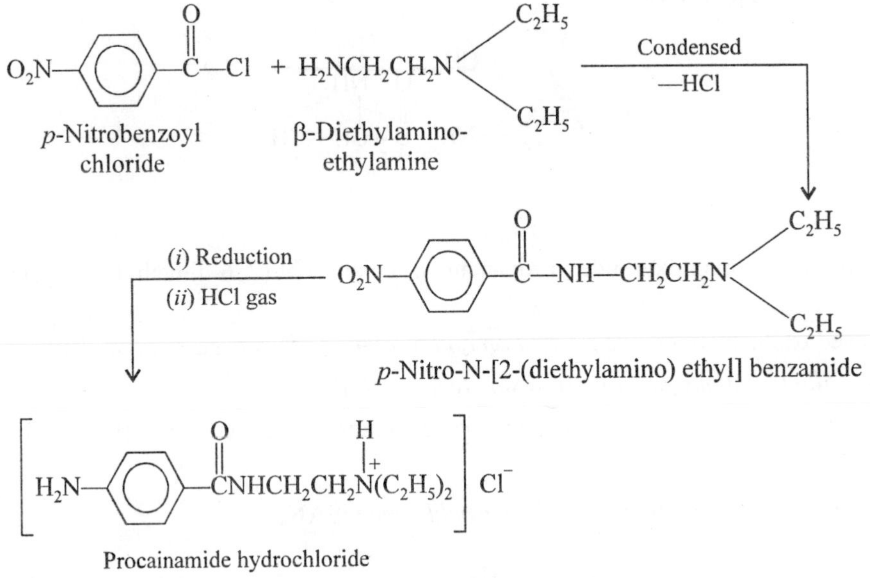

p-Nitrobenzoyl   β-Diethylamino-
   chloride   ethylamine

p-Nitro-N-[2-(diethylamino) ethyl] benzamide

Procainamide hydrochloride

   Procainamide is a class I anti-arrhythmic agent with actions similar to those of quinidine. It depresses myocardial automaticity and excitability and increases the effective refractory period of the atrium. It finds its application *towards the suppression of ventricular extrasystoles and paroxysmal ventricular tachycardia. It is also useful in the control and management of atrial fibrillation and premature atrial contractions.*

   Dose : Initial, adult, oral, for atrial dysrhythmias, 1.25 g, then 750 mg after 1 hour, followed by 500 mg to 1 g every 2 hours as required ; for ventricular dysrhythmias, 1 g followed by 250 to 500 mg every 3 hours.

   Lignocaine has already been discussed under **'local anaesthetics'** and it is also found to possess anti-arrhythmic actions. There are a few other synthetic compounds which exhibit lignocaine-like properties, *e.g.,* mexiletine hydrochloride ; tocainide which shall be discussed as under :

E. *Mexiletine* INN, *Mexiletine Hydrochloride* BAN

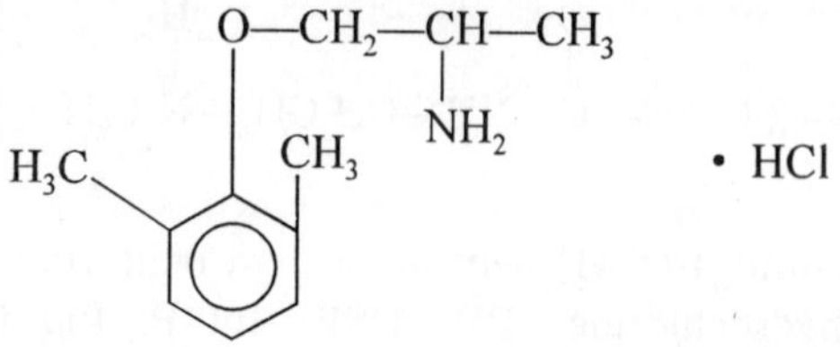

$$O—CH_2—CH—CH_3$$

with NH_2 and H_3C, CH_3 substituents · HCl

1-Methyl-2-(2, 6-xylyloxy) ethylamine hydrochloride ; Mexitil[R] (Boehringer Ingelheim, U.K.)

It belongs to the Class I anti-arrhythmic agent having properties very much alike to those of lignocaine. *Unlike lignocaine it may be administered orally. It is employed for the management and control of ventricular arrhythmias.*

Dose : Initial, oral, 400 to 600 mg, followed by 200 to 250 mg 3 or 4 times per day, starting 2 hours after the loading dose ; maintenance dose 600 to 800 mg daily in divided doses.

F. *Tocainide* INN, BAN, USAN,

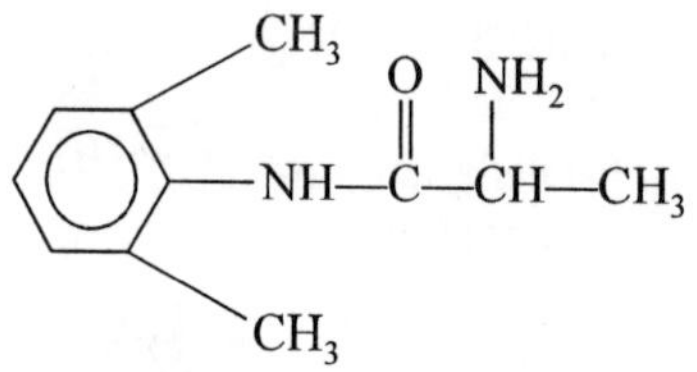

$$—NH—\overset{O}{\underset{\|}{C}}—\overset{NH_2}{\underset{\|}{CH}}—CH_3$$

with two CH_3 groups on the ring

Amino-2', 6'-propionoxylidide ; Propanamide, 2-amino-N-(2, 6-dimethyl-phenyl) ; Tonocard[R] (Merck)

It is used exclusively *for the prevention and treatment of ventricular arrhythmias.*

Dose : 500 to 750 mg administered slowly through IV.

There are a few compounds which may be conveniently grouped together under the miscellaneous category.

G. *Dexpropranolol* INN, *Dexpropranolol Hydrochloride* USAN,

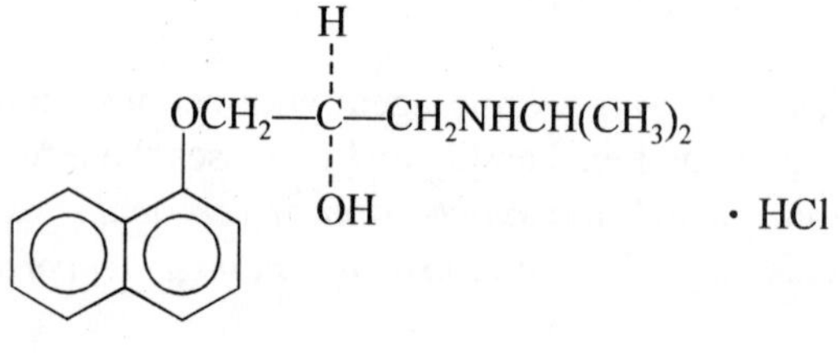

$$OCH_2—\overset{H}{\underset{OH}{C}}—CH_2NHCH(CH_3)_2$$

· HCl

(+)-(R)-1-Isopropylamino-3-(1-naphthyloxy) prcpan-2-ol hydrochloride ; 2-Propanol, 1-[(1-methylethyl)-amino]-3-(1-naphthalenyloxy)-, hydrochloride ; Dexpropranolol Hydrochloride[R] (ICI Pharmaceuticals, U.K.) ;

Besides possessing similar membrane-stabilizing effects like propranolol, it also exerts little beta-adrenoreceptor blocking activity.

5.1.1. Mechanism of Action of Membarne-Stabilizing Agents

The mechanism of action of certain membrane-stabilizing agents are described here under :

5.1.1.1. Quinidine

The bioavailability of quinidine seems to be governed solely on a combination of metabolism and P-gp* efflux. It has been demonstrated that the bioavailabilities of quinidine gluconate and quinidine sulphate are 70-75% and 80-85%, respectively. Interestingly, once the 'drug' gets absorbed, it is subjected to the *hepatic first-pass metabolism* and is found to be plasma-protein bound to the extent of almost 85%, having an elimination half-life of nearly 6 hours. The 'drug' is largely metabolized in the liver, and the renal excretion of the *'unchanged drug'* is also substantially appreciable *i.e.,* nearly 10%-15%. The *two* predominant metabolites are the corresponding hydroxylated derivatives, namely : (*a*) *o*-Demethylquinidine ; and (*b*) Oxydihydroquinoline, as given below :

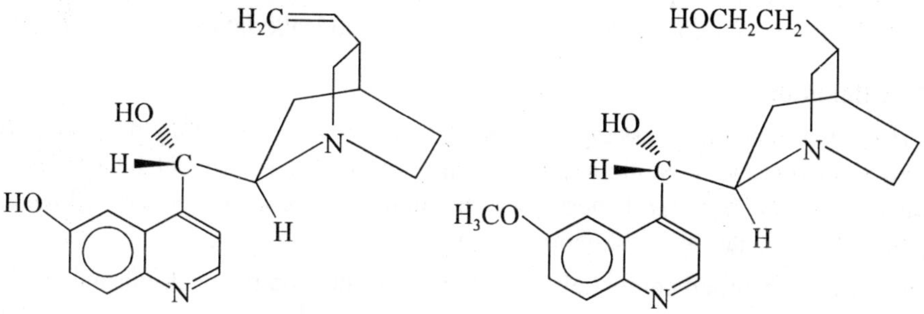

o-Demethylquinidine (*a*)

[Obtained by hydroxylated structural

analogues at the **'Quinoline Ring'**]

Oxydihydroquinidine (*b*)

[Obtained by oxidation of the *'vinyl'*

function' at the **'Quinuclidine Ring'**]

Interestingly, the aforesaid metabolites (*a*) and (*b*) essentially retain only about 33% of the pharmacological activity in comparison to that of *quinidine*. Furthermore, it has been observed that the quinidine, which being a P-gp substrate, invariably checks the renal tubular secretion of digoxin *via* the P-gp efflux pump route, thereby giving rise to an enhanced plasma concentration for digoxin.

5.1.1.2. Disopyramide

Though its antiarrythmic features are quite identical to those of *quinidine* and *procainamide*, yet there exist certain exceptions with regard to its specific antimuscarinic activities which are found to be much more marked and pronounced ; and strategically manifested at the two prominent extracardiac and intracardiac sites.

Salient Features. The various salient features of the 'drug' are :

1. Minimises cardiac automaticity in non nodal cells.

2. Enhances the functional refractory period and minimises the relative refractory period in atrial as well as ventricular cells.

3. Lowers the responsiveness of particularly the myocardial cells to the electrical stimulation.

4. Minimises the ensuing conduction velocity and enhances the stimulus threshold.

*****P-gp Complex** : P-Glycoprotein ;

5. Both at the SA-node and A V-node, its inherent direct myocardial depressant pharmacological actions are adequately opposed by its antimuscarinic property. Hence, at a dosage regimen varying between low to intermediate doses, specfically, it can caused essentially *sinus tachycardia* in certain subjects ; besides, decreasing A V-nodal capability to afford a *second-degree block* of considerably high frequency atrial impulses that eventually pass through directly to the ventricle. Perhaps, that is why such patients that do have particularly *supraventricular tachyarrythmias*, are normally **digitalized** before being treated with disopyramide.

The drug is practically absorbed completely when administered orally, having the bioavailability extending upto 90%. About 50% of the drug gets usually excreted parctically unchanged in the urine having a biological half-life of 5 to 7 hours in subjects with both adequate cardiac output and almost normal renal function. A small fraction $\simeq$ 10% is usually secreted into bile. Its thereapeutic plasma levels normally vary between 2-4 mcg. mL^{-1}. Lastly, it has been demonstrated with adequate evidence that a substantial portion of a dose gets eliminated by N-monodealkylation.

5.1.1.3. Procainamide

The '*drug*' is found to depress myocardial contractility and hence, may produce hypotension ; it must be given to patients very cautiously those who are having a clear cut history of *heart-failure, valvular disease* or *aortic stenosis.* It possesses an antimuscarinic action on the atrioventricular node which may ultimately negate its direct depressant action on that node.

It is practically absorbed completely from the oral route, and the peak-plasma levels are accomplished within a span of 1-2 hours, where it is protein bound to the extent of 15%. Its volume of distribution is approximately 2 mL . g^{-1}. About 50 to 60% of the drug gets eliminated by the renal excretion (along with the tubular secretion). Importantly, it gets metabolized to N-acetylprocainamide, which is an **active metabolite,** and can be accumulated.

5.1.1.4. Tocainide Hydrochloride

It has been established that the total body clearance of tocainide hydrochloride stands at 166 mL. min^{-1} only thereby suggesting that the hepatic clearance is not to an appreciable extent. By virtue of the fact that the drug has very low hepatic clearance, the prevailing hepatic extraction ratio should be quite small ; and hence, the drug is most unlikely to afford a sizable first pass effect. It gets hydrolyzed in a manner very much akin to lidocaine ; and as such its metabolites are **active.**

5.2. Antisympathetic Drugs

The therapeutic agents that possess antisympathetic characteristic features is of the type propranolol.

Propranolol INN, *Propranolol Hydrochloride* BAN, USAN,

1-(Isopropylamino)-3-(1-naphthyloxy)-2-propanol hydrochloride ; 2-Propanol, 1-[(1-methylethyl) amino]-3-(1-naphthalenyloxy-, hydrochloride ; BP ; USP ; Inderal[(R)] (Ayerst)

Synthesis

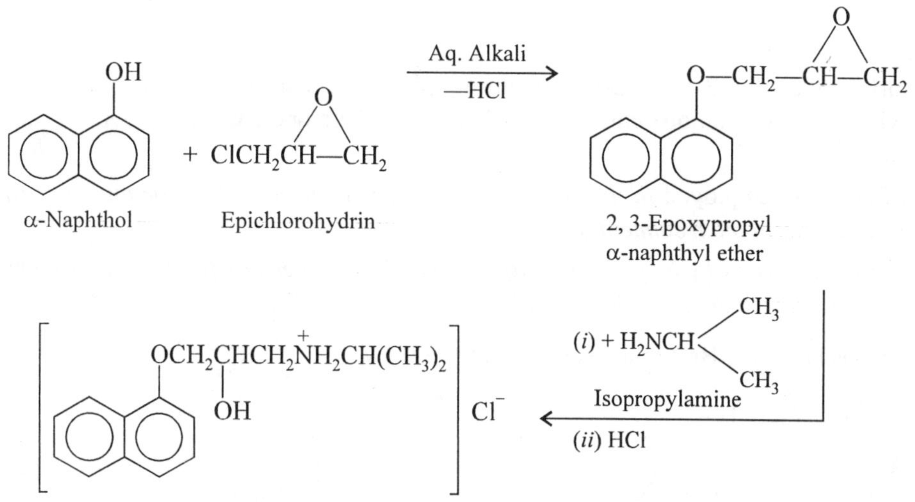

α-Naphthol   Epichlorohydrin   2, 3-Epoxypropyl α-naphthyl ether

Propranolol hydrochloride

The interaction of α-naphthol and epichlorohydrin in the presence of aqueous alkali gives 2, 3-epoxypropyl-α-naphthyl ether which upon treatment with isopropylamine affords the rupture of the epoxy ring thereby forming the propranolol base and this ultimately yields the official compound with HCl.

In addition to its prominent β-adrenergic blocking effect, it does possess independent quinidine-like anti-arrhythmic actions. Hence, it finds its utility *to minimise ventricular and atrial extrasystoles, digitalis-induced tachyarrhythmias, ventricular tachycardia and above all paroxysmal atrial tachycardia.*

Dose : Adult, i.v., for arrhythmias, 1 to 3 mg at a rate not exceeding 1 mg per minute.

5.2.1. Mechanism of Action

The '*drug*' exerts its actions by penetrating into the CNS and thereby causes the central effects. Though the precise mechanism for this specific action has not yet been established, it has been suggested that the β-blockers usually lower blood pressure in one of the following several methods, namely :

    (*i*) a direct effect on the heart and blood vessels,

    (*ii*) minimising sympathetic outflow from the CNS, and

    (*iii*) affecting the renin-angiotensin-aldosterone system.

5.3. Prolonging Cardiac Action

These are a few compounds which specifically prolong the cardiac action potential, such as : amiodarone and bretylium tosylate.

A. Amiodarone *INN, BAN,*

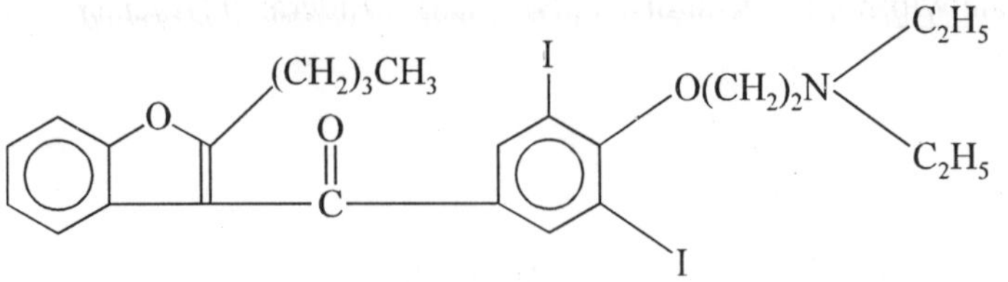

2-Butylbenzofuran-3-yl 4-(2-diethylaminoethoxy)-3, 5-diiodophenyl ketone ; 2-Butyl-3-benzofuranyl 4-[2-(diethylamino) ethoxy]-3, 5-diiodophenyl ketone ; Cardarone X[(R)] (Labaz Sanofi, U.K.)

It is frequently employed *in the control and management of ventricular and supraventricular arrhythmias, and also in the treatment of angina pectoris.*

Dose : Initial (as its hydrochloride salt), 200 mg 3 times per day for a week or more and then reduced to a maintenance dose of 200 mg daily.

B. Bretylium Tosilate *INN,* Bretylium Tosylate *BAN, USAN,*

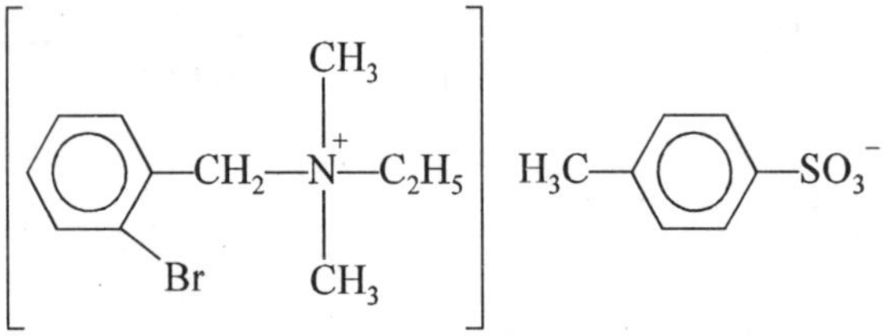

[2-Bromo-N-ethyl]-N, N-dimethylbenzene-methanaminium 4-methyl-benzene sulphonate ; (*o*-Bromobenzyl) ethyldimethyl-ammonium *p*-toluenesulphonate ; Bretylol[(R)] (American Critical Care) ; Bretylate[(R)] (Wellcome, U.K.)

Synthesis

It may be prepared by the interaction of *o*-bromobenzyl bromide and dimethyl ethylamine and teh resulting product on quaternization with *p*-toluenesulphonic acid yields the desired compound.

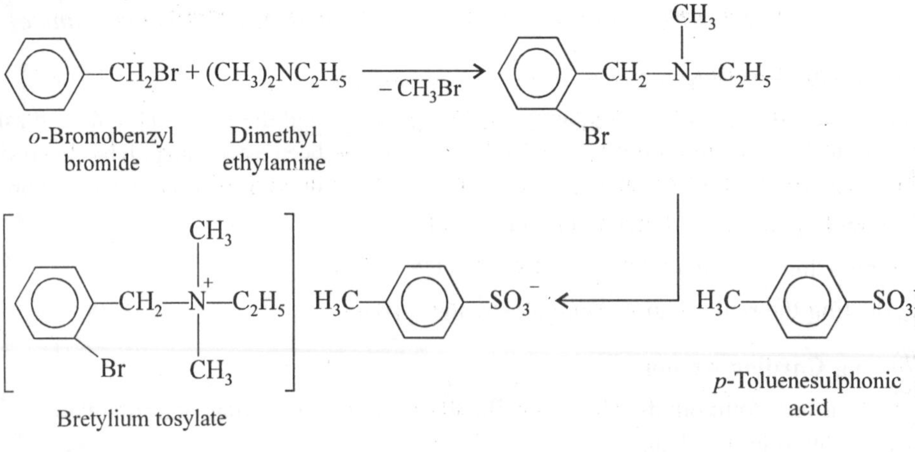

o-Bromobenzyl   Dimethyl
bromide   ethylamine

p-Toluenesulphonic
acid

Bretylium tosylate

Bretylium tosylate is frequently used in the *treatment of ventricular arrhythmias which are refractory to other anti-arrhythmic drugs. It is specifically useful in the diagnosis of ventricular tachycardia.*

Dose : Initial, adult, i.m., 5 to 10 mg per kg, repeated after 1 or 2 hours, then 5 to 10 mg per kg every 6 to 8 hours for maintenance.

5.3.1 Mechanism of Action

The mechanism of action of amiodarone and bretylium tosylate are as stated below :

5.3.1.1. Amiodarone

It is categorized as *'class-III antidysrythmic drug'* that has been exclusively approved for *life-threatening recurrent ventricular arrythmias* which do not respond to other **antiarrythmic drugs.** It has a biological half-life ranging between 25 to 100 days. It inhibits metabolism of many drugs usually cleared by oxidative-microsomal enzymes.

5.3.1.2. Bretylium Tosylate

The *'drug'* is found to afford postural decrease in arterial pressure.* Nowadays, bretylium has been solely reserved for usage in the *ventricular arrythmias* which are observed to be resistant to other therapy. It fails to suppress phase-4-depolarization specifically, which action is generally found quite common in other antiarrhythmic agents. It is also categorized as a **'class-III antiarrythmic agent'** by virtue of the fact that it frequently prolongs the effective refractory period with respect to the action potential duration regardless the effect on conduction time.

However, the exact mechanism of antiarrhythmic action is still remain to be resolved.

5.4. Interference with Calcium Conductance

This particular class of compounds directly interfere with calcium conductance, such as verapamil hydrochloride.

Verapamil INN, *Verapamil Hydrochloride* BAN, USAN,

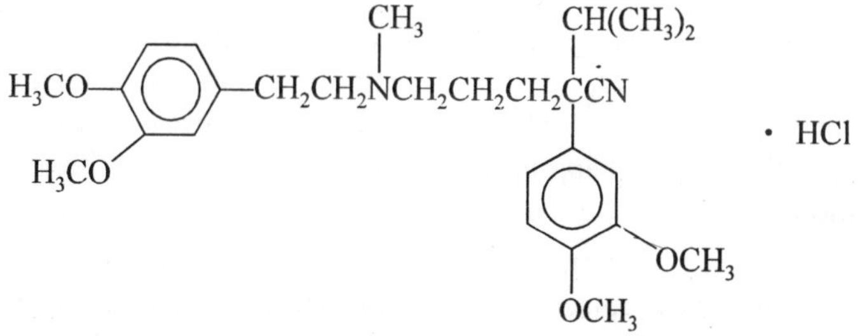

5-[N-3, 4-Dimethoxyphenethyl)-N-methylamino]-2-(3, 4-dimethoxy-phenyl)-2-isopropylvale-ronitrile monohydrochloride ; Benzene-acetonitrile, α-[3-[[2-(3, 4-dimethoxyphenyl)-ethyl] methylamino] propyl]-3, 4-dimethoxy, [-α-(1-methylethyl)-. monohydrochloride ; BP ; Calan[R] (Searle) ; Isoptin[R] (Knoll)

Verapamil hydrochloride is mostly advocated in the *control and management of supraventricular arrhythmias and angina pectoris. It also finds its usefulness in the treatment of supraventircular tachycardias.*

Dose : Initial, oral, 40 to 120 mg 3 times per day according to the severity of the condition of the patient and his response.

---

*Nadermanee K *et al. Circulation,* **66** : 202, 1982

5.4.1. Mechanism of Action

The plasma half-life of this **'drug'** is unable to predict most precisely the actual duration of action on account of the presence of **active metabolites.** As it possesses an appreciably higher *first-pass metabolism*, nearly 100% of the *'drug'* gets ultimately excreted in the urine as its corresponding metabolites.

In general, the anti-arrhythmic drugs may be categorized judiciously into *four* distinct classes I through IV that may be described as given below :

Class I anti-arrhythmic drugs normally prolong the refractory period of cardiac muscle, reduce its excitability, and above all minimise the conduction velocity for example, quinidine sulphate, disopyramide phosphate, tocainide hydrochloride, procainamide hydrochloride, mexiletine hydrochloride, aprindine, etc.

Class II anti-arrhythmic drugs essentially improve myocardial cell responsiveness such as propranolol hydrochloride.

Class III anti-arrhythmic drugs usually prolong the cardiac action, for instance : amiodarone and bretylium tosylate. *The latter depresses automaticity and increases the threshold with respect to fibrillation-inducing electrical stimulation in either normal or infarcted myocardium. It has also been found to enhance the functional refractory period thereby shortening the relative refractory period, thus the overall effect being to discourage re-entry.*

Class IV anti-arrhythmic drugs usually interfere with calcium conductance such as verapamil hydrochloride. Verapamil inhibits the action potential of the upper and middle nodal regions of the heart where the slow inward calcium-ion-mediated current contributes to depolarisation. This is responsible for the blockade of slow-channel conduction in the atrioventricular node. It has been found to inhibit one limb of the re-entry circuit which is assumed to underlie most paroxysmal supraventricular tachycardias, thereby causing the reduction of ventricular rate in atrial flutter and fibrillation.

6. VASOPRESSOR DRUGS

Vasopressor drugs (vasodilators) are chiefly those employed for angina pectoris, cerebral or peripheral vascular disorders. Besides, there are a plethora of *'drug substances'* belonging to certain other classes, treated elsewhere in this text, do also exert *vasoconstrictor* as well as cardiostimulator activity which may be judiciously and appropriately employed to increase the prevailing blood pressure under suitable conditions.

In typical physiological conditions wherein the *'plasma volume'* gets diminished sharply, as in *hypovolemic shock*, the immediate replacement tends to restore the blood pressure. The use of *'plasma-extenders'* in such conditions to restore the blood pressure are **not** regarded as the true **'vasopressor drugs'** for obvious reasons as they fail to affect vasoconstriction.

Over the years, there is much substantial evidence that explicitly prove that vasoconstriction particularly enhances the **ischemic damage** already caused on account of the improper and inadequate blood circulation prevailing in the body. As a result of these startling findings, the main emphasis got ligitimately shifted to the α-*adrenerging blocking drugs* in 1960s along with a host of *vasodilators*, and importantly to the *cardiostimulants.*

A few important members of this category shall be discussed in this section, such as : Buphenine, Isoxsupurine and Prenylamine.

A. Buphenine *INN,* Buphenine Hydrochloride *BAN,* Nylidrin Hydrochloride *USAN,*

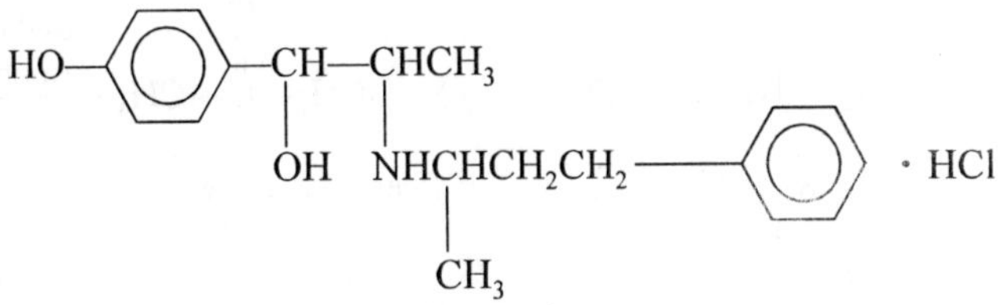

1-(4-Hydroxyphenyl)-2-(1-methyl-3-phenylpropylamino) propan-1-ol-hydro-chloride ; *p*-Hydroxy-α-[1-[(1-methyl-3-phenylpropyl) amino] ethyl] benzyl alcohol hydrochloride ; USP ; Arlidin[R] (USV Pharmaceutical)

Buphenine is used in the *treatment of peripheral vascular disease. It has also been employed in the treatment of Meniere's disease and similar disorders of the internal ear.*

Dose : Usual, initial, oral 6 mg thrice daily, which may be enhanced to 36 or 48 mg per day in divided doses.

B. Isoxsupurine *INN,* Isoxsupurine Hydrochloride *BAN, USAN,*

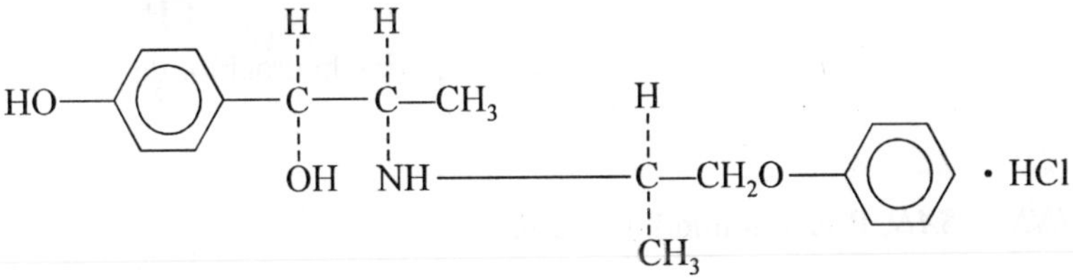

1-(4-Hydroxyphenyl)-2-(1-methyl-2-phenoxy-ethylamino) propan-1-ol hydrochloride ; Benzenemethanol, 4-hydroxy-α-[1-[(1-methyl-2-phenoxyethyl)-amino] ethyl]-, hydrochloride, stereo-isomer ; BP ; USP ; Vasodilan[R] (Mead Johnson)

Synthesis

Sodium phenoxide and chloroacetone reacts to give a mole of phenoxyacetone with the elimination of a mole of NaCl, which on reductive amination in the presence of ammonia and hydrogen yields an amine. This amine on treatment with *p*-hydroxy-propiophenone gives an intermediate that on further treatment with sodium borohydride yields isoxupurine which when treated with hydrochloric acid affords the official compound.

It is used in the *treatment of cerebral and peripheral vascular disease.*

Dose : Oral, 20 mg 4 times per day ; i.v. infusion as a solution containing 100 mg in 500 ml of sodium chloride solution.

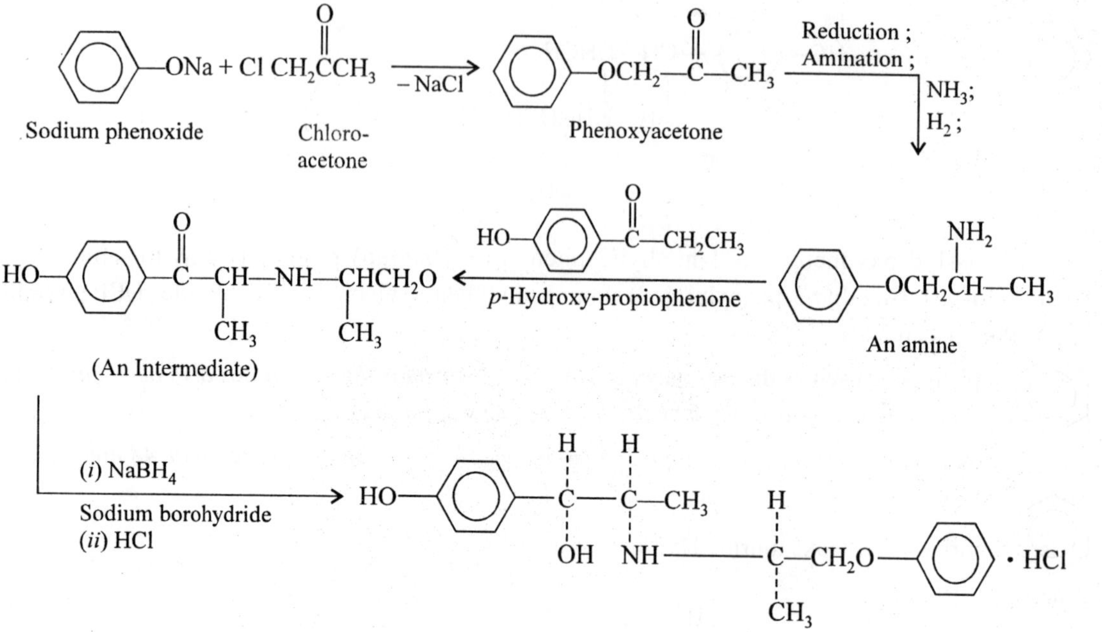

Sodium phenoxide   Chloro-acetone   Phenoxyacetone

(An Intermediate)   *p*-Hydroxy-propiophenone   An amine

Isoxsupurine hydrochloride

C. Prenylamine *INN, USAN*, Prenylamine Lactate *BAN,*

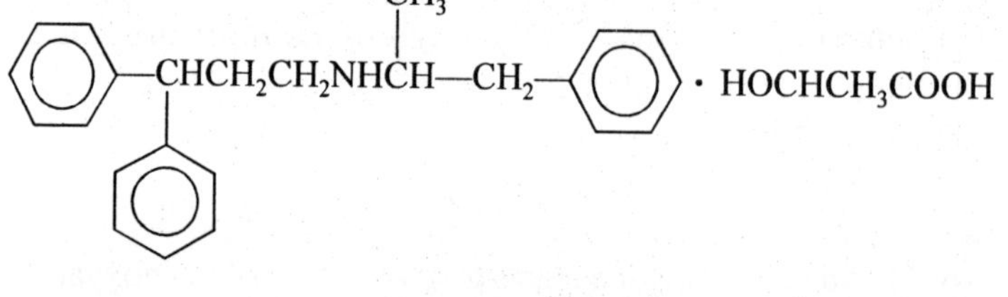

N-(2-Benzhydrylethyl)-α-methylphenethylamine lactate ; Benzenepropanamine, N-(1-methyl-2-phenylethyl)-γ-phenyl- ; BP ; Synadrin[R] (Hoechst, U.K.)

Synthesis

Benzaldehyde and malonitrile undergoes Knoevnagel condensation to give an unsaturated cyanoester which on reaction with phenyl magnesium bromide yields an intermediate. This on hydrolysis and subsequent decarboxylation yields a nitrile which on reduction gives an amine. The resulting amine on alkylation with 2-chloro-1-phenylpropane yields the prenylamine base which on treatment with lactic acid affords the official product.

It is used prophylactically in the *treatment of angina pectoris.*

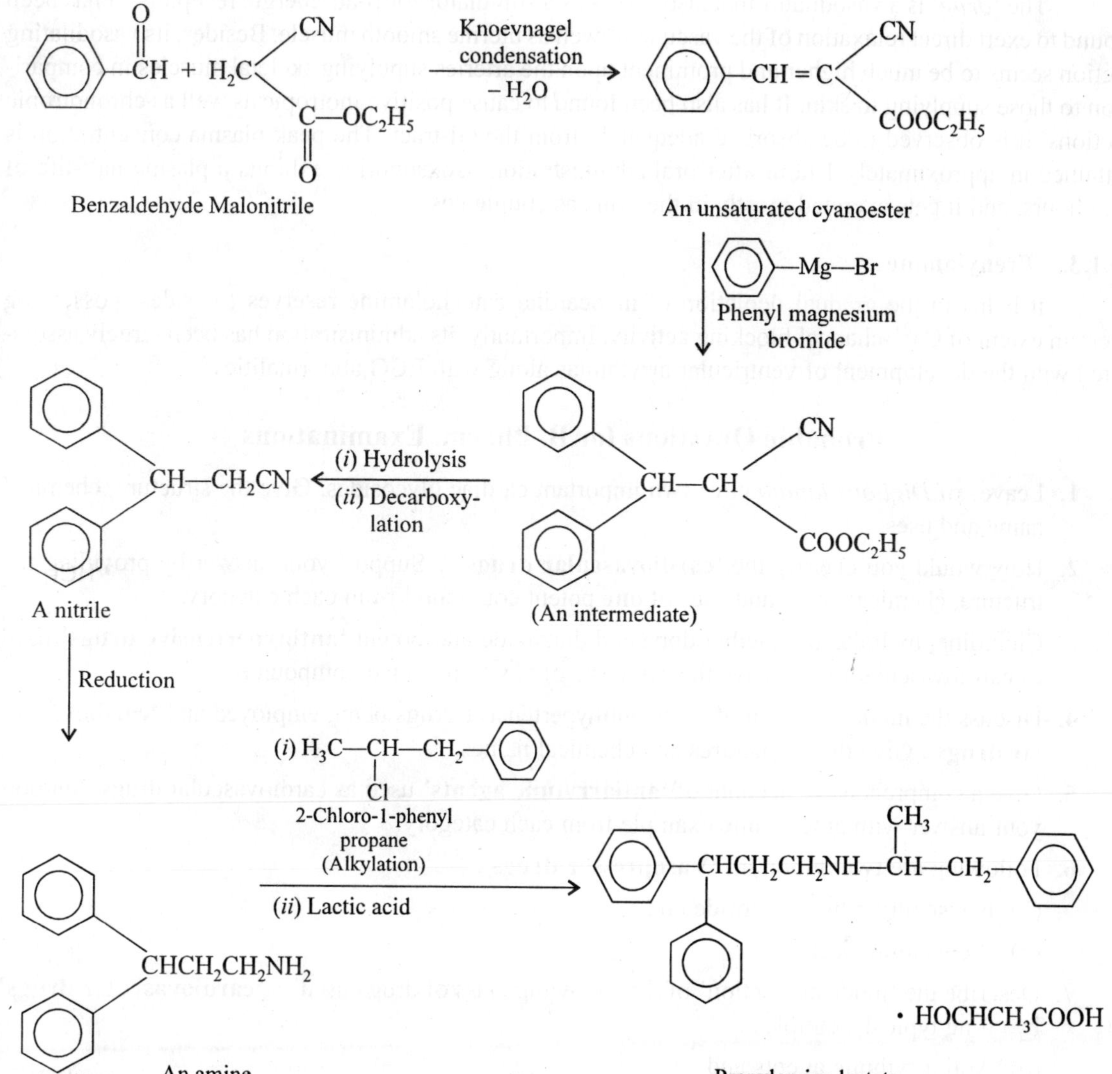

Benzaldehyde Malonitrile

An unsaturated cyanoester

Phenyl magnesium bromide

A nitrile

(An intermediate)

Reduction

2-Chloro-1-phenyl propane (Alkylation)

(ii) Lactic acid

An amine

Prenylamine lactate

Dose : Usual, initial, 180 mg (of base) per day in 3 divided doses.

6.1. Mechanism of Action

The mechanism of action of the aforesaid *'drug substances'* shall be described in the section that follows :

6.1.1. Buphenine (Nylidrin Hydrochloride)

It exhibits β_2-activity more predominantly for the skeletal muscle. It has been used extensively for the treatment of skeletal muscle with a vasospastic component.

6.1.2. Isoxsupurine

The *'drug'* is a vasodilator that also serves as a stimulator for β-adrenergic receptors. It has been found to exert direct relaxation of the vascular as well as uterine smooth muscle. Besides, its vasodilating action seems to be much higher and prominent upon the arteries supplying skeletal muscles in comparison to those supplying to skin. It has also been found to cause positive inotropic as well as chronotropic actions. It is observed to be absorbed adequately from the GI-tract. The peak plasma concentration is attained in approximately 1 hour after oral administration. Isoxsupurine exhibits a plasma half-life of 1.5 hours, and it gets excreted mostly in the urine as conjugates.

6.1.3. Prenylamine

It helps in the gradual depletion of myocardial catecholamine reserves ; besides, possessing certain extent of Ca^{2+}-channel blocking activity. Importantly, its administration has been largely associated with the development of ventricular arrythmias along with ECG abnormalities.

Probable Questions for B. Pharm. Examinations

1. Leaves of *Digitalis lanata* gave **two** important cardiac glycosides. Give the structure, chemical name and uses.

2. How would you classify the **'cardiovascular drugs'** ? Support your answer by providing the tructure, chemical name and uses of **one** potent compound from each category.

3. Clonidine, hydralazine, methyl dopa and diazoxide and potent **'antihypertensive drugs'** used as cardiovascular drugs. Give the synthesis of any **two** named compounds.

4. Discuss the mode of action of some antihypertensive drugs being employed and **'cardiovascular drugs'**. Give their structures and chemical names.

5. Give a comprehensive account of **'antiarrytmic agents'** used as cardiovascular drugs. Support your answer with at least **one** example from each category.

6. Following are **two** important **'Vasopressor drugs'**.
   (*a*) Isoxsupurine hydrochloride and
   (*b*) Prenylamine lactate.

7. Describe the 'mode and action' of the following class of drugs used as **'cardiovascular drugs'** by citing typical examples :
   (*a*) Anti-arrythmic agents and
   (*b*) Vasopressor drugs

8. Discuss the synthesis of any **one** of the following membrane-stabilizing agents usually employed as **'cardiovasculawr drugs'**.
   (*a*) Disopyramide phosphate
   (*b*) Procainamide hydrochloride

9. Discuss the Antisympathetic drugs **Propranolol Hydrochloride**. Give its synthesis from alpha naphthol and epichlorohydrin.

10. Discuss the synthesis of Verapamil hydrochloride and describe its mode of action.

RECOMMENDED READINGS

1. NV Costrini and W M Thomson *Manual of Medical Therapeutics* (22nd edn) Little Brown and Co. Boston (1977).

2. ME Wolff (*Ed*) : *Burger's Medicinal Chemistry and Drug Discovery* (5th edn), John Wiley & Sons Inc., New York, (1995).

3. LS Goodman and A Gilman *The Pharmacological Basis of Therapeutics* (9th edn) Macmillan Co. London (1995).

4. WO Foye (*Ed*) *Principles of Medicinal Chemistry* Lea and Febiger, Philadelphia (1989).

5. DLednicer and L A Mitscher *The Organic Chemistry of Drug Synthesis* John Wiley and Sons, New York (1995).

6. EF Reynolds (*Ed*) *Martindale the Extra Pharmacopoeia* (30th edn) The Pharmaceutical Press London (1992).

7. MC Griffiths (*Ed*) *USAN and the USP Dictionary of Drug Names* United States Pharmacopeial Convention Inc. Rockville (1985).

12 Autonomic Drugs

1. INTRODUCTION

The drugs which act on the **'autonomic nervous system'** (ANS) and control the vital internal processes which ordinarily, are not under volition, are known as **autonomic drugs.**

Noradrenaline is released particularly at the target organs and ultimately gives rise to the ensuing contraction of the cardiac muscles ; and ultimately an enhancement in the heart rate. Besides, it is also helpful in relaxing the smooth muscles and thereby causes a visible reduction in the contractions of the GI-tract as well as the urinary tract. It also minimises a distinct reduction in salivation and lowers dilatation of the peripheral blood vessels particularly.

Broadly speaking, the sympathetic (autonomic) nervous system aids categorically the *'fight or flight'* response by closing down the body's housekeeping mechanisms, such as : *digestion, defecation, urination* and the like, and thereby stimulating the heart ultimately. It has been observed that the '*adrenal medulla*' eventually helps in the release of the hormone *adrenaline*, that reinforces the action of noradrenaline finally.

It is, however, pertinent to mention here that the effects of autonomic (sympathetic) nervous system activation and, therefore, the effects of sympathomimetic drugs are evaluated mostly by the specific type and ultimately the localization of the post synaptic receptor to which the released neurotransmitter or exogenous sympathomimetic binds finally.

ANS is comprised of *two* divisions, namely : *sympathetic* and *parasympathetic.* Acetylcholine (ACh) invariably acts as a neurotransmitter at both sympathetic and parasympathetic nerve endings, postganglionic nerve fibers in the parasympathetic zone, and certain postganglionic fibers, such as : salivary and sweat glands, in the sympathetic division of ANS. Interestingly, ANS modulates the different types of activities of both the smooth muscle and the glandular secretions. Generally, all these activities, as a rule, exert their actions much below the level of consciousness, for instance ; *circulation, respiration, body temperature, digestion* and *metabolism.* It is worth while to mention here that the said two functional divisions afford almost contrasting effects upon the internal environment of the human body. Explicitly, the sympathetic division invariably shows its effect as a unit, particularly during conditions of fit, anger or shock (fright), and hence, expends energy. Interestingly, the parasympathetic division is organized for its absolutely localized and discrete discharge and, therefore, not only conserves but also stores energy as a potential reserve in the body.

2. CLASSIFICATION

The autonomic drugs may be classified into the following categories, namely :

(*a*) Sympathomimetic Drugs

(*b*) Antiadrenergic Drugs

(*c*) Cholinomimetic Drugs

(*d*) Antimuscarinic Drugs

(*e*) Ganglionic Blocking Agents

(*f*) Adrenergic Neurone Blocking Agents.

The above categories of autonomic drugs have been treated separately, including typical examples from each group.

3. SYMPATHOMIMETIC DRUGS

Sympathomimetic drugs usually mimic stimulation of the peripheral endings of the sympathetic or 'adrenergic' nerves, the action being exerted on the effector cells supplied by postganglionic endings. There is now enough evidence to show that the neurohormone directly concerned with such an action is noradrenaline.

It is, however, interesting to observe that a good number of sympathomimetics in fact do not really mimic the actions of noradrenaline or adrenaline at the effector receptor. They merely induce the release of noradrenaline from the sympathetic postganglionic adrenergic nerves. Such sympathomimetics which exert their action indirectly are comparatively less effective in patients treated with noradrenaline depleting drugs, for instance, the rauwolfia alkaloids, or other adrenergic neuron blockers.

A few important compounds used as *sympathomimetic drugs* are discussed below : ephedrine : epinephrine ; adrenaline ; isoprenaline ; methoxamine hydrochloride ; metarminol bitartrate ; naphazoline hydrochloride ; oxymetazoline hydrochloride ; phenylpropanolamine hydrochloride ; xylometazoline hydrochloride, etc.

A. *Ephedrine* BAN, USAN,

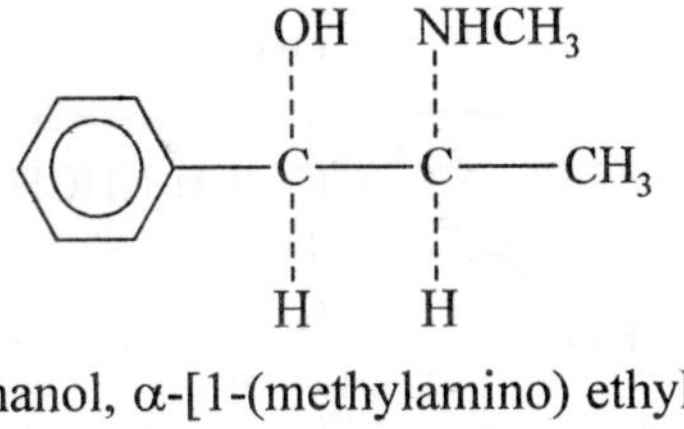

(-)-Ephedrine ; Benzenemethanol, α-[1-(methylamino) ethyl)]-*R*-(R^*, S^*)- ; BP ; USP ; Eur. P., Ind. P.

Synthesis

Nagai first isolated ephedrine in 1887 from a well-known Chinese herb, *ma huang* ; by moistening the powdered drug with either aqueous sodium carbonate or with lime water and subsequently extracting it with ethanol or benzene.

Neuberg's synthetic method is the most ideal one for the commercial production of ephedrine :

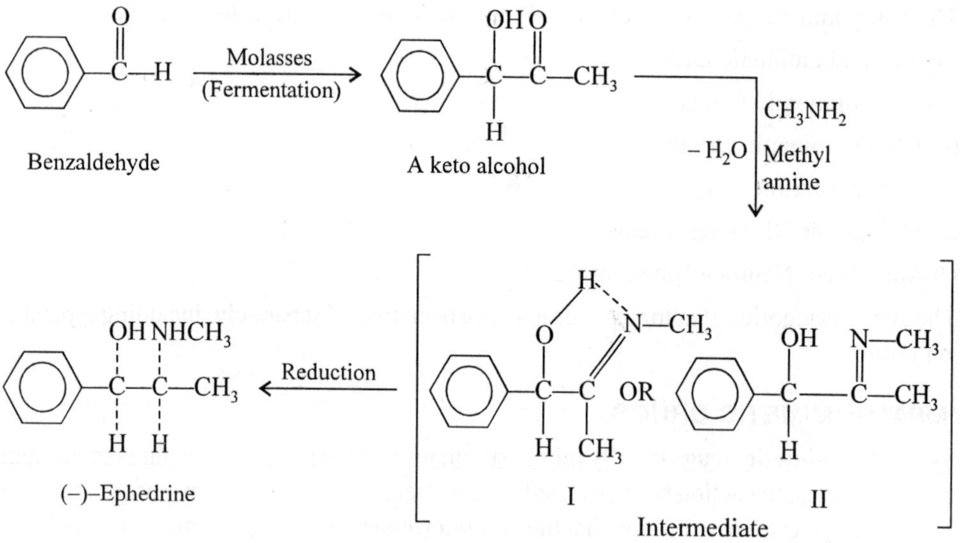

Benzaldehyde → Molasses (Fermentation) → A keto alcohol → CH₃NH₂ Methyl amine, −H₂O → Intermediate I or II → Reduction → (−)-Ephedrine

Fermentation of benzaldehyde with either molasses or a mixture of glucose and yeast yields a keto alcohol which on reaction with methyl amine gives rise to an intermediate which may be depicted either as I or II with the loss of a mole of water. This on reduction yields the official compounds.

Ephedrine helps to increase the blood pressure at a therapeutic dose level in *two* different ways ; first by enhancing peripheral vasoconstriction. It is found to exert a stimulant action on the respiratory centre. It shows a wide spectrum of actions, namely : *reduction in the activity of the uterus, bronchodilatation and lowering of intestinal tone and motility. Ephedrine is also used in postural hypotension and in subjects having more or less complete heart block.*

Dose : 10 to 25 mg every 3 to 4 hours.

B. *Epinephrine* INN, USAN ; *Adrenaline* BAN,

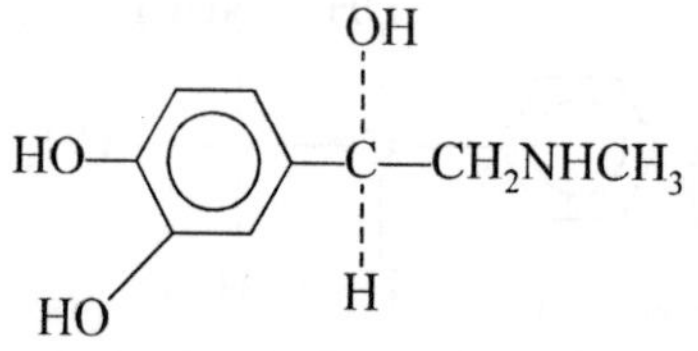

(-)-3, 4-Dihydroxy-α-[(methylamino) methyl] benzyl alcohol ; 1, 2-Benzenediol, 4-[1-hydroxy-2-(methylamino) ethyl]-, (*R*)- ; Epinephrine (USP) ; Adrenaline BP ; Int. P., Ind. P., ; Adrenaline[R] (Parke-Davis).

Synthesis

Abel and Fuerth isolated independently the active principle from the *suprarenal glands* which were later on established as the same compound now known as **adrenaline.** It may be synthesized conveniently by several methods, but a very common process is described on next page :

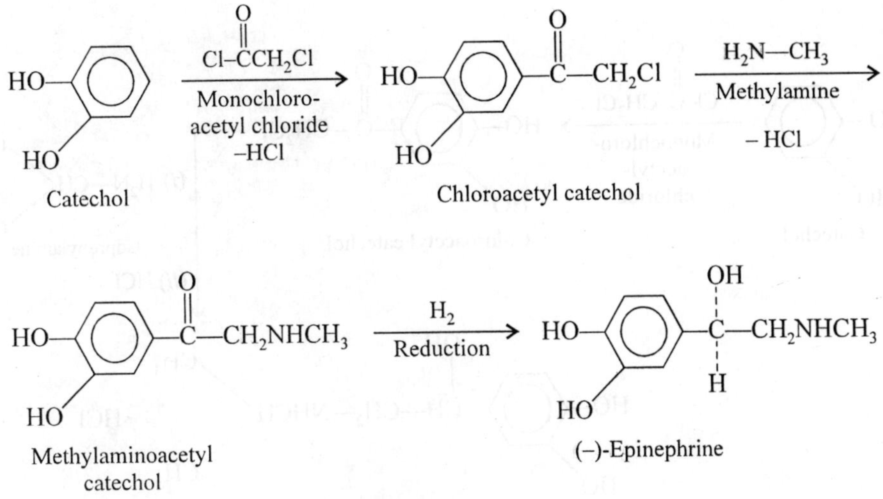

Catechol — Chloroacetyl catechol

Methylaminoacetyl catechol — (−)-Epinephrine

The interaction between catechol and monochloracetyl chloride gives chloroacetyl catechol with the elimination of a mole of hydrogen chloride which on subsequent reaction with methylamine yields methylaminoacetyl catechol. This on reduction gives rise to racemic epinephrine, which may be re-solved conveniently with *d*-tartaric acid.

Epinephrine, the sympathomimetic adrenal hormone, is found to act on smooth muscles, heart and the gland cells thereby causing a similar pattern of actions as may have been produced by the stimulation of the respective adrenergic nerves. Hence, it had been invariably *used to stimulate the heart, enhances the heart rate, tones up the blood pressure and above all affords relaxation of the musculature of the intestine and bronchi.* It is usually the drug of choice in acute allergic disorders and histamine reactions.

Dose : Subcutaneous, 0.2 to 0.5 mg in 0.1% solution ; intramuscular, 1 to 3 mg in a 0.2% oil suspension, repeated as required.

C. *Isoprenaline* INN, BAN, *Isoproterenol Hydrochloride* USAN,

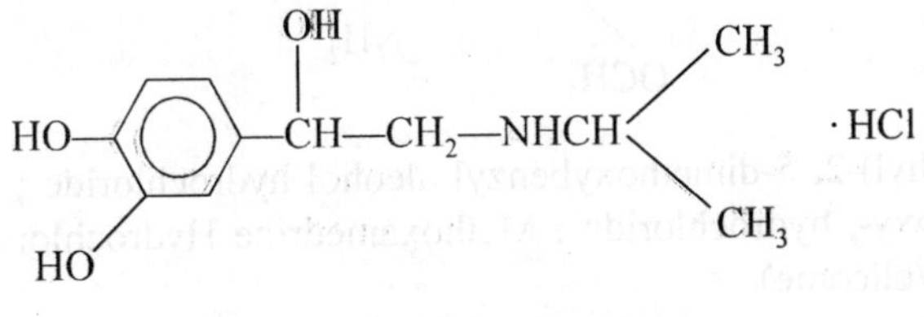

3, 4-Dihydroxy-α-[(isopropylamino) methyl] benzyl alcohol hydrochloride ; 1, 2-Benzenediol, 4-[1-hydroxy-2-[(methyl-ethyl)-amino] ethyl]-, hydrochloride ; Isoprenaline Hydrochloride BP ; Int. P., Isoproterenol Hydrochloride USP ; Norisodrine Aerotrol[R] (Abbott) ; Vapo-Iso[R] (Fisons) ;

Synthesis

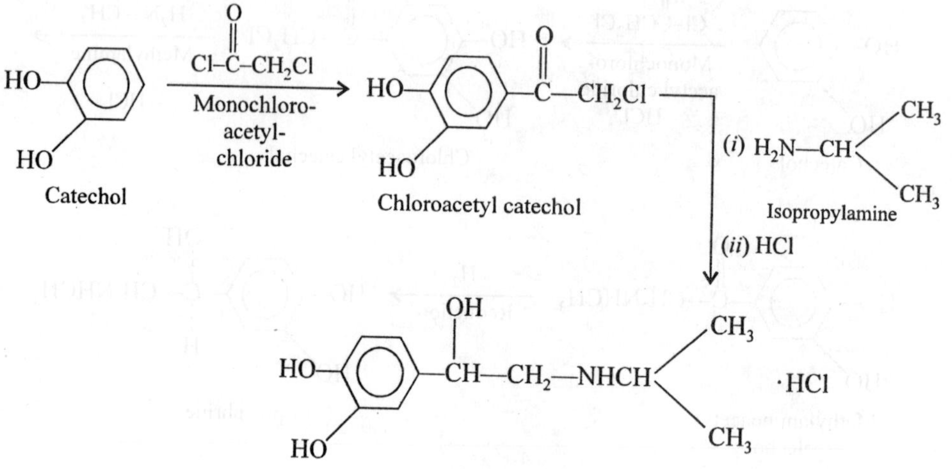

Isoproterenol hydrochloride

Chloroacetyl catechol may be prepared by the interaction of catechol and monochloroacetyl chloride, which on reaction with isopropylamine and hydrochloric acid yields the official product.

It exerts a stimulating action on the heart thereby enhancing cardiac output, rate and excitability. It is also *employed in the management and treatment of bradycardia, as a stimulant following cardiac arrest and prevention for attacks of Strokes-Adams Syndrome. It is an effective bronchodilator in asthma.*

Dose : Sublingual, 10 to 15 mg 3 to 4 times daily ; intramuscular or subcutaneous, 0.01 to 0.2 mg ; repeated as necessary ; infusion, 1 to 2 mg per 500 ml of 5% dextrose infusion at such a rate so as to maintain blood pressure.

D. *Methoxamine* INN, BAN, *Methoxamine Hydrochloride* USAN,

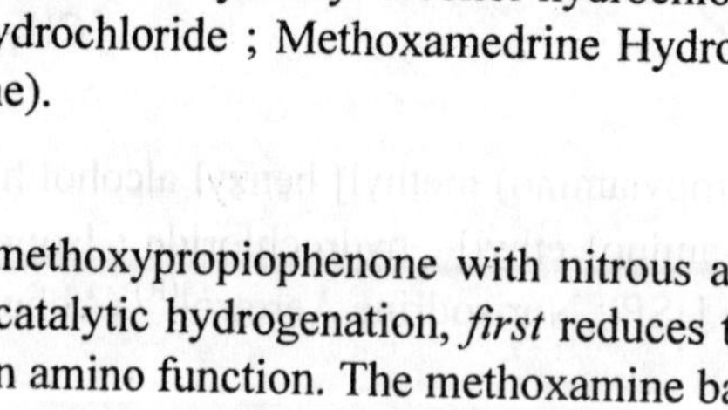

(±)-α-(1-Aminoethyl)-2, 5-dimethoxybenzyl alcohol hydrochloride ; Benzene-methanol, α-(1-aminoethyl)-2, 5-dimethoxy-, hydrochloride ; Methoxamedrine Hydrochloride : BP ; USP ; Int. P. ; Vasoxyl[(R)] (Burroughs Wellcome).

Synthesis

The reaction of 2′, 5′-dimethoxypropiophenone with nitrous acid gives the corresponding 2-isonitroso derivative, which on catalytic hydrogenation, *first* reduces the keto function, and *secondly* converts the nitroso group into an amino function. The methoxamine base when dissolved in an appropriate solvent and subjected to a stream of hydrogen chloride gas yields the official compound.

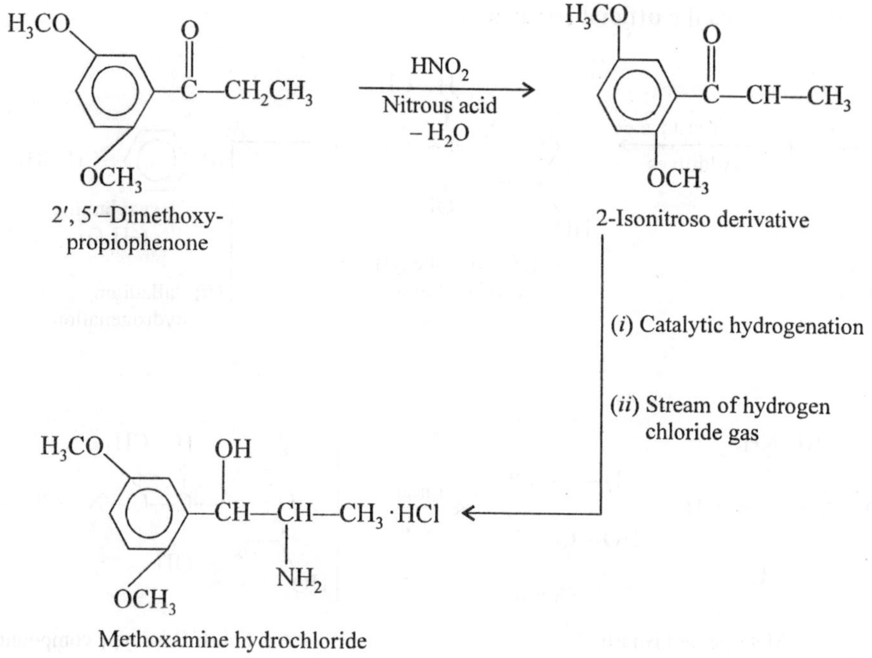

2', 5'–Dimethoxy-
propiophenone

$\xrightarrow[\substack{\text{Nitrous acid} \\ -H_2O}]{HNO_2}$

2-Isonitroso derivative

(*i*) Catalytic hydrogenation

(*ii*) Stream of hydrogen
chloride gas

Methoxamine hydrochloride

Methoxamine hydrochloride affords an increase in arterial blood perssure through peripheral vasoconstriction. It is the drug of choice for the treatment of hypotensive conditions when it is required to boost-up blood pressure without any cardiac stimulation.

Dose : Usual, intramuscular, 5 to 20 mg ; intravenous, 3 to 10 mg administered slowly in divided doses.

E. *Metaraminol* INN, BAN, *Metaraminol Bitartrate* USAN,

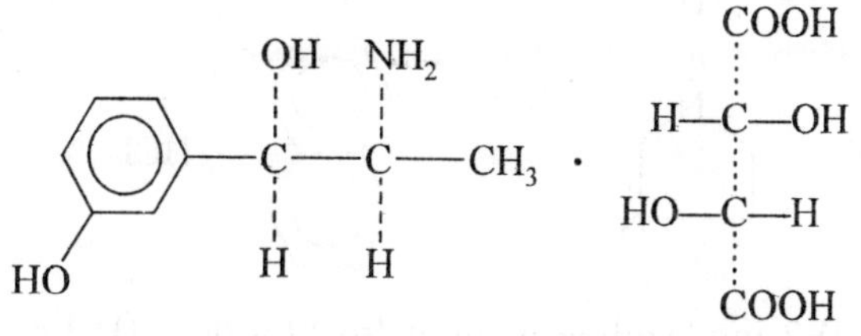

(-)-α-(1-Aminoethyl)-*m*-hydroxybenzyl alcohol tartrate (1 : 1) ; 1-*m*-Hydroxy-norphedrine Bitartrate ; Metaraminol Tartrate B.P., Metaraminol Bitartrate U.S.P. ; Aramine[R] (Merck)

Synthesis

m-(Hydroxyphenyl) acetyl carbinol is obtained by the fermentation of *m*-hydroxy benzaldehyde in the presence of yeast and a few additives, which on subsequent treatment with benzyl amine followed by hydrogenation with palladium yields an intermediate which is considered to be a *Schiff's Base type compound.* This product undergoes cleavage at $NH-CH_2$, the hydrogenated residue of C—N linkage, to result the desired metaraminol base which on treatment with an equimolar proportion of tartaric acid in an alcoholic medium gives the official compound.

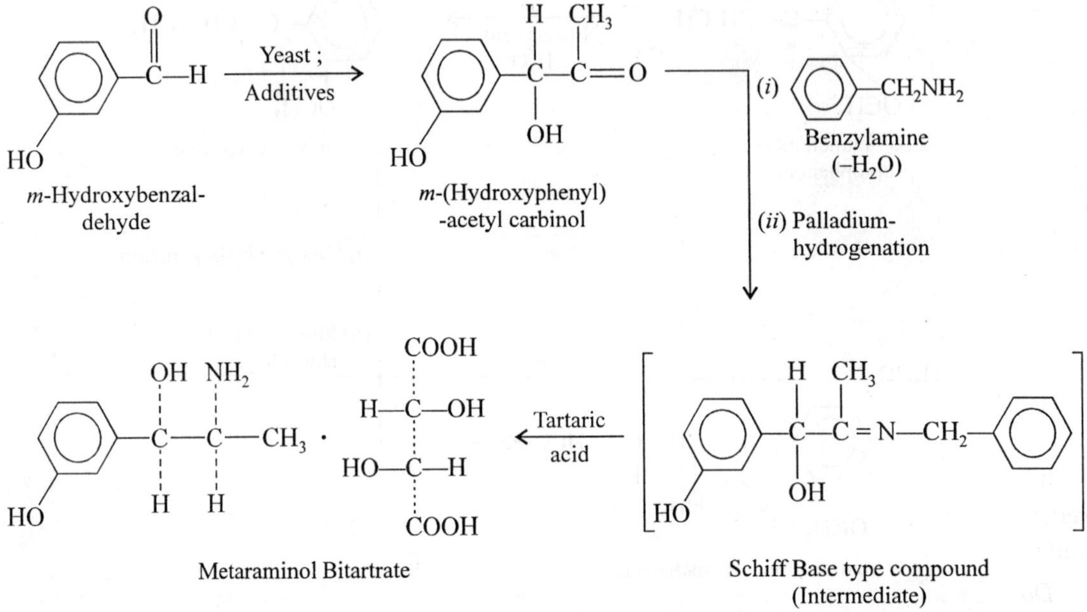

It enhances cardiac output, peripheral resistance, and blood pressure. It helps to increase the coronary blood flow thereby decreasing the heart-rate. The drug is *employed frequently in actue hypotensive states such as anaphylactic shock or shock secondary to myocardial infraction and trauma.*

    Dose : Intravenous, 0.5 to 5 mg in an emergency ; by infusion, 15 to 100 mg/500 ml of dextrose injection or sodium chloride injection ; intramuscular, 2 to 12 mg.

F. *Naphazoline* INN, BAN, *Naphazoline Hydrochloride* USAN,

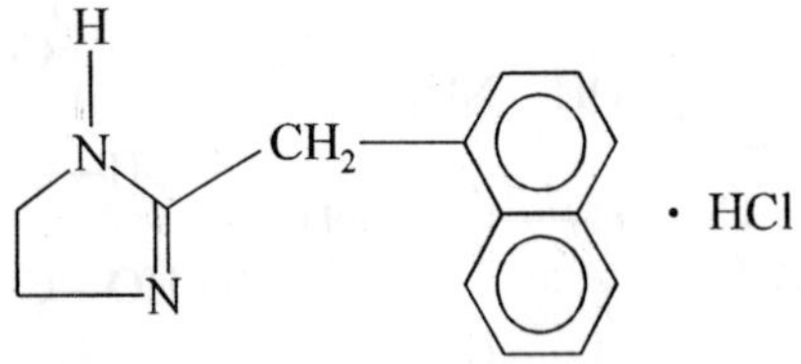

    2-(1-Naphthylmethyl)-2-imidazoline monohydrochloride ; 1H-Imidazole, 4, 5-dihydro-2-(1-naphthylmethyl)-, monohydrochloride ; BP ; (1968), USP ; Clear Eyes[R] (Abbott) ; Privine Hydrochloride[R] (Ciba-Geigy).

Synthesis

    Chloromethylation of naphthalene with formaldehyde and hydrogen chloride yields 1-naphthyl methyl chloride, which on reaction with KCN gives 1-naphthyl acetonitrile. This product on condensation with ethylene diamine monohydrochloride between 175-200°C for 1 hour yields the desired official compound with the elimination of a mole of ammonia.

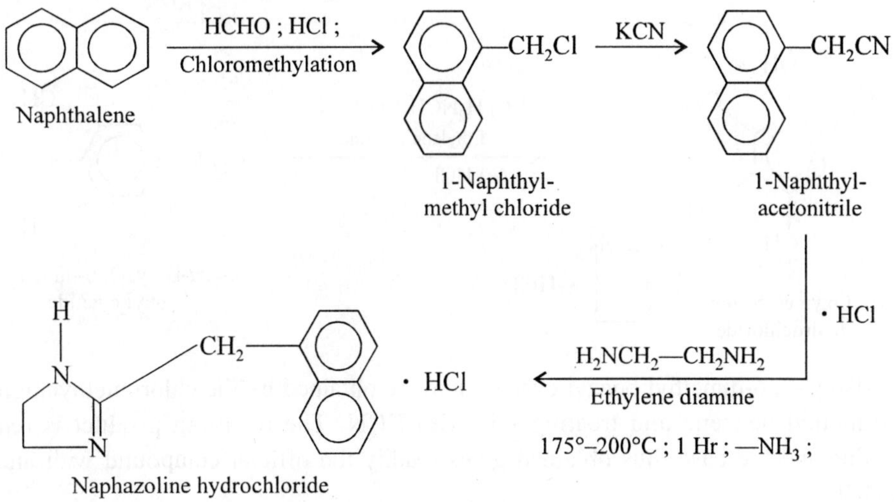

Naphthalene

1-Naphthyl-
methyl chloride

1-Naphthyl-
acetonitrile

Naphazoline hydrochloride

It is a directly acting sympathomimetic drug which is mostly *used as a local vasoconstrictor for the relief of nasal congestion due to allergic or infectious manifestations. It is also employed as an ophthalmic solution for the relief of ocular congestion and blepharospasm.*

Dose : For nasal mucosa, 2 drops of 0.05% solution ; for conjunctivita, 1 to 2 drops of a 0.1% solution after every 3 to 4 hours.

G. *Oxymetazoline* INN, BAN, *Oxymetazoline Hydrochloride* USAN,

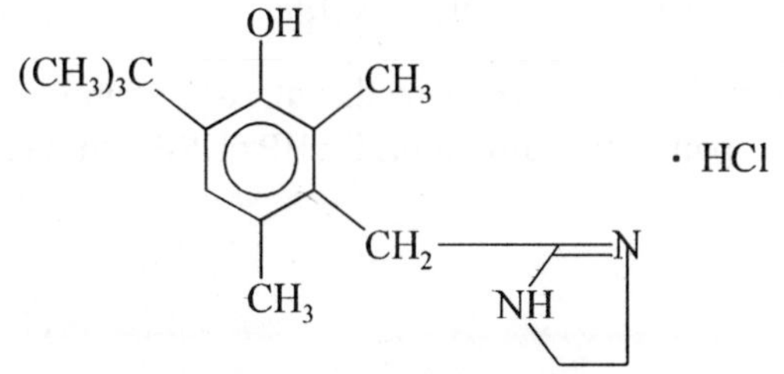

· HCl

6-*tert*-Butyl-3-(2-imidazolin-2-ylmethyl)-2, 4-dimethylphenol monohydrochloride ; 2-(4-*tert*-Butyl-3-hydroxy-2, 6-dimethyl-benzyl)-2-imidazoline hydrochloride ; USP ; Daricon[R] (Pfizer) ; Iliadin-Mini[R] (E. Merck, U.K.)

Synthesis

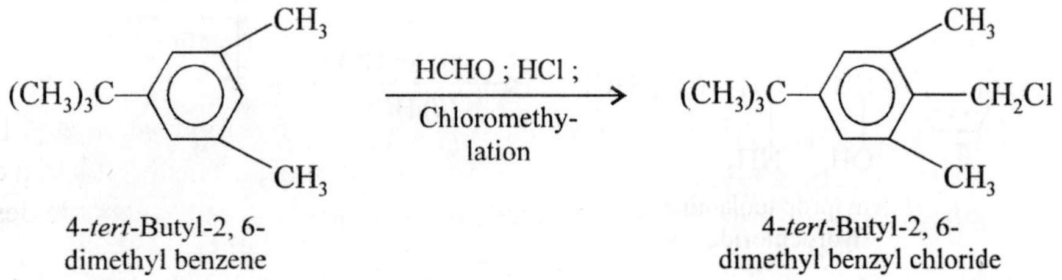

4-*tert*-Butyl-2, 6-
dimethyl benzene

4-*tert*-Butyl-2, 6-
dimethyl benzyl chloride

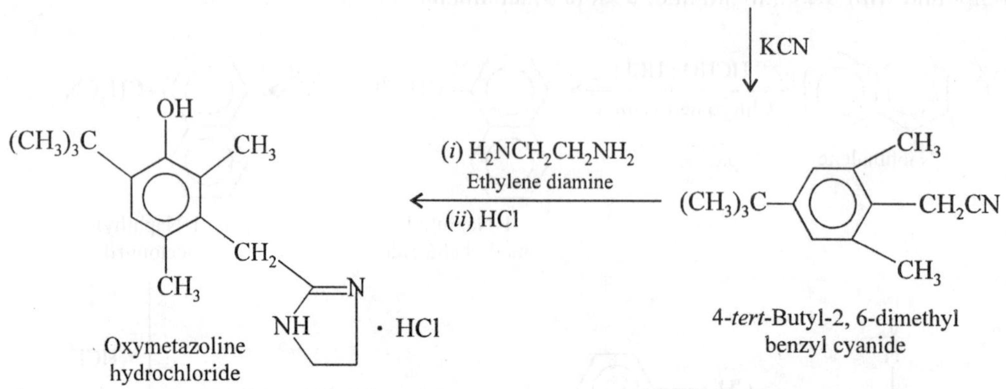

4-*tert*-Butyl-2, 6-dimethyl benzyl cyanide may be obtained by the chloromethylation of 4-*tert*-butyl-2, 6-dimethyl benzene and treating this with KCN. The resulting product is reacted with ethylenediamine and the base thus obtained gives readily the official compound with an equimolar quantity of HCl.

It is employed exclusively topically to *decongest the nasopharyngeal membranes in sinusitis, rhinitis and otitis media. Its on-set of action is several minutes but its action lasts up to several hours.*

Dose : *Intranasal, 2 to 4 drops of a 0.05% solution.*

H. *Phenylpropanolamine* BAN, *Phenylpropanolamine Hydrochloride* USAN,

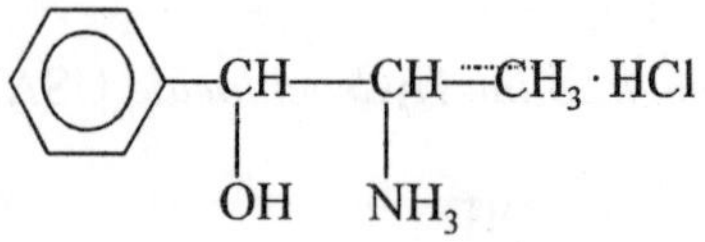

(±)-Norephedrine hydrochloride ; Benzenemethanol, α-(1-amino-ethyl)-, hydrochloride, (R^*, S^*)-, (±) ; (±)-2-Amino-1-phenyl-propan-1-ol hydro-chloride ; BP ; USP ; Propadrine Hydrochloride[R] (MSD) ;

Synthesis

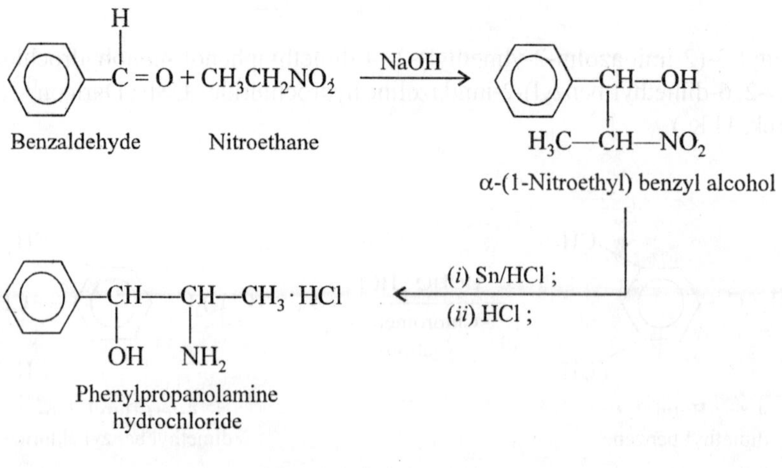

α-(1-Nitroethyl) benzyl alcohol is obtained by the interaction of benzaldehyde and nitroethane in the presence of sodium hydroxide. This product when subjected to reduction in the presence of tin and hydrochloric acid, followed by dissolution of the phenylpropanolamine base in an appropriate solvent and passing a stream of HCl gas yields the final product.

It is administered orally for the symptomatic relief of nasal congestion. It has also been used for the reduction of appetite and to monitor urinary incontinence. It also finds its use as a bronchodilator and bronchial decongestant in asthma.

Dose : Usual, oral, 25 to 30 mg three or four times per day ; Topical, 0.1 to 3% aqueous solution.

I. *Xylometazoline* INN, BAN, *Xylometazoline Hydrochloride* USAN,

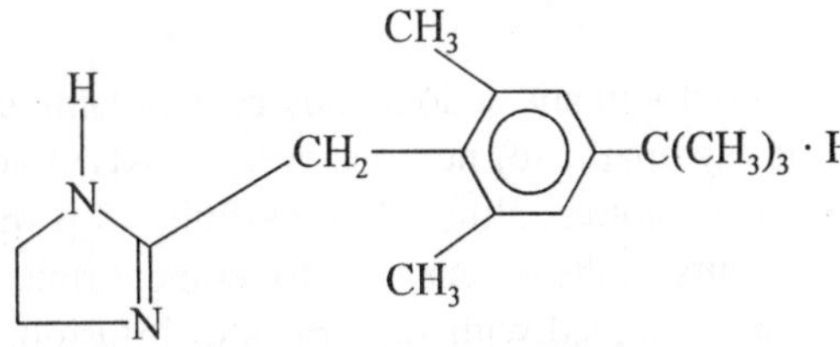

2-(4-*tert*-Butyl-2, 6-dimethylbenzyl)-2-imidazoline monohydrochloride ; 1H-Imidazole, 2-[[4-(1, 1-dimethylethyl)-2, 6-dimethylphenyl] methyl]-4, 5-dihydro-, monohydrochloride ; BP ; USP ; Otrivin Hydrochloride[R] (Ciba-Geigy)

Synthesis

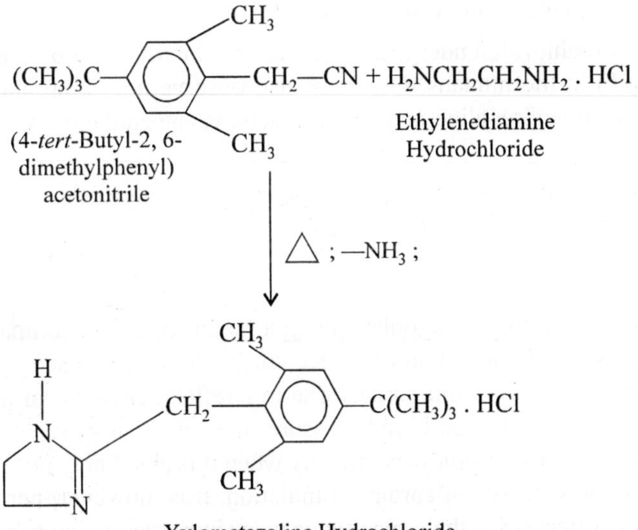

Xylometazoline Hydrochloride

It may be prepared by the interaction of (4-*tert*-butyl-2, 6-dimethyl-phenyl) acetonitrile with ethylenediamine hydrochloride at an elevated temperature with the loss of a mole of ammonia.

It is frequently employed as a *local vasoconstrictor for nasal congestion caused by sinusitis or rhinitis.*

Dose : Intranasal, 1 drop of a 0.1% solution in adult ; or a spray of 0.05% solution.

3.1. Mechanism of Action

The mechanism of action of certain '*sympathomimetic drugs*' discussed under section 12.3 are dealt with in the sections that follows :

3.1.1. Ephedrine

The '*drug*' acts as a direct and indirect agonist *i.e.,* it helps in the release of norepinephrine. Besides, it also exerts CNS-stimulatory actions. It has been observed that the ephedrine stereoisomer having essentially the (1R, 2S) absolute configuration exhibits direct activity on the receptors, both α and β, as well as an indirect component. It is worthwhile to state here that the (1S, 2R) entantiomer has primarily an indirect activity.

3.1.2. Epinephrine

The '*drug*' happens to be predominant endogenous catecholamine released from the adrenal medulla in response to autonomic (sympathetic) nervous system (ANS) activation. It is believed that epinephrine acts on all the α- and β-receptors, although the affinity of β-receptors for it is found to be much higher as compared to the affinity of the α-receptors for epinephrine. As a result, with the administration of relatively low doses amalgamated with slow rates of infusion, epinephrine may minimise diastolic blood pressure by virtue of β_2-*receptor-regulated vasodilation* and a corresponding enhancement of heart rate through the activation of β_1-receptors. Besides, the systolic blood pressure may have been enhanced on account of enhanced cardiac output. Furthermore, it has been observed that with regular increment in doses, α_1-regulated vasoconstriction occurs promptly, having an overall net enhancement in vascular resistance followed by blood pressure.

3.1.3. Isoprenaline (Isoproterenol Hydrochloride)

Isoproterenol hydrochloride undergoes appreciable *o*-methylation by the aid of catechol-*o*-methyl transferase (COMT) in the humans. One may also observe a striking and pronounced identical structural features between isoprenaline and the endogenous catecholamines, namely : *dopamine* and *norepinephrine* (NE).

It is a prototypic , non selective β-agonist invariably employed to stimulate heart rate in heart block, bradycardia, and *torsades de pointes*.

3.1.4. Methoxamine

The '*drug*' is a direct-acting α_1-agonist having a rather rapid and comparatively longer pressor action. It is found to possess certain extent of β-receptor-blocking pharmacological characteristic features. In fact, it gives rise to bradycardia, largely due to a reflex activation of the strategically located vagus nerve secondary to the increased blood pressure. It is invariably employed for the control and management of *critical hypotensive states* specifically when it is absolutely necessary to raise the blood pressure without causing any degree of cardiac stimulation. It is, however, pertinent to state here that this '*drug*' has very little effect upon the *capacitance veins* in particular, so as to accomplish a reasonable compromise with its usefulness, especially in the treatment of various types of shock. The reflex bradycardia produced by methoxamine is used juidiciously to terminate effectively *paroxysmal supraventricular tachycardia*. Besides, it is found to have very little effect on the bronchial muscles, thereby affording neither any central stimulation nor an enhancement of the irritability of the **anaesthetized-sensitized heart.**

3.1.5. Metaraminol

The '*drug*' essentially has the characteristic features, such as : *meta*—OH of a catechol, an α-CH$_3$ of an amphetamine, and the attenuating (3-OH for attributing the classic sympathomimetic activity. Thus, an overall pressor activity is predicted gainfully. It has an ability to enter the neuronal vescicles followed by displacement of NE gives rise to an indirect activity. Besides, it is also observed to stimulate the α-adrenoreceptors directly. Indeed, both of these activities are absolutely necessary for the pressor activity.

3.1.6. Naphazoline

The '*drug*' essentially has an imidazole nucleus and exhibits *peripheral α-adrenoceptor stimulant characteristic features* ; however, surprisingly no β-effects have been observed. Importantly, very much in contrast with other several direct and indiect agonists, there is no appreciable neuronal uptake mechanism involved. It is found to exert major action as decongestants upon the nasal mucosa particularly in addition to ocular membranes. It is, however, pertinent to mention here that a certain extent of α$_2$ central stimulation most likely puts forward a plausible explanation with regard to a certain degree of sedation encountered by this drug substance.

3.1.7. Phenylpropanolamine (PPA)

It is employed frequently as a nasopharynegeal and bronchial decongestant. It also finds its usage in the treatment of urinary incontinence and retrograde ejaculation. One may accomplish a partial success in the moderate release of histamine by coadministering sympathomimetic decongestant agents, for instance : ephedrines and PPA, that might relieve both nasal congestion as well as counteract drowsiness to a certain degree by virtue of their inherent CNS stimulant side effects.

3.1.8. Xylometazoline

It is a potent sympathomimetic agent having marked and pronounced α-adrenergic pharmacologic profile. However, it is found to act as a vasoconstrictor when applied topically to mucous membranes particularly ; and, therefore, retards both swelling and congestion to a considerable degree.

3.2. Structure Activity Relationships (SARs)

The SARs of the sympathomimetic drugs are as enumerated under :

(*i*) Aliphatic amines longer than *n*-butylamine [CH$_3$ (CH$_2$)$_3$] prove to be adequate for pressor activity.

(*ii*) Introduction of aromatic rings significantly increase potency.

(*iii*) Presence of one —OH group at C-3 or C-4 of the aromatic rings enhances vasoconstrictor activity, *e.g.*, metraminol bitartrate, oxymetazoline hydrochloride.

(*iv*) Introduction of two —OH groups at C-3 and C-4 invariably enhance the tendency to induce vasodilation in the presence of other preferred molecular substituents, *e.g.*, adrenaline, isoprenaline hydrochloride, etc.

(*v*) N-substituents favour vasodilation, *e.g.*, epinephrine (—CH$_3$) ; isoprenaline [–CH(CH$_3$)$_2$].

(*vi*) Enzymatic degradation of the drug molecule may be prevented by affording substitution on the side chain immediately adjacent to the amino function *e.g.*, methoxamine hydrochloride, isoprenaline hydrochloride. Such an innovation also accomplishes prolonged duration of action along with oral efficacy.

4. BETA ADRENERGIC RECEPTOR STIMULANTS

As stated earlier there are two types of beta-receptors, viz : β_1 and β_2. Agents that stimulates β_2-receptors selectively are very effective in relaxing the smooth muscles of the bronchi and uterus. They are, therefore, useful in asthma and in the management of uncomplicated premature labour in the trimester of pregnancy.

These are also known as the β-*adrenergic receptor agonists.*

A few typical examples are as stated below : Salbutamol ; Terbutaline ; Pirbuterol hydrochloride ; Salmetrol xinafoate ;

A. *Salbutamol* INN, BAN, *Albuterol* USAN,

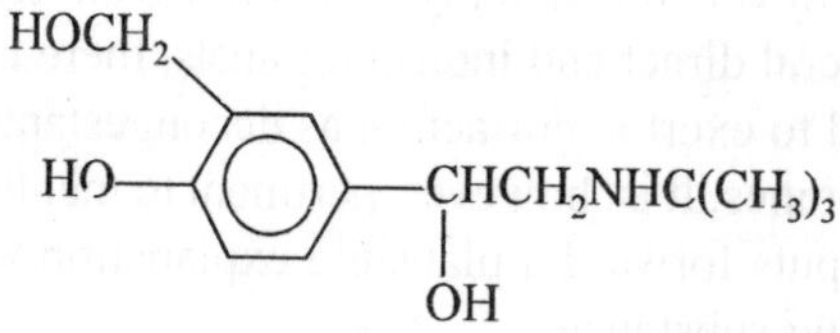

HOCH$_2$

HO— —CHCH$_2$NHC(CH$_3$)$_3$
OH

1, 3-Benzenedimethanol, α'-[[(1, 1-dimethylethyl) amino] methyl]-4-hydroxy- ; α'-[(*tert*-Butylamino) methyl]-4-hydroxy-*m*-xylene-α-α'-diol ; BP ; Ventolin Inhaler[R] (Glaxo, Inc.)

Salbutamol is used as *a bronchodilator.* Its bronchodilating property being relatively more marked and pronounced than its effect on the heart.

Dose : Oral, inhalation, adult, 100 mcg, followed by a second dose after 5 minutes, if required.

B. *Terbutaline* USAN ;

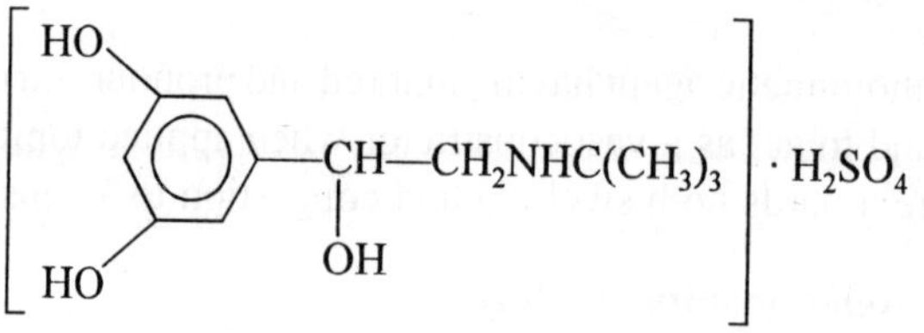

HO

—CH—CH$_2$NHC(CH$_3$)$_3$ · H$_2$SO$_4$
OH

HO

1, 3-Benzenediol, 5-[2-[(1, 1-dimethylethyl) amino]-1-hydroxy-ethyl]- ; Bricanyl[R] ; Brethaire[R] ;

It is used for treating premature labour. It in indicated parenterally for the emergency treatment of *status asthmaticus.*

C. *Pirbuterol Hydrochloride* USAN ;

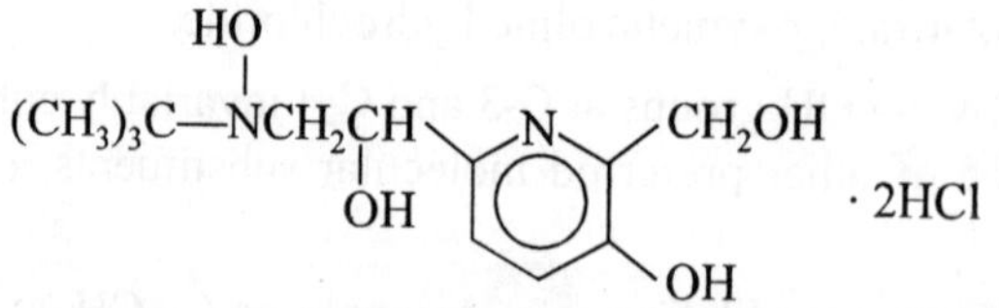

HO

(CH$_3$)$_3$C—NCH$_2$CH CH$_2$OH
OH N
· 2HCl
OH

2, 6-Pyridine dimethanol, α^6-[[(1, 1-dimethylethyl) amino] methyl]-3-hydroxy-, hydrochloride, Maxair[R] ;

It is imployed as a β_2-selective bronchodilator.

D. *Salmetrol Xinafoate* USAN ;

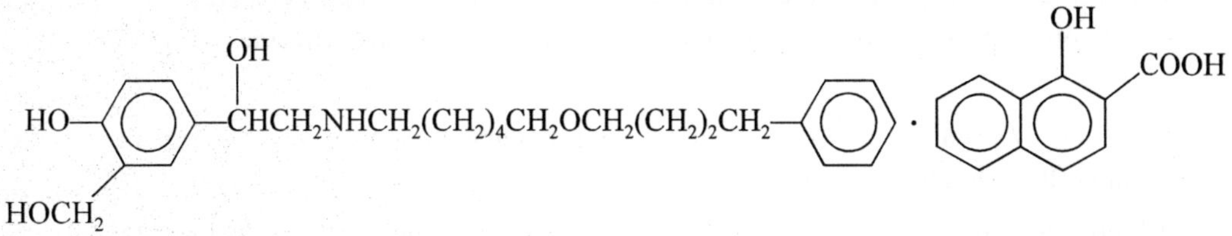

HO—⟨ring⟩—CHCH₂NHCH₂(CH₂)₄CH₂OCH₂(CH₂)₂CH₂—⟨ring⟩ · ⟨naphthalene ring⟩

(structure with OH, HOCH₂ substituents and OH, COOH on naphthalene)

1, 3-Benzenedimethanol, (±)-4-hydroxy-α^1-[[[6-(4-phenylbutoxy)hexyl] amino]-methyl-, 1-hydroxy-2-naphthalenecarboxylate (salt) ; Serevent[R] ;

It is invariably used as a rather more lipophilic β_2 agonist recommended for long-term *bid* maintenance treatment of asthma ; it has a duration of action even after inhalation of 12 hr.

4.1. Mechanism of Action

The possible mechanism of action of these drugs shall be described as under :

4.1.1. Salbutamol (Albuterol)

It is a direct-acting sympathomimetic agent with predominantly β-adrenergic activity together with a very selective action on the β_2-receptors (*i.e.*, β_2-agonist). It is pertinent to emphasize here that this preferential activity for the specific β_2-receptor stimulation gives rise to its spectacular bronchodilating action being comparatively more marked and pronounced than its effect on the heart in particular.

4.1.2. Terbutaline

The '*drug*' exhibits a direct-acting sympathomimetic agent having predominantly β-adrenergic activity plus a selective action on the β_2-receptors (*i.e.*, (β_2-agonist). This is, therefore, recommended for the treatment of severe and acute forms of bronchospasm *via* IM, IV, sub-cutaneous routes.

4.1.3. Pirbuterol Hydrochloride

It is closely related structurally to salbutamol, the only difference being that the former has a basic-pyridine nucleus while the latter has a benzene ring (nucleus). It is also a direct-acting sympathomimetic agent having a most prominent β-adrenoreceptor stimulant action, and a selective action on β_2-receptors.

4.1.4. Salmetrol Xinafoate

Its mechanism of action is very much akin to salbutamol, terbutaline and pirbuterol.

5. ADRENERGIC RECEPTOR BLOCKING AGENTS

Adrenergic receptor blocking agents are broadly sub-divided into *two* heads, namely :

5.1. α-Adrenoceptor Blocking Agents

They usually exert a generalised direct vasodilator effect on all muscular walled vessels. They are found to reduce vasoconstriction and enhance tissue perfusion. They are also termed as either α-*adrenoreceptor antagonists* or *nonselective α-antagonists*.

However, it has been observed that the blockade of α_1-*adrenoreceptors* invariably produces prompt apparent actions ; whereas the blockade of α_2-*receptors* gives rise to rather subtle effects.

Effects of α_1-Antagonism.   The various effects are as follows :

(*a*) The impulses to the arterioles minimises vascular resistance appreciably, which consequently tends to reduce blood pressure, produce a pink-warm skin, ptosis, and nasal, scleroconjunctival congestion.

(*b*) It not only enhances venous capacitance at the 'capacitance vessels (*i.e.,* venules) that essentially necessites fluids-loading, but also gives rise to postural hypotension.

(*c*) It produces mild to moderate miosis and serious interference with ejaculation (sexual debility).

(*d*) It causes *palpitations, tachycardia,* and enhanced *secretion of renin.*

[Note. The actions depicted in (d) above are essentially caused due to β_1-adrenoreceptor responses which are obviously not suppressable by α-blockade.]

(*e*) It is now well established that there exists a simultaneous enhanced quantum of norepinephrine (NE) released from the adrenergic nerve-endings (*i.e.,* the transmitter '*overflow*') on account of concurrent blockade of α_2-adrenoreceptors that subsequently affords a negative-feedback mechanism thereby to lower the release of transmitter significantly. As a result, *tachycardia, palpitations* and enhancement of *plasma-renin-levels* may take place even when BP falls to a very little extent. Furthermore, the aforesaid overflow/reflex actions are, in reality, counter productive with respect to the broad-spectrum applications of non-selective α-blocking drugs.

Only *three* non selective α-adreno receptor antagonists, such as : phenoxybenzamine, phentolamine and tolazoline, are presently being used in the US. These drugs shall now be discussed below :

A. *Phenoxybenzamine* INN, BAN, *Phenoxybenzamine Hydrochloride* USAN,

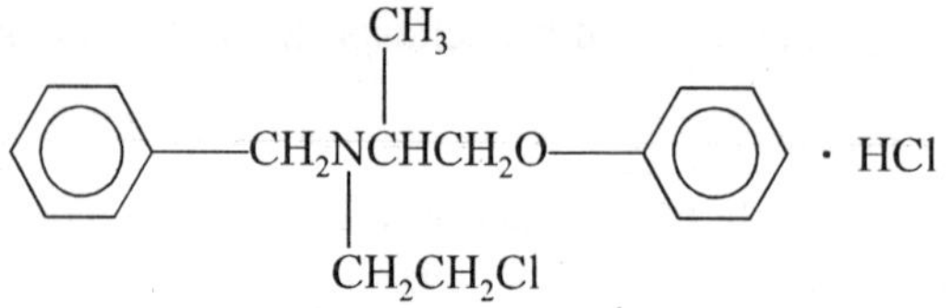

N-(2-Chloroethyl)-N-(1-methyl-2-phenoxyethyl)-benzylamine   hydrochloride   ; Benzenemethanamine, N-(2-chloroethyl)-N-(1-methyl-2-phenoxyethyl)-, hydrochloride ; SKF 688A ; USP Dibenzyline[R] (Smith Kline and French)

Synthesis

Condensation of phenol with propylene chlorohydrin gives an alcohol which upon treatment with thionyl chloride yields the corresponding chloride. This on alkylation with enthanolamine affords a secondary amine which on treatment with benzyl chloride followed by chlorination and lastly with hydrochloric acid yields the official compound.

It is a potent alpha-adrenoceptor (α_1 and α_2) blocking agent with a prolonged duration of action. Phenoxybenzamine hydrochloride has been *employed intravenously in the treatment of shock. It also finds its application in the treatment of pulmonary oedema.*

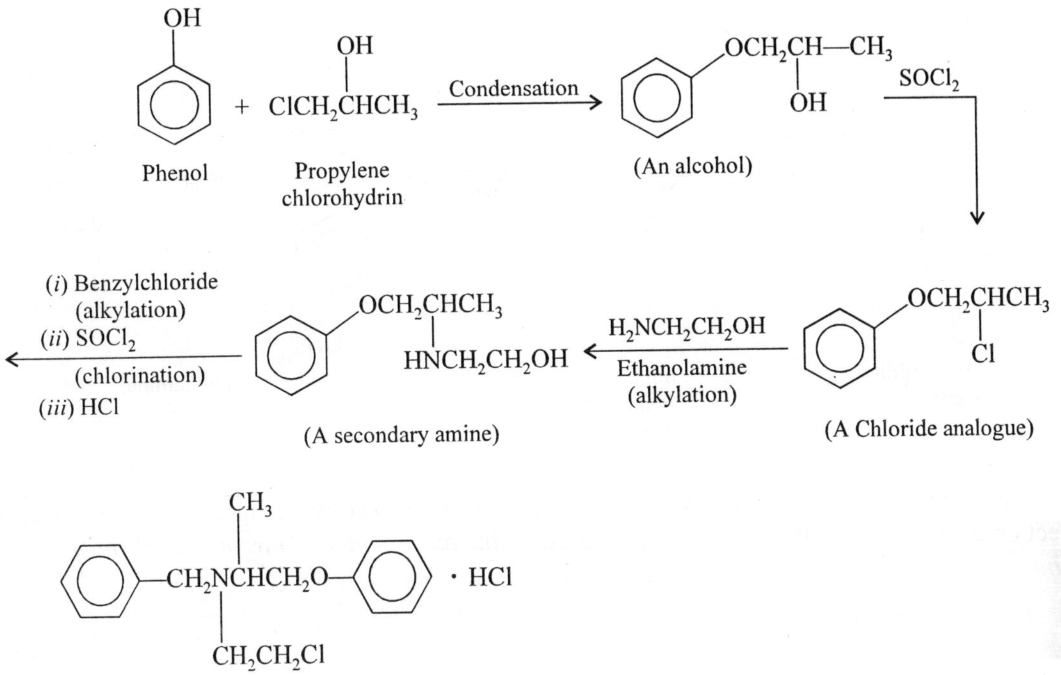

Phenoxybenzamine hydrochloride

Dose : Usual, initial, 10 mg per day, increased gradually to 60 mg per day in divided doses.

B. *Phentolmine* INN, *Phentolamine Hydrochloride* BAN,

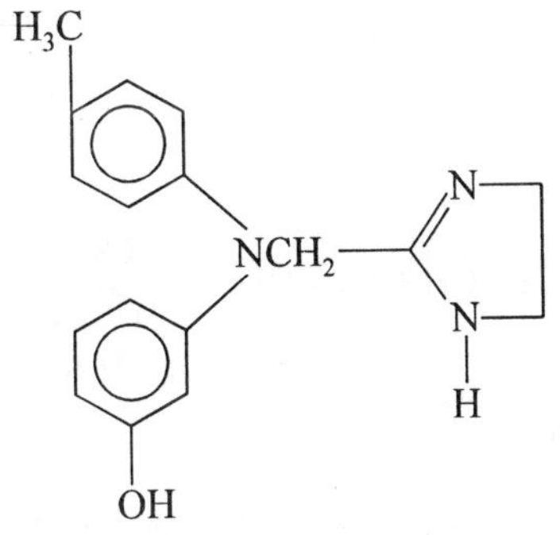

3-[N-(2-Imidazolin-2-ylmethyl)-*p*-toluidino] phenol hydrochloride ; BP, (1963), USP, Int. P. ; Regitin[R] (Ciba, U.K.)

Synthesis

Phentolamine may be prepared by the condensation of 3-(4-toluidino) amino phenol and 2-imidazolin-2-ylmethyl chloride with the elimination of a mole of hydrogen chloride.

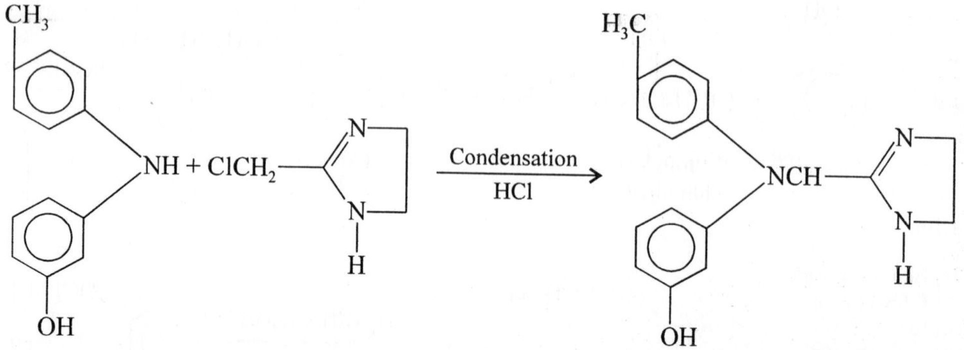

| 3-(4-Toluidino) amino phenol | 2 Imidazolin- 2 ylmethyl- chloride | Phentolamine |

It is a short-acting α-adrenoceptor blocking agent which also possesses an overall direct vasolidator effect on all muscular walled vessels. *It is of great value as an adjunct in treatment of shock or heart failure.*

C. *Tolazoline Hydrochloride* USAN,

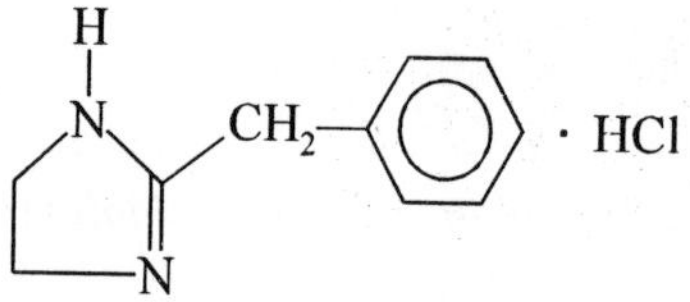

1 *H*-Imidazole, 4, 5-dihydro-2-(phenylmethyl)-monohydrochloride ; Priscoline[R] (Ciba-Geigy).

Synthesis

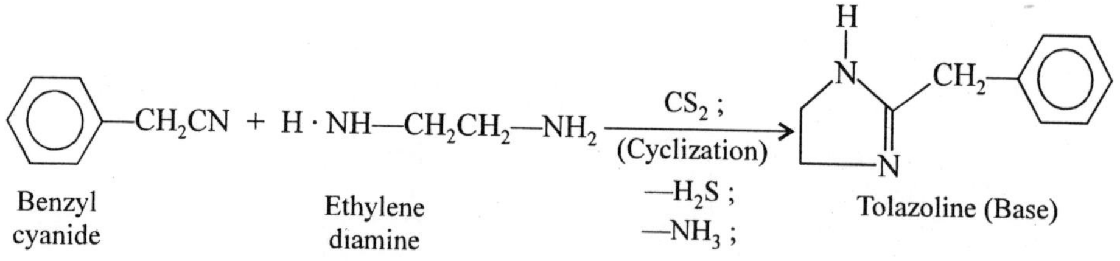

| Benzyl cyanide | Ethylene diamine | | Tolazoline (Base) |

The reaction between a mole each of benzyl cyanide and ethylene diamine in the presence of carbon disulphide as a medium gives rise to the formation of the desired tolazoline (base) through cyclization ; and subsequent elimination of a mole each of hydrogen sulphide and ammonia that are liberated from the reaction mixture. The base thus obtained is treated with a HCl in molar concentration to obtain the corresponding salt.

It is mostly used as a vasodilator with α-adrenergic blocking activity and direct vasodilator actions. It exhibits a sympathomimetic effect to stimulate the heart, in doing so the BP sometimes gets increased moderately, despite vasodilation. It is found to be an effective α$_2$-antagonist having a free access to the CNS ; and, therefore, it has been employed efficaciously to antagonize the overdoses of clonidine besides other centrally acting α_2-agonists.

5.1.1. Mechanism of Action

The mechanism of action of the *three* aforesaid drug substances shall now be discussed in the sections that follows :

5.1.1.1. Phenoxybenzamine Hydrochloride

The '*drug*' exerts a quite unpredictable effectiveness and thereby produces an unduly excessive tachycardia ; besides, causing orthostatic hypotension in patients having essential hypertension because of its apparent competitiveness with newer breed of drugs. As an adjuvant with a β-antagonist it is specifically beneficial in the control and management of *inoperable pheochromocytoma** by virtue of its prolonged duration of action.

5.1.1.2. PHENTOLAMINE HYDROCHLORIDE

The '*drug*' is a non selective α-adrenoreceptor antagonist ; and the blockade caused is reversible. Besides, exhibiting α-blocking activity it possesses a plethora of other pharmacological actions, namely :

- Mild to moderate sympathomimetic-like mydriatic as well as '*cardiostimulant*' (*viz.*, rate, force of contraction and dysrythmias) activity,
- Weak *muscarinic activity* in the GI-tract, and weak to mild *histaminergic activity* in the stomach (*viz.*, acid secretion), and
- Flushing and slight fall in BP due to *arterioles.*

5.1.1.3. Tolazoline Hydrochloride

Interestingly, tolazoline distinctly possesses a histamine-like effect to cause stimulation of the gastric secretion ; and also on ACh-like activity to enhance GI-motility. It is also observed to produce mydriasis by means of a definite sympathomimetic action. It is found to be an excellent and effective α_2-**antagonist** having an easy access to the CNS ; and, therefore, has been used judiciously to negate (antagonize) overdoses of clonidine and a host of other centrally acting α_2-agonists.

5.2. β-Adrenoceptor Blocking Agents

β-**Adrenoceptor blocking agents** or **antagonists** usually inhibit the actions of catecholamines at the β-adrenergic receptor sites competitively. They are also frequently termed as β-**adrenoreceptor** or β-**adrenergic blocking agents.** These agents normally retard the cardiac activity by preventing β-adrenoceptor stimulation. The effect on the heart may be viewed through different angles, *viz* : minimising its rate and force of contraction, reducing its reaction to stress and exercise and lastly, reducing the rate of conduction of impulses through the conducting system. All these remarkable characteristics are vital for their numerous applications in the therapeutic armamentarium, *e.g.*, in the treatment of angina pectoris and cardiac arrhythmias. These are also used in the control and treatment of hypertension.

The β-adrenoceptor blocking agents (or β-blockers) have received an overwhelming cognizance in the therapeutic armamentarium in the past four decades that they have been judiciously classified into the followed *three* categories, namely :

---

*A chromaffin cell tumour of the sympathoadrenal system that produces catecholamines (*i.e.,* EP = epinephrine ; and NE = norepinephrine).

(*i*) First-generation β-blockers,

(*ii*) Second-geneartion β-blockers, and

(*iii*) Third-generation β-blockers.

These *three* specific types of β-blockers shall now be treated individually in the sections that follows :

5.2.1.   First Generation β-Blockers

The first main objective towards the exploratory development of these agents was to accomplish selectivity for β-receptors with respect to α-receptors.

Salient Features for the Development of Propranolol

The various stages that were essentially followed in a sequential manner for the development of propranolol, a predominant candidate drug of this category are as stated under :

(1) Isoprenaline [I] (see section 12.3.C) was specifically picked up as the '**lead compound**', which was proved to be an '*agonist*'* and not an '*antagonist*',** besides, being active at β-*receptors* and not α-*receptors*. Interestingly, the cardinal objective was to take advantage of the inherent specificity on one hand, and to modify the molecule meticulously to convert it from an '*agonist*' to an '*antagonist*' on the other.

(2) '*Phenolic functional moieties*' are found to be absolutely necessary for the '*agonist activity*'-profile. Nevertheless, it does not imply that the '*phenolic groups*' are essential for antagonist activity, because antagonists mostly block receptors by binding in different manners from the agonist. Thus, the two phenolic functions in isoprenaline were skilfully replaced by chloro-functions to yield dichloro-isoprenaline (DCI) [II], which proved to be a '*partial agonist*'. Compound [II] was capable of blocking the binding ability to the '*natural messangers*'. Thus, it could be regarded as an antagonist because it lowered the adrenergic activity appreciably.

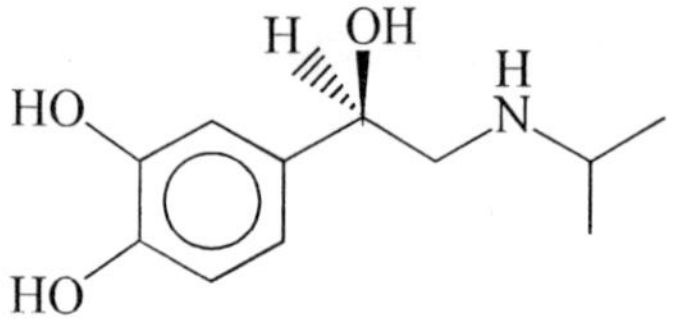

[I] 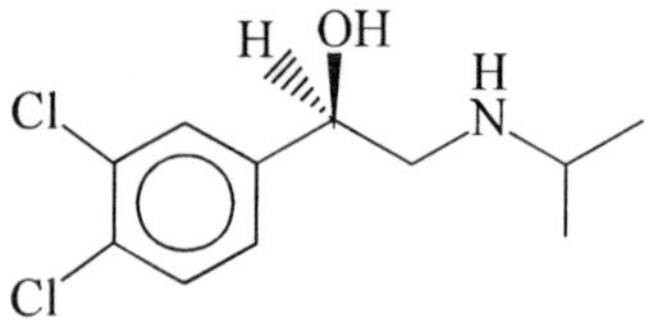   [II] 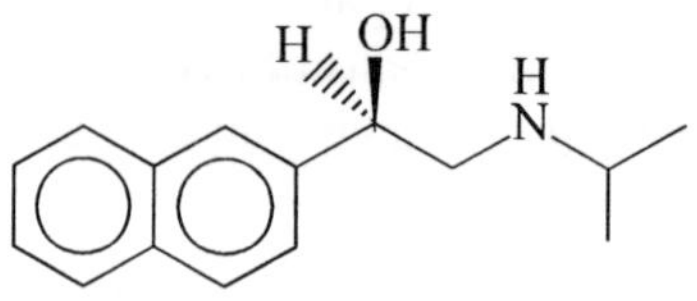   [III]

Partial β-Agonists-Development

(3) The next vital and crucial step was to get rid of the partial agonist activity.

Medicinal chemists usually convert an '*agonist*' into an '*antagonist*' by introducing an additional '*aromatic ring*' *i.e.,* changing benzene with a naphthyl ring. The said proposed modification invariably give rise to an altogether different induced fit existing between the *ligand* and the *binding site*—, thereby without activating the receptor. Thus, the two adjacent '*chloro functional moieties*' of [II] were removed and an additional benzene ring introduced to obtain **pronethalol** [III]. Compound [III] was still acted as a partial agonist, but ultimately recognized as the very first and foremost β-

---

*Drugs that mimic the body's own regulatory function are known as '**agonist**'.

**That which counteracts the action of something else *e.g.,* drug substance.

blocker to be employed profusely as an wonderful drug for the control, management and treatment of angina, high BP, and arrythmias.

(4) Further structural modifications were carried out with respect to :

- Extension of the chain-length,
- Connection of the aromatic ring, and
- Joining to the amine function.

As evidenced profusely in the literature most of the *'drug discoveries'* were more or less accidental. The same was the fate in the synthesis of **propranolol**, for which α-naphthol was used in the reaction mixture instead of the β-naphthol, as the latter was not readily available in the laboratory to arrive at the predetermined **'target structure'** as illustrated below :

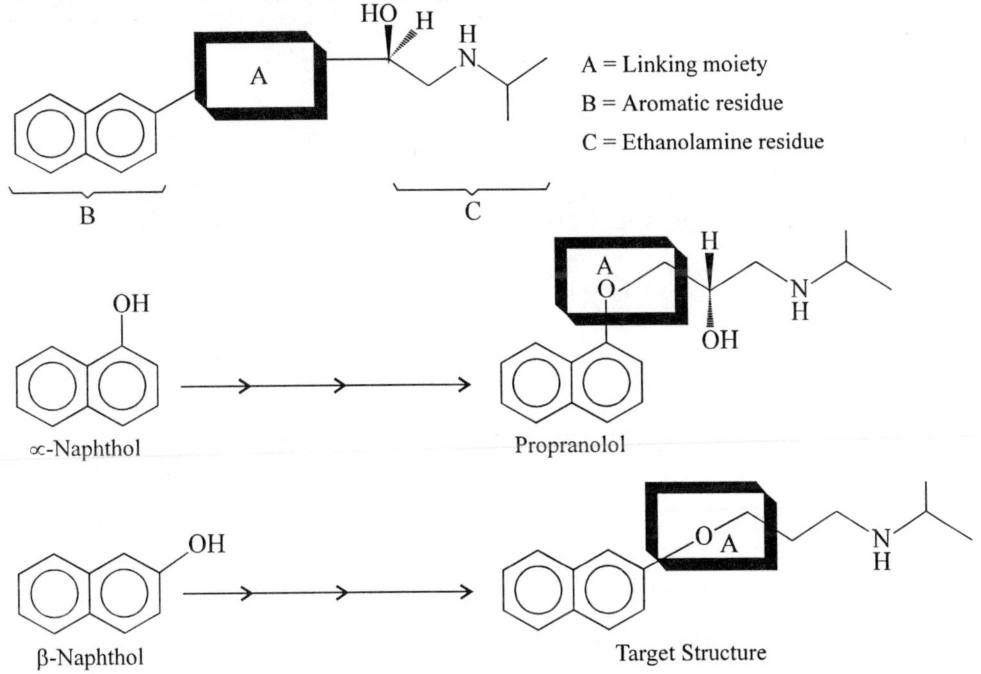

Explanations :

(a) It vividly displays the academic as well as the professional challenge adopted usually by the *'medicinal chemist'* in the accomplishment of **chain extension**.

(b) **Propranolol** was synthesized by using α-naphthol, extending the linking moiety with an ether linkage (—O—), and an ethanolamine residue.

(c) The *'targetted drug molecule'* was synthesized by using β-naphthol, extending the residue with an ethereal linkage (—O), and an ethanolamine portion.

The net result of the entire exercise was the epoch-making discovery of **propranolol,** that was observed to be a pure *antagonist* and which was approximately 20 times more potent in comparison to **pronethalol** (*i.e., the original 'targetted-drug molecule'*).

The most potent drug substance belonging to the *first generation* β-*blockers*, propranolol, shall be discussed here under :

A. *Propranolol* INN, *Propranolol Hydrochloride* BAN, USAN,

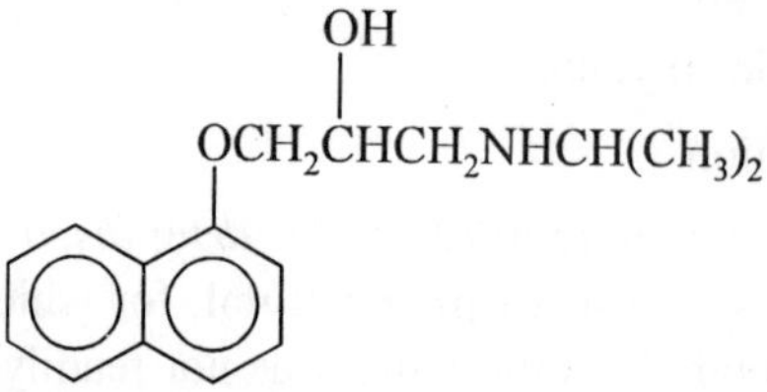

OH
|
OCH₂CHCH₂NHCH(CH₃)₂

HCl

(±)-1-Isopropylamino-3-(1-naphthyloxy) propan-2-ol-hydrochloride : 2-Propanol, -1-[(1-methylethyl) amino]-3-(1-naphthalenyloxy)-hydrochloride ; BP., USP. ; Inderal[(R)] (Ayerst)

Synthesis

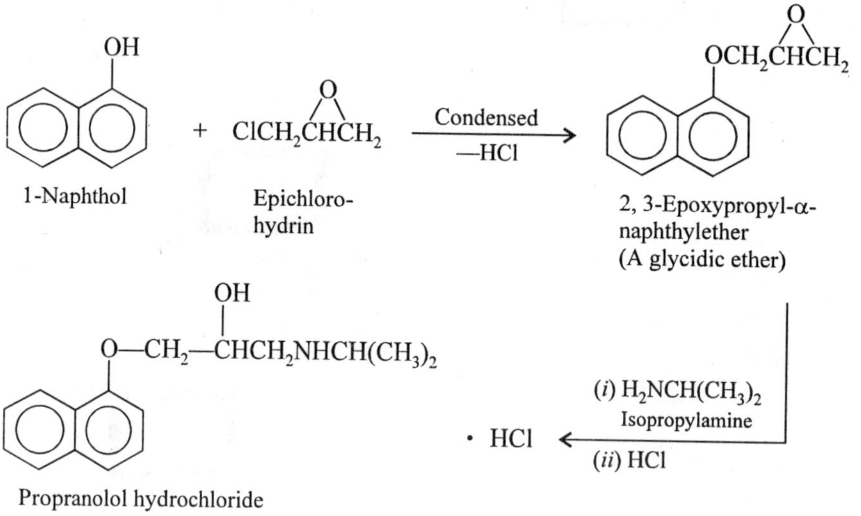

OH
1-Naphthol

+

O
CICH₂CHCH₂
Epichloro-hydrin

→ Condensed
—HCl

OCH₂CHCH₂
2, 3-Epoxypropyl-α-naphthylether
(A glycidic ether)

OH
|
O—CH₂—CHCH₂NHCH(CH₃)₂

· HCl ←

(i) H₂NCH(CH₃)₂
Isopropylamine
(ii) HCl

Propranolol hydrochloride

Interaction of 1-naphthol with epichlorohydrin affords a glycidic ether which upon treatment with isopropylamine aids in the opening of the oxirane ring yielding the propranolol base and this on being treated with a known quantity of hydrochloric acid gives the official compound.

Propranolol has been reported to exhibit quinidine-like antiarrhythmic actions which are quiet independent of beta-adrenergic blockade. Hence, these pharmacological properties are usually *employed to suppress ventricular tachycardia, digitalis-induced tachyarrhythmias, paroxysml atrial tachycardia, and lastly ventricular and atrial extra-systoles. It is also currently receiving a lot of attention in the treatment and management of essential hypertension.*

Dose : Oral, adult, for arrhythmias, 10 to 30 mg 3 to 4 times daily.

5.2.1.1. Structure Activity Relationships (SARs) of Aryloxypropanolamines

After the most successful synthesis of *propranolol*, a good number of **aryloxypropanolamines** have been synthesized in various laboratories ; and, therefore, the SARs of these drug substances have been summarized as stated below :

1. The 'branched and bulky N-alkyl functional moieties', such as : tert-butyl, iso-propyl etc., proved to be extremely vital for attributing the β-antagonist activity, thereby suggesting a possible interaction taking place with a hydrophobic pocket strategically located in the binding site.

2. It is, however, feasible to afford a variation of the aromatic ring system as well as heteroaromatic rings into the drug-molecules e.g., timolol, pindolol etc.

3. The probable substitution of the two methylene moieties present in the 'side-chain' enhances the **metabolic stability** at the expense of **therapeutic potency** (lowering of activity).

4. The **'alcoholic function'** on the side-chain is an absolute necessary requirement for its activity.

5. Isosteric replacement of the ethereal linkage (—O—) with such moieties as : CH_2, S or NCH_3 is found to be more or less detrimental ; however, a tissue-selective β-blocker has been synthesized by replacing NH for O.

6. The introduction of relatively longer alkyl substituents in comparison to 'isopropyl' or 'tert-butyl' are found to be much less therapeutically potent and efficient.

7. The addition of an arylethyl functional moiety, for instance :
$CH(CH_3)$—CH_2—C_6H_5 or $CH(CH_3)_2$—CH_2—C_6H_5 has proved to be useful in having better efficacious drug substances.

8. The 'amine nitrogen' should always be a secondary in character with regard to the optimum activity.

The various aspects described under SARs of aryloxypropanolamines may be summarized in the following expression more categorically and rather explicitly :

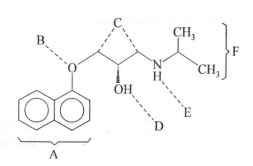

A = Variable with heteroaromatic rings

B = Intimately engaged to H-bonding to receptor site

C = Substitution with functional groups lowers therapeutic efficacy

D = Absolutely essential for H-bonding interaction

E = Essential ionic-bonding interaction and must be **secondary in nature.**

F = Branching and extension both useful ; and fits into **hydrophobic pockets.**

5.2.1.2. Mechanism of Action

The 'drug' penetrates into the CNS and thereby affords the predominant central effects. β-Antagonists are invariably employed in the treatment of essential hypertension. In reality, the exact and precise mechanism for this specific therapeutic effect has not yet been established, it has been adequately advocated that the β-blockers (e.g. propranolol) cause an effective decrease in BP by one of the three following manners, namely :

(a) Exerting a direct effect on the heart and the blood vessels.

(b) Minimising sympathetic outflow from the CNS, and

(c) Affecting the renin-angiotensin-aldosterone system.

However, the probable best usage of propranolol shall be its combination with an antihypertensive vasodilators *e.g.* hydralazine, minoxidil etc., to preferentially check and prevent the **reflex tachycardia***.

5.2.2. Second Generation β-Blockers (Selective β₁-Blockers)

The genesis of the selective β₁-blockers or the second generation β-blockers received its legitimate cognizance by virtue of the stark reality that propranolol (see Section 12.5.2.1.A) happens to be a *non-selective β-antagonist* that eventually acts as an antagonist at both *β₁-receptors* and *β₂-receptors*. Hence, it does give rise to a very serious problem if the patient suffers from '*asthma*', because the very administration of propranolol (*i.e.,* first-generation β-blocker) would not only initiate but also precipitate an '*asthmatic attack*' by sharply antagonizing the prevailing β₂-receptors in the bronchial smooth muscle. Ultimately, it would give rise to sudden contraction of bronchial smooth muscle followed by an eventual closure of the airways.

A typical example of a potent drug belonging to this particular class is *practolol* which shall be described as given below :

A. *Practolol* INN

N-[4-[2-Hydroxy-3-[(1-methylethyl)-amino] propoxy] phenyl] acetamide ; Eraldin[(R)] ;

Coleman (1979)** synthesized for the first time a series of twelve *para*-acylphenoxyethanol- and propanolamines ; of which, only one of them, **practolol**, was subjected to both extensive and intensive clinical trials.

Practolol was found to be less potent than propranolol. It also exhibited intrinsic sympathomimetic activity (ISA) to a certain extent. It is pertinent to state here that one particular property not earlier observed with the β-blockers : the ability predominantly to inhibit *isoproterenol (IPR)-induced tachycardia,* while exhibiting a minimal or no effect on the IPR-depressor (*i.e.,* hypotensive) response. In other words, practolol displayed cardioselectivity. Hence, for the first ever instance it was revealed and evidenced that the inhibition of β-adrenergic response may be confined significantly to certain sites.

Note. Practolol was eventually withdrawn due to relatively rare, but of course extremely serious dermatologic as well as ophthalmic toxicological actions that ultimately led to total blindness and fatalities.

A comprehensive and intensive research of *practolol* was accomplished in a systematic manner, and it was proved amply that the **'acetamido functional moiety'** had to be strategically located at the *para* position of the aromatic phenyl ring rather than the corresponding *ortho-* or *meta-*positions if the structure was to retain specifically the desired selectivity for the **cardiac β₁-receptors.** Fig. 12.1(*a*) and

---

*Tachycardia (i.e., an abnormal rapidity of heart action, usually heart rate more, than 100 beats per minute in adults) resulting from stimuli outside the heart, reflexly accelerating the heart-rate or depressing vagal tone.

Coleman AJ *et al. Biochem. Pharmacol.,* **28, 10, 1979.

(b) evidently depicts that in the particular instance of *para*-substitution (*practolol*) there exists an **extra H-bonding interaction** in the β₁-**receptors** but not the β-**receptors** ; whereas, in the *meta*-substitution the above mentioned criteria (*i.e.*, physical characteristic feature of an extra H-bonding interaction) is absolutely non-existence.

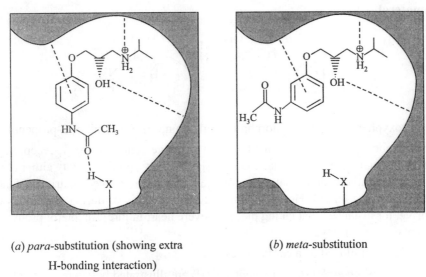

(a) *para*-substitution (showing extra   (b) *meta*-substitution
       H-bonding interaction)

Fig. 12.1.   (*a*) and (*b*) Bonding Interactions with β₁-Receptors.

[Adapted from : Patrick GL, *An Introduction to Medicinal Chemistry*, Oxford University **Press,** Oxford UK, 2nd edn., 2001]

Special Note. The subsequent replacement of the acetamido functional moiety with other groups that are certainly capable of hydrogen bonding ultimately gave rise to the synthesis of a series of highly specific cardioselective β₁-blockers that were eventually employed as potent drugs, namely : Atenolol ; Betaxolol ; Metoprolol.

5.2.3.   Third Generation β-Blockers

It has been stated earlier that the *first generation* and the *second generation* β-blockers essentially possess either **isopropyl** or *tertiary*-**butyl** functional moieties attached to the N-alkyl groups.

The wisdom and skill of the '*medicinal chemist*' further, promulgated the on-going research through the **'extension tactics'** that essentially incorporated the strategical addition of **arylalkyl functional moieties** attached to the terminal N-atom thereby producing a new series of *third generation β-blockers* that get ultimately bound to the β₁-receptor by means of an additional H-bonding interaction.

A few typical examples of the third generation β-blockers are discussed below ; namely : Epanolol :

5.2.3.1.   Epanolol

N-[2-[[3-(2-Cyanophenoxy)-2-hydroxy-propyl] amino] ethyl]-4-hydroxybenzeneacetamide ; Visacor$^{(R)}$;

Epanolol is a cardioselective β-blocker. It possesses intrinsic sympathomimetic activity to a certain extent. It finds its application as an antihypertensive as well as antianginal agent.

5.2.3.2.   Xamoterol

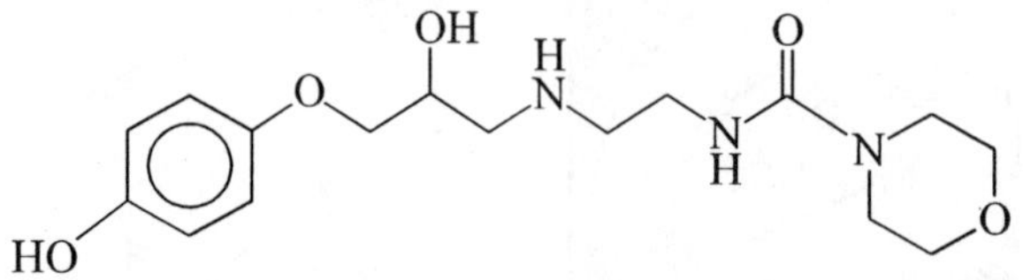

1-(4-Hydroxyphenoxy)-3-[2-(4-morpholinocarboxamido) ethylamino]-2-propanol ;

It is a β-adrenoceptor partial agonist having a selective action on the β_1-receptors. It has been observed that as a partial agonist it normally causes significantly agonist activity either at rest or under conditions of low sympathetic drive that may ultimately result in improved ventricular function and an obvious enhanced cardiac output. However, either during exercise or during conditions of enhanced sympathetic drive, for instance : that taking place in severe heart failure, xamoterol produces distinct β-adrenoceptor antagonist activity.

It is invariably employed as a cardiotonic.

The structures of *epanolol* (I) and *xamoterol* (II) are illustrated below respectively, wherein the specific moieties particularly involved in the additional hydrogen bonding have been demarcated explicitely as given below :

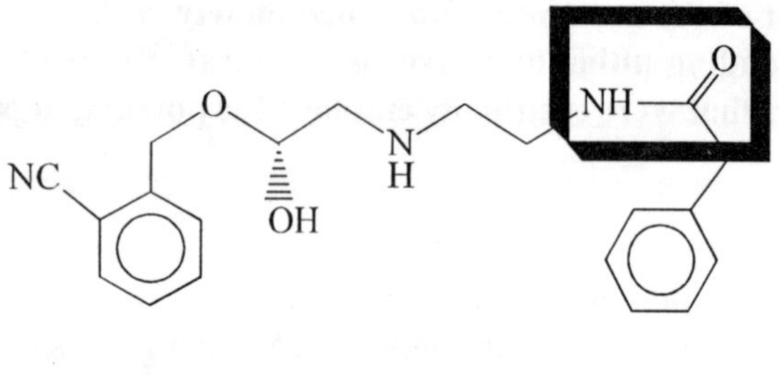

Epanolol (I) 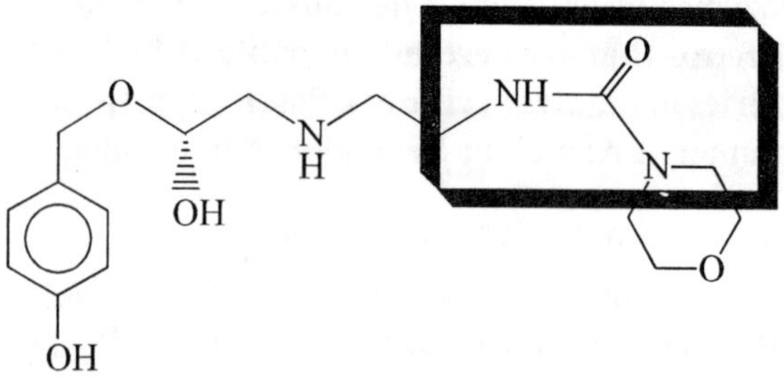   Xamoterol (II)

It has been observed that xamoterol is highly selective as a β_1-partial agonist ; and has been indicated largely in the control, management and treatment of mild heart failure. Interestingly, it emphatically acts as an *agonist* and it augments cardiac stimulation when the subject is usually taking rest ; however, (II) serves as a β-blocker in the course of a strenuous exercise *i.e.,* a situation when excessive quantum of epinephrine (EP) and norepinephrine (NE) are being produced simultaneously *in vivo.*

5.3. Alpha- and Beta-Adrenergic Receptor Blocking Agent

A. *Labetalol* INN, *Labetalol Hydrochloride* BAN, USAN,

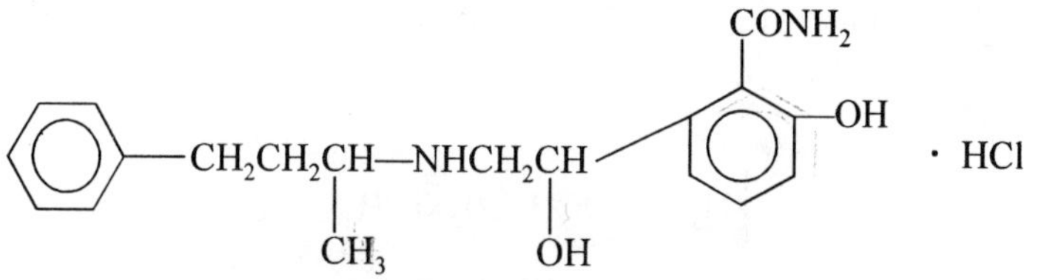

2-Hydroxy-5 [1-hydroxy-2-(1-methyl-3-phenylpropylamino) ethyl] benzamide hydrochloride ; Trandate[R] (Glaxo Inc.)

It is an antihypertensive agent with β-adrenoceptor blocking actions similar to those of propranolol hydrochloride. Besides, it possesses α-adrenoceptor blocking properties which lower blood pressure by drecreasing peripheral vascular resistance.

Dose : Usual, initial, 100 or 200 mg twice per day with food.

6. CHOLINOMIMETIC (PARASYMPATHOMIMETIC) DRUGS

The **cholinomimetic or parasympathomimetic or cholinergic drugs** are those which cause a muscarinic action on the receptors of the effector organs provided by the post-ganglionic cholinergic nerves. Invariably, these drugs exert their action in two different ways, namely : *direct action,* whereby they act on the cholinoceptive receptors like acetylcholine ; *indirect action,* by rendering the cholinesterase enzymes inactive and preserving endogenously secreted acetylcholine, *e.g.,* anticholinesterase drugs like physostigmine (naturally occurring, neostigmine and pyridostigmine (synthetic).

Cholinomimetic drugs may be broadly classified under the following *two* categories. They are :

(*a*) Directly Acting

(*b*) Indirectly Acting (Anticholinesterase Drugs).

6.1. Directly Acting

A few typical examples of medicinal compounds bleonging to this category are discussed below : acetylcholine chloride ; bethanechol chloride ; carbachol ; methacholine chloride ; pilocarpine nitrate, etc.

A. *Acetylcholine* INN, *Acetylcholine Chloride* BAN, USAN,

$$\left[H_3CCOOCH_2CH_2\overset{\oplus}{N}(CH_3)_3 \right] \cdot Cl^{\ominus}$$

Choline chloride acetate ; (2-Acetoxyethyl) trimethyl-ammonium chloride ; (2-Hydroxyethyl) trimethylammonium chloride, acetate ; Ethanaminium, 2-(acetyloxy)-N, N, N-trimethyl-, chloride ; USP. Miochol[R] (Cooper Vision Pharm.) ;

Synthesis

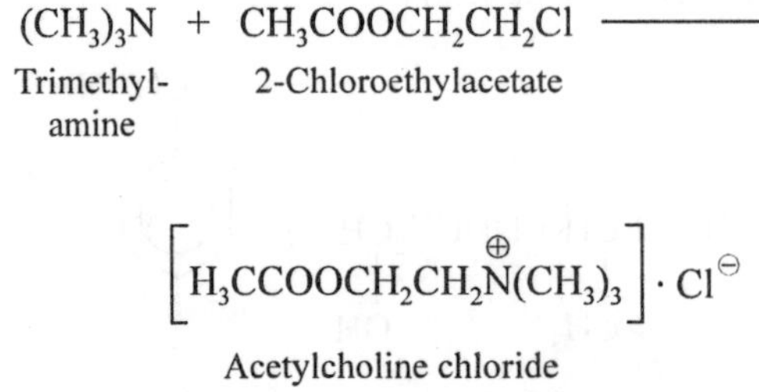

$$(CH_3)_3N \ + \ CH_3COOCH_2CH_2Cl$$

Trimethyl-   2-Chloroethylacetate
amine

$$\left[H_3CCOOCH_2CH_2\overset{\oplus}{N}(CH_3)_3 \right] \cdot Cl^{\ominus}$$

Acetylcholine chloride

Acetylcholine chloride may be prepared by the interaction of trimethylamine and 2-chloroethyl acetate.

It is a potent quaternary ammonium parasympathomimetic agent. Its transient action is due to its destruction by cholinesterase. It is a vasodilator and cardiac depressant. The vasodilator action caused by acetylcholine is found to be most prominent in the peripheral vascular areas. It has been *used in a wide range of conditions such as cataract surgery, iridectomy, trophic ulcers, paroxysmal tachycardia, gangrene and Raynaud's disease.*

Dose : Topical, as a 1% solution.

B. *Bethanechol Chloride* BAN, USAN,

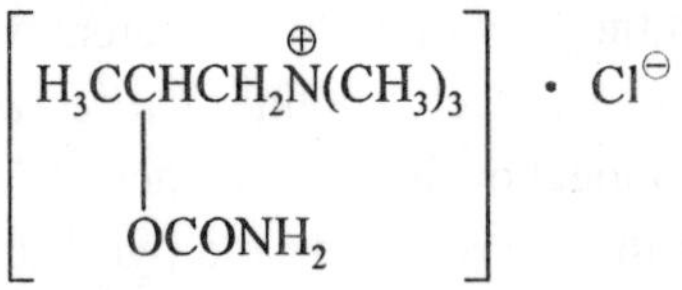

$$\left[H_3CCHCH_2\overset{\oplus}{N}(CH_3)_3 \right] \cdot Cl^{\ominus}$$
$$\underset{OCONH_2}{|}$$

(2-Hydroxypropyl) trimethylammonium chloride carbamate ; Carbmylmethyl-choline chloride ; 1-Propanaminium, 2-[(aminocarbonyl) oxy]-N, N, N-trimethyl-, chloride ; USP. NF/. ; Urecholine[R] (Merck) ; Myotonachol[R] (Glenwood)

Synthesis

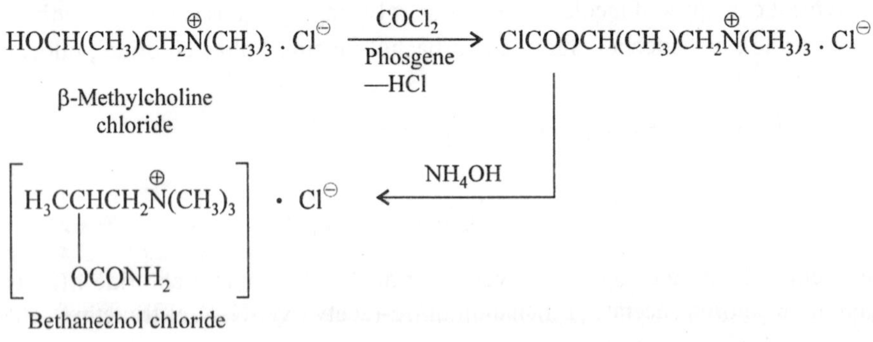

$$HOCH(CH_3)CH_2\overset{\oplus}{N}(CH_3)_3 \cdot Cl^{\ominus} \quad \xrightarrow[\substack{Phosgene \\ -HCl}]{COCl_2} \quad ClCOOCH(CH_3)CH_2\overset{\oplus}{N}(CH_3)_3 \cdot Cl^{\ominus}$$

β-Methylcholine
chloride

$$\left[H_3CCHCH_2\overset{\oplus}{N}(CH_3)_3 \right] \cdot Cl^{\ominus} \quad \xleftarrow{NH_4OH}$$
$$\underset{OCONH_2}{|}$$

Bethanechol chloride

Bethanechol chloride is perpared by the interaction of β-methylcholine chloride with phosgene in chloroform solution followed by treatment of the resulting product with ammonium hydroxide.

It is not promptly inactivated by hydrolysis in the presence of enzyme cholinesterase, thereby enabling it to exert comparatively prolonged parasympathomimetic action. It is usually employed in the *treatment of functional urinary retention and postvagotomy gastric atony.*

Dose : Oral, 5 to 30 mg 3 or 4 times per day ; subcutaneous, 2.5 to 10 mg 3 or 4 times daily.

C. *Carbachol* INN, BAN, USAN,

$$[H_2NCOO . CH_2CH_2N^{\oplus} (CH_3)_3] . Cl^{\ominus}$$

Choline chloride carbamate ; Carbamoylcholine chloride ; Ethanaminium, 2-[(aminocarbonyl) oxy]-N, N, N-trimethyl-chloride ; BP. (1973), USP ; Ind. P., Int. P. ; Carcholin[R] (MSD) ; Moistat[R] (Alcon)

Synthesis

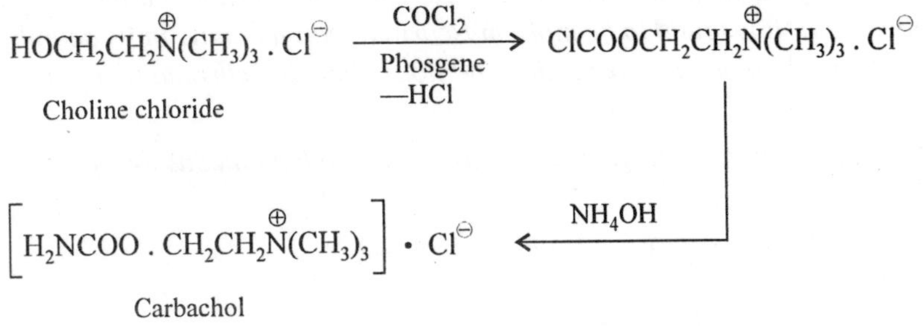

$$\underset{\text{Choline chloride}}{HOCH_2CH_2\overset{\oplus}{N}(CH_3)_3 . Cl^{\ominus}} \xrightarrow[\substack{\text{Phosgene} \\ -HCl}]{COCl_2} ClCOOCH_2CH_2\overset{\oplus}{N}(CH_3)_3 . Cl^{\ominus}$$

$$\underset{\text{Carbachol}}{\left[H_2NCOO . CH_2CH_2\overset{\oplus}{N}(CH_3)_3 \right] . Cl^{\ominus}} \xleftarrow{NH_4OH}$$

Carbachol may be prepared by reacting choline chloride with phosgene in chloroform solution followed by treatment of the product with ammonium hydroxide.

It possesses both muscarinic and nicotinic actions of acetylcholine. It is *used for its miotic actions in the treatment of primary glaucoma. It is employed invariably in urinary retention, peripheral vascular disease and intestinal paresis.*

Dose : Topical, 0.1 ml of a 0.75 to 3% solution.

D. *Methacholine* INN, *Methacholine Chloride* BAN, USAN,

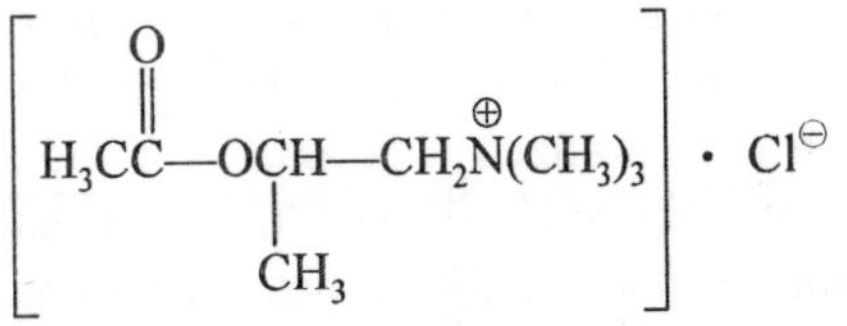

$$\left[\underset{\underset{CH_3}{|}}{\overset{\overset{O}{\parallel}}{H_3CC}-OCH-CH_2\overset{\oplus}{N}(CH_3)_3} \right] . Cl^{\ominus}$$

(2-Hydroxypropyl) trimethylammonium chloride acetate ; Acetyl-β-methylcholine chloride ; (2-Acetoxypropyl) trimethyl-ammonium chloride ; BPC. (1973) ; USP., Ind. P. ; Provocholine[R] (Hoffman La Roche)

Synthesis

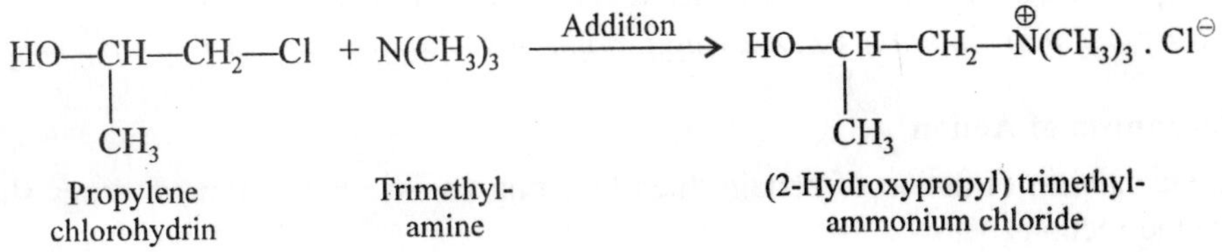

$$\underset{\substack{\text{Propylene} \\ \text{chlorohydrin}}}{HO-\underset{\underset{CH_3}{|}}{CH}-CH_2-Cl} + \underset{\substack{\text{Trimethyl-} \\ \text{amine}}}{N(CH_3)_3} \xrightarrow{\text{Addition}} \underset{\substack{\text{(2-Hydroxypropyl) trimethyl-} \\ \text{ammonium chloride}}}{HO-\underset{\underset{CH_3}{|}}{CH}-CH_2-\overset{\oplus}{N}(CH_3)_3 . Cl^{\ominus}}$$

$$\left[H_3CC\overset{\overset{\displaystyle O}{\|}}{} -O-\underset{\underset{\displaystyle CH_3}{|}}{C} -CH_2\overset{\oplus}{N}(CH_3)_3 \right] \cdot Cl^{\ominus} \xleftarrow{\quad\quad} \begin{array}{c} (CH_3CO)_2O \\ \text{Acetic anhydride} \end{array}$$

Methacholine chloride

(2-Hydroxypropyl) trimethylammonium chloride may be prepared by the addition of propylene chlorohydrin to trimethylamine, which on acetylation with acetic anhydride yields the official compound.

Its actions on the cardiovascular system are more marked and pronounced than on the genitourinary and gastro-intenstinal systems. It has been *used successfully to terminate attacks of supraventricular paroxysmal tachycardia. It is found to be more muscarinic than nicotinic in its actions. It is frequently employed to afford vasodilation in vasospastic conditions, namely : chronic varicose ulcers, cold exposure and phlebits.*

Dose : Usual paroxysmal tachycardia, 10 to 25 mg ; subcutaneous for peripheral vascular disease, 10 to 25 mg.

E. *Pilocarpine Nitrate* BAN, USAN,

Pilocarpine mononitrate ; 2(3H-Furanone, 3-ethyldihydro-4-[(1-methyl-1H-imidazol-5-yl) methyl]-, (3S-*cis*)-, mononitrate ; BP., USP., Eur. P., Ind. P., Int. P. ; PV. Carpine(R) (Allergan)

Preparation

The dried and powdered leaves of *Pilocarpus microphyllus* is subjected to extraction for total alkaloids with ethanol acidified with hydrochloric acid. The solvent is removed under reduced pressure and the resulting aqueous residue is neutralized with ammonia and kept aside till all the resins settle down completely. It is subsequently filtered and the filtrate concentrated by evaporation to a small volume, made alkaline with ammonia and finally extracted with chloroform. The solvent is removed under reduced pressure and the contents dissolved in a minimum quantity of dilute nitric acid and the product is crystallized.

Pilocarpine nitrate is a parasympathomimetic agent possessing muscarinic effects of acetylcholine. *It is mostly used as a solution (1 to 5%) to exert an action on the eye to cause miosis and retard intraocular tension in the treatment of open-angle glaucoma.* Pilocarpine nitrate being less hygroscopic than its corresponding hydrochloride and hence it is more easy to handle.

Dose : Topical, 0.1 ml of 0.5 to 6% solution into the conjunctival sac 1 to 5 times in a day.

6.1.1. Mechanism of Action

The mechanism of action of certain directly acting parasympathomimetic drugs shall now be discussed in the sections below :

6.1.1.1. Acetylcholine Chloride

The '*drug*' is invariably employed as a main topical opthalmological agent to *induce miosis* during certain intraocular surgical operations, namely : cataract surgery, iridectomy, penetrating keratoplasty, and other anterior segment surgery. Importantly, when applied to the intact cornea, ACh penetrates rather too sluggishly to be a clinically efficient miotic. As ACh gets rapidly destroyed by acetylcholin-esterases ; therefore, it has hardly any systemic usages.

6.1.1.2. Bethanechol chloride

It evidently possesses somewhat prominent and apparently stronger muscarinic activity for the GI-tract and the urinary tracts in comparison to the cardiovascular system ; and, therefore, is extensively used systemically exclusively for the *gasteroenterological* and *genitourinary* uses. It is found to be resistant to hydrolysis by the cholinesterases, and perhaps that is why it has a comparatively prolonged duration of action.

6.1.1.3. Carbachol

As on date, the '*drug*' finds its enormous use in ophthalmology, mostly for the control, management and treatment of narrow-angle glaucoma ; besides, to *induce miosis* just prior to ocular surgery. It is worthwhile to mention here that it does not undergo hydrolysis by cholinesterase ; and, hence, exhibits a much longer span of activity in comparison to ACh.

6.1.1.4. Methacholine Chloride

The '*drug*' shows a highly selective muscarinic activity. Interestingly, weak nicotinic actions are manifested usually at the neuromuscular junction specifically in *mysthenic patients*, and also at adrenal medullary tumours in pheochromocytoma*. However, it distinctly exhibits a tendency toward false positives amongst the nonasthmatic smokers and relatives of asthmatics ; there is also a relatively smaller percentage of '*false negatives*'. The *drug* is found to be contraindicated in the presence of β-adrenoreceptor-blocking drugs.

6.1.1.5. Pilocarpine Nitrate

The '*drug*' serves as a potential *muscarinic agonist* which is found to be totally devoid of the nicotinic activity but is found to be nonselective with regard to the muscarinic targets. Pilocarpine nitrate is observed to penetrate membranes in a much better and efficient manner than do quaternary ammonium cholinomimetics perhaps due to its basic tertiary amine characteristic feature.

Interestingly, the free base, pilocarpine, is used frequently in the *ocular controlled-release system*, because the nonionized form may be able to diffuse readily and exclusively through the specific *hydrophobic* membrane.

6.2. Indirectly Acting (Anticholinesterase) Drugs

Following are a few examples of indirectly acting cholinomimetics (anticholinesterase drugs) : demecarium bromide ; edrophonium chloride ; physostigmine salicylate and pyridostigmine bromide.

---

*A well encapsulated, lobular, vascular tumour of chromaffin tissue of the adrenal medulla or sympathetic paraganglia.

A. *Demecarium Bromide* INN, BAN, USAN,

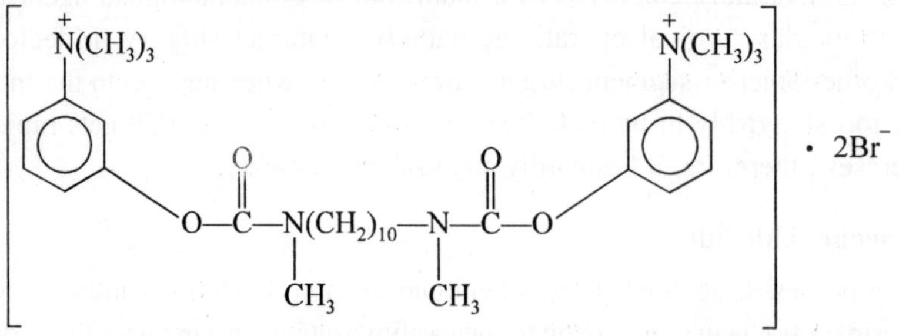

(*m*-Hydroxyphenyl) trimethylammonium bromide decamethylenebis (methyl-carbamate) (2 : 1) ;
3, 3'-[N N'-Decamethylenebis (methyl-carbamoyloxy] *bis*-(NNN-trimethylanilinium) dibromide ; USP.,
NF. Humorsol[R] (Merck)

Synthesis

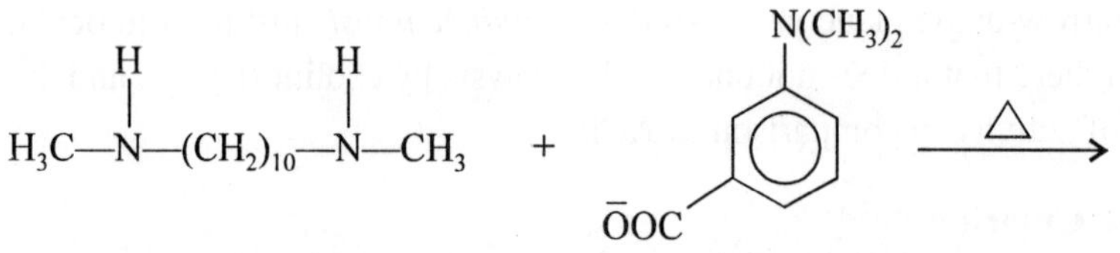

N, N'-Dimethyl-1, 10-
decamethylene diamine

3-(Dimethylamino)-
phenyl carbonate

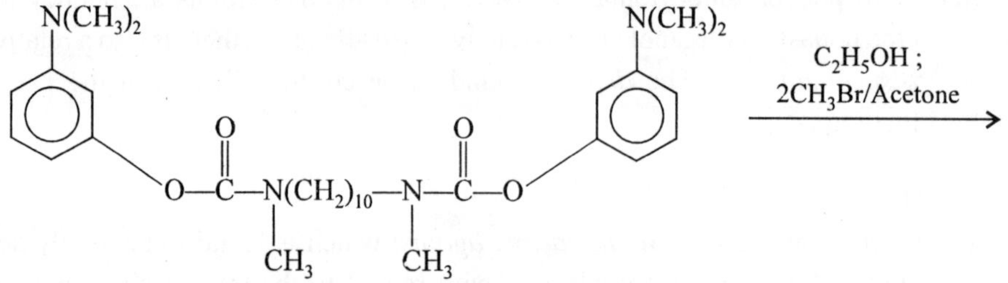

C$_2$H$_5$OH ;
2CH$_3$Br/Acetone

1, 10-Decamethylenebis [3-(dimethylamino)-
phenyl N-methylcarbamate]

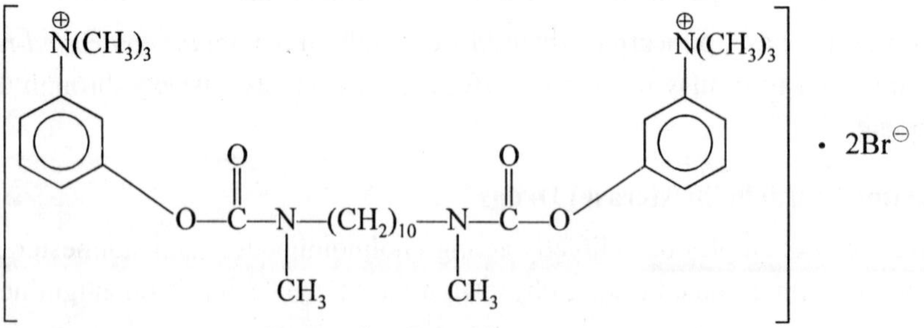

Demecarium Bromide

The interaction of N, N′-dimethyl-1, 10-decamethylene diamine and molten 3-(dimethylamino) phenyl carbonate yields 1, 10-decamethylenebis [3-(dimethylamino)-phenyl N-methyl-carbamate] which on subsequent dissolution in ethanol and treatment with methylbromide in acetone gives the doubly quaternized official compound.

Demecarium bromide is a quaternary ammonium anticholinesterase drug which *possesses a very high degree penetrability in the eye. It is employed in the treatment of open-angle glaucoma and accomodative convergent strabismus (esotropia) by means of local instillation into the affected eye.*

Dose : Topical, to the conjunctiva, 1 to 2 drops of a 0.125 to 0.25% solution twice weekly to 1 or 2 times per day.

B.   *Edrophonium* INN, *Edrophonium Chloride* BAN, USAN,

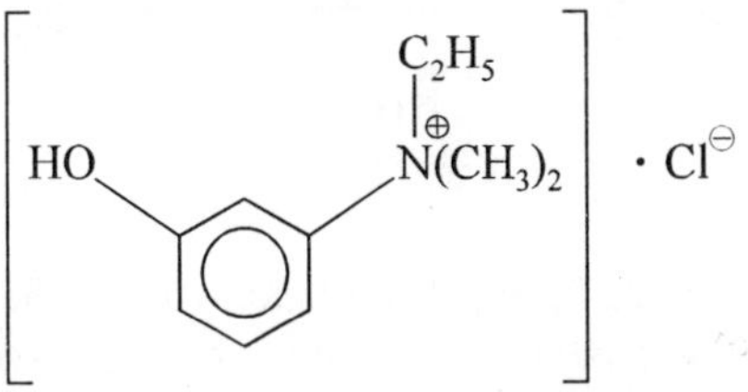

Ethyl-(*m*-hydroxyphenyl) dimethylammonium chloride ; Benzenaminium, N-ethyl-3-hydroxy-N, N-dimethyl-, chloride ; BP., USP., Int. P. ; Tensilon[R] (Hoffmann La Roche)

Synthesis

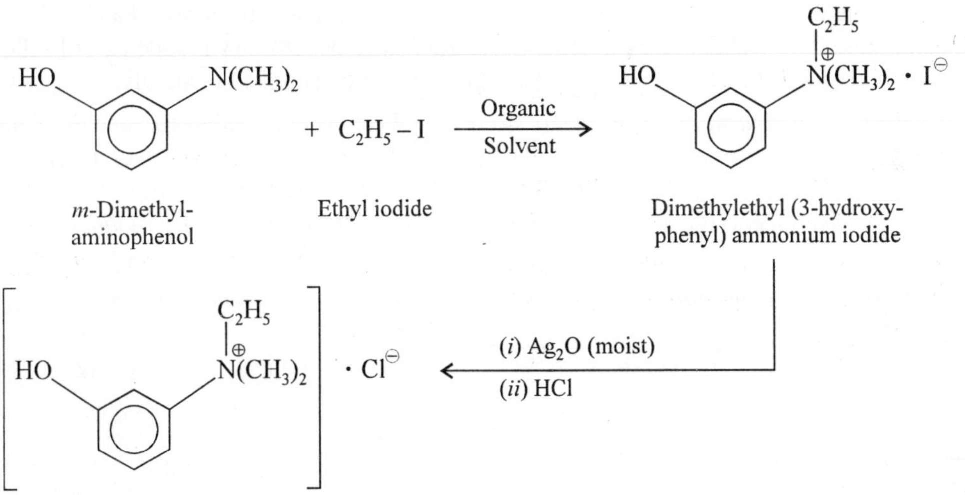

Dimethylethyl (3-hydroxyphenyl) ammonium iodide is prepared by quaternization of *meta*-dimethylaminophenol with ethyl iodide in a suitable organic solvent. Edrophonium chloride may now be obtained *via* treatment with moist silver oxide followed by neutralization with HCl.

It is of particular *utility in the diagnosis of myasthenia gravis.* It may also be *employed to make a clear distinction between a myasthenic crisis and a cholinergic crisis, because in the first instance an improvement of neuromuscular function is usually observed while in the second it worsens it further.*

Dose : Intravenous, 2 to 10 mg ; usually 2 mg is injected initially and if no adverse reaction takes place within 30 seconds, the remaining 8 mg is injected.

C. *Physostigmine Salicylate* USAN

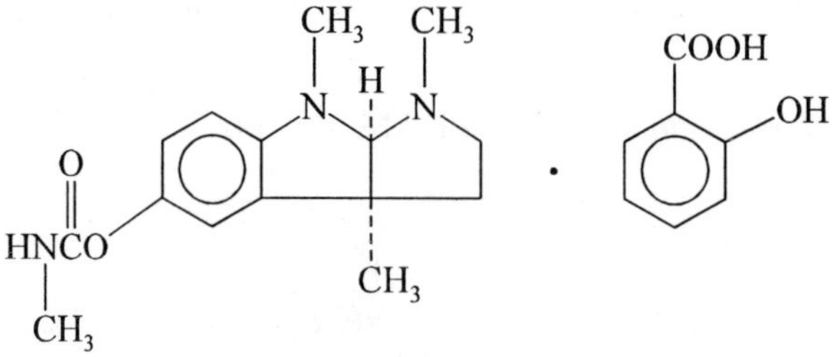

Pyrrol [2, 3-*b*] indol-5-ol, 1, 2, 3, 3a, 8, 8a-hexahydro-1, 3a, 8-trimethyl-, methyl-carbamate (ester), (3a S-*cis*), mono (2-hydroxybenzoate) ; Eserine Salicylate ; BP ; USP ; Eur. P., Int. P., Ind. P. ; Isopto-Eserine[(R)] (Alcon)

Preparation

The powdered seeds of *Physostigma venenosum* Balfour (*Family : Leguminoseae*) is extracted with hot alcohol. The excess of alcohol is distilled off under reduced pressure, the residue made alkaline with sodium carbonate and extracted with solvent ether. From the resulting solution physostigmine is removed with the aid of sulphuric acid. The base is liberaetd again from an alkaline medium. The official compound is now obtained by treating two parts of physostigmine with one part of salicylic acid in 35 parts of boiling distilled water and finally allowing it to crystallize out slowly.

It is *used chiefly as a miotic.* The constriction of pupil commences within 10 minutes of application and the effect lasts up to 12 hours. It is also *employed to decrease intra-ocular pressure in glaucoma.* The salicylate is comparatively less deliquescent than the sulphate. The drug is invariably *recommended for marginal corneal ulcers.* It also finds its seldom *use for atony of the urinary bladder.*

Dose : Topical, for open-angle glaucoma, 0.1 ml of a 0.25 to 5% solution instilled into the conjunctival sac 2 or 4 times daily :

D. *Pyridostigmine* INN, *Pyridostigmine Bromide* BAN, USAN,

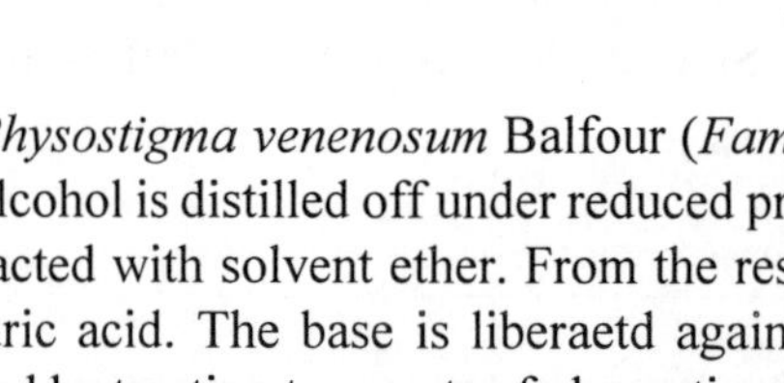

3-Hydroxy-1-methylpyridinium bromide dimethylcarbamate ; Pyridinium, 3-[[(dimethylamino)-carbonyl] oxy]-1-methyl-, bromide ; BP ; USP ; Int. P. ; Mestinon[R] (Hoffmann La Roche) ; Regonol[R] (Organon)

Synthesis

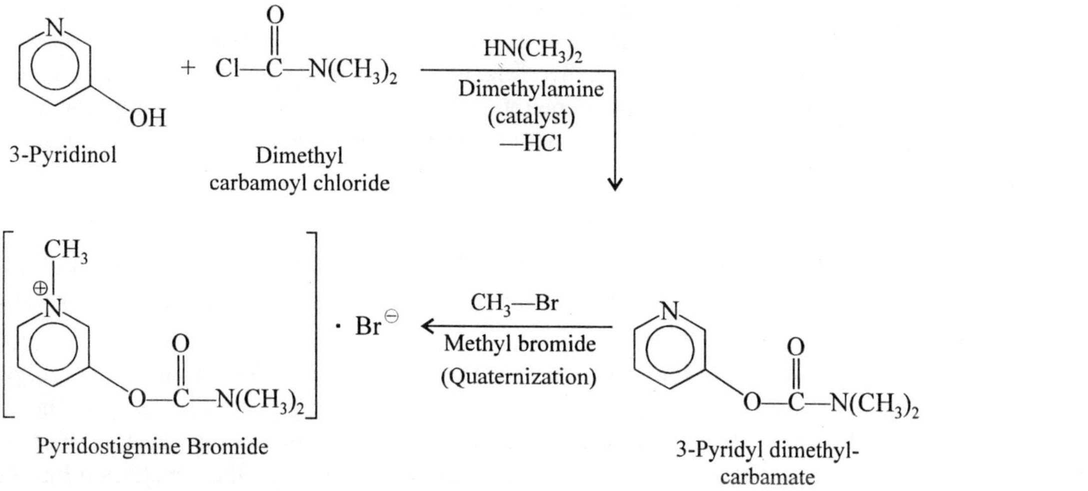

3-Pyridyl dimethyl carbamate is prepared by the interaction of 3-pyridinol with dimethyl carbamoyl chloride in the presence of a basic catalyst like dimethyl amine with the loss of a mole of HCl. The resulting product is quaternized and methyl bromide to yield the official compound.

Pyridostigmine bromide is abundantly employed in the treatment of myasthenia gravis. It is also used in the *treatment of paralytic ileus or postoperative urinary retention.*

Dose : Initially, 60 mg every 4 to 8 hours, but 120 to 300 mg 6 times daily is the usual dose.

6.2.1.　Mechanism of Action

The mechanism of action of the *four* indirectly acting (anticholinesterase) drugs discussed in the section 12.6.2. are described below :

6.2.2.　Demecarium Bromide

It is a quaternary ammonium anticholinesterase '*drug*' which shows relatively high topical penetrability into the eye. Importantly, the inhibition of acetylcholinesterase has several consequences depending upon exactly where the enzymes are inhibited. It has, however, been observed that neither the butyrylcholinesterase located in plasma nor the acetylcholinesterase present in erythrocyets do possess any well-defined functions ; and, hence, their inhibition has no known physiological consequences. Interestingly, such an inhibition may ultimately result moderate increase in the plasma halflife and the concentration of ACh together with some other *hydrolyzable choline esters.* However, the most pivotal and critical effects accrue to inhibition at the specific sites of cholinergic neuroeffector transmission. Nevertheless, the actual and realistic preservation of ACh at such sites not only prolongs but also intensifies the cholinergic activity precisely and effectively.

Consequently, bradycardia, partial heart block, miosis, enhanced gastric secretion and motility and tendency to urinate all result from the appreciable anticholinesterase profile.

6.2.3. Physostigmine Salicylate

It enjoys the reputation of being one of the oldest anticholinesterases. It has been observed that the '*drug*' invariably combines with the said enzyme particularly at the *esteratic site* to give rise to the corresponding *methylcarbamoyl enzyme*, which is **inactive**. Importantly, it is found to share with neostigmine marked and pronounced stimulatory activities on the bowel but produces excessive functions, such as : secretion of glands, effect on BP, constriction of the pupil, and relatively lesser action on the skeletal muscle in particular. By virtue of its inherent characteristic feature as a tertiary amine, it gets penetrated into the nervous system, and, therefore, may cause central activities on being administered in sufficiently large doses. It is found to penetrate rapidly into the eye. Interestingly, the salicylate salt exerts a dual purpose *viz., first*, its specific action on CNS ; and *secondly*, its ability to cause ophthalmological activities.

6.2.4. Pyridostigmine Bromide

The '*drug*' is a quaternary ammonium salt which is observed to be a reversible inhibitor of cholinesterase activity having activities very much akin to those of neostigmine, but is definitely much slower in onset and of longer duration.

Physostigmine is very sluggishly absorbed from the GI-tract. The drug gets hydrolyzed by the enzymes cholinesterases, and is also metabolized in the liver. It is excreted mostly in the urine as its metabolites and partly as unchanged drug. The '*drug*' happens to cross the placental barrier and only very small quantum are usually excreted in the breast-milk. However, the penetration into the CNS is comparatively very slow.

Interestingly, the '*drug*' is approximately $\frac{1}{4}$ th as potent as neostigmine at the *neuromuscular junction* ; and nearly $\frac{1}{8}$ th as potent on the *exocrine glands, bowel*, and *gentiurinary tract*. Finally, the '*drug*' is frequently employed to antagonize competitive neuromuscular-blocking drugs.

7. ANTIMUSCARINIC (ANTICHOLINERGIC) AGENTS

The term **'antimuscarinic'** is derived from the action of acetylcholine at the postganglionic synapse which is closely imitated by the alkaloid, muscarine.

Antimuscarinic agents like atropine chiefly exert their action by blocking the normal responses to excitation of postganglionic parasympathetic (cholinergic) nerves that stimulate both smooth muscle as well as exocrine glands. Sometimes they are also termed as parasympatholytic and as anticholinergic agents. They usually act by preventing the normal effect of acetylcholine on the receptor cells.

The antimuscarinic effects mainly include an elevation of heart-rate. Besides, a diminution in the production of bronchial, lachrymal, gastric, nasal, intestinal, sweat and saliva secretions are observed together with a reduction in intestinal motility.

It has been established beyond any reasonable doubt that specifically the cholinergic transmission invariably takes place either at the neuroeffectors innervated by the parasympathetic and some sympathetic postganglionic nerves or mostly at all autonomic ganglia *viz.,* the somatic neuromuscular junction, besides certain central synapses. Broadly speaking, the '*anti-muscarinic agents*' behave as **competitive antagonists** which exert their action exclusively upon the cholinergic receptors at smooth muscle, secretory cells, and some central synapses. There are several synonymous terminologies for the **anti-muscarinic agents**, namely : *anticholinergic, cholinolytic, parasympatholytic*, and *parasympathetic blocking drugs.*

As a host of cholinergic, ganglionic and neuromuscular blocking drugs do commonly antagonize the activity of ACh ; it may, therefore, be expected that a few of these drug substances would help in blocking at more than one type of cholinergic receptor.

Antimuscarinic agents may be classified on the basis of their chemical structures under the following heads :

(*i*) Aminoalcohol Esters.

(*ii*) Aminoalcohol Ethers.

(*iii*) Aminoalcohol Carbamates.

(*iv*) Aminoalcohols.

(*v*) Aminoamides.

(*vi*) Diamines.

(*vii*) Miscellaneous Amines.

7.1. Aminoalcohol Esters

Aminoalcohol esters, having the following general formula, have emerged from the outstanding pharmacological actions displayed by atropine :

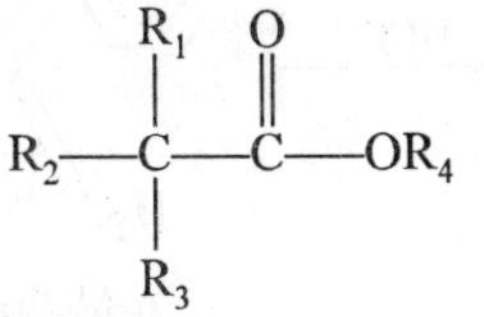

A follow-up from this generalised structure resulted into the formation of an altogether new breed of '*tropeines*' that possessed remarkable antimuscarinic activity. A few typical examples of this category of compounds are discussed below, namely :

A.   *Atropine* BAN, USAN,

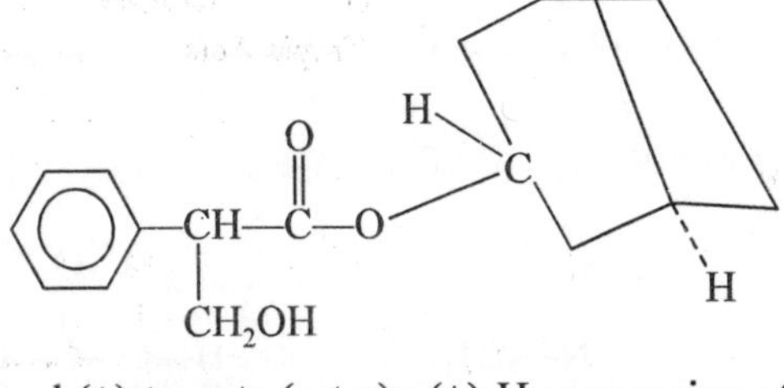

1 αH, 5αH-Tropan-3α-ol (±)-tropate (ester) ; (±)-Hyoscyamine ; (1R, 3R, 5S)-Tropan-3-yl (±)-tropate ; B.P.C. (1973) ; USP ; NF ; Ind. P. ; Atropinol[(R)] (Winzer, Germany)

Synthesis

The synthesis essentially consists of the following *three* parts :

(*a*) synthesis of tropic acid ;

(*b*) synthesis of tropine ; and

(*c*) condensation of tropic acid and tropine.

(a) *Synthesis of Tropic Acid.* Methyl phenyl ketone on treatment with HCN yields methylphenyl nitrile carbinol which on acidification gives the corresponding α-hydroxy-α-phenyl-propionic acid. The resulting product loses a molecule of water upon heating under reduced pressure to give α-methylene-α-phenyl acetic acid which on treatment with HCl undergoes addition reaction according to the Markownikoff's rule and yields α-chloro-methyl-α-phenyl acetic acid. This product on reaction with K_2CO_3 gives tropic acid.

(a) *Synthesis of Tropic Acid*

Methylphenyl ketone

Methylphenyl
nitrile carbinol

α-Hydroxy-α-
phenylpropionic acid

α-Chloromethyl-α-
phenyl acetic acid

α-Methylene-α-
phenyl acetic acid

Tropic Acid

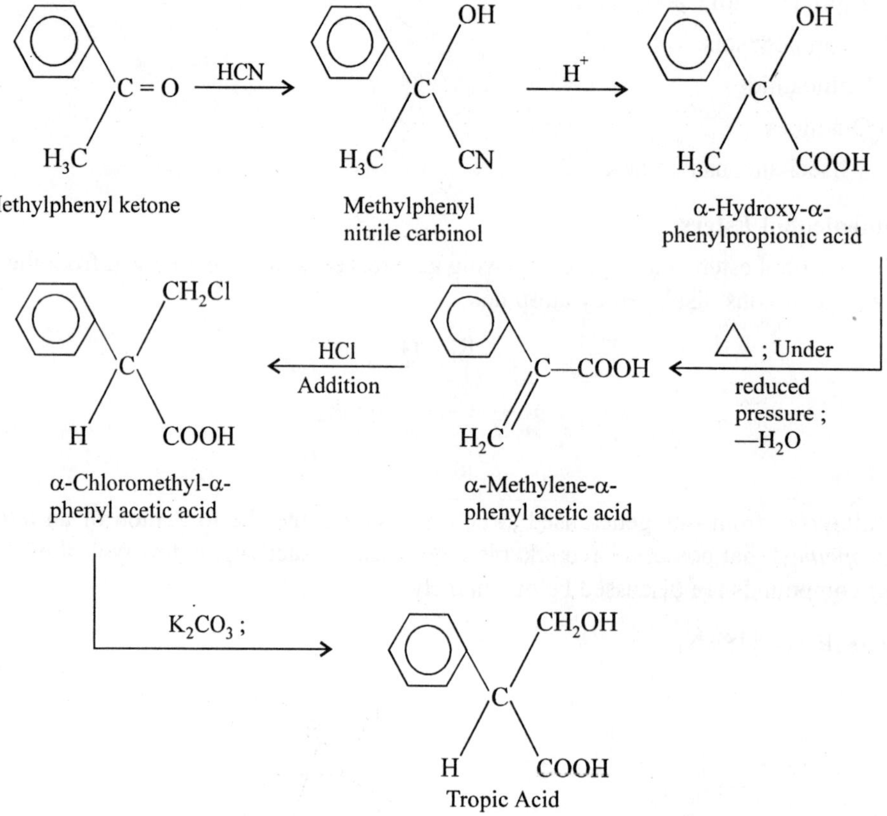

(b) *Synthesis of Tropine*

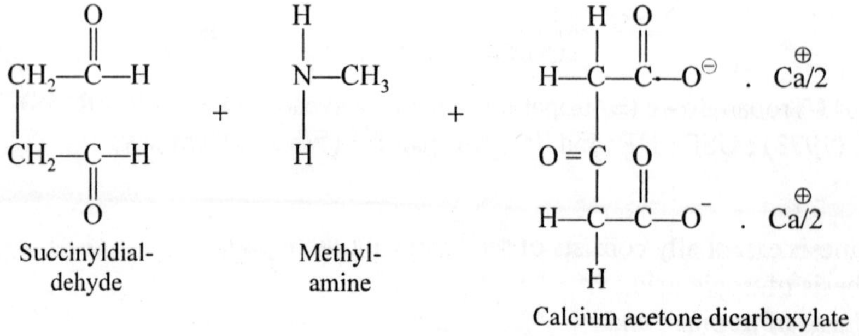

Succinyldial-
dehyde

Methyl-
amine

Calcium acetone dicarboxylate

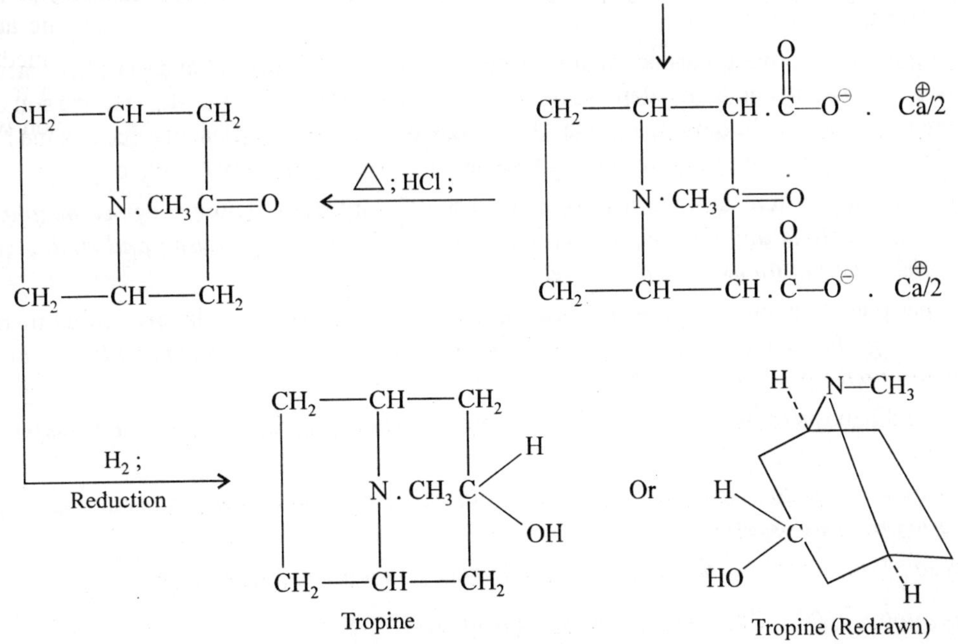

Tropine

Tropine (Redrawn)

The interaction of succinyldialdehyde, methylamine and calcium acetone dicarboxylate yields the corresponding calcium salt which on treatment with HCl and subsequent reduction gives rise to tropine.

(c) *Condensation of Tropic Acid and Tropine*

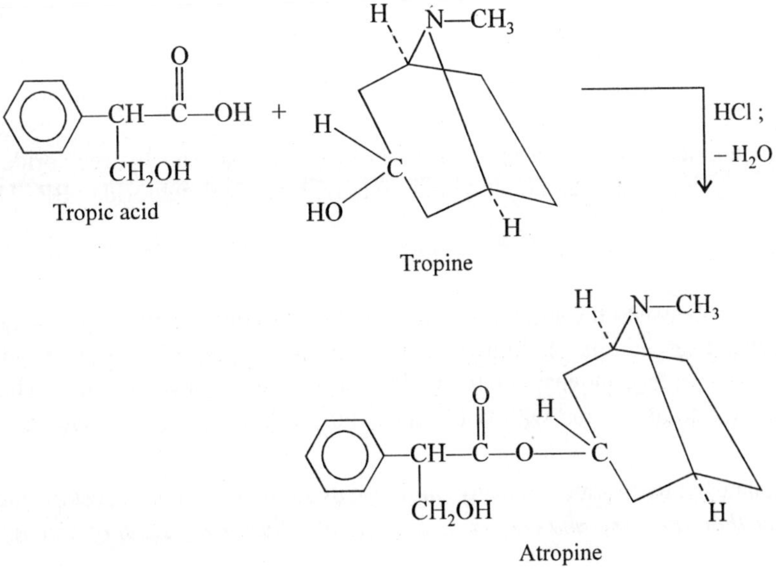

Tropic acid

Tropine

Atropine

A molecule each of tropic acid and tropine in the persence of HCl loses a molecule of water to yield the official product.

Atropine is an antimuscarinic alkaloid which possesses both central and peripheral actions. It exerts first a stimulating and then a depressing action on the central nervous system (CNS) and exhibits antispasmodic actions on the smooth muscle. On account of its broad spectrum of effects in the body the therapeutic applications are numerous, but unfortunately it lacks selectivity of action.

Its centrally potent actions are used in various conditions, namely : *postencephphalitic parkinsonism, paralysis agitans, respiratory stimulation, some types of spastic and rigid states and rarely in schizophrenia (at its toxic dose level).*

Its peripheral actions are so widespread as could be observed from the diversified therapeutic applications, *e.g., in ophthalmology to dilate the pupil and to paralyze accommodation, in bronchial asthma to dry up the bronchial secretions.*

In the diagnosis of heart diseases atropine finds its use to stop *extrasystoles and complete heart block.*

Atropine helps to control the spastic conditions of the bowel, for instance, spastic colitis, cardiospasm and pylorospasm.

Dose : Usual, 0.25 mg thrice daily normally taken 30 minutes before meals.

B. *Cyclopentolate INN, Cyclopentolate Hydrochloride BAN, USAN,*

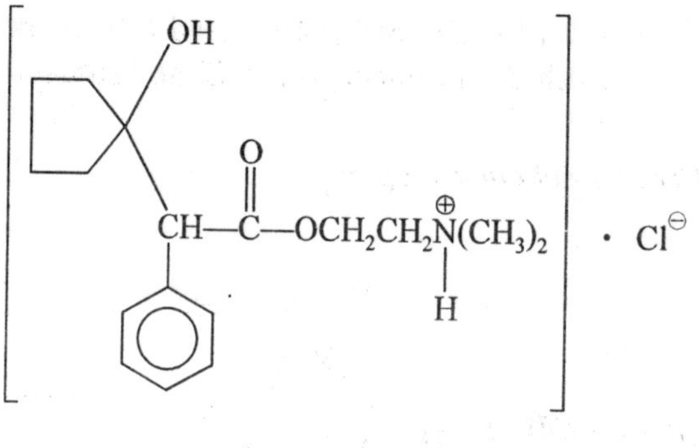

2-(Dimethylamino) ethyl-1-hydroxy-α-phenylcyclopentane-acetate hydrochloride ; Benzeneacetic acid, α-(1-hydroxy-cyclopentyl)-, 2-(dimethylamino) ethyl ester, hydrochloride ; BP ; USP ; Cyclogy[(R)] (Alcon)

Synthesis

The interaction between the sodium salt of phenylacetic acid and isopropyl magnesium bromide results into a doubly charged species known as the *Ivanov reagent.* This product on treatment with cyclopentanone affords aldol condensation to yield the corresponding hydroxy acid. The resulting product on being subjected to alkylation with N-(2-chloro-ethyl)-dimethyl-amine gives the desired official compound.

Cyclopentolate hydrochloride is usually *employed as eye drops to cause cycloplegia and mydriasis. It acts much faster than atropine and possesses a relatively shorter duration of action.*

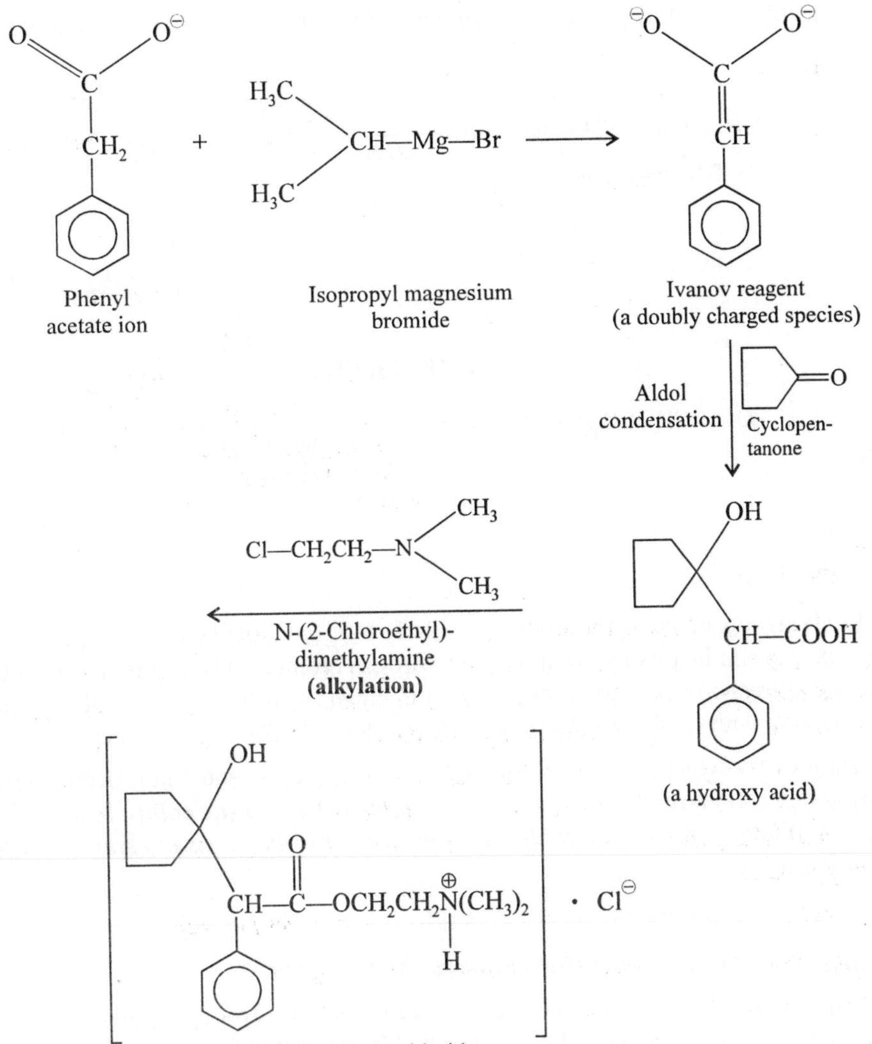

Phenyl acetate ion + Isopropyl magnesium bromide ⟶ Ivanov reagent (a doubly charged species)

Aldol condensation | Cyclopentanone

(a hydroxy acid)

N-(2-Chloroethyl)-dimethylamine (**alkylation**)

Cyclopentolate hydrochloride

Dose : Topical, adult, 1 drop of 1 or 2% solution to the conjuctiva ; for refraction, 1 drop of a 0.5% solution repeated after 5 to 15 minutes.

C. *Dicycloverine* INN, *Dicyclomine* BAN, *Dicyclomine Hydrochloride* USAN,

2-(Diethylamino) ethyl [bicyclohexyl]-1-carboxylate hydrochloride ; Bicyclo-hexyl]-1-carboxylic acid, 2-(diethylamino) ethyl ester, hydrochloride ; BP ; USP ; Bentyl[R] (Merrell Dow) ; Merbentyl[R] (Merrell, U.K.)

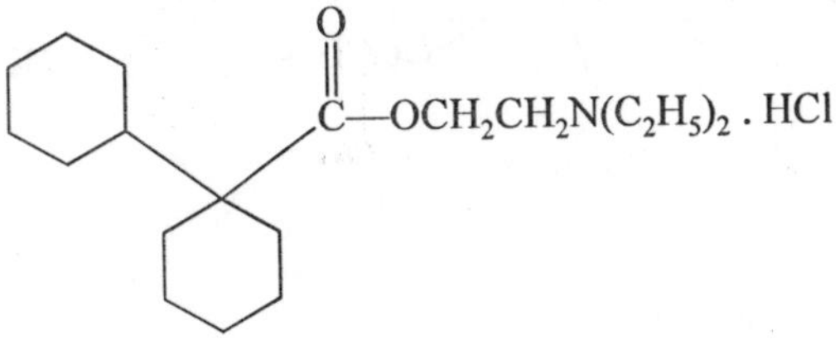

Synthesis

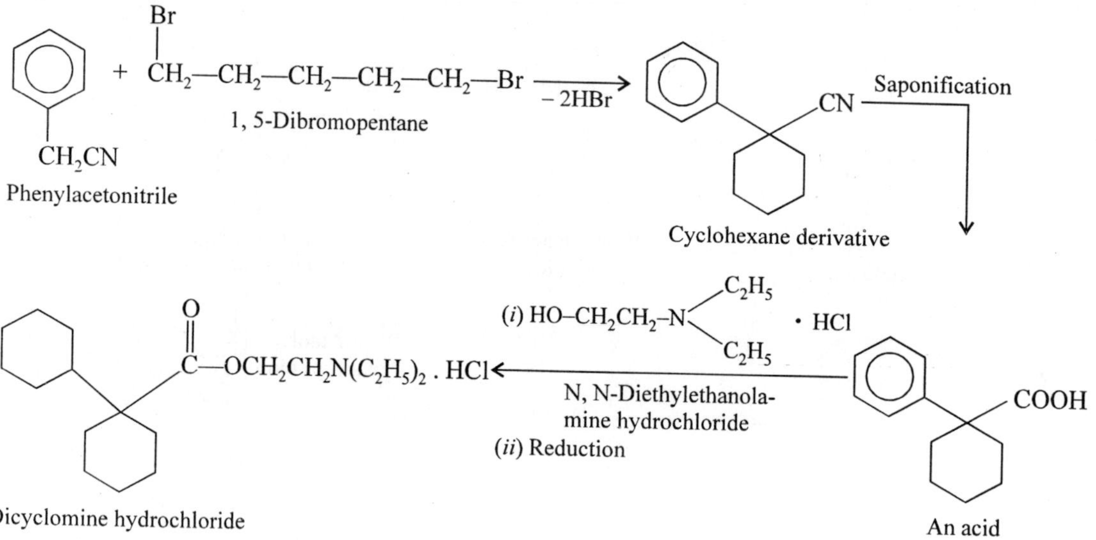

Dicyclomine hydrochloride   An acid

Double alkylation of phenylacetonitrile with 1, 5-dibromopentane yields the corresponding cyclohexane, with the elimination of two moles of hydrogen bromide. The resulting product on saponi-fication gives the corresponding acid, which on *first* treatment with N, N-diethylethanolamine hydro-chloride and *secondly* with catalytic reduction yields the desired official compound.

Dicyclomine hydrochloride behaves both as an antimuscarinic and a nonspecific antispasmodic agent. It is frequently employed in the *treatment of irritable colon, spastic colitis, mucous colitis, spas-tic constipation and biliary dyskinesia. It also finds its use in the diagnosis of peptic ulcer by delaying gastric emptying process.*

Dose : Oral or intramuscular, 10 to 20 mg after 4 to 6 hours per day.

D.   *Homatropine* INN, *Homatropine Hydrobromide* BAN, USAN,

1αH, 3αH-Tropan-3α-ol mandelate (ester) hydrobromide ; Tropyl mandelate hydrobromide ; Benzeneacetic acid, α-hydroxy-, 8-methyl-8-azabicyclol [3, 2, 1] oct-3yl ester hydrobromide, endo- (±) ; BP ; USP ; Eur. P., Int. P. ; Ind. P., Isopto Homatropine[(R)] (Alcon)

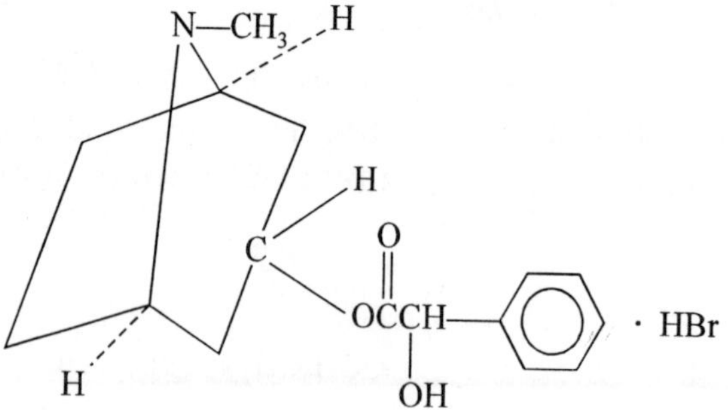

Synthesis

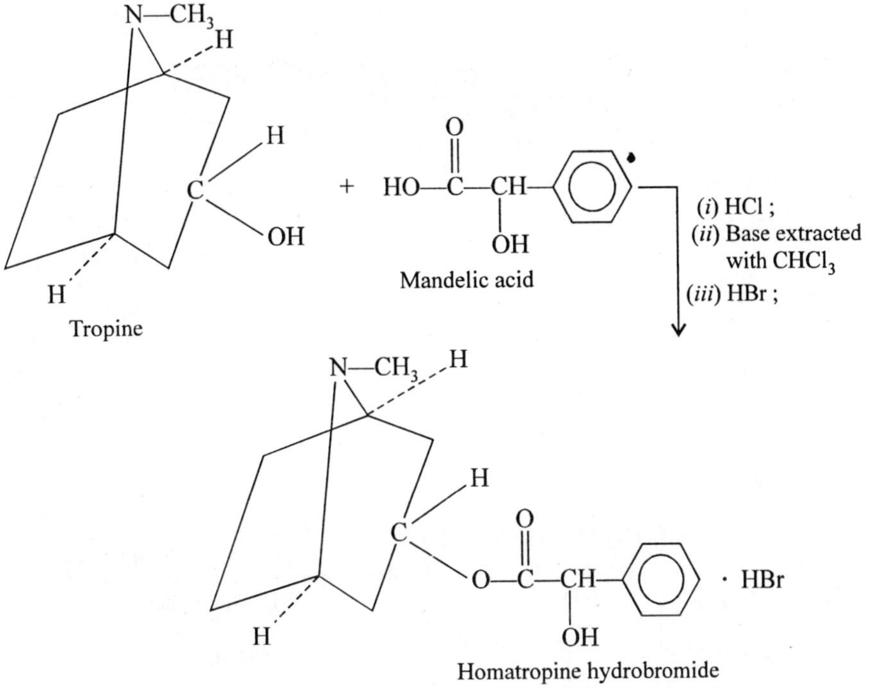

Tropine

Mandelic acid

(i) HCl ;
(ii) Base extracted
    with CHCl₃
(iii) HBr ;

Homatropine hydrobromide

It may be prepared by the interaction of tropine and mandelic acid in the presence of hydrochloric acid. The base is liberated on alkalization with ammonia, extracted with chloroform, the solution is evaporated and treated with hydrobromic acid to obtain the desired compound.

Homatropine hydrobromide is employed as *a mydriatic. It is used preferentially to atropine because of its three vital reasons, namely ; more rapid action ; less prolonged action and conveniently controlled by physostigmine.*

Dose : *Topical, adult, to the conjuctiva, 1 drop of a 2-5% solution given 3 times at 10 minutes intervals.*

E.   *Propantheline* INN, *Propanetheline Bromide* BAN, USAN,

Di-isopropylmethyl [2-(xanthen-9-ylcarbonyloxy) ethyl] ammonium bromide ; (2-Hydroxyethyl) diisopropylmethylammonium bromide xanthene-9-carboxylate ; BP ; USP ; Int. P. ; Pro-Banthine[R] (Searle) ; SK-Propentheline Bromide[R] (SK & F)

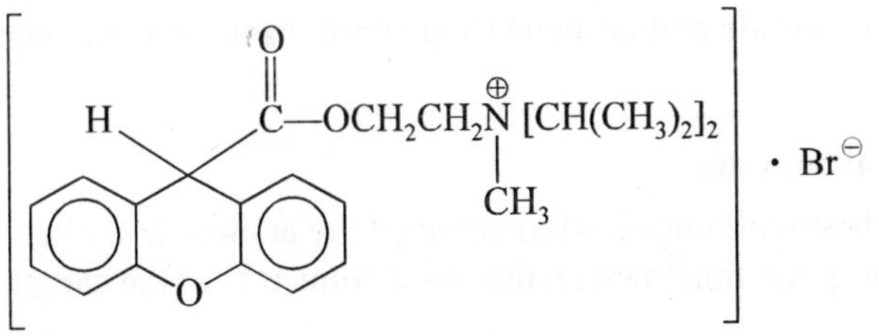

Synthesis

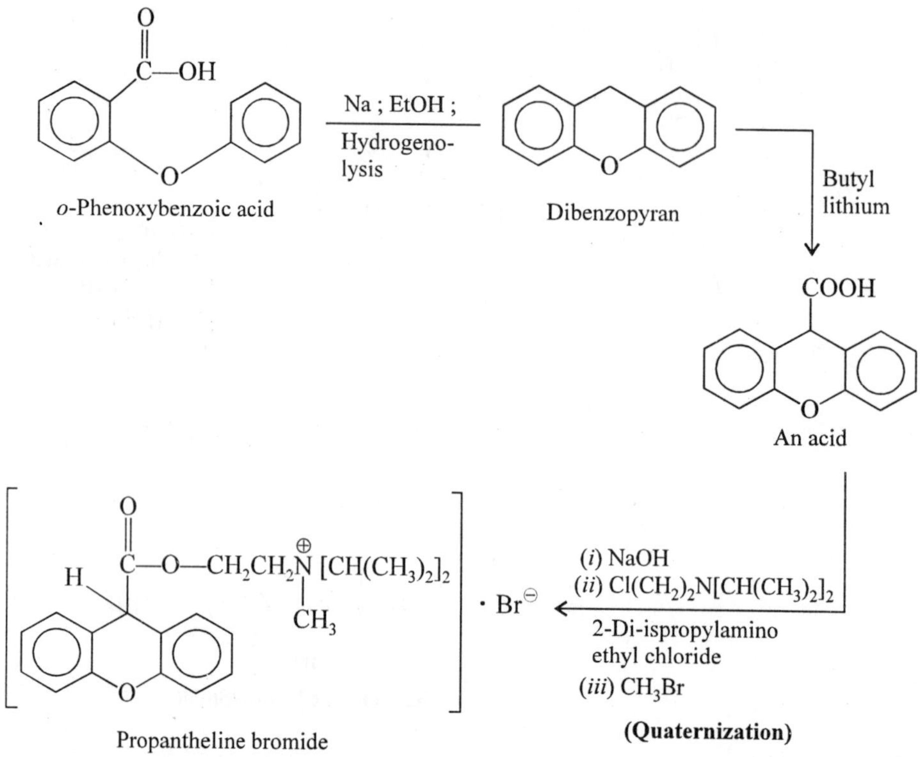

o-Phenoxybenzoic acid   Na ; EtOH ;
Hydrogeno-
lysis

Dibenzopyran   Butyl
lithium

COOH

An acid

Propantheline bromide

(i) NaOH
(ii) Cl(CH$_2$)$_2$N[CH(CH$_3$)$_2$]$_2$
2-Di-ispropylamino
ethyl chloride
(iii) CH$_3$Br

(Quaternization)

     o-Phenoxybenzoic acid undergoes Friedel-Crafts cyclization in the presence of sodium and alcohol to form dibenzopyran, which on treatment with butyl lithium yields the corresponding acid. The resulting compound *first* converted to its sodium salt, *secondly* treated with 2-di-isopropylamino ethyl chloride, and *thirdly* subjected to quaternization with methylbromide to give the desired official compound.

     Propantheline bromide is used in the treatment of acute and chornic pancreatitis, pylorospasm, gastritis, and ureteric and biliary spasm. It also finds its application as an adjunct *in the treatment of gastric and duodenal ulcer.*

     Dose : Usual, initial, 15 mg 3 times per day before meals ; and 30 mg at bed-time.

7.2. Aminoalcohol Ethers

     Generally, the aminoalcohol ethers have been more widely employed as anti-parkinsonism agents rather than as usual antimuscarinic drugs. A few such compounds, namely ; benztropine mesylate, chlorphenoxamine hydrochloride and orphenadrine citrate have been adequately discussed under anti-parkinsonism drugs.

7.3. Aminoalcohol Carbamates

     The concept of structural modification amongst the amino-alcohol carbamates has crept up due to the introduction of non-polar functions on the unsubstituted N of the carbamoyl moiety in the following general formula :

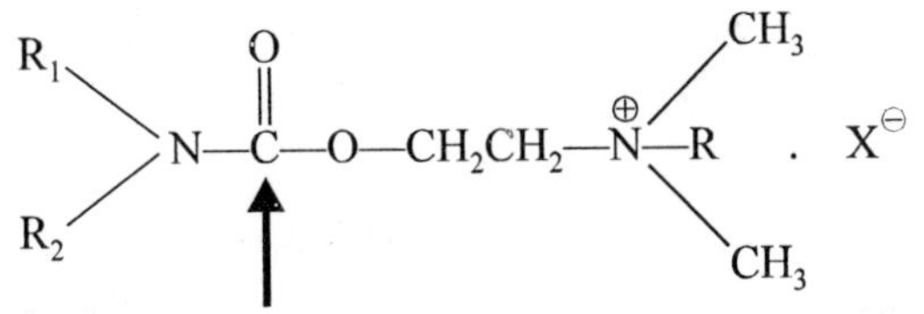

Only one such compound belonging to this category is discussed here.

A. *Fencarbamide* INN, *Phencarbamide* USAN,

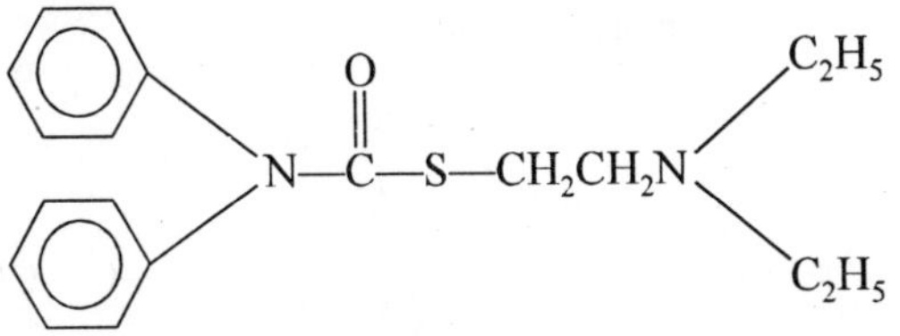

S-[2-(Diethylamino) ethyl] diphenylthiocarbamate ; Carbamothionic acid, diphenyl-1, S-[2-(diethylamino) ethyl] ester ; Escorpal$^{(R)}$ (Farbenfabriken Bayer A.G., W. Germany)

7.4. Aminoalcohols

Most of the aminoalcohols have been developed in the past two decades. These compounds have gained their prominence as antiparkinsonism agents, such as biperiden hydrochloride, procyclidine hydrochloride, trihexyphenidyl hydrochloride, etc., (see chapter on **'antiparkinsonism agents'**).

7.5. Aminoamides

The aminoamides essentially differ from the aminoalcohols whereby the polar hydroxyl group in the latter is replaced by the corresponding polar amide function. However, the bulky structural features found commonly at one terminal end of the molecule remain the same in both the species stated above.

Examples : Isopropamide iodide ; Tropicamide.

A. *Isopropamide* INN, *Isopropamide Iodide* BAN, USAN,

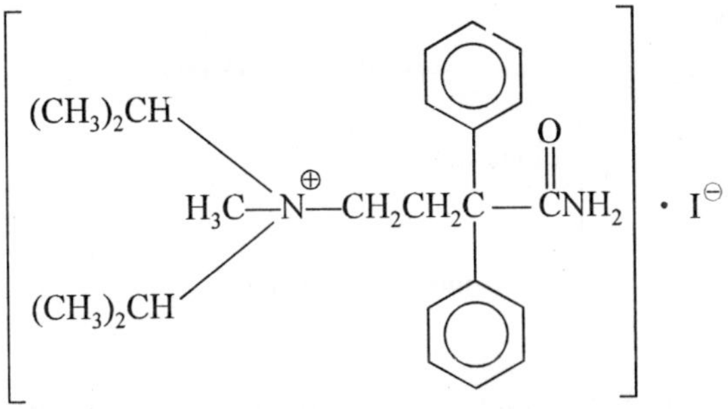

(3-Carbamoyl-3, 3-diphenylpropyl) diisopropylmethylammonium iodide ; Benzenepropana-minium, γ-(aminocarbonyl)-N-methyl-N, N-*bis* (1-methyl-ethyl)-γ-phenyl-, iodide ; USP ; Darbid$^{(R)}$ (SK & F)

Synthesis

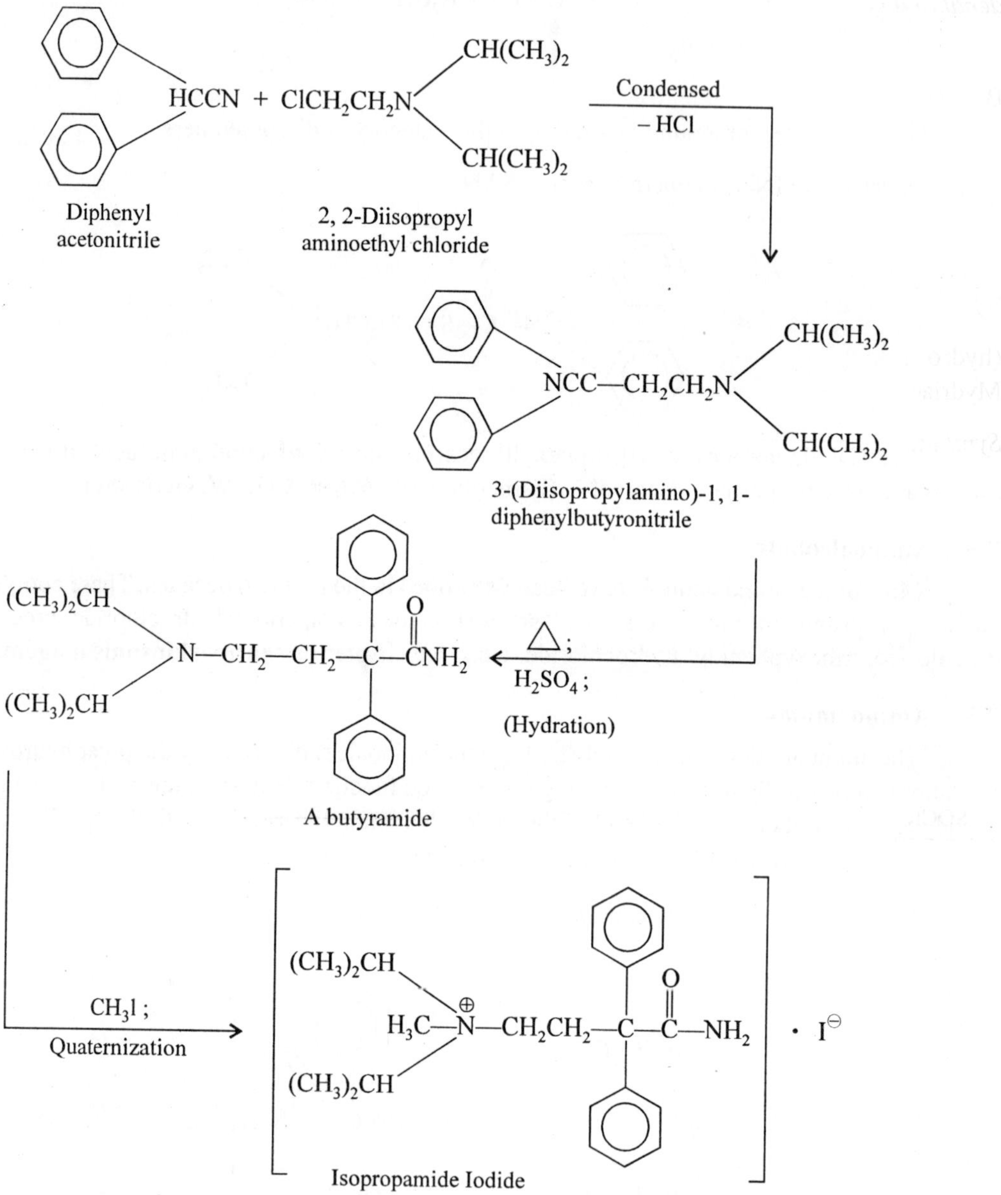

Diphenyl
acetonitrile

2, 2-Diisopropyl
aminoethyl chloride

Condensed
– HCl

3-(Diisopropylamino)-1, 1-
diphenylbutyronitrile

$\triangle$;
H_2SO_4 ;
(Hydration)

A butyramide

CH_3I ;
Quaternization

Isopropamide Iodide

3-(Diisopropylamino)-1, 1-diphenylbutyronitrile is obtained by the condensation of diphenyl acetonitrile with 2, 2-diisopropylamino ethyl chloride, which undergoes hydration by heating in the presence of sulphuric acid to yield the corresponding butyramide. Quaternization of the butyramide with methyl iodide results into the formation of the official compound.

Isopropamide iodide is recommended for *use in the treatment of peptic ulcer and various other states of gastrointestinal hyperactivity. It has also been advocated as an adjunct in the therapy of duodenal and gastric ulcer and invariably in the relief of visceral spasms.*

Dose : Initial, adult, oral, 5 mg twice a day.

B.   *Tropicamide* INN, BAN, USAN,

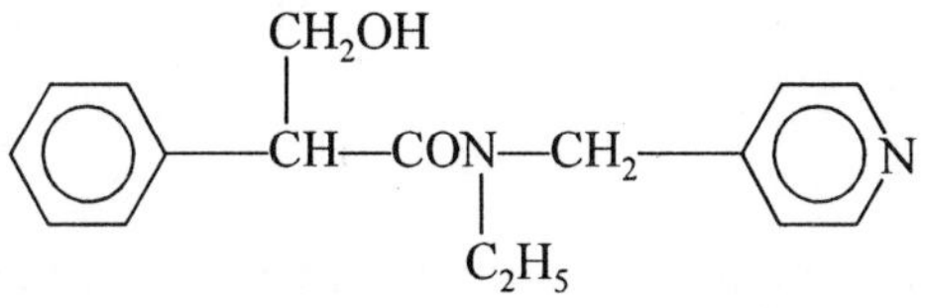

N-Ethyl-2-phenyl-N-(4-pyridylmethyl)-hydracrylamide ; Benzeneacetamide, N-ethyl-α-(hydroxymethyl)-N-(4-pyridinyl-methyl)- ; N-Ethyl-N-(4-pyridyl-methyl) tropamide ; B.P. U.S.P., Mydriacyl[R] (Alcon) ;

Synthesis

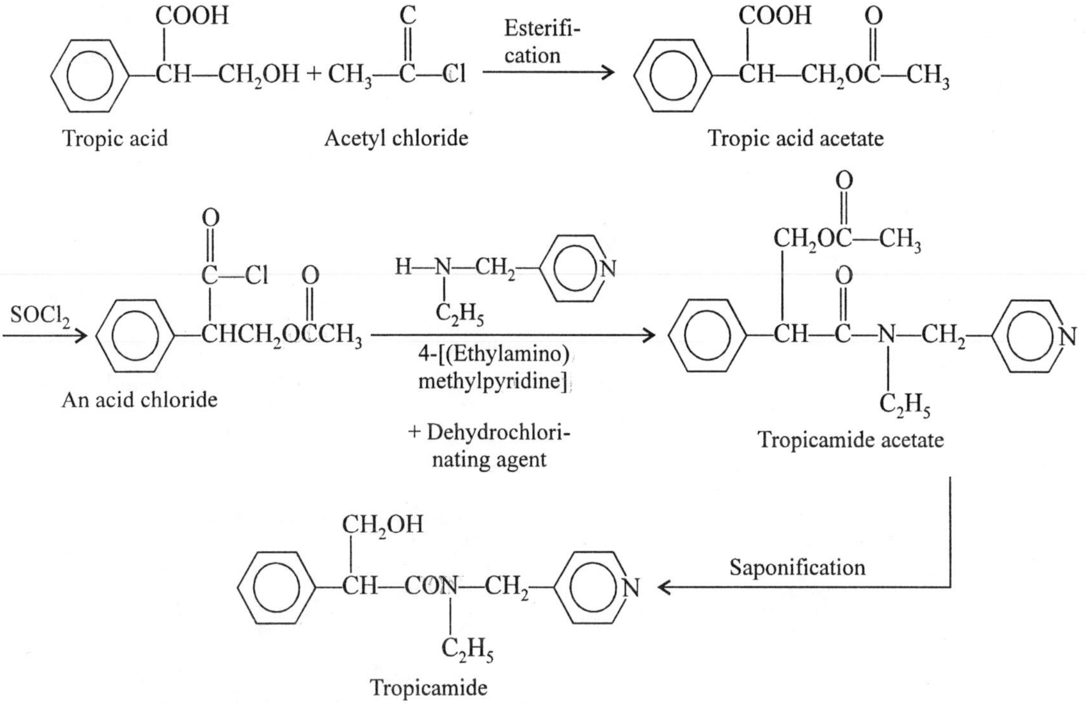

Esterification of tropic acid with acetyl chloride gives tropic acid acetate which on chlorination with thionyl chloride yields the corresponding acid chloride. The resulting product on treatment with 4-((ethylamino) methyl-pyridine and subsequently with an appropriate dehydrochlorinating agent gives rise to tropicamide acetate. Saponification of this compound affords the desired official product.

Tropicamide is *frequently used in the ophthalmologic practice to induce mydriasis and cycloplegia.*

Dose : Usual, adult, topical, 1 to 2 drops of a 0.5 or 1% solution to the conjunctiva ; for mydriasis 0.5% solution is employed and for cycloplegia 1% solution.

7.6.   Diamines

The *two* derivatives of phenothiazine belonging to the classification of diamines are diethazine and ethopropazine hydrochloride which are broadly grouped together under antiparkinsonism agents. The former being more toxic than the latter and hence has been withdrawn. The latter has been discussed under **'antiparkinsonism agents'.**

7.7.   Miscellaneous Amines

The miscellaneous amines essentially consists of two potent compounds both of which possess characteristic bulky group that attribute to the typical features of the usual antimuscarinic molecule.

Examples : Diphemanil methylsulphate, Methixene hydrochloride, Glycopyrronium bromide and Pirenzepine.

A.   *Diphemanil Metisulfate* INN, *Diphemanil Methylsulphate* BAN,

   Diphemanil Methylsulfate USAN,

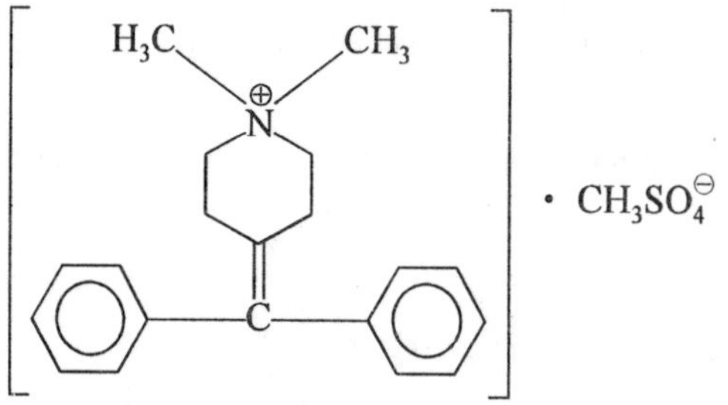

4-(Diphenylmethylene)-1, 1-dimethylpiperidinium methyl sulphate ; 4-Benz-hydrylidene-1, 1-dimethyl piperidinium methyl sulphate ; Piperidinium, 4-(diphenylmethylene)-1, 1-dimethyl, methyl sulphate ; USP ; Prantal[R)] (Schering-Plough)

Synthesis

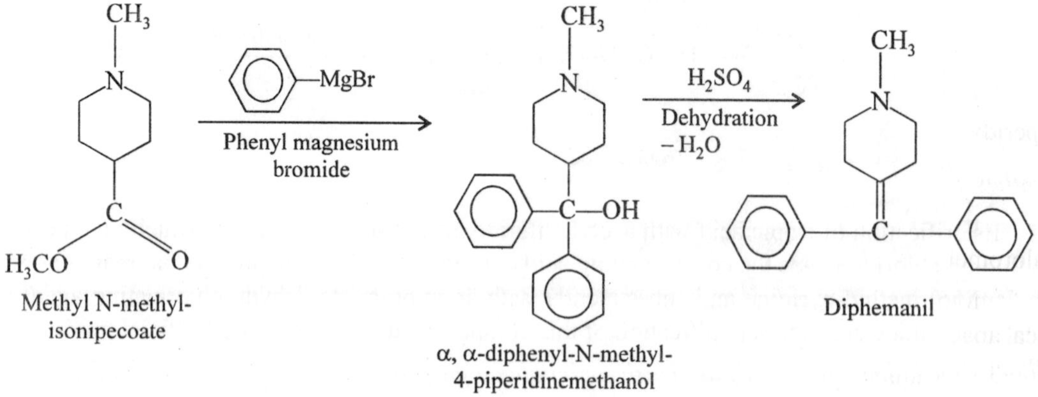

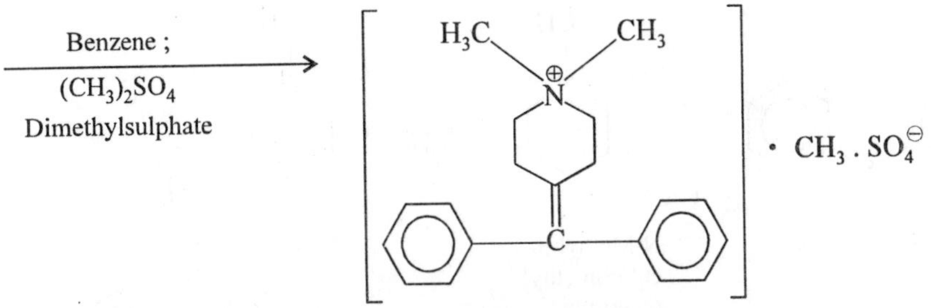

Benzene ;

$(CH_3)_2SO_4$

Dimethylsulphate

$\cdot$ $CH_3 . SO_4^{\ominus}$

Diphemanil methylsulphate

Grignardization of methyl N-methyl isonipecoate with phenyl magnesium bromide yields α, α-diphenyl-N-methyl-4-piperidine methanol which on dehydration with concentrated suplhuric acid gives the diphemanil base. Quaternization of the purified base in benzene with dimethyl sulphate yields the desired official compound.

Diphemanil methylsulphate is the drug of choice in causing the relief of pylorospasm. At comparatively low doses it exerts its action in two ways, namely, by effectively suppressing sweating and by causing bronchodilation. Hence, it finds its wide application in the *treatment of gastric hyperacidity, hypermotility, hyperhidrosis and peptic ulcer. It is also recommended for gastric and duodenal ulcer, and to relieve visceral spasms.*

Dose : Initial, oral, 100 mg 4 to 6 times daily

B. *Metixene INN, Methixene Hydrochloride BAN, USAN,*

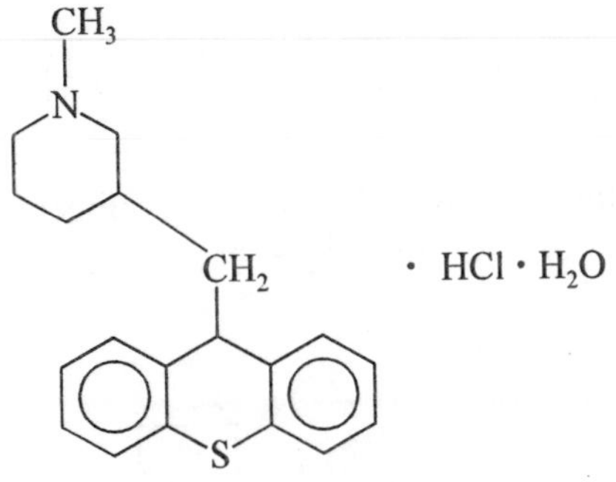

$\cdot$ HCl $\cdot$ H_2O

1-Methyl-3-(thioxanthen-9-ylmethyl)-piperidine hydrochloride monohydrate ; 9-(1-Methyl-3-piperidylmethyl) thioxanthene hydrochloride monohydrate ; Trest[R] (Dorsey Pharm.)

Synthesis

It may be prepared by the alkylation of the sodium salt of thioxanthene with N-methyl-3-chloromethyl piperidine.

Methixene hydrochloride besides possessing antimuscarinic properties also exhibit antihistaminic, local anaesthetic and antispasmodic actions. It is invariably employed in the *management of pylorospasm, biliary dyskinesia, spastic colon, gastritis and also in duodenal ulcer.*

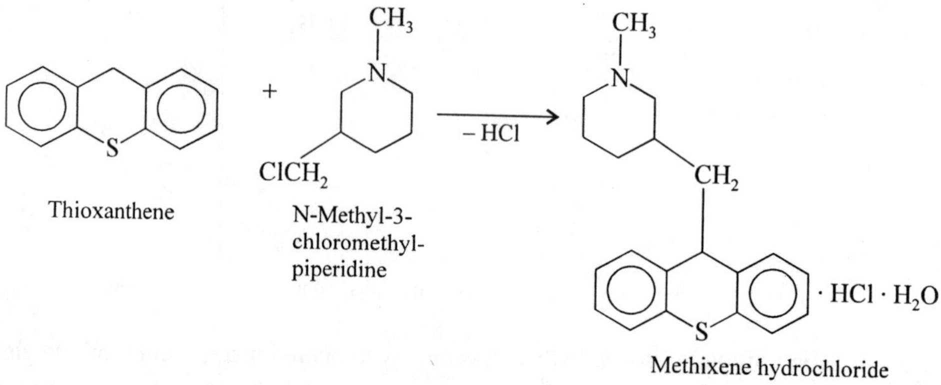

Thioxanthene   N-Methyl-3-
                      chloromethyl-
                      piperidine

Methixene hydrochloride

Dose : 1 to 2 mg 3 times daily.

C. *Glycopyrronium Bromide* INN, BAN, *Glycopyrrolate* USAN,

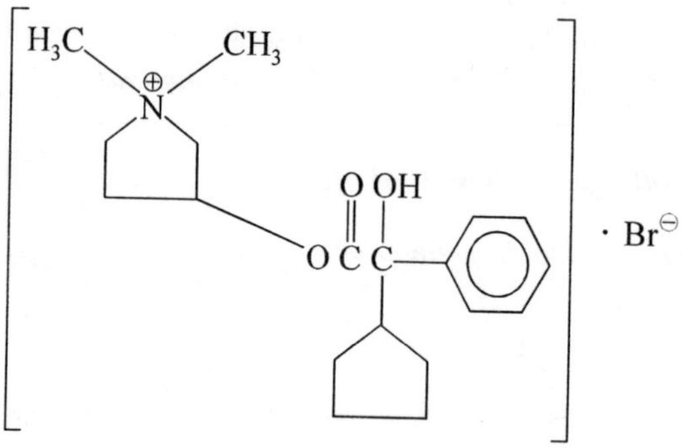

3-Hydroxy-1, 1-dimethylpyrrolidinium bromide α-cyclopentyl-mandelate ; Pyrrolidinium, 3-(cyclopentylhydroxyphenylacetyl) oxy-1, 1-dimethyl-, bromide ; U.S.P. ; Robinul[(R)] (Robins)

Synthesis

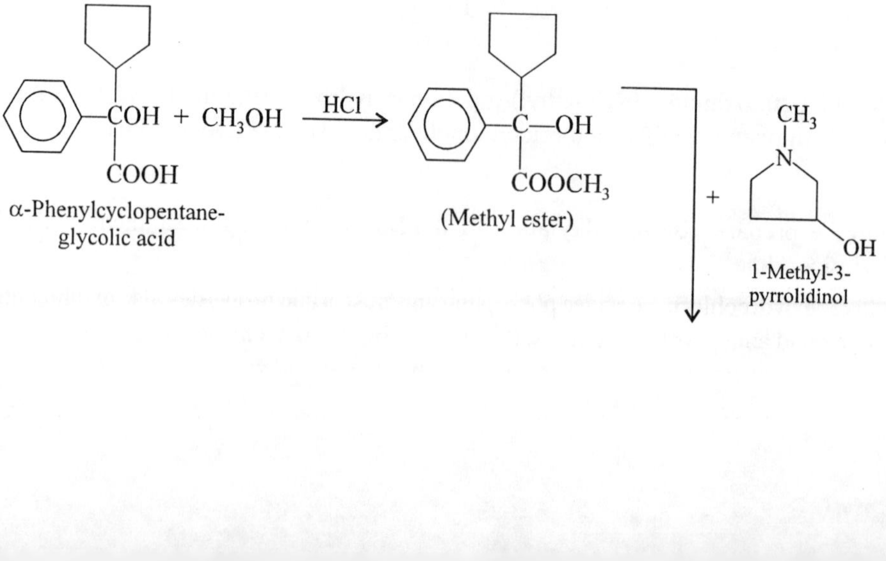

α-Phenylcyclopentane-   (Methyl ester)
glycolic acid

1-Methyl-3-
pyrrolidinol

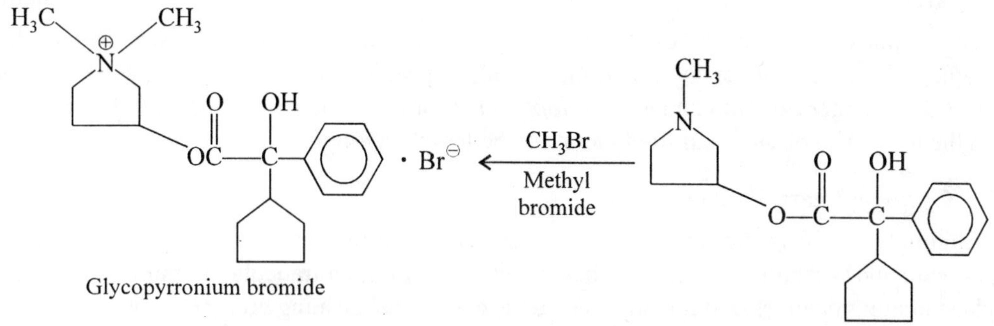

Glycopyrronium bromide

A transester

Esterification of α-phenylcyclopentane glycolic acid with methanol in the presence of hydrochloric acid yields the corresponding methyl ester which on further treatment with 1-methyl-3-pyrrolidinol undergoes transesterification and this on subsequent reaction with methyl bromide gives the desired official compound.

Glycopyrrolate is one of the four choicest drugs used in the management and control of gastric secretion. *It exerts a two-fold action, first by prolonging the gastric emptying time and secondly by decreasing the gastric acid production. Thus it generously favours the retention of antacids in acute cases of peptic ulcer. Besides, it also finds its usefulness in the treatment of colitis, biliary spasm, spastic colon, spastic duodenum.*

Dose : Initial, oral, 1 to 3 mg, 6 to 8 hours per day.

D. *Pirenzepine* INN, BAN,

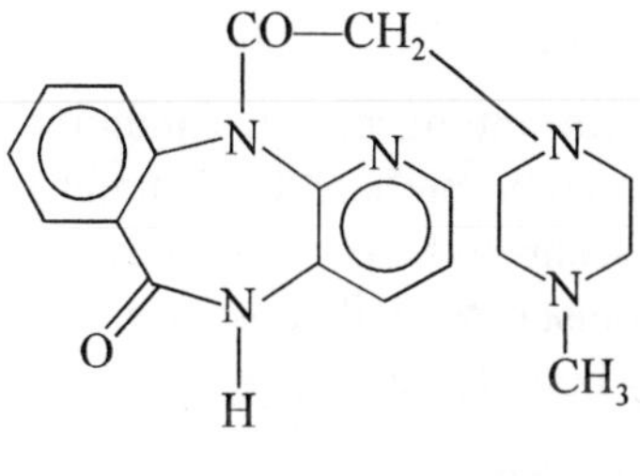

5, 11-Dihydro-11-[(4-methyl-1-piperazinyl) acetyl]-6H-pyridol [2, 3-β] [1, 4] benzodiazepine-6-one ; Gastrozepin[R] (as dihydrochloride) [Boehringer Ingelheim] ;

Pirenzepine hydrochloride is found to retard gastric secretion appreciably with minimal systemic anticholinergic effects. Hence it has been successfully employed for the *management and treatment of hyperacidity and peptic ulcer.*

7.8. Mechanism of Action

The mechanism of action of the various categories of '*muscarinic agents*' described under section 7.1 through 7.7 shall be discussed individually in the sections that follows :

7.8.1. Aminoalcohol Esters

The *five* potent compounds discussed under section 7.1 are as follows :

7.8.1.1. Atropine

The antimuscarinic activity of atropine largely resides in the *l-isomer* (*l-hyoscyamine*). It is absorbed rapidly from the GI-tract and is distributed rather speedily throughout the body. It also produces *mydriasis** for a longer duration and also *cycloplegia*** for more than upto 7 days. It gets metabolized solely in the liver. The plasma half life is found to be less than 4 hours.

7.8.1.2. Cyclopentolate Hydrochloride

Although, the '*drug*' does not seem to affect intraocular tension appreciably, yet it must be used with great care and percautions in patients having either very high intraocular pressure or with unrecognized glaucomatous changes. After being used to the cornea, the ensuing *cycloplegia* is almost complete in 25 to 75 minutes. A necessary neutralization with a few drops of pilocarpine nitrate (1—2%) solution, invariably affords almost complete recovery in a span of 6 hours ; otherwise it would take a relatively longer duration of nearly 6—24 hours for its complete recovery.

7.8.1.3. Dicyclomine Hydrochloride

It possesses approximately 1/8th the neurotropic activity of *atropine,* and nearly double the musculotropic activity of *papaverine.* It has substantially minimised the undesirable side effect intimately linked with the *atropine-type compounds.* It exerts its spasmolytic effect on various smooth-muscle spasms, specifically those associated with the GI-tract.

7.8.1.4. Hematropine Hydrobromide

The '*drug*' a tertiary amine hydrobromide, causes both *mydriasis* and *cycloplegia.* However, its duration of action seems to be much shorter and rapid when compared to those of atropine ; and perhaps it may be the reason why it is preferred for such purposes.

7.8.1.A.   SAR of Amino-alcohol Esters

A plethora of extremely potent antimuscarinic medicinal compounds essentially possess an esteratic functional moiety, and this could perhaps be a major contributing feature for the effective binding. However, this explanation seems to be not only reasonable but also plausible by virtue of the fact that *acetylcholine* (ACh) the agonist essentially possesses an almost identical characteristic feature and function for getting strategically bound to the same site.

7.8.2.   Aminoalcohol Ethers

The *three* medicinal compounds cited under this category, namely : benzotropine mesylate, chlorphenoxamine hydrochloride and orphenadrine citrate have been discussed under **'antiparkinsonism drugs'.**

7.8.3.   Aminoalcohol Carbamates

The only compound dealt with in this section is phencarbamide.

7.8.3.1.   Phencarbamide

The '*drug'* is reported to possess appreciable antimuscarinic activities. It finds its usage as its napadisylate salt (derivative) exclusively in the control, management and treatment of gastro-intestinal disorders.

---

*Pronounced or abnormal dilation of the pupil.

**Paralysis of the ciliary muscle.

7.8.3.A.   SAR of Aminoalcohol Carbarmates

The introduction of non-polar bulky functional moieties *e.g.,* phenyl ring, on the unsubstituted N-atom of the carbamoyl group perhaps render the molecule capable of van der Waal's interactions to the receptor surface to a certain extent.

7.8.4.   Aminoalcohols

The compounds stated under this specific section, such as biperiden hydrochloride, procyclidine hydrochloride, trihexyphenidyl hydrochloride, have been dealt with under the chapter on **'antiparkingonism agents'.**

7.8.5.   Aminoamides

There are *two* drugs discussed under this section *viz.,* isopropamide iodide and tropicamide which would be treated as under :

7.8.5.1.   Isopropamide Iodide

The *'drug'* exerts a potent anticholinergic effect and causes atropine-like effects peripherally. It has been observed that only at high dose levels it affords sympathetic blockade at the ganglionic sites. It can provide a long extendable action lasting upto almost 12 hours to contain the antispasmodic and antisecretory effects.

7.8.5.2.   Tropicamide

The *'drug'* possesses an obvious advantage over the belladona alkaloids in two aspects, namely : *first,* its evident shorter duration of action ; and *secondly,* the over homatropine in its ability to induce cycloplegia. It is found to cause little enhancement of the intraocular pressure in the normal subjects ; but it may do so in such patients having either glaucoma or suffering from some sort of structural deformities in the anterior chamber of the eye.

7.8.5.A. SAR-of Aminoamides

In fact, the aminoamide type of anticholinergic usually designates almost the same type of molecule as the aminoalcohol group ; however, it has an apparent and absolutely distinguishable characteristic feature duly represented by the replacement of the corresponding polaramide group with the corresponding polar hydroxyl moiety. Nevertheless, the aminoamides still retain the same bulky structural moieties as are commonly seen at one end of the molecule.

7.8.6.   Diamines

The *two* drugs sited under this category *viz.,* diethazine and ethopropazine hydrochloride have been dealt with adequately under the chapter on **'antiparkinsonism agents'.**

7.8.7.   Miscellaneous Amines

The *four* compounds discussed under this category, namely : diphemanil methylsulphate, methixene hydrochloride, glycopyrronium bromide and pirenzepine, will be described as under :

7.8.7.1.   Diphemanil Methylsulphate

It acts as a potent parasympatholytic by blocking the nerve impulses at the site of parasympathetic ganglia ; however, it fails to invoke a sympathetic ganglionic blockade. It has been observed to be highly specific in its activity upon those innervations which categorically activate both functionalities, namely : (*a*) GI-motility ; and (*b*) gastric secretion. Interestingly, based on the extremely specific nature of its pharmacological activity on the gastric functions actually renders diphemanil particularly beneficial in the control, management and treatment of peptic ulcer. Besides, its inherent lack of atropine-like effects makes this specific usage relatively less painful in comparison to several other antispasmodic drugs.

Note. The methylsulphate salt is considered to be the best in terms of its stability in comparison to its corresponding halides e.g., Cl⁻ (hygroscopic) ; Br⁻ and I⁻ (toxic reactions).

7.8.7.2.   Methixene Hydrochloride

The '*drug*' essentially exhibits both antihistaminic and direct antispasmodic activities. It has been duly observed that the drug is invariably absorbed from the GI-tract. It subsequently gets excreted through the urine, partly as its corresponding isomeric sulphoxides or their metabolites ; and a reasonable quantum as unchanged product.

Methixene hydrochloride has been profusely indicated in such preparations that may ultimately relieve gastro-intestinal spasms.

7.8.7.3.   Glycopyrronium Bromide

It is a quaternary ammonium antimuscarinic agent having marked and pronounced peripheral effects that are quite akin to those of atropine. It has been found that approximately 10-25% of the '*drug*' gets absorbed from the GI-tract when administered orally. It penetrates the blood brain barrier (BBB) rather very poorly. It gets ultimately excreted in bile and urine.

7.8.7.4.   Diphemanil Methylsulphate

The '*drug*' is quite active for the symptomatic management and subsequent control of visceral spasms. It may also cause enough relief from the ensuing painful and distressing spasms of the biliary and the genitro-urinary systems. Importantly, it also finds its adequate usage for the treatment of symptomatic bradycardia.

8.   GANGLIONIC BLOCKING AGENTS

Langley first studied the action of drugs on the autonomic ganglia with the alkaloid nictoine obtained from *Nicotiana tabacum*. Subsequently, some other alkaloids were also found to possess similar effects, namely ; **coniine** from the poison Hemlock obtained from *Conicum maculatum*, **gelsemine** from the yellow *jasmine* and *lobeline* from the lobelia, a native of America.

Generally, ganglionic blocking agents normally exert their action by competing with acetylcholine from the cholinergic receptors present in the autonomic postganglionic neurons. In fact, the ganglia of either of the sympathetic and parasympathetic nervous systems are cholinergic in character, these therapeutic agents interrupt the outflow through both these systems. Therefore, it is not practically possible to accomplish a complete therapeutic block of the autonomic outflow to a specific locus without encountering a few undesirable but unavoidable '*side–effects*' that may arise as a consequence from the blockade of other ensuing autonomic nerves. Thus, the resulting blockade of the sympathetic outflow to the blood vessels may ultimately lead to either a distinct hypotension or an enhanced blood flow (having a pink or warm skin).

In general, the ganglionic blocking agents must be employed with great caution particularly when such drugs as : hypotensives, antihypertensives or anaesthetic drugs are used concomitantly, by virtue of the fact that the '*hypotension*' may be aggrevated to such an enormous degree that would cause complete jeopardization of normal blood flow through the brain, heart, or kidney.

The ganglionic blocking agents are usually used in hypertension, peripheral vascular disease, vasopastic disorders—thereby lowering the blood pressure and increasing the peripheral blood flow. Occasionally these agents are also employed for their interruption of parasympathetic nervous outflow, as is observed in peptic ulcer and intestinal hypermotility.

Such ganglionic blocking agents that interfere with the nervous transmission of sympathetic as well as para-sympathetic ganglia may include : hexa-methonium bromide ; mecamylamine hydrochloride ; pempidine ; pentolinium tartrate, trimetaphan camsylate, etc.

A.   *Hexamethonium Bromide* INN, BAN,

$$(CH_3)_3 \overset{\oplus}{N}(CH_2)_6 \overset{\oplus}{N}(CH_3)_3 \cdot 2\overset{\ominus}{Br}$$

Hexamethylenebis (trimethylammonium bromide) ; Hexonium bromide ; BPC. (1968) ; Bistrium Bromide[R] (Squibb)

Hexamethonium bromide is a quaternary ammonium ganglion blocking agent which inhibits the transmission of nerve impulses in both sympathetic and parasympathetic ganglia. It is usually employed in the *treatment of severe or malignant hypertension.*

Dose : Initial, subcutaneously or intramuscularly 5 to 15 mg ; administration after each 4 to 6 hours, if necessary.

B.   *Mecamylamine* INN, *Mecamylamine Hydrochloride* BAN, USAN,

N, 2, 3, 3-Tetramethyl-2-norbornanamine hydrochloride ; Bicyclo [2, 2, 1] heptan-2-amine, N, 2, 3, 3-tetramethyl-, hydrochloride ; BP ; (1968), USP ; Int. P., Ind. P. ; Inversine[R] (Merck)

Synthesis

dl-Camphene on treatment with an excess of hydrogen cyanide in a rather strongly acidic medium to 5°C gives rise to 2-formamido-2, 3, 3-trimethyl-norcamphane. The course of reaction may be considered as due to the addition of HCN to the 2, 8-double bond present in the camphene moiety to yield the corresponding 2-isocyanate which on subsequent hydration gives the formamido derivative. Mecamylamine base is obtained by the reduction of the formyl function in the formamido derivative with lithium aluminium hydride which may be converted into its hydrochloride salt by dissolving the base in an appropriate solvent and passing a stream of HCl gas.

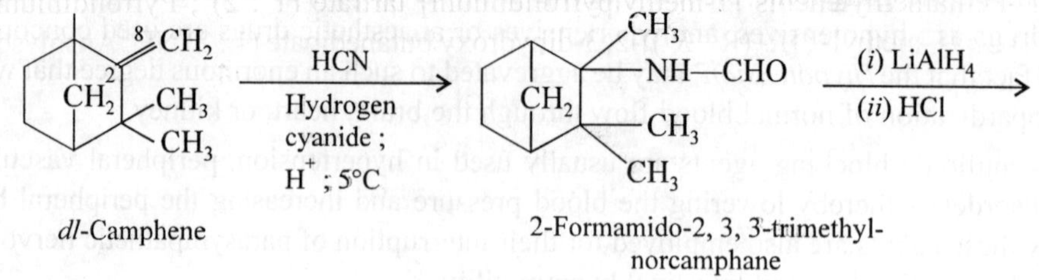

dl-Camphene → 2-Formamido-2, 3, 3-trimethyl-norcamphane

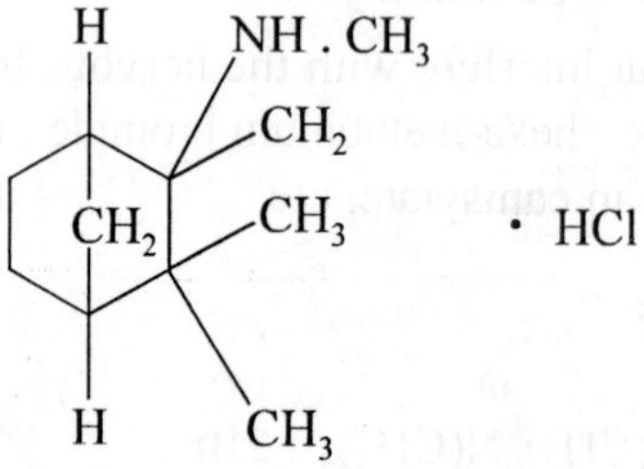

(i) LiAlH$_4$
(ii) HCl

Mecamylamine hydrochloride

A non-quaternary ammonium compound with very poor ionizability in the small intestine thus rendering it absorbable more easily. It gets excerted through the kidney comparatively slowly thereby enhancing its duration of action.

Dose : Usual, initial, 2.5 mg twice daily ; maintenance, 7.5 mg thrice daily.

C.   Pempidine *INN, BAN*

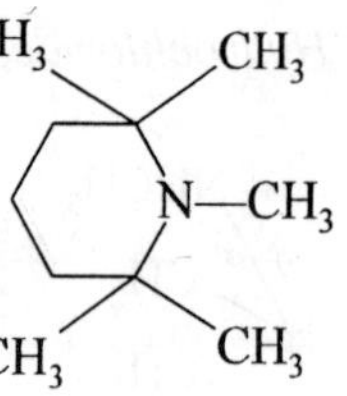

1, 2, 2, 6, 6-Pentamethyl-piperidine ; Pempidine Tartrate (BP ; 1968) ; Perolysen[R] (May & Baker)

It is a tertiary amine ganglion-blocking agent which essentially blocks the transmission of nerve impulses in both sympathetic and parasympathetic ganglia. It is used in the *managemnet and treatment of severe or malignant hypertension.*

Dose : Usual, initial, 2.5 mg of pempidine tartrate orally 3 to 4 times daily ; maintenance dose, 10 to 80 mg per day in 4 divided doses.

D.   Pentolonium Tartrate *INN,* Pentolinium Tartrate *BAN,*

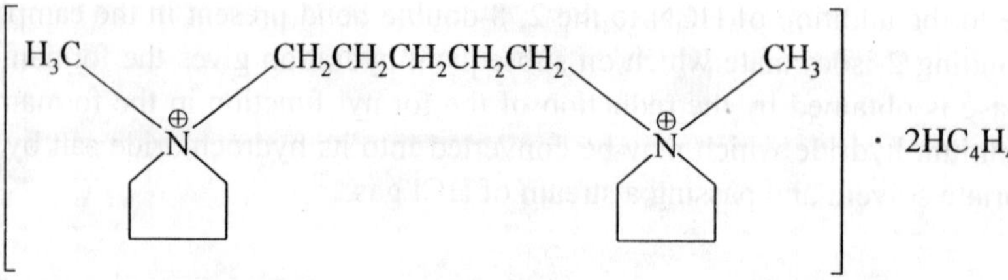

1, 1'-Pentamethylenebis [1-methylpyrrolidinium] tartrate (1 : 2) ; Pyrrolidinium, 1, 1-(1, 5-pentanediyl) *bis* [1-methyl- ; [*R*- (R*, *R**)]-2, 3-dihydroxy butanedioate (1 : 2) NF ; Ansolyson[R] (Wyeth)

Synthesis

N-Methylpyrrolidine and pentamethylene diiodide on heating with ethanol yields the quaternary pyrrolidinium base which upon treatment with moist silver oxide and double equimolar proportion of tartaric acid gives the official product.

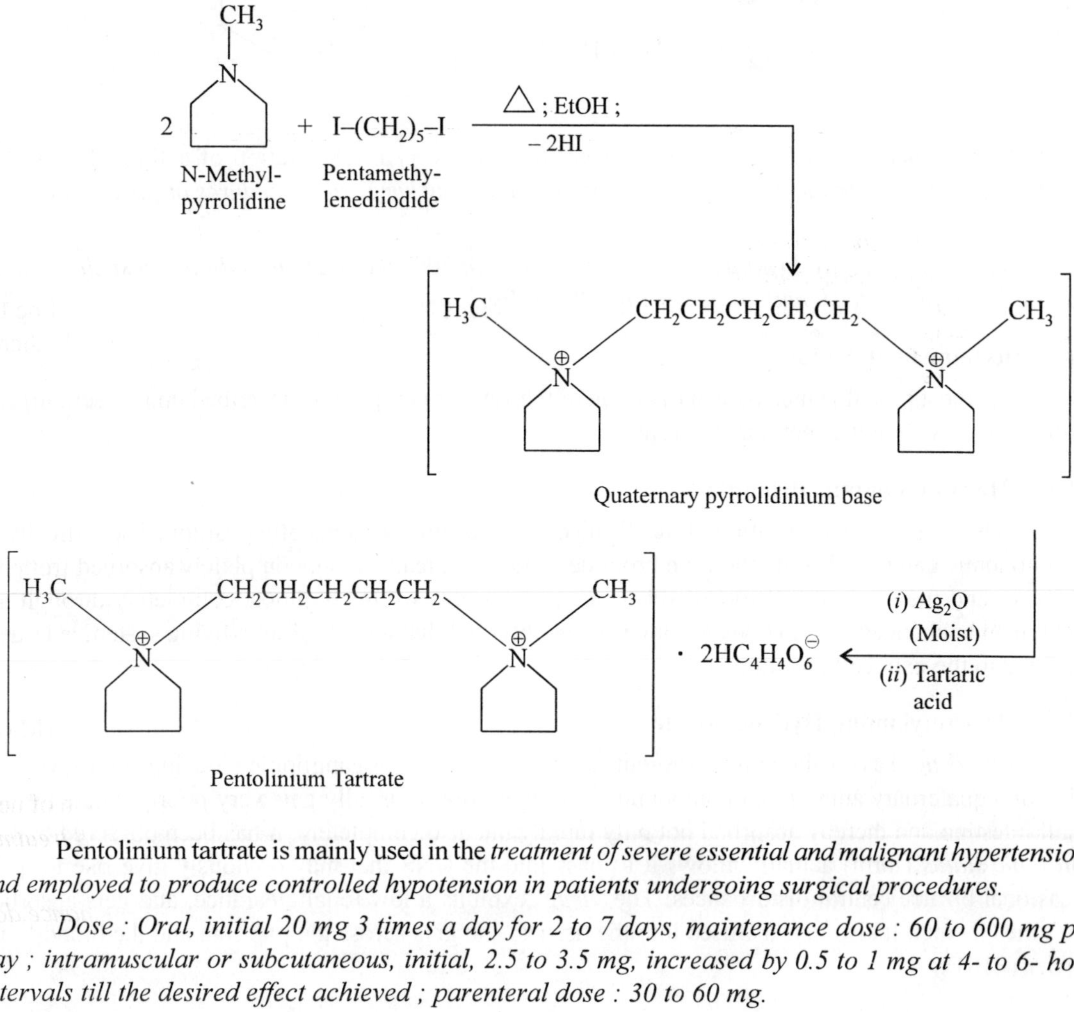

Quaternary pyrrolidinium base

Pentolinium Tartrate

Pentolinium tartrate is mainly used in the *treatment of severe essential and malignant hypertensions and employed to produce controlled hypotension in patients undergoing surgical procedures.*

Dose : Oral, initial 20 mg 3 times a day for 2 to 7 days, maintenance dose : 60 to 600 mg per day ; intramuscular or subcutaneous, initial, 2.5 to 3.5 mg, increased by 0.5 to 1 mg at 4- to 6- hour intervals till the desired effect achieved ; parenteral dose : 30 to 60 mg.

E. Trimetaphan Camsilate *INN,* Trimetaphan Camsylate *BAN,* Trimethaphan Camsylate *USAN,*

(+)-1, 3-Dibenzyldecahydro-2-oxoimidazo [4, 5-*c*]-thieno [1, 2- α]-thiolium 2-oxo-10-boranesulphonate (1 : 1) ; BP ; (1968), USP ; Int. P. ; Arfonad[R] (Roche, U.K.)

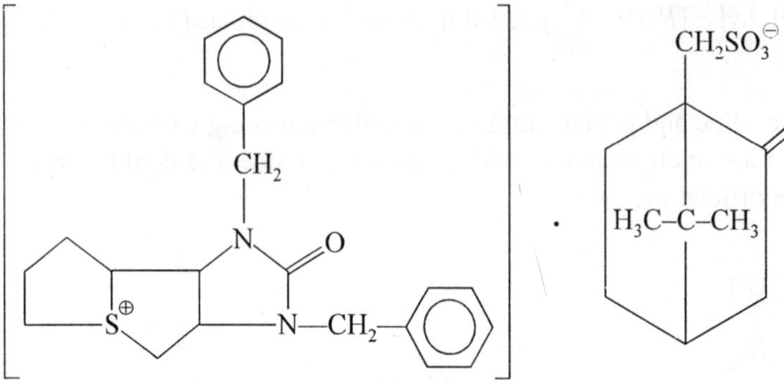

Trimetaphan is ganglionic-blocking agent having a very brief duration of action. Hence, it is *chiefly used either for inducing controlled hypotension during surgical procedures or for certain diagnostic procedures.*

Dose : Intravenous, by infusion, 0.2 to 5 mg/min. in 500 ml of isotonic solution at such a rate so that the blood pressure should not fall below 60 mm Hg.

8.1. Mechanism of Action

The most probable mechanism of action of the various compounds described under section 12.8. shall be dealt with in the sections that follows :

8.1.1.   Hexamethonium Bromide

The '*drug*' blocks transition at the N_2 nicotinic receptors strategically positioned specifically in the autonomic ganglia. Hexamethonium bromide is both erratically and incompletely absorbed from the GI-tract ; and therefore, the absorbed substance is invariably excreted almost completely through the urine in an unchanged form. However, it has been observed that the '*drug*' in a living system is largely confined to the extracellular fluid.

8.1.2.   Mecamylamine Hydrochloride

The '*drug*' has an altogether different status from most other ganglionic blocking agents wherein it is not a quaternary ammonium compound ; and, therefore, it usually gets very poorly ionized in the small intestine and thereby absorbed not only rapidly but also completely. It has been observed that its *non ionic status* (form) actually allows it to pass into the CNS that may obviously give rise to quite occasional bizarre central disturbances. The '*drug*' exhibits a low renal clearance, and gets absorbed from the GI-tract. It also gets diffused into the tissues, and gets across the placenta and the blood-brain barrier (BBB).

8.1.3.   Pempidine

Pempidine exerts its actions in two ways : *first*, the sympathetic block causes peripheral vasodilation thereby affecting enhanced blood flow, raised skin temperature, and reduction blood pressure ; *secondly*, the parasympathetic block causes reduction of gastric and salivary secretion together with retarded motility of the gastro-intestinal tract and bladder.

8.1.4. Pentolinium Tartrate

The two quaternary N-atoms separated by a shorter distance *i.e.,* five CH_2 groups, caused primarily ganglionic blockade and a clinically appreciable fall in BP. Importantly, this *bis*-quaternary drug substance was much less satisfactory. Sympathetic blockade, however, did afford the well predicted pharmacological actions, such as : vasodilation and fall in BP. It is found to be absorbed from the GI-tract in an irregular and incomplete manner.

8.1.5. Trimethaphan Camsylate

Though the 'drug' is usually classified as a ganglionic blocking agent, but it only blocks ganglia in the therapeutic dosage regimen not to an appreciable extent. It has been observed that to a certain extent its hypotensive activities are caused exclusively due to a direct peripheral vasodilator action.

Note. The '*drug*' very often causes the release of histamine ; and, therefore, its usage as a ganglionic blocking agent in subjects having a history of *allergy* and *asthma* should be avoided as far as possible.

9. ADRENERGIC NEURONE BLOCKING AGENTS

Adrenergic neurone blocking agents usually interfere with postganglionic sympathetic nervous transmission but they do not exert any effect on the parasympathetic nervous system.

It is, however, pertinent to state here that the adrenergic neurotransmitter, norepinephrine (NE) is invariably synthesized in the adrenergic neurons. Interestingly, 3, 4-dihydroxyphenylalanine (DOPA), that is usually formed in the adrenergic neuron by means of the hydroxylation of tyrosine, subsequently undergoes decarboxylation (by the aid of aromatic amino acid decarboxylase) to yield the corresopnding catecholamine dopamine (3, 4-dihydroxyphenylethylamine). NE is produced normally within the adrenergic neuron in the viccinity of the nerve endings being *granular organelles* essentially comprising the enzyme *dopamine β-hydroxylase* that subsequently inducts the desired *side-chain-hydroxyl moiety* to dopamine and then to NE. Ultimately, the resulting NE thus generated gets stored in the *granular organelles.*

In short, the adrenergic neuron blocking agents help distinctly in reducing the delivery of catecholamines *viz.,* NE, to the adrenergic receptors genuinely, which may eventually take place by disallowing the synthesis of catecholamines and their subsequent storage or release.

The following *two* compounds belonging to this category shall be discussed, namely : Guanethidine monosulphate ; Bethanidine sulphate :

A.   Guanethidine *INN,* Guanethidine Monosulphate *BAN, USAN,*

[2-(Hexahydro-1 (2H)-azocinyl) ethyl] guanidine sulphate (1 : 1) ; 1-[2-(Perhydroazocin-1-yl) ethyl] guanidine monosulphate ; BP ; USP ; Ismelin[R] (Ciba-Geigy)

Synthesis

Alkylation of saturated azocine with chloroacetonitrile yields an intermediate and a mole of hydrogen chloride gets eliminated. Catalytic reduction of the intermediate affords the diamine. The resulting product on condensation with 2-methyl-2-thiopseudourea (*i.e.,* S-methyl ether of thiourea) and the subsequent treatment with sulphuric acid yields the official compound by the elimination of a mole of thiomethanol.

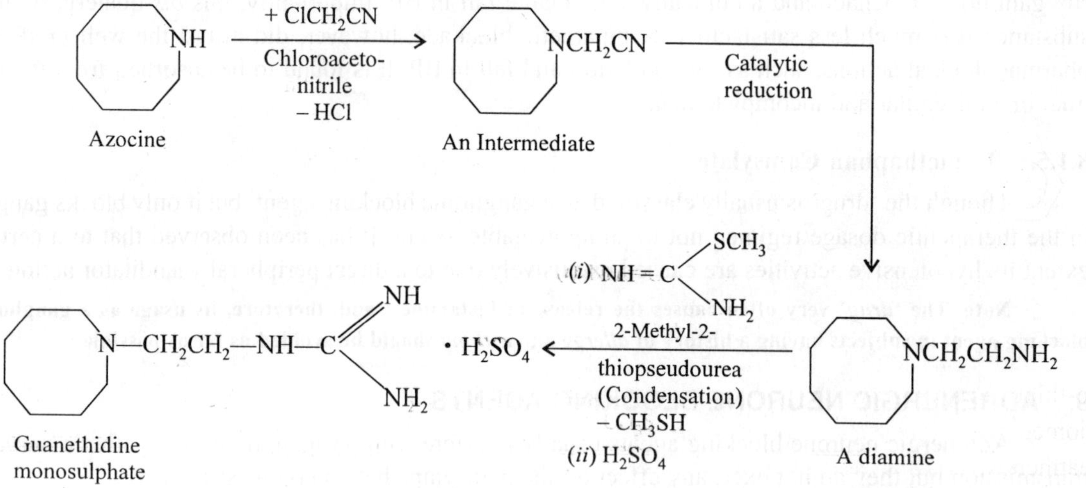

Azocine

An Intermediate

Guanethidine monosulphate

2-Methyl-2-thiopseudourea (Condensation)

A diamine

It is an antihypertensive agent which is *frequently employed in the treatment of moderate and severe hypertension, or for mild hypertension. It has a rather slow onset of action, the full effect may range from several hours to 2 or 3 days, and its duration of action may extend from 4 or more days.*

Dose : Usual, initial 10 to 20 mg per day ; usual maintenance dose 30 to 100 mg per day as a single dose but up to 300 mg daily may be given.

B.   Bethanidine *INN*, Bethanidine Sulphate *BAN, USAN,*

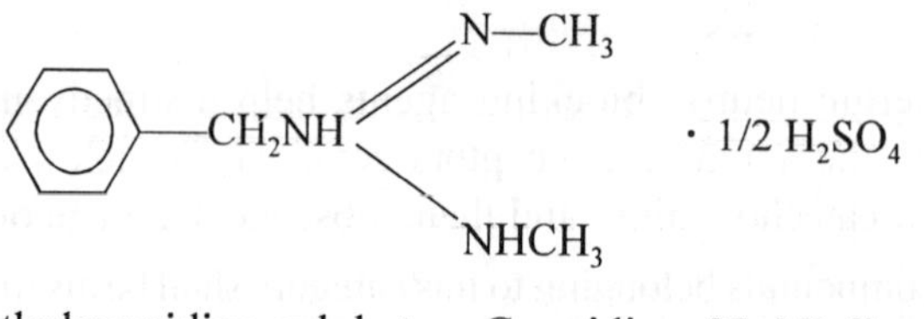

1-Benzyl-2, 3-dimethylguanidine sulphate ; Guanidine, N, N'-dimethyl-N''-(phenylmethyl)-, sulphate ; BP ; Esbatal[R] (Calmic, U.K.)

Synthesis

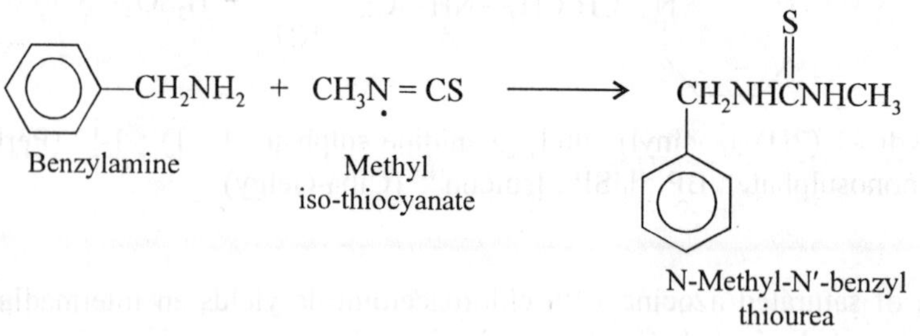

Benzylamine

Methyl iso-thiocyanate

N-Methyl-N'-benzyl thiourea

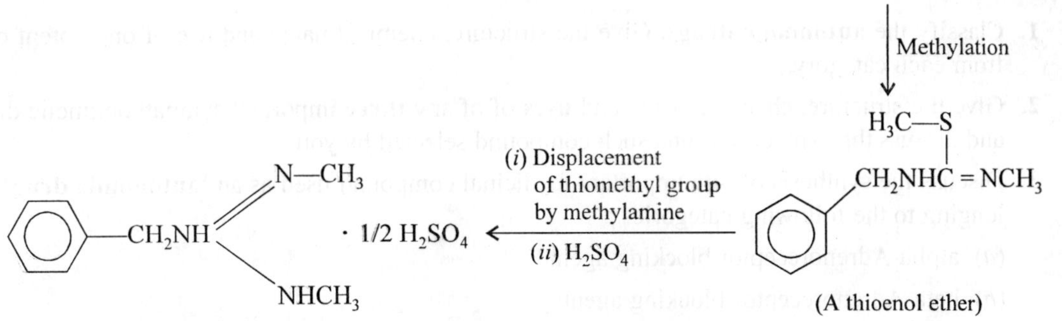

Methylation

$H_3C—S$

(i) Displacement
of thiomethyl group
by methylamine

$CH_2NHC = NCH_3$

(ii) H_2SO_4

(A thioenol ether)

N—CH₃ ... · 1/2 H_2SO_4 ← CH₂NH ... NHCH₃

Bethanidine sulphate

N-Methyl-N′-benzyl thiourea may be obtained by the interaction of benzylamine and methyl iso-thiocyanate, which on methylation yields a thioenol ether. Displacement of the corresponding thiomethyl group by methylamine *via* an addition elimination process affords bethanidine which on treatment with an aliquot of sulphuric acid gives the official product.

It is usually employed in the *treatment of moderate and severe hypertension or for the treatment of mild hypertension in patients.*

Dose : Usual, initial dose 10 mg 3 times per day. Maintenance dose ranges from 20 to 200 mg per day.

9.1. Mechanism of Action

The specific mechanism of action of the selected compounds under section 12.9.1 are enumerated below :

9.1.1. Guanethidine Monosulphate

The '*drug*' inherently contains the '*guanido moiety*' that essentially inducts an appreciable degree of high basicity (pKa = 12) ; and hence, is 99.99% protonated at the physiologic pH. It is worthwhile to mention here that the drug does not pass through the BBB and, unlike reserpine, influences absolutely little CNS-mediated sedation.

Guanethidine distinctly produces vasodilation and positively enhances the venous capacitance significantly. It is observed to bring down the blood pressure to such a dangerously low levels in certain subjects, the drug invariably and importantly is used in *submaximal dose levels ;* and, therefore, administered in conjunction with either thiazides (diuretics) or hydralazine (hypotensive), to allow certain adrenergic function. Thus, the '*drug*' is **not** usually used in the treatment of mild to moderate hypertension, but exclusively moderately severe to severe hypertension.

9.1.2. Bethanidine Sulphate

The mechanism of action of the '*drug*' is very much akin to guanethidine. It has almost identical actions to the ganglionic-blocking drugs, but there is no parasympathetic blockade whatsoever.

Probable Questions for B. Pharm. Examinations

1. Classify the **autonomic drugs.** Give the structure, chemical name and use of one potent drug from each category.

2. Give the structure, chemical name and **uses** of of any **three** important sympathomimetic drugs and discuss the synthesis of one such compound selected by you.

3. Discuss the synthesis of **one** important medicinal compound used as an **'autonomic drug'** belonging to the following categories :

    (*a*) alpha-Adrenoreceptor blocking agent

    (*b*) beta-Adrenoreceptor blocking agent

4. Name **one** compound each belonging to the cateogry of **'chlinomimetic drugs'** that are either *directly acting* or *indirectly acting.* Describe their synthesis.

5. Give a comprehensive account of **'antimuscarinic agents'** by giving the structure, chemical name and uses of **one** important representative from each cateogry.

6. (*a*) Structure modification amongst the amino-alcohol carbamates by introducing non-polar functions on the unsubstituted N of the carbonyl moiety gave rise to **Phenyl carbamide.** Discuss this medicinal compound.

    (*b*) Aminoalchols have gained their cognizance as **antiparkinson agents.** Discuss the synthesis of isopropamide iodide from diphenyl acetonitrile.

7. Give the structure, chemical name and uses of the following **'autonomic drugs'** :

    (*a*) Diphemanil methylsulphate

    (*b*) Methixene hydrochloride

    (*c*) Glycopyrronium bromide

    Describe the synthesis of any **one** compound.

8. (*a*) Enumerate the importance of **'ganglionic blocking agents'** as autonomic drugs.

    (*b*) Describe the synthesis of Mecamylamine hydrochloride.

9. Describe the **'mode of action'** of the following :

    (*a*) Ganglionic blocking agents

    (*b*) Antimuscarinic agents

10. Give a comprehensive account of the **'adrenergic neurone blocking agents'** with specific reference to :

    (*a*) Bethanidine sulphate

    (*b*) Guanethidine monosulphate.

RECOMMENDED READINGS

1. ME Wolff (*Ed*) : *Burger's Medicinal Chemistry and Drug Discovery* (5th edn), John Wiley and Sons Inc, New York, (1996).

2. LS Goodman and A Gilman *The Pharmacological Basis of Therapeutics* (9th edn) Macmillan Co. London (1995).

3. WO Foye (*Ed*) *Principles of Medicinal Chemistry* (5th edn) Lippincott Williams and Wilkins, Philadelphia, 2002.

4. JEF Reynolds (*Ed*) *Martindale the Extra Pharmacopoeia* (31st edn) Royal Pharmaceutical Society The Pharmaceutical Press London (1996).

5. MC Griffiths (*Ed*) *USAN and the USP Dictionary of Drugs Names* United States Pharmacopoeial Convention Inc Rockville (1985).

6. D Lednicer and L A Mitscher *The Organic Chemistry of Drug Synthesis* John Wiley and Sons, New York (1995).

7. Gennaro AR, Remington : *The Science and Practice of Pharmacy*, Lippincott Williams and Wilkins, Philadelphia, Vol. II, 20th edn, 2000.

8. Gringauz A, *Introduction to Medicinal Chemistry,* Wiley-VCH, New York, 1997.

13 Diuretics

1. INTRODUCTION

Diuretics are drugs that promote the output of urine excreted by the kidneys.

The increased excretion of water and electrolytes by the kidneys is dependent on three different processes, *viz.*, glomerular filtration, tubular reabsorption (active and passive) and tubular secretion.

Every normal human being essentially bears a daily rhythm in the excretion of water and electrolytes, being minimum during night and maximum in the morning. This may be a reflection of intra-cellular metabolism. Alteration prevailing in the diurnal rhythm is normally characterized by initial symptoms of disturbance of fluid balance of the body as evidenced in heart failure (Addison's disease), hepatic failure and renal diseases.

Diuretics are very effective in the *treatment of cardiac oedema, specifically the one related with congestive heart failure.* They are employed extensively in various types of disorders, for example, *nephrotic syndrome, diabetes insipidus, nutritional oedema, cirrhosis of the liver, hypertension, oedema of pregnancy and also to lower intraocular and cerebrospinal fluid pressure.* In some instances where oedema is not present, the diuresis may be specifically indicated and effected by certain highly specialized diuretics as in hypertension, epilepsy, migraine, glaucoma, anginal syndrome and bromide intoxication.

In its simplest explanation the formation of urine from the blood mainly comprises of *two* cardinal processes taking place almost simultaneously, namely : (*a*) glomerular filtration ; and (*b*) selective tubular reabsorption, and subsequent secretion. It has been duly observed that as the *'glomerular filtrate'* gets across through the tubules, substances that are absolutely essential to the blood and tissues, such as: water, salts, glucose, and amino acids are reabsorbed eventually.

However, under perfect normal physiologic circumstances the glomerular filtration rate is approximately 100 mL min^{-1}. And from this volume about 99 mL of the fluid is sent back to the blood pool, and thus only 1 mL is excreted as urine. From these critical and vital informations one may infer that the *'diuretics'* may enhance the rate of urine-formation by either of the *two* following phenomena, *viz.*,

(*a*) Increasing glomerular filtration, and

(*b*) Depressing tubular reabsorption.

2. CLASSIFICATION

Diuretics may be broadly classified under the following *two* categories :

(*a*) Mercurial Diuretics

(*b*) Non-mercurial Diuretics.

2.1. Mercurial Diuretics

The mercurial diuretics essentially contain Hg^{++} in an organic molecule. They usually inhibit sodium reabsorption in the proximal tubuler and ascending loop of Henle. There may be slight effect in the distal tubule where inhibition of chloride reabsorption also occurs. The mercurials have been found to enhance K^+ excretion though potassium loss is less than that produced by many other diuretics. However, the overall action of mercurial diuretics is invariably increased by acidification of urine. The mercurial diuretics are not very much used in clinical practices due to their pronounced and marked side-effects *viz.,* mercurialism, hypersensitivity and excessive diuresis which may lead to electrolyte depletion and vascular complications. Most of the mercurials are administered by intramuscular route and the availability of orally active diuretics has limited their use.

The mercurial diuretics has the following general formula :

$$Y—CH_2—CH—CH_2—Hg—X$$
$$|$$
$$OR$$

where X = OH, halide, or heterocyclic moiety,

Y = substituted side chain or substituted aromatic function

and R = Methyl group.

Examples : Chlormerodrin Hg 197 ; Meralluride ; Mercaptomerin sodium ; Merethoxylline procaine ; Mersalyl ; and Mercumatilin sodium.

A. *Chlormerodrin Hg 197* USAN. *Chlormerodrin* BAN, *Chlormerodrin* (197 Hg) INN,

$$H_2NCONH—CH_2CHCH_2—HgCl$$
$$|$$
$$OCH_3$$

Chloro (2-methoxy-3-uriedopropyl) mercury^{-197} Hg ; Mercury^{197} Hg, [3-(aminocarbonyl) amino] -2-methoxypropyl] chloro-USP ; BPC (1959) ; Neohydrin-197$^{(R)}$ (Abbott)

Synthesis :

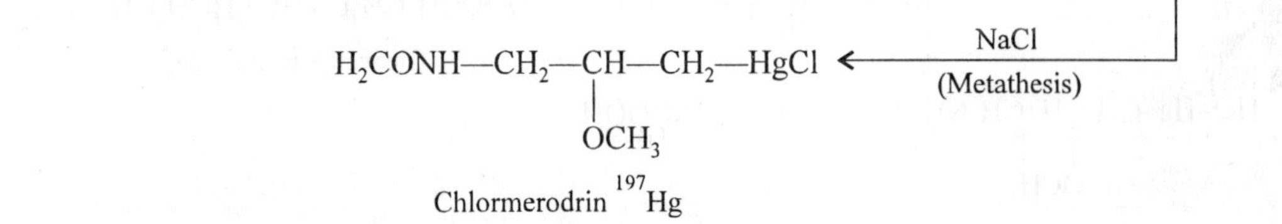

H_2N\
          >C = O + (CH_3COO)_2Hg —MeOH/Reflux;→ H_2N—CONH—CH_2—CH—CH_3
(H_2C = CHCH_2)NH/ |
                                                                      OCH_3

(H_2C = CHCH_2)NH
N-Allylurea   Mercuric acetate   2-Methoxy-N-propyl urea

H_2CONH—CH_2—CH—CH_2—HgCl ←—NaCl/(Metathesis)—
          |
          OCH_3
Chlormerodrin ^{197}Hg

It may be prepared by the interaction of N-allyl urea with mercuric acetate in the presence of methanol when the former gets acetoxymercurated. The saturation also takes place simultaneously when the methoxy group is introduced at position 2. Metathesis occurs on the addition of aqueous NaCl resulting in the precipitation of chlormerodrin which is subsquently filtered, washed and dried.

Chlormerodrin[197] Hg is used in the *treatment of oedema of congestive heart failure. It has also been employed in the management of chronic nephritis, ascites of liver disease and nephrotic oedema.*

Dose : Usual, oral, 18.3 to 73.2 mg per day ($\equiv$ to 10 to 40 mg of mercury per day)

B. *Meralluride* INN, BAN, USAN,

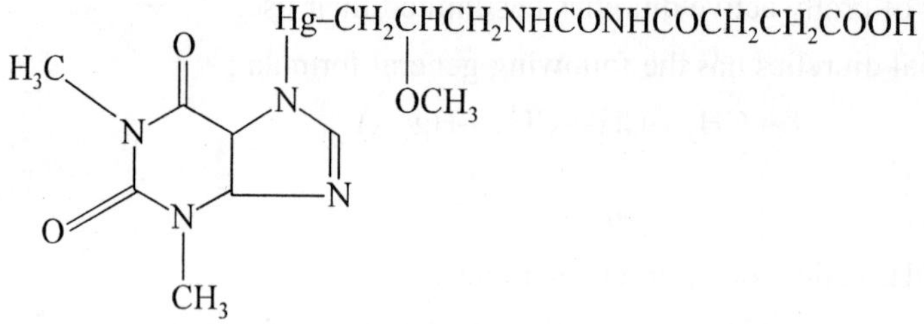

$$Hg-CH_2CHCH_2NHCONHCOCH_2CH_2COOH$$
$$OCH_3$$

[3-[3-(3-Carboxypropionyl) ureido]-2-methoxypropyl]-hydroxy-mercury mixture with theophylline ; Mercury [3[[[(3-carboxy-1-oxopropyl) amino] carbonyl] amino]-2-methoxypropyl] (1, 2, 3, 6-tetrahydro-1, 3-dimethyl-2, 6-dioxo, 7H-purin-7-yl)-; BPC (1959) ; NFXIV ; Mercuhydrin[(R)] (Merrell Dow)

Synthesis :

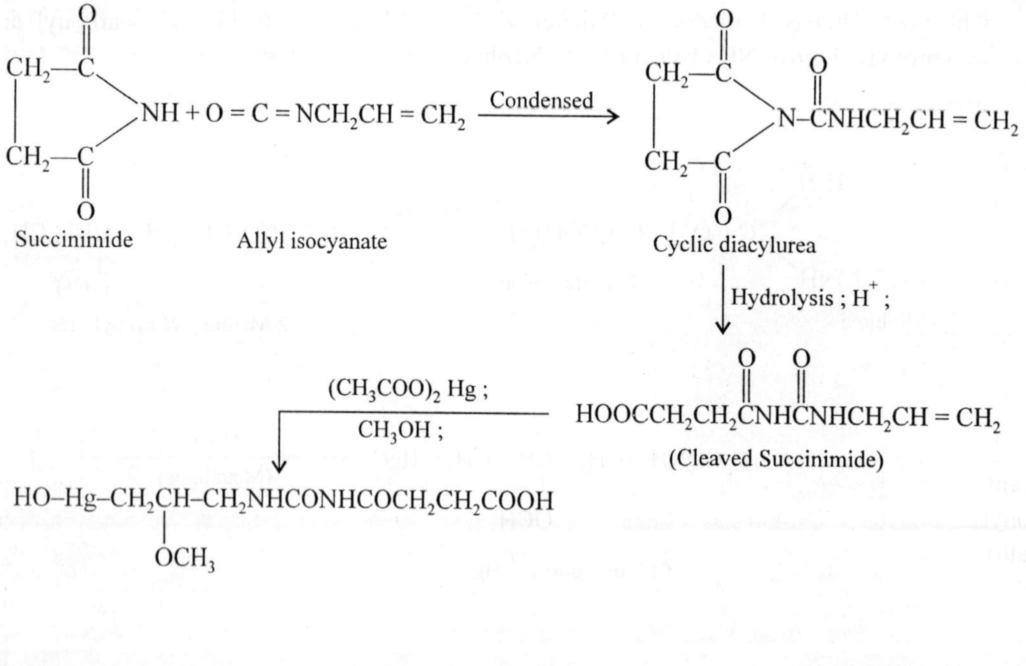

Succinimide   Allyl isocyanate   Cyclic diacylurea

Hydrolysis ; H$^+$;

$$HOOCCH_2CH_2\overset{O}{\overset{\|}{C}}NH\overset{O}{\overset{\|}{C}}NHCH_2CH = CH_2$$

(Cleaved Succinimide)

$(CH_3COO)_2$ Hg ;
CH_3OH ;

$$HO-Hg-CH_2CH-CH_2NHCONHCOCH_2CH_2COOH$$
$$OCH_3$$

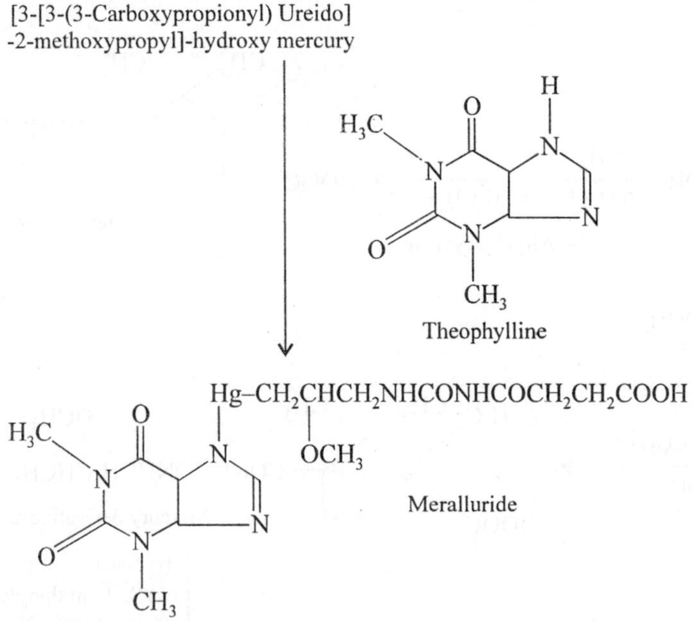

[3-[3-(3-Carboxypropionyl) Ureido]
-2-methoxypropyl]-hydroxy mercury

H_3C

O

H

N

N

N

N

O

CH_3

Theophylline

H_3C

O

N

N

N

O

CH_3

Hg–CH₂CHCH₂NHCONHCOCH₂CH₂COOH

OCH₃

Meralluride

Cyclic diacylurea is prepared by the condensation of succinimide and allyl isocyanate which upon acid hydrolysis affords the cleavage of the ring of the succinimide. Oxymercuration of the terminal olefin bond in the presence of mercuric acetate in methanol solution gives [3-[3-(3-caboxy-propionyl) ureido]-2-methoxypropyl]-hydroxy mercury. This on condensation with an equimolar portion of theophylline gives the official compound.

Meralluride is employed for the *treatment of oedema secondary to congestive heart failure, the nephrotic state of glomerulonephritis and hepatic cirrhosis.*

Dose : Usual, 1 ml (≡ to 39 mg of Hg and 43.6 mg of anhydrous theophylline) 1 or 2 times a week, parenteral 1 to 2 ml.

C. *Mercaptomerin Sodium* BAN, USAN, *Mercaptomerin* INN,

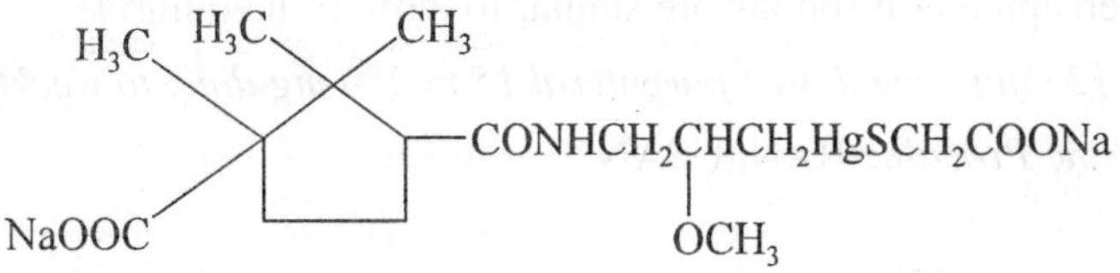

H_3C H_3C CH_3

—CONHCH₂CHCH₂HgSCH₂COONa

NaOOC

OCH₃

[3-(3-Carboxy-2, 2, 3-trimethylcyclopentane-carboxamide)-2-methoxypropyl] (hydrogen mercaptoacetate)-mercury disodium salt ; Mercury, [3-[[(3-carboxy-2, 2, 3-trimethylcyclopentyl) carbonyl] amino]-2-methoxypropyl] (mercapto-acetate-S)-, disodium salt ; USP. ; Ind. P. ; Thiomerin[R] (Wyeth)

Synthesis :

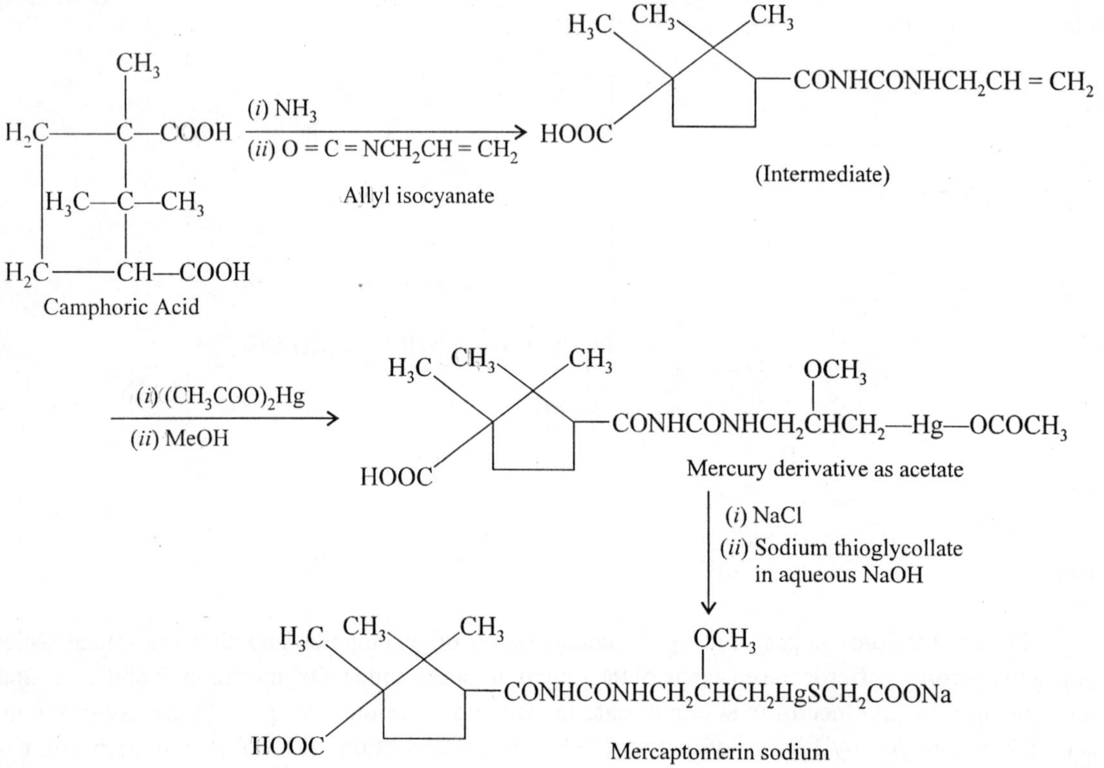

(Intermediate)

Mercury derivative as acetate

(*i*) NaCl
(*ii*) Sodium thioglycollate
in aqueous NaOH

Mercaptomerin sodium

Camphoric acid on condensation with ammonia and subsequent treatment with allyl isocyanate affords an intermediate which on reaction with mercuric acetate in methanol gives rise to the corresponding mercury derivative as acetate. This on treatment with sodium chloride followed by sodium thioglycollate in aqueous NaOH solution yields the official compound which may be obtained either by evaporation or by precipitation with an appropriate solvent.

The uses of mercaptomerin sodium are similar to those of meralluride.

Dose : Usual, 125 mg once daily ; parenteral 15 to 250 mg daily to weekly

D. *Merethoxylline Procaine* BAN, USAN,

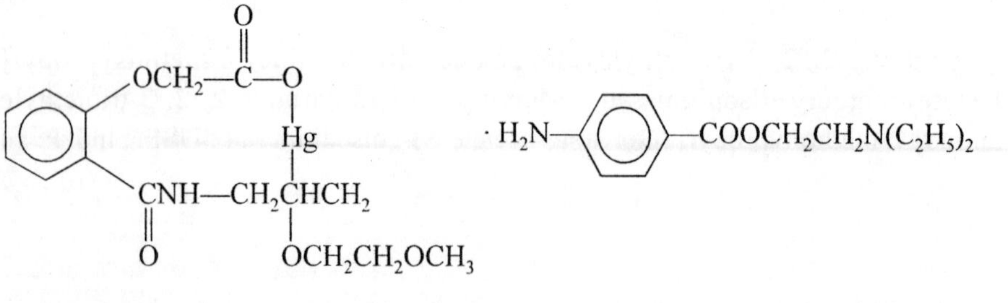

The procaine merethoxylline is an equimolar mixture of procaine and merethoxylline, the latter being the inner salt of [*o*-[[3-(hydroxymercuri)-2-(2-methoxyethoxy)-propyl]-carbamoyl] phenoxy] acetic acid. A mixture of the procaine merethoxylline and theophylline in the molecular proportion of 1 : 1.4 is available as a solution. Dicurin Procaine[(R)] (Lilly, USA.)

It has been used effectively in the *treatment of oedema and ascites in cardiac failure and also in ascites due to cirrhosis of the liver. The procaine component helps in reducing the discomfort of local irritation which may be caused by the mercurial compound when injected into tissues.*

Dose : Usual i.m., subcutaneous, daily 0.5 to 2.0 ml (containing 100 mg of merethoxylline procaine and 50 mg of theophylline per ml)

E. *Mersalyl* INN, USAN, *Mersalyl Sodium* BAN,

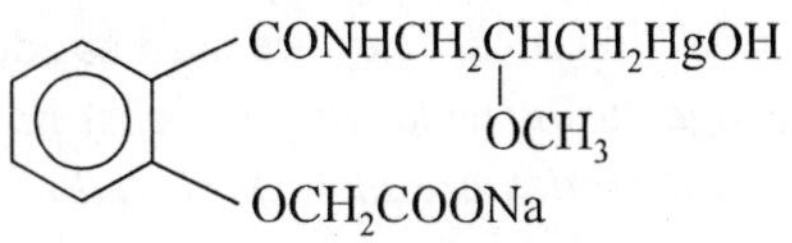

$$\text{CONHCH}_2\text{CHCH}_2\text{HgOH}$$
$$\text{OCH}_3$$
$$\text{OCH}_2\text{COONa}$$

Sodium salt of *o*-[(3-hydroxymercuri-2-methoxypropyl) carbamoyl]-phenoxy-acetic acid ; BPC. 1959 ; NF. XI ; Salygran[(R)] (Winthrop)

Mersalyl is *used to increase the output of oedema fluid in such typical conditions as renal disease, heart failure etc. It is also employed in the treatment of nephrotic oedema and in ascites due to cirrhosis of the liver.*

Dose : After assessing patient's tolerance by giving i.m. injection of 0.5 ml (10% m/v) ; 0.5 to 2 ml i.m. on alternate days.

2.1.1.   Mechanism of Action : The mechanism of action of the **'mercurial diuretics'** described under section 13.2.1 are stated as under :

2.1.1.1.   Chlormerodrin : Mercury-197 has been used in the form of chlormerodrin ([197]Hg), but has been largely superseded by other agents, such as : **Technetium-99 m.** It mainly survived as the [197]Hg isotope ($t_{1/2}$ = 64 hr.) employed for the exact visualization of renal parenchyma.

2.1.1.2.   Meralluride : Organic mercurial diuretics were widely employed prior to the introduction of *'thiazides'* and a host of other potent non-mercurial diuretics, but now have been virtually superseded by these orally active drugs that are found to be both potent and less toxic.

2.1.1.3.   Mercaptomerin Sodium : The statement given under section 2.1.1.2. also holds good for this *'drug'.*

2.1.1.4.   Merethoxylline Procaine : The statement provided under section 2.1.1.2. also holds good for this *'drug'.*

2.1.1.5.   Mersalyl Sodium : The *'drug'* is a powerful diuretic that acts on the renal tubules specifically, thereby enhancing the excretion of Na^+ and Cl^- ions, in almost equal amounts, and of water.

Salient Features of Organomercurials : The exact mechanism of action pertaining to the *organomercurials* is still quite a mystery. However, a few important salient features are as stated under :

(1) They breakdown usually to ionic mercury at the acidic urinary pH.

(2) Bonding of a Hg-atom to the organic residue overwhelmingly lowered the degree of toxicity of the corresponding **'inorganic compounds'** to an appreciable *'acceptable'* levels *in vivo.*

(3) Besides, it may be suggested that as an *'organic ligand'* the chances of legitimate cellular penetration to the specific **sulfhydryl enzymes** present in the proximal tubules is significantly improved thereby inactivating the renal enzymes directly involved with the tubular reabsorption processes, causing *diuresis* ultimately.

2.2.   Non-Mercurial Diuretics

The **non-mercurial diuretics** usually are predominant in terms of their significant clinical effectiveness and wider applications. They, in general, possess fewer side-effects and are much less toxic than the corresponding mercurial diuretics. They are used as adjunct specifically in the *treatment of either poisoning or drug over-dosage during which they increase the process of elimination of poisons or drugs through the kidneys. These diuretics are also employed to counter water and salt retention caused by various drug treatments.*

The most commonly used **diuretics** are invariably classified by their respective chemical class, mechanism of action, site of action, or effects on the *urine* contents. Nevertheless, these drugs normally exert their action rather widely with regard to their prevailing efficacy as well as their definite site of action located within the nephron. The real efficacy of a diuretic is often measured by its ability to enhance the rate of excretion of Na^+ ions filtered usually at the glomerulus (*i.e.,* the filtered load of sodium) and hence, must not be misunderstood with the potency, that is the actual amount of the *'diuretic'* essentially needed to cause a specific diuretic response. In other words, the efficacy of a diuretic is invariably estimated in portion by the site of action of the diuretic.

The **non-mercurial diuretics** may be classified on the basis of their chemical structures together with their physical characteristics as follows :

      1. Thiazides (Benzothiadiazines)

      2. Carbonic-Anhydrase Inhibitors

      3. Miscellaneous Sulphonamide Diuretics

      4. Aldosterone Inhibitors

      5. 'Loop' or 'High-Ceiling' Diuretics

      6. Purine or Xanthine Derivatives

      7. Pyrimidine Diuretics

      8. Osmotic Diuretics

      9. Acidotic Diuretics and

    10. Miscellaneous Diuretics

2.2.1.   Thiazides (Benzothiadiazines)

A major breakthrough in diuretic therapy was the introduction of chlorothiazide as a reliable, oral and non-mercurial diuretic in 1955 by Nouello and Spagne in the research laboratories of Merck, Sharp and Dohme. A number of benzothiadiazines (I), having the following general chemical formula :

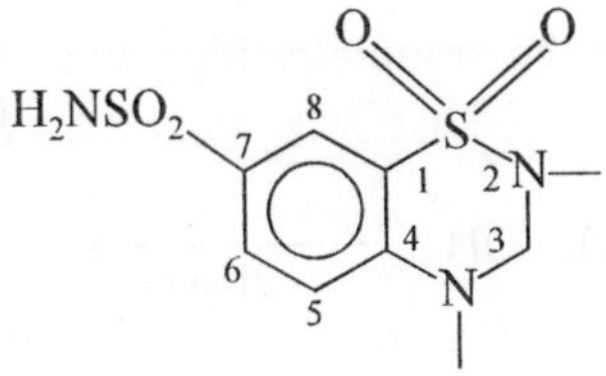

(I)

were subsequently synthesized and found to possess varying degree of diuretic actions. The benzothiadiazines are frequently known as **thiazides** or **benzothiazides.**

It has been amply observed that the thiazide diuretics enhance urinary excretion of both Na and H_2O by specifically inhibiting Na reabsorption located in the cortical (thick) portion of the ascending limb of Henle's loop* and also in the early distal tubules. Besides, they also progressively cause an increase in the excretion of Cl^-, K^+ and HCO_3^- (to a lesser extent) ions. However, the latter effect is predominantly by virtue of their mild carbonic anhydrase-inhibitory action. Importantly, due to their site of action, they invariably interfere with the dilution ; whereas, the concentration of urine is not affected appreciably.

In general, the thiazide diuretics minimise the glomerular filtration rate. Furthermore, this specific action fails to contribute to the diuretic action of such drugs, and this would perhaps put forward a logical explanation of their observed lower efficacy in instances having *impaired-kidney function.*

Examples : Chlorothiazide ; Hydrochlorothiazide ; Hydroflumethiazide ; Bendroflumethiazide ; Benzthiazide ; Cyclothiazide ; Cyclopenthiazide ; Methylclothiazide ; Trichlormethiazide ; Polythiazide ; Altizide

A. *Chlorothiazide* INN, BAN, USAN,

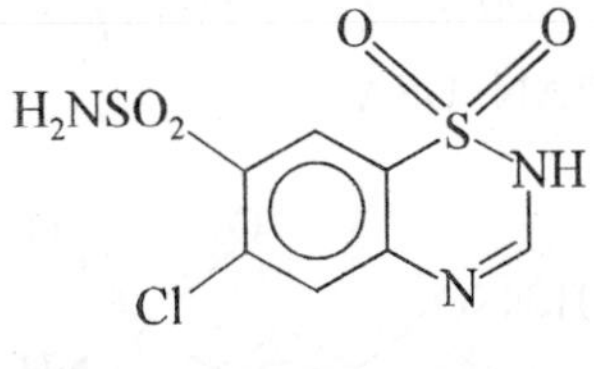

6-Chloro-2H-1, 2, 4-benzothiadiazine-7-sulphonamide 1, 1-dioxide ; 2H-1, 2, 4-Benzothiadiazine-7-sulphonamide, 6-Chloro-1, 1-dioxide ; BP., USP., Int. P. Diuril[R] (Merck Sharp and Dohme) ; SK-Chlorothiazide[R] (Smith Kline and French)

Synthesis :

It may be prepared by the chlorination of 3-chloroaniline with chlorosulphonic acid to yield 3-chloroaniline-4, 6-disulphonyl chloride, which is then amidated with ammonia to give the corresponding 4, 6-disulphonamide analogue. This on heating with formic acid affords cyclization through double condensation.

---

*The U-shaped portion of a renal tubule lying between the proximal and distal convoluted portions. It consists of a thin-descending limb and a thicker ascending limb.

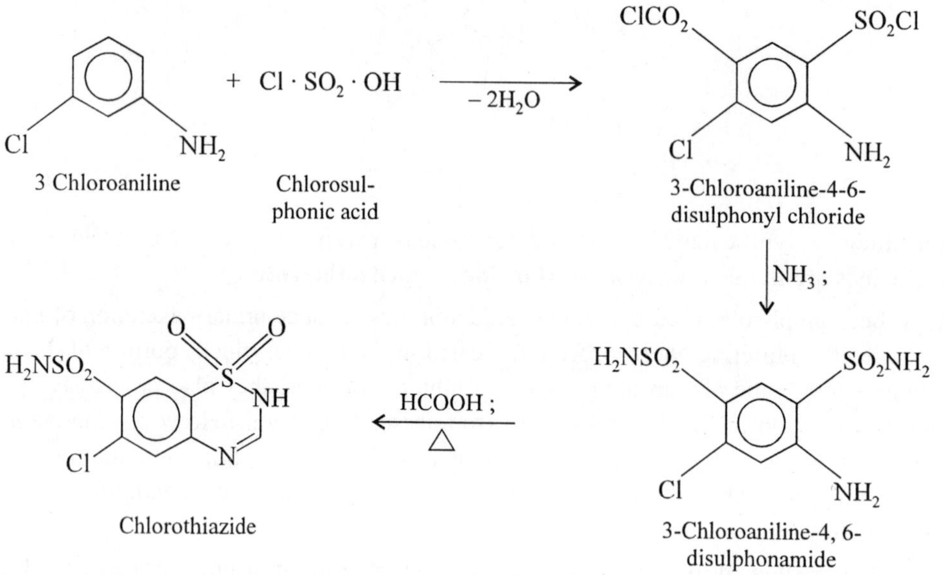

3 Chloroaniline   Chlorosul-   3-Chloroaniline-4-6-
                     phonic acid   disulphonyl chloride

Chlorothiazide   3-Chloroaniline-4, 6-
                                               disulphonamide

Chlorothiazide is used in the *treatment of oedema associated with congestive heart failure and renal and hepatic disorders. It is also employed in hypertension, either alone or in conjunction with other antihypertensive agents. It is also used in oedema associated with corticosteroid therapy thereby increasing the potassium-depleting action of the latter.*

Dose : Antihypertensive, 250 to 500 mg ; usual, antihypertensive, 250 mg 3 times per day ; diuretic 500 mg to 1g ; usual, diuretic, 500 mg 1 or 2 times per day.

B. *Hydrochlorothiazide* INN, BAN, USAN,

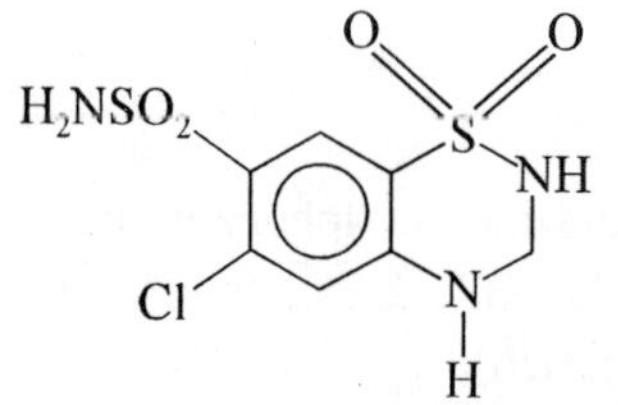

6-Chloro-3, 4-dihydro-2H-1, 2, 4-benzothiadiazine-7-sulphonamide 1, 1-dioxide ; 2H-1, 2, 4-Benzothiadiazine-7-sulphonamine, 6-chloro-3, 4-dihydro-1, 1-dioxide ; BP ; USP ; Int. P ; Esidrix[R] (Ciba-Geigy) ; Hydro DIURIL[R] (MS & D) ; Thiuretic[R] (Parke-Davis) ; Oretic[R] (Abbott)

Synthesis :

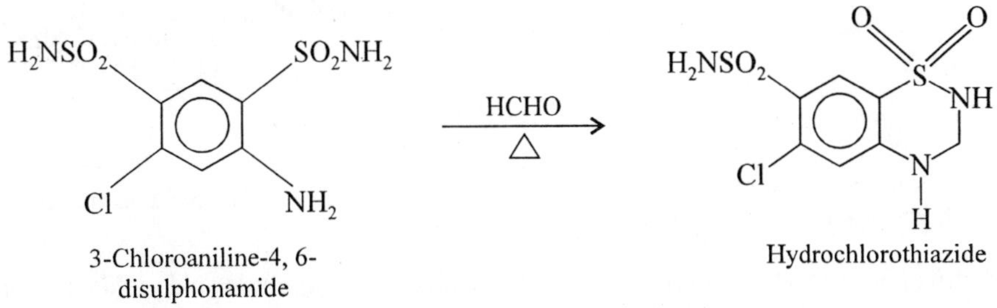

| 3-Chloroaniline-4, 6- disulphonamide | Hydrochlorothiazide |

The route of synthesis is more or less identical with that for chlorothiazide described earlier except that formaldehyde is used instead of formic acid in the final cyclization step from 3-chloroaniline-4, 6-disulphonamide.

Its diuretic actions are similar to those of chlorothiazide but it is ten times more potent than the latter. However, when the treatment is prolonged loss of K^+ causes hypokalemia which may be prevented by supplementation with potassium salts.

Dose : 25 to 200 mg per day ; usual, 50 mg 1 or 2 times daily.

C. *Hydroflumethiazide* INN, BAN, USAN,

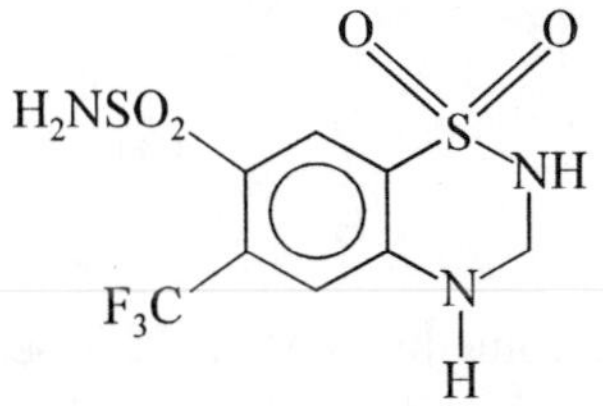

3, 4-Dihydro-6-(trifluoromethyl)-2H-1, 2, 4-benzothiadiazine-7-sulphonamide, 1, 1-dioxide ; 2H-1, 2, 4-Benzothiadiazine-7-sulphonamide, 3, 4-dihydro-6-(trifluoromethyl)-1, 1-dioxide ; Trifluoro-methylhydrothiazide ; BP. USP ; Int. P ; Diucardin[R] (Ayerst) ; Saluron[R] (Bristol)

Synthesis :

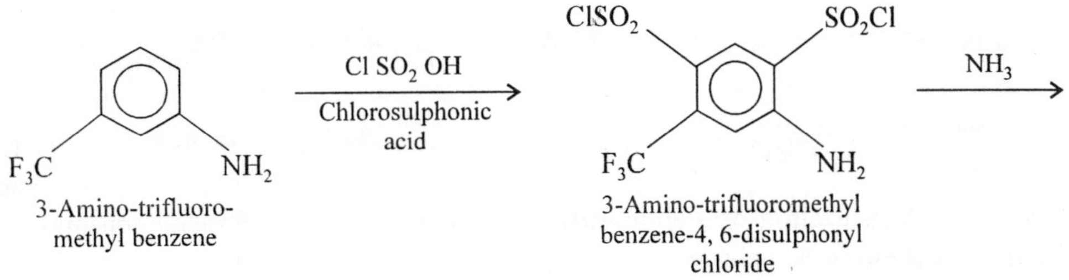

| 3-Amino-trifluoro-methyl benzene | 3-Amino-trifluoromethyl benzene-4, 6-disulphonyl chloride |

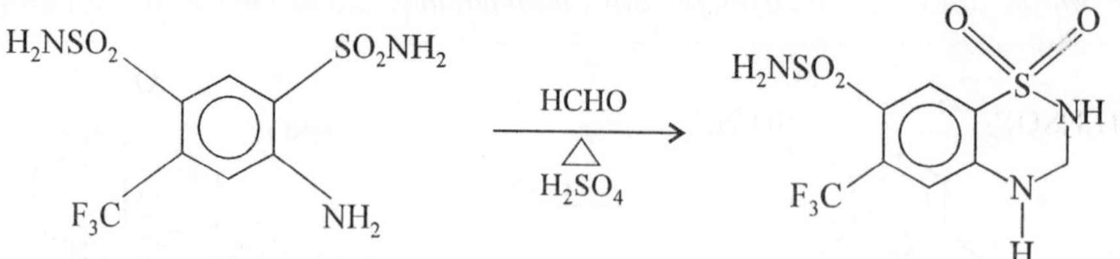

3-Amino-trifluoromethyl
benzene 4, 6-disulphonamide

Hydroflumethiazide

Treatment of 3-amino-trifluoromethyl benzene with chlorosulphonic acid yields the corresponding 4-disulphonyl chloride derivative which on reaction with ammonia gives rise to 3-amino-trifluoro methyl benzene 4, 6-disulphonamide. This on heating with formaldehyde in an environment of sulphuric acid affords a concomitant condensation and finally cyclization to the official compound.

Hydroflumethiazide is a potent diuretic employed in the management of oedema associated with cardiac failure, steroid administration, premenstrual tension and hepatic cirrhosis.

Dose : 25 to 200 mg ; usual, 50 to 100 mg daily.

D. *Bendroflumethiazide* INN, USAN, *Bendrofluazide* BAN,

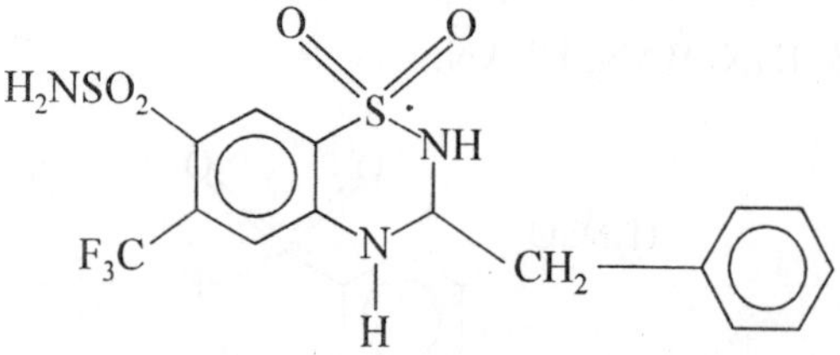

3-Benzyl-3, 4-dihydro-6-(trifluoromethyl)-2H, 1, 2, 4-benzo-thiadiazine-7 sulphonamide 1, 1-dioxide ; 2H, 1, 2, 4-Benzothiadiazine-7-sulphonamide, 3, 4-dihydro-3-(phenylmethyl)-6-(trifluoromethyl)-, 1, 1-dioxide ; BP., USP., Int. P. ; Naturetin[R] (Squibb) ; Neo-Naclex[R] (Glaxo)

Synthesis :

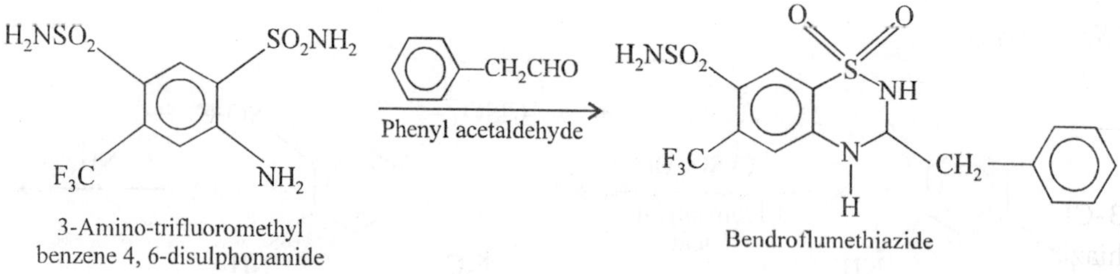

3-Amino-trifluoromethyl
benzene 4, 6-disulphonamide

Bendroflumethiazide

It consists of cyclization of 3-amino-trifluoromethyl benzene 4, 6-disulphonamide through condensation with phenylacetaldehyde.

Bendroflumethiazide is used in the *control and management of oedema, nephrosis and nephritis, cirrhosis and ascites, congestive heart failure, and other oedematous states. It is also employed as an antihypertensive agent.*

Dose : Initial, diuretic, 5 to 20 mg per day ; maintenance, 2.5 to 5 mg daily ; as antihypertensive, initial, 5 to 20 mg per day, maintenance, 2.5 to 15 mg per day.

E. *Benzthiazide* INN, BAN, USAN,

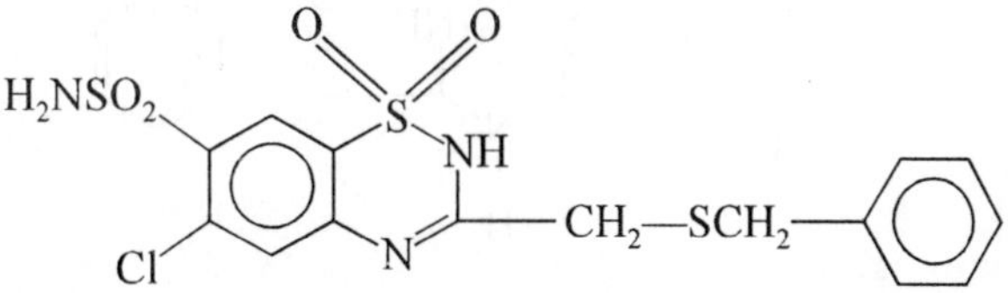

3-[(Benzylthio) methyl]-6-chloro-2H-1, 2, 4-benzothiadiazine-7-sulphonamide 1, 1-dioxide ; 2H-1, 2, 4-Benzothiadiazine-7-sulphonamide-6-chloro-3-[[(phenylmethyl) thio] methyl]-, 1, 1-dioxide ; BPC ; (1963), USP ; Exna[R] (Robins) ; Aquatag[R] (Tutag)

Synthesis :

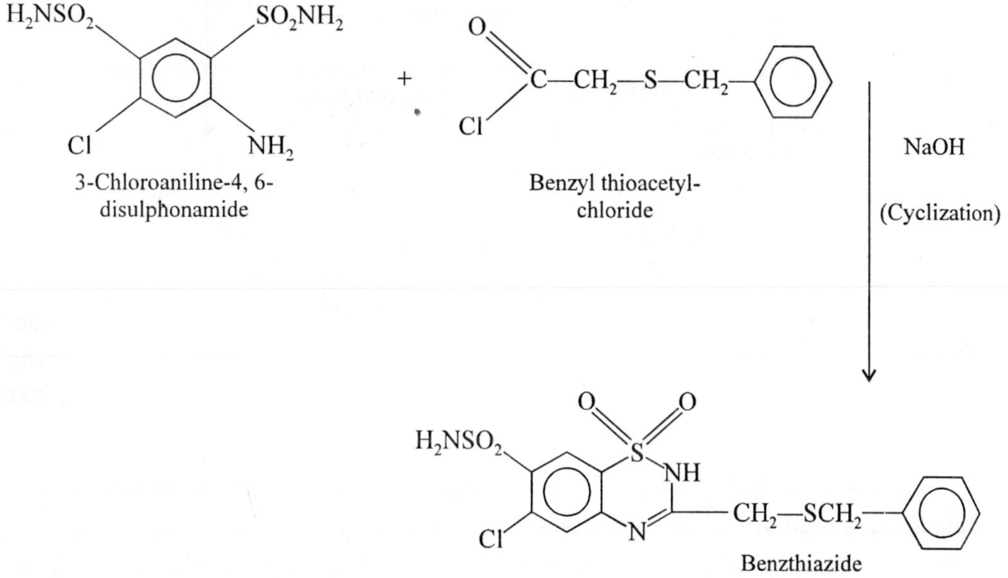

3-Chloroaniline-4, 6-
disulphonamide

Benzyl thioacetyl-
chloride

NaOH

(Cyclization)

Benzthiazide

3-Chloroaniline-4, 6-disulphonamide is prepared in the same manner as described for chlorothiazide which is then made to condense and cyclize by treatment with benzyl thioacetyl chloride in the presence of sodium hydroxide to yield benzthiazide.

It is used as a diuretic and an antihypertensive agent with pharmacological actions similar to those of chlorothiazide.

Dose : Usual, diuretic, initial, 50 to 200 mg per day ; maintenance, 50 to 150 mg per day ; usual, antihypertensive, initial, 50 mg 2 times a day ; maintenance, maximal dose of 50 mg 3 times daily.

F. *Cyclothiazide* INN, BAN, USAN,

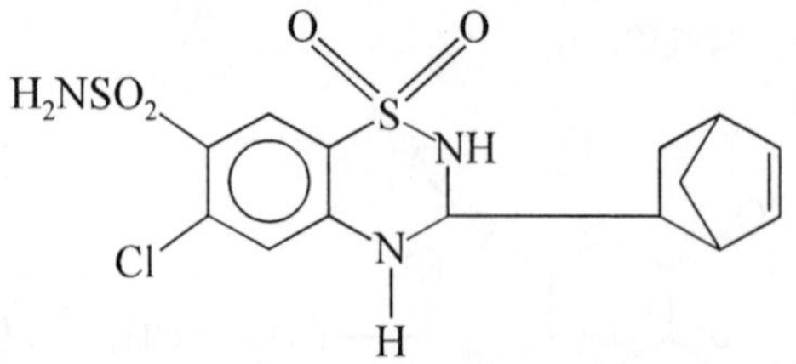

6-Chloro-3, 4-dihydro-3-(5-nor-bornen-2-yl)-2H-1, 2, 4-benzothiadiazine-7-sulphonamide 1, 1-dioxides ; 2H-1, 2, 4-Benzothiadiazine-7-sulphonamide, 3-bicyclol [2, 2, 1] hept-5-en-2-yl-6-chloro-3, 4-dihydro-, 1, 1-dioxide ; USP ; NF ; Anhydron[R] (Lilly) ; Fluidil[R] (Adria)

Synthesis :

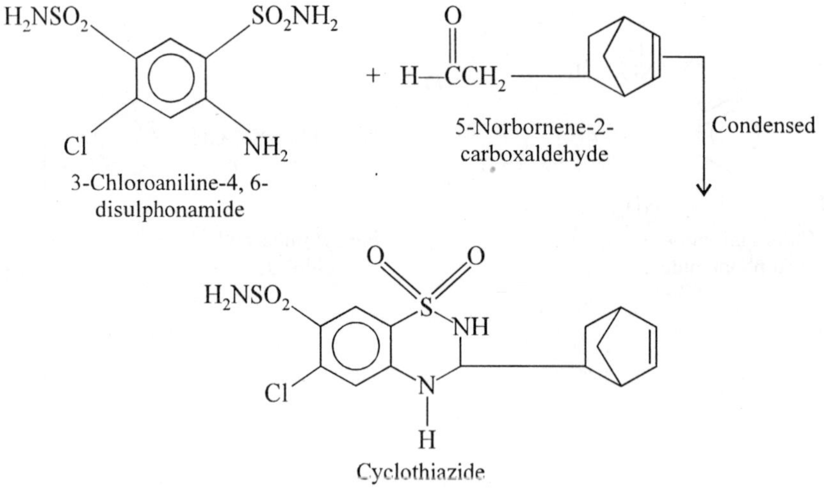

5-Norbornene-2-carboxaldehyde

3-Chloroaniline-4, 6-disulphonamide

Condensed

Cyclothiazide

The synthesis of cyclothiazide is analogous to that for chlorothiazide, except that 5-nonbornene-2-carboxaldehyde is used in the cyclization process in place of formic acid.

It possesses both diuretic and antihypertensive actions. It is often *used as an adjunct to other antihypertensive drugs, such as reserpine and the ganglionic blocking agents.*

Dose : Usual, initial, diuretic, 1 to 2 mg per day ; maintenance, 1 to 2 mg on alternate days, or 2 or 3 times per week ; usual, antihypertensive, 2 mg 1 to 3 times daily

G. *Cyclopenthiazide* INN, BAN, USAN,

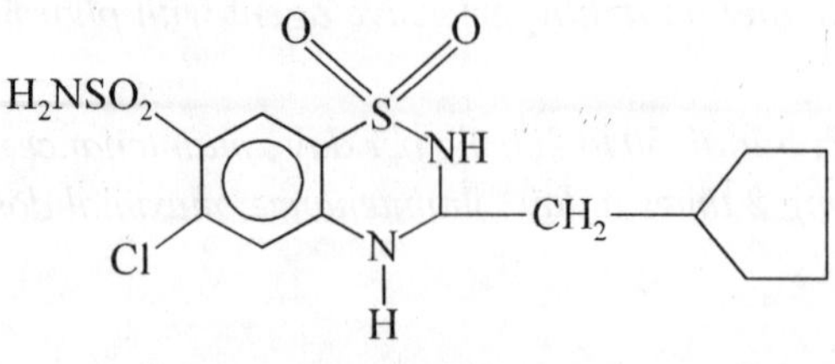

6-Chloro-3-(cyclopentylmethyl)-3, 4-dihydro-2H, 1, 2, 4-benzothiadiazine-7-sulphonamide 1, 1-dioxide ; 2H-1, 2, 4-Benzothiadiazine-7-sulphonamide, 6-chloro-3-(cyclopentylmethyl)-3, 4-dihydro, 1, 1-dioxide ; BP ; Navidrex-K[(R)] (Ciba, U.K.)

Synthesis :

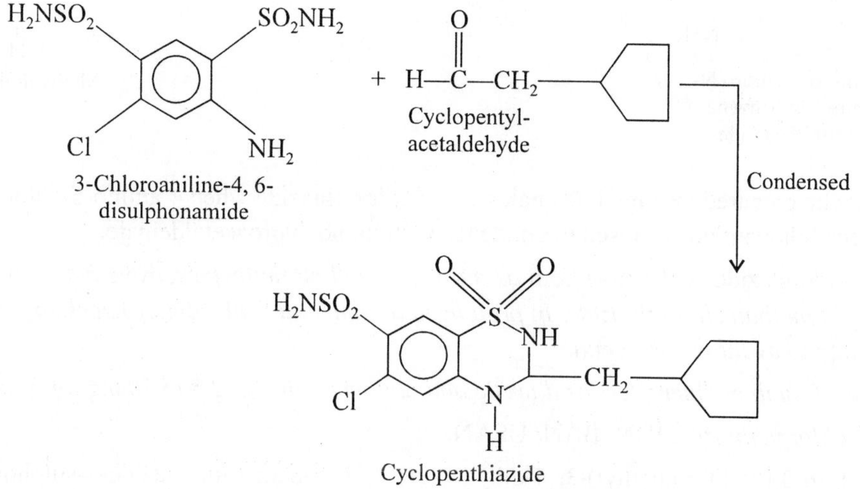

3-Chloroaniline-4, 6-disulphonamide

Cyclopentyl-acetaldehyde

Condensed

Cyclopenthiazide

The process is similar to that for chlorothiazide, except that cyclopentyl acetaldehyde is used in the cyclization to yield the official compound.

Cyclopenthiazide possesses actions and uses similar to those of chlorothiazide.

Dose : For oedema, usual, initial, 0.5 to 1 mg per day, reduced to 250 to 500 mcg per day or 500 mcg on alternate days ; For hypertension, usual, 250 to 500 mcg per day either alone, or in conjunction with other antihypertensive agents.

H. *Methyclothiazide* INN, BAN, USAN,

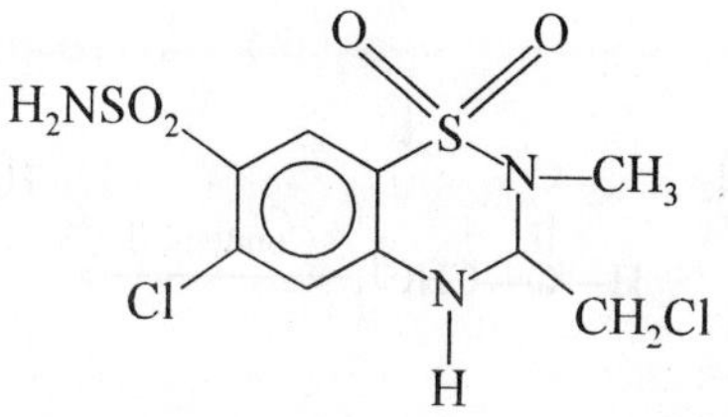

6-Chloro-3-(chloromethyl)-3, 4-dihydro-2-methyl-2H-1, 2, 4-benzothiadiazine-7-sulphonamide-1, 1-dioxide ; 2H-1, 2, 4-Benzothiadiazine-7-sulphonamide, 6-chloro-3-(chloromethyl)-3, 4-dihydro-2-methyl, 1, 1-dioxide ; USP ; Enduron[(R)] (Abbott)

Synthesis :

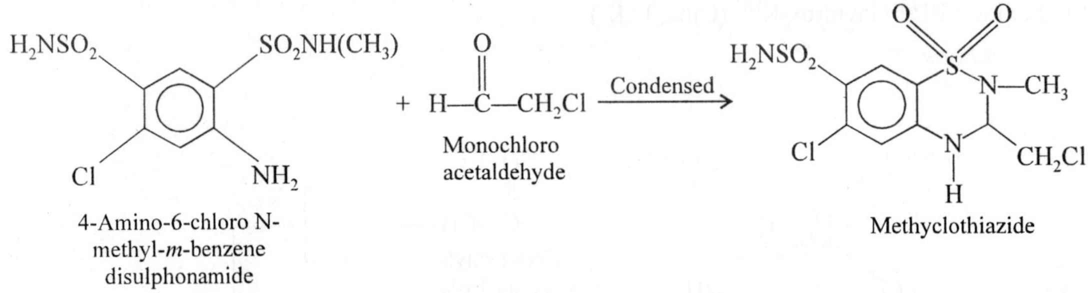

4-Amino-6-chloro N-
methyl-*m*-benzene
disulphonamide

Monochloro
acetaldehyde

Methyclothiazide

It may be prepared by a method analogous to chlorothiazide when 4-amino-6-chloro-N^3-methyl-*m*-benzenedisulphonamide is caused to condense with monochloroacetaldehyde.

Methyclothiazide is effective both *as a diuretic and an antihypertensive agent. It is about 100 times more potent than chlorothiazide. In prolonged treatment it is absolutely necessary to supplement with potassium to avoid hypokalemia.*

Dose : Usual, maintenance, as diuretic and antihypertensive, 2.5 to 10 mg once per day.

I. *Trichlormethiazide* INN, BAN, USAN,

6-Chloro-3-(dichloromethyl)-3, 4-dihydro-2H-1, 2, 4-benzo-thiadiazine-7-sulphonamide 1, 1-dioxide ; 2H-1, 2, 4-Benzothiadiazine-7-sulphonamide, 6-chloro-3-(dichloromethyl)-3, 4-dihydro-1, 1-dioxide USP ; Metahydrin[R] (Merrell Dow) ; Naqua[R] (Schering)

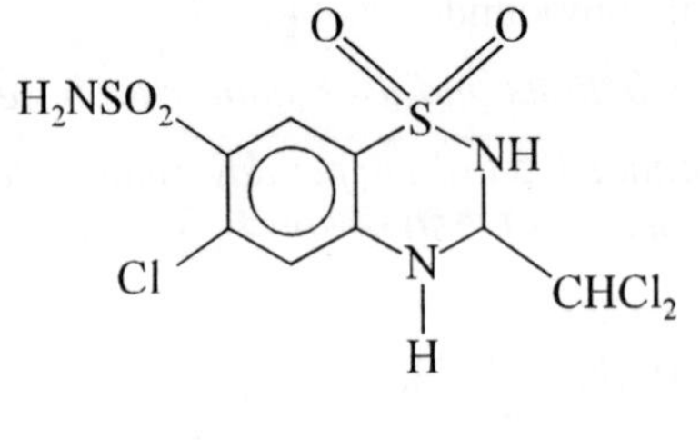

Synthesis :

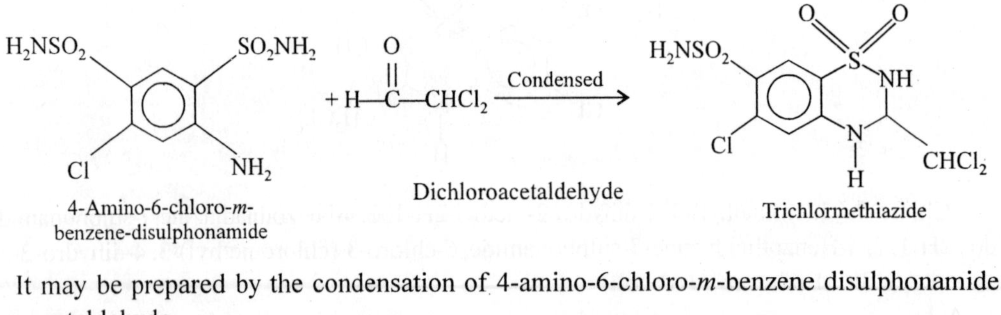

4-Amino-6-chloro-*m*-
benzene-disulphonamide

Dichloroacetaldehyde

Trichlormethiazide

It may be prepared by the condensation of 4-amino-6-chloro-*m*-benzene disulphonamide with dichloroacetaldehyde.

Trichlormethiazide belongs to the class of *long-acting diuretic and antihypertensive thiazide.*

Dose : Usual, 2 to 4 mg twice daily ; maintenance 2 to 4 mg once per day.

J. *Polythiazide* INN, BAN, USAN,

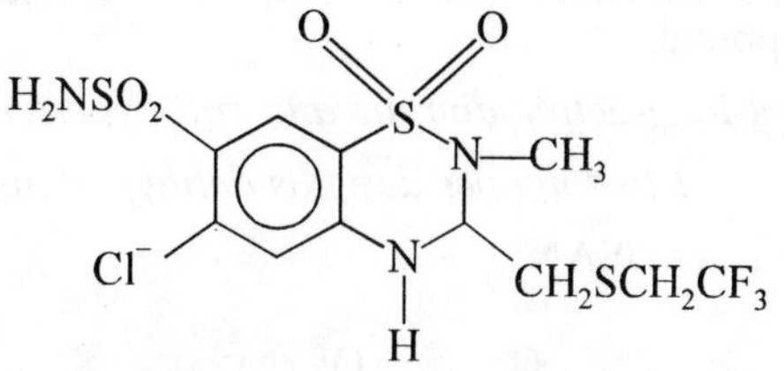

6-Chloro-3, 4-dihydro-2-methyl-3-[[2, 2, 2-trifluoroethyl) thio] methyl]-2H-1, 2, 4-benzothia-diazine-7-sulphonamide 1, 1-dioxide ; 2H-1, 2, 4-Benzothiadiazine-7-sulphonamide, 6-chloro-3, 4-dihydro-2-methyl-3-[[(2, 2, 2-trifluoroethyl) thio] methyl]-, 1, 1-dioxide ; BP ; USP ; NF ; Renese[R] (Pfizer)

Synthesis :

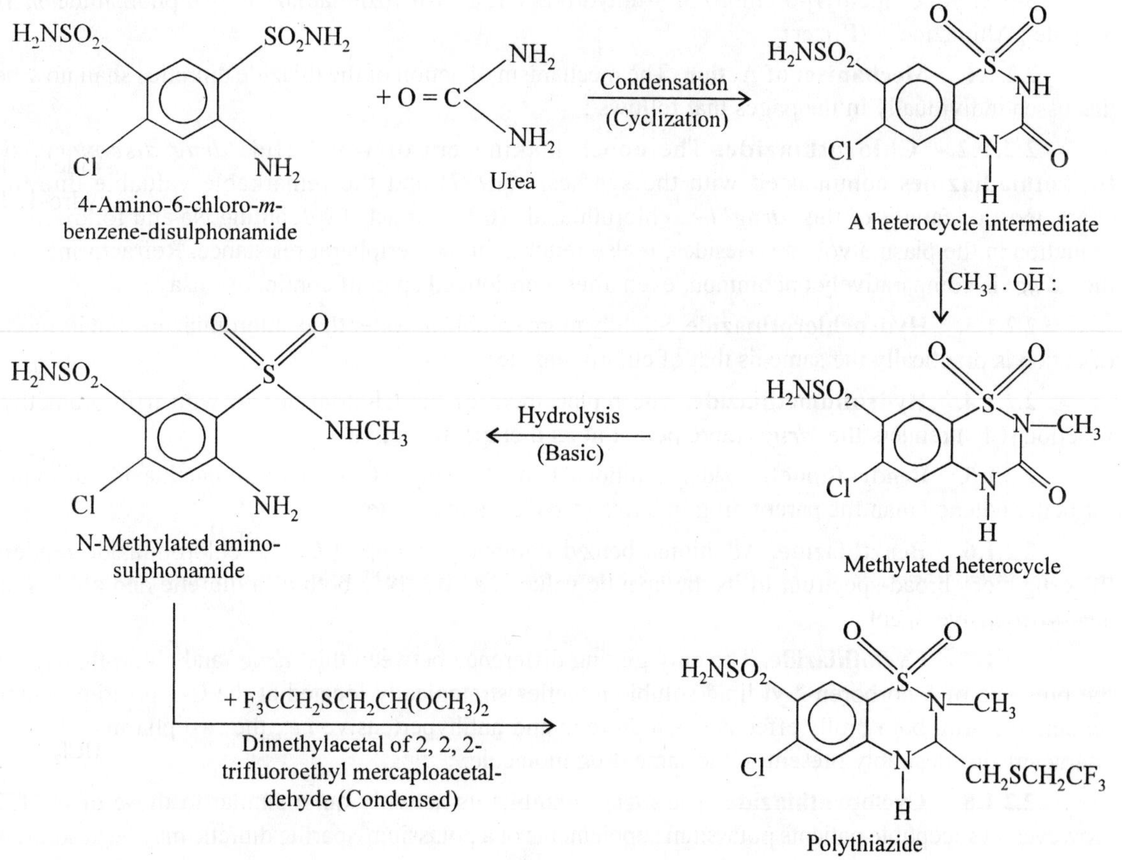

A heterocycle intermediate is prepared by the condensation of 4-amino-6-chloro-*m*-benzene disulphonamide with urea which on treatment with methyl iodide in a basic medium yields the corresponding methylated heterocycle. This on hydrolysis in the presence of a base affords N-methylated aminosulphonamide which on condensation with dimethylacetal of 2, 2, 2-trifluoroethylmercaptoacetaldehyde yields the official compound.

Polythiazide is a *potent long-acting diuretic and anti-hypertensive agent.*

Dose : As diuretic, usual, 1 to 4 mg per day ; as antihypertensive, 2 to 4 mg as required.

K. *Altizide* INN, *Althiazide* USAN,

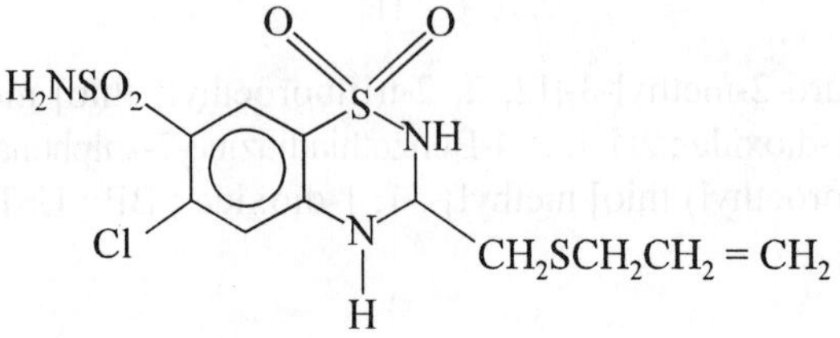

3-[(Allythio) methyl]-6-chloro 3, 4-dihydro-2H-1, 2, 4-benzothiadiazine-7-sulphonamide 1, 1-dioxide ; Althiazide[(R)] (Pfizer)

2.2.1.1.   Mechanism of Action. The mechanism of action of the thiazide diuretics shall now be discussed individually in the pages that follows :

2.2.1.2.   Chlorothiazide. The epoch making era of wonderful *'drug discovery'* of **benzothiadiazines** commenced with the synthesis (1957) and the remarkable valuable diuretic characteristic features of this *'drug' i.e.,* chlorothiazide (CTZ). It acts by depliting Na and followed by reduction in the plasma volume. Besides, it also reduces in the peripheral resistance. Refractoriness of the *'drug'* is comparatively uncommon, even after a prolonged span of continuous usage.

2.2.1.3.   Hydrochlorothiazide. Slightly more soluble in water than chlorothiazide, but its mode of action is practically the same as that of chlorothiazide.

2.2.1.4.   Hydroflumethiazide. The replacement of the Cl-atom at C-6 with trifluoromethyl function (CF_3) renders the *'drug'* more potent in its therapeutic activity.

2.2.1.5.   Bendroflumethiazide. Additional benzyl moiety at C-3 of hydroflumethiazide attributes far better potency than the parent drug in terms of its diuretic profile.

2.2.1.6.   Benzthiazide. Additional benzyl thiomethyl group at C-3 of chlorothiazide renders the drug more broad-spectrum in its therapeutic values *i.e.,* it serves both as a diuretic and also as an antihypertensive agent.

2.2.1.7.   Cyclothiazide. The only glaring difference between this *'drug'* and chlorothiazide is the presence of 5-norboren-2-yl lipid-soluble moieties strategically located at the C-3 position which renders the drug both orally effective as a diuretic and antihypertensive *i.e.,* the two pharmacological characteristics desirably present in the same drug molecule.

2.2.1.8.   Cyclopenthiazide. The *'drug'* exhibits its activitiy quite similar to those of HCTZ. However, in suceptible patients potassium supplements or a potassium-sparing diuretic may be absolutely important and necessary.

2.2.1.9.   Methyclothiazide. The dosage regimen of clinically used compounds invariably ranges between 1 to 2000 mg. Besides, there exists one more important and clinically useful variable within which a choice is obviously preferable is the duration of action. Methyclothiazide possesses a range of 24+ hours in comparison to CTZ having as much a low range of 6 hours.

2.2.1.10. Trichlormethiazide. The *'drug'* is an orally effective as well as long-acting thiazide diuretic and antihypertensive. It resembles CTZ with respect to its pharmacologic actions, therapeutic uses and untoward effects.

2.2.1.11. Polythiazide. It is also a long-acting diuretic and antihypertensive agent that causes diuresis within 2 hr, attains a peak in 6 hr and lasts 24 to 48 hr. The mean plasma half-lives for absorption and elimination are 1.2 and 25.7 hour respectively. It has been observed that nearly 20% of the drug gets excreted unchanged in the urine. On being compared on a milligram basis, 2 mg of polythiazide has almost nearly the same diuretic activity as produced by 500 mg of CTZ.

2.2.1.12. Altizide (Althiazide). It is a thiazide diuretic having action very similar to hydrochlorothiazide (HCTZ). It is invariably administered in combination with spironolactone.

2.2.1.13. SARs of Thiazide Diuretics. The SARs of these benzene disulphonamide structural analogues yielded a broad-spectrum of compounds having a relatively high degree of diuretic activity, which are summarized as stated under :

(1) Thiazide diuretics are found to be weakly acidic in nature having a benzothiadiazine 1, 1-dioxide nucleus.

Benzothiadiazine 1, 1-dioxide

(2) Chlorothiazide (CTZ) being the simplest member of this series of structural analogues having two pKa (dissociation constant) values of 6.7 and 9.5. The two acidic zones in CTZ are virtually due to the presence of : (*a*) presence of a H-atom at the 2-N that essentially attributes the most acidic character by virtue of the influence of the prevailing electron withdrawing effects of the neighbouring sulphone moiety ; and (*b*) presence of the sulphonamide ($-SO_2NH_2$) functional moiety strategically located at C-7 position which affords an additional environment (zone) of creating acidity in the molecule ; however, its acidic influence is much less than the 2-N proton. Importantly, these acidic protons enable the formation of the corresponding water-soluble sodium salt which may be gainfully used for IV-administration of the diuretics as shown below :

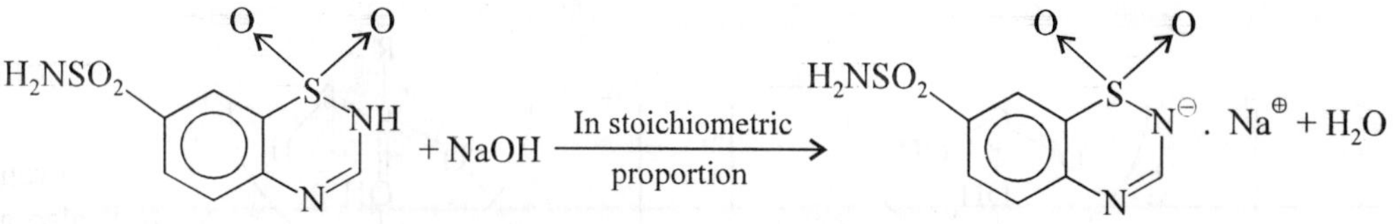

(3) Presence of an electron-withdrawing moiety at C-6 is an absoluble necessity for the diuretic activity. A few important and vital observations are as enumerated below :

(*a*) Practically negligible diuretic activity is obtained by having a H-atom at C-6 ;

(*b*) Substitution with a chloro cr trifluoromethyl moiety at C-6 are quite active pharmacologically.

(*c*) Further, the CF_3 moiety renders the resulting diuretic compound more lipid-soluble and also with a much longer duration of action in comparison to its chloro-substituted derivatives ;

(*d*) Presence of electron-releasing moieties, namely : *methyl* or *methoxyl* at C-6 position attributes significantly reduced diuretic activity.

(4) Removal or possible replacement of the sulphonamide function at C-7 results into such compounds possessing either little or almost no diuretic activity.

(5) Saturation of the prevailing double-bond between 3 and 4 positions to give rise to a corresponding 3, 4-dihydro structural analogue which is observed to be having nearly 10 times more diuretic activity than the unsaturated analogue.

(6) Introduction of a lipophilic functional moiety at C-3 position renders a marked and pronounced enhancement in the diuretic potency. For instance : aralkyl, haloalkyl, or thioether substitution, enhances the lipoidal solubility of the molecule to a considerable extent thereby producing compounds with a much longer duration of action.

(7) Alkyl substitution on the N-2 position is observed to lower the polarity and ultimately enhancing the duration of the ensuing diuretic action.

2.2.2. Carbonic Anhydrase Inhibitors

In early 1940s, attempts were made towards the synthesis and subsequent screening of sulphonamides possessing carbonic anhydrase inhibitory characteristics of sulphanilamide which resulted in the production of a variety of heterocyclic sulphonamides. When the enzyme is inhibited, the generation of carbonic acid (H_2CO_3) that usually dissociated into HCO_3^- and H_3O^+, is also inhibited. Thus in glomerular filtrate, a deficiency of H_3O^+ which normally exchanges for Na^+ occurs. The Na^+ remains in the renal tubule together with the HCO_3^- plus an osmotic equivalent of water, which ultimately results in the excretion of a large quantity of urine and hence diuresis. These compounds which inhibit carbonic anhydrase, besides acting as diuretics, also cause acidosis because of the elimination of HCO_3^- and Na^+ ions. The acidosis tends to limit its diuretic activity.

A most logical H-bonding mechanism which is believed to act competitively perhaps explains the action of some sulphonamide carbonic anhydrase inhibitors which predominantly exhibit both diuretic and antiglaucoma activities. It is, however, assumed that carbonic acid being the normal substrate which not only fits into a cavity but also complexes with the corresponding enzyme carbonic anhydrase (CA) as illustrated in Fig. 12.1(A). Consequently, this complex is strongly stabilized by **four** H-bonds.

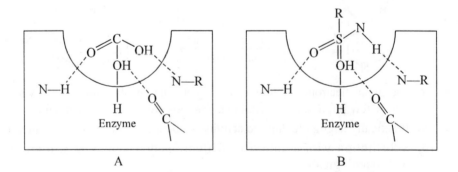

Fig. 12-1. Interactions Occurring at the Hypothetical Reactive Sites
of Enzymes Carbonic Anhydrase

Interestingly, the *sulphonamide moiety* which essentially possesses a *geometric structural entity* thus allowing an equally perfect fit compatible into the cavity of the enzyme CA also get bound quite securely and effectively, perhaps to the same four areas by H-bonds as depicted in Fig. 12.1(B).

Therefore, one may safely draw an inference that these sulphonamide structural analogues competitively prevent the carbonic acid from getting bound at this specific site. Consequently, such an action shall obviously inhibit the prvailing action of the enzyme CA, thereby giving rise to an apparent acid-base imbalance that would ultimately cause diuresis.

It is, however, pertinent to state here that the sulphonamides of the type wherein the possibilities for H-bonding having been lowered from *four* to *three* usually render the compounds **inactive.**

A few typical examples of this class of diuretics are described here.

Examples : Acetazolamide ; Methazolamide ; Ethoxzolamide ; Diclofenamide ; Disulfamide

A. *Acetazolamide* INN, BAN, USAN,

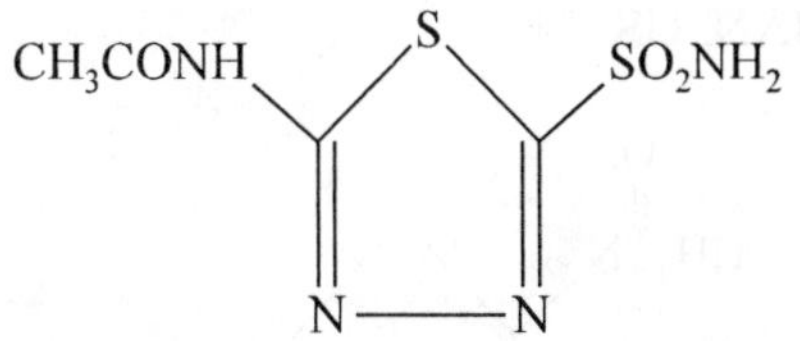

N-(5-Sulfamoyl-1, 3, 4-thiadiazol-2-yl) acetamide ; Acetamide, N-[5-(amino-sulphonyl)-1, 3, 4-thiadiazol-2-yl]- ; BP ; USP ; Ind. P. ; Diamox[R] (Lederle)

Synthesis :

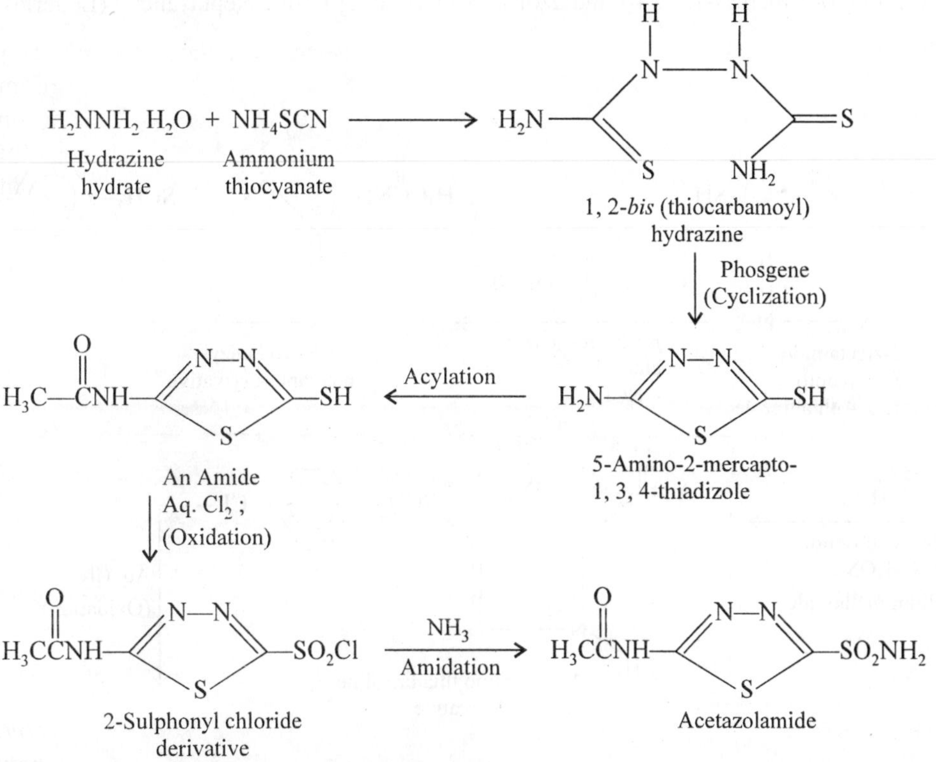

$H_2NNH_2 H_2O$ + NH_4SCN $\longrightarrow$

Hydrazine
hydrate

Ammonium
thiocyanate

1, 2-*bis* (thiocarbamoyl)
hydrazine

Phosgene
(Cyclization)

An Amide

Acylation

5-Amino-2-mercapto-
1, 3, 4-thiadizole

Aq. Cl₂ ;
(Oxidation)

2-Sulphonyl chloride
derivative

NH_3
Amidation

Acetazolamide

Reaction between hydrazine hydrate and ammonium thiocyanate yields 1, 2-*bis* (thiocarbamoyl) hydrazine which on treatment with phosgene undergoes molecular rearrangement through loss of ammonia to yield 5-amino-2-mercapto-1, 3, 4-thiadiazole. This on acylation gives a corresponding amide which on oxidation with aqueous chlorine affords the 2-sulphonyl chloride. The final step essentially consists of amidation by treatment with ammonia.

Acetazolamide is employed effectively for adjunctive treatment of drug-induced oedema, oedema caused by congestive heart failure, petit mal and other centrencephalic epilepsies. It has also been used to lower the intraocular pressure prior to surgery in acute conditions of angle-closure glaucoma, besides open-angle and secondary glaucoma.

Dose : Usual, 250 mg 2 to 4 times per day.

B. *Methazolamide* INN, BAN, USAN,

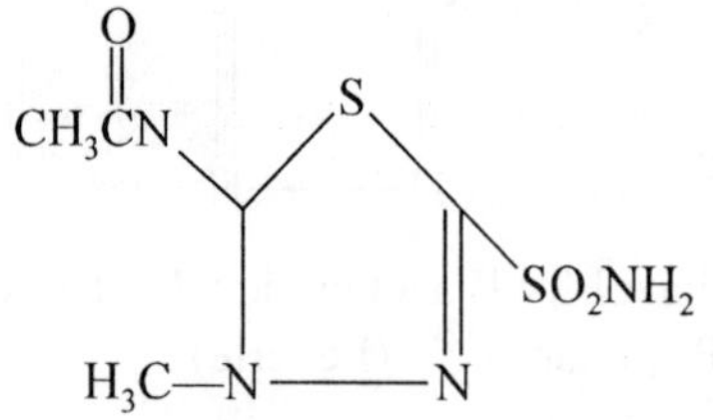

N-(4-Methyl-2-sulphamoyl-Δ^2-1, 3, 4-thiadiazolin-5-ylidene) acetamide ; Acetamide, N-[5-(aminosulphonyl)-3-methyl-1, 3, 4-thiadiazol-2 (3H)-ylidene] USP ; Neptazane[R] (Lederle)

Synthesis :

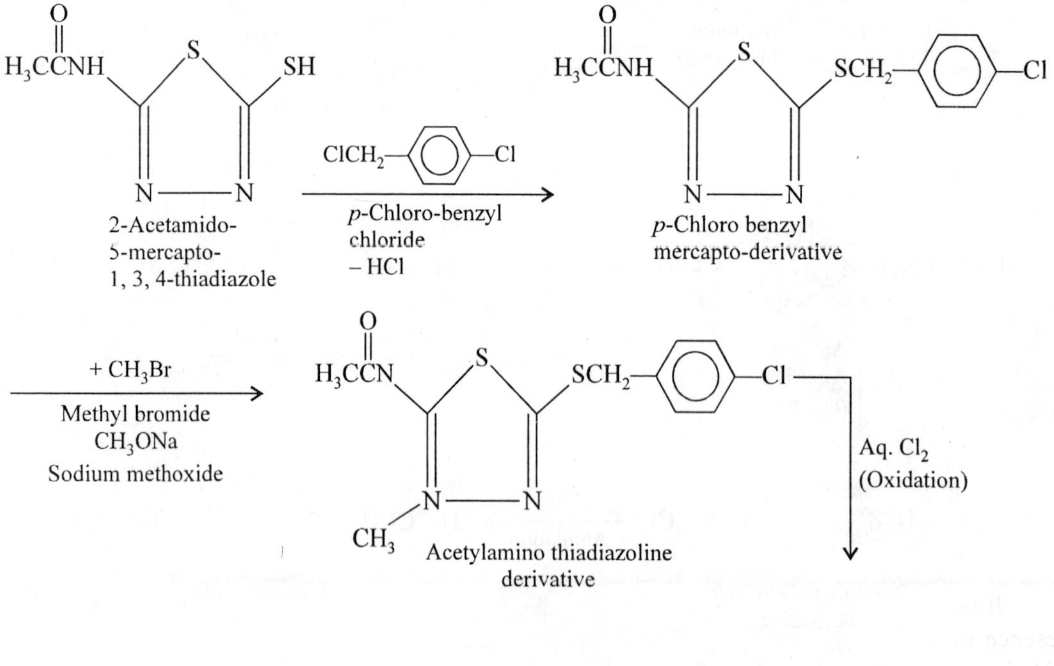

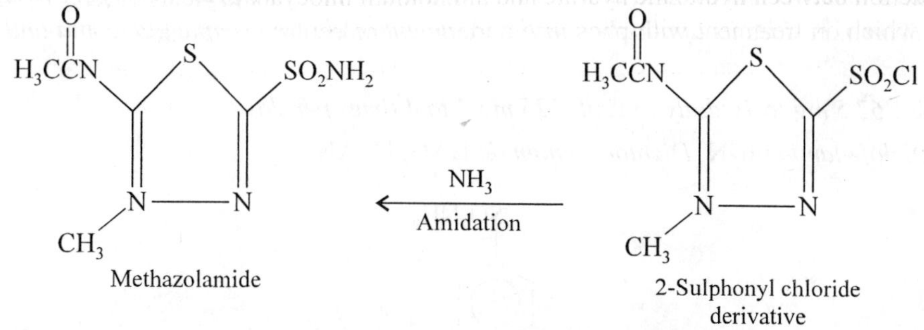

Methazolamide   2-Sulphonyl chloride derivative

2-Acetamido-5-mercapto-1, 3, 4-thiadiazole is prepared as described under acetazolamide. This on treatment with *p*-chlorobenzyl chloride forms the corresponding *p*-chloro benzyl mercapto derivative, which when reacted with methyl bromide in the presence of sodium methoxide yields the acetylamino thiadiazoline derivative. On oxidation with aqueous chlorine it gives rise to the 2-sulphonyl chloride derivative which finally yields methazolamide on amidation with ammonia.

Its actions and uses are similar to those of acetazolamide. However, its action has been found to be *relatively less prompt but of definitely longer duration than that of the latter, lasting for 10 to 18 hours.*

Dose : 100 to 600 mg per day ; usual, 50 to 100 mg 2 to 3 times per day.

C. *Ethoxzolamide* BAN, USAN,

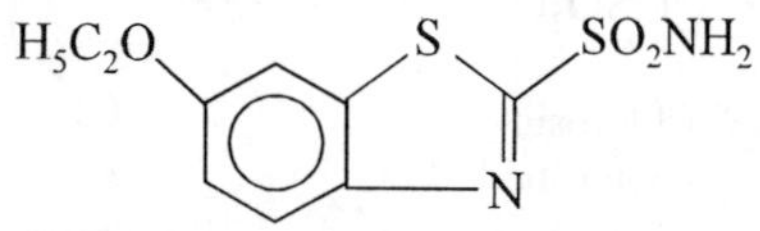

6-Ethoxy-2-bezothiazolesulphonamide ; 2-Benzothiazolesulphonamide, 6-ethoxy- ; Ethoxyzolamide ; USP ; Cardrase (Upjohn)

Synthesis :

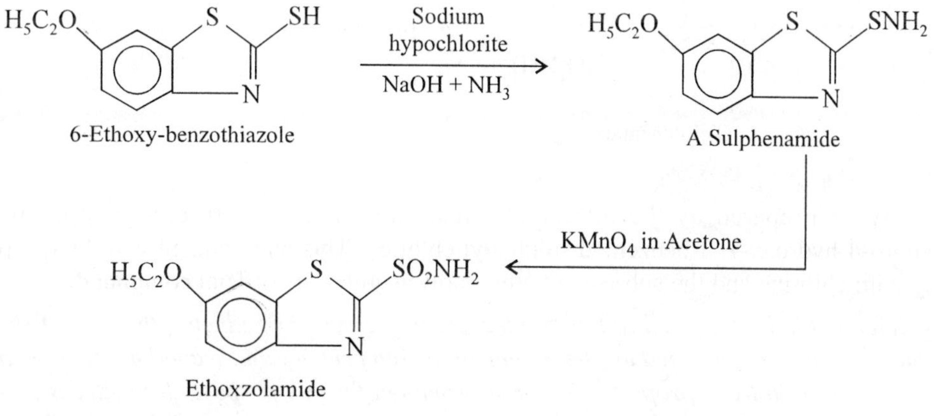

Ethoxzolamide

It may be prepared by the reaction of 6-ethoxy-benzothiazole with sodium hypochlorite in the presence of sodium hydroxide and ammonia to yield the corresponding sulphenamide, which upon oxidation with potassium permanganate in acetone forms the official compound.

Ethoxzolamide is mainly used to lower the intraocular pressure prior to surgery in acute angle-closure glaucoma, besides its application in the treatment of chronic simple glaucoma and secondary glaucoma.

Dose : 62.5 mg to 1g daily ; usual, 125 mg 2 to 4 times per day.

D. *Diclofenamide* INN, *Dichlorphenamide* BAN, USAN,

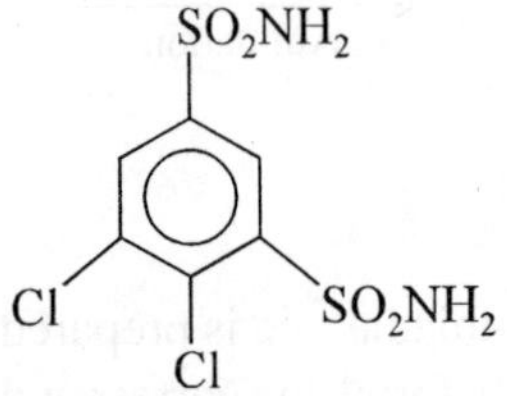

$$SO_2NH_2$$

4, 5-Dichloro-*m*-benzenedisulphonamide ; 1, 3-Benzenedisulphonamide, 4, 5-dichloro- ; 4, 5-Dichlorobenzene-1, 3-disulphonamide ; BP ; USP ; Daranide[R] (Merck Sharp & Dohme) ; Oratrol[R] (Alcon)

Synthesis :

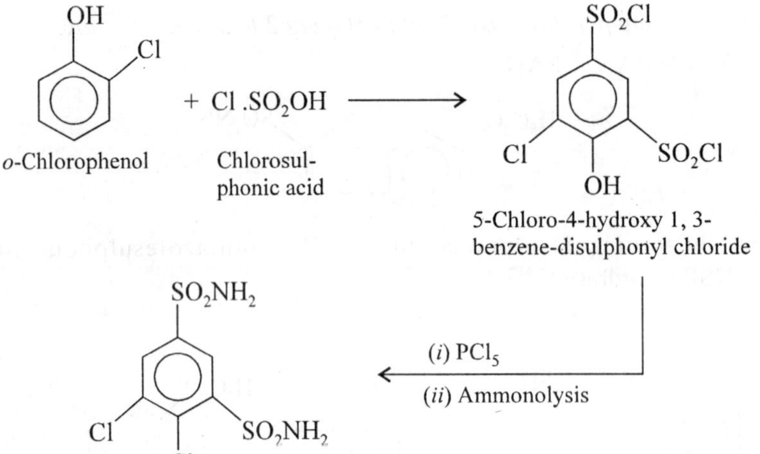

5-Chloro-4-hydroxy 1, 3-benzene-disulphonyl chloride

(*i*) PCl$_5$
(*ii*) Ammonolysis

Diclofenamide

It may be prepared by the interaction of *o*-chlorophenol with chlorosulphonic acid to yield 5-chloro-4-hydroxy-1, 3-benzene-disulphonyl chloride. This on treatment with PCl$_5$ replaces the 4-hydroxy with chlorine and the subsequent ammonolysis gives the official compound.

Diclofenamide is employed to lower intraocular pressure by reducing the rate of secretion of aqueous humor. It is recommended for the treatment of both primary and secondary glaucoma. Though it possesses inherent diuretic properties it is not promoted for this purpose. It produces less acidotic refractoriness to diuretic action than acetazolamide.

Dose : 50 to 300 mg per day ; usual ; 25 to 50 mg 1 to 3 times daily.

E. *Disulfamide* INN, USAN, *Disulphamide* BAN,

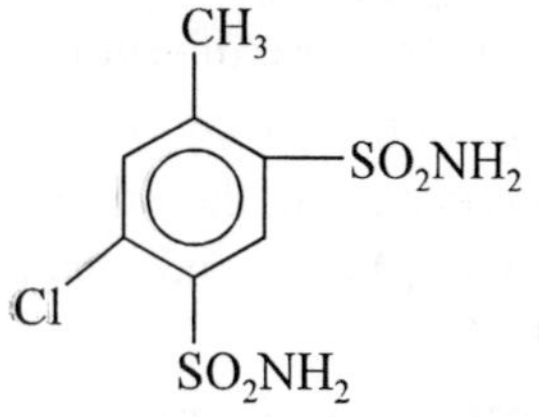

5-Chlorotoluene-2, 4-disulphonamide ; BPC (1968) ; Diluen$^{(R)}$ (Libra, Italy).

Synthesis :

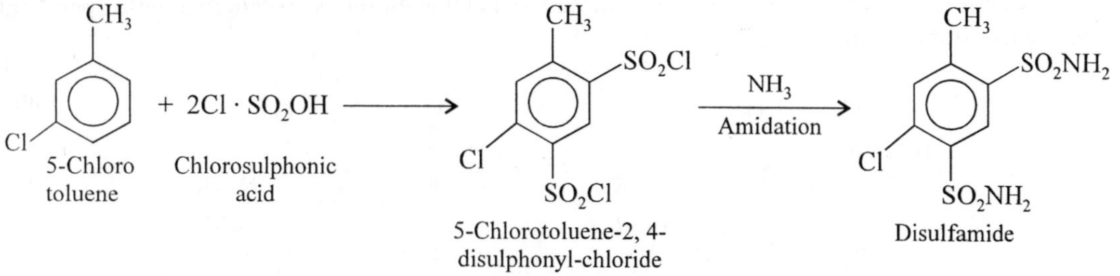

5-Chlorotoluene-2, 4-disulphonyl chloride is prepared by the interaction of 5-chlorotoluene and chlorosulphonic acid, which on amidation gives rise to disulfamide.

It actions and uses are simialr to those of chlorothiazide. It is invariably employed for the *treatment of oedema.*

Dose : For oedema, usual, initial, 200 mg per day for 5 days a week or on alternate days, reduced to 100 mg per day.

2.2.2.1.   Mechanism of Action : The mechanism of action of certain carbonic anhydrase inhibitors used as diuretics shall be discussed in the sections that follows :

2.2.2.1.1. Acetazolamide. The *'drug'* still remanis the most vital carbonic anhydrase inhibitor and being regarded as the prototype member of this specific category. It gets absorbed appreciably from the GI-tract, bound extensively to the plasma proteins, and does not undergo biotransformation. It is eliminated almost completely from the plasma by the kidneys within a span of 24 hr. The drug is subjected to filtration at the glomeruli, and viable tubular secretion in the proximal tubule. Importantly, it also invariably affords a varying range of pH-dependent non-ionic back diffusion taking place particularly in the distal segments of the nephron.

2.2.2.1.2. Methazolamide. If has been amply demonstrated *in vitro* that the *'drug'* is definitely has an edge over the prototype acetazolamide with regard to its potency as CA inhibitor. Besides, it is also observed to exhibit an improved penetration into the eye*, which action strongly recommends its usage in the treatment of glaucoma.

2.2.2.1.3. Ethoxzolamide. The *'drug'* is a carbonic anhydrase inhibitor which is found to exert its action by lowering the intraocular pressure prior to surgery when employed preoperatively in acute angle-closure glaucoma.

---

*Sprague JM : *Advances in Chemical Series,* American Chemical Society, Washington DC., **45**, 87–101, 1964.

2.2.2.1.4. Dichlorphenamide. The *'drug'* acts by lowering the intraocular pressure just like several other CA-inhibitors ; and hence, may be beneficial in the control, management and treatment of glaucoma *i.e.,* in primary as well as the acute phase of secondary glaucoma.

Interestingly, the major importance of this *'drug'* is that it ultimately served as stepping stone far away from the *'pure'* CA-inhibiting diuretics ; and, therefore, paved the way towards the development of the **'thiazides'** that proved to be extremely useful and effective Na$^+$ and Cl$^-$ depleting agents having almost negligible CA-inhibitory activity.*

2.2.2.1.5. Disulphamide. Its mechanism of action is almost similar to that of chlorothiazide (CTZ) in the relief of fluid retention in the body.

2.2.3.   Miscellaneous Sulphonamide Diuretics

The actions of these drugs are very similar to the thaizide diuretics, except that these specifically possess longer duration of action.

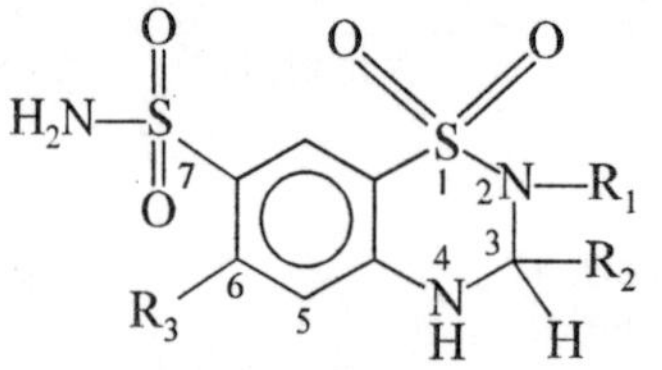

Benzothiadiazines
(Thiazides)

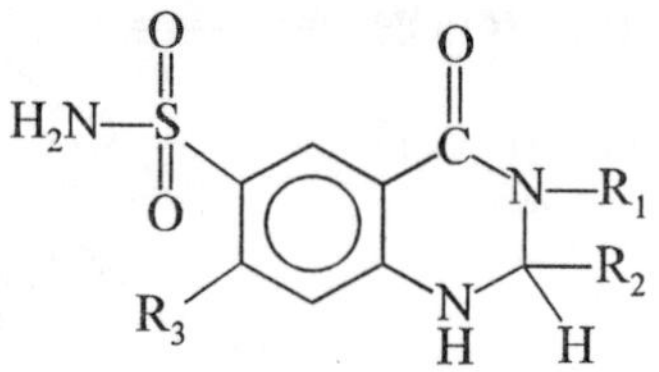

Thiazide Isosteres

In the above benzothiadiazine (thiazides) moiety the 'SO$_2$' at position 1 has been duly changed

to (—C—) carbonyl function *i.e.,* the sulphonyl moiety replaced with carbonyl moiety, thereby resulting into the formation of a series of *'thiazide isosteres'*, namely : *quinethazone, chlorthalidone, metolazone* and *indapamide.*

All these compounds grouped together under **'miscellaneous sulphonamide diuretics'** shall be treated individually as under :

A. *Quinethazone* INN, BAN, USAN,

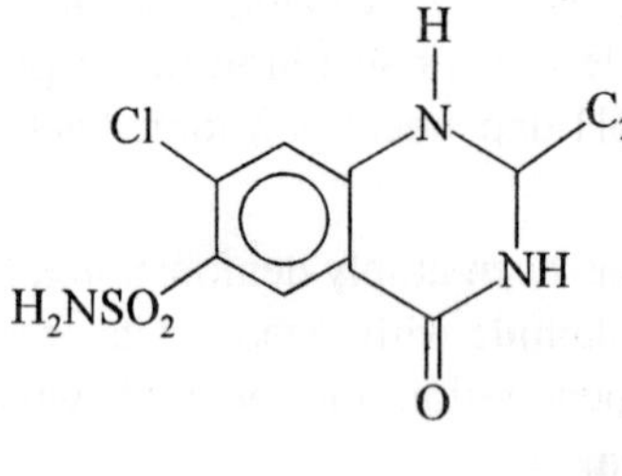

7-Chloro-2-ethyl-1, 2, 4-tetrahydro-4-oxo-6-quinazolinesulphonamide ; 6-Quinazolinesulphonamide, 7-chloro-2-ethyl-1, 2, 3, 4-tetrahydro-4-oxo- ; USP. ; Hydromox$^{(R)}$ (Lederle)

---

*Allen RC : In : Cragoe EJ (ed.) : *Diuretics-Chemistry Pharmacology and Medicine,* John Wiley and Sons, New York, pp-49–200, 1983.

Synthesis :

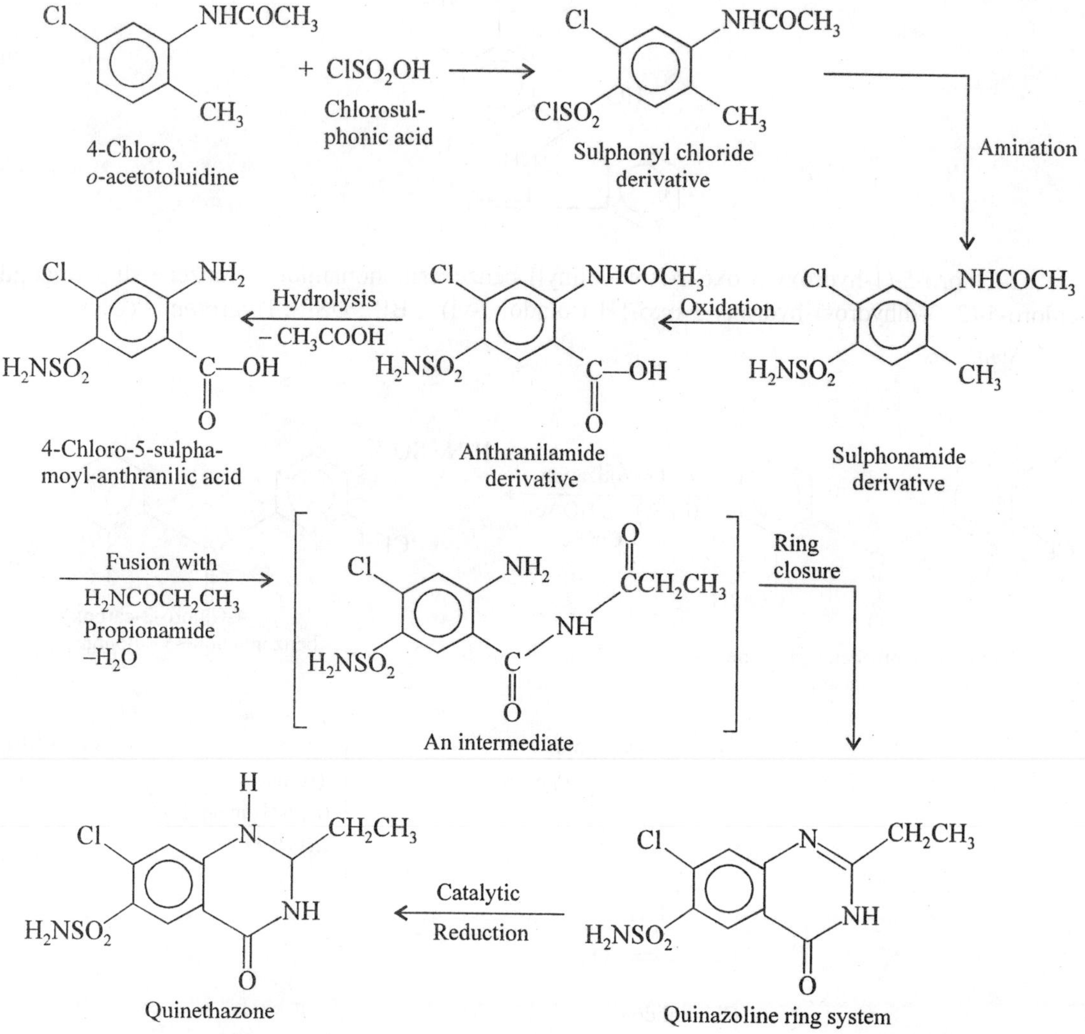

Chlorosulphonation of 4'-chloro-*o*-acetotoluidine yields the corresponding sulphonyl chloride derivative which on amination forms the sulphonamide derivative. Oxidation of the methyl moiety gives the respective anthranilamide derivative which on hydrolysis eliminates the acetyl group to yield the substituted anthranilic acid. Fusion of this amino acid with propionamide first gives rise to an intermediate by the loss of a mole of water and ultimately helps in the closure of the ring to generate the quinazoline ring system. Catalytic reduction of this finally produces the official compound.

Quinethazone essentially differs from the benzothiazide type of diuretics only in the replacement of a sulphur atom by a carbon at position 4.

It possesses both diuretic and antihypertensive properties similar to those of the thiazides.

Dose : 50 to 200 mg per day ; usual, 50 to 100 mg once daily.

B. *Chlortalidone* INN, *Chlorthalidone* BAN, USAN,

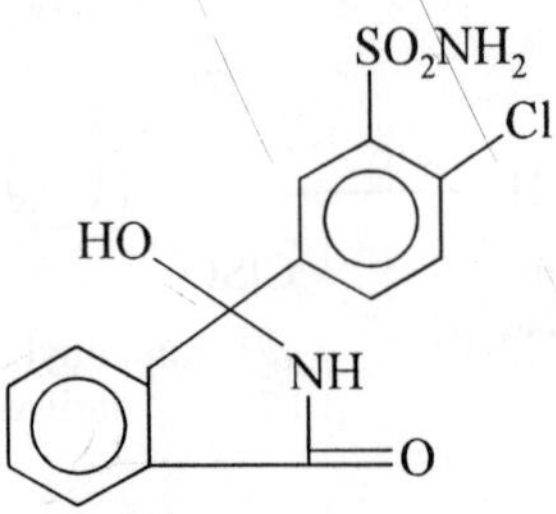

2-Chloro-5-(1-hydroxy-3-oxo-1-isoindolinyl) benzenesulphonamide ; Benzene sulphonamide, 2-chloro-5-(2, 3-dihydro-1-hydroxy-3-oxo-1H-isoindol-1-yl)- ; BP ; USP ; Hygroton[(R)] (USV)

Synthesis :

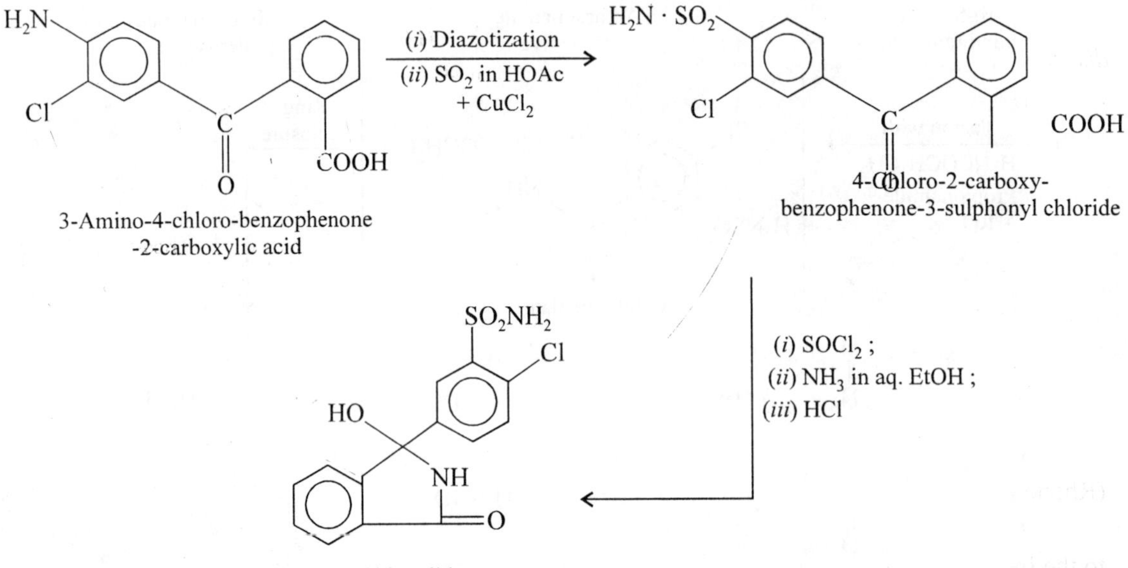

Chlortalidone is a thiazide-like diuretic agent *which essentially contains an isoindole ring.*

It may be prepared by the diazotization and subsequent treatment with sulphur dioxide in glacial acetic acid in the presence of cupric chloride of 3-amino-4-chloro-benzophenone-2-carboxylic acid to yield 4-chloro-2′-carboxy-benzophenone-3-sulphonyl chloride. This on treatment with thionyl chloride followed by amidation in aqueous ethanol and finally with HCl gives crude chlortalidone which is recrystallized from aqeous ethanol.

Chlortalidone is employed in the treatment of oedema associated with obesity, pregnancy, renal disease, hepatic cirrhosis, premenstrual syndrome and above all the congestive heart failure.

Dose : As diuretic, 50 to 200 mg per day or alternate day ; usual, 100 mg once daily ; As antihypertensive, 100 mg alternate day or 50 mg every day.

C. *Metolazone* BAN, USAN

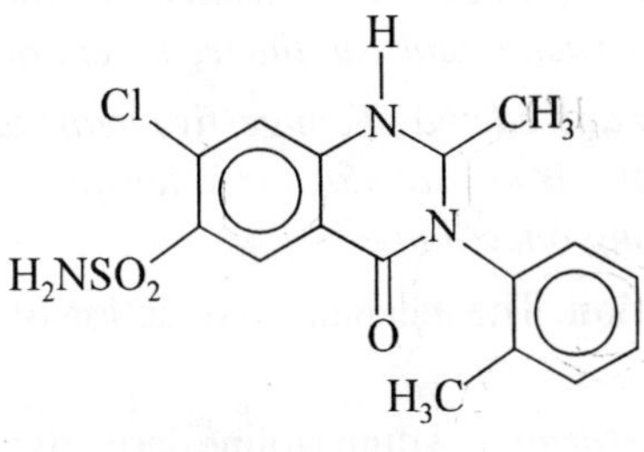

6-Quinazolinesulphonamide, 7-chloro-1, 2, 3, 4-tetrahydro-2-methyl-3-(2-methylphenyl)-4-oxo- ; Diulo[R] ; Zaroxolyn[R] ;

It is a quinazoline-derived nonthiazide diuretic. It is found to be more effective in comparison to the thiazide-like diuretics in the treatment of edema in such patients who have a history of compromised renal function. It is extensively indicated for *hypertension, edema accompanying congestive heart failure, renal disease* including the *nephrotic syndrome* and other *conditions of* retarded renal function.

Dose : Usual, adult, oral, edema of cardiac failure, 5 to 10 mg once daily ; edema of renal disease, 5 to 20 mg once daily ; mild essential hypertension, 2.5 to 5 mg once daily.

D. *Indapamide* BAN, USAN,

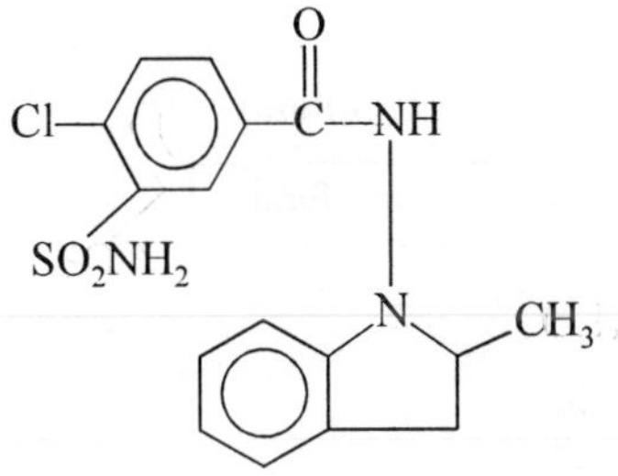

Benzamide, 3-(aminosulphonyl)-4-chloro-N-(2, 3-dihydro-2-methyl-1H-indol-1-yl)- ; Lozol[R] (Rhone Poulenc Rorer) ;

It is an orally active and effective diuretic acid and anithypertensive drug closely related chemically to the **indolines.**

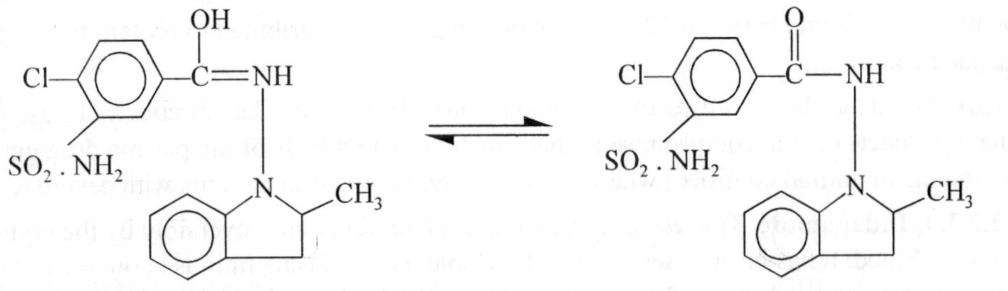

Indapamide (acid)
(*enol*-form)

Indapamide
(*keto*-form)

The *'drug'* undergoes *keto-enol* tautomerism and the *'acid form'* is the active one.

It is used for the control and management of edema associated with congestive heart failure and, alone or in combination with other such agents, in the treatment of hypertension.

Dose : Usual, hypertension and edema of congestive heart failure, 2.5 mg as a single daily dose taken in the morning ; if the response is not satisfactory after one (edema) to four (hypertension) week, the dose is usually increased to 5 mg once daily.

2.2.3.1. Mechanism of Action. The mechanism of action of the above *'drug'* shall be described individually as under :

2.2.3.1. Quinethazone. The *'drug'* is a quinazoline derivative having 6-thiazide-like effect. Based on the available clinical evidence one may suggest that its site, mechanism of action, electrolyte excretion pattern and above all the therapeutic activities are very much similar to those of CTZ.

2.2.3.2. Chlorthalidone. The biochemical studies carried out with the *'drug'* suggest that the prolonged duration of action is solely on account of the slow gastrointestinal absorption, enterohepatic recirculation and above all the critical binding to RBCs in the body. It has been observed that nearly 30–60% of the *'drug'* gets excreted almost unchanged by the kidney.

SAR of Chlorthalidone. The *enol*-form (*i.e.,* the acid form) of chlorthalidone is the **'active drug'** as depicted below :

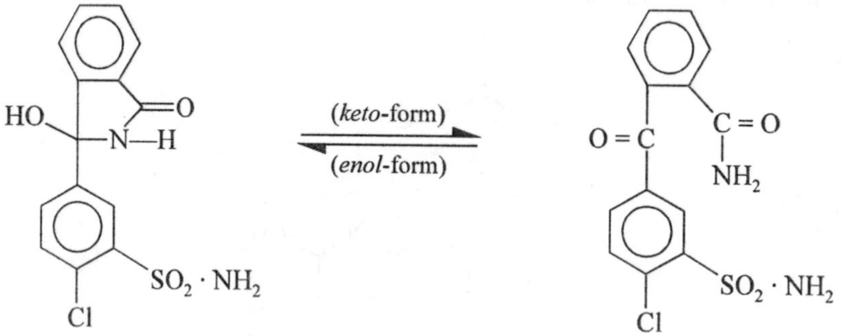

Predominant & Active Drug   Inactive form of Drug

It is not strictly speaking a thiazide.

2.2.3.3. Metolazone. The *'drug'* exerts its inhibition of Na^+ (and Cl^-) reabsorption in early *distal tubule* and the *ascending limb of loop of Henle*. It is also demonstrated to show its action primarily to inhibit Na^+ reabsorption both at the *cortical diluting site* and in the *proximal convoluted tubule*. Its long duration of action ranging between 12 to 24 hours is appreciably attributed to protein binding as well as enterohepatic recycling.

However, it may be more effective in comparison to the thiazide like diuretics in the usual treatment of edema in subjects having compromised renal function. About 95% of the plasma drug gets bound to plasma proteins in normal controls ; whereas, about 90% is bound in patients with severe renal failure.

2.2.3.4. Indapamide. The *'drug'* is taken up preferentially and reversibly by the erythrocytes in the peripheral blood. It has been observed that the whole blood/plasma ratio is about 6 : 1 at the time of peak concentration and reduces to 3.5 : 1 after a lapse of 8 hr. It has been found that 71 to 79% of the drug gets bound to plasma proteins. It gets metabolized extensively *in vivo* ; and only 7% of the unchanged form of the drug is excreted by the kidneys.

SAR of Indapamide. It apparently differs from the thiazides structurally. However, it may be viewed chemically as comprising of a **polar** *sulphamoylchlorobenzamide* and a highly **lipoidal** *methylindolyl* functional moiety.

2.2.4. 'Loop' and 'High-Ceiling' Diuretics

These are a group of diuretics which essentially contain carboxylic acid moieties. They usually produce an intense diuresis of relatively short duration (4-6 hrs) with rapid onset (30 min). They have been found to act mainly on the ascending limb of the loop of Henle (hence often referred to as **loop diuretics**), besides exerting some effect on both the proximal and distal tubules. They seem to act by inhibiting th reabsorption of Cl⁻ (and therefore of NaCl). They cause loss of Cl⁻, Na⁺ and K⁺ ions to a considerable extent.

The **'loop diuretics'** usually possess a much greater diuretic profile in comparison to the *'thiazides'*; and are observed to be even more potent and effective in a situation having electrolyte as well as acid-base disturbances concurrently. Besides, the time of onset and duration of action of the *'high-ceiling diuretics'* are emuch shorter than those with the *thiazides*.

Interestingly, there exists a little controversy with regard to the *relative superiority* of the **'loop diuretics'** in a specific situation of hypertension intimately associated with renal insufficiency than the *thiazides*. Furthermore, the former tend to *enhance* renal blood flow, whereas the latter tend to *minimise* renal blood flow, and thereby lead to further compromise to renal function.

As a point of caution it may, be added that a very **'close monitoring is absolutely warranted'** to avoid severe ensuing electrolyte imbalances in patients being treated with the *'loop diuretics'* by virtue of the fact that they normally possess much greater potency in comparison to the *thiazides*.

Examples : Bumetanide ; Furosemide ; Etacrynic acid.

A. *Burmetanide* INN, BAN, USAN,

3-Butylamino-4-phenoxy-5-sulphamoylbenzoic acid ; Benzoic acid, 3-(amino-sulphonyl)-5-(butylamino)-4-phenoxy- ; Bumex[R] (Hoffman-La Roche) ; Burinex[R] (Leo, U.K.)

Bumetanide is used in the treatment of renal insufficiency and, in conditions which warrant forced diuresis regimens for the control and management of acute drug poisoning e.g., barbiturate poisoning in attempted suicide cases. It is also employed in the treatment of oedema.

Dose : For oedema, usual, oral 1mg once in the morning followed by another similar dose after 6–8 hours if necessary.

B. *Furosemide* INN, USAN *Frusemide* BAN,

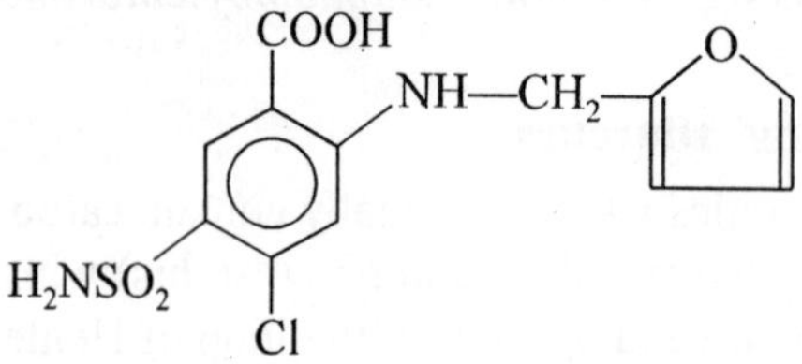

4-Chloro-N-furfuryl-5-sulphamoylanthranilic acid ; Benzoic acid, 5-(amino-sulphonyl)-4-chloro-2-[(2-furanylmethyl) amino]- ; Frusemide (BP ; Eur. P.,) ; Furosemide (USP.) ; Lasix[R] (Hoechst) ; SK-Furosemide[R] (SK & F)

Synthesis :

2, 4-Dichloro-5-sulphamoyl benzoic acid may be prepared by reacting 2, 4-dichlorobenzoic acid with chlorosulphonic acid at an elevated temperature and then carrying out the amidation. This on treatment with furfuryl amine in the presence of sodium bicarbonate, affords nucleophilic aromatic displacement of the highly activated chlorine at C-2, thereby yielding furosemide. However, the protection of the chlorine atom at C-4 may be achieved by regulating the temperature of the furfurylamination.

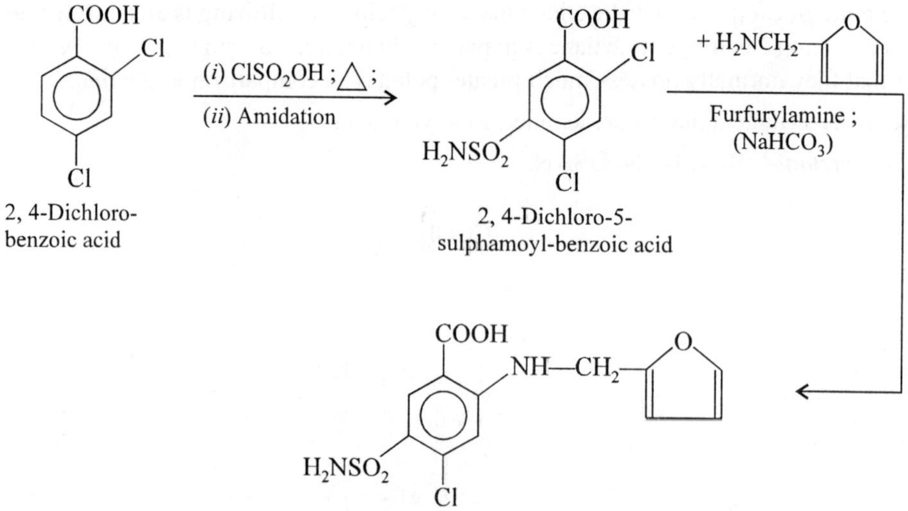

Furosemide possesses relatively high efficacy, rapid onset of action, short duration of action, and 1:10 ratio between the minimum and maximum diuretic dose. It is used for the *treatment of oedema associated with renal disease, nephrotic syndrome, cirrhosis of the liver and congestive heart failure. It has an edge over other commonly used diuretic agents specifically when a greater diuretic potential is required. It may also be employed towards the management of hypertension.*

Dose : Oral, 40 to 600 mg per day ; usual, 40 to 80 mg per day ; i.m. or i.v., 20 to 40 mg.

C. *Etacrynic Acid* INN, *Ethacrynic Acid* BAN, USAN,

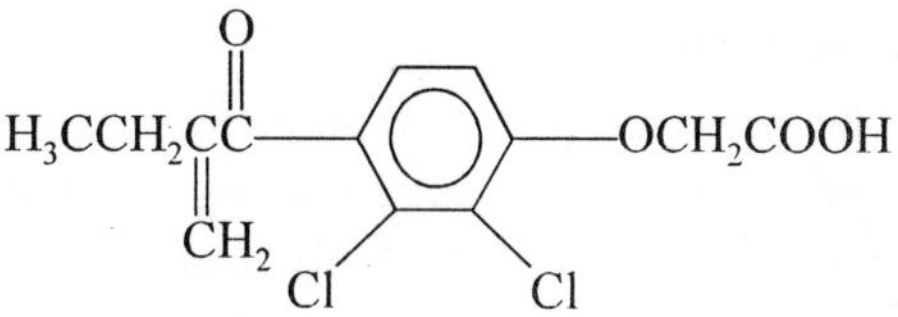

[2, 3-Dichloro-4-(2-methylenebutyryl) phenoxy] acetic acid ; Acetic acid, [2, 3-dichloro-4-(2-methylene-1-oxobutyl) phenoxyl]- ; Ethacrynic acid (BP., USP.) ; Etacrynic Acid (Eur. P.) ; Edecrin[(R)] (Merck Sharp & Dohme)

Synthesis :

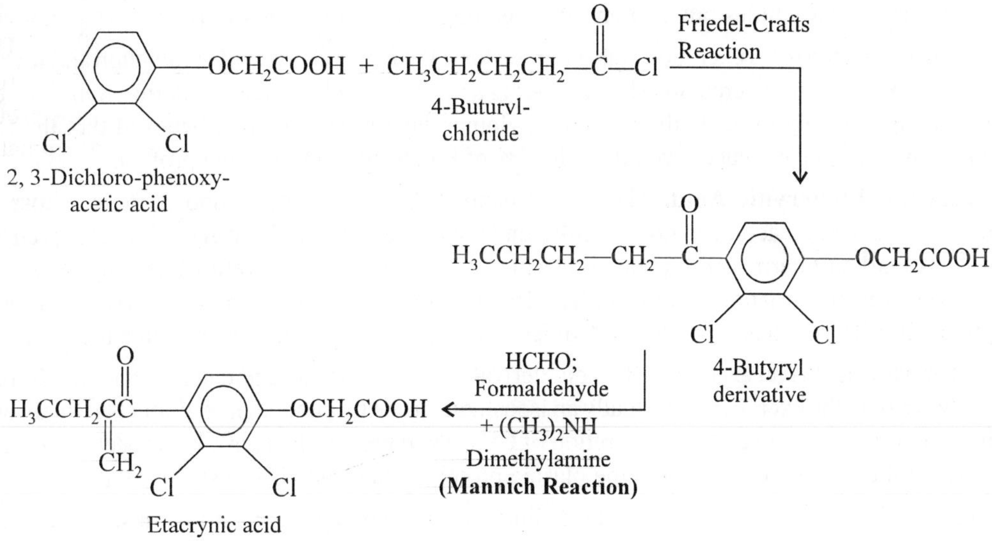

2, 3-Dichloro-phenoxy acetic acid undergoes Friedal-Craft's reaction with 4-butyryl chloride to yield the corresponding 4-butyryl analogue. This is subsequently subjected to Mannich reaction with formaldehyde and dimethylamino thereby introducing the methylene group caused by thermal decomposition, yields the official compound.

Ethacrynic acid is normally used in the *treatment of fluid retensive conditions due to congestive heart failure, cirrhosis of the liver, renal disease, and the nephrotic syndrome. It is invariably employed for the control and management of ascites due to lymphoedema, idiopathic oedema and malignancy. It is also recommended through i.v. in an emergency situation of acute pulmonary oedema.*

Dose : 50 to 200 mg per day ; 50 mg 2 times daily or 2 times every alternate day ; i.v. 100 mg per day in divided doses.

2.2.4.1. Mechanism of Action. The mechanism of action of the various **'loop diuretics'** discussed in the previous section shall be dealt with in the sections that follows :

2.2.4.1.1. Bumetanide. The *'drug'* is found to inhibit both chloride and sodium reabsorption in the ascending limb of the loop of Henle. Besides, it is somewhat little more *chloruretic* than *natriuretic*. It markedly affords dilation of renal vasculature and enhances the renal blood flow. It gets bound to protein to an extent of 95%, and the volume of distribution ranges between 12–35 L. Nearly 45% of an oral dose gets excreted almost unchanged. The biological half-life varies between 1–1.5 hr. and is usually prolonged in patients having renal failure.

SAR of Bumetanide. The presence of a 3-aminobenzoic acid along with the sulphonamido moiety at C-5 renders the drug significantly potent (1 mg ≡ 40 mg Furosemide). Furthermore, the presence of the phenoxy functional group at C-4 may substantially account for this portion through markedly enhanced lipophilicity. Interestingly, a rather newer structural analogue **azosemide,** is of great therapeutic advantage because of its logical as well as successful replacement of the COOH moiety with the corresponding isosteric tetrazolyl moiety.

2.2.4.1.2. Furosemide. The *'drug'* is found to be slightly more potent than the organomercurial agents (see section 13.2.1.), is orally effective ; and its diuretic action is independent of possible changes taking place in body acid-base balance. It has been demonstrated that it acts predominantly not only on the proximal and distal tubules but also on the ascending limb of the loop of Henle.

Furthermore, the renal excretion was observed to be the major channel of elimination and invariably averaged 92% of the administered dose levels, having a mean renal clearance of 149 mL. min^{-1}. Because, this quantum appreciably exceeds the prevailing glomerular filtrate rate, it is believed that the tubular secretion of this drug takes place, even though 95% of it gets bound to plasma protein.

2.2.4.1.3. Ethacrynic Acid. The *'drug'* happens to be powerful loop diuretic whose exact molecular mechanism of action is not yet fully understood. Interestingly, it has been observed that it does possess marked pharmacodynamic similarities of the mercurial diuretics like *'mersalyl'*, which being a phenoxyacetic structural analogue. Besides, it exhibits both *in vivo* and *in vivo* compatibility in its reaction with SH moieties. Moreover, it logically competes with them for the same receptors.

Ethacrynic acid is an aryloxyacetic acid derivative which acts as a potent short-acting diuretic. It actually gives rise to the excretion of virtually an isoosmotic urine by altogether stopping Na^+ reabsorption from the loop of Henle ; however, the excretion of Cl^- is even greater than Na^+. It has been observed that nearly 95% of the *'drug'* gets bound to the plasma proteins. Plasma half-life stands at about 1 hr.

The maximum water as well as sodium diuresis is very much identical to that with *furosemide* ; but largely exceeds that with the thiazides.

SARs of Aryloylphenoxyacetic acids. The comparative study of nine aryloylphenoxyacetic acids revealed that the very presence of an activated double bond susceptible to a nucleophilic attack is almost imperative to cause an effective diuresis. Furthermore, the plausible reduction of the double bond, thereby making 1, 4-addition of an SH moiety practically impossible, ultimately lowered appreciably but failed to eliminate **saluretic activity.** It may be suggested that there exists no definite evidence to demonstrate the sulphhydryl binding in the mechanism of diuretic action of ethacrynic-type drugs.

Ethacrynic acid

2.2.5. Aldosterone Inhibitors

Aldosterone is one of the most important members amongst the corticosteroids secreted by the adrenal cortex. It promotes the retention of Na^+, Cl^- and water *via* distal tubular reabsorption.

Aldosterone inhibitors (antihormone diuretics) are agents which particularly compete with aldosterone at the specific receptor site located in the distal tubule, thereby reversing the electrolyte

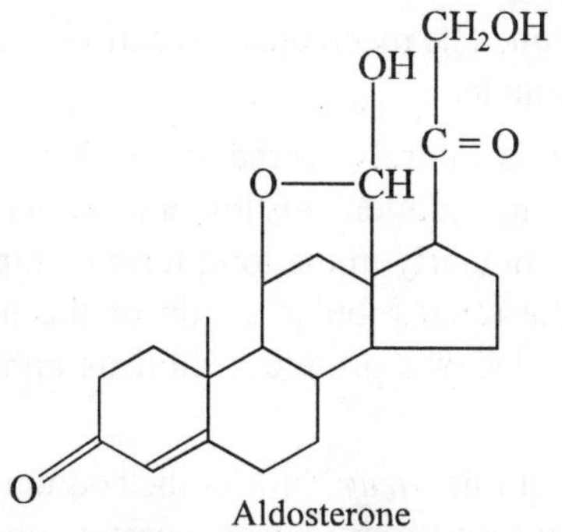

Aldosterone

actions of this naturally occurring hormone, and causing diuresis. The following are the two members of this class of compounds, namely ; spironolactone and metyrapone.

A. *Spironolactone* INN, BAN, USAN,

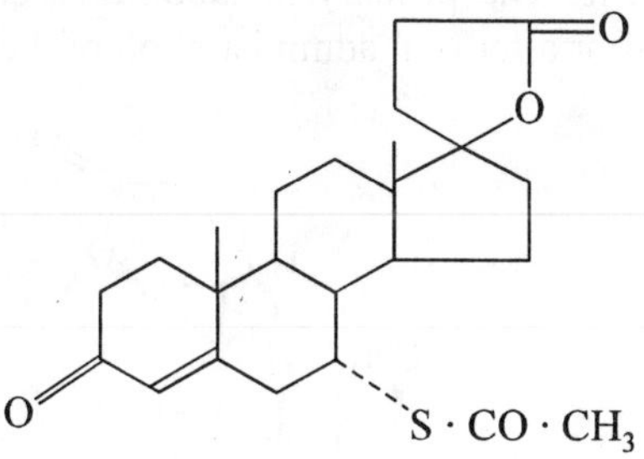

17-Hydroxy-7α-mercapto-3-oxo-17α-pregn-4-ene-21-carboxylic acid, γ-lactone acetate ; Pregn-4-ene-21-carboxylic acid, 7-(acetylthio)-17-hydroxy-3-oxo-γ-lactone, (7α, 17α)- ; 7α-Acetylthio-3-oxo-17α-pregn-4-ene-21, 17β-carbolactone acid γ-lactone ; Spirolactone ; BP ; USP ; Aldactōne(R) (Searle)

It acts both as a diuretic and as an antihypertensive drug. It is mostly employed in the *treatment of refractory oedema associated with congestive heart failure, nephrotic syndrome or cirrhosis of the liver in which secretion of aldosterone plays a part. It has also been used successfully in the treatment of primary hyperaldosteronism.*

Dose : Usual, initial, 100 mg per day in divided doses ; in hyperaldosteronism 400 mg per day.

B. *Metyrapone* INN, BAN, USAN,

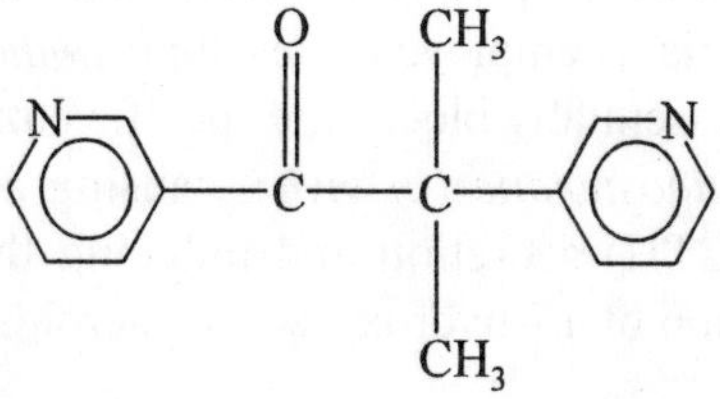

2-Methyl-1, 2-di-3-pyridyl-1-propanone ; 1-Propanone, 2-methyl-1, 2-di-3-pyridinyl- ; BP ; USP ; Metopirone$^{(R)}$ (Ciba-Geigy).

Metyrapone inhibits the synthesis of aldosterone which has been used in the treatment of some cases of resistant oedema. It is necessary to administer a glucocorticoid (cortisone) along with metyrapone because the latter also exerts an inhibitory effect on the former.

Dose : 2.5 to 4.5 mg per day in divided doses.

2.2.5.1. Mechanism of Action. The mechanism of action of the two compounds discussed in the previous section shall be treated as under :

2.2.5.1.1. Spironolactone. It is a purely synthetic steroid which essentially exerts its action as a *'competitive antagonist'* of the potent, endogeneous mineral-corticosteroid, aldosterone. Its *natriuretic* action seems to be slightly more particularly in the long-term therapy. In other words, it reverses these electrolyte changes by blocking the renal tubular action of the hormone. Importantly, by critically inhibiting Na$^+$ reabsorption spironolactone produces diuresis and simultaneously minimises the K$^+$ excretion.

It has been duly observed that this *'drug'* blocks the sodium-retaining effects of aldosterone on the distal convoluted tubule, in doing so it particularly corrects one of the most cardinal mechanisms solely responsible for causing edema ; however, spironolactone is effective only in the presence of aldosterone.

It is metabolized rapidly after the oral administration. It is found that metabolites are excreted mostly in the urine, but also in bile. The primary metabolite is, *canrenone,* which attains the peak plasma levels within a span of 2–4 hr after oral administration of the drug.

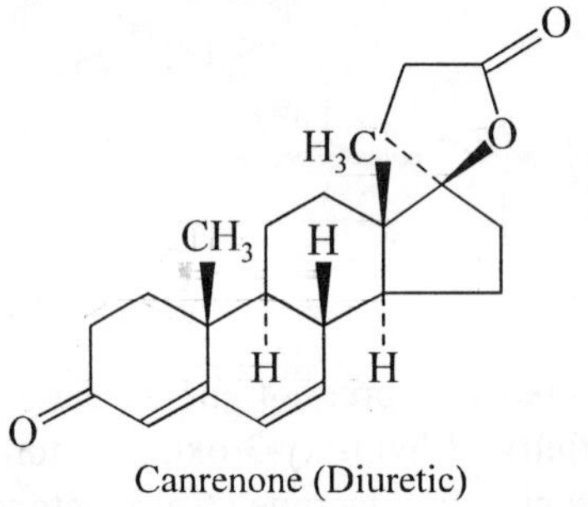

Canrenone (Diuretic)

The half-life of canrenone, following multiple doses of the drug is 13 to 24 hour. It has been observed that both spironolactone and carnenone are usually get bound to the plasma proteins even more than 90%.

Note : The *'drug'* is particularly ineffective in such clinical situations that are known to have high circulating aldosterone levels (*e.g.*, cirrhosis with ascites).

2.2.5.1.2. Metyrapone. The *'drug'* is a purely synthetic compound which possesses a distinct unique characteristic feature of inhibiting 11-β-hydroxylation in the biosynthesis of cortisol, corticosterone and aldosterone. Therefore, it is invariably employed to test for *hypothalamic-pituitary* function. However, in the normal individual, the drug essentially blocks the specific enzymatic step that ultimately leads to the synthesis of *cortisol* and *corticosterone* (*in vivo*), causing an absolute intense stimulation of adrenocorticotropic hormone (ACTH) secretion and inducing thereby a marked and pronounced enhancement in the urinary excretion of 17-hydroxy-corticosteroids.

2.2.6. Purine or Xanthine Diuretics

These are the structural analogues of the parent compounds of unsubstituted purine (7-imidazo [4, 5-*d*] pyrimidine) and xanthine.

Purine   Xanthine

(7-Imidazo [4, 5-*d*] pyrimidine)   (3, 7-Dihydro-1H-purine-2, 6-dione)

Caffeine : 1, 3, 7-Trimethyl xanthine ;

Theophylline : 1, 3-Dimethyl xanthine ;

Theobromine : 3, 7-Dimethyl xanthine ;

Caffeine, theophylline and theobromine are the three important members of the xanthine diuretics, which are commonly found in the common beverages *viz.,* coffee (*Coffee arabica*), coca-cola (*Cola acuminata*) and cocoa (*Theobroma cacao*) contain caffeine ; tea (*Thea sinensis*) contains caffeine and theophylline ; and cocoa (*Theobroma cacao*) contains theobromine.

1. *Caffeine* BAN, USAN,

1, 3, 7-Trimethylxanthine ; 1H-Purine-2, 6-dione, 3, 7-dihydro-1, 3, 7-trimethyl- ; Guranine ; Methyl-theobromine ; Caffeine (BP ; USP ;) ; Coffeinum (Eur. P.).

General Method of Extraction

The crude milled natural product is usually moistened with an aqueous alkali, for instance Na_2CO_3, $NaHCO_3$ or line so as to release the alkaloids from their respective salt and subsequently percolated with benzene, ether, or some other appropriate water-immiscible solvent. The solvent layer is extracted with dilute mineral acid to convert the alkaloids into their corresponding salts and also to push them into the aqueous phase. The free alkaloids may be precipitated by the addition of alkali and finally separated by suitable means.

The diuretic effects of caffeine are less than those of theobromine and theophylline. Caffeine may enhance renal blood flow and glomerular filtration rate, but its main action may be attributed to the reduction of the normal tubular reabsorption.

Dose : 100 to 300 mg.

2. *Theophylline* BAN, USAN,

1, 3-Dimethylxanthine ; 3, 7-Dihydro-1, 3-dimethylpurine-2, 6 (1H)-dione ; BP ; USP ; Eur. P., Int. P., Ind. P. ; Constant-T[(R)] (Ciba-Geigy) ; Duraphyl[(R)] (McNeil) ; Elixicon[(R)] (Berlex) ;

It may be extracted from the leaves of tea by the general method described under caffeine.

Dose : Oral, 60 to 200 mg.

3. *Theobromine* BAN, USAN,

3, 7-Dimethylxanthine ; 3, 7-Dihydro-3, 7-dimethylpurine-2, 6 (1H)-dione ; Santheose ;

It is extracted from cocoa by adopting the general method discussed under caffeine.

It possesses a weaker diuretic activity than theophylline.

Dose : 300 to 600 mg.

2.2.6.1. Mechanism of Action. The mechanism of action of the three well-known *'purine diuertics'* shall be discussed as under :

2.2.6.1.1. Caffeine. It is a well-recognized CNS-drug which action is solely attributed on account of its inhibition of the enzyme phosphodiesterase in the brain and the ultimate accumulation and actions of cyclic 3', 5'-adenosine monophosphate (C-AMP).

Caffeine stimulates the voluntary skeletal muscle, thereby enhancing the requisite force of contraction and minimising the ensuing muscular fatigue. Besides, it is found to stimulate parietal cells, increasing gastric juice (acid) secretion ; it also induces a mild diuretic action by aggravating renal blood flow and glomerular filtration rate and lowering proximal tubular reabsorption of Na^+ and H_2O significantly.

2.2.6.1.2. Theophylline. The *'drug'* produces CNS stimulation and skeletal muscles but to a much lesser extent as compared to caffeine ; however, it exhibits a greater effect on the coronary dilatation, smooth muscle relaxation, diuresis and cardiac stimulation than caffeine.

In general, it possesses relatively more pharmacologic activity practically in all categories than *'theobromine'.*

2.2.7. Pyrimidine Diuretics

The display of diuretic properties by the methylated xanthines, stimulated enough interest in research to establish and ascertain the fact whether or not the pyrimidine analogues, which incidentally are closely related biochemically to the purine derivatives *in vivo,* also possess diuretic activity, This paved the way towards the synthesis of two uracil analogues, namely : aminometradine-having 6-amino group and amisometradine-having 1, 3-diaklyl substituents.

A. *Aminometradine* INN, BAN, USAN,

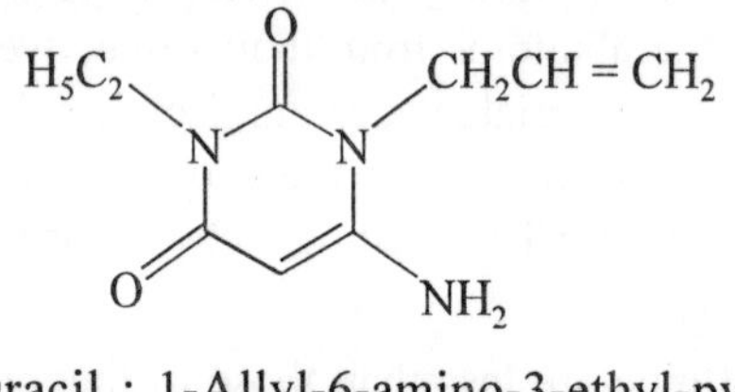

1-Allyl-6-amino-3-ethyluracil ; 1-Allyl-6-amino-3-ethyl-pyrimidine-2, 4 (1H, 3H)-dione ; Aminometramide ; BPC (1959) ; Minacard$^{(R)}$ (Searle)

It is a relatively weak diuretic which has been employed in the control of oedema in subjects having mild congestive heart failure. It is rarely used now.

Dose : 200 to 800 mg per day in divided doses on 3 days a week, or on alternate days.

2.2.7.1. Mechanism of Action. The *'drug'* essentially has a pyrimidine nucleus that possesses an almost similar activity as those of the purine derivatives, such as : caffeine, theophylline etc. Besides, aminometradine happens to be intimately related biochemically to the xanthine analogues.

2.2.8. Osmotic Diuretics

The functional capacity and reabsorption capability of the renal tubule towards various electrolytes and nonelectrolytes are restricted to a limited extent only, and this vary with respect to each ionic species. A large intake of any of these substances by an individual, may enhance its concentration in the body fluids and will ultimately afect the glomerular filtration rate and the reabsorption capacity of the tubule. The substance will finally appear in the urine with an increased volume of water. Such a substance which increases the output of urine in this fashion is called **osmotic diuretics.**

Osmotic diuretics may be classified into *two* sub-groups, *viz.,*

(*a*) *Osmotic electrolytes, e.g.,* potassium and sodium salts, and

(*b*) *Osmotic nonelectrolytes, e.g.,* urea, sucrose, mannitol, trometamol.

A. *Sodium Acid Phosphate* BAN,

$$NaH_2PO_4 . H_2O$$

Sodium acid phosphate (BP ; Int. P ; Ind. P ;) ; Sodiumbiphosphate USP ;

It is used quite often as a urinary acidifier, for instance, during therapy with methenamine.

Dose : 500 mg to 1 g to 6 times daily ; usual, 600 mg 4 times per day.

B. *Potassium Acetate* BAN, USAN,

$$CH_3COOK$$

Acetic acid, potassium salt ; BP ; USP ; Ind. P ;

It has been used as a diuretic.

C. *Urea* BAN, USAN,

$$H_2NCONH_2$$

Carbamide ; BP ; USP ; Ind. P. ; Ureaphil[R] (Abbott) ; Elaqua XX[R] (Elder)

It is an osmotic diuretic with a low renal threshold. It is also *administered to maintain the output of urine during surgical procedures. It is also recommended to decrease intra-ocular pressure in acute glaucoma.*

Dose : Oral, up to 20 g from 2 to 5 times per day ; as a 40% solution in water or carbonated beverages, has been given as maintenance therapy after i.v. application for the relief of cerebral oedema.

D. *Mannitol* BAN, USAN,

$$HOCH_2-\overset{\overset{\displaystyle H}{|}}{\underset{\underset{\displaystyle OH}{|}}{C}}-\overset{\overset{\displaystyle H}{|}}{\underset{\underset{\displaystyle OH}{|}}{C}}-\overset{\overset{\displaystyle OH}{|}}{\underset{\underset{\displaystyle H}{|}}{C}}-\overset{\overset{\displaystyle OH}{|}}{\underset{\underset{\displaystyle H}{|}}{C}}-CH_2OH$$

D-Mannitol ; BP ; USP ; Osmitrol[R] (Travenol, U.K.)

Preparation

On commercial scale it is produced by the catalytic or electrolytic reduction of certain monosaccharides, for instance, glucose and mannose.

It is a diuretic and a diagnostic agent for the kidney function test. The osmotic and diuretic is usually initiated by the administration of a hypertonic solution of mannitol. It is also employed to lower the intraocular pressure and cerebrospinal fluid pressure, before, during the after surgical procedures. Mannitol is considered to be superior to dextrose because of the fact that it is both metabolized in vivo and reabsorbed by the renal tubule to a negligible extent.

Dose : Usuaul, i.v. infusion, 50 to 200 g per day ; as diuretic, 50 to 100 g, administered as a 5 to 20% solution.

E. *Trometamol* INN, BAN, *Tromethamine* USAN,

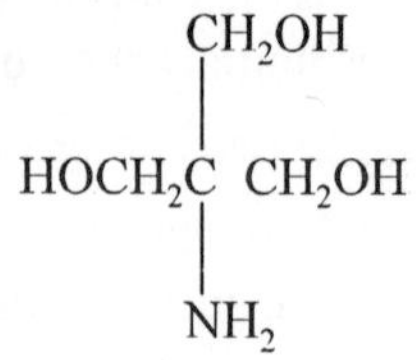

$$CH_2OH$$
$$|$$
$$HOCH_2C\ CH_2OH$$
$$|$$
$$NH_2$$

2-Amino-2-(hydroxymethyl)-1, 3-propanediol ; 1, 3-Propanediol, 2-amino-2-(hydroxymethyl)- ; Trihydroxymethylaminomethane ; *Tris*-(hydroxymethyl) aminomethane ; 2-Amino-2-(hydroxymethyl) propane-1, 3-diol ; Tromethamine USP. ; THAM[R] (Abbott)

It is an organic amine base that reacts with cations of fixed or metabolic acids and also combines with H$^+$ ions from H_2CO_3 to yield bicarbonate as well as a cationic buffer. *An intravenous infusion usually affords an osmotic diuresis.*

Dose : Usual, 300 mg/kg body weight administered i.v. as a 0.3 M solution stretched over a period of not less than 60 minutes.

2.2.8.1. Mechanism of Action. The mechanism of action of a few osmotic diuretics are enumerated in the sections that follow :

2.2.8.1.1. Sodium Acid Phosphate. The inorganic salt when employed in large doses usually cause short-term diuresis besides affording acidification.

2.2.8.1.2. Urea. Simply by employing large amounts (*e.g.,* 15 g or an adult) for the water-soluble and also nonmetabolizable compounds to afford a hypertonic condition *i.e.,* high osmolarity, water content is eventually withdrawn from tissues for instance, the eye-ball, thereby lowering pressure in it appreciably (*i.e.,* intraocular pressure). It also helps in reducing the intracranial pressure using almost the same mechanism.

2.2.8.1.3. Mannitol. The IV administration of the hypertonic solutions of the *'drug'*, which is a sugar alcohol, is usually employed to promote an *osmotic diuresis.* It invariably exerts its action because of the glaring fact that the drug is not obsorbed significantly from the GI-tract ; and if administered orally, it gives rise to definite osmotic diarrhea.

It is found that when this *'drug'* is administered parenterally it gets distributed adequately in the extracellular space. Furthermore, only a small portion ranging between 7–10% usually gets metabolized to glycogen ; and remaining quantity is excreted in the urine. Plasma half-life after single IV dose is only 15 minutes having normal renal function.

2.2.8.1.4. Tromethamine (Trometamol). The *'drug'* happens to be a weak amine base having pKa value of 7.8 at the normal body temperature (98.4°F). Hence, it is almost very close to plasma pH (7.4) ; and, therefore, well-acceptable for the preparation of a buffer mixture for controlling the extracellular pH.

It is, however, pertinent to sate here that at pH 7.4 (plasma pH) it is almost 30% non ionized ; and, therefore, it slowly penetrates the cells, where it would buffer the intracellular contents. Under the prevailing circumstances it is able to react with any proton donor, and the usual notion that it reacts first and foremost with carbonic acid (H_2CO_3) or CO_2 is absolutely erroneous. In this manner protons are removed from the H_3O^+ ions, whereby the ionization of H_2CO_3 is shifted so as to minimise pCO_2 and also to enhance the concentration of HCO_3^-. Thus, the excess quantum of HCO_3^- gets excreted slowly through the kidney. This is, therefore, an extremely beneficial manner by which the level of high pCO_2 may be managed conveniently in various conditions, namely : *respiratory acidosis* (*e.g.,* drug intoxication, asphyxia neonatorum, status asthmaticus etc.) where in the pulmonary ventilation is quite insufficient.

2.2.9. Acidotic Diuretics

The acidotic diuretics are essentially the inorganic compounds having a cation function. Examples are-ammonium or calcium, combined with a fixed anion *viz.,* chloride ion, which causes two vital, actions, namely ; systemic hyperchloremic acidosis and weak diuretic effect. These compounds (*e.g.,* ammonium chloride, calcium chloride) invariably potentiate the diuretic action of mercurial diuretics and hence may be administered at least 48–72 hours prior to the treatment of a mercurial compound so as to facilitate hyperchloremic acidosis. Recently, insoluble cation exchange resins have been used to act as diuretics by this mechanism.

A. *Ammonium Chloride* BAN, USAN,

$$NH_4Cl$$

Sal ammoniac ; Muriate of ammonia ; BP ; USP ; Eur. P., Int. P ; Expiger[R] (Pharmacia, Denm.)

Ammonium chloride causes diuresis by inducing mild acidosis. The acid-forming property is due to the conversion of NH_4^+ ion in the liwe to urea, which leaves the Cl^- ion free to combine with the available cation, liberated from the elastic HCO_3^- ion. This eventually upsets the $BHCO_3 : H_2CO_3$ ratio thereby causing acidosis, thus :

$$2NH_4Cl + CO_2 \longrightarrow H_2NCONH_2 + H_2O + 2HCl$$
$$NaHCO_3 + HCl \longrightarrow NaCl + H_2CO_3$$

The liberated acid may be buffered by the phosphates as follows :

$$Na_2HPO_4 + H_2CO_3 \rightleftharpoons NaH_2PO_4 + NaHCO_3$$

| Disodium hydrogen | Sodium acid |
| phosphate | phosphate |

The net result in that Cl^- ion displaced HCO_3^- ion and the latter is converted to CO_2. This phenomenon appreciably enhances the Cl^- load to the kidneys thereby allowing a substantial amount of it to escape unabsorbed with a matching amount of Na^+ ion along with an isoosmotic amount of water. The overall effect being the net loss of extracellular fluid thereby helping the mobilization of oedema fluid.

Dose : Oral, 4 to 12g per day ; usual, oral, 1 to 2 g 4 times daily ; i.v. 100 to 100 ml of 2% solution ; usual, i.v., 500 ml of a 2% solution infused over a period of 3 hours.

2.2.9.1. Mechanism of Action. The mechanism of action of ammonium chloride is as follows :

2.2.9.1.1. Ammonium Chloride. The *'drug'* is indeed a combination of a labile cation (NH_4^+) and a rather fixed anion (Cl^-). The ammonium ion on being converted to urea usually liberates H^+ that instantly reacts with HCO_3^- and other body buffers. The ultimate product is that the Cl^- ion displaces HCO_3^- ; and the latter is subsequently converted to CO_2. In doing so, the Cl^- load to the kidney is enhanced appreciably, and an adequate quantum escapes reabsorption together with an equivalent quantity of cation (mostly Na^+) and an isoosmatic quantum of water. In fact, this is the fundamental mechanism by which NH_4Cl brings forth a net loss of the extracellular fluid and thereby augments the actual mobilization of edema fluid from the body.

2.2.10. Miscellaneous Diuretics

There are a few potent diuretics which could not be accommodated conveniently into any of the classifications made so far (A-I), hence they have been grouped together under this head. *Examples* : Triamterene ; Muzolimine.

A. *Triamterene* INN, BAN, USAN,

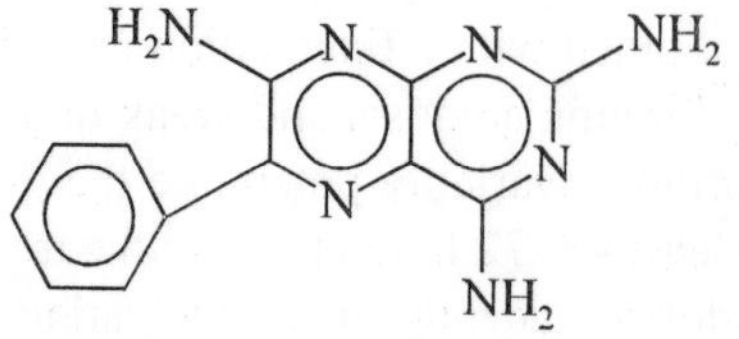

2, 4, 7-Triamino-6-phenylpteridine ; 2, 4, 7-Pteridinetriamine, 6-phenyl- ; BP ; USP ; Dyrenium[R] (SK & F)

Synthesis

Tautomerism of 2, 4, 5, 6-tetraaminopyrimidine gives the "nonenolized" amine which renders the amino moiety at C-5 more basic in character as compared to the remaining amine functions. Therefore,

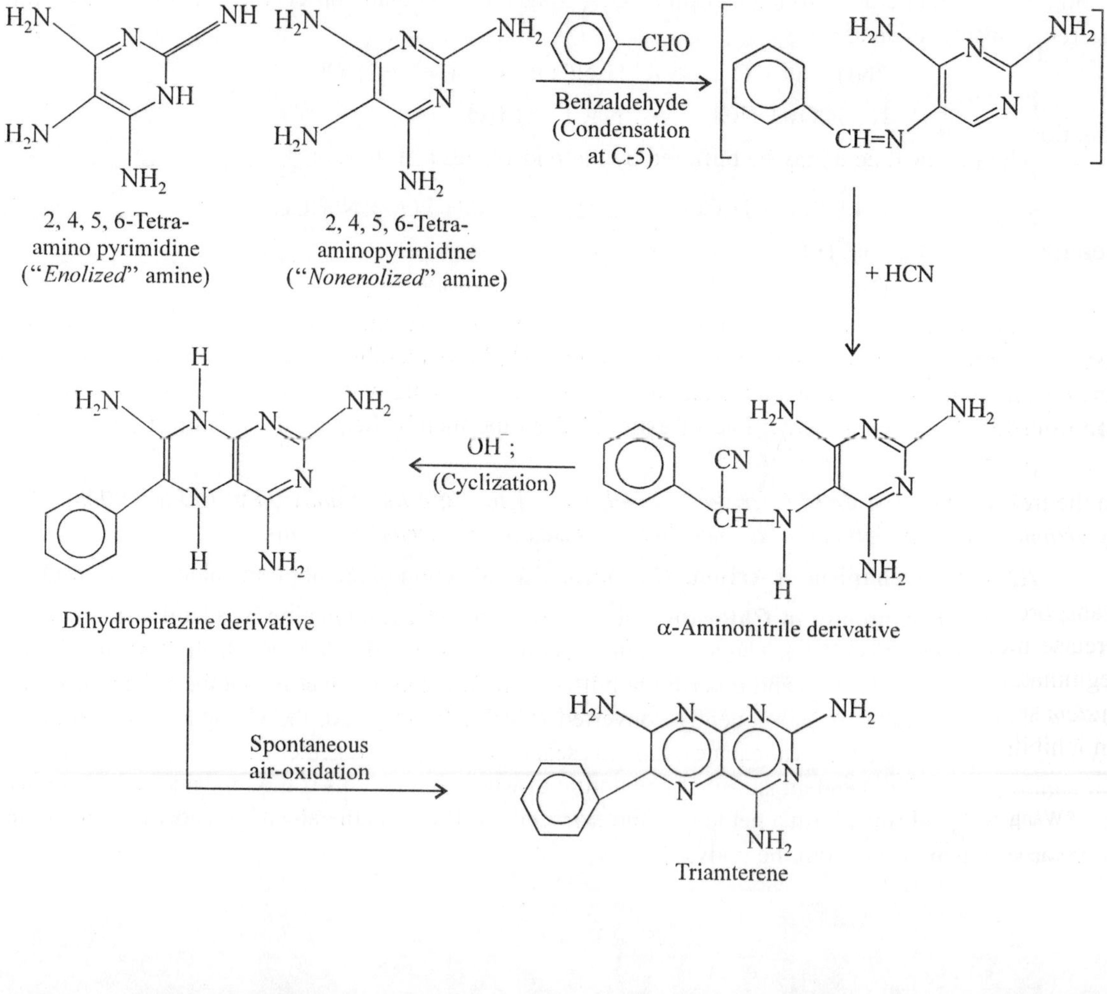

2, 4, 5, 6-Tetra-amino pyrimidine ("*Enolized*" amine)

2, 4, 5, 6-Tetra-aminopyrimidine ("*Nonenolized*" amine)

Benzaldehyde (Condensation at C-5)

+ HCN

OH⁻ ; (Cyclization)

Dihydropirazine derivative

α-Aminonitrile derivative

Spontaneous air-oxidation

Triamterene

on condensation with benzaldehyde, the union takes place at the most preferred basic nitrogen to form a benzylidene analogue as an intermediate with the loss of a mole of water. Treatment with hydrogen cyanide yields the corresponding α-aminonitrile derivative, which when subjected to a basic medium undergoes cyclization to form the dihydropirazine derivative. This undergoes spontaneous air-oxidation to yield the official compound.

Triamterene is usually recommended in the treatment of oedema associated with nephrotic syndrome, cirrhosis of liver, and congestive heart failure. It has also been used for the control and management of idiopathic oedema, steroid-induced oedema, oedema caused by hyperaldosteronism and in such oedematus patients who fail to respond to other therapy. It is usually used in conjunction with other diuretics like thiazides.

Dose : 100 mg every alternate day to 300 mg per day ; usual, 100 mg once daily.

B. *Muzolimine* INN, BAN, USAN,

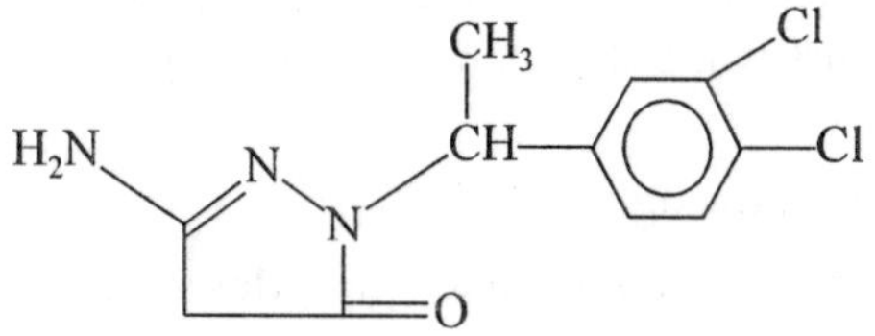

3-Amino-1-(3, 4-dichloro-α-methylbenzyl)-2-pyrazolin-5-one ; 3H-Pyrazol-3-one, 5-amino-2 [1-(3, 4-dichlorophenyl) ethyl]-2, 4-dihydro- ;

Its actions and uses have been reported to be similar to those of frusemide. It has a prolonged duration of action.

Dose : For oedema, 40 to 80 mg per day.

2.2.10.1. Mechanism of Action. The mechanism of action of the *'miscellaneous diurects'* are described as under :

2.2.10.1.1. Triamterene. The *'drug'* gets metabolized primarily by the liver in the form of *hydroxy triamterene sulphate,* which is an **active metabolite.** It has been observed that 3–5% of the drug gets usually excreted unchanged in the urine. Although it is known to promote the adequate excretion of Na^+ and Cl^-, it is also believed to conserve K^+ by significantly retarding the usual transport of this specific ion from the tubular cell to the tubular lumen.

In usual practice, however, it is given in conjunction with hydrochlorothiazide (HCTZ) particularly in the treatment of edema linked with *congestive heart failure, cirrhosis of the liver,* and the *nephrotic syndrome.*   .

2.2.10.1.2. Muzolimine. Muzolimine possesses diuretic activity on account of its action on the transport mechanism pertaining to the cells of the thick ascending limb of Henle's loop. However, the precise mechanism(s) by which muzolimine causes a diuretic action still remains to be established legitimately[*]. Furthermore, it has been proposed that muzolimine, in fact, inhibits the K^+/Cl^- *contransport system* specifically upon the basolateral membrane of the thick ascending limb cells, whereby it affords an inhibition of the 1 Na^+/1 K^+/2 Cl^- contransport system[**].

---

[*]Wangemann PH *et al. Pflugers Arch.* **410**, 674, 1987.

[**]Landan RL *et al. Annu. Rev. Med.,* **41**, 265, 1955.

Probable Questions for B. Pharm. Examinations

1. (a) What are **'diuretics'** ? Classify diuretics by citing the structure, chemical name and uses of at least **one** compound from each categroy.

   (b) Why do the **'non-mercurical diuretics'** have an edge over the **'mercurial diuretics'** ?

2. Discuss **Benzothiadiazines (Thiazides)** as an important class of diuretics. Give the structure, chemical name and uses of any **four** official compounds.

3. Acetazolamide, Methazolamide, Diclofenamide and Disulfamide and potent carbonic anhydrase inhibitors employed as diuretics. Discuss the structural difference amongst these drugs and give the synthesis of any **two** compounds.

4. How would you synthesize **chlorthalidone,** a sulphonamide diuretic, from 3-amino-4-chloro benzophenone-2-carboxylic acid ?

5. 'High-ceiling diuretics exert an intense diuresis of relatively short duration (4-6 hours) with a rapid onset (30 mins)'. Justify the statement with the help of at least **two** important members of this class of drugs.

6. Write a brief note on the following :

   (a) Aminometradine   (b) Sapironolactone   (c) Metyrapone   (d) Caffeine.

7. How would you synthesize :

   (a) Furosemide from 2, 4-dichlorobenzoic acid

   (b) Disulphamide from 5-chlorotoluene

   (c) Methixene hydrochloride from thioxanthene.

8. With the help of some specific examples give an account of the following :

   (a) Osmotic diuretics   (b) Acidotic diuretics.

9. **Trimterene** and **Muzoimine** are two diuretics belonging to the 'miscellaneous group'. Give their structures and the synthesis of the former drug.

10. Discuss the **mode of action** of the following class of diuretics :

    (a) Thiazides

    (b) Carbonic anhydrase inhibitors

    (c) Mercurial diuretics

    Support your answer with typical examples.

RECOMMENDED READINGS

1. RW Berliner and J Orlaff Carbonic Anhydrase Inhibitors, *Pharmacy Rev,* June (1956).

2. J Merrill Use and Abuse of Diuretics *Med Clin N Amer* 44, 1155 (1960).

3. TC Daniels and EC Jorgensen, Diuretics in : Text book of Organic Medicinal and Pharmaceutical Chemistry, Eds. CP Wilson, O. Gisvold and RF Doerge (10th end.) J B Lippincott Co. Philadelphia (1998).

4. J M Spragne ''Diuretics in : *Molecular Modification in Drug Design (Ed)* R F Gould *Adv Chem Ser* No. 45, American Chemical Society Washington, D.C. (1964).

5. JM Spragne Diuretics, in : *Topics of Medicinal Chemistry,* Vol. 2 (*Eds*) J L Rabinowitz and R M Myerson, Interscience Publishers New York (1968).

6. D Lednicer and LA Mitscher *The Organic Chemistry of Drug Synthesis* John Wiley and Sons, New York (1995).

7. V Papsech and EF Schroeder Non-mercurial diuretics in : *Medicinal Chemistry (Ed)* F F Blicke Vol. III J Wiley, New York (1956) 175.

8. ME Wolff (*Ed*) : *Burger's Medicinal Chemistry and Drug Discovery* (5th edn), John Wiley & Sons Inc, New York, (1996).

9. EF Reynolds (*Ed*) *Martindale the Extra Pharmacopoeia* (31st end), Royal Pharmaceutical Society, London, (1996).

10. Nogrady T, *'Medicinal Chemistry : A Biochemical Approach',* Oxford University Press, New York, 2nd. edn, 1988.

14

Antihistaminics

1. INTRODUCTION

Histamine, 4-(2-aminoethyl) imidazole as such could not for many years attract enough attention from physiologists, pharmacologists and biochemists alike because of the absence of any therapeutic utility. It occurs in many storage sites in the body in varying amounts. It is present in the mast cells of many body organs, in blood basophils, the mucosal cells of the gastrointestinal tract especially the acid-secreting parietal cells, in the hypothalamus and area postrema in the central nervous system.

In the living organism, histamine is synthesized from the naturally occurring α-amino acid, histidine, by the loss of a carboxyl group through bacterial or enzymatic decarboxylation as stated below :

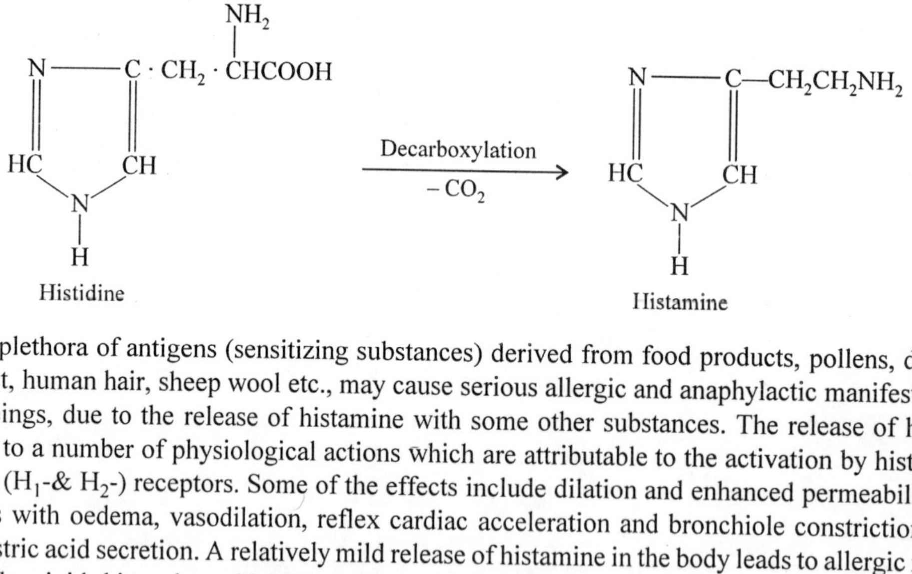

Histidine   Histamine

A plethora of antigens (sensitizing substances) derived from food products, pollens, dust mite, house dust, human hair, sheep wool etc., may cause serious allergic and anaphylactic manifestations in human beings, due to the release of histamine with some other substances. The release of histamine gives rise to a number of physiological actions which are attributable to the activation by histamine of histamine (H_1-& H_2-) receptors. Some of the effects include dilation and enhanced permeability of the capillaries with oedema, vasodilation, reflex cardiac acceleration and bronchiole constriction. It also causes gastric acid secretion. A relatively mild release of histamine in the body leads to allergic reactions displayed by vivid skin rashes with itching, whereas in extreme instances it may result in an anaphylactic shock which may be fatal. The actions of histamine can be antagonized chemically using histaminase or formaldehyde but this is of no practical value. The actions are best modified by the use of substances that block competitively the histamine sensitive receptors. Such substances are known as **antihistamines (antihistaminics).** The term antihistamine is traditionally used to refer to drugs that block the H_1-receptors.

Best *et al.** (1927) made an epoch making observation that **'histamine'** was present in relatively high concentration in the lungs, which eventually gave rise to vasoconstriction, anaphylactic shock-like syndrome and acute respiratory distress in laboratory animals treated with IV administration. Subsequently, the histamine's specific role in the particular pathogenesis of the ensuing anaphylactic reaction was duly demonstrated and substantially established.

In the early 1930s, an amalgamation of sequential events and circustancial evidences almost established the various deleterious and harmful effects caused by an excess of *'histamine'* ; and these findings were exclusively based on both *in vitro* and *in vivo* methods that were meticulously developed so as to screen the chemical effects upon the various physiological aspects, namely : bronchial, gastro-intestinal and other smooth muscle tissues. Thus, a broad platform was made available for the extensive as well as intensive screening of several synthesized drug molecules with respect to their possible viable *'histaminic activity'*, in addition to their anticipated *anticholinergic* and *antiserotoninergic* pharmacological profile.

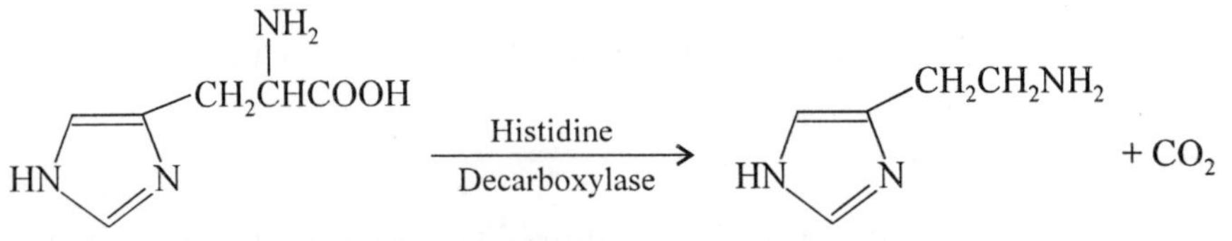

L-Histidine → (Histidine Decarboxylase) → Histamine + CO_2

Antihistaminics are widely used in the *palliative treatment in allergic conditions like hay fever, urticaria, some forms of pruritus, rhinitis, conjunctivitis, nasal discharge, mild asthma etc. A few antihistaminics possess potent antiemetic action and hence are frequently employed in the prevention and treatment of irradiation sickness, motion sickness (air, sea, road), nausea in pregnancy and post-operative vomiting.*

In general, the most common side-effect of antihistaminics is sedation which may be followed by drowsiness, impaired alertness and retarded ability to perform jobs which need high precision and concentration.

2. CLASSIFICATION

The commonly used antihistaminics may be classified on the basis of their chemical structures and these all are of the type histamine H_1-receptor antagonists. They are :

2.1. Histamine H_1-Receptor Antagonists

(*i*) *Aminoalkylethers* : *Examples*-Diphenhydramine Hydrochloride ; Bromodiphenhydramine Hydrochloride ; Dimenhydrinate ; Doxylamine Succinate ; Diphenylpyraline Hydrochloride.

(*ii*) *Ethylenediamines* : *Examples*-Mepyramine Maleate ; Tripelennamine Hydrochloride, Thonzylamine Hydrochloride ; Zolamine Hydrochloride.

(*iii*) *Thiophene Derivatives* : *Examples*-Methapyrilene Hydrochloride ; Methaphenilene Hydrochloride, Thenyldiamine Hydrochloride ; Chlorothen Citrate.

(*iv*) *Cyclic Basic Chain Analogues : Examples*

(*a*) Imidazoline Derivatives, *e.g.,* Antazoline Hydrochloride ; (*b*) Piperazine Derivatives, *e.g.,* Cyclizine Hydrochloride ; Chlorcyclizine Hydrochloride ; Meclizine Hydrochloride ; Buclizine Hydrochloride ; (*c*) Piperidine Derivativs, *e.g.,* Thenalidine Tartrate.

---

*Best CH *et al. J. Physiol*, **62** : 397, 1927.

(v) *Phenothiazine Derivatives* : *Examples*-Promethazine Hydrochoride ; Promethazine Teoclate ; Trimeprazine Tartrate ; Methdilazine Hydrochloride.

(vi) *Second-generation Non Sedating Antihistamines* : *Examples :* Terfenadine ; Astemizole ; Loratadine ; Acrivastine ;

(vii) *Miscellaneous Agents* : *Examples*-Phenindamine Tartrate ; Triprolidine Hydrochloride ; Chlorpheniramine Maleate ; Cyproheptadine Hydrochloride.

2.1.1. Aminoalkylethers

These are also referred to as basic ether group because of the presence of an alkyl amino function in the drug molecule.

A. *Diphenhydramine Hydrochloride* BAN, USAN, *Diphenhydramine* INN,

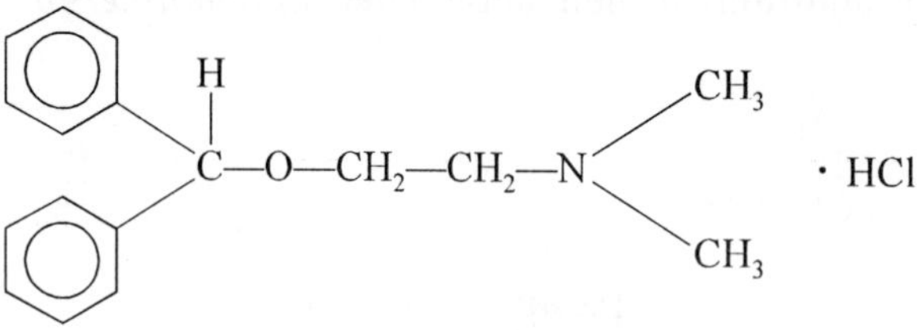

2-(Diphenylmethoxy)-N, N-dimethylethylamine hydrochloride ; Ethanamine, 2-(diphenylmethoxy)-N, N-dimethyl-, hydrochloride ; BP ; BPC ; USP ; Benadryl[R] (Parke-Davis) ; Bendylate[R] (Reid-Provident) ; SK-Diphenhydramine[R] (Smith Kline & French)

Synthesis

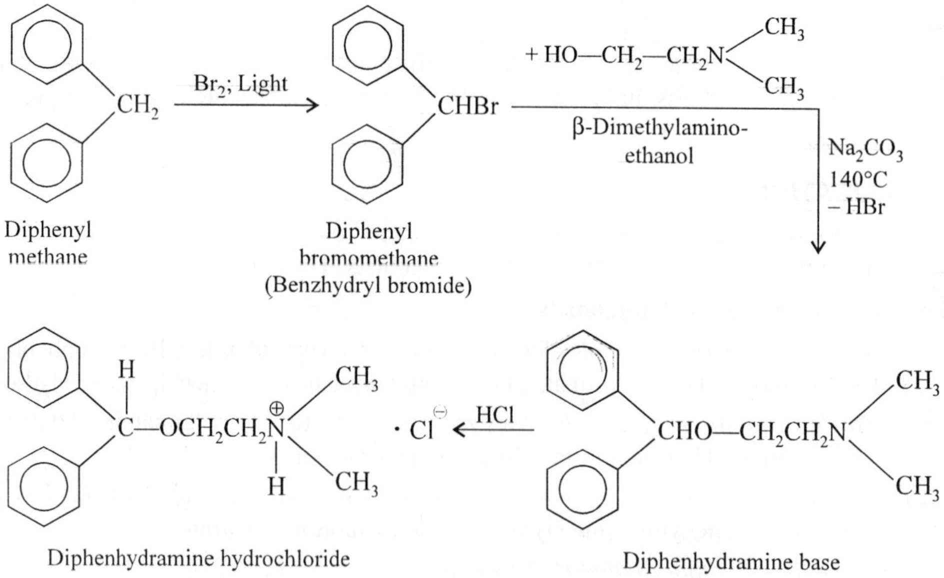

Diphenyl methane

Diphenyl bromomethane (Benzhydryl bromide)

Diphenhydramine hydrochloride

Diphenhydramine base

Diphenylbromomethane is first prepared by the bromination of diphenylmethane in the presence of light. Subsequently diphenhydramine base is obtained by heating diphenylbromomethane, β-dimethyl-amino-ethanol, and sodium carbonate is toluence. After distilling off toluene, the purified diphenhydramine is converted to the hydrochloride with hydrogen chloride.

It is *frequently used in mild, local allergic reactions due to insect bites. It possesses sedative, antiemetic and anti-tussive properties and can be used in seasonal allergic rhinitis, allergic manifestations due to urticaria and allergic conjunctivitis of inhalant allergens.*

Dose : 25–50 mg, usual, adult, oral dose 3 to 4 times a day, with maximum of 400 mg daily ; topical to skin 2% cream 3 or 4 times a day.

B. *Bromodiphenhydramine Hydrochloride* BAN, USAN, *Bromazine* INN,

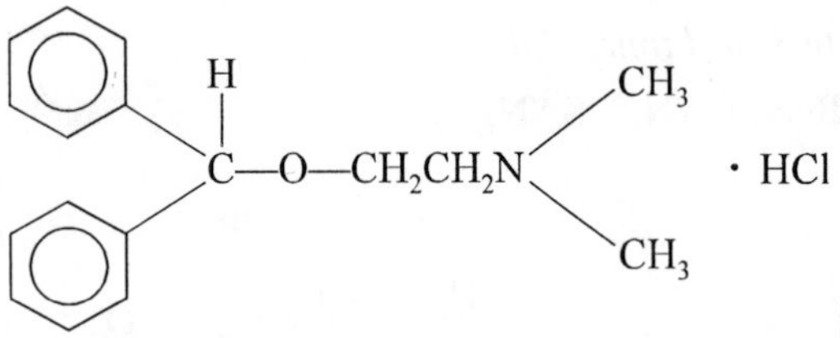

2-[(*p*-Bromo-α-phenylbenzyl) oxy]-N, N-dimethylamine hydrochloride ; Ethanamine, 2-[(4-bromophenyl) phenylmethoxy]-N, N-dimethyl-, hydrochloride ; Bromazine hydrochloride ; BP ; USP ; Ambodryl Hydrochloride[R] (Parke-Davis)

Synthesis :

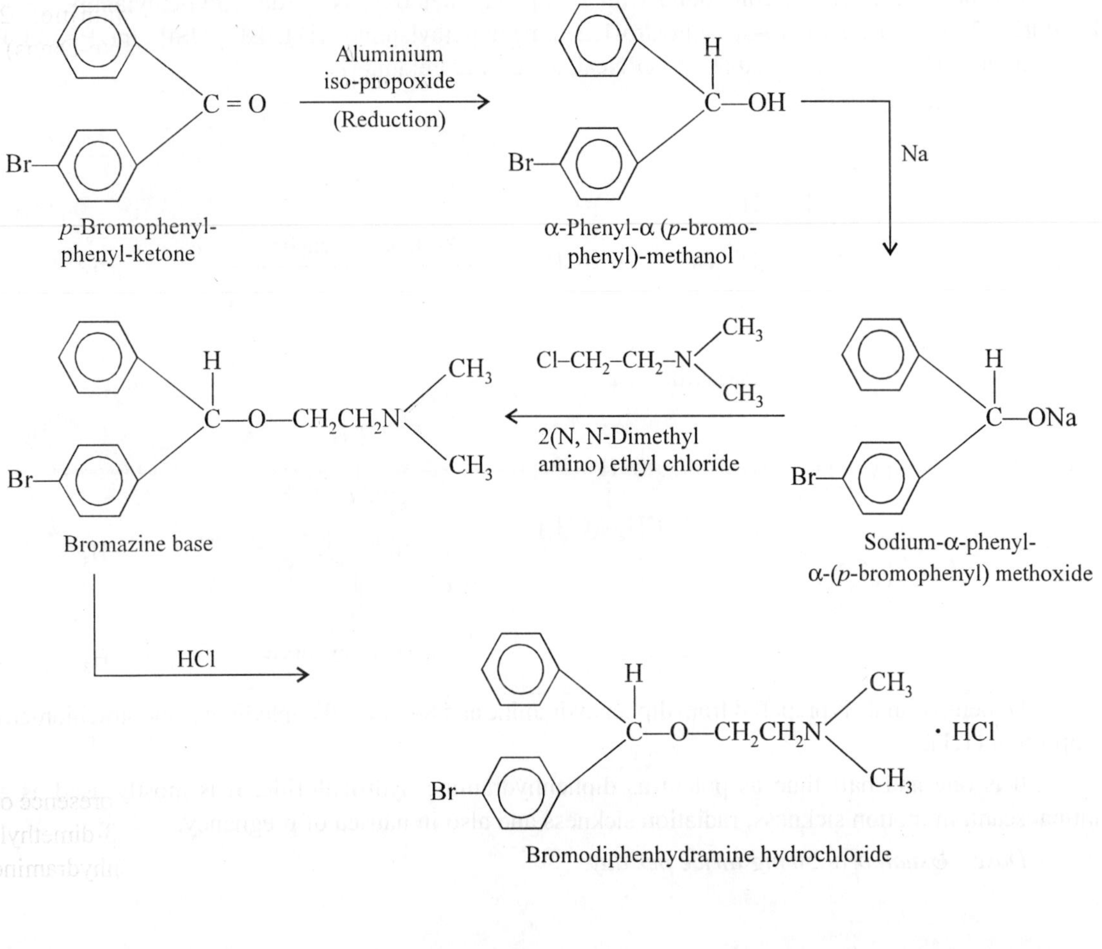

p-Bromophenyl-phenyl-ketone

α-Phenyl-α (*p*-bromophenyl)-methanol

Sodium-α-phenyl-α-(*p*-bromophenyl) methoxide

Bromazine base

Bromodiphenhydramine hydrochloride

α-Phenyl-α-(*p*-bromophenyl) methanol is first prepared by the reduction of *p*-bromophenyl-phenyl ketone with aluminium isopropoxide, which on treatment with sodium metal results into the corresponding mono-sodium salt. This on reaction with 2(N, N-dimethyl amino) ethyl chloride loses a molecule of sodium chloride and provide the bromazine base which on neutralization with hydrogen chloride gives the bromodiphenhydramine hydrochloride.

It is probably effective for mild, local allergic reactions to insect bites, physical allergy, and *for minor drug reactions characterised by pruritis.*

Dose : 25 mg, usually 3 or 4 times daily.

C. *Dimenhydrinate* INN, BAN, USAN,

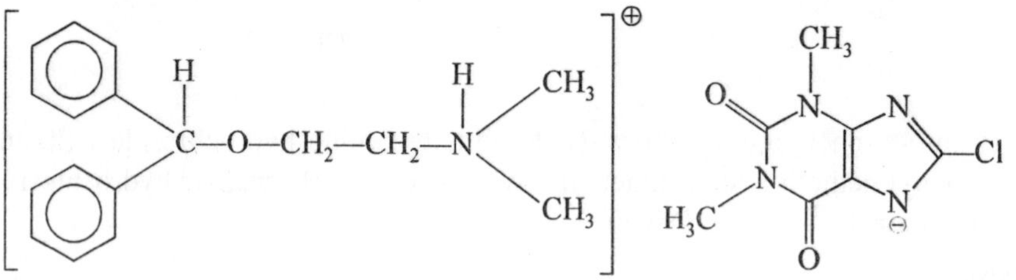

8-Chlorotheophylline, compound with 2-(diphenylmethoxy)-N, N-dimethyl-ethylamine (1:1) ; 1H-Purine-2, 6-dioine, 8-chloro-3, 7-dihydro-1, 3-dimethyl-ethylamine (1:1) ; BP ; USP ; Int. P ; Ind. P ; Dramamine[R] (Searle) ; Dommanate[R] (O'Neal, Jones and Feldman)

Synthesis :

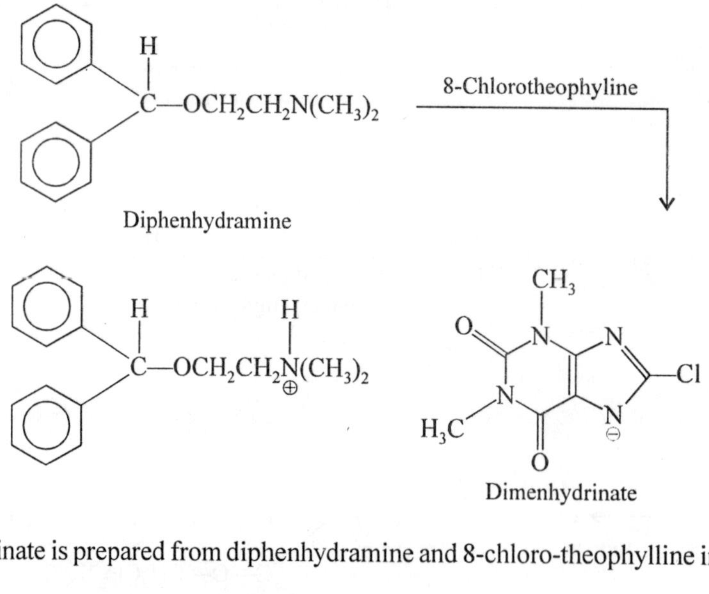

Diphenhydramine

Dimenhydrinate

Dimenhydrinate is prepared from diphenhydramine and 8-chloro-theophylline in the stoichiometric proportion (1:1).

It is one and half time as potent as diphenhydramine hydrochloride. It is mostly used as an antinauseant, in motion sickness, radiation sickness and also in nausea of pregnancy.

Dose : Usual, oral 50 mg thrice per day.

D. *Doxylamine Succinate* BAN, USAN,

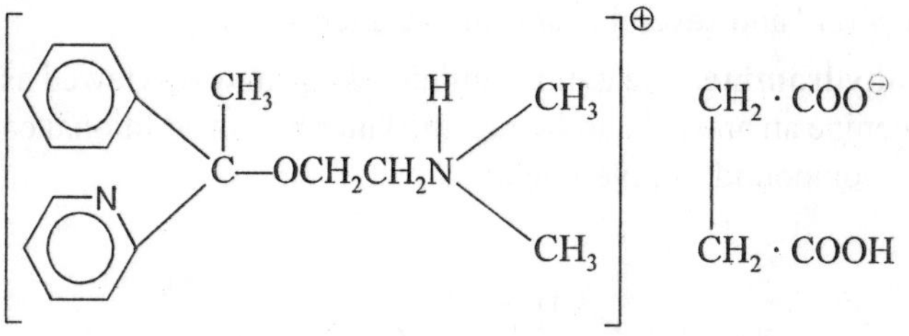

2-[α-[2-(Dimethylamino) ethoxy]-α-methylbenzyl] pyridine succinate (1:1) ; Ethanamine, N, N-dimethyl-2-[1-phenyl-1-(2-pyridinyl) ethoxy]-butanedioate (1:1) USP ; Decapryn Succinate[R] (Merrell Dow) ; Unisom[R] (Pfizer)

It may be *used for allergic conjunctivitis due to inhalant allergens (like pollens and dust), seasonal and perennial allergic rhinitis, and uncomplicated allergic skin manifestations of urticaria.*

Dose : 12.5 to 25 mg ; Usual, adult, oral 4 to 6 times a day.

E. *Diphenylpyraline Hydrochloride* BAN, USAN, *Diphenylpyraline* INN,

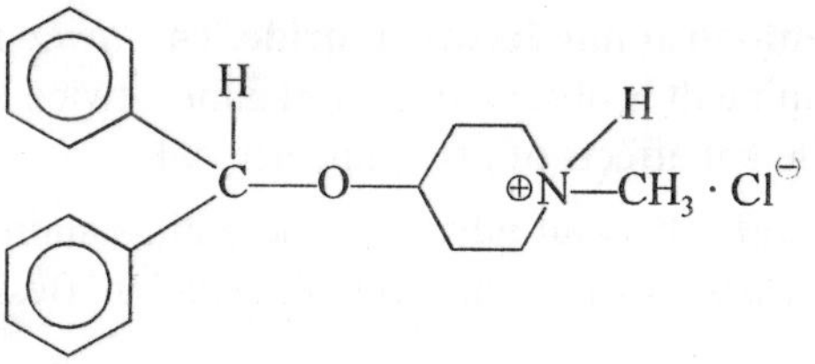

4-(Diphenylmethoxy)-1-methylpiperidine hydrochloride ; Piperidine, 4-(diphenylmethoxy)-1-methyl-, hydrochloride ; BP ; USP ; Diafen[R] (Riker) ; Hispril[R] (Smith, Kline & French)

It may be employed for the *treatment of angioedema, dermographism and amelioration of reactions to blood or plasma. It is also effective for use in seasonal and perennial allergic rhinitis, vasomotor rhinitis, allergic conjunctivitis due to inhalant allergens and foods.*

Dose : Usual, adult, oral 5 mg, 2 times a day.

2.1.1.1. Mechanism of Action. The mechanism of action of aminoalkylethers dealt with in the previous section shall be discussed as under :

2.1.1.1.1. Diphenhydramine Hydrochloride. The *'drug'* is found to be well-absorbed after the oral administration ; and the first-pass metabolism is notedly so predominant that only 40-60% virtually reaches systemic circulation almost unchanged. It has been observed that the peak-plasma concentrations are invariably accomplished within a span of 1 to 4 hr, 80 to 85% gets bound to plasma protein and the elimination half-life varies between 2.4 to 9.3 hour.

As the *'drug'* has an atropine-like specific action, it needs to be administered with great caution and supervision in subjects having a history with asthma.

In reality, quite a few of the so called *'first-generation'* H_1-antihistaminics are believed to antagonize ACh. Nevertheless, the parasympatholytic activity could be regarded as the major undesirable side effects because it essentially gives rise to dry mouth and voiding difficulties. Interestingly, the ability to minimise nasal discharges (secretions) may be loohed upon as a positive clinical attribute, when included in *'hay fever'* and several *'cold'* medicaments.

SAR of Diphenhydramine. The antihistaminic *'drug'* may be viewed as possessing an isosteric relationship with adiphenine an antispasmodic *i.e.,* the latter has an additional carbonyl moiety to give it the status of the *'ester'* compound as given below :

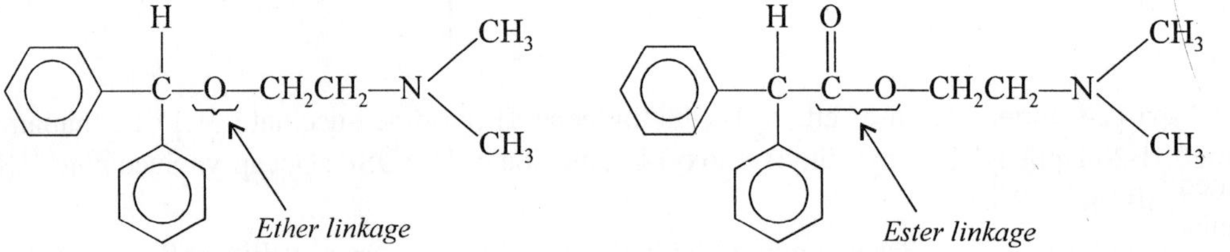

Diphenhydramine   Adiphenine
(*Antihistaminic*)   (*Antispasmodic*)

The *'etiology'* of this drugs anticholinergic activity may be explained explicity if one takes into consideration the two functional moieties *viz.,* ether and ester as isosterically related to one another.

2.1.1.1.2. Bromodiphenhydramine Hydrochloride. The *'drug'* is found to be more lipid soluble in comparison to diphenhydramine. It is observed to exert almost twice its effective activity particularly in the guinea pigs against the lethal effects of histamine aerosols.

2.1.1.1.3. Dimenhydrinate. It is an ethanolamine antihistaminc agent belonging to the first-generation H_1-antagonist drug classes causing appreciable sedation. Besides, it also exhibits significant anticholinergic activity.

2.1.1.1.4. Doxylamine Succinate. The *'drug'* is an ethanolamine antihistamine agent having appreciable sedative pharmacologic acitivity ; and, generally, listed in OTC* *sleep aids.* It also possesses significant anticholinergic activity.

2.1.1.1.5. Diphenylpyraline Hydrochloride. It exerts its antihistaminic properties along with antimuscarinic and central sedative pharmacologic activities. Though it may be applied topically but it has a risk of sensitization. Therefore, it finds its enormous application for the relief of hypersensitivity reactions, such as : *urticaria, angiodema, rhinitis, conjunctivitis,* and in pruritic skin disorders.

2.1.2. Ethylenediamines

This class of compounds essentially have the following general structure :

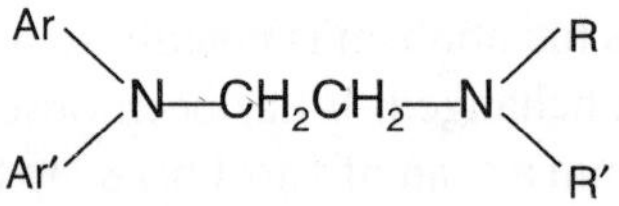

where Ar and Ar′ are aromatic or heteroaromatic moieties, and *R* and *R′* are small alkyl entities. A few classical members of this category shall be discussed here.

---

*OTC = Over the Counter.

A. *Mepyramine Maleate* BAN, *Pyrilamine Maleate* USAN, *Mepyramine* INN,

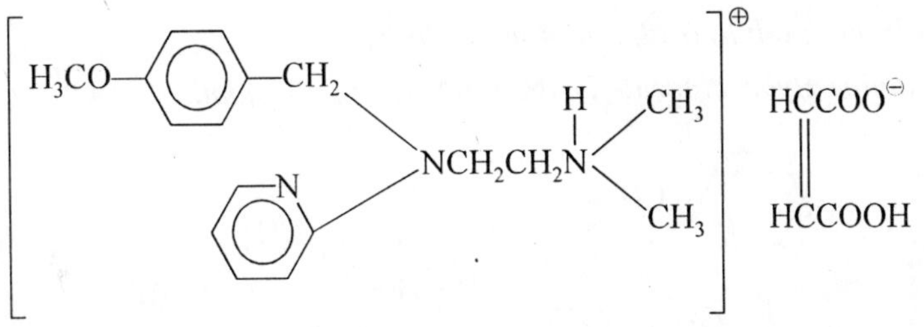

2-[[(2-Dimethylamino) ethyl] (*p*-methoxybenzyl) amino] pyridine maleate (1:1) ; 1, 2-Ethanediamine, N-[(4-methoxy-phenyl) methyl]-N′, N′-dimethyl-N-2-pyridinyl-, (Z)-2-butanedioate ; Pyranisamine maleate ; Mepyramine Maleate BP ; Int. P., Ind. P. ; Purilamine Maleate U.S.P., Anthisan[R] (May & Baker) ; Dorantamin[R] (Dorsey Lab.) ; Minihist[R] (Ives) ; Pymafed[R] (Hoechst-Roussel)

Synthesis :

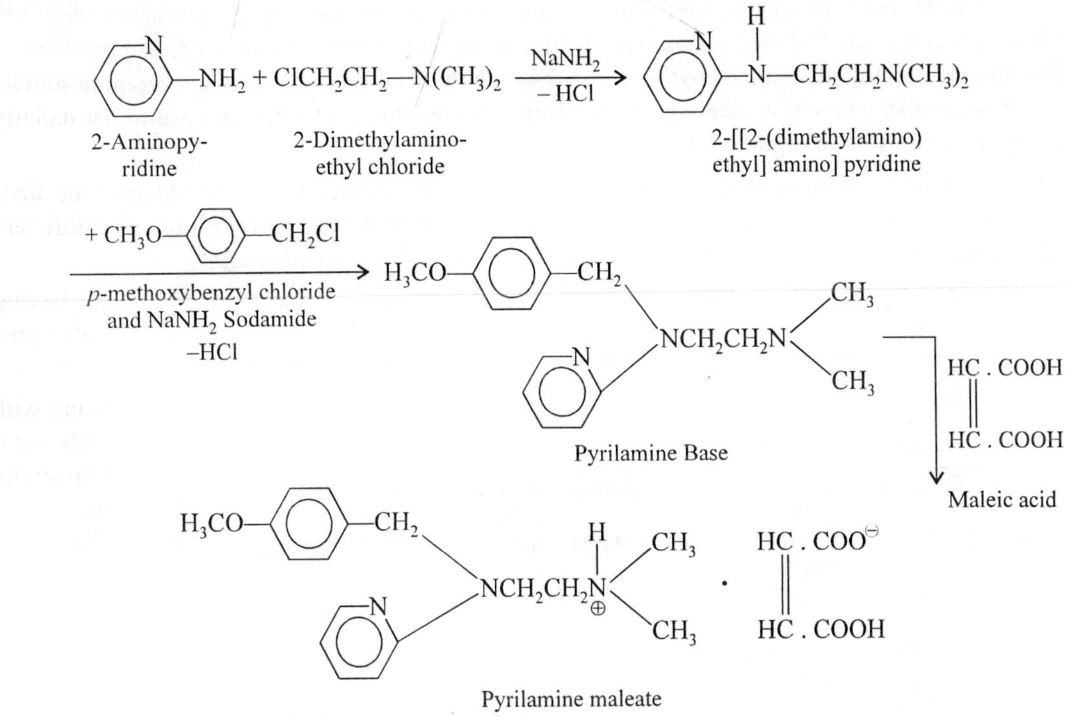

In the first step, 2-[[2-(dimethylamino) ethyl] amino] pyridine is prepared by the condensation of 2-aminopyridine with 2-dimethylamino ethyl chloride with the elimination of a molecule of hydrogen chloride in the presence of sodamide. The resulting product on further condensation with *p*-methoxy benzyl chloride in the presence of sodamide yields the pyrilamine base which on neutralization with maleic acid gives rise to the desired product pyrilamine maleate.

It is a *potent antihistaminic agent with a low incidence of sedative effects*. It has acclaimed a legitimate entrance as component in a number of proprietary antitussive formulations.

Dose : 25 to 50 mg ; adult, oral, 3 to 4 times daily.

B. *Tripelennamine Hydrochloride* BAN, USAN, *Tripelennamine* INN,

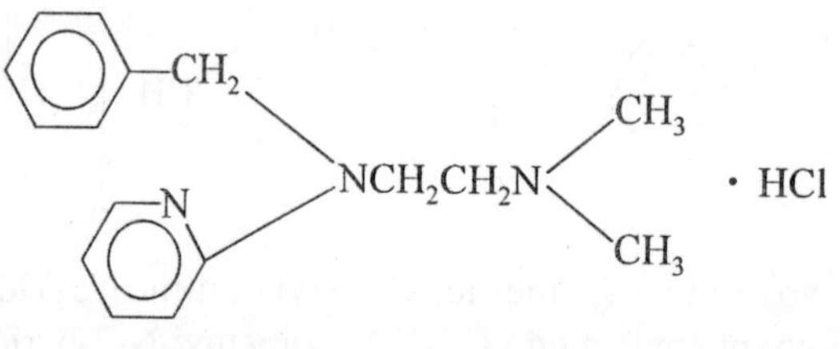

2-[Benzyl [2-(dimethylamino) ethyl] amino] pyridine monohydrochloride ; 1, 2-Ethanediamine, N, N-dimethyl-N'-(phenylmethyl)-N'-2-pyridinyl-, monohydrochloride ; BP 1963 ; USP ; Int. P., Ind. P. ; Pyribenzamine Hydrochloride[(R)] (Ciba-Geigy)

Synthesis :

Tripelennamine can be prepared as follows : 2-aminopyridine, prepared by the action of sodamide on pyridine, is reacted with β-dimethylaminoethyl chloride in the presence of sodamide, and the resulting 2-[2-(dimethylamino) ethylamino] pyridine is subsequently condensed with benzyl bromide in the presence of sodamide. The corresponding hydrochloride salt is obtained from the base by treatment with hydrogen chloride in an organic solvent.

It is frequently employed in the *treatment of perennial and seasonal allergic rhinitis, allergic conjunctivitis due to inhalant allergens and foods, simple allergic skim manifestations of urticaria and angioedema, dermographism and anaphylactic reactions as an adjunct to adrenaline.*

Dose : 25 to 50 mg ; Usual, adult, oral 4 to 6 times a day.

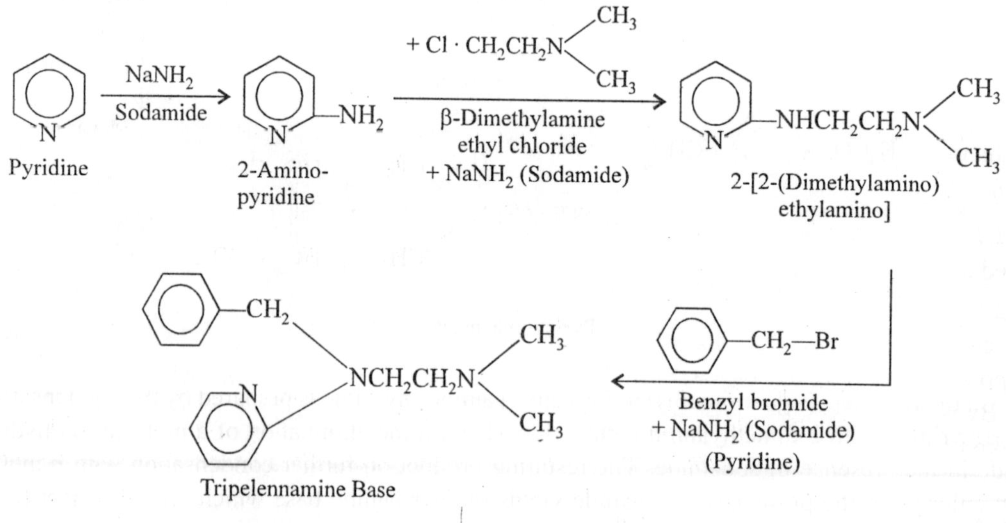

Tripelennamine Base

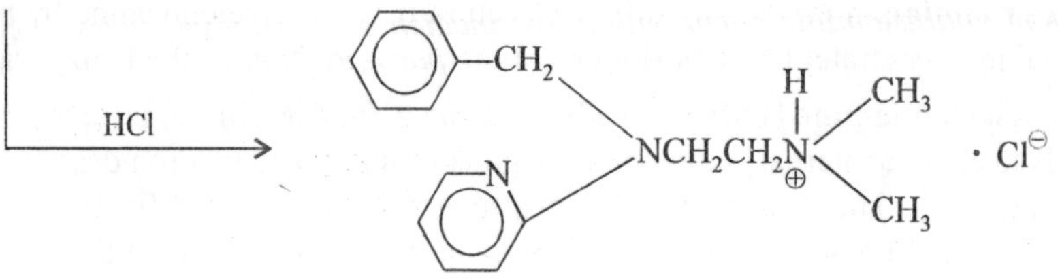

Tripelennamide Hydrochloride

C. *Thonzylamine Hydrochloride* INN, BAN, USAN,

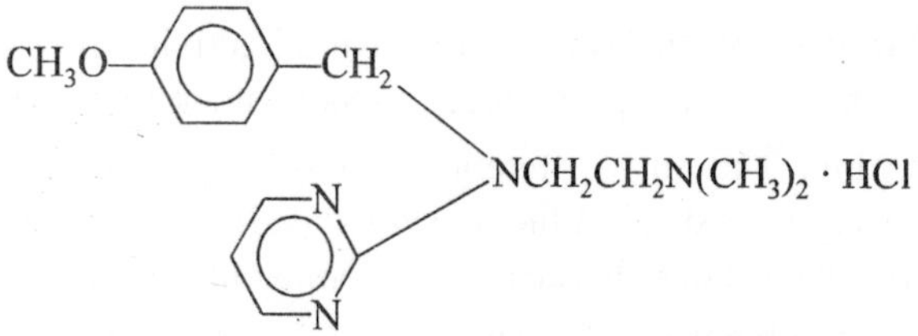

2-{[2-(Dimethylamino) ethyl]-(*p*-methoxybenzyl) amino} pyrimidine hydrochloride ; NF XII ; Resistab[R] (Bristol-Myers) ; Neohetramine Hydrochloride[R] (Nepera)

It is recommended for use with streptomycin in exudative human tuberculosis. It is used in treating the symptoms of hay fever, urticaria, drug reactions and other mild allergic conditions.

Dose : 50 mg ; Usual, adult, oral up to 4 times a day.

D. *Zolamine Hydrochloride* USAN, *Zolamine*, INN,

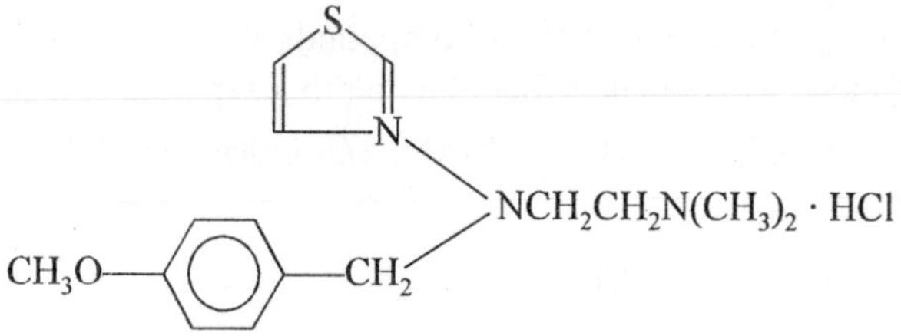

2-[[2-(Dimethylamino) ethyl]-(*p*-methoxybenzyl) amino] thiazole monohydrochloride ; 1, 2-Ethanediamine, N-[(4-methoxyphenyl) methyl]-N', N'-dimethyl-N-2-thiazolyl-, monohydrochloride.

It is used both as an antihistaminic and anaesthetic (topical) agent.

2.1.2.1. Mechanism of Action. The mechanism of action of some of the ethylenediamines are described as under :

2.1.2.1.1. Mepyramine Maleate (Pyrilamine Maleate). The *'drug'* and *tripelennamine* are both pronounced clinically to belong to the category of less potent antihistaminics. It has been reported to be highly potent particularly in antagonizing the histamine-induced contractions produced in guinea-pig ileum*. By virtue of this marked and pronounced local anaesthetic action, the *'drug'* is recommended to be taken along with food and **not be chewed prohibitively.**

---

*Casy AF : **Chemistry of H₁-histamine antagonists** : In : Rochae Silva M (ed.). **Handbook of Experimental Pharmacology,** Vol. 18.2, p-175, Springer-Verlag, New York. 1978.

SAR of Pyrilamine. It essentially differs structurally from tripelennamine by havig a methoxy (OCH$_3$) functional moiety strategically positioned at the *para*-position of the benzyl radical.

2.1.2.1.2. Tripelennamine Hydrochloride. The *'drug'* (and its citrate) seem to be get metabolized almost completely to either quaternary ammonium N-glucuronide or *O*-glucuronides of the corresponding hydroxylated metabolites. The drug also undergoes several chemical modifications *in vivo*, namely : (*a*) ring hydroxylation ; (*b*) N-oxidation ; and (*c*) N-demethylation. In view of these glaring evidences the *'drug'* gets excreted principally in the urine as its conglomerate of metabolites.

Incidentally, tripelennamine enjoys the reputation of being the first and foremost ethylenediamine ever developed in the American Laboratories.

2.1.2.1.3. Thonzylamine Hydrochloride. The overall activity of this *'drug'* seems to be very much identical to tripelennamine but it is pronounced to have much less toxicity.

SAR to Thonzylamine. The only difference between this **'drug'** and tripelennamine is the presence of a *'pyrimidine nucleus'* instead of the *'pyridine nucleus'* in the latter, which perhaps retards its toxicity profile to an appreciable extent because of its comparatively faster metabolism *in vivo.*

2.1.2.1.4. Zolamine Hydrochloride. The *'drug'* possesses less toxicity in comparison to tripelennamine and thonzylamine. It exhibits both antihistaminic and anaesthetic pharmacological profile.

SAR of Zolamine. It essentially contains the *'thiazole moiety'* instead of the *pyridine* and *pyrimidine* groups present in tripelennamine and a 5-thonzylamine respectively. Being, a 5 membered ring (thiazole) which is certainly much more compact than the 6-membered heterocyclic ring present in the other two compounds.

2.1.3. Thiophene Derivatives

The thiophene moiety is an essential component of this group of compounds which exhibit significant antihistaminic properties. Though these compounds also possess an ethylene-diamine nucleus yet the presence of a thiophene group makes them belong to a separate category altogether.

A. *Methapyrilene Hydrochloride* BAN, USAN, *Methapyrilene* INN,

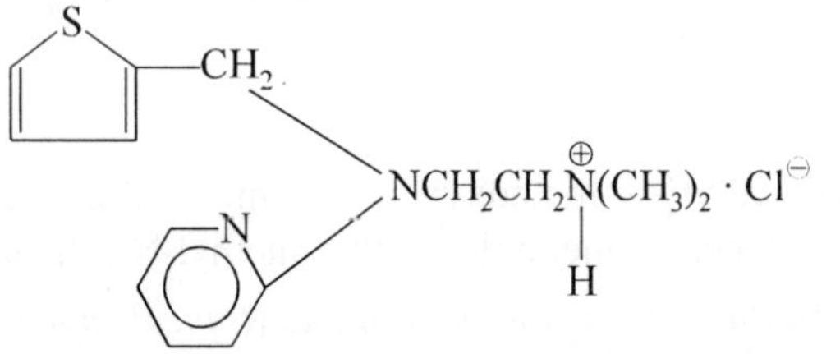

2-[[2(Dimethylamino) ethyl]-2-thenylamino] pyridine monohydrochloride ; 1, 2-Ethanediamine, N, N-dimethyl N'-2-pyridinyl-N'-(2-thenylmethyl)-, monohydrochloride ; USP ; Histadyl[R] (Lilly) ; Semikon Hydrochloride[R] (Beecham) ; Thenylene Hydrochloride[R] (Abbott)

Synthesis :

N, N-Dimethyl-N'-(2-pyridyl)-ethylene diamine is prepared by the condensation of 2-amino pyridine and 2-dimethylamino ethyl chloride in the presence of sodamide. The resulting product is further condensed with 2-thenyl chloride, using sodamide as a catalyst, to obtain the methapyrilene base, which on neutralization with hydrogen chloride yields the methapyrilene hydrochloride.

It essentially differs from tripelennamine in having a 2-thenyl (*i.e.,* thiophene-2-methylene) moiety instead of the benzyl group.

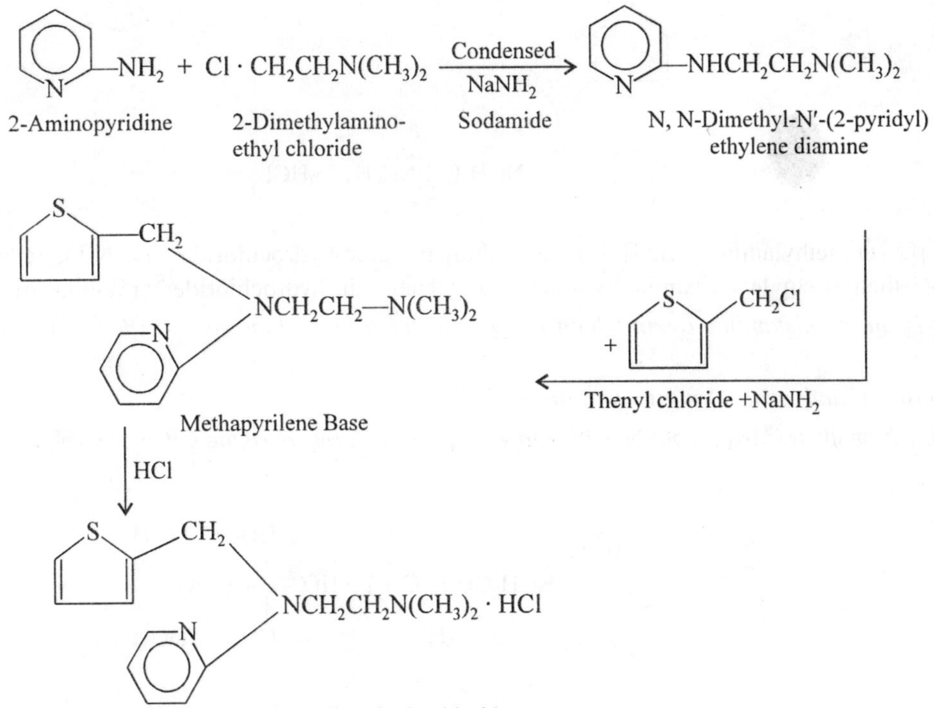

NH2 + Cl · CH2CH2N(CH3)2
Condensed
NaNH2
NHCH2CH2N(CH3)2
2-Aminopyridine
2-Dimethylamino-ethyl chloride
Sodamide
N, N-Dimethyl-N'-(2-pyridyl) ethylene diamine
CH2
NCH2CH2—N(CH3)2
CH2Cl
Methapyrilene Base
Thenyl chloride +NaNH2
HCl
CH2
NCH2CH2N(CH3)2 · HCl

Methapyrilene hydrochloride

It possesses a low incidence of side-effects and may be employed in the treatment of all types of suspected allergies, namely—hay fever, chronic urticarias, allergic rhinitis and allergic dermatitis.

Dose : 50 to 100 mg, 3 to 4 times a day.

B. *Methaphenilene Hydrochloride* USAN, *Methaphenilene* INN,

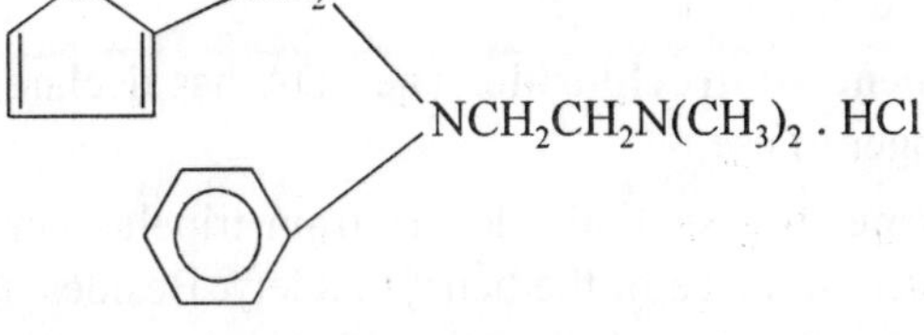

CH2
NCH2CH2N(CH3)2 . HCl

N, N-Dimethyl-N'-(α-thenyl)-N'-phenethylenediamine hydrochloride ; NF X ; Diatrine Hydrochloride(R) (Warner Chilcott)

It is an effective antihistaminic agent. It induces slow incidence of side reactions but possesses a moderate tendency to cause gastro-intestinal irritation. It is a drug of choice for the symptomatic relief of upper respiratory infections.

Dose : Usual, 25–50 mg 4 times per day.

C. *Thenyldiamine Hydrochloride* BAN, *Thenyldiamine* INN, USAN,

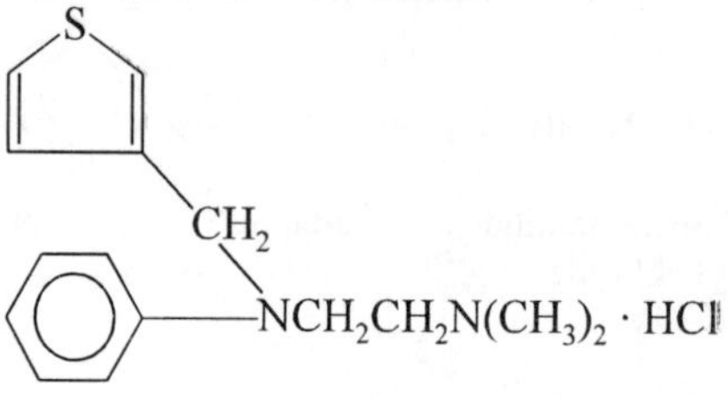

2-[[2-(Dimethylamino) ethyl]-3-thenylamino] pyridine-hydrochloride ; N, N-Dimethyl-N'-(2-pyridyl)-(3-thenyl)-ethylene-diamine hydrochloride ; Thenfodil hydrochloride[R] (Winthrop) ;

It is an antihistaminic agent which is recommended in comparatively milder type of allergic conditions.

Dose : Usual, oral, 15 mg up to 6 times per day.

D. *Chlorothen Citrate* USAN, *Chloropyrilene* INN, *Chloropyrilene Citrate* BAN,

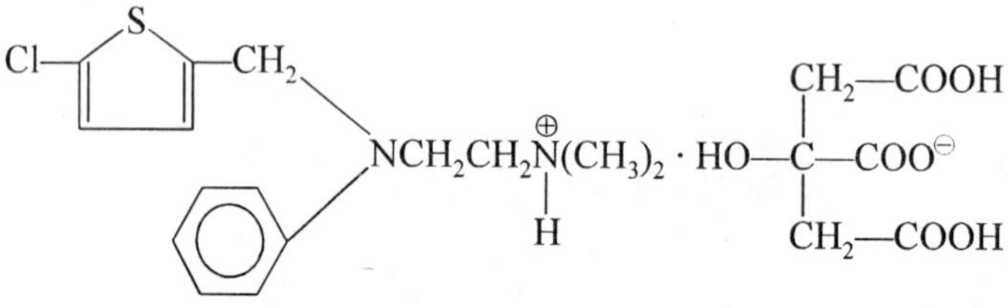

2-[(5-Chloro-2-thenyl [2-(dimethylamino) ethyl] amino] pyridine dihydrogen citrate ; Chloromethapyrilene citrate ; N.F. XIII ; Panta[R] (Valeas, Italy)

Chlorothen is similar in structure to tripelennamine, the only difference being that the benzyl group in the latter is substituted by the 5-halothenyl moiety. It has been observed that the halogen-substitution enhances the antihistaminic activity and renders the compound less toxic than its corresponding non-halogenated version.

Dose : 25 mg ; every 3 to 4 hours.

2.1.3.1. Mechanism of Action. The mechanism of action of the thiophene structural analogoues of histamine shall be discussed briefly as under :

2.1.3.1.1. Methapyrilene Hydrochloride. The FDA has declared it a potential carcinogen in 1979, and hence, it is no longer in use.

SAR of Methapyrilene. It essentially differs from tripelennamine in possessing a 2-thenyl (thiophene-2-methylene) moiety in place of the benzyl moiety. Besides, the thiophene ring is regarded to be isosteric with the benzene ring ; and, therefore, the isosteres found to display almost identical activity. An exhaustive study with respect to the *'solid-state conformation'* of this *'drug'* evidently showed that the geometrical *trans*-conformation is obviously the most preferred one for the two ethylene N-atoms.

2.1.3.1.2. Methaphenilene Hydrochloride. It is a potent antihistaminic agent that may give rise to rather mild type of gastro-intestinal irritation.

2.1.3.1.3. Thenyldiamine Hydrochloride. The *'drug'* is regarded as a traditional antihistaminic which action is usually associated with both troublesome sedative and antimuscarinic effects.

2.1.3.1.4. Chlorothen Citrate (Chloropyrilene Citrate). It is invariably associated with an antibacterial formulation that is indicated mostly for the treatment of vasomotor rhinitis and other hypersensitivity reaction of the upper respiratory tract (URT) complicated by bacterial infections.

2.1.4. Cyclic Basic Chain Analogues

A variety of more potent and less toxic antihistaminic agents have been tailored by effecting molecular modifications of the general ethylenediamine structure whereby the dimethylamino function is essentially replaced by a small compact heterocyclic ring.

Thus, the cyclic basic chain analogs may be further sub-divided into three categories, namely :

(*a*) *Imidazoline Derivatives*

A. *Antazoline Hydrochloride* BAN, USAN, *Antazoline* INN,

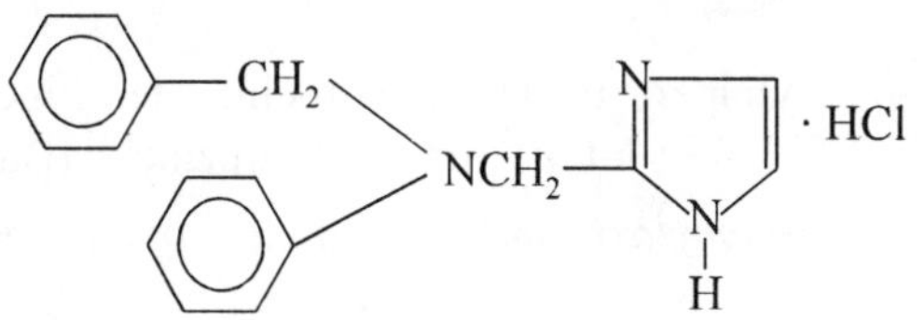

2-(N-Benzylanilino) methyl-2-imidazoline hydrochloride ; N-Benzyl-N-(2-imdazoline-2-yl-methyl) aniline hydrochloride ; BP ; 1973, USP ; XV, Int. P., Ind. P., Antistine[R] (Ciba) ; Histostab[R] (Boots)

Synthesis :

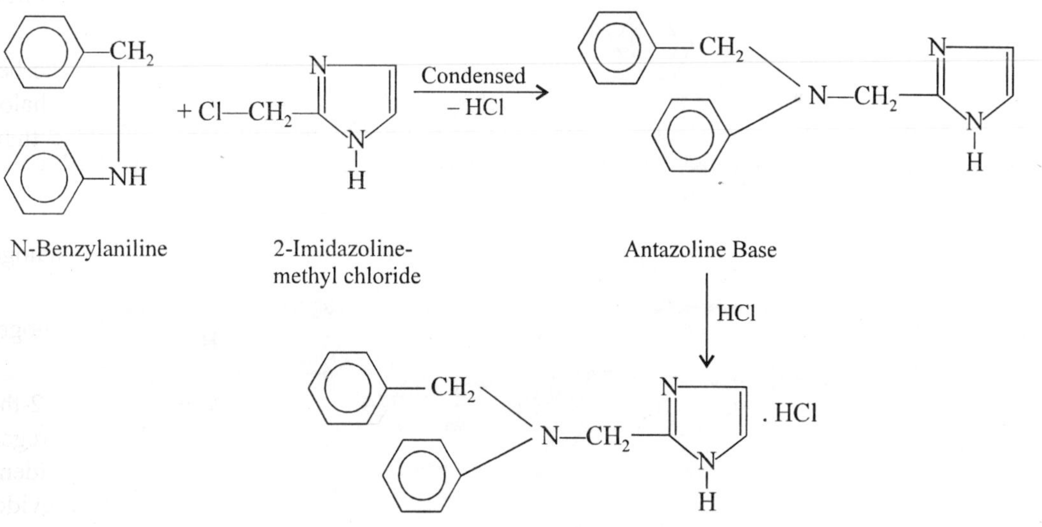

N-Benzylaniline   2-Imidazoline-methyl chloride   Antazoline Base

Antazoline hydrochloride

Antazoline base is prepared by the interaction of N-benzyl aniline with 2-imidazoline methyl chloride with the elimination of hydrogen chloride. The base is neutralized with hydrogen chloride to yield the desired antazoline hydrochloride.

It is less active than most of the other antihistaminic drugs, but has the advantage of being devoid of local irritant characteristics.

Dose : 50 to 100 mg.

(*b*) *Piperazine Derivatives*

A. *Cyclizine Hydrochloride* BAN, USAN, *Cyclizine* INN,

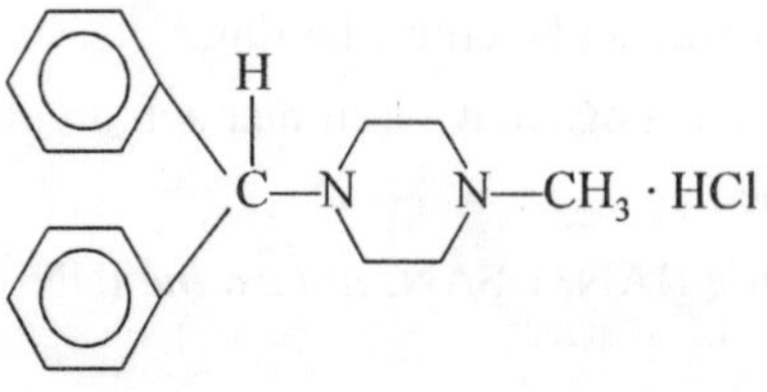

1-(Diphenylmethyl)-4-methylpiperazine monohydrochloride ; Piperazine, 1(diphenylmethyl)-4-methyl-, monohydrochloride ; BP ; USP ; Int. P., Ind. P., Marezine[R] (Burroughs Wellcome)

It is mostly employed as a prophylaxis and for treatment of motion sickness.

Dose : 25 to 50 mg.

B. *Chlorcyclizine Hydrochloride* BAN, USAN, *Chlorcyclizine* INN,

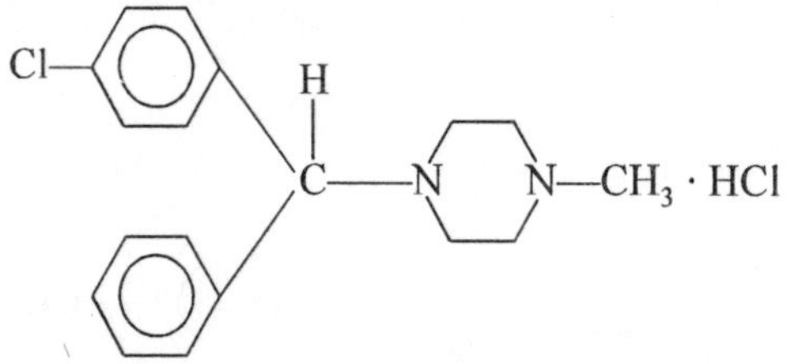

1-(*p*-Chloro-α-phenylbenzyl)-4-methylpiperazine monohydrochloride ; Piperazine, 1-[(4-chlorophenyl) phenylmethyl]-4-methyl-, monohydrochloride ; BP ; USP ; Int. P., Ind. P., Di-Paralene[R] (Abbott) ; Perazil[R] (Burroughs Wellcome)

Synthesis :

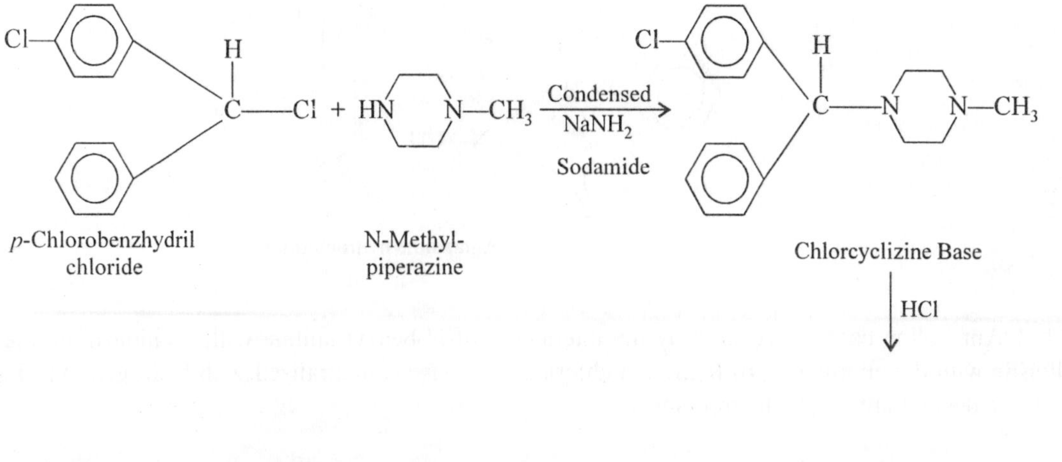

p-Chlorobenzhydril chloride	N-Methyl-piperazine	Chlorcyclizine Base

HCl

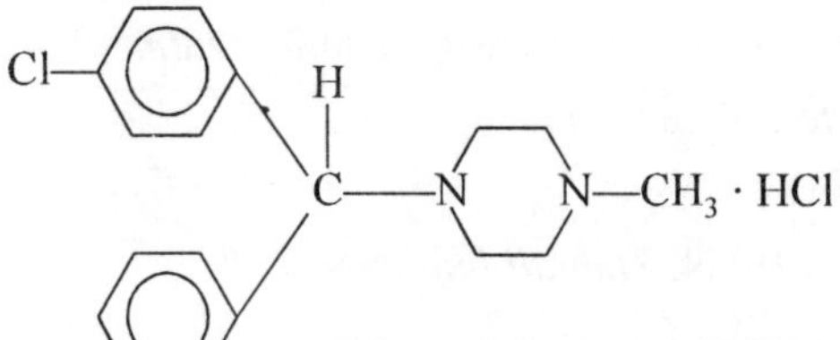

Chlorcyclizine hydrochloride

Chlorocyclizine base is first prepared by condensing together *p*-chlorobenzhydril chloride with N-methylpiperazine in the presence of sodamide. The resulting base is neutralized with hydrogen chloride to obtain the required chlorcyclizine hydrochloride.

Though it is less potent, it possesses a prolonged antihistaminic action of similar duration to that of promethazine hydrochloride. It has local anaesthetic, antiemetic and anticholinergic characteristics.

Dose : 50 to 200 mg.

C. *Meclizine Hydrochloride* USAN, *Meclozine* INN, BAN,

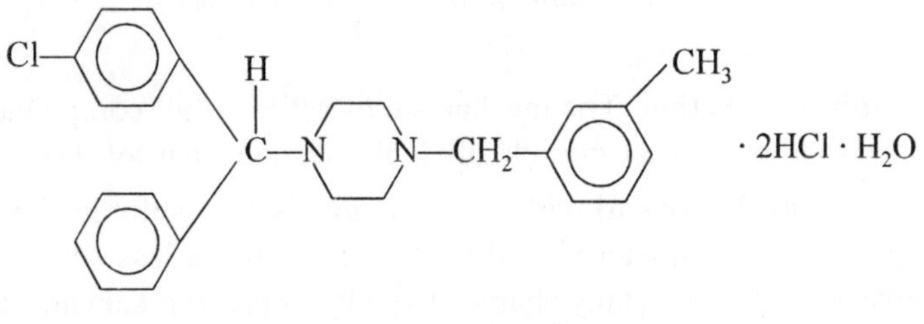

1-(*p*-Chloro-α-phenylbenzyl)-4-(*m*-methylbenzly) piperazine dihydrochloride monohydrate ; Piperazine, 1-[(4-chlorophenyl) phenylmethyl]-4-[(3-methylphenyl)-methyl]-, dihydrochloride, monohydrate ; BP ; USP ; Int. P., Ind. P., Antivert[R] (Roerig) ; Bonine[R] (Pfizer)

It is mostly employed for its inherent antiemetic action which is quite marked and pronounced and lasts for up to 24 hours. It has also been used for the prevention and treatment of motion sickness and also for the relief of allergic conditions.

Dose : 25 to 50 mg.

D. *Buclizine Hydrochloride* BAN, USAN, *Buclizine* INN,

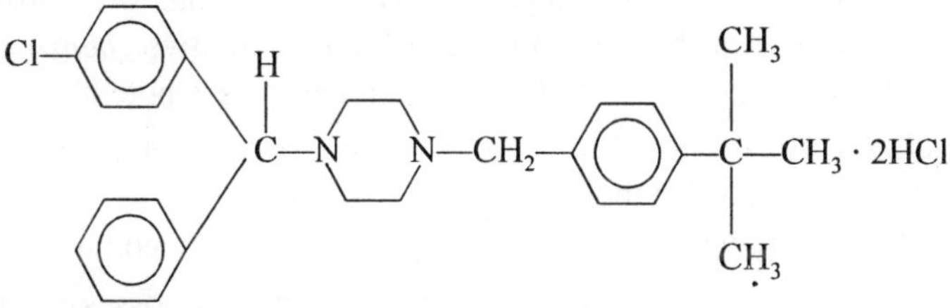

1-(*p-tert*-Butylbenzyl)-4-(*p*-chloro-α-phenylbenzyl) piperazine dihydrochloride ; Piperazine, 1-[(4-chlorophenyl) phenylmethyl]-4-[(1, 1-dimethylethyl) phenyl] methyl]-, dihydrochloride ; Bucladin-S[R] (Stuart) ; Vibazine[R] (Pfizer)

It is chiefly used for its antiemetic properties. It possesses less pronounced sedative effects than promethazine. *It is also recommended for the symptomatic treatment of allergic conditions and vertigo.*

Dose : 25 to 50 mg, 2 to 3 times a day.

(c) *Piperidine Derivatives*

A. *Thenalidine Tartrate* BAN, *Thenalidine* INN,

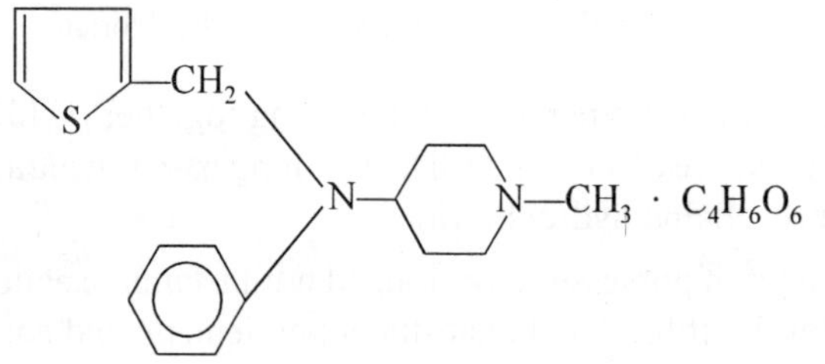

1-Methyl-4-(N-then-2-ylanilino) piperidine tartrate ; Thenophenopiperidine Tartrate ; Thenopiperidine Tartrate ; Sandosten[(R)] (Sandoz)

It is used for the prevention and treatment of allergic conditions.

Dose : 100 to 150 mg per day.

2.1.4.1. Mechanism of Action. The mechanism of action of all compounds enumerated under sections (a) through (c) shall be treated individually in the sections that follows :

2.1.4.1.1. Antazoline Hydrochloride. The *'drug'* is less soluble than the corresponding phosphate salt and is mostly administered orally. It is found to be less active than a host of other antihistaminics ; however, it has been duly characterized by its predominant absence of local irritation.

Besides, it exhibits more than double than local anaesthetic potency of *'procaine'* and also exhibits the anticholinergic properties.

SAR of Antazoline. Just like the ethylenediamines, it also essentially comprises of an N-benzylamino function directly attached to a basic N-atom through a 2-carbon chain.

2.1.4.1.2. Cyclizine Hydrochloride. The *'drug'* is basically employed as a potent prophylaxis and also for the control, management and treatment of motion sickness.

2.1.4.1.3. Chlorcyclizine Hydrochloride. The *'drug'* is used invariably in the symptomatic relief of urticaria, hay fever, and a few other allergic manifestations.

SAR of Chlorcyclizine. It has been adquately demonstrated that the distribution or substitution of halogen either at the *ortho*-or at the *meta*-position of any of the two *'benzhydryl functional rings'* mostly gives rise to such compounds that do possess appreciably less potent acitivity.

2.1.4.1.4. Meclizine Hydrochloride. The *'drug'* is effective in vertigo intimately associated with such ailments that essentially affect the vestibular system. As the *'drug'* also exhibits anticholinergic activity, it may be employed in patients having a history of asthma, glaucoma, or prostatic enlargement.

SAR of Meclizine. It apparantly differs form chlorcyclizine by possessing an N-*m*-methylbenzyl functional moiety instead of the prevailing N-methyl moiety. Thus, it exhibits moderately potent antihistaminic profile.

2.1.4.1.5. Buclizine Hydrochloride. The *'drug'* acts by exerting appreciable anticholinergic and antihistaminic activities. Besides, it possesses CNS-depressant profile. Hence, indicated for the control and management of nausea, vomitting and dizziness closely related to motion sickness.

SAR of Buclizine. Importantly, buclizine-a member of the piperazine class of antihistaminics are very much structurally related to both the **ethylenediamines** as well as the **benzyhydryl ethers of ethanolamines.** Its structure essentially include the 2 carbon separation existing between the N-atoms, that forms a part of the piperazine ring.

2.1.4.1.5. Thenalidine Tartrate. The *'drug'* possesses the actions and uses of antihistaminics ; and has been employed parenterally for the symptomatic relief of hypersensitivity reactions.

2.1.5. Phenothiazine Derivatives.

Since 1945, plethora of antihistaminics have come into existence as a consequence of bridging the aryl functional moieties of agents that were intimately related to the ethylene-diamines. **Phenothiazines** essentially possess S-atom as the bridging entity.

It is, however, pertinent to state here that the *phenothiazines* which predominantly exhibit therapeutically potential and useful antihistaminic activities should essentially contain the following characteristic features, namely :

• at least 2 to 3 C-atom

• branched alkyl chain between the prevailing ring system

• terminal N-atom

Thus, the significant point of difference between the *phenothiazine antihistaminics* and the *phenothiazine antipsychotics* is that the latter should have an **unbranched propyl chain.** However, there are *two* most important aspects that are most essential for the phenothiazine antihistaminics, namely :

(*a*) 3 C-bridge between N-atoms are more potent *in vitro,* and

(*b*) heterocyclic ring of the antihistamines should be unsubstituted (*i.e.,* unlike the phenothaizine antipsychotics*.)

Toldy *et al.** (1959) resolved the two enantiomers of promethazine and observed identical antihistaminic and a number of other pharmacologic activities, such as : antiemetic, anticholinergic, and sedating agent. Importantly, this specific feature in *promethazines* is found to be in absolute contrast with regard to investigative studies carried out with *pheneramines* and *carbinoxamines,* wherein the strategically located chiral centre is located quite closer to the aromatic feature of the drug molecule.

Promethazine

A number of phenothiazine derivatives have evolved into potent antihistaminic agents ; a few important ones are described below :

A. *Promethazine Hydrochloride* BAN, USAN, *Promethazine* INN,

$CH_2CH(CH_3)N(CH_3)_2 \cdot HCl$

---

*Toldy L et al. Acta. Chim. Acàd. Sci. Hung., **19** : 273, 1959.

10-[2-(Dimethylamino) propyl] phenothiazine monohydrochloride ; 10H-Phenothiazine-10-ethanamine, N, N, α-trimethyl-, monohydrochloride ; BP ; USP ; Eur. P., Int. P., Ind. P ; Phenergan[R] (Wyeth) ; Remsed[R] (Endo) ; Zipan[R] (Savage) ; Ganphen[R] (Reid-Provident) ; Fellozine[R] (O'Neal, Jones & Feldman) ; Tixylix[R] (May & Baker, U.K.)

Synthesis :

Phenothiazine is first prepared by fusing together diphenylamine and sulphur in the presence of iodine or aluminium trichloride. Promethazine base may be prepared by reacting the resulting phenothiazine

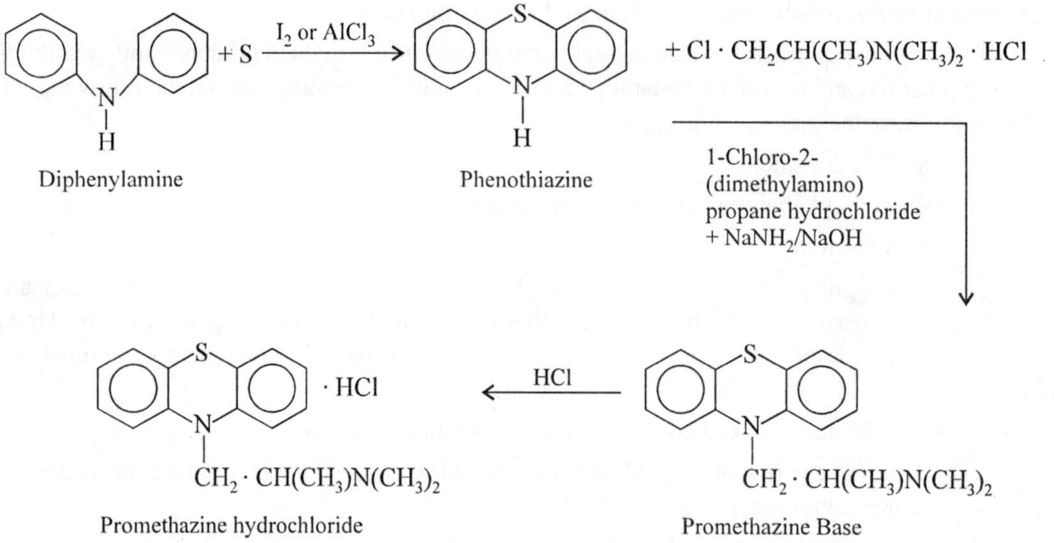

Diphenylamine   Phenothiazine

1-Chloro-2-(dimethylamino) propane hydrochloride + NaNH$_2$/NaOH

Promethazine hydrochloride   Promethazine Base

with 1-chloro-2-(dimethylamino) propane hydrochloride in the presence of sodamide and sodium hydroxide in xylene. The corresponding base is extracted, purified and converted to the hydrochloride.

It has a prolonged duration of action. It may be used effectively in perennial and seasonal allergic rhinitis, vasomotor rhinitis, allergic conjunctivitis due to inhalant allergens and foods ; and certain milder type of skin manifestations of urticaria. It also possesses some anticholinergic, antiserotoninergic, and marked local anaesthetic properties.

Dose : 20 to 50 mg per day (B.P.) ; 12.5 to 150 mg per day (USP).

B. *Promethazine Teoclate* INN, USAN, *Promethazine Theoclate* BAN,

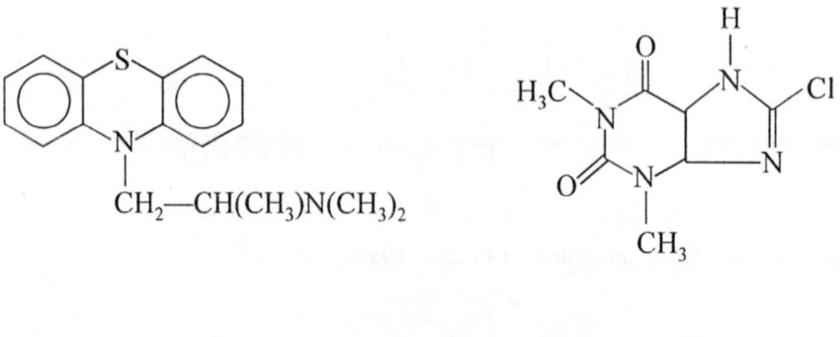

10-(2-Dimethylaminopropyl) phenothiazine compound of 8-chlorotheophylline ; Promethazine chlorotheophyllinate ; BP ; BPC ; USP ; Avomine[R] (May & Baker)

It is mainly used as an antiemetic in the prevention and treatment of motion sickness. It may also be used in post-operative vomiting, the nausea and vomiting of pregnancy, drug-induced nausea and vomiting and in irradiation sickness.

Dose : 25 to 50 mg per day.

C. *Trimeprazine Tartrate* BAN, USAN, *Alimemazine* INN,

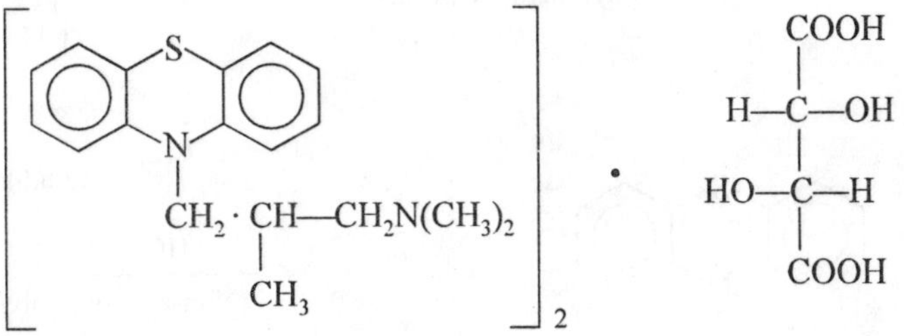

10-[3-(Dimethylamino)-2-methylpropyl] phenothiazine tartrate (2:1) BP ; USP ; Temaril[R] (Smith, Kline & French)

It is used mainly for its marked and pronounced effect in the *relief of pruritus*. Its overall pharmacological characteristic lie between that of promethazine and chlorpromazine.

Dose : 10 to 40 mg ; adult, oral per day.

D. *Methdilazine Hydrochloride* BAN, USAN, *Methdilazine* INN,

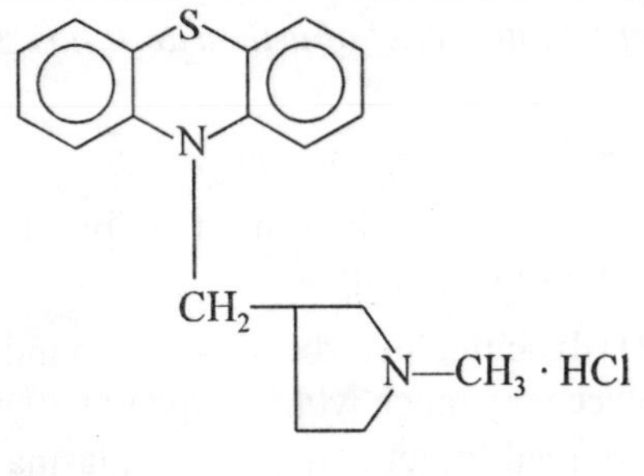

10-[(1-Methyl-3-pyrrolidinyl)-methyl phenothiazine monohydrochloride ; 10H-Phenothiazine, 10-[(1-methyl-3-pyrrolidinyl) methyl]-, monohydrochloride ; USP ; Tacaryl Hydrochloride[R] (Westwod)

Synthesis :

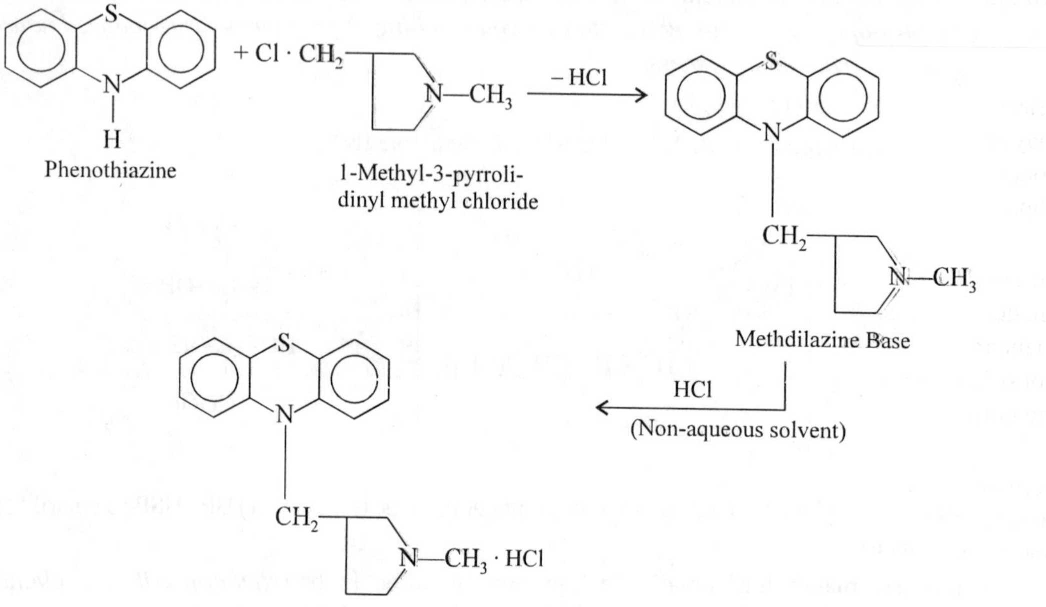

Methdilazine is prepared by reacting together phenothiazine and 1-methyl-3-pyrrolidinyl methyl chloride ; the resulting base is treated with equimolar quantity of hydrogen chloride in a non-aqueous solvent.

It may be used for the symptomatic relief of urticaria. It has also been used successfully for the treatment of migraine headache.

Dose : 8 mg usual, adult, oral 2 to 4 times a day.

2.1.5.1. Mechanism of Action. The mechanism of action of certain members of this particular category of antihistaminics shall be discussed below :

2.1.5.1.1. Promethazine Hydrochloride. The *'drug'* is found to be well absorbed, and the peak optimum effects invariably take place very much within a span of 20 minutes after adequate oral, rectal, or IM administration. It is also observed to get bound to the plasma proteins to an extent of 76-80%. However, the *'drug'* gets excreted gradualy both in the urine and faeces, primarily in the form of its corresponding **inactive metabolites** *viz., sulphoxides,* and *glucuronic'es.*

2.1.5.1.2. Trimeprazine Tartrate. The *'drug'* has an additio al methylene (–CH_2–) unit in the side-chain of *promethazine,* which renders it more active than the latter, and interestingly less active than CPZ in histamine-induced bronchospasm in the guinea pigs. Hence, it is mainly used as an antipuritic drug.

2.1.5.1.3. Methdilazine Hydrochloride. The replacement of the moderately longer side chain of promethazine (PMZ) at position 10 has been meticulously substituted by a methylene-linked N-methyl pyrrolidine nucleus, which being rather compact and small in its dimensions, enables the *'drug'* to exert its effect solely for the symptomatic relief of *urticaria.* It is also indicated invariably in very seriously ill or dehydrated children on account of the significantly greater susceptibility of *dystonias** with the phenothiazines.

2.1.6. Second-generation Nonsedating Antihistamines

Since 1980s an enormous impetus has been geared towards the development of improved H_1 selectivity so that the new breed of antihistaminics should bear practically no sedative properties, and may also possess adequate antiallergic activities. In fact, the outcome of such an overwhelming rigorous concerted research activities towards producing an altogether new class of antihistaminics have been baptised as the **second generation antihistamines.**

It has been critically observed that most of these newer compounds belonging to this category do possess a wide variation in their structural profile ; however, their pharmacologic characteristic features are not so variant in nature, as they invariably exert their action principally in the periphery. It is pertinent to mention here that the structural resemblance to the *first generation H_1-antagonists* (see section 2.1) is not strictly adhered to by virtue of the fact that most of these drug substances came into existence first in the normal process of investigation pertaining to several other diversified pharmacologic targets.

In general, the second-generation non-sedating antihistamines essentially possess selective peripheral H_1-antagonism activities, and the specifically exhibit much less anticholinergic activity. Besides, they are associated with lowered affinity for adrenergic and serotonergic receptors, and usually possess very limited CNS-effects.

The mechanism of action of these agents show that they do not penetrate the blood brain barrier (BBB) appreciably most probably on account of the following cardinal factors, such as :

(*a*) Amphoteric nature (*i.e.,* majority of them are usually zwitter ionic at the prevailing physiologic pH) ;

(*b*) Partitioning properties ; and

(*c*) Behave as substractes for the drug efflux of either **P-glycoprotein transporter** or organic anion transporter protein.**

A few typical members of this category of antihistaminics shall now be discussed as under :

A. **Terfenadine** USAN, BAN,

Alpha-[4-(1, 1-Dimethylethyl) phenyl]-4-(hydroxydiphenylmethyl)-1-piperidinebutanol ; Seldane[R] ;

SAR : Terfenadine was discovered in the course of an extensive and intensive search for new butyrophenone antipsychotic drugs as could be seen by the presence of the N-phenylbutanol substituent. It is also studded with a diphenylmethyl piperidine group structural analogous as is normally observed in the piperazine antihistaminics.

---

*Prolonged muscle contractions that may cause twisting and repititive movements or abnormal posture.

Cvetkovic M *et al. Drug Metab. Dispos.* **27, 866–871, 1999.

Although terfenadine once enjoyed the very popular nonsedating antihistamines, but the extensive clinical experience pronounced it to be an altogether dangerous drug causing serious cardiac arrythmias, taking place very often in the event when certain other drugs were administered concomitantly. Such effects are caused due to the blockade of delayed rectifier K^+ channels in cardiac tissue, and hence are intimately related to the parent molecule.

Mechanism of Action. The *'drug'* undergoes rapid oxidation *in vivo* that finally gives rise to the formation of the corresponding carboxylic acid metabolite,* which is presently marketed as **fexofenadine** as given below :

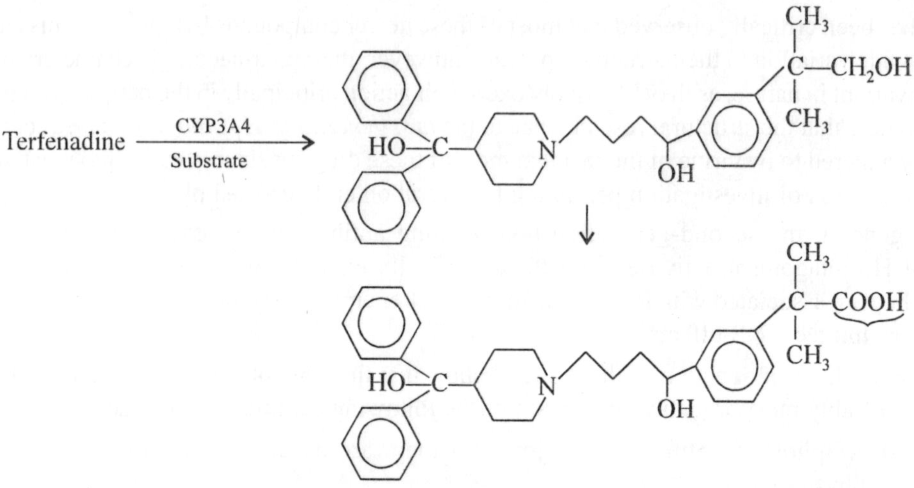

Fexofenadine

In fact, the acid metabolite is ultimately responsible for the antihistaminic properties of *terfenadine* in humans because the parent compound is readily metabolized *via* CYP3A4 substrate catalyzed processes in due course.

Nevertheless, the histamine receptor affinity of terfenadine are supposed to be associated primarily to the presence of the respective diphenylmethyl piperidine functional group. The actual cause of its prolonged action is solely on account of its slow dissociation from these receptors**.

B. **Astemizole** BAN, USAN,

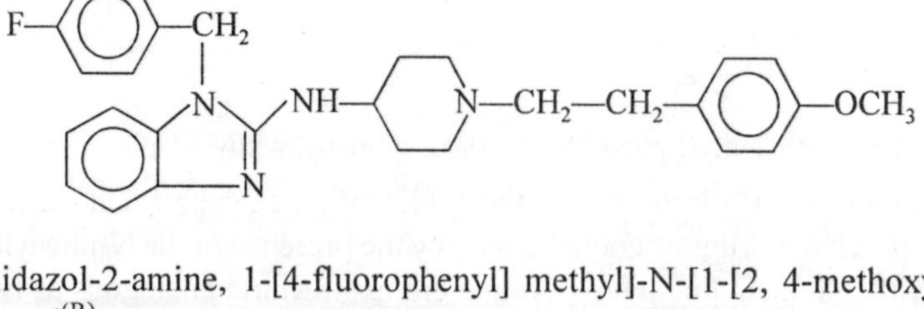

1H-Benzimidazol-2-amine, 1-[4-fluorophenyl] methyl]-N-[1-[2, 4-methoxyphenyl) ethyl]-4-piperidinyl- ; Hismanal[R] ;

---

*Lalonde RL *et al. Pharm. Res.* **13,** 832–8, 1996.

**Facts and Comparison pp. 188-194C, 1993.

Astemizole is the creative product by the medicinal chemists of an extensive as well as intensive search of a series of *benzimidazoles.** These new breed of the synthesized products may be regarded as the 4-aminopiperidines wherein the *para*-amino functional moiety essentially holds the *two aromatic rings viz., first,* present in the benzimidazol structure itself ; and *secondly,* as the *para*-fluorophenyl moiety linked at one of the N-atoms.

The drug is found to be more potent and possesses longer duration of action than the terfenadine. It is a slow-onset, long acting and nonsedating piperidine antihistaminic having practically little anticholinergic activity. It is indicated for seasonal allergic rhinhitis and chornic urticaria.

Mechanism of Action. At least two *active metabolites,* namely : (*a*) *o*-desmethylastemizole ; and (*b*) norestemizole, as shown belew are obtained :

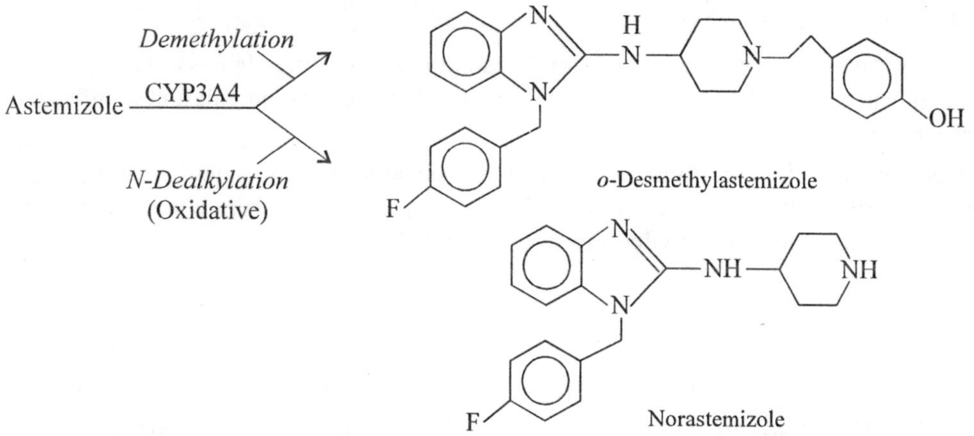

Kamei *et al.*** (1991) confirmed the presence of a *'third metabolite'* that may also contribute to the effects of astemizole to a certain extent.

Astemizole gets largely distributed in the peripheral tissues having the highest concentration found in the *liver, pancreas,* and *adrenal glands.* The *'drug'* is observed to undergo substantial first-pass metabolism involving such processes as : oxidative dealkylation, aromatic hydroxylation, and glucuronidation. Interestingly, one of the active metabolites *i.e, o*-desmethylastemizole essentially possesses antihistaminic activity comparable to the parent drug, and hence helps to enhance the therapeutic activity.

Besides, astemizole is highly protien bound (96%) and has a plasma half-life of 1–6 days. However, the metabolite *desmethylastemizole* shows a half-life ranging between 10 to 20 days, that solely depends on the dosage regimen and its frequency.

SAR of Astemizole. The piperidino-amino-benzimidazol group seems to be absolutely an essential requisite for the H_1-receptor affinity ; besides, helping appreciably to the persistent receptor binding which ultimately gives rise to the prolonged action.

---

*Janssens F *et. al. J. Med. Chem.,* **28,** 1943–7, 1985.

Kamei C *et. al Arzneim Forsch,* **41, 932–6, 1991.

C. **Loratadine** BAN, USAN,

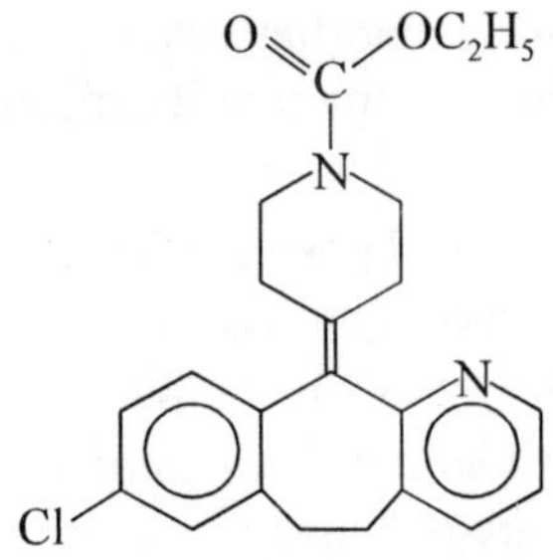

1-Piperidinecarboxylic acid, 4-(8-chloro-5, 6-dihydro-11H-benzo [5, 6]-cycloheptal [1, 2-*b*] pyridin-11-ylidene]- ; Claritin[R] ;

It is a long-acting, nonsedating tricyclic antihistaminic drug. It possesses practically little anticholinergic activity. It shows potency that is fairly comparable to astemizole and significantly greater than terfenadine.*

Mechanism of Action. Loratadine undergoes metabolic conversion to the corresponding major metabolite decarboethoxyloratadine (desloratadine) that specifically occurs *via* an **oxidative process** rather than *via* a direct means of **hydrolysis** as depicted below :

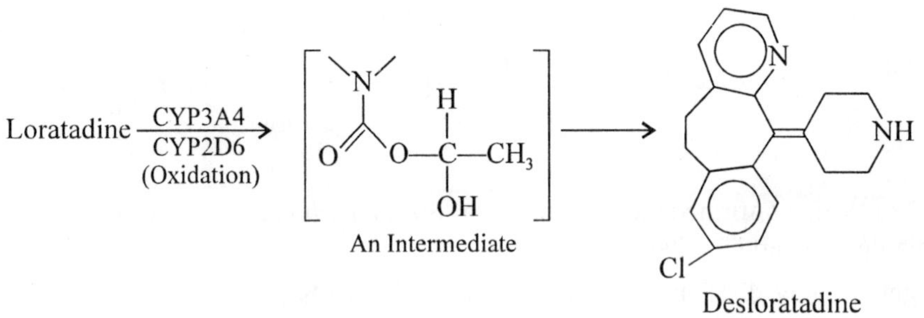

It has been observed that both CYP2D6 and CYP3A4 seem to be the perspective CYP450 isoenzymes that particularly help in the process of catalysis of this oxidative metabolic phenenomenon**. Interestingly, the ensuing metabolite (desloratadine) fails to gain its entry into the CNS in appreciable concentrations. It is, however, pertinent to state here that apparently among the *nonsedating second-generation antihistaminics,* this specific metabolite is found to be the only **nonzwitterizonic species.** Simons *et al.**** (1999) made a critical observation that while on one hand the failure of zwitterionic concentrations to have an easy access to the respective CNS sites in reasonably appreciable concentrations may be rationalized promptly, while on the other an identical explanation is not probably apparant for either the parent drug *loratadine* or its corresponding metabolite *desloratadine.*****

---

*Ahn HS *et. al. Eur. J. Pharmacol,* **127,** 153, 1986.

Yumibe N *et. al. Biochem Pharmacol,* **51 : 165–72, 1996.

***Simons FE and Simons KJ, *Clinical Pharmacokinetics,* **36,** 329–52, 1999.

****Smith SJ, *Cardiorascular toxicity of antihistamines : Otolaryngol Head Neck Surg.* **III,** 348–54, 1994.

Besides, the competitive substrates for CYP3A4 do not significantly give rise to drug-drug interaction, as could be seen with astemizole and terfenadine, by virtue of the fact that the parent molecule (*loratadine*) overwhelmingly lacks effect on K^+ rectifying channels located in the cardiac tissue.

SAR of Loratadine. The *'drug'* is intimately related to the *first generation tricyclic antihistaminics* and also to the *antidepressants.*

D. **Acrivastine** BAN, USAN,

(E, E)-3-[6-[1-(4-methylphenyl)-3-(1-pyrrolidinyl)-1-propenyl-2-puridinyl]-2-propenoic acid ; (Semprex)$^{(R)}$;

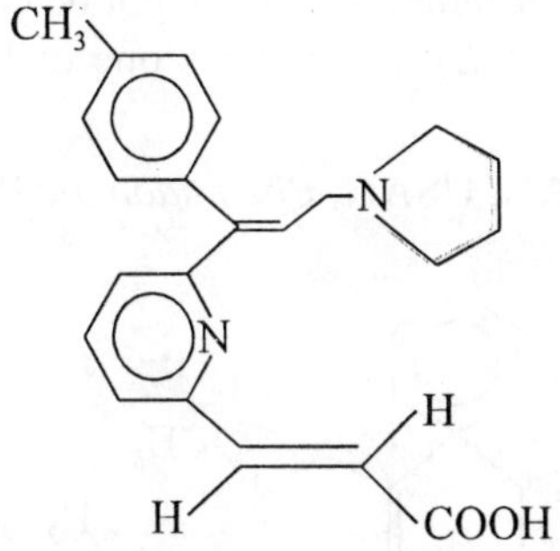

Acrivastine displays antihistaminic potency as well as the duration of action fairly comparable to *tripolidine* ; however, unlike latter the former fails to exhibit appreciable anticholinergic activity at the therapeutic concentrations. Besides, the obvious increase in the polarity of acrivastine on account of the strategically positioned carboxyethyl actually limits the BBB penetration significantly which ultimately allows this *'drug'* to cause less sedation in comparison to tripolidines.

SAR of Acrivastine. The *'drug'* is of specific interest solely from a *'drug design'* standpoint. In fact, making an intensive search for new molecular entities, it does not bring forth any **'new chemistry'.** Interestingly, the already known old compound, *tripolidine,* has been restructured by enhancing the hydrophilicity *via* strategically introducing an acrylic acid functional moiety that ultimately yielded acrivastine as shown below :

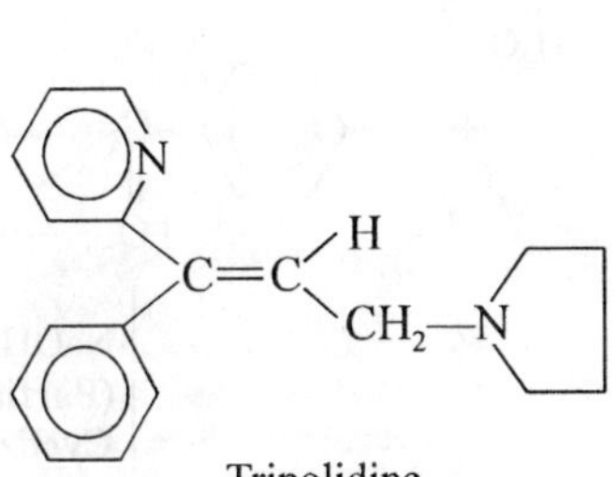

Tripolidine
[Actidil$^{(R)}$]
(An 'Alkenylamine')

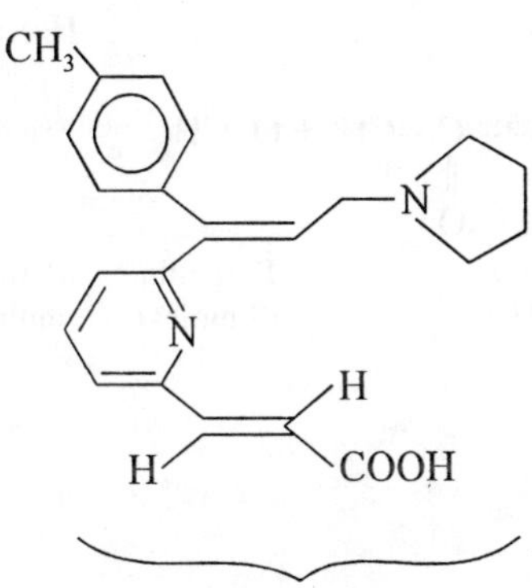

Acrylic acid
Acrivastine
[Semprex$^{(R)}$]

The lowering of the lipoidal solubility characteristic feature in the drug molecule still retained an effective H_1-antagonism peripherally. It also drastically squeezed in the $t_{1/2}$ to 1.7 hours only when compared to 4.6 hours for the parent molecule tripolidine.

Mechanism of Action. The *'drug'* has a mean peak plasma concentration varying too widely ; and it seems to penetrate the CNS quite sluggishly. However, the metabolic fate of the *'drug'* is yet to be established.

Usual adult dose : Oral, 8 mg/60 mg/3–4 times per day.

2.1.7. Miscellaneous Agents

A few medicinally potent antihistaminics that cannot be conveniently accommodated under the above-mentioned categories (*viz : A* to *E*), but possess one or two nitrogen atoms in a heterocyclic system are discussed below :

A. *Phenindamine Tartrate* BAN, USAN, *Phenindamine* INN,

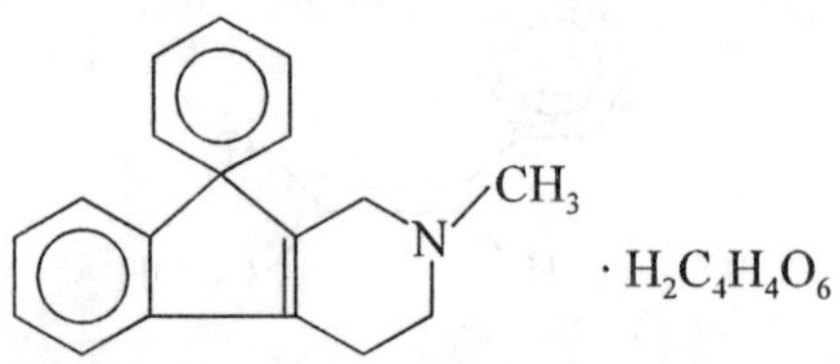

1, 2, 3, 4-Tetrahydro-2-methyl-9-phenyl-9-azafluorene hydrogen tartrate ; 2, 3, 4, 9-Tetrahydro-2-methyl-9-phenyl-1H-indenol [2, 1-C] pyridine hydrogen tartrate ; BP ; NF XIV, Int. P., Ind. P., Theophorin[R] (Hoffmann-La Roche)

Synthesis :

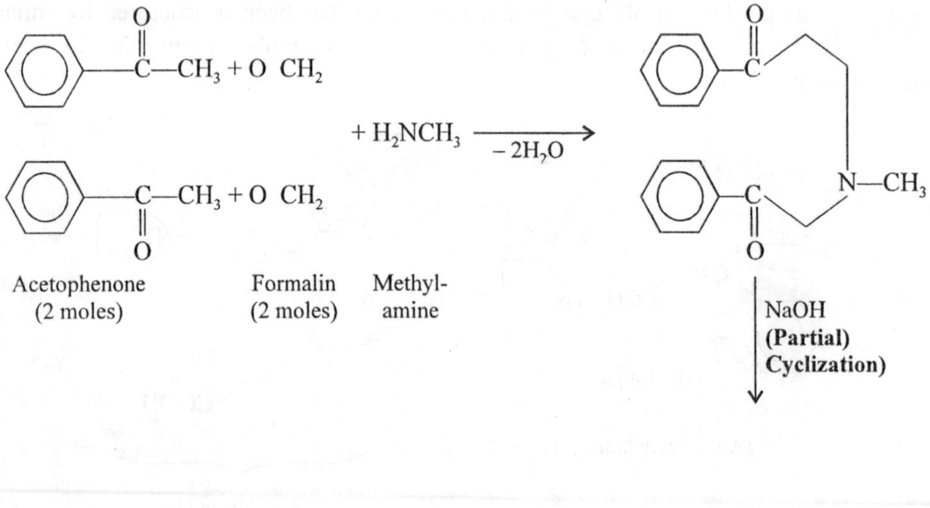

| Acetophenone | Formalin | Methyl- |
| (2 moles) | (2 moles) | amine |

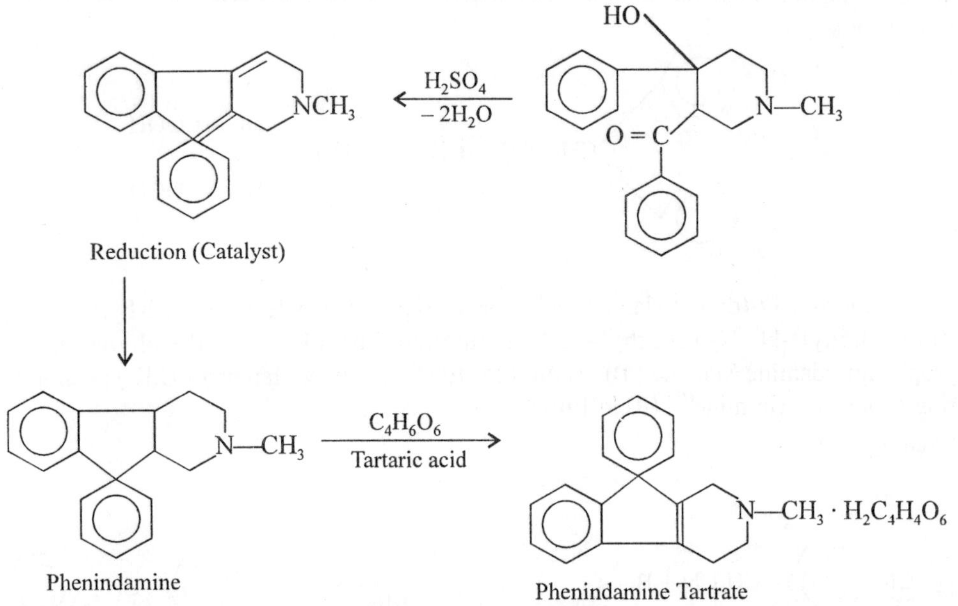

Reduction (Catalyst)

Phenindamine

Phenindamine Tartrate

Two moles each of acetophenone, formaldehyde and one mole of methylamine are condensed to form an intermediate compound with the loss of two moles of water. This intermediate product on treatment with sodium hydroxide undergoes partial cyclization. On further treatment of the resulting product with concentrated sulphuric acid the complete cyclization takes place thereby losing two moles of water. This compound on catalytic reduction yields phenindamine which on treatment with an equimolar proportion of tartaric acid yields phenindamine tartrate.

It is less effective than promethazine but it does not generally produce drowsiness and may even cause a mild stimulation.

Dose : 75 to 150 mg per day.

B. *Triprolidine Hydrochloride* BAN, USAN, *Triprolidine* INN,

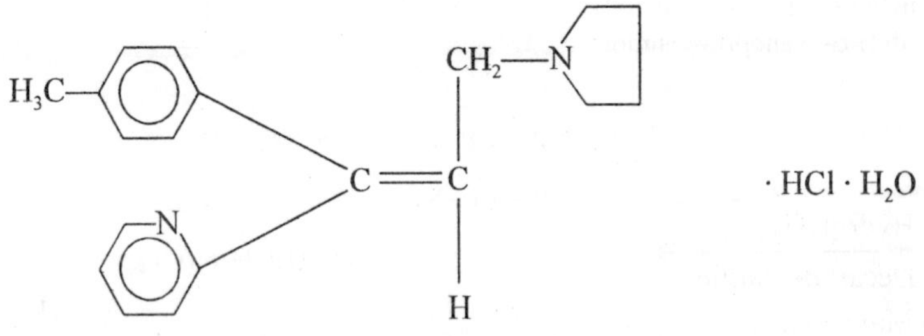

(E)-2 [3-(1-Pyrrolidinyl)-1-*p*-tolylpropenyl] pyridine monohydrochloride monohydrate ; BP ; USP ; Actidil[(R)] (Burroughs Wellcome)

It is one of the more potent of the antihistaminics and its action last for up to 12 hours.

Dose : 5 to 7.5 mg per day.

C. *Chlorpheniramine Maleate* BAN, USAN, *Chlorphenamine* INN,

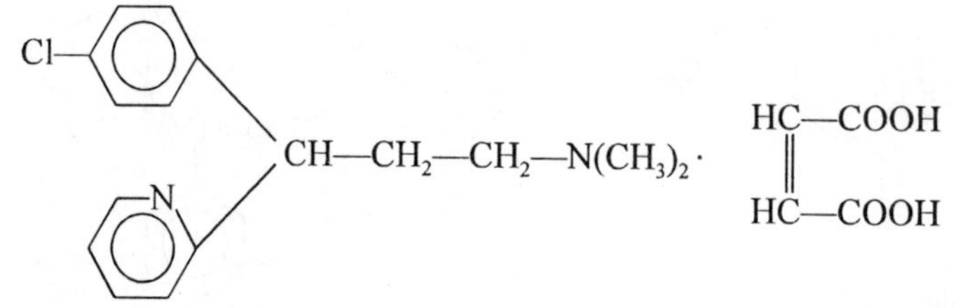

2-[*p*-Chloro-α-[2-(dimethylamino) ethyl] benzyl] pyridine maleate (1:1) ; 2-Pyridinepropanamine, γ-(4-chlorophenyl)-N, N-dimethyl-, (Z)-2-butenedioate (1:1) ; Chlorphenamine Maleate ; Chlorprophenpyridamine Maleate ; BP ; USP ; Piriton[R] (Allen & Hanburys U.K.) ; Chlor-Trimeton[R] (Schering-Plough) ; Alermine[R] (Reid-Provident)

Synthesis :

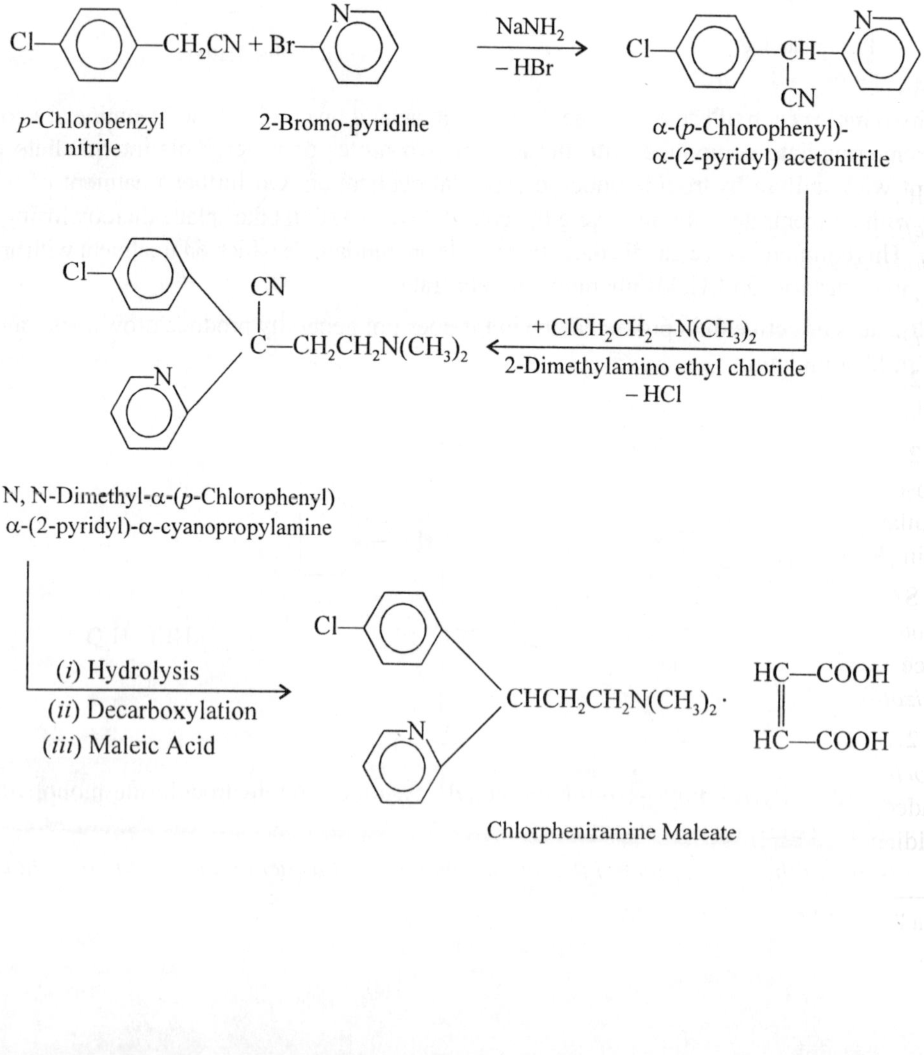

Chlorpheniramine Maleate

α-(p-Chlorophenyl)-α-(2-pyridyl) acetonitrile is prepared by the interaction of p-chlorobenzyl nitrile and 2-bromopyridine in the presence of sodamide with the elimination of a molecule of hydrogen bromide. This on treatment with 2-dimethylamino ethyl chloride abstracts the lonely active hydrogen atom present in the acetonitrile function as hydrogen chloride and yields N, N-dimethyl-α-(p-chlorophenyl)-α-(2-pyridyl)-α-cyano-propyl amine. The resulting product when subjected to hydrolysis, followed by decarboxylation and finally heated with maleic acid gives rise to the official compound.

It is one of the most potent of the antihistaminics which generally causes less sedation than promethazine.

Dose : Usual, oral, 4 mg 3 or 4 times per day.

D. *Cyproheptadine Hydrochloride* BAN, USAN, *Cyproheptadine* INN,

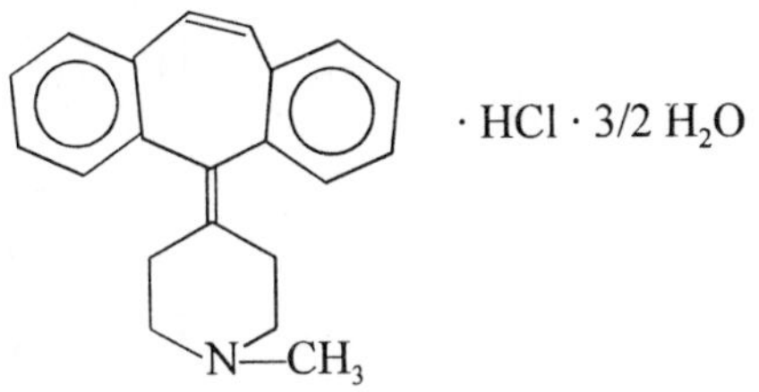

· HCl · 3/2 H$_2$O

4-(5H-Dibenzo [a, d] cyclohepten-5-ylidene)-1-methylpiperidine hydrochloride sequihydrate ; Piperidine, 4-(5H-dibenzo [a, d]-cyclohepten-5-ylidene)-1-methyl-, hydrochloride, sesquihydrate ; BP ; USP ; Periactin[R] (Merck Sharp & Dohme)

It possesses antiserotonin, anticholinergic and antialdosterone characteristics. It is *highly potent and is effective in smaller doses than promethazine hydrochloride though the effect lasts for a short duration. It helps to stimulate the appetite in under-weight patients and those suffering from anorexia nervosa.*

Dose : Usual, adult, oral, 4 mg 3 to 4 times a day.

2.1.7.1. Mechanism of Action. The mechanism of action of certain important members of this particular class of compounds shall now be discussed as under :

2.1.7.1.1. Phenindamine Tartrate. The *'drug'* exerts its action by temporarily relieving running nose and also sneezing related to the common cold. However, it may cause drowsiness sleepiness, just similar to most of the antihistaminics ; but at the same time it may also give rise to a mild stimulating action in patients and may cause insomnia when taken prior to going to bed.

SAR of Phenindamine. Phenindamine may be structurally related to an unsaturated propylamine analogue wherein the rigid ring system essentially embedded with a distorted *trans*-alkene system. The presence of either oxidizing substances or an exposure to heat may render this compound to undergo *isomerization* which ultimately leads to an **inactive** form.

2.1.7.1.2. Triprolidine Hydrochloride. The *'drug'* is shown to exhibit both high *activity* and *superiority* of its (E)-isomer with respect to its corresponding (Z)-isomer as the H$_1$-antagonists.* It has been adequately demonstrated using the guinea pig ileum sites that the actual prevailing affinity of triprolidine for the H$_1$-receptors was found to be higher more than 1,000 times the affinity exhibited

---

*Ison RR *et. al. J. Pharm. Pharmacol,* **25**, 887, 1973.

by its (Z)-isomer. However, the overall relative potency of this *'drug'* is very much comparable to that of dexchlorpheniramine.

SAR of Triprolidine. It has been established that the pharmacoligic activity solely resides in the geometric isomer wherein the pyrrolidinomethyl moiety is present as *trans*-to the corresponding 2-pyridyl functional moiety as given below :

Triprolidine	Triprolidine
(*trans*-Form)	(*cis*-Form)
Active	**Inactive**

2.1.7.1.3. Chlorpheniramine Maleate. The *'drug'* is widely used as an essential component of a plethora of *antitussive formulations*. It is found to attain appreciable first-pass metabolism ranging between 40–55%. The peak plasma levels of 5.9 and 11 ng. mL^{-1} are accomplished within a span of 2–6 hours.

SAR of Chlorpheniramine. The presence of the strategically positioned chloro moiety at the *para*-position of the benzene ring affords a 10-times enhancement in its potency without making an significant alteration in its toxicity. Besides, it has been observed that the maximum activity of the **'drug'** resides in the *dextro*-enantiomorph exclusively.

2.1.7.1.4. Cyproheptadine Hydrochloride. The *'drug'* is an antihistamine having serotonin-antagonist, and calcium channel blocking activities. Besides, it also exhibits antimuscarinic and central sedative actions.

2.1.7.2. STRUCTURE ACTIVITY RELATIONSHIPS (SARs) AMONGST H_1-RECEPTOR BLOCKERS

The large number of potent **antihistaminic agents** used in the therapeutic armamentarium belong to various defined chemical categories, namely : aminoalkylethers, ethylenediamines, thiophene analogs, cyclic basic chain analogs and phenothiazine derivatives. However, it is now possible to derive some important conclusions with respect to their structural requirements for optimal activity and pharmacological actions, namely :

1. In all derivatives the terminal N atom must be tertiary amine so as to exhibit maximum activity.

2. The terminal N atom may constitute part of a heterocyclic structure, *e.g.,* pyribenzamine hydrochloride, thenylene hydrochloride, etc.

3. For maximum activity the carbon-chain between the O and N atoms or the N and N atoms must be the ethylene moiety, *i.e.,* —CH_2CH_2—. However, a long or branched chain combination gives rise to a less potent analog.

4. It is interesting to observe that in the promethazine hydrochloride molecule the two carbon chain is linked with an iso-propyl moiety, but the presence of the phenothiazine group might exert better therapeutic effect on the molecules as such.

5. Introduction of a halogen atom *viz,* Cl, Br at the *para*-position of the phenyl function improves the antihistaminic activity of the parent molecule, *e.g.,* pheniramine compared with, chloropheniramine and brompheniramine.

6. Amongst the ethylenediamine analogs many potent compounds have evolved due to the inclusion of various groups on the second N of the chain. Such groups may be either heterocyclic aromatic rings or isocyclic group. Hydrogenation of such ring(s) leads to loss in activity.

7. The nucleus of an antihistaminic must bear a minimum of two aralkyl or aryl functions or an equivalent embeded in a polycyclic ring.

8. Antihistaminics exhibiting optical isomerism revealed that the *dextro*-isomer supersedes the *levo*-in their potency, *e.g.,* dexchlorpherniramine, dexbrompheniramine, triprolidine etc.

9. For enhanced effectiveness is antihistaminics it is essential that one of the aromatic moieties is α-pyridyl while the second substituent on the N atom could be either a benzyl function of a substituted benzyl group or one of the isosteres of the benzyl moiety, *e.g.,* thenyldiamine, pyrilamine, etc.

10. Introduction of basic-cyclic ring system by altering the position of dimethyl amino group also enhances the antihistaminic activity, *e.g.,* cyclizine, chlorcyclizine, meclizine etc.

2.2. Prevention of Histamine Release

The release of histamine *in vitro* may be prevented by the aid of certain medicinally acive compounds. Such substances are not usually absorbed by the gastro-intestinal tract. An attempt has been made to find a similar substance which might be absorbed when administered orally and which possesses anti-allergic properties. Two such compounds are disucssed below namely : Sodium cromoglycate and Ketotifen fumarate.

A. *Sodium Cromoglycate* BAN, *Cromolyn Sodium* USAN, *Cromoglicic Acid* INN

Sodium 5, 5′-(2-hydroxytrimethylenedioxy) *bis* (4-oxo-4H-chromene-2-carboxylate ; Disodium 4, 4′-dioxo-5, 5′-(2-hydroxytrimethylenedioxy) di (4H-chromene-2-carboxylate) ; Sodium Cromoglycate BP ; BPC ; Cromolyn Sodium USP ; Intal[R] (Fisons ; U.K.) ; Aarane[R] (Syntex)

Sodium cromoglycate is a chromono derivative which inhibits the release of histamine and SRS-A in allergic reactions. It is mostly employed in the prophylactic treatment of asthma. It is administered only by inhalation, either alone or in conjunction with a small quantity of isoprenaline to prevent bronchospasm caused by the inhalation of the fine powder.

Dose : 20 mg, by inhalation 2 to 6 times a day.

B. *Ketotifen Fumarate* USAN, *Ketotifen* INN, BAN,

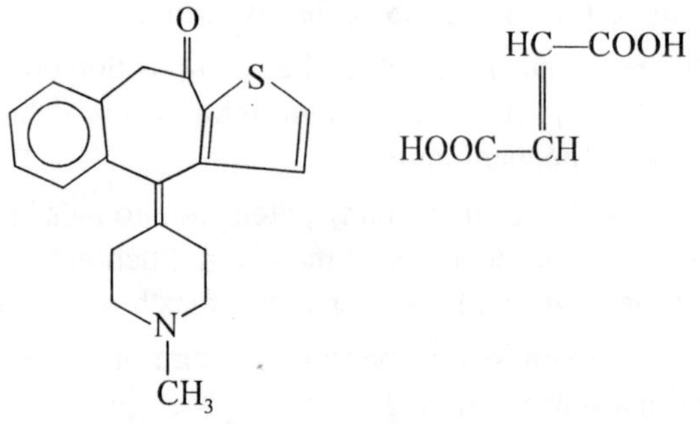

4, 9-Dihydro-4-(1-methyl-4-piperidylidene)-10H-benzo [4, 5] cyclohepta [1, 2-*b*] thiophen-10-one fumarate (1:1) ; 10H-Benzo [4, 5]-cyclohepta [1, 2-*b*] thiophen-10-one, 4, 9-dihydro-4-(1-methyl-4-piperidinylidene)-, (E)-2-butenedioate (1:1) ; USP ; Zaditen[R] (Sandoz)

It possesses anti-allergic properties similar to those of sodium cromoglycate and is used in the prophylactic treatment of asthma.

Dose : Usual, oral, equivalent to 1 mg of ketotifen 2 times per day.

2.2.1. Mechanism of Action. The mechanism of action of these compounds shall be described as under :

2.2.1.1. Sodium Cromoglycate. Though the precise mechanism of action of this *'drug'* remains uncertain, it is believed to exert its action primarily by preventing the release of mediators of inflammation from the sensitised mast cells through the stabilization of mast-cell membranes.

2.2.1.2. Ketotifen Fumarate. The *'drug'* is an antihistamine which also inhibits release of inflammatory mediators. It also exerts stabilizing action on the mast cells analogous to that of sodium cromoglycate.

2.3.   Histamine (H$_2$) Receptor Blockers

At stated earlier, histamine activates two types of receptors *viz.*, H$_1$ and H$_2$ receptors. The activation of H$_2$ receptors leads to increased gastric acid secretion, increased contraction of the isolated atria and inhibition of isolated uterus. These effects are blocked by H$_2$-receptor antagonists which are now being used for the treatment of peptic ulcer.

Examples : Cimetidine ; Ranitidine ; Oxmetidine Hydrochloride

A. *Cimetidine* INN, BAN, USAN,

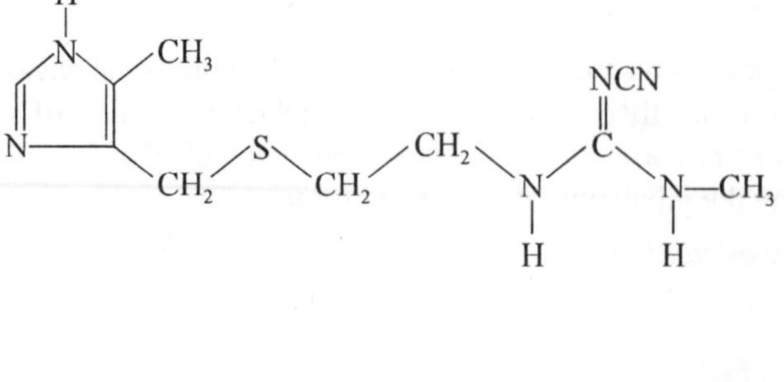

2-Cyano-1-methyl-3-[2-[[(5-methylimidazol-4-yl) methyl]-thio] ethyl] guanidine ; Guanidine, N''-cyano-N-methyl-N'-[2-[[(5-methyl-1H-imidazol-4-yl) methyl] thio] ethyl]- ; SKF 92334 ; BPC ; USP ; Tagamet$^{(R)}$ (Smith Kline & French)

Cimetidine being a histamine H_2-receptor antagonist not only inhibits gastric acid secretion but also prevents other actions of histamine mediated by H_2-receptors. *It is employed in gastric and duodenal ulcer, and in all other situations where an inhibition of the secretion of gastric juice is considered to be useful.*

Dose : Usual, oral, 200 mg 3 times per day with meals and 400 mg at night.

B. *Ranitidine* INN, BAN, USAN,

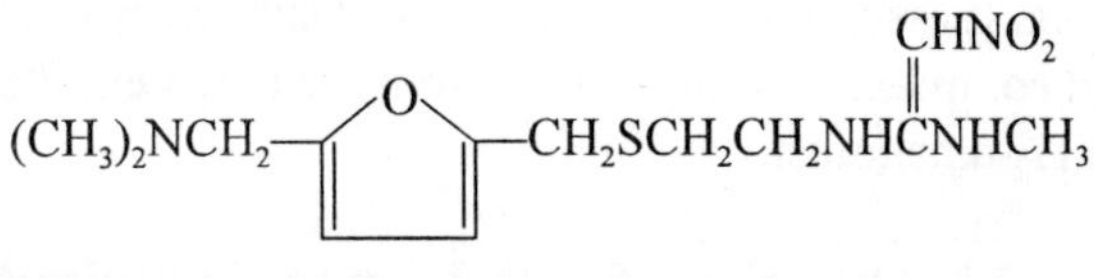

$$(CH_3)_2NCH_2 - \underset{O}{\fbox{ }} - CH_2SCH_2CH_2NHC(=CHNO_2)NHCH_3$$

N-[2-[[5-[(Dimethylamino) methyl] furfuryl] thio] ethyl]-N'-methyl-2-nitro-1, 1-ethenediamine ; 1, 1-Ethenediamine, N-[2-[[[5-[(dimethylamino) methyl]-2-furanyl] methyl] thio] ethyl]-N'-methyl-2-nitro ; Zantac$^{(R)}$ (Glaxo, U.K.)

The actions and uses of ranitidine are similar to cimetidine.

Dose : Usual, 150 mg (as the hydrochloride) 2 times a day.

C. *Oxmetidine Hydrochloride* BAN, USAN, *Oxmetidine* INN,

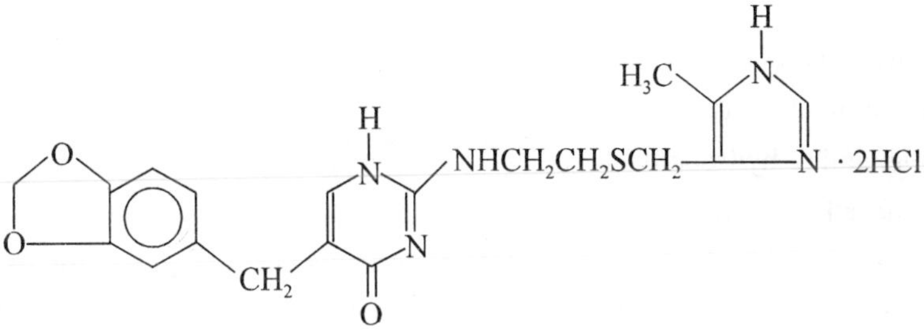

2-[[2-[[(5-Methylimidazol-4-yl) methyl] thio] ethyl] amino]-5-piperonyl-4-(1H)-pyrimidinone dihydrochloride ; SK & F 92994-A$_2^{(R)}$ (Smith Kline & French)

It is reported to have histamine H_2 receptor blocking activity.

2.3.1. Mechanism of Action. The mechanism of action of the compounds enumerated under section 14.2.3. are described as under :

2.3.1.1. Cimetidine*. The *'drug'* is observed to minimise the hepatic metabolism of such drugs that are eventually biotransformed by the cytochrome P-450 mixed oxidase system by way of either delaying the elimination or enhancing the serum levels of these pharmacologic agents.

Importantly, it exhibits relatively higher oral bioavailability (60–70%), and a plasma half-life of – 2 hour that gets enhanced particularly either in agedsubjects or those having a renal and hepatic impairment. However, it has been observed that nearly 30 to 40% of a cimetidine dose gets metabolized either as **S-oxidation** or a **5-CH$_3$ hydroxylation.** Finally, the parent drug and the metabolites are virtually eliminated primarily by the renal excretion.

---

*Introduced in 1977 and enjoyed the reputation of being one of the most profusely prescribed drugs in history for several years.

2.3.1.2. Ranitidine. Zantac gives rise to *three* known metabolites, namely : (*a*) *ranitidine N-oxide ;* (*b*) *ranitidine S-oxide ;* and (*c*) *desmethyl ranitidine.* It is observed to be only a weak inhibitor of the *hepatic cytochrome P-450* mixed function oxidation system. The plasma half-life ranges between 2 to 3 hours ; and it usually gets excreted together with its metabolites in the urine.

Note : *It is established that some 'antacids' may afford a reduction in the ranitidine absorption ; and, therefore, must not be taken at least within a span of one hour of the administration of this H_2-blocker particularly. In fact ranitidine is a stronger base (pKa 8.44).*

2.3.1.3. Oximetidine Hydrochloride. The *'drug'*, which being a 3-pyridyl structural derivative soon gained recognition by virtue of the fact that it predominantly exhibited certain H_1-antagonism.

SAR of Oximetidine. It essentially possesses a 5-substituted isocytosine functional moiety and also shows the latitude and complexitiy of digression from the urea permitted that renders the *'drug'* to become a potent selective H_2-antagonist.

Probable Questions for B. Pharm. Examinations

1. What are allergens ? What is the importance of **'antihistaminics'** in combating various types of allergic conditions ? Give suitable examples to support your answer.

2. Classify the Histamine **H_1-Receptor Antagonists.** Give the structure, chemical name and uses of **one** compound from each cateogry.

3. Name any **three 'aminoalkylethers'** being used as antihistaminics. Discuss the synthesis of **one** of them.

4. Give the structure, chemical name and uses of :
   (*a*) Mepyramine maleate and
   (*b*) Tripelenamine hydrochloride
   Describe the synthesis of any **one** drug.

5. **'Thiophene** derivatives have low side-effects and may be employed in the treatment of all types of suspected allergies'. Justify the statement with the help of at least **two** typical examples.

6. Cyclic basic-chain analogues essentially having
   (*a*) Imidazoline,
   (*b*) Piperazine and
   (*c*) Piperidine
   the above **three** types of heterocyclic nucleus give rise to potent 'antihistaminics'. Explain.

7. Discuss the synthesis oif :
   (*a*) Promethazine hydrochloride
   (*b*) Methdilazine hydrochloride
   Elaborate their applications separately.

8. Phenindamine tartrate and chlorpheniramine mealeate are two important antihistaminics. Describe the synthesis of **one** drug in detail.

9. Give a brief account of :
   (*a*) Drugs used in the 'Prevention of Histamine Release'
   (*b*) Histamine (H_2) Receptor Blocksers

10. Give a comprehensive account of :

(a) SAR-amongst H_1 -receptor blockers

(b) Mode of action of antihistaminics.

RECOMMENDED READINGS

1. RP Orange, MA Kaliner and KF Austen in : *Biochemistry of the Acute Allergic Reaction (Eds)* KF Austen and EL Becker Oxford Blackwell (1971).

2. JH Burn Antihistamine B M J ii (1955).

3. RB Hunter and DM Dunlop A Review of Antihistamine *Drugs Q J Med 25* (1948).

4. KLandsteiner in : *The Specificity of Serological Reactions,* Dover Publications, Inc. New Tork (1962).

5. JW Black and KEV Spencer in : *Cimetidine, International Symposium on Histamine H_2 Receptor Antagonists* (Eds. Cl Wood and MA Simkins) Smith Kline and French Laboratories Ltd. Welwyn Garden City, Dasprint Limited Lodon (1973).

6. DT Witiak *Antiallergic Agents in Medicinal Chemistry and Drug Discovery,* M E Wolff *(Ed)* 5th Edn. Wiley & Sons Inc., New York (1995).

7. B Idson Antihistamine Drugs *Chem Rev* 47 (1950).

8. DT Witiak Antiallergic Agents, in : *Principles of Medicinal Chemistry (Ed)* W O Foye Lea & Febiger Philadelphia (1974).

9. D Lednicer and LA Mitscher *The Organic Chemistry of Drug Synthesis* John Wiley and Sons New York (1995).

10. USAN and the USP *Dictionary of Drug Names (Ed)* M C Griffiths United States Pharmacopeial Convention Inc. Rockville (1985).

15

Non-Steroidal Anti-Inflammatory Drugs (NSAIDs)

1. INTRODUCTION

Inflammation may be defined as the series of changes that occur in living tissues following injury. The injury which is responsible for inflammation may be brought about by a variety of conditions such as : physical agents like mechanical trauma, ultra-violet or ionizing radiation ; chemical agents like organic and inorganic compounds, the toxins of various bacteria ; intracellular replication of viruses ; hypersensitivity reactions like reaction due to sensitized lymphocytes with antigenic material *viz.*, inhaled organic dusts or invasive bacteria ; and necrosis of tissues whereby inflammation is induced in the surrounding tissues.

Almost three decades ago, steroids namely : prednisolone, dexamethasone, betamethasone, triamcinoline and hydrocortisone were considered to be the drug of choice as anti-inflammatory agents. Owing to the several adverse effects caused by either short-term or long-term steroid therapy, these have been more or less replaced by much safer and better tolerated **non-steroidal anti-inflammatory drugs (NSAIDs).**

The seriousness and enormous after effects of steroid therapy necessitated an accelerated research towards the development of non-steroidal anti-inflammatory drugs since the past three decades. A good number of these agents have been put into clinical usage widely and confidently thereby exhibiting positive therapeutic efficacy accompanied with fewer untoward reactions.

The mechanism of action principally responsible for most of the NSAIDs seems to be inhibition of prostaglandin synthesis by causing almost complete blockade of the activity of the precursor enzyme, *cyclogenase.* In fact, there are two *isozymes* that have been duly recognized for the cyclo-oxygenase enzyme (*viz.*, COX-1 and COX-2). However, both *isozymes* practically perform the same reactions, but COX-1 is the isozyme that is found to be *active* under normal healthy conditions. Importantly, in rheumatoid arthritis, COX-2, which is usually found to be quite dormant, gets duly activated and yields a substantial quantum of inflammatory prostaglandins. Based on these critical facts and observations a vigorous concerted effort is being geared up to develop such newer drug substances that are specifically selective for the COX-2 isozyme, with a view to arrest particularly the production of the inflammatory prostaglandins.

In general, there exists virtually very little difference between the therapeutic efficacy of different NSAIDs, as certain patients would respond to one *'drug'* better than another. In reality, it is almost difficult to predict the best suitable drug for a patient ; thus, it invariably necessitates to arrive at the *best-fit-drug via* trial and error only.

Keeping in view the innumerable adverse side effects caused by the NSAIDs their clinical usefulness are restricted drastically. Therefore, patients who are taking such drugs for a relatively longer periods should have periodic white-blood cell counts as well as determinations of serum creatinine levels, besides hepatic enzyme activities.

2. CLASSIFICATION

NSAIDs may be classified on the basis of their basic chemical structures as described below along with various classical examples belonging to each category, namely :

1. Heteroarylacetic acid analogues 2. Arylacetic acid analogues
3. Arylpropionic acid analogues 4. Naphthalene acetic acid analogues
5. Gold compounds 6. Miscellaneous anti-inflammatory drugs
7. Salicylic acid analogues and 8. Pyrazolones and pyrazolodiones.

2.1. Heteroarylacetic Acid Analogues

This constitutes an important class of non-steroidal anti-inflammatory drugs which have gained recognition in the recent past. A few classical examples of this group are, indomethacin ; sulindac ; tolmetin sodium ; zomepirac sodium ;

A. *Indomethacin* BAN, USAN, *Indomethacin* INN,

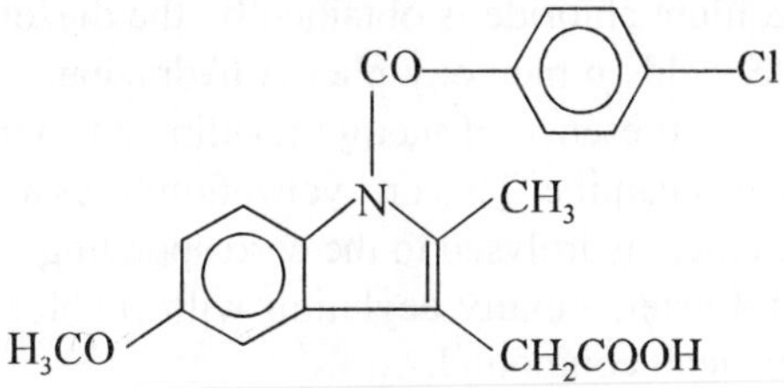

1-(*p*-Chlorobenzoyl)-5-methoxy-2-methylindole-3-acetic acid ; 1H-Indole-3-acetic acid, 1-(4-chlorobenzoyl)-5-methoxy-2-methyl- ; BP ; USP ; Indocin[(R)] (MSD)

Synthesis

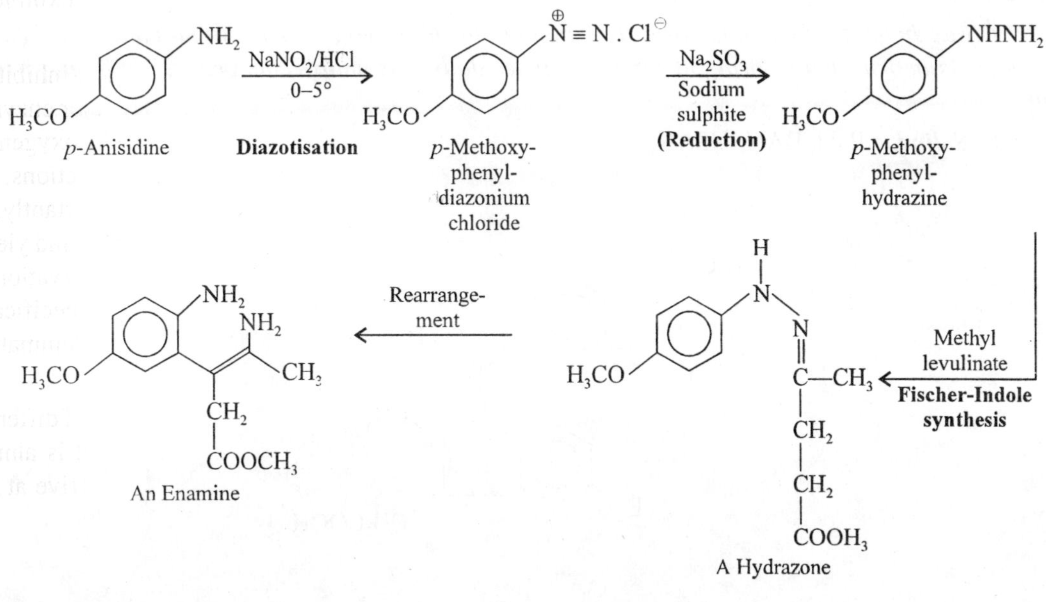

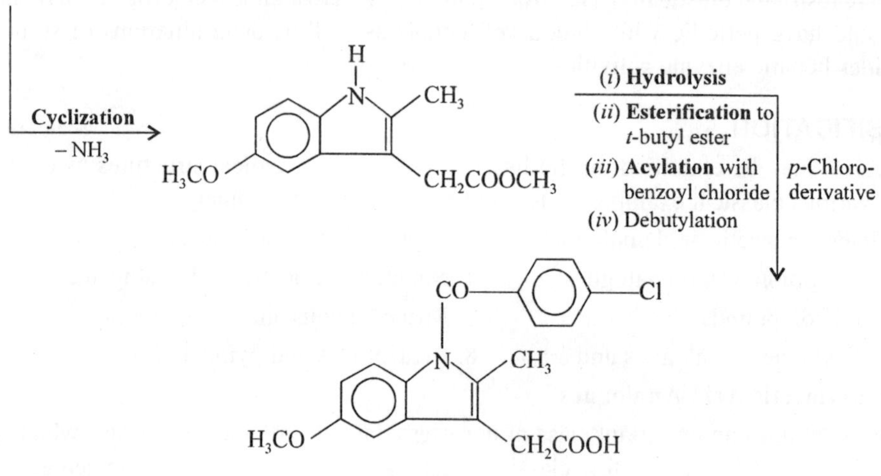

Indomethacin

p-Methoxy phenyl diazonium chloride is obtained by the diazotization of p-anisidine which on reduction with sodium sulphite yields p-methoxy phenyl hydrazine. The resulting product undergoes the Fischer-indole synthesis in the presence of methyl levulinate to form a hydrazone which on intra-molecular rearrangement gives an enamine. This on cyclization loses a molecule of ammonia and forms an intermediate compound. It is then hydrolysed to the corresponding acid which is re-esterified *via* the anhydride to give the tert-butyl ester. Finally acylation with p-chlorobenzoyl chloride followed by debutylation gives rise to the official compound.

It is a non-steroid drug possessing anti-inflammatory, antipyretic and analgestic properties. *It is usually used for the treatment of rheumatoid arthritis, ankylosing (rheumatoid) spondylitis, gouty arthritis and osteoarthritis.* It is not an ordinary simple analgesic and owing to its reasonaly serious untoward effects should be used with great *caution.*

Dose : In gout, usual, adult, oral, 100 mg initially, followed by 50 mg 3 times daily until pain is relieved ; As antirheumatic, oral, 50 mg 2 or 3 times daily ; As antipyretic, oral, 25 to 50 mg 3 times daily.

B. *Sulindac* INN, BAN, USAN,

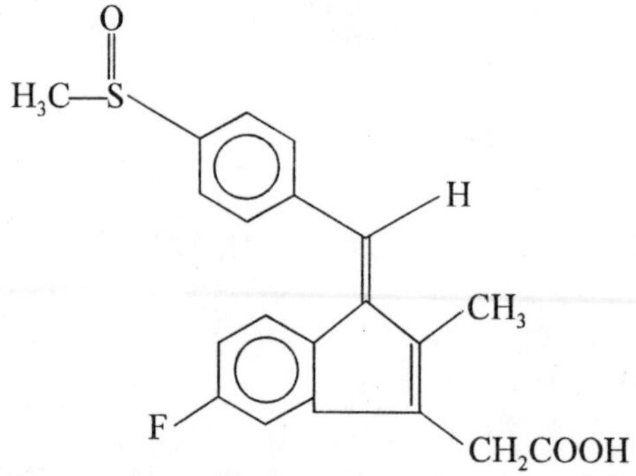

cis-5-Fluoro-2-methyl-1-[(*p*-methylsulfinyl) benzylidene] indene-3-acetic acid ; 1H-Indene-3-acetic acid, 5-fluoro-2-methyl-1-[[4-(methylsulfinyl) phenyl] methylene]-, (Z)-; USP ; Clinoril[R] (MSD).

It is a fluorinated indene with a structural resemblance to indomethacin. It has anti-inflammatory, analgesic and antipyretic properties. *It is usually employed in the treatment of rheumatic and musculoskeletal disorders ; and also for severe and long-term relief of signs and symptoms of acute painful shoulder, acute gouty arthritis and osteoarthritis.*

Dose : Usual, adult, oral, 150 mg twice a day with food.

C. *Tolmetin Sodium* BAN, USAN, *Tolmetin* INN,

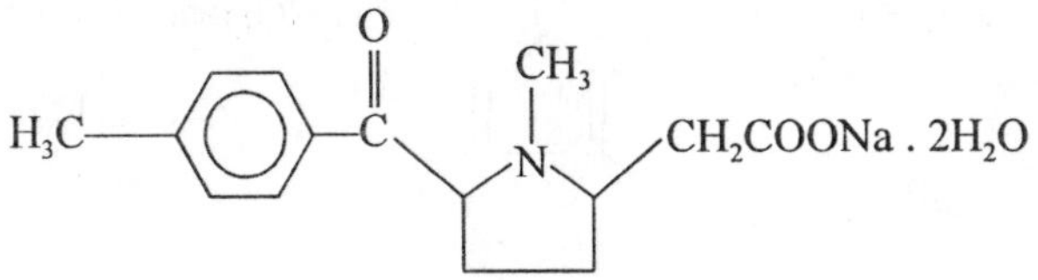

Sodium 1-methyl-5-*p*-toluoylpyrrole-2-acetate dihydrate ; 1H-Pyrrole-2-acetic acid, 1-methyl-5-(4-methylbenzoyl)-, sodium salt, dihydrate ; USP ; Tolectin[R] (McNeil).

Synthesis

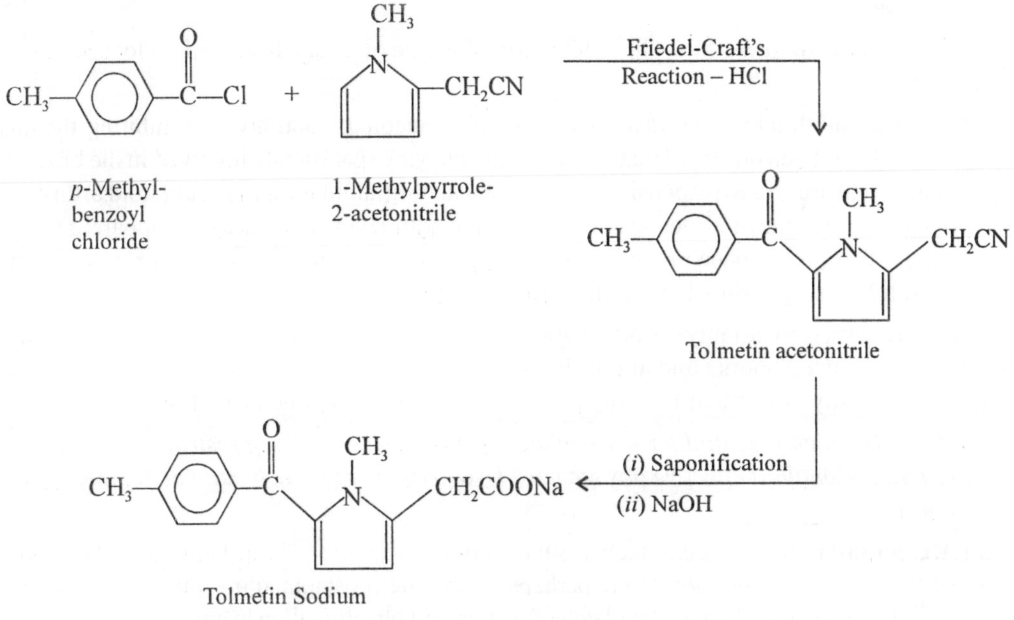

p-Methyl-
benzoyl
chloride

1-Methylpyrrole-
2-acetonitrile

Tolmetin acetonitrile

(i) Saponification
(ii) NaOH

Tolmetin Sodium

Tolmetin acetonitrile may be prepared by the Friedel-Craft's reaction between *p*-methylbenzoyl chloride and 1-methylpyrrole-2-acetonitrile with the elimination of a mole of hydrochloric acid. The resulting product after appropriate separation from its 4-aroyl isomer is finally subjected to saponification followed by conversion to its sodium salt.

It has antipyretic, analgesic and anti-inflammatory actions. It is employed in the *treatment of rheumatic and other musculoskeletal disorders.* The drug is, however, comparable to indomethacin and aspirin in the control and management of disease activity.

Dose : (Equivalent to tolmetin) Adult, oral, initial, 400 mg 3 times per day, subsequently adjusted to patient's response.

D. *Zomepirac Sodium* BAN, USAN, *Zomepirac* INN,

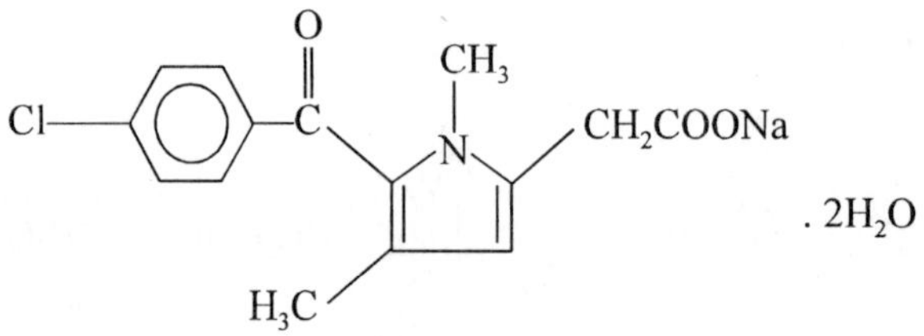

Sodium 5-(*p*-chlorobenzoyl)-1, 4-dimethylpyrrole-2-acetate dihydrate ; 1H-Pyrrole-2-acetic acid, 5-(4-chlorobenzoyl)-1, 4-dimethyl-, sodium salt, dihydrate ; USP ; Zomax[(R)] (McNeil).

It is an analgesic and anti-inflammatory drug structurally very similar to tolmetin sodium. It is normally used in mild to moderate pain, including that of musculoskeletal disorders.

Dose : (Zomepirac sodium 1.2g is approximately equivalent to 1g of zomepirac) ; 400 to 600 mg of zomepirac daily.

2.1.1. Mechanism of Action. The mechanism of action of drugs discussed under section 15.2.1. are as under :

2.1.1.1. Indomethacin. The *'drug'* exerts its pharmacologic activity by inhibiting the enzyme *cyclo-oxygenase.* It has been observed that this aforesaid enzyme specifically involved in the biosynthesis of prostaglandins that are solely responsible for the pain and inflammation of rheumatoid arthritis ; and, therefore, inhibiting the 'enzyme' decreases the prostaglandin levels and eases the apparent symptoms of the disease. Besides, it has been proved beyond any reasonable doubt that the *'drug'* also inhibits the synthesis of useful prostaglandins both in the GI-tract and kidney.

Indomethacin, is invariably absorbed quite rapidly after oral administration ; peak plasma levels are accomplished in just 2 hours ; and almost 97% of the drug is protein bound. It has a half-life of 2.6 to 11.2 hours ; and only 10-20% of the drug gets excreted practically unchanged in the urine.

Caution. *The high potential for dose-related adverse reactions both restrains as well as makes it imperative that the smallest effective dosage must be determined for each individual patient carefully and meticulously.*

2.1.1.2. Sulindac. The precise mechanism of action of the *'drug'* is still unknown. However, it is believed that the *'sulphide metabolite'* may perhaps inhibit the prostaglandin synthesis. Interestingly, it exerts appreciably much less effect on the platelet function and bleeding time in comparison to *'aspirin',* it must be used with great caution in such patients who could be affected quite adversely by this sort of action.

The *'drug'* gets absorbed invariably to the extent of 90% after the oral administration. The peak plasma levels are usually accomplished in about 2 hour in the fasting patient and may be extended between 3-4 hours when given with food. It has been duly observed that the mean half-life of sulindac is 7 = 8 hours ; and the mean half-life of the corresponding sulphide-metabolite is 16.4 hour (almost double than the parent drug).

2.1.1.3. Tolmetin Sodium. The exact mechanism of action of the *'drug'* is not yet established, although inhibition of prostaglandin synthesis most probably contributes heavily to its anti-inflammatory activity. It has been observed adequately that in such patients having rheumatoid arthritis different types of manifestations of its anti-inflammatory and analgesic actions do occur, but there exists little distinct proof of alteration of the progressive course of the prevailing disease.

The *'drug'* is usually absorbed not only rapidly but also completely having peak-plasma levels being attained within a span of 30-60 minutes after an oral therapeutic after a dosage regimen (40 mcg. mL^{-1}) after a 400mg dosage). Besides, it gets bound nearly to 99% to the plasma proteins ; whereas, the mean plasma-life is almost 1 hour. Importantly, all of a dose gets excreted in the urine within 1 day, either as conjugates of the parent drug *'tolmetin'* or as an *inactive oxidative metabolite.*

2.1.1.4. Zomepirac Sodium. The *'drug'* happens to be the *chloro*-derivative of tolmetin ; and, therefore, it predominantly shows appreciably longer plasma levels nearly 7 hours*, thereby attributing much lesser dosing frequency. It has been demonstrated adequately that a dose ranging between 25-50mg of this *'drug'* gives relief almost equivalent to that produced by aspirin, 650mg. Interestingly, in advanced cancer subjects, oral doses of 100-200mg were as effective the moderate parenteral doses of morphine.**

Note. The *'drug'* is presently not marketed because of its severe anaphylactoid reactions.***

2.2. Arylacetic Acid Analogues

It has been observed that organic compounds which bear some sort of resemblance either with respect to their structural features or functionally often display similar biological actions. However, it may be noted with interest that by contrast there exists no such common goals between arylacetic and arylpropionic acids.

The early 1970s have withnessed the introduction of arylacetic acids into numerous beneficial antiarthritic-analgesic agents ; however, their various structural parameters are still being explored exhaustively. A few *salient features* are enumerated below :

• Indole and pyrrole acetic acid, that are also aromatic in character, are regarded as a subgroup.

• Acidic heterocyclic sulphonamide compounds also afford clinically important NSAIDs.

Interestingly, all these compounds additionally show useful antipyretic activities. They all share a more or less common mechanism of action.

A few potent analogues belonging to this class of compounds are described below : ibuprofen ; ibufenac ; diclofenac sodium.

A. *Ibuprofen* INN, BAN, USAN,

$$CH_3CHCH_2-\!\!\!\bigcirc\!\!\!-CHCOOH$$
$$\overset{|}{CH_3} \qquad\qquad \overset{|}{CH_3}$$

---

*O' Neill PJ *et al. J Pharmacol Exp Therap.,* **209**, 366, 1979.

**Wallenstein SL, *Unpublished Report.*

***An agent producing anaphylactic reactions.

(±)-*p*-Isobutylhydratropic acid ; Benzeneacetic acid, α-methyl-4-(2-methyl-propyl), (±)-; BP ; USP ; Motrin[R] (Upjohn) ; Brufen[R] (Boots) ; Nuprin[R] (Bristol-Myers) ; Advil[R] (American Home Prod.).

Synthesis

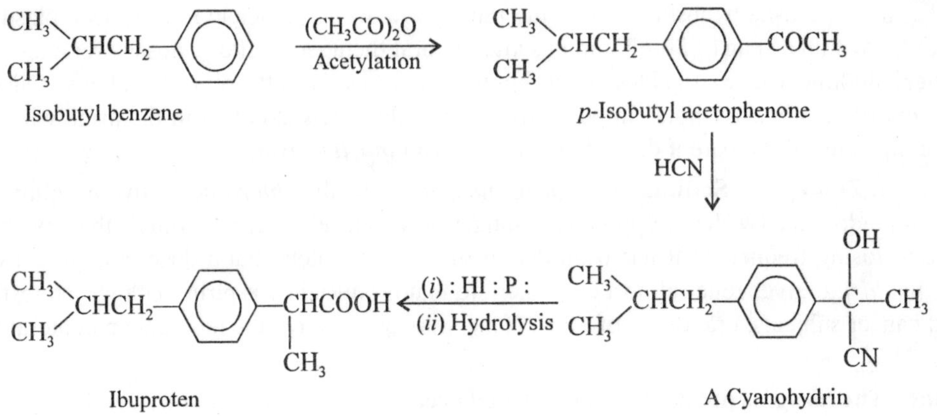

Isobutyl benzene   → (Acetylation, (CH₃CO)₂O)   *p*-Isobutyl acetophenone   → (HCN)   A Cyanohydrin   → ((i) : HI : P : (ii) Hydrolysis)   Ibuproten

p-Isobutyl acetophenone is prepared by the acetylation of isobutyl benzene which upon treatment with hydrocyanic acid yields the corresponding cyanohydrin. This on heating with hydrogen iodide in the presence of red phosphorous helps to reduce the benzylic hydroxyl moiety ; further hydrolysis of the nitrile groups gives the official compound.

It is an anti-inflammatory drug that possesses anti-pyretic and analgesic actions. It is indicated for the *treatment of rheumatoid arthritis and osteoarthritis*. It is also recommended *to arrest acute flares and in the long-term management of these diseases.*

Dose : Usual, oral adult, analgesia (dysmenorrhea), 200 to 400mg 4 to 6 times per day ; in rheumatoid arthritis, osteoarthritis, 300 to 400mg 3 oe 4 times daily.

B. *Ibufenac* INN, BAN, USAN,

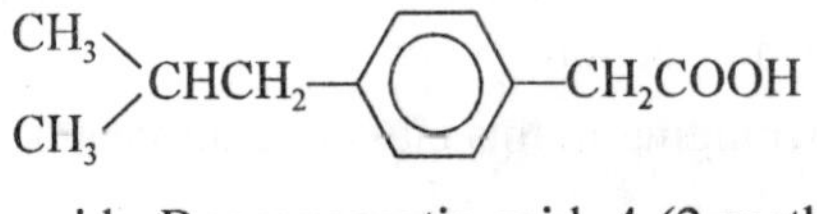

(*p*-Isobutyl-phenyl) acetic acid ; Benzeneacetic acid, 4-(2-methylpropyl); Dytransin[R] (Boots).

Synthesis

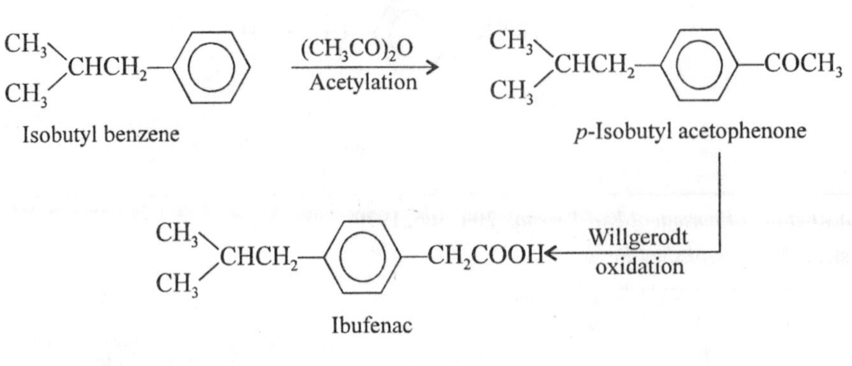

Isobutyl benzene   → (Acetylation, (CH₃CO)₂O)   *p*-Isobutyl acetophenone   → (Willgerodt oxidation)   Ibufenac

The *p*-isobutyl acetophenone obtained by the acetylation of isobutylbenzene is subjected to Wilgerodt oxidation to yield ibufenac.

It has anti-inflammatory, analgesic and antipyretic actions. It was formerly employed in the rheumatic conditions but was found to cause hepatotoxicity.

C. *Diclofenac Sodium* BAN, USAN, *Diclofenac* INN,

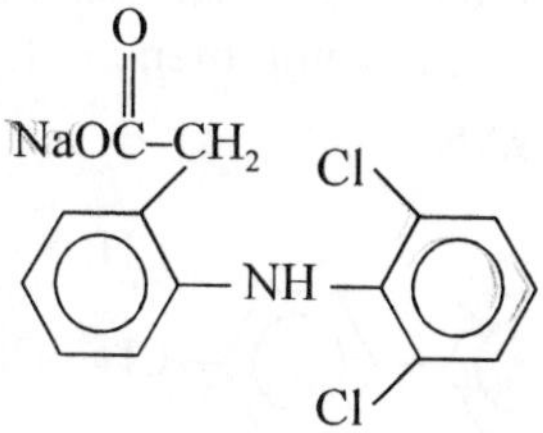

Sodium [*o*-(2, 6-dichloroanilino) phenyl] acetate ; Benzene-acetic acid, 2-[(2, 6-dichlorophenyl) amino]-, monosodium salt ; Dichlorophenac sodium ; Voltaren[R] (Ciba-Geigy) ;

It is a phenylacetic acid derivative which has analgesic, antipyretic and anti-inflammatory actions. It is mostly employed in the *treatment of rheumatoid arthritis and other rheumatic disorders.*

Dose : 20 to 50mg 3 times day. It is also given as a suppository.

2.2.1. Mechanism of Action. The mechanism of action of compounds described under section 15.2.2 shall be dealt with in the sections that follows :

2.2.1.1. Ibuprofen. The *'drug'* seems to be fairly comparable to *'asprin'* in the control, management and treatment of rheumatoid arthritis having a distinct and noticeable lower incidence of side effects.* It has been amply proved and established that the pharmacologic activity of this *'drug'* exclusively resides in the S–(+)-isomer not only in *ibuprofen* but also throughout the arylacetic acid series. Nevertheless, these strategic isomers are exclusively responsible for causing more potent inhibition of the prostaglandin synthetase.

The *'drug'* gets absorbed quite fast after the oral administration ; and evidently the peak plasma serum levels generally are attainable within a span of 1 to 2 hour. It is usually metabolized rapidly and eliminated in the urine. The serum half-life is quite transient ranging between 1.8 and 2.0 hour.

Note. The inclusion of '*ibuprofen*' as an OTC-drug (*i.e.*, non prescription drug) in the United States is solely based on its lack of any serious untoward problems stretched over a decade of meticulous clinical observation and experience.

2.2.1.2. Ibufenac. The *'drug'* is a precursor in which the α-methyl benzeneacetic acid (ibuprofen) is replaced with simple benzeneacetic acid function, that was abandoned on account of its severe hepatotoxicity and was found to be less potent.

2.2.1.3. Diclofenac Sodium. The *'drug'* is believed to exert a wide spectrum of its effects as a consequence of its ability to inhibit the prostaglandin synthesis noticeably. However, its anti-inflammatory activity is very much akin to other members of NSAIDs having a potency, *on weight basis,* which is nearly 2.5 times that of **indomethacin.** Likewise, *on weight basis,* its analgesic potency is about 8-16 times than that of **ibuprofen.**

---

*Dorman J *et. at. Can Med. Assoc J,* **110,** 1370, 1974.

Note. *The corresponding **potassium salt** (Voltaren$^{(R)}$; Cataflam$^{(R)}$, which is proved to be faster acting, is invariably indicated for the management of acute pain and primary dysmenorrhea. It is also specifically recommended for patients having a history of high BP.*

2.3. Arylpropionic Acid Analogues

Like the arylacetic acids the arylpropionic acid analogues also exhibit potent anti-inflammatory properties besides usual antipyretic and analgesic characteristics. A few examples of this category of compounds are discussed here, flurbiprofen ; ketoprofen ; indoprofen ; fenoprofen calcium.

A. *Flurbiprofen* INN, BAN, USAN,

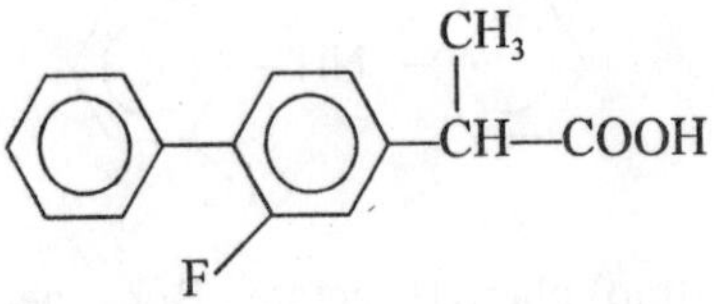

2-(2-Fluorobiphenyl-4-yl) propionic acid ; [1, 1'-Biphenyl]-4-acetic acid, 2-fluoro-α-methyl-, (±)-; (±)-2-Fluroro-α-methyl-4-biphenylacetic acid ; (±)-2-(2-Fluoro-4-biphenylyl) propionic acid ; Ansaid$^{(R)}$ (Upjohn).

Synthesis

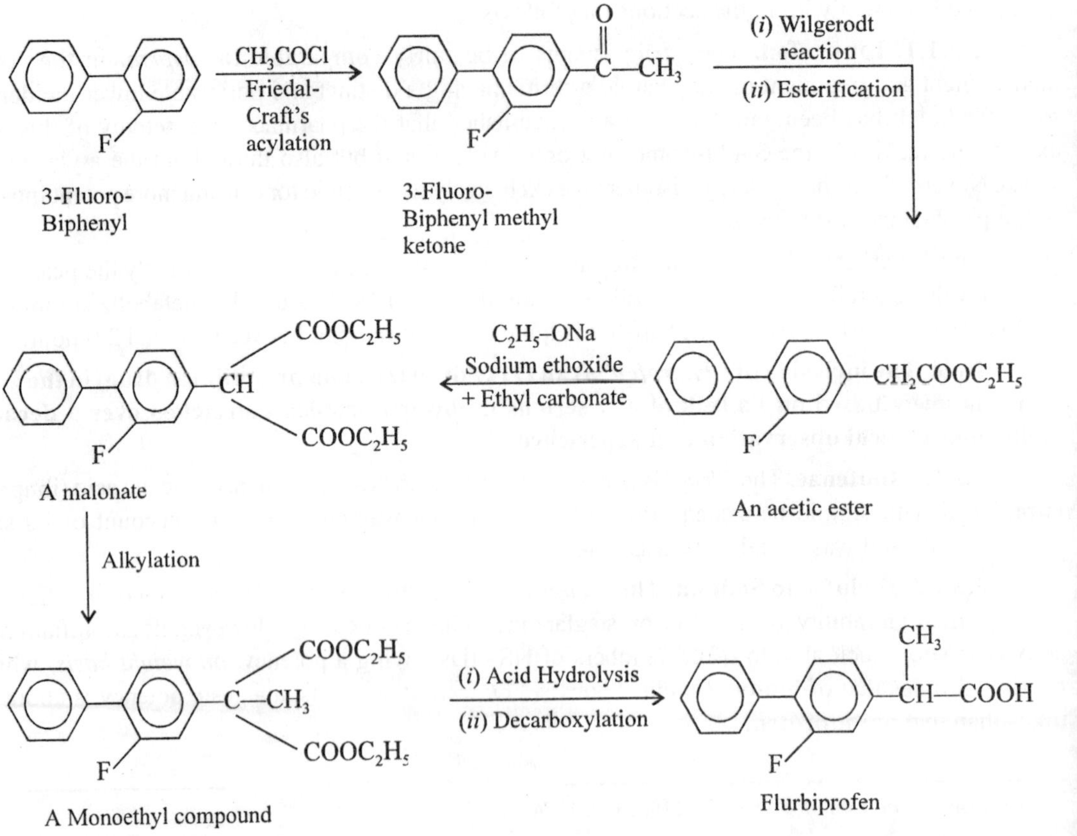

3-Fluoro biphenyl methyl ketone may be prepared by the Friedal-Craft's acylation of 3-fluorobiphenyl with acetyl chloride which upon Wilgerodt reaction followed by esterification yields the corresponding acetic ester. This on treatment with sodium ethoxide and ethyl carbonate yields a malonate which on alkylation forms a monoethyl compound. The resulting product on subsequent hydrolysis and concomitant decarboxylation yields flurbiprofen.

It is a phenylpropionic acid analogue which *possesses analgesic, anti-inflammatory and antipyuretic actions.* It is generally employed in the *treatment of rheumatoid arthritis and other rheumatic disorders.*

Dose : Usual, adult, 150 to 200mg per day in 3 to 4 divided doses.

B. *Ketoprofen* INN, BAN, USAN,

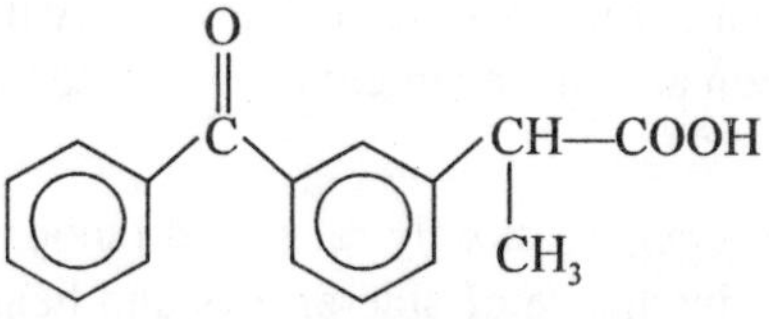

2-(3-Benzoylphenyl) propionic acid ; *m*-Benzoylhydratropic acid ; Benzeneacetic acid, 3-benzoyl-α-methyl- ; BP ; Alrheumat[R] (Bayer, U.K.) ; Orudis[R] (May & Baker, U.K.)

It is used in the *treatment of rheumatoid arthritis and osteoarthritis.*

Dose : 50 to 100mg twice daily with food.

C. *Indoprofen* INN, BAN, USAN,

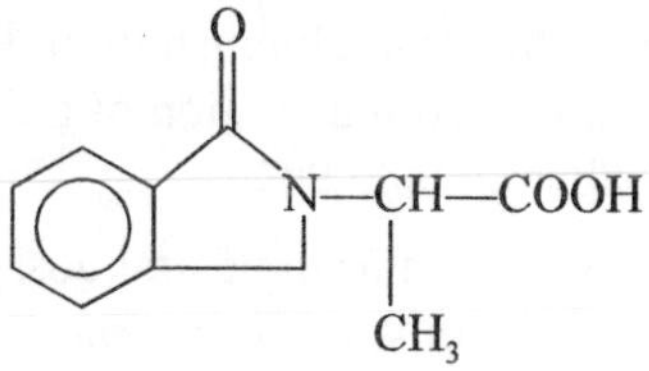

2-[4-(1-Oxoisoindolin-2-yl) phenyl] propionic acid ; *p*-(1-Oxo-2-Isoindolinyl) hydratropic acid ; Benzeneacetic acid, 4-(1, 3-dihydro-1-oxo-2H-isoindol-2-yl)-α-methyl- ; Endyne[R] (Adria).

It is generally used for the relief of various types of pain. It is also employed in the treatment of rheumatoid arthritis and osteoarthrosis.

Dose : 600 to 800mg per day in divided doses.

D. *Fenoprofen Calcium* BAN, USAN, *Fenoprofen* INN,

Calcium (±)-2-(3-phenoxyphenyl) propionate dihydrate ; Calcium (±)-*m*-phenoxyhydratropate dihydrate ; Benzeneacetic acid, α-methyl-3-phenoxy-, calcium salt dihydrate, (±)- ; BP ; USP ; Nalfon[R] (Lilly).

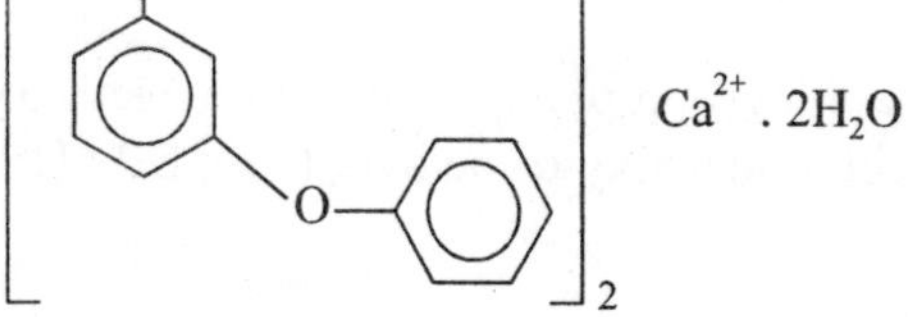

Fenoprofen calcium has *anti-inflammatory, (antiarthritic), and analgesic properties.* It has beem shown to inhibit prostaglandin synthetase. It is known to *reduce joint-swelling, decrease the duration of morning stiffness and relieve pain.* It is also indicated *for acute flares and exacerbations and in the long-term management of osteoarthritis and rheumatoid arthritis.*

Dose : (Fenoprofen equivalent) Usual, adult, oral, rheumatoid arthritis, 600mg 4 times daily ; osteoarthritis, 300 to 600mg 4 times per day.

2.3.1. Mechanism of Action. The mechanism of action of the compounds discussed under section 15.2.3. shall now be dealt with in the sections that follows :

2.3.1.1. Flurbiprofen. The *'drug'* is structurally and pharmacologically related to *fenoprofen, ibuprofen* and *ketprofen.* It is used for its specific ocular effects ; and therefore, is administered topically to the eye just before certain ocular surgeries so as to prevent any intra operative miosis. However, the exact mechanism for the prevention and management of the postoperative ocular inflammation is yet to be established.

2.3.1.2. Ketoprofen. The *'drug'* is closely related to fenoprofen in its structure and properties. Besides, it has shown a very low incidence of side-effects and hence, has been approved as an OTC-drug in US.

2.3.1.3. Indoprofen. It is a NSAID now rarely used because of its adverse rections. The *'drug'* shows carcinogenicity in *animal* studies ; and, therefore, it has been withdrawn from the market completely.

2.3.1.4. Fenoprofen Calcium. It is a propionic acid structural analogue closely related to *ibuprofen* and *naproxen.* The mechanism of action most probably relates to its inhibition of prostaglandin synthesis. The *'drug'* gets rapidly absorbed after the oral administration. Peak plasma-levels (of about 50 mcg. mL^{-1}) are attained within 2 hour after oral administration of a 600 mg dosage. The plasma half-life is nearly 3 hour. It is highly bound to albumin upto 90%.

It has been observed that nearly 90% of a single oral dosage gets eliminated within a span of 24 hours mostly as *fenoprofen glucoronide* and *4'-hydroxy fenoprofen glucuronide,* the obvious major-urinary metabolites of the *'drug'.*

2.4. Naphthalene Acetic Acid Analogues

The recent intensive quest for non-steroid anti-inflammatory drugs and arylacetic acids in particular offer a brighter scope that the naphthalene acetic acid analogues might turn out to be the leading compounds of an extensive series of promising clinical agents. *Example* : Naproxen.

A. *Naproxen* INN, BAN, USAN,

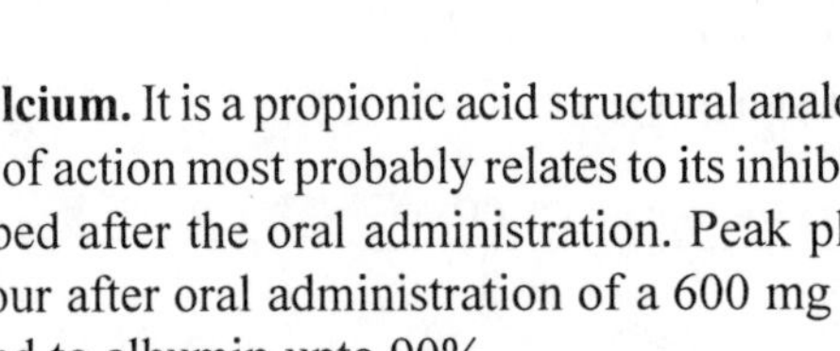

(±)-2-(6-Methoxy-2-naphthyl)-propionic acid ; (+)-6-Methoxy-α-methyl-2-naphthaleneacetic acid ; 2-Naphthaleneacetic acid, 6-methoxy-α-methyl-, (+)- ; BP ; USP ; Naprosyn(R) (Syntex).

Synthesis

2-Acetyl-6-methoxy-naphthalene may be prepared by the acylation of 6-methoxynaphthalene. The resulting product is then subjected to a series of reactions, namely ; Wilgerodt-Kindler reaction, esterification, alkylation and hydrolysis ultimately yields *DL*-Naproxen. Resolution of the resulting racemic mixture is caused through precipitation of the more potent *D*-enantiomer as the cinchonidine salt.

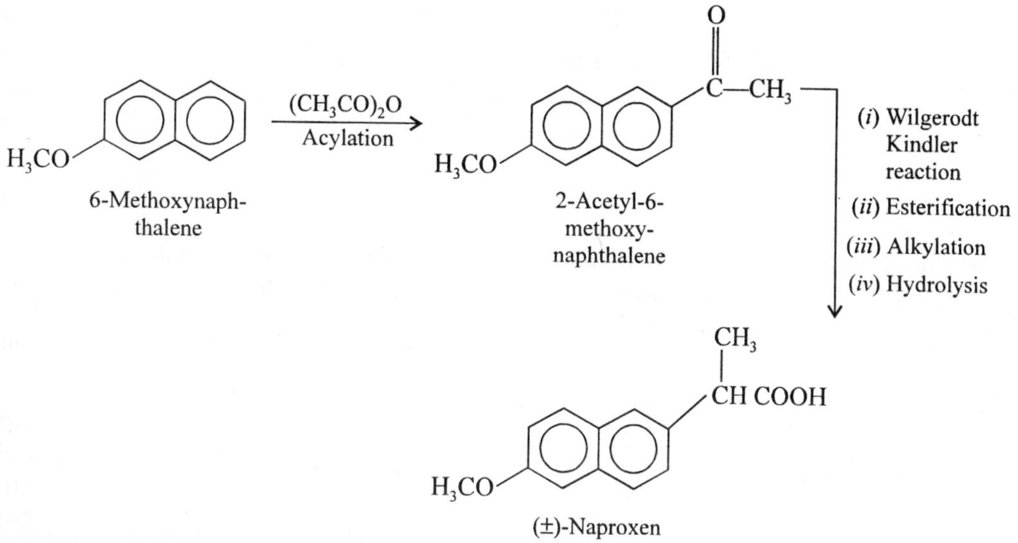

6-Methoxynaph-
thalene

2-Acetyl-6-
methoxy-
naphthalene

(i) Wilgerodt
Kindler
reaction

(ii) Esterification

(iii) Alkylation

(iv) Hydrolysis

(±)-Naproxen

It possesses analgesic, anti-inflammatory and anti-pyretic actions. It is normally used in the *treatment of rheumatic or musculoskeletal disorders, rheumatoid arthritis, dysmenorrhea, and acute gout.* However, the sodium salt is mostly employed as an analgesic for a variety of other painful conditions.

Dose : Adult, in rheumatoid arthritis, 250 to 375mg as initial dose 2 times per day ; in acute gout, 750mg as loading dose followed by 250mg 3 times a day until relieved.

2.4.1. Mechanism of Action. The mechanism of action of naproxen is described below :

2.4.1.1. Naproxen. It is a naphthalene acetic acid structural analogue available commercially as the acid and its sodium salt and is sold OTC. The *'drug'* is fairly comparable to *aspirin* both in the management and control of disease symptoms. Nevertheless, it has relatively lesser frequency and severity of nervous system together with milder GI-effects.

It is absorbed almost completely from the GI-tract after an oral administration. Peak plasma levels ($\simeq$ 55 mg.mL^{-1}) are accomplished after 4 to 5 doses at an internal of 12 hours. It has been observed that more than 99% gets bound to serum albumin. The mean plasma half-life is nearly 13 hour. About 95% of a dose gets excreted in the urine, largely as *conjugates of naproxen* and its corresponding **inactive** *metabolite 6-demethyl-naproxen.*

2.5. Gold Compounds

In general, gold compounds either suppress or prevent, but do not cure arthritis and synovitis. The use of organic gold derivatives for the treatment of rheumatoid arthritis was first reported in 1927.

However, the monovalent gold compounds bring symptomatic relief to rheumatoid arthritis in patients. A few classical examples of this class of compounds are discussed below. *Examples* : auranfin ; aurothioglucose ; aurothioglycanide ; sodium aurothiomalate.

A. *Auranofin* INN, BAN, USAN,

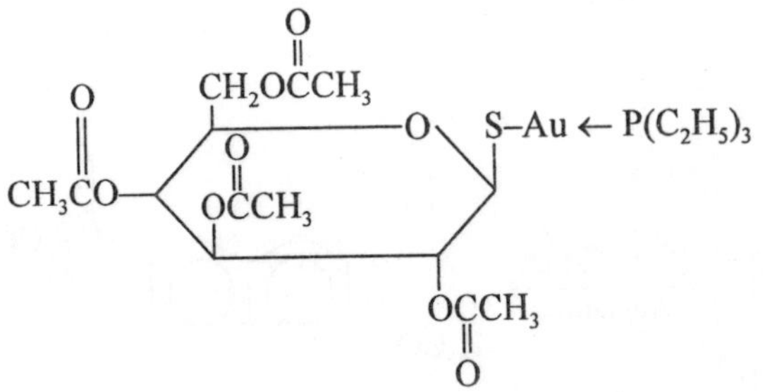

(1-Thio-β-D-glucopyranosato) (triethylphosphine) gold 2, 3, 4, 6-tetra-acetate ; Gold, (2, 3, 4, 6-tetra-*o*-acetyl-1-thio-β-D-glucopyranosato-S) (triethylphosphine)- ; Ridaura[(R)] (SK & F).

Synthesis

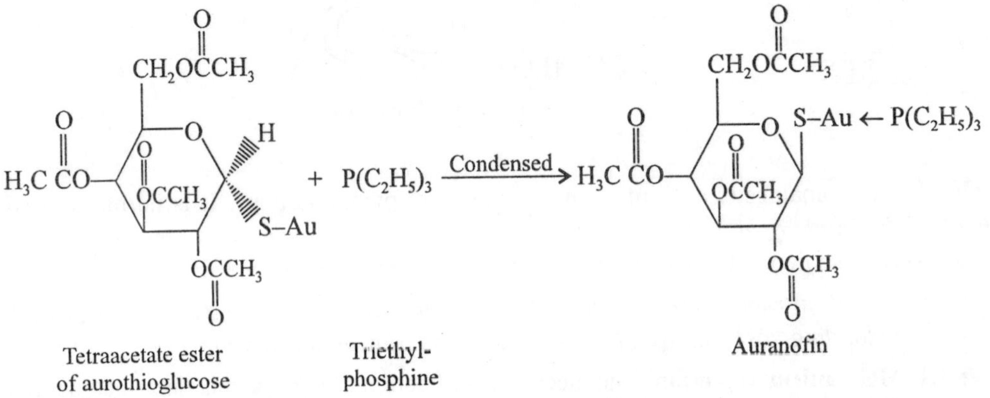

| Tetraacetate ester of aurothioglucose | Triethyl-phosphine | Auranofin |

It may be prepared by the condensation of the tetraacetate ester of aurothioglucose with triethylphosphine to yield the co-ordination complex, auranofin.

Auranofin is administered orally and is used chiefly for its *anti-inflammatory action in the cure of rheumatoid arthritis.*

Dose : *Usual, adult, oral 3mg 2 times daily.*

B. *Aurothioglucose* BAN, USAN,

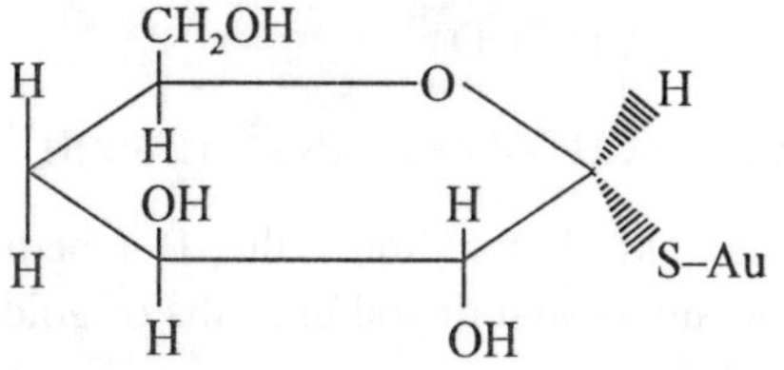

(1-Thio-D-glucopyranosato) gold ; Gold, (1-thio-D-glucopyranosato)- ; Gold Thioglucose ; (D-Glucosylthio) gold ; USP ; Solganal$^{(R)}$ (Schering-Plough).

Synthesis :

Aurothioglucose is prepared by refluxing together an aqueous solution of thioglucose and gold tribromide in the presence of sulphur dioxide. The resulting compound is precipitated, and is purified by dissolving in water and after which it is reprecipitated by the addition of alcohol.

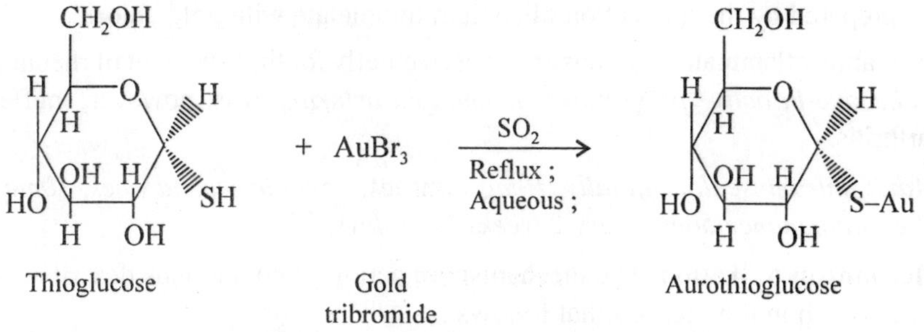

Thioglucose	Gold tribromide	Aurothioglucose

It is an antirheumatic drug employed for *treatment of active and progressing rheumatoid arthritis and nondisseminated lupus erythematosus. It has been reported that no other antirheumatic drug possesses the capability of arresting the progression of the disease, as gold can do in some cases.*

Dose : Intramuscular, administration as a suspension in oil for adult in an usual weekly dose of 10mg increasing gradually to 50mg ; children between 6 to 12 years, may be given one quarter the usual dose.

C. *Aurothioglucanide* INN, BAN, USAN,

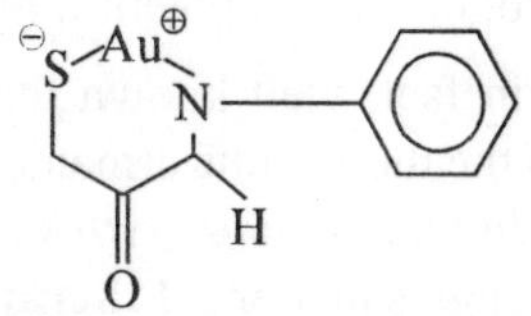

S-Gold derivative of 2-mercaptoacetanilide ; α-Auromercaptoacetanilide ; 2-Aurothio-N-phenylacetamide ;

It is used mainly for its *anti-inflammatory effect in the treatment of rheumatoid arthritis.* Being practically insoluble in water it is more gradually released and subsequently absorbed than the other water-soluble gold compounds.

D. *Sodium Aurothiomalate* INN, BAN, *Gold Sodium Thiomalate* USAN,

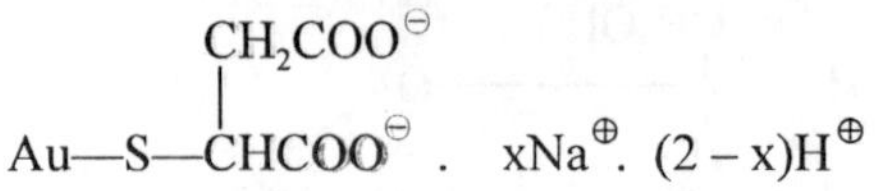

$$CH_2COO^{\ominus}$$
$$|$$
$$Au—S—CHCOO^{\ominus} \quad . \quad xNa^{\oplus}. \quad (2-x)H^{\oplus}$$

Mercaptosuccinic acid, monogold (1+) sodium salt ; Butanedioic acid, mercapto-, monogold (1+) sodium salt ; (A mixture of the mono- and di-sodium salts of gold thiomalic acid) ; Gold Sodium Thiomalate USP ; Myochrysine[R] (MSD).

Synthesis

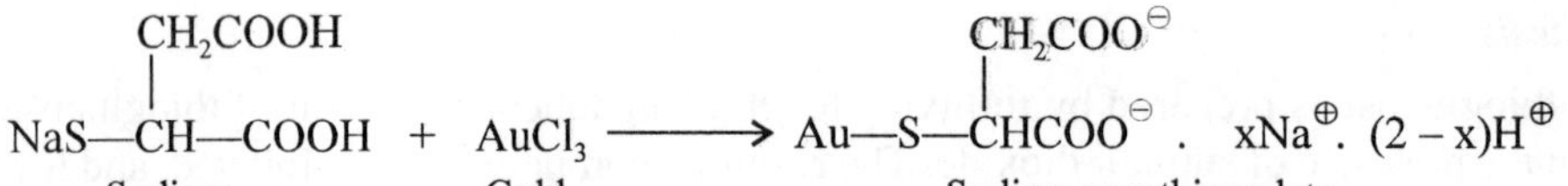

$$CH_2COOH \qquad\qquad\qquad\qquad CH_2COO^{\ominus}$$
$$| \qquad\qquad\qquad\qquad\qquad\qquad\qquad |$$
$$NaS—CH—COOH \; + \; AuCl_3 \longrightarrow Au—S—CHCOO^{\ominus} \; . \; xNa^{\oplus}. \; (2-x)H^{\oplus}$$

Sodium   Gold   Sodium aurothiomalate
thiomalate   chloride

It may be prepared by the interaction of sodium thiomalate with gold chloride.

It possesses anti-inflammatory actions and is used chiefly for the treatment of rheumatoid arthritis. It is extremely *effective in active pregressive rheumatoid arthritis*. It is, however, ineffective against other types of arthritis.

Dose : Adult, intramuscular, initially, 10mg 1st week, 25mg in second week, 50mg per week for 20 weeks, and for maintenance 50mg every 2 weeks for 4 days.

2.5.1. Mechanism of Action. The mechanism of action of compounds described under section 15.2.5 shall be dealt with in the sections that follows :

2.5.1.1. Auranofin. The value of gold salts in the rheumatoid arthritis is fairly well established ; except for this drug, all available gold preparations should be IM administered. Nearly 25% of the gold content in the *'drug'* gets absorbed. The mean terminal body half-life varies between 21 to 31 days. It has been observed that nearly 60% of the *'absorbed gold'* gets exereted in the urine ; while the remainder is excreted in the faeces.

However, the exact mechanism by which this *'drug'* exhibits its therapeutic effect in rheumatoid arthritis is still not properly understood, although there are ample evidences whereby the *'drug'* does affect a plethora of cellular processes directly linked with inflammation. Importantly, in contrast to the parenteral gold preparations, it is not recognized as a potent inhibitor of sulphydryl moiety reactivity.

2.5.1.2. Aurothioglucose. It is, in fact, well known that once the *'adrenal steroids'* mostly displaced for the *'gold compounds'* from the therapeutic armamentarium for the treatment of active and progressive rheumatoid arthritis and disseminated *lupus erythematosus*. * However, bearing in mind the recognition of the numerous hazardous dangers of **steroid therapy** and the potential curative properties has virtually restored the usage of gold compounds. It has been duly demonstrated that no other *'antirheumatic drug'* is as capable of arresting the progression of the disease, as gold compounds can do in certain instances.

---

*A chronic autoimmune inflammatory disease involving multiple organ systems and marked by periods of exacerbation and remission.

The best therapy normally takes place when the *'drug'* is employed almost in the early active stages of the disease, and also it is solely based on the daily excretion rate of gold in an individual patient.

It has been observed that the *'drug'* invariably comprises of 50% gold, time to peak effect is 4-6 hours, almost 95-99% gets bound to plasma protein, plasma half-life after a single dose varies from 3 to 27 days ; and finally about 70% is excreted in the urine and 30% in the faeces.

2.5.1.3. Aurothioglycanide. It is one of the sulphur containing gold compounds with a heterocyclic moiety in which the gold (Au) is imbeded strategically. The *'drug'* gets absorbed *in vivo* rather slowly by virtue of its poor solubility in water.

2.5.1.4. Sodium Aurothiomalate. The *'drug'* gets absorbed rapidly after the intramuscular injection and 85 to 95% becomes bound to plasma proteins. It is widely distributed to body tissues and fluids, including synovial fluid, and hence accumulates in the body. The serum half-life of gold is nearly 5/6 days ; however, it increases after successive doses and after a complete course of treatment, gold may be seen in the urine even upto 1 year or more due to its presence in deep body compartments. It is mainly excreted in the urine, with similar quantum in the faeces.

2.6. Miscellaneous Anti-Inflammatory Drugs

There are a number of compounds which incidentally do not fall into any of the categories mentioned so far but they possess anti-inflammatory actions. A few such compounds are described here.

2.6.1. Antimalarial Agents

Chloroquine and hydroxychloroquine belonging to the class of 4-amino-quinoline anthmalarials are being used in clinical practice in the cure and treatment of rheumatoid arthritis since 1957. However, the two important disadvantageous factors, namely : slow onset of therapeutic effect and significant ocular toxicity seemed to have shadowed the clinical supremacy of these drugs.

2.6.2. Uricosuric Agents

Such drugs that help in the enhanced excretion of excess uric acid through urination and thus reduce the urea concentration in the plasma are known as **uricosuric agents.** There are two important agents which are frequently used in hyperuricemia *viz.,* sulfinpurazone and probenecid both of which enhance the level of penicillin in plasma by inhibiting its secretion. The former agent has already been dealt under antipyretic analgesics in pyrazolones and pyrazolodiones ; the latter will be discussed here.

A. *Probenecid* INN, BAN, USAN,

p-(Dipropyl-sulfamoyl) benzoic acid ; Benzoic acid, 4-[(dipropylamino) sulfonyl]-; BP ; USP ; Int. P., Benemid[R] (MSD) ; SK-Probenecid[R] (SKF).

$$(CH_3CH_2CH_2)_2NSO_2-\text{\Large\bigcirc}-COOH$$

Synthesis

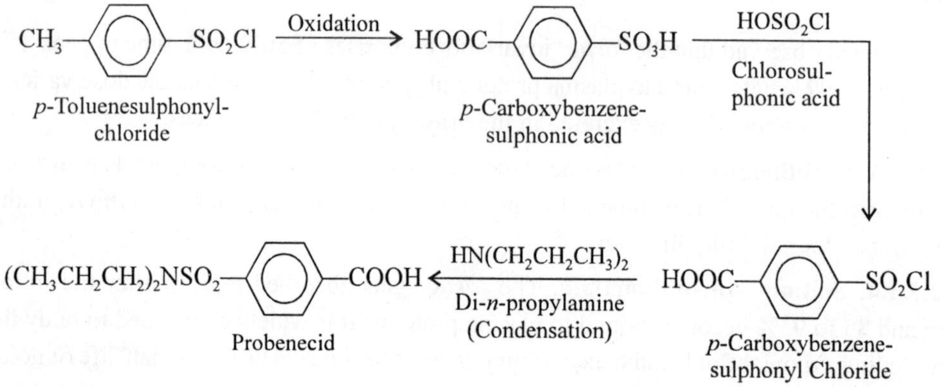

CH$_3$—⟨O⟩—SO$_2$Cl $\xrightarrow{\text{Oxidation}}$ HOOC—⟨O⟩—SO$_3$H $\xrightarrow{\substack{\text{HOSO}_2\text{Cl} \\ \text{Chlorosul-} \\ \text{phonic acid}}}$

p-Toluenesulphonyl-
chloride

p-Carboxybenzene-
sulphonic acid

(CH$_3$CH$_2$CH$_2$)$_2$NSO$_2$—⟨O⟩—COOH $\xleftarrow{\substack{\text{HN(CH}_2\text{CH}_2\text{CH}_3)_2 \\ \text{Di-}n\text{-propylamine} \\ \text{(Condensation)}}}$ HOOC—⟨O⟩—SO$_2$Cl

Probenecid

p-Carboxybenzene-
sulphonyl Chloride

p-Carboxybenzenesulphonic acid is obtained by the oxidation of the methyl group present in *p*-toluenesulphonyl chloride which on further treatment with chlorosulphonic acid yields the corresponding *p*-carboxybenzene sulphonyl chloride. Condensation with di-*n*-propylamine gives rise to the official compound.

Probenecid inhibits renal tubular reabsorption of water and by this mechanism enhances the urinary excretion of uric acid. This lowers the level of urate in the serum. It thus serves as a potent uricosuric agent in the treatment of gout. Probenecid also blocks the renal tubular secretion of penicillins and cephalosporins. It is, therefore, used as *an adjuvant therapy with penicillin V or G, ampicillin, cloxacillin, oxacillin, methicillin and naficillin to increase and prolong their plasma levels.* Besides it also *enhances the plasma levels of anti-inflammatory agents like naproxen and indomethacin,* and a host of medicinal compounds such as sulphonamides, sulphonylureas, dapsone, etc.

Dose : Adult, oral, 500mg to 2g per day ; usual, 250mg 2 times daily for one week, then 500mg twice a day thereafter.

B. *Allopurinol* INN, BAN, USAN,

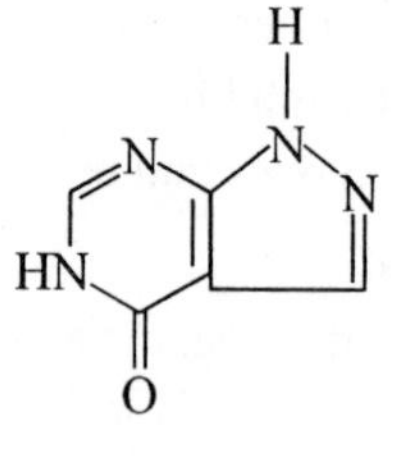

1H-Pyrazolo [3, 4-*d*] purimidin-4-ol ; 1, 5-Dihydro-4H-pyrazolo [3, 4-*d*] pyrimidin-4-one ; BP ; USP ; Zyloprim$^{(R)}$ (Burroughs Wellcome).

Synthesis

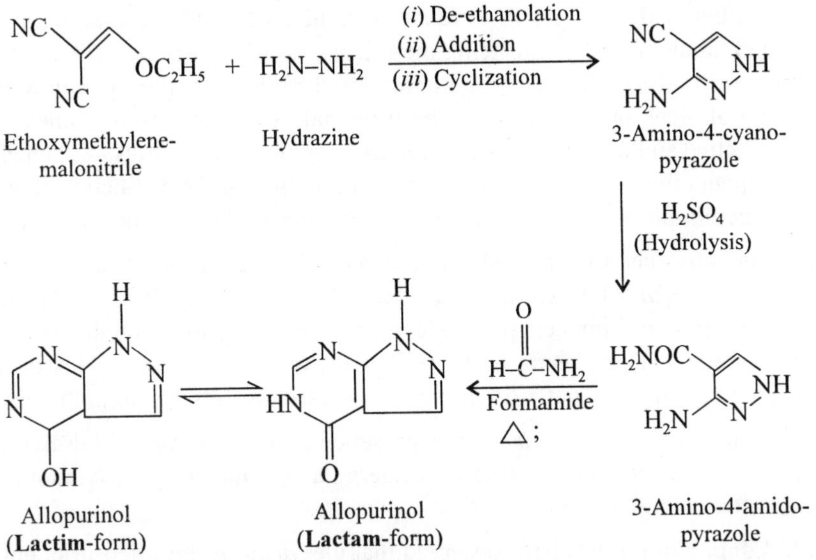

Condensation of ethoxymethylenemalonitrile with hydrazine *via* deethylation, addition and cyclization gives rise to 3-amino-4-cyanopyrazole which upon hydrolysis in the presence of sulphuric acid yields 3-amino-4-amino pyrazole. This on heating with formamide inserts the last carbon atom to afford allopurinol which exhibits tautomerism.

It is a structural analogue of hypoxanthine and is classified as xanthine oxidase inhibitor. It is administered for an indefinite duration in the *treatment of chronic gout*. It helps to decrease the concentration of uric acid in plasma by blocking the conversion of hypoxanthine and xanthine to uric acid and by reducing purine synthesis. Thus it *causes gradual resolution of tophi and minimises the risk of the formation of uric acid calculi.*

Dose : Usual, adult, oral, antigout, 100 to 200mg 2 or 3 times a day.

C. *Piroxicam* INN, BAN, USAN,

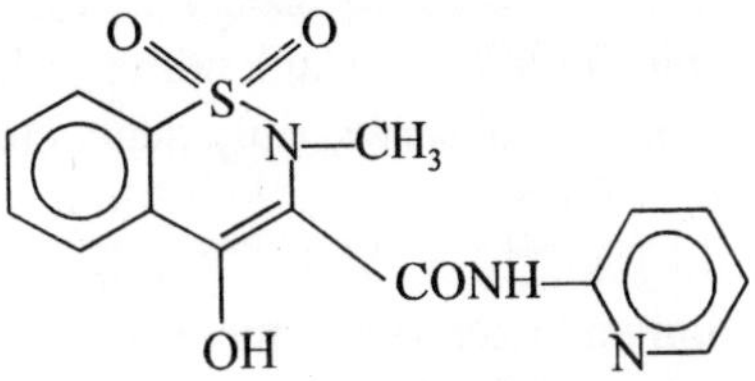

4-Hydroxy-2-methyl-N-2-pyridyl-2H-1, 2-benzothiazine-3-carboxamide 1, 1-dioxide ; 2H-1, 2-benzothiazine-3-carboxamide, 4-hydroxy-2-methyl-N-2-pyridinyl-, 1, 1-dioxide ; Feldene[(R)] (Pfizer)

It is employed for acute and long-term therapy for the *relief of symptoms of osteoarthritis and rheumatoid arthritis*. It also possesses uricosuric actions and has been used in the *treatment of acute gout*.

Dose : Usual, adult, oral, 20mg daily.

2.6.2.1. Mechanism of Action. The mechanism of action of some of the typical compounds described under section 15.2.6.2. are treated in the sections that follows :

2.6.2.1.1. Probenecid. The *'drug'* is found to inhibit its tubular reabsorption of urate at the proximal convoluted tubule, thereby enhancing the urinary excretion of uric acid and mimising serum uric acid levels. Interestingly, with respect to the **'outward renal transport phenomenon'** the *'drug'* blocks the secretion of weak organic acids at the proximal as well as distal tubules. Therefore, it is overwhelmingly effective and useful as an *'adjuvant therapy'* with such drugs as : pencillin G, O, or V, or with ampicillin, methicillin, oxacillin, cloxacillin, or naficillin for the distinctive elevation as well as prolongation of penicillin plasma levels by whatever route the antibiotic is actually administered.

The *'drug'* get absorbed rather rapidly and completely after an oral administration. It has been observed that plasma levels of 100 to 200 mcg. mL^{-1} are almost necessary for an adequate and sufficient uricosuric effect ; whereas, an equivalent plasma levels of 40-60 mcg. mL^{-1} produce maximal inhibition of the pencillin excretion. The plasma half-life varies from 4 to 17 hour. However, at a plasma concentration of 14 mcg. mL^{-1}, about 17% of the drug invariably gets bound to the plasma protein.

SAR of Probenecid. In this *'drug'* the presence of its electron withdrawing carboxy and sulphonamido-moieties, has not been reported to undergo any aromatic hydroxylation, which explains its fast absorption after an oral administration.

2.6.2.1.2. Allopurinol. It has been observed that the *'drug'* is not uricosuric, but it does inhibit the production of uric acid by way of blocking categorically the biochemical reactions that are essentially involved immediately preceding uric acid formation. Hence, it also inhibits **xanthine oxidase** (enzyme), which is exclusively responsible for the conversion of hypoxanthine to xanthine and of xanthine ultimately to uric acid.

Besides, allopurinol, inhibits *de novo* purine synthesis *via* a feedback mechanism, that specifically provides another benefit to the subject. It is found to get metabolized by xananthine oxidase to *oxypurinol,* that also invariably inhibits xanthine oxidase. However, oxypurinol possesses a much longer half-clearance time from plasma than allopurinol.

2.6.2.1.3. Piroxicam. The *'drug'* represents a class of *acidic inhibitors* of prostaglandin synthesis, although it fails to antagonize PGE_2 directly.* It is found to be exerting a rather long duration of action having a plasma half-life of 38 hour, which remarkably pegs a dosage of only 20 to 30mg once daily. Besides, its overall pharmacologic activity has been determined to be almost equivalent to either 400mg of *ibuprofen* or 25mg of *indomethacin* 3-times daily.**

Like other NSAIDs, the *'drug'* inhibits prostaglandin synthesis chemotaxis and the release of liposomal enzymes (from liver). It has been observed duly that a chronic administration with 20mg per day causes steady state plasma levels of 3-5 mcg. mL^{-1} within a span of 7 to 12 days. The volume of distribution is found to be 0.12–0.14 $L.kg^{-1}$; mean half life is ~ 50 hour (range, 30-86 hour). It gets metabolized mostly *via* hydroxylation and excreted in the urine ultimately.

---

*Wiseman EH : *R Soc Med Int Congr Ser,* **1,** 11, 1978.

Balogh Z *et al. Curr Med Res Opin,* **6, 148, 1979.

Piroxicam
(*keto*-form)

Piroxicam
(*enol*-form) **(Acidic)**

2.7 Salicylic Acid Analogues

A good number of salicylic acid analogues have also been found to possess anti-inflammatory actions, *e.g.,* aspirin, salol, salsalate, sodium salicylate, salicylamide, benorilate, choline salicylate, flufenisal etc., in addition to their antipyretic analgesic property. These compounds have been individually treated in Chapter 9.

2.8 Pyrazolones and Pyrazolodiones

Drugs like phenazone, aminophenazone (aminopyrine), dipyrone, phenylbutazone, oxyphenbutazone, sulfinpyrazone, etc., belonging to this category, besides their antipyretic-analgesic action, have also been reported to exhibit anti-inflammatory properties. These compounds have been dealt separately in the chapter on 'antipyretic-analgesics'.

Probable Questions for B. Pharm. Examinations

1. What are the advantages of NSAID(s) over the steroidal drugs used as anti-inflammatory drugs ? Support your answer with the suitable examples.

2. Classify NSAID based ont heri chemical structures. Give examples of **one** potent drug from each category.

3. Indomethacin and Tolmetin Sodium are two typical examples of heteroarylacetin acid analogue of NSAID. Give the synthesis of one of them while differentiating their chemical structures.

4. Give the structure, chemical name and uses of **three** important members of arylacetic acid analogues employed as NSAID. Discuss the synthesis of any **one** drug selected by you.

5. 'The arylpropionic acid analogue also exhibits potent anti-inflammatory propertise besides analogesic and antipyretic activities'. Justify the statement with suitable examples of NSAID.

6. Naproxen derived from **naphthalene acetic acid analogue** proved to be the leading compound of an extensive series of promising clinical agents. Describe its synthesis from 6-methoxy naphthalene.

7. Discuss the monovalent gold compounds as NSAID. Give the synthesis of **auranotin** and **aurothioglucose** along with their usage.

8. Give the structure, chemical name and uses of the following uricosuric agents :

   (*a*) Allopurinol   (*b*) Probenecid   (*c*) Piroxicam

   Describe the synthesis of any **one** drug.

9. Give a comprehensive account of the following categories of drugs used profusely as NSAID(*s*) :

   (*a*) Salicylic acid analogues,

(*b*) Pyrazolones and pyrazolodiones and

(*c*) Antimalarial agents

Support your answer with appropriate examples

10. Discss the mode of action of NSAID(*s*) by citing the examples of typical representative drugs.

RECOMMENDED READINGS

1. D Lednicer and LA Mitscher *The Organic Chemistry of Drug Synthesis* John Wiley and Sons New York (1995).

2. E Arrigoni-Martelli, *Inflammation and Anti-inflammatories,* Spectrum Publications, inc. New York (1977).

3. WE Coyne, *Nonsteroidal Anti-inflammatory Agents and Antipyretics,* in : *Burger's Medicinal Chemistry and Drug Discovery,* ME Wolff *(Ed)* 5th edn., John Wiley and Sons inc., New York (1995).

4. RA Scherrer and MW Whitehouse, *Anti-inflammatory Agents,* Academic Press, New York (1974).

5. TY Shen, *Perspectives in Non-steroidal Anti-inflammatory Agents,* Chem. (Internal, edn.), (1972).

6. Eds M H J Smith and PK Smith, *The Salicylates,* Interscience Publishers New York (1966).

7. CA Winder, *Nonsteroid Anti-inflammatory Agents,* in Progress in Drug Research, Ed. E Jucker Vol. 10, Birkhauser, Basel, (1966).

8. MC Griffiths, *USAN and the USP Dictionary of Drug Names,* United States Pharmacopoeial Convention, Inc., Rockwille Md. (1985).

9. JEF Reynolds, *Martindale-The Extra Pharmacopoeia*, (31st edn.) The Royal Pharmaceutical Society, London, 1996.

10. Patrick GL, *An Introduction to Medicinal Chemistry,* Oxford University Press, Oxford (UK), 2nd, end., 2001.

16   Antiparkinsonism Agents

1. INTRODUCTION

Parkinson's disease or *Paralysis agitans* was first described as early as 1817 by James Parkinson, a London doctor, as consisting essentially of *'Involuntary tremulous motion with decreased muscular power in parts not in action and even when supported, with a propensity to bend the trunk forwards and to pass from a walking to a running pace, the senses and intellect being uninjured'*.

Parkinsonism is usually idiopathic but can arise from ischaemic changes in the brains as in arteriosclerotic and postencephatic parkinsonism.

Various drugs are invariably used in Parkinsonism for the following effects ; *first,* to lower abnormal reflex rigidity and tremor, and restore normal motor activity *e.g.,* natural atropine group of alkaloids, synthetic atropine substituted, antihistaminics ; *secondly,* to minimise mental depression-*e.g.,* analeptics ; and *thirdly,* to allay restlessness, tension and anxiety-*e.g.,* sedatives and tranquillizers.

Parkinsonism is a vivid example of a disorder that lends itself to such particular treatment as : disorders within separate nervous structures which essentially comprise of neurons predominately of one or two transmitter types. In reality, however, the *'antiparkinsonism agents'* are basically not interneuron depressants.

It has been duly demonstrated and established that the disorder in parkinsonism invariably occurs very much within the *substantia nigra** and *corpus striatum.***

1.1. Etiology. It has been observed that although the neuropathology is well understood, the actual cause of Parkinson's disease is not yet known. In order to understand the etiology of the disease comprehensively one may have to take into consideration the development of both effective *pharmacotherapeutic* and *prophylactic therapy* adequately.

There are several *'theories'* that have been duly put forward with respect to the actual cause of the Parkinson's disease, namely :

    (*a*) endogenous and/or environmental neurotoxicants,

    (*b*) mitochondrial dysfunction, and

   **Substantia nigra :* The black substance in a section of the *erus cerebri.*

  ***Corpus striatum :* A structure in the cerebral hemispheres consisting of two basal ganglia and the fibres of the internal capsules that separate them.

(*c*) oxidative metabolism : all of which may ultimately lead to a distinct *'oxidative stress'*.

(*d*) neurodegenerative disorders (*e.g.,* movement disorder Huntington's disease*) have been established genetically and that the researchers also proved a possible link and influence in the Parkinson's disease.

(*e*) epidemiological investigative researchers have adequately revealead that, besides the age-factor involved, — a family history of Parkinson's disease bears almost the strongest predictor of an enhanced possible risk of the ensuing disorder.**

(*f*) α-synucle in protein *i.e.,* a distinct mutation observed in the α-synuclein gene located on the choromosome 4*q*, happens to be a highly conserved and abundant 140-amino acid protein having quite unknown function (*modus operandi*) which is invariably expressed largely in the presynaptic nerve terminals in the brain.***

(*g*) there exists almost little evidence thereby suggesting that Parkinson's disease is virtually *autoimmune related*****; and it further proves that there exists neither a prevalent communicable infectious etiology nor a genetic etiology.

(*h*) the most striking and interesting characterized epidemiologic findings in Parkinson's disease is its remarkable *lower incidence in cigarrette smokers* than the corresponding nonsmokers.*****

(*i*) **dopamine** – imiplicates itself in the disease process by means of the production of *chemically reactive oxidation products* which suggest evidently that the endogenously liberated products may be the etiologic factors in Parkinson's disease.******

(*j*) MAO – catalyzed oxidation of the **monoamine neurotransmitters** (*e.g.,* dopamine, norepinephrine, serotonin) produces H_2O_2 (Eqn. '*a*'), that may subsequently undergo a redox reaction with superoxide in the *Haber-Weiss reaction******* to give rise to the formation of the highly cytotoxic hydroxy radical as depicted in (Eqn. '*b*') below :

$$RCH_2CH_2NH_2 + O_2 + H_2O \longrightarrow RCH_2CHO + NH_3 + H_2O_2 \qquad ...(a)$$

$$H_2O_2 + O_2 \bullet \longrightarrow O_2 + OH^{\ominus} + OH \bullet \qquad ...(b)$$

Furthermore, the subsequent auto-oxidation of dopamine to the corresponding electrophilic **semiquinone** and **quinone** analogues has also received considerable cognizance because these *'oxidation products'* are also cytotoxic in nature.******* Besides, Mn^{2+}, is observed to catalyze oxidation of dopamine, and interestingly the resulting species *viz.,* semiquinone and quinone have been duly implicated in the Mn-neurotoxicity.********

---

*An inherited disease of the CNS that usually has its onset between ages 25 and 55. Degeneration in the cerebral cortex and basal ganglia causes chronic progressive chorea and mental deterioration, ending in dementia.

Semchuk KM *et al. Neurology,* **43, : 1173–1180, 1993.

***Kruger R *et al. Nature Genetics,* **18** : 106–108, 1998.

****Duvoisin RC : In Marsden CD, Fahn S eds., **Movement Disorders,** Butterworth Scientific London, 8–24, 1982.

*****Kessler II *et al. Am. J. Epidemiol,* **94**, 16-25, 1971.

******Haber F, Weiss J : *Naturwissenschaften,* **5**, 45-92, 1932.

*******Graham DG *et al. Mol. Pharmacol,* **14** : 644-653, 1978.

********Graham DG, *Neurotoxicology,* **5** : 83-95, 1984.

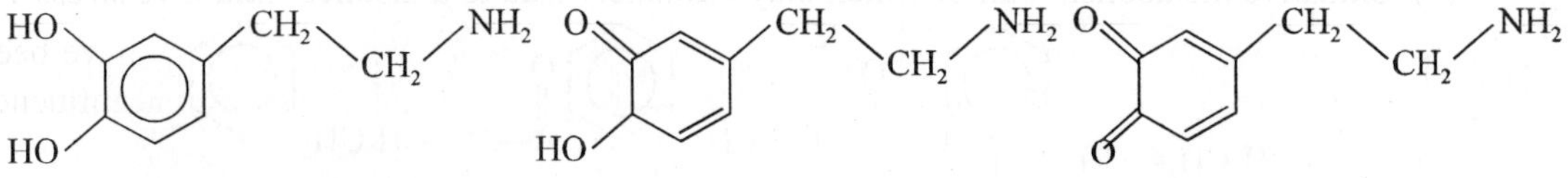

Dopamine Semi-Quinone Quinone

(k) various theories put forward till date suggest amply that the Parkinson's disease could be the consequence of normal aging phenomenon adequately superimposed on a lesion strategically located in the *substantia nigra* that might have occurred much earlier in one's life span.*

(l) symptoms of Parkinson's disease become distinctly apparant when the striatal dopamine levels invariably gets declined to about 80%.**

(m) imaginatively, the various apparent/visible symptoms of parkinsonism might be caused by *two* distinct phenomena, namely :

(i) a particular disease-linked episode amalgamated with certain observed pathological changes by virtue of normal aging process ; and

(ii) wonderful discovery of the selective as well as potent **dopaminergic neurotoxicant** N-methyl-4-phenyl-1, 2, 3, 6-tetrahydropyridine (MPTP) has largely helped scientists carrying out intensive and extensive researches to establish the etiology of Parkinson's disease.

1.2. Parkinsonism Produced by MPTP :

It has been reported that the **cyclic tertiary amine MPTP** *i.e.,* N-methyl-4-phenyl-1, 2, 3, 6-tetrahydropyridine, has caused an induction of a specific type of *parkinsonism* both in humans as well as monkeys that are found to be virtually identical equally in *neuropathology* and *motor abnormalities* to the resulting idiopathic Parkinson's disease.***

Interestingly, MPTP was actually obtained during the course of synthesis of MPPP (as *by-produds*) *i.e.,* the reverse ester of the narcotic analgesic meperidine termed as '**MPPP**' (N-methyl-4-propionoxy-4-phenylpiperidine), also commonly known as '**designer-heroin**' or '**synthetic heroin**'. *MPPP*-is invariably regarded as a structural analogue of another narcotic analgesic α-**prodine**. The *three* vital phenylpiperidine synthetic analgesics. *viz.,* MPPP, α-prodine, and meperidine are illustrated as under :

Importantly, the neuropathological and clinical characteristic features of MPTP-induced parkinsonism invariably resemble idiopathic Parkinson's disease rather more intimately than anyother previous human or experimented animal disorder exhibited by toxins, viruses, metals, or other modes. In short, the molecular pathophysiology of the ensuring MPTP neurotoxicity has virtually decephered the mystery surrounding the *neurodegenerative mechanisms* particularly associated with the **idiopathic parkinsonism.**

---

*Calne DB *et al. Nature,* **317** : 246-248, 1985

Riederer *et al. J Neural Transmission,* **38 : 277-301, 1976.

***Davis GC *et al. Psychiatric Res.* **1** : 249-254, 1979.

Burns RS *et al. Proc Natl Acad Sci* USA, **80** : 4546-4550, 1983.

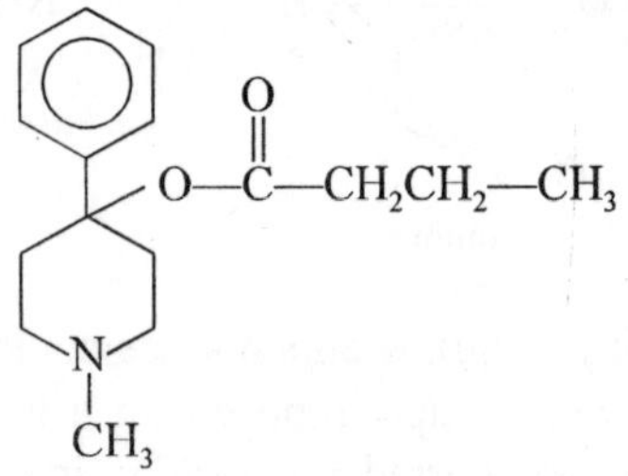

| "Designer Heroin" (MPPP) | α-Prodine [Nisentil[R]] | Meperidine [Demerol[R]] | N-Methyl-4-phenyl 1, 2, 3, 6-tetrahydropyridine (MPTP) |

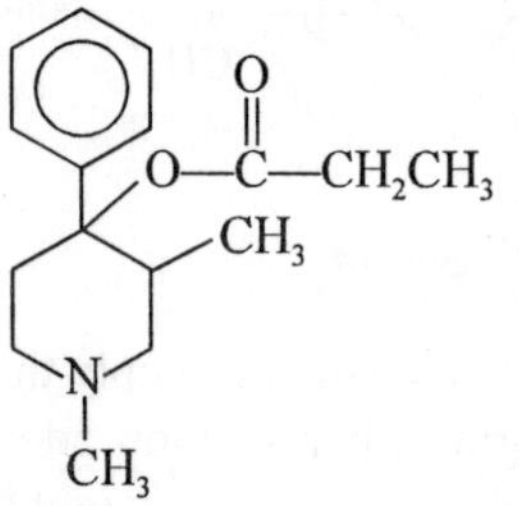

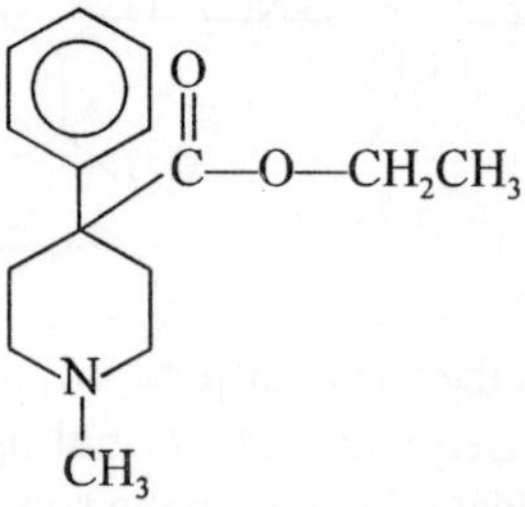

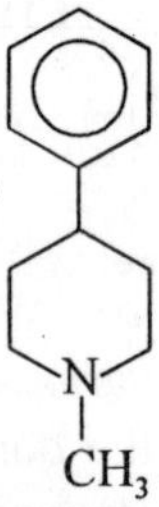

2. CLASSIFICATION

Antiparkinsonism agents may be classified on the basis of their chemical structures as follows :

1. Piperidine analogues

2. Pyrrolidine analogues

3. Phenothiazine analogues, and

4. Miscellaneous drugs.

2.1. Piperidine Analogues

A few strucutral analogues of piperidine proved to be potent antiparkinsonism agents. A few examples belonging to this class of compound is given below, namely : Biperiden hydrochloride ; Cycrimine hydrochloride and Trihexyphenidyl hydrochloride.

A. *Biperiden Hydrochloride* BAN, USAN, *Biperiden* INN,

α-5-Norbornen-2-yl-α-phenyl-1-piperidinepropanol hydrochloride ; Piperidinepropanol, α-bicyclo [2, 2, 1] hept-5-en-2-yl-α-phenyl-, hydrochloride ; USP ; NF ; Akineton Hydrochloride[R] (Knoll).

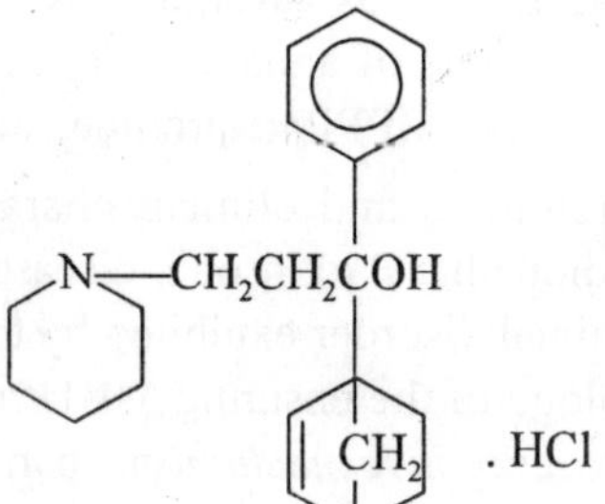

Synthesis :

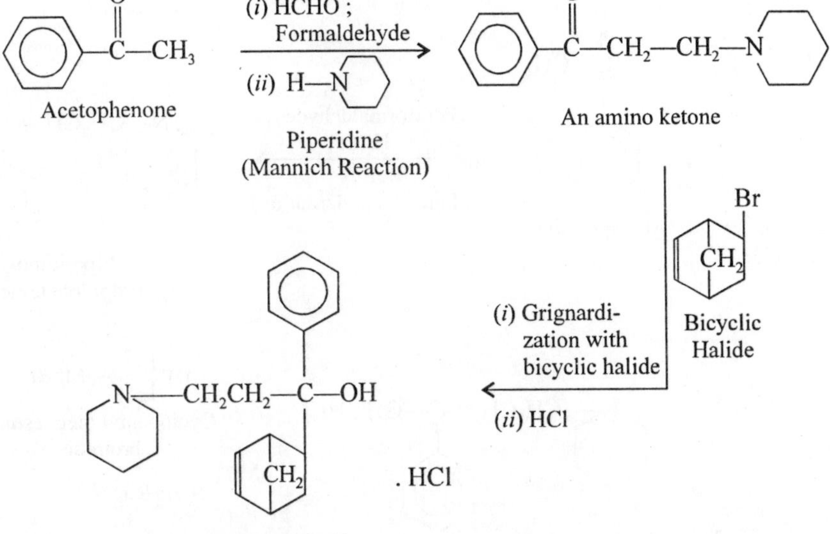

Biperiden Hydrochloride

Mannich reaction of acetophenone with formaldehyde and piperidine results into the formation of an aminoketone, which on grignardization with bicyclic halide and subsequent neutralization with HCl affords the official compound biperiden hydrochloride.

Biperiden is used in the *treatment of parkinsonism, muscle rigidity, akinesia and drooling*. It is also employed in the *acute crises due to oculogyration (movement of eyeball)*. It also finds its use in lowering spasticity in pyramidal tract disorders.

Dose : For parkinsonism, 2mg 3 or 4 times daily.

B. *Cycrimine Hydrochloride* BAN, USAN, *Cycrimine* INN,

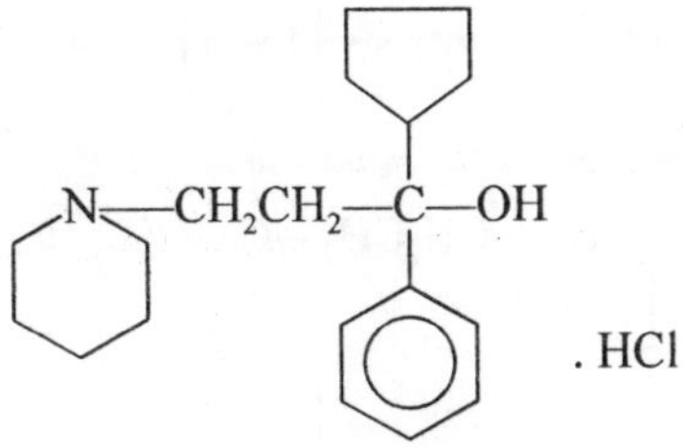

α-Cyclopentyl-α-phenyl-1-piperidinepropanol hydrochloride ; 1-Piperidine-propanol, α-cyclopentyl-α-phenyl-, hydrochloride ; USP ; NF ; Pagitane Hydrochloride[(R)] (Lilly).

Synthesis

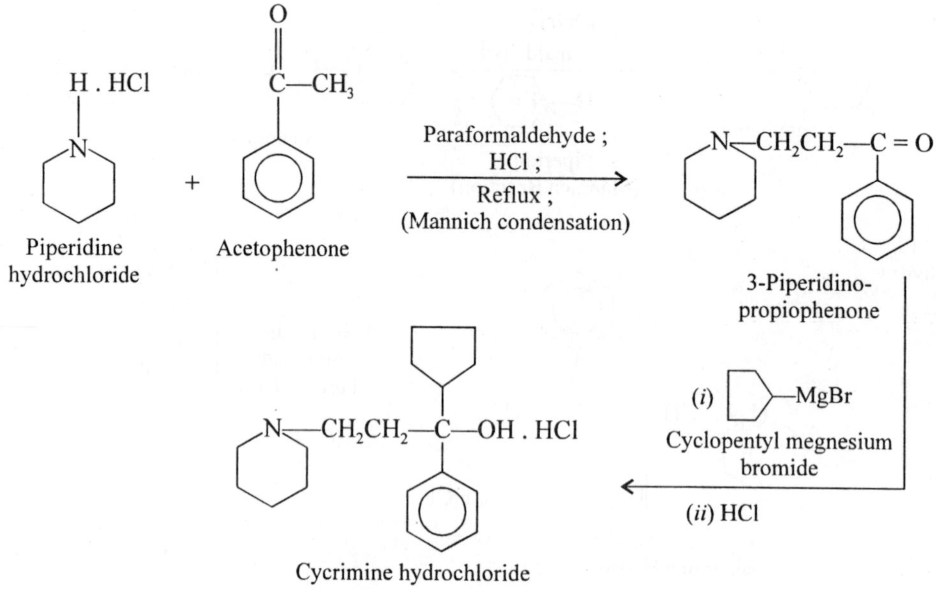

Cycrimine hydrochloride

Mannich condensation takes place when piperidine hydrochloride, acetophenone, paraformal-dehyde and HCl are refluxed for several hours to yield 3-piperidino propiophenone, which upon grignardization with cyclopentyl magnesium bromide followed by neutralization of the cycrimine base forms the official compound.

It is mostly employed in *all forms of paralysis agitans (parkinsonism).*

Dose : Initial, oral, 12.5 mg 2 to 3 times daily ; maintenance dose, 3.75 to 15 mg per day.

C. *Trihexyphenidyl Hydrochloride* USAN, *Trihexyphenidyl* INN, *Benzhexol Hydrochloride* BAN,

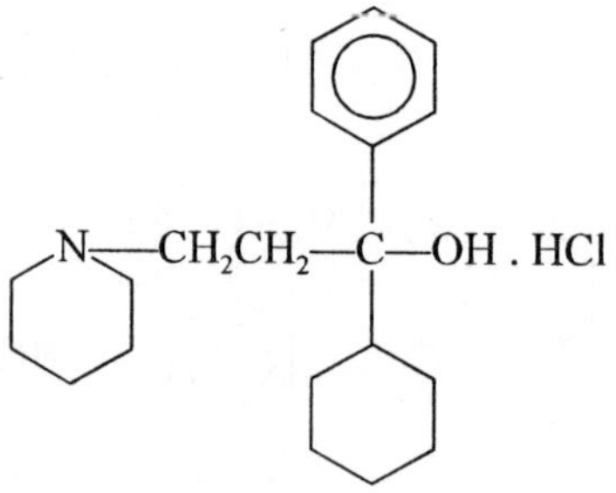

α-Cyclohexyl-α-phenyl-1-piperidinepropanol hydrochloride ; 1-Piperidine-propanol, α-cyclohexyl-α-phenyl-, hydrochloride ; USP ; Benzhexol Hydrochloride BP ; Artane[R] (Lederle) ; Pipanol[R] (Winthrop) ; Tremin[R] (Schering-Plough).

Synthesis

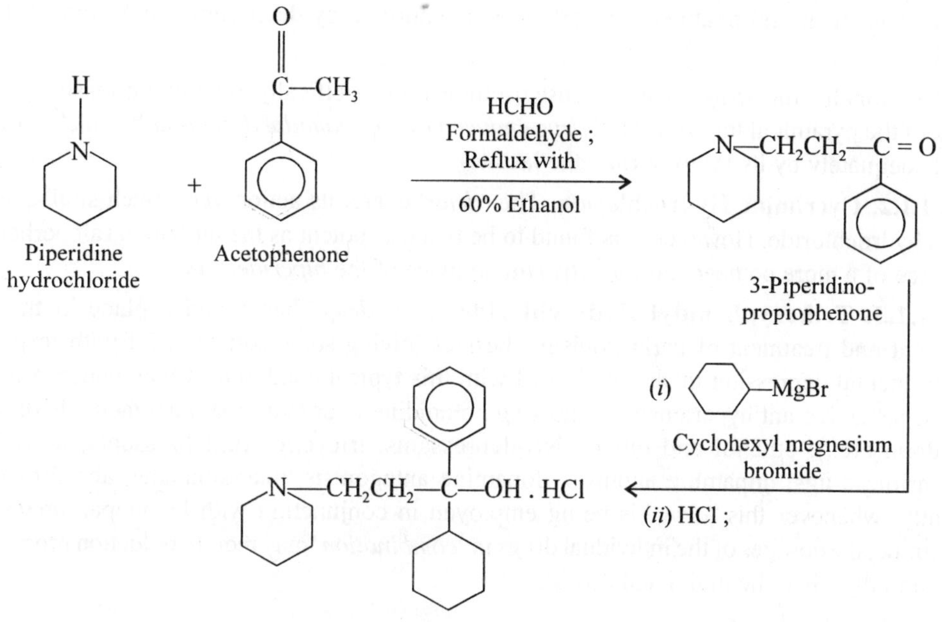

3-Piperidinopropiophenone is prepared by the Mannich condensation of acetophenone and piperidine with formaldehyde by refluxing the reactants in 60% ethanol. The resulting product is subjected to Grignard reaction with cyclohexyl-magnesium bromide to yield the corresponding base which is subsequently precipitated by passing a stream of hydrogen chloride through a solution of the base in a suitable solvent.

It is an antiparkinsonism drug possessing relatively weaker antispasmodic and antimuscarinic properties. In fact, *it is the drug of choice for the treatment of parkinsonism.* It is also employed *for the relief of akinesia, muscular rigidity, tremor and oculogyria.* It is also found to be useful in the *treatment of drug-induced extrapyramidal symptoms.*

Dose : Initial, oral, 1mg on 1st day, followed by 2mg daily after 3 to 5 days ; maintenance dose, 6 to 10mg per day in 3 to 4 divided doses but not exceeding 20mg per day.

2.1.1. Mechanism of Action. The mechanism of action of *'drugs'* discussed under section 16.2.1. are dealt with in the pages that follows :

2.1.1.1. Biperiden Hydrochloride. The *'drug'* possesses a comparatively weak visceral anticholinergic, but a strong nicotinolytic, action with regard to its ability to *block nicotine-induced convulsions.* Perhaps it amply expatiates its neurotropic activity being distinctly low on the intestinal musculature together with the corresponding blood vessels. Besides, it has been demonstrated that the *'drug'* exhibits a comparatively stronger musculotropic action, that is almost equivalent to that of *papaverine,* in comparison to several synthetic anticholinergic drugs, such as : atropine, tropicamide, methixene hydrochloride, glycopyrrolate etc. It is, however, pertinent to state here that the action of biperiden on the eye, although mydriatic, is found to be much lower relative to that of atropine. Interestingly, these prevailing inherent *weak anticholinergic actions* essentially add to its utility in the control, management and treatment of Parkinson's syndrome by decreasing the ensuing side effects appreciably.

Occasionally, the *'drug'* is of immense usefulness in specifically minimizing spasticity in certain disorders of the pyramidal tract, such as : *drug-induced extrapyramidal dyskinesia**, which is eventually managed adequately by its IV form (*i.e.,* the lactate).

2.1.1.2. Cycrimine Hydrochloride. The *'drug'* exerts its action very much similar to that of biperiden hydrochloride. However, it is found to be twice as potent as the biperiden salt perhaps due to the presence of a more compact *cyclopentyl ring* in place of the *biperiden ring.*

2.1.1.3. Trihexyphenidyl Hydrochloride. The *'drug'* has found a place in the control, management and treatment of parkinsonism ; besides, giving some sort of relief with respect to the prevailing mental depression invariably linked with this typical condition. It may undergo interaction with the CNS-active antihypertensive drugs (*e.g.,* clonidine hydrochloride, guanfacine hydrochloride, methyl dopa etc.,), ethanol and other CNS-depressants, tricyclic antidepressants, MAOIs, other antimuscarinic drugs, dopamine agonists, dopamine antagonists, phenothiazine, and procainamide. Importantly, whenever this *'drug'* is being employed in conjunction with levodopa, amantadine, or bromocriptine, the dosages of the individual drugs in *'combination'* may require reduction proportionately and substantially (than the individual doses).

2.2 Pyrrolidine Analogue

The introduction of a 5-membered heterocyclic ring, *i.e.,* pyrrolidine instead of the 6-membered piperidine ring also gave rise to important antiparkinsonism agent.

Example : Procyclidine Hydrochloride.

A. *Procyclidine Hydrochloride* BAN, USAN, *Procyclidine* INN,

α-Cyclohexyl-α-phenyl-1-pyrrolidinepropanol hydrochloride ; 1-Pyrrolidine-propanol, α-cyclohexyl-α-phenyl-, hydrochloride ; 1-Cyclohexyl-1-phenyl-3-(pyrrolidin-1-yl) propan-1-ol hydrochloride ; BP ; USP ; Int. P., N.F. ; Kemadrin[R] (Burroughs Wellcome).

---

*Outside the pyramidal tracts of the CNS-; a defect in the ability to perform voluntary movement.

Synthesis

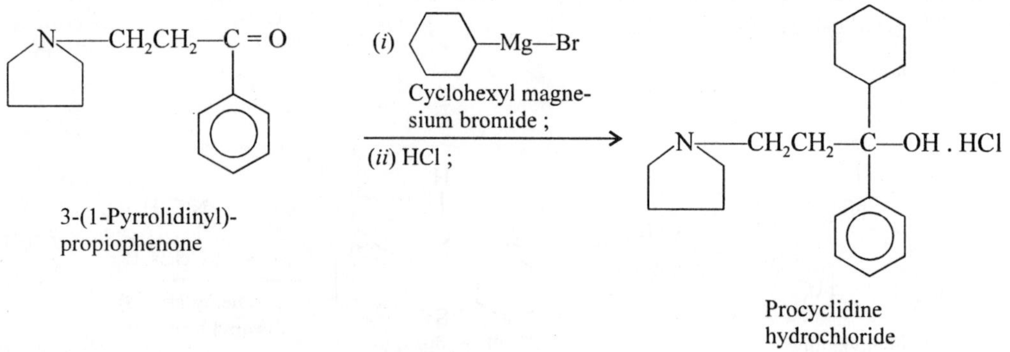

3-(1-Pyrrolidinyl)-
propiophenone

(i) Cyclohexyl magne-
sium bromide ;
(ii) HCl ;

Procyclidine
hydrochloride

It may be prepared by reacting 3-(1-pyrrolidinyl) propiophenone with a Grignard reagent cyclohexyl magnesium bromide to yield the procyclidine (base) which is subsequently treated with HCl to form the official compound.

Procyclidine being a structurally similar chemical congener of trihexyphenidyl possesses also similar properties. It is used for the *symptomatic treatment of postencephalitic parkinsonism. It has also been employed successfully to alleviate the extrapyramidal syndrome induced by such drugs as reserpine and phenothiazine analogues.*

Dose : Initial, oral, 7.5mg per day in 3 or 4 divided doses after meals ; maintenance dose usually 20 to 30mg per day.

2.2.1. Mechanism of Action. The mechanism of action of procyclidine hydrochloride is discussed as under :

2.2.1.1. Procyclidine Hydrochloride. The *'drug'* exerts its potent peripheral anticholinergic effects that happen to be quite akin to its corresponding analogue methochloride (*i.e.,* tricyclamol chloride) ; however, its pivotal clinical usefulness solely lies in its ability to cause substantial relief to the *voluntary muscle spasticity* by virtue of its central activity precisely. Hence, it has been employed both successfully and efficaciously in the treatment of Parkinson's syndrome. Importantly, its ensuing activity on tremor is not quite anticipated ; and, therefore, must be supplemented by usual combination with other identical drug substances.

2.3 Phenothiazine Analogue

The only phenothiazine analogue official in USP and used in parkinsonism is Ethopropazine hydrochloride.

A. *Ethopropazine Hydrochloride* BAN, USAN, *Profenamine* INN,

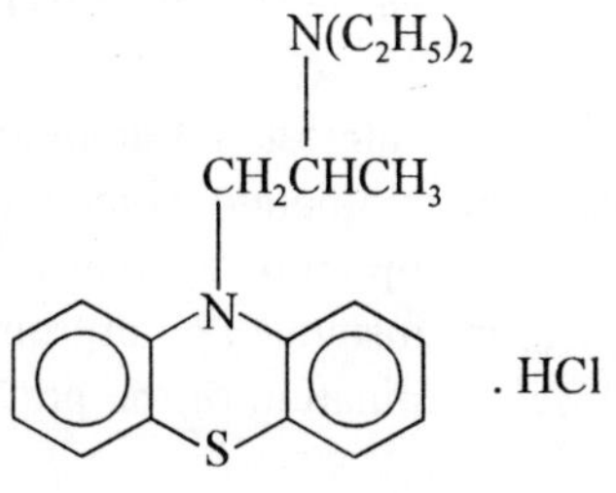

10-[2-(Diethylamino) propyl] phenothiazine monohydrochloride ; 10H-Phenothiazine-10-ethanamine, N, N-diethyl-α-methyl-, monohydrochloride ; Isothazine Hydrochloride ; BP ; USP ; Int. P. ; Parsidol$^{(R)}$ (Parke-Davis).

Synthesis

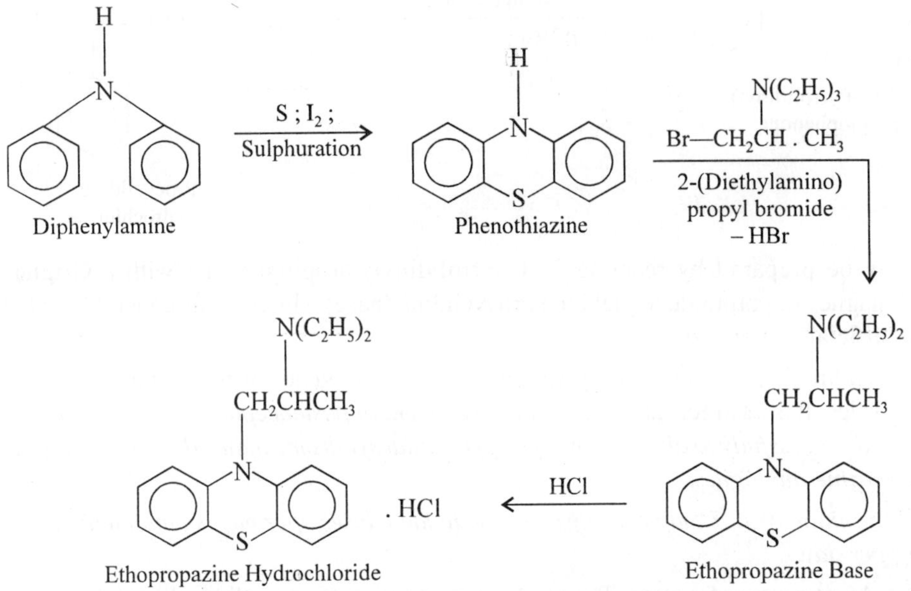

Phenothiazine nucleus may be prepared by the sulphuration of diphenyl amine in the presence of sulphur and iodine which is then treated with the corresponding Grignard complex of 2-(diethylamino) propyl bromide to yield the ethopropazine (base). The resulting base is dissolved in an appropriate solvent and treated with an equimolar quantity of HCl to form the official compound.

It is employed in the *management of parkinsonism, particularly for the control of rigidity*. It has also been used to *reduce the spasm, tremor and oculogyration. It also possesses anticholinergic, adrenergic blocking, mild anti-histaminic, local anaesthetic and ganglionic blocking actions.*

Dose : Usual, oral, initial, 50mg per day, slowly increased to 500mg per day in divided doses.

2.3.1. Mechanism of Action. The mechanism of action of ethopropazine hydrochloride is described as under :

2.3.1.1. Ethopropazine Hydrochloride. The *'drug'* possesses antimuscarinic activity. It is particularly beneficial in the symptomatic treatment of parkinsonism. It finds its enormous therapeutic value in the control of rigidity ; besides, having a favourable response in oculogyric crises, tremor and sialorrhea.*

SAR of Ethopromazine. Another wonderful and dramatic appropriate example of appreciable pharmacologic differences existing between compound which are *'practically-look-alikes'* is the usage of *phenothiazine structural analogue* **'ethopropazine'** to control and treat the extrapyramidal Parkinson like syndrome caused by antipsychotics, for instance : chlorpromazine — the well-known neuroleptic chlorpromazine (Chapter : 23). A close examination of the prevailing chemical structures of the two types

---

*Excessive secretion of saliva.

of drugs vividly shows that the *ring-N-atom* in the antiparkinsonism drug (*i.e.,* ethopropazine) is actually separated from the *chain-N-atom* by *2-C-atoms* ; whereas, the antipsychotic (*i.e.,* chlorpromazine) or tranquillizer essentially has a *3-C-atom* distance — indeed a *'small'* difference. Thus, the former *'drug'* exerts a antiparkinsonism action ; whereas, the latter is employed invariably as an antipsychotic having an overwhelming and appreciable ability to induct tremors.

2.4 Miscellaneous Drugs

There exist a good number of potent antiparkinsonism agents that do not fall into any of the classifications discussed above (A through C) ; and hence, they have been grouped together under this head. *Examples :* Benztropine mesylate ; Orphenadrine citrate ; Chlorphenoxamine hydrochloride ; Levodopa and Amantadine hydrochloride.

A. *Benztropine Mesylate* BAN, USAN, *Benzatropine* INN,

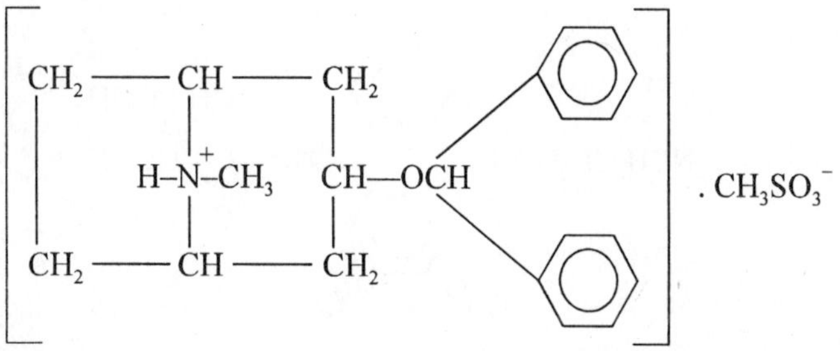

3α-(Diphenylmethoxy)-1αH-5αH-tropane methanesulphonate ; 8-Azabicyclo [3.2.1] octane, 3-(diphenylmethoxy)-, *endo*, methanesulphonate ; Benzotropine Methanesulphonate ; BP ; USP ; Cogentin$^{(R)}$ (MS & D).

Synthesis

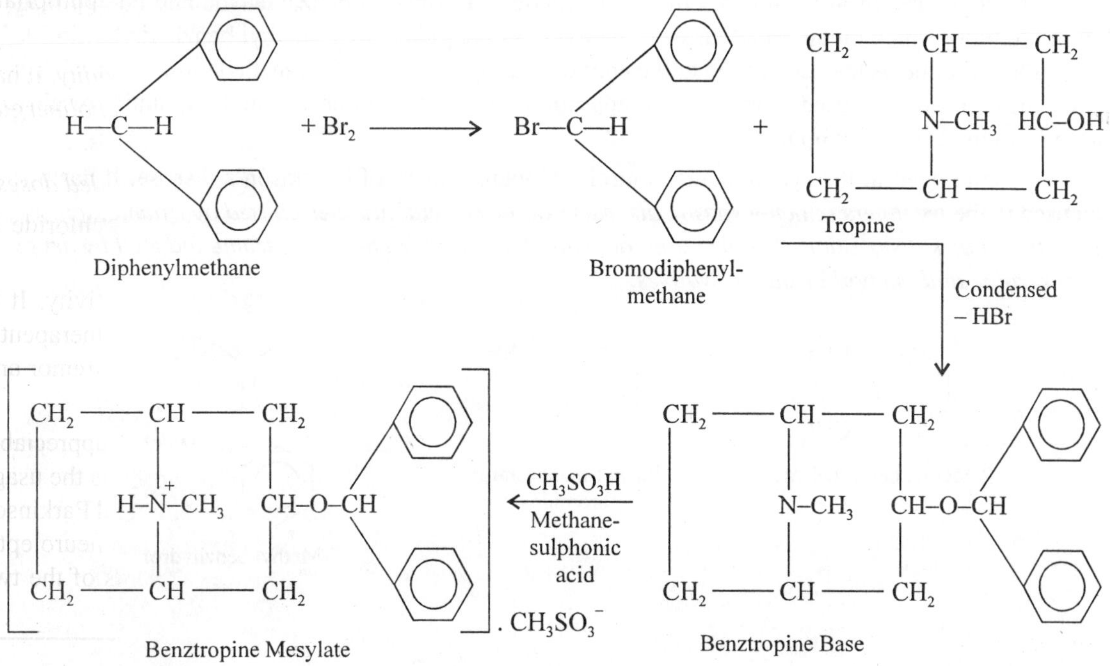

Diphenylmethane

Bromodiphenyl-methane

Tropine

Condensed – HBr

Benztropine Mesylate

Benztropine Base

CH_3SO_3H
Methane-sulphonic acid

Bromodiphenylmethane may be prepared by the direct bromination of diphenylmethane which is then condensed with tropine, through the Williamson ether synthesis (making use of sodium alkoxide derivative of tropine), to yield the benztropine base. This is dissolved in an appropriate solvent and precipitated by treating it with an equimolar quantity of methanesulphonic acid.

It has a mixed chemical features of both diphenhydramine class of antihistaminics and atropine. It has been used successfully in the *treatment of parkinsonism, to arrest tremor and rigidity, oculogyric crises and pain secondary to muscular spasm.* It is also employed to *control extrapyramidal dyskinesia caused by tranquillizers, namely, chlorpromazine or reserpine.* Its actions and uses are similar to those of benzhexol and is preferred to that of the later due to its inherent sedative effective at its normal dose.

Dose : Usual, initial, oral, 0.5 to 1mg, slowly increased to 500mcg after every 5 to 6 days till optimal dose is achieved ; i.m. or i.v. 1 to 2mg.

B. *Orphenadrine Citrate* BAN, USAN, *Orphenadrine* INN,

$(CH_3)_2NCH_2CH_2OCH$

$$\begin{array}{c} CH\ COOH \\ | \\ HO—C—COOH \\ | \\ CH\ COOH \end{array}$$

N, N-Dimethyl-2[o-methyl-α-phenylbenzyl) oxy] ethylamino citrate (1:1) ; Ethanamine, N, N-dimethyl-2-[(2-methylphenyl) phenylmethoxy]-, 2-hydroxy-1, 2, 3-propanetricarboxylate (1:1) Mephenamine Hydrochloride ; Orphenadin Hydrochloride ; BP ; USP ; Norflex[R] (Riker).

Synthesis

2-Methylbenzhydrol is prepared by the interaction of 2-methyl benzaldehyde and phenyl magnesium bromide, which on chlorination with thionyl chloride yields the 2-methyl benzhydryl chloride. The resulting product is then converted to the amino ether by reaction with dimethylamino ethanol. The orphenadrine (base) is caused to react, in an appropriate solvent, with an equimolar quantity of citric acid to form the official compound.

It is employed in the symptomatic control and management of Parkinson's disease. It has also been used in the *treatment of acute spastic disorders of the skeletal muscles caused by trauma, tension, and vertebral disk dislocation.* It is also used *alleviate the extrapyramidal syndrome induced by drugs, e.g., reserpine and phenothiazine derivatives.*

| 2-Methyl benzaldehyde | Phenyl-magnesium bromide | 2-Methyl benzhydrol |

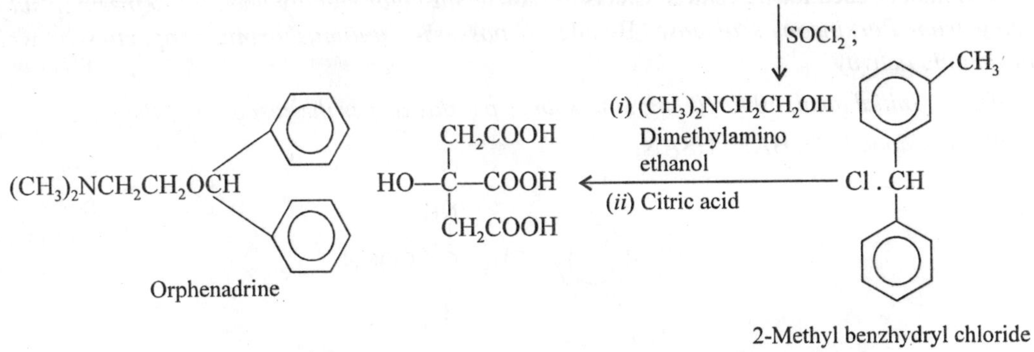

SOCl$_2$;

(i) (CH$_3$)$_2$NCH$_2$CH$_2$OH
Dimethylamino
ethanol

(ii) Citric acid

(CH$_3$)$_2$NCH$_2$CH$_2$OCH

Orphenadrine

CH$_2$COOH
|
HO—C—COOH
|
CH$_2$COOH

Cl . CH

2-Methyl benzhydryl chloride

Dose : Initial, oral, 100mg 2 times per day ; i.m. or i.v. 60mg every 12 hours.

C. *Chlorphenoxamine Hydrochloride* BAN, USAN, *Chlorphenoxamine* INN,

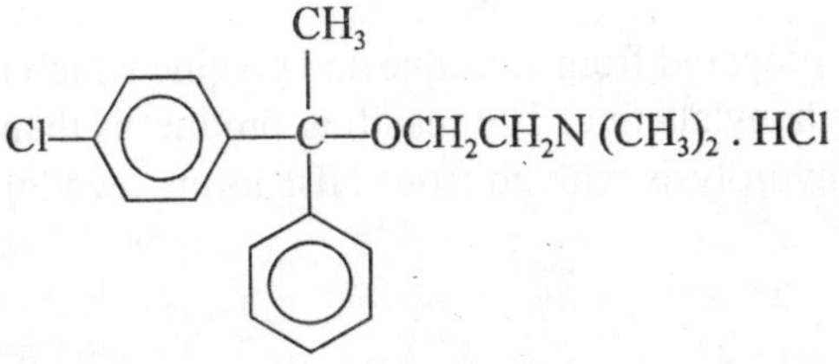

CH$_3$
|
Cl—C—OCH$_2$CH$_2$N (CH$_3$)$_2$. HCl

p-Chloro-α, α-dimethylphenethylamine hydrochloride ; Benzene-ethanamine, 4-chloro-α, α-dimethyl-, hydrochloride ; USP ; NF ; Phenoxene$^{(R)}$ (Merrell Dow).

Synthesis

p-Chloro-α-methyl-α-phenyl benzyl alcohol is prepared by the grignardization of *p*-chloro-acetophenone with phenyl magnesium bromide. This on etherification by treatment with N-(2-chloroethyl) dimethyl amine yields the chlorphenoxamine (base) which is then dissolved in an appropriate solvent and converted to the hydrochloride by a stream of hydrogen chloride to form the official compound.

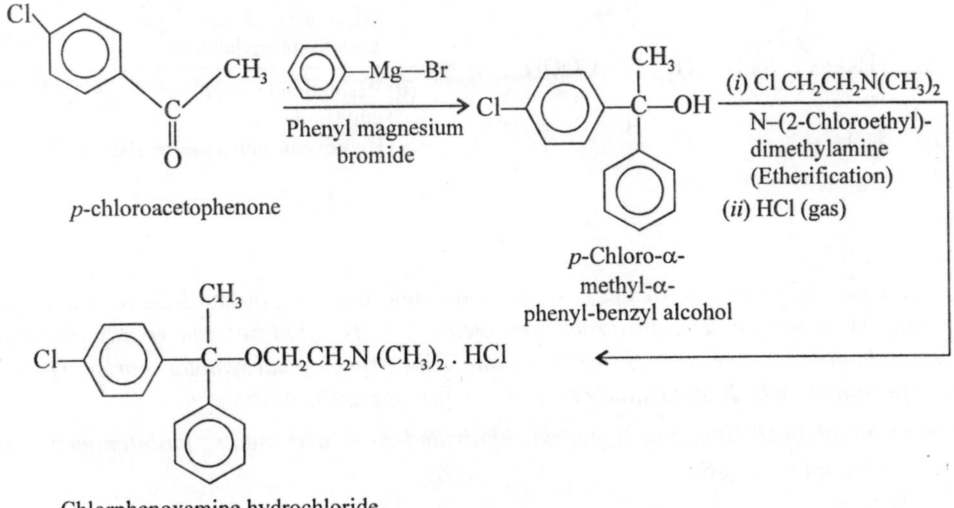

Cl

CH$_3$
|
C
‖
O

p-chloroacetophenone

Mg—Br

Phenyl magnesium
bromide

CH$_3$
|
Cl—C—OH

p-Chloro-α-
methyl-α-
phenyl-benzyl alcohol

(i) Cl CH$_2$CH$_2$N(CH$_3$)$_2$

N–(2-Chloroethyl)-
dimethylamine
(Etherification)

(ii) HCl (gas)

CH$_3$
|
Cl—C—OCH$_2$CH$_2$N (CH$_3$)$_2$. HCl

Chlorphenoxamine hydrochloride

It is mainly used for its central effects to *reduce muscular rigidity and also akinesia in subjects suffering from Parkinsons's disease.* Besides it *possesses antimuscarinic properties as well as antihistaminic activity.*

Dose : Initial, oral, 50 to 100mg 3 or 4 times per day as per the response of the patient.

D. *Levodopa* INN, BAN, USAN,

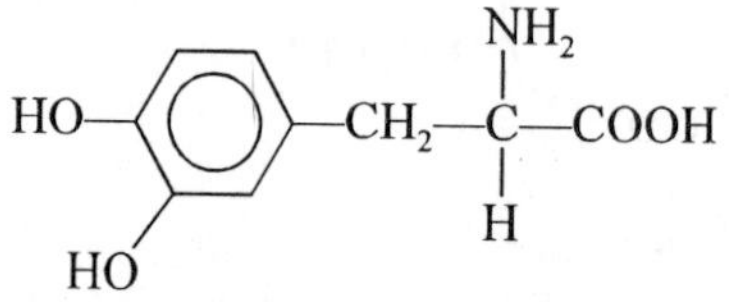

(–)-3-(3, 4-Dihydroxphenyl)-L-alanine ; L-Tyrosine, 3-hydroxy- ; L-Dopa ; BP ; USP ; Larodopa[R] (Roche) ; Levopa[R] (SK & F) ; Bendopa[R] (ICN) ; Dopar[R] (Norwich Eaton).

Synthesis

DL-Dopa may be first prepared from vanilline and glycine which is then converted to the DL-N-acetyl-3-methoxy-4-acetoxy phenylalanine. The resulting product is then resolved by means of α-phenyl-ethyl amine which upon hydrolysis with aqueous HBr forms levodopa.

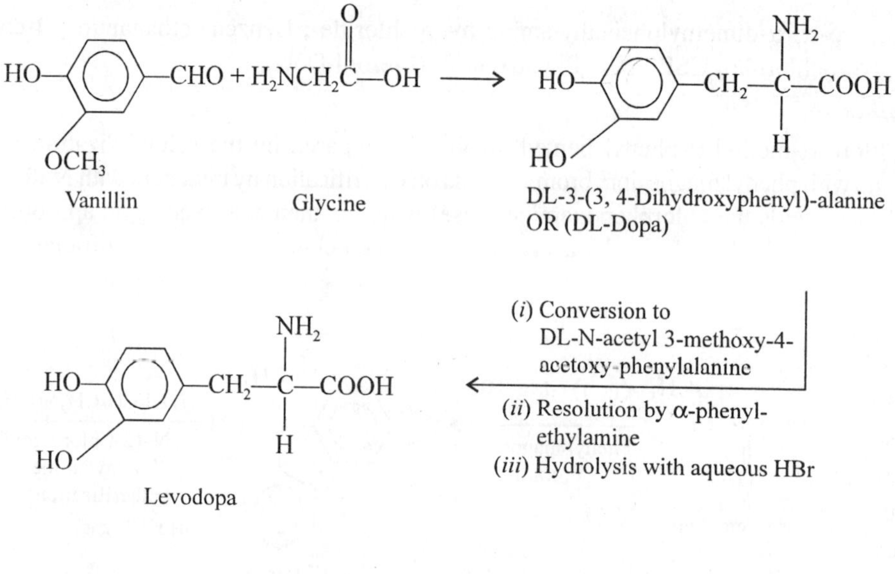

Vanillin   Glycine   DL-3-(3, 4-Dihydroxyphenyl)-alanine
                                       OR (DL-Dopa)

Levodopa

(*i*) Conversion to DL-N-acetyl 3-methoxy-4-acetoxy-phenylalanine
(*ii*) Resolution by α-phenyl-ethylamine
(*iii*) Hydrolysis with aqueous HBr

It is considered to be one of the costliest and single most important drug for the treatment of incapacitating parkinsonism. *The maximum therapeutic effects could be seen vividly on rigidity and hypokinesia. It has also been used successfully to control the neurological symptoms arising from chronic manganese poisoning, which incidentally resemble those of parkinsonism.*

Dose : Initial, oral, 100mg to 1g per day in divided doses with meals ; maintenance dose, 2.5 to 6g daily and must not exceed 8g.

E. *Amantadine Hydrochloride* BAN, USAN, *Amantadine* INN,

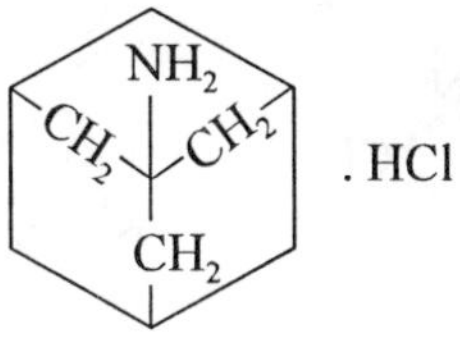

. HCl

1-Adamantanamine hydrochloride ; Tricyclo [3, 3, 1, $1^{3, 7}$] decan-1-amine, hydrochloride ; USP ; Symmetrel$^{(R)}$ (Endo).

Amantadine has been found to potentiate dopaminergic activity and hence it finds its use in the treatment of parkinsonism usually in conjunction with other therapy. It helps to improve hypokinesia and rigidity but usually displays relatively less effect on tremor.

Dose : In parkinsonism, initial, 100mg per day, increased to 100mg twice daily, after one week.

2.4.1. Mechanism of Action. The mechanism of action of all the *'drugs'* described under section 16.2.4. shall now be treated individually in the sections that follows :

2.4.1.1. Benztropine Mesylate. The *'drug'* exerts its distinct central actions to suppress tremor as well as rigidity that are gainfully employed therapeutically in Parkinsonism. It is also employed for the treatment of *extrapyramidal dyskinesia ;* but certainly not *tardive dyskinesia* which is caused due to the administration of various potent tranquillizers, namely : reserpine, chlorpromazine (CPZ) etc. It may be worthwhile to mention here that the *'drug'* fails to cause any sort of central stimulation (*i.e.,* a definite plus point), and instead produces the characteristic sedative effect normally seen amongst the antihistaminics.

SAR of Benztropine. Interestingly, this *'drug'* resembles as well as possesses the dual-characteristic features of **atropine**–an *antimuscarinic drug* (*i.e.,* having potency equivalent to almost 1/4th to that of atropine) ; and **pyrilamine maleate**-an *antihistaminic drug* (*i.e.,* having potency practically equal to that of pyrilamine maleate). One may look at this *'drug'* as an wonderful fusion of the two potential drug molecules namely : atropine and pyrilamine.

2.4.1.2. Orphenadrine Citrate. The *'drug'* is observed to minimise voluntary muscle spasm by a central effect. However, the indications for the citrate are considered usually as an adjunct for the relief of discomfort amalgamated with severe painful musculoskeletal conditions, which is not yet vividly understood, but that may be associated with the *'analgesic characteristics'* of the drug substance. It has been duly observed that it does not exert its action by relaxing directly the tense skeletal muscles in man. However, the observed peripheral atropine-like actions are relatively mild in nature. Importantly, it helps to minimise voluntary muscle spasm by virtue of its central inhibitory activity specifically on the *cerebral motor areas* ; of course, an apparent central effect very much identical to that of atropine.

SAR of Orphenadrine. The *'drug'* is very much closely related to *diphenhydramine*–an *'aminoalkylether'* antihistaminic structurally but possesses much **higher anticholinergic profile** and much **lower antihistaminic activity.**

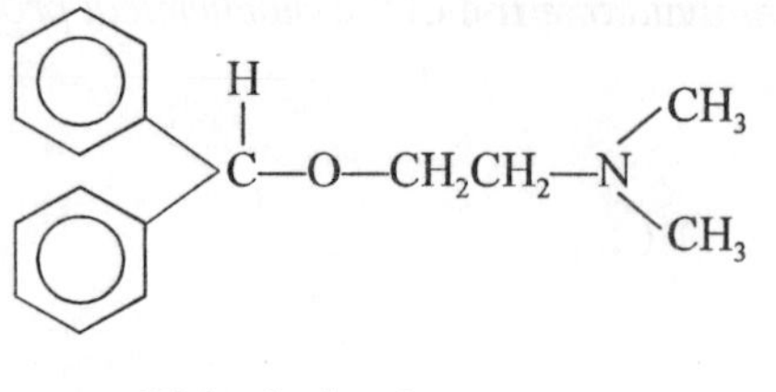

Diphenhydramine 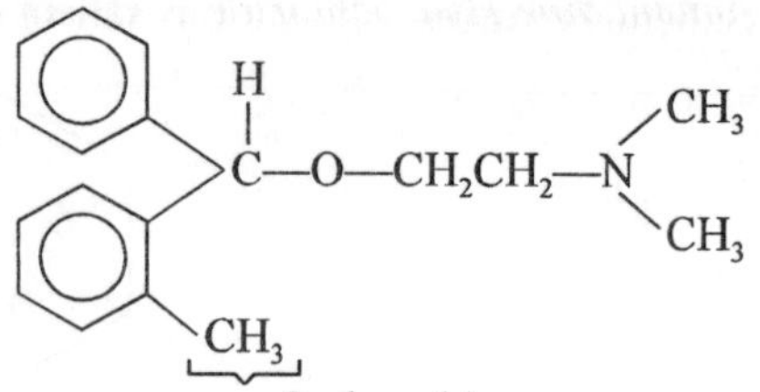   Orphenadrine

The strategical positioning of an additional *o*-methyl group at 2-α-phenylbenzyloxy moiety in orphenadrine makes the glaring difference between the two compounds as shown above.

2.4.1.3. Chlorphenoxamine Hydrochloride. The *'drug'* exerts its action in causing an appreciable reduction in the *muscular rigidity* besides akinesia in patients having Parkinson's disease through the central effects exclusively. It also displays both **antimuscarinic** and antihistaminic activities that makes the *'drug'* a little sensitive and hence may be used with caution in patients.

2.4.1.4. Levodopa. The *'drug'* is usually absorbed between 40-70% through the oral administration. It has been found that even less than 1% gets penetrated into the brain. Besides, it has been observed to undergo decarboxylation resulting to **'dopamine'** to the maximum extent of 99%. Interestingly, the concurrent administration of *carbidopa* normally checks peripheral decarboxylation and increases availability to the brain significantly. Peak concentrations of dopamine normally in the brain take place 1-2 hour after administration. The plasma half-life of levodopa above is 0.5 to 1 hour ; and in combination with carbidopa is 1.2 to 2.3 hour.

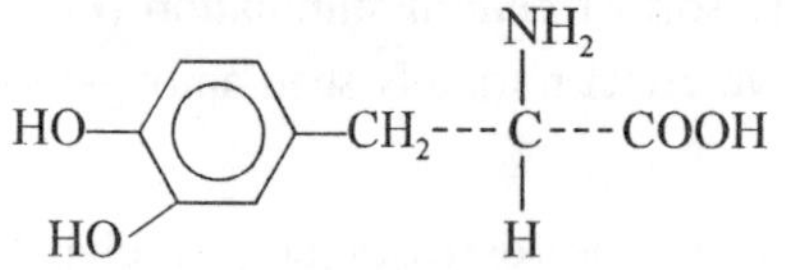

Levodopa 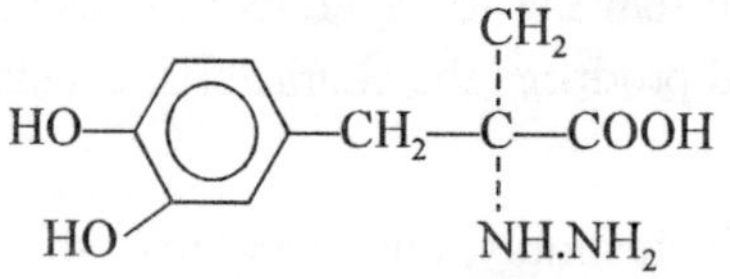   Carbidopa 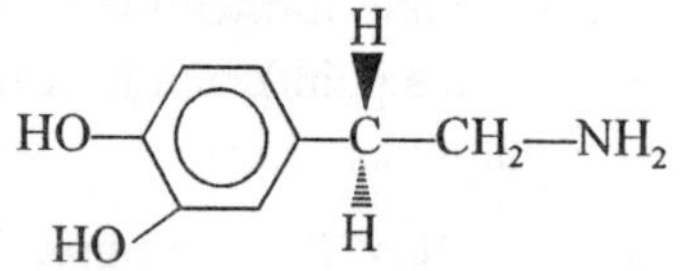   Dopamine

Interestingly, **pyridoxine** is known to antagonize *'levodopa'*, perhaps by promoting *possible premature decarboxylation* (as a coenzyme to dopa decarboxylase) before the *'drug'* has gained entry into the brain. It has been observed that *carbidopa* curtails antagonism by pyridoxine to a certain extent.

(Note : Patients must not take multivitamin supplements containing pyridoxine during the course of levodopa therapy).

Furthermore, both methyldopa and reserpine, that essentially interfere with the *catecholamine synthesis* as well as its storage, exacerbate the Parkinson syndrome and, therefore, antagonize the activity of levodopa. The pharmacological profile of levodopa actually synergized by antimuscarinics.

2.4.1.5. Amantadine Hydrochloride. The *'drug'* is found to inhibit the replication phenomenon of the *influenza type A viruses* specifically at low concentrations. The above admantanamine essentially possesses *two* vital mechanisms, namely :

(*a*) it particularly inhibits an initial *preliminary step* in the viral replication process—most probably **viral uncoating,*** and

(*b*) in certain strains they invariably affect a *later step* which most likely involves the **viral assembly,** perhaps by direct interference with *hemagglutinin processing.*

*Hay AJ, : *Semin Virol,* **3** : 21, 1992.

Salient Features. The various salient features that comprise of the *biochemical processes* are :

(*i*) the major biochemical locus of action is the typical influenza type A virus M2 protein, which happens to be an integral membrane protein component that virtually serves as an ion-channel,

(*ii*) the M2 channel is recognized as a *'proton transport system'*,

(*iii*) the actual interference with the *transmembrane proton pumping,* thereby sustaining a high intracellular proton concentration with respect to the extracellular concentration,

(*iv*) increasing the *p*H-induced conformational changes in the hemagglutinin content in the course of its intracellular transport at a later stage, and

(*v*) the prevailing comformational modifications taking place in the hemagglutinin content thereby check the transferance of the nascent virus particles specifically to the cell-membrane for causing exocytosis.

Special Note. The *'drug'* shows the following *three* distinct characteristic noteworthy features :

(*a*) *it exerts absolutely very little effect on influenza type B,*

(*b*) *it affords seasonal prophylaxis within a range of 70-90% protective against influenza type A,* and*

(*c*) *its primary side effects are very much associated with CNS, and are also dopaminergic.*

Probable Questions for B. Pharm. Examinations

1. What are **Antiparkinsonism Agents** ? How would you classify them ? Give the structure, chemical name and uses of at least **one** compound from each category.

2. Give the structure and uses of the following drugs :

   (*a*) Biperiden hydrochloride,

   (*b*) Benzhexol hydrochloride.

3. Procyclidine—a structurally similar congener of trihexyphenidyl, exhibits similar properties. Discuss its synthesis from 3-(1-pyrrolidinyl)-propiophenone.

4. Describe how would you synthesize ethopropazine hydrochioride—a phenothiazine analogue from diphenylamine.

5. Explain the sequential steps adopted for the synthesis of :

   (*a*) Benztropine mesylate from diphenylmethane,

   (*b*) Levodopa from vanilline.

6. Name the two compounds that are used as **antiparkinsonism agents** and may be syntheized from :

   (*a*) *p*-Chloroacetophenone,

   (*b*) 2-methylbenzaldehyde,

   Give the detailed synthesis of **one** drug.

---

*Douglas RG, *N Engl J Med.* **332**, 443, 1990.

7. Give a comprehensive account of the *mode of action* of various poetnt **antiparkinsonism agents** with the help of some typical examples.

8. Explain the following :

   (*a*) Piperidine hydrochloride and acetophenone yields cycrimine hydrochloride,

   (*b*) Piperidine and acetophenone yields trihexyl-phenidyl hydrochloride.

RECOMMENDED READINGS

1. RC Schwab and AC England, *'Newer Preparations in the Treatment of Parkinsonism'*, *Med. Clin. N. America,* March (1957).

2. AC England and RC Schwab, 'Parkinson Disease Management, *Arch. Int. Med.* 104, 111 and 439 (1959).

3. DLednicer and LA Mitsher, *'The Organic Chemistry of Drug Synthesis,* John Wiley and Sons New York (1995).

4. JEF Reynolds, Ed. *'Martindale : he Extra Pharmacopoeia* (31st Edn.), The Royal Pharmaceutical Society London (1996).

5. PJ Roberts, GN Woodruff and LL Iverson, Eds., *Dopamine, Adv. Biochem'*, *Psychopharmacol.,* Raven Press, New York (1978).

6. DB Calne, *'Developments in the Pharmacology and Therapeutics of Parkinsonism'*, *Ann Neurol.* 1, 111, (1977).

7. JH Block and JM Beale, *'Wilson and Gisvold's Textbook of Organic Medicinal and Pharmaceutical Chemistry,* Lippincott Williams and Wilkins, New York, 11th edn., 2004.

8. Gringauz A, *'Introduction to Medicinal Chemistry',* Wiley-VCH, New York, (1997).

17 Expectorants and Antitussives

1. INTRODUCTION

Expectorants are drugs employed to aid in the relief of congestion in the lower respiratory tract below the epiglottis, in the trachea, bronchi, or lungs and, therefore, they are helpful in the treatment of cough. Expectoration may be caused by (*i*) enhancing brochial secretion, (*ii*) making secretion less viscous, or (*iii*) suppressing cough. Besides expectorants may also possess antiseptic, anaesthetic, or other pharmacological activity.

Release and clearance of sputum may be helped by humidifying the respiratory tract with luke-warm beverages (*e.g.,* tea, coffee) or by inhalation of sodium chloride aerosols. Sometimes the inhalation of a surface-active agent like tyloxapol may find its useful application. Mucolytic agents like (-) bromhexine hydrochloride [Bisolvon$^{(R)}$ (Boehringer Ingelheim)], acetylcystein, trypsin and chymotrypsin have also been found to help the excretion of sputum by changing its structure.

Another school of thought designates the **expectorants** as — *'drugs that are proved to be beneficial in loosening and liquefying mucous, in soothening irritated bronchial mucosa, and in making coughs more productive'*. In reality, these agents are believed to afford an appreciable affect upon the respiratory tract in *two* different manners, namely :

- by minimising the viscosity of the bronchial secretions, which in turn remarkably stimulates their elimination thereby helping in the removal of the local irritants ; and thus, the ineffective coughing is either alleviated significantly or rendered more productive.
- by enhancing the quantum of respiratory tract fluid thereby producing a demulscent action on the dry mucosal lining ; and thus, relieving the unproductive cough considerably.

Antitussives are agents that are employed invariably in the symptomatic control of cough by way of depressing the cough-centre strategically situated in the medulla. Interestingly, these are also commonly known as *anodynes, cough suppressants,* and *centrally acting antitussives.*

Importantly, the *'narcotic analgesic agents'* (see chapter - 10) retained an almost comfortable status in this are upto the recent times, such as : morphine, hydromorphone, codeine, hydrocodone, methadone and levorphanol.

It is, however, pertinent to mention here that in the recent years, a good number of newer drug substances have been synthesized meticulously which remarkably exhibited significant antitussive characteristic feature absolutely devoid of the absurd and most discouraging **'addiction liabilities'** of the aforesaid narcotic drugs. Surprisingly, quite a few of these agents usually exert their activity very much identical fashion *via* a central effect.

1.1. Hypotheses suggested for relief of cough. In early 1960s, *two* befitting hypotheses were put forward to expatiate most logically and convincingly the probable mechanism for the ultimate relief of cough, namely :

(*a*) Salem and Aviado's hypothesis, and

(*b*) Chappel and Seemann's hypothesis.

These hypothesis shall now be discussed briefly as under :

1.1.1. Salem and Aviado's Hypothesis. According to their considered proposal and suggested explanation — **'bronchodilation is a critically essential and important mechanism exclusively responsible for the relief of cough'.** It further gives a plausible account that irritation of the mucosa first and foremost gives rise to bronchoconstriction which subsequently excites the **'cough receptors'.**

1.1.2. Chappel and Seemann's Hypothesis. As per the hypothesis put forward by these researchers — a large variety of the antitussives belonging to this category normally fall into *two* structural variants, namely :

(*a*) **Large structural moieties.** They have a close resemblance to that of **methadone** (see Chapter : 10). *Noscapine* is a suitable example of this type, as shown below :

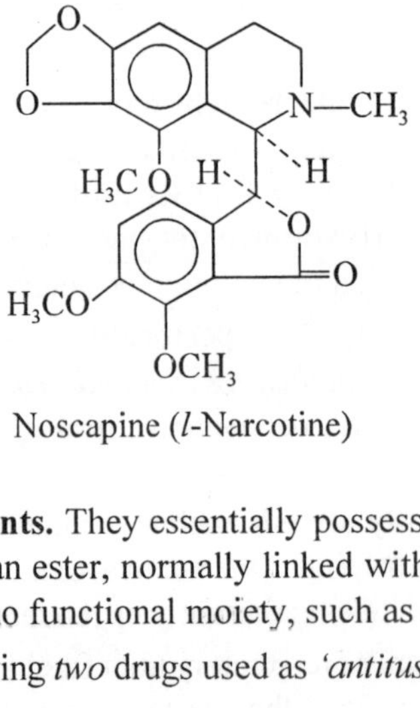

Noscapine (*l*-Narcotine)

(*b*) **Large bulky substituents.** They essentially possess large bulky substituents particularly on the acid residue of an ester, normally linked with the help of a long, ether containing chain to a *tertiary* amino functional moiety, such as : *carbetapentane citrate.*

Exceptions. The following *two* drugs used as *'antitussives'* are exceptions as given below :

(*i*) **Sodium Dibunate :**   (*ii*) **Benzonatate :**

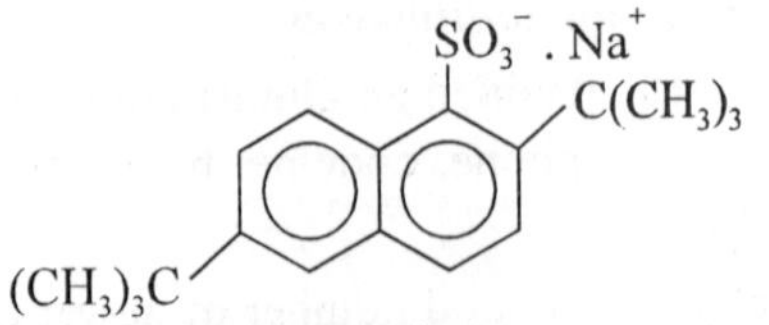

(Antitussive)

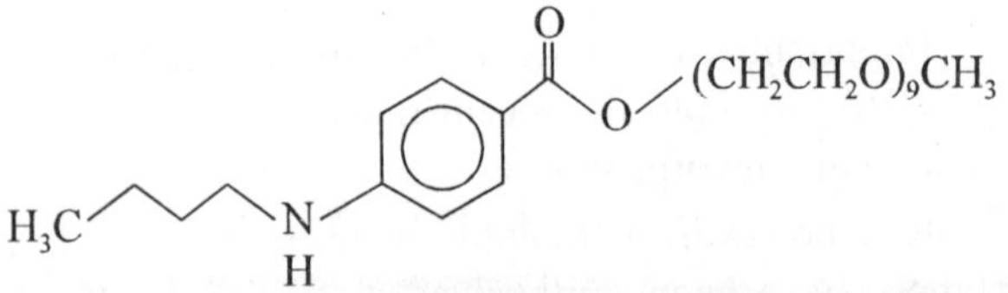

(Antitussive)

2. CLASSIFICATION

Expectorants and antitussive agents may be broadly classified into the following *three* categories :

  (*i*) Sedative Expectorants

  (*ii*) Stimulant (Irritant) Expectorants

  (*iii*) Centrally Acting Antitussive Agents.

2.1. Sedative Expectorants

These group of drugs specifically help in the secretion of a protective mucous film that covers up inflamed membranes and increases the efficiency of the removal of slimy exudates by coughing.

Sedative expectorants may be further sub-divided into *three* different classes :

(*a*) *Saline Expectorants* : These usually enhance bronchial secretion and help to "loosen" the cough, *e.g.,* ammonium carbonate, ammonium chloride, ammonium acetate, alkali citrates (Na or K) and inorganic iodides (KI or NaI).

(*b*) *Nauseant Expectorants* : These act as expectorants in small doses and nauseant and emetic in large doses, *e.g.,* tartar emetic, ipecac, etc. These are usually mixed with sweet-tasting cough syrups that help to cure *croupous bronchitis* in children.

(*c*) *Demulcent Expectorants* : These agents are normally mucilaginous in nature and serve to coat and protect the mucous membrane of the upper respiratory tract, *e.g.,* syrup of acacia, ginger, glycyrrhiza (liquorice) tolubalsam and the like.

A few examples belonging to the class of sedative expectorants will now be discussed here. *Examples* : Acetylcysteine ; Bromhexine hydrochloride ; Ammonium chloride ; Prepared Ipecacuanha ; Liquorice ; Cocillana ; Potassium iodide.

  A. *Acetylcyteine* INN, BAN, USAN,

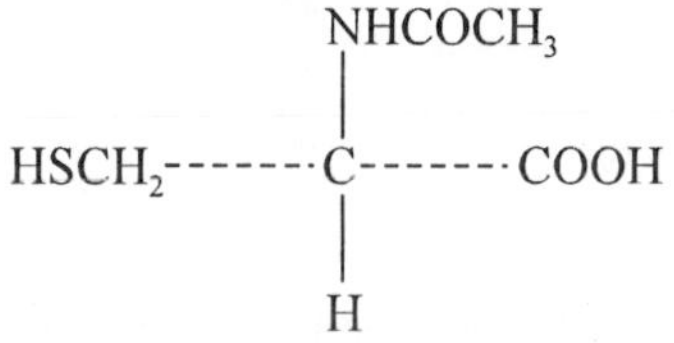

$$
\text{HSCH}_2 \text{------} \underset{\underset{H}{|}}{\overset{\overset{NHCOCH_3}{|}}{C}} \text{------COOH}
$$

N-Acetyl-L-cysteine ; L-Cysteine, N-acetyl-; USP ; Mucomyst[R] (Mead Johnson) ; Aribron[R] (Duncan, Flockhard, U.K.) ;

Synthesis

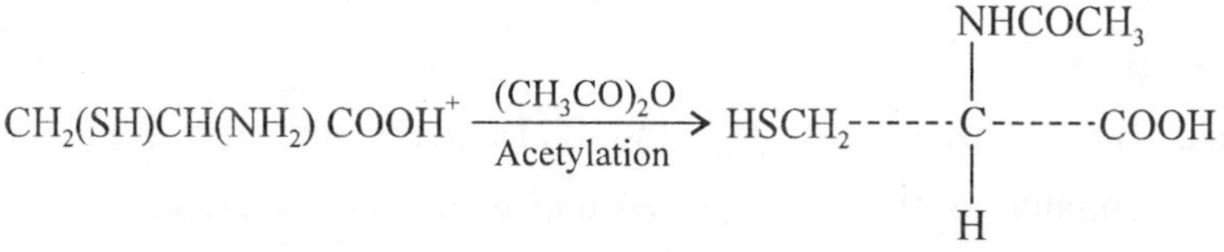

$$
\text{CH}_2(\text{SH})\text{CH}(\text{NH}_2)\,\text{COOH}^{+} \xrightarrow[\text{Acetylation}]{(\text{CH}_3\text{CO})_2\text{O}} \text{HSCH}_2\text{------}\underset{\underset{H}{|}}{\overset{\overset{NHCOCH_3}{|}}{C}}\text{------COOH}
$$

          L-Cysteine   Acetylcysteine

It may be prepared by carrying out the direct acetylation of naturally occurring L-cysteine.

Acetylcysteine is used to *reduce the viscosity of pulmonary secretions and thereby help their removal. Therefore, it is usually incorporated as an adjuvant in preparations meant for bronchopulmonary disorders when mucolysis is sought for. Lowering of viscosity of the mucous may be due to the opening of the disulphide bond present in the sulphydryl moiety in the mucous medium.*

Dose : Direct instillation, 1 to 2 ml of a 10 or 20% solution every 1 to 4 hours ; inhalation of nebulized solution, 3 to 5 ml of a 20% solution daily.

B. *Bromhexine Hydrochloride* BAN, USAN, *Bromhexine* INN,

. HCl

3, 5-Dibromo-Nα-cyclohexyl-Nα-methyltoluene-α, 2-diamine monohydrochloride ; Benzenemethanamine, 2-amino-3, 5-dibromo-N-cyclohexyl-N-methyl-, monohydrochloride ; BP ; Bisolvon[R] (Boehringer Ingelheim, U.K.).

Bromhexine hydrochloride has been reported to alter the structure of bronchial secretion, besides playing the dual role of enhancing the volume and reducing the viscosity of sputum.

Dose : Usual, adult, 8 to 16mg 3 or 4 times per day ; For children below 5 years, 4mg 2 times per day ; and for 5 to 10 years, 4mg 4 times per day ; i.m. or i.v. 8 to 24mg per day.

C. *Ammonium Chloride* BAN, USAN,

NH₄Cl

BP ; USP ; Eur. P., Int. P.; Expigen[R] (Pharmacia, Denm).

It is used as an ingredient of expectorant cough mixtures.

Dose : 300mg to 1g.

D. *Prepared Ipecacuanha* BAN,

BP ; Eur. P., Int. P., Ind. P., Powdered Ipecac USP ;

It is the finely powdered ipecacuanha, the dried root or rhizome and roots of *Cephaelis ipecacuanha* [(*i.e., Uragoga ipecacuaha* (*Rubiaceae*)] adjusted with powdered ipecacuanha of lower alkaloidal content or powdered lactose to contain 1.9 to 2.1% of total alkaloids, calculated as emetine.

Prepared ipecacuanha is employed in smaller doses as an expectorant.

Dose : 25 to 100mg (approximately 0.5 to 2mg of total alkaloid).

E. *Liquorice* BAN,

Glycyrrhiza ; Glycyrrhizae Radix ; BP ; Ind. P., Liquorice root (Eur. P.).

Liquorice essentially consists of the dried unpeeled root and stolons of *Glycyrrhiza glabra (Leguminosae)* containing not less than 25% of watersoluble extractive, having approximately 7% of glycyrrhizin consisting of the potassium and calcium salts of glycyrrhizinic acid (a glucoside of glycyrrhetinic acid). It possesses a characteristic odour and a slightly aromatic sweet taste.

Liquorice is used as a *demulcent and expectorant*. It is mostly used as a *flavouring agent in cough mixtures specifically containing nauseous components like alkali iodides, ammonium chloride, creosote, and cascara liquid extract.* In grandmother's prescription a decoction of liquorice and linseed has long been used reputably as a cure for cough and bronchitis. Deglycyrrhizinised liquorice has been used in the treatment of peptic ulcer owing to its **reduced mineral corticoid activity.**

Dose : Liquorice Extract (BPC 1973), 0.6 to 2g.

F. *Cocillana* BAN,

Grape bark ; Guapi bark ; BP ; Cocillana Liquid Extract (BPC ; 1973).

Cocillana is the dried bark of *Guarea rusbyi (Meliaceae)* containing not less than 3.5% of alcohol soluble extract.

It has been employed as a *substitute for ipecacuanha in the treatment of coughs.* It usually forms an ingredient as the liquid extract along with other expectorants in cough mixtures.

Dose : 0.5 to 1ml.

G. *Potassium Iodide* BAN, USAN,

KI

BP ; USP ; Eur. P ; Ind. P., Ind. P. ; SSKI[R] (Upsher-Smith, U.S.A.).

Potassium iodide is used as an expectorant and also in the treatment of cutaneous lymphatic sporotrichosis.

Dose : 250 to 500mg.

2.1.1. Mechanism of Action. The mechanism of action of a few compounds discussed under section 17.2.1 shall be treated with in the sections that follows :

2.1.1.1. Acetylcysteine. The *'drug'* is usually employed as adjuvant therapy specifically in *bronchopulmonary disorders* when mucolysis is to be accomplished. It is, however, believed that the *sulphhydryl functional moiety* present in the molecule helps to **open** the disulphide bondages in the mucous whereby the viscosity is lowered. Importantly, the mucolytic activity of the *'drug'* is directly associated with the *p*H ; and an appreciable mucolysis takes place within a *p*H range of 6 and 9.

Note. The *'drug'* **when administered either orally or parenterally serves as a potential 'antidote' to check or lower the hepatotoxicity caused due to acetaminophen (paracetamol) overdosage.**

2.1.1.2. Bromhexine Hydrochloride. It has been duly observed that the *'drug'* virtually brings about a change in the structure of the ensuing bronchial secretion. In addition to this, it also plays the dual role of increasing not only the actual volume but also minimising the viscosity of the sputum to a considerable extent.

2.1.1.3. Ammonium chloride. The *'drug'* exerts its action as a **'saline expectorant'.** The basic mechanism by which it brings forth a net loss of the *extracellular fluid* is due to the fact that the NH_4^+ ion gets changed to urea, and the liberated H^+ ion reacts with HCO_3 and similar body buffers. The ultimate outcome being that Cl^- ion displaces HCO_3^- ion ; and the latter eventually gets converted to CO_2. In this way, a significant quantum of Cl^- — load practically excapes reabsorption along with an equivalent amount of cation (mostly Na^+) together with an isoosmotic quantum of water. In short, NH_4Cl releases the secretion of relatively viscous cough and promotes relief to the patient.

2.1.1.4. Prepared Ipecacuanha. The total alkaloids (as *'emetine'*) varying between 1.9 to 2.1% (*w/w*) serve as an expectorant.

2.1.1.5. Liquorice. The *'drug'* contains 7% of *glycyrrhizin* that exerts its action as a demulcent and expectorant.

2.1.1.6. Potassium Iodide. The *'drug'* is found to exert its action by liquefying the thick and tenacious sputum in chronic bronchitis, bronchietasis, bronchial asthma, and pulmonary emphysema. In reality, the precise therapeutic value of KI as an expectorant has not yet been proved substantially and convincingly.

2.2 Stimulant (Irritant) Expectorants

Drugs in this category usually induce healing in chronic inflammatory processes of the mucous membranes of the respiratory tract. These mostly comprise of mildly irritating volatile terpenoid oils and phenolic compounds based on creosote that may be inhaled and augment repair in the inflammed areas of the bronchis.

Some official preparations belonging to this classification are : Creosote ; Guaifensin ; Eucalyptol ; Terpin hydrate ; Sulfogaiacol.

A. *Creosote* BAN,

Wood Creosote ; BPC ; (1959) ; N.F. XII ; Ind. P.

A mixture of phenols obtained from fractional distillation of wood tar consisting mainly of cresol, guaicol, phlorol ($C_8H_{10}O$) and methylcresol.

Creosote has expectorant actions and been used in bronchitis and bronchiectasis.

Dose : 0.12 to 0.6ml.

B. *Guaifensin* INN, USAN, *Guaiphensin* BAN,

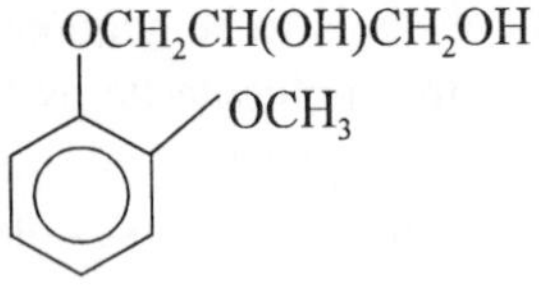

$$OCH_2CH(OH)CH_2OH$$

3-(*o*-Methoxyphenoxy)-1, 2-propanediol ; 1, 2-Propanediol, 3-(2-methoxyphenoxy)-; Glyceryl guaicolate ; Guaiphensin BP ; Guaifensin, USP ; Glyceryl Guaicolate N.F.

Synthesis

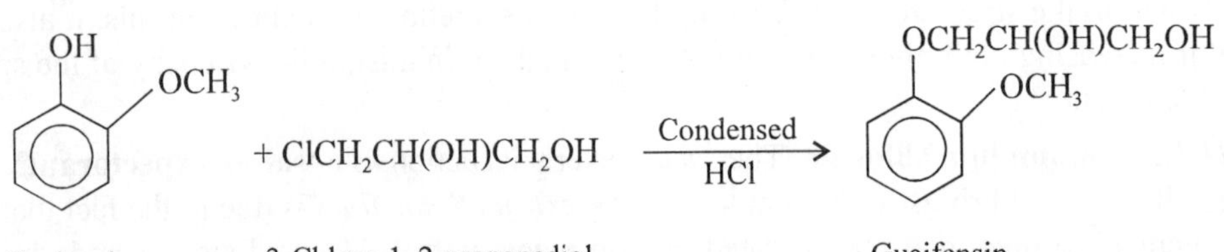

| Guaiacol | 3-Chloro-1, 2-propanediol | Guaifensin |

It may be prepared by the condensation of guaiacol and 3-chloro-1, 2-propanediol through the elimination of a molecule of hydrogen chloride. The reaction proceeds by warming the reactants with a base.

Guaifensin finds its extensive use as an expectorant. It is reported to *lower the viscosity of the tenacious secretions by enhancing the volume of the respiratory tract fluid.*

Dose : Usual, 100mg every 3 or 4 hours.

C. *Eucalyptol* USAN, *Cineole* BAN,

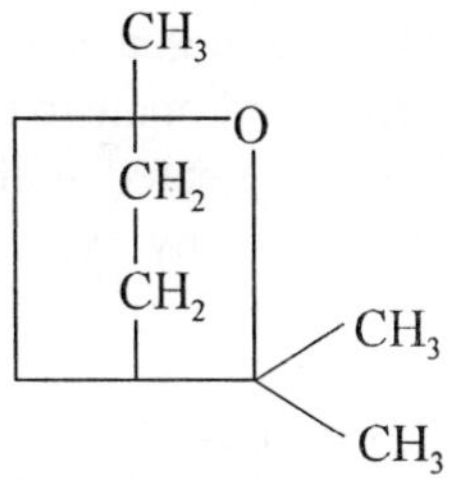

1, 8-Epoxy-*p*-menthane ; 1, 3, 3-Trimethyl-2-oxabicyclo [2, 2, 2] octane ; NF XII ; Cineole BPC (1973).

Preparation : It may be conveniently separated from the purified volatile oils by making use of its salient feature of forming crystals on being subjected to a low temperature.

It is used locally for its antiseptic action in inflammations of nose and throat. It is also used by inhalation in bronchitis.

D. *Terpin Hydrate* BAN, USAN,

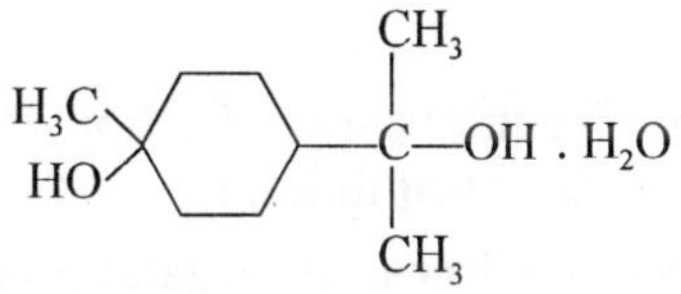

p-Menthane-1, 8-diol monohydrate ; Cyclohexanemethanol, 4-hydroxy-α, α-4-trimethyl-, monohydrate ; Terpine, Terpinol ; BPC ; 1968, USP ; Ind. P.; Terpin Hydrate and Dextromethorphan Hydrobromide Elixir (USP); Terpin Hydrate, Codeine Phosphate, Cineole and Menthol as Tercoda[R] (Sinclair U.K.).

Terpin hydrate may be separated from a good quality terpentine oil (or pine oil) by stirring it with 2 to 3 times its volume of 30% sulphuric acid at a temperature ranging between 20-30°C. The stirring is usually prolonged continuously for 4 to 6 days with intermittent passage of air blowing through the mixture so as to assure thorough contact. The crude crystals of terpin hydrate separate which may be recrystallized from alcohol.

It is frequently employed as an *expectorant in bronchitis, in combination with other components like dextromethorphan hydrobromide and codeine phosphate for cough mixtures.*

Dose : Usual, 125 to 300mg every 6 hours.

E. *Sulfogaiacol* INN, *Potassium Guaiacosulphonate* BAN, *Potassium Guiacosulfonate* USAN,

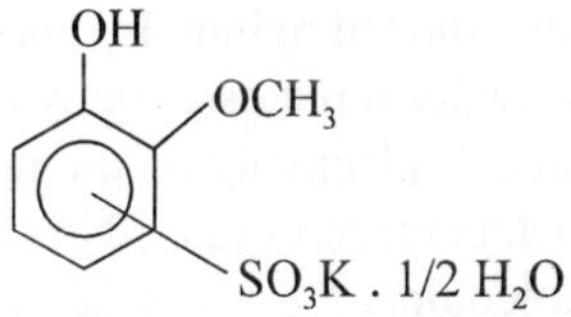

Potassium hydroxymethoxybenzenesulphonate hemihydrate ; Benzenesulphonic acid, hydroxymethoxy-, monopotassium salt, hemihydrate ; Broncovanil[R] (Scharper, Italy).

Synthesis

It is reported to be a mixture of 3- and 5-sulphonate isomers. It may be prepared by the interaction of guaiacol with sulphuric acid at 70 to 80°C to form guaiaco sulphonic acid. After adequate dilution the reaction mixture is *first* neutralized with barium carbonate to remove excess of H_2SO_4 as a precipitate of barium sulphate ; and *secondly*, the filtrate is treated with potassium carbonate to neutralize the guaicosulphonic acid itself and also to precipitate any excess barium. The reaction mixture is filtered and the filtrate concentrated to crystallization.

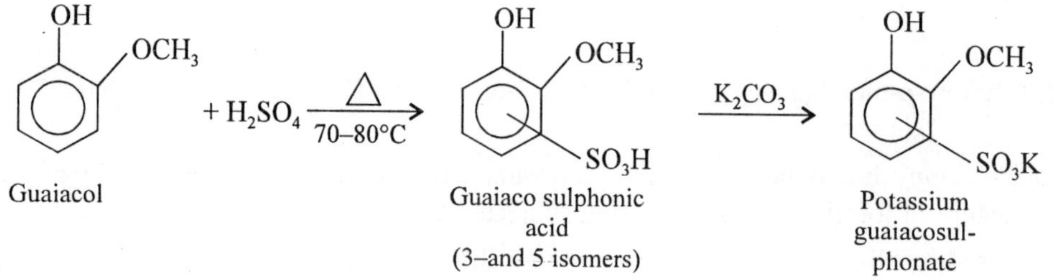

Guaiacol   Guaiaco sulphonic   Potassium
                                  acid   guaiacosul-
                            (3–and 5 isomers)   phonate

Potassium guaiacosulphonate is employed both as an *expectorant in bronchitis and also as an intestinal antiseptic.*

Dose : 0.5 to 1g.

2.2.1. Mechanism of Action. The mechanism of action of certain compounds described under section 17.2.2. are dealt with in the sections that follows :

2.2.1.1. Guaifensin. It has been established adequately through various subjective clinical studies that the action of guaifensin essentially makes better the *'dry unproductive cough'* by (*a*) distinctly lowering the sputum viscosity ; (*b*) difficulty encountered in expectoration ; and (*c*) enhancing the sputum volume considerably. However, experimentally, it simply augments the respiratory tract secretions significantly ; and that too when given in doses much higher than those usually employed therapeutically.

2.2.1.2. Eucalyptol (Cineol ; Cajeputol). The *'drug'* exerts its *'antiseptic effect'* in the inflammations of the throat and nose. It is quite often employed by inhalation in bronchitis.

2.2.1.3. Terpin Hydrate. The *'drug'* exhibits its therapeutic action in *bronchitis* as an expectorant. It has been duly observed that the **'terpin hydrate elixir'** as such normally contains too little of the active compound to make it effective single ; and, therefore it is invariably used mainly as a vehicle for the cough mixtures, namely : (*a*) **Terpin Hydrate and Dextromethorphan Elixir ;** and (*b*) **Terpin Hydrate and Codeine Elixir.**

2.3 Centrally Acting Antitussive Agents

These drugs specifically act by depressing the meduallary cough centre in the central nervous system (CNS) to suppress cough reflex. They are mostly narcotics. The centrally acting antitussives consist primarily of the phenanthrene alkaloids of opium. In this classification are other synthetic agents that are not derived from the opium derivatives but essentially exhibit antitussive action. These drugs, like the opium alkaloids, are thought to act selectively on the medullary centres to suppress the cough reflex. *Examples* : Benzonatate ; Carbetapentane citrate ; Noscapine ; Levopropoxyphene napsylate ; Dextromethorphan hydrobromide ; Pholcodine.

A. *Benzonatate* INN, BAN, USAN,

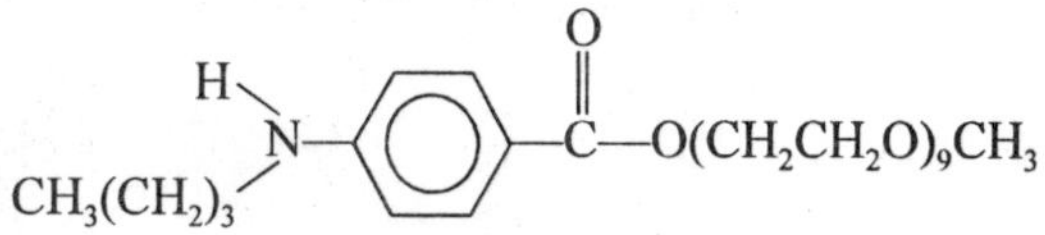

2, 5, 8, 11, 14, 17, 20, 23, 26-Nonaoxaoctacosan-28-yl p(butyl-amino) benzoate ; Benzoic acid, 4-(butylamino)-, 2,5,8,11,14,17,20,23,26-nonaoxaoctacos-28-yl-ester ; Benzonatine ; USP ; Tessalon[R] (Endo).

Synthesis

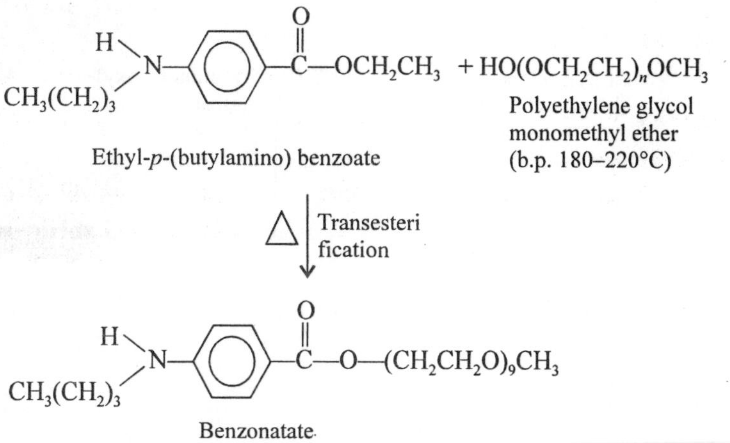

Ethyl-*p*-(butylamino) benzoate

Polyethylene glycol monomethyl ether (b.p. 180–220°C)

Transesterification

Benzonatate

Benzonatate may be prepared by the transesterification of ethyl-*p*-(butylamino) benzoate with a polyethyleneglycol monomethyl ester (b.p. 180-220°C) at 1 mm Hg. The reaction is carried out *in vacuo* whereby a thin stream of xylene is made to pass through it. After complete removal of the traces of moisture and volatile components, a solution of sodium methoxide in methanol is added to the reaction mixture. The contents are heated *under vacuo,* after addition of xylene, for 2 to 3 hours at 100°C. The crude benzonatate thus obtained may be purified by suitable means.

It is a potent antitussive agent. *It usually acts by inhibiting transmission of impulses of the cough reflex in the vagal nuclei of the medulla and predominantly depresses polysynaptic spinal reflexes. It is regarded as a cough suppressant acting both centrally and peripherally.*

Dose : 100 to 200mg ; usual, 100mg 3 times daily.

B. *Carbetapentane Citrate* BAN, USAN, *Pentoxyverine* INN,

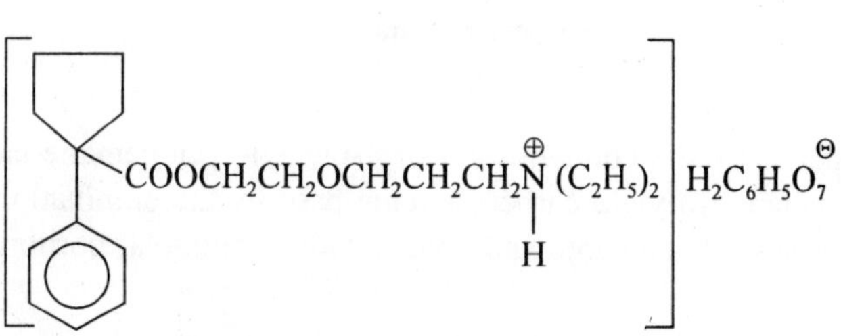

2-[2-(Diethylamino) ethoxy] ethyl 1-phenylcyclopentane-carboxy citrate (1:1) ; NF XIII ; Toclase[R] (Pfizer).

Synthesis

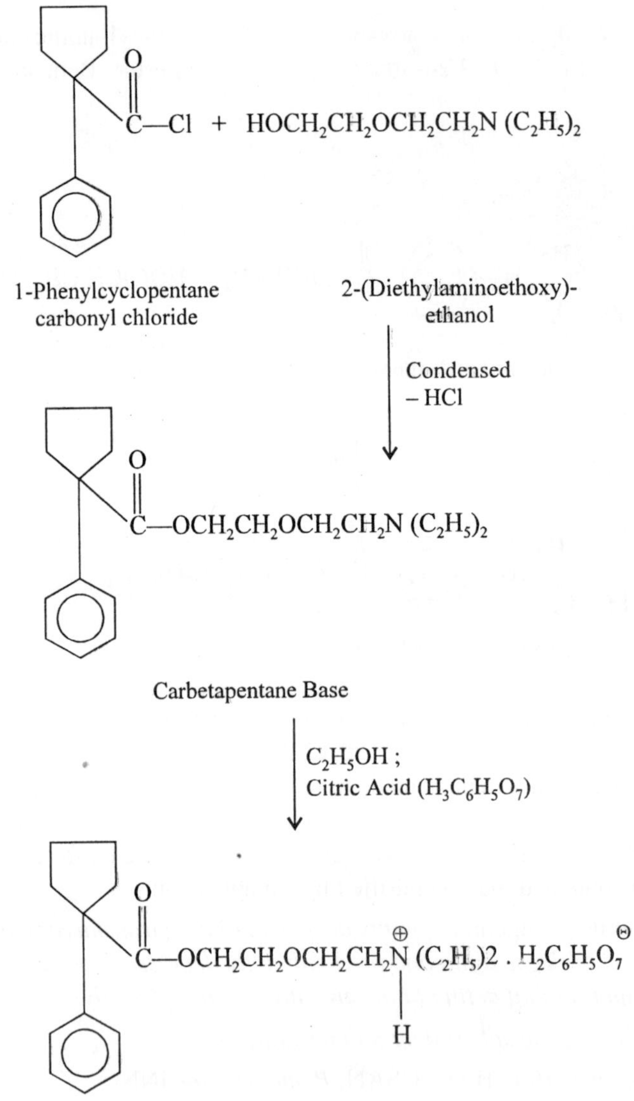

1-Phenylcyclopentane
carbonyl chloride

2-(Diethylaminoethoxy)-
ethanol

Condensed
– HCl

Carbetapentane Base

C_2H_5OH ;
Citric Acid $(H_3C_6H_5O_7)$

Carbetapentane Citrate

It may be prepared by the condensation of 1-phenyl-cyclopentane carbonyl chloride with 2-(diethylaminoethoxy) ethanol to yield carbetapentane base by the elimination of a mole of hydrogen chloride. The base is dissolved in ethanol and treated with a equimolar portion of citric acid to give the official compound.

Carbetapentane citrate is a cough suppressant and is reported to reduce bronchial secretions. It is found to be *effective in acute coughs associated with common upper respiratory infections.*

Dose : 25 to 150mg per day in divided doses.

C. *Noscapine* INN, BAN, USAN,

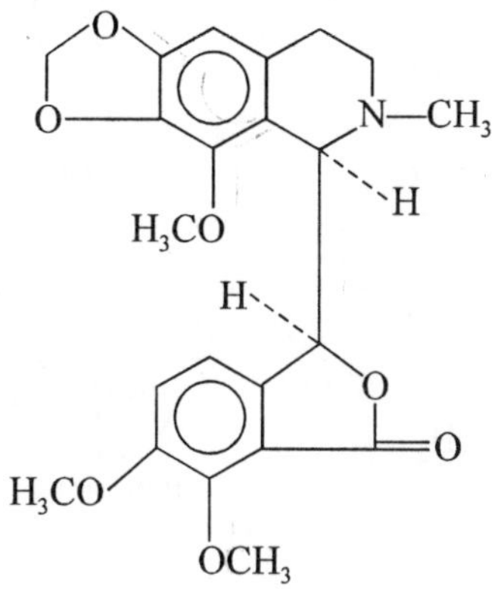

(3S)-6, 7-Dimethoxy-3-[(5R)-5, 6, 7, 8-tetrahydro-4-methoxy-6-methyl-1, 3-dioxolo [4, 5-g] isoquinolin-5-yl] phthalide ; Narcotine ; L-α-Narcotine ; BP ; USP ; Eur. P ; Int. P ; Tusscapine[R] (Fisons).

It is isolated from opium in which noscapine concentration ranges from 3 to 10%.

Noscapine is invariably employed in the control and management of cough due to bronchial asthma and pulmonary emphysema. It remakably reduces both the frequency and intensity of coughing paroxyms. Besides, it possesses weak bronchodilator actions and stimulates the respiration. It has no analgesic activity.

Dose : Usual, 15 to 30mg 3 or 4 times per day.

D. *Levopropoxyphene Napsylate* BAN, USAN, *Levopropoxyphene* INN,

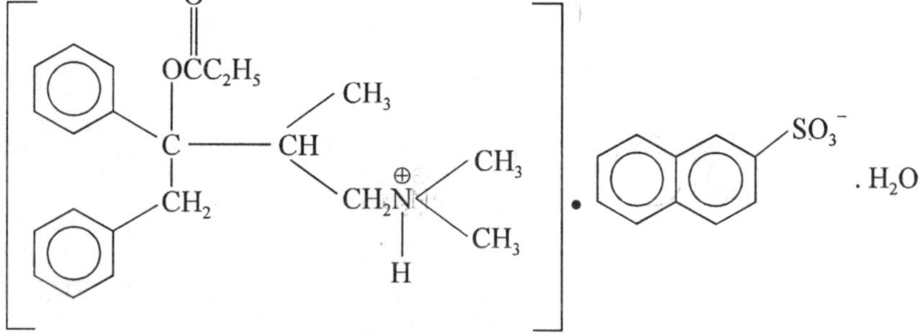

2-Naphthalenesulphonic acid compound with (—)-α-[2-(dimethylamino)-1-methylethyl]-α-phenylphenethyl propionate (1:1) monohydrate ; Benzeneethanol, α-[2-dimethyl-amino)-1-methylethyl]-α-phenyl-, propanoate (ester), [R-(R*, S*)]-, compound with 2-naphthalenesulphonic acid (1:1); USP ; Novrad[R] (Lilly).

Synthesis

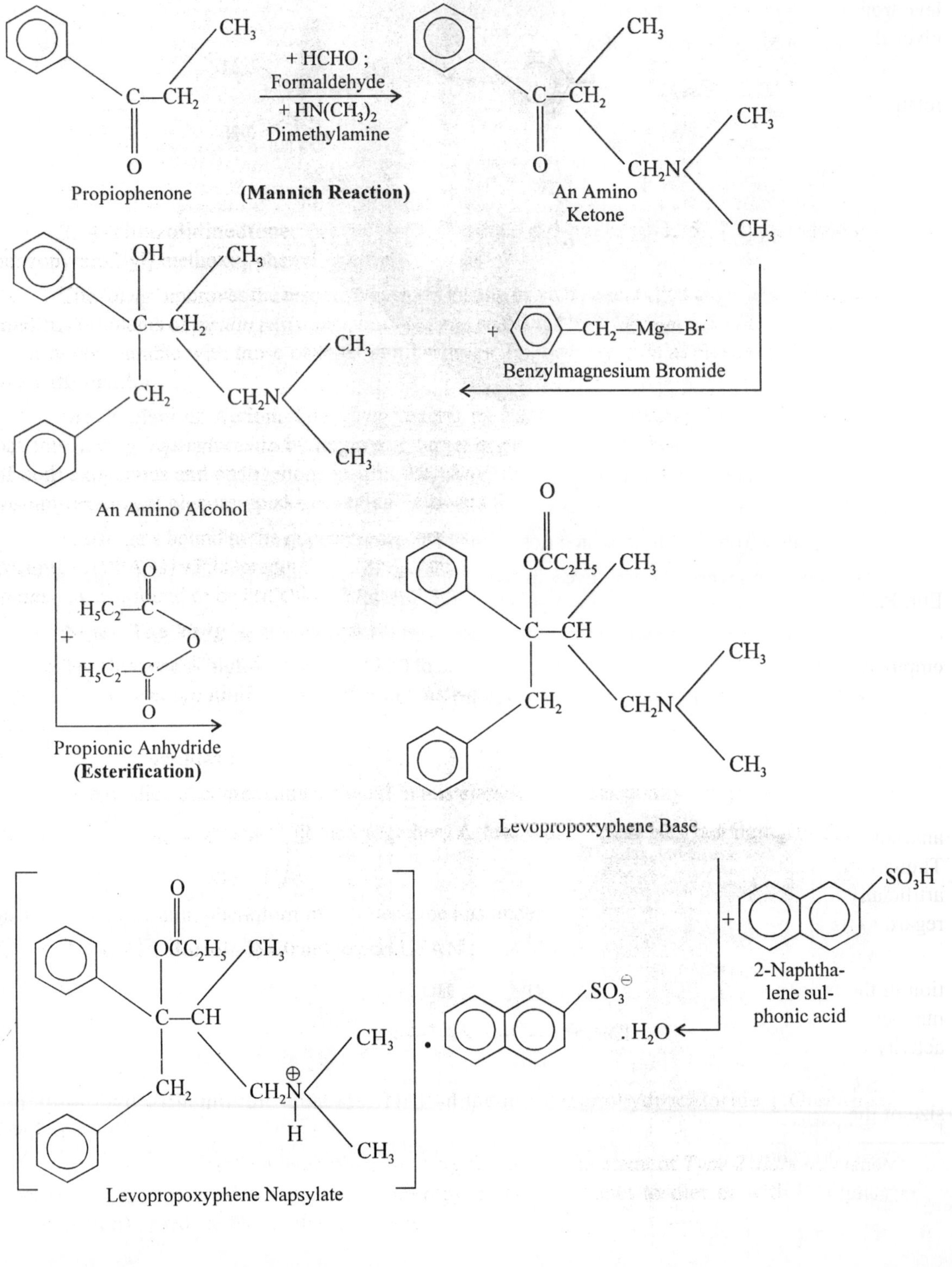

Propiophenone   **(Mannich Reaction)**   An Amino
                                                                    Ketone

An Amino Alcohol

Propionic Anhydride
(Esterification)

Levopropoxyphene Base

2-Naphtha-
lene sul-
phonic acid

Levopropoxyphene Napsylate

Mannich reaction of propiophenone with formaldehyde and dimehtylamine yields the corresponding amino ketone, which on treatment with benzylmagnesium bromide gives rise to the corresponding amino alcohol. Esterification of this alcohol with propionic anhydride forms the levopropoxyphene base, which on reaction with an equimolar quantity of 2-naphthalene sulphonic acid gives the official compound.

Levopropoxyphene napsylate is an antitussive and a cough suppressant. It has been found to be relatively less potent than codeine in the treatment of cough reflexes.

Dose : Usual, 50 to 100mg (base equivalent) after every 4 hours.

E. Dextromethorphan Hydrobromide

This compound has already been discussed in details under 'narcotic analgesics'.

F. *Pholcodine* INN, BAN, USAN,

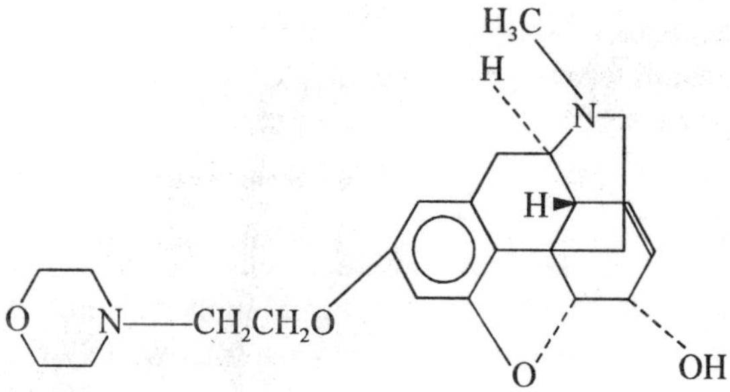

Morpholinylethylmorphine ; O^3-(2-Morpholinoethyl) morphine monohydrate ; Pholcod ; BP ; Eur. P., Int. P.; Ethnine Simplex[R] (Purdue Frederick).

It is a cough suppressant with mild sedative but practically negligible analgesic action. It is employed for the relief of unproductive cough.

Dose : Adult, 5 to 15mg ; Children over 2 years 5mg ; children below 2 years 2.5mg.

2.3.1. Mechanism of Action. The mechanism of action of the compounds discussed under section 17.2.3. shall now be treated individually in the sections that follows :

2.3.1.1. Benzonatate. The *'drug'* exerts its action by reducing the cough reflex at its source by anaesthetizing the *stretch receptors* strategically located in the respiratory passages, lungs and pleura. Though its antitussive potency and profile is fairly comparable to that of *codeine* when evaluated against artificially (experimentally) induced cough in man as well as in animals, yet it is slightly inferior with regard to its effect in comparison to codeine against cough associated with clinical illness.

2.3.1.2. Carbetapentane Citrate. The *'drug'* is found to act by causing an appreciable reduction in the bronchial secretion. Morren*(1957) and Levis *et al.*** (1955) synthesized and evaluated a number of closely related compounds, of which carbapentane was found to be the most active. Its activity was one and a half times more potent than *codeine.*

SAR of Carbetapentane. It has been duly established that either by increasing or decreasing the size of the hydrocarbon ring of carbetapentane remarkably lowers its therapeutic activity.

---

*Morren HG, *Chem. Abstr.* **51,** 7443, 1957.

Levis S *et al. Arch. Int. Pharmacodyn.* **103, 200, 1995.

2.3.1.3. Noscapine [(–)-Narcotine]. Noscapine* usually makes up 0.75—9% of opium alkaloids. It is found to exert a marked and pronounced action both on the frequency as well as the intensity of the coughing paroxyms.

2.3.1.4. Levopropoxyphene Napsylate. The (–)-isomer of the *'drug'* (*i.e.,* levopropoxyphene) showed greater antitussive activity in comparison to either the (+)-isomer or the racemic mixture.

SAR of Levopropoxyphene. In a series of congeners of the esters of 1, 2-diphenyl-4-dialkylamino-2-butanol which were duly synthesized and tested for their corresponding *antitussive* and *analgesic* activity the following compound exhibited an active **antitussive property** without any **analgesic activities.**

(Antitussive Activity)

2.3.1.5. Dextromethorphan Hydrobromide. The *'drug'* acts by controlling cough spasms by depressing the cough centre in the medulla. It has been amply demonstrated in man that it exhibits a cough depression potency almost one-half that of *codeine.* Interestingly, the *'drug'* does not afford any **addiction,** whatsoever, even after the usage of large doses for prolonged durations. Besides, it possesses the antitussive characteristics of codeine, without having any analgesic, central depressant, and constipating features. However, it legitimately provides an enormous opportunity to register the ensuing specificity of action displayed by quite intimately related molecules.

In this particular instance, the (+) and (–) isomers both should get attached to the definitive receptors actually responsible for the suppression of the cough reflex. In fact, the prevailing (+) form (isomer) bears a steric relationship which essentially precludes being attached to the definitive receptors particularly associated in the various therapeutic activities, such as : analgesic, constipative, addictive, and other properties displayed by the corresponding (–) form (isomer).

2.3.1.6. Pholcodine. The *'drug'* exerts its action as an effective cough suppressant, having almost minimal side effects and practically very little physical-dependence liability.**

Probable Questions for B. Pharm. Examinations

1. Give the structure and chemical name of the **five** important drugs that are used abundantly in :
    (*a*) Enhancing bronchial secretion,
    (*b*) Making secretion less viscous, and
    (*c*) Suppressing cough.

---

*An opium alkaloid first isolated in 1817 by Robiquet, rather easily from the drug by ether extraction.

May AJ *et al. Brit J Pharmacol,* **9, 335, 1954.

2. How would you classify the **expectorants** and **antitussive agents ?** Support your answer with the help of least one example from each category.

3. What are **sedative expectorants ?** Classify them and give the structure chemical name and uses of one compound from each group.

4. Give a brief account of the **'stimulant expectorants'.** Discuss the synthesis of guaiphensin and potassium guaiacosulphonate from guaiacol with 3-chloro-1, 2-propanediola nd sulphuric acid respectively under different experimental parameters.

5. Give the structure, chemical name and uses of **two** important drugs obtained from the natural products *viz* ; eucalyptus oil and pine oil (turpentine oil).

6. 'Centrally acting antitussive agents act by depressing the medullawry cough centre in the CNS to suppress cough reflexes'. Justify the statement by citing at least **three** potent drugs.

7. Discuss the synthesis of :

   (*a*) Benzonatate

   (*b*) Carbetapentane citrate.

8. Name **two** drugs that act as centrally acting antitussive agents and are analogues of morphine. Discuss the synthesis of **one** such compound.

9. Mannich reaction of propiophenone yields levopropoxyphene napsylate. Describe its course of reaction to the final product.

RECOMMENDED READINGS

1. EM Boyd, *'Expectorants and Respiratory Tract Fluids'*, *Pharmacological Reviews,* Dec. (1954).

2. MC Griffiths, *'USAN and the USP Dictionary of Drug Names',* United States Parmacopeial Convention, Inc. Rockville (1985).

3. *Remington : The Science and Practice of Pharmacy,* Vol. I and II (20th, edn.), Lippincot Williams and Wilkins , New York (2000).

4. ME Wolff (*Ed*) : *Burger's Medicinal Chemistry and Drug Discovery*, (5th edn), John Wiley and Sons. Inc., New York, (1995).

5. Block JH and Beale JM, *'Wilson and Gisvold's Textbook of Organic Medicinal and Pharmaceutical Chemistry*, Lippincott Williams and Wilkins, New York, 5th edn, (2004).

18 Sulphonamides

1. INTRODUCTION

A considerable progress had been made before 1930 towards the development of externally applicable bactericides, but most bactericides that could be administered internally having reasonably safety margin unfortunately lost their activity in the presence of blood serum. There are limited agents available for the treatment of most diseases of bacterial origin. Diseases like pneumonia, meningitis, dysentry etc., could not be treated effectively until the epoch-making discovery of **sulpha drugs**.

The compound *p*-aminobenzenesulphonamide, now known as **sulphanilamide,** was first synthesized by Gelmo in 1908 as an intermediate in the study of azo dyes. Surprisingly it was many years before its therapeutic value was actually ascertained. Gerhard Domagk* in 1935 screened a number of these azo-dyes for their antibacterial effects and observed that they were active against *streptococci*. In 1935, a German firm prepared a red dye 4-sulphonamide-2', 4'-diamino-benzene or *p'*-sulphonyl chrysoidine, and after three years Domagk suggested significant curative properties of this compouhd and named it *Prontosil*.

Trefouel *et al.*** (1935) at Pasteur Institute discovered that *Prontosil* breaks down in the tissues to *p*-aminobenzenesulphonamide, now known as **sulphanilamide,** and suggested that the antibacterial characteristics of the drug resided in this part of the molecule.

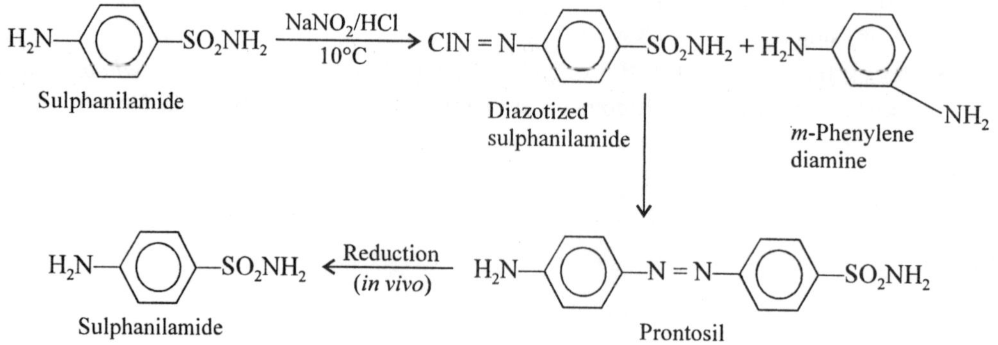

*Domagk, G : *Dtsch. Med. Wochenschr.* **61**, 250, 1935.

Trefouel J *et al.* : *CR Senaces Soc. Biol.* **120, 756, 1935

Prontosil may be prepared by the diazotization of sulphanilamide and subsequent reaction with *m*-phenylene-diamine.

Interestingly, it has been observed that *Prontosil* is absolutely **inactive** *in vivo* but possesses superb and excellent antimicrobial activity *in vivo*. In fact, this specific characteristic property of the drug eventually gained overwhelming recognition and stimulated an astronomical extensive and intensive research activity focussed onto the **sulphonamides.**

Fuller* (1937) further substantiated and confirmed by isolating *'free sulphonamide'* from the blood and urine of subjects being treated with *Prontosil*. Nevertheless, copius clinical findings were adequately reported with *Prontosil* and its **active metabolite**, sulphanilamide, in the control, management and treatment of *puerperal sepsis*** and meningococcal infections. In reality, these critical findings and observations legitimately and judiciously opened the prevailing *'modern era of chemotherapy'*; besides, afforded a tremendous push towards the very concept and ideology of the **'prodrug'.**

In 1937, two British researchers prepared **'sulphapyridine'** that was indeed the first and foremost structural analogue of **'sulphanilamide'.** This particular compound proved to be a grand and tremendous success in curing pneumonia. This magnificent discovery, in fact, paved the flood gates for the synthesis and screening of hundreds of derivatives of *sulphanilamide,* but only a few have retained the glory of being potent medicinal compounds.

As on date, there exist a few *typical sulphonamides* and particularly the *sulphonamide-trimethoprim combinations* which find their applications exclusively and most extensively for the management and treatment of the **opportunistic infections** in humans having AIDS.*** A few typical examples are as illustrated below :

S.No.	Sulphonamide Commonly Used	Disease/Infection
1.	Trimethoprim + Sulphamethoxazole	• Treatment/prophylaxis of *Pneumocystis carinii* pheumonia. • First attack of Urinary Tract Infections (UTIs).
2.	Pyrimethamine + Sulphadiazine	• Treatment and prophylaxis of cerebral taxoplasmosis.
3.	Silver sulphadiazine + Mafenide	• Burn therapy : prevention and treatment of bacterial infection.

The two N-atoms present in the sulphanilamide molecule have been designated N_1 and N_4 as shown below :

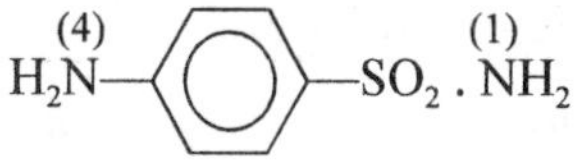

$$\overset{(4)}{H_2N}-\langle\bigcirc\rangle-SO_2 . \overset{(1)}{NH_2}$$

Many structural modifications of sulphanilamide were made by the substitution of heterocyclic aromatic nuclei at N_1 which yielded highly potent compounds. The substitution at N_4 is comparatively rare. It must be noted that the *ortho-* and *meta-*isomers are valueless therapeutically, and any substitution on the aromatic ring either destroys or reduces the activity of the drug.

Since the past four decades sulphonamides have been extensively used against many common Gram-positive bacterial infections.

---

*Fuller AT : *Lancet,* **1** : 194, 1937.

**Septicemia following child birth.

***McDonald L *et al. Formulary,* **31** : 470, 1996.

It is, however, pertinent to state here that one may take cognizance of the various cardinal factors while selecting *'systemic antimicrobial agents'* for therapy in subjects must include :

- identification of probable or specific microorganisms
- antimicrobial susceptibility
- bactericidal *Vs* bacteriostatic activity
- status of host
- allergy history, age, pharmacokinetic factors, renal and hepatic function, pregnancy status, genetic or metabolic abnormalities
- anatomical site of infection and host defenses, particularly neutrophil function.

2. CLASSIFICATION

Sulphonamides *i.e.,* the *systemic antibacterial drugs* may be classified broadly on the basis of their **site of action** as described in the sections that follows :

2.1. Sulphonamides for General Infections

These sulphonamides are invariably employed against the streptococcal, meningococcal, gonococcal, staphylococcal and pneumococcal infections.

Examples : sulfanilamide, sulfapyridine, sulfathiazole, sulfadiazine, sulfamerazine, sulfadimidine, sufalene, sulfamethizole etc.

A. *Sulfanilamide* INN,

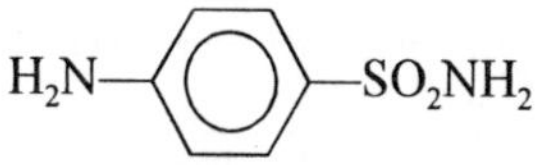

$$H_2N-\langle\bigcirc\rangle-SO_2NH_2$$

p-Aminobenzenesulfonamide ; Sulphanilam ; Solfammide ; BPC ; 1968 ; Int. P., Rhinamid[(R)] (Bengué U.K.).

Synthesis :

Sulfanilamide may be prepared by any one of the *three* following methods, namely :

Method-I (From Benzene)

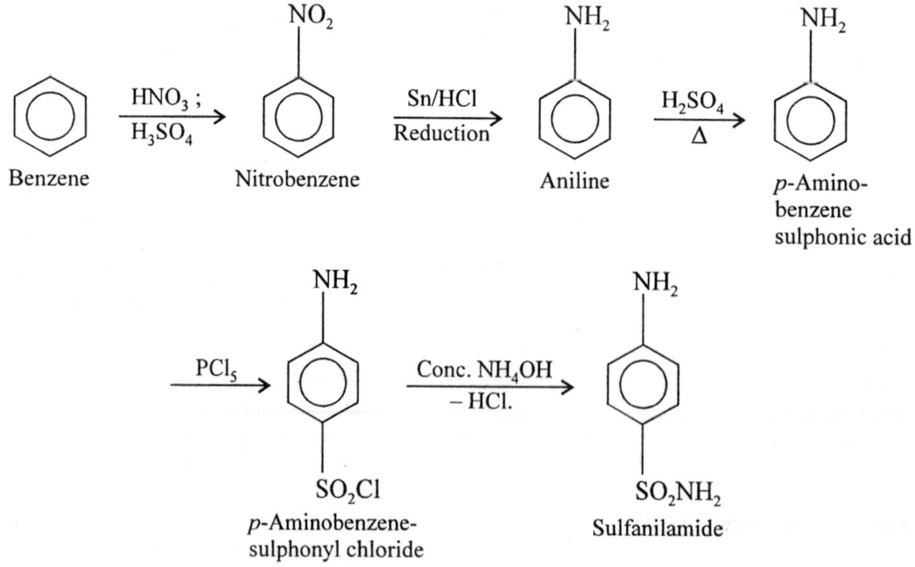

| Benzene | $\xrightarrow[\text{H}_3\text{SO}_4]{\text{HNO}_3 ;}$ | Nitrobenzene | $\xrightarrow[\text{Reduction}]{\text{Sn/HCl}}$ | Aniline | $\xrightarrow[\Delta]{\text{H}_2\text{SO}_4}$ | *p*-Amino-benzene sulphonic acid |

$\xrightarrow{\text{PCl}_5}$ *p*-Aminobenzene-sulphonyl chloride (SO_2Cl) $\xrightarrow[\text{– HCl.}]{\text{Conc. NH}_4\text{OH}}$ Sulfanilamide (SO_2NH_2)

Benzene on nitration yields nitrobenzene which on reduction gives aniline. *p*-Amino benzene sulphonic acid is obtained by treating aniline with hot concentrated sulphuric acid which on chlorination with phosphorus pentachloride gives *p*-aminobenzene sulphonyl chloride ; and this on amination with concentrated ammonia solution yields sufanilamide.

Method-II (From Sulphanilic Acid)

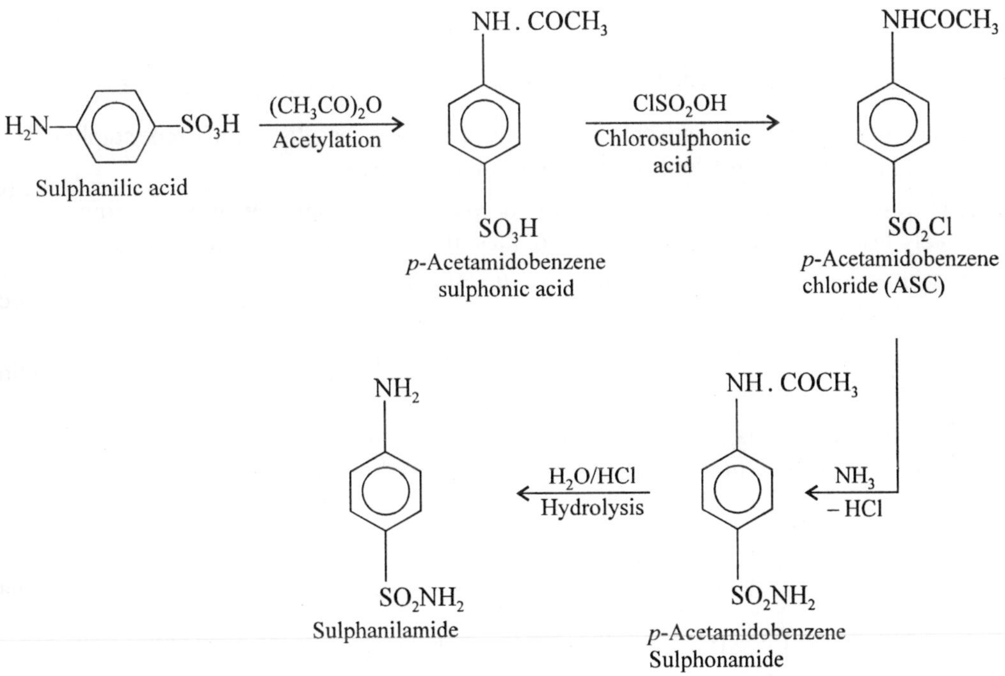

The free and active amino function in sulphanilic acid is first protected by acetylation and the resulting *p*-acetamido-benzene sulphonic acid is chlorinated with chlorosulphonic acid to obtain *para*-acetamido benzene sulphonyl chloride. This on amination with concentrated ammonia solution changes into its corresponding sulphonamide analog, which on hydrolysis results into the formation of sulfanilamide.

Method-III (From Acetanilide)

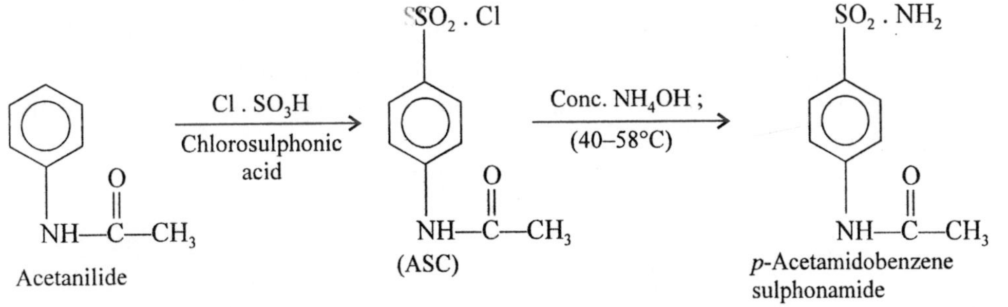

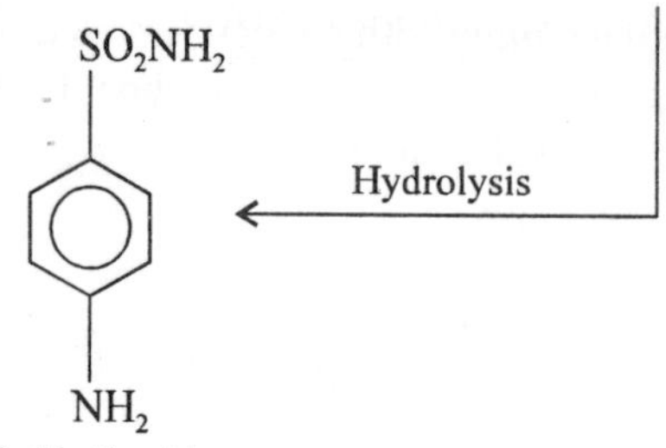

Sulfanilamide

Acetanilide on treatment with chlorosulphonic acid gives *p*-acetamidobenzene sulphonyl chloride which on amination and further hydrolysis yields sufanilamide.

It is now used very rarely because of its high toxicity and hence now being replaced by comparatively less toxic sulpha drugs and antibiotics. It is still used in veterinary medicine.

B. *Sulfapyridine* INN, USAN, *Sulphapyridine* BAN,

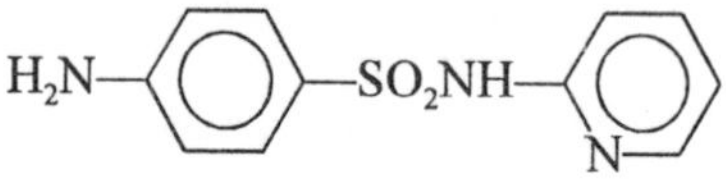

N^1-2-Pyridylsulfanilamide ; Benzene sulfonamide, 4-amino-N-2-pyridinyl-; Sulphapyrid ; Sulphapyridine BP ; BPC ; Sulfapyridine USP ; M & B 693$^{(R)}$ (May & Baker).

Synthesis :

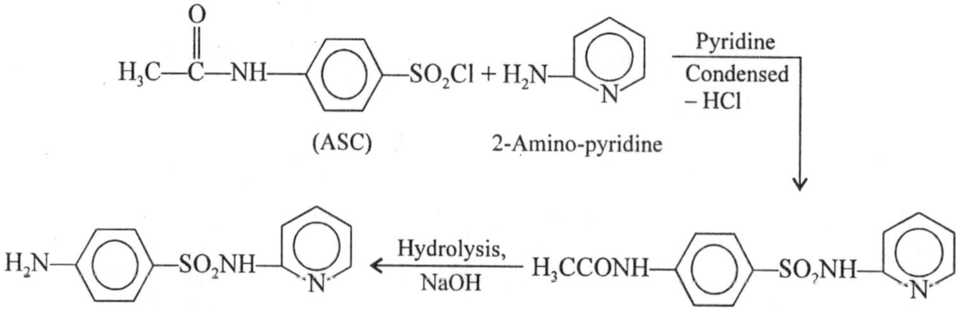

p-Acetamidobenzene sulphonyl chloride (ASC) is condensed with 2-aminopyridine using pyridine as a solvent, followed by alkaline hydrolysis of the resulting product to yield sulfapyridine.

It is mainly used in the treatment of *dermatitis herpetiformis for such patients who do not give positive response to dapsone.* It is effective in pneumonia. Though more potent than sulfanilamide, it is more toxic and has been replaced by sulfadiazine.

Dose : 0.5 to 3g daily ; USP dose range 0.5 to 6g daily.

C. *Sulfathiazole* INN, USAN, *Sulphathiazole* BAN,

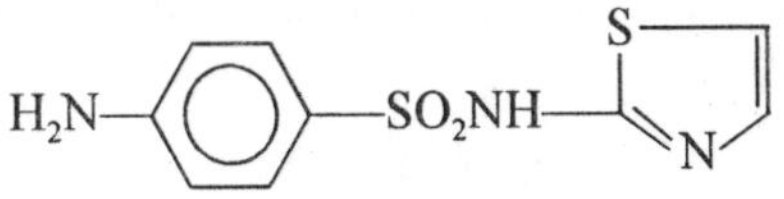

N^1-2-Thiazolylsulfanilamide ; Benzenesulfonamide, 4-amino-N-2-thiazolyl ; Norsulfazolum ; M & B 760 ; USP ; BP ; Int. P., Cerazole$^{(R)}$ (Beecham) ; Sulfex$^{(R)}$ (Smith Kline & French) ; Thiazamide$^{(R)}$ (May & Baker) ; Cibazol$^{(R)}$ (Ciba),

Synthesis :

It may be prepared by :

(*i*) Preparation of *p*-acetamidobenzene sulphonyl chloride (ASC)

(*ii*) Preparation of 2-aminothiazole

(*iii*) Condensation of (*i*) and (*ii*) above

(*a*) *Preparation of ASC :* It can be prepared by a method described under sulfanilamide.

(*b*) *Preparation of 2-aminothiazole :*

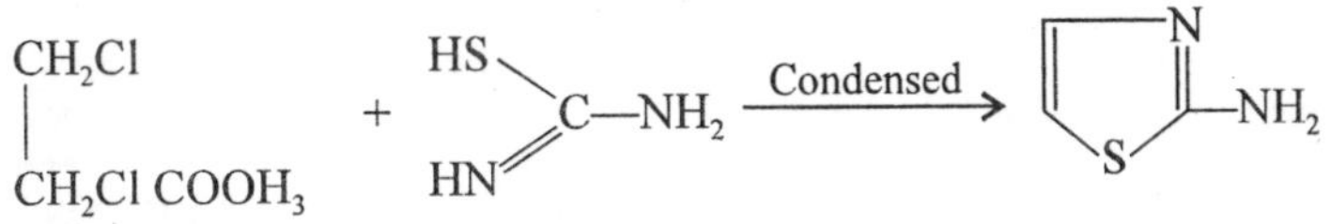

| 1, 2-Dichloro-ethyl acetate | Thiourea | 2-Aminothiazole |

It may be prepared by the condensation of ASC and 2-aminothiazole in pyridine and subsequently hydrolysing the product in the presence of sodium hydroxide.

It is occasionally used, as an adjunct to antibiotics, in severe staphylococcal infections. It is of general use, but *specifically useful against staphylococcal infections and also in bubonic plague.*

Dose : 3g initially ; subsequent doses, 1g every 4 hours.

D. *Sulfadiazine* INN, USAN, *Sulphadiazine,* BAN,

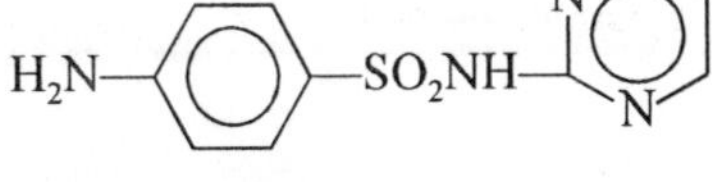

N^1-2-Pyrimidinylsulfanilamide ; Benzenesulfonamide, 4-amino-N-2-pyrimidinyl ; Sulfapyrimidine ; USP ; BP ; Eur. P. ; Int. P ; Ind. P ; Codiazine$^{(R)}$ (Beecham) ; Coco-Diazine$^{(R)}$ (Lillly) ; Eskadiazine$^{(R)}$ (SK & F) ; Diazyl$^{(R)}$ (Abbott).

Synthesis :

It may be prepared by the condensation of :

(*i*) *p*-Aminobenzene sulphonyl chloride (ASC) and

(*ii*) 2-Aminopyrimidine

(*a*) *Preparation of ASC :* It can be prepared as described under sulfanilamide.

(*b*) *Preparation of 2-amino-pyrimidine*

(i) Formyl Acetic Acid

It is prepared by the interaction of fuming sulphuric acid on malic acid followed by its dehydration and decarboxylation :

$$\underset{\text{Malic acid}}{HOOC.CH(OH)CH_2COOH} \xrightarrow[\substack{(ii) \text{ Dehydration and} \\ \text{decarboxylation}}]{(i) \text{ Fuming } H_2SO_4} \underset{\text{Formyl acetic acid}}{HOC = C.COOH} + H_2O + CO$$

(ii) Condensation of Formyl Acetic Acid with Guanidine

Formyl acetic acid and guanidine undergo cyclization after condensation in the presence of fuming sulphuric acid with the loss of two moles of water. The cyclized product undergo *keto-enol* tautomerism, when the *enol*-form, *i.e.,* 4-hydroxy-2-amino pyrimidine is subsequently chlorinated with either phosphorus oxychloride ($POCl_3$) or chlorosulfonic acid ($ClSO_2OH$) and finally reduced with zinc metal and ammonium hydroxide to obtain 2-amino-pyrimidine.

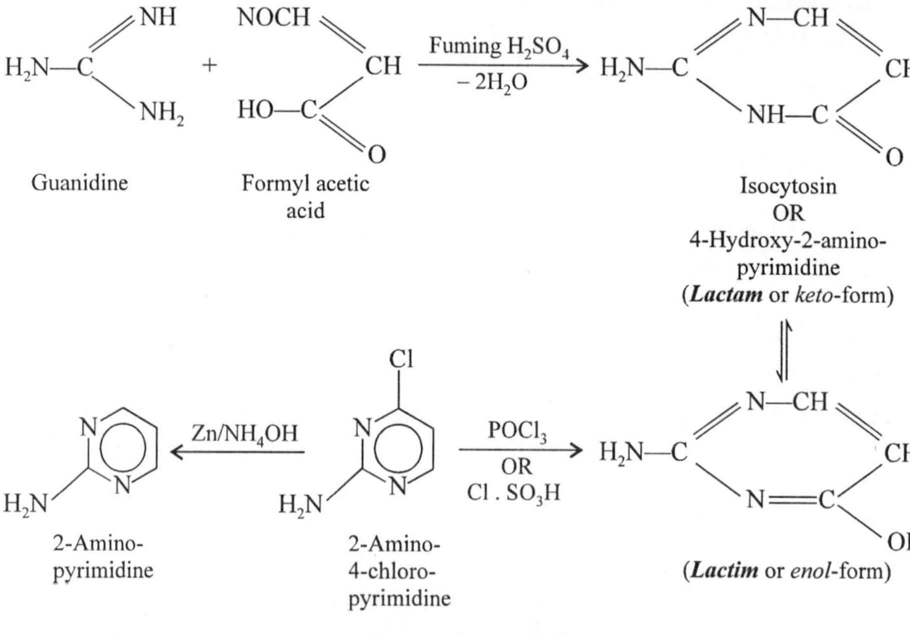

(iii) Condensation of ASC with 2-Aminopyrimidine

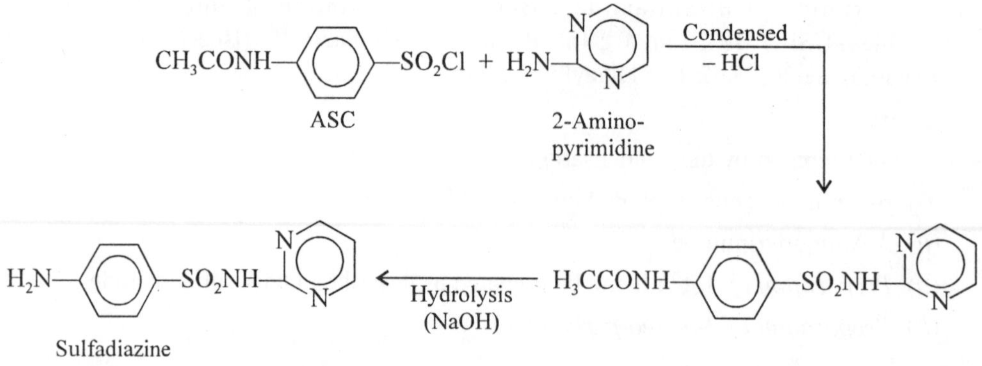

It is effective against most coccus infections. It is *better tolerated than sulfanilamide or sulfathiazole*. In fact, it is as potent as sulfathiazole and possesses fewer side reactions. It has also been used in the *treatment of chancroid due to Haemophilus ducreyi*. It may also be used for the prophylaxis of recurrences of rheumatic fever.

Dose : 3g initially ; USP dose range 2 to 8g daily.

E. *Sulfamerazine* INN, USAN, BAN,

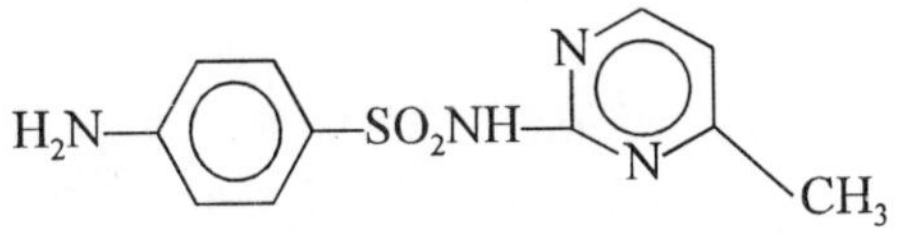

N^1-(4-Methyl-2-pyrimidinyl) sulfanilamide ; Benzenesulfonamide, 4-amino-N-(4-methyl-2-pyrimidinyl)-; Sulfamethyl-pyrimidine ; Sulfamethyldiazine ; BP ; USP ; Int. P ; Solumedine[R] (Specia).

Synthesis :

It is obtained by the condensation of :

(*a*) *p*-aminobenzene sulphonyl chloride (ASC) and

(*b*) 2-amino-4-methyl pyrimidine

(*c*) alkaline hydrolysis

(*a*) *Preparation of ASC :* It has been discussed under sulfanilamide.

(*b*) *2-Amino-4-methyl pyrimidine :* It is achieved by the interaction of sodium derivative of formyl acetone with guanidine as follows : Guanidine

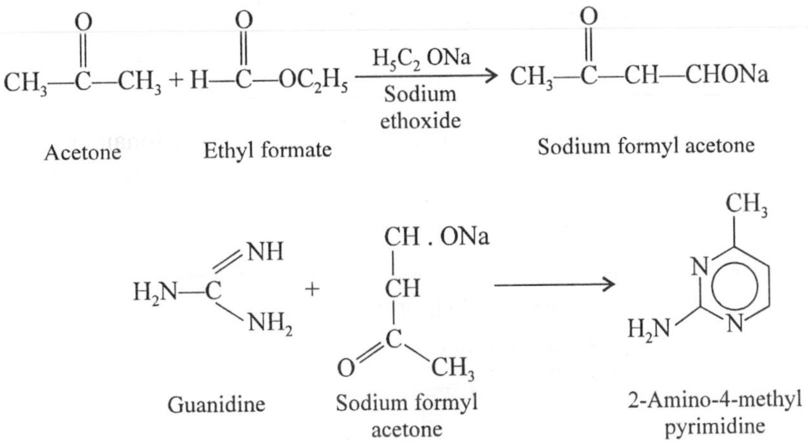

Acetone   Ethyl formate   Sodium formyl acetone

Guanidine   Sodium formyl   2-Amino-4-methyl
                 acetone   pyrimidine

(c) Condensation of (a) and (b) followed by alkaline hydrolysis :

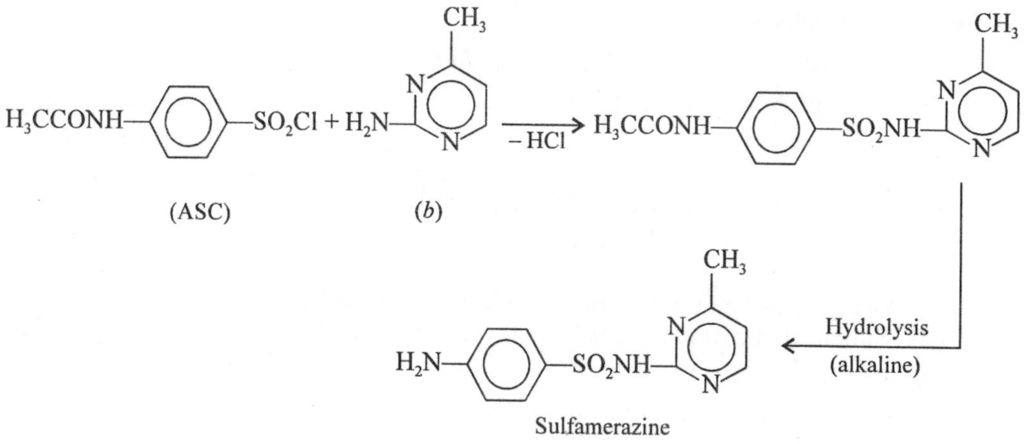

It has the general properties of the sulphonamides. It is mostly used in conjunction with other sulphonamides.

Dose : 4g initially ; subsequent doses 1g every 6 hours.

F. *Sulfadimidine* INN, *Sulphadimidine* BAN, *Sulfamethazine* USAN,

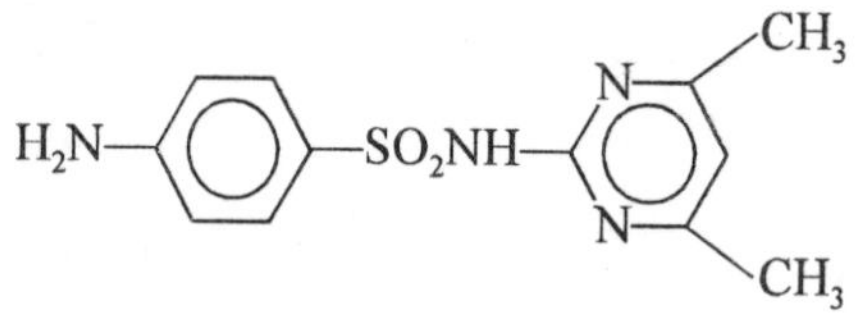

N^1-(4, 6-Dimethyl-2-pyrimidinyl) sulfanilamide ; Benzenesulfonamide, 4-amino-N-(4, 6-dimethyl-2-pyrimidinyl)-; Sulphadimethylpyrimidine ; USP ; BP ; Eur. P ; Int. P ; Ind. P ;

Synthesis :

It may be prepared from

    (*i*) *p*-Acetamido beznene sulphonyl chloride (ASC)

    (*ii*) 2-Amino-4, 6-dimethylpyrimidine

    (*iii*) Condensation of (*i*) and (*ii*) and

    (*iv*) Hydrolysis in alkaline medium

    (*a*) *Preparation of ASC* : It has been described under sulfanilamide.

    (*b*) *Preparation of 2-amino-4, 6-dimethyl pyrimidine* : It is prepared by reacting together the *Lactim*-form of acetyl acetone and guanidine as follows :

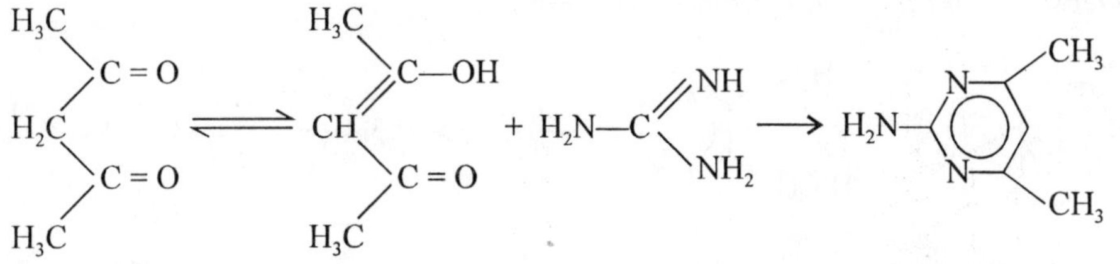

Acetyl acetone (**Lactam**-form)   Acetyl acetone (**Lactim**-form)   Guanidine   2-Amino-4, 6-dimethyl pyrimidine

(c) *Condensation of (i) and (ii)* :

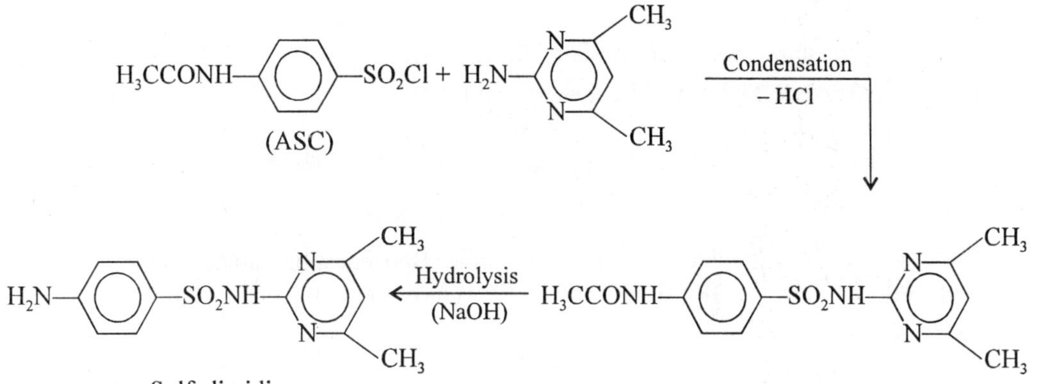

Sulfadimidine

It is comparatively less effective than sulfadiazine in meningal infections because of its poor penetration into the cerebrospinal fluid. However, for other infections *it is often regarded as the choicest sulphonamide. It is readily absorbed from the gastro-intestinal tract, hence desired concentration in blood may be achieved with regular oral doses.*

Dose : 3g initially ; subsequent doses up to 6g per day in divided doses.

G. *Sulfalene* INN, USAN, *Sulfametopyrazine* BAN,

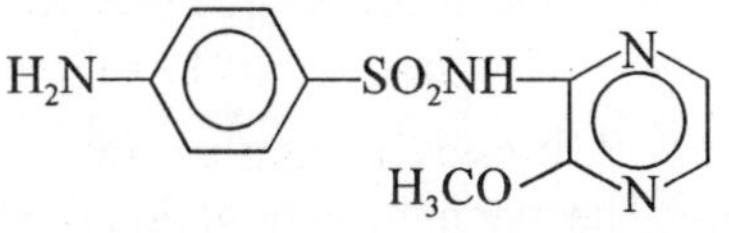

N^1-(3-Methoxypyrazinyl) sulfanilamide ; Benzenesulfonamide, 4-amino-N-(3-methoxypyrazinyl)-; Sulfametopyrazine ; Sulfamethoxypyrazine ; Int. P ; Kelfizina$^{(R)}$ (Abbott).

Synthesis :

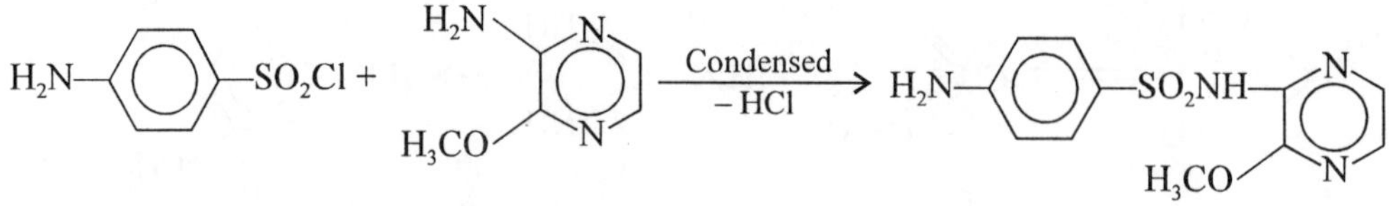

p-Aminobenzene	2-Amino	Sulfalene
sulphonyl chloride	3-methoxy-	
	pyrazine	

It is prepared by the condensation of *p*-aminobenzene sulphonyl chloride with 2-amino-3-methoxy-pyrazine when a molecule of hydrogen chloride is eliminated.

It is recommended for the *treatment of chronic bronchitis, urethritis, malaria and respiratory tract infections.*

Dose : 800mg ; initial dose, followed by 200mg daily.

H. *Sulfamethizole* INN, USAN, *Sulfamethizole* BAN,

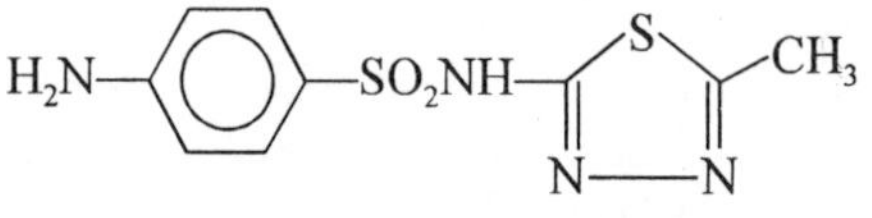

N[1]-(5-Methyl-1, 3, 4-thiadiazol-2-yl) sulfanilamide ; Benzenesulfonamide, 4-amino-N-(5-methyl-1, 3, 4-thiadiazol-2-yl) ; USP ; BP ; BPC ; Thiosulfil[(R)] (Ayerst) ; Ultrasul[(R)] (Alcon) ;

Syntehsis :

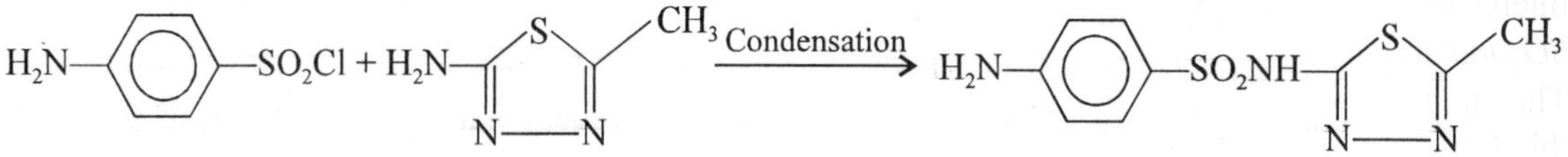

p-Aminobenzene	2-Amino-5-methyl-	Sulfamethizole
sulphonyl	1, 3, 4-thiadiazole	
chloride		

It may be prepared by the condensation of *p*-aminobenzene sulphonyl chloride with 2-amino-5-methyl-1, 3, 4-thiadiazole.

It has the general properties of sulphonamides. It is also employed in the *treatment of coliform infections of the urinary tract.*

Dose : 2 to 4g initially, followed by 2 to 4g per day in 3 to 6 divided doses.

2.1.1. Mechanism of Action. The mechanism of action of various drugs discussed under section 18.2.1. shall now be treated individually in the sections that follows :

2.1.1.1. Sulfanilamide. The *'drug'* has been largely superseded in medicine by other qualified structural analogues that are either less toxic or are for individual purposes preferable. It has been duly observed that the overall net result remains the same, but the prevailing molecular basis of the effect is rather different in these strains *e.g.,* streptococcal, pneumococcal, meningococcal, gonococcal and *E. coli.* However, the bacteria that are capable of taking up the preformed **folic acid** into their cells are found to be intrinsically resistant to the sulphonamides.

In fact, sulphanilamdie gives rise to a *'false metabolite'* that eventually prevents its ultimate conversion to tetrahydrofolic acid (THFA) as shown below :

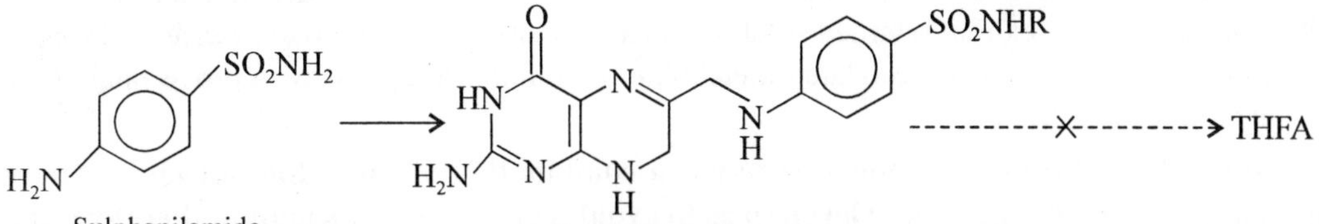

| Sulphanilamide | A False Metabolite |

SAR of Sulphanilamide

The fundamental basis of the prevailing structural resemblance of sulphanilamide-in particular and sulphonamides-in general to *para*-aminobenzoic acid (PABA) which is so destructive to these microorganisms is found to be quite evident.

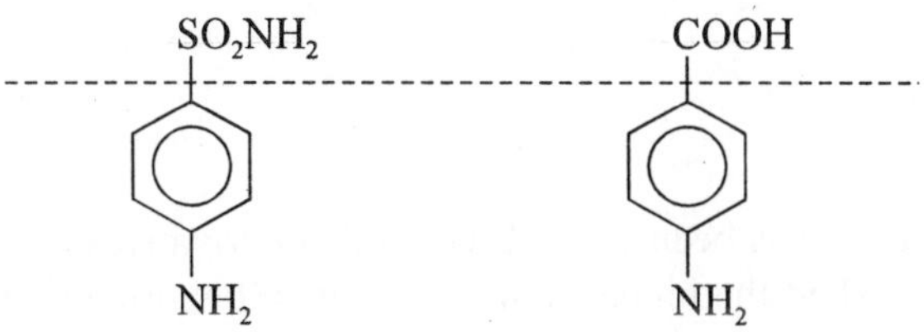

Sulphanilamide   PABA

The most glaring difference between the functional group which essentially differs in the two above cited molecules is the **carboxyl** of PABA and the **sulphonamide** moiety of sulphanilamide. Interestingly, the latter contains an evidently strong electron withdrawing feature by virtue of the *aromatic SO_2 moiety* that renders the N-atom to which it is directly linked partially elecropositive in character. This, in turn, eventually enhances the acidity of the two H-atoms linked to the N-atom thereby making this functional moiety (*i.e., sulphonamide moiety*) slightly acidic in nature (*pKa* 10.4). In contrast, the *pKa* value of the former, due to the presence of the carboxyl moiety of PABA, stands at 6.5. Based on this wonderful theory, immediately persued by a vigorous synthetic drive programme, that specifically involved the critical replacement of one of the H-atoms attached to the sulphonamido N-atoms by an **electron-withdrawing heteroaromatic nucleus** (ring) remarkably gave rise to the following *two* advantageous and useful features, namely :

(*a*) consistently improved antimicrobial activity, and

(*b*) appreciably acidified the remaining H-atom and substantially increased potency.

The wisdom and the skill of the *'medicinal chemist'* in designing newer drug entities brought about the following apparent changes :

- *pKa* value came down to almost very close to that of PABA (*i.e.,* 6.5) ; *e.g.,* sulfisoxazole *pKa* : 5.0,

- Significant enhancement of the *'antibacterial potency'* of the newer product, and

- dramatically potentiated the water solubility under the prevailing physiologic conditions.

2.1.1.2. Sulfapyridine. The *'drug'* enjoys the reputation of being the first agent to exhibit a remarkable significant curable action on pneumonia by Whitby.

SAR of Sulfapyridine. The *'drug'* afforded a tremendous impetus to the study of the whole class of N[1]-heterocyclically modified and substituted structural analogues of sulphanilamide.

2.1.1.3. Sulfathiazole. The primary amino moiety of this *'drug'* gets acetylated *in vivo,* and the resulting amides usually exhibit retarded solubility that may ultimately lead to toxic effects. Perhaps the poor solubility of the insoluble acetylated metabolite of sulfathiazole may even prove fatal in case it blocks the kidney tubules.

Note : It is always recommended to use a combination of the sulpha-drugs, rather than using a single one with higher dose level, so as to avoid 'crystal-urea' formation thereby reducing the chances of kidney blockade.

The possible metabolism of sulphathiazole is depicted as under :

Sulfathiazole   Metabolite
(Insoluble)

2.1.1.4. Sulfadiazine. It has been duly observed that derivatives of *'pyrimidine'* have enjoyed the most effort and accomplished the highest clinical success. Importantly, the effect of this *'drug'* is almost equal to 2-sulfapyrazine (*i.e.,* the 1, 4-diazine derivative) both *in vitro* and *in vivo* studies against a host of organisms.*

Comparison of Sulphadiazine *Vs* Sulphanilamide

It has been proved adequately that *sulphanilamide* essentially needs **142 times** the actual concentration of *sulfadiazine* so as to prevent the growth of *Escherichia coli.* In reality, the lipid solubility of the latter (sulphadiazine) is first **2.5 times** greater than that of the former (sulphanilamide) that would not be a strong enough evidence to explain explicitly the vast difference in their therapeutic potency. In case, one takes into consideration the prevailing *pKa* values of the two drug susbtances calculated at pH 7.4 it may be observed that only 0.03% of sulphanilamide shall undergo ionization ; whereas, sulfadiazine would get ionized upto 80% under such experimental conditions.

It is, however, pertinent to state here that even though sulphanilamide has sufficient lipid solubility (10.5%) to cross the bacterial membrane, 99.7% of the total available molecules would usually remain in the molecular **(inactive)** state once inside. Based on the above statement of facts the existing difference in the *'lipoidal solubility'* between the two aforesaid drugs would certainly permit the scope of prediction that **'sulfadiazine'** should possess a distinct **longer biological half-life.**

SAR of Sulfadiazine. The 5-sulphanilamido isomer together with various methoxy and methyl structural analogues are found to be comparatively less active *in vitro* than **sulfadiazine.** However, the *para*-isomer is nearly equivalent to sulfadiazine *in vitro* but very poorly active *in vivo.* It has been observed that the corresponding 4- and 5-isomers are relatively less active orally against streptococci in mice.**

*Bell PH *et al. J Am Chem Soc,* **64**, 2905, 1942

Redin GS *et al. Chemotherapia,* **11, 309, 1966.

The replacement of the *thiazole ring* with a more *electron withdrawing pyrimidine nucleus* enhanaces the acidity of the NH proton by stabilizing the anion ultimately. Consequently, sulphadiazine and its metabolite are appreciably ionized at pH of blood thereby rendering them **more soluble** and **less toxic** as illustrated below :

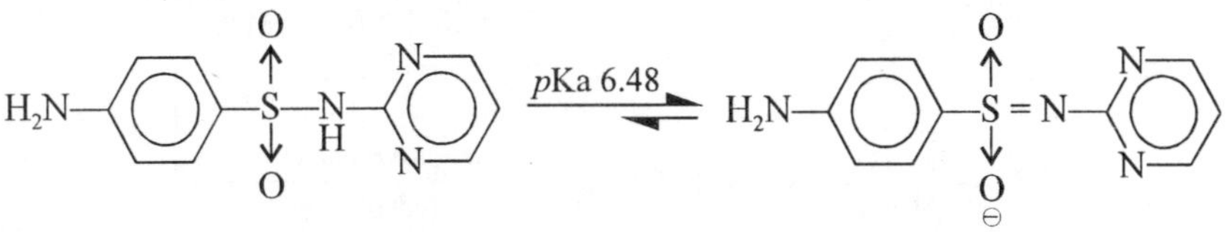

| Sulphadiazine | Sulphadiazine |
| (Unionized) | (Ionized 86%) |

2.1.1.5. Sulfamerazine. The *'drug'* gets easily absorbed from the GI-tract after oral administration ; and, therefore, the desired concentration in blood is accomplished with usual dosage regimen employed. As it exhibits relatively poor penetration right into the **cerebrospinal fluid** ; hence, it is much less effective in the treatment of meningal infections in comparison to the congener *sulfadiazine.*

2.1.1.6. Sulfamethazine. Because the *'drug'* is more soluble in acidic urine in comparison to sulfamerazine, therefore, its chances and scope of *'kidney damage'* by its usage is lowered to a great extent. Its *p*Ka value stands at 7.2. Its plasma half-life is 7 hours.

SAR of Sulfamethazine. The *'drug'* possesses almost similar chemical properties to those of *sulfadiazine* and *sulfamerazine,* but does show greater water-solubility than either.

2.1.1.7. Sulfalene. A combination of the *'drug'* with trimethoprim has exhibited an almost spectacular effect in clinical trials against the resistant falciparum malaria contracted in Vietnam, when the American soldiers were fighting there in the late sixties.*

2.1.1.8. Sulfamethiazole. A *'drug'* almost acts in the same manner as the sulphanilamide. However, its action is specifically useful in the treatment of UTIS. It has a plasma half-life of 2.5 hours.

2.2. Sulphonamides for Urinary Infections

A number of sulphonamides have been used extensively for the prevention and cure of urinary-tract infections over the past few decades. They are used sometiems as a prophylactic before and after manipulations on the urinary tract. A few such sulphanilamide analogues belonging to this category shall be dealt with here. *Examples :* sulfacetamide, sulfafurazole, sulfisoxazole acetyl, sulfacitine, etc.

A. *Sulfacetamide* INN, USAN, *Sulphacetamide* BAN,

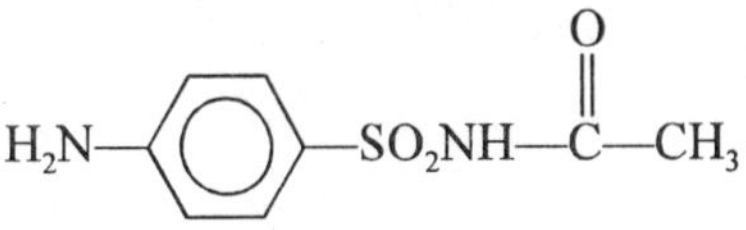

N-Sulfanilylacetamide ; Acetamide, N-[(4-aminophenyl) sulfonyl]-; BPC ; 1959 ; Ind. P ;

Synthesis :

It is prepared by the selective and partial hydrolysis of the N^1, N^4-diacetyl derivative of sulphanilamide which is obtained by the acetylation of sulfanilamide.

It possesses general characteristics of a sulphonamide and was formerly used in the treatment of bacterial infections of the urinary tract.

---

*Modell W, *Science,* **162,** 1346 (1968).

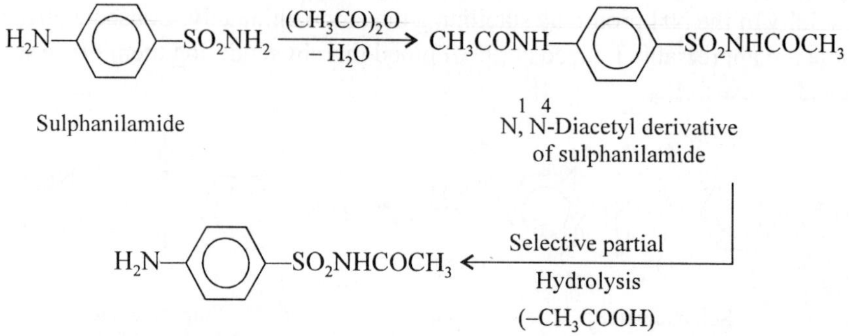

$$H_2N-\langle\bigcirc\rangle-SO_2NH_2 \xrightarrow[-H_2O]{(CH_3CO)_2O} CH_3CONH-\langle\bigcirc\rangle-SO_2NHCOCH_3$$

Sulphanilamide

1 4
N, N-Diacetyl derivative
of sulphanilamide

$$H_2N-\langle\bigcirc\rangle-SO_2NHCOCH_3 \xleftarrow[\text{Hydrolysis}]{\text{Selective partial}}$$
$(-CH_3COOH)$

Dose : 4g initially ; 1g every 4 hours.

B. *Sulfafurazole* INN, *Sulfisoxazole* USAN, *Sulphafurazole* BAN,

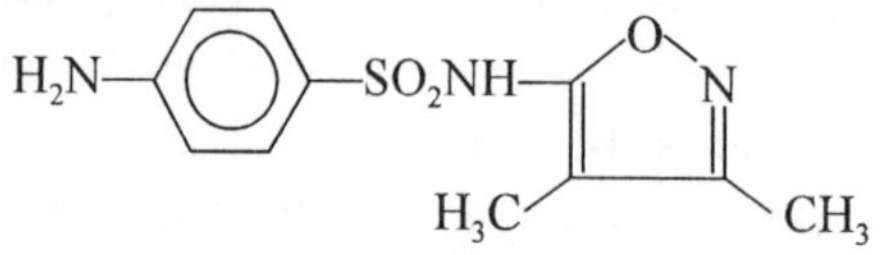

$$H_2N-\langle\bigcirc\rangle-SO_2NH-\text{[isoxazole ring]}$$

with H_3C and CH_3 substituents

N^1-(3, 4-Dimethyl-5-isoxazolyl) sulfanilamide ; Benzenesulfonamide, 4-amino-N-(3, 4-dimethyl-5-isoxazolyl)-; BP ; USP ; Ind. P ; Int. P ; Gantrisin[R] (Roche) ; SK-Soxazole[R] (SK & F) ; Soxomide[R] (Upjohn) ; Sulfalar[R] (Parke-Davis).

Synthesis :

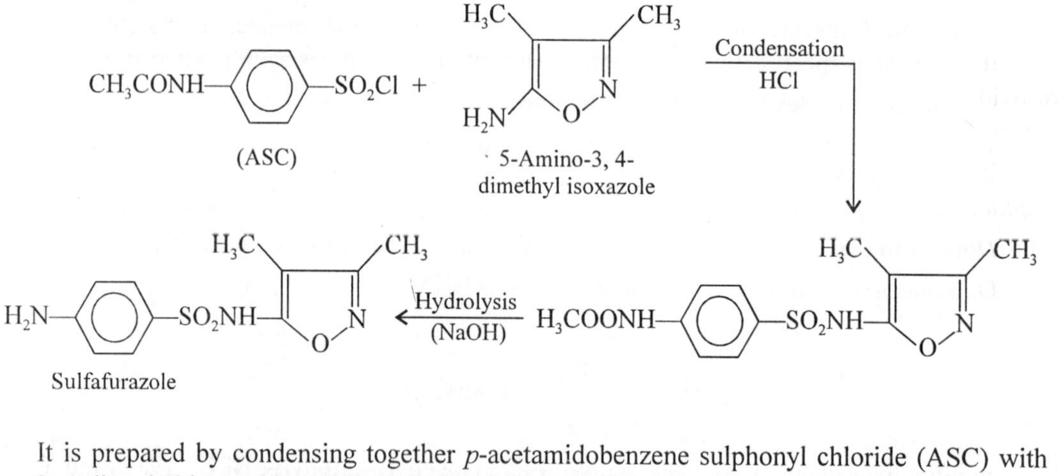

$$CH_3CONH-\langle\bigcirc\rangle-SO_2Cl + \text{[5-Amino-3,4-dimethyl isoxazole]} \xrightarrow[\text{HCl}]{\text{Condensation}}$$

(ASC)

· 5-Amino-3, 4-
dimethyl isoxazole

$$H_2N-\langle\bigcirc\rangle-SO_2NH-\text{[ring]} \xleftarrow[\text{(NaOH)}]{\text{Hydrolysis}} H_3COONH-\langle\bigcirc\rangle-SO_2NH-\text{[ring]}$$

Sulfafurazole

It is prepared by condensing together *p*-acetamidobenzene sulphonyl chloride (ASC) with 5-amino-3, 4-dimethyl isoxazole and hydrolysing the resulting product in an alkaline medium.

Its characteristics and therapeutic utilities are almost similar to those of sulfadiazine. It finds favour in the treatment of various urinary-tract infections. It may be used in the form of topical preparations for the *treatment of some infections, such as vaginitis caused by Hemophilus vaginalis.*

Dose : 2 to 4g initially ; oral, followed by 4 to 8g a day in 4 to 6 divided doses.

C. *Sulfisoxazole Acetyl* INN, USAN, *Acetyl Sulphafurazole* BAN,

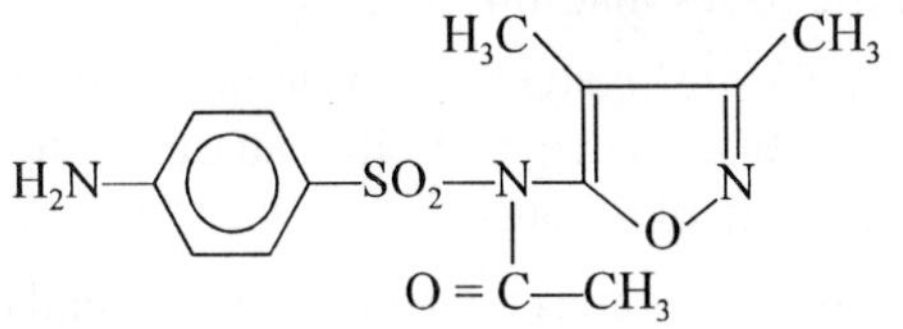

H₂N—⟨benzene⟩—SO₂—N—[3,4-dimethyl-5-isoxazolyl]
                    |
               O = C—CH₃

N-(3, 4-Dimethyl-5-isoxazolyl)-N-sulfanilyl-acetamide ; Acetamide, N-[(4-aminophenyl) sulfonyl]-N-(3,4-dimethyl-5-isoxazolyl)-; USP ; Lipo Gantrisin[R] (Roche).

Synthesis :

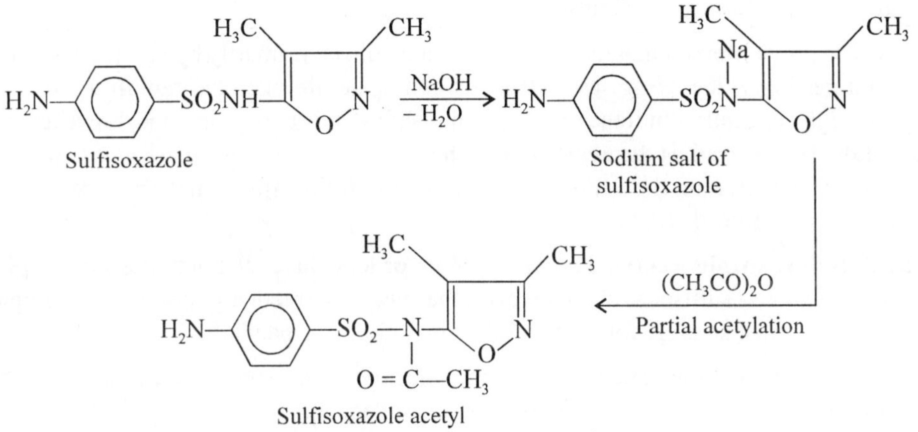

H₂N—⟨benzene⟩—SO₂NH—[3,4-dimethyl-5-isoxazolyl] $\xrightarrow[-H_2O]{NaOH}$ H₂N—⟨benzene⟩—SO₂N—[3,4-dimethyl-5-isoxazolyl with Na]

Sulfisoxazole   Sodium salt of sulfisoxazole

H₂N—⟨benzene⟩—SO₂—N—[3,4-dimethyl-5-isoxazolyl] $\xleftarrow{(CH_3CO)_2O}$
                    | Partial acetylation
               O = C—CH₃

Sulfisoxazole acetyl

It is prepared by first converting sulfisoxazole into its sodium salt by treatment with sodium hydroxide and then carrying out the selective acetylation at N^1 with an equimolar quantity of either acetic anhydride or acetyl chloride.

It is tasteless, unlike its parent compound, hence it is more suitable for liquid oral preparations. Its *renal toxicity is lower than that of sulfadiazine.*

Dose : 2 to 4g oral, adult and initial dose ; maintenance dose 4 to 8 g per day in divided doses.

D. *Sulfacitine* INN, *Sulfacytine* USAN, BAN,

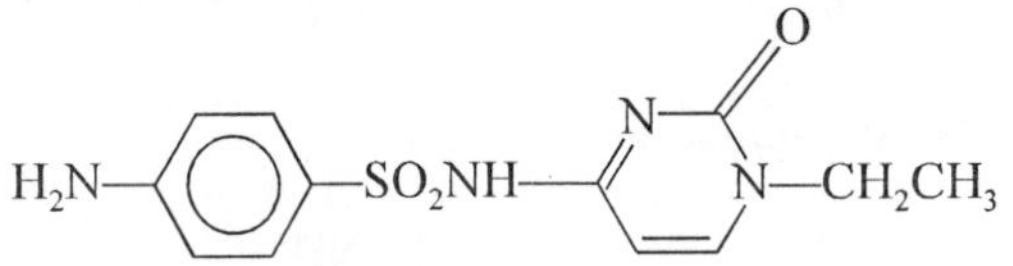

H₂N—⟨benzene⟩—SO₂NH—[1-ethyl-2-oxo-pyrimidinyl]—N—CH₂CH₃

N^1-(1-Ethyl-1, 2-dihydro-2-oxo-4-pyrimidinyl)-sulfanilamide ; Benzenesulfonamide, 4-amino-N-(1-ethyl-1, 2-dihydro-2-oxo-4-pyrimidinyl)-; 1-Ethyl-N-sulfanilylcytosine ; Renoquid[R] (Parke-Davis).

Sulfacitine is a short-acting sulphonamide which is employed likewise to sulfafurazole in the *treatment of acute urinary-tract infections.*

Dose : Initial loading dose 500mg ; maintenace dose 250mg 4 times per day up to 10 days.

2.2.1. Mechansim of Action. The mechanism of action of the *four* compounds described under section 18.2.2 are dealt with in the pages that follows :

2.2.1.1. Sulfacetamide. The *'drug'* exerts its action topically in conjunction with *sulphabenzamide* and *sulfathiazole* for the control and treatment of vaginitis caused due to the microorganism *Gardnerella (Hemophilus) vaginalis.* It has half-life of 7 hours.

2.2.1.2. Sulfisoxazole. The *'drug'* does not penetrate cells and happen to pass through barriers as well as most sulphonamides. Therefore, it is found to be not always effective against the systemic infections which are particularly sensitive to other sulphonamides. Importantly, in the specific genitourinary tract infections wherein penetration right into the involved tissues is required essentially, it may not prove to be as useful as **sulphadiazine.** It is known that the *'drug'* gets secreted into the *prostatic fluid ;* however, it has not yet been ascertained whether it gets secreted likewise into other genitourinary fluids.

It has been duly observed that the *'drug'* gets metabolized primarily by *acetylation* and *oxidation* in the liver. Interestingly, the *'drug'* as well as its conjugate are excreted rapidly by the kidney and thereby attains high concentration it the urine. The half-life stands at 6 hours. It is, however, pertinent to state here that the **free** as well as the **acetylated** forms of the drug substance are *highly soluble, even in an acidic urine ;* and, therefore, the adjuvant follow-up **'alkali therapy'** is not absolutely necessary and also fluids need not be forced.

2.2.1.3. Sulfisoxazole Acetyl. The *'drug'* more or less shares the activities and applications of the parent compound *i.e.,* sulfisoxazole. However, the *'drug'* is practically tasteless as compared to the parent drug ; hence, most suitable for liquid oral formulations for paediatric usage.

The acetyl compound is usually split in the *intestinal tract* and subsequently gets absorbed as **'sulfisoxazole'**, *i.e.,* it is a befitting **'prodrug'** for sulfisoxazole.

2.2.1.4. Sufacytine. The *'drug'* exerts its action for a relatively shorter duration ; and is employed invariably for the management and treatment of severe UTIs very much akin to **sulphafurazole.**

2.3. Sulphonamides for Intestinal Infections

A plethora of insoluble sulphonamide analogues, for instance phthalylsulfathiazole and succinylsulfathiazole, are not readily absorbed from the gastrointestinal tract. However, the release of active sulphonamide in high concentration, obtained due to hydrolysis in large intestine, enables their application for intestinal infections and also for pre-operative preparation of the bowel for surgery. A few examples of sulphonamides belonging to this specific use will be discussed here.

Examples : sulfaguanidine, phthalylsulfathiazole, succinylsulfathiazole, phthalylsulfacetamide, salazosulfapyridine, etc.

A. *Sulfaguanidine,* INN, *Sulphaguanidine* BAN,

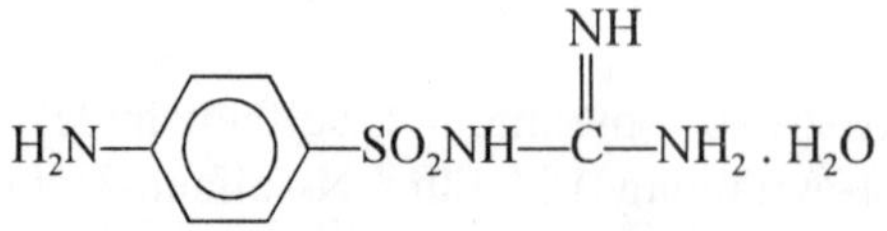

$$H_2N-\langle\bigcirc\rangle-SO_2NH-\overset{\overset{\displaystyle NH}{\|}}{C}-NH_2 \cdot H_2O$$

N^1-(Diaminomethylene) sulfanilamide ; Benzenesulfonamide, 4-amino-N-(diaminomethylene)-; N-*p*-Aminobenzenesulfonyl guanidine monohydrate ; Sulphaguanid ; BPC ; 1973 ; NF ; XI ; Int. P.; Ind. P ;

Synthesis :

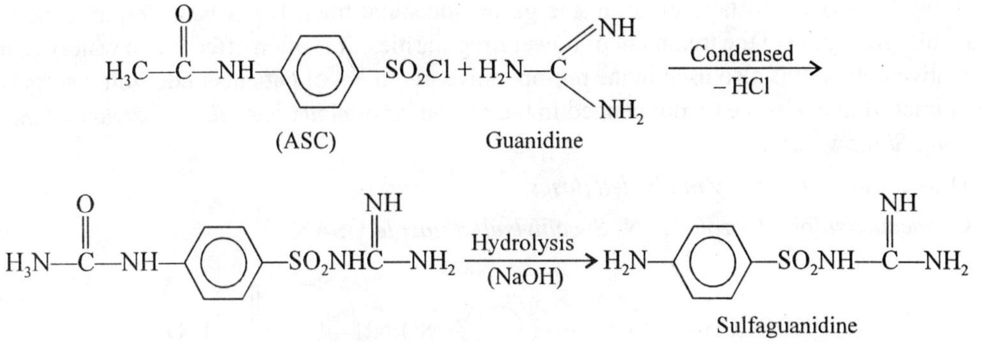

(ASC)   Guanidine   Condensed – HCl

Sulfaguanidine

It may be prepared by condensing *p*-amino benzene sulfonyl chloride with guanidine and hydrolysing the resultant in the presence of sodium hydroxide.

It has been used for the *treatment of local intestinal infections, specifically bacillary dysentry, but it has mostly been replaced by comparatively less toxic analogues, namely ; phthalylsulphatiazole and succinylsulphathiazole.*

Dose : 3g, 3 to 4 times per day for 3 days.

B. *Phthalylsulfathiazole,* INN, USAN, *Phthalylsulphathiozole* BAN,

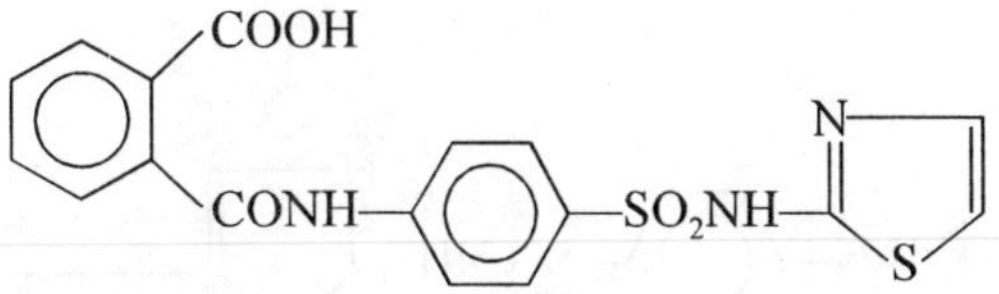

4′-(2-Thiazolylsulfamoyl) phthalanilic acid ; 2*p*-(*o*-Carboxy-benzamido) benzene sulfonamido thiazole ; Sulfaphtalyl-thiazole ; USP ; BP ; Ind. P ; Int. P ; Thalazole[R] (May and Baker, U.K.).

Synthesis :

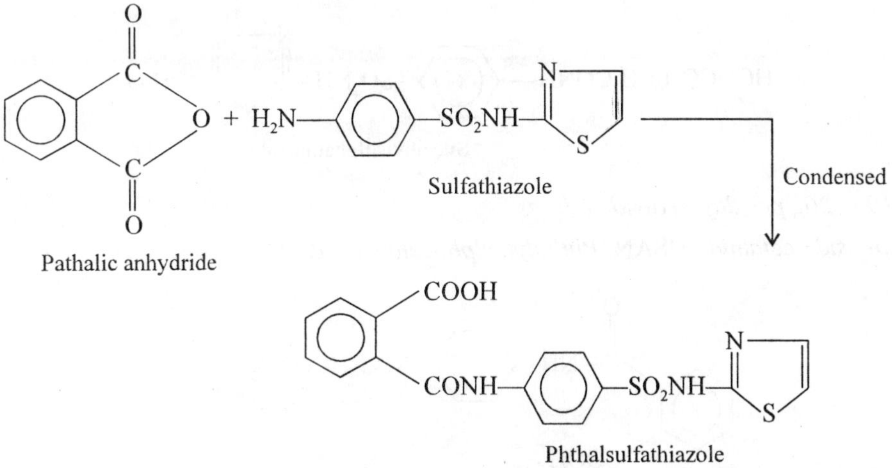

Pathalic anhydride   Sulfathiazole   Condensed

Phthalsulfathiazole

It may be prepared by the interaction of sulfathiazole and phthalic anhydride in equimolar proportions.

It exerts its bacteriostatic effect in the gastro-intestinal tract. It has been found to be twice as active as sulfaguanidine in the treatment of bowel irregularities. It is often effective in watery diarrhoeas and ulcerative colitis. It is also used in the pre-operative treatment of patients undergoing surgery of the intestinal tract. It may also be recommended in the *treatment of acute bascillary dysentry of the Sonne, Flexner and Shiga species.*

Dose : 5 to 10g per day in divided doses.

C. *Succinylsulphathiazole* BAN, *Succinylsulfathiazole* USAN,

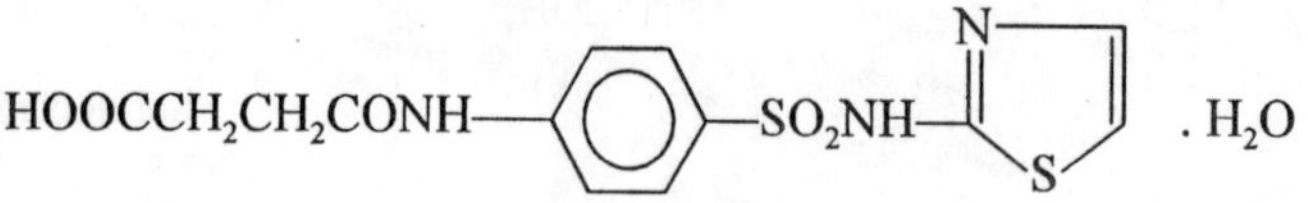

$$\text{HOOCCH}_2\text{CH}_2\text{CONH}-\langle\bigcirc\rangle-\text{SO}_2\text{NH}-[\text{thiazole}] \ . \ H_2O$$

4'-(2-Thiazolylsulfamoyl-succinanilic acid monohydrate ; 2-*p*-(3-Carboxy-propionamide) benzene sulfonamido thiazole monohydrate ; USP ; XVIII, BP ; Eur. P ; Ind. P ; Int. P ; Sulfauxidine[R] (Merck Sharp and Dohme).

Synthesis :

It can be prepared by heating together succinic anhydride and sulfathiazole.

It is used in bacillary dysentry and cholera. Its other uses are more or less the same as that of phthalylsulfathiazole.

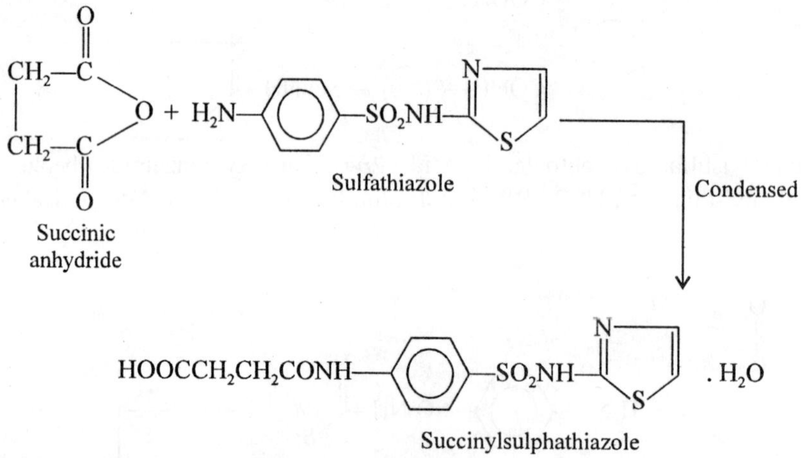

Succinic anhydride + Sulfathiazole → Condensed → Succinylsulphathiazole . H_2O

Dose : 10 to 20g per day in divided doses.

D. *Phthalylsulfacetamide* USAN, *Phthalylsulphacetamide* BAN,

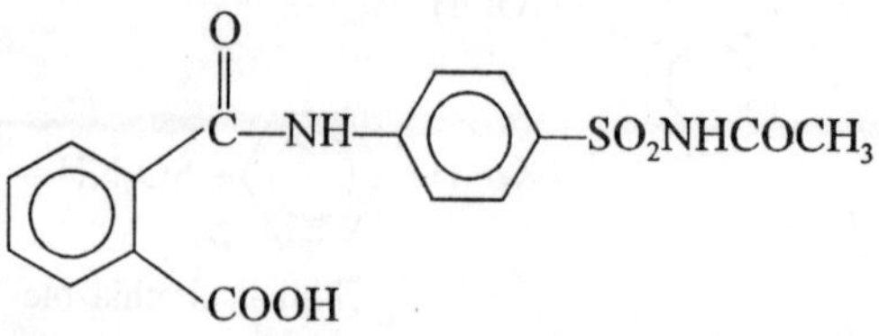

4'-(N-Acetylsulfamoyl) phthalanilic acid ; N-[p(o-Carboxybenzamido) benzene sulfonyl] acetamide ; NF ; XIII, Ind. P., Talsigel[R] (Squibb) ; Thalisul[R] (Beecham) ; Thalamyd[R] (Schering-Plough).

Synthesis :

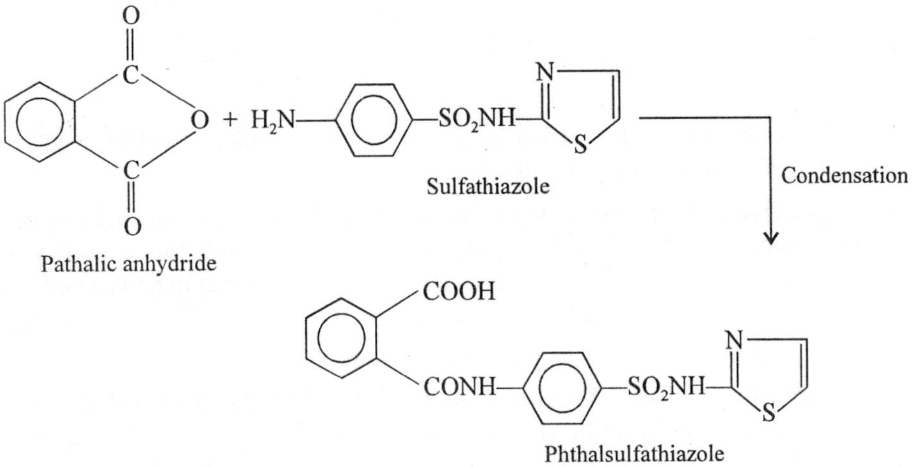

Pathalic anhydride

Sulfathiazole

Condensation

Phthalsulfathiazole

It may be prepared by the interaction of phthalic anhydride and sulfacetamide.

The therapeutic uses of this compound is very much similar to phthalylsulfathiazole.

Dose : 6g per day in divided doses.

E. *Salazosulfapyridine* INN, *Sulfasalazine* USAN, *Sulphasalazine* BAN,

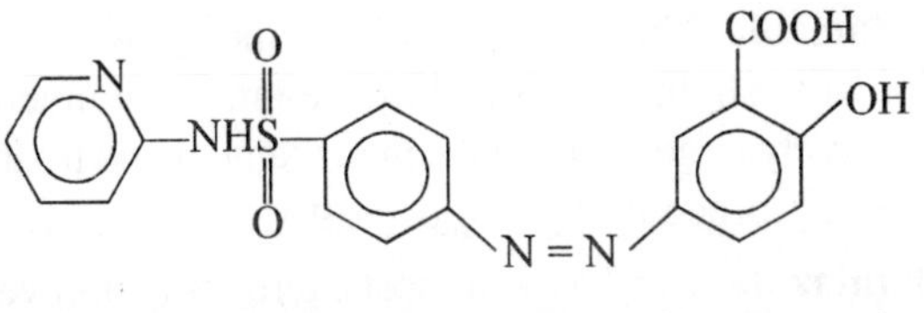

5-[[p-(2-Pyridylsulfamoyl) phenyl]azo] salicylic acid ; Benzoic acid, 2-hydroxy-5-[[4-(2-pyridinylamino) sulfonyl]-phenyl] azo-; USP ; NF ; Azulfudine[R] (Lederle) ; Salazopyrin[R] (Pharmacia U.K.) ; SAS–500[R] (Rowell).

Synthesis :

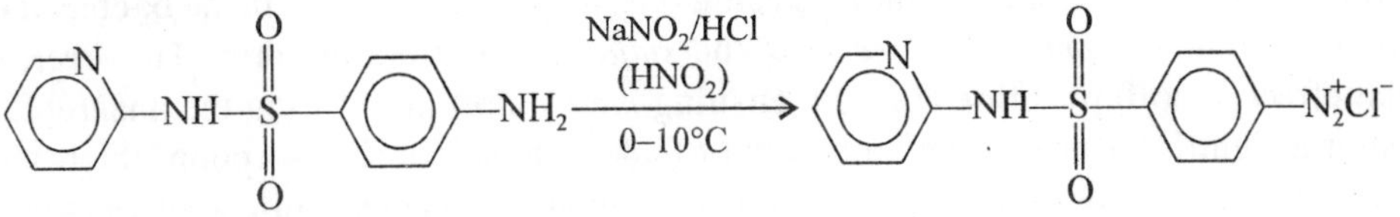

N'-2-Pyridylsulphani-
lamide

Diazotisation

Diazonium salt

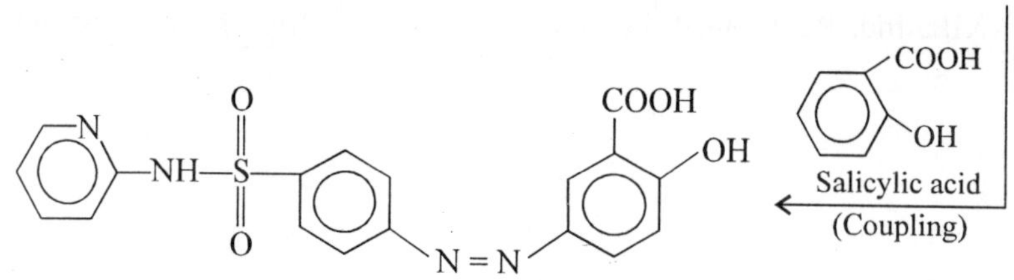

Salazosulfapyridine

It may be prepared *first* by diazotising N^1-2-pyridyl-sulfanilamide at 0-10°C and *secondly* by coupling the resulting diazonium salt with salicylic acid.

It has a suppressive effect on ulcerative colitis. This therapeutic action may perhaps be attributed to a local immunosuppressive effect. Salazosulfapyridine ultimately releases sulfapyridine and 5-aminosalicyclic acid in the colon with the aid of bacterial enzymes. It often imparts a yellow colour to alkaline urine.

Dose ; 2 to 8g per day preferably in 4 to 8 divided doses.

2.3.1. Mechanism of Action. The mechanism of action of the compounds discussed in section 18.2.3. are now treated individually in the sections that follows :

2.3.1.1. Sulphaguanidine. The percentage unbound *i.e.,* the protein-unbound portion which is predominantly significant for the ensuing activity, toxicity, metabolism and glomerular filtration of the *'drug'*, ranges from 95% for *sulfaguanidine* to 0.2—0.5% for *4-sulfa-2, 6-dimethoxyprimidine*. From the above observations one may infer that the criterion of *'protein-binding'* would lead ultimately one to believe that an extensively bound *'drug entity'* is quite undesirable not only from the antibacterial but also from the pharmacological stand point crtically. It also gets very slightly absorbed from the intestinal mucosa.

2.3.1.2. Phthalylsulfathiazole. The *'drug'* is of low inherent toxicity. Besides, it also enjoys an additional plus point for being only slightly absorbed by the intestinal mucosa ; and, therefore, may be safely administered in comparatively large doses in the management and treatment of bacillary infections of the intestine. The drug is not absorbed orally and is mostly employed for ulcerative colitis.

2.3.1.3. Succinylsulfathiazole. The *'drug'* almost exerts its action very much identical to that of phthalylsulfathiazole discussed earlier.

2.3.1.4. Phthalylsulfacetamide. The mechamism of action of this *'drug'* is very much akin to phthalylsulfathiazole.

2.3.1.5. Sulfasalazine. The *'drug'* gets poorly absorbed from the small intestine, so as to enable a major portion of the drug to penetrate into the **colon** (*i.e., large intestine*) where the **bacterial enzymes** strategically release both *5-aminosalicylic acid* and *sulfapyridine* from the drug. However, the local antibacterial effect of **sulfapyridine** (*i.e.,* the **ensuing** *metabolite*) in lowering the anaerobic bacteria may not be of a significant magnitude on account of adequate systemic absorption. Interestingly, the first metabolite *i.e.,* 5-aminosalicylic acid specifically inhibits the **arachidonic acid casade** *i.e.,* both *cyclooxygenase* and *lipooxygenase* pathways, effectively. Perhaps the most prominent and important would be the legitimate inhibition of leukotriene B_4 production by PMNs.

2.4. Sulphonamides for Local Infection

There are some sulphonamides which are used exclusively for certain local applications. A few such typical sulphonamides shall be discussed below :

Examples : Sulfacetamide sodium, Mafenide, etc.

A. *Sulfacetamide Sodium* USAN, *Sulphacetamide Sodium* BAN,

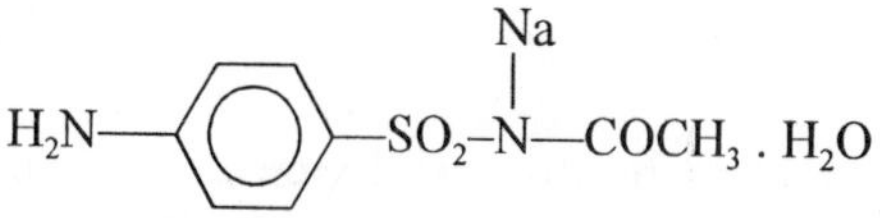

$$H_2N\text{---}\langle\bigcirc\rangle\text{---}SO_2\text{--}\underset{\underset{Na}{|}}{N}\text{---}COCH_3 \ . \ H_2O$$

N-Sulfanilylacetamide monosodium salt monohydrate ; Acetamide ; N-[(4-Aminophenyl) sulfonyl]-, monodium salt, monohydrate ; Soluble sulphacetamide ; USP ; BP ; Int. P ; Ind. P ; Albucid[R] (Nicholas U.K.) ; Cetamide[R] (Alcon) ; Sulf-10[R] (Cooper Vision).

Synthesis :

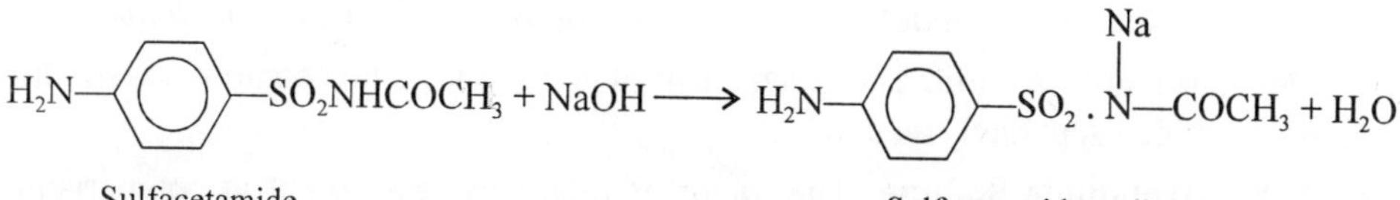

$$H_2N\text{---}\langle\bigcirc\rangle\text{---}SO_2NHCOCH_3 + NaOH \longrightarrow H_2N\text{---}\langle\bigcirc\rangle\text{---}SO_2 \ . \ \underset{\underset{Na}{|}}{N}\text{---}COCH_3 + H_2O$$

Sulfacetamide   Sulfacetamide sodium

It may be prepared by heating together sulfacetamide and sodium hydroxide in *equimolar concentrations.*

It is chiefly employed by local application in injuries or infections of the eyes at various strengths ranging from 10 to 30%. It is also used in the *treatment of acute conjunctivitis and in the prophylaxis of ocular infections after injuries or burns.*

Dose : In eye, drops 10%, 15%, 20% and 30% ; In ointments 2.5% and 6%.

B. *Mafenide* INN, BAN, USAN,

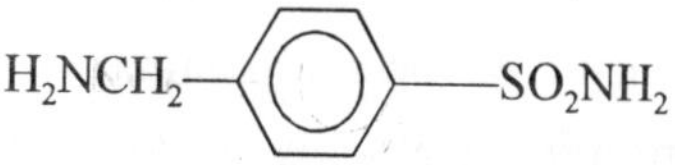

$$H_2NCH_2\text{---}\langle\bigcirc\rangle\text{---}SO_2NH_2$$

α-Amino-*p*-toluenesulfonamide ; Benzenesulfonamide, 4-(amino-methyl)-; Marfanil ; Mafenide Acetate, USP ; Mafenide Hydrochloride, Jap. P ; BPC ; 1949 ; Mafenide Propionate, Sulfonyl[R] (Winthrop).

Synthesis :

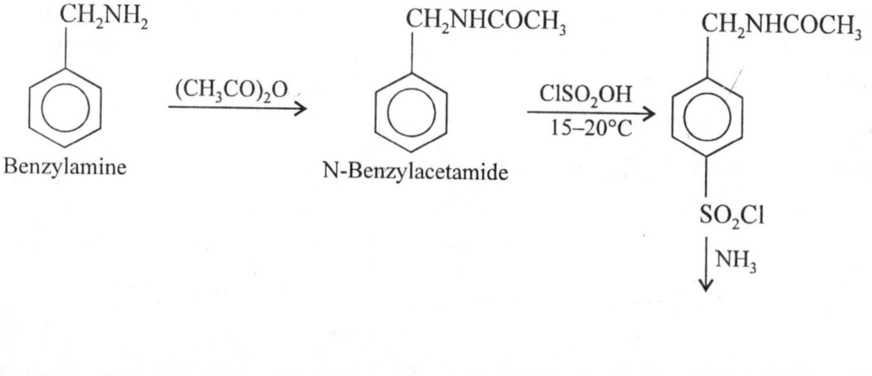

Benzylamine $\xrightarrow{(CH_3CO)_2O}$ N-Benzylacetamide $\xrightarrow[15-20°C]{ClSO_2OH}$

CH$_2$NH$_2$ CH$_2$NHCOCH$_3$ CH$_2$NHCOCH$_3$

SO$_2$Cl $\downarrow$ NH$_3$

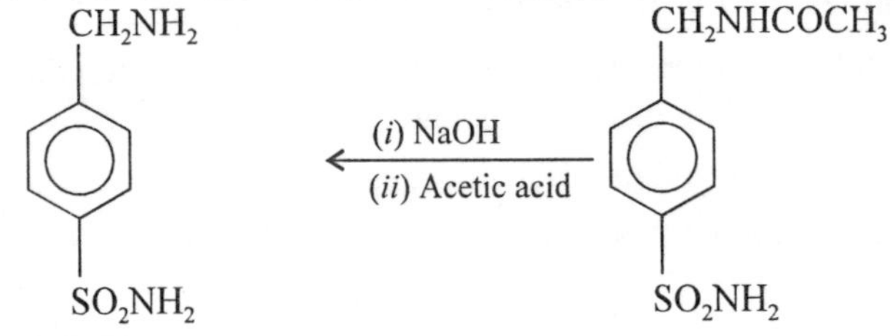

CH$_2$NH$_2$ CH$_2$NHCOCH$_3$

(i) NaOH
(ii) Acetic acid

SO$_2$NH$_2$ SO$_2$NH$_2$

Mafenide

N-Benzylacetamide is prepared by the acetylation of benzylamine, which on treatment with chlorosulfonic acid at 15-20°C yields *p*-benzylacetamido sulphonyl chloride. This on amination gives the corresponding sulfonamide derivatve, which upon hydrolysis with sodium hydroxide and subsequent neutralization with acetic acid yields mafenide.

It is used in the treatment and cure of gas gangrene. It is also used for the *treatment of infection especially that caused by Pseudomonas alruginosa, in second-and third-degree burns.*

Dose : 5% solution of mafenide hydrochloride or mafenide propionate for topical use.

2.4.1. Mechanism of Action. The mechanism of action of the *two* compounds described under section 18.2.4 are treated separately as under :

2.4.1.1. Sulfacetamide Sodium. The *'drug'* is relatively less potent in comparison to *'other sulphonamides'.* This retardation of therapeutic value is perhaps due to the poor penetration into both *tissues* and *bacteria.* However, if used in high concentration by means of local application, it is found to be of great utiliry in different types of *ophthalmologic infections,* especially those produced by pyogenic cocci, gonococcus, *E. coli* and Koch-Week's bacillus.

As the *'drug'* is obviously nonirritating in nature even at a high dosage regimen, it may be employed safely in sufficient concentration to accomplish adequate penetration of the ocular tissues with much ease and fervour.

2.4.1.2. Mafenide. It is **not a true sulfanilamide-type** drug substance, because it is *not* inhibited by PABA. Evidently, its antibacterial activity predominantly involves a mechanism which essentially differs from that of the true sulphanilamide-type compounds.

Interestingly, the *'drug'* is specifically effective against *Clostridium welchii* as a prophylaxis of wounds in the form of topical medicaments *viz.,* lotions, ointments, or dusting powder.

Note. The corresponding acetate salt *i.e.,* mefenide acetate used in an ointment base proved to be the most efficacious agent devoid of any untoward metabolic acidosis.

2.5. Sulphonamide Related Compounds

There are some sulphonamides which essentially differ from the basic sulphonamide nucleus, but do possess anti-bacterial properties. A few such typical examples belonging to this type of compounds are dealt with below :

Examples : Nitrosulfathiazole, dapsone, silver sulfadiazine, etc.

A. *Nitrosulfathiazole* INN, *Para-Nitrosulfathiazole* USAN, *Paranitrosulphathiazole* BAN,

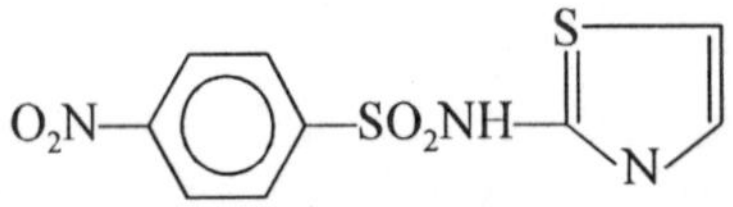

O$_2$N—⟨O⟩—SO$_2$NH—

p-Nitro-N-2-thiazolylbenzenesulfonamide ; 4-Nitro-N-(thiazol-2-yl) benzenesulphonamide ; NFXI ; Nisulfazole[R] (Breon) ;

It is only administered as a rectal injection as an adjunct in the local treatment of non-specific ulcerative colitis.

Dose : 10ml of a 10% suspension after each stool and at bed time.

B. *Dapsone* BAN, USAN,

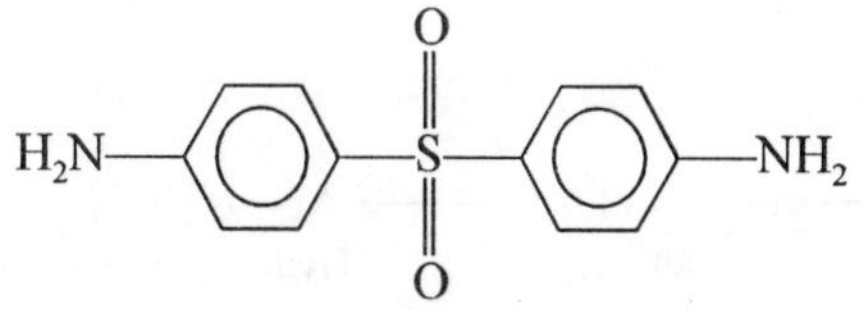

4, 4'-Sulfonyldianiline ; Benzenamine, 4, 4'-Sulfonyl *bis*- ; Diaphenylsulfone ; Disulone ; BP ; USP ; Int. P ; Ind. P ; Avlosulfon[R] (Ayerst).

Synthesis :

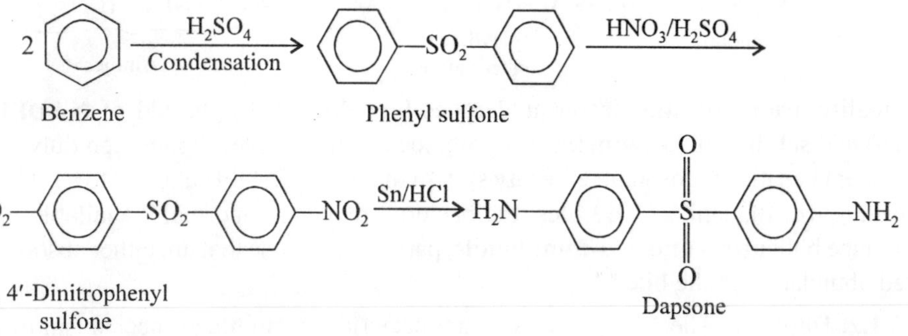

Benzene is made to condense with sulphuric acid to give phenyl sulfone, which is then nitrated and subsequently reduced to obtain dapsone.

It exhibits an antibacterial spectrum and mechanism of action similar to that of sulphanilamide. *It is the drug of choice in the chemotherapy of leprosy and also for the treatment of nocardiosis.* It has also been used successfully as a suppressant in the treatment of *dermatitis herpetiformis.*

Dose : As leprostatic, 25mg twice a week initially for one month followed by 25mg per day each month ; As suppressant for dermatitis herpetiformis-100 to 200mg per day.

C. *Silver Sulfadiazine* USAN, *Silver Sulphadiazine* BAN,

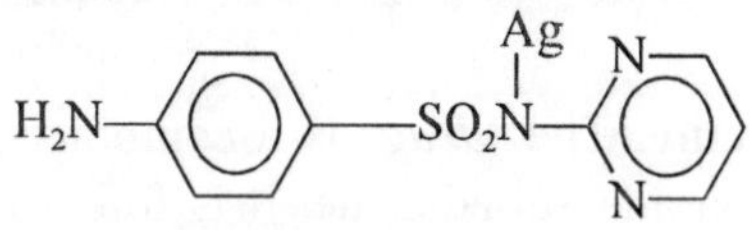

N^1-2-Pyrimidinylsulfanilamide monosilver (1+) salt ; Benzene-sulfonamide, 4-amino-N-2-pyrimidinyl-, mono-silver (1+) salt ; Flint SSD[R] (Flint) ; Silvadene[R] (Marion).

It is an effective topical antimicrobial agent, especially against *Pseudomonas species.* It finds its extensive use in burn-therapy because it attacks the *pseudomonas* radically which is perhaps considered to be the ultimate cause of failures in the treatment of burn cases.

2.5.1. Mechanism of Action. The mechanism of action of compounds discussed under section 18.2.5. shall be dealt with individually in the sections that follows :

2.5.1.1. Paranitrosulfathiazole. In general, the reduction of both aromatic into and azo xenobiotics ultimately gives rise to the corresponding **primary amine metabolites.*** These reactions may be summarized explicitly in the following *two* steps :

(*a*) *Conversion of nitro-to an amine group.* An aromatic nitro compound usually get reduced initially to the nitroso and hydroxylamine intermediates, as illustrated in the following metabolic sequence :

$$Ar—N{\overset{\oplus}{\underset{O^{\ominus}}{\diagup^O}}} \longrightarrow \underset{Nitroso}{Ar—N = O} \longrightarrow \underset{Hydroxylamine}{Ar—NHOH} \longrightarrow \underset{Amine}{Ar—NH_2}$$
$$\underset{Nitro}{}$$

(*b*) *Conversion of azo-to amines.* It is, however, believed that an azo reduction normally proceeds *via* a hydrazo intermediate (—NH—NH—) which subsequently gets cleaved by undergoing reduction to give rise to the corresponding aromatic amines, as depicted under :

$$\underset{Azo\ compound}{Ar—N = N—Ar'} \longrightarrow \underset{\substack{Hydrazo\\Derivative}}{Ar—NH—NH—Ar'} \longrightarrow \underset{Amines}{\underbrace{Ar—NH_2 + H_2N—Ar'}}$$

In reality, paranitrosulfathiazole undergoes bioreduction by the aid of **NADPH-dependent microsomal** and **soluble nitro reductases** usually located in the liver. It has been duly observed that a multicomponent hepatic microsomal reductase system requiring NADPH appears to be solely responsible for the ensuing azo reduction.** Besides, the bacterial reductases normally available in the intestine may also reduce both **nitro- and azo-compounds,** particularly those that are either absorbed very poorly or excreted abundantly in the bile.***

2.5.1.2. Dapsone. The *'drug'* exerts its antibacterial spectum and mechanism of action almost identical to those of sulfanilamide. It is duly absorbed by the oral route. It has been observed that the absorption is more efficient with comparatively low than with high doses. It gets eliminated in the liver by acetylation. The half-life is 10 to 50 hours ; and at least 8 days are normally needed to accomplish *plateau concentrations.*

Combinations with other drugs	*Therapeutic Usages*
Dapsone + Rifampin ⎫ Dapsone + Clofazimine ⎭	Prevention of **multibacillary leprosy**
Dapsone + Pyrimethamine	Prevention/treatment of **Malaria**
Dapsone + Trimethoprim	Treatment of *Pneumocystis carinii pneumonia* (PCP)

2.5.1.3. Sodium Sulfadiazine. The *'drug'* is a combination in *'one single compound'* the **antibacterial** properties of Ag^+ ion and sulfadiazine. It is found to be specifically effective against *Pseudomonas aeruginosa.* Although sulfadiazine gets absorbed systematically to a certain extent ;

---

*Gillette JR : In Brodie BB and Gillette JR (eds.) *Concepts in Biochemical Pharmacology,* Part 2, Springer-Verlag, Berlin, 349, 1971.

Hernandez PH *et al. Biochem. Pharmacol* **16 : 1877, 1967 ; Gillette JR *et al. Mol. Pharmacol,* **4 :** 541, 1968.

***Scheline RR : *Pharmacol Rev.* **25** : 451, 1973.

however, it is not sufficient to afford *'crystalluria formation'*. Importantly, Ag^+ ion usually inactivates proteolytic enzymes employed for *debridement* (*i.e.,* removal of foreign material and dead or damaged tissue, especially in a wound).

3. IONIZATION OF SULPHONAMIDES

It is squarely demonstrated and established that the sulfonamide functional moiety (SO_2NH_2), has an apparent tendency to gain stability provided it happens to lose a proton, which fact is adequately substantiated as the resulting negative charge gets *resonance stabilized* ultimately as shown under :

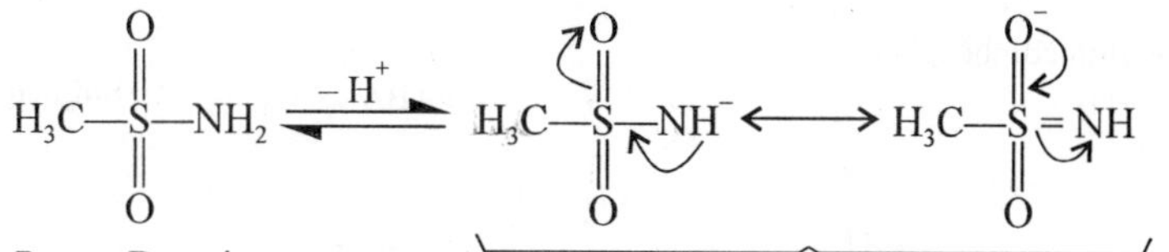

Proton Donating
(Sulphonamide Moiety)

Resonance Stabilized
(Sulphonamide Moiety)

Furthermore, it can be expatiated by considering the fact that because the proton-donating form of the functional moiety (*i.e.,* sulphonamide) bears absolutely no charge, one may even characterize the same as an HA acid, just in the same vein as *phenols, thiols* and *carboxyl* groups.

Consequently, the loss of a *'proton'* may be directly linked to a *pKa* value of the *'drug'* under investigation ; and, therefore, it also applies to all the structural analogoues (*i.e.,* congeners or series).

Example : Sulfisoxazole (see under section 18.2.2.B) has a *pKa* value 5.0 which evidently shows that this specific sulphonamide is a slightly weaker acid in comparison to the acetic acid (pKa 4.8) as shown below :

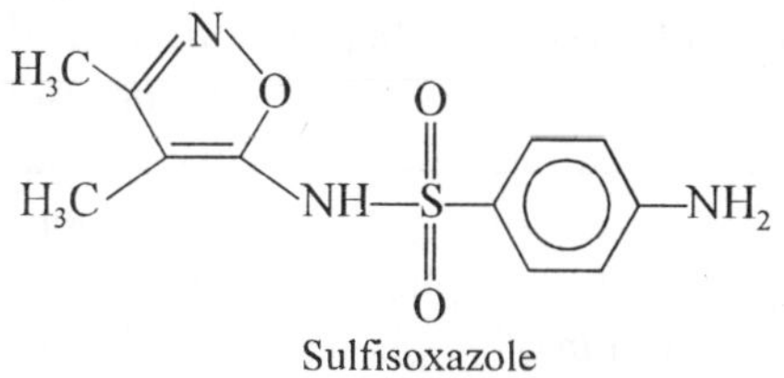

Sulfisoxazole
(*p*Ka 5.0)

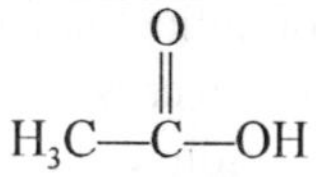

Acetic Acid
(*p*Ka 4.8)

4. SULPHONAMIDE INHIBITION AND PROBABLE MECHANISMS OF BACTERIAL RESISTANCE TO SULPHONAMIDES

The coupling of dihydropteridinediphosphate and PABA to yield pteroic acid is catalyzed by an enzyme known as dihydropteroate synthetase, which is, in fact, inhibited by sulphanilamide.

The varying degree of response of organisms to sulphanilamide-type antibacterial action may be attributed to various factors, namely : *first*, being the different biological nature of organisms which essentially involves a quantitative difference in the capacity of the enzyme, dihydropteroate synthetase, to result folic acid from PABA in the presence of sulphanilamide. *Secondly*, it is the biological difference towards preferential permeability of the cell for some specific sulphonamides. Thirdly, it is due to electronic configuration and steric factors of the drugs causing the individual folic acid synthesizing together with the relative permeability of bacterial cell walls to these drugs.

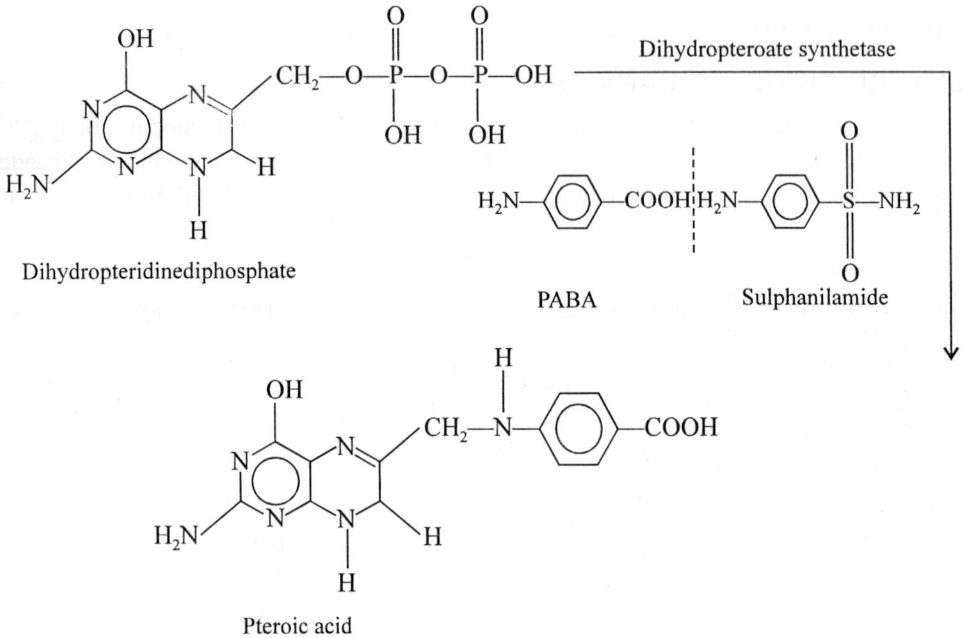

Dihydropteridinediphosphate

PABA   Sulphanilamide

Pteroic acid

5. CHEMOTHERAPEUTIC CONSIDERATION

In general, the sulphonamides are converted *in vitro* to N^4-acetyl analogues, a portion of which is excreted as such. It is, however, pertinent to observe here that these acetylated products have a lower pKa value and lower solubility than the corresponding unacetylated sulphonamides. These N^4-acetylated products have a tendency to crystalize in the renal tubules, thereby causing obstruction in the kidney and ultimately may lead to kidney damage. The degree of crystalluria formation is solely dependent on certain cardinal factors, namely : solubility of the free sulphonamide in the urine, degree of acetylation, solubility of the acetylated product, rate of excretion of sulphonamide and its metabolites and lastly the pH and volume of the urine excreted.

The crystalluria formation can be minimised by several ways : *first*, by increasing the intake of water specifically during sulphonamide therapy ; *secondly*, alkalinization of the urine by administering sodium bicarbonate or other alkaline formulations ; and *thirdly*, by administering a mixture of two or more sulphonamides, belonging to the same category, in such quantities that none of the drugs may reach a concentration which would otherwise cause crystalluria.

Probable Questions for B. Pharm. Examinations

1. (*a*) Explain how an 'azo-dye' breaks down *in vivo* to yield sulphanilamide ?

   (*b*) N1-substitution in sulphanilamide is more effective and useful than N4-substitution. Explain.

2. Classify sulphonamide on the basis of their site of action. Give the structure, chemical name and uses of **one** potent drug each class.

3. How would you synthesize sulphanilamide from :

(*a*) Acetanilide (*b*) Benzene (*c*) Sulphanilic acid.

4. Give the structure, brand name, official status and uses of :

(*a*) Sulphapyridine (*b*) Sulphathiazole (*c*) Sulphadiazine (*d*) Sulfamerazine.

Discuss the synthesis of any **two** drugs.

5. How would you synthesize the following drugs employed for urinary-tract infections :

(*a*) Sulphacetamide from sulphanilamide

(*b*) Sulfafurazole from *p*-acetamidobenzene sulphonyl chloride

(*c*) Sulfisoxazole acetyl sulfisoxazole

6. Give the structure, chemical name the uses of the following branded drugs :

(*a*) Thalazole® (*b*) Sulfauxidine® (*c*) Thalisul®.

7. Describe the synthesis of a potent sulphonamide used mostly in :

(*a*) Gas-gangrene

(*b*) Second-and third degree burns.

8. Name any **two** important members of sulphonamide related compounds that are used in chemotherapy of leprosy and burn-therapy. Discuss the synthesis of **one** medical compound.

9. Give a comprehensive account of the **'mode of action of sulphonamides'**. Support your answer with suitable examples.

10. (*a*) Enumearte briefly sulphonamide inhibition and probable mechanisms of bacterial resistance to sulphonamides.

(*b*) Write a short note on **'chemotherapeutic consideration of sulphonamides'**.

RECOMMENDED READINGS

1. EN Northey : Sulphonamides and Allied Compounds, *Am. Chem. Soc. Monograph.*, (1948).

2. A Goldstein : Antibacterial Chemotherapy, *New England J Med.* (240) (1949).

3. H Busch and M Lane, *Chemotherapy,* Yearbook Medical Publishers Chicago (1967).

4. ME Wolff (*Ed.*) : *Burger's Medicinal Chemistry and Drug Discovery*, (5th edn), John Wiley and Sons. Inc., New York, (1995).

5. F Hawking and JS Lawrence : *The Sulfonamides,* Grune and Stratton New York (1961).

6. FW Schuler : *Molecular Modification in Drug Design, Am. Chem. Soc.* Advanced in Chemistry Series No. 45, Washington. D.C. (1964).

7. WO Foye : *Principles of Medicinal Chemistry,* (5th Edn.) Lippincott Williams and Wilkins, New York, 2002.

8. JN Delgado and WA Remers : Wilson dn Gisvold's Textbook of Organic and Medicinal Chemistry, (10th edn.), Philadelphia, J B Lipincott Company (1998).

9. JK Seydel : Molecular basis for the Action of Chemotherapeutic Drugs, Structure-activity Studies of Sulfonamides, Proc. III International Pharmacology Congress, Sao Paulo, Pergamon Press, New York (1966).

10. DLednicer and LA Mitscher, *The Organic Chemistry of Drug Synthesis,* John Wiley and Sons. New York (1995).

19

Antimalarials

1. INTRODUCTION

Antimalarials are chemotherapeutic agents which are used for the prevention and treatment of malaria.

Malaria is still one of the most dreadful protozoal diseases affecting man. Until recently it was of a world-wide spread. However, the disease occurs now mainly in tropical countries of the world. It is quite painful and distressing to know that more than one million people die of malaria each year and most of these people belong to the third world countries. Malaria has been eradicated from the developed countries as a result of the Malaria Eradication Programme of the World Health Organization (WHO). This was achieved through improved living conditions, use of insecticides, destruction of breeding places for mosquitoes and use of antimalarials as prophylactic agents. Similar efforts in developing countries have not yielded good results.

The causal organisms responsible for malaria belong to the genus *plasmodium* which is of the class of protozoa known as **sporozoa.** There are four different species which are accepted as being responsible for human malaria. These are *Plasmodium malariae,* the parasite of *quartan malaria ; Plasmodium vivax,* the parasite of benign *tertian malaria, Plasmodium falciparum,* the parasite of *malignant* or *subtertian malaria,* and *Plasmodium ovale,* the parasite that causes a mild type of *tertian malaria.*

These protozoa have complex life cycles embodying both the female anopheles mosquito and the liver and the erythrocyte of the human host. Hence, an ideal antimalarial must be able to exert an effect on two fronts simultaneously, namely : to eradicate the microzoan from the blood and also from the tissues, in order to produce an effectice *'radical cure'*. It has been established beyond reasonable doubt that the various antimalarials differ essentially in their point of interruption of the cycle of the parasite and, therefore, the stages of the infection that is effected.

In actual practice, there are *three* well recognized and predominant manners to control malaria effectively, namely ;

(*a*) Elimination of the vector*,

(*b*) Drug therapy, and

(*c*) Vaccination.

---

*A carrier, usually an insect or other arthropod, that transmits the causative organisms of disease from infected to non infected individuals, especially one in which the organism goes through one or more stages in its life-cycle.

532

Elimination of the Vector. Currently, the elimination of the vector is considered to be one of the easiest and most cost-effective measure adopted across the globe.

In fact, there are *two* different ways to control the *'mosquito carrier'*. *First,* being check and prevent the usual contact usually taking place between the insect and the human beings. It is, however, pertinent to observe here that the **Anopheles mosquito** happens to be a *nocturnal feeder* ; and, therefore, it is relatively much easier to control than the corresponding **Aedes aegypti mosquito** which is a *day feeder* and is responsible solely for carrying **dengue** as well as **yellow fever** (prevalent in the African continent). Simply by installing nylon or iron screens on widows and using mosquito netting (at night while sleeping) in bed-rooms may provide an effective measure of prevention. *Secondly,* the elimination of the *Anopheles mosquito*, normally by total erradication by the application of *insecticide* and drastically destroying the breeding hide-outs, is believed to be one of the most practical ways to eliminate (as opposed to control) malaria.

Examples : **Dichloro diphenyl trichloroethane (DDT)** an insecticide discovered by the Nobel laureate Dr. Muller (1948), almost kills the malaria-carrying *Anopheles mosquito* completely. Though its *long lasting* effect is eventually very much beneficial from the standpoint of Anopheles mosquito control, but it gets accumulated in the environment unfortunately that may ultimately gain entry into the **food chain** and can affect both **humans** and **animals** equally. Hence, the use of DDT has been banned completely by FDA, WHO and other law enforcing authorities ; and duly replaced by other *'safer insecticides'*.

$$Cl-\langle\bigcirc\rangle-\overset{\overset{\displaystyle CCl_3}{|}}{CH}-\langle\bigcirc\rangle-Cl$$

DDT

Drug Therapy. A host of *'drug substances'* either isolated from the *plant sources*, such as : quinine ; quinidine, cinchonine, cinchonidine, artimisinine ; or *synthetic compounds* for instance : chloroquine, paludrine, pamaquine etc., are being used profusely in topical countries to fight the menace of malaria as the *'life-saving drugs'*. It is worth noting at this juncture that the currently employed **'antimalarials'**, while being effective against certain species, also exhibit certain adverse reactions, and noticeable resistance is enhancing also progressively.

Vaccination. Inspite of the tremendous impetus and thrust instituted legitimately by WHO, the dream of developing an effective, safer, economically viable **'antimalarial vaccine'** is yet to discovered to combat the human sufferings, more specifically the rate of infant mortality in economically less privileged and developed countries of the world.

It has been duly observed that the *malaria parasite* does elicit obviously an immune-response, demonstrated by virtue of the fact that usually the children, in particular, having an initial exposure are more prone to die than the adults who have since experienced/exposed to several recurring attacks.

Besides, a T-cell response which essentially comprise of both **CD4$^+$ and CD8$^+$ T-cells,** production of **interferon gamma,** and **nitric oxide synthase** induction serves as an additional proof of evidence that the **human-immune system** is *able to detect the parasite and hence responds accordingly.*[*]

A plethora of chemotherapeutic agents having divergent chemical structures have been introduced clinically since the mid-twenties, *e.g.,* pamaquine (1926), quinacrine (1930). Although historical evidence of cinchona dates back to 1638 when it was used to treat countess of Cinchona, wife of the then governor

---

*Pombo DJ *et al. Lancet,* **360** : 610, 2002.

of Peru, and hence the name. Later on, in 1820, Polletier and Caventou succeeded in the isolation of quinine from the circhona bark. Now, more than twenty-five alkaloids from the circhona bark have been characterized, out of which only a few are useful clinically, *viz.*, quinine, quinidine, cinchonine, cinchonidine.

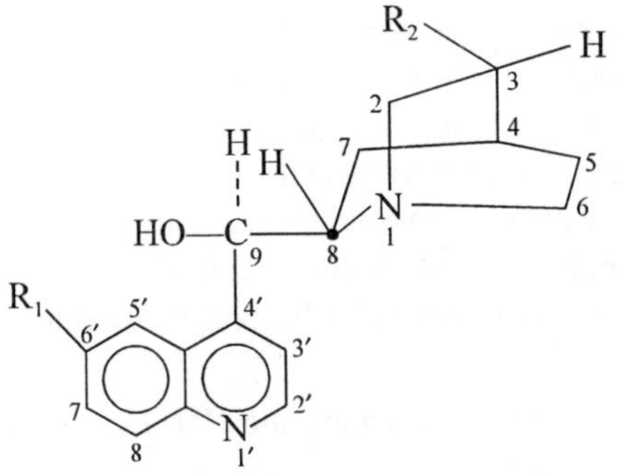

Quinine :　　　　　$R_1 = OCH_3$; $R_2 = - CH = CH_2$; (–) 8S : 9R isomer

Quinidine :　　　 $R_1 = OCH_3$: $R_2 = - CH = CH_2$; (+) 8R : 9S isomer

Cinchonine :　　 $R_1 = H$; $R_2 = - CH = CH_2$; (+) 8R : 9S isomer

Cinchonidine :　 $R_1 = H$; $R_2 = - CH = CH_2$; (–) 8S : 9R isomer

All these alkaloids bearing the same substitution at R_1 and R_2 are essentially **diastereoisomers,** only having different configuration at the *third and fourth chiral centres (C-8 and C-9).*

Soon after the Second World War (1943), a large number of compounds were synthesized and tested for antimalarial actions, and these eventually gave birth to a host of potent drugs like, chloroquine, proguanil, comoquine and amodiaquine, etc.

2. CLASSIFICATION

The antimalarials may be classified on the basis of their basic chemical nucleus as stated below and some representative examples belonging to each class are given.

2.1. 4-Aminoquinoline Analogues

In 1942, a group of German researchers first reported that 4-, 6- and 8-aminoquinolines when duly substituted produced antimalarial agents. An extensive research in East and West was augmented due to the acute shortage of cinchona bark during the Second World War. These drugs are found to be active against the erythrocytic form of most malarial parasites ultimately affecting a clinical cure. They do not cause prevention of the disease, and they are inactive against the liver-infecting forms.

A. *Chloroquine Phosphate* BAN, USAN, *Chloroquine* INN,

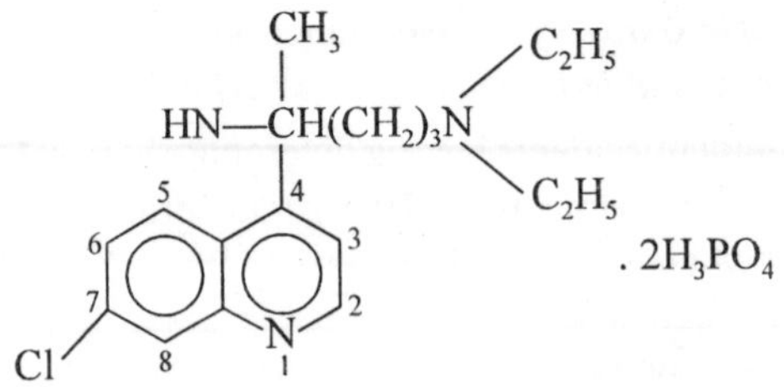

7-Chloro-4-[[4-(diethylamino)-1-methyl] butyl] amino]-quinoline phosphate (1:2) ; 1, 4-Pentanediamine, N^4-(7-chloro-4-quinolinyl)-N^1, N^1-diethyl-, phosphate (1:2) ; Resochin ; BP ; USP ; Int. P., Ind. P. ; Aralen[R] (Winthrop).

Synthesis :

It is prepared by adopting the following *four* steps *viz.,*

(*a*) Preparation of 4, 7-Dichloroquinoline (*i.e.,* the nucleus)

(*b*) Preparation of 2-amino-5-diethyl amino pentane, or 1-diethylamino-4-amino pentane (*i.e.,* the side chain).

(*c*) Condensation of '*a*' and '*b*'.

(*d*) Addition of concentrated phosphoric acid to a hot ethanolic solution of the condensed product.

(*a*) *Preparation of Nucleus :*

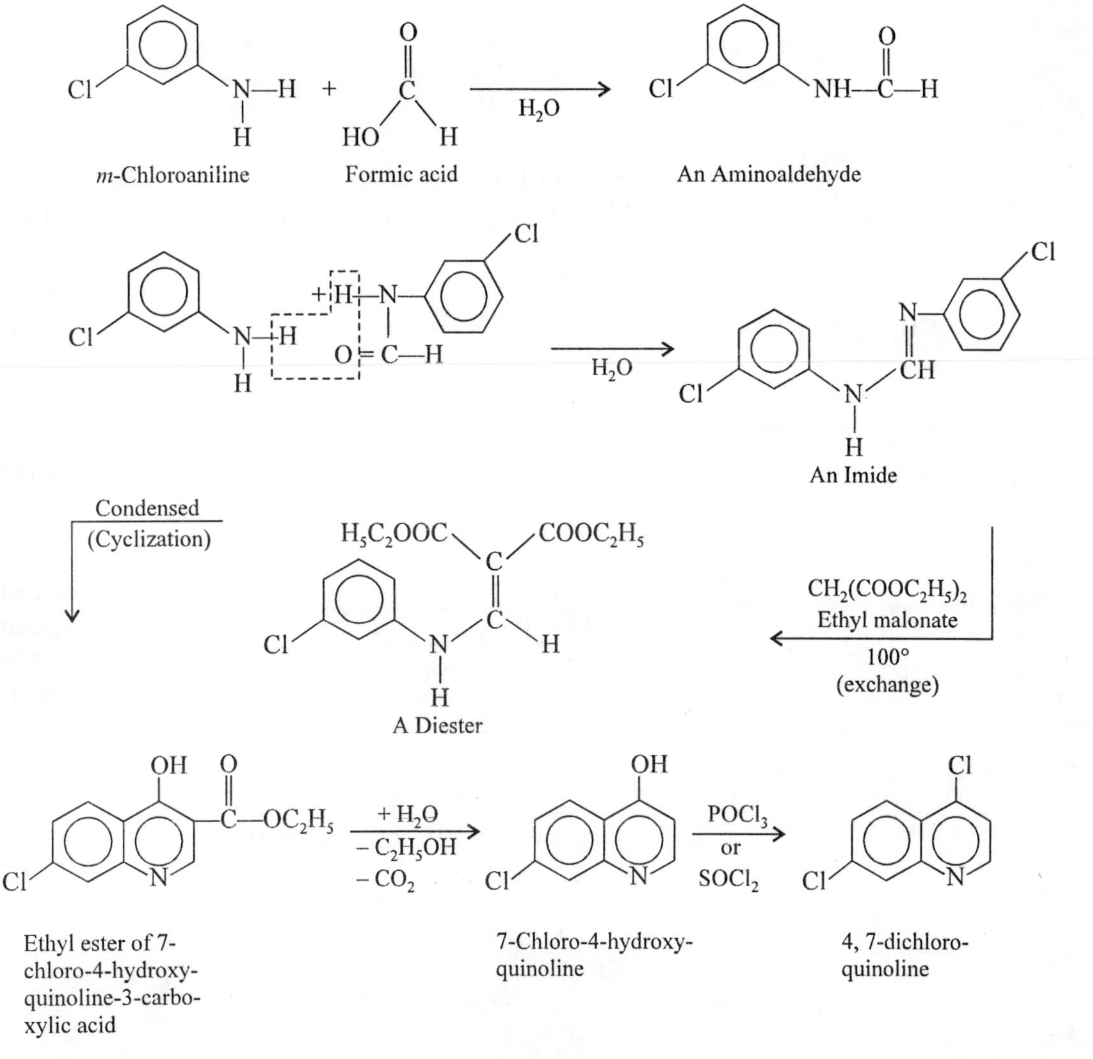

m-Chloroaniline   Formic acid   An Aminoaldehyde

An Imide

Condensed (Cyclization)

A Diester

$CH_2(COOC_2H_5)_2$
Ethyl malonate
100°
(exchange)

Ethyl ester of 7-chloro-4-hydroxy-quinoline-3-carboxylic acid

7-Chloro-4-hydroxyquinoline

4, 7-dichloroquinoline

$+H_2O$
$-C_2H_5OH$
$-CO_2$

$POCl_3$
or
$SOCl_2$

(*b*) *Preparation of Side Chain :*

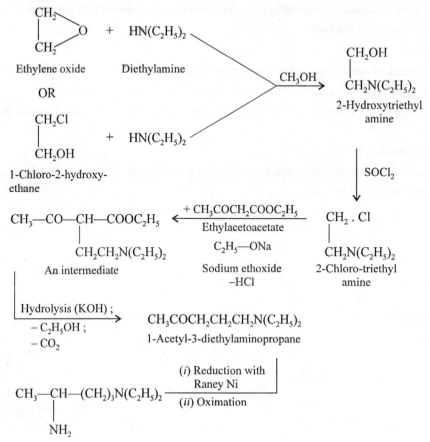

(*c*) *Condensation of (a) and (b) :*

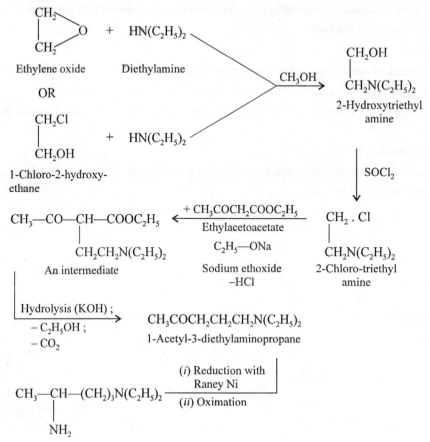

(*d*) *Preparation of Phosphate Salt :*

$$\text{Chloroquine Base} + 2H_3PO_4 \rightarrow \left[\begin{array}{c} \overset{CH_3}{\underset{|}{}} \\ \overset{+}{NH_2}-CH(CH_2)_3\overset{+}{NH}(C_2H_5)_2 \\ \\ \text{Chloroquine Phosphate} \end{array} \right] . 2H_2PO_4^-$$

Reaction between two moles of *m*-chloroaniline and a mole of formic acid yields an imidine which on treatment with ethyl malonate at 100° undergoes exchange and results into a diester. This further undergoes cyclization through condensation to yield the corresponding ethyl ester of 7-chloro-4-hydroxy-quinoline-3-carboxylic acid which on hydrolysis gives 7-chloro-4-hydroxy quinoline. Finally, on chlorination with either phosphorus oxychloride or thionyl chloride yields the nucleus 4, 7-dichloro quinoline.

The side chain is prepared by the interaction between either ethylene oxide or 1-chloro-2-hydroxy ethane with diethyl amine in methanol yields 2-hydroxy-triethyl hydrochloride. This on chlorination with thionyl chloride yields 2-chloro-triethyl amine which on treatment with ethyl aceto-acetate in the presence of sodium ethoxide gives an intermediate compound. Alkaline hydrolysis produces 1-acetyl-3-diethylamino propane which on reduction with Raney nickel followed by oximation yields 4-amino-1-diethylamino pentane.

Condensation of the nucleus and the side chain gives rise to the chloroquine base which on treatment with hot phosphoric acid in an ethanolic solution yields the official compound.

Chloroquine is extensively employed for the suppression and treatment of malaria. *It has been found to exert a quick schizonticidal effect and seems to affect cell growth by interfering with DNA.* The overall activity of chloroquine appears to be partially dependent on the preferential accumulation in the infected erythrocyte. It has been observed that *it specifically kills the erythrocytic forms of all malaria parasites at all states of development, but has no effect on the malaria parasite in the human liver cells. Hemce, chloroquine produces complete cure of malaria caused by P. falciparum.* It fails to check the relapse caused by the secondary exoerythrocytic phase of *P. malariae, P. ovale and P. vivax.*

Dose : As prophylactic, suppressive, 500mg once per week ; therapeutic, initially, 1g, followed by 500mg in 6 hours, and 500mg on the 2nd and 3rd days.

Chloroquine sulphate is another salt of chloroquine [Nivaquin[(R)], May & Baker] which possesses almost similar actions to those of resochin.

B. *Amodiaquine Hydrochloride* BAN, USAN, *Amodiaquine* INN,

4-[(7-Chloro-4-quinolyl) amino]-α-(diethylamino)-*o*-cresol dihydrochloride dihydrate ; Phenol, 4-[(7-Chloro-4-quinolyl)-amino]-2-[(diethylamino) methyl]-dihydrochloride, dihydrate, Amodiachin Hydrochloride ; BP ; USP ; Int. P ; Ind. P ; Camoquin[(R)] (Parke-Davis).

Synthesis :

It consists of the preparation of :

   (*i*) 4, 7-Dichloroquinoline (*i.e.,* nucleus),

   (*ii*) 4-Amino-2-diethylaminomethyl phenol,

   (*iii*) Condensation of (*i*) and (*ii*),

   (*iv*) Formation of hydrochloride salt.

   (*a*) *Preparation of 4, 7-dichloroquinoline nucleus.* It is prepared as described under chloroquine phosphate earlier.

   (*b*) *Preparation of the side chain 4-amino-2-diethylamino-ethyl phenol* : It may be prepared by two methods as described below :

Method-I : From p-Acetamido phenol

   o-Diethylaminomethyl-*p*-acetamidophenol is prepared by the interaction of *p*-acetamidophenol, formaldehyde and diethyl amine with the elimination of a molecule of water. Hydrolysis of this product yields 4-amino-2-diethylamino ethyl phenol with the elimination of a mole of acetic acid.

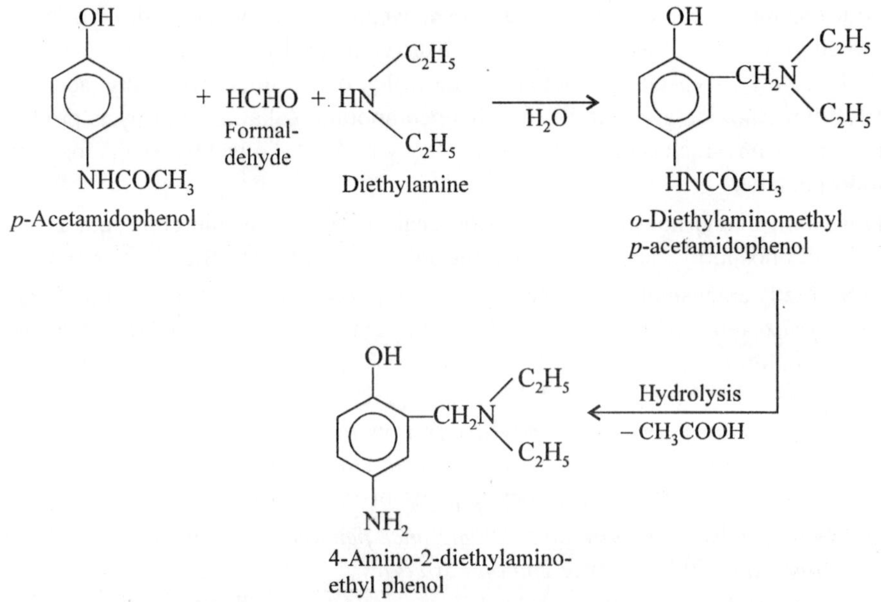

Method-II : From 2-Hydroxy-5-nitrobenzyl chloride

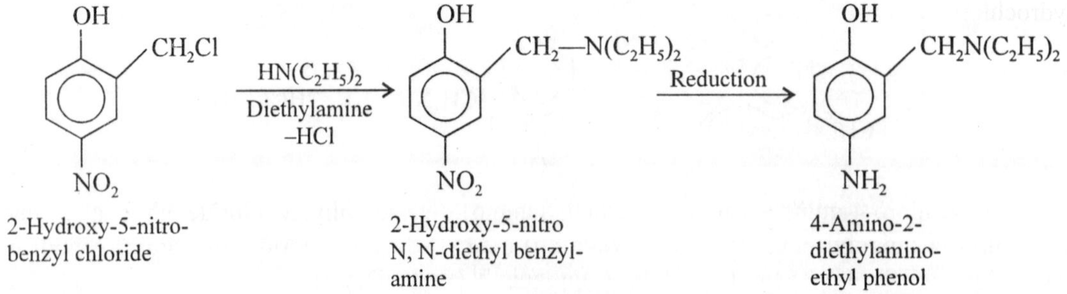

The reaction between 2-hydroxy-5-nitro-benzyl chloride with diethylamine yields 2-hydroxy-5-nitro-N, N, diethyl benzyl amine, which on reduction gives 5-amino-2-diethylamino ethyl phenol.

(c) *Condensation of (a) and (b),* we have :

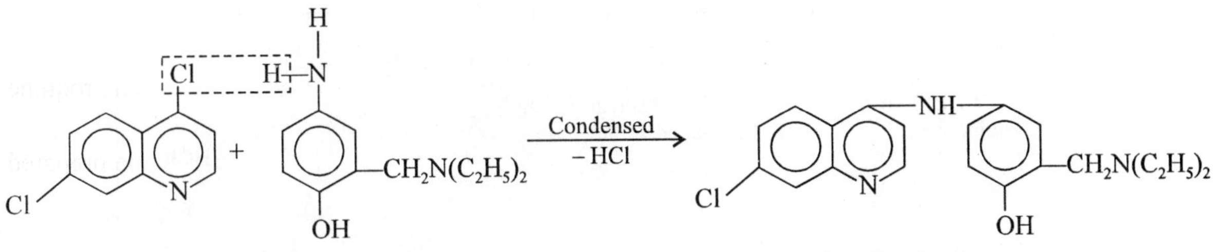

4, 7-Dichloro-   4-Amino-2-diethylamino-   Amodiaquine Base
quinoline   ethyl phenol

(d) *Preparation of Hydrochloride Salt :*

$$\text{Amodiaquine Base} \xrightarrow{\text{HCl}} \text{Amodiaquine Hydrochloride}$$

The condensation of the nucleus and the side chain prepared above in (a) and (b) gives the amodiaquine base which on neutralization with hydrochloric acid yields the official compound.

Its antimalarial action is very much similar to that of chloroquine and hence may be used alternatively for the same purpose.

Dose : Initially, 600mg of base followed by 300mg doses 6, 24, 48 hours later.

C. *Hydroxychloroquine Sulphate* BAN, *Hydroxychloroquine Sulfate* USAN,

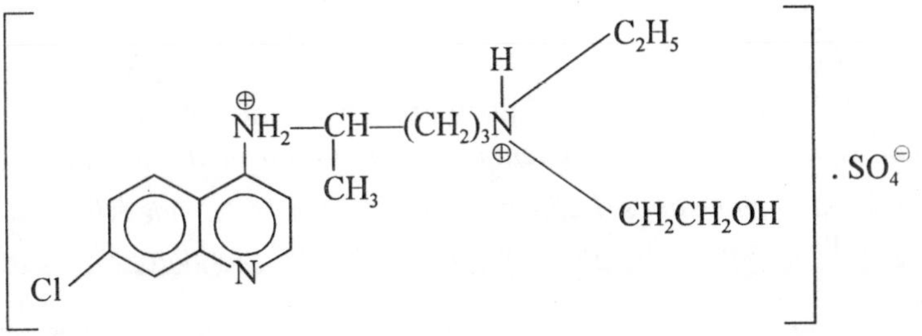

2-[[4-(7-Chloro-4-quinolyl] amino] pentyl] ethylamino] ethanol sulphate (1:1) (salt) ; Ethanol, 2-[[4-(7-chloro-4-quinolyl) amino]-pentyl] ethylamino]-, sulphate (1:1) salt ; Oxichlorochin Sulphate ; Hydrochloroquine Sulphate BP ; Hydroxychloroquine Sulfate USP ; Plaquenil Sulfate[R] (Winthrop).

Synthesis :

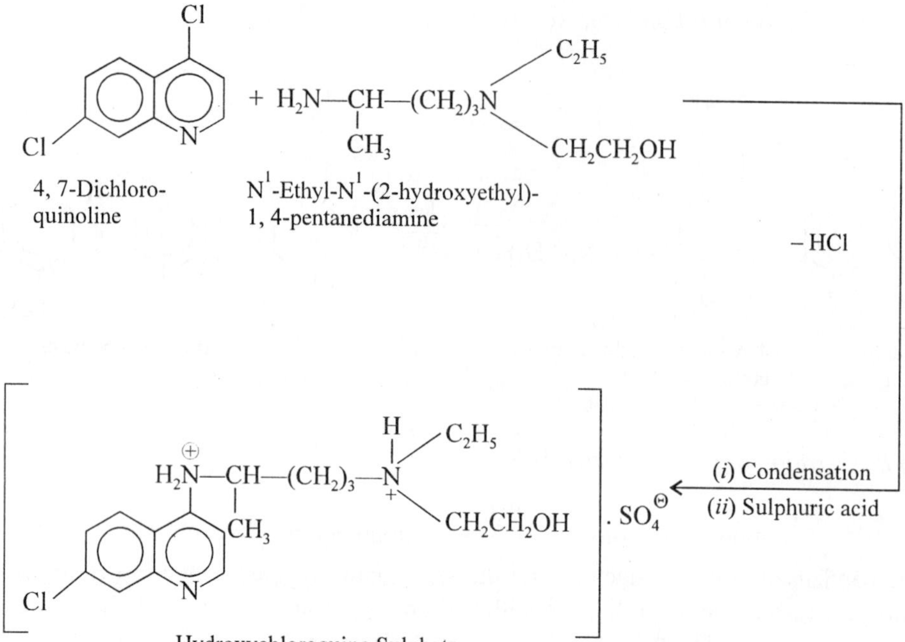

4, 7-Dichloro-
quinoline

N^1-Ethyl-N^1-(2-hydroxyethyl)-
1, 4-pentanediamine

– HCl

(*i*) Condensation
(*ii*) Sulphuric acid

Hydroxychloroquine Sulphate

It may be prepared by the condensation of 4, 7-dichloro quinoline and N^1-ethyl-N^1-(2-hydroxyethyl)-1, 4-pentanediamine and dissolving the resulting hydroxychloroquine base in absolute ethanol. The coversion to the corresponding sulphate is achieved by treating with an equimolar portion of sulphuric acid.

Its actions and uses are similar to those of chloroquine. Owing to its specific action on the erythrocytic phase of the malaria parasite it fails to serve as a 'radical cure' for *P. vivax* infections. *It also finds its clinical usefulness in the treatment of rheumatoid arthritis and lupus erythematosus.*

Dose : In P. falciparum infections, 1.25g in a single dose or in 2 divided doses at 6-hour intervals ; in rheumatoid arthritis, 400mg daily ; in lupus erythematosus, 200 to 400mg 1 or 2 times daily.

D. *Santoquin*

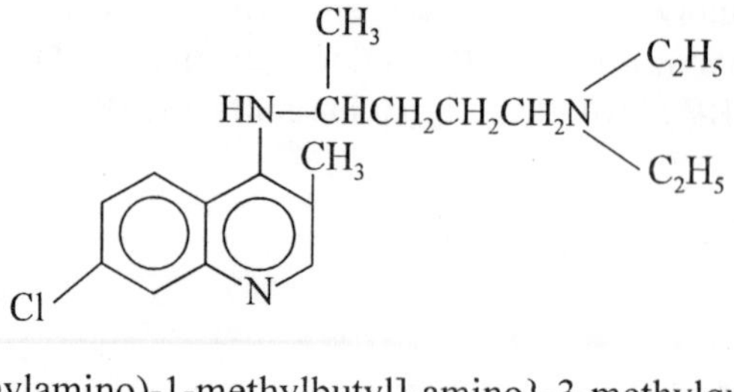

7-Chloro-4-{[4-(diethylamino)-1-methylbutyl]-amino}-3-methylquinoline.

Synthesis :

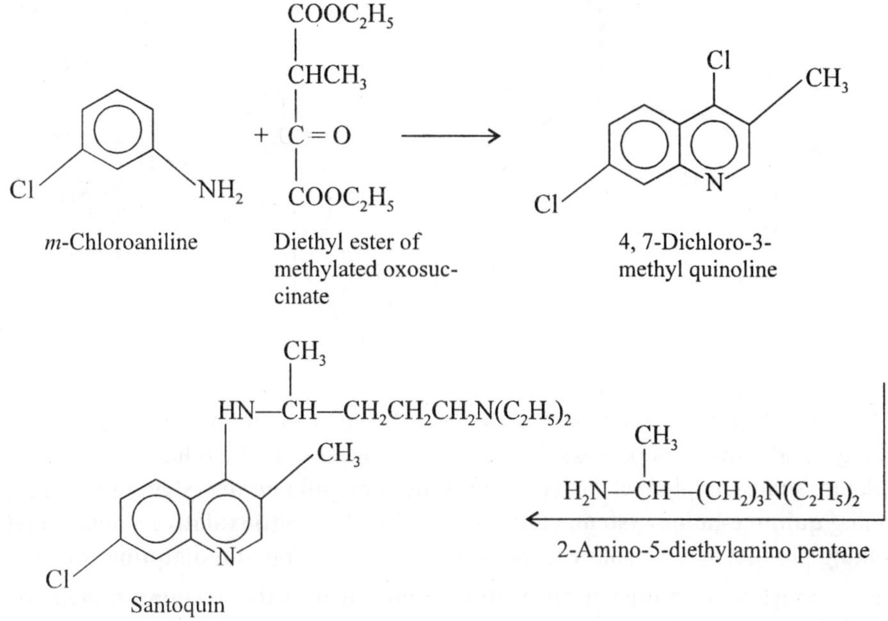

m-Chloroaniline	Diethyl ester of methylated oxosuccinate	4, 7-Dichloro-3-methyl quinoline

2-Amino-5-diethylamino pentane

Santoquin

4, 7-Dichloro-3-methylquinoline is prepared by the interaction of *m*-chloroaniline and diethyl ester of methylated oxosuccinate which on treatment with 2-amino-5-diethyl-amino pentane affords santoquin.

It has an additional methyl group at C-3 in the quinoline nucleus of chloroquine. It is found to be less reactive than chloroquine.

2.1.1. Mechanism of Action. The mechanism of action of the four compounds described under section 19.2.1 shall be dealt with individually in the sections that follows :

2.1.1.1. Chloroquine Phosphate. The *'drug'* particularly causes significant dysfunction of the *acid phagosomes in plasmodia* and also in human leukocytes and macrophages. It is neither a prophylactic nor a radical curative agent in vivax malaria. Interestingly, in regions wherein *Plasmodium falciparum* is invariably sensitive to chloroquine, it is specifically and predominantly effective in terminating acute attacks of **non-resistant falciparum malaria ;** and, therefore, normally affords total cure in this type of malaria. As chloroquine is well tolerated, it has been recommended to be used rountinely in **amebiasis** *without any demonstrable hepatic involvement.*

The *'drug'* usually gets absorbed almost completely from the GI tract when administered orally. The parenteral IM administraction is usually given with its HCl-salt. It has also been observed that the tissues bind the drug, but not to the same extent as that of *quinacrine.* It gets degraded in tissues to unknown products.

SAR of chloroquine. It may be regarded as the **prototypical** structure which overwhelmingly succeeded **'quinine'** and recognized as a potential *'synthetic antimalarial drug'* since the mid-1940s, as shown below :

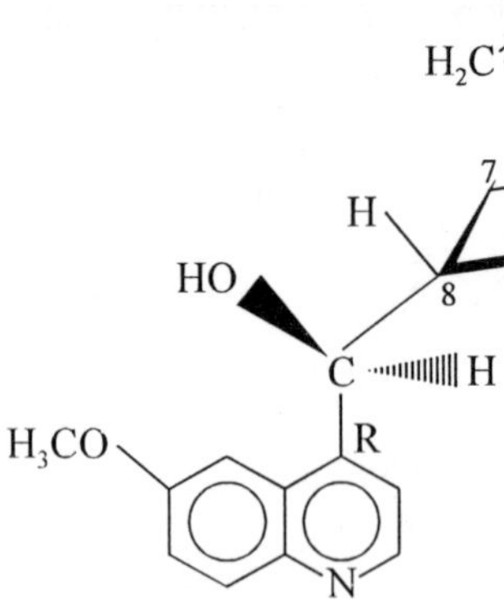

Quinine

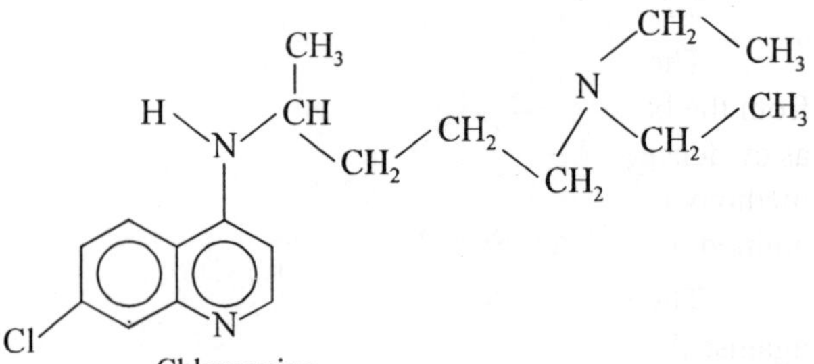

Chloroquine

2.1.1.2. Amodiaquine Hydrochloride. The *'drug'* resembles very similar to chloroquine mechanistically ; and it does not possess any added advantages over the other *4-aminoquinoline drugs*. It has been demonstrated amply that the **hydroquinone (phenol) amine system** rapidly gets oxidized to a corresponding **quinone-imine system,** either accomplished *via* **antioxidatively** and/or **metabolically ;** and the resulting product may be solely responsible for the ensuing **amodiaquine toxicity.**

Note : The quinine-imine system is almost identical to the acetaminophen (paracetamol) toxic metabolite.

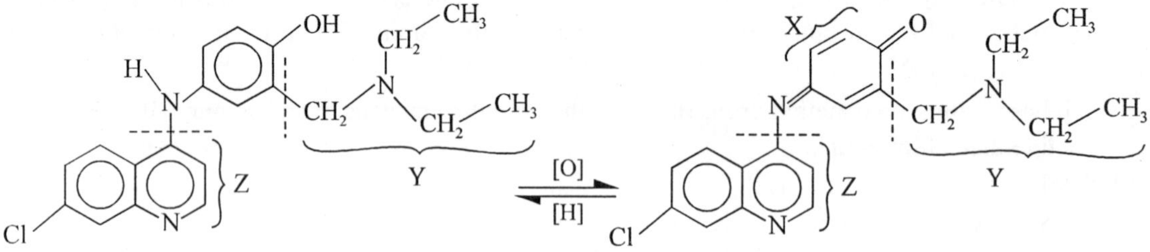

Amodiaquine

Amodiaquine Iminoquinone

$$\begin{bmatrix} X = \text{Quinonoid/Benzenoid Form} \\ Y \& Z = \text{No Change Moieties} \end{bmatrix}$$

In other words, amodiaquine upon oxidation gets converted to its *ketone-form* termed as **amodiaquine iminoquinone** which essentially embodies in it a quinonoid/benzenoid moiety.

2.1.1.3. Hydroxychloroquine Sulfate. The *'drug'* exerts its action exclusively in the *suppressive treatment* of **autoimmune inflammatory** diseases, for instance : *rheumatoid arthritis* (RA) and *systemic lupus erythromatosus* (SLS). Just like chloroquine (CQ), this *'drug'* (HCQ) is found to remain in the body for over a month and the prophylactic dosage is once-*a*-week only. It is somewhat less toxic than CQ.

SAR of HCQ. Structurally, it essentially differs exclusively in having an additional hydroxy (OH) moiety strategically attached to one of the terminal N-ethyl functions which eventually renders it less toxic than CQ perhaps due to H-bonding *in vivo.*

2.1.1.4. Sontoquine. The *'drug'* is found to be effective against **amaebic hepatitis** in *man* as well as *hamsters*. Unfortunately, the drug fails to show any specifically promising activity against the intestinal infections, most probably by virtue of the fact that it gets rapidly absorbed and do not reach the *lower intesine zone* in an effective therapeutic concentrations.

2.2. 8-Aminoquinoline Analogues

These structural analogues, unlike 4-aminoquinolines offer a rather more significant derivative from the basic quinine moiety. From the structural aspect these drugs seem to be optimally substituted as evidenced by the presence of a side-chain consisting of 4 to 6-carbon atoms as well as the location of methoxy group at C-6. In general, the 8-aminoquinoline analogues are relatively more toxic than the 4-aminoquinoline counterparts.

They are active against the pre- or exoerythrocytic form of the malarial parasite, but lack activity against the erythrocytic forms. The 8-amino-quinolines possess gametocidal activity. They are used mainly for the radical cure of relapsing malaria like vivax malaria.

A few classical examples belonging to this category are discussed below :

Pamaquine ; Primaquine phosphate ; Pentaquine phosphate ; Isopentaquine and Quinocide hydrochloride.

A. *Pamaquine* INN, *Pamaquin* BAN, *Pamaquine Naphthoate* USAN,

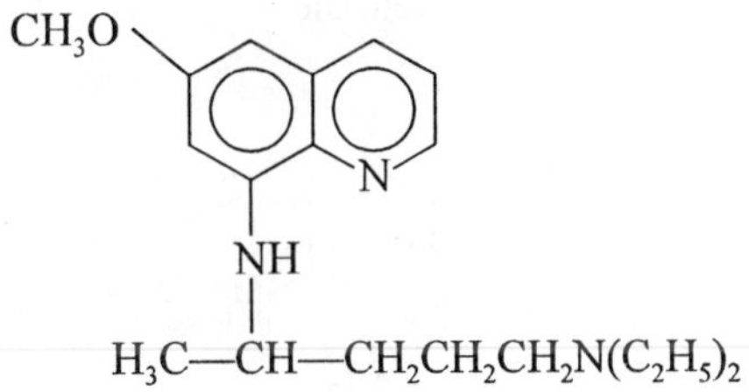

$$CH_3O$$

$$NH$$

$$H_3C-CH-CH_2CH_2CH_2N(C_2H_5)_2$$

8-(4-Diethylamino-1-methylbutylamino)-6-methoxyquinoline ; Pamachin ; Pamaquine Embonate ; Plasmoquinum ; BP ; 1953, NF IX.

Synthesis :

It consists of the preparation of :

(*a*) 4-amino-diethylamino pentane, *i.e.,* side-chain,

(*b*) 8-amino-6-methoxy quinoline, *i.e.,* quinoline nucleus,

(*c*) Condensation of (*a*) and (*b*).

(*a*) *Preparation of Side-Chain*

It has been described earlier under chloroquine phosphate.

(*b*) *Preparation of Quinoline Nucleus*

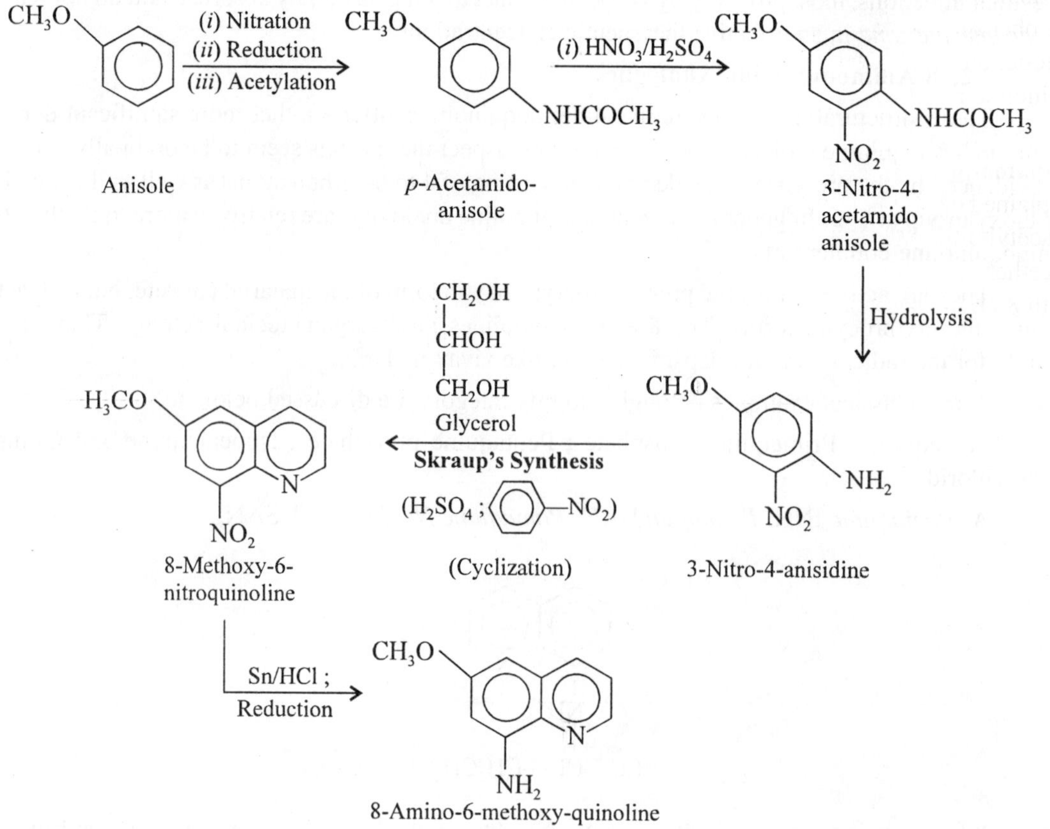
CH3O
(i) Nitration
(ii) Reduction
(iii) Acetylation
CH3O
NHCOCH3
(i) HNO3/H2SO4
CH3O
NHCOCH3
NO2
Anisole
p-Acetamido-anisole
3-Nitro-4-acetamido anisole
Hydrolysis
CH2OH
CHOH
CH2OH
Glycerol
Skraup's Synthesis
(H2SO4 ; —NO2)
(Cyclization)
H3CO
N
NO2
8-Methoxy-6-nitroquinoline
CH3O
NH2
NO2
3-Nitro-4-anisidine
Sn/HCl ;
Reduction
CH3O
N
NH2
8-Amino-6-methoxy-quinoline

(*c*) *Condensation of (a) and (b)*

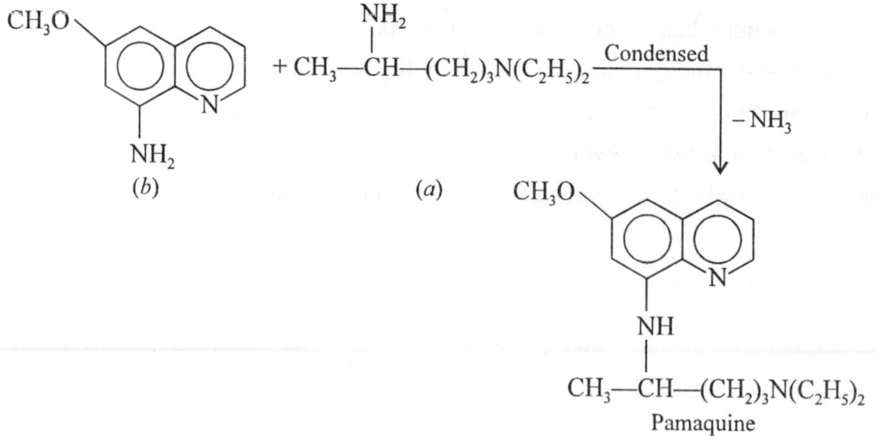
CH3O
N
NH2
(b)
NH2
+ CH3—CH—(CH2)3N(C2H5)2
(a)
Condensed
– NH3
CH3O
N
NH
CH3—CH—(CH2)3N(C2H5)2
Pamaquine

p-Acetamido anisole may be prepared by the sequential nitration, reduction and acetylation of anisole which on further nitration yields 3-nitro-4-acetamido anisole. This on hydrolysis gives 3-nitro-4-anisidine which on treatment with glycerol in the presence of concentrated sulphuric acid and nitrobenzene undergoes cyclization through Skraup's synthesis to yield 8-methoxy-6-nitro quinoline. Reduction of the resulting product gives rise to 8-amino-6-methoxy quinoline. Condensation of this quinoline residue with 4-amino-1-diethylamino pentane forms pamaquine.

Craig (1944) first discovered the presence of an isomeric form of pamaquine known as **isopamaquine,** in the commercial sample of pamaquine. The evolution of isopamaquine may be logically explained on the basis of the fact that the oximation of the amino alcohol obtained from the reduction of 1-acetyl-3-diethylamino propane actually gives rise to 4-amino-1-diethylamino pentane as major product together with 3-amino-1-diethylamino pentane as minor product. The latter product then condenses with 8-amino-6-methoxy quinoline to yield *isopamaquine* as given below :

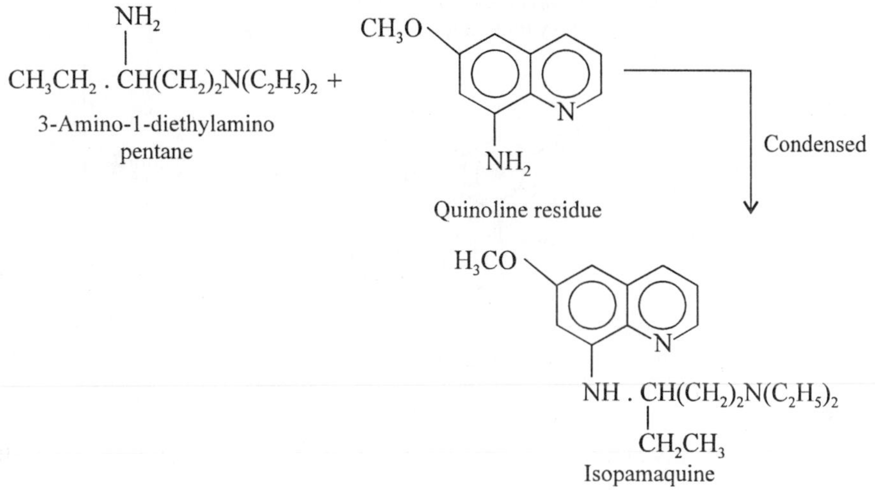

$$NH_2$$
$$CH_3CH_2 . CH(CH_2)_2N(C_2H_5)_2 +$$

3-Amino-1-diethylamino
pentane

CH_3O

Quinoline residue

NH_2

Condensed

H_3CO

$NH . CH(CH_2)_2N(C_2H_5)_2$
CH_2CH_3

Isopamaquine

The antimalarial activity of isopamaquine and pamaquine are fairly identical.

Toptchiev and Braude, in 1947, put forward a modified synthesis for pamaquine which consists of the following steps, namely :

(*a*) Preparation of a ketal from 1-acetyl-3-diethyl amino propane

(*b*) Preparation of 8-amino-6-methoxy quinoline

(*c*) Condensation of (*a*) and (*b*)

(*d*) Reduction of the condensed product.

(*a*) *Preparation of a Ketal*

$$OCH_3$$
$$CH_3CO(CH_2)_3N(C_2H_5)_2 \xrightarrow{\text{Methyl orthoformate}} CH_3C(CH_2)_3N(C_2H_5)_2$$
$$OCH_3$$

1-Acetyl-3-diethyl
amino propane

A ketal

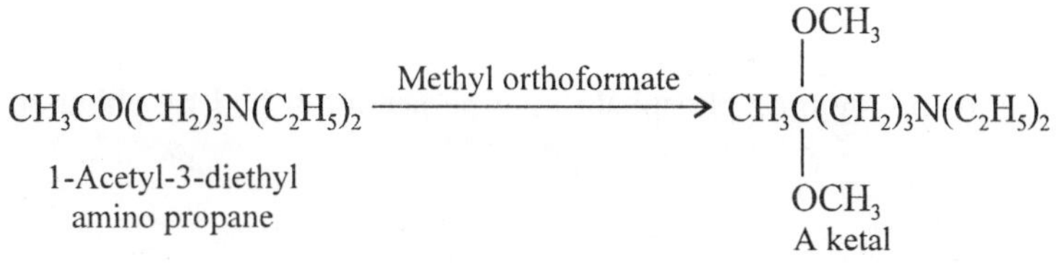

The starting compound is prepared by the method described earlier for the preparation of the side chain, which on treatment with methyl orthoformate yields a ketal.

(b) *Preparation of quinoline residue*

It is same as described under pamaquine.

(c) *Condensation of (a) and (b) ; and (d) Reduction.*

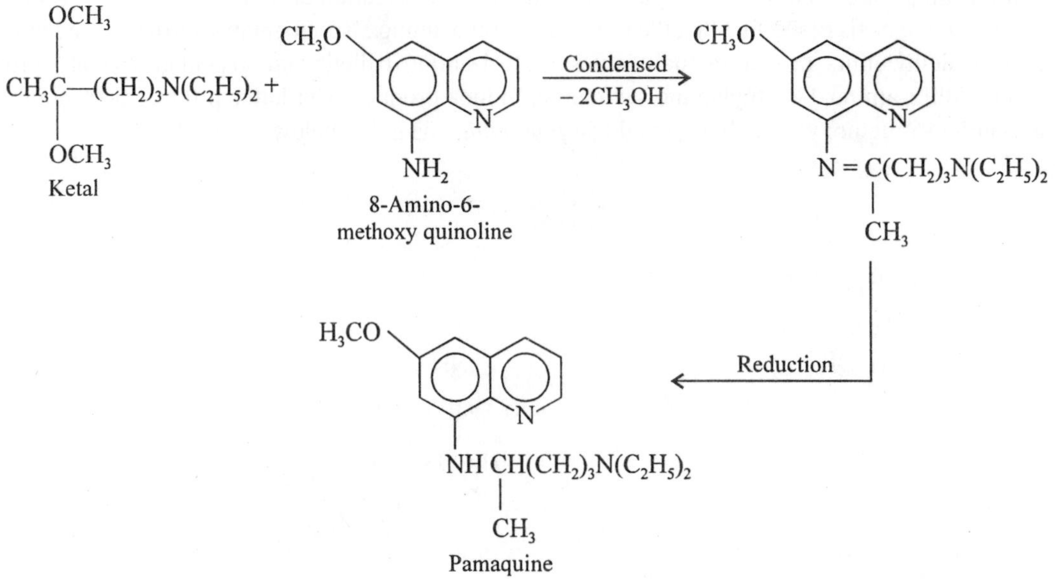

8-Amino-6-methoxy quinoline

Pamaquine

Pamaquine was initially employed for the treatment of malaria but has since been superseded by primaquine phosphate.

B. *Primaquine Phosphate*, BAN, USAN, *Primaquine* INN,

8-[(4-Amino-1-methylbutyl) amino]-6-methoxy quinoline phosphate (1:2) ; 1, 4-Pentanediamine, N^4-(6-methoxy-8-quinolyl)-, phosphate (1:2) ; Primachin phosphate ; BP ; USP ; Int. P. ; Primaquine Phosphate[R] (ICI Pharmaceuticals, U.K.)

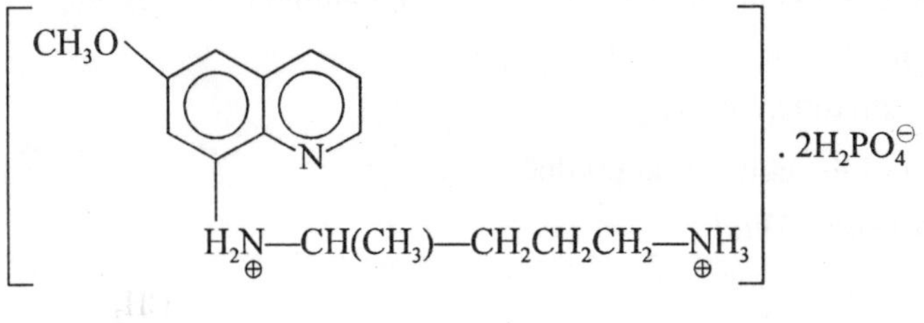

Synthesis :

The synthesis of primaquine phosphate may be accomplished by either of the *two* following methods :

Method-1 ; Elderfield's Method from 1, 4-Dibromopentane

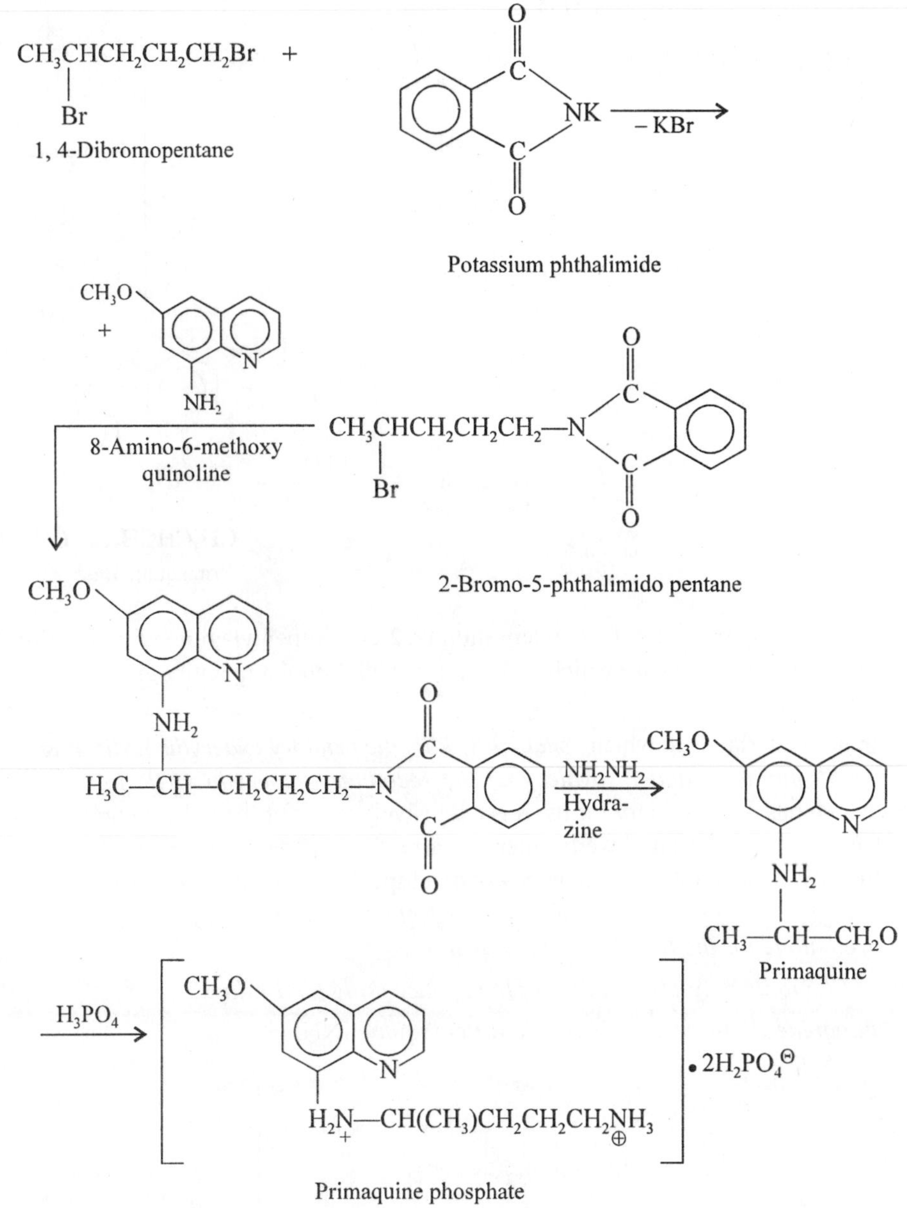

Potassium phthalimide

8-Amino-6-methoxy
quinoline

2-Bromo-5-phthalimido pentane

Primaquine

Primaquine phosphate

2-Bromo-5-phthalimido pentane is prepared by the interaction of 1, 4-dibromopentane with potassium phthalimide, which on reaction with 8-amino-6-methoxy quinoline yields the condensed product. Further treatment with hydrazine eliminates the phthalimido residue and yields the primaquine base which on reaction with a double molar quantity of phosphoric acid forms the official compound.

Method-II. From 2-Chloropentylamine

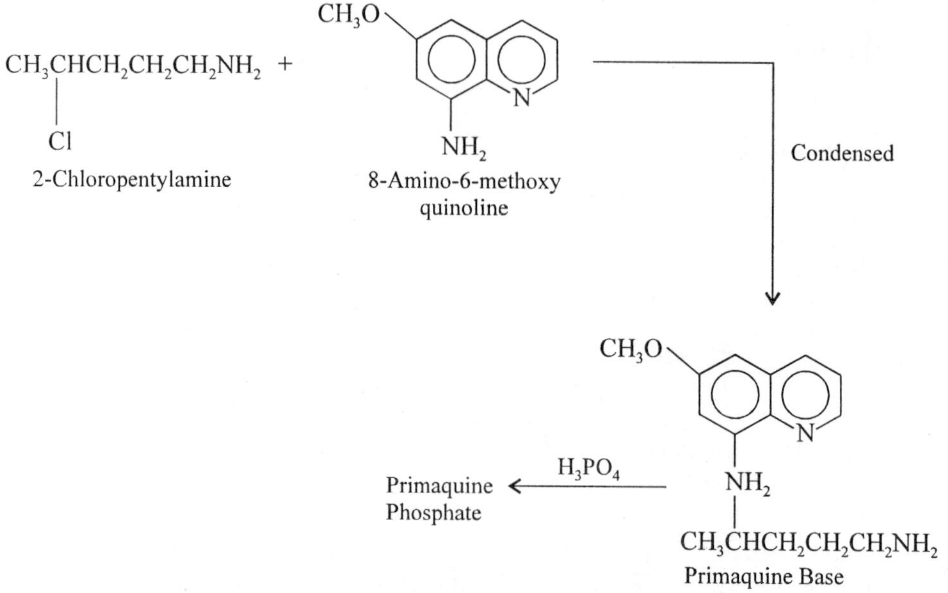

$CH_3CHCH_2CH_2CH_2NH_2$ +

|
Cl

2-Chloropentylamine

8-Amino-6-methoxy
quinoline

Condensed

Primaquine ← H_3PO_4
Phosphate

$CH_3CHCH_2CH_2CH_2NH_2$
Primaquine Base

It may also be prepared by the condensation of 2-chloro-pentylamine with 8-amino-6-methoxy quinoline to obtain primaquine base which on treatment with bimolar quantity of phosphoric acid yields primaquine phosphate.

It is an antimalarial drug which *specifically kills the primary exoerythrocytic stages of P. vivax, P. falciparum, P. malariae and P. ovale, and the secondary exoerythrocytic form of all except P. falciparum,* which has no secondary forms. It is extensively used for the radical cure of relapsing vivax malaria, but is not normally employed either for arresting the severe attacks of the disease or for suppressive therapy. It invariably kills gametocytes of all species, or inhibits their growth and development in the mosquito. *It fails to produce any significant effect on other erythrocytic stages and hence it must not be employed alone for the treatment of malaria.*

Dose : 17.5 to 26.3mg (10 to 15mg of base) once daily for 14 days.

C. *Pentaquine Phosphate* BAN, USAN, *Pentaquine* INN,

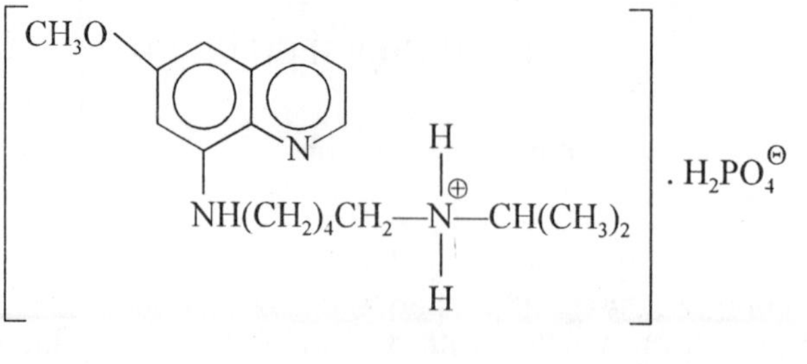

8-(5-Isopropylaminoamylamino)-6-methoxy quinoline phosphate ; 8-(5-Isopropylaminopentyl-amino)-6-methoxyquinoline phosphate. USP ; XIV.

Synthesis :

It consists of the preparation of :

(*a*) 8-Amino-6-methoxy quinoline

(*b*) 1-Chloro-5-isopropylamino pentane

(*c*) Condensation of (*a*) and (*b*) and

(*d*) Phosphate salt.

(*a*) *Preparation of 8-amino-6-methoxy quinoline*

It is prepared as described under pamaquine.

(*b*) *Preparation of 1-chloro-5-isopropylamino pentane*

Tetrahydrofurfuryl   A pyran

$$\text{HOCH}_2(\text{CH}_2)_4\text{NHCH(CH}_3)_2 \xrightarrow{\text{SOCl}_2} \text{ClCH}_2(\text{CH}_2)_4\text{NHCH(CH}_3)_2$$

5-Isopropyl-aminopentanol   1-Chloro-5-isopropyl-amino pentane

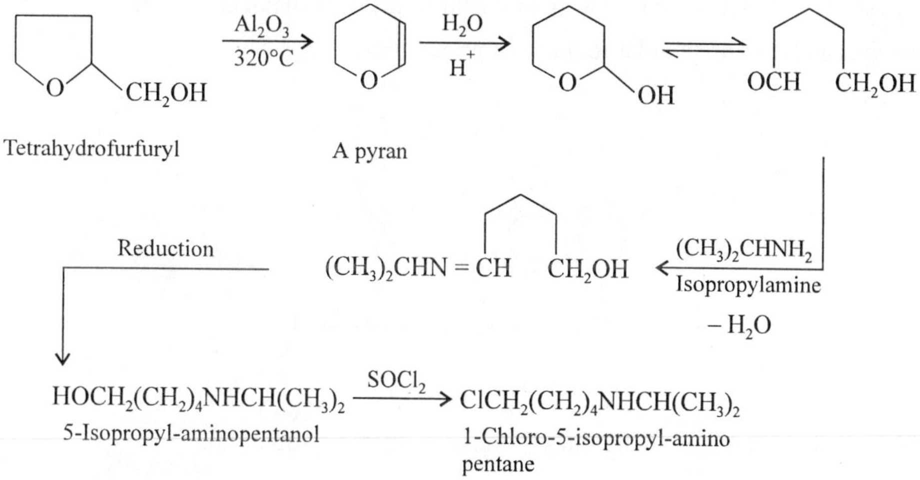

Tetrahydrofurfuryl alcohol on heating with aluminium oxide at 320° forms a partially saturated pyran which upon hydrolysis in an acidic medium yields a hydroxy analogue of pyran. This undergoes cleavage and the cleaved product on treatment with isopropyl amine forms an intermediate which on reduction gives rise to 5-isopropyl amino pentanol. Chlorination with thionyl chloride yields-1-chloro-5-isopropyl amino pentane.

(*c*) *Condensation of (a) and (b), (d) Treatment with* H_3PO_4

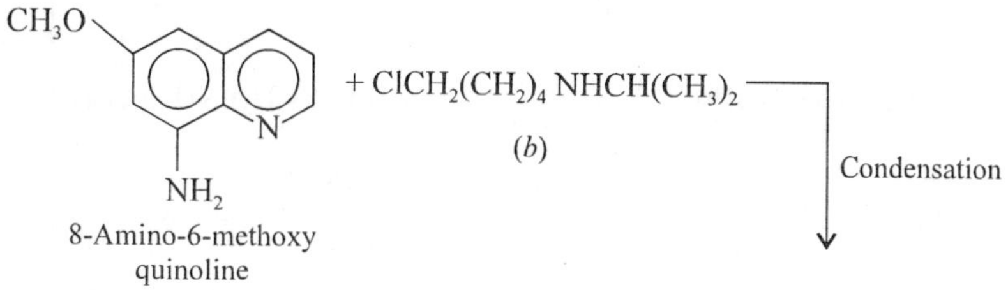

8-Amino-6-methoxy
quinoline

$+ \text{ClCH}_2(\text{CH}_2)_4\ \text{NHCH(CH}_3)_2$

(*b*)

Condensation

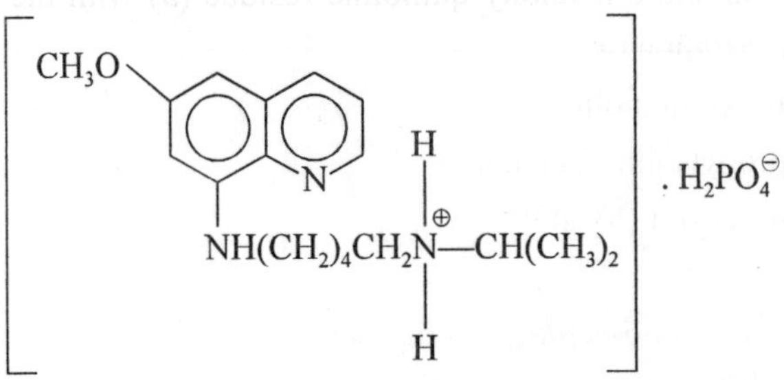

$$\left[\text{CH}_3\text{O} \underset{\text{N}}{\bigotimes} \text{H} \atop \text{NH(CH}_2)_4\text{CH}_2\overset{\oplus}{\text{N}}\text{—CH(CH}_3)_2 \right] . \text{H}_2\text{PO}_4^{\ominus}$$

Pentaquine Phosphate

Condensation of the quinoline residue (*a*) and the side chain (*b*) yields the pentaquine base which on treatment with one mole of phosphoric acid forms the official compound.

Its actions and uses are similar to those of primaquine.

Dose : 100mg per day.

D. *Isopentaquine*

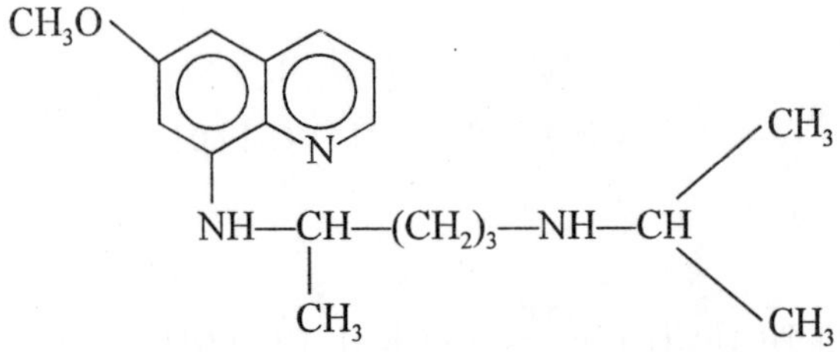

8-[[4-(Isopropylamino)-4-methylbutyl] amino]-6-methoxy-quinoline.

Synthesis :

It consists of the preparation of the side chain 2-bromo-5-isopropylaminopentane as given below :

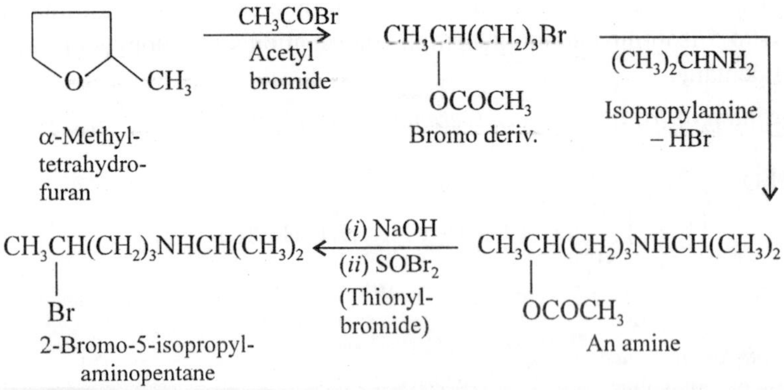

Interaction between α-methyl-tetrahydrofuran and acetyl bromide yields bromo derivative which on treatment with isopropyl amine forms a corresponding amine derivative. This on treatment with aqueous NaOH and thionyl bromide gives the side chain.

Treatment of the 8-amino-6-methoxy quinoline residue (*a*) with the side chain (*b*) yields the isopentaquine as shown below :

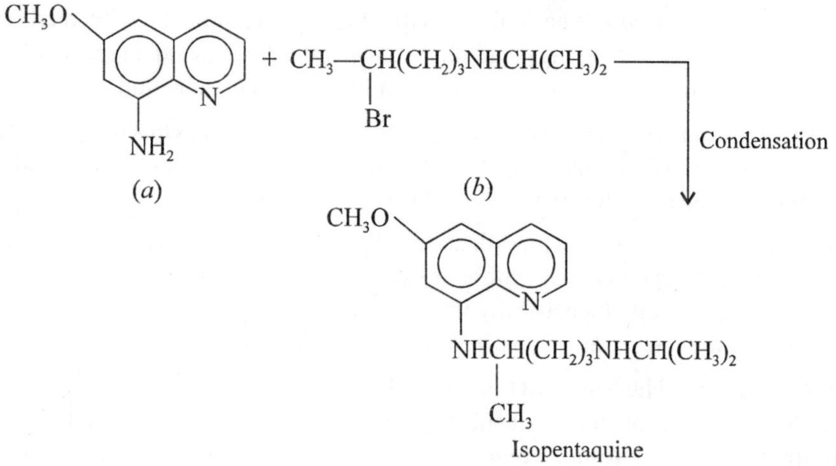

(*a*)   (*b*)

Isopentaquine

Isopentaquine is an isomer of pentaquine, and reported to be more active than the later as an antimalarial agent. It is also less toxic than pentaquine.

E. *Quinocide Hydrochloride* BAN, *Quinocide* INN, USAN,

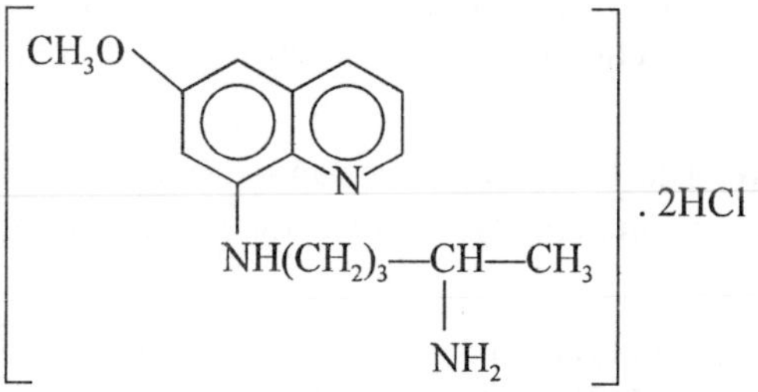

8-(4-Aminopentylamino)-6-methoxyquinoline dihydrochloride.

It is a structural isomer of primaquine, has actions and uses resembling to those of primaquine phosphate.

Dose : 30mg per day.

2.2.1. Mechanism of Action. The mechanism of action of the various compounds discussed under section 19.2.2. are dealt with separately in the pages that follows :

2.2.1.1. Pamaquine. The *'drug'* exerts its action against the exoerythrocytic stages of *P. ovale* and primary exoerythrocytic stages of *P. falciparum.* It has also been observed that it particularly inhibits the **gametocyte stage,** that essentially helps to eliminate the form required to infect the *'mosquito carrier'.* It also appears to disrupt and destabilize the parasite's mitochondria *via* several processes that include maturation into the subsequent resulting forms. The glaring advantage being the destruction of the exoerythrocytic forms before the parasite may actually infect the erythrocytes *i.e.,* the specific stage in the *'infectious process'* which ultimately renders malaria so weakening.

SAR of Pamaquine. It is indeed structurally related to the cinchona alkaloids essentially having a 6-methoxy group like quinine, but the various substituents on the *'quinoline nucleus'* are strategically positioned at C-8 rather than C-4 as found on the cinchona alkaloids. It has a *four-carbon alkyl linkage* or *bridge between the two N-atoms.* It has only one **chiral centre.** Though it has been critically observed that there exists certain differences in the metabolism of each stereoisomer and type of adverse response, there is hardly any difference in the antimalarial action based on the pamaquin's stereochemistry.

2.2.1.2. Primaquine. Its mechanism of action is very much similar to that of *'pamaquin'*. However, its spectrum of activity is regarded to be one of the narrowest of all the currently employed antimalarials ; and is recommended exclusively for exoerythrocytic *P. vivax* malaria.

SAR of Primaquine. Structural modifications of *pamaquine* produced the *unsubstituted primary aminoalkyl derivative i.e., primaquine,* whose relatively more predominant therepeutic activity and significantly much lower toxicity (specifically the tendency for causing hemolysis) essentially replaced pamaquine virtually as the most well recognized **tissue schizonticide** of choice.

2.2.1.3. Pentaquine. The degree of toxicity in the 8-amino quinoline structural analogues appears to be directly associated with the degree of substitution at the terminal amino function.* Using the said criterion **pamaquine,** having a *tertiary amino moiety,* happens to be **more toxic** than **primaquine,** having a *primary terminal nitrogen ;* whereas, **pentaquine** and **isopentaquine,** having *secondary terminal amino moieties,* are found to be **intermediate in toxicity.**

SAR of Pentaquine. It may be observed that with the exception of pentaquine, the other *three* **8-aminoquinolines** *viz., pamaquin, primaquine* and *isopentaquine* have only one **chiral centre** (*i.e.,* asymmetric carbon). In fact, certain differences do take place in the actual metabolism of individual stereoisomer, but there exists practically little difference in the antimalarial profile based on the compound's stereochemistry.

2.2.1.4. Isopentaquine. The *'drug'* possesses an intermediate degree in toxicity because it has an essential secondary terminal amino moiety. Besides, the two N-atoms are duly separated by a chain of four C-atoms.

2.2.1.5. Quinocide (Chinocide). The *'drug'* is an isomer of primaquine, has been studied extensively by Russian researchers.** It has been used widely in the Eastern Europe, but despite claims to the contrary its chemotherapeutic index is appreciably lower in comparison to primaquine.***

2.3. 9-Aminoacridines

The earlier hypothesis put forward by Ehrlich that methylene blue exerts antimalarial activity paved the way for the discovery of a number of acridine analogues. The 9-aminoacridine analogues, however, are found to be extremely toxic in nature and, therefore, they have been successfully replaced by the 4-aminoquinoline analogues to a great extent. A few typical examples of this category are discussed below :

A. *Mepacrine Hydrochloride* BAN, *Quinacrine Hydrochloride* USAN, *Mepacrine* INN,

6-Chloro-9-[[4-(diethylamino)-1-methylbutyl] amino]-2-methoxy-acridine dihydrochloride dihydrate ; 1, 4-Pentane-diamine, N^4-(6-chloro-2-methoxy-9-acridinyl)-N^1, N^1-diethyl-, dihydrochloride,

---

*Edgeombe JH *et al. J Natl Malaria Soc,* **9,** 285 (1950).

Lysenko AJ *et al. Med Parasitol Parasit Dis* (USSR) **24, 132, 137, (1955).

***Powell RD : *Clin Pharmacol Therap.* **7,** 48, (1966).

dihydrate ; Acrinamine ; Mepacrine Hydrochloride BP ; Eur. P ; Int. P ; Ind. P ; Quinacrine Hydrochloride USP ; Atabrine Hydrochloride[(R)] (Winthrop).

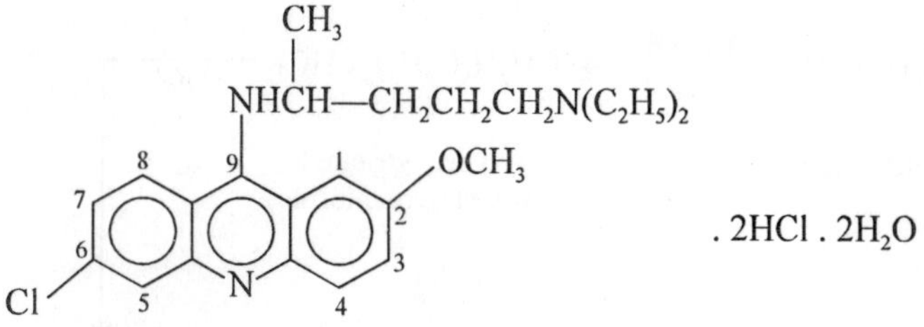

$$. 2HCl . 2H_2O$$

Synthesis :

It essentially consists of the following steps :

(*i*) Preparation of the side chain :

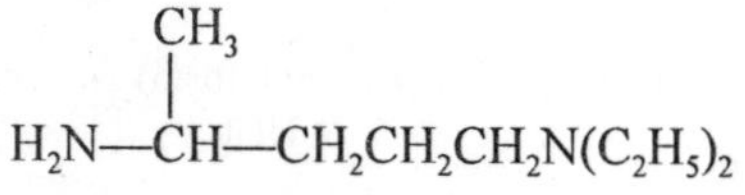

4-Diethylamine-1-methylbutylamine ;

4-Diethylamine-1-methylbutylamine ;

(*ii*) Preparation of the acridine nucleus :

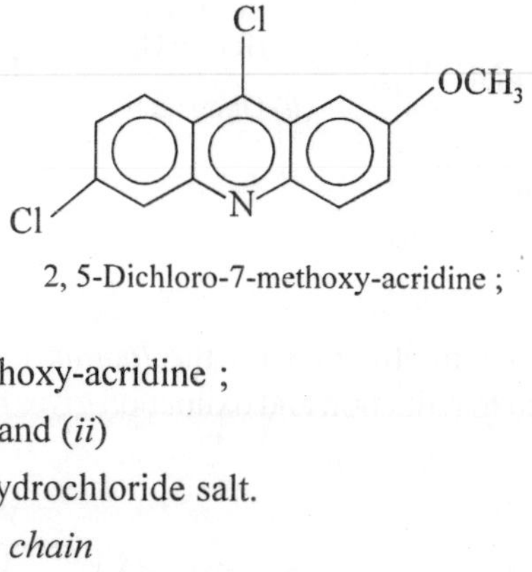

2, 5-Dichloro-7-methoxy-acridine ;

2, 5-Dichloro-7-methoxy-acridine ;

(*iii*) Condensation of (*i*) and (*ii*)

(*iv*) Preparation of the hydrochloride salt.

(*a*) *Preparation of the side chain*

2-Hydroxy triethylamine hydrochloride is obtained by the interaction of ethylene oxide and diethylamine in the presence of methanol and hydrochloric acid which on chlorination with thionylchloride yields 1-chloro-triethylamine.

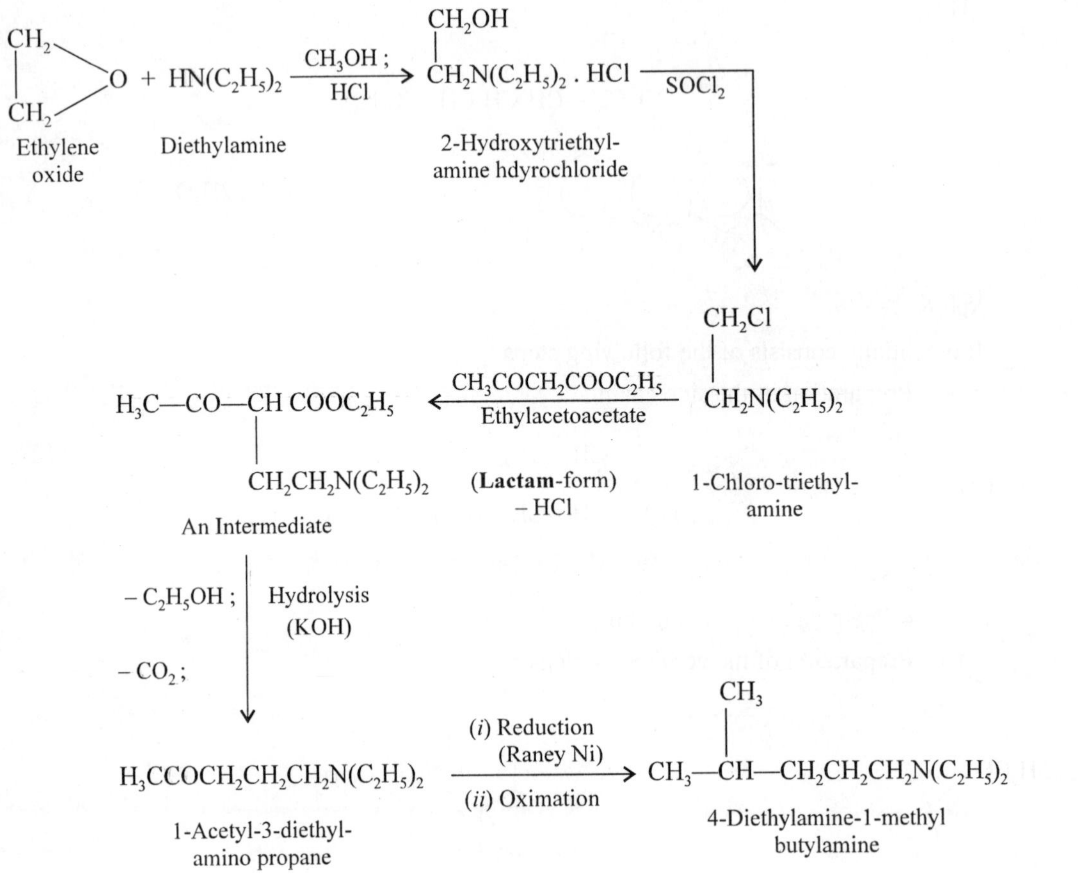

$$\begin{array}{c}CH_2\\ |\\ CH_2\end{array}\!\!\!\!>\!\!O + HN(C_2H_5)_2 \xrightarrow[HCl]{CH_3OH;} \begin{array}{c}CH_2OH\\ |\\ CH_2N(C_2H_5)_2 . HCl\end{array} \xrightarrow{SOCl_2}$$

Ethylene   Diethylamine   2-Hydroxytriethyl-
oxide   amine hdyrochloride

$$\begin{array}{c}CH_2Cl\\ |\\ CH_2N(C_2H_5)_2\end{array}$$

$$H_3C\!-\!CO\!-\!CH\ COOC_2H_5 \xleftarrow[\text{Ethylacetoacetate}]{CH_3COCH_2COOC_2H_5}$$

$$\begin{array}{c}|\\ CH_2CH_2N(C_2H_5)_2\end{array}$$

An Intermediate   (**Lactam**-form)   1-Chloro-triethyl-
                         – HCl   amine

$$\begin{array}{l} - C_2H_5OH;\ \Big|\ \text{Hydrolysis}\\ \qquad\qquad\quad (KOH)\\ - CO_2;\ \ \Big\downarrow \end{array}$$

$$H_3CCOCH_2CH_2CH_2N(C_2H_5)_2 \xrightarrow[(ii)\ \text{Oximation}]{(i)\ \text{Reduction (Raney Ni)}} \begin{array}{c}CH_3\\ |\\ CH_3\!-\!CH\!-\!CH_2CH_2CH_2N(C_2H_5)_2\end{array}$$

1-Acetyl-3-diethyl-   4-Diethylamine-1-methyl
amino propane   butylamine

The resulting compound on treatment with the *lactam*-form of ethylacetoacetate forms an intermediate which when subjected to reduction and oximation gives 4-diethylamine-1-mthyl butylamine.

(*b*) *Preparation of the acridine nucleus*

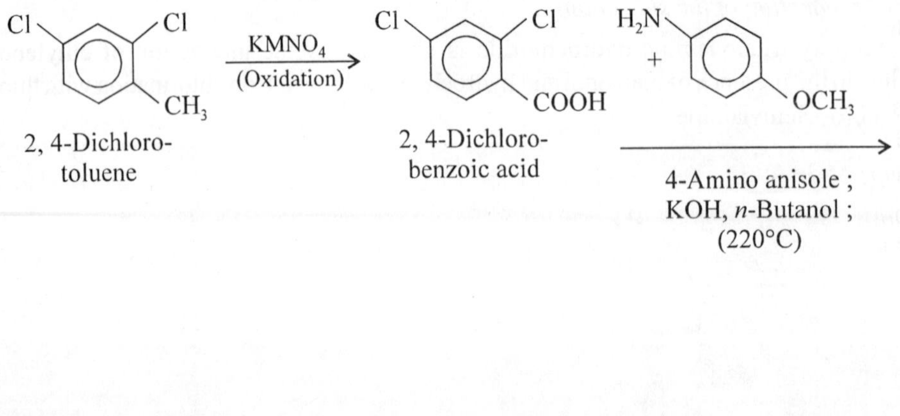

2, 4-Dichloro-   2, 4-Dichloro-
toluene   benzoic acid

4-Amino anisole ;
KOH, *n*-Butanol ;
(220°C)

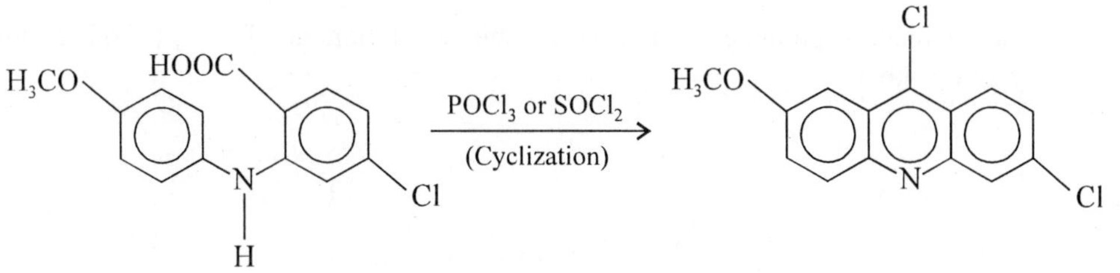

4-Chloro-2-(4-methoxy-
anilino) benzoic acid

2, 5-Dichloro-7-methoxy
acridine

(c) Condensation of (a) and (b) above ; and (d) Treatment with Hydrochloric Acid

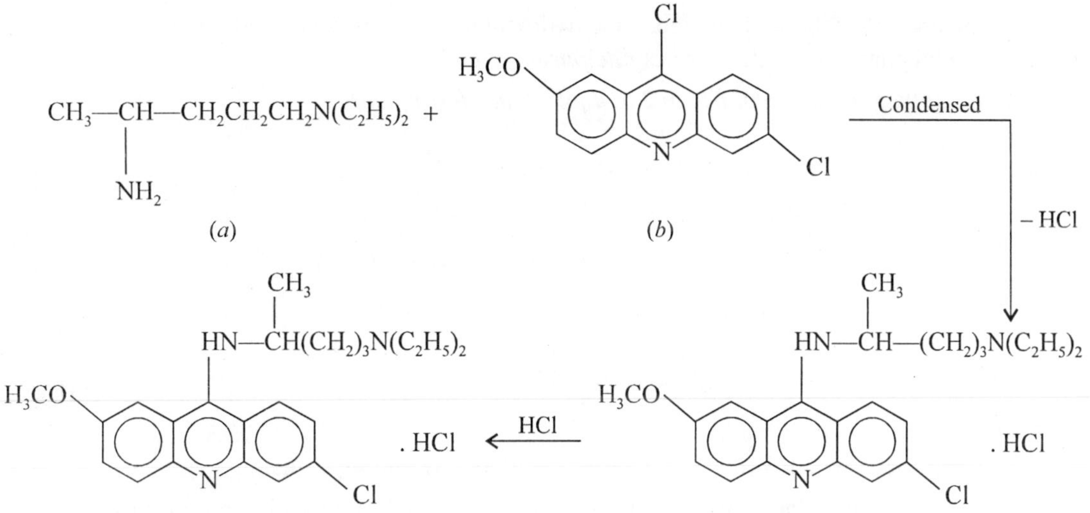

Mepacrine Hydrochloride

Mepacrine Base

2, 5-Dichloro-7-methoxy acridine may be prepared by the oxidation of 2, 4-dichloro toluene and treating the resulting acid with 4-amino anisole at 220°C in the presence of KOH and *n*-butanol ; the additional compound when reacted with either POCl$_3$ or SOCl$_2$ undergoes cyclization. One mole each of the side chain and the acridine residue get condensed to yield the mepacrine base which on treatment with hydrochloric acid gives the official compound.

Mepacrine hydrochloride inhibits the erythrocytic state of development of the malarial parasite. It is considered neither as a causal prophylactic nor as a radical curative agent. It is found to be more toxic and less effective than chloroquine. *Besides, it has also been used in giardiasis, amebiasis, tapeworm and pinworm infestations.*

Dose : As therapeutic, 200mg with 300mg of sodium bicarbonate each 6 hours up to 5 doses, followed by 100mg 3 times per day for 6 days ; as suppressive, 100mg once daily.

B. *Mepacrine Mesylate* BAN,

Mepacrine Methanesulphonate ; Quinacrine methanesulphonate ; BPC ; (1963) ; Quinacrine Soluble[R] (May and Baker).

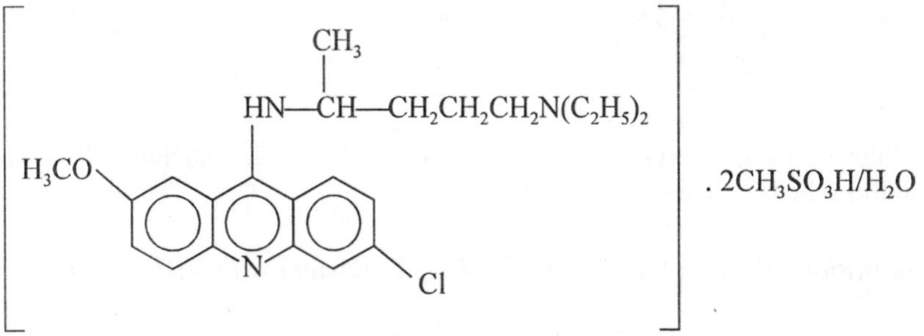

. $2CH_3SO_3H/H_2O$

It is much more soluble than mepacrine hydrochloride. It has been used more conveniently for parenteral administration in acute cases of *falciparum* malaria.

Dose : 360mg intramuscularly in 2 to 4 ml 'Water for Injection'.

C. *Aminoacrichin*

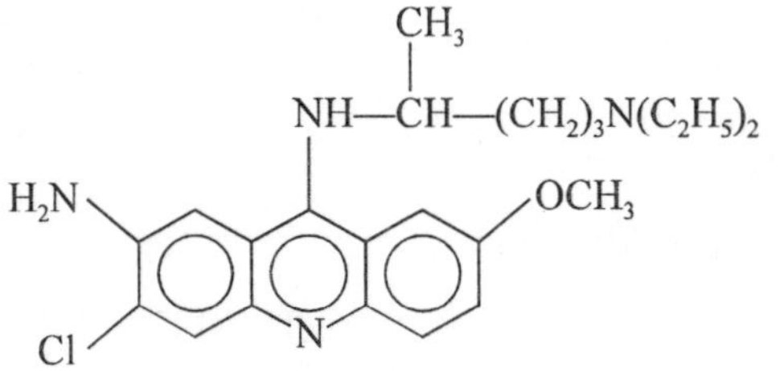

7-Amino-6-chloro-9-[[4-(diethylamine)-1-methylbutyl] amino] 2-methoxy-acridine ;

Its use as an antimalarial drug has been discontinued and replaced by more effective and less toxic agents.

2.3.1. Mechanism of Action. The mechanism of action of the antimalarial agents described under section 19.2.3. shall now be treated individually as under :

2.3.1.1. Quinacrine Hydrochloride. The *'drug'* almost exhibits the same effects as those caused by the 4-aminoquinolines. The GI irritancy is registered to be much higher than the 4-aminoquinolines ; and, therefore, it is a common practice to administer sodium bicarbonate concomitantly.

It is absorbed quite rapidly from the GI-tract and also from IM and intracavitary sites of injection. The *'drug'* gets excreted very gradually in the urine and gets accumulated in tissues on chronic administration.

Interestingly, the *'drug'* is believed to act at several sites within the cell, including **intercalation of DNA strands, succinic dehydrogenase, mitochondrial electron transport system,** and **cholinesterase.** It may serve as *tumorigenic* and *mutagenic,* and hence, has been employed profusely as a *sclerosing agent.**

---

*An agent that causes or developes *sclerosis* (*i.e.,* hardening or induration of an organ or tissue, especially that due to excessive growth of fibrous tissue.

2.3.1.2. Mepacrine Mesylate. The *'drug'* is a methanesulphonate salt of mepacrine (or quinacrine) whose therapeutic potency is relatively higher than its corresponding HCl-salt.

2.3.1.3. Aminoacrichin. The *'drug'* along with its two other acridine structural analogues, namely : *acriquine* and *azacrin* were introduced based on a combined structural features of 8-aminoquinoline and 4-aminoquinoline, but were not so successful due to their high toxicity.

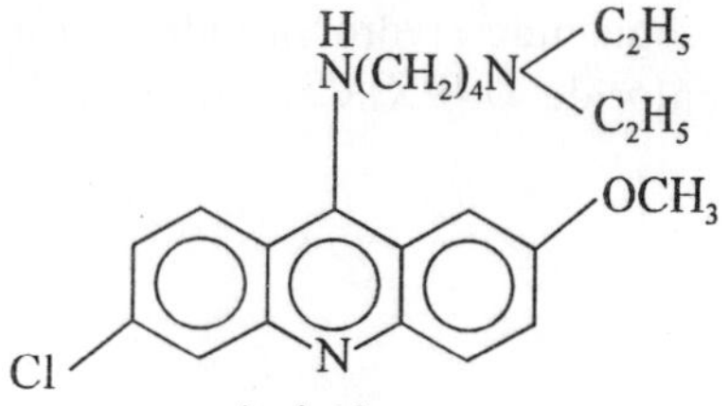

Acriquine 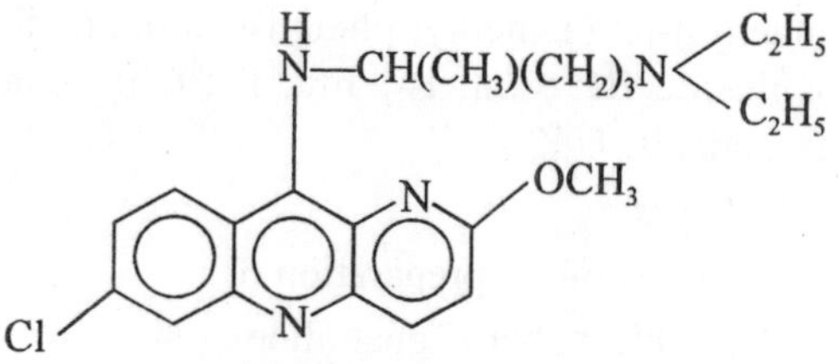   Azacrin

2.4. Guanidine Analogues (Biguanides)

The guanidine analogues, in general, are not found to be active unless and until they get cyclized metabolically to a dihydro-s-triazine analogue having a close resemblance either to the pteridine moiety of folic acid or pyrimethamine as shown below :

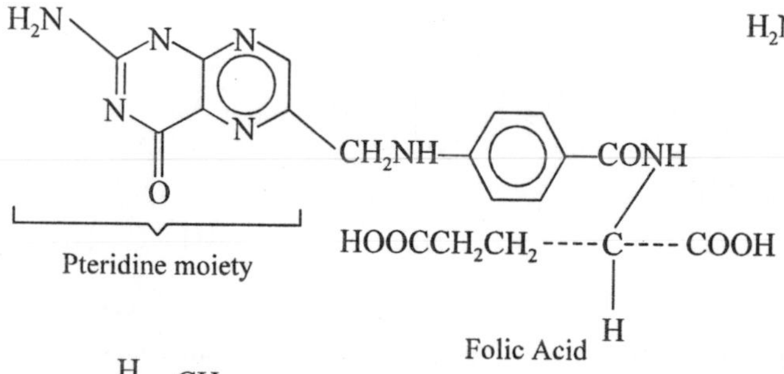

Pteridine moiety

Folic Acid

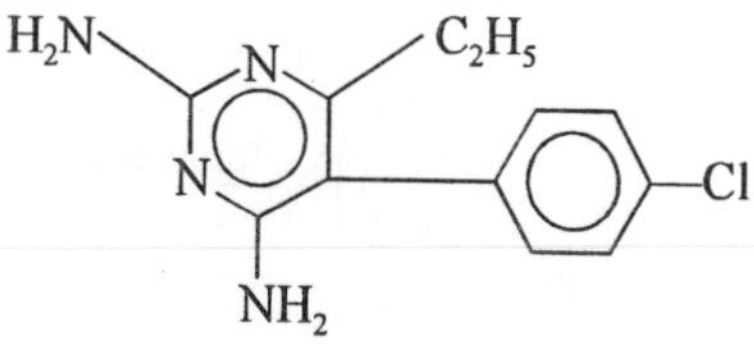

Pyrimethamine

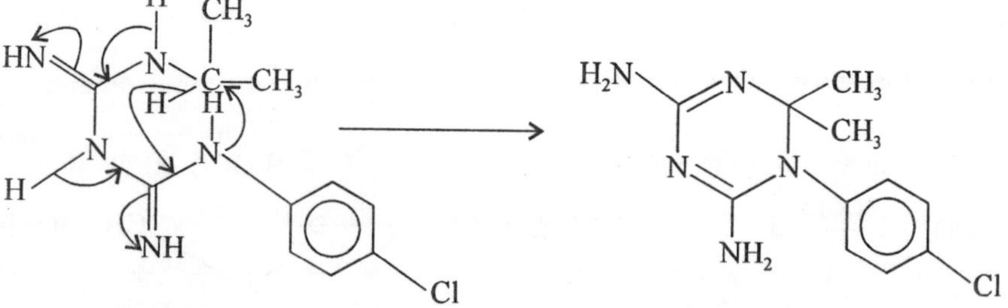

Proguanil

Cycloguanil
(Active metabolite)

The other structural analogues of guanidine are also metabolised in a similar fashion.

A few members of this class of compounds are described below, *viz.,* Proguanil hydrochloride ; Cycloguanil embonate ; Chlorproguanil ; Bromoguanil.

A. *Proguanil Hydrochloride* BAN, *Proguanil* INN, *Chlorguanide Hydrochloride* USAN,

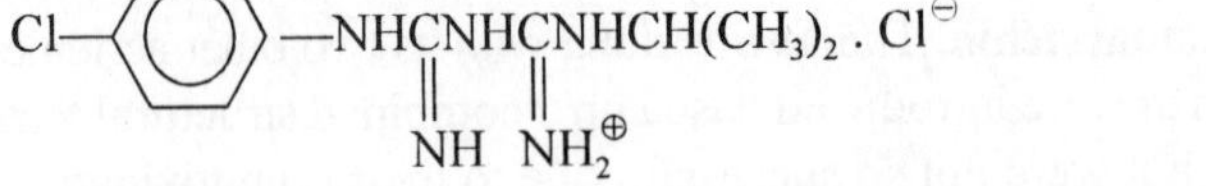

$$Cl-\langle\bigcirc\rangle-NHCNHCNHCH(CH_3)_2 \cdot Cl^{\ominus}$$
$$\qquad\qquad \underset{NH}{\overset{\|}{}}\ \underset{NH_2^{\oplus}}{\overset{\|}{}}$$

1-(*p*-Chlorophenyl)-5-isopropylbiguanide hydrochloride ; Imidodicarbonimidic diamide, N-(4-chlorophenyl)-N′-(1-methyl-phenyl)-, monohydrochloride ; Proguanide Hydrochloride ; Proguanil Hydrochloride BP ; Int. P ; Ind. P ; Chlorguanide Hydrochloride USPXIV ; Paludrine[R] (ICI Pharmaceuticals, U.K.).

Synthesis :

It consists of the preparation of :

(*a*) *p*-Chlorophenyl guanidine

(*b*) Iso-propyl cyanamide

(*c*) Condensation (*a*) and (*b*)

(*d*) Hydrochloride salt.

(*a*) *Preparation of p*-Chlorophenyl guanidine

p-Chloro-nitrobenzene is subjected to reduction, treatment with cyanobromide and amination to yield *p*-chlorophenyl guanidine.

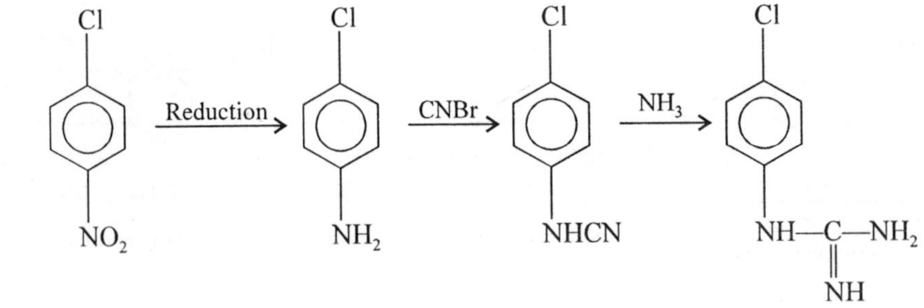

| p-Chloro-nitro-benzene | p-Chloro-aniline | p-Chlorophenyl aminocyanide | p-Chlorophenyl guanidine |

(*b*) *Preparation of iso-propyl cyanamide*

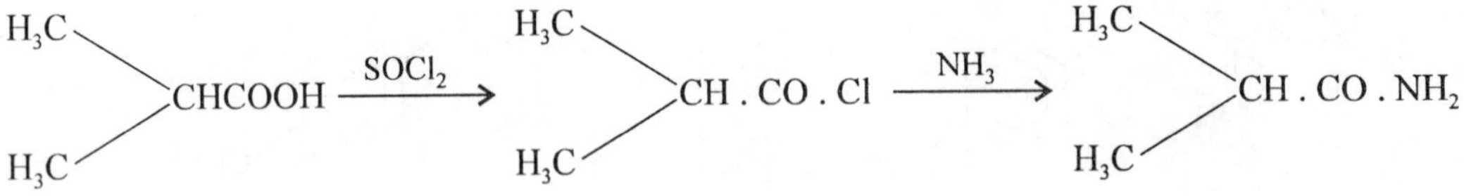

Iso-propanoic acid   Iso-propylchloride   Iso-propionamide

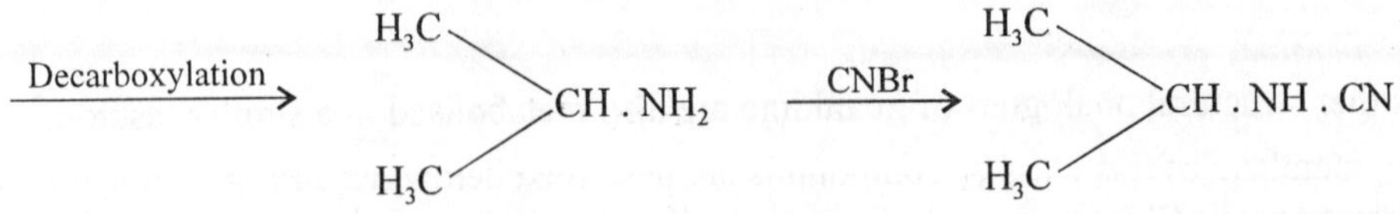

Iso-propylamine   Iso-propyl cyanamide

Iso-propyl cyanamide may be prepared by the chlorination of iso-propionic acid followed by amination, decarboxylation and finally treating with cyanogen bromide.

(c) *Condensation of* (a) *and* (b) ; (d) Formation of Hydrochloride Salt

Condensation of *p*-chlorophenyl guanidine with iso-propyl cyanamide gives the proguanil base which on treatment with one mole of hydrochloric acid yields proguanil hydrochloride.

It is an antimalarial drug whose metabolite is a potent dihydrofolate reductase inhibitors. It is active against the pre-erythrocytic (liver) forms of malaria. It is also active against the erythrocytic forms but their activity is slow.

Hence, proguanil is used mainly for prophylactic treatment of malaria.

Dose : As prophylactic and suppressant, 100 to 200mg per day in non-imune subjects ; 300mg per week or 200mg 2 times per week in semi-imune subjects ; in acute vivax malaria, initial loading dose

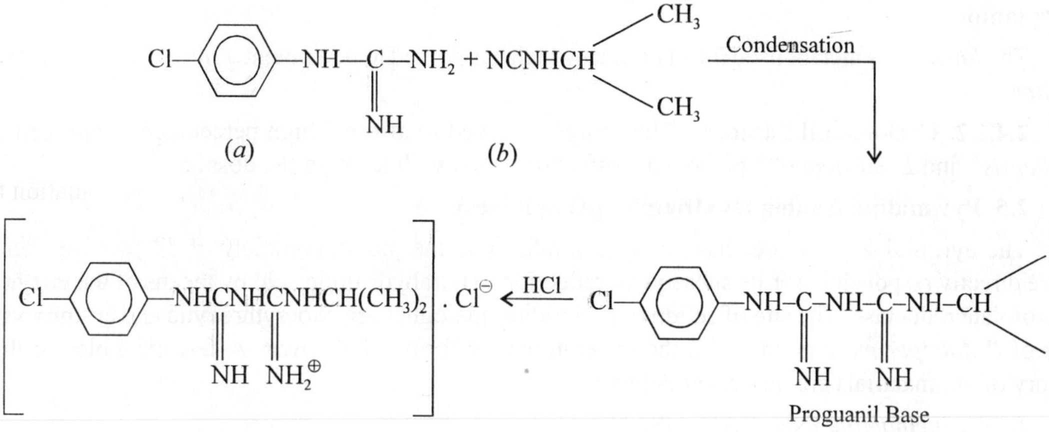

(a)   (b)

Proguanil Hydrochloride   Proguanil Base

300 to 600mg followed by 300mg per day for 5 to 10 days ; in falciparum malaria, 300mg 2 times daily for 5 days.

B. *Cycloguanil Embonate* INN, BAN, *Cycloguanil Pamoate* USAN,

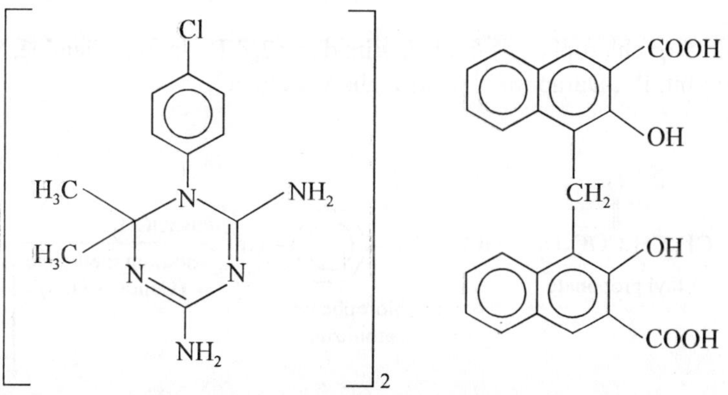

4, 6-Diamino-1-(*p*-chlorophenyl)-1, 2-dihydro-2, 2-dimethyl-s-triazine compound (2:1) with 4, 4′ methylene-bis [3-hydroxy-2-naphthoic acid] ; Camolar(R) (Parke-Davis).

Cycloguanil is the active metabolite of proguanil as shown earlier. Its actions and uses are similar to paludrine. It has been recognized as a dihydrofolate reductase inhibitor and employed for the suppression of malaria, but failed to achieve a wide acceptance. It exerts little therapeutic value in such cases where resistance to either proguanil or pyrimethamine is prevalent. *To attain prolonged immunization in areas infested with hyperendemic malaria, administration of cycloguanil and amodiaquine every 4 months is recommended.*

Dose : Usual, adult, intramuscular, 350mg of cycloguanil base every 4 months.

2.4.1. Mechanism of Action. The mechanism of action of two drug substances discussed under section 19.2.4. shall be dealt with separately as under :

2.4.1.1. Chlorguanide Hydrochloride (Proguanil HCl). British scientists during World War II had adopted an altogether different line of action in breaking away from the normal quinoline and acridine types of structure, and eventually paved the way in the epoch making discovery of the biguanide, **chlorguanide.**

The *'drug'* gets metabolised into a product which has proved to be a potent *dihydrofolate reductase inhibitor.*

2.4.1.2. Cycloguanil Pamoate. The *'drug'* is proved to afford a high percentage of cures in *L. brasiliensis** and *L. mexicana*** pathogenic infections even with a single IM dosage.

2.5 Pyrimidine Analogues (Diaminopyrimidines)

The pyrimidine analogues have a close similarity to the pteridine moiety of dihydrofolic acid, and are directly responsible for its subsequent reduction to tetrahydrofolic acid by means of the enzyme dihydrofolate reductase. The site of action of pyrimidine analogues are exoerythrocytic and erythrocytic forms of *P. falciparum,* together with the exoerythrocytic forms of *P. vivax.* A few examples of this category of antimalarials are described below :

A. *Pyrimethamine* INN, BAN, USAN,

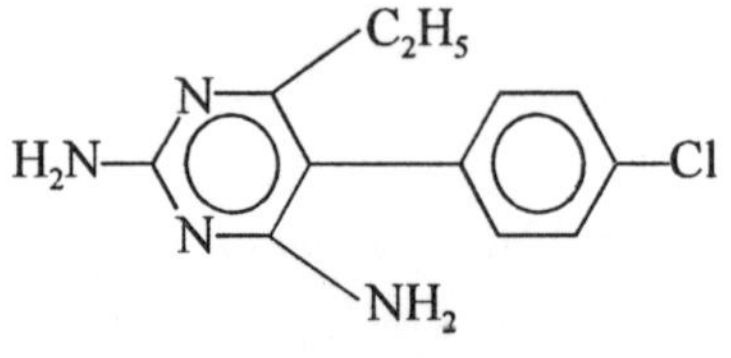

2, 4-Diamino-5-(*p*-chlorophenyl)-6-ethylpyrimidine ; 2, 4-Pyrimidinediamine, 5-(4-chlorophenyl)-6-ethyl-; BP ; USP ; Int. P. ; Daraprim[R] (Burroughs Wellcome).

Synthesis :

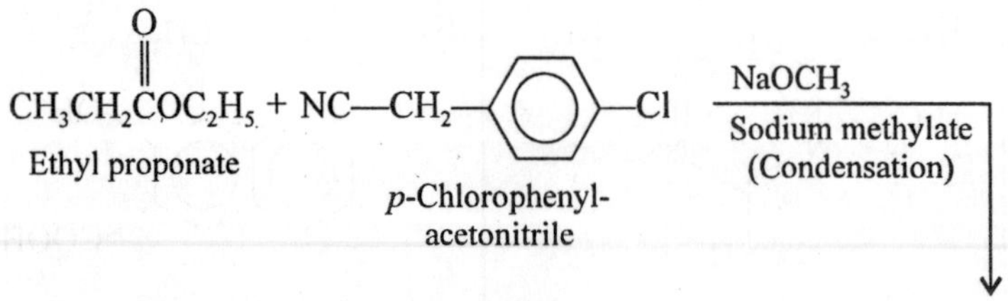

---

*Pena-Chavarria *et al. J Am Med Assoc,* **194,** 1142 (1965).

Beltran F *et al. Prensa Med Mex,* **31, 365 (1966).

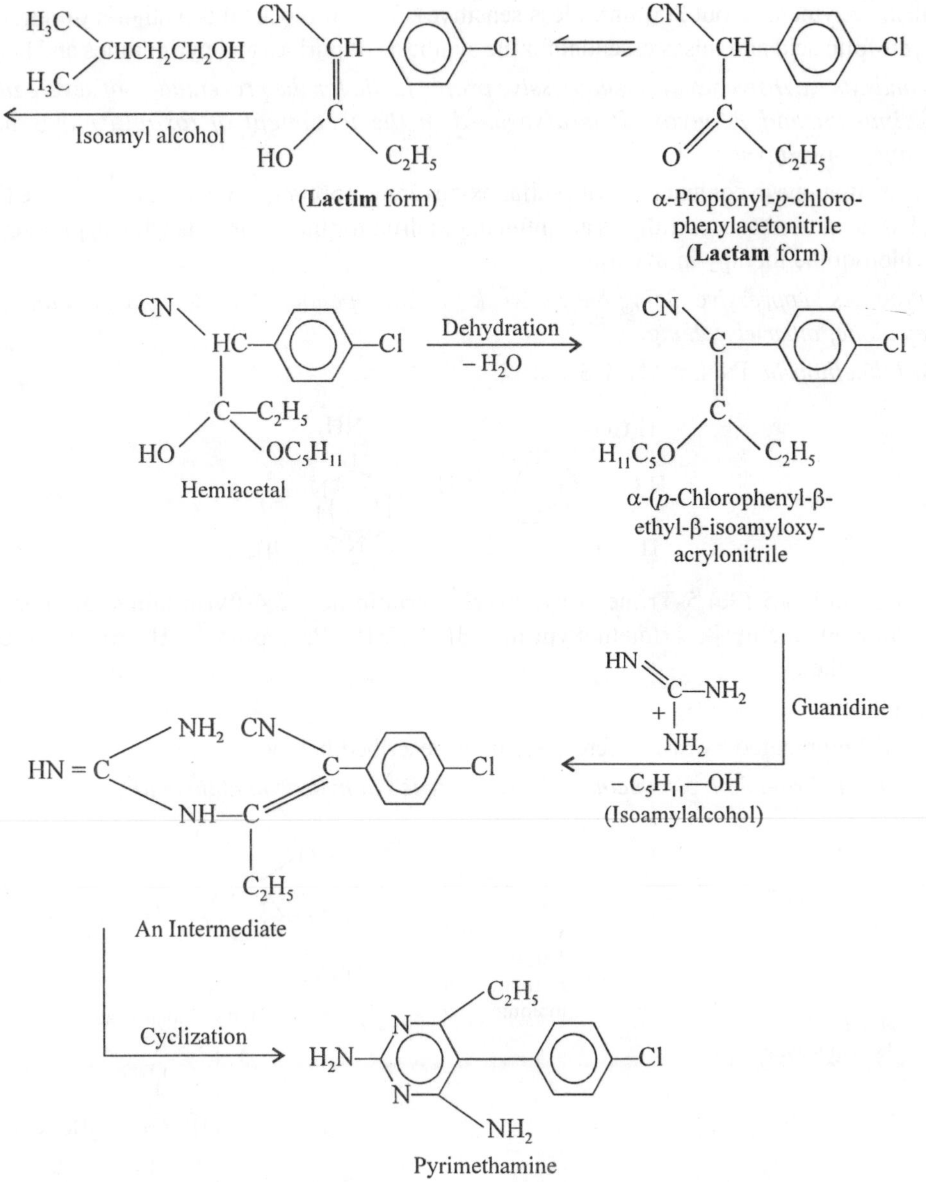

α-Propionyl-*p*-chloro-phenylacetonitrile
(**Lactam** form)

Hemiacetal

α-(*p*-Chlorophenyl-β-ethyl-β-isoamyloxy-acrylonitrile

Guanidine

An Intermediate

Pyrimethamine

α-Propionyl-*p*-chlorophenylacetonitrile (*lactum*-form) is prepared by the condensation of ethyl propionate and *p*-chlorophenylacetonitrile which undergoes *tautomerism* to form the corresponding *lactim*-form. This on reaction with isoamyl alcohol forms the hemiacetal which upon dehydration yields α-(*p*-chlorophenyl)-β-ethyl-β-isoamyloxyacrylo-nitrile. The resulting product on treatment with guanidine affords cyclization *via* two different steps : *first*, elimination of a mole of isoamyl alcohol by condensation involving the imino hydrogen of guanidine, and *secondly*, an addition reaction between an amino group of guanidine and the nitrile group of the intermediate compound.

Like the biguanides it is a potent inhibitor of dihydrofolate reductase of the plasmodium (mammalian enzyme is about 200 times less sensitive). Thus it blocks the synthesis of tetrahydrofolic from dihydrofolic acid and this is essential for the synthesis of purines and pyrimidines and hence DNA.

It finds its extensive use as a suppressive prophylactic for the preventation of severe attacks due to P. falciparum and P. vivax. It is also used in the treatment of toxoplasmosis and as an immunosuppressive agent.

Pyrimethamine in conjunction with sulfadoxine (25mg : 500mg), under the brand name Fansidar[R] (Roche), has been used successfully as an antimalarial drug for those subjects who display sensitization towards chloroquine therapy in malaria.

Dose : As suppressive, 25mg once a week ; as therapeutic, 50 to 75mg once a day for 2 days when used alone, otherwise 25mg.

B. *Trimethoprim* INN, BAN, USAN,

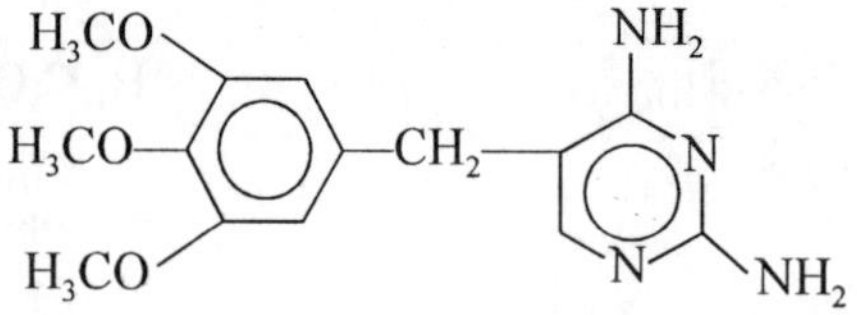

2, 4-Diamino-5-(3,4,5-Trimethoxybenzyl) pyrimidine ; 2,4-Pyrimidinediamine, 5-[(3,4,5-trimethoxyphenyl) methyl]-; Trimethoxyprim ; BP ; USP ; Proloprim[R] (Burroughs Wellcome) ; Trimpex[R] (Roche).

Synthesis :

It may be prepared by *two* different methods described below :

Method-1. From 3, 4, 5-trimethoxy benzaldehyde via hydrocinnamic acid

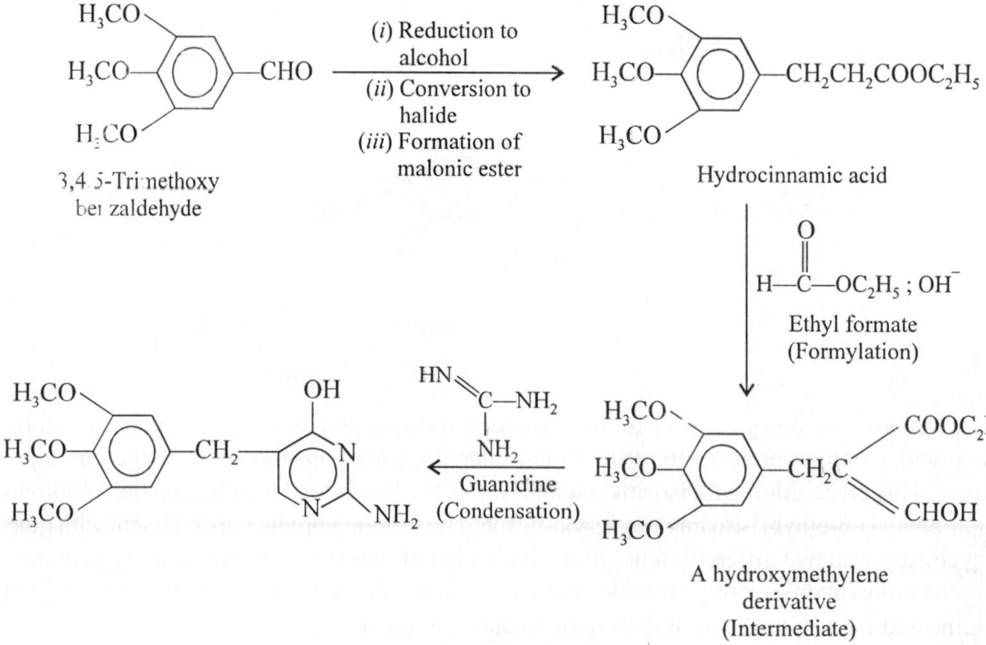

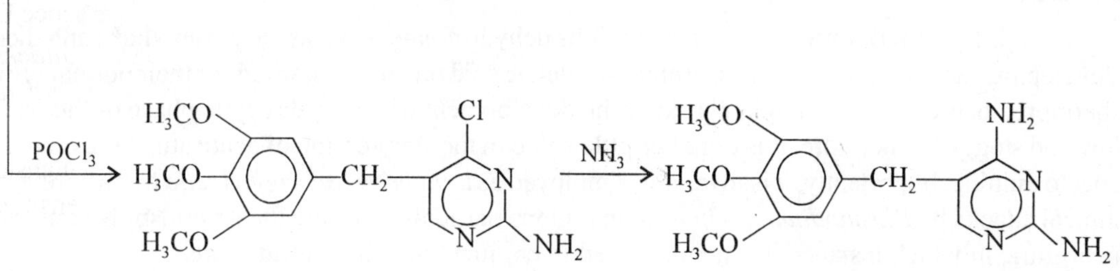

Trimethoprim

Hydrocinnamic acid is prepared by the bishomologation of 3, 4, 5-trimethoxy benzaldehyde, *i.e.,* subjecting the later to reduction forming an alcohol, coversion to halide and finally formation of the malonic ester. This is then subjected to formylation with ethyl formate and base to yield the corresponding hydroxymethylene derivative. Condensation of this intermediate with guanidine gives the pyrimidine residue, by a scheme very similar to the one discussed under pyrimethamine. The hydroxyl moiety present in the pyrimidine nucleus is converted to the chloro group by treatment with phosphorus oxychloride and finally amination leads to the formation of the official compound.

Method-II. From 3, 4, 5-trimethoxybenzaldehyde via cinnamonitrile

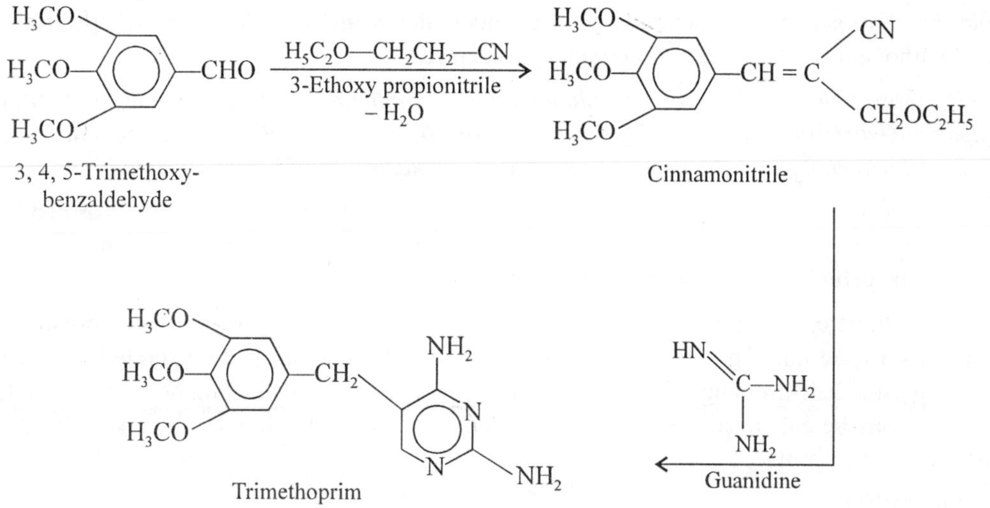

3, 4, 5-Trimethoxy-
benzaldehyde

Cinnamonitrile

Trimethoprim

Guanidine

It is comparatively a shorter course of reaction whereby cinnamonitrile is prepared by the interaction of 3, 4, 5-trimethoxy-benzaldehyde with 3-ethoxy propionitrile with the elimination of a mole of water. The resulting product on treatment with guanidine affords the formation of trimethoprim directly.

Like pyrimethamine, *trimethoprim is a potent inhibitor of dihydrofolate reductase.* It has been employed in conjunction with sulfametopyrazine in the treatment of chloroquine-resistant malaria but unfortunately could not attain wide acceptance. *It has also been used in conjunction with sulphonamides in the treatment of bacterial infections viz.,* trimethoprim with sulphamethoxazole.

Dose : 1.5g with 1g of sulfametopyrazine per day for 3 days.

2.5.1. Mechanism of Action. The mechanism of action of the compounds described under section 19.2.5. are dealt with individually in the sections that follows :

2.5.1.1. Pyremethamine. The *'drug'* inhibits dehydrofolate reductase in plasmodia* ; and thereby the developing parasite cannot synthesize and use nucleic acid precursors needed for their normal growth. Furthermore, its prevailing action in checking the development of the erythrocytic phase of the parasite is slow and sluggish ; therefore, it is of rather little value in the suppression of acute attacks, except as an adjunct to quinine. Importantly, it is invariably employed as a *suppressive prophylactic* for the prevention of clinical attacks by *Plasmodium falciparum* in regions particularly where the **organism is resistant to chloroquine,** in which instance it is administered in conjunction with **sulfadoxine.**

Besides, it also helps in rendering the *'parasite'* incapable of sporulating in the mosquito whereby the **'life-cycle of the parasite'** is disrupted squarely.

It has been duly reported that success rate of the *'drug'* is almost 90% in certain regions, which may be increased to even 95% by the addition of **quinine.** However, in the control, management and treatment of *toxoplasmosis*** it is usually combined with **trisulfapyrimidines.**

2.5.1.2. Trimethoprim. The *'drug'* also shows its action by the inhibition of dihydrofolate reductase, though its potency is appreciably lower. However, it is found to be most important as an **'antibacterial agent'.** It is worthwhile to mention here that the *bacterial dihydrofolate reductases* are invariably more susceptible in comparison to the plasmodial ones. Hence, the *'drug'* is observed to be extremely effective against all bacteria which should exclusively synthesize their own **folinic acid** (*leucovorin*). This specific characteristic profile renders the *'drug'* to acclaim a broad spectrum against a host of pathogenic (causative) microorganisms, such as :

Streptopyrogenes, viridans, and *pneumoniae* ; *Staphylococcus aureus* and *epidermidis ; H. influenzae ; Klebsiella-Enterobacter Serratia, E. coli,* different *Shigella* and *Salmonella, Bordetella pertussis ; Vibrio cholerae ; Pneumocystis carinii, Toxoplasma gondii ;* and *Plasmodia.*

It is, however, pertinent to mention here that the **mammalian dihydrfolate reductase** is approximately 1 : 10,000 to 1 : 50,000 as sensitive to it as the bacterial enzymes, so that there prevails almost **little interference with folate metabolism in humans.**

The volume of distribution is nearly 1.8 mL g^{-1}. The concentration in the cerebrosphinal fluid (CSF) attains a level ranging between 30-50% of the drug in plasma. It gets excreted mostly into the urine. The plasma half-life ranges between 9 to 12 hr. in normal adults having normal kidney-function ; however, it may be enhanced even upto 2 to 3 times in a situation whereby the creatinine clearance falls below 10 mL . min^{-1}.

2.6. Sulfones

A large number of diphenylsulfone analogues have been developed for the treatment of leprosy. Incidentlly one such member chemically known as 4, 4'-diaminodiphenyl sulfone (dapsone) exhibited prophylactic activity against resistant *P. falciparum.* Dapsone in conjunction with pyrimethamine has been effectively used in the treatment of malaria due to chloroquine resistant *P. falciparum.*

---

*Med. Lett. **29,** 53, 1987.

**A disease caused by infection with the protozoan *Toxoplasma gondii.*

A. *Dapsone* INN, BAN, USAN,

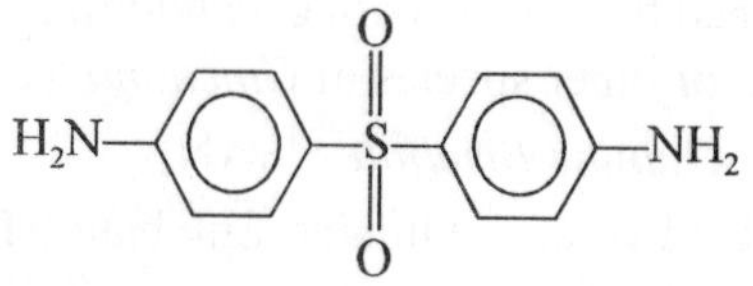

4, 4'-Sulfonyldianiline ; Benzeneamine, 4, 4'-sulfonylbis-; 4, 4'-Diaminodiphenyl sulfone ; Diaphenylsulfone ; Disulone ; BP ; USP ; Int. P ; Ind. P. ; Avlosulfon[R] (Ayerst).

Synthesis :

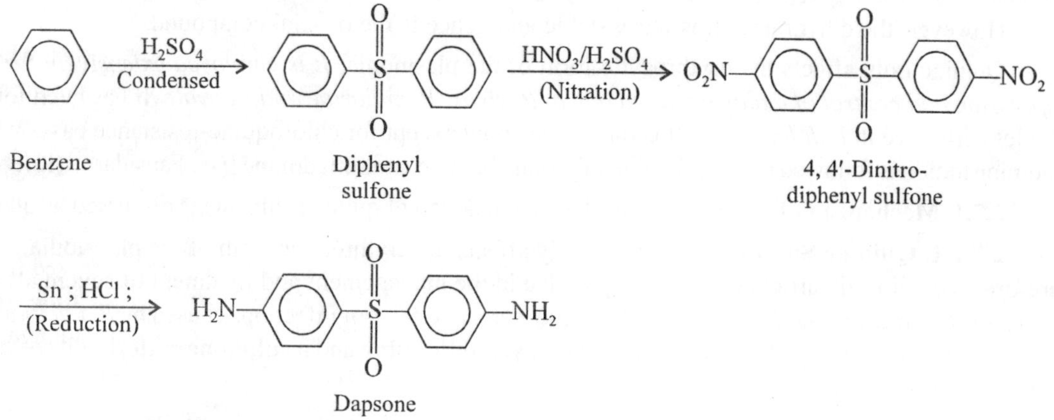

Diphenyl sulfone is prepared by the condensation of benzene with sulphuric acid. Nitration is afforded by treatment with a mixture of nitric acid and sulphuric acid to yield 4, 4'-dinitrodiphenyl sulfone which on reduction with tin and hydrochloric acid gives the official compound.

Dapsone possesses limited therapeutic value in the treatment of malaria, except when combined with other agents for the treatment of chloroquine-resistant cases.

2.6.1. Mechanism of Action. The mechanism of action of *'dapsone'* shall be discussed as under :

2.6.1.1. Dapsone. Its mechanism of action is very much similar to that of *sulphanilamide*. It is employed profusely in the treatment of both *lepromatous* and *tuberculoid* types of leprosy. However, in combination with **rifampin,** it is regarded as the *'drug of choice'* in the chemotherapy of leprosy. Besides, the combination with **clofazimine** affords a similar therapeutic effect. The *'drug'* is the most preferred **'sulfone'** because of the two cardinal facts, such as : (*a*) cost-effective ; and (*b*) equally efficacious to other sulfones.

Interestingly, when combined with *trimethoprim,* it is found to exert almost identical activity as trimethoprim-sulfamethoxazole in the plausible treatment of *Pneumocystis carinii pneumonia.* Also used with **pyrimethamine** for treatment of malaria.

It is most absorbed by the oral administration. Absorption is more efficient at low than high dosage regimen. Finally, it gets eliminated in the liver by acetylation. Patients may respond to this *'drug'* as 'slow' and 'fast' acetylators. The plasma half-life ranges between 10 to 50 hours ; and at least 8 hours are needed to accomplish plateau concentrations.

2.7. Quinine Analogues

Quinine is an alkaloid obtained from the bark of *Cinchona officinalis* Linne (*C. ledgeriana* Moens) belonging to the family *Rubiaceae* or other species of *Cinchona*.

A. *Quinine Sulphate* BAN, *Quinine Bisulfate* USAN,

Quinine sulphate (2:1) salt dihydrate ; Quinine Sulphate BP ; Quinine sulfate USP ; Quinine Bisulate NFXI ; Kinine[R] (ICN, Canada) ;

Preparation

The quinine is isolated from the bark of *Cinchona* sp., after recrystallization several times from mildly acidified (H_2SO_4) hot water. Quinine sulphate obtained after recrystallization retains up to seven moles of water, but undergoes efflorescence in dry environment to lose up to five moles of water.

However, the dihydrate salt is fairly stable and hence is the official compound.

Quinine only affects the erythrocytic form of the plasmodia. *It is employed extensively for the suppression and control of malaria caused due to P. vivax, P. malariae and P. ovale.* It has been found to be less effective in *P. falciparum*. It is rarely used now except for chloroquine-resistance cases when its administration is followed by combination of purimethamine and sulfadoxine [*i.e.,* Fansidar[R] (Roche)].

2.7.1. Mechanism of Action. The mechanism of action of quinine sulphate is discussed as under :

2.7.1.1. Quinine Sulfate : The *'drug'* only affects the erythrocytic form of the plasmodia ; and, therefore, is employed particularly as a suppressive in the management and treatment of severe attacks of *P. vivax. P. malariae* and *P. ovale* malaria. It may cure upto 50% of infections caused by *P. falciparum*. The *'drug'* may be employed in combination with **pyrimethamine** and a **sulphonamide,** but it seems to be antagonized by **chloroquine.**

Choice of combinations with other drugs : A few typical examples are as follows :

    (*i*) **Quinine-pyrimethamine-sulfadiazine (or sulfadoxine).** In the treatment of choice for infections caused by chloroquine-resistant *Plasmodium falciparum ;*

    (*ii*) **Quinine-tetracycline.** In infections produced by chloroquinine-resistant *P. falciparum.*

    (*iii*) **Quinine-Clindamycin.** In the treatment of choice for *babesiosis.*

The *'drug'* has a tendency to suppress neuromuscular transmission, and hence used in **myotonia congentia** or **Thomsen's disease.**

Note. The '*drug*' is mostly given orally after meals to minimize gastric irritation.

2.8. New Antimalarial Drugs. A few important newer antimalarial drugs are discussed as under :

2.8.1. Artemisinin. The marked and pronounced antimalarial activity of **'Quinghausu'** as the constituent of a traditional Chinese medicinal herb *Artemisia annuna* L., (sweet wormwood) has been known in China for over 200 years. However, the active principle was first isolated in 1972 and found to be a sesquiterpene lactone with a peroxy moiety.

The following *four* chemical structures, namely : (*i*) artemisinin ; (*ii*) dihydroartemisinin ; (*iii*) artemether (oil-soluble) ; and artemotil (oil soluble) ; and artesunate (water soluble) are found to be active against the entire *Plasmodium* genera that cause malaria predominantly across the tropical regions of the globe, such as : Africa, Indian sub-continent, South East Asia and the like.

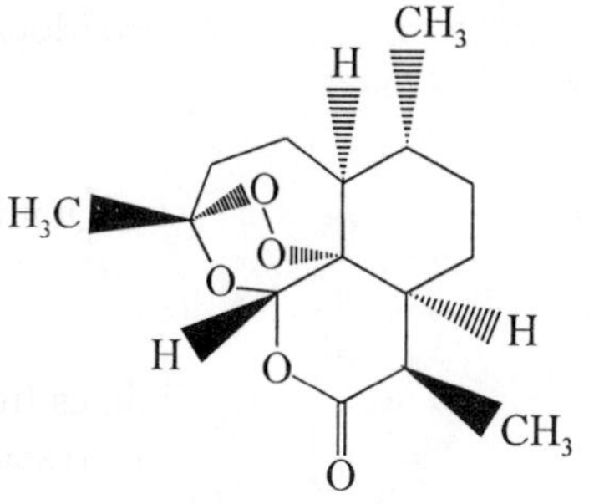

Artemisinin*

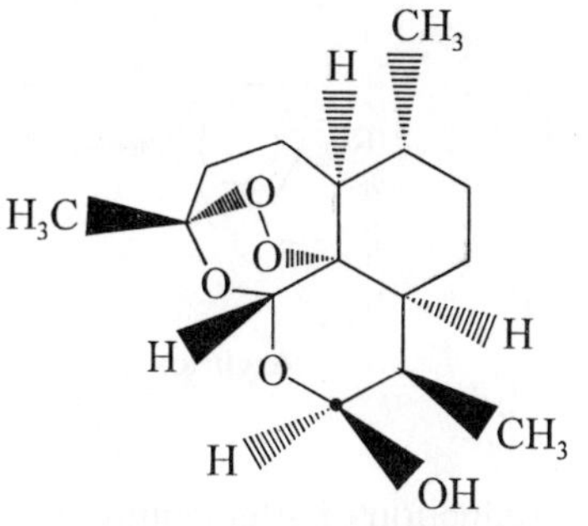

Dihydroartemisinin

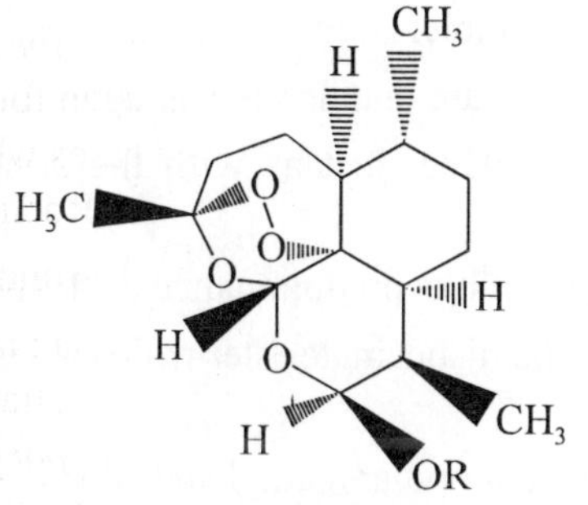

Artemether [R = CH₃]
Artemotil [R = C₂H₅]

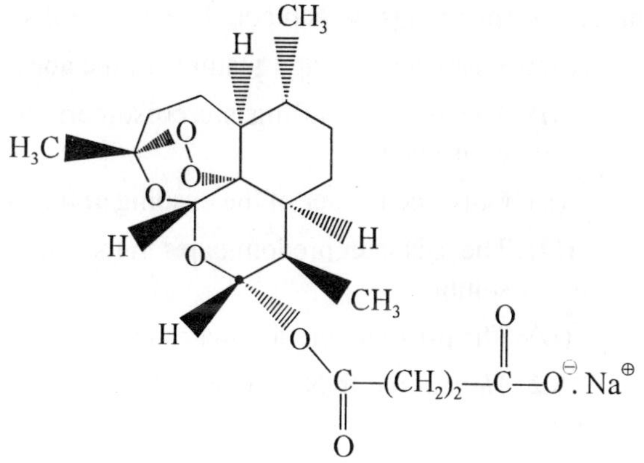

Artesunate

SAR of Artemisinin. The most important, critical and key structure of the *'drug'*, artemisinin, is the presence of a **'trioxane'** moiety which essentially consists of the **endoepoxide** and **doxepin oxygens** that is evidently displayed by a rather simplified versions of *3-aryltrioxanes* as shown on the next page, which are responsible for exerting the antimalarial activity against the parasite.

It is, however, pertinent to state here that the prevailing stereochemistry at C-12 is not so critical and vital.**

Mechanism of Action. In humans (*i.e.,* the host) erythrocyte, it has been observed that the *malaria parasite* actually consumes the haeomoglobin comprising mainly of Fe^{2+} iron, thereby changing it to the corresponding *toxic hematin* consisting of Fe^{3+} iron, subsequently get reduced to heme with its Fe^{2+} iron. Later on, the resulting **'heme iron'** eventually interacts with the prevailing *trioxane moiety,* thereby releasing the **'reactive oxygen' carbon radicals** and the extremely reactive $Fe^{IV} = O$ species. It has been established that the latter is proved to be lethal to the parasite.***

---

*Chemical name of Artemisinin :

(3α, 5αβ, 6β, 8αβ, 9α, 12β, 12aR·)- (+)-Octahydro-3, 6, 9-trimethyl-3, 12-epoxy-12H-pyrano [4, 3-j]-1, 2-benzodioxepin-10 (3H)-one ; ($C_{15}H_{22}O_5$).

Posner GH *et al. J Med Chem* **44, 3054, 2001.

***Posner GH *et al. J Am Chem Soc*, **118**, 3537, 1996.

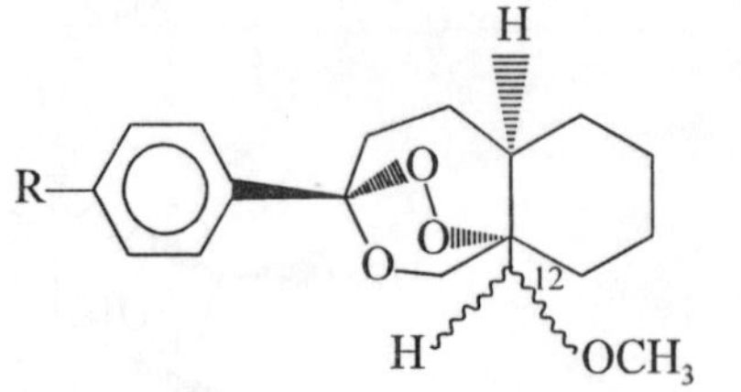

Aryltrioxanes [R = – F or —COOH]
(Simplified Versions)

Interestingly, the reduction of artemisinin to dihydroartemisinin gives rise to a **chiral centre,** as shown by a bold black spot in the structure of dihydro artemisinin that may ultimately lead to the formation of **'prodrugs'** which could be either oil soluble or water soluble.

A few characteristic vital features of the above cited *'prodrugs'* are enumerated as under :

(*i*) The two prevailing stereoisomers are found to be **active,** just as with the simpler aryltrioxanes.

(*ii*) Only one isomer of the ensuing **artemisinin prodrug** exhibits predominance exclusively.

(*iii*) The α-isomer predominates in forming the subsequent hemisuccinate ester which is water-soluble.

(*iv*) The β-isomer predominates in producing the subsequent nonpolar methyl and ethyl ethers.

2.8.2. *Mefloquine* INN, USAN, *Mefloquine Hydrochloride* BAN,

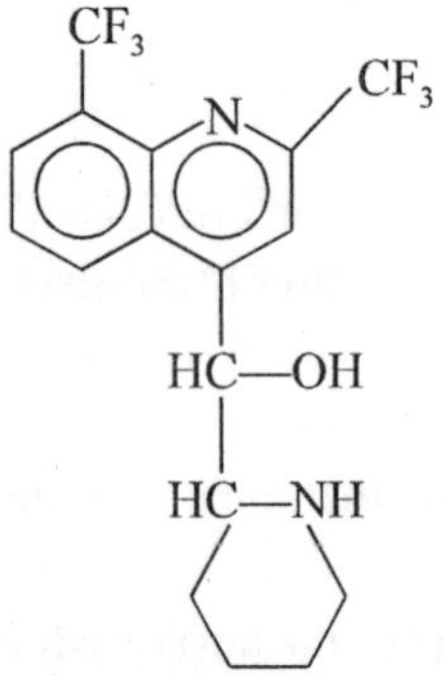

DL-*erythro*-α-2-Piperidyl-2-, 8-*bis* (trifuoromethyl)-4-quinolinemethanol ; 4-Quinolinemethanol, α-2-piperidinyl-2, 8-*bis* (trifuoromethyl)-, (R·, S·)-(±)-;

This is the outcome of many years of research by the United States department of the Army. It belongs to the 4-quinoline methanol series, several of which were found to have potent schizonticidal activity but could not be used clinically, because they possessed photosensitizing activity in man. Mefloquine is devoid of this effect.

It is very effective against the erythrocytic forms of malaria. However, its use is restricted to cases of chloloquine-resistant falciparum malaria in order to prevent the emergence of parasites that are resistant to it.

Dose : Oral single dose, 0.4 to 1.5g.

Probable Questions for B. Pharm. Examinations

1. (*a*) What are the causal organisms responsible for malaria ? How do the antimalarials affect the life cycle cycle of mosquito ? Explain.

   (*b*) With the help of a 'General Structure' give the status of four important alkaloids isolated from cinchona bark.

2. Classify the synthetic antimalarials based on their basic chemical nucleus. Give examples of at least **one** compound from each class.

3. Modifications of the side-chain at C-4 on the 4-amino-7-chloro quinoline nucleus give rise to the following drugs :

   (*a*) Chloroquine phosphate

   (*b*) Amodiaquine hydrochloride

   (*c*) Santoquin

   Give their structures and the synthesis of any **one** drug.

4. Name **three** important antimalarials derived from **8-amino-6-methoxy quinoline nucleus.** Give their structure, chemical name, uses and the synthesis of any **one** drug.

5. Elaborate the synthesis of **mepacrine hydrochloride** by adopting the following steps sequentially :

   (*i*) **Side chain :** 4-Diethylamine-1-methylbutyl amine.

   (*ii*) **Nucleus :** 2, 5-Dichloro-7-methoxy acridine

   (*iii*) Condensation of (*i*) and (*ii*) and

   (*iv*) HCl.

6. Discuss the synthesis of **'Primaquine Phosphate'** :

   (*a*) Elderfield's method—from 1, 4-dibromopentane

   (*b*) From 2-chloropentylamine.

7. Interaction of **8-amino-6-methoxy quinoline** with the following side chains :

   (*a*) 1-chloro-5-isopropylamine pentane

   (*b*) 2-bromo-5-isopropylamino pentane

   yield **two** potent antimalarials. Discuss their synthesis in details.

8. Ehrlich's hypothesis that methylene blue exerts antimalarial activity led to the discovery of **Mepacrine Mesylate.** Describe its synthesis sequentially.

9. Paludrine (proguanil hydrochloride) the wonder drug for malaria gets metabolizsed to its active form cycloguanil *in vivo.*

   (*a*) Explain its biotransformation

   (*b*) Give its synthesis from *p*-chlorophenyl guanidine and *iso*-propyl cyanamide.

10. (*a*) Discuss the synthesis of **one** important antimalarial drug belonging to the class :

    (*i*) Diaminopyrimidines   (*ii*) Sulfones.

    (*b*) Give a brief account of the **'Mode of Action'** of antimalarials.

                                                 Or

    Give a comprehensive account of 'ARTEMISININ'

RECOMMENDED READINGS

1. EA Stec, *The Chemotherapy of Protozoan Diseases,* Vols. I-IV, Walter Reed Army Institute of Research Washington D.C. (1971).

2. WC Cooper, *'Summary of Antimalarial Drugs'*, Report No. 64, U.S. Public Health Service (1949).

3. PE Thompson and LM Werbal, *'Antimalarial Agents'* Chemistry and Pharmacology, Academic Press New York (1972).

4. RM Pinder, Antimalarials, in *Burger's Medicinal Chemistry and Drug Discovery*, M.E. Wolff (*Ed*) (5th edn), John Wiley and Sons Inc., New York (1995).

5. GM Findlay, *'Recent Advances in Chemotherapy'*, (2nd edn), Vol. 2, Philadelphia, Blakiston, (1951).

6. WB Pratt, *'Fundamentals of Chemotherapy'*, Oxford University Press London (1973).

7. DLednicer and LA Mitscher, *'The Organic Chemistry of Drug Synthesis'*, John Wiley and Sons, New York (1995).

8. JEF Reynolds *(Ed)*, *'Martindale : the Extra Pharmacopoeia* : 31st Edn., The Pharmaceutical Press London (1997).

9. Co-ordinating Group for Research on the Structure of Quing Hau Sau, *'A New Type of Sesquiterpene Lactone'*, K'O Hsuch Tung Pao22 (3), 142 (1977), *In Drugs of the Future,* Vol. VI, No. 1, 37, (1981).

10. World Health Organization : *The World Health Report 2002*—available as a PDF file at : *http//www.who.int/whr/en/.*

11. Honingsbaum M : *The Fever Trail : In search for the Cure for Malaria,* Farrar, Straus and Girous, New York, 2001.

20 | Anthelmintics

1. INTRODUCTION

The restrictive application of the terminology **anthelmintic** is invariably meant for such drugs exerting their action locally to expel parasites from the GI tract exclusively. Nevertheless, there exists several varieties of *worms* which are able to penetrate other tissues as well ; therefore, the *'drugs'* that predominantly act on these *parasitic infections* are frequently termed as **anthelmintics.**

At this juncture one may come across *two* more terminologies, namely ;

(*a*) **Vermicides.** *i.e.,* the *'drugs'* that solely kill worms are called *vermicides,* and

(*b*) **Vermifuges.** *i.e.,* the *'drugs'* that specifically affect the worm in such a fashion that either the **peristaltic activity** or **catharsis** expels it from the intestinal tract are commonly known as *vermifuges.*

Importantly, such absolute arbitraty categorization actually affords no useful and gainful objective as a host of **anthelamintics** have been recognized that particularly mainfest both actions equally, as per the strength of dosages employed. Therefore, in a broader sense and perspective the **anthelmintics** are defined more appropriately as—*'drugs used to combat any type of helminthiasis'.*

1.1. Types of Worm Parasites. In fact, the worm parasites of man actually belong to *two* phyla, namely ; (*a*) **Nemathelminthes** (*roundworms*) ; and (*b*) **Platyhelminthes** (*flat worms*).

1.1.1. Roundworms. The roundworms essentially comprise of the following *seven* species which shall be discussed briefly *vis-a-vis* the disease they produce in humans in the sections that follows :

(*a*) *Hookworm* : These are of *two* types, namely :

    (*i*) **American Variety.** *Necator americanus,* and

    (*ii*) **European Variety.** *Ancylostoma duodenale.*

They are found to attach themselves to the mucosa of the duodenum and subsequently obtain their nourishment (for survival) by sucking blood from the surrounding blood vessels.

(*b*) *Roundworm.* The most prevalent human helminths belonging to this category is *Ascaris lumbricoides.* It is observed to inhabit in the upper segment of the small intestine ; and, therefore, it is vomitted up quite often.

(*c*) *Whipworm.* It exactly resembles a tiny whip and the causative species in *Trichuris trichiura.* It is mostly inhabited in the **cecum,** but is also locateed in the lower segment of the ileum and the appendix.

571

(d) *Pinworm (Threadworm).* It is only 1.5 to 3mm long, *Enterobius vermicularis,* and resides mostly in the small intestine, cecum and colon.

(e) *Strongyloides stercoralis.* It inhabits in the duodenum mostly but may also be located in various other parts, for instance : biliary passages, pancreatic ducts, stomach, various segments of the intestinal passage.

(f) *Trichinella spiralis.* The infection usually caused with *T. spiralis* is known as **trichinosis** *i.e.,* a condition which comes into being due to the ingestion of partially cooked pork meat profusely infested with the larvae of the worm. The intake of such meat allows the cysts to dissolve, the parasites get matured which eventually gives rise to a new crop of larvae that not only develops but also penetrates right into the intestinal mucosa and ultimately lodge in the muscles.

(g) *Wuchereria bancrofti.* It is one of the most vital filarial worms that is particularly transmitted by the bite of the mosquito. The prevailing symptoms are the blocking of the lymphatic. ducts with the adult worms.

1.1.2. Flat worms. The flatworms are normally of *two* kinds, namely : (*a*) Segmented (*cestodes*) ; and (*b*) non segmented (*trematodes*).

(*i*) *Cestodes.* They include the **'tapeworms',** which are of *four* categories commonly found in humans *viz.,* beef tapeworm (*Taenia saginata*) ; pork tapeworm (*Taenia solium*) ; fish tapeworm (*Diphyllobothrium latum*) ; and dwarf tapeworm (*Hymenolepis nana*). In reality, the larval stage of all the four tapeworms is invariably spent in the muscles of the intermediate host, and human infection usually takes place by means of eating partially (improperly) cooked meat and fish.

(*ii*) *Trematodes.* They mostly include the **flukes ;** and in man they occur in *three* varities that solely inhabit the blood stream thereby causing prominently **schistosomiasis.** These blood flukes are, namely : *S. haematobium ; S. mansoni ; S. mekongi ;* and *S. japonicum.* Importantly, all these human parasites predominantly produce epigastric distress, abdominal pain, anorexia, diarrhea with blood and mucus in the stools, pyrexia, enlarged and tender liver, and ascites. It has been established that the intermediate host is either a *freshwater snail* or a *freshwater mollusk.* The usual mode of transmission in humans is on account of drinking contaminated water.

2. CLASSIFICATION :

The **anthelmintics** are classified based upon their chemical structures invariably, whereas a few *'natural products'* also are used to combat the infections in humans. They may be classified as :

(*i*) Piperazines

(*ii*) Benzimidazoles

(*iii*) Heterocycles

(*iv*) Antimalarials

(*v*) Natural Products

These different categories of *'anthelmintics'* shall now be treated with specific examples in the sections that follows :

2.1. Piperazines. In general, *piperazine* and a good a number of its salts *e.g.,* adipate, calcium edetate, citrate, phosphate and tartrate—have been employed profusely in therapeutic treatment of roundworm and pinworm infections. Piperazine itself is a representative example of this class of compounds :

2.1.1. Piperazine BAN, USAN

Hexahydropyrazine ; Diethylenediimine ; Arthriticine[R] ; Dispermin[R] ;

Synthesis :

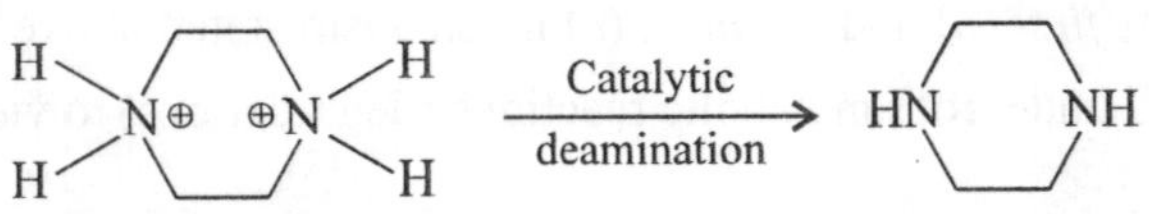

Piperazine may be prepared by the catalytic deamination of diethyldiamine (US Pat 2, 267, 686).

It is used as an anthelmintic for the management and treatment of *pinworm* and *roundworm* infestations.

Mechanism of Action. The *'drug'* blocks the response of the ascaris muscle to ACh, thereby affording flaccid paralysis in the worm, which is eventually dislodged from the intestinal inner lumen and ultimately get expelled in the faeces.

2.1.2. Diethylcarbamazine Citrate BAN, USAN,

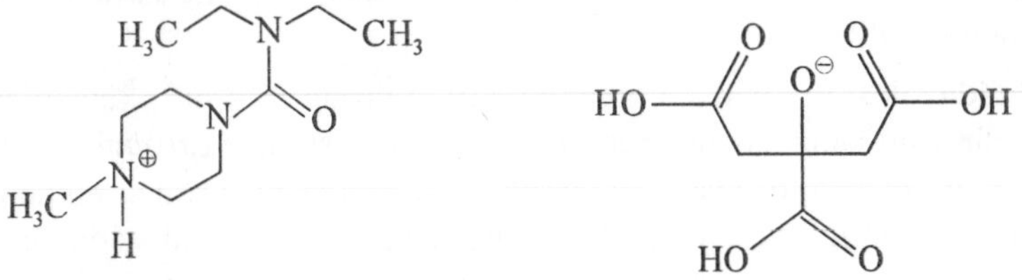

N, N-Diethyl-1, 4-methyl-1-piperazinecarboxamide citrate ; 1-Diethylcarbamyl-4-methyl-piperazine dihydrogen citrate ; USP ; Hetrazan[R] ;

It is effective against various forms of *filariasis,* including *Bancroft's onchocerciasis,* and *laviasis.* It is also found to be active against *ascariasis.*

Mechanism of Action. The explicit mechanism of action of diethylcarbamazine (DEC) is not yet known. It has been observed that DEC seems to be the **'active form'** of the drug having a very fast onset of action ; however, interestingly the drug is absolutely **inactive** *in vitro* thereby affirming the glaring fact that activation of a cellular component is a must for the ensuing **filaricidal action.**

In a broader perspective the following *three* most probable mechanisms have been proposed, namely :

(*a*) direct involvement of blood platelets triggered by the action of filarial excretory antigens, thereby accomplishing a rather **'complex reaction'** taking place amongst the *'drug',* antigen and *platelets,* *

---

*Cesbron JY *et al.* *Nature,* **325** : 533-536, 1987.

(b) **inhibition** of *microtubule polymerization* and **disruption** of *preformed microtubules,* *
and

(c) interference with *arachadonic acid metabolism.* **

Besides, DEC also exerts antiinflammatory activity which action is caused solely due to the bockade occurring at *cyclooxygenase* and *LTA₄ synthase* **(leukotriene synthesis),** whereby two predominant activity takes place :

- To change vascular and cellular adhesiveness, and

- To alter cell activation.

Perhaps the latter biological action may propose a plausible suggestion that a possible relationship would prevail between the *first (a)* and the *third (c)* mechanism stated above.

Metabolism. DEC undergoes metabolic reactions (degradations) to yield *three* products *viz.,* A, B and C as shown below :

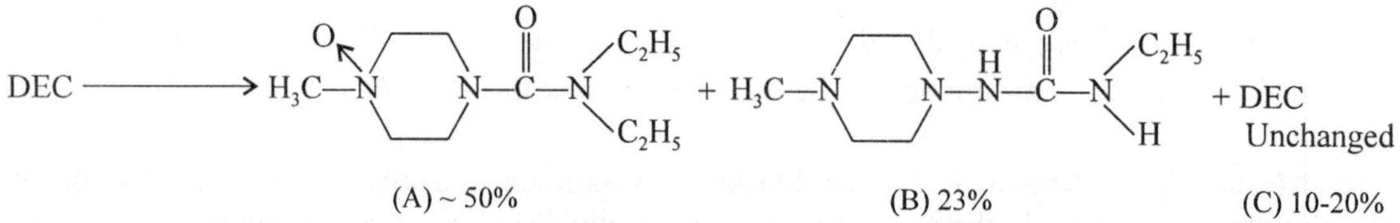

(A) ~ 50%   (B) 23%   (C) 10-20%

Besides, the traces of **piperazine** and **methylpiperazine** are also obtained. In fact, all these metabolites are finally excreted through the urine. Based on the fact that the *'drug'* possesses a very rapid onset of action one may believe that none of these metabolites are virtually involved in exerting the therapeutic action of DEC.

2.2. Benzimidazoles. In 1960s, a broad-spectrum group of drugs, konwn as **benzimidazoles,** were discovered with a big-bang having specific activity against the *gastrointestinal helminths.* In fact, out of several thousand benzimidazoles synthesized and evaluated for their anthelmintic profile only **three** members of this family have gained enormous recognition and wide acceptance, namely : Albendazole, Mebendazole and Thiabendazole. These drugs shall now be discussed individually in the sections that follows :

2.2.1. Albendazole BAN, USAN, INN,

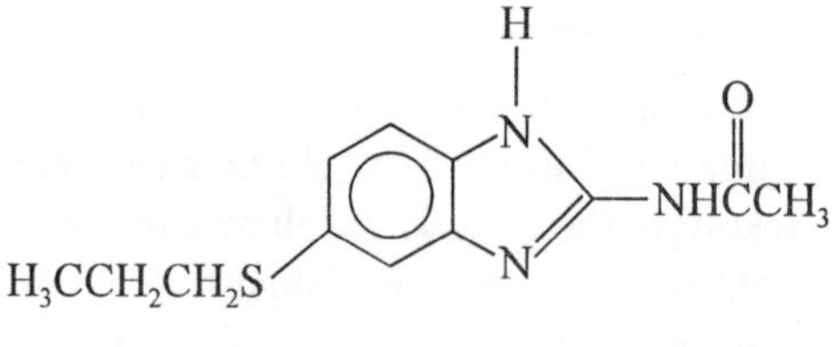

Carbamic acid, [5-(propylthio)-1H benzimidazol-2yl-] methyl ester ; USP ; IP ; BP ; Albenza[R] ; Eskazole[R] ; Zentel[R] ;

---

*Fujimaki Y *et al. Biochem Phamacol,* **39,** 851-856, 1992.

Maizels RM *et al. Parasitol,* **105, S49-S60, 1992.

Synthesis :

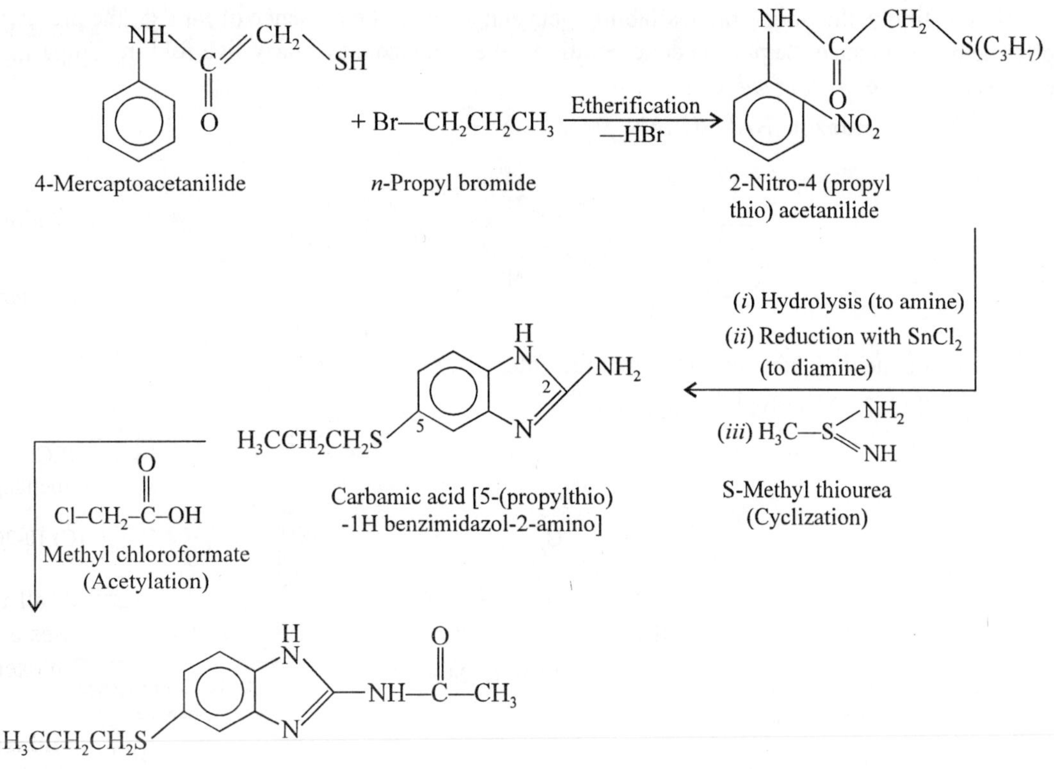

4-Mercaptoacetanilide   *n*-Propyl bromide

Etherification / —HBr → 2-Nitro-4 (propyl thio) acetanilide

(*i*) Hydrolysis (to amine)
(*ii*) Reduction with $SnCl_2$ (to diamine)
(*iii*) H_3C-S S-Methyl thiourea (Cyclization)

Carbamic acid [5-(propylthio) -1H benzimidazol-2-amino]

$Cl-CH_2-C-OH$
Methyl chloroformate (Acetylation)

Albendazole

The etherification of 4-mercaptoacetanilide with *n*-propyl bromide gives rise to the formation of 2-nitro-4 (propylthio) acetanilide with the elimination of one mole of HBr. The resulting product upon hydrolysis converts it to an amine, reduction with $SnCl_2$ to a diamine, and finally interaction with S-methyl thiourea affords cyclization to yield carbamic acid [5-(propylthio)-1H benzimidazol-2-amino]. This upon acetylation with methyl chloroformate affords the official compound albendazole.

It is widely used across the globe for the management and treatment of *intestinal nematode infection.* It is also quite effective as a single-dose-treatment for *ascariasis, New and Old World hookworm infections,* and *trichuriasis.* It has been observed that a recommended multi-dose therapy with albendazol may help in the complete eradication of pinworm, threadworm, capillariasis, chlonorchiasis, and hydated disease as well. However, the overall observed effectiveness of albendazole against tapeworms (cestodes) is obviously more variable and less impressive apparently.

Mechanism of Action. The precise mechanism of action of the *'drug'* is not propely understood ; however, it seems to afford its primary anthelmintic effect by binding to the free (3-tubulin present in the parasite cells, thereby causing a more or less selective **inhibition of parasite micotubule**

polymerization, and inhibition of **micotubule-dependent glucose-up-take** significantly. Besides, the effective inhibition of parasite β-tubulin usually takes place at rather lower strengths of the *'drug'* than those that are normally needed to check and suppress human microtubule polymerization.

Interestingly, the drug's bioavailability gets enhanced in the presence of fat *e.g.,* the presence of 40g fat helps to enhance the plasma concentrations of albendazole to nearly five fold in comparison to that observed in the *'fasting subjects'.*

2.2.2. Mebendazole BAN, USAN, INN

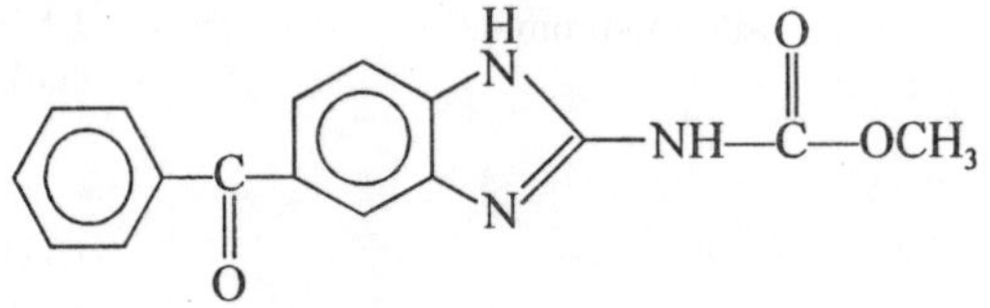

Carbamic acid (5-benzoyl-1H-benzimidazol-2-yl), methyl ester ;

IP., BP, USP ; Vermox[R] ; Antiminth[R] ;

Synthesis :

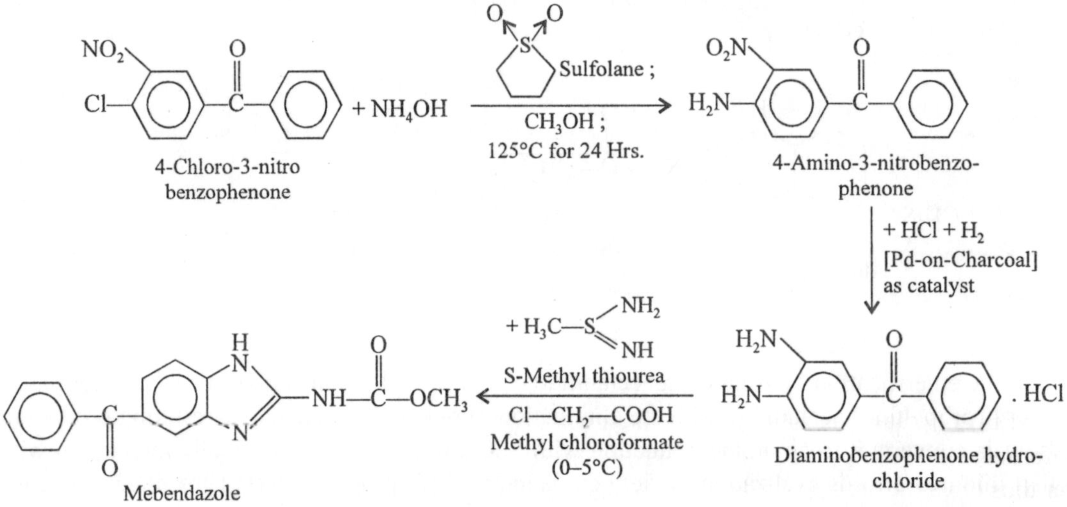

By heating together 4-chloro-3-nitro benzophenone and ammonia at 125°C for 24 hours in the presence of sulfolane yields 4-amino-3-nitrobenzophenone. The resulting product on being treated with hydrochloric acid and hydrogenation wiht *Pd-on-charcoal* as a catalyst yields diaminobenzophenone hydrochloride. This on being treated with S-methyl thiourea in the presence of methyl chloroformate at 0-5°C gives rise to the desired official drug.

It is the anthelmintic drug of first choice in hookworm, pinworm, roundworm, whipworm, guinea worm, in filariasis, and also as an alternative drug for *Visceral Larva Migrans.* It is also employed as an adjunct to steroids for curing *trichinosis.*

Mechanism of Action. The *'drug'* specifically blocks the glucose uptake by susceptible heliminths, thereby depleting the stored glycogen within the parasite. Obviously, the glycogen depletion invariably causes in an actual decreased generation of *adenosine triphosphate* (ATP), the latter is essentially needed for the survival and reproduction of the helminth. Besides, it inhibits cell-division in nematodes.*

2.2.3. Thiabendazole USAN, BAN, INN

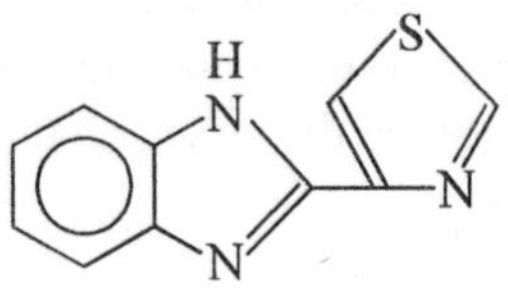

2-(4-Thiazolyl) benzimidazole ; 1H-Benzimidazole, 2-(4-thiazolyl)- ; USP ; IP ; Mintezol[R] ; Thibenzole[R] ;

Synthesis :

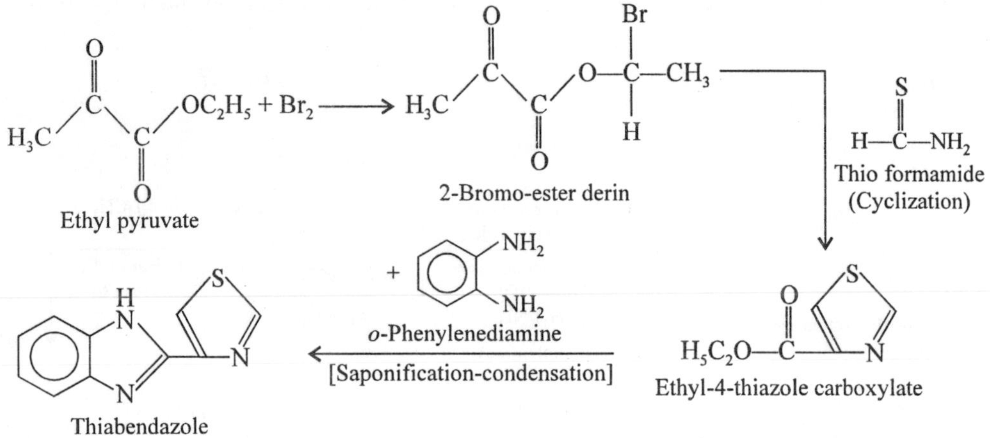

Ethyl pyruvate is first brominated, and the resulting 2-bromo ester derivative is treated with thioformamide when cyclization takes place with the formation of ethyl-4-thiazole carboxylate. The ester thus obtained is further saponified and condensed with *o*-phenylenediamine so as to introduce the benzimidazole heterocyclic nucleus and obtain the official drug.

The *'drug'* exhibits broad-spectrum anthelmintic activity. It is mostly employed for the management and treatment of enterobiasis, strongloidiasis (causing threadworm infection) ; ascariasis, uncinariasis (causing hookworm infection) ; and trichuriasis (causing whipworm infection). Besides, it also finds its usefulness to get rid of symptoms associated with cutaneous larva migrans (*i.e.,* the creeping eruption**), and the ensuing invasive phase of trichinosis.

Mechanism of Action. The *'drug'* exerts its anthelmintic action by inhibiting the *helminth-specific enzyme* **fumarate reductase.***** However, it has not yet been fully established whether metal

---

*Dessan A *et al. Science,* **267** : 1638, 1995.

**Creeping eruption caused by *Angiostrongylus costaricensis.*

***Prichard RK, *Nature,* **228** : 684, 1970.

ions are involved in the inhibition mechanism or if the inhibition of the enzyme is exclusively associated with the anthelmintic effect of thiabendazole. It has also been established beyond any reasonable doubt that, in general, the benzimidazole anthelmintic drugs *viz.,* thiabendazole, mebendazole in helping to arrest totally the **nematode-cell division** particularly in the *metaphase state* by directly interfering with the microtubule assembly.* In fact, the two aforesaid compounds are responsible for exhibiting a high affinity for **tubulin**** *i.e.,* the well-known **precursor protein** essential for the *microtubule synthesis.*

2.3. Heterocyclics. There are some compounds structurally based upon the heterocyclic necleus *viz.,* pyridine (oxamniquine), pyrimidine (pyrantel pamoate) which are also found to exert anthelmintic activities. These *two* compounds shall now be discussed as under.

2.3.1. Oxamniquine USAN, INN,

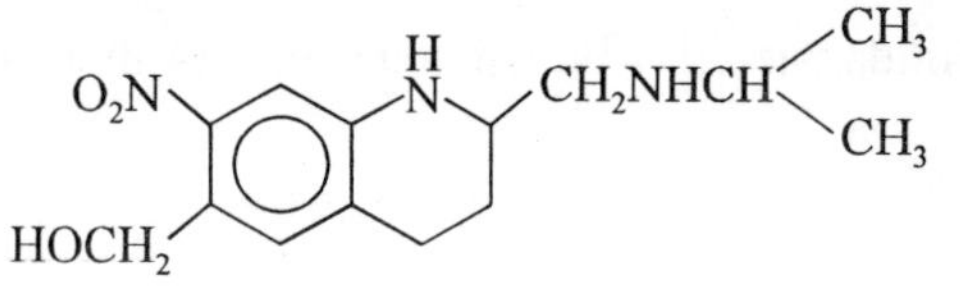

6-Quinolinemethanol, 1, 2, 3, 4-tetrahydro-2-[[(1-Methylethyl) amino]-methyl]-7-nitro ; USP ; Vansil[R] ;

Synthesis :

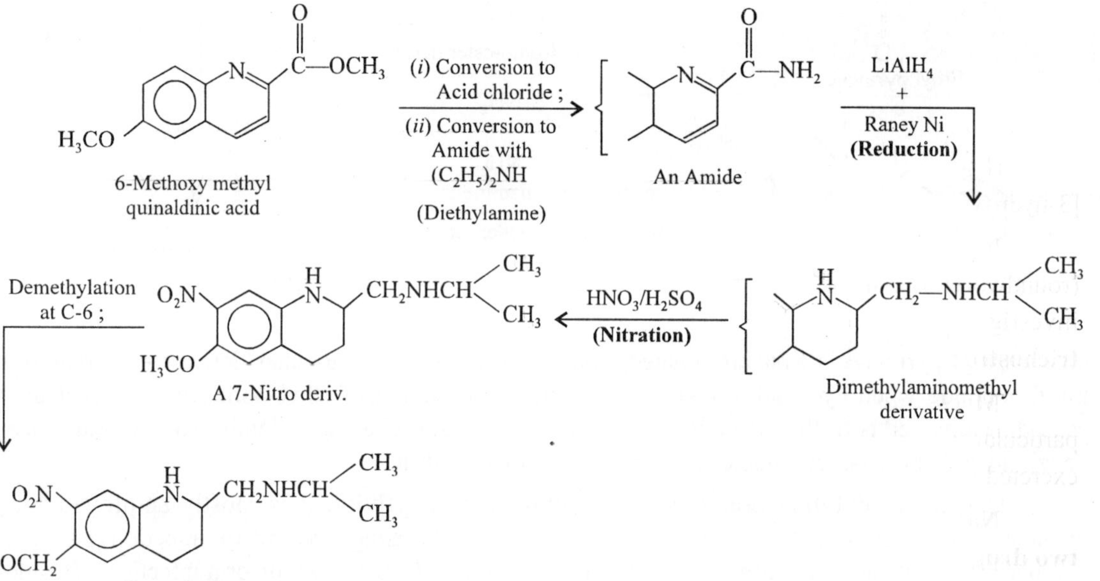

First of all the 6-methoxymethyl quinaldinic acid is converted to an acid chloride, followed by its conversion to an amide with diethylamine to produce an *'amide'* derivative, which upon reduction with

---

*Friedman PA *et al. Biochim. Biophys Acta,* **544** : 605, 1978.

**A colchicine-binding protein.

LiAlH$_4$ and Raney Nickel to yield the dimethylaminomethyl derivative wherein the pyridine ring also gets saturated. The resulting product is subjected to nitration in the presence of HNO$_3$/H$_2$SO$_4$ to give rise to the formation of the corresponding 7-nitro derivative ; and finally the demethylation at C-6 yields the desired product oxamniquine.

Oxamniquine is a potent antischistosomal agent which is specifically indicated for the treatment of *intestinal schistosomiasis* caused by *S. mansoni,* including the severe and the chronic phase with hepatosplenic involvement. It is found to minimize appreciably the *egg-load* of *Schistosoma mansoni.*

Mechanism of Action. The *'drug'* is observed to critically cause inhibition of DNA, RNA and protein synthesis schistosomes. The oral bioavailability of oxamniquine is fairly good, and effective plasma levels are accomplished within a span of 1 to 1.5 hours. Interestingly, the *'drug'* gets metabolized to its corresponding inactive metabolites predominantly, of which the major component is the 6-carboxy derivative.

SAR of Oxamniquine. The most critical and vital entity present in this *'drug'* is the presence of the 6-hydroxymethyl moiety ; and the subsequent metabolic activation of the precursor 6-methyl derivatives is equally critical in nature.

2.3.2. Pyrantel Pamoate USAN, BAN,

(E)-1, 4, 5, 6-Tetrahydro-1-methyl-2-[2-(2-thienyl) ethenyl]-, compound with 4, 4'-methylenebis [3-hydroxy-2-naphthalenecarboxylic acid] (1:1) ; USP ; Antiminth$^{(R)}$;

It enjoys the reputation of being one of the anthelmintics of choice in the treatment of *ascariasis* (roundworm infection), *enterobiasis* (pinworm infection). Recently, it is under intensive and extensive investigation with regard to this potential for the treatment of **hookworm, moniliformis,** and **trichostrongylus** infections.

Mechanism of Action. The *'drug'* exerts its action as a depolarizing blocking agent that particularly affords spastic paralysis in susceptible helminths. More than half of the oral dosage gets excreted in the faeces.

Note. As the action of pyrantel pamoate is just the reverse of piperiazine ; therefore, these two drug substances must not be administered simultaneously to a patient.

2.4. Antimalarials. Some of the *'antimalarials',* discussed in Chapter-19, are also used as *'anthelmintics'* at different dosage regimen.

2.5. Natural Products. There are quite a few natural products that are used extensively as **anthelmintics** *i.e.,* as anti-infective agents. A few important as well as typical examples shall be discussed in the pages that follows :

2.5.1. Invermectin USAN, BAN, INN

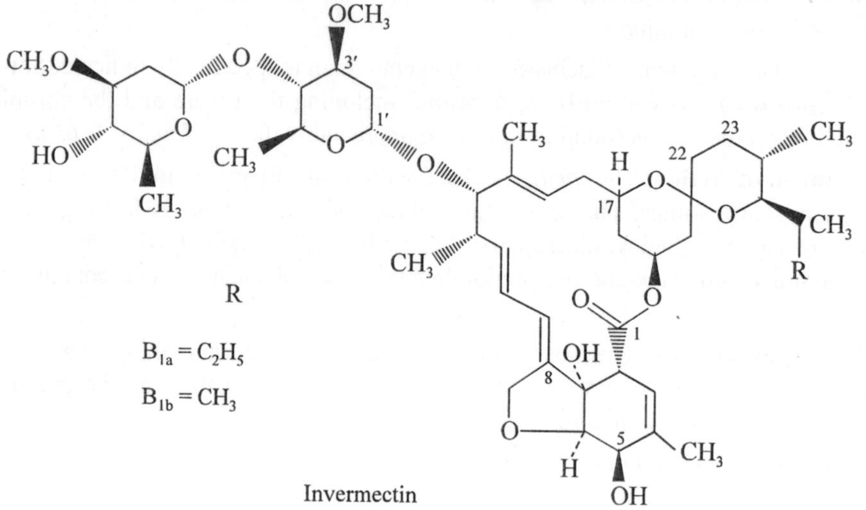

Invermectin

R

$B_{1a} = C_2H_5$

$B_{1b} = CH_3$

USP ; Int. P. ; BP ; Invomec[(R)] ; Cardomec[(R)] ; Equalan[(R)] ; Mectizan[(R)] ;

Invermectin is usually extracted from the soil of actinomycete *Streptomyces avermitilis,* the **natural avermectins** are 16-membered macrocyclic lactones and is found to be a mixture of 22, 23-dihydro structural analogues of avermectins B_{1a} and B_{1b} prepared by catalytic hydrogenation (reduction). In reality, *avermectins* are members of a family of rather structurally **complex antibiotics** obtained by fermentative process with the pure isolated strain of *S. avermitilis.* An intensive screening of cultures for the anthelmintic drugs exclusively from the *'natural products'* ultimately gave birth to this wonderful drug.*

It has been amply demonstrated that the natural avermectins invariably exhibit **minimal biologic profile of activity,** whereas **invermectin** has proven to be extermely useful and hence recognized for the management and treatment of good number of nematode infections. Besides, it is found to be active against arthropods that usually parasitize the animal folks.**

Mechanism of Action. There are *two* different modes of *'mechanism of action'* for invermectin have been suggested, namely :

(a) **Indirect Action.** In this particular instance the motility of *microfalaria* is minimized appreciably which subsequently permits the cytotoxic cells of the host to enable them adhere to the parasite thereby causing an elimination from the host finally. This specific action may be afforded due to the ability of invermectin to either exhibit its action a **GABA agonist** or as an **inducer of Cl⁻ ion influx** that may ultimately cause *hyperpolarization* and *muscle paralysis.* However, the latter mechanism *i.e.,* the Cl⁻ ion influx seems to be the more logical and plausible explanation.* The overall net result of this action is a rapid lowering in the prevailing *microfilarial concentrations.*

---

*Burg RW *et al. Antimicrob. Agents Chemother* : **15** : 361, 1979.

Campbell WC, *Science,* **221, 823, 1983.

(*b*) **Degeneration of Microfilaria *in uterio*.** This specific action essentially would give rise to relatively fewer microfilaria being released from the female worms and normally extends over a longer duration of time interval. The overall effect caused due to the presence of the *degenerated microfilaria' in uterio* directly prevents **fertilization phenomenon** and the **production of microfilaria.**

Metabolism. The *'drug'* gets absorbed rapidly, bound to an appreciable extent to *plasma protein ;* and excreted ultimately either through the urine or faeces in *two* forms, namely : (*i*) unchanged invermectin ; and (*ii*) 3′-O-demethyl-22, 23-dihydro-avermectin B_{1a} or as dihydroavermectin B_{1b} monosaccharide. Ethanol is found to aggravate the absorption of the *'drug'* even upto 100%.

2.5.2. Pyrethrum and Pyrethroids. In fact, a plethora of naturally occurring pyrethrums have been used quite extensively as viabally potent insecticides since the 1800s. A number of potent chemical entities have been successfully isolated from the extract of the flowering portion of the **Chrysanthemum** plant. Importantly, the plants grown in Kenya (East Africa) contain upto **1.3% pyrethrins.** The *'pyrethrum extracts'* earn a sizable agricultural revenue for the country.

The **Chrysanthemum Extract** comprise of a mixture of *ester e.g.,* chrysanthemic and pyrethric acids ; *alcohols e.g.,* cinerolone and pyrethrolone. As the *'esters'* are usually more prone to get *hydrolyzed* and *oxidized,* hence it must be stored in sealed light-proof containers in a cool place.

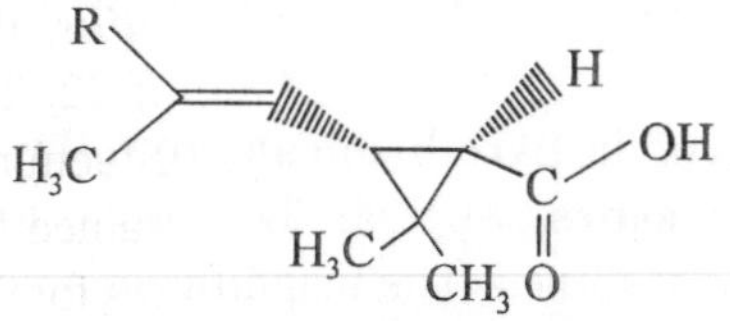

$R_1 = CH_3$; Chrysanthemic Acid
$R_1 = COOCH_3$; Pyrethric Acid

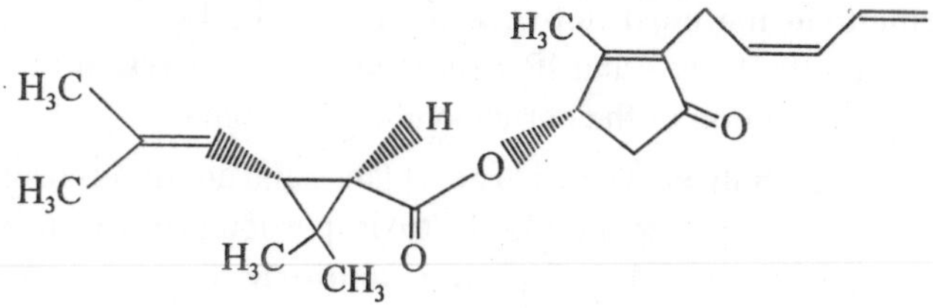

Pyrethrin I

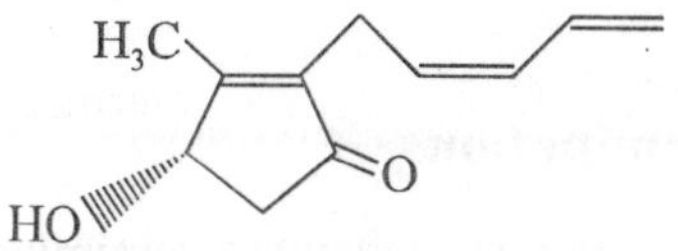

Pyrethrolone

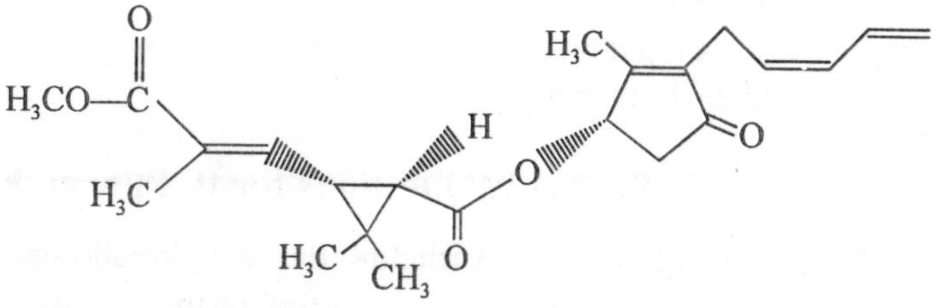

Pyrethrin II

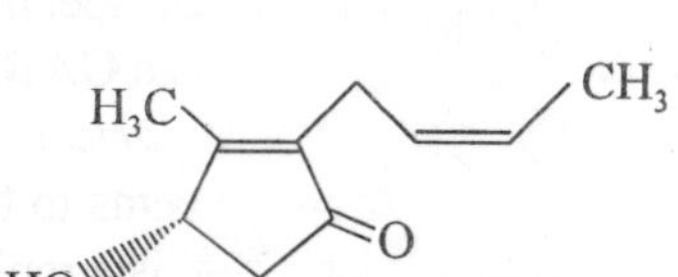

Cinerolone

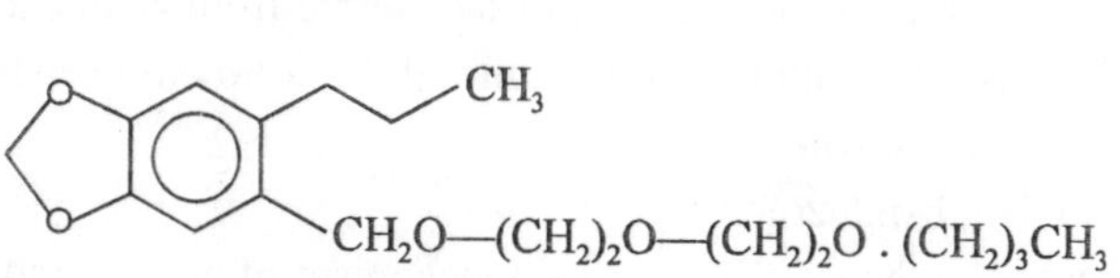

Piperonyl butoxide

Mechanism of Action. The mechanism of action of **pyrethrins** and **pyrethroids (permethrin)** are due to their inherent characteristic feature as *nerve membrane sodium channel toxins* that fail to exert any action upton the potassium channels. In reality, most of these chemical entities get bound to specific sodium-channel proteins and thereby slow down the rate of inactivation of the sodium current elicited by membrane depolarlization.. The net overall affect being the prolongation of the 'open time' of the sodium channel.

However, at low concentrations the pyrethroids (permethrin) is observed to display **repititive action**

Permethrin

potentials and also afford **neuron firing ;** whereas, at relatively higher concentrations the nerve membrane gets depolarized almost completely thereby causing a blockade of excitation.

Stereospecific Aspects. It has been well established that the ensuing receptor interaction of the pyrethrums with the sodium channel complex is absolutely **stereospecific ;** and, therefore, solely dependent on the stereochemistry of the carboxylic acid in question. Interestingly, in the case of **permethrin** the **most active isomers** are the IR, 3-*cis*- and IR, 3-*trans*-cyclopropane-carboxylates. However, the IS *cis*- and IR *trans*-isomers are **inactive ;** and are found to serve as antagonists to the therapeutic action of the corresponding IR-isomers.

Metabolism. The wide acceptance and enormous usefulness of the **pyrethrum and pyrethroids** are that they pose to be highly toxic particularly to the **ectoparasites,** whereas they prove to be comparatively much less toxic (*i.e.,* **nontoxic**) to mammals in case absorbed. The magnificent notoxic characteristic feature is associated with the excellent and rapid metabolism of these drug substances either **via** hydrolysis or oxidation. More specifically, the extent of either hydrolysis or oxidation is exclusively dependent upon the structure of the prevailing **pyrethrins or pyrethroids.**

Besides, the rapid breakdown of these drug substances also accounts for their low persistence in the surrounding environment.

Probable Questions for B. Pharm. Examinations

1. Give a brief account on 'Anthelmintics' and provide suitable examples wherever necessary.
2. How would you classify 'Anthelonintics' on the basis of chemical structures ? Give the structure, chemical name and uses of **one** example from each category.
3. Discuss the synthesis of any **one** of the following drugs :
    (*a*) Albendazole
    (*b*) Thiabendazole.
4. Describe the synthesis, uses, mechanism of action, and SAR of **oxamniquine.**
5. Write a short note on any **one** of the following potent 'Anthelmintics' :
    (*i*) Pyrantel Pamoate
    (*ii*) Pyrethrum and Pyrethroids.
6. What is Invermectin ? Discuss its metabolism and mechanism of action.

RECOMMENDED READINGS

1. Freeman CD *et al.* : **Metronidazole : A Therapeutic Review and Update,** *Drugs,* **54** : 679-708, 1997.

2. Wilson JD *et al.* (eds), *Harrison's Principles of Internal Medicine,* McGraw Hill, New York, 12th edn., 772, 1992.

3. Williams DA and Lemke TL, *Foye's Principles of Medicinal Chemistry,* Lippincott Williams and Wilkins, New York, 5th edn, 2002.

4. Block JH and Beale JM Jr., (eds), *'Wilson and Gisvold's Textbook of Organic Medicinal and Pharmaceutical Chemsitry',* Lippincott Williams and Wilkins, New York, 5th edn, 2004.

5. Gennaro AR : *Remington : The Science and Practice of Phramacy,* Vol. I and II, Lippincott Williams and Wilkins, New York, 20th edn., 2000.

6. Mandel GL *et al.* (eds). *Principles and Practices of Infectious Diseases,* Vol. I, Churchill-Livingstone, New York, 4th edn, 1995.

7. Yamaguchi H *et al.* (eds). *Recent Advances in Antifungal Chemotherapy,* Mercell Dekker, New York, 1992.

8. Testa B (ed.), *Advances in Drug Research,* Vol. 21, Academic Press, New York, 1991.

21 Insulin and Oral Hypoglycemic Agents

1. INTRODUCTION

A major portion of the *pancreas** essentially comprises of glandular tissue which specially contains acinar cells that predominantly gives rise to the secretion of certain **digestive enzymes.** Besides, there also exist some '*isolated groups of pancreatic cells*' commonly known as the **islets of Langerhans** which usually made up of *four* cell types, each of which generates a *distinct polypeptide hormone,* namley :

(*a*) **Insulin** — in the beta (β) cells,

(*b*) **Glucagon** — in the alpha (α) cells,

(*c*) **Somatostatin** — in the delta (δ) cells, and

(*d*) **Pancreatic polypeptide** — in the PP or F cell.

Interestingly, the β-**cells** made up 60-80% of the **islets of Langerhans** most predominantly and distinctly.

Diabetes — a general term for diseases marked by excessive urination ; and is usually refers to *diabetes mellitus.*

However, the *clinical diabetes mellitus* invariably occurs in *two* forms, associated with different causes and methods of therapy.

Type 1 Diabetes : The *insulin-dependent diabetes mellitus* (IDDM), normally takes place when the β-cells of the prevailing pancreatic *islets of Lanherhans* are destroyed, perhaps by an **autoimmune, mechanism,** as a consequence of which the 'insulin production' *in vivo* is overwhelmingly insufficient. Subjects undergoing such abnormalities in biological functions may show appreciable metabolic irregularity that may ultimately lead to develop **diabetic β-ketoacidosis** together with other manifestations of acute diabetes. Therapeutically Type-I diabetes is largely treated with **insulin.**

Type 2 Diabetes : The *noninsulin-dependent diabetes mellitus* (NIDDM), *i.e.,* type 2 diabetes, is most abundantly linked with obesity in its adult patients largely. In such a situation, the insulin levels could be either elevated or normal ; and therefore, in short, it is nothing but a disease of abnormal **'insulin resistance'.** However, it has been duly observed that the impact of the disease is relatively

---

*Both an *exocrine* and *endocrine* orgin ; a compound acinotubular gland situated behind the stomach in front of the first and second lumbar vertebrae in a horizontal position, its head attached to the duodenum and its tail reaching to the spleen.

milder, occassionally leaving to β-ketoacidosis and may also be accompanied by certain other degenerative phenomena *in vivo*. The etiology of the condition bears a *strong genetic hereditory ;* and, hence, *insulin therapy* may not prove to be quite effective.

2. INSULIN-PRIMARY STRUCTURE

Sanger (in 1950s) put forward the primary structure of insulin as illustrated below in Fig. 1.

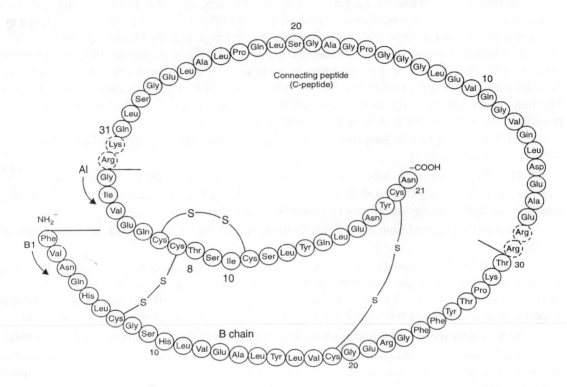

Fig. 1. Primary structure of proinsulin, depicting cleavage sites to produce insulin.

[Adapted from : *Foye's Principle of Medicinal Chemistry*, 5th Internationl Student Edition, Lippincott Williams and Wlikin, New York, 2001]

The above Fig. 1 has the following **Salient Features,** namely :

(1) **Proinsulin** is the immediate precursor to insulin in the single-chain peptide.

(2) **Proinsulin** folds to adopt the 'correct orientation of the prevailing *'disulphide bonds'* plus other relevant conformational constraints whatsoever on account of its primary structure exclusively.

(3) **Proinsulin** in reality, has a precursor of its own, *preproinsulin*–a peptide, that essentially comprises of hundreds of **'additional residues'.**

(4) At an emerging critical situation the **insulin** gets generated from *proinsulin* due to the ensuing cleavage of *proinsulin* at the *two points indicated.* This eventually produces **insulin,** that comprises of a **21-residue A chain** and strategically linked with **two disulphide bonds** ultimately to a **30-residue B chain**. Interestingly, these bondages between the two aforesaid residual chains 'A' and 'B' are invariably oriented almost perfectly and correctly by virtue of the prempted nature of proinsulin folding.

2.1. Variants of Insulin Products. There are a number of variants of insulin products that are available as a *'drug'*, namely :

2.1.1. Insulin Injection. [*Synonyms* : Regular Insulin ; Crystalline Zinc Insulin ;] :

It is available as a sterile, acidified or neutral solution of insulin. The solution has a potency of 40, 80, 100 or 500 USP Insulin Units in each ml.

Mechanism of Action. It is a rapid-action insulin. The time interval from a hypodermic injection of this 'drug' until its action may be observed ranges between 1/2 to 1 hour. It has been observed that the duration of action is comparatively short but evidently a little longer than the plasma half-life that stands at nearly 9 minutes. Importantly, the duration of action is not linearly proportional to the size of the dose, but it is a simple function of the logarithm of the dose *i.e.,* if 1 unit exerts it action for 4 hours then 10 units will last 8 hours. In usual practice the duration is from 8 to 12 hour after the subcutaneous injection, which is particularly timed a few minutes before the ingestion of food so as to avoid any possible untoward fall in the prevailing blood-glucose level.

2.1.2. Isophane Insulin Suspension. [*Synonyms* : Isophane Insulin ; Isophane Insulin Injection ; NPH Insulin ; NPH Iletin ;] :

The *'drug'* is a sterile suspension of Zinc-insulin crystals and protamine sulphate in buffered water for injection, usually combined in such a fashion that the *'solid phase of the suspension'* essentially comprises of crystals composed of *insulin, protamine*,* and *zinc.*

Each mL is prepared from enough insulin to provide either 40, 80, or 100 USP Insulin units of insulin activity.

Mechanism of Action. The *'drug'* exerts its action as an intermediate-acting insulin for being insoluble and obtained as repository form of insulin. In reality, the action commences in 1–1.5 hour, attains a peak-level in 4 to 12 hour, and usually lasts upto 24 hours, with an exception that *'human isophane insulin'* exerts a rather shorter duration of action. It is, however, never to be administered IV.

Note : Incidence of occasional hypersensitivity may occur due to the presence of 'protamine'.

2.1.3. Insulin Zinc Suspension. It is invariably obtained as a sterile suspension of insulin in buffered water for injection, carefully modified by the addition of zinc chloride ($ZnCl_2$) in such a manner that the *'solid-phase of the suspension'* comprises of a mixture of **crystalline** as well as **amorphous** insulin present approximately in a ratio of 7 portions of crystals and 3 portions of amorphous substance. Each mL is obtained from sufficient insulin to provide either 40, 80, or 100 USP Insulin Units of the **Insulin Activity.**

Mechanism of Action. It has been duly observed that the *'amorphous zinc-insulin component'* exerts a duration of action ranging between 6–8 hours, whereas the *'crystalline zinc-insulin component'* a duration of action more than 36 hour, certainly due to the sluggishness and slowness with which the larger crystals get dissolved. However, am appropriate dosage of the 3 : 7 mixture employed usually displays an onset of action of 1 to 2.5 hour and an intermediate duration of action which is very near to that of *'isophane insulin suspension'* (24 hour), with which preparation this *'drug'* could be employed interchangeably without any problem whatsoever. However, it must not be administered IV.

Note : The major advantage of 'zinc insulin' is its absolute freedom from *'foreign proteinous matter'*, such as : globin, or protamine, to which certain subjects are sensitive.

---

*The protamine sulphate is usually prepared from the sperm or from the mature testes of fish belonging to the genera **Oncorhynchus** Suckley, or **Salmo** Linne (*Family : Salmonidae*).

2.1.4. Extended Insulin Zinc Suspension : [*Synonyms* : Ultra-Lente Iletin ; Ultralente Insulin/ Ultratard] :

Mechanism of Action : The actual **'crystalline profile'** in this specific form are of sufficient size to afford a slow rate of dissolution. It is found to exert its *long-acting action* having an onset of action ranging between 4 to 8 hours, an optimal attainable peak varying between 10-30 hours, and its overall duration of action normally in excesss of 36 hours, which being a little longer than that of **Protamine Zinc Insulin.**

Note : Because the *'drug'* is free of both protamine and other foreign proteins, the eventual incidence of allergic reactions gets minimized to a significant extent.

2.1.5. Prompt Insulin Zinc Suspension : [*Synonyms :* Semi-Lente Iletin ; Semitard] :

The *'drug'* is usually a sterile preparation of insulin in **'buffered water for injection'**, strategically modified by the addition of zinc chloride ($ZnCl_2$) in such a manner that the **'solid phase of the prevailing suspension'** is rendered amorphous absolutely.

Each mL of this preparation provides sufficient insulin either 40, 80, or 100 USP Insulin Units.

Mechanism of Action. The zinc-insulin in this particular form is a mixture of amorphous and extremely fine crystalline materials. As a result, the *'drug'* serves as a rapid-acting insulin with an onset of 1 to 1.5 hour, an attainable peak of 5-10 hours, and a duration of action ranging between 12-16 hours.

Note : Since this specific form of insulin is essentially free of any foreign proteins, the incidence of allergic reactions is found to be extremely low.

2.1.6. Lispro Insulin. [*Synonyms* : Human Insulin Analog ; Humalog] : It is a human insulin analogue of *r* DNA origin meticulously synthesized from a special nonpathogenic strain of *E. coli,* genetically altered by the addition of the gene for insulin lispro ; Lys (B28), Pro (B29). In fact, the prevailing amino acids at position 28 and 29 of human insulin have been reversed altogether.

Mechanism of Action. The *'drug'* is a very **rapid-acting insulin** which may be injected conveniently just prior to a meal. It exhibits an onset of action within a short span of 15 minutes besides having a relatively much shorter peak ranging between 0.5 to 1.5 hour, and having duration of action varying between 6 to 8 hours in comparison to the *'regular insulin injection'.*

2.1.7. Protamine Zinc Insulin Suspension. [*Synonyms :* Zinc Inuslin ; Protamine Zinc Insulin Injection ; Protamine Zinc and Iletin ;] :

The *'drug'* is a sterile suspension of insulin in buffered water for injection, that has been adequately modified by the addition of zinc chloride ($ZnCl_2$) and protamine sulphate. The protamine sulphate is usually prepared from the sperm or from the mature testes of fish belonging to the genus *Oncorhynchus* Suckley or *Salmo* Linne (Family : *Salmonidae*). Each mL of the suspension prepared from sufficient insulin to provide wither 40, 80, or 100 USP Insulin Units.

Mechanism of Action. The *'drug'* exerts a long-acting action having an onset of action of 4 to 8 hour, a peak at 14 to 24 hour, and a duration of action nearly 36 hour. As a result this *'drug'* need not be administered with any definite time relation frame to the corresponding food intake. Besides, it should not be depended upon solely when a very prompt action is required, such as : in **diabetic acidosis and coma.** Since the *'drug'* possesses an inherent prolonged action, it must not be administered more frequently than once a day. It has been duly observed that *'low levels'* invariably persists for 3 o 4 days ; and, therefore, the dose must be adjusted at intervals of not less than 3 days. It is given by injection, normally into the **loose subcutaneous tissue.**

Note : The *'drug'* should never be administered IV.

3. ORAL HYPOGLYCEMIC AGENTS :

The synthetic oral hypoglycemic agents have been added to the therapeutic armamentarium over the last five decades in lieu of the various **'insulin variants'** discussed earlier. In this particular section the focus shall be made on the different categories of synthetic oral hypoglycemic agents based on their chemical structures, namely :

    (*i*) Sulfonylureas,

    (*ii*) Non sulfonylureas,

    (*iii*) Thiazolindiones,

    (*iv*) Bisguanides, and

    (*v*) α-Glucosidase Inhibitors

The important *'drugs'* belonging to each of the above categories shall now be discussed individually in the sections that follows :

3.1. Sulfonylureas. The sulfonylurea hypoglycemic agents are basically sulphonamide structural analogues but they do not essentially possess any *'antibacterial activity'* whatsoever. In fact, out of 12,000 sulfonylureas have been synthesized and clinically screened, and approximately 10 compounds are being used currently across the globe for lowering blood-sugar levels significantly and safely. The sulfonylureas may be represented by the following general chemical structure :

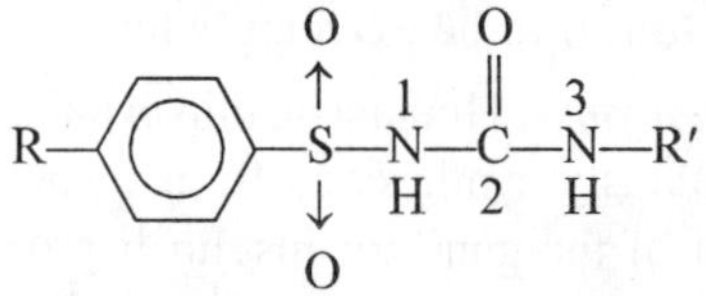

$$R-\!\!\bigcirc\!\!-\overset{\displaystyle O}{\underset{\displaystyle O}{\overset{\uparrow}{\underset{\downarrow}{S}}}}-\overset{1}{\underset{H}{N}}-\overset{\overset{\displaystyle O}{\|}}{\underset{2}{C}}-\overset{3}{\underset{H}{N}}-R' \, ,$$

Salient Features : The salient features of the **'sulfonylureas'** are as given below :

    (1) These are urea derivatives having an arylsulfonyl moiety in the 1 position and an aliphatic function at the 3-position.

    (2) The aliphatic moiety, R', essentially confers lipophilic characteristic properties to the newer drug molecule.

    (3) Optimal therapeutic activity often results when R' comprises of 3 to 6 carbon atoms, as in acetohexamide, chlorpropamide and tolbutamide.

    (4) Aryl functional moieties at R' invariably give rise to toxic compounds.

    (5) The R moiety strategically positioned on the *'aromatic ring'* is primarily responsible for the duration of action of the compound.

However, these agents are now divided into *two* sub-groups, namely :

(*a*) First-generation sulfonylureas, and

(*b*) Second-generation sulfonylureas.

These two aforesaid classes of sulfonylureas will be dealt with separately as under :

3.1.1. First-Generation Sulfonylureas. The various important drugs that belong to this category are, namely : Acetohexamide ; Chlorpropamide ; Tolazamide ; and Tolbutamide. These drugs shall be treated individually as under :

3.1.1.1. Acetohexamide BAN, USAN, INN

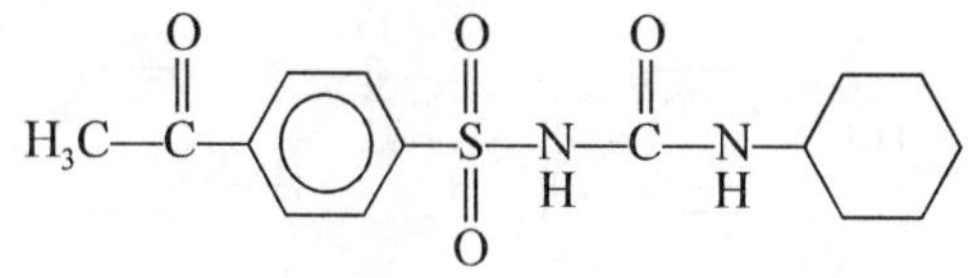

1-[(p-Acetylphenyl) Sulfonyl]-3-cyclohexyl urea ; USP ; Dymelor[R] ;

It lowers the blood-sugar level particularly by causing stimulation for the release of endogenous insulin.

Mechanism of Action. The *'drug'* gets metabolized in the liver solely to a reduced entity, the corresponding α-hydroxymethyl structural analogue, which is present predominantly in humans, shares the prime responsibility for the ensuing hypoglycemic activity.

SAR of Acetohexamide. It is found to be an intermediate between *'tolbutamide'* and *'chlorpropamide' i.e.,* in the former the cyclohexyl ring is replaced by butyl moiety and *p*-acetyl group with methyl group ; while in the latter the cyclohexyl group is replaced by propyl moiety and the *p*-acetyl function with chloro moiety.

3.1.1.2. Chlorpropamide USAN, BAN, INN,

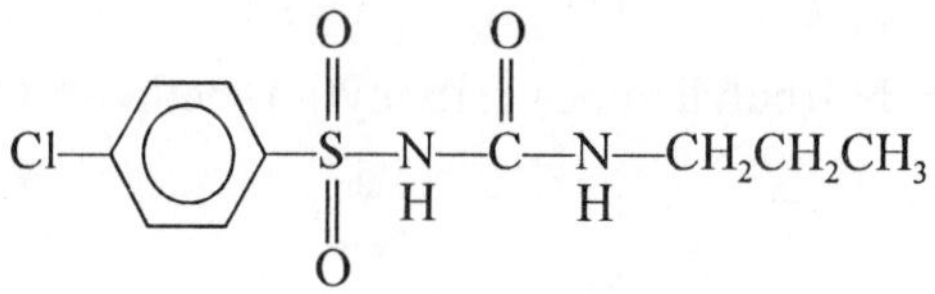

1-[(p-Chlorophenyl)-Sulphonyl]-3-propyl urea ; Diabinese[R] ;

Synthesis :

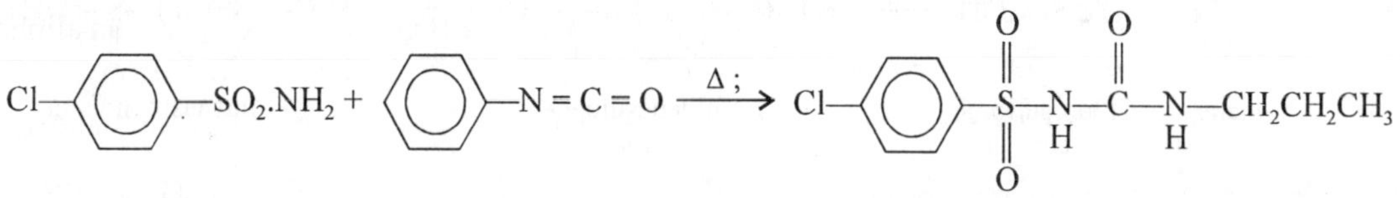

p-Chlorobenzene   Phenyl isocyanate   Chlorpropamide
Sulphonamide

The interaction between *p*-chlorobenzenesulphonamide and phenyl isocyanate in equimolar concentrations under the influence of heat undergoes *addition reaction* to yield the desired official compound.

The therapeutic application of this *'drug'* is limited to such subjects having a history of stable, mild to mderately severe diabetes melitus who still retain residual pancreatic β-cell function to a certain extent.

Mechanism of Action. The *'drug'* is found to be more resistant to conversion to its corresponding **inactive metabolites** than is **'tolbutamide'** ; and, therefore, it exhibits a much longer duration of action. It has also been reported that almost 50% of the *'drug'* gets usually excreted as metabolites, with the principal one being hydroxylated at the C-2 position of the *propyl-side chain.* *

---

*Thomas RC *et al. J Med Chem,* **15**, 964, 1972.

3.1.1.3. Tolazamide USAN, BAN, INN ;

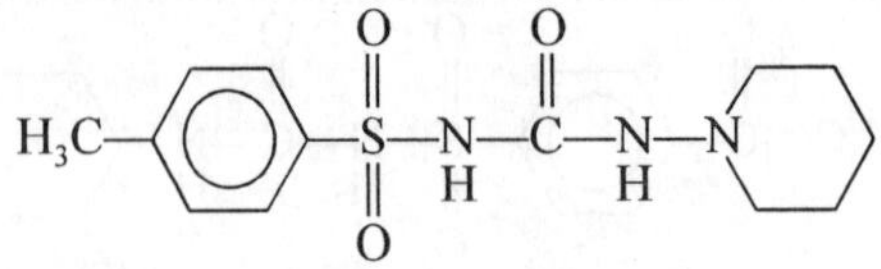

1-(Hexahydro-1H-azepin-1-yl)-3-(*p*-tolylsulphonyl) urea ; Tolinase$^{(R)}$;

It is found to be more potent in comparison to '*tolbutamide*', and is almost equal in potency to *chlorpropamide*.

Mechanism of Action. Based on the radiactive studies it has been observed that nearly 85% of an oral dose usually appears in the urine as its corresponding metabolites which were certainly more water-soluble than the parent tolazamide itself.

3.1.1.4. Tolbutamide :

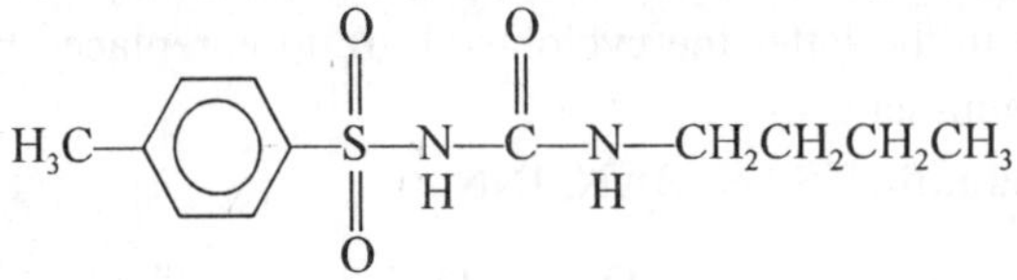

Benzenesulphonamide, N-[(butylamino) carbonyl]-4-methyl- ; Orinase$^{(R)}$;

Synthesis :

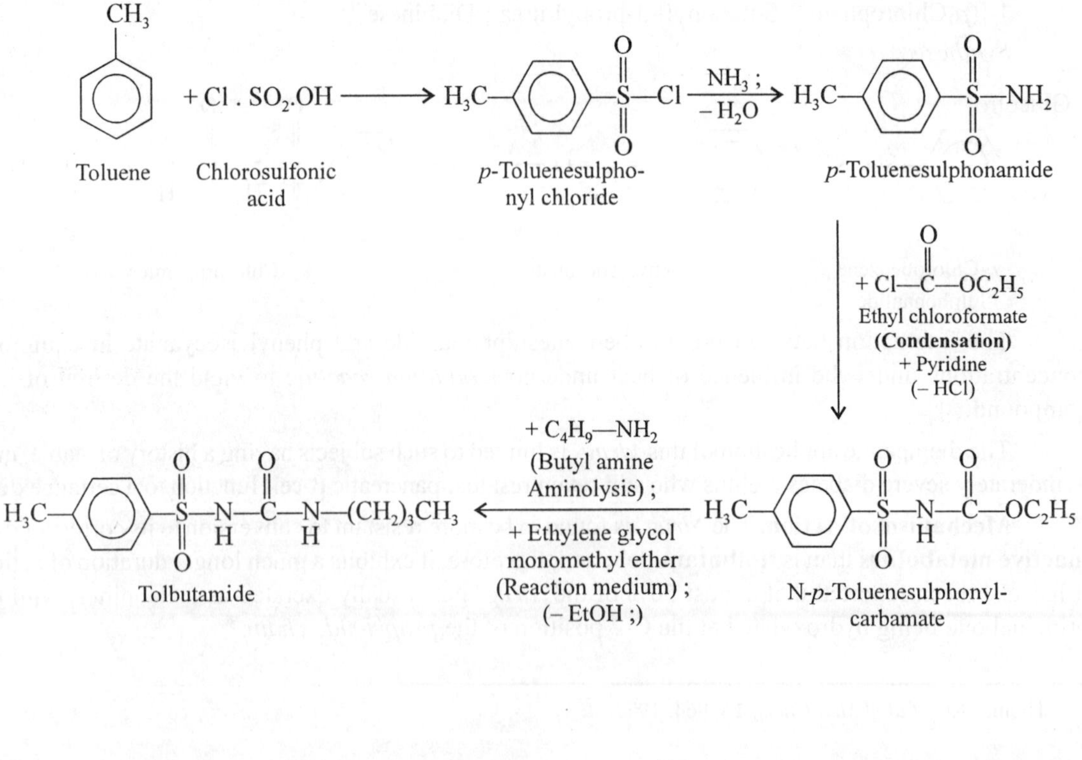

Toluene   Chlorosulfonic   *p*-Toluenesulpho-   *p*-Toluenesulphonamide
                 acid   nyl chloride

+ Cl—C—OC$_2$H$_5$
Ethyl chloroformate
(Condensation)
+ Pyridine
(– HCl)

Tolbutamide   + C$_4$H$_9$—NH$_2$
                     (Butyl amine
                     Aminolysis) ;
                     + Ethylene glycol
                     monomethyl ether
                     (Reaction medium) ;
                     (– EtOH ;)

N-*p*-Toluenesulphonyl-
carbamate

First of all toluene is treated with chlorosulfonic acid to yield *p*-toluenesulphonyl chloride, which on treatment with ammonia gives rise to the formation of *p*-toluenesulphonamide. The resulting product on condensation with ethyl chloroformate in the presence of pyridine produces N-*p*-toluenesulphonyl carbamate with the loss of a mole of HCl. Further aminolysis of this product with butyl amine using ethylene glycol monomethyl ether as a reaction medium loses a mole of ethanol and yields tolbutamide.

It is mostly beneficial in the treatment of selected cases of non-insulin-dependent diabetes melitus. Interestingly, only such patients having *some residual functional islet β-cells* which may be stimulated by this drug shall afford a positive response. Therefore, it is quite obvious that such subjects who essentially need more than 40 Units of insulin per day normally will not respond to this drug.

Mechanism of Action. The *'drug'* usually follows the major route of breakdown ultimately leading to the formation of butylamine and *p*-toluene sulphonamide respectively.

Importantly, the observed hypoglycemia induced by rather higher doses of the *'drug'* is mostly not as severe and acute as can be induced by **insulin ;** and, therefore, the chances of severe hypoglycemic reactions is quite lower with tolbutamide ; however, one may observe acute refractory hypoglycemia occasionally does take place. In other words, refractoriness to it often develops.

3.1.2. Second-Generation Sulfonylureas. The vital and important members of this class of compounds are, namely : Glipizide ; Glyburide ; and Glumepiride. These drug substances will be dealt with separately in the sections that follows :

3.1.2.1. Glipizide USAN, INN,

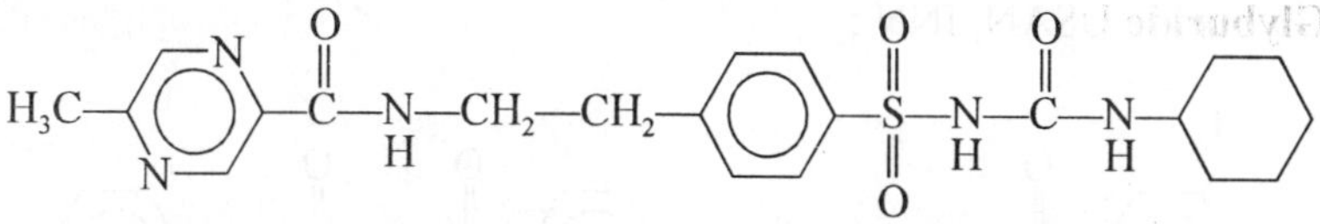

Pyrazinecarboxamide, N-[2-[4-[[[(cyclohexylamino) carbonyl] amino] sulfonyl] ethyl]-5-methyl-; Glucotrol$^{(R)'}$;

Synthesis :

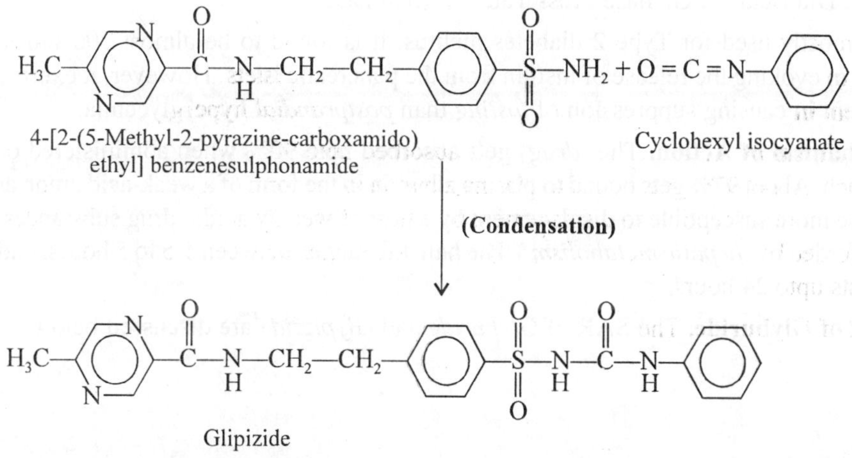

4-[2-(5-Methyl-2-pyrazine-carboxamido) ethyl] benzenesulphonamide

Cyclohexyl isocyanate

(Condensation)

Glipizide

Glipizide may be prepared by the condensation of 4-[2-(5-methyl-2-pyrazine-carboxamido)-ethyl] benzenesulphonamide with cyclohexylisocyanate in equimolar proportions.

It is employed for the treatment of **Type 2 diabetes mellitus** which is found to be 100 folds more potent than *tolbutamide* in evoking the pancreatic secretion of insulin. It essentially differs from other oral hypoglycemic drugs wherein the ensuing tolerance to this specific action evidently does not take place.

Mechanism of Action. The primary hypoglycemic action of this *'drug'* is caused due to the fact that it upregulates the insulin receptors in the periphery. It is also believed that it does not exert a direct effect on **glucagon secretion.**

The *'drug'* gets metabolized *via* oxidation of the cyclohexane ring to the corresponding *p*-hydroxy and *m*-hydroxy metabolites. Besides, a *'minor metabolite'* which occurs invariably essentially involves the N-acetyl structural analogue that eventually results, from the acetylation of the primary amine caused due to the hydrolysis of the amide system exclusively by **amidase enzymes.**

Note : The 'drug' enjoys *two* special status, namely :

(*a*) **Treatment of non-insulin dependent diabetes mellitus (NIDDM) since it is effective in most patients who particularly show resistance to all other hypoglycemic drugs ; and**

(*b*) **Differs from other oral hypoglycemic drug because it is found to be more effective during eating than during fasting.**

3.1.2.2. Glyburide USAN, INN ;

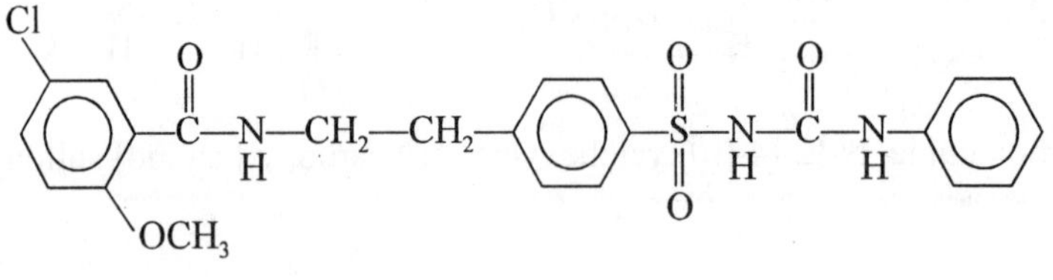

Benzamide, 5-chloro-N-[2-[4-[[[(Cyclohexylamino) carbonyl] amino] sulphonyl] phenyl] ethyl]-2-methoxy- ; Dia Beta[R] ; Glynase Press Tab[R] ; Micronase[R] ;

It is mostly used for Type 2 diabetes melitus. It is found to be almost 200 times as potent as *tolbutamide* in evoking the release of insulin from the pancreatic islets. However, it exerts a rather more effective agent in causing suppression of *fasting* than *postprandial* hyperglycemia.

Mechanism of Action. The *'drug'* gets absorbed upto 90% when administered orally from an empty stomach. About 97% gets bound to plasma albumin in the form of a weak-acid anion and, therefore, is found to be more susceptible to displacement by a host of weakly acidic drug substances. Elimination is mostly afforded by *'hepatic metabolism'*. The half-life ranges between 1.5 to 5 hours, and the duration of action lasts upto 24 hours.

SAR of Glyburide. The SAR of *Glyburide* and *Glypizzide* are discussed below :

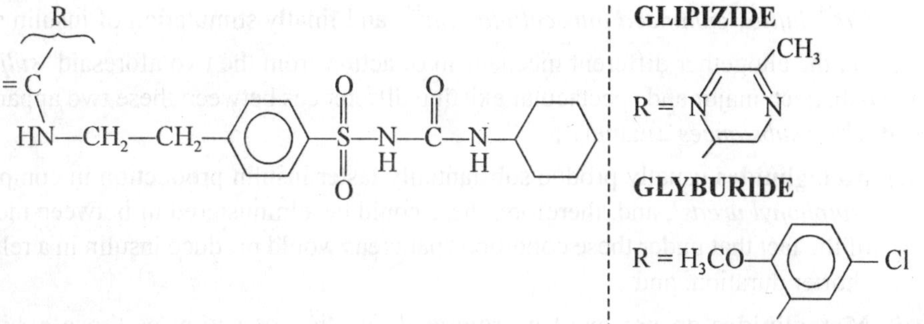

GLIPIZIDE

$R = $ (pyrazine ring with CH_3)

GLYBURIDE

$R = H_3CO$—(benzene ring)—Cl

DRUG	pKa	Potency Compared to Tolbutamide
Glipizide	5.9	100 times more potent
Glyburide	5.3	200 times more potent

Obviously the presence of 'R' in glyburide potentiates the hypoglycemic activity 200 times, whereas the heterocylic nucleus in glipizide potentiates 100 times in comparison to tolbutamide.

3.1.2.3. Glimepiride USAN, INN,

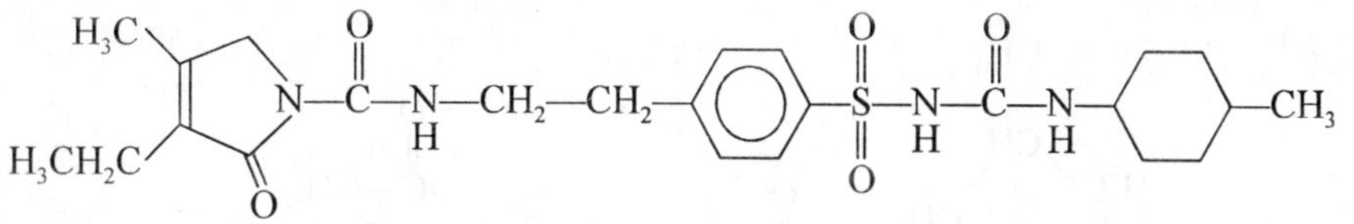

1-[[p-[2-(3-Ethyl-4-methyl-2-oxo-3-pyrroline-1-carboxamido) ethyl] phenyl] sulphonyl]-3 (*trans*-4-methylcyclohexyl) urea ; Amaryl[R] ;

Its hypoglycemic activity is very much akin to glipizide.

Mechanism of Action. The *'drug'* is found to be metabolized primarily through oxidation of the alkyl side chain attached to the pyrrolidine nucleus *via* a minor metabolic path that essentially involves acetylation of the amine function.

SAR of Glimepiride. The only major distinct difference between this *'drug'* and **glipizide** is that the former contains a five-membered **'pyrrolidine ring'** whereas the latter contains a six-membered **'pyrazine ring'**.

3.2. Non Sulfonylureas-Metaglinides

Metaglinides are nothing but non sulphonylurea oral hypoglucemic agents normally employed in the control and management of type 2 diabetes (*i.e.,* non-insulin-dependent diabetes mellitus, NIDDM). Interestingly, these agents have a tendency to show up a quick and rapid onset and a short duration of action. Just like the *'sulphonylureas'*, they also exert their action by inducing insulin-release from the prevailing functional pancreatic β-cells.

Importantly, the mechanism of action of the **'metaglinides'** is observed to differ from that of the *'sulphonylureas'*. In fact, the mechanism of action could be explained as under :

    (*a*) through binding to the particular receptors in the β-cells membrane that ultimately lead to the closure of ATP-dependent K^+ channels, and (*b*) K$^+$ channel blockade affords depolarizes the β-cell membrane, which is turn gives rise to *Ca^{2+} influx, enhanced intracellular Ca^{2+}*, and finally stimulation of insulin secretion.

Based on the altogether different mechanism of action from the two aforesaid *'sulphonylureas'* there exist *two* distinct, major and spectacular existing differences between these two apparantly similar categories of *'drug substances'*, namely :

    (*i*) **Metaglinides** usually produe substantially faster insulin production in comparison to the *'sulphonyl ureas'*, and, therefore, these could be administered in-between meals by virtue of the fact that under these conditions pancreas would produce insulin in a relatively much shorter duration, and

    (*ii*) **Metaglinides** do not exert a prolonged duration of action as those exhibited by the *'sulphonylureas'*. Its effect lasts for less than 1 hour whereas sulphonylureas continue to cause insulin generation for several hours.

Note : The glaring advantage of short duration of action by the metaglinides being that they possess comparatively much lesser risk of hypoglycemia in patients.

A few typical examples from this category are : repaglinide, nateglinide which would be treated as under :

3.2.1. Repaglinide USAN, INN,

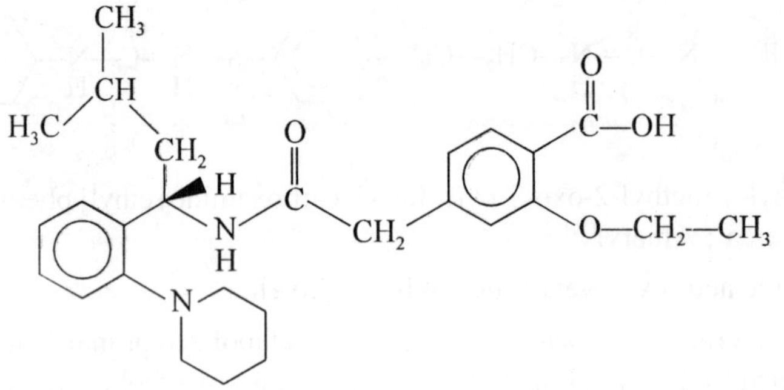

p-Tolu.ic acid, (+)-2-ethoxy-α-[[(S) α-isobutyl-*o*-piperidino-benzyl] carbamoyl]- ; Prandin$^{(R)}$;

It is used in the control and management of Type-2 diabetes mellitus. It must be taken along with meals.

Mechanism of Action. The *'drug'* is found to exert its action by stimulating insulin secretion by binding to and inhibiting the ATP-dependent K$^+$ channels in the β-cell membrane, resulting ultimately in an opening of Ca^{2+} channels. It gets absorbed more or less rapidly and completely from the GI tract ; and also is exhaustively metabolized in the liver by *two* biochemical phenomena, such as : (*a*) glucuronidation ; and (*b*) oxidative biotransformation. Besides, it has been established that the **hepatic cytochrome P-450 system 3A4** is predominantly involved in the ultimate metabolism of *repaglinide*. However, this specific metabolism may be reasonably inhibited by certain drug substances', for instance : *miconazole, ketoconazole, and erythromycin.*

3.2.2. Nateglinide

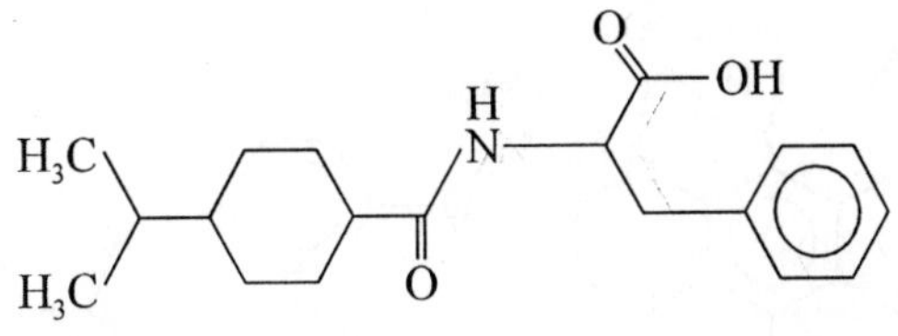

N-(4-Isopropylcyclohexanecarbonyl)-D-phenylalanine ; Starlix[R] ;

It is a phenylalanine structural analogue and belongs to the class of *'metaglinides'*. It is mostly employed in the control and management of type 2 diabetes.

3.3. Thiazolindiones

The **thiazolindiones** exclusively designate a distinct and novel nonsulphonylurea group of potent hypoglycemic agents that are used invariably for the treatment of NIDDM. However, these *'drugs'* essentially needs a **'functioning pancreas'** which may give rise to the reasonably adequate secretion of insulin from β-cells, very much akin to the sulphonylureas. It has been observed duly that insulin may be released in *'normal levels'* from the β-cells ; however, the peripheral sensitivity to this particular hormone may be lowered appreciably. It has been amply established that **'thiazolindiones'** are highly selective agonists for the peroxisome proliferator-activated receptor-*r* (PPAR*r*), that is primarily responsible for improving *'glycemic control'* exclusively *via* the marked and pronounced efficacy of **insulin sensitivity** in the *adipose tissue* and *muscles*. Besides, they also prevent and inhibit the prevailing **hepatic gluconeogenesis.** In short, one may add that **thiazolindiones** invariably help to normalize blood-sugar level in two ways : (*a*) through glucose metabolism ; and (*b*) through reduction of the amount of insulin required to accomplish glycemic control.

Note : These agents are effective exclusively in the presence of 'insulin'

A few typical examples belonging to this class of compounds shall be discussed in the sections that follows :

3.3.1. Rosiglitazone USAN,

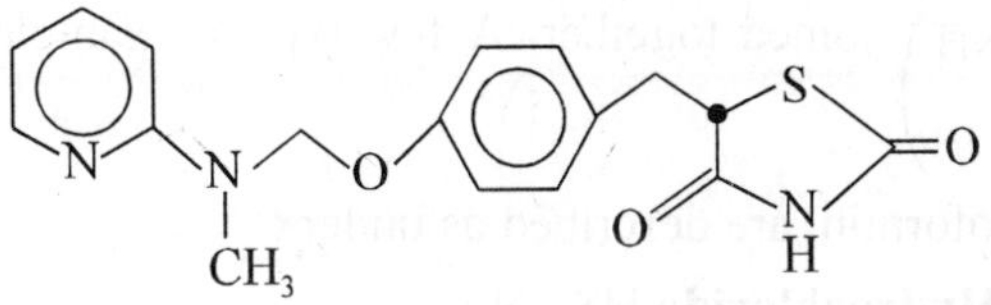

(±)-5-[[4-[2-(Methyl-2-pyridinylamino) ethoxy] phenyl] methyl]-2, 4-thiazolidinedione ; Avandia[R] ;

The *'drug'* has a single chiral centre (marked ●) and, therefore, exists as a racemate. Importantly, the enantiomers are found to be **'absolutely indistinguishable'** by virtue of their rapid *interconversion.*

3.3.2. Troglitazone :

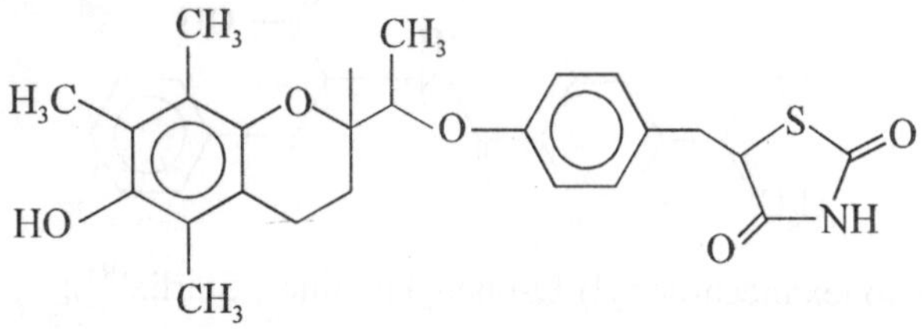

2, 4-Thiazolidinedione, (±)-5-[[4-[3, 4-dihydro-6-hydroxy-2, 5, 7, 8-tetramethyl-2H-1-benzopyran-2-yl) methoxy] phenyl] methyl]- ; Rezulin[R] ;

The *'drug'* improves the responsiveness to insulin in such patients that experience **Type 2 diabetes mellitus** problems of *insulin resistance* initiated and sustained by a *'unique mechanism of action'* which is fairly comparable with those of other similar drugs. Importantly, it is at present **only approved for use with insulin.**

Mechanism of Action. The *'drug'* exerts its action by decreasing blood glucose in diabetic patients having *hyperglycemia* by improving target organ response to insulin. Besides, in the presence of both exogenous and endogenous insulin the *'drug'* minimizes the hepatic glucose output, enhances insulin-dependent glucose uptake, and finally lowers fatty acid output in adipose tissue.

It also gets bound to the nuclear receptors usually termed as **peroxisome proliferator-activated receptors** (PPARs) which predominantly regulate solely the transcription of a host of **insulin-responsive genes** that are found to be critical to *'glucose'* and *'lipid'* metabolism.

Note : The *'drug'* is not an insulin secretagogue.

Troglitazone is highly bound (> 99%) to serum albumin. It gets metabolized solely in the liver to several **inactive compounds,** including a *sulphate-conjugate*—a major metabolite, and mostly excreted in the faeces.

3.4. Bisguanides :

The medicinal compounds included in this classification essentially comprise of two **'guanidine residues'** $\left(i.e., \text{H}_2\text{N}-\underset{\underset{\text{NH}_2}{|}}{\text{C}}=\text{NH} \right)$ joined together. A few typical examples belonging to this category, namely ; metoformin, phenoformin, are described as under :

3.4.1. Metoformin Hydrochloride USAN ;

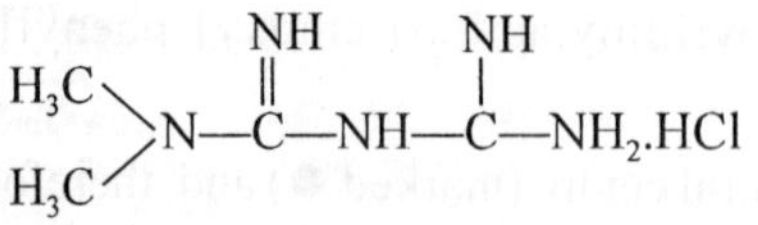

Imidodicarbenimidic diamide, N, N-dimethyl-, monohydrochloride ; Glucophage[R] ; Metiguanide[R] ;

It is used as an oral antihyperglycemic drug for the management of *Type 2 diabetes mellitus*. It is invariably recommended either as monotherapy or as an adjunct to diet or with a sulphonylurea (combination) to reduce blood-glucose levels.

Mechanism of Action. The *'drug'* is found to lower both basal and postprandial glucose. Interestingly, its mechanism of action is distinct from that of sulphonylureas and does not cause hypoglycemia. However, it distinctly lowers hepatic glucose production, reduces intestinal absorption of glucose, and ultimately improves **insulin sensitivity** by enhancing appreciably peripheral glucose uptake and its subsequent utilization. The *'drug'* is mostly eliminated unchaged in the urine, and fails to undergo hepatic metabolism.

3.4.2. Phenoformin

The *'drug'* is obsolete nowadays.

3.5. α-Glucosidase Inhibitors

It is quite well-known that the specific enzyme α-glucosidase is strategically located in the *brush-border* of the small intestine ; and, is exclusively responsible for affording cleavage of the dietary carbohydrates and thereby augmenting their rapid absorption into the body. Therefore, any mean by which the inhibition of this enzyme is affected would certainly permit less-dietary carbohydrate to be available for absorption ; and, hence, less available in the blood-stream soon after ingestion of an usual meal. It has been observed that the prevailing inhibitory characteristic features of such agents are maximum for **glycoamylase,** followed by sucrose, maltase and dextranase respectively.

A. few classical examples are discussed below :

3.5.1. Acarbose USAN, INN,

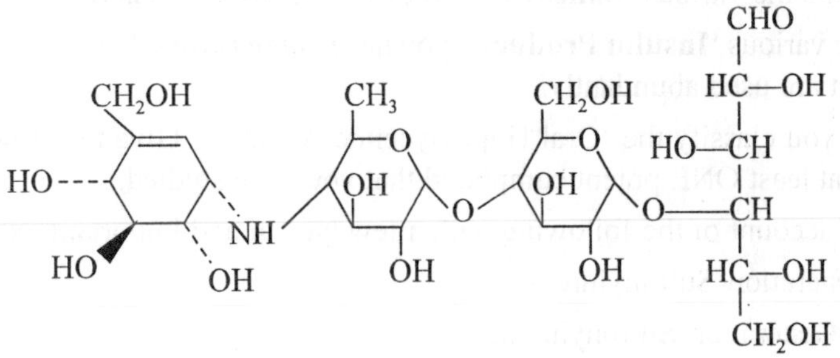

Glucose, *o*-4, 6-dideoxy-4-[[[15-(1α, 4α, 5(3, 6α)]-4, 5, 6-trihydroxy-3-(hydroxymethyl)-2-cyclohexen-1-yl] amino]-α-*o*-glucopyranosyl-(1-4)-*o*-α-D-glucopyranosyl-(1-4)- ; Precose[R] ;

It is used in the control and management of **Type 2 diabetes mellitus.**

Mechanism of Action. The *'drug'*, which is obtained from the microorganism *Actinoplanes utahensis,* is found to a complex oligosaccharide that specifically delays digestion of indigested carbohydrates, thereby causing in a smaller rise in blood glucose levels soonafter meals. It fails to increase insulin secretion ; and its antihyperglycemic action is usually mediated by a sort of competitive, reversible **inhibition of pancreatic α-amylase membrane-bound intestinal α-glucosidase hydrolase enzymes.**

The *'drug'* is metabolized solely within the GI tract, chiefly by intestinal bacteria but also by diagestive enzymes.

3.5.2. Miglitol USAN, INN

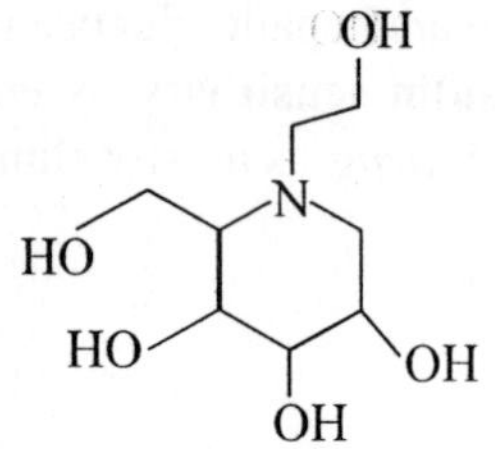

1-(2-Hydroxyethyl)-2-(hydroxy-methyl)-[2R-(2α, 3β, 4α, 3β)]-piperidine ; Glyset$^{(R)}$;
It also lowers blood-glucose level.

Mechanism of Action. It resembles closely to a sugar, having the heterocyclic nitrogen serving as an isosteric replacement of the *'sugar oxygen'*. The critical alteration in its structure enables its recognition by the α-glycosidase as a substrate. The ultimate outcome is the overall competitive inhibition of the enzyme which eventually delays complex carbohydrate absorption from the ensuing GI tract.

Probable Questions for B. Pharm. Examinations

1. (*a*) What are type-I and type-II 'Diabetes' ? Explain with some typical examples.

   (*b*) Enumerate the various **'Salient Features'** of the *Insulin-Primary Structure.*

2. What are the various **'Insulin Products'** you have come across ? Discuss briefly any FIVE such products that are used abundantly.

3. How would you classify the 'Oral Hypoglycemic Agents' ? Give the structure, chemical name and uses of at least ONE potent compound that you have studied.

4. Give a brief account of the following with a few typical and important examples :

   (*a*) First-Generation Sulfonylureas

   (*b*) Second-Generation Sulfonylureas

5. How would synthesize the following 'Drugs' ? Explain the course of reaction(s) involved in the synthesis.

   (*i*) Chlorpropamide

   (*ii*) Tolbutamide

   (*iii*) Glipizide

6. Explain the following :

   (*i*) **Glyburide** is 200 times more potent than Tolbutamide.

   (*ii*) **Glipizide** is 100 times more potent than Tolbutamide.

   (*iii*) SAR of Glymepiride

   (*iv*) Mechanism of action of Tolbutamide.

7. (*a*) Discuss the **'Metaglinides'** with regard to their specific 'mechanism of actions'.

   (*b*) Give the structure and uses of any ONE of the following drugs :

   (*i*) Repaglinide

   (*ii*) Nateglinide.

8. Give a comprehensive account on **'Thiazolindiones'** with specific reference to the following potent **drugs** :

   (*a*) Rosiglitazone

   (*b*) Troglitazone.

9. Write a short note on the following **'oral hypoglycemic agents'** :

   (*a*) Bisguanides ; and

   (*b*) α-Glucosidase Inhibitors.

RECOMMENDED READINGS

1. Cook NS : *Potassium Channels : Structure, classification, Function and Therapeutic Potential,* John Wiley and Sons, New York, 1990.

2. Meisheri KD : *Direct Acting Vasodilators : In Singh BJ et al. (eds.) Cardiovascular Pharmacology,* Churchill Livingstone, New York, 1994.

3. Gennaro AR : *Remington : The Science and Practice of Pharmacy,* Lippincott Williams and Wlikins, Vol. II, 20th edn., 2004.

22 Steroids

1. INTRODUCTION

The **steroids** constitute a group of structurally related compounds that are widely distributed both in the plant and the animal kingdom. The basic nucleus of these physiologically potent an biochemically dynamic medicinal compounds do possess a more or less similar stereochemical relationship. The *steroids*, in genreal, have been found to contain either the partly or completely hydrogenated 17H-cyclopenta-phenanthrene nucleus.

The **steroids** include a broad-spectrum of important compounds which exhibit remarkable pharmacodynamic properties, namely : adrenal cortical hormones, sex hormones, cardiac glycosides, antirachitic vitamins (Vitamin D), toad poisons, saponins, bile acids and some alkaloids.

Broadly speaking both steroid hormones and related structural analogues constitute and designate one of the most abundantly employed categories of pharmacologically active and potént agents. These *'medicinal compounds'* are invariably used as first in importance in the control and management of birth control, inflammatory conditions, hormone-replacement theraphy (HRT), and above all in the treatment of neoplastic diseases (cancer). Interestingly, the plethora of these agents are exclusively based on a specific common structural nucleus usually termed as the'**steroid backbone'**. However, the different steroidal variants essentially attribute to the specific and unique molecular targets.

2. STEROID NOMENCLATURE, NUMBERING, DOUBLE BONDS AND STEREOCHEMISTRY

The general formula for the basic structure of the above cited compounds may be represented as follows :

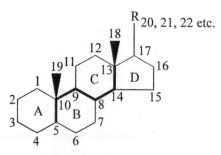

Fig. 1. General Steroid Formula.

600

The rings are conventionally lettered and numbered as indicated above. However, in actual conformation the basic structure of steroid is not planar. It has also been observed that in the naturally occurring steroidal compounds the substitutions in the rings usually occur at C-3, C-7 and C-11 positions.

According to the standard convention the direction of projection from the plane of the ring system of substituting groups located at centres of asymmetry is usally designated by the Greek letters α and β.

The α-substituting group is viewed as projecting beneath the ring plane and is conventionally represented by a broken line (dotted line).

The β-substituting groups is viewed as projecting above the ring plane and is normally represented by a solid line.

It has been observed that all the steroids on dehydrogenation with selenium at 360°C usually yield Diel's hydrocarbon, i.e., 3′-methyl-1:2-cyclopentanophenanthrene, whereas at 420°C, the steroids give mainly chrysene and a small amount of picene.

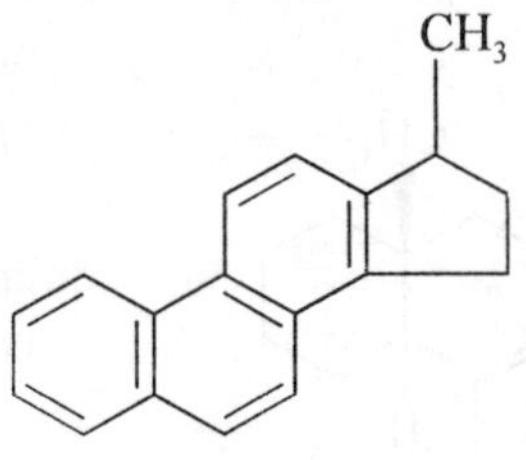

Diel's
hydrocarbon

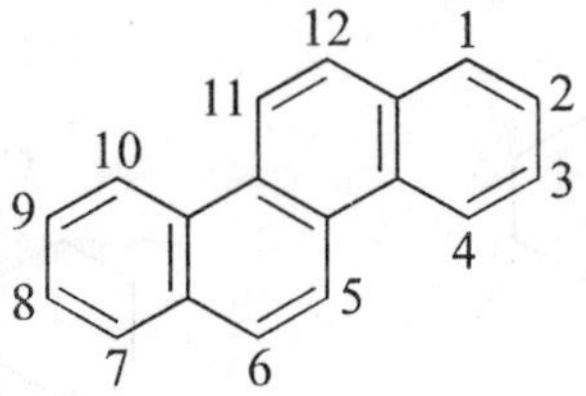

Chrysene
Or
1 : 2-Benzphenanthrene

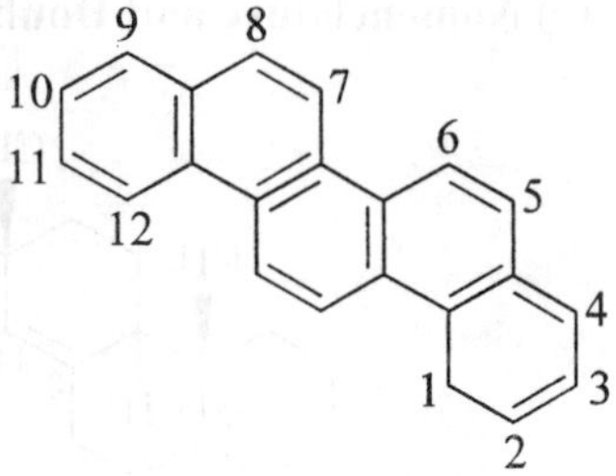

Picene
Or
1 : 2 : 7 : 8-Dibenzphen-
anthrene

A few typical examples of 'steroidal drugs' together with their *nomenclature* and *numbering* are illustrated below :

(*a*) **Common and Systematic Nomenclature :**

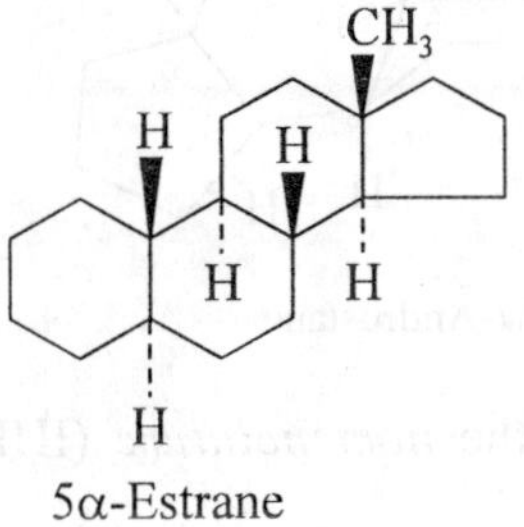

5α-Estrane

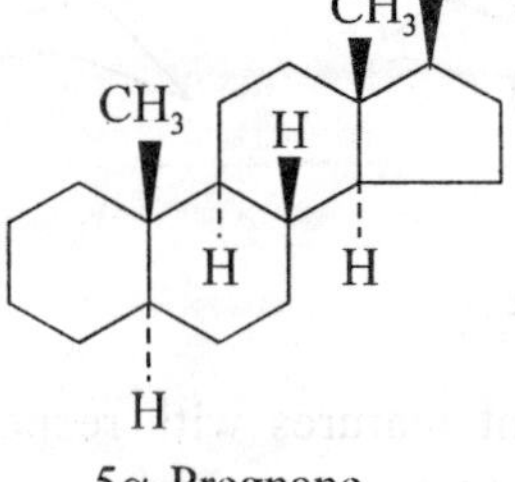

5α-Pregnane

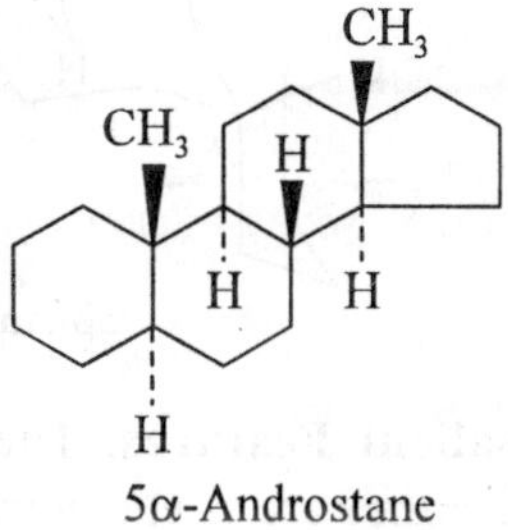

5α-Androstane

(b) Nomenclature and Numbering :

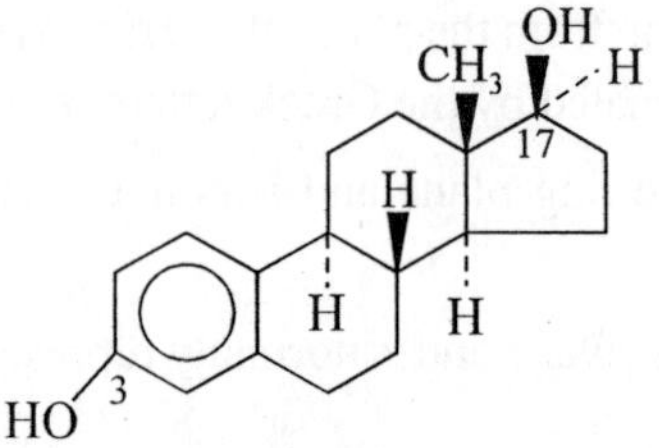

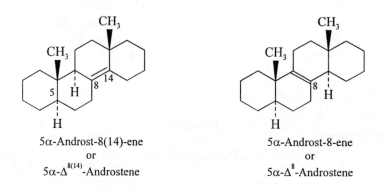

17β-Estradiol
[Estra-1, 3, 5(10)-triene-
3, 17β-diol]

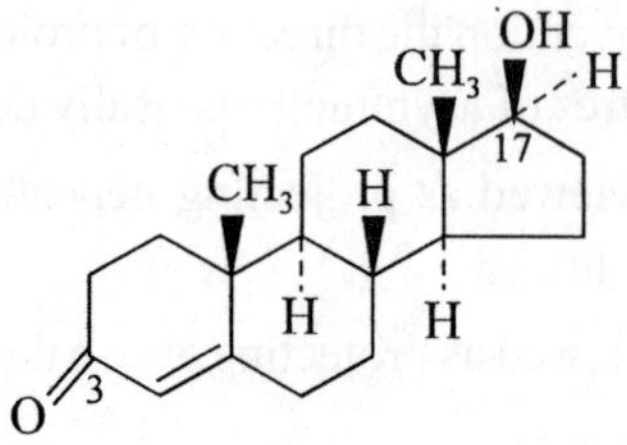

Testosterone
[17β-Hydroxyandrost-
4-en-3-one]

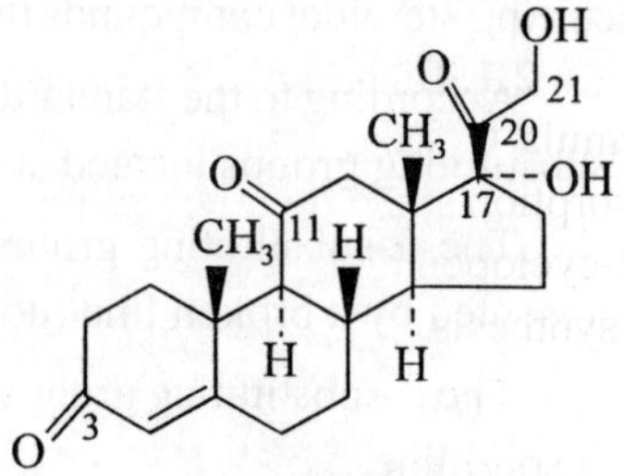

Cortisone
[17, 21-Dihydroxypregn-4-ene
− 3, 11, 20-trione]

(c) Nomenclature and Double Bonds :

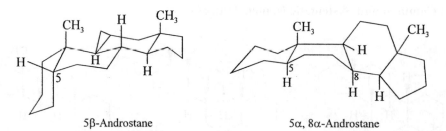

5α-Androst-8(14)-ene
or
5α-Δ$^{8(14)}$-Androstene

5α-Androst-8-ene
or
5α-Δ^{8}-Androstene

(d) Nomenclature and Stereochemistry :

5β-Androstane

5α, 8α-Androstane

Salient Features. The salient features with respect to the *nomenclature* (IUPAC) and *stereochemistry* are as enumerated below :

(1) Stereochemistry of the H-atom at C-5 is invariably incorporated in the 'name' itself,

(2) Stereochemistry of other H-atoms is **not** usually indicated unless and until it essentially happens to differ from **5 α-cholestane,** and

(3) Altering the stereochemistry at any of the '*ring-juncture*' with a heavy-dark line (see 'general steroid formula') changes immensely the prevailing '**shape of the steroid**', as may be observed in the above cited examples of 5β-androstane and 5α, 8α-androstane.

2.1. Diel's Hydrocarbon. It is a solid substance having a melting point 126-127°C and a molecular formula $C_{18}H_{16}$. Based on the results of oxidation reactions, X-rays crystal analysis coupled with absorption spectrum measurements it was revealed that the hydrocarbon in question could be 3′-methyl-1:2-cyclopentanophenanthrene. The next essential step was to establish the structure of this compound by synthesis, *e.g.*, that of Harper *et al* (1934) who used the Bogert-Cook method commencing from :

    (*i*) 2-(1-naphthyl)-ethyl-magnesium bromide

    (*ii*) 2:5-dimethylcyclopentanone

2-(1-Naphthyl)-ethyl-magnesium bromide and 2:5-dimethyl-cyclopentanone react to give a condensed product which on oxidation with phosphorus pentoxide at 140°C and subsequent distillation under reduced pressure yields an intermediate. This undergoes cyclization first and later on when distilled with selenium gives the Diel's hydrocarbon.

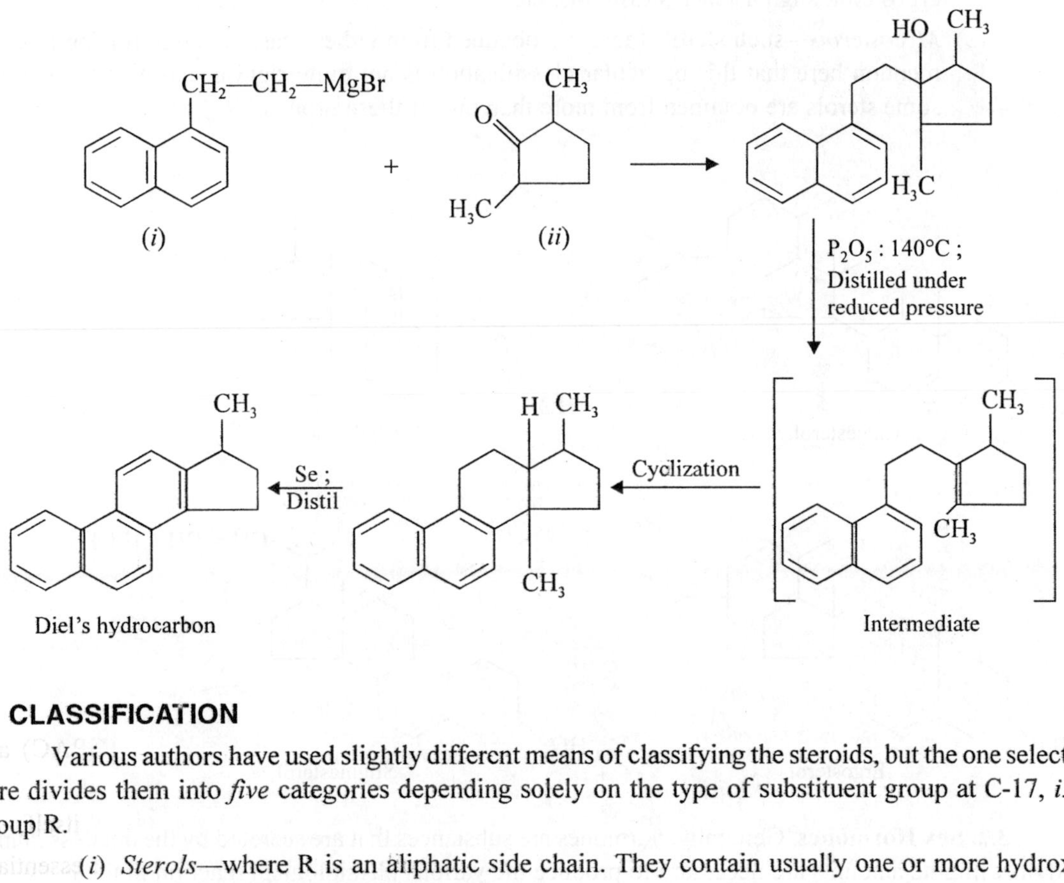

Diel's hydrocarbon   Intermediate

3. CLASSIFICATION

Various authors have used slightly different means of classifying the steroids, but the one selected here divides them into *five* categories depending solely on the type of substituent group at C-17, *i.e.*, group R.

    (*i*) *Sterols*—where R is an aliphatic side chain. They contain usually one or more hydroxyl groups attached in alicyclic linkage.

    (*ii*) *Sex Hormones*—where R bears a ketonic or hydroxyl group and mostly possesses a two-carbon side chain.

(*iii*) *Cardiac Glycosides*—where R is a lactone ring. The glycosides also contain sugars linked through oxygen in other parts of the molecule. Normally on hydrolysis it yields this sugar together with the cardiac aglycone.

(*iv*) *Bile Acids*—where R is essentially a five-carbon side chain ending with a carboxylic acid moiety.

(*v*) *Sapogenins*—where R contains an oxacyclic (ethereal) ring system.

3.1. Sterols. The term sterols has been coined from the words '*ster*oidal alc*ohols*'. They have been found to occur both in animal and plant oils and fats. These are usually crystalline compounds and mostly bear an alcoholic group. They may occur either as free or as esters of the higher fatty acids, and are isolated from the unsaponifiable fraction of oils and fats.

The sterols may be further sub-divided into the following three categories, namely :

(*a*) *Zoosterols*—such sterols those are obtained from the animal kingdom only, *e.g.*, cholesterol, cholestanol, coprostanol (coprosterol), etc.

(*b*) *Phytosterols*—such sterols those are derived exclusively from the plant sources, *e.g.*, ergosterol, stigmasterol, sitosterols, etc.

(*c*) *Mycosterols*—such sterols those are obtained from either yeast or fungi. It is pertinent to mention here that this particular classification is not quite rigid because of the fact that some sterols are obtained from more than one of these groups.

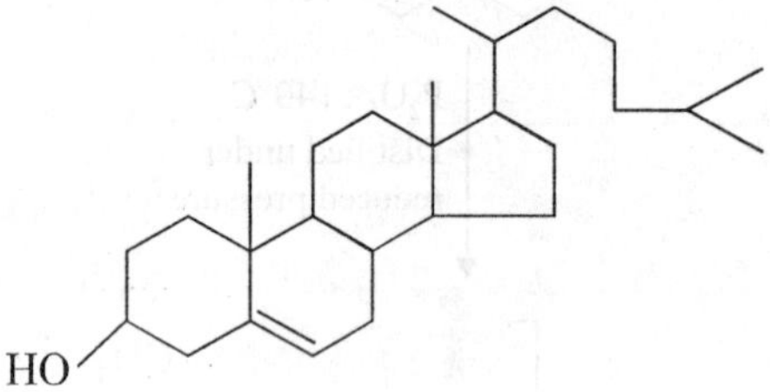

Cholesterol

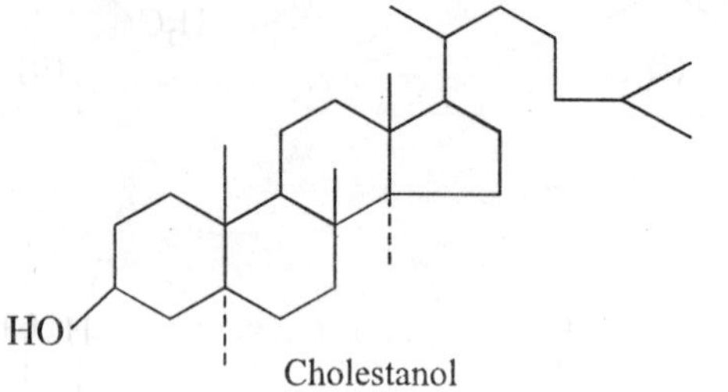

Cholestanol

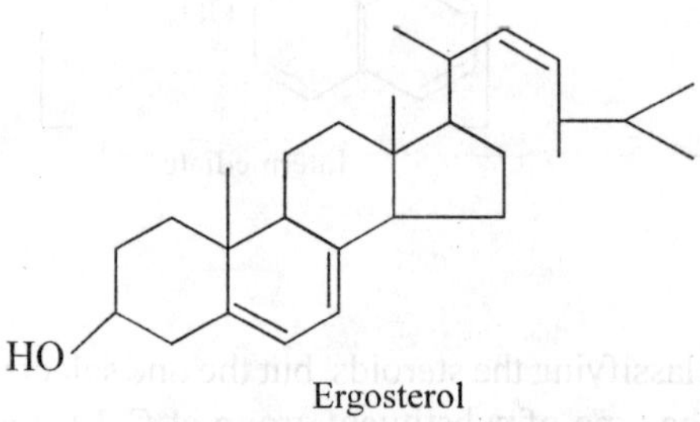

Ergosterol

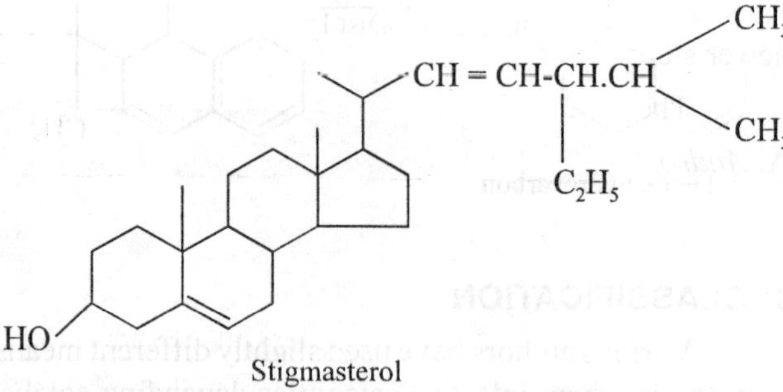

Stigmasterol

3.2. Sex Hormones. Generally, hormones are substances that are secreted by the ductless glands, and only minute amounts are necessary to produce the various physiological reaction in the body.

However, the sex-hormones belong to the steroid class of compounds and are produced in the gonads, *i.e.,* testes in the male and ovaries in the female. In fact, their activity seems to be controlled and monitored by the hormones that are produced in the anterior lobe of the pituitary glands. Perhaps because of this inherent characteristies the sex hormones are invariably termed as the secondary sex hormones and the hormones of the anterior lobe of the pituitary are called the primary sex-hormones.

A general survey of the literature stretching over the past three decades would reveal that a vast number of structural modifications of the steroid hormones have taken place. These newer compounds have been prepared with a view to enhance their biological activities, oral activity and duration of action, besides attributing better solubility properties, minimising the requirement for some essential perimeter functional group of the parent hormone and lastly to effect a marked separation of their biological activities.

These modifications have been duly accomplished through a number of means, for instance, protecting some vital moieties against the metabolic attack or attack by intestinal bacteria, prevention of the conversion *in vivo* of one steroid hormone into another steriod and lastly through alteration of physical properties by preparing their respective 19-nor analogues, ester derivatives, enol ethers, acetals and ketals, bringing about conformational changes and electron attracting effect.

3.2.1. Classification. Sex-hormones are usually classified under the following *three* heads, namely :

(*i*) Androgens (Male Hormones) *e.g.,* androsterone, testosterone.

(*ii*) Oestrogens (Female or Follicular Hormones), *e.g.,* oestrone, oestriol, oestradiol, stilbesterol, hexesterol.

(*iii*) Gestogens (The Corpus Luteum Hormones) *e.g.,* progesterone.

3.2.2. Androgens. Experiments with testicular extract has more or less enjoyed a chequered career. Veronoff successfully transplanted tested from monkeys into elderly men and claimed to rejuvenate them. Likewise, ligature of *vas deferens* which is known to cause atrophy of spermatogenic tissue and indirectly hypertrophy of intestenal tissue which secrete testosterone. Another researcher Steinach carried out similar studies by ligaturing the *vas deferens* and obtained identical results.

Androgens besides showing a specific action on gonads, also stimulate production of elements that are absolutely essential for all tissue growth. This characteristic which the androgens share with other steroidal hormones such as corticosterones and oestrogens has paved the way towards synthesis of newer steroids that possess mainly metabolic acitivity without androgenic effect.

The *two* most important androgens are androsterone and testosterone.

A. *Androsterone*

3α-Hydroxy-5α-androstan-17-one.

Androsterone (m.p. 185°C) is a naturally occurring androgen that may be isolated from male urine. It can also be synthesized from epi-cholestanyl acetate as given below :

Synthesis :

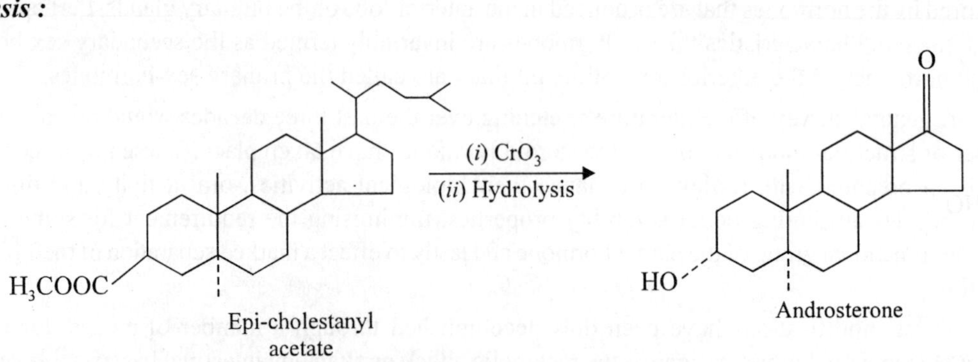

(*i*) CrO₃
(*ii*) Hydrolysis

H₃COOC

Epi-cholestanyl
acetate

HO

Androsterone

It may be synthesized from epi-cholestanyl acetate at two stages : *first,* by its oxidation with chromium-6-oxide and *secondly,* by subjecting the resulting product to hydrolysis to yield the desired product.

Butanandt and co-workers (1931) first isolated androsterone (15 mg) from 15,000 litres of urine.

B. *Testosterone* INN, BAN, USAN,

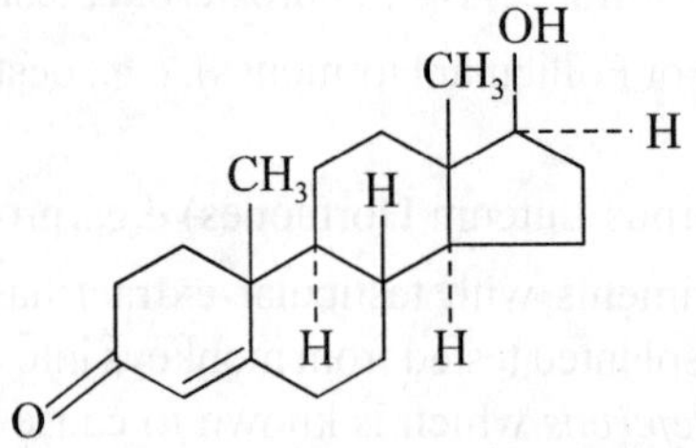

Testost. 17β-Hydroxyandrost-4-en-3-one ; Androst-4- en-3-one, 17-hydroxy-, (17β)- ; BP, USP ; Synadrol F⁽ᴿ⁾ (Pfizer) ; Mertestate⁽ᴿ⁾ (Sterling).

Synthesis :

Testosterone may be synthesized from the following two starting materials, namely ;

    (*i*) From Cholesterol, Butenandt (1935) ; Ruzica (1935) ; Oppenauer (1937) ;

    (*ii*) From Dehydroepiandrosterone, Mamoli (1938).

(*a*) *From Cholesterol*

Cholesteryl acetate dibromide is first prepared by the acetylation of chloesterol and its subsequent bromination. This on oxidation with chromium-6-oxide reduces the 8-carbon side chain at C-17 to a mere CO moiety, which on reduction followed by hydrolysis yields dehydroepiandro-sterone. The resulting product on acetylation protects the acetyl moiety at C-3 and treatment with sodium propoxide introduces a hydroxy group at C-17. Benzoylation followed by mild hydrolysis causes the reappearances of free OH moiety at C-3 and a benzoxy function at C-17. Oppeanauer oxidation cuased by refluxing the resulting secondary alcohol with aluminium tertiary butoxide in excess of acetone affords a ketonic function at C-3, which upon hydrolysis in an alkaline medium yields the official compound.

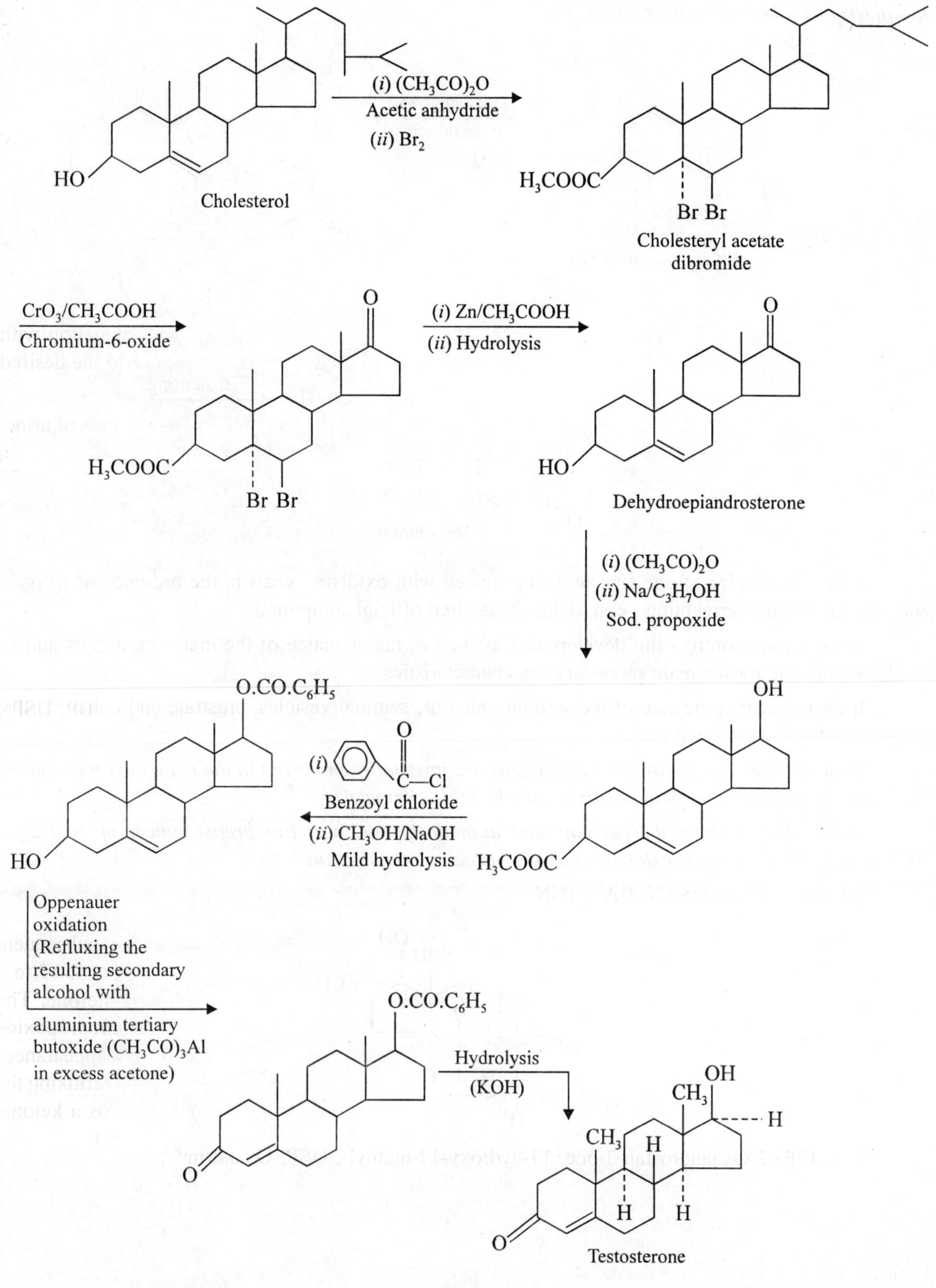

HO
Cholesterol
(i) (CH3CO)2O
Acetic anhydride
(ii) Br2
H3COOC
Br Br
Cholesteryl acetate dibromide
CrO3/CH3COOH
Chromium-6-oxide
O
H3COOC
Br Br
(i) Zn/CH3COOH
(ii) Hydrolysis
O
HO
Dehydroepiandrosterone
(i) (CH3CO)2O
(ii) Na/C3H7OH
Sod. propoxide
OH
H3COOC
(i) Benzoyl chloride
(ii) CH3OH/NaOH
Mild hydrolysis
O.CO.C6H5
HO
Oppenauer oxidation
(Refluxing the resulting secondary alcohol with aluminium tertiary butoxide (CH3CO)3Al in excess acetone)
O.CO.C6H5
O
Hydrolysis
(KOH)
OH
CH3
H
CH3
H
H
H
O
Testosterone

(b) From Dehydroepiandrosterone

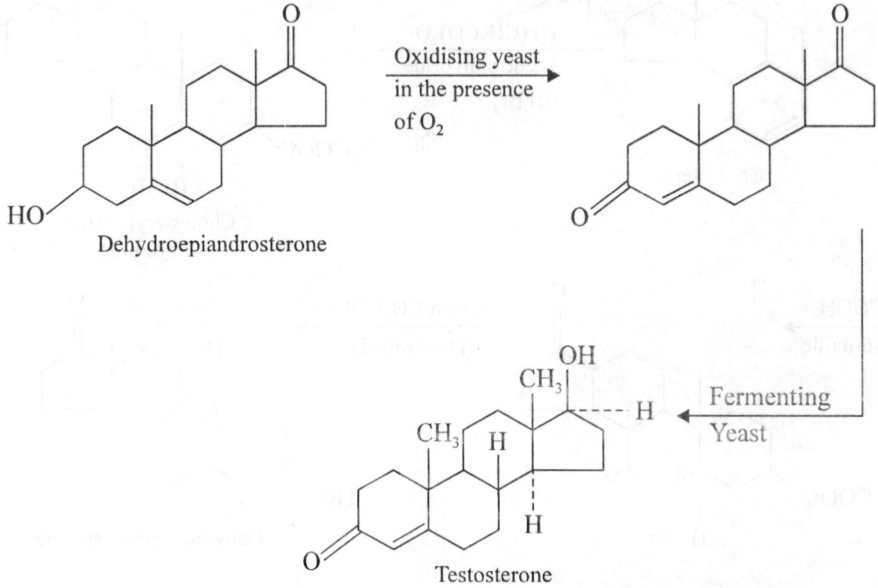

Dehydroepiandrosterone

Oxidising yeast
in the presence
of O_2

Fermenting
Yeast

Testosterone

Dehydroepiandrosterone first on being treated with oxidising yeast in the presence of oxygen and secondly by the fermenting yeast yields the desired official compound.

Testosterone controls the development as well as maintenance of the male sex organs and is solely responsible for the male secondary sex characteristics.

It also increases the size of the serotum, phallus, seminal vesicles, prostrate and enhances the sexual activity in adolescent males.

Testosterone along with other androgens are invariably employed in the male for replacement therapy in hypogonadism, eunuchoidism, and the male climacteric.

Dose : For prolonged treatment, subcutaneously, 600 mg; For breast cancer up to 1.5g ; Alternatively 10 to 30 mg per day through the buccal administration.

(c) Oxandrolone USAN, BAN, INN

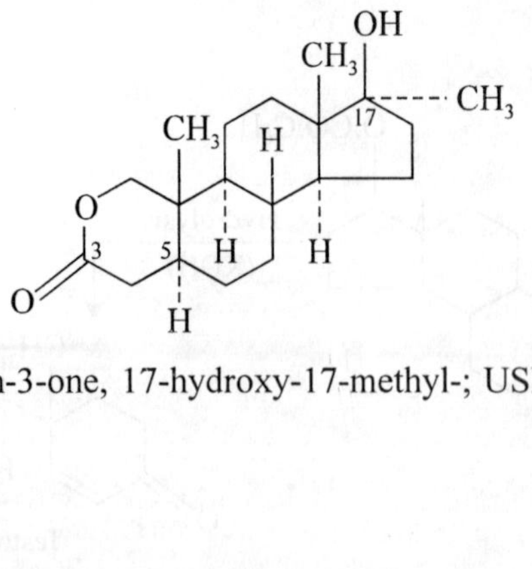

(5α, 17β)-2-Oxaandrostan-3-one, 17-hydroxy-17-methyl-; USP; Oxandrin[R];

Synthesis :

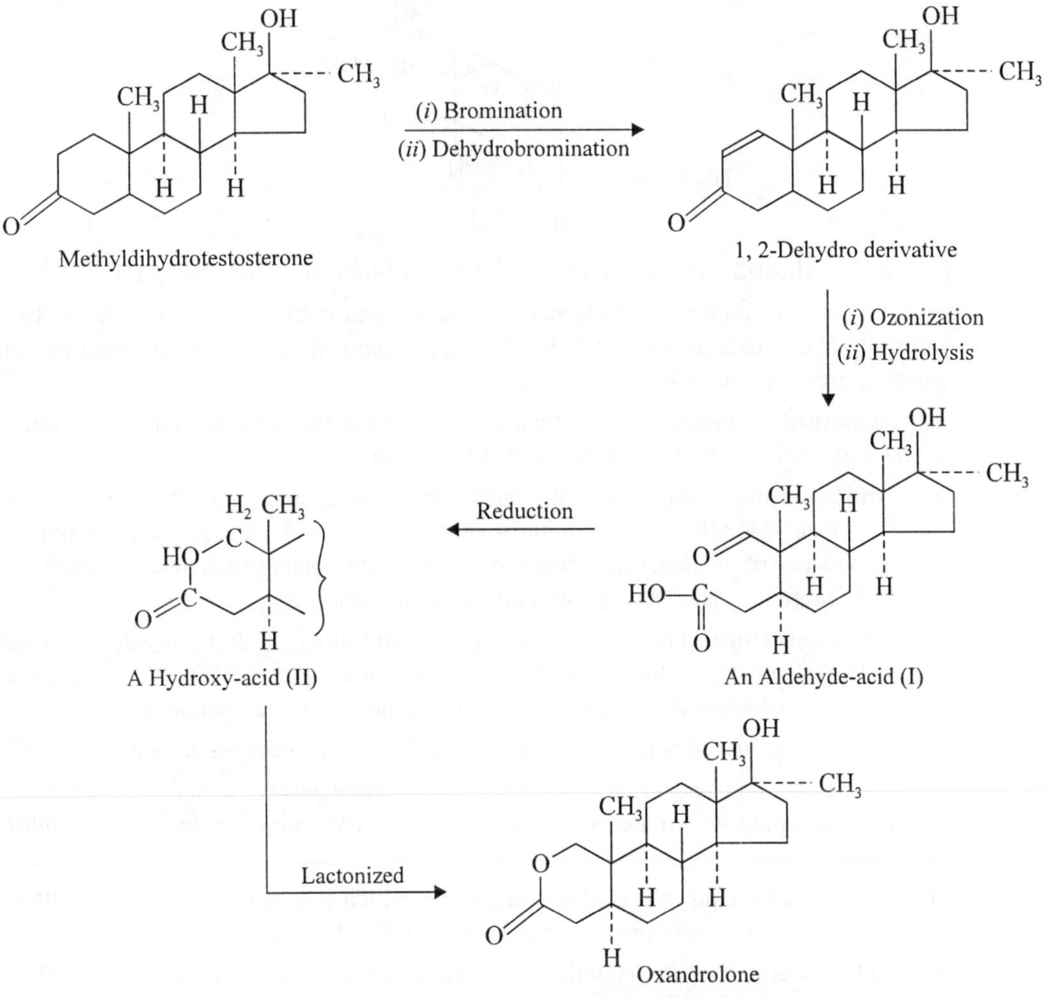

Methyldihydrotestosterone

(i) Bromination
(ii) Dehydrobromination

1, 2-Dehydro derivative

(i) Ozonization
(ii) Hydrolysis

Reduction

A Hydroxy-acid (II)

An Aldehyde-acid (I)

Lactonized

Oxandrolone

Methyldihydrotestosterone on being subjected to bromination followed by dehydrobromination gives rise to the formation of 1, 2-dehydro derivative, which upon ozonization and hydrolysis yields an aldehyde-acid (I). The resulting acid (I) on reduction produces the corresponding hydroxy-acid (II) which when lactonized produces the desired compound, oxandrolone.

It is an androgenic steroid having comparatively higher *anabolic activity* in relation to the *androgenic activity.* Hence, it is used mostly to promote nitrogen anabolism (protein synthesis) and weight-gain in cachexia* and other debilitating diseases and after serious infections, burns, trauma or surgical procedures. It may also be employed to relieve pain in some types of **osteoporosis** thereby augmenting Ca^{2+} retention and hence improving the condition of bone. It also finds its application for its predominant erythropoetic effects in the treatment of both *hypoplastic* and *aplastic anemias.***

*A state of ill-health, malnutrition, and wasting (*e.g.,* chronic diseases, certain malignancies and advanced pulmonary tuberculosis.

**Hypoplastic i.e.,* aplastic anemia ; *Aplastic :* Anemia caused by deficient red cell production due to bone-marrow disorders.

D. *Stanozolol* USAN, BAN, INN

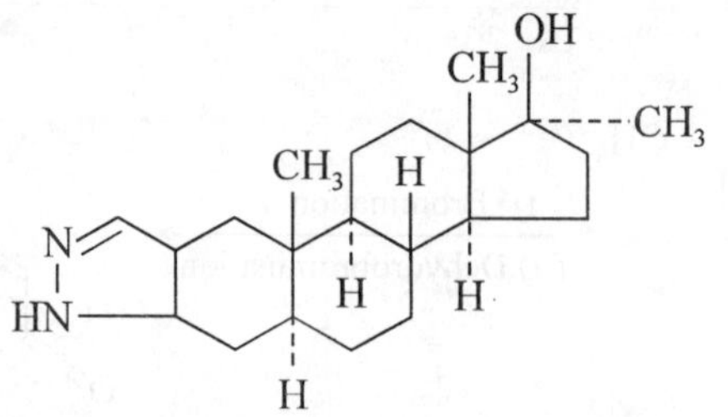

(5α, 17β)-2′H-Androst-2-enol [3, 2-C] pyrazol-17-ol, 17-methyl ; USP ; Winstrol[R] ;

It is an androgen having comparatively *strong anabolic* and weak *androgenic activity.* Its uses are almost identical to that of oxandrolone. Besides, it is also employed in the prophylaxis of hereditary angiodema, which is presently the only approved use.

3.2.2.1. Mechanism of Action. The mechanism of action of the various compounds described under section 22.3.2.2. shall now be treated individually as under :

3.2.2.1.1. Androsterone. Being one of the naturally occurring androgens it exerts widespread anabolic effects. The '*drug'* also affects hypopituitarism and with Addison's disease, relief of impotence not associated with evidence of testicular underactivity, pituitary dwarfism to accelerate growth, and in functional dysmenorrhea giving relief through an antiestrogenic action.

3.2.2.1.2. Testosterone. The '*drug'* undergoes metabolism that may lead to either pharmacologically **active steroids** *e.g.,* estradiol, 5α-dihydrotestosterone (OR 5α-DHT), and androsterone ; or to **inactive steroids** *e.g.,* 6α-hydroxytestosterone, epitestosterone, and etiocholanolone.*

However, the enzyme **5α-reductase** brings about the following changes, namely :

(*a*) In prostate gland (an androgen target tissue) testosterone gets converted to 5α-DHT, which enjoys the reputation of being the *most potent endogenous androgen metabolite of testosterone, and*

(*b*) It helps to catalyze an irreversible reaction for which it essentially needs NADPH as a cofactor that strategically provides the H-atom at C-5,**

It is found to be not effective when administered orally as it almost gets destroyed in the liver on absorption. Its plasma half-life ranges between 10-20 minutes.

3.2.2.1.3. Oxandrolone. The '*drug'* exerts its action by virtue of its inherited protein catabolism associated with long-term usage of corticosteroid. Besides, it is also indicated in HIV wasting syndrome and alcoholic hepatitis.

Note. Strictly speaking it is not a 'steroid', and its configuration is that of a 17-methyl androgenic steroid.

3.2.2.1.4. Stanzolol. The '*drug'* acts by significantly lowering the frequency and severity of attacks in angioedema ; and it is now the only approved application.

3.2.2.2. Derivatives of Testosterone. There are, in fact, quite a few improtant derivatives of testosterone that have been used extensively in therapy ; and these are summarized in Table 1.

---

*RW Brueggemier, Burger's Medicinal Chemistry, 5th edn, Vol : 3, ME Wolff ed., John Wiley & Sons, NewYork, pp 445-510, 1996.

P Ofner, *Vit Horm.,* **26, 237, 1968.

Table 1. Derivatives of Testosterone

Approved Names	Official Status	Proprietary Names	Dose
Testosterone Acetate	—	Cetovister(R) (Substancia, Spain)	—
Testosterone Cypionate	USP ;	dep Andro 100(R) (Forest)	50 to 200 mg/ml in oil solution
Testosterone Decanoate	BP ;	—	—
Testosterone Enanthate	BP ; USP ;	Delatestryl(R) (Squibb)	100 to 400 mg every 2 to 4 weeks
Testosterone Isocaproate	BP ;	—	—
Testosterone Ketilaurate	—	—	—
Testosterone Phenylacetate	—	—	—
Testosterone Phenylpropionate	BP ;	Tess PP(R) (Organon)	—
Testosterone Propionate	BP ; USP ; Eur. P. ; Int. P. ; IP. ;	Synadrol(R) (Pfizer)	5 to 20 mg daily as buccal tablets
Testosterone Undecanoate	—	Restandol(R) (Organon, UK)	40 to 160 mg daily

3.2.3. Oestrogens

The oestrogens are mainly concerned with growth and function of the sex organs.

In general, they are classified under *two* sub-heads, namely :

   (*a*) Steroidal Oestrogens

   (*b*) Non-steroidal Oestrogens

(*a*) *Steroidal Oestrogens*

All of them essentially possess a steroidal nucleus and attribute oestrogenic activity. *Examples :* Oestrone, oestriol, oestradiol.

A. Estrone INN, USAN *Oestrone,* BAN,

   3-Hydroxyestra-1, 3, 5 (10)-trien-17-one ; Estra-1-3, 5(10)-trien-17-one, 3 hydroxy-; Oestrone Eur. P. ; Estrone USP ; Theelin(R) (Parke-Davis).

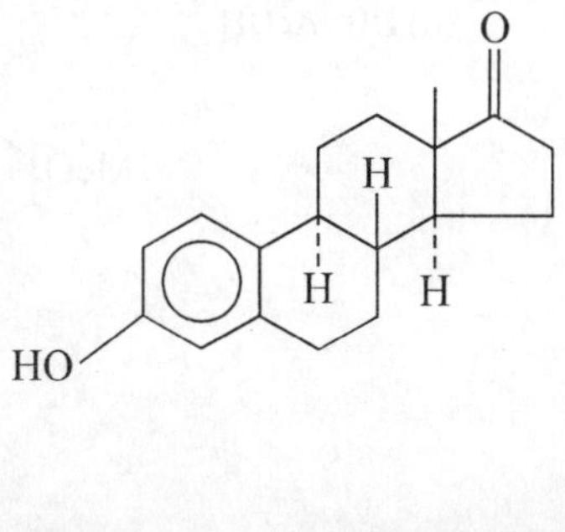

Synthesis :

Johnson and co-workers (1958, 1962) have carried out a total synthesis of oestrone ; each step in their synthesis was *stereoselective*, but Hughs and co-workers (1960) have put forward a total synthesis of oestrone which appear to be comparatively simpler than any other previous method.

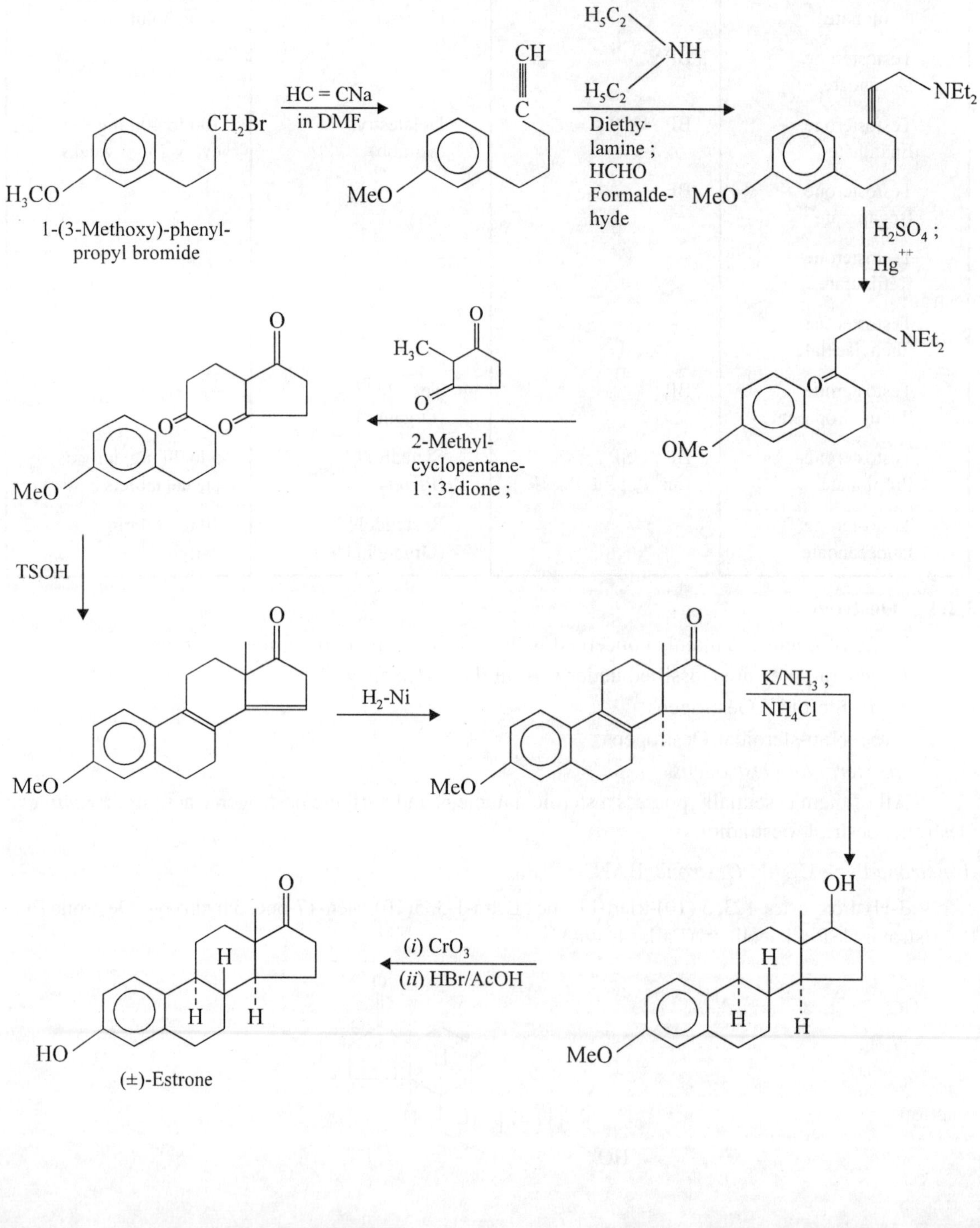

1-(3-Methoxy)-phenyl-propyl bromide

$HC \equiv CNa$ in DMF

H_5C_2—NH / H_5C_2 Diethylamine ; HCHO Formaldehyde

H_2SO_4 ; Hg^{++}

2-Methyl-cyclopentane-1 : 3-dione ;

TSOH

H_2-Ni

K/NH_3 ; NH_4Cl

(i) CrO_3
(ii) HBr/AcOH

(±)-Estrone

It is mainly used for the replacement therapy in deficiency states, e.g., primary amenorrhoea, delayed onset of puberty, control and management of menopausal syndrome, malignant neoplasms of the prostate.

Dose : 0.1 to 5mg per day.

B. *Estriol* INN, USAN, *Oestriol,* BAN,

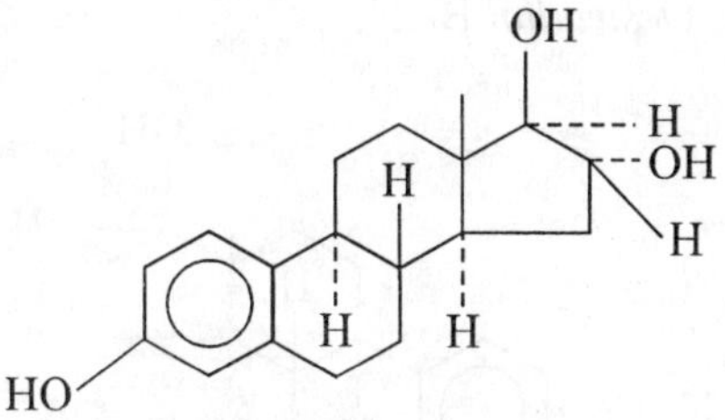

Estra-1,3,5 (10-triene-3,16α,17β- triol ; Estriol USP ; Ovestin(R) (Organon, UK).

Soon after the discovery of oestrone *two* other hormones were isolated, namely : oestriol and oestradiol. Oestriol was first isolated from human pregnancy urine.

Synthesis :

Leeds et al. (1954) have converted oestrone into oestriol by a simple method as discussed below.

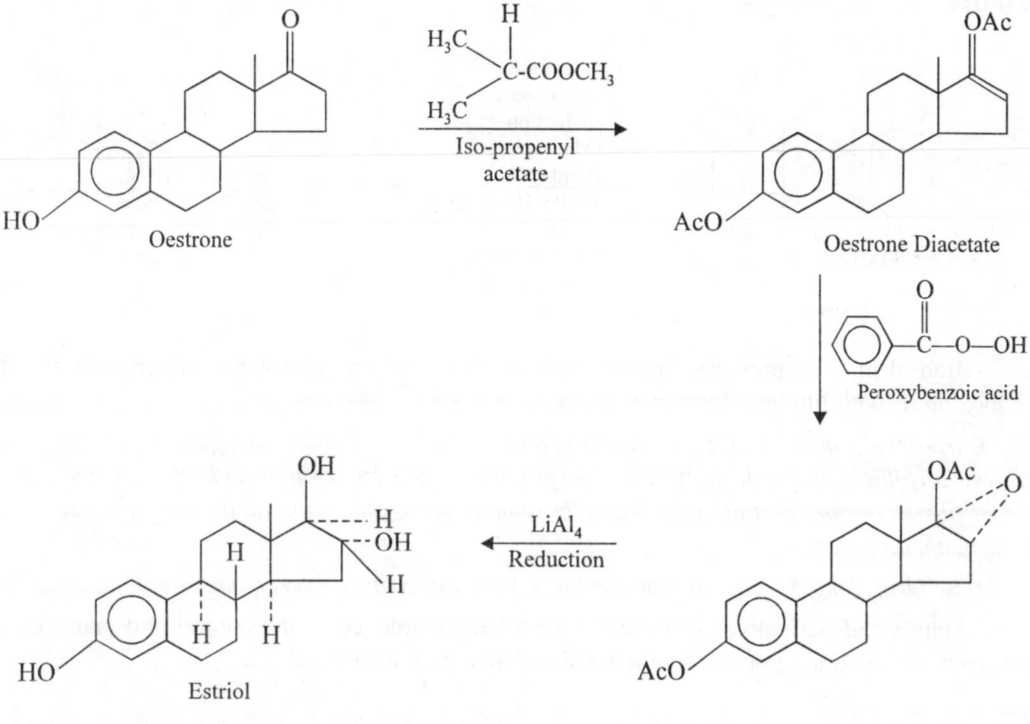

Oestrone on treatment with iso-propenyl acetate yields the corresponding diacetate which on reaction with peroxybenzoic acid removes the double bond between C-16 and C-17 and introduces an oxygen bridge having alpha configuration between the said two carbon atoms. This on reduction with lithium aluminium hydride yields the official compound.

Oestriol is more potent than either oestrone or oestradiol in its oestrogenic activity when administered orally. It is reported to possess a selective action on the vagina and cervix.

Dose. For menopausal symptoms, 250 to 500 mcg per day.

C. *Estradiol* INN, USAN, *Oestradiol* BAN,

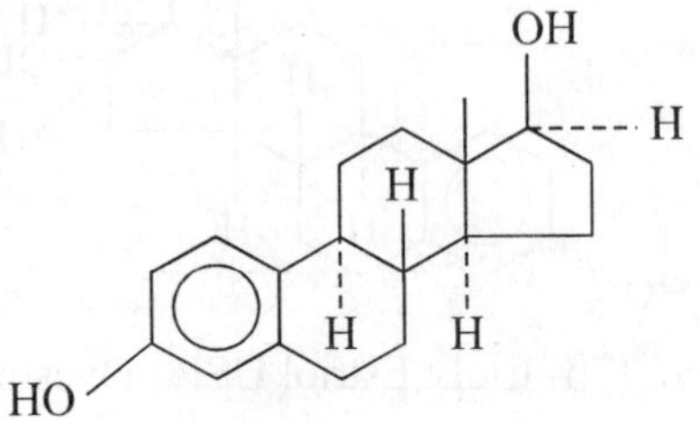

Estra-1, 3, 5 (10)-triene-3, 17β-diol ; Beta-oestradiol; Estradiol USP ; Oestradiol (BPC1968) ; Diogyn[R] (Pfizer) ; Oestradiol Implants[R] (Organon, U.K.).

Estradiol was first obtained by the reduction of oestrone, but later it was isolated from the ovaries of cows.

Synthesis :

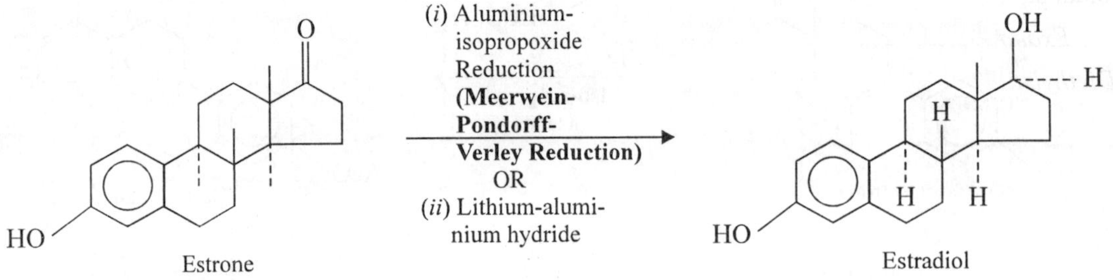

Estradiol may be prepared conveniently by the reduction of estrone either with aluminium isopropoxide or with lithium-aluminium-hydride.

Estradiol is found to be the most active of the naturally occurring oestrogenic hormones produced in the ovarian follicles under the influence of the pituitary. It helps to regulate and subsequent maintenance of the female sex organs, certain functions of the human uterus and above all the secondary sex features, and the mammary glands.

Dose. Oral, 2 mg per day ; intramuscular, 1.5 mg 2 or 3 times weekly ; implantation, 20 to 100 mg.

A number of derivatives of estradiol have been employed in the control and management of oestrogenic activity and these are summarized in Table 2 on next page.

Table 2. Derivatives of Estradiol

Approved Names	Official Status	Proprietary Names	Dose
Oestradiol Benzoate	BP; Eur. P. ;	Benzotrone[R] (Paines & Byrne, U.K.)	1 to 5 mg daily
Estradiol Cypionate	USP ;	Depo-Estradiol Cypionare[R] (Upjohn, USA)	1 to 5mg intramuscular every 3 to 4 weeks
Oestradiol Dipropionate	BPC (1954) ; Ind. P. ;	Ovocyclin[R] (Ciba-Geigy, Switz),	1 to 5mg i.m. every 1 to 2 weeks
Estradiol Enanthate	—	—	10 mg
Oestradiol Undecanoate	—	Primogyn Depot[R] (Schering)	100 to 200 mg every 2 to 3 weeks
Estradiol Valerate	USP ;	Delestrogen[R] (Squibb) ;	5 to 40 mg every 1 to 3 weeks

(b) *Non-Steroidal Oestrogens*

A large number of medicinal compounds possessing remarkable oestrogenic activity, but not of steroidal structure (nucleus), have been prepared *synthetically.*

Examples. Diethylstibesterol : Hexestrol ; Dienestrol.

A. *Diethylstilbesterol* INN, USAN, *Stilbesterol* BAN,

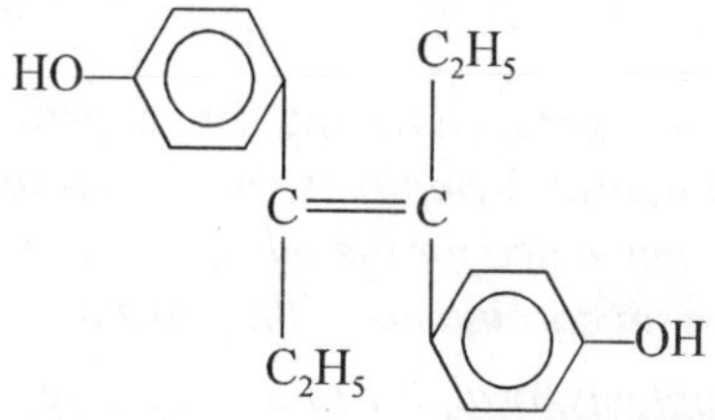

(E)-αβ-Diethylstilbene-4-4′ diol ; Phenol 4,4′-(1,2-diethyl-1,2-ethenediyl) *bis*-(E)- ; Diethylstilbesterol (USP) ; Stilboesterol (BP ; Eur. P. ; Int. P. ; Ind. P.) ; Stilbetin[R] (Squibb).

Synthesis :

Diethylstilbesterol may be synthesized by *two* different methods. *First*, from anisaldehyde (Dodds and Lawson, 1939) ; and *secondly*, from anethole (Kharasch *et al.* 1943). The latter shall be discussed here.

Anethole on treatment with hydrogen bromide undergoes **Markownikoff's addition** to yield anethole hydrobromide. The resulting product in the presence of sodamide and liquid ammonia gives an intermediate product which on subsequent demethylation followed by isomerization in alkali yields the official compound.

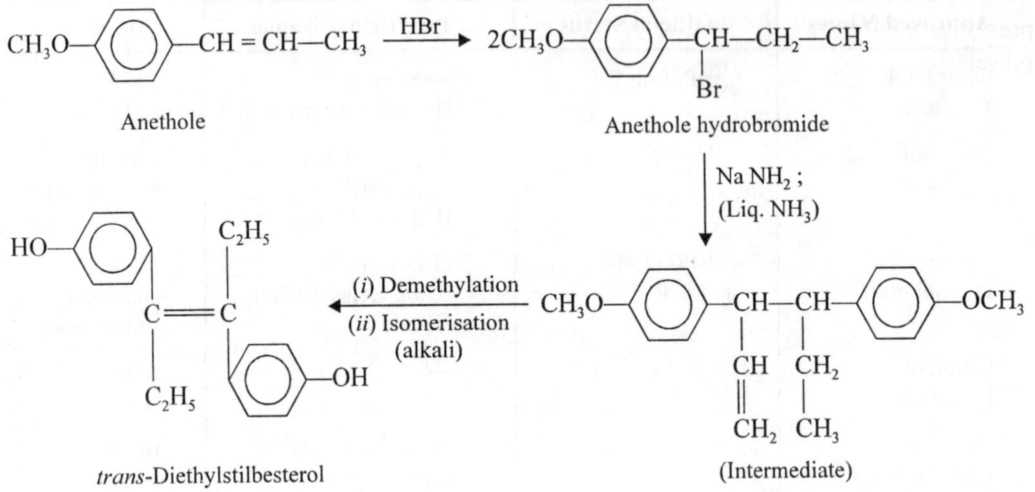

$$CH_3O-\langle\bigcirc\rangle-CH=CH-CH_3 \xrightarrow{HBr} 2CH_3O-\langle\bigcirc\rangle-\underset{\underset{Br}{|}}{CH}-CH_2-CH_3$$

Anethole   Anethole hydrobromide

NaNH$_2$;
(Liq. NH$_3$)

HO$-\langle\bigcirc\rangle$

C$_2$H$_5$

$$\underset{\underset{C_2H_5}{|}}{C}=\underset{\underset{\langle\bigcirc\rangle-OH}{|}}{C}$$

(i) Demethylation
(ii) Isomerisation
(alkali)

$$CH_3O-\langle\bigcirc\rangle-\underset{\underset{CH_2}{\overset{CH}{||}}}{CH}-\underset{\underset{CH_3}{\overset{CH_2}{|}}}{CH}-\langle\bigcirc\rangle-OCH_3$$

trans-Diethylstilbesterol   (Intermediate)

There exists *two geometrical isomeric* forms of diethyl-stilbesterol ; *cis-* and *trans-*out of which only the latter exhibits potent oestrogenic activity.

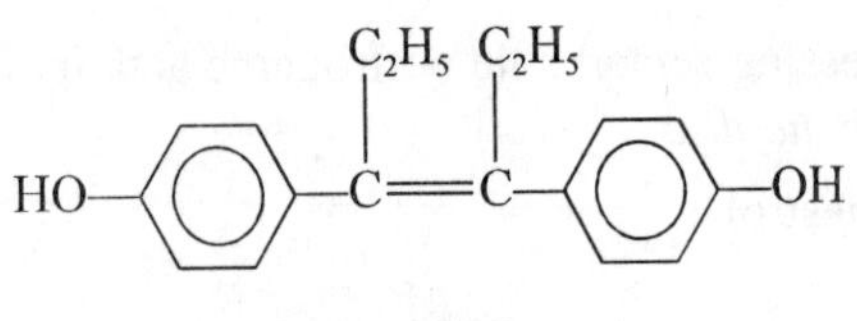

C$_2$H$_5$ C$_2$H$_5$

$$HO-\langle\bigcirc\rangle-\underset{}{C}=\underset{}{C}-\langle\bigcirc\rangle-OH$$

C$_2$H$_5$

$$HO-\langle\bigcirc\rangle-\underset{}{C}=\underset{\underset{C_2H_5}{|}}{C}-\langle\bigcirc\rangle-OH$$

cis-Stilbesterol 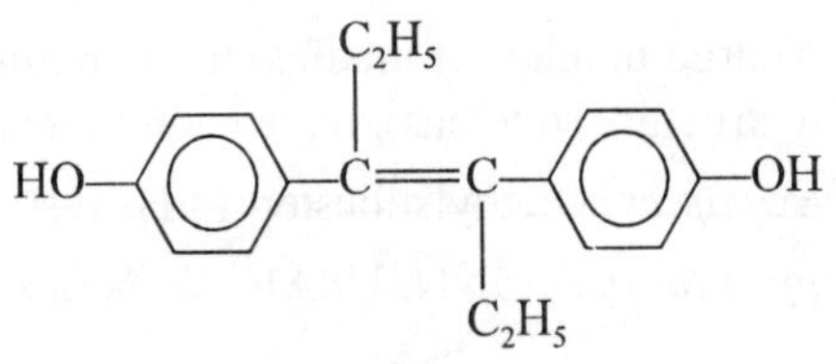   *trans*-Stilbesterol

It is a synthetic non-steroidal oestrogen having similar actions and uses to those of oestradiol. It is used in the *treatment of menopausal symptoms and in secondary amenorhoea due to ovarian insufficiency. It has also been recommended for the inhibition of lactation, in the palliative treatment of malignant neoplasms of the breast, in carcinoma of the prostate and for postcoital contraception.*

Dose. For menopausal symptoms, oral, 0.1 to 2 mg; for secondary amenorrhoea, 0.2 to 0.5 mg ; for carcinoma of prostate, 3 mg per day.

B. *Hexestrol* INN, *Hexoestrol* BAN,

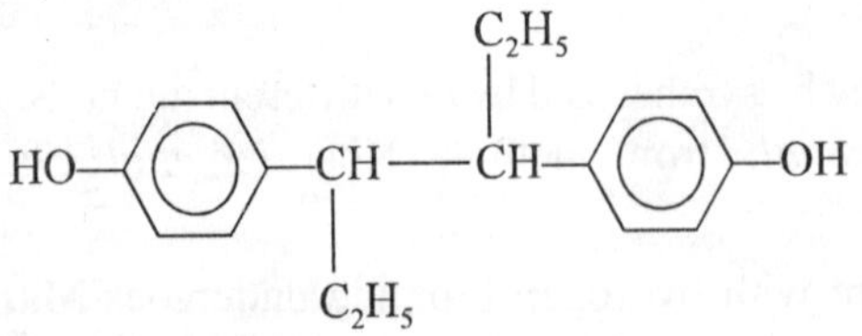

C$_2$H$_5$

$$HO-\langle\bigcirc\rangle-\underset{\underset{C_2H_5}{|}}{CH}-\underset{}{CH}-\langle\bigcirc\rangle-OH$$

*meso-*4, 4'-(1,2-Dimethylethylene) diphenol ; Dihydrostilboestrol ; Hexanoestrol ; BPC (1968) ; Ind. P. ; Hormoestrol[(R)] (Siegfried, Switz).

Synthesis :

Hexestrol is prepared by subjecting anethole hydrobromide to the **Wurtz-Fittig reaction** in the presence of sodium to get the corresponding diethyl derivative. This on further demethylation in the presence of alcoholic potassium hydroxide yields the official compound.

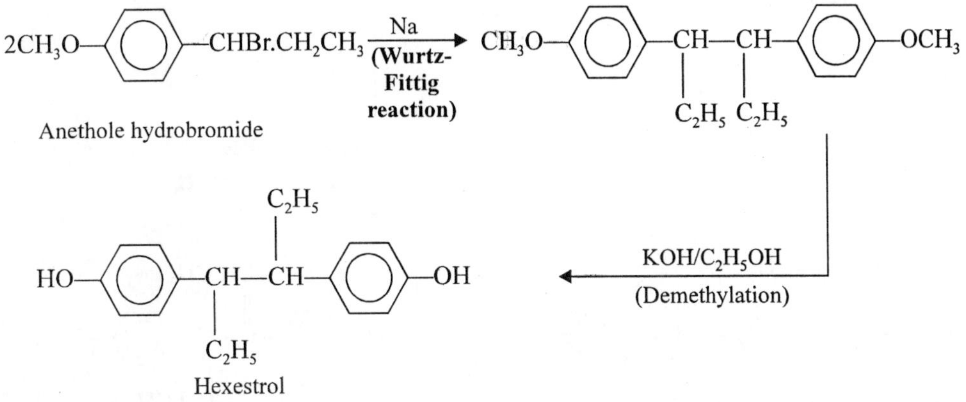

Anethole hydrobromide

Hexestrol

It is used for menopausal symptoms and also for the treatment of neoplasms of the breast and prostate.

Dose. Oral, usual, 1 to 5 mg.

C. *Dienestrol* INN, USAN, *Dienoestrol* BAN,

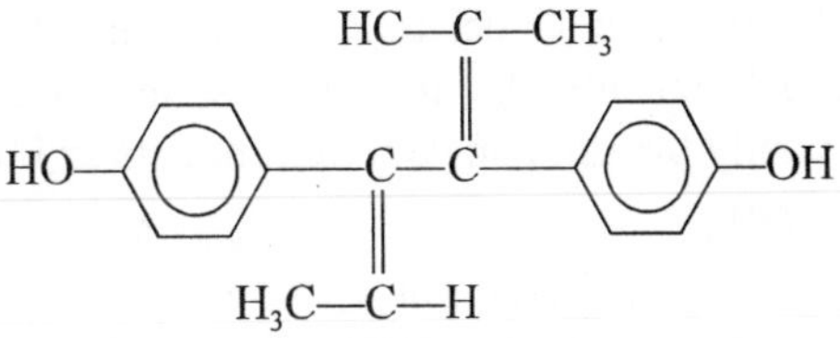

(E.E)-4, 4'-Di (ethylidene) ethylene diphenol ; Phenol, 4, 4'-(1, 2-diethyl-idene-1, 2-ethanediyl) *bis*-, (E,E)- ; Dienoestrol (BP ; Eur.P., Int. P ; Ind. P ; Dienoestrol (USP) ; Estraguard[R] (Reid-Provident).

Synthesis

Dienoestrol may be synthesized by various methods. The synthesis put forward by Dodds *et al.* is described here.

Reduction of *para*-hydroxypropiophenone yields a diphenol derivative which upon benzoylation with benzoyl chloride followed by acetylation with a mixture of acetylchloride and acetic anhydride gives the dienestrol dibenzoate. This on treatment with alcoholic KOH yields the official product.

Its actions and uses are similar to those of oestradiol. Besides, it is also employed by local application in creams.

Dose. For menopausal symptoms, 0.5 to 5 mg per day ; For mammary or prostatic carcinoma, 15 to 30 mg per day.

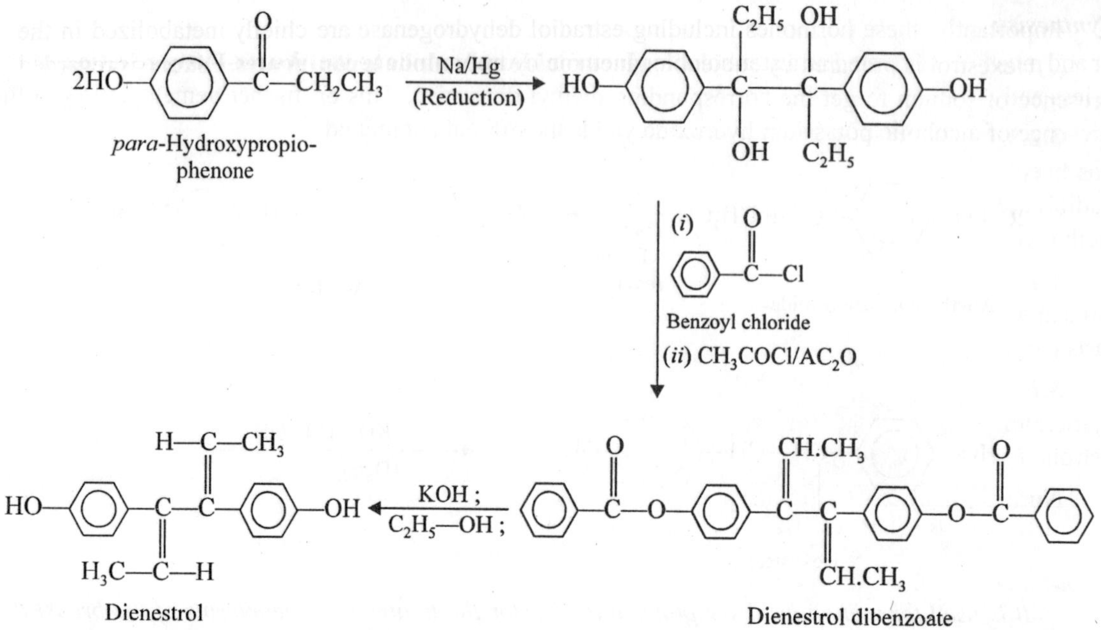

para-Hydroxypropio-
phenone

(i) Benzoyl chloride

(ii) CH_3COCl/AC_2O

Dienestrol   Dienestrol dibenzoate

3.2.3.1. Mechanism of Action.

The mechanism of action of 'oestrogens' shall be dealt with separately under the following *two* heads, namely :

A. Steroidal Oestrogens

3.2.3.1.A-1. Estrone.

The endogenous oestrogens, for instance : *estrone* and *17β-estradiol* are observed to be interconvertible biochemically in the presence of the specific enzyme **estradiol dehydrogenase** and yield practically the same metabolic products as illustrated below :

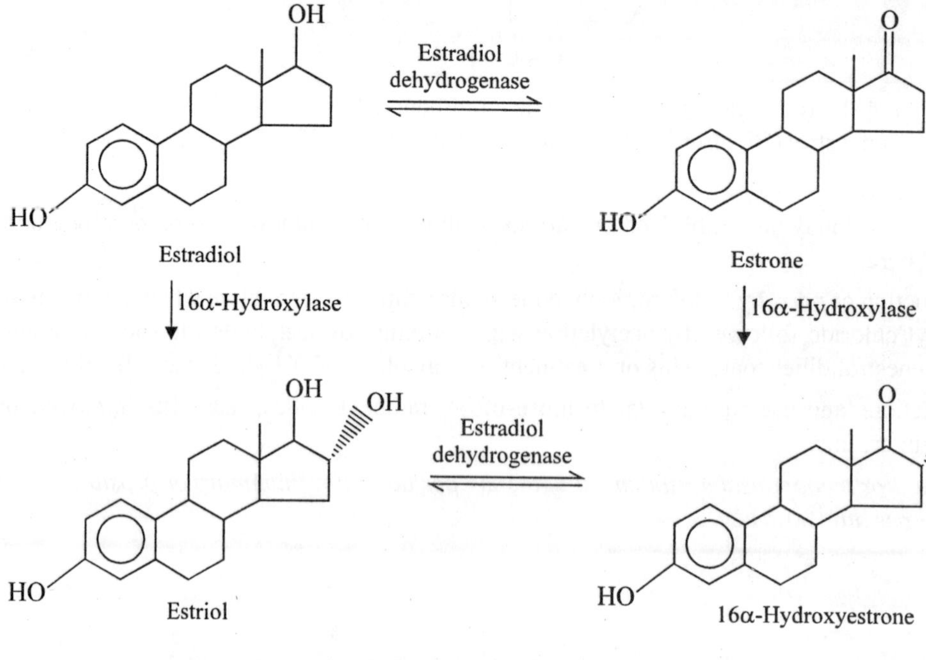

Estradiol

Estrone

Estriol

16α-Hydroxyestrone

Estrogen Metabolism

Importantly, these hormones including estradiol dehydrogenase are chiefly metabolized in the liver and mostly get excreted as water-soluble **glucuronide** and **sulphate conjugates.** Estrone is regarded as a less active (1/12) metabolite of estradiol.

3.2.3.1.A-2. Estriol. This specific steroid is most abundantly synthesized in the human placenta. It has been observed that in both pregnant and nonpregnant women estriol (along-with estrone and estradiol) are duly metabolized to small quantum of other structural analogues *viz.,* 2-hydroxyestrone ; 2-methoxyestrone ; 4-hydroxyestrone ; and 16β-hydroxy-17β-estradiol.

The '*drug*' affords a proliferation of the breast ductile system. It also stimulates the development of lipid and other tissues which essentially contributes to breast shape and function. Fluid retention in the breasts particularly in the later-stages of the menstrual cycle is found to be a common feature of estriol.

3.2.3.1.A-3. Estradiol. The '*drug*' distinctly possesses a **high presystemic elimination rate;** and, therefore, gives rise to a low bioavailability by the oral route. The drug gets appreciably converted to estrone *in vivo.* The plasma half-life stands at 1 hour.

> **Note. Both transdermal and micronized preparations are employed effectively for the replacement therapy.**

B. Non-Steroidal Oestrogens

3.2.3.1.B-1. Diethylstilbesterol (DES). The '*drug*' has an advantage over the other estrogens (*e.g.,* estrone, estriol and estradiol) by virtue of the fact that it gets absorbed quite effectively through the oral administration. Besides, its *rate of inactivation* is very slow and sluggish. DES was found to be appreciably cheaper in comparison to the naturally occurring estrogens and still can produce all the same pharmacological estrogenic activities. Interestingly, DES has exhibited 10 fold the estrogenic potency of its corresponding *cis*-isomer due to the fact that the *trans*-isomer bears a close relationship to **estradiol.***

> **Note. Due to the high incidence of 'uterine cancers' as replacement therapy in menopausal women its usage in women has been banned. However, its use in men for the treatment of 'prostatic cancer' still continues.**

SAR of DES. One may consider DES as another form of estradiol wherein the two 6-membered rings 'B' and 'C' open up and a 6-membered aromatic ring 'D' introduced in place of the cyclopentane ring. It was further suggested that the actual distance prevailing between the two DES phenol OH moieties was virtually the same as the C-3 OH to C-17 OH distance existing in estradiol ; and, hence, these two entities may prove to be a '*perfect fit*' to the same receptor site. *Recently, with the advent of latest computer softwares the medicinal chemist has established the distance between the two OH moieties in DES to be 12.1Å and in estradiol 10.9Å.*

The following Figure 22.1 is the '**computer generated graphics**' illustrating explicitly the *top-view* and the *side-view* of the actual superimposition of **estradiol** $(H_2O)_2$ shown by **dark-lines** with **DES** represented by **light-lines.** It is, however, pertinent to state here that in an aqueous medium *estradiol* essentially has two water moles which are hydrogen-bonded to the 17-OH moiety. In case, one of the two water moles is considered in the distance measurement of the hydroxyl groups, there exists a '*perfect fit*' associated with the two OH moieties of DES as may be observed in Fig. 1. Hence, it may be implied juistifiably that water may play a vital pivotal role for *estradiol* in its *receptor site.*

---

*UV Solmssen, *Chem Rev.,* **37**, 481, 1945.

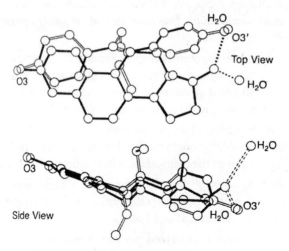

ESTRADIOL (H₂O)₂ : DARK LINES
DIETHYLSTILBESTEROL (DES) : LIGHT LINES

Figigure 22.1. Computer Generated Graphics Showing Superimposition of Estradiol and DES.

[*Adapted from Wilson and Gisvold's Textbook of Organic Medicinal and Pharmaceutical Chemistry, Lippincott Williams and Wiltzns, New York, 5th edn. 2004.*]

3.2.3.1. B-2. Hexestrol. The '*drug*' represents the *meso* form of 3, 4-*bis* (*p*-hydroxyphenyl)-*n*-hexane that distinctly possesses the greatest estrogenic potency of the three stereoisomers belonging to the corresponding dihydro analogue of DES. Interestingly, it is found to be less potent than DES.

3.2.3.1. B-3. Dienestrol. The '*drug*' is a potent estrogen which finds its abundant use only topically, for the treatment of *atrophic vaginitis* and *kraurosis vulvae.*

CAUTION. Not recommended in patients with known or suspected cancer of the breast ; known or suspected estrogen-dependent neoplasia; undiagnosed abnormal genital bleeding ; thrombophlebitis or thromboembolic disorders or a previous case-history of such typical conditions ; and hypersensitivity to the ingredients of the cream or suppositories of dienestrol or during pregnancy.

3.2.4. Gestogens

Gestogens or **corpus luteum hormones** are mostly secreted by the corpus luteum portion of the ovary and the metabolized to various inactive products, *e.g.,* pregnanediol. The metabolities are esentially excreted through urine.

Example : Progesterone.

A. *Progesterone* INN, BAN, USAN,

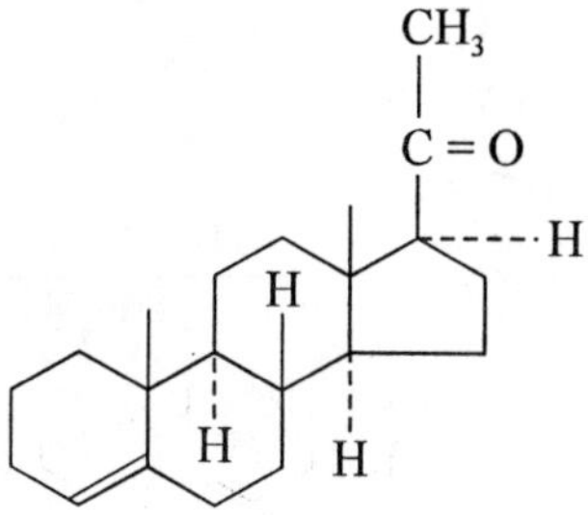

Pregn-4-ene-3, 20-dione ; BP ; USP ; Eur. P. ; Int. P. ; Ind. P. ; Syngesterone[R] (Pfizer) ; Gesterol 50[R] (Forest).

Synthesis :

Progesterone has been synthesized by various researchers from different starting materials as indicated below :

(*i*) From Pregnanediol (Butenandt *et al.* 1930)

(*ii*) From Stigmasterol (Butenandt *et al.* 1934)

(*iii*) From Cholesterol (Butenandt *et al.* 1939)

(*iv*) From Ergosterol (Shephard *et al.* 1955)

It has also been synthesized from diosgenine by Marker *et al.* (1940-1941) which will be discussed here.

Acetylation of diosgenin at 200°C gives the corresponding diosgenyl diacetate which upon oxidation with chromium-6-oxide removes the side-chain at C-17 and the resulting product on reduction followed by hydrolysis yields pregnenolone. This on being subjected to Oppenauer oxidation affords the official compound.

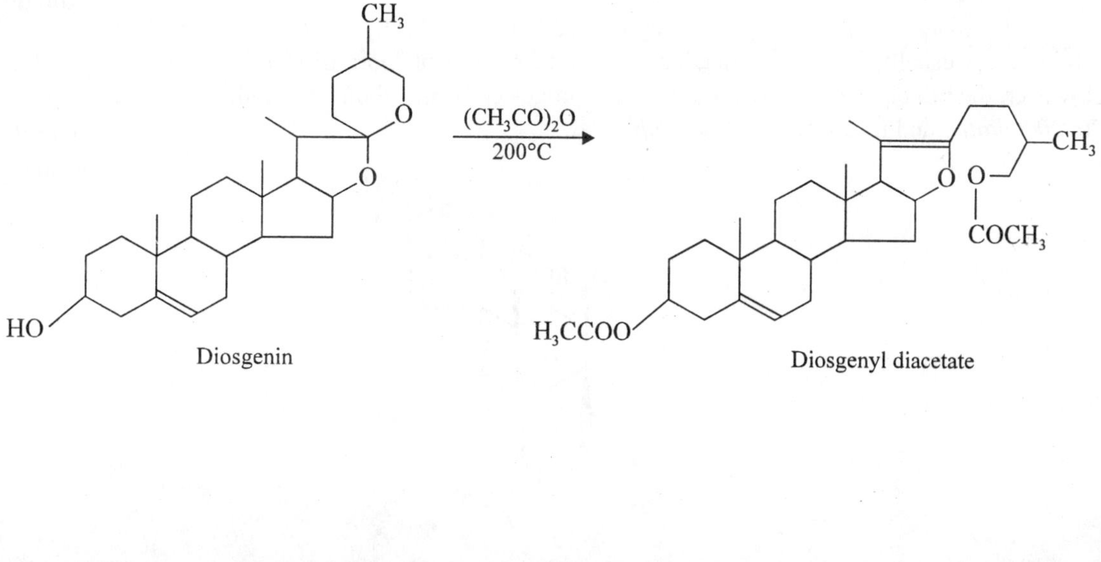

Diosgenin

Diosgenyl diacetate

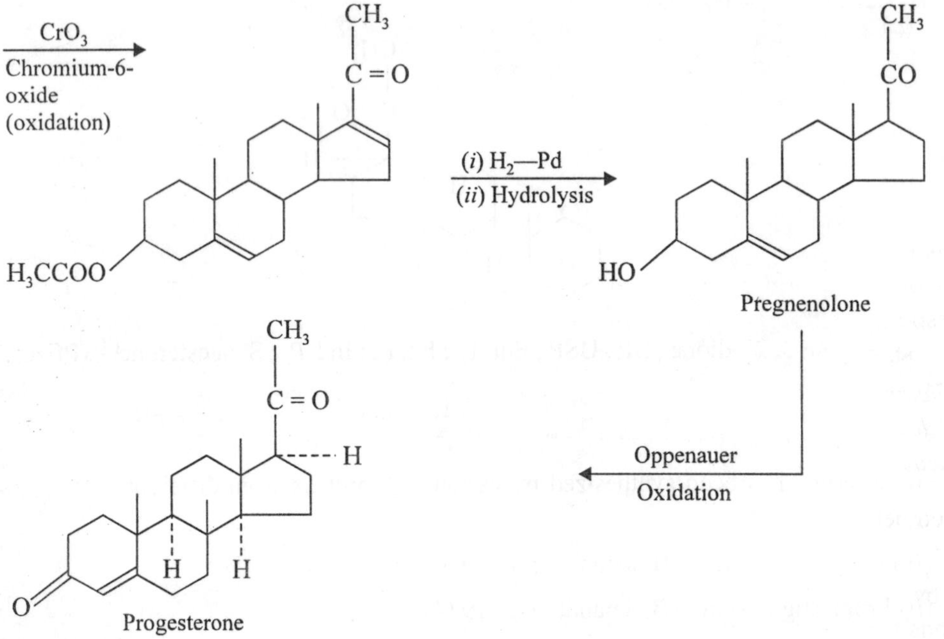

Progesterone

It is employed in the treatment of functional uterine bleeding. It is also used in conjuction with an oestrogen in the treatment of menstrual disorders, neoplasms of the breast and endometrium. Sometimes it also finds its use in habitual and threatened abortion.

Dose. For uterine bleeding, 5 to 10 mg injected per day up to 5 to 10 days ; For habitual abortion, 5 to 20 mg twice or thrice per week by intramuscular injection.

3.2.4.1. Mechanism of Action. The mechanism of action of progesterone shall now be discussed as under :

3.2.4.1.1. Progesterone. One school of thought considered it to be the '*drug of choice*' specifically in the **luteal-phase dysfunction,** a disorder that gives rise to either *infertility* or *repititive early* abortion.

The '*drug*' gets metabolized rapidly when adminstered orally showing a plasma half-life of only 5 minutes. It usually undergoes transformation leading to a plethora of steroidal metabolic products. However, the principal excretory product of the progesterone metabolism is nothing bout *5β-pregnane-3α-20α-diol(I)* and its *corresponding conjugates.*

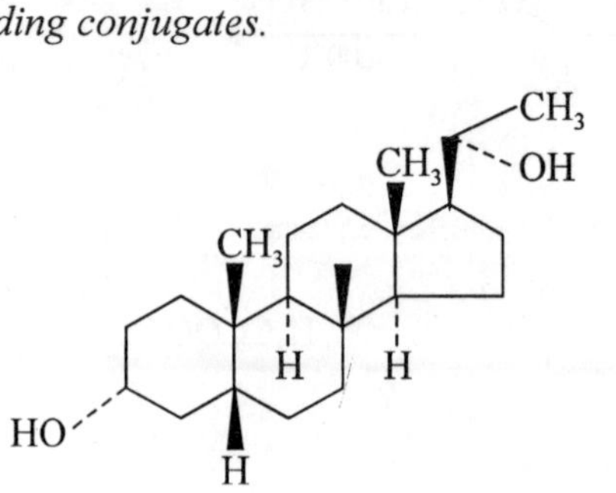

(I)

A few salient-features of the aforesaid metabolism are :

(*a*) reduction of the double bond between C-4 and C-5,

(*b*) reduction of the ketone (—$\overset{\overset{\textstyle O}{\|}}{C}$—) function at C-3 giving rise to 3α-ol, and

(*c*) reduction of the ketone moiety at C-20 to provide the 20α-ol.

It has been duly observed that the prevailing reduction invariably taking place at C-5 must precede the reduction of the C-3 ketone. Besides, the characteristic structural features which may specifically cause blockade of the reduction either at C-5 or C-20 have enormously enhanced the half-lives of the corresponding progesterone derivatives.

Note. [*Progestasert, Alza*[R] *an intrauterine contraceptive device consists of 38mg of progesterone in silicone oil. In this instance, the 'drug' is blieved to increase the contraceptive effectiveness of the said device by a local effect on the endometrium followed by effects upon the motility of sperm, capacitation and metabolism.*]

3.3. CARDIAC GLYCOSIDES

Plant extracts containing cardiac glycosides were invariably employed as poisons in the medieval trial by both African and South American natives for the preparation of their lethal arrow and spear poisons for use in fighting as well as hunting.

'**Digitalis**' a preparation made by extraction of dried seeds and leaves of the purple foxglove *Digitalis purpurea,* found certain application in the control and management of dropsy. Later on, in 1785, a noted Scottish physician William Withering first introduced the use of *'digitalis'* in heart therapy and this became a spectacular success and tremendous achievement for curing heart patients.

The active components of digitalis are glycosides of digitoxigenin, digoxigenin and gitoxigenin and afford these glycones on acid hydrolysis. The free *aglycones* or *genins* are only weekly active.

Digitoxigenin   Digoxigenin   Gitoxigenin

The salient features of the abvoe *three* genins are enumerated below :

   (*i*) All the above three genins have an α, β-unsaturated five-membered lactone ring.

   (*ii*) All of them have 3β and 14β hydroxy groups ; digoxigenin has an additional 12β-hydroxy group and gitoxigenin a 16β-hydroxy group.

   (*iii*) The unsaturated lactone ring and the 14β-hydroxy group are both essential to cardiac activity.

The corresponding glycosides digitoxin and digoxin are all **triosides** of comparable high cardiotonic activity. They are described briefly here :

A. *Digitoxin* INN, BAN, USAN,

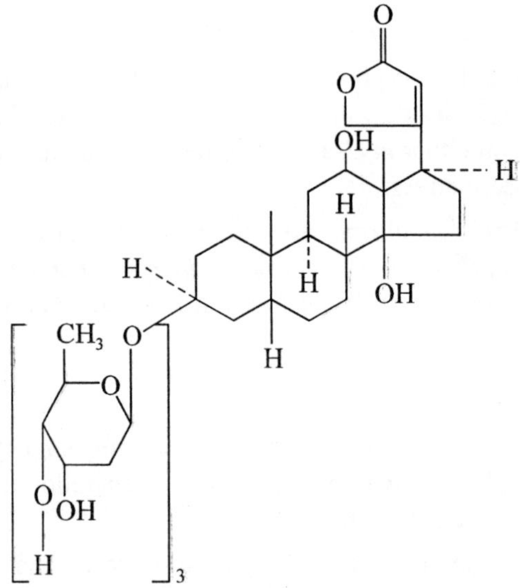

Card-20 (22)-enolide, 3-[(o-2, 6-dideoxy-β-D-*ribo*-hexopyranosyl-(1 → 4)-o-2, 6-dideoxy-β-D-*ribo*-hexopyranosyl-(1 → 4)-2, 6-dideoxy-β-D-hexopyranosyl) oxy]- 14-hydroxy, (3β, 5β)- ; BP ; USP ; Eur. P ; Ind. P ; Crystodigin(R) (Lilly).

Digitoxin is the most potent of the digitalis glycosides besides being the most cumulative in action.

It enhances the force of myocardial contraction and in the case of heart failure this dominating inotropic effect results in a much modified cardiac output with regard to more complete emptying of the ventricle at systole, an apparent decrease in the elevated end-diastolic ventricular pressure, and above all a positive reduction in the size of the dilated heart. It is used in the treatment of congestive heart failure.

Dose. Adult, initial, 600 mcg, followed by doses of 200 to 400 mcg every 6 hours as necessary; For slow digitalisation, 300 mcg has been given twice daily for 4 days ; Maintenance dose ranges from 50 to 200 mcg per day.

B. *Digitoxin* INN, BAN, USAN,

3β-[o-2, 6-Dideoxy-β-D-*ribo*-hexopyranosyl-(1 → 4)-o, 2, 6-dideoxy-β-D-*ribo*-hexopyranosyl-(1 → 4)-2, 6-dideoxy-β-*ribo* hexopyranosyl) oxy]-12β,-14-dihydroxy-5β-card-20 (22)-enolide ; BP ; USP ; Eur. P. ; Int. P ; Ind. P ; Lanoxicaps(R) (Burroughs Wellcome) ;

Its uses and actions are very similar to those of digitoxin.

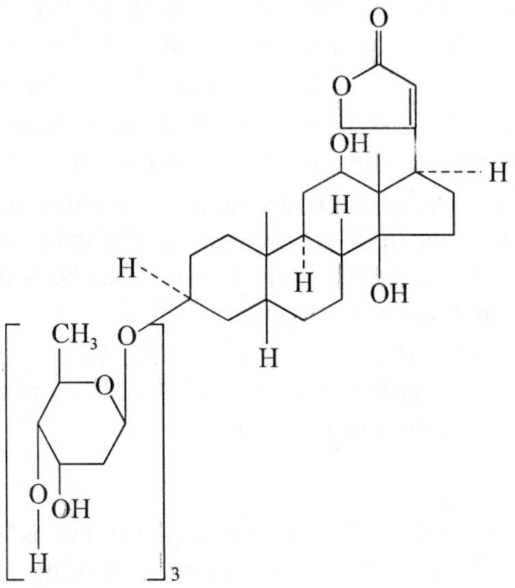

Dose. For rapid digitialization, 0.75 to 1.5 mg orally followed by 250 mcg, every 6 hours until the desired therapeutic effect is achieved.

3.3.1. Mechansim of Action. The mechansim of action of digitoxin and digoxin are treated in the sections that follows :

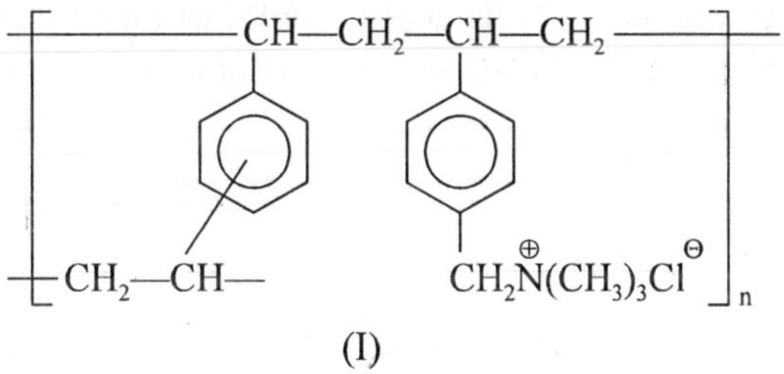

(I)

3.3.1.1. Digitoxin. The '*drug*' usually gets absorbed almost completely after oral administration ; of course, with an exception when **cholestyramine(I)** is used concomitantly. It is found to exhibit its optimal activity within a span of 4 to 12 hour. However, after **full digitalization,** the duration of action extends upto 14 days. It gets protein bound in plasma upto almost 97%. Its volume of distribution (v_d^{ss}) is approximately 0.6 mL g^{-1}. It has been duly established that a plasma concentration of 15-25mg mL^{-1} are regarded to be therapeutic range ; whereas, 35-40 ng mL^{-1}, or even more to be toxic. However, significant variation in the plasma concentration may be afforded by plasma K^+ and Ca^{2+} levels along with other such factors. It has been osberved that the ensuing '*hepatic metabolism* usually accounts for 52-70% of the entire elimination of this '*drug*'. The β-half-life varies between 2.4 to 9.6 (average 7.6) days.

CAUTION. Phenytoin. *(anticonvulsant) and* **phenobarbital** *(longacting barbiturate) can induce hepatic microsomal enzymes and thereby retard the half-life significantly ; and, therefore, ultimately interfering with the prevalent efficacy ofthe 'drug'.*

3.2.1.2. Digoxin. The '*drug*' is invariably used IV for accomplishing rapid digitalization because of its high degree of purity ; and its action becomes manifest within a span of 15-30 minutes, eventually attaining its peak in 2-5 hours. However, after full digitalization its duration of action extends upto 6 days (unlike 14 days for digitoxin). The '*drug*' is bound to protein in plasma between 20-30%. Its volume of distribution (v_d^{ss}) stands at 5.1 L. kg^{-1} in normal adults ; whereas, in patients with a history of renal failure v_d^{ss} is nearly 3.3 L. kg^{-1}. It has been observed that an observed **extensive intracellular binding** is usually responsible for the large volume of distribution.

The renal exeretion in adults normally accounts for 60-90% of its total elimination ; a small quantum gets converted in the liver to *dihydrodigoxin*. The elimination half-life in adults (normal) ranges between 29 to 135 hours (normally 36-41 hour). It has been observed that an enhanced GI motility lowers and decreased motility augments absorption.

CAUTION. *Bioavailability of 'digoxin' gets altered due to the presence of such drugs as : antacids, antineoplastic agents, cholestyramine resins, dietary fibre, erythromycin, neomycin, tetracyclines, metoclopramide, sulphasalazine and propantheline.*

3.4. BILE ACIDS

The liver secretes a clear, golden yellow viscous liquid known as 'bile'. It is stored in gall bladder and is solely usuful for the digestive system. It mainly consists of the inorganic ions like HCO_3^-, Cl^- Na^+, K^+, etc., in addition to organic compounds such as bile acids, bile pigments, liquid fatty acids and cholesterol. Cholic Acid Deoxycholic Acid Chenodeoxycholic Acid

The bile acids are usually present as the salt of amide with either glycine or taurine, for instance ; sodium glycocholate (glycine + cholic acid), and sodium taurocholate (taurine + cholic acid).

In all twelve natural bile acids have been identified and characterised duly. Of these the most abundant bile acids in human bile are : cholic acid (26-60% of total bile acids) ; deoxycholic acid (5-25%), and chenodeoxycholic acids (30-35%), whose structures and chemical names are stated below :

3α, 7α, 12α-Trihydroxy
cholanic acid
OR
Cholic Acid

3α, 12α-Dihydroxy
cholanic acid
OR
Deoxycholic Acid

3α, 7α-Dihydroxy
cholanic acid
OR
Chenodeoxycholic Acid

The bile acids may be isolated from the bile by cleaving the peptide linkage present in them by hydrolysis with alkali. From the resulting solution the bile acids are conveniently isolated either by crystallization from organic solvent or by treating the ethereal solution of the acids with various concentration of hydrochloric acid, for instance ; the trihydroxy, dihydroxy and the monohydroxy acids may be ioslated by treating the ethereal solution with 15%, 25% and concentrated hydrochloric acid respectively.

3.5. SAPOGENINS

Saponins are the plant glycosides that have the characteristic of forming colloidal aqueous solutions which normally foam upon shaking. Like other glycosides, saponins usually vary in their chemical structures. Saponins, in general possess an unique property to effect hydrolysis of red-blood cells (RBC) even in high dilutions. In this respect, they are very toxic to cold-blooded animals. In general, saponins have a bitter taste and are very irritating to the eyes and the nose. The more commonly and abundantly occurring saponins are those found in soap bark, soap root, snake root, similax and cacti.

Saponins on hydrolysis yield sugars such as glucose, galactose, rhamnose and xylose together with an aglycone (sapogenin) *i.e.,* the non-sugar moeity. They have been used extensively in medicine, as foaming agents in fire extinguishers and as fish poisons.

Following are a few examples of steroidal saponins with their respective sources :

Source	Saponin	Sapogenin	Sugars
Digitalis purpurea or *Digitalis lanata*	Digitonin	Digitogenin $(C_{27}H_{44}O_5)$	Glucose, Galactose
Trillium erectum	Trillin	Diosgenin $(C_{27}H_{42}O_3)$	Glucose

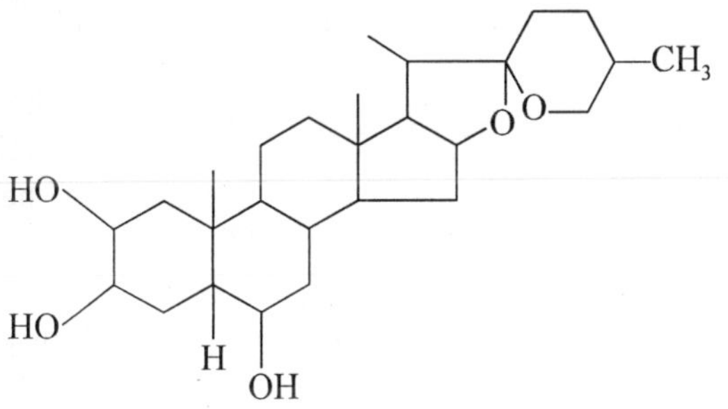

Digitogenin

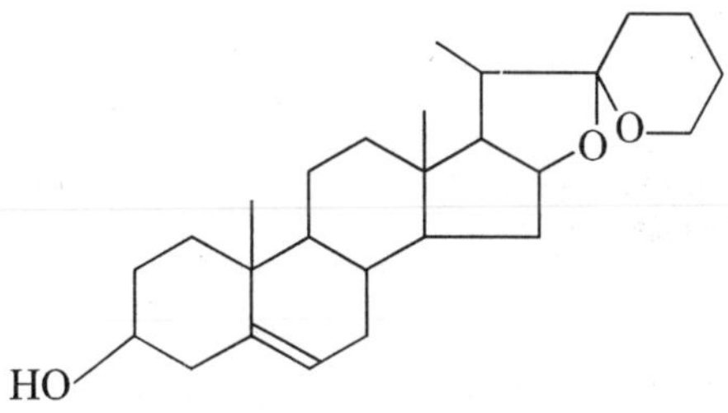

Diosgenin

Probable Questions for B. Pharm. Examinations

1. Write short notes on the folloiwng :
   (*a*) Nomenclature of Steroids
   (*b*) Diel's hydrocarbon
   (*c*) Sterols.

2. Give a brief account of the ANDROGENS. How would you synthesize Testosterone from :
   (*a*) Cholesterol
   (*b*) Dehydroepiandrosterone.

3. What are Follicular Hormones ? Classify them and describe the synthesis of one potent drug from each class.

4. Name a prominent **Corpus Luteum Hormone** and discuss its synthesis from a glycoside obtained from *Digitalis lanata.*

5. Discuss **'Cardiac Glycosides'** by giving its plant source, **three** important known genins, structure of the corresponding glycosides and their uses.

6. Give a comprehensive account of the **'Bile Acids'**. How are they isolated from the natural bile ? Support your answer with the structure of known bile acids.

7. Naturally occurring plant sources yield **'Sapogenins'.** Discuss their importance and usage in medicine and steroidal chemistry.

8. Hugh's total synthesis of **OESTRONE** from 1-(3-methoxy)-phenyl propyl bromide offers a comparatively simpler method than others. Explain.

9. Give the names and official status of at least **five** derivatives of :

   (*a*) Testosterone

   (*b*) Estradiol

   which are used in medicine.

RECOMMENDED READINGS

1. JEF Reynolds (Ed), '*Martindale the Extra Pharmacopoeia',* (31st edn.), The Pharmaceutical Press London (1996).

2. ME Wolff, *Burger's Medicinal Chemistry and Drug Discovery,* (5th edn), John Wiley and Sons, Inc., New York, (1995).

3. MC Griffiths (Ed), 'USAN and the USP Dictionary of Drug Names', United States Pharmacopoeial Convention, Inc. Rockville (1985).

4. 'British Approved Names 1986', British Pharmacopoeia Commission London, Her Majesty's Stationary Office.

5. 'International Nonproprietary Names (INN) for Pharmaceutical Substances', World Health Organisation Geneva (Cumulative List No. 6), (1982)

6. A W Norman and Litwack, G., *Hormones,* Academic Press, San Diego, 2nd edn, 1997.

7. JG Hardman and Limbird LE (eds): *Goodman and Gilman's The Pharmacological Basis of Therapeutics,* MC Graw Hill, New York, 10th edn., 2001.

8. Duax WL *et al.* **Biochemical Actions of Hormones,** Academic Press, New York, Vol. II, 1984.

9. Bhatnagar A, Brodie AMH *et al.* Eds. *Fourth International Anrnatase conference, J. Steroid Biochem Mol Biol, 61,* 107-426, 1997.

10. Zeelen FJ, *Medicinal chemsitry of Steroids*, Elsevier, Amsterdam, 1990.

23   Antibiotics

1. INTRODUCTION

The term **"antibiotic"** was put forward by Vuillemin in 1889, *to designate the active component involved in the process of 'antibiosis' or to the opposition of one living micro-organism to another.* According to another school of thought—*'antibiotics are nothing but the microbial metabolites which in relatively high dilution may inhibit the growth of micro-organisms'.*

Waksman proposed the widely cited definition that—*'an antibiotic or an antibiotic substance is a substance produced by the microorganisms, which has the capacity of inhibiting the growth and even of destroying other micro-organisms'.*

However, the restriction that an antibiotic must be a product of a micro-organism is not in keeping with common use.

Later on Benedict and Langlykke coined a more general and acceptable definition of an **antibiotic** which states that—*'a chemical compound derived from or produced by a living organism, which is capable, in small concentrations, of inhibiting the life processes of micro-organisms.'*

Therefore, a substance may be classified as an **antibiotic** provided it meets the following *four* cardinal requirements :

    (*a*)   that it is a product of metabolism

    (*b*)   that it is a synthetic product produced as a structural analogue of a naturally occurring antibiotic.

    (*c*)   that it antagonizes the growth and/or the survival of one or more species of microorganisms

    (*d*)   that it is effective in low concentrations.

In another latest version '**antibiotics**' may be defined as—*'microbial metabolites or synthetic structural analogues inspired by them which, in small dosage regimens, inhibit the growth and survival of microorganisms without any serious toxicity whatsoever to the parent host.'*

Importantly, selective toxicity happens to be the '*key concept*' amongst the antibiotics. There are several vital and glaring instances whereby the '*clinical utility of natural antibiotics*' has been enormously augmented *via.* critical medicinal chemical manipulation of the *mother structure* that would ultimately give rise to not only broader antimicrobial spectrum but also higher potency, lower toxicity and more convenient way of administration.

It is pertinent to state at this point in time that in the modern era the enormous and wide application of '*antibiotics*' in animal nutrition and disease has eventually resulted in the overwhelming sensitization of a comparatively huge number of the susceptible people across the globe, most of whom have developed serious reactions upon contact with such type of drug substances. In the same vein, the most frequent and wide agricultural usage have also made a significant contribution to the ever increasing *'pool of antibiotic resistant bacteria'* in a community.

2. CLASSIFICATION

In this chapter the **antibiotics** will be discussed explicitely under the following *four* main heads, namely :

(*a*) β-Lactam antibiotics,

(*b*) Aminoglycoside Antibiotics,

(*c*) Chloramphenicol, and

(*d*) Tetracyclines.

These *four* types of antibiotics shall be treated in an elaborated manner in the pages that follows :

3.   β-LACTAM ANTIBIOTICS

The β-lactam antibiotics may be further sub-divided into *two* categories, namely :

(*a*) Penicillins, and   (*b*) Cephalosporins.

3.1. Penicillins

Penicillin is the name assigned to the mixture of natural compounds having the molecular formula $C_9H_{11}O_4N_2SR$, and differing only in the nature of 'R'.

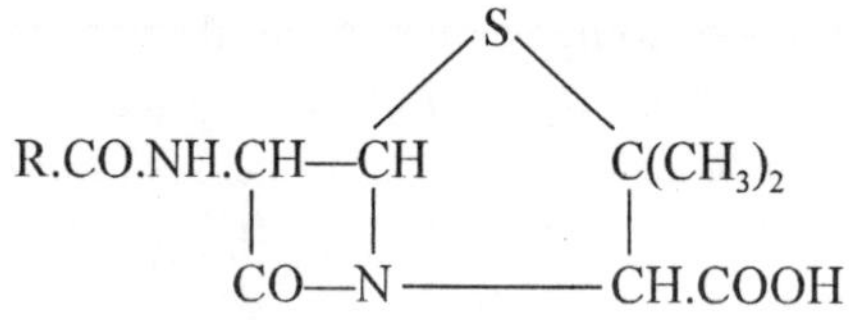

$$R.CO.NH.CH\text{---}CH \quad\quad C(CH_3)_2$$
$$CO\text{---}N\text{--------}CH.COOH$$

Penicillin

These are mainly produced by various strains of *Penicillium notatum* and *Penicillium chrysogenum.* There are at least **six naturally occurring penicillins,** whose chemical names, other names and the nature of 'R' are given in the following table :

3.1.1. Naturally Occurring Penicillins

S. No.	Chemical Name	Other Names	—R
1.	Pent-2-enylpenicillin	Penicillin-1 or F	$—CH_2.CH = CH.CH_2CH_3$
2.	Benzylpenicillin	Penicillin-II or G	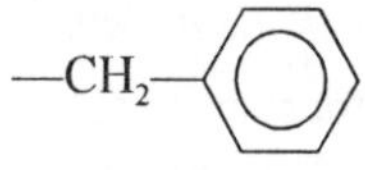 $—CH_2$—
3.	*p*-Hydroxybenzyl-penicillin	Penicillin-III or X	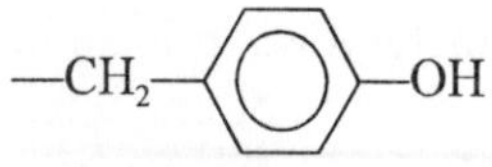 $—CH_2$——OH
4.	*n*-Heptylpenicillin	Penicillin-IV or K	$—(CH_2)_6.CH_3$
5.	*n*-Amylpenicillin	Dihydro-F-penicillin	$—(CH_2)_4.CH_3$
6.	Phenoxymethyl-penicillin	Penicillin-V	$—CH_2—O—C_6H_5$

ANTIBIOTICS

631
3.1.2. Structure of the Penicillins

Following are the various salient features which ultimately determine the general structure of all the penicillins :

1. The penicillins are all strong monobasic acids, *i.e.,* they form salts.

2. The penicillins are hydrolysed by hot dilute inorganic acids ; one carbon atom is eliminated as carbon dioxide (CO_2) and two products are obtained in equimolecular proportions, one being an **amine,** *Pencillamine* and the other an **aldehyde,** *Penniloaldehyde.*

$$C_9H_{11}O_4N_2SR + 2H_2O \xrightarrow{\text{HCl}} CO_2 + C_5H_{11}O_2NS + C_3H_4O_2NR$$
$$\text{Amine} \qquad \text{Aldehyde}$$

All the penicillins give the same amine, but different aldehydes, because it bears the variable component 'R' in it.

3. D-Penicillamine ($C_5H_{11}O_2NS$)

Penicillamine instantly gives the indigo colour reaction with ferric chloride, a test characteristic of cysteine, thereby suggesting that the amine is probably a substituted cysteine. The structure of penicillamine was later on proved to be D-β: β-dimethylcysteine by synthesis as described below :

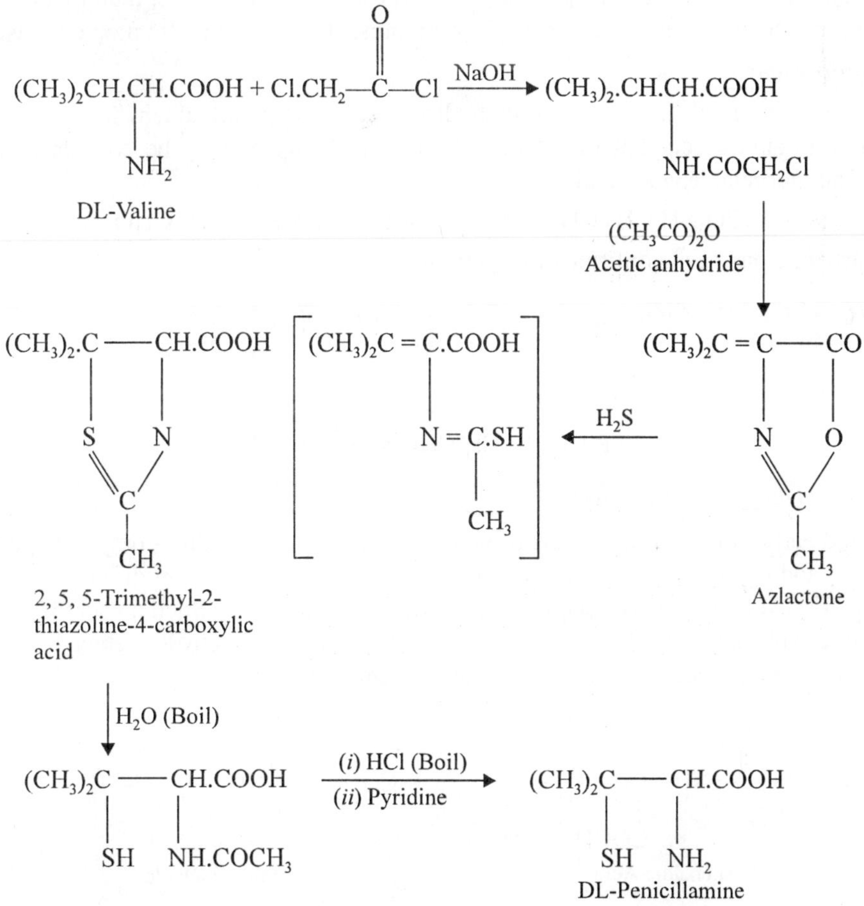

2, 5, 5-Trimethyl-2-thiazoline-4-carboxylic acid

Azlactone

DL-Penicillamine

Resolution of the racemic amine

The resulting racemic mixture of penicillamine was *first* converted into the formyl derivative ; and *secondly* it was resolved by means of brucine and thirdly, the formyl group was removed by hydrolysis, thus :

$$(CH_3)_2C\!-\!CH.COOH \quad \xrightarrow[-H_2O]{H.COOH} \quad (CH_3)_2C\!-\!CH.COOH$$
$$\qquad\quad | \quad\ | \qquad\qquad\qquad\qquad\qquad\qquad | \quad\ |$$
$$\qquad\quad SH \ \ NH_2 \qquad\qquad\qquad\qquad\qquad\quad SH \ \ NH.CHO$$

DL-Penicillamine   DL-Form

$$(CH_3)_2C\!-\!CH.COOH \quad \xleftarrow{\qquad} \begin{array}{l} (i) \text{ Brucine} \\ (ii) \text{ HCl} \\ (iii) \text{ Pyridine} \end{array}$$
$$\qquad\quad | \quad\ |$$
$$\qquad\quad SH \ \ NH_2$$

D-Penicillamine

D-Penicillamine was found to be identical with the natural penicillamine. When treated with diazomethane ($CH_2\!=\!N^+\!=\!N^-$), penicillin is converted into its methyl ester and this, on treatment with an aqueous solution of mercuric chloride, gives the methyl ester of pencillamine, thereby proving that the carboxyl group in penicillamine is the carboxyl group present in the penicillin moelcule itself.

4. *Penilloaldehyde*

It has been observed that on vigorous hydrolysis, all the penilloaldehydes give a substituted acetic acid and an aminoacetaldehyde. Hence, the penilloaldehydes may be considered as acylated derivatives of aminoacetaldehyde. Thus

$$R.CO.NH.CH_2.CHO + H_2O \rightarrow R.COOH + H_2N.CH_2.CHO$$

This structure has been confirmed by synthesis :

$$R.CO.Cl + H_2N.CH_2CH(OC_2H_5)_2 \xrightarrow{\qquad\qquad} R.CONHCH_2CH(OC_2H_5)_2$$

$$R.CO.NH.CH_2CHO \xleftarrow{\quad HCl \quad}$$

5. *Carbon Dioxide (CO$_2$) Molecule*

As stated earlier the acid hydrolysis of penicillin yields three products only *viz.*, penicillamine, penilloaldehyde and carbon dioxide. The liberation of a molecule of carbon dioxide gave rise to the belief that it is formed by the ready decarboxylation of an unstable acid. Such an acid is a β-keto acid. Hence, a possible explanation may be put forward that perhaps a penilloaldehyde-carboxylic acid (penaldic acid) is formed as an intermediate in the hydrolysis of penicillin, thus

$$\begin{array}{ccccc} O & & O & & \\ \| & & \| & & \\ R\!-\!C\!-\!NH\!-\!CH\!-\!C\!-\!H & \longrightarrow & CO_2 + R\!-\!CNH\ CH_2\!-\!C\!-\!H \\ & | & & \| \qquad\qquad\quad \| \\ & COOH & & O \qquad\qquad\quad O \end{array}$$

Penaldic Acid   Penilloaldehyde

6. *Combination of Penicillamine and Penilloaldehyde in Penicillin*

It has been observed that the hydrolysis of penicillin with dilute alkali or with the enzyme (*penicillinase*) yields penicilloic acid (a dicarboxylic acid), which readily eliminates a molecule of carbon dioxide to form penilloic acid, thereby suggesting that a carboxyl group is present in the β-position with regard to a negative group.

7. *Presence of Thiazolidine Ring*

It has been established experimentally that penilloic acid upon hydrolysis with aqueous mercuric chloride yields penicillamine and penilloaldehyde respectively. This type of hydrolysis is characteristic of compounds containing a thiazolidine ring, *i.e.,*

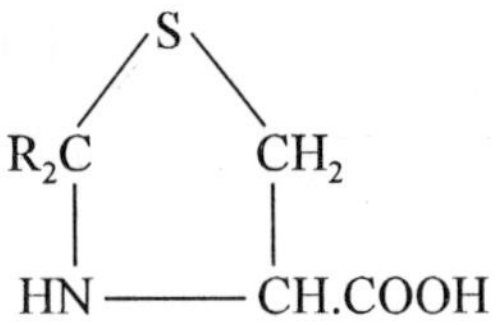

Thiazolidine

Hence, penilloic acid could be (I), because this particular structure would give the above required products. Thus

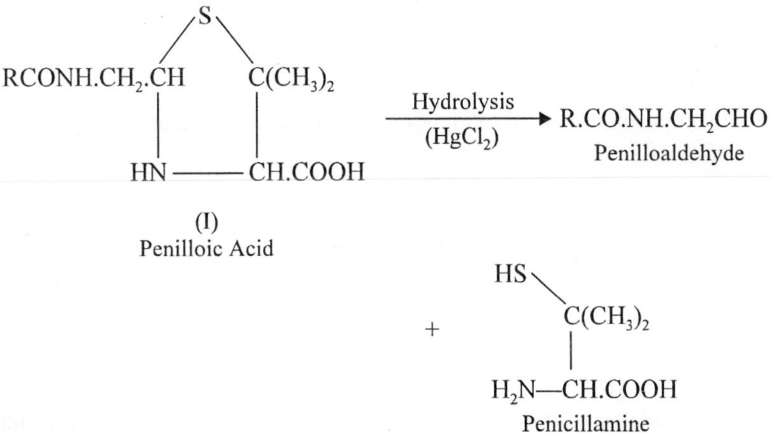

(I)
Penilloic Acid

$$\text{R.CO.NH.CH}_2\text{CHO}$$
Penilloaldehyde

+

Penicillamine

Therefore, if (I) is penilloic acid, then penicilloic acid would be (II)-

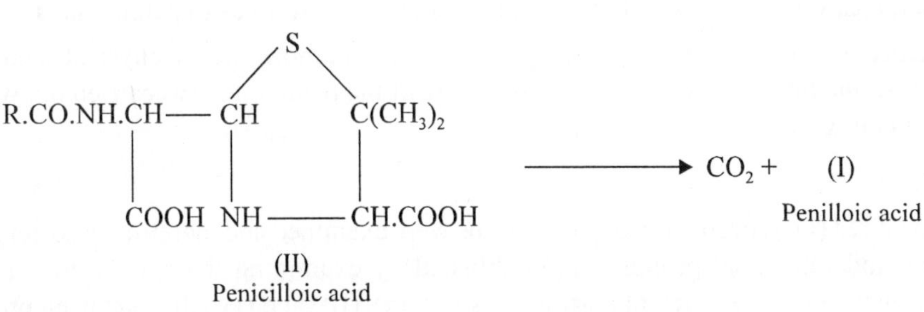

(II)
Penicilloic acid

$$\longrightarrow \text{CO}_2 + \quad \text{(I)}$$
Penilloic acid

8. *Evidence for Structure (II)*

The treatment of penicillin with methanol yields the corresponding ester methyl penicilloate which, on hydrolysis with aqueous mercuric chloride, gives methyl penaldate and penicillamine. Thus

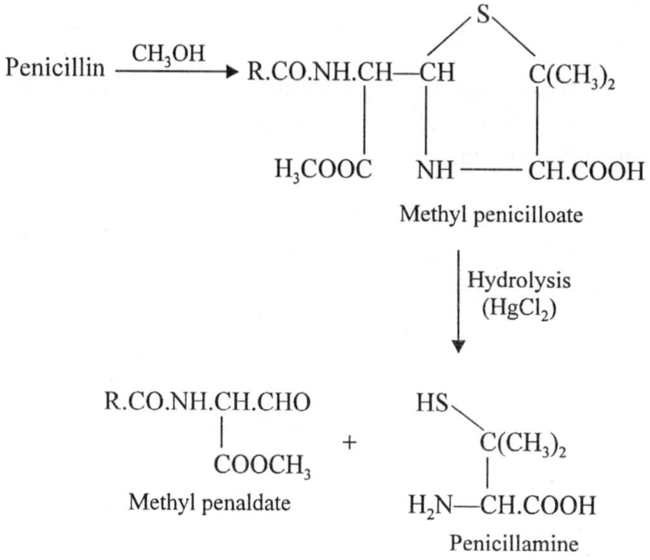

Methyl penicilloate

Hydrolysis (HgCl$_2$)

R.CO.NH.CH.CHO
|
COOCH$_3$

Methyl penaldate

+

HS—C(CH$_3$)$_2$
|
H$_2$N—CH.COOH

Penicillamine

9. *Probable Structures for Penicillin*

Based on the foregoing chemical evidences two probable structures for penicillin have been put forward *viz* ; (III) and (IV) ;

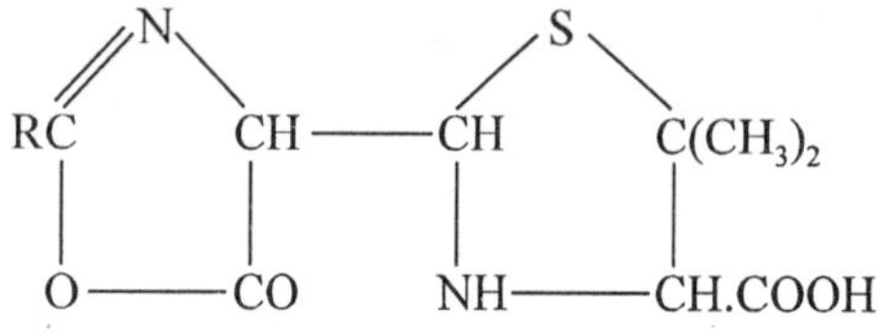

(III)
Oxazolone Structure

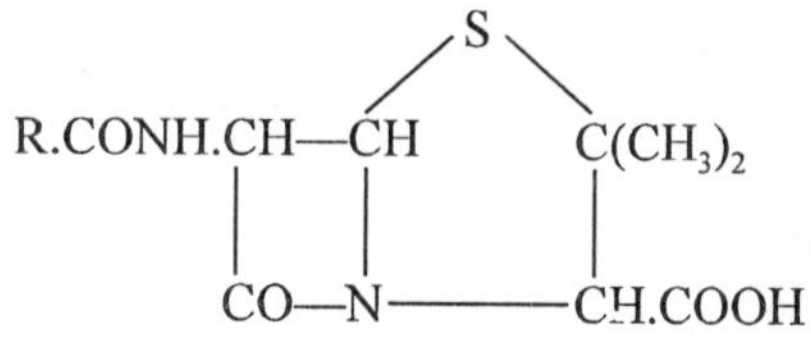

(IV)
β-Lactam Structure

At this juncture, however, it was not quite possible to decide between the two structures (III) and (IV) on the ground of chemical evidence alone, since penicillin is prone to undergo abrupt molecular rearrangement, *e.g.,* on treatment with dilute acid, penicillin rearranges to penillic acid.

Therefore, it was absolutely necessary to examine the molecule by physical methods (thereby leaving the molecule intact). In fact, an intensive study of the penicillins was carried out with respect to their infra-red and X-ray diffraction analysis.

(a) Infra-Red Analysis

The infra-red (i.r.) spectra of many penicillins were examined and therefrom a correlation between various bands and functional groups was established by examining the spectra of synthetic model compounds which contained different portions of structures (III) and (IV) that had been proposed above on the basis of chemical evidence.

This fact may be further illustrated by taking into consideration the methyl ester as well as the sodium salt of benzyl penicillin that exhibited the following characteristic peaks of all the penicillins in these regions :

Penicillin (as)	Characteristic Peaks (cm^{-1})
Methyl Ester :	3333, 1770, 1748, 1684, 1506
Sodium Salt :	3333, 1770, 1613, 1681, 1515

The band at 3333 cm^{-1} in both compounds was due to the NH group (str.)

The 1748 cm^{-1} band of methyl ester and the 1613 cm^{-1} band of the corresponding salt were assigned to the carbonyl group (str.) in the carboxyl group (as ester or salt respectively).

Further, model oxazolones were studied that showed two characteristic bands ; one at 1825 cm^{-1} for the carbonyl group and the other at 1675 cm^{-1} for the C = N moiety.

However, the absence of the first band but possible presence of the second in the benzylpenicillin derivatives would not allow an ultimate decision to be reached between (III) and (IV).

On specifically examining a large number of thiazolidines in the double bond region down to 1470 cm^{-1}, only the carbonyl bond was revealed to be present (~ 1748 and 1613 cm^{-1}).

$$[Z = - COOCH_3 ; - COO^{\ominus} \text{ etc.}]$$

Oxazolones   Thiazolidines

A number of 1°, 2° and 3° amides were studied extensively. It was found that all the three types showed a characteristic band close to 1670 cm^{-1} which may be attributed to the carbonyl group ; but in the case of the primary amides there was an additional band at 1613 cm^{-1} and with the secondary amides the band was found to be close to 1515 cm^{-1}. These findings reveal that the penicillins possess the secondary amide structure (IV), because the secondary amide band at 1670 cm^{-1} was almost equal to 1684 and 1681 cm^{-1}, besides the band at 1515 cm^{-1} was equivalent to 1506 and 1515 cm^{-1}. Thus, in all, four out of five bands have been accounted for duly.

Finally, a large number of β-lactams and fused thiazolidine-β-lactams were studied intensively. The former category of compounds did not display a band near 1770 cm^{-1}, but all the latter were found to exhibit a band at 1770 cm^{-1}. This ultimately accounts for the fifth band, and hecne it follows that (IV) is the structure of the penicillins.

(b) X-Ray Diffraction Analysis

The **X-ray diffraction analysis** of the sodium, potassium and the rubidium salts of the benzylpenicillin showed the presence of a β-lactam ring, thereby further supporting the fact that structure (IV) is the probable structure of penicillin.

Based on this structure (IV) and also the various chemical reactions studied so far, one may summarise them as described below :

3.1.3. Chemical Reactions of the Penicillins

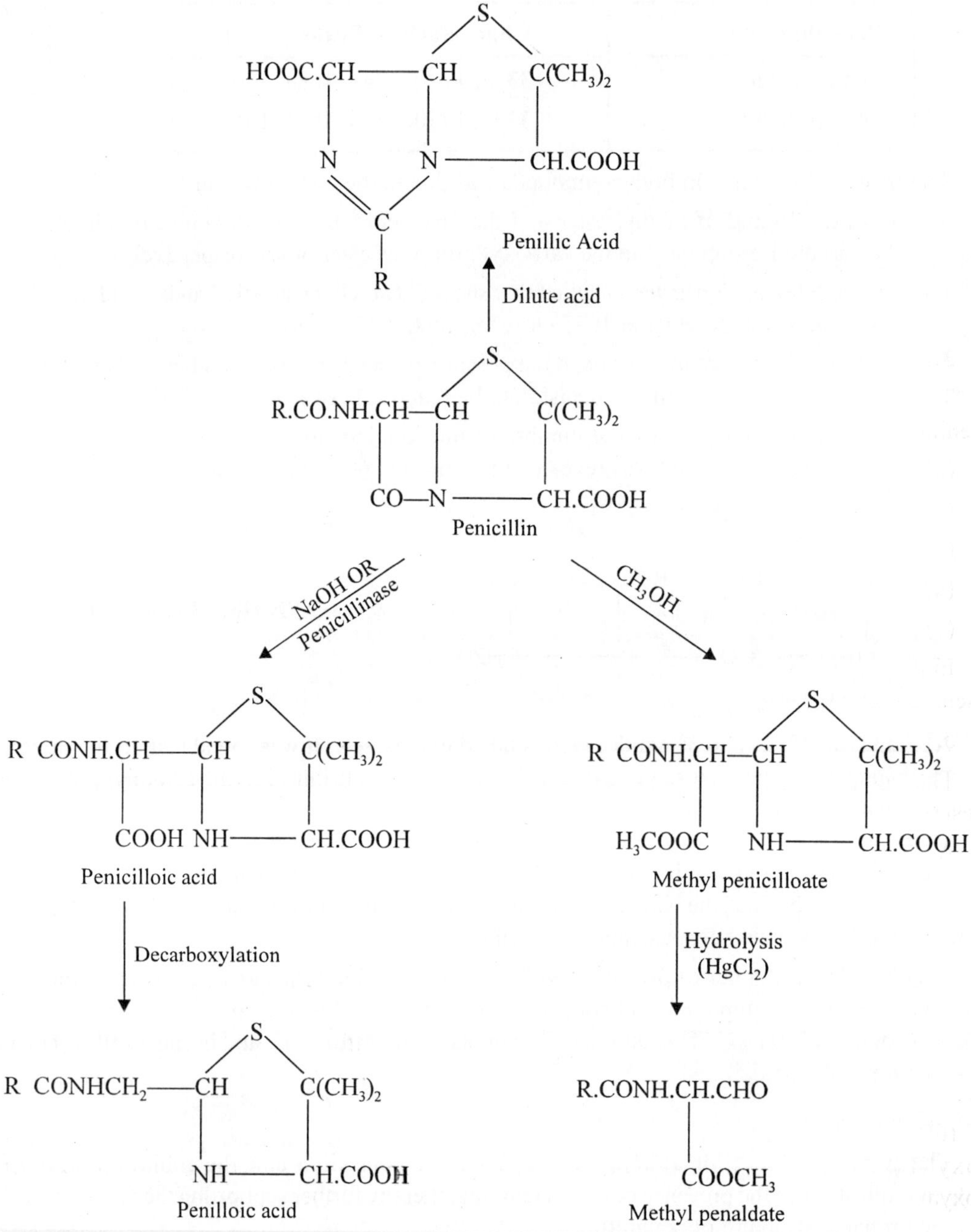

HOOC.CH————CH   C(CH$_3$)$_2$

N   N————CH.COOH

C

R

Penillic Acid

↑ Dilute acid

R.CO.NH.CH——CH   C(CH$_3$)$_2$

CO—N————CH.COOH

Penicillin

NaOH OR
Penicillinase

CH$_3$OH

R CONH.CH————CH   C(CH$_3$)$_2$

COOH NH————CH.COOH

Penicilloic acid

R CONH.CH——CH   C(CH$_3$)$_2$

H$_3$COOC   NH————CH.COOH

Methyl penicilloate

↓ Decarboxylation

↓ Hydrolysis
(HgCl$_2$)

R CONHCH$_2$————CH   C(CH$_3$)$_2$

NH————CH.COOH

Penilloic acid

R.CONH.CH.CHO

COOCH$_3$

Methyl penaldate

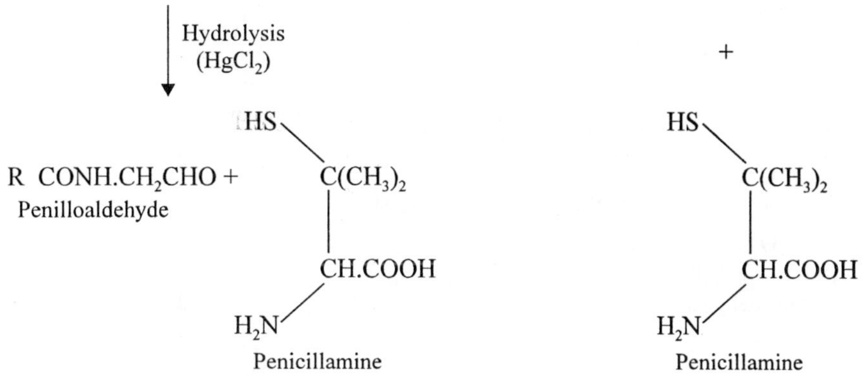

3.1.4. The Penicillin Variants

The advent of latest developments in 'medicinal chemistry', in fact, put forward the following *five* penicillin variants, namely :

(*a*) Natural Penicillins (best streptococcal and narrow spectrum)

(*b*) Penicillinase-resistant Penicilins (antistaphylococcal)

(*c*) Aminopenicillins (improved Gram – ve : *H-influenzae, Enterococcus, Shigella, Salmonella)*,

(*d*) Extended-spectrum (antipseudomonal) penicillins, and

(*e*) β-Lactamase combinations (expand spectrum to staph, β-lactamase producers).

In this particular section a few typical examples from each category of *penicillin variants* shall be discussed comprehensively.

3.1.4.1. Natural Penicillins (best streptococcal and narrow spectrum)

The first successful synthesis of penicillin was carried out by Sheehan *et al.* (1957, 1959) who synthesized Penicillin-V (*i.e.,* phenoxymethyl-penicillin).

A. *Phenoxymethylpenicillin* INN, BAN *Penicillin*-V USAN,

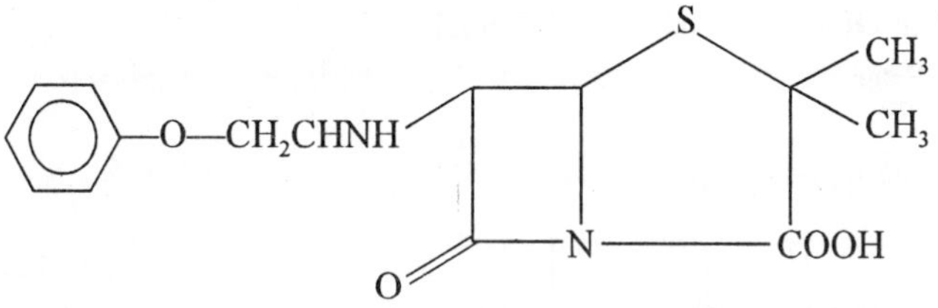

(6R)-6-2-(-2-Phenoxyacetamido) pencillanic acid ; 4-Thia-1-azabicyclol [3,2,0]-heptane-2-carboxylic acid, 3, 3-dimethyl-7-oxo-6- [(phenoxyacetyl) amino]-, [2S-(2α, 5α, 6β)]- ; Phenoxymethylpenicillin (BP ; Eur. P ; Int. P.,) ; Penicillin-V (USP) ; V-Cillin[(R)] (Lilly).

Synthesis :

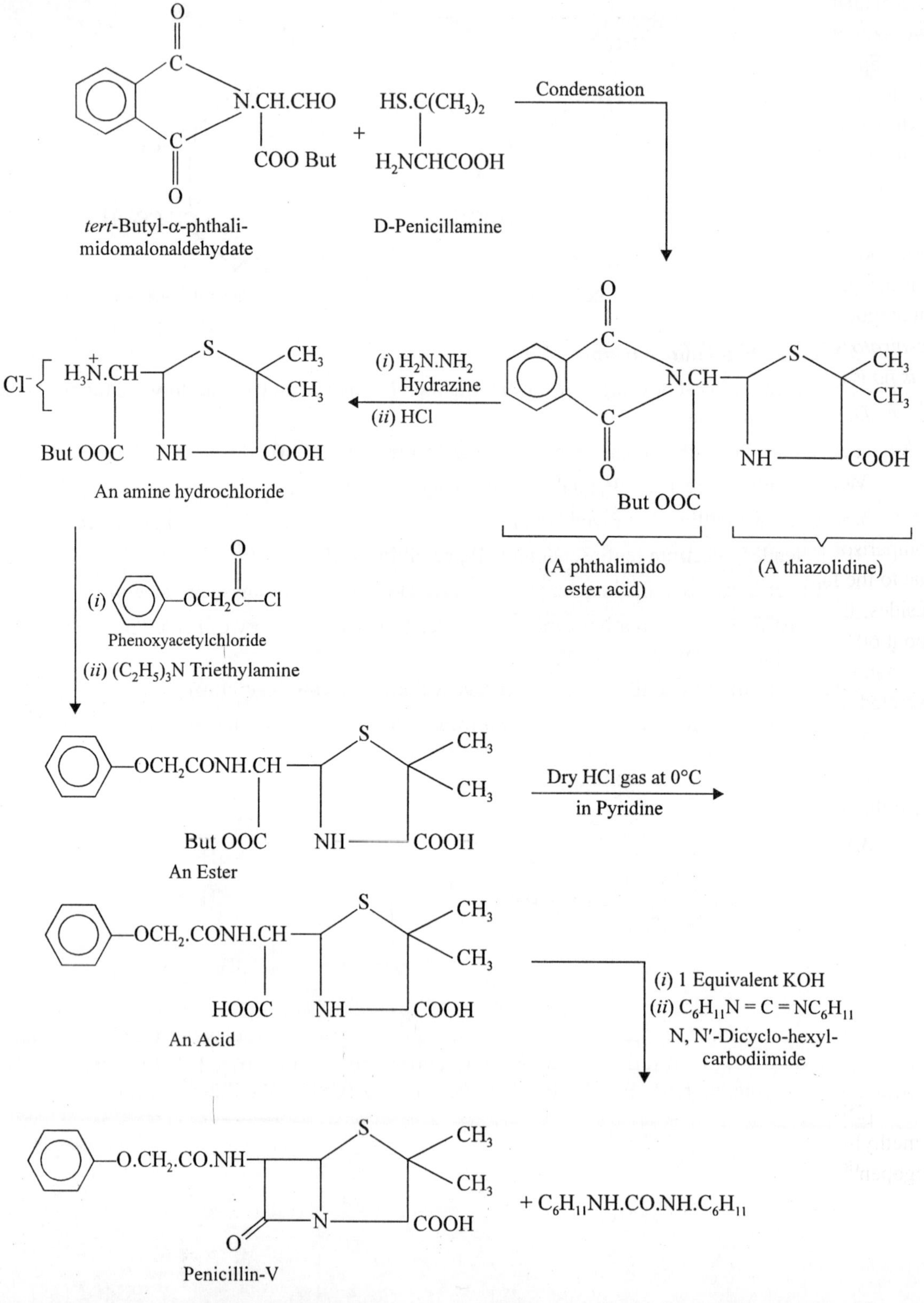

tert-Butyl-α-phthali-
midomalonaldehydate

D-Penicillamine

Condensation

(A phthalimido ester acid)

(A thiazolidine)

(*i*) $H_2N.NH_2$
Hydrazine
(*ii*) HCl

Cl^- { $H_3\overset{+}{N}.CH$ — S — $C(CH_3)_2$ ...

An amine hydrochloride

(*i*) $\langle\bigcirc\rangle$—$OCH_2\overset{O}{C}$—Cl
Phenoxyacetylchloride
(*ii*) $(C_2H_5)_3N$ Triethylamine

An Ester

Dry HCl gas at 0°C
in Pyridine

An Acid

(*i*) 1 Equivalent KOH
(*ii*) $C_6H_{11}N = C = NC_6H_{11}$
    N, N'-Dicyclo-hexyl-
    carbodiimide

Penicillin-V

$+ C_6H_{11}NH.CO.NH.C_6H_{11}$

Condensation of D-Penicillamine and *tert*-butyl-α-phthalimidomalonaldehydate yields an intermediate embedded with a phthalimido ester acid and a thiazolidine ring. Treatment of this with hydrazine followed by hydrochloric acid helps the removal of phthaloyl moiety as phthalhydrazide and gives an amine hydrochloride. The resulting product on further treatment with phenoxyacetylchloride in triethylamine introduces the side chain to yield an ester. This product when subjected to a stream of hydrogen chloride gas at 0°C in pyridine helps to remove the blocking *tert*-butyl function thereby yielding penicilloic acid. The resulting acid undergoes cyclization to afford penicillin-V by stirring for 20 minutes with a solution of N, N′-dicyclohexyl carbodiimide in dioxane.

Penicillin-V is paritcularly effective in the management and control of infections caused by gram-positive bacteria, namely ; streptococcal, staphylococcal, pneumococcal and clostridial infections. It is also the drug of choice in the treatment of a number of gram-positive bacteria *viz.*, gonococcal and meningococcal infections. Besides, it is also used exclusively in the *treatment of pneumonia and other respiratory tract infections caused by Staphylococcus aureus, B. anthracis and Streptomyces pyrogenes. It is now solely used in the treatment of both syphillis and gonorrhea.*

Dose. Oral, adults and children over 12 years of age or older, usually 125 to 500 mg 3 to 4 times a day.

Mechanism of Action. The mechanism of action of penicillin-V shall be discussed as under :

3.5.1.1. Penicillin-V. The '*drug*' gets inactivated to a relatively lesser extent by gastric juice in comparison to penicillin G. It is, however, the *most preferred oral penicillin* for *less serious infections* due to the fact that serum-levels are found to be 2-5 times higher than matching doses of penicillin G ; besides, there exists relatively less variability in its degree of absorption. The oral bioavailability is about 60% at the most. It gets bound to plasma proteins between 75-80%. The volume of distribution v_d^{ss} stands at 0.73 mL. g^{-1}, which is significantly much higher than that of penicillin G. The '*drug*' gets excreted unchanged in the urine between 20-40% . The plasma half-life is nearly 0.5 to 1. hour.

3.1.4.2. Penicillinase-resistant Penicillins (antistaphylococcal) :

A few typical examples belonging to this class of penicillins are, namely : cloxacillin, methicillin, oxacillin etc., which would be treated individually in the sections that follows :

3.1.4.2.1. Cloxacillin Sodium USAN, INN,

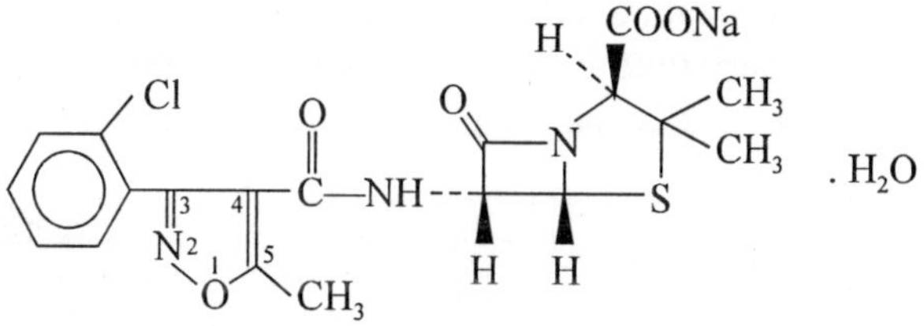

[2S-(2α, 5α, 6β)-4-Thia-1-azabicyclo [3, 2, 0] heptane-2-carboxylic acid 6-[[[3-(2-chlorophenyl)-5-methyl-4-isoxazolyl] carbonyl] amino]- 3, 3-dimethyl-7-oxo-, monosodium salt, monohydrate ; Tegopen$^{(R)}$; Cloxapen$^{(R)}$; USP ;

Synthesis :

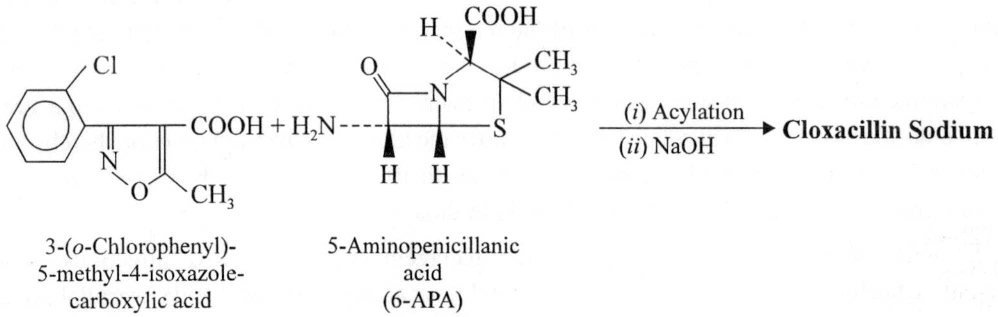

| 3-(*o*-Chlorophenyl)-
5-methyl-4-isoxazole-
carboxylic acid | 5-Aminopenicillanic
acid
(6-APA) | | |

6-APA is duly acylated with 3-(*o*-chlorophenyl)-5-methyl-4-isoxazole carboxylic acid and the resulting cloxacillin base is adequately purified by recrystallization. The base thus obtained is converted to the corresponding sodium salt by treating with an equimolar concentration of NaOH.

It is a pencillinase-resistant penicillin (antistaphylococcal) usually administered orally.

Potency : Equivalent of not less than 825 mcg of cloxacillin per mg.

3.1.4.2.2. Methicillin Sodium. USAN, INN,

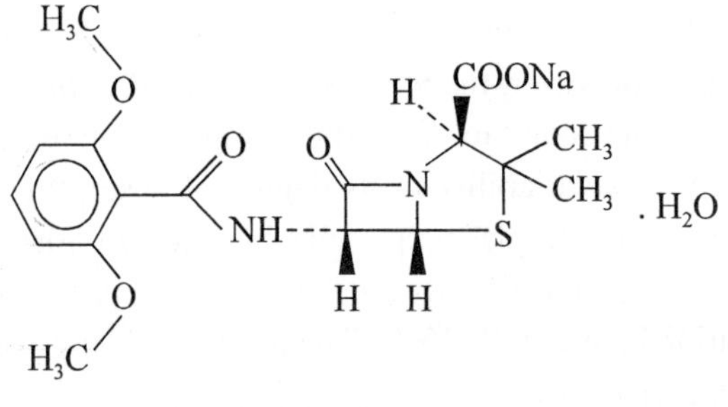

4-Thia-1-azabicyclo [3, 2, 0] heptane-2-carboxylic acid, 6-[(2, 6-dimethoxybenzoyl) amino]-3, 3-dimethyl-7-oxo-, monosodium salt, monohydrate, [2S-(2α, 5α, 6β)]- ; USP ; Staphcillin[R] ;

The official compound may be prepared by condensing the fermentation-produced 6-APA in an appropriate solvent with 2, 6-dimethoxybenzoyl chloride ; and the resulting methicillin is subsequently precipitated as its corresponding sodium salt by the addition of sodium acetate.

It is usually indicated in the treatment of staphylococcal infections caused by strains resistant to other penicillins. As a precautionary note it is recommended that this '*drug*' should **not** be used in general treatment and theraphy so as to avoid the possibility of widespread development of organisms resistant to it.

Mechansim of Action. The '*drug*' is specifically resistant to inactivation by the presence of the enzyme **penicillinase** found in staphylococci. It has been observed to induce penicillinase formation which specifically restrains its usage in the control, management and treatment of penicillin G-sensitive infections.

SAR of Methicillin. The steric hindrence afforded by the presence of 2, 6-dimelthoxy moieties categorically renders the '*drug*' resistant to enzymatic hydrolysis.

Note. Methicillin gives rise to higher incidence of intestinal nephritis which is reportedly higher in comparison to other penicillins.

3.1.4.2.3. Oxacillin Sodium USAN, INN,

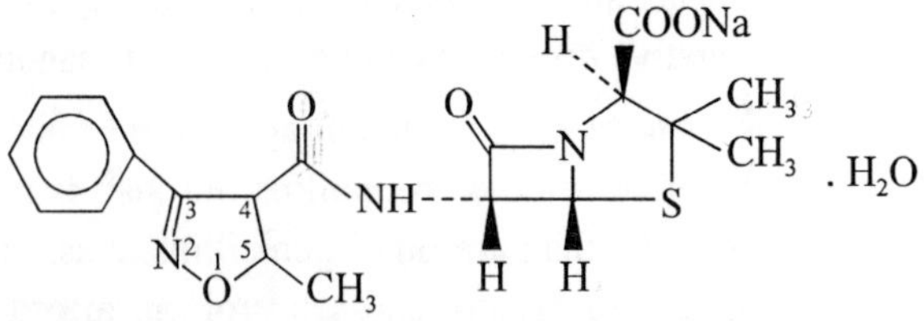

[2S-(2α, 5α, 6β)]-4-Thia-1-azabicyclo [3, 2, 0] heptane-2-carboxylic acid, 3, 3-dimethyl-6-[[5-methyl-3-methyl-4-isoxazolyl) carbonyl]-amino]-7-oxo-, monosodium salt, monohydrate ; USP ; Bactocill[R] ; Prostaphlin[R] ;

6-APA produced by fermentation is subjected to condensation with 5-methyl-3-phenyl-4-isoxazolyl chloride in an appropriate organic solvent. The resulting oxacillin base is precipitated as the sodium salt by the addition of requisite quantum of pure sodium acetate.

The use of the '*drug*' must be restricted to the treatment of infections produced by staphylococci resistant to penicillin G.

Mechanism of Action. The '*drug*' is reasonably well absorbed from the GI-tract, specifically in fasting subjects. It has been observed that the effective plasma levels of it are easily achievable in about 1 hour ; however, its extensive plasma-protein binding, it gets excreted quickly through the kidneys. The '*drug*' affords first pass metabolism in the liver to the corresponding 5-hydroxymethyl derivative. Interestingly, the resulting '**metabolite**' exhibits antibacterial acitivity fairly comparable to that of oxacillin, but observed to be less intimately protein bound and more readily excreted.

SAR of Oxacillin. Evidently, the steric effects due to the presence of inherent 3-phenyl and 5-methyl moieties of the isoxazolyl ring not only prevent the binding of this penicillin to the β-**lactamase active site** but also afford protection to the ensuing lactam ring from degradation. The '*drug*' is found to be comparatively resistant to acid hydrolysis and ; hence, could be given orally with good pharmacologic effect.

3.1.4.3. Aminopenicillins (improved Gram –ve : H. influenzae, Enterococcus, Shigella, Salmonella) :

The medicinal compounds that are to be discussed in this category are, namely : ampicillin and bacampicillin.

3.1.4.3.1. Ampicillin USAN, BAN, INN,

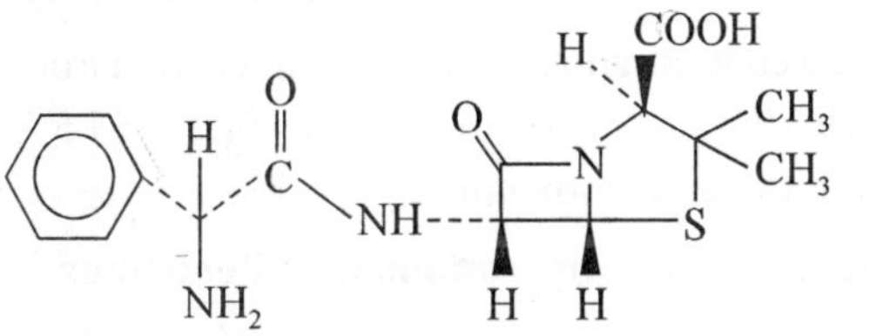

[2S-[2α, 5α, 6β (S*)]]-4-Thia-1-azabicyclo [3,2,0] heptane-2-carboxylic acid, 6[(aminophenylacetyl) amino]-3, 3-dimethyl-7-oxo ; USP ; Penbriten[R] ; Polycillin[R] ; Omnipen[R] ; Amcill[R] ; Principen[R] ;

Ampicillin has an antibacterial spectrum broader than that of pencillin G. It is active against the same Gram-positive organisms which are susceptible to other penicillins. Besides, it is also found to be more active against certain Gram-negative organisms and enterococci than are other penicillins.

Mechanism of Action. The *'drug'* exerts its action against the same species of Gram-positive organisms which are not only susceptible to other penicillins but also is more active against certain Gram-negative organisms and enterococci than are other penicillins. It is, however, quite evident that the α-amino functional moiety does play a pivotol and vital role in affording its wider activity, but unfortunately the exact mechanism for its action is not yet established. It has been proposed that the α-amino moiety inducts its ability to cross cell wall barriers which are otherwise believed to be impenetrable to other penicillins.

SAR of Ampicillin. The D-(–) ampicillin is found to be more active appreciably in comparison to its isomer L-(+) ampicillin. The α-amino function in ampicillin gets protonated extensively in an acidic media that perhaps legitimately offers a satisfactory explanation with respect to ampicillin's stability to acid hydrolysis, and observed instability to alkaline hydrolysis.

3.5.3.2. Bacampicillin Hydrochloride USAN, BAN, INN

4-Thia-1-azabicyclo [3,2,0] heptane-2-carboxylic acid, [2S-[2α, 5α, 6β(S*)]]-6-[(aminophenylacetyl) amino]-3,3-dimethyl-7-oxo-,1-[(ethoxycarbonyl) oxyethyl] ester, monohydrochloride ; USP ; Spectrobid[R] ;

It is an *aminopenicillin oral prodrug* which gets converted to ampicillin *in vivo* after undergoing hydrolysis rapidly in the presence of esterases in plasma.

It has been established beyond any reasonable doubt that its oral absorption is both rapid and virtually complete in comparison to ampicillin and also less affected by food.

Interestingly, the plasma concentrations of ampicillin released from bacampicillin evidently exceed from those of oral pure *ampicillin* or *amoxicillin* for the first 2 to 5 hours but after that the pattern remains the same as for ampicillin and amoxicillin.*

3.1.4.4. Extended-Spectrum (Antipseudomonal) Penicillins

The potent penicillin variants belonging to this category which shall be described here are : piperacillin and ticarcillin.

---

*Neu HC : *Rev. Infect. Dis.* **3** : *110, 1981.*

3.1.4.4.1. Piperacillin Sodium. USAN, INN,

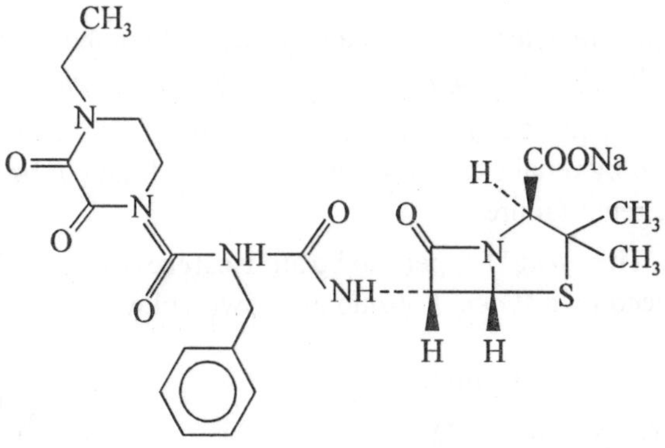

[2S-[2α, 5α, 6β (S*)]]-4-Thia-1-azabicyclo [3,2,0] heptane-2-carboxylic acid, 6-[[[[4-ethyl-2, 3-dioxo-1-piperazinyl) carbonyl] amino]-phenylacetyl] amino]-3, 3-dimethyl-7-oxo-, monosodium salt ; Pipracil[R]; USP;

It is found to be the most active penicillin variant against *Ps aeruginosa,* having a potency almost matching to that of gentamycin. It distinctly shows more activity against *Klebsiella* and several other enteric bacilli in comparison to either *carbenicillin* or *ticarcillin.* It is invariably employed as an '*alternative drug*' for specific use against infections caused by various pathogenic strains, such as : *Acinetobacter, Bacteroides fragilis* (GI-strains), Enterobacter, *E. coli, Kl pneumoniae, Morganella morganii, Pr mirabilis or vulgaris, Providencia rettgeri or stuartii, Ps aeruginosa* (UTIs) or *Serratia.* Besides, it exhibits a low efficacy against penicillinase—and other members of the β-lactamase-producing organisms.

Mechanism of Action. The '*drug*' gets destroyed readily by stomach acid (HCl) ; and, therefore, it is only active when adminstered either by IM or IV. It has been observed that the β-lactamase susceptibility of this '*drug*' is not '*absolute*' by virtue of the fact that β-lactamase producing, ampicillin-resistant strains of *N-gonorrhoeae* and *H. influenzae* are found to be susceptible to it.

3.1.4.4.2. Ticarcillin Disodium. USAN, BAN, INN,

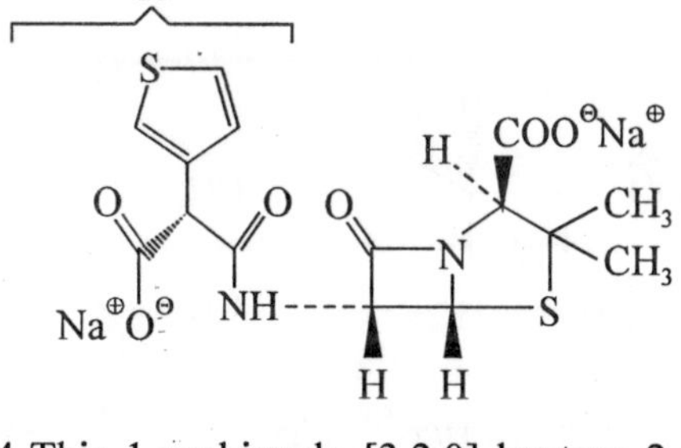

[2S-[2α, 5α, 6β (S*)]]-4-Thia-1-azabicyclo [3,2,0] heptane-2-carboxylic acid, 6-[(carboxy-3-thienylacetyl) amino]-3, 3-dimethyl-7-oxo-, disodium salt ; USP ; Ticar[R] ;

The '*drug*' possesses an almost similar antibacterial profile and pharmacokinetic characteristics as those of *carbenicillin (I).* Because, of the following *two* positive advantages ticarcillin proved to be highly crucial and vital in the treatment of serious infections (*e.g., Ps aeruginosa*) that may require a *high-dose therapy* :

(a) Possesses a slightly better pharmacokinetic characteristics *viz.,* elevated serum levels, and longer duration of action.

(b) Higher *in vitro* antibacterial activity against a good number of species belonging to Gram-negative bacilli *e.g., Ps aeruginosa* and *Bacteroides fragilis.*

In plasma, the '*drug*' is protein-bound to the extent of 55-65%. The volume of distribution is 0.22 mL. g^{-1}. It is eliminated by renal excretion. The half-life is found to be 0.5 to 1 hr, except 15 hour in the instance of complete renal failure.

SAR of Ticarcillin. The '*drug*' happens to be an '**isostere**' of carbenicillin wherein the '*phenyl*' moiety has been duly replaced by a '*thienyl*' group as shown below :

Carbenicillin (I)   Ticarcillin

3.1.4.5. β-Lactamase Combinations (expand spectrum to staph-, β-lactamase producers) :

In general, it has been observed that the **penicillinase—resistant penicillins** normally get bound to the *penicillinases* ; however, the actual dissociation of the '*drug*'-*enzyme complex* is rather quite rapid. In actual practice, they have been successfully supplanted by *three* substances, namely : *clavulanic acid, sulbactam* and *tazobactam.* In fact, all these are regarded as newer breeds of β-lactamase inhibitors that specifically acylate the enzymes by creation of a '*double-bond*' (greater electronic bondage) and consequently afford dissociation very slowly, thereby significantly enhancing the potency of the penicillins against certain organisms and ultimately increase their therapeutic efficacy*.

The combination of β-lactam inhibitors with other antibiotics helps to expand the spectrum of the antibiotic to a significant extent which may be observed evidently by carrying out the *in vitro* studies.

However, there are *three* important β-lactamase inhibitors duly recognized, namely : clauvulanic acid, sulbactam, and tazobactam as given under : Clavulanic acid Sulbactam Tazobactam

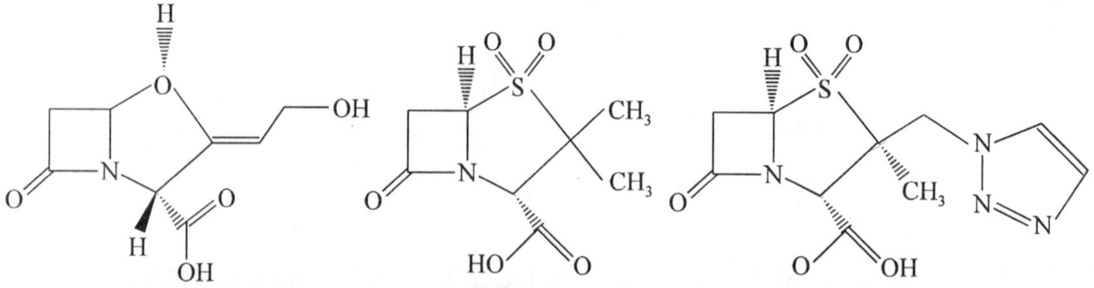

Clavulanic acid   Sulbactam   Tazobactam

A few typical examples of these β-*lactamase inhibitors* in combination with other '*antibiotics*', which are available commercially, shall now be discussed individually as under :

---

*Kar A., *Pharmacognosy and Pharmacobiotechnology,* New Age International, New Delhi, p. 736, 2003.

3.1.4.5.1. Clavulanate-Amoxicillin. A combination administered orally. It causes *more diarrhea than amoxicillin.*

3.1.4.5.2. Clavulanate-Ticarcillin. A combination given IV. It is active versus *more Gram-negative bacilli.*

3.1.4.5.3. Sulbactam-Ampicillin. A combination given IV. It is active versus *Staphylococcus* and β-lactamase producing *H. influenzae* and *Streptococcus pneumoniae.*

3.1.4.5.4. Tazobactam-Piperacillin. A combination given IV. It is active versus *more Gram-negative bacilli.*

3.1.5. Other Clinically Useful Derivatives of Penicillin-V

Name	Official Status	Brand Name(s)	Dose
Penicillin V Benzathine	USP,	—	—
Phenoxymethylpenicillin Calcium	BP, Int. P.;	—	—
Penicillin V Hydrabamine	USP (XX)	—	—
Penicillin V Potassium	BP, USP, Int Int. P., Ind. P.;	Pfizerpen[R] (Pfizer) ; Panapar VK[R] (Parke-Davis)	—

3.1.6. Structures of Some Cinically Useful Penicillins

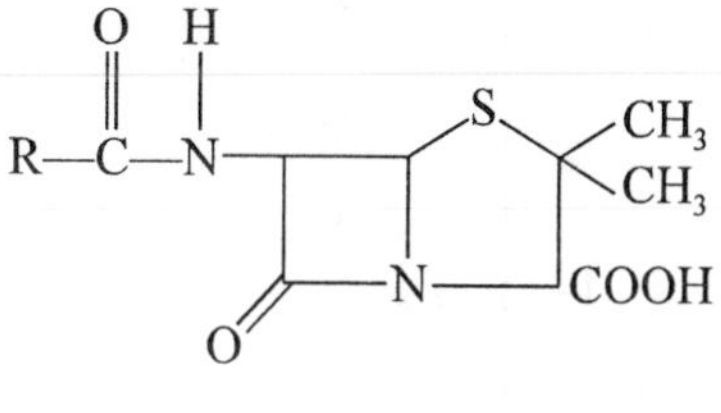

Name	R	Official Status	Brand Name(s)	Dose
Cloxaxillin Sodium	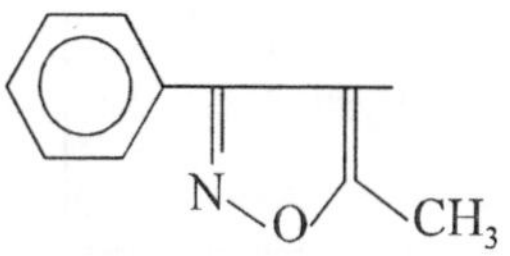	BP; USP; Int. P.;	Cloxapen[R] (Beecham)	500 mg 4 times daily
Methicillin Sodium	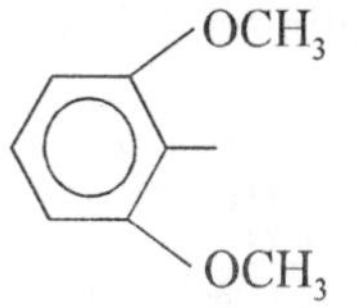	B.P. (1973), USP ; Int. P.	Azapen[R] (Pfizer) ; Celbenin (Beecham) ;	1g IM every 4 to 6 hours

| Nafcillin Sodium | 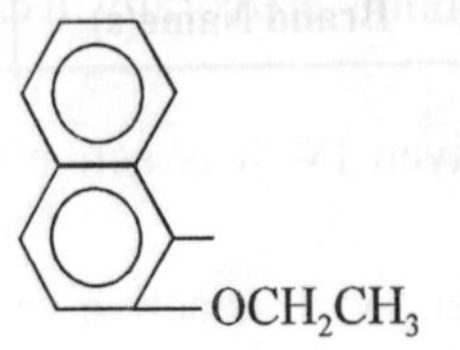 | USP ; Int. P. ; | Unipen(R) (Wyeth) | 500 mg (*e.g.,* of Nafcillin) every 4 to 6 hours |
| Mecillinam Sodium | 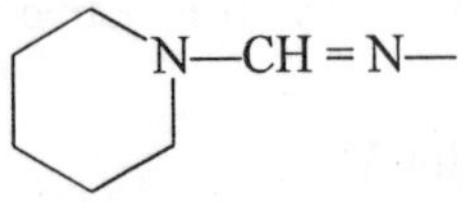 | | Coactin[R] (Hoffmann-La Roche) | 12.5 to 15 mg per kg every 6 hours |

3.1.7. Structures of Some Clinically Useful Penicillins Related to Ampicillin

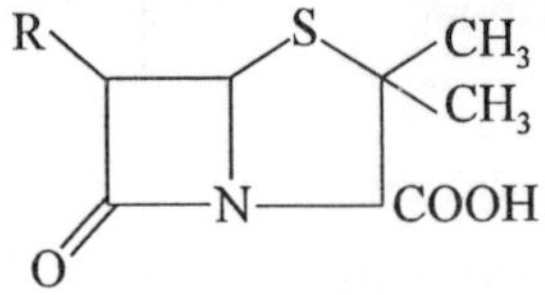

Name	R	Official Status	Brand Name(s)	Dose
Ampicillin	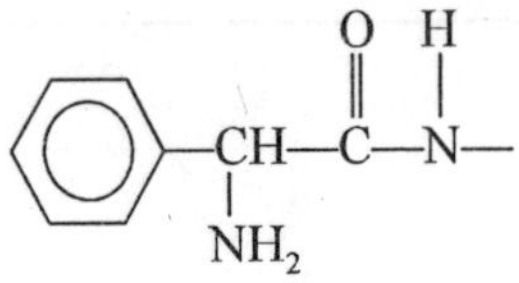	BP; USP; Int. P.;	Amcil[R] (Parke-Davis) ; Omnipen[R] (Wyeth) ; Pfizerpen A[R] (Pfizer)	150-300 mg per kg body weight daily
Amoxicillin Trihydrate	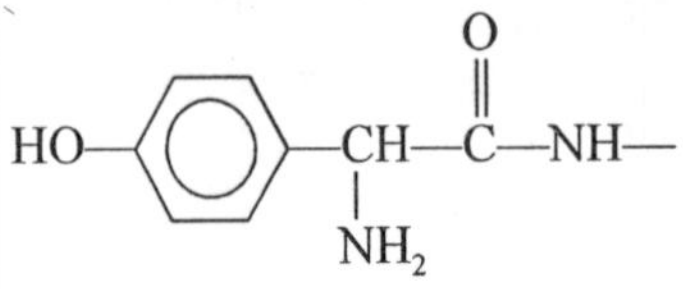	BP ; USP ;	Amoxil[R] (Bencard, U.K.) ; Utimox[R] (Parker-Davis)	500 mg IM every 8 hours
Ciclacillin	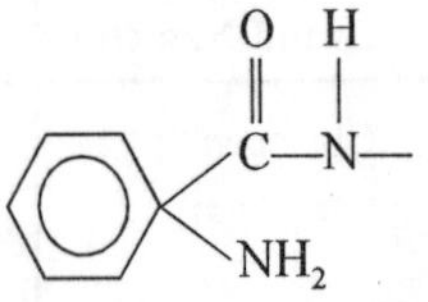	USP ;	Cyclapen-W[R] (Wyeth)	150-500 mg 4 times daily

3.1.8. Structures of some Cinically Useful Ester of Ampicillin

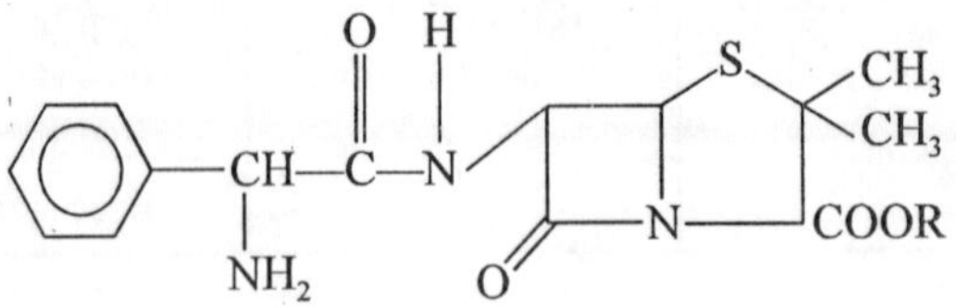

Name	R	Official Status	Brand Name(s)	Dose
Pivampicillin Hydrochloride	$-CH_2O-\overset{\overset{O}{\|\|}}{C}-C(CH_3)_3$	—	Pondocillin(R) (Burgers, U.K.) ;	500 mg 3 to 4 times daily
Talampicillin Napsylate	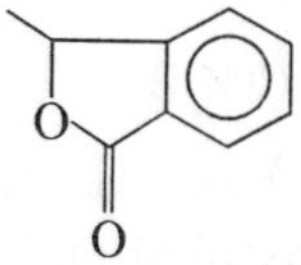	—	Talpen(R) (Beecham Research, U.K.)	250 to 500 mg 3 times daily

3.2. CEPHALOSPORINS

After the spectacular world-wide recognition and tremendous success of the *penicillins*, the best known family of β-lactams are termed as the *cephalosporins*, wherein the β-lactam ring is strategically fused to a 6-membered *dihydrothiazine* ring system as shown below :

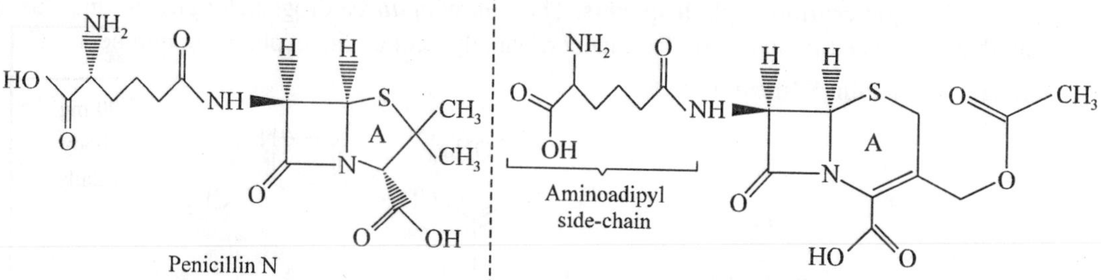

Penicillin N
[**Ring A : 5-membered**]

Caphalosporin C
[**Ring A : 6-membered**]

Giuseppe Brotzu's epoch making discovery, in 1945, in the species *cephalosporium* fungi obtained from *C. acremonium* showed a remarkable inhibition in the growth of a rather wide spectrum of both Gram-positive and Gram-negative organisms. Abraham and Newton (1961) at Oxford for the first time not only isolated successfully but also characterized cephalosporin C.* However, the confirmation of its structure was ascertained by X-ray crystallography.**

Inspite of the glaring evidence that cephalosporin C was resistant to *S. aureus* β-lactamase, besides its prevailing antibacterial activity was inferior in comparison to penicillin N and other penicillin structural analogues.

It has been observed critically that the *natural products* usually exhibit a relatively *lower level of antibacterial activity.* Therefore, the articulate and judicial *'cleavage'* of the amide bond of the **aminoadipyl side-chain** present in cephalosporin C provides **7-amino-cephalosporanic acid (7-ACA)**, which is most ideally suitable for the synthesis of a wide range of semisynthetic cephalosporins *via* acylation of the C(7)-amino functional moiety*** as depicted under :

*EP Abraham and GGF Newton, *Biochem J.*, **79**, 377, (1961)

DC Hodgkin and EN Masien, *Biochem. J.*, **79, 393, (1961).

***Fechtig B *et al.*, *Helv. Chim Acta.* **51**, 1108, 1968.

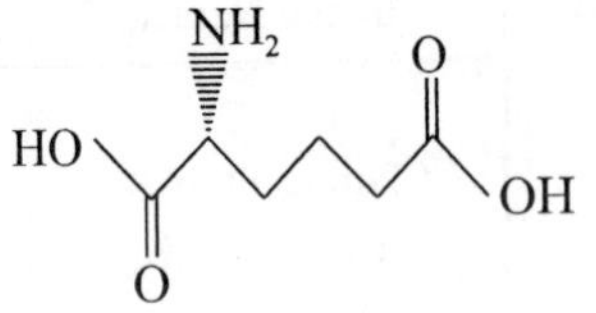

Aminoadipyl Side-chain
(Present in Cephalosporin C)

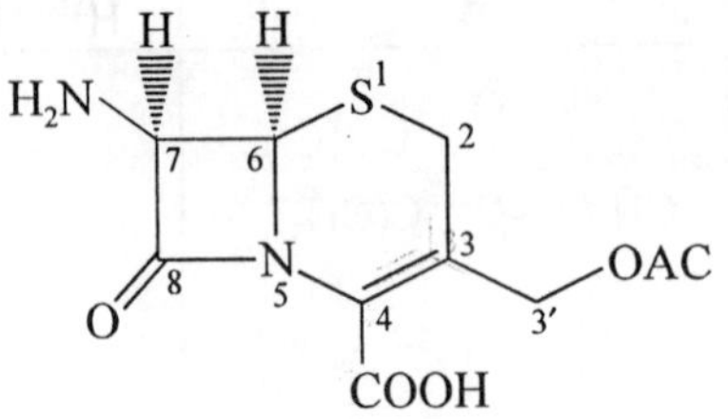

7-ACA

3.2.1. Classification. The 'cephalosporins' may be classified under the following *four* categories : Aminoadipyl

(*a*) First generation (staph, some enteric Gram-negative, bacilli)

(*b*) Second generation (more active *Vs* Gram-negative, some active *Vs H. influenzae* and anaerobes)

(*c*) Third generation (best Gram-negative spectrum, β-lactamase resistant, poor *Vs* staph.)

(*d*) Fourth generation.

A few typical examples of '*cephalosporins*' belonging to each of the above *four* generation shall now be discussed more explicitly in the sections that follows :

3.2.1.1. First generation cephalosporins. The following *three* drugs belonging to this class of compounds shall be treated in an elaborated manner, namely : cefazolin, cephalexin, and cephradine.

3.2.1.1.1. Cefazolin Sodium USAN ;

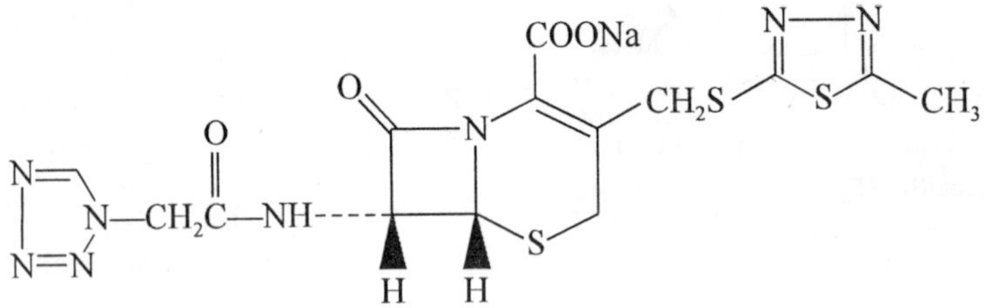

5-Thia-1-azabicyclo [4,2,0] oct-2-ene-2-carboxylic acid, 3-[[(5-methyl-(6R-*trans*)-1,3,4-thiadiazol-2-yl) thio] methyl]-8-oxo-7-[[(1H-tetrazol-1 yl) acetyl]-amino]-, monosodium salt, USP ; Ancef[(R)]; Kefzol[(R)];

Synthesis :

The acylation of the sodium salt of 7-aminocephalosporanic acid (*i.e.,* 7-ACA) with 1H-tetrazole-1-acetyl chloride gives rise to the formation of an intermediate with the elimination of a mole of HCl. The resulting product on being treated with 5-methyl-1, 3, 4-thiadiazole-2-thiol affords the displacement of the acetoxy moiety which upon treatment with an equimolar concentration of NaOH yields the official compound.

It is a *first-generation cephalosporins* given IM or IV. The '*drug*' may be employed to treat infections of the skin, bone, soft tissues, respiratory tract, urinary tract, and endocarditis and septicemia caused by susceptible organisms. It has been observed that amongst UTIs, cystitis responds much predominantly and better in comparison to *pyelonephritis.** It is regarded to be the preferred cephalosporin for most surgical prophylaxis due to its inherent long half-life.

*Inflammation of kidney and renal pelvis.

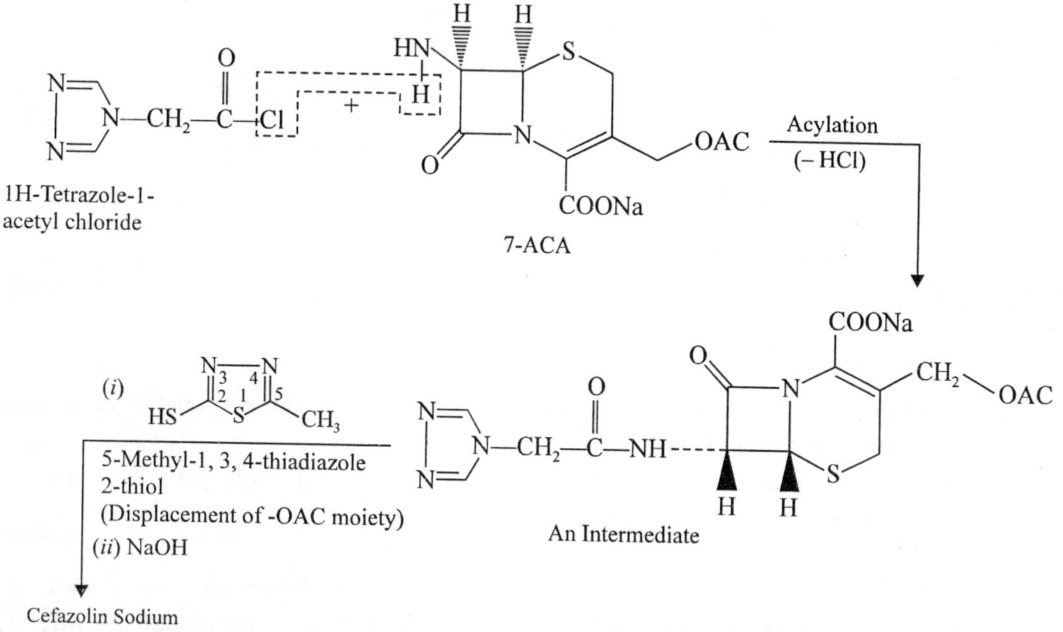

1H-Tetrazole-1-
acetyl chloride

7-ACA

(i) 5-Methyl-1, 3, 4-thiadiazole
2-thiol
(Displacement of -OAC moiety)
(ii) NaOH

An Intermediate

Cefazolin Sodium

Mechanism of Action. The '*drug*' possesses activity against Gram-positive organism, but exhibits a relatively narrow spectrum against Gram-negative strains due in part to their susceptibility to the β-lactamases. However, the Gram-negative activity essentially confined to *E. coli, Klebsiella* and *Pr mirabilis*. It has also been observed that certain *Gram-negative organisms* and *penicillinase-producing staphylococci* which are resistant to both *penicillin G* and *ampicillin* are evidently sensitive to **cefazolin**.

3.2.1.1.2. Cephalexin USAN, INN, BAN

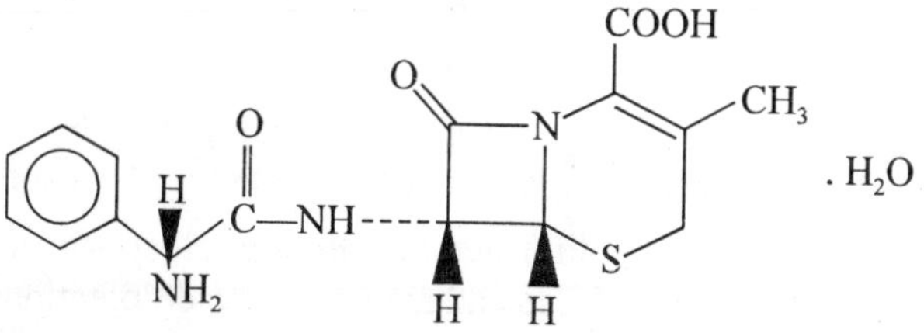

5-Thia-1-azabicyclo [4,2,0] oct-2-ene-2-carboxylic acid, [6R-(6α,7β (R*)]]-7-[(aminophenylacetyl) amino]-3-methyl-8-oxo-, monohydrate ; Keflex[R] ;

It is approved and recommended for use against respiratory infections caused by pneumococcus together with β-hemolytic streptococci ; otitis media by *H. influenzae, Branhamella catarrhalis*, pneumococcus, staphylococci ; skin and soft tissue infections by staphylococci and streptococci ; bone and joint infections by *Pr mirabilis* and staphylococci ; and above all the UTIs produced by *E. coli, Klebsiella* and *Pr mirabilis*.

Mechanism of Action. The '*drug*' has been specifically designed as an **orally active semisynthetic cephalosporin**. Importantly, there are *two* vital reasons that are solely responsible for the oral inactivation of cephalosporins, namely :

(*a*) β-Lactam ring's instability to acid hydrolysis, such as : **cephalothin** (I) and **cephaloridine** (II) :

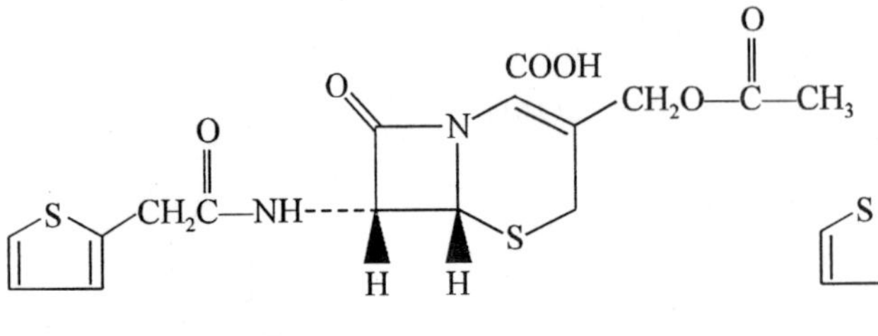

(I)

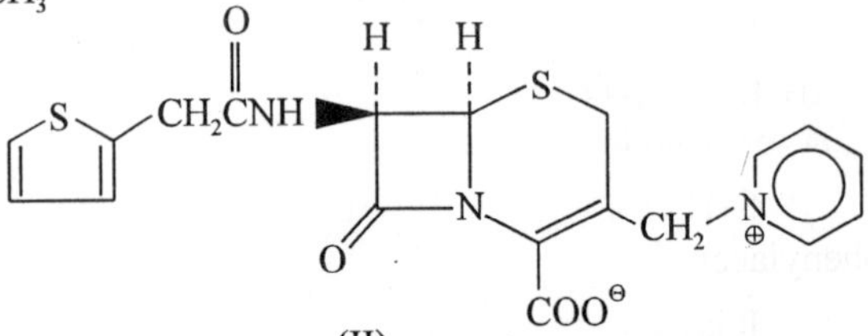

(II)

(*b*) Solvolysis or microbial transformation of the 3-methylacetoxy moiety, for instance : **cephalothin** and **cephaloglycin**.

The presence of α-amino moiety of cephalexin makes the drug '**acid stable**', whereas the reduction of the 3-acetoxymethyl to the methyl function helps in a big way to circumvent profusely the reaction taking place at the specific desired site.

Note. The '*drug*' is significantly much less potent than cephalothin and cephaloridine ; and, hence, is grossly inferior to both of them for the treatment of systemic infections of a vary serious nature.

3.2.1.1.3. Ceptradine USAN BAN, INN,

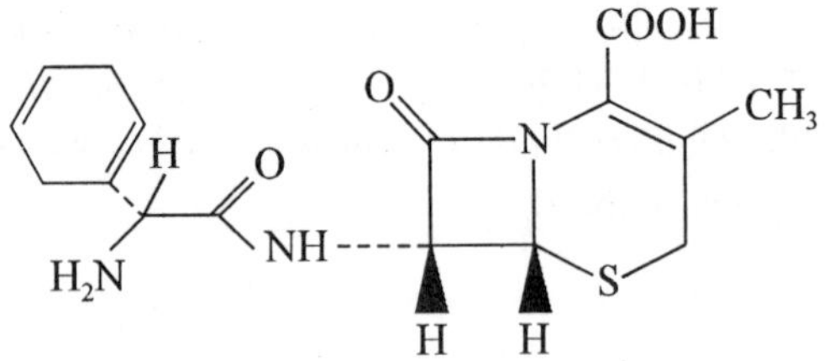

5-Thia-1-azabicyclo [4,2,0] oct-2-ene-2-carboxylic acid, [6R-[6α, 7β-(R*)]]-[(amino-1,4-cyclohexadien-1-ylacetyl] amino]-3-methyl-8-oxo-; USP ; Anspor[R] ; Velosef[R] ;

A is recognized as a short-acting first-generation cephalosporin administered IM or IV. Recommended usually for the treatment of UTIs and respiratory tract infections.

Mechanism of Action. The '*drug*' is minimally protein bound and gets excreted almost 100% after oral administration *via* the kidneys. It is, however, found to be fairly stable in an acidic media (*e.g.*, gastric juice).

SAR of Cephradine. The '*drug*' has a close resemblance to '*cephalexin molecule*' chemically ; and the former may be viewed as a partially hydrogenated derivative of the latter. Therefore, perhaps cephradine possesses quite similar antibacterial as well as pharmacokinetic characteristics.

Note. The secondary pharmaceutical products (*i.e.*, dosage forms) contain usually a non-stoichiometric hydrate essentially containing upto 16% water ; and, therefore, all such products must indicate explicitly by the labeling on the package itself.

3.2.1.2. Second Generation Cephalosporins. In this particular class of compounds the following typical examples shall be treated individually in delails *e.g.,* cefamandole, cefoxitin, and cefuroxime.

3.2.1.2.1. Cefamandole Nafate USAN, INN,

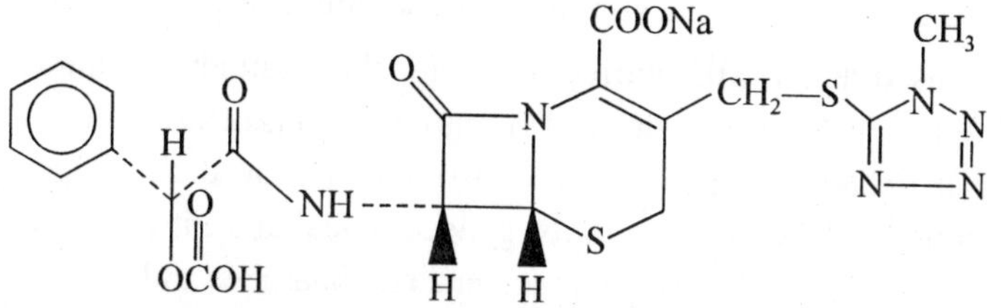

5-Thia-1-azabicyclo [4,2,0] oct-2-ene-2-carboxylic acid, [6R- [6α, 7β(R*)]]-7-[[(formyloxy) phenylacetyl] amino]-3-[[(1-methyl 1H-tetrazol-5-yl)-thio]methyl]-8-oxo-, monosodium salt ; Mandol[(R)];

It is a short-acting *second-generation cephalosporins* normally administered IM or IV. It exhibits a broader spectrum of activity with an increased activity against *Haemophilus influenzae*, besides the *enterobacteriaceae* produced as a result from a overwhelmingly enhanced stability to the β-lactamases.

SAR of Cefamandole. The various salient features are :

(*i*) A formate ester of cefamandole, a semisynthetic cephalosporin which essentially inducts D-mandelic acid as the '**acyl portion**' ; and a sulphur-containing heterocycle (*e.g.*, 5-thio-1,2,3,4-tetrazole) instead of the acetoxy moiety positioned on the C-3 methylene C-atom.

(*ii*) Esterification of the α-hydroxyl function of the D-mandeloyl moiety eventually circumvents the instability function of this '*drug*' particularly in solid-state dosage forms.* This important salient feature caters for the satisfactory concentrations of the parent antibiotic *in vivo via* spontaneous hydrolysis of the prevailing ester between a neutral to alkaline pH range.

(*iii*) D-Mandeloyl functional group present in this '*drug*' seems to afford noticeable resistance to a few β-lactamases, by virtue of the fact that certain β-lactamase-producing Gram-negative organisms (specifically Enterobacteriaceae) which display obvious resistance of **cefazolin** and other first generation cephalosporins (see section 3.2.1.1.) are found to be sensitive to cefamandole,

(*iv*) Besides, the '*drug*' is also active against a few ampicillin-resistant strains of *Neisseria* and *Haemophillus* species ; and

(*v*) Permeability and intrinsic acitivity along with viable resistance to the β-lactamases are the glaring factors that establishes the ensuing sensitivity of individual bacterial strains to this '*drug*'**.

3.2.1.2.2. Cefoxitin Sodium USAN, INN,

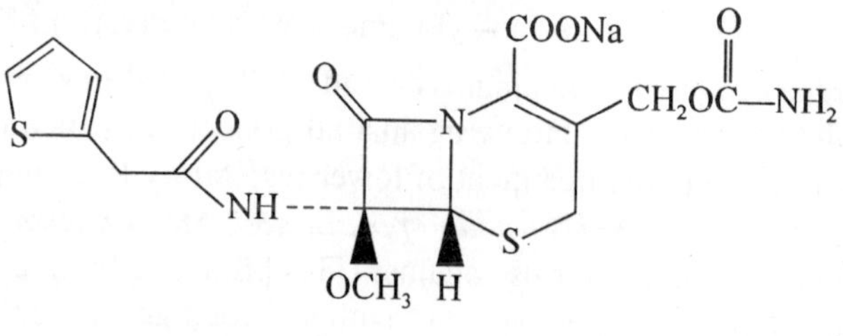

---

*Indelicato JM *et al.* J Pharm Sci*, **65** : 1175, 1976.

Ott JL *et al.* Antimicrob Agents Chemother*, **15 : 14, 1979.

(6R-*cis*)-5-Thia-1-azabicyclo [4,2,0] oct-2-ene-2-carboxylic acid, 3-[[(aminocarbonyl) oxy] methyl]-7-methoxy-8-oxo-7-[(2-thienylacetyl) amino]-, sodium salt ; USP ; Mefoxin[R] ;

It is invariably used as an '**alternative drug**' for the treatment of intra-abdominal infections, colorectal surgery or appendectomy and ruptured viscus beause it is active against most enteric anaerobes including the organism *Bacteroides fragilis*. It is also indicated in the management and treatment of bone and joint infections produced by *S. aureus*, gynecological and intra-abdominal infections caused by *Bacteroides* species together with other common enteric anaerobes and Gram-negative bacilli ; lower respiratory tract infections produced by *Bacteroides* species, *E.coli*, *H. influenzae*, *Klebsiella* spp. *S. aureus* or *Streptococcus* spp. (except enterococci) ; septicemia caused by *Bacteroides* spp., *E. coli*, *Klebsiella* spp., *S. aureus* or *Strep pneumoniae* ; skin infections produced by *Bacteroides* spp., *E. coli*, Klebsiella spp., *S. aureus* or *epidermidis* or *Streptococcus* spp. (except enterococci) or UTIs by *E. coli*, *Klebsiella* spp. or indole positive *Proteus*, and for preoperative prophylaxis.

Mechanism of Action. The '*drug*' is found to be resistant to certain β-lactamases which are responsible for the hydrolysis of cephalosporins. It has been duly observed that cefoxitin helps to antagonize the action of cefamandole (see section 3.2.1.2.1) against *E. cloacae* and also that of carbenicillin against *P. aeruginosa*. As the half-life is comparatively of shorter duration ; therefore, the drug must be administered 3 to 4 times per day.

Note. Solutions of its sodium salt stable for 24 hours at an ambient temperature and lasts upto 1 week when refregerated (0-10°C). However, 7α-methoxyl solutions helps to stabilize the β-lactam to alkaline hydrolysis to a certain extent.

3.2.1.2.3. Cefuroxime Sodium USAN, BAN, INN,

5-Thia-1-azabicyclo [4, 2, 0] oct-2-ene-2-carboxylic acid, (6R, 7R)-7-[2-(2-furyl) glyoxylamido]-3-(hydroxymethyl)-8-oxo-, 7-(Z)-mono (*o*-methyloxime) carbamate (ester) ; Kefurox[R] ; Zinacef[R] ;

It is approved for the treatment of meningitis caused by *Streptococcus pneumoniae*, *N. meningitidis* and *S. aureus*. It exhibits an excellent activity against all gonococci, hence is also employed to treat gonorrhea. It also finds it usage in the treatment of lower respiratory tract infections invariably caused by *H. influenzae* and *parainfluenzae*, *Klebsiella spp.*, *E. coli*, *Strep pneumoniae*, and pyrogens and *Staph aureus*. It is also recommended for use against UTIs produced by *E. coli* and *Klebsiella*, which designates a more limited approval in comparison to other second generation cephalosporins. It is also indicated in bone infections, septicemias, and above all the surgical prophylaxis.

Mechanism of Action. The '*drug*' exerts its activity against *H. influenzae*, besides its inherent ability to penetrate directly into the cerebrospinal fluid (CSF) which renders it specifically beneficial for the treatment of *meningitis* caused by that susceptible organism. Cefuroxime gets distributed evenly throughout the entire body segments. It has been observed that almost 85% of the '*drug*' gets eliminated in the urine. Its half-life range between 1.3—1.7 hour but may get extended upto even 24 hours in the specific instance of renal failure.

Note. *The sodium salt of cefuroxime is poorly absorbed by the oral route. However, its corresponding 'axetil ester' is also available for the oral administration of* **otitis media, pneumonia and UTIs**.

SAR of Cefuroxime. The presence of a *syn*-alkoxiamino substituent in the drug molecule is closely associated with the prevalent β-**lactamase activity** in these cephalosporins. Perhaps its inclusion into the '*second generation cephalosporins*' is duly justified due to the fact that its antibacterial spectrum bears a close similarity to that of **cefamandole** (see section 3.2.1.2.1).

3.2.1.3. Third Generation Cephalosporins

Though there are several drugs that are approved and marketed belonging to the '*third generation cephalosporins*', but only *three* such compounds shall be discussed in this particular section, namely : cefixime, ceftazidime, and ceftibuten.

3.2.1.3.1. Cefixime USAN, INN,

5-Thia-azabicyclo [4,2,0] oct-2-ene-2-carboxylic acid, [6R-[6α, 7β(Z)]]-7-[[(2-amino-4-thiazolyl) [(carboxymethyoxy) imino] acetyl] amino]-3-ethenyl-8-oxo- ; USP ; Suprax(R) ;

It is a well-known orally active *third generation cephalosporin* having superb and excellent therapeutic profile against a plethora of *E. coli, Klebsiella, H. influenzae, Branhametla catarrhalis, N. gonorrhoeae* and *meningitidis,* besides including β-*lactamase producing strains*. It is found to be active against certain common **streptococci** spp. whereas **staphylococci** are genuinely resistant. It is invariably recommended for the respiratory infections*, otitis media and uncomplicated UTIs; however, its actual therapeutic role is yet to be understood exhaustively.

Mechanism of Action. The '*drug*' gets absorbed gradually and rather incompletely from the GI tract and exhibits a bioavailability ranging between 40-50%. Importantly, the apparent appreciable good oral absorption of this drug substance is due to its facilitated and augmented transport across the **intestinal brush-border membranes** that essentially implicate the ensuing carrier system for the '**dipeptides**'.** However, this result was not quite expected by virtue of the fact that the prevailing '*drug*' predominantly is devoid of the **ionizable α-amino moiety** either present in the '**dipeptides**' or the 'β-**lactams**'previously known to be transported by the aforesaid carrier system.***

---

*Infections due to acute bronchitis, pharyngitis, and tonsititis.

Tsuji A *et al. J. Pharm. Pharmacol.,* **39, 272, 1987.

***Westphal JP *et al. Clin. Pharmacol. Ther.* **57**, 257, 1995.

3.2.1.3.2. Ceftazidime Sodium USAN, INN,

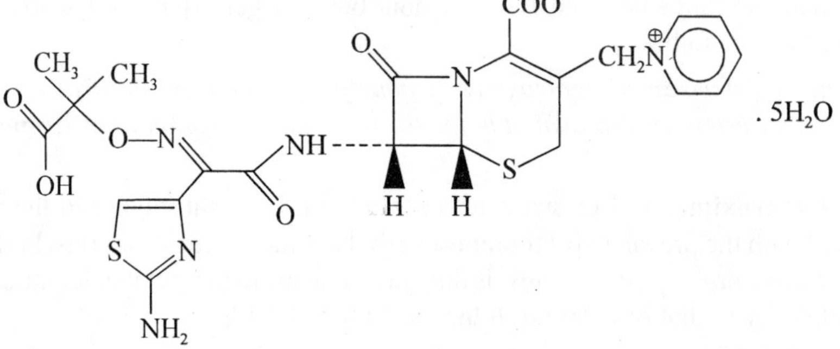

Pyridinium, [6R (6α, 7β(Z)]]-1-[[7-[[(2-amino)-4-thiazolyl)-I(1-carboxy-1-methylethoxy) imino] acetyl] amino]-2-carboxy-8-oxo-5-thia-1-azabicyclo [4,2,0] oct-2-ene-3yl]-, hydroxide, inner salt ; USP ; Fortaz[R] ; Tazicef[R] ; Tazidime[R] ;

The '*drug*' displays its special interest due to its inherent high activity against the *Pseudomonas* and *Enterobacteriaceae* but fails to do so for *enterococci*. It is well recognized widely as an '**alternative drug**' specifically for the management and treatment of hospital-acquired Gram-negative infections. However, a combination with **amikacin** in the treatment of infections in immunocompromised patients when *Ps aeruginosa* happens to be a causative organism.

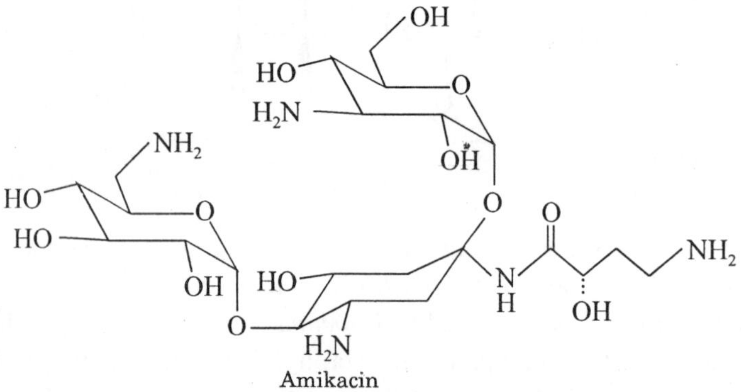

Amikacin

Ceftazidime is profusely recommended for use in the treatment of bone and joint infections, CNS-infections, gynecological infections, lower respiratory tract infections, septicemia, skin and UTIs.

Mechanism of Action. The mechanism of action of this '*drug*' is solely attributed by the presence of *two* characteristic structural features, namely :

  (*i*) a **2-methylpropionicoxaminoacyl moiety** which exclusively confers the β-lactamase resistance and, perhaps is responsible for an enhanced permeability *via* the porin channels of the cell envelope, and

  (*ii*) a **pyridinium functional moiety** strategically positioned at the 3'-position which essentially affords the Zwitterionic characteristic features on the '*drug molecule*'.

The '*drug*' is eliminated upto 80-90% in the urine. The plasma half-life in normal healthy persons is almost 2 hours, but may be prolonged in patients having renal failure.

3.2.1.3.3. Ceftibuten INN

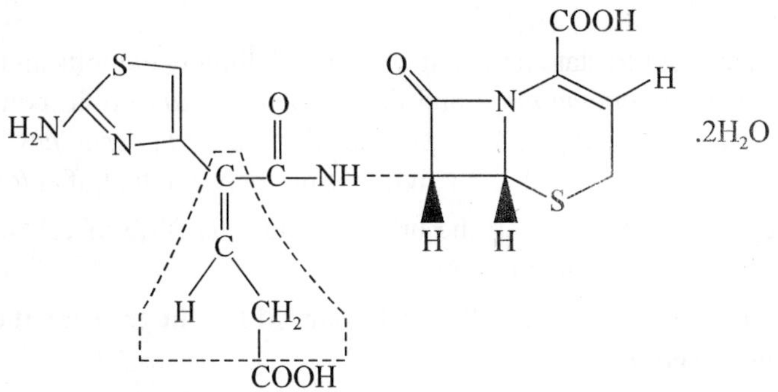

5-Thia-1-azabicyclo [4, 2, 0] oct-2-ene-2-carboxylic acid, [6R-[6α, 7β (Z)]]-7-[[2-(2-amino-4-thiazolyl)-4-carboxy-1-oxo-2-butenyl] amino]-8-oxo-, dihydrate ; Cedax$^{(R)}$;

The '*drug*' has excellent potency against a majority of the members of the *Enterobacteriaceae* family, *H. influenzae, Neisseria* spp., and *M. catarrhalis.* It is found to be **not** active against *S. aureus* or *P. aeruginosa* and exhibits mild streptococcal activity. It is invariably indicated in the management and control of community-acquired respiratory tract, urinary tract and above all the gynecological infections.

SAR of Ceftibuten. It is one of the most recent chemically novel structural analogues of the **oximinocephalosporins** wherein an *olefinic methylene moiety* (C = CHCH$_2$—), as shown by the dotted line in the above structure,with Z stereochemistry has virtually replaced the *syn*-oximino (C = NO—) moiety (which is present in *ceftazidime).*

Mechanism of Action. The aforesaid isosteric replacement gives rise to a compound which predominantly retains the resistance to hydrolysis catalyzed by a host of β-lactamases, has not only increased chemical stability but also rendered the '*drug*' orally active. It enjoys the glory of being the '*drug*' having the highest oral bioavailability amongst the prevailing third-generation cephalosporins.*

The '*drug*' gets excreted mostly unchanged in the urine. It possesses a half-life of about 2.5 hours. Ceftibuten has been found to have the plasma protein binding upto 63%.

3.2.1.4. Fourth Generation Cephalosporins. The only approved drug substance that belongs to the *fourth generation cephalosporins* is cefepime which will be discussed as under :

3.2.1.4.1. Cefepime Hydrochloride USAN, INN,

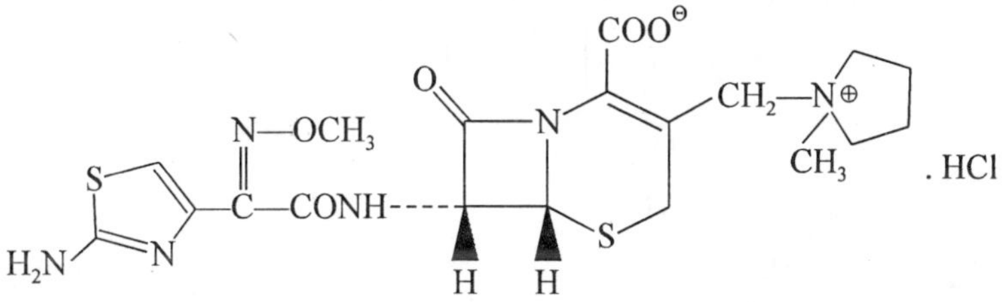

Pyrrolidinium, [6R-(6α, 7β(Z)]]-1-[[7-(2-amino-4-thiazolyl) (methoxy-imino) acetyl] amino-2-carboxy-8-oxo-5-thia-1-azabicyclo [4,2,0] oct-2-ene-3yl]-methyl]-1-methyl-, hydroxide, inner salt hydrochloride ; Maxipime$^{(R)}$; Axepin$^{(R)}$;

It is profoundly recognized as an altogether new approved *fourth-generation cephalosporin* which essentially possesses an extended Gram-negative spectrum against Gram-negative aerobic bacilli usually

*Fassbenden M *et al. Clin. Infect. Dis.,* **16,** 646, 1993.

covered by **cefotaxime** and **ceftazidime** including certain strains that are found to be resistant to these *third-generation cephalosporins.*

It is, however, pertinent to state here that cefepime definitely exhibits an improved antibacterial profile against *Streptococcus pneumoniae* and *Staphylococcus aureus* in comparison to the *third-generation cephalosporins.* Interestingly, its specific activity against *P. aeruginosa* is found to be variable just like other antibiotics ; and the profile of activity resides between that of *ceftazidime* and *ceftoxime.*

The '*drug*' gets excreted mostly in the urine having a half-life of 2.1 hours. It is found to be bound almost minimally to the plasma proteins.

Note. Cefepime HCl may be administered IVor IM for the treatment of UTIs, pneumonias and skin infections.

4. AMINOGLYCOSIDE ANTIBIOTICS

The aminoglycoside antibiotics constitute an important category of antibacterial agents in the therapeutic armamentarium, *e.g.,* streptomycins, neomycins, paramomycins, kanamycins, gentamycins and the corresponding derivatives of these antibiotics.

These are a bunch of closely related chemically basic carbohydrates that are mostly water-soluble. Their respective hydrochlorides and sulphates are crystalline in nature. They are found to be effective in inhibiting the growth of gram-positive as well as gram-negative bacteria. They are also effective to a great extent against mycobacteria.

In general, *they are prepared biosynthetically exclusively from an admixture of carbohydrate components of the fermentation media.*

They usually act by causing interference with the '**reading**' of the genetic code.

A few typical examples cited earlier shall be discussed below :

A. *Streptomycin* INN, *Streptomycin Sulphate* BAN, *Streptomycin Sulfate* USAN,

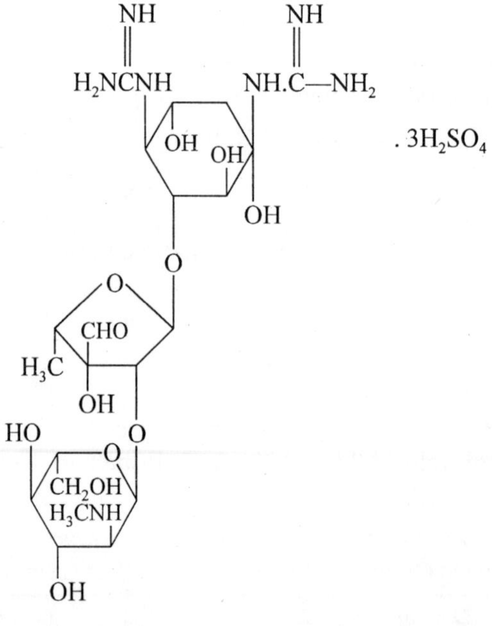

BP ; USP ; Eur. P ; Int. P ; Ind. P ; Isoject Streptomycin Injection[R] (Pfizer) ; Streptomycin Sulphate[R] (Glaxo, U.K.).

Streptomycin is chiefly employed in the *treatment of tuberculosis in conjunction with other drugs such as isoniazid and rifampicin.*

Streptomycin and penicillin exert a synergistic action against bacteria and are usually employed together in the treatment of subacute bacterial endocarditis caused by *Streptococcus faecalis*

It exerts **bacteriostatic action** in low concentrations and **bactericidal** in high concentrations against a plethora of Gram-negative and Gram-positive organisms. The only infection wherein this '*drug*' alone is the '*drug of choice*' are **tularemia*** and **bubonic plague.**** A combination with a *tetracycline* it may be employed in the treatment of **brucellosis*** and infections produced by *Pseudomonas mallei.* It is also an alternative drug of choice in the treatment of chancroid, rat-bite fever and tuberculosis.

Dose. For non-tuberculosis infections, usual, 1g per day up to5 to 10 days.

Mechansim of Action. The '*drug*' exerts its maximum effectiveness against the organism *Mycobacterium tuberculosis.* Interestingly, the antibiotic is not a cure itself but has proved to be an excellent and valuable adjunct to other modalities of therapeutic treatment for tuberculosis. It acquires a rapid development with respect to certain strains of microorganisms. The combined administration of *streptomycin* and *penicillin* has been suggested to combat infections which may be due to organisms that are sensitive to both these antibiotics. The '*drug*' is neither obsorbed nor destroyed appreciably in the GI tract.

SAR of Streptomycin. The '*drug*' serves as a triacidic base due to the presence of *two* characteristic chemical entities, namely : (*a*) two strongly basic guanido moieties ; and (*b*) rather weakly basic methylamino function. Furthermore, hydroxy-streptomycin differs from streptomycin in essentially having a strategically positioned OH moiety in place of one of the H-atoms of the streptose methyl function. Besides, streptomycin B(*i.e.,* mannisido streptomycin) possesses a **mannose residue** attached to a glycosidic linkage *via* a OH moiety at C-4 of the N-methyl-L-glucosamine functional group. The designated stereochemical structure of the '*drug*' has been reconfirmed *via* the total synthesis.****

B. *Neomycin* INN, *Neomycin Sulphate* BAN, *Neomycin Sulphate* USAN,

Fradiomycin Sulphate; BP ; USP ; Eur. P ; Int. P ; Ind. P ; Neobiotic[R] (Pfizer) ; Mycifradin[R] (Upjohn).

Neomycin is mostly used in a wide variety of local infection such as burns, ulcers, wounds, impetigo, infected dermatoses, furunculosis, conjunctivitis, etc. It is also employed as an adjuvant in topical steroid preparations to control secondary infections in the case of inflammatory disorders.

The '*drug*' is employed to produce intestinal antisepsis prior to large bowel surgery, for the treatment of gastroenteritis produced by toxigenic *E. coli*, and also to afford suppression of ammonia producing bowel flora in the management of hepatic coma. As it causes a rapid overgrowth of nonsusceptible organism, including staphylococci, oral therapy must not be prolonged in any case for more than 3 days. It displays broad-spectrum activity against a good number of pathogenic organisms. Besides, it demostrates a low incidence of toxic and hypersensitivity reactions.

---

* An acute plaguelike infectious disease caused by *Francisella tularensis.*

**It is caused by *Yersinia pestis* usually found in infected rats, ground squirrels and gets transmitted to humans by the bite of rat flies.

***A widespread infectious febrile disease affectiing humans and cattle (also called **Malta Fever).**

****Umezawa S *et al. J. Antibiotic* (Tokyo) **27**, 997, 1974.

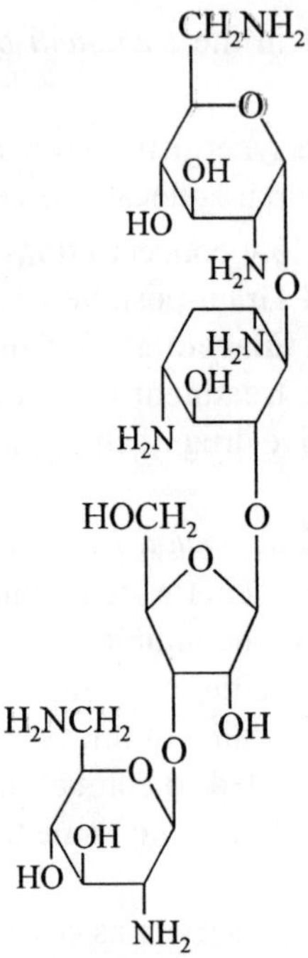

Neomycin

Dose: *Topical, to the skin, as 5% solution, aerosol or ointment 2 to 3 times a day.*

Mechanism of Action. The '*drug*' usually gets absorbed very rarely from the digestive system ; therefore, its oral administration primarily fails to produce any substantial systemic effect.

SAR of Neomycin. The structures of neomycin A(neamine), neomycin B and neomycin C have been established ; besides, the absolute configurational structures of *neomycin* and *neamine* have been reported. It has been demonstrated that neamine could be obtained by the methanolysis of neomycin B and C respectively, whereby the glycosidic linkage existing between D-ribose and deoxystreptamine undergoes cessation.

Note. The '*drug*' is invariably combined with other '*antibiotics*' namely : *gramicidin,*
**   *polymyxin B sulfate* and *bacitracin*.**

C. *Kanamycin* INN, *Kanamycin Sulphate* BAN, *Kanamycin Sulfate* USAN,

D-streptamine, *o*-3-amino-3-deoxy-α-D-glucopyranosyl (1 → 6)-*o*- [6-amino-6-deoxy-α-D-glucopyranosyl (1 → 4)]-2-deoxy-, sulphate ; BP ; USP ; Kelbcil[R] (Beecham) ; Kantrex[R] (Bristol).

It is effective against some Mycoplasma and gram-positive bacteria, for instance, *Staphylococcus Pyogenes* and *Staphylococcus epidermidis*. Along with penicillin it is found to be effective against *Streptomyces fecalis*. It is used invariably either alone or in combination with other drugs for a variety of disorders, namely : *acute staphylococcal infections, gonorrhea, tuberculosis, acute urinary tract infections, for bowl sterilization in hepatic coma and also prior to bowl surgery.*

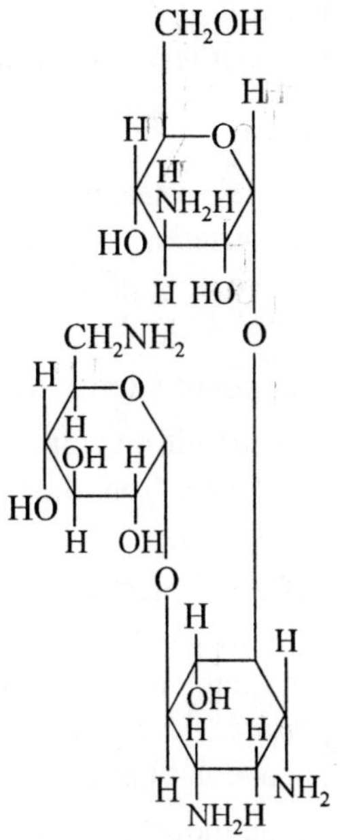

CH$_2$OH

H

H O

H'
NH$_2$H

HO

H HO

CH$_2$NH$_2$ O

H   O

H
OH H

HO

H   OH

O

H   H

OH
H   H

H   NH$_2$
NH$_2$H

Kanamycin

In US the use of kanamycin is normally restricted to the infections related to the *intestinal tract*, such as : bacillary dysentry; *systemic infections* caused due to Gram-negative bacili, such as : *Klebsiella*, *Proteus*, *Enterobacter*, and *Serratia spp.*, which have developed resistance to some other antibiotics. It has also been indicated for preoperative antisepsis of the bowel. However, this '*drug*' could not be useful in tuberculosis perhaps due to the fact that it develops resistance to mycoorganism rather rapidly.

Dose : (Base equivalent)-Oral, adult, for intestinal infection, 1g after every 8 hours for 5 to 7 days ; For preparative preparations, 1g every hour for 4 doses followed by 1g every 6 hours for 36 to 72 hours.

Mechanism of Action. Based on both clinical experience and experimental demonstration* it has been duly observed that the '*drug*' develops cross-resistance overwhelmingly in the **tubercle bacilli** specifically along with some other medicinal entities, such as : vincomycin, dihydrostreptomycin and antitubercular drug substances.

SAR of Kanamycin. Kanamycins A, B and C *i.e.*, the three closely related analogues of kanamycin have been duly established by the aid of chromatography. Kanamycin A is the '*drug*' available for therapeutic usage. It has been proved that the vital point of difference amongst the kanamycins resides solely in the **suger residuces** strategically linked to the *glycosidic oxygen* at the C-4 position of the central deoxystreptamine. Interestingly, the kanamycins do **not** essentially possess the **D-ribose residue** as is present in *neomycins* and *paromomycins*. In all the three structural variants of kanamycin the presence of **kanosamine entity** is found to be attached glycosidically at the C-6 position of deoxystreptamine *i.e.*, 3-D-glucosamine. They also differ in the substituted D-glucoses which are observed to be attached glycosidically at the C-4 position of the inherent deoxystreptamine ring.

---

*Morikubo Y, *J Antibiot* [A] : **12,** 90, 1959.

CAUTION : (1) *The 'drug' causes either retarded or impaired loss of hearing.*

(2) *Kanamycin and penicillin salts must not be combined in the same solution perhaps due to the possible inactivation of either agents significantly.*

5. CHLORAMPHENICOL

Chloramphenicol (chloromycetin) is a levorotatory broadspectrum antibiotic originally produced from several streptomycetes, namely : *S. venezualae, S. omiyamensis* and *S. phacochromogenes var.* chloromyceticus. It has been reported to be the drug of choice for the treatment of typhus and typhoid fever.

However, chloramphenicol is of paramount interest owing to the following *three* reasons :

(*a*) It is a naturally occurring aromatic nitro compound of which there is only one previously recorded example of *hiptagin,* obtain from the root bark of *Hiptage madablota* Gaertn is noteworthy.

(*b*) It is capable of exerting its effect against viral diseases as well as those due to bacterial invasion and opens up the whole field of the chemotherapy of *virus* and *rickettsial* infections in man including typhus, undulant fever, *Salmonella septicaemia,* whooping cough, gastroenteritis, *lymphogranuloma inguinale,* typhoid and paratyphoid. So far, chloramphenicol-fast strains have not been isolated.

(*c*) It is amenable to synthesis on an industrial scale.

5.1. Structure of Chloramphenicol

The structure of chloramphenicol has been established on the basis of the following vital chemical evidences. They are :

(*i*) The molecular formula of chloramphenicol is $C_{11}H_{12}O_5N_2Cl_2$.

(*ii*) Its absorption spectrum is similar to that of nitrobenzene.

(*iii*) The presence of a nitro group was revealed by the reduction of chloramphenicol with tin (Sn) and hydrochloric acid, followed by diazotization and then coupling to yield an orange precipitate with β-naphthol (Rebstock *et al.* 1949).

(*iv*) When reduced catalytically (with palladium, Pd) it gives a prodcut which has an absorption spectrum very similar to that of *para*-toluidine and the resulting solution gives a positive test for ionic chlorine.

(*v*) Hydrolysis of chloramphenicol with either acid or alkali produces dichloroacetic acid together with an optically active base $C_9H_{12}O_4N_2$. Thus :

$$C_{11}H_{12}O_5N_2Cl_2 \xrightarrow{\text{H}_2\text{O}} Cl_2CH.COOH + C_9H_{12}O_4N_2$$

Chloramphenicol   Dichloro-acetic acid   (Base)

(*vi*) The resulting base was shown to contain a primary amino group, and on being treated with methyl dichloroacetate, the base regenerated chloramphenicol (Rebstock *et al.* 1949).

(*vii*) Chloramphenicol is converted into a diacetyl derivative on treatment with acetic anhydride in pyridine ; whereas the base obtained from chloramphenicol yields a triacetyl derivative on similar treatment thereby suggesting that chloramphenicol probably contains two-OH groups.

(*viii*) When the chloramphenicol base is treated with periodic acid (HIO_4) two molecules of the latter are consumed with the formation of one molecule each of ammonia, formaldehyde and *para*-nitrobenzaldehyde respectively.

However, these products may be accounted for provided the base is assumed to be 2-amino-1-nitrophenyl propane-1, 3-diol (Rebstock *et al.* 1949). Thus :

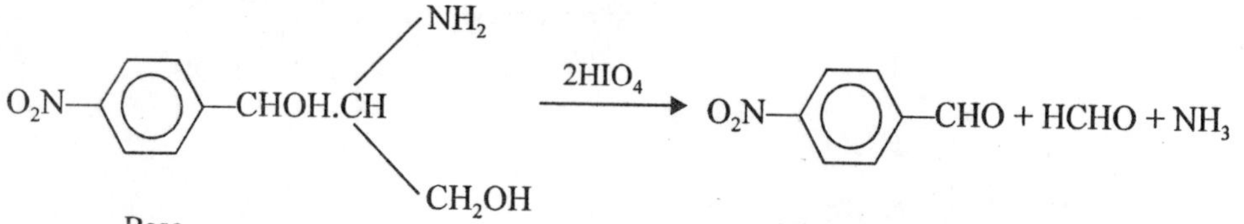

| Base | | *p*-Nitrobenzalde-hyde | Formaldehyde |

Hence, chloramphenicol may be written as :

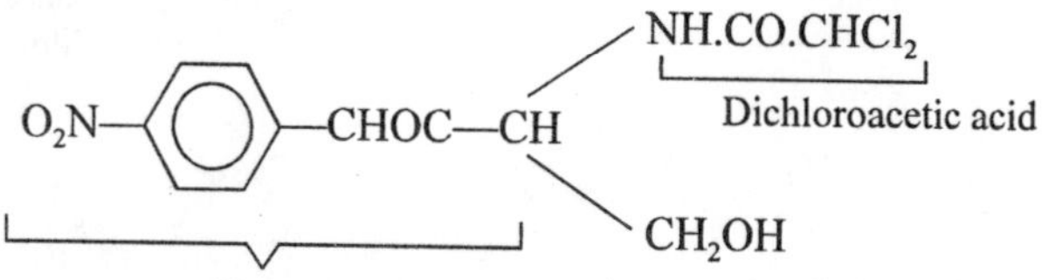

D-(−)-*Threo*-2-dichloroacetamide-1-*p*-nitrophenylpropane-1, 3-diol.

A. *Chloramphenicol* INN, BAN, USAN,

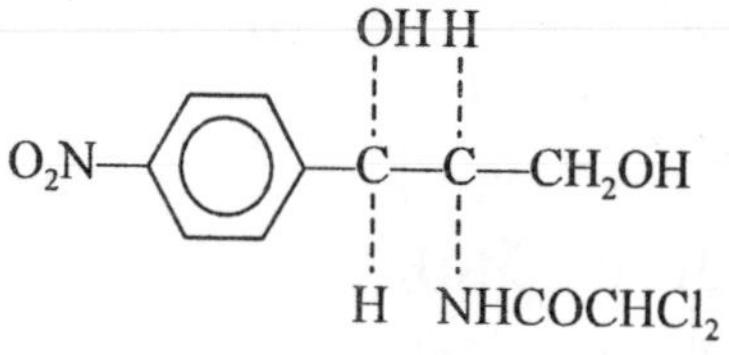

D-*threo*-(−)-2, 2-Dichloro-N-[β-hydroxy-α-(hydroxy methyl)-*p*-nitrophenyl] acetamide ; Acetamdie, 2,z-dichloro-N-[2-hydroxy-1-(hydroxymethyl)-2-(4-nitrophenyl) etheryl]-, [R-(R*, R*)]-; BP ; USP ; Eur. P ; Int. P ; Ind. P ; Chloromycetin[R] (Parke-Davis).

5.2. Synthesis of Chloramphenicol

Chloramphenicol has been successfully synthesized by different methods and the present global demand of this drug is adequately met exclusively by chemical synthesis. The synthesis put forward by Long *et al.* (1949) is discussed below :

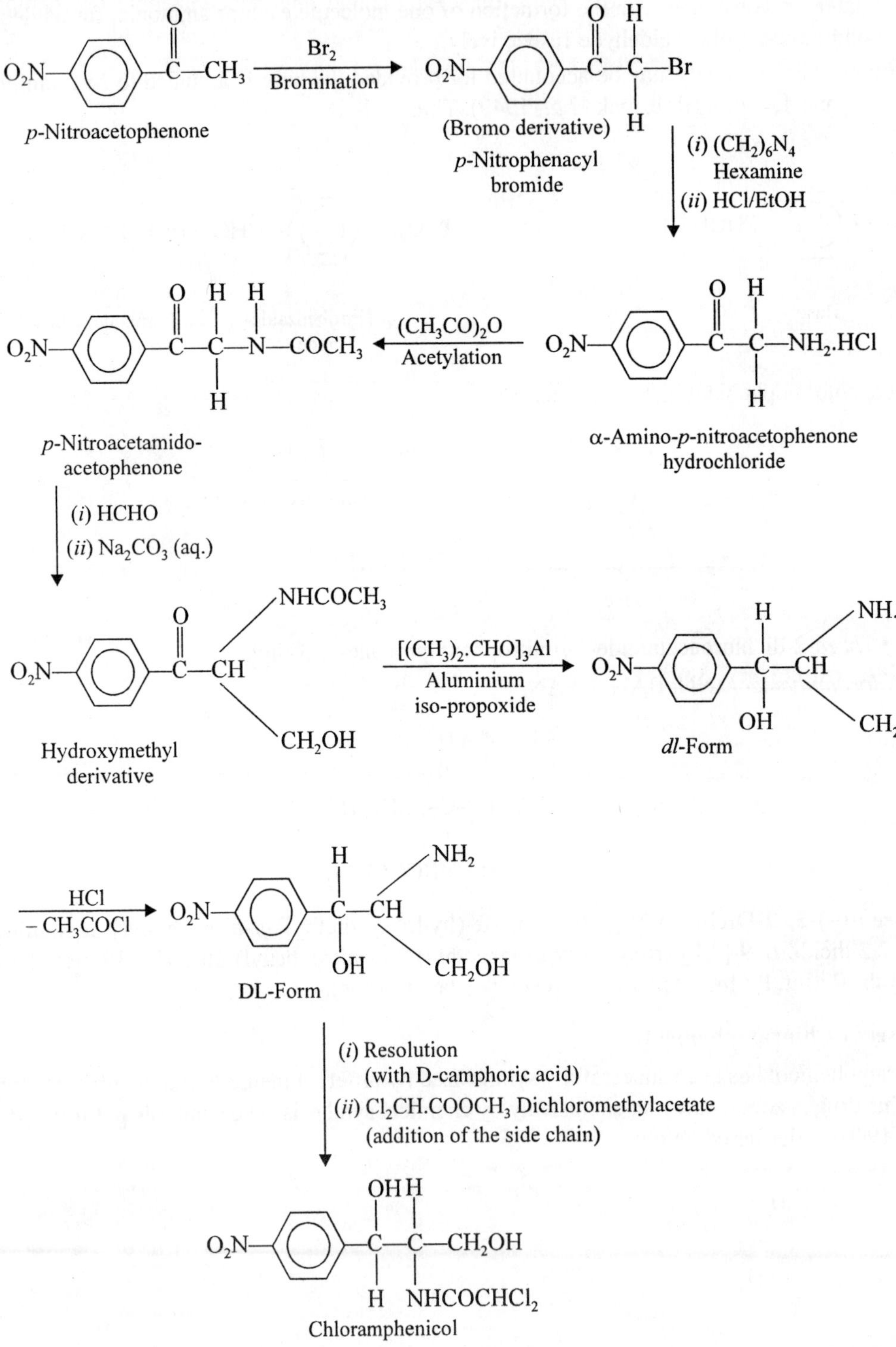

O_2N—⬡—$\overset{\overset{O}{\|}}{C}$—$CH_3$ $\xrightarrow[\text{Bromination}]{Br_2}$ O_2N—⬡—$\overset{\overset{O}{\|}}{C}$—$\overset{\overset{H}{|}}{\underset{H}{C}}$—$Br$

p-Nitroacetophenone

(Bromo derivative)
p-Nitrophenacyl
bromide

$\downarrow$ (*i*) $(CH_2)_6N_4$
Hexamine
(*ii*) HCl/EtOH

O_2N—⬡—$\overset{\overset{O}{\|}}{C}$—$\overset{\overset{H}{|}}{\underset{H}{C}}$—$N$—$COCH_3$ $\xleftarrow[\text{Acetylation}]{(CH_3CO)_2O}$ O_2N—⬡—$\overset{\overset{O}{\|}}{C}$—$\overset{\overset{H}{|}}{\underset{H}{C}}$—$NH_2.HCl$

p-Nitroacetamido-
acetophenone

α-Amino-*p*-nitroacetophenone
hydrochloride

$\downarrow$ (*i*) HCHO
(*ii*) Na_2CO_3 (aq.)

O_2N—⬡—$\overset{\overset{O}{\|}}{C}$—$CH$⟨$\overset{NHCOCH_3}{CH_2OH}$ $\xrightarrow[\text{Aluminium iso-propoxide}]{[(CH_3)_2.CHO]_3Al}$ O_2N—⬡—$\overset{\overset{H}{|}}{\underset{OH}{C}}$—$CH$⟨$\overset{NH.COCH_3}{CH_2OH}$

Hydroxymethyl
derivative

dl-Form

$\xrightarrow[\text{– } CH_3COCl]{HCl}$ O_2N—⬡—$\overset{\overset{H}{|}}{\underset{OH}{C}}$—$CH$⟨$\overset{NH_2}{CH_2OH}$

DL-Form

$\downarrow$ (*i*) Resolution
(with D-camphoric acid)
(*ii*) $Cl_2CH.COOCH_3$ Dichloromethylacetate
(addition of the side chain)

O_2N—⬡—$\overset{\overset{OH}{|}}{\underset{H}{C}}$—$\overset{\overset{H}{|}}{\underset{NHCOCHCl_2}{C}}$—$CH_2OH$

Chloramphenicol

para-Nitroacetophenone on bromination gives the corresponding bromo derivative which on treatment with hexamine followed by acidic ethanol yields α-amino-*p*-nitroacetophenone hydrochloride. This on acetylation gives the acetamido derivative which on treatment with formaldehyde followed by aqueous sodium carbonate affords the corresponding hydroxy methyl analogue. Reduction of the keto moiety is effected by treatment with aluminium iso-propoxide to give the product in DL-form which on reaction with HCl removes the acetyl function to yield the chloramphenicol base in its DL-form. The resulting product is first subjected to resolution with α-camphoric acid and secondly with dichloromethyl acetate to afford the addition of the side chain to yield chloramphenicol.

Typhoid fever and similar salmonellal infections are usually considered the prime indications for the use of chloramphenicol. It is also employed *in acute infections due to Heamophilus influenzae, including meningitis attributed to ampicillin-resistant strains.* It also find its enormous applications in *topical infections of eye and skin.* It has also been used *to eradicate vibrios from patients with cholera.* It is employed *for rickettsial infections like typhus and Rocky Mountain spotted fever.*

Chloramphenicol is particularly recommended for the management and treatment of serious infections produced by the strains of both Gram-positive and Gram-negative organism that have developed eventually resistance to either **ampicillin** or **penicillin G,** for instance : *H. influenzae, Salmonella typhi, S. pneumoniae, B. fragilis,* and *N. meningitidis.*

It is used topically extensively for the superficial *conjunctival infections* and *blepharitis* essentially caused by *E. coli, H. influenzae, Moraxella lacunata, Streptococcus hemolyticus,* and *S. aureus.* However, it is still the drug of choice for the **typhoid fever.**

Dose. Usual, adult, 500 mg every 6 hours.

Mechanism of Action. The '*drug*' has specifically the ability to penetrate right into the central nervous system (CNS) ; therefore, it is still an important alternative therapy for meningitis. The major course for the metabolism of chloramphenicol essentially involves the formation of the 3-*o*-glucuronide. Howerver, the minor reactions necessarily include : (*i*) reduction of the inherent *para*-nitro moiety to the corresponding '*amine*' function ; (*ii*) hydrolysis of the amide moiety ; (*iii*) hydrolysis of α-chloroacetamido group ; and (*iv*) reduction to yield α-hydroxyacetyl analogue.*

Chloramphenicol gets absorbed very fast from the GI tract, having a bioavailability of almost 90%. It has been observed that about 60% of the drug in blood is bound to serum albumin. It is biotransformed in the liver within a range of 85-95%. The volume of distribution v_d^{ss} stands at 0.7 mL. g^{-1}. The plasma half-life varies between 1.5–5 hours, except over 24 hours in neonates 1-2 days old, and 10 hours in infants 10-16 days old. The '*drug*' may cross the placental barrier and in turn intoxicate the fetus ; therefore, it must be avoided as far as possible in pregnant women.

Note. The '*prodrug*' of chloramphenicol viz., chloramphenicol palmitate (USP), which is a tasteless product, is solely intended for pediatric usage profusely, because the parent drug has a distinct bitter taste.

5.3. Structure Activity Relationship

Chloramphenicol possesses *two chiral (asymmetric) carbon atoms* in the '**acylamino-propanediol chain**' as shown below :

---

*Glazko A : *Antimicrob Agents Chemother.*, 655, 1966.

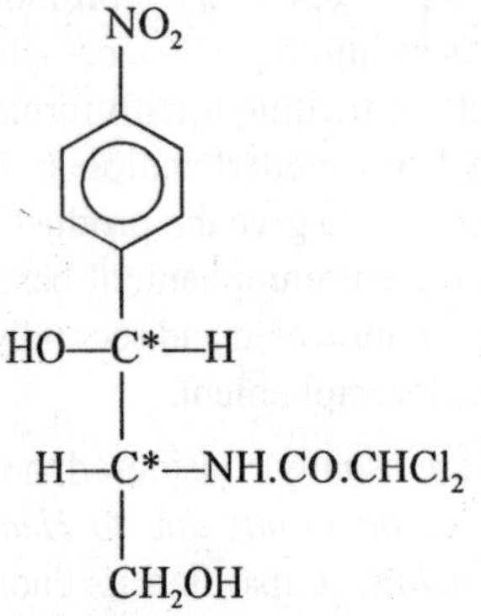

	NO_2			NO_2

HO—C*—H   H—C*—OH

H—C*—NH.CO.CHCl$_2$ H—C*—NHCOCHCl$_2$

CH$_2$OH   CH$_2$OH

(*Threo*-form) 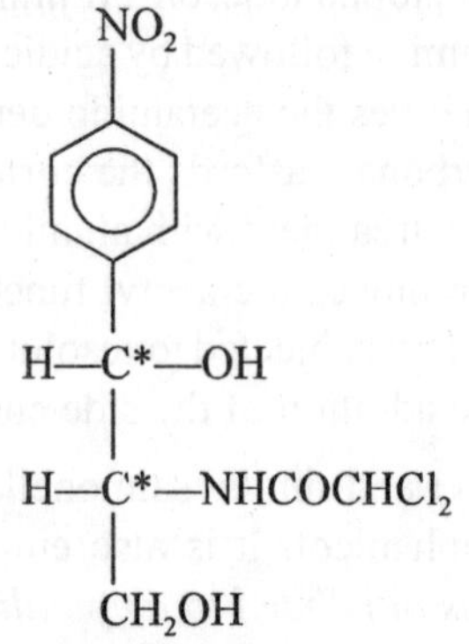   (*Erythro*-form)

Thus there are *two* possible pairs of enantiomorphs.

It has been observed that the biological activity resides almost exclusively in the '**D-Threo-isomer**' whereas the L-*Threo*, and D- and L-*Erythro* isomers are virtually inactive.

A large number of structural analogues of chloramphenicol have been prepared on the basis of the following themes : removal of the chlorine atom, transference of chlorine atom to the aromatic nucleus, transference of the nitro moiety to the *ortho-* or *meta*-position, esterification of the hydroxyl function(s), replacement of the phenyl ring with furyl, naphthyl and xenyl rings respectively, addition of alkyl or alkoxy substituents to the aryl ring and lastly replacement of the inherent nitro group by a halogen atom. It is, however, pertinent to mention here that none of these structurally modified analogues showed an activity approaching to that of chloramphenicol towards *Shigella paradysenteriae*

6. THE TETRACYCLINES

The epoch-making discovery of chlortetracycline (aureomycin) in 1947 by Duggar paved the way for a number of structural analogues used as broad-spectrum antibiotics that belong to the tetracycline family. The tetracyclines which are found to be effective therapeutically are listed in the following table.

6.1. Salient Features of the Tetracyclines

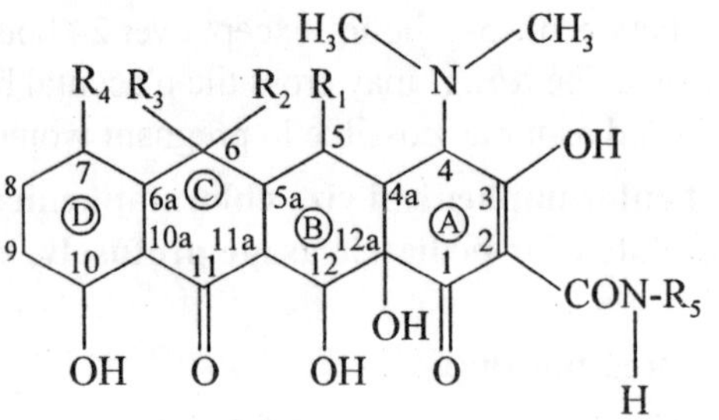

Name of Compound	Official Status	Brand Name(s)	R_1	R_2	R_3	R_4	R_5
Tetracycline	BPC ; (1973); USP ;	Tetracyn[R] (Pfizer) ; SK-Tetracycline[R] (SK & F)	H	OH	CH_3	H	H
Oxytetracycline	USP ;	Terramycin[R] (Pfizer)	OH	OH	CH_3	H	H
Chlortetracycline HCl	BP, USP ; Eur. P.; Int. P.; Ind. P.;	Aureomycin[R] (Lederle)	H	OH	CH_3	Cl	H
Demeclocycline HCl	BP, USP ; Eur. P.;	Ledermycin[R] (Lederle, UK)	H	OH	H	Cl	H
Methacycline HCl	BP (1973); USP ;	Rondomycin[R] (Wallace)	OH	=	CH_2	H	H
Doxycycline	USP ;	Vibramycin[R] (Pfizer)	OH	H	CH_3	H	H
Rolitetracycline	USP ;	Syntetrin[R] (Bristol)	H	OH	CH_3	$H—CH_2—N$	

6.2. Nomenclature

Based on the above conventional numbering of various carbon atoms and subsequent labelling of the **four** aromatic rings present in the tetracycline nucleus, oxytetracycline is chemically designated as :

"4-Dimethylamino-1, 4, 4a, 5, 5a, 6, 11, 12a-octahydro-3, 6, 10, 12, 12a-penta-hydroxy-6-methyl-1, 11-dioxo-2-naphthacenecarboxamide".

Some other members of the tetracycline family may conveniently be named as follows :

Methacycline : 6-Methylene-5-oxytetracycline ;

Doxycycline : α-6-Deoxy-5-oxytetracycline ;

Rolitetracycline : N-(Pyrrolidinomethyl)-tetracycline.

6.3. General Chracteristics of the Tetracyclines

Following are the *general characteristic features* of all the members of the tetracycline family :

(a) The tetracyclines are obtained by fermentation procedures from streptomyces species or by the chemical transformations of the natural products.

(b) The important members of this family are essentially derivatives of an octahydron-aphthacene, *i.e.*, a hydrocarbon made up of a system of four-fused rings.

(c) The antibiotic spectra and the chemical properties of these compounds are quite similar but not identical.

(d) The tetracyclines are amphoteric compounds, *i.e.*, forming salts with either acids or bases. In neutural solutions these substances exist mainly as zwitter ions.

(e) The acid salts of the tetracyclines that are formed through protonation of the dimethylamino group of C-4, usually exist as crystalline compounds which are found to be very much soluble in water. However, these amphoteric antibiotics will crystallize out of aqueous solutions of their salts unless they are duly stabilized by an excess of acid.

(f) The corresponding hydrochloride salts are used most commonly for oral administration and are usually encapsulated owing to their bitter taste.

(g) The water soluble salts are obtained either from bases such as sodium/potassium hydroxides or formed with divalent/polyvalent metals, e.g., Ca^{++}. The former ones are not stable in aqueous solutions, while the latter ones, e.g., calcium salt give tasteless products that may be employed to prepare suspensions for liquid oral dosage forms.

(h) The unusual structural features present in the tetracyclines afford three acidity constants (pKa values) in aqueous solutions of the acid salts. The thermodynamic pKa values has been extensively studied by Lesson et al. and discussed in the chapter on 'Physical-chemical factors and biological activities'.

(i) An interesting property of the tetracyclines is their ability to undergo epimerizaton at C-4 in solutions having intermediate pH range. These isomers are called epitetracyclines.

The **four epi-tetracyclines** have been isolated and characterized. They exhibit much less, activity than the corresponding **'natural' isomers** ; thus accounting for an apparent decrease in the therapeutic value of aged solution.

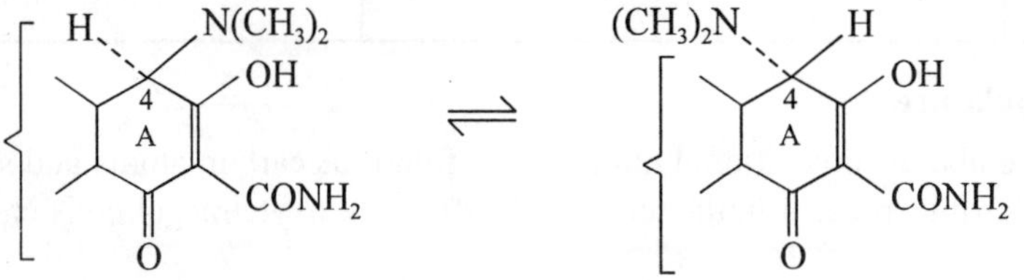

epi (less active)   **Natural** (more active)

(j) It has been observed that the strong acids and bases attack the tetracyclines having a hydroxy moiety at C-6, thereby causing a considerable loss in activity through modification of the C-ring as shown below :

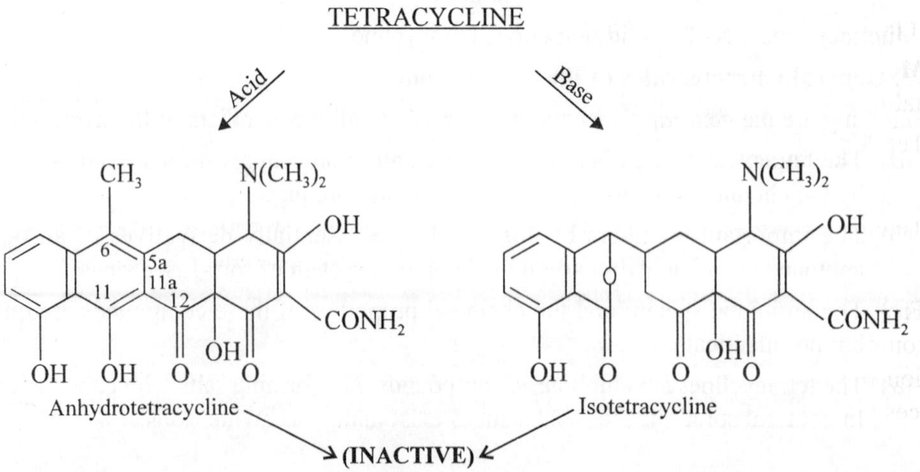

TETRACYCLINE

Anhydrotetracycline   Isotetracycline

(INACTIVE)

Strong acids produce a dehydration through a reduction involving the OH group at C-6 and the H atom at C-5a. The double bond thus generated between positions C-5a and C-6 induces a shift in the position of the double bond between the carbon atoms C-11 and C-11a thereby forming the relatively more energetically favoured resonant system of the naphthalene group found in the *inactive* anhydrotetracyclines.

The strong bases on the other hand promote a reaction between the hydroxyl group at C-6 and the carbonyl moiety at C-11, thereby causing the bond between C-11 and C-11a atoms to cleave and eventually form the lactone ring found in the *inactive* isotetracyclines.

(*k*) The tetracyclines form stable chelate complexes with many metals, *e.g.,* Ca^{++}, Mg^{++}, Fe^{++}, etc.

A few typical examples of the tetracyclines shall be dealt with in the sections that follows :

6.3.1. Tetracycline USAN, BAN, INN,

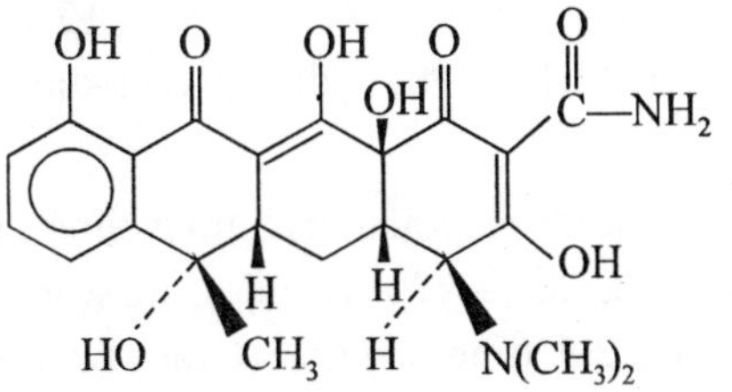

2-Naphthacenecarboxamide [4S-(4α, 4aα, 5aα, 6β, 12aα)]-4-(dimethylamino)-1, 4, 4a, 5, 5a, 6, 11, 12a-octahydro-3,6,10,12,12a-pentahydroxy-6-methyl-1,11-dioxo-; USP; Achromycin[R]; Cyclopar[R]; Panmycin[R]; Tetracyn[R] ;

The '*drug*' is the durg of choice in the treatment of chloera, relapsing fever, granuloma inguinale and infections produced by rickettsia, *Borrelia, Mycobacterium fortuitum* and *marinum,* and *Chlamydia psittaci* and *trachomatis* (except pneumonia and inclusion conjunctivitis).

It may be employed as an '**alternative drug**' in the following *two* situations, namely :

(*a*) For silver nitrate in the prevention of neonatal ocular prophylaxis of chlamydial and gonococcal cojunctivitis, and

(*b*) For treatment of actinomycosis, anthrax, chancroid, mellioidosis, plague, rat-bite fevers, syphilis and yaws.

It has also been reported to be beneficial in the treatment of *toxoplasmosis.*

Mechanism of Action. The mechanisms of action of its combination with other agents have been established adequately, such as :

Tetracycline + $MgCl_2$. $6H_2O$—**Panmycin (R)**—Enhances the rate and peak of plasma concentration.

Tetracyclines + Aluminium/Calcium gluconates—Observed enhanced plasma levels in experimental animals.

From the above two cited examples one may evidently conclude that the tetracyclines may form stable complexes with bivalent metal ions (*e.g.,* Mg^{2+}, Ca^{2+};) that would appreciably minimize the absorption from the GI-tract. In reality, these '**adjuvants**' seen to compete with the tetracyclines for substances present in the GI-tract which might otherwise be free to complex with these antibiotics, and thus ultimately retard their absorption significantly. Of course, there is no concrete evidence which may suggest that the metal ions (Mg^{2+}, Ca^{2+}) *per se* serve as '*buffers*'; a theoretical explanation quite often put forward in the literature.

6.3.2. Minocycline Hydrochloride USAN, BAN, INN,

. HCl

2-Napththacenecarboxamide, [4S-(4α,4aα, 5aα, 12aα)]-4,7-*bis* (dimethylamino)-1,4,4a,5,5a,6,11,12a-octahydro-3,10,12,12a-tetrahydroxy-1,11-dioxo-, monohydrochloride ; USP ; Minocin[R] ; Vectrin[R] ;

The '*drug*' is found to be 2-4 folds as potent as tetracycline ; however, it essentially shares an equally low potency against *Enterococcus fecalis*. Besides, it is observed to be 8 times more potent against *Streptococcus viridans,* and 2-4 times against Gram-positive organisms in comparison to tetracyclines. It is the drug of choice for the treatment of infections caused by *Mycobacterium marinum.* It remarkably differs from the other structural analogues of tetracyclines wherein the observed bacterial resistance to the drug stands at a *low ebb and incidence* ; it is particularly true for *Staphylococci,* in that the prevailing cross-resistance is only upto 4%.

Minocycline has been indicated for the management and treatment of *chronic bronchitis* and othe upper respiratory tract infections (URTs). Though it essentially possesses comparatively low renal clearance, which is partially compensated for by means of its high serum and tissue levels, it has been duly recommended for the treatment of urinary tract infections (UTIs). The '*drug*' has been equally useful in the virtual erradication of *N. meningitidis* in specific asymptomatic carriers.

Mechanism of Action. The '*drug*' is usually absorbed by the oral route upto 90-100%. However, its absorption is predominantly diminished to a small extent milk and food intake ; and appreciably by the presence of '*iron preparations*' and '*nonsystemic antacids*'. It is protein-bound in plasma between a range of 70-75%. The volume of distribution v_d^{ss} stands at $0.14 - 0.7$ mL. g^{-1}. The plasma half-life ranges between 11–17 hours. It gets excreted unchanged in urine upto 10%; however, its biological half-life is usually prolonged chiefly in the incidence of renal failure.

6.4. Structure Activity Relationship (SAR)

The structure activity relationship amongst the various members of the tetracycline family has ben studied extensively.

The high level of antimicrobial activity of tetracycline established earlier reveal that the substitutions on the C-5 and C-7 were not an essential requirement.

The activity of 6-dimethyltetracycline (demecycline) and demeclocycline has established that the methyl function at C-6 may be replaced by hydrogen.

The activity of deoxycycline and 6-deoxy-6-demethyltetracycline (minocycline) shows that the presence of hydroxy moiety at C-6 is not essential either.

The 6-deoxy-6-methylenetetracyclines and their corresponding mercaptan adducts possess typical characteristics tetracycline activity and illustrate further the level of modification feasible at C-6 with the possible retention of biologic activity.

It is, however, interesting to observe that the subsequent removal of the 4-dimethylamino function affords a loss of about 75% of the antibiotic effect of the parent tetracyclines.

The X-ray diffraction studies reveal that the following stereochemical formula represents the orientations, as observed in the natural tetracyclines :

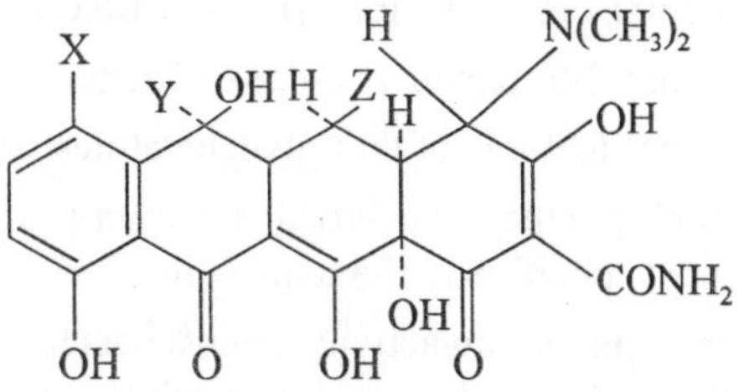

Tetracycline : X, Z = H ; Y = CH$_3$;

Chlortetracycline : X = Cl ; Y = CH$_3$; Z = H ;

Oxytetracycline : X = H ; Y = CH$_3$; Z = OH ;

Demeclocycline : X = Cl ; Y = Z = H ;

X-ray diffraction studies further reveal that the 4-dimethylamino function is placed in a *trans*-orientation rather than the *cis*-form as inferred earlier by chemical investigations. It further establishes the presence of a conjugated system existing in the structures of tetracycline from C-10 through C-12.

6.5. Newer Tetracyclines

Since 1992, several newer breeds of '**tetracyclines**' have emerged that were exclusively based on the recent researches focussed on the following aspect, namely :

(*a*) superb broad spectrum antimicrobial profile of the '*tetracyclines*', and

(*b*) recent astronomically broad emergence of bacterial **genes** and and **plasmids** encoding tetracycline resistance.

Therefore, keeping in view of the stringent limitations imposed on the '**tetracyclines**' as a class has caused the researchers at the Lederle Laboratories to augment extensive and intensive studies to rediscover SARs of tetracyclines with strategical substitutions in the **aromatic ring 'D'** in a meaningful and sincere effort to lay hand on to certain newer breeds of tetracyclines that might give rise to such drug substances which are specifically effective against the resistant strains.*

---

* Tally FT *et al. J Antimicrob. Chemother*, **35**, 449, 1995.

The concerted efforts ultimately gave birth to a few newer tetracyclines as illustrated below :*

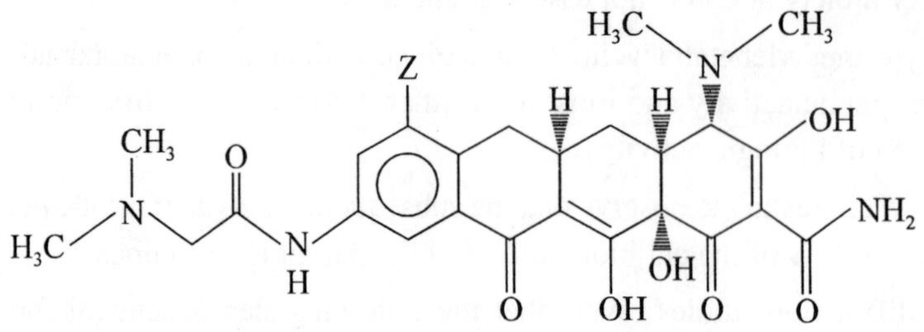

Examples :

    (*a*) 9-(Dimethylglycylamino) minocycline : [DMG-MINO] ; Z = N(CH$_3$)$_2$;

    (*b*) 9-(Dimethylglycylamino)-6-demethyl-6-deoxytetracycline [DMG-DMDOT] ; Z = H ;

Salient Features. The salient features of the '**glycylcyclines**' are as stated under :

    (*i*) retain essentially both potency and broad spectrum profile as displayed by the '**parent tetracyclines**' against specifically the tetracycline-sensitive microbial strains, and

    (*ii*) exhibit predominantly maximum activity against bacterial strains which show tetracycline resistance either through the ribosomal protecting determinants or afford mediation by efflux.

    The future prospects of a possible '**second generation tetracyclines**' are almost written on the wall provided the meaningful and fruitful clinical trials of the ongoing **glycylcyclines** do emerge both favourable *pharmacokinetic* and *toxicological* profiles for such '**medicinal compounds**' in the near future.

Probable Questions for B. Pharm. Examinations

1. (*a*) What are the four cardinal requirements of a substance to be called an '**antibiotic**' ?
    (*b*) Give the structure, chemical name and other names of the **six** naturally occurring **Penicillins.**
2. How would you establish the structure of the Penicillins as per the following steps ?
    (*a*) Hydrolysis by hot dilute inorganic acid
    (*b*) D-Penicillamine
    (*c*) Penilloaldehyde
    (*d*) Presence of CO$_2$ molecule
    (*e*) Combination of Penicillamine and Penilloaldehyde in Penicillins
    (*f*) Presence of Thiazolidine ring
    (*g*) Evidence for Penicilloic acid
    (*h*) Probable structure for Penicillin.
3. Discuss the synthesis of phenoxy methyl penicillin from *tert*-butyl-alpha-*phthalimidom-alonalhydate and D-penicillamine.*
4. Give the structure, chemical name, official status of at least **two** clinically useful :
    (*a*) Penicillins
    (*b*) Penicillins related to Ampicillin
    (*c*) Ester of Ampicillin.

5. (*a*) What are **'Aminoglycoside Antibiotics'** ?

   (*b*) Give the structure, official status and uses of any **three** potent drugs.

6. Based on vital chemical evidences, how will one establish the structure of an **'antibiotic'** produced from *Streptomyces venezualae.*

7. (*a*) What are the **three** reasons of paramount interest of **chloramphenicol** for its cognizance as a potent antibiotic ?

   (*b*) Discuss the SAR and stereochemistry of **chloramphenicol.**

   (*c*) Describe the synthesis of **chloramphenicol** from *p*-nitroacetophenone.

8. (*a*) Discuss the salient features of the **'Tetracylines'**.

   (*b*) Give a brief account of the SAR of **'Tetracylines'**.

9. Elaborate the characteristics of the 'Tetracylines' with specific reference to :

   (*a*) pKa values

   (*b*) Epimerization

   (*c*) Effect of strong acids and bases.

10. Give a comprehensive account of a 'CEPHALOSPORINS' and provide appropriate examples.

OR

Write short note on the following :

(*a*) Minocycline Hydrochloride

(*b*) Newer Tetracyclines.

RECOMMENDED READINGS

1. W O Foye, (*Ed.*) *'Principles of Medicinal Chemistry'*, (5th edn.), Lippincott, Williams & Wilkins, New York, 2002.

2. J E F Reynolds, (*Ed.*) *'Martindale : The Extra Pharmacopoeia'*, (31st edn.), The Pharmaceutical Press London (1997).

3. M C Griffiths (*Ed.*), 'USAN and the USP Dictionary of Drug Names 1987, United States Pharmacopoeial Convention, Inc. Rockville (1985).

4. M E Wolff, *Burger's Medicinal Chemistry and Drug Discovery,* (5th edn.), John Wiley and Sons, Inc., New York, (1995).

5. L S Goodman and A Gilman *'The Pharmacological Basis of Therapeutics'*, (10th edn.), Macmillan Co. London (1995).

6. Havaka JJ and Boothe JH (eds.) : *The Tetracyclines,* springer Verlag, New York, 1985.

7. Coute JE : *Manual of Antibiotics and Infectious Diseases,* Baltimore, Williams & Wilkins, 8th ed., 1995.

8. Mandell GL *et al.* (eds.) : *Priniciples and Practice of Infectious Diseases,* Churchill-Livingstone, New York, Vol. I, 4th edn. 1995.

9. Mitscher LA : *The Chemistry of Tetracycline Antibiotics,* Marcell Dekker, New York, 1978.

10. Gennaro AR : *Remington : The Science and Practice of Pharmacy,*Lippincott Williams & Wilkins, New York, Vol. II, 20th edn., 2004.

24 Antineoplastic Agents

1. INTRODUCTION

The past three decades have witnessed a remarkable revolution in the field of tumour chemotheraphy. A spectacular wealth of basic knowledge with regard to molecular and cellular biology, better understanding of mechanisms of cellular division, tumour immunology, fundamental factors involved in both viral and chemical carcinogenesis and above all the improved investigative techniques have ultimately led to the introduction of a substantial number of newer **antineoplastic agents.**

A few years ago significant palliative results were obtained by chemotherapy in a number of human neoplasma. Today it is, however, possible to list at least certain neoplastic diseases that can be associated with a normal life expectancy after treatment with drugs alone or in combination with other modalities. These neoplasms essentially include : carcinoma in women, acute leukemia, Burkitt's lymphoma, Ewing's sarcoma, retinoblastoma in children, lymphosarcoma, Hodgkin's disease, rhabdomyosarcoma, mycosis fungoides and testicular carcinoma.

A **neoplasm,** or **tumour** is an abnormal mass of tissue, the growth of which exceeds and is uncoordinated with that of the normal tissue and continues in the same manner after cessation of the stimuli which have initiated it.

A **malignant tumour** grows rapidly and continuously, and even when it has improverished its host and source of nutrition, it still retains the potentiality for further proliferation. Besides, malignant tumours invade and destroy neighbouring tissues and possess no effective capsule, a malignant tumour readilty ulcerate and tend sooner or later to disseminate and form metastases.

The **causation of neoplasms** are many, for instance : the *genetic factors e.g.,* retinoblastoma is determined by a Mendelian dominant factor and so are the multiple benign tumours ; the *chemical carcinogens e.g.,* arsenic, soot, coal tar, petroleum lubricating oil ; the *polycyclic hydrocarbon carcinogens e.g.,* 1, 2, 5, 6-dibenzanthracene, 3, 4-benzpyrene.

There has been a tremendous growth in different aspects of cancer research, cancer chemotheraphy *vis-a-vis* a better understanding of the intricacies of the '**tumour biology**' that has ultimately led to not only the legitimate evolution but also the explicit elucidation of the probable mechanisms of action for the **antineoplastic agents.** In fact, the various strategies involved to augment the speedy as well as meaningful progress in the develoment of **antineoplastic agents** may be accomplished as follows :

(*a*) Fundamental basis for the more rational approach in the design of newer drugs,

(*b*) Large collaborative investigations, concerted integrated research based on recent developments and advances in the clinical techniques, and

672

(c) Combination of such privileged advantages with improved preliminary screening methodologies.

As on data nearly ten differnet types of '**neoplasms**' may be 'cured'* with the aid of chemotheraphy in patients quite satisfactorily, namely : leukemia in children, Hodgakin's disease, Burkitt's lymphoma, Ewing's sarcoma, choriocarcinoma in women, lymphosarcoma, mycosis fungoides, rhabdomyosarcoma, testicular carcinoma, and retinoblastoma in children.

It is pertinent to raise a vital question at this point in time—'*why cancer is rather difficult to cure in comparison to other microbial infections*'. One may put forward the following plausible explanations as :

(*i*) Qualitative differences existing between the human and bacterial cells. It is well known that the bacterial cells. possess distinctive cell walls ; besides, the ribosomes also differ entirely from those of 'human cells',

(*ii*) Quantitative differences do prevail between normal and neoplastic human cells, and

(*iii*) Body's immune mechanisms and other host defenses play a vital role in killing bacteria (*i.e.,* bactericidal) plus other susceptible foreign cells ; whereas they are not so prevalent in destroying cancerous cells.

Evidences of quantitative differences do exist in the natural characteristics of *proteins* observed in monitoring various essential pathways which in turn control *three* major operations, namely : (*a*) cell proliferation ; (*b*) cell differentiation ; and (*c*) induction of programmed cell death (*i.e.,* **apoptosis**)— also necessarily catering for much desired '**targets for antineoplastic agents.**** In a situation, whenever the cancerous cells overcome the '*body's suveillance mechanism*', the chemotherapeutic agents (*i.e.,* antineoplastic agents) should be able to destroy, kill and thus erradicate completely each and every residual **clonogenic malignant cell,** since even one cell may refurbish and reestablish the cancerous tumour.

Incidence of Tumors :

The **incidence of tumours** vary from age, sex, geographical, ethnic, environmental, virus, radiation and hormone factors as stated below :

(*a*) *Age Incidence : e.g., embryonic mesenchymoma* group originate and disseminate even before birth ; *sarcoma* arises in adolescence ; *carcinoma* takes place after the age of 40 years and increases with advancing years ; *bone sarcoma* occurs between 10-12 years ; *cancer of prostrate* becomes active in old age.

(*b*) *Sex Incidence : e.g., post-cricoid cancer* is found 90% in young women ; *cancer of lower part of oesophagus* occurs in elderly men.

(*c*) *Geographical Incidence : e.g., nasopharyngeal cancer* is common among Chinese and rare in other races ; *Cancer of mouth and tongue* is common in India ; *Cancer of bladder* is common in Egypt ; *Cancer of liver* is common in Central Africa.

(*d*) *Ethnic Incidence : e.g.,* uncircumsised males suffer from *penile carcinoma* and their wives often suffer from carcinoma of cervix.

---

*Cure means-an expectation of normal longevity.

**Dorr RT and Von Hoff DD (eds) : *Cancer Chemotheraphy Handbook,* Appleton and Lange, Norwalk CT, 2nd edn., pp3-14, 1994.

(*e*) *Environmental Incidence : e.g.,* bronchogenic carcinoma is found mostly among cigarette smokers and people in industrialised areas due to air pollution and asbestos fibre inhalation.

(*f*) *Virus Incidence : e.g., polyoma virus* when gets in contact with host cell, it destroys it by feeding on it and releasing its DNA. Consequently, when this DNA gets in contact with host DNA, a new DNA with different genetic (genotype) material is formed. As this genotype is different, it grows differently from the normal cell leading to cancerous cells.

(*g*) *Radiation Incidence : e.g., osteosarcoma* is found in subjects handling paints containing radium ; radiologists mostly suffer from leukemia.

(*h*) *Hormone Incidence : e.g.,* breast cancer in mice is produced by administration of large doses of *oestrogens.*

1.1. Chemotherapeutic Intervention :

The various aspects of **chemotherapeutic intervention** may be discussed in an elaborated manner under the following defined categories, such as :

(*i*) Phase specificity,

(*ii*) Tumour selectivity and response,

(*iii*) Determinants of sensitivity and selectivity,

(*iv*) Requirements for 'kill',

(*v*) Combination chemotheraphy,

(*vi*) Log cell-kill principle, and

(*vii*) Drug resistance.

Each of the above aspects shall now be treated individually in the sections that follows :

1.1.1. Phase Specificity. Broadly speaking the *'antineoplastic drugs'* may be categorized under *two* heads, namely :

(*a*) *Phase nonspecific drugs.* These drugs have an ability to act on the cell throughout the cell-cycle, and

(*b*) *Phase specific drugs.* The drugs act **preferentially** during one or more of the nonresting phases. In other words, they prove to be *'absolutely ineffective'* when delivered to the cell specifically during the *wrong phase.*

Fig. 22.1 and 22.2 illustrate the cell-life cycle and the cell-cycle specificity respectively as given below :

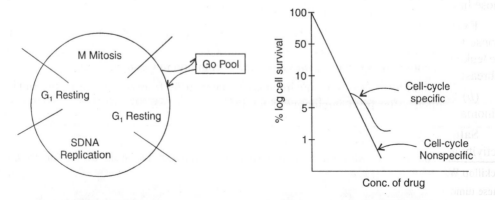

Fig. 22.1 Cell-Life cycle.   Fig. 22.2. Cell-cycle specificity.

Fig. 22.1 : evidently shows a circular pictorial model actually obtained for the clockwise progression of the cell-cycle. In actual practice, however, both the duration of individual phase in the cell cycle alters appreciably guided by the cell type, and also within a single tumour. Following are some of the '**typical durations**', for instance :

S—DNA : replication phase	= 10-20 hours ;
G_2—Resting phase	= 2-10 hours ;
G_1—Resting phase	= highly variable due to another phase ;
M—Mitosis phase	= 0.5 – 1 hours ;
G_o—Pool	= Cell not active during cell division.

Salient-Features of Antineoplastic Drugs

There are as follows :

(*i*) Block the biosynthesis or transcription of nucleic acids to check cell-division through direct interference with mitotic spindles,

(*ii*) Both mitosis phases and cells that are engaged in DNA-synthesis are found to be highly susceptible to these antineoplastic agents, and

(*iii*) In the resting state the not-so-fast growing tumours invariably possess good number of cells.*

Fig. 22.2 explicitely represents the overall effects of antineoplastic agents upon the cell survival which is exponentially related to dose. Now, if a plot is made between **log cell survival** along the Y-axis and the **drug concentration** along the X-axis one would get a '**straight line**'. Nevertheless, these drugs usually display their cytotoxicity irrespective of the cell-cycle-phase ; and, hence, are known as the '**non-cell cycle phase specific drugs**'. Importantly, such other drugs *viz.,* mitotic inhibitors and antimetabolites, that particulary act at one phase of the cell cycle only, normally display a distinct **plateau** soonafter a preliminary low-dose exponential region.

1.1.2. Tumour selectivity and Response. It has been duly observed that particularly for '*phase-specific drugs*' (see section 1.1.1.*b*), the probability apprehended for a lethal action on a cancerous tumour cell (or nomal cell) is usually directly proportional to the actual percentage of time consumed in the '*vulnerable phase.*' In other words, one may ascertain that the real percent of time spent specifically in the vulnerable phase appears to be an important '*determinant factor*'for the susceptibility of tumours belonging to different cell types. Not withstanding to any particular growth phase one may safely generalize that such tumours having a large growth fraction one prone to chemotherapy in comparison to those having a low fraction ; and this constitutes a very ideal and equally important percept.

Examples : (*a*) Cancerous tumors having high growth fractoins which are found to give adequate response to chemotheraphy are, namely : Hodgkin's disease, Burkitt's lymphoma**, Wilm's tumour, acute leukemia in children, choriocarcinoma**, chronic myelogenous leukemia**, lymphocytic leukemia, and breast cancer.

(*b*) Neoplasms (*i.e.,* malignant tumours) which afford a very poor response are, for instance : carcinoma of the GI-tract, malignant melanoma, and tumours of the uterus and cervix.

Salient Features. Following are some of the cardinal salient features with regard to tumour selectivity and response :

*Mackillop WJ *et. al. J. Natl. Cancer Inst.,* **70** : 9, 1983.

**These tumours are now considered curable to a great extent.

(1) Efficacy of antineoplastic drugs is increased significantly in early treatment of *newly developed small cancerous tumours* having relatively higher growth fractions.

(2) Most effective antineoplastic drug should invariably be expected to be of such kind which is rather specific to the phase with the longest duration *e.g.,* S-DNA : replication phase (10-20 hours), and G_2-resting phase (2-10 hours).

(3) Recently, investigation of the possibility of *synchronizing cancerous tumor cells* is gaining momentum so that most likely **all cells are in the same phase of the cycle.** In case, such in '*ideal situation*' may be accomplished then :

(*a*) Cancerous tumor might become more vulnerable to the suitable drugs administered at the right-time, and

(*b*) Therapeutic index of the '*drug*' may be enhanced appreciably.

(*c*) *Synchronization* is achieved by a *holding pulse* of a *mitostatic drug* which essentially holds the cells in a specified phase till such time the *out-of-phase* cells also come into that phase.

(*d*) Sudden or planned discontinuation of the '*synchronizing antineoplastic drug*' at the same time releases the cancerous cells to resume their own specific cycle *i.e.,* all commencing afresh from the same phase.

(*e*) Combination chemotherapy, the antineoplastic drugs are frequently administered in a particular '*sequence*', instead of simultaneously ; however, in usual practice the **first-administered antineoplastic drug** invariably stands for a *synchronizing drug.*

1.1.3. Determinants of Sensitivity and Selectivity

There are certain pivotal factors that essentially help in determining the *selectivity* of antineoplastic drugs required for some definite cell-types. Besides, the actual demand for *various nutrients* also varies significantly amongst different tumor types, as do they differ frequently between the **tumor cells** and the **normal cells.**

Example. A plethora of malignant tumors of need much more **asparagine** (a nonessential amino acid) in comparison to normal cells ; therefore, if by any manner the plasma asparagine gets destroyed (enzymatically), the cancerous tumor cells in turn are selectively *starved* to death.

It has been observed duly that some '*drugs*' either get metabolized in the *liver* or the *perpheral cells* thereby the various cell types substantially differ in their respective ability to metabolize these drug substances.

Example. **Bleomycin**—an antineoplastic drug usually gets metabolized much less in the *susceptible tumor cells* in relation to other cells, thereby allowing distinctly higher local concentrations. In addition to this there are a host of other anineoplastic drugs which are converted to the *active metabolites* by the aid of the prevailing '**target-cells**' (also termed as '*lethal synthesis*') ; and the ensuing differences in the conversion rates ultimately contribute directly to **selectivity.**

Degree of variance in penetrance also account for certain critical apperent differences amongst the antineoplastic drugs.

Example. (*i*) Neoplasms in the CNS are more effectively curable by lipid-soluble drugs than the water-soluble ones,

(*ii*) Some durgs exhibit *greater active transport* into cancerous tumor cells than into normal cells, (*iii*) Certain drugs show differences in '*outward transport*' as well,

(*iv*) Selectivity is also governed by tumor-cell attacking **killer T-cells, suppressor T-cells**, and blocking factors from **B-cells** which specifically guard and protect some cancerous cells from the prevailing immune attack, and

(*v*) Generally, the immune cells are established to be the most suppressed ones ; and thus two situations may arise : *first*, antineoplastic drugs which augment response to malignant cells and *secondly,* which antagonize it squarely.

1.1.4. Requirements for 'kill'

Ideally, a remission (*i.e.,* reduction in intensity) normally may be accomplished with a '*kill*' ranging between 90-99% of the neoplastic cells. Interestingly, a '*kill*' amounting to 99% is supposed to leave a bear minimum of 107-108 surviving neoplastic cells to continue tumor growth, and consequently the remission would stay on 3-4 doubling times only. From these observations one may safely infer that such neoplasms against whcih the immune-system is absolutely ineffective, a **100% kill'** is not only a prerequisite but also necessary to cause a '**true cure'.**

1.1.5. Combination Chemotherapy

It has been amply tested, tried and established beyond any reasonable doubt that one may enhance the '**percent of kill'** by employing the **combination therapy** of two or even more antineoplastic agents judiciously. Of course, the usage of *radiation therapy* can also be effectively used with durgs. In reality, there exits *four* cardinal factors that may optimize such '*combinations*', namely :

(*a*) Each component drug should have certain degree of efficacy by itself.

(*b*) Each component drug must have an altogether different mechanism of '**cytotoxic profile'** and, preferably, command *phase specificity.*

(*c*) Each component drug should have a distinct and different spectrum of toxicity in comparison to the other components, so as to avid specifically any overwhelming ensuing toxicity of a given type.

(*d*) The '*mechanism of resistance*' to each component must be invariably different to that of the other components.

1.1.6. Log Cell-kill Principle

It has been well-defined that the efficiency of anitneoplastic drugs may be characterized by their inherent **log cell-kill index.** In other words, the *negative log of the fraction* of the cancerous tumor cell population which essentially survives a *single-course of treatment.*

Example. A neoplastic drug that eventually kills 99.9% of the malignant tumor cell population, *i.e.,* leaves 0.0001 (or 1/104) of the population is usually termed as a **4-log drug** ; whereas, a second drug which kills 99.9% is known as a **3-log drug.**

However, the *log cell-kill index* represents a very thin (tenuous) number, but it definitely serves a tremendous usefulness in rightly predicting the effects of combinations which essentially fulfil criteria (*a*) and (*b*) above (section 1.1.5). Thus, the very close predicted effect of a combination is usually accomplished by the simple addition of the various indices obtained from the component drugs.

Example. Theoretically, a *4-log drug* together with a *3-log drug* must provide ordinarily a **7-log combination** *i.e.,* almost kills 99.99999% or leaves 1/107 of the population. Now, at this juncture a 3**rd** drug which essentially kills 99% (2-log-drug) may further minimize the remaining population to the extent of 1/109, that ultimately comes close to the complete eradication of a cancerous tumor noticed at an early stage.

1.1.7. Drug Resistance

It has been obseved that there are certain tumour populations that seem to be heterogeneous in nature by the time the cancerous tumor is discovered after the usual **biopsy examination' ;** whereas a few of the cells being resistant to some antineoplastic agents right at the very outset of the recommended treatment. The said findings hold good for certain well-established organs of the human body, such as : colon, jejunal, adrenal, kidney and liver carcinomas. A maximum of *four* different malignant cell-types have been duly recognized and identified in a single tumor.

It is, however, pertinent to mention here that a certain degree of resistance appears to be acquired in much as the same manner as in *microbial resistance,* such as :

(*a*) resistance granting '*genetic change'* taking place during treatment, and

(*b*) resistant '*daughter cells'* consequently proliferate in the prevailing environment of the antineoplastic agent.

In a nut shell, irrespective of the actual cause, the prevailing resistance invariably negates the usefulness of an antineoplastic agent to an appreciable extent.

In fact, there are **ten** different mechanisms of resistance that have been duly identified, namely :

(*i*) Complete loss of the '*transport system*' required essentially for the permeation of the drug into the tumor cell *e.g.,* methotrexate.

(*ii*) Disappearance of the enzyme necessary for the intratumor *lethal systhesis* of an essential active metabolite.

(*iii*) An enhancement in the production of the *target enzyme e.g.,* methotrexate.

(*iv*) Retardation in the *affinity for* or the *quantum of the target enzyme e.g.,* methotrexate, fluorouracil and topoisomerase inhibitors.

(*v*) *Pleiotropic drug-resistance i.e.,* an enhancement in the outward active transport of the antineoplastic drug, whereby the effective intracellular concentrations cannot be accomplished or maintanced.

(*vi*) *Overexpression of metallothionine* in resistance to Pt-containing drug and some alkylating antineoplastics.

(*vii*) *Formation of antibodies e.g.,* interferons.

(*viii*) *Membrance of antibodies* which essentially afford resistance to natural killer (NK) cells.

(*ix*) Enhance *glutathione synthesis* in malignant cells being treated with anthracyclinedione cells.

(*x*) Repair of potentially *lethal DNA damage.*

2. CLASSIFICATION

Antineoplastic agents are classified under the following *seven* categories, namely :

(*i*) Alkylating Agents   (*ii*) Antimetabolites

(*iii*) Antibiotics   (*iv*) Plant products

(v) Miscellaneous compounds   (vi) Hormones

(vii) Immunotherapy.

2.1. Alkylating Agents

Alkylating agents are chemically reactive compounds that combine most readily with nucleophilic centres a fully saturated carbon atom of the alkylating group becoming attached to the nucleophile.

The term 'alkylating agents' is applied to compounds which, in a sense, *alkylate* the substance with which they react, by joining it through a *covalent bond,* although a strong polar bond is not excluded from this general definition. Any 'antineoplastic agent' whose activity is explained by such a mechanism is called an **alkylating agent**.

These are further sub-divided into *four* categories, namely :

(i) Mustards

(ii) Methanesulphonates

(iii) Ethylenimines

(vi) Nitrosoureas

2.1.1. Mustards

After the discovery of the antileukemic activity of mustard gas : (Cl CH_2 CH_2)$_2$S Bis-β-chloroenthyl sulphide (Mustard Gas) in human being, its clinical application for the treatment of neoplasms could not be persued further due to its high toxicity, low solubility in water, oily nature and blister-producing properties.

Nitrogen mustards were selcted for the clinical application for the treatment of neoplasms because they presented fewer problems in handling, besides their respective hydrochlorides and other salts are generally stable solids having low vapour pressure and high solubility in water.

A few important nitrogen mustards used as antineoplastic agents are discussed below, for instance : Mechlorethamine hydrochloride, Mephalan, Cyclophosphamide and Chlorambucil.

A. Mechlorethamine Hydrochloride USAN

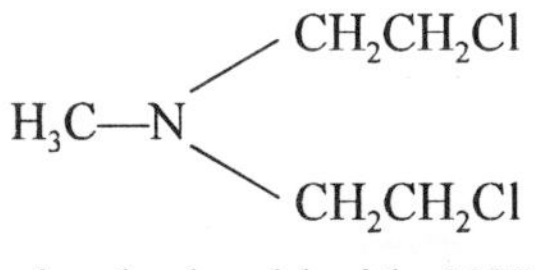

$$H_3C-N \begin{cases} CH_2CH_2Cl \\ CH_2CH_2Cl \end{cases}$$

2, 2-Dichloro-N-methyldiethylamine hydrochloride USP ;

Mustine Hydrochloride BP ;

Mustargen$^{(R)}$ (Merck Sharp & Dohme) ;

Synthesis :

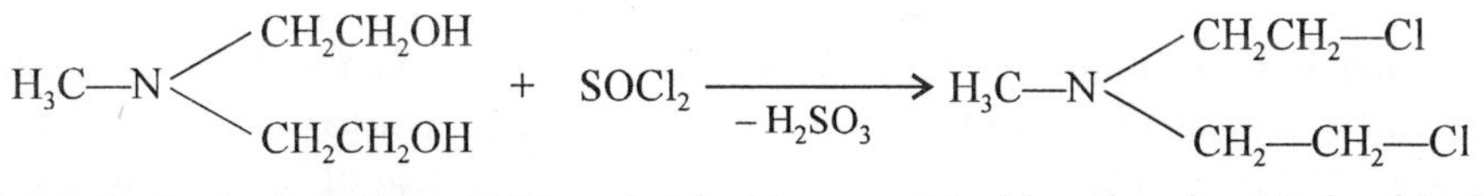

$$H_3C-N \begin{cases} CH_2CH_2OH \\ CH_2CH_2OH \end{cases} + SOCl_2 \xrightarrow[-H_2SO_3]{} H_3C-N \begin{cases} CH_2CH_2-Cl \\ CH_2-CH_2-Cl \end{cases}$$

2, 2'-(Methylimino) diethanol   Thionyl chloride   Mechlorethamine Hydrochloride.

Chlorination of 2, 2'-(methylimino) diethanol with thionyl chloride gives rise to the desired official compound, with the elimination of sulphurous acid.

It is effective in Hodgkin's disease. Usual practice is to administer mechlorethamine with other antineoplastic agents such as vincristine, prednisone etc. It is the drug of choice for the treatment of mycosis fungoides and lymphomas.

Dose. Single doses of 400 mcg per kg body weight or a course of 4 daily doses of 100 mcg per kg are normally administered by iv injection in a strength of 1 mg per ml in sodium chloride injection.

B. *Melphalan* USAN,

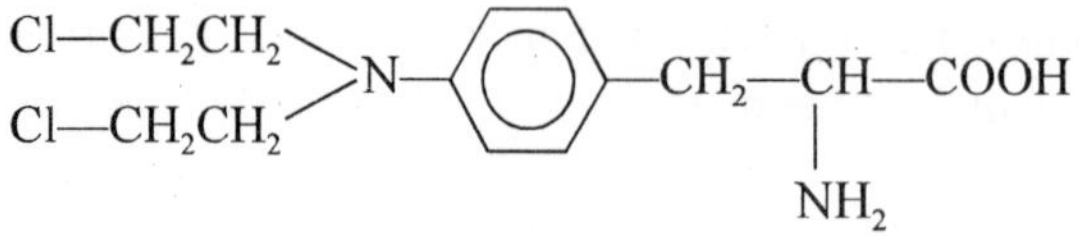

4-[Bis (2-chloroethyl) amino]-L-phenylalanine ; L-Mustard ; L-Sarcolysin ; USP ; BP ;

Alkeran$^{(R)}$ (Burroughs Wellcome) ;

Synthesis :

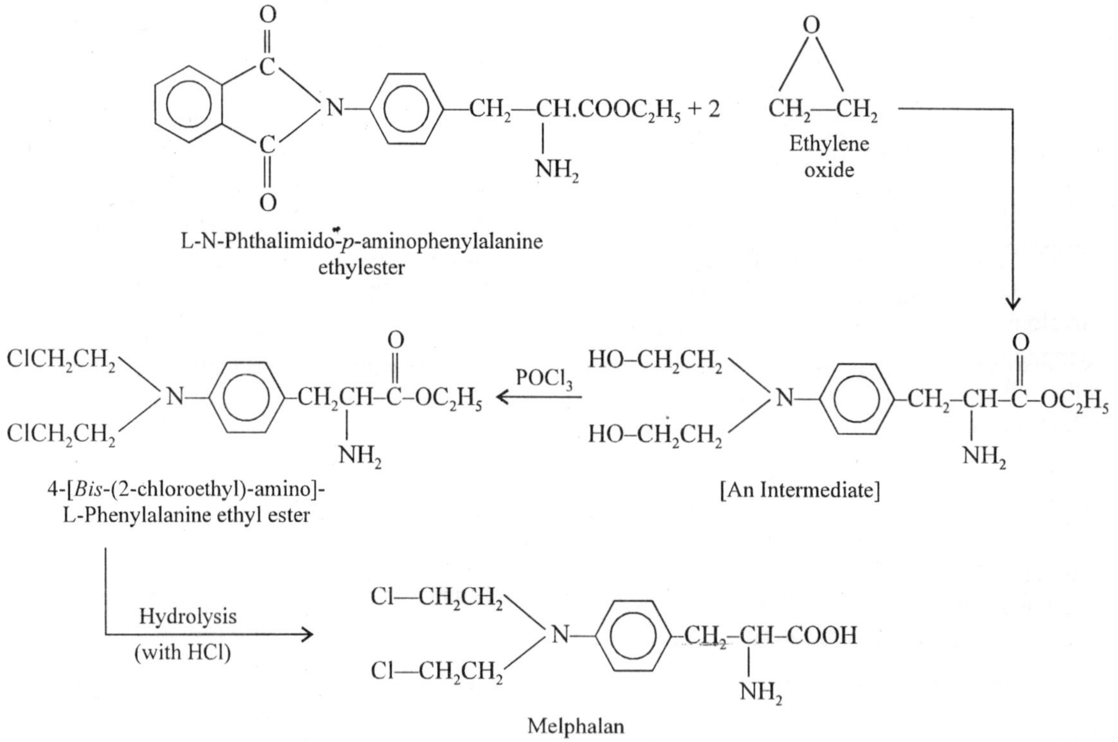

L-N-Phthalimido-*p*-aminophenylalanine
ethylester

Ethylene
oxide

4-[*Bis*-(2-chloroethyl)-amino]-
L-Phenylalanine ethyl ester

[An Intermediate]

Hydrolysis
(with HCl)

Melphalan

L-N-Phthalimido-*p*-aminophenylalanine ethyl ester when reacted with ethylene oxide yields an intermediate which on treatment with phosphorus oxychloride gives rise to 4-*Bis*-(2-chloroethyl)-amino-L-phenylalanine ethyl ester. This on hydrolysis with hydrochloric acid offers the desired compound.

Melphalan is very effective in preventing the recurrence of cancer in premenopausal women who have undergone radical mastectomy.

Dose. Oral, 150 mcg per kg body weight daily for 4 to7 days, combined with prednisone 40-60 mg dialy ; 250 mg per kg daily for 4 to 5 days ; or 6 mg daily by 2 to 3 weeks.

C. *Cyclophosphamide* BAN, USAN,

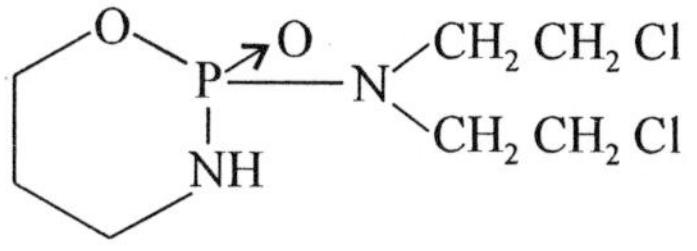

N, N-Bis (2-chloroethyl) tetrahydro-2H-1, 1, 3, 2-oxazaphosphorin-2-amine-2-oxide ; BP ; USP ; Cytoxan$^{(R)}$ (Mead-Johnson) ;

Synthesis :

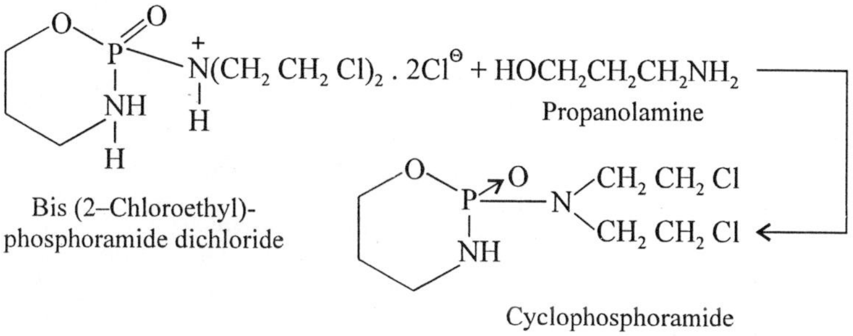

Bis (2–Chloroethyl)-
phosphoramide dichloride

Cyclophosphoramide

It is prepared by the interaction of bis-(2-chloroethyl) phosphoramide dichloride with propanolamine.

Cyclophosphamide is effective against acute leukemia, chronic lymphocytic leukemia and multiple myeloma. In combination with other chemotherapeutic agents it is found to cause radical cure in acute lymphoplastic leukemia in children and also in Burkitt's lymphoma. It has a positive advantage over other alkylating agents because of its activity both parenterally and orally besides its tolerance over prolonged periods in divided doses.

Dose. Intitial, adult dose of 40-50 mg per kg, given intravenously in divided doses over 2 to 5 days ; Children : 2-8 mg per kg daily iv injection.

Cyclophosphamide is one of the most useful antineoplastic agents and substitution at C-4 position has led to 4-phenyl cyclophosphamide (*a*) and 4-methyl cyclophosphamide (*b*) The most commonly used analogues of cyclophosphamide are Ifosfamide (*c*) and Trofosfamide (*d*) :

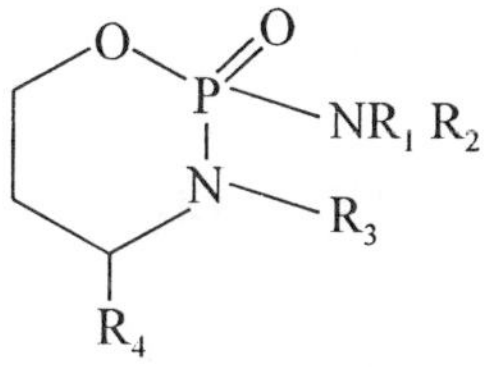

(*a*) : $R_1 = R_2 =$ —$CH_2 CH_2 Cl$; $R_3 = H$; $R_4 = C_6H_5$;

(*b*) : $R_1 = R_2 =$ —$CH_2 CH_2 Cl$; $R_3 = H$; $R_4 = CH_3$;

(*c*) : $R_1 = R_3 =$ —$CH_2 CH_2 Cl$; $R_2 = R_4 = H$;

(*d*) : $R_1 = R_2 = R_3 =$ —$CH_2 CH_2 Cl$; $R_4 = H$.

Enzyme catalyzed oxidation of cyclophosphamide yields 4-hydroxy cyclophosphamide (*e*) and subsequent formation of aldophosphamide (*f*) that leads to phosphamide mustard (*g*) which is considered to be the ultimate alkylating agent.

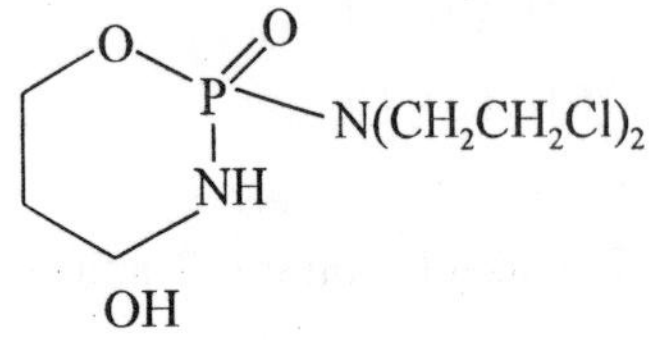

4-Hydroxy cyclophosphamide
(*e*)

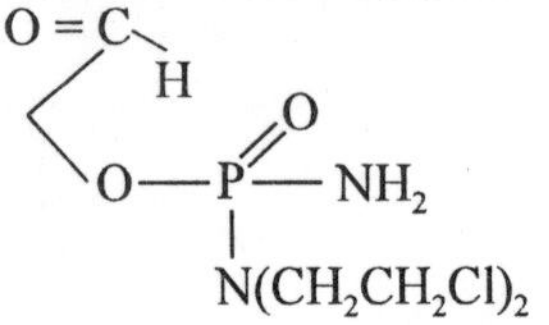

Aldophosphamide
(*f*)

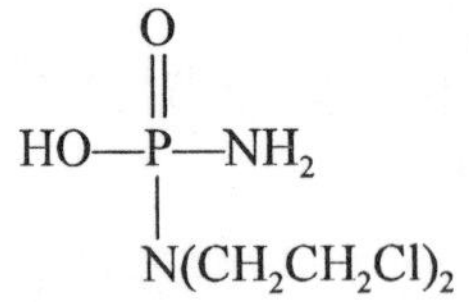

Phosphoramide Mustard
(*g*)

D. *Chlorambucil* USAN

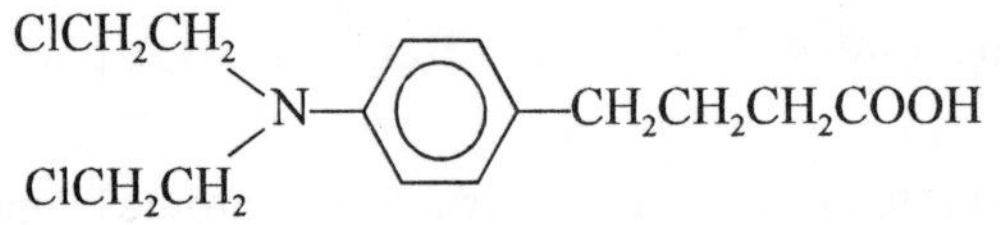

4-[*p*-Bis (2-chloroethyl) amino] phenyl] butyric acid ; Chloraminophene ; USP ;

Leukeran[R] (Burroughs Wellcome) ;

Synthesis :

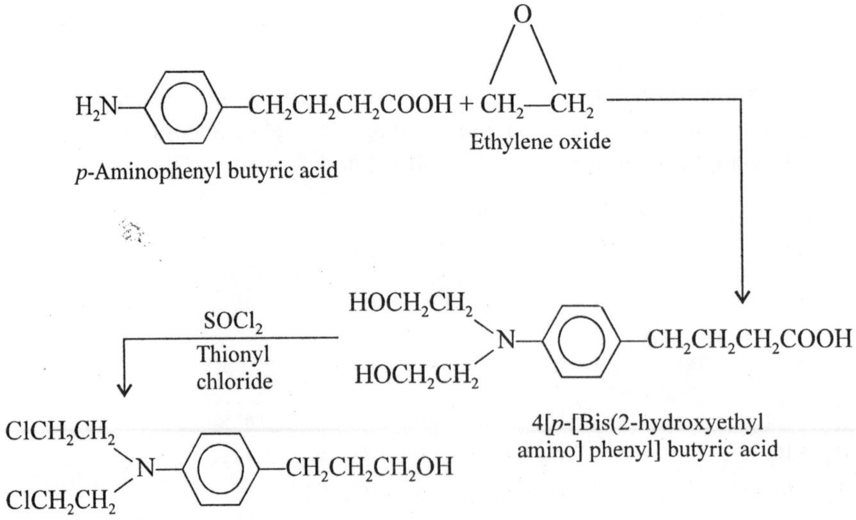

p-Aminophenyl butyric acid

Ethylene oxide

SOCl$_2$
Thionyl chloride

4[*p*-[Bis(2-hydroxyethyl amino] phenyl] butyric acid

Chlorambucil

Chlorambucil is prepared by treating *p*-aminophenyl butyric acid with enthylene oxide to yield 4-[*p*-[Bis (2-hydroxyethyl) amino] phenyl] butyric acid which on chlorination with thionyl chloride offers the desired product.

It is indicated in treatment of Hodgkin's disease, lymphosarcoma, primary microglobulinemia and chronic lymphocytic leukemia. It has an edge over other nitrogen mustards because of its least toxicity and slowest activity.

Dose. Usual, oral, 100 to 200 mcg per kg body weight daily (usually 4 to 10 mg as a single daily dose) for 4 to 8 weeks.

2.1.1.1. Mechanism of Action

The mechanism of action of the drugs described under section 2.1.1 shall be dealt with in the sections that follows :

2.1.1.1.1. Mechlorethamine Hydrochloride. The β-chloroethyl moieties lose Cl⁻ ions to generate carbonium and azardium (ethylerimonium ions), that are found to be extremely reactive ; and are capable of alkylating many biologically vital chemical moieties. It has been observed that in DNA they alkylate guanine moieties ; if one arm alkylates *one guanine group* and the second arm another guanine on the opposing strand of prevailing double-stranded DNA, the DNA turns into irreversibly cross-linked. It ultimately gives rise to inhibition of mitosis, besides causing chromosomal breakage. Importantly,some undifferentiated germinal cells are nonproliferative and hypertrophied during exposure to the '*drug*', whereas the rather more differentiated germinal cells usually disintegrate. Besides, some malignant growths, specifically of the lymph modes and bone marrow, seem to be more sensistive to the drug in comparison to the normal more slowly proliferative tissues.

2.1.1.1.2. Melphalan. The '*drug*' serves as a primary immunosuppressive drug. It is found to be well absorbed *via* the oral route, being also equally efficacious as administered by the IV route. The '*drug*' gets transformed into active metabolites in probably all tissues. The elimination half-life ranges between 1 to 3 hours.

2.1.1.1.3. Cyclophosphamide. The '*drug*' behaves unlike other β-chloroethylamino alkylators, and hence fails cyclize rapidly to the corresponding active ethylene imonium form unless and until activated by the hepatic enzymes. Importantly, the liver is protected by the further metabolism of activated metabolites into the corresponding inactive end products. Therefore, the '*drug*' is fairly stable in the GI-tract, well tolerated, and quite efficacious both by the oral and parenteral routes. It fails to produce any sort of '*local vasication*', necrosis, phlebitis, or even pain.

It is distributed to the tissues having volume of distribution v_d^{ss} more than the total body water. The '*drug*' gets metabolized by the hepatic microsomal system to the corresponding alkylating metabolites, which in turn eventually are duly converted to phosphoramide mustard and acrolein (an aldehyde). However, the relatively high doses readily induce the metabolism of cyclophosphamide. The plasma half-life ranges between 4 to 6 hours.

2.1.1.1.4. Chlorambucil. The '*drug*' is one of the slowest-acting and also least toxic currently employed nitrogen mustards. Importantly, its toxicity is manifested chiefly as bone-marrow depression ; however, in the prevailing therapeutic doses it is observed to be fairly moderate and reversible. The '*drug*' gets adsorbed well *via* the oral administration. It generally is degraded extensively *in vivo*. The elimination half-life is nearly 1.5 hour.

2.1.2. Methanesulphonates

From a mechanistic point of view the methanesulphonates (or the methanesulphonate esters) are specially interesting since the long alkylene chains separating the reductive ester groups virtually exclude the possibility of the formation of reactive intermediate ring structures. Thus, these ester groups constitute a level of direct alkylating ability whcih need not be mediated by cyclization. The methanesulphonate ion is a weakly nucleophilic group which is displaced from carbon by a more strongly nucleophilic group that is present in the biological system acted upon by the drug.

The most important alkylating agent in this group is Busulfan :

A. *Busalfan* USAN

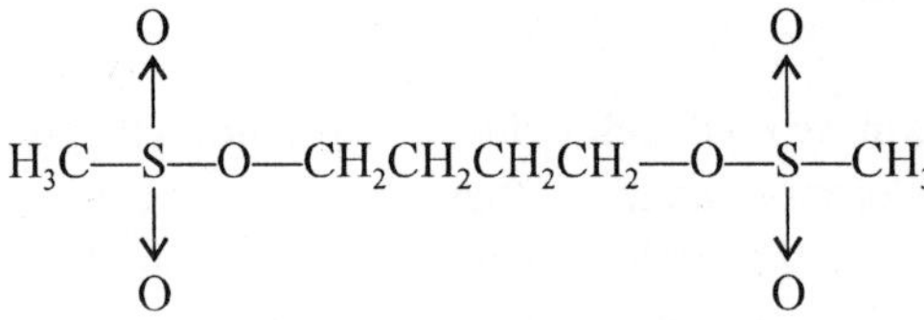

$$H_3C-\underset{\underset{O}{\downarrow}}{\overset{\overset{O}{\uparrow}}{S}}-O-CH_2CH_2CH_2CH_2-O-\underset{\underset{O}{\downarrow}}{\overset{\overset{O}{\uparrow}}{S}}-CH_3$$

1, 4-Butanediol dimethanesulphonate ; 1, 4-Di (methanesulfonyloxy) butane ; BP, USP,

Mylearn[k] (Burroughs Wellcome) ;

Synthesis :

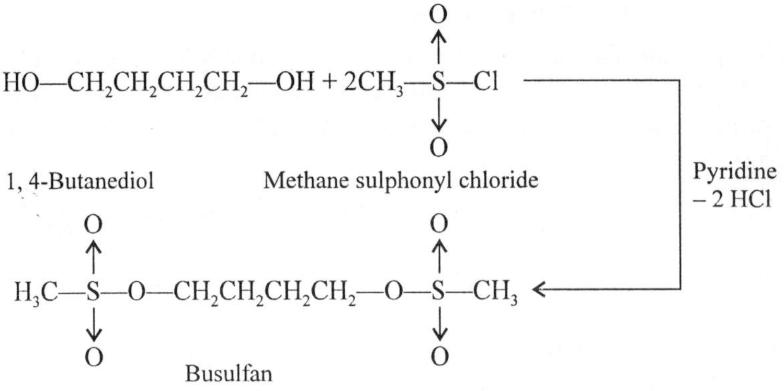

$$HO-CH_2CH_2CH_2CH_2-OH + 2CH_3-\underset{\underset{O}{\downarrow}}{\overset{\overset{O}{\uparrow}}{S}}-Cl$$

1, 4-Butanediol   Methane sulphonyl chloride

Pyridine
– 2 HCl

$$H_3C-\underset{\underset{O}{\downarrow}}{\overset{\overset{O}{\uparrow}}{S}}-O-CH_2CH_2CH_2CH_2-O-\underset{\underset{O}{\downarrow}}{\overset{\overset{O}{\uparrow}}{S}}-CH_3$$

Busulfan

It is prepared by the interaction of 1, 4-butanediol with two moles of methane sulphonyl chloride in the presence of pyridine, when the final product obtained is recrystallised either from acetone or alcohol. Busulfan is broadly used in the treatment of granulocytic leukemia.

Dose. For granulocytic leukemia : 60 mcg per kg body weight daily orally, upto a maximum single daily dose of 4 mg and to be continued till the white-cell count falls between 15000 to 25000 per mm³.

2.1.2.1. Mechanism of Action. The mechanism of action of busulfan shall be discussed as under :

2.1.2.1.1. Busulfan. The '*drug*' is phase nonspecific. It exerts almost negligible action on rapidly proliferative tissues other than the bone marrow. However, at relatively lower dose levels granulocytopoesis may be suppressed quite selectively without causing any effect on erythropoises. As the '*drug*' has little effect on lymphopoesis, it is of no value in lymphocytic leukemia and malignant lymphoma. It is found to be not immunosuppressive. Its elimination half-life ranges between 2-3 hours.

2.1.3. Ethylenimines

Clossley first reported the inhibition of experimental tumours in mice by treatment with triethylene melamine (TEM) and this resulted in the discovery of another drug triethylenethio phosphoramide.

A. *Triethylenemelamine*

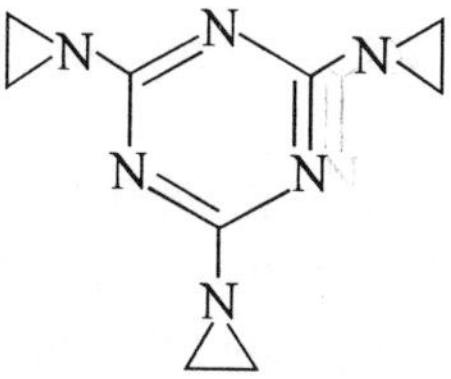

2, 4, 6-Tri (1-azridinyl)-S-triazine ; 2, 4, 6-Tris (ethyleneimino)-S-triazine ; 2, 4, 6-Triethyleneimino-1, 3, 5-triazine.

It is used as an adjuvant to radiation therapy of retinoblastoma and injected into the carotid artery. It is used in the palliative treatment of malignant neoplasms.

B. *Triethylenethio Phosphoramde* BAN

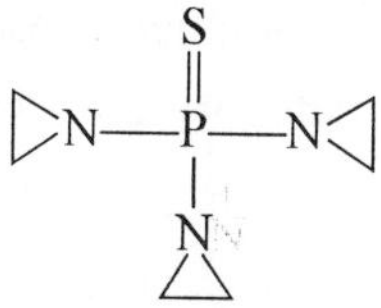

Tris (1-aziridinyl) phosphine sulphide ; N, N', N''-Triethylenethio-phosphoramide ; Thiotepa USP ; BP,

Ledertepa(R) (Lederle) ; Thiofosyl(R) (Astra) ;

Synthesis :

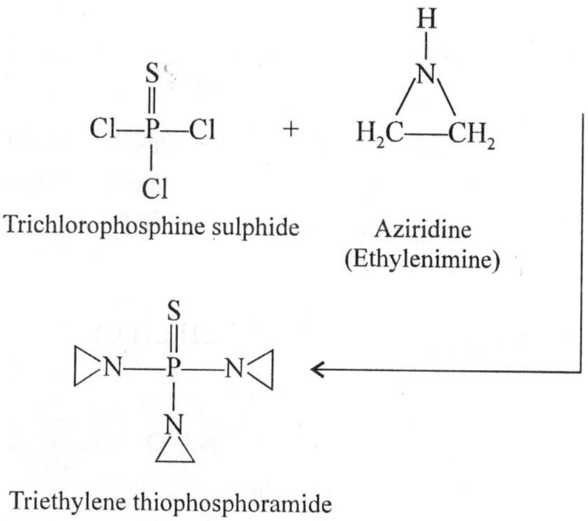

Trichlorophosphine sulphide   Aziridine
(Ethylenimine)

Triethylene thiophosphoramide

It is prepared by teating trichlorophosphine sulphide with aziridine and recrystalizing the official product from water.

It is of value in the treatment of carcinoma of breast, ovaries, colon-rectum and rectum. It is also found to be useful in the treatment of malignant lymphomas and bronchogenic carcinomas.

Dose. Upto 60 mg in single or divided doses may be given by im injection or by instillation in adults and children over 12 years.

2.1.3.1. Mechanism of Action

The mechanism of action of the '*drugs*' discussed under section 2.1.3 shall now be treated individually.

2.1.3.1.1. Triethylenemelamine (Tretamine). Its action and properties are very much akin to thiotepa (triethylenethio phosphoramide. It happens to cross blood brain barrier (BBB).

2.1.3.1.2. Triethylenethio Phosphoramide (Thiotepa). The '*drug*' exerts its action due to its alkylating characteristics. It is extensively metabolized ; and traces of unchanged drug substance are excreted in the urine, along with a large proportion of metabolites. The '*drug*' also crosses the blood-brain barrier (BBB). It has been observed that it undergoes absorption through *serous membrances*, for instance : bladder and pleura, to a certain degree.

2.1.4. Nitrosoureas

Nitrosoureas are having both practical and theoretical interest. They are very highly lipid soluble antineoplastic compounds first synthesized at Southern Research Institute, Birmingham.

A few important members of this category are discussed below, namely : Carmustine and Lomustine.

A. *Carmustine* BAN, USAN,

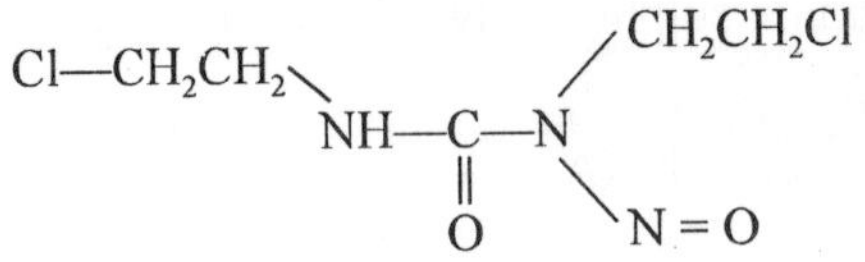

$$Cl-CH_2CH_2 \diagdown NH-\underset{\underset{O}{\|}}{C}-N \diagup \overset{CH_2CH_2Cl}{\diagdown} N=O$$

1, 3-Bis (2-chlorethyl)-1-nitrosourea ; BCNU ; Carmubris(R) (Bristol),

Synthesis :

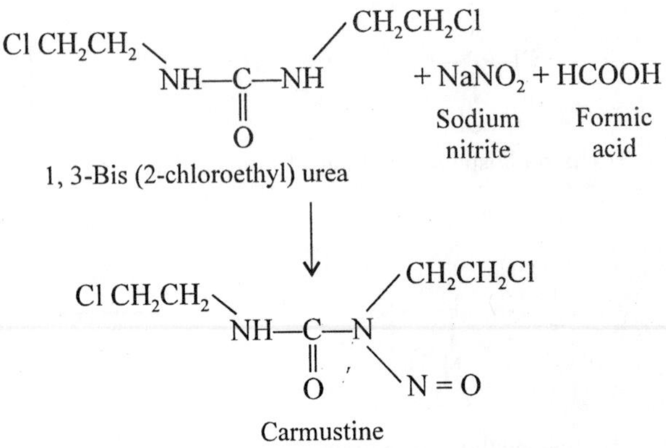

$$Cl\,CH_2CH_2 \diagdown NH-\underset{\underset{O}{\|}}{C}-NH \diagup CH_2CH_2Cl$$

1, 3-Bis (2-chloroethyl) urea

$+ NaNO_2 + HCOOH$

Sodium   Formic
nitrite   acid

$$Cl\,CH_2CH_2 \diagdown NH-\underset{\underset{O}{\|}}{C}-N \diagup \overset{CH_2CH_2Cl}{\diagdown} N=O$$

Carmustine

Carmustine may be prepared byinteracting 1, 3-bis (2-chloroethyl) urea with sodium nitrite and formic acid. It is a low-melting white powder that undergoes decomposition at 27°C and hence it is supplied as a lypholized powder.

As it possesses the potential to cross the blood-brain-barrier, carmustine is employed specifically for brain tumours and other tumours, for instance leukemias, which have metastasized to the brain. A combination of carmustine and prednisone is used for the treatment of multiple mycloma. As a secondary therapy it is frequently employed in conjunction with other antineoplastic agents for lymphomas and Hodgkin's disease.

Dose. A single dose by IV injection at 100 to 200 mg/m².

B. *Lomustine USAN*

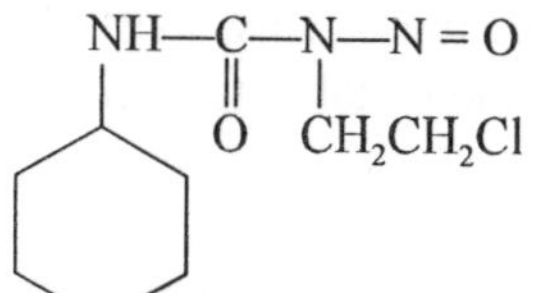

1-(2-Chloroethyl)-3-cyclohexyl-1-nitrosourea ; CCNU ; CeeNU(R) (Bristol) ; CINU(R) (Bristol-Myers) ;

Synthesis

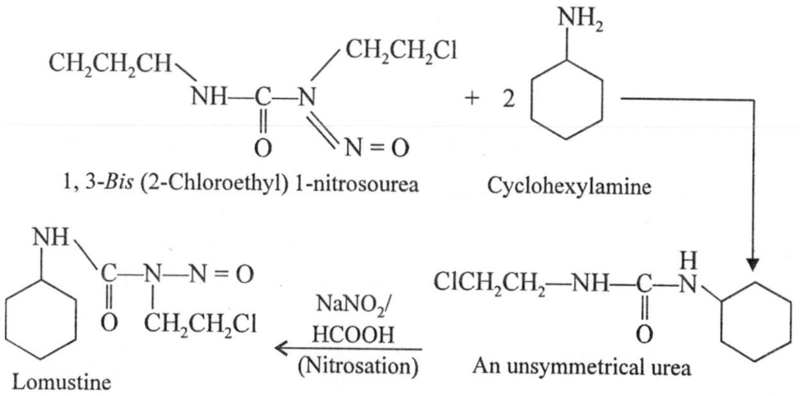

It is prepared by the decomposition of 1, 3-bis (2-chlorethyl)-1-nitrosourea in the presence of two equivalents of cyclohexylamine to yield an unsymmetrical urea which on nitrosation with sodium nitrite and formic acid offers the desired compound.

It is employed effectively in the treatment of primary and metastatic brain tumours. It is also used as secondary therapy in Hodgkin's disease.

Dose. Usual dosage : 130 mg/m² orally every six weeks.

2.1.4.1. Mechanism of Action. The mechanism of action of the two medicinal compounds described under section 2.1.4. shall now be treated individually as under :

2.1.4.1.1. Carmustine. The '*drug*' most probably exerts its action due to the ability to cross-like cellular DNA. Thus the very synthesis of both DNA and RNA is ihibited. If is specifically phase nonspecific. The '*drug*' gets metabolized, *via* oral administration, practically 100% as it happens to pass through the liver ; therefore, it should be given IV. It has been observed that after IV administration, its plasma half-life is of rather shorter duration ranging between 3-30 minutes. By virtue of the fact that the '*drug*' has a high lipid solubility profile which renders it to pass through the *blood-brain barrier* (BBB) rather swiftly. Besides, the prevailing concentrations in the *cerebrospinal fluid* (CSF) varies from approximately 50-115% of those in plasma.

2.1.4.1.2. Lomustine. The '*drug*' is a chemical congener of **Carmustine** and, therefore, almost possesses the same mechanisms of action. Just like carmustine, it accomplishes maximum concentrations in the CSF ; and, hence, shares with carmustine a **first choice status** for the treatment of *glioblastoma.* *

The '*drug*' is found to be well absorbed orally and thereby survives the first pass through the liver to be effective by the oral administration. Besides, it gets distributed evenly amongst the various tissues having a volume of distribution (v_d^{ss}) much higher than the total body water content. However, in the CSF the concentration of metabolites attains almost 150% of that normally present in plasma. It has been observed that the biotransformation usually takes place throughout the body ; the half-life is nearly 15 minutes, and the half-lives of the metabolites are 48 hour.

2.2. Antimetabolites

Antimetabolites are such compounds which essentially prevent the biosynthesis of normal cellular metabolites. They generally possess close structural resemblance to the metabolite which is ultimately antagonized. Thus they have a tendency to unite with the active site, as if they are the actual substrate.

In general, following are the various classes of antimetaboites usually employed in the treatment of cancer. They are namely :

(*a*) Antifolic acid compounds   (*b*) Analogues of Purines

(*c*) Analogues of Pyrimidines   (*d*) Amino acid antagonists

2.2.1. Antifolic Acid Compounds

Antifolic acid compounds are also referred to as '*Antifolics*' or '*Folate Antagonists*' Drugs belonging to this category act by preventing the synthesis of folic acid which is required by the tissues. They bind strongly to dihydrofolate reductase (DHFR) thereby inhibiting the conversion of dihydrofolic acid to tetrahydrofolic acid and thus inhibit the synthesis of purines and thymidines. Antifolics kill cells by inhibiting DNA synthesis in the S phase of the cell cycle. Therefore, they are found to be most effective in the log growth phase.

The most important drug in this group is methotrexate.

A. *Methotrexate* BAN, USAN, INN

*A neuroglia (*i.e.,* cells and fibers forming the interstitial elements of CNS) cell tumour.

N-[4-[[(2, 4-Diamino-6-pteridinyl) methyl] methylamino] benzoyl]-L-glutamic acid ;

BP (1973) ; USP ;

Amethopterin[(R)] (Lederle) ;

Synthesis :

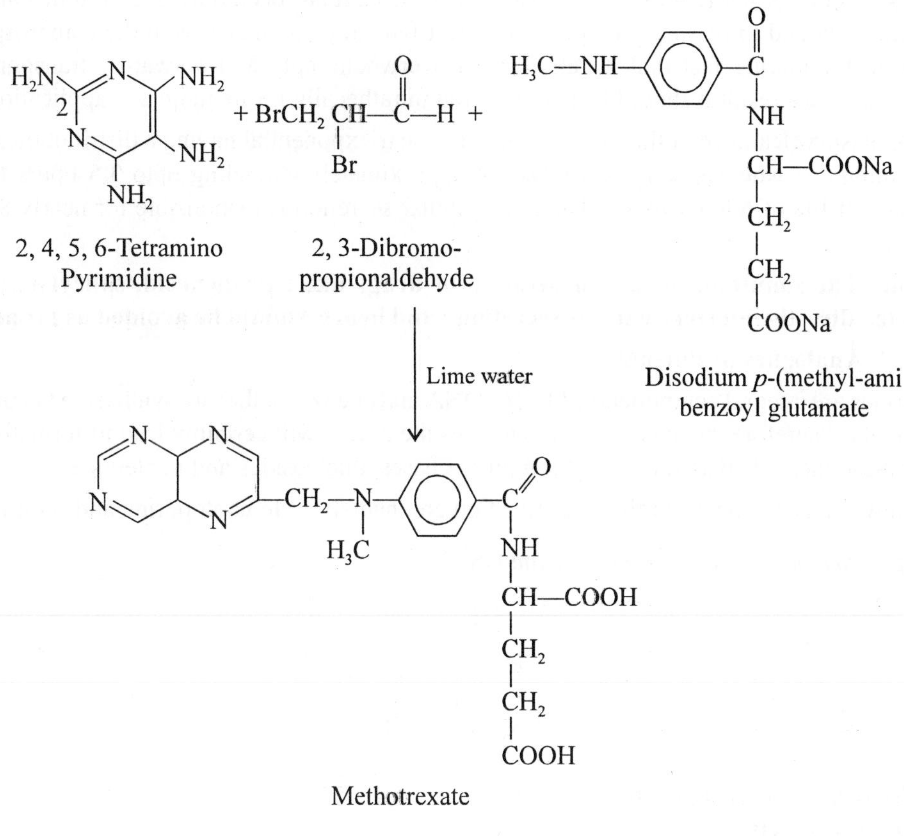

2, 4, 5, 6-Tetramino
Pyrimidine

2, 3-Dibromo-
propionaldehyde

Lime water

Disodium *p*-(methyl-ami
benzoyl glutamate

Methotrexate

It is prepared by treating together 2, 4, 5, 6-tetraaminopyrimidine ; 2, 3-di-bromopropionaldehyde, disodium *p*-(methylamino)-benzoylglutamate, iodine and potassium iodide and subsequently followed by heating with lime water.

It is the first ever antineoplastic agent that produced appreciable remissions in leukemia. It is extensively employed for the treatment of acute lymphoblastic leukemia. It is invariably used in combination cheotherapy for palliative management of lung cancer, breast cancer and epidermoid cancers of the head. It is frequently recommended for the treatment and prophylaxis of meningeal leukemia based on its ability to penetrate the central nervous system. It is also of value in choricarcinoma and related trophoblastic tumours of women.

Dose. For maintenance therapy of acute lymphoblastic leukemia is 15-30 mg per m² body surface once or twice weekly, either orally or intramuscularly, with other agents such as mercaptopurine.

2.2.1.1. Mechanism of Action

The mechanism of action of methotrexate shall be discussed as under :

2.2.1.1.1. Methotrexate. The '*drug*' exerts its action by inhibiting the enzyme dihydrofolate reductase (DHFR), and thus prevents effectively the conversion of *deoxyuridylate* to *thymidylate*, that ultimately blocks the synthesis of new DNA required urgently for the cellular replication.

Interestingly, the '*drug*' in doses less than 30 mg/m^2 usually gets absorbed well by the oral administration ; however, nearly 1/3rd of an oral dose is metabolized by both *intestinal organisms* and *antibiotics* that ultimately affect the quantam absorbed. Furthermore, in doses greater than 80 mg/m^2 the amount absorbed is further reduced to the extent of 30-50%. It has been duly observed that almost 50% of the '*drug*' is bound to plasma-protein, however, it fails to gain an access to the cerebrospinal fluid (CSF) due to the glaring fact that it gets ionized overwhelmingly and outwardly transported at the *choroid plexus*. As a result it should be administered intrathecally for its judicious application in CNS.

The plasma clearance of the '*drug*' is found to be triexponential having a distribution half-life of nearly 45 minutes ; whereas, a *second-phase* of approximately extending upto 3.5 hours.* It has an elimination half-life of 6 to 69 hours. The renal tubular secretion is responsible for nearly 80% of the elimination.

Note. The simultaneous administration of drugs like : probenecid, salicylate and other NSAIDs etc., directly interfere with its secretion ; and hence, should be avoided as far as possible.

2.2.2. Analogues of Purines

Purines are integral components of RNA, DNA and coenzyme that are synthesized in proliferaton of cancer cells. Therefore, an agent that antagonizes the purine will certainly lead to formation of false DNA and these include analogues of natural purine bases, nucleosides and nucleotides.

A few drugs belonging to this classification are, namely : Mercaptopurine and Azathiopurine :

A. *Mercaptopurinum* BAN ; *Mercaptopurine* USAN ;

o-Mercapto-6-purine ; 6 MP ;

BP(1973) ; USP ;

Purinethol$^{(R)}$ (Burroughs Wellcome) ;

Synthesis :

| Hypoxanthine | Phosphorus pentasulfide | | Mercaptopurine |

It may be prepared by the interaction of hypoxanthine with phosphorus pentasulphide.

---

*Perhaps due to an enterohepatic component—as about 10% of the '*drug*' is secreted into the **bile**.

Mercaptopurine is found to inhibit experimental orthoimmune encephalomyletis and thyroiditis and hence used in combination with vincristine, methortrexate and prednisone in the treatment of childhood leukemia. As such 6-MP may cause hyperuricamia but it is usually administered with *allopurinol*—an analogue of hypoxanthine which blocks the conversion of 6-MP to uric acid and hence the dose of 6-MP is reduced and still the desired response is obtained.

Dose. Oral, usual, initial for children and adults : 2.5 mg per kg body weight daily, but the dosage varies as per individual response and tolerance.

B. *Azathiopurine* BAN, USAN,

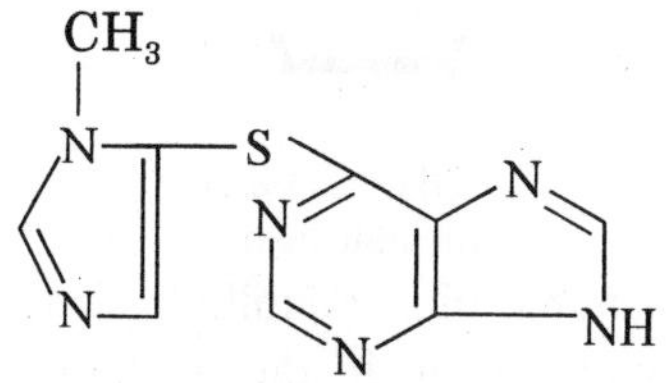

6-[1-Methyl-4-nitromidazole-5 yl] thio] purine ;

BP ; USP ;

Imuran[R] (Burroughs Wellcome) ;

Synthesis :

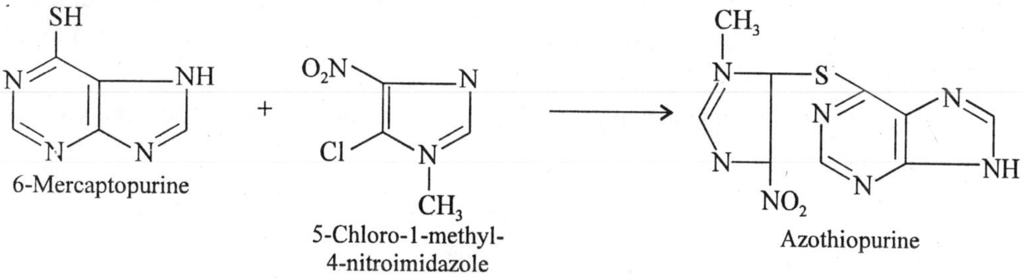

It is prepared by treating 6-mercaptopurine with 5-chloro-1-methyl-4-nitroimidazole.

The main use of azathiopurine is as an adjunct for the management and prevention towards the rejection of renal homotransplants.

Dose. Usual, adult, and children : 1 to 25 mg per kg body weight daily by mouth.

2.2.2.1. Mechanism of Action. The mechanism of action of the two medicinal compounds described under section 2.2.2 shall be dealt with in the sections that follows :

2.2.2.1.1. Mercaptopurine. The '*drug*' gets converted to 6-thioinosinic acid that predominantly serves as an antimetabolite to inhibit synthesis of *adenine* and *guanine* ; besides, it also presents conversion of *purine bases* into the corresponding *nucleotides.*

It also mimics inosinic acid thereby causing a negative feedback suppression of the synthesis of inosinic acid. It has been observed taht a portion of the '*drug*' gets converted to *thioguanine,* which is ultimately incorporated into both DNA and RNA to give rise to the formation of defective nucleic acids.

In this manner the synthesis and functionalities of the resulting nucleic acid are impaired in various ways. It finally helps in the *inhibition of cell mitosis.*

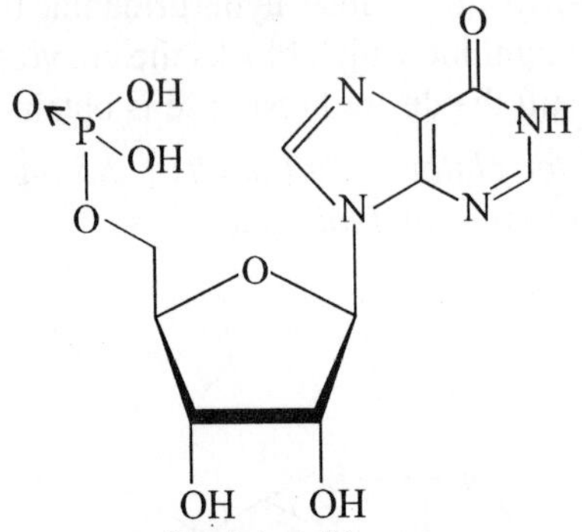

Inosinic acid

It has been found that the systemic bioavailability of mercaptopurine *via* the oral route varies from 5-37%, due to its first-pass metabolism in the intestinal mucosa and liver, wherein the two biochemical reactions usually take place, namely : (*a*) oxidation by *xanthine oxidase* ; and (*b*) S-methylation. The '*drug*' gets bound to plasma protein to nearly 20%. Its volume of distribution (v_d^{ss}) is much higher in comparison to the extracellular space ; however, the access to CSF is minimal. The half-life in children is 21 minutes and in adults 47 minutes.

2.2.3. Analogues of Pyrimidines

Pyrimidine analogues have the capacity to interfere with the synthesis of pyrimidine nucleoside and hence the DNA synthesis. Aside from their antineoplastic effects they are also found to be equally effective in psoriasis and fungal infections.

A few characteristic compounds of this category are, namely : Fluorouracil and Cytarabine.

A. *Fluorouracil* BAN, USAN,

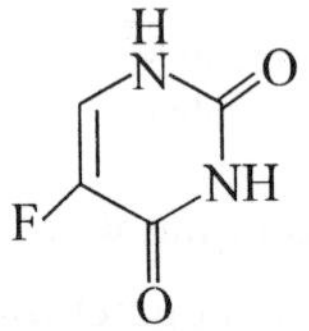

5-Fluoro-2, 4 (1H, 3H)-pyrimidinedione ; 2, 4-Dioxo-5-fluoropyrimidine ;

USP ; Efudex$^{(R)}$ (Roche) ; Fluoroplex(R) (Allergan) ;

Synthesis :

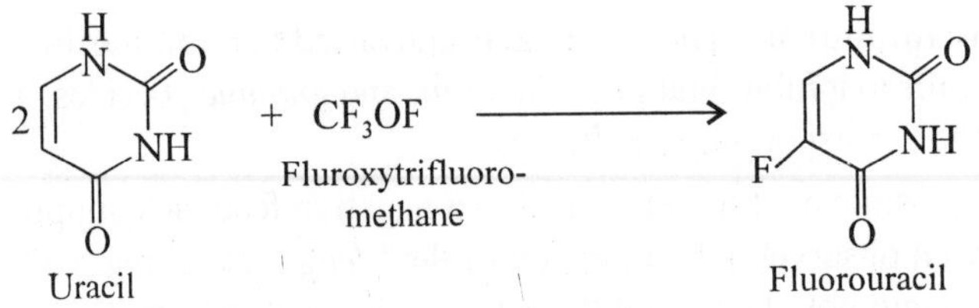

Uracil   Fluroxytrifluoro-   Fluorouracil
                              methane

This official compound is prepared by the direct fluorination of uracil with fluoroxytri-fluoromethane.

It is used in the palliative treatment of carcinoma of the breast, pancreas, prostrate, colon and hepatoma for which surgery or irradiation is not possible. It is also found to be beneficial in tropical treatment of premalignant solar keratosis.

Dose. Usual, iv injection : 1.2 mg per kg body weight daily to a maximum of 1 g daily for 3 or 4 days.

B. *Cytarabine* BAN, USAN,

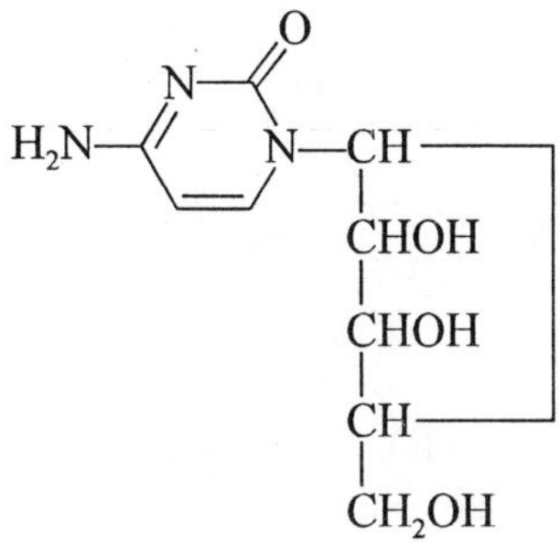

o-Amino-4-arabinofurannosyl-1-oxo-2dihydro-1, 2-pyrimidine ; Cytosine arabinoside ; USP ; Aracytin[R] (Upjohn) ; Cytosar-U[R] (Upjohn).

Synthesis :

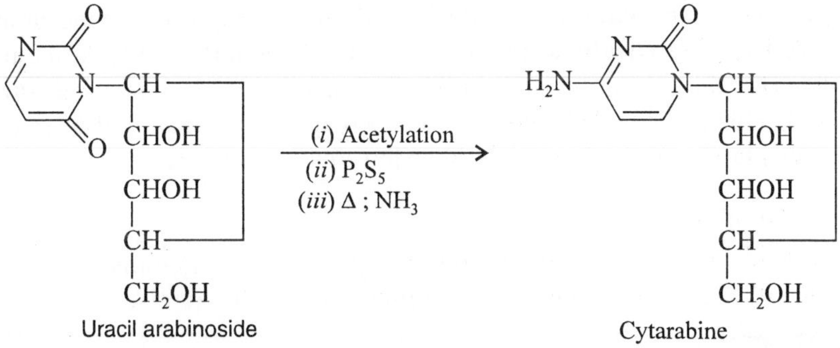

Cytarabine may be synthesized by the acetylation of uracil arabinoside followed by treatment with phosphorus pentasulphide and subsequent heating with ammonia.

It is indicated in both adult and childhood leukemia. It is specifically useful in acute granulocytic leukemia and found to be more effective when combined with thioguanine and daunorubacine.

Dose. Usual adult, and children for leukemia : 2 mg per kg body weight intravenously per day for 0 days.

2.2.3.1. Mechanism of Action. The mechanism of action of fluorouracil and cytarabine discussed under section 2.2.3 shall now be treated individually as under :

2.2.3.1.1. Fluorouracil. The '*drug*' is a congener of uracil which eventually serves both as a surrogate and as an antimetabolite of the nueleotide. Interestingly, its metabolite, 5-fluorodeoxyuridine-5′-monoplosphate (FUMP), blocks the synthesis of *thymidylic* and hence of *deoxyribonucleic acid* (DNA).

It also gets incorporated into the RNA directly. The '*drug*' is poorly absorbed orally and hence shows variable first-pass metabolism of the drug but the gut and the liver ; and hence, IV administration is an absolute necessity. It has been observed that nearly 60% of it gets metabolized to CO_2 ; however, more than 15% is excreted through the urine. The '*drug*' gains entry into the CSF and effusions. The plasma half-life is nearly 10 minutes ; however, the **active metabolite FUMP,** may be detectable for quite a few days at a stretch.

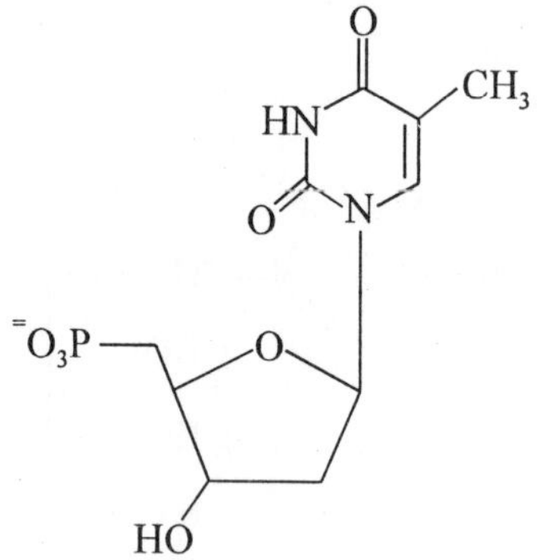

Thymidylic acid

2.2.3.1.2. Cytarabine. The '*drug*' is a **pyrimidine nucleoside antimetabolite** which is cytotoxic to a plethora of cell-types. Precisely the induction of the enzyme **nucleotidase** into DNA inhibits polymerization *via* termination of strand synthesis. It is S-phase specific.

As the '*drug*' is not absorbed quite effectively by oral administration, hence its oral bioavailability is merely 0.2. Nevertheless, it penetrates right into the CSF and accomplishes a concentration upto 40% in plasma. It gets destroyed *in vivo* to an extent of 90% by **deamination**. Its plasma half-life ranges between 1-3 hours. The elimination half-life in the CSF stands at 3.5 hours. The '*drug*' undergoes **detoxification** through the entire body ; and, therefore, perhaps it may be administered even in patients with renal impairment, however, the dosage could be lowered accordingly.

2.2.4. Amino Acid Antagonists

The amino acid antagonists broadly act as a glutamine antagonists in the synthesis of formylglycinamidine ribotide from glutamine and formylglycinamide ribotide.

Example. Azaserine ;

A. *Azaserine* USAN,

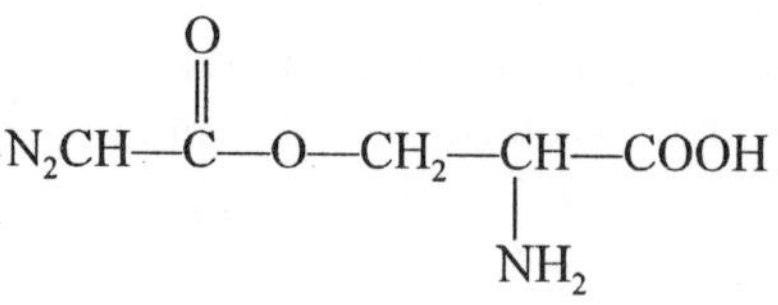

o-Diazoacetyl-L-serine ;

CI-337[R] (Parke Davis) ;

Azaserine inhibits the growth of sarcoma 180 and several leukemias. In clinical trials, although there was improvement in some cases of Hodgkin's disease, acute leukemia in children and chronic lymphocytic leukemia, the results in general were not very encouraging.

2.2.4.1. Mechanism of Action

The mechanism of action of azaserine is discussed as under :

2.2.4.1.1. Azaserine. The '*drug*' is believed to be a glutamine antagonist that specifically inhiits purine biosynthesis and thus may exert antitumour activity.

2.3. Antibiotics

The recognitiion of antibiotics as an important class of antineoplastic agents is quite recent. Consequently, the production of antineoplastic agents through proper strain selection and controlled microbial fermentation conditions may ultimately optimize the formation of a particular component in an antibiotic mixture.

A few important members of this category are described below, namely ; Dactinomycine ; Daunorubicin ;

A. *Dactinomycine* USAN,

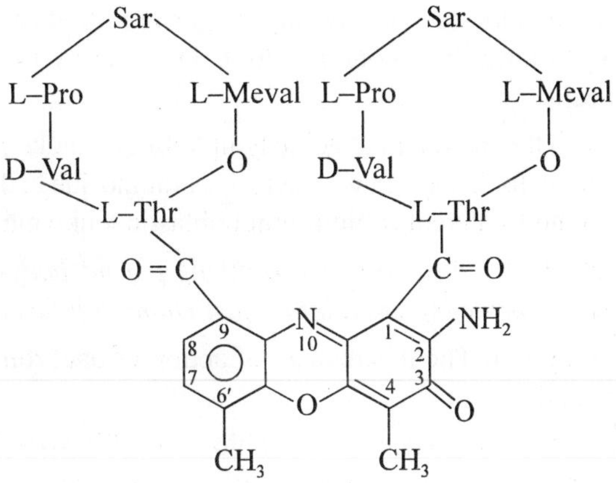

Actinomycin D ;

USP ; Cosmegen[R] (Merck, Sharp & Dohme).

The first antibotic to be isolated from a species of *Streptomyces* was Actinomycin A and many related antibiotics including Actinomycin D were latter obtained. Actinomycin C was the first to be tried on neoplastic diseases. Actinomycin D is commercially available as Dactinomycine. It is found to the acitve against L-1210, P-1534, P-388 and adenocarcinoma strains. It binds to DNA thereby preventing DNA transcription.

It is used in the treatment or rhabdomyosarcoma in children and methotrexate-resistant choricarcinoma in women. It has also been used to inhibit immunoligical response particularly the rejecton of renal transplants.

Dose. Adults, iv, 0.01 mg (10 mcg) per kg body weight ; Children : 0.015 mg (15 mcg) per kg body weight for not more than 5 days.

B. *Daunorubicin* BAN ; *Daunorubicin Hydrochloride* USAN

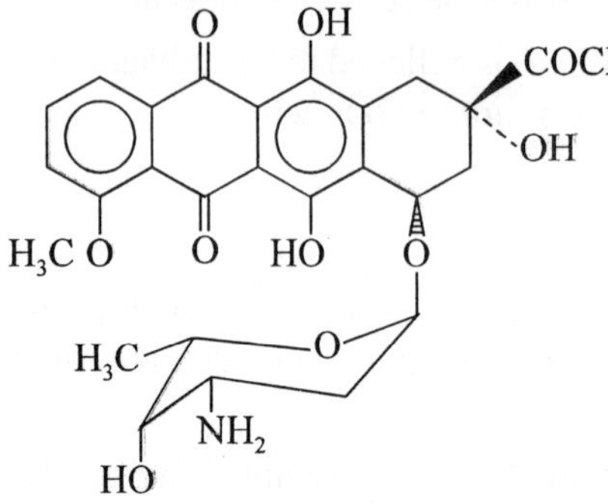

5,12-Naphthacenedione,   (8S-*cis*)-8-acetyl-10-[(3-amino-2,3,6-trideoxy)-α-1-lyxo-hexanopyranosyl) oxy]-7, 8, 9, 10-tetrahydro-6, 8, 11-trihydroxy-10-methoxy, hydrochloride ;

Ondena[R] (Bayer).

Anthracyclines constitute another complex and bigger family of antibiotics. They mostly occur as glycosides of the anthracyclinones (aglycone residue). They act by intercalation with the DNA in both normal and neoplastic cells.

Daunorubicin is useful in the treatment of acute lymphoblastic leukemia in children. It is normally employed in combination therapy, for instance : with cytosine arabinoside in the treatment of myclogenous leukemia ; with cytarabine in the treatment of non-lymphoblastic leukemia in adult.

Dose. For acute mycloblastic leukemia : 45 to 60 mg per m² body-surface daily for 3 days by injecting a solution in sodium chloride injection into a fast-running infusion of sodium chloride.

2.3.1. Mechanism of Action. The mechanism of action of dactiromycin and daunorubicin will be dealt with individually as under :

2.3.1. Dactinomycin. The '*drug*' specifically inhibits the **DNA-dependent RNA-polymerase.** Interestingly, the drug also significantly potentiates *radiation recall* (otherwise known as 'radiotherapy'). It also serves as a *secondary (efferent) immunosuppressive agent.* It has been demonstrated that almost 50% of the dose is excreted in fact into the bile and 10% into the urine ; the half-life is nearly 36 hour. The drug does not pass the blood-brain barrier (BBB).

2.3.2. Daunorubicin Hydrochloride. The '*drug*' intercalates into DNA, inhibits topoisomerase II, yields oxygen radicals, and ultimately inhibits DNA synthesis. It can invariably prevent and check cell division in doses that virtually fail to interfere directly with the nucleic acid synthesis.

It has been observed that the oral absorption is reasonably poor ; and, therefore, it must be administered IV. The half-life of distribution is about 45 minutes and of elimination, nearly 19 hours. The active metabolite, *daunorubicinol,* has a half-life of almost 27 hours. The '*drug*' gets metabolized largely in the liver, and also secreted right into the bile (ca 40%).

CAUTION : *Dosage should be lowered in instances where liver or renal insufficiencies occur.*

2.4. Plant Products

Plant products have been used extensively in the treatment of malignant disease since thousands of years, but the studies of Dustin in 1938 on the cytotoxicity of colchicin heralded the start of the search for natural antineoplastic drugs. Today a large number of chemical constituents isolated from naturally occurring plant products have proved to the quite efficacious as antitumour agents.

An attempt is made here to reivew the action, clinical usefulness, their sources and the classification is done based on their chemical nucleus ; *viz.,*

(*a*) Imides and Amides

(*b*) Tertiary Amines

(*c*) Heterocyclic Amines

(*d*) Lactones

(*e*) Glycosides

2.4.1. Imides and Amides

Examples. Colchicine ; Narciclasine ;

A. Colchicine

Colchicine occurs as the major alkaloid of the autumn crocus, *Colchicum autumnale* and the African climbing Lily, *Gloriosa superba* Linn., (Family : *Liliaceae*).

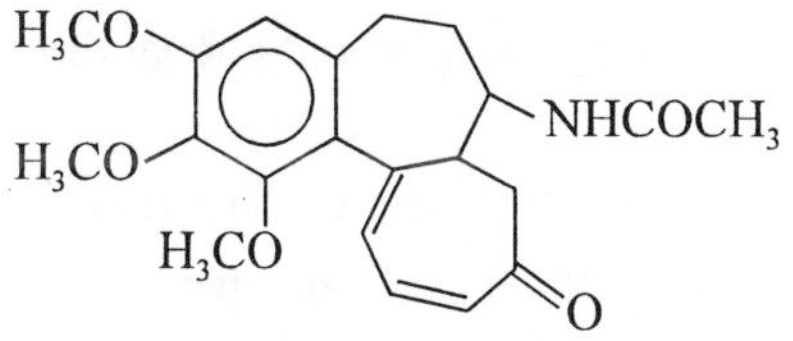

Colchicine

It arrests mitosis at the metaphase preventing anaphase and telophase. It was observed that colchicine diminishes deoxy-cytidylate aminohydrolase activity in Ehrlich ascites cells suggesting thereby that its action on mitosis and DNA synthesis could be by this method only. It is mainly used in terminating acute attacks of gout.

However, its derivative, demecoloine (Colcemid) is found to be active against myelocytic leukemia.

B. *Narciclasine*

Narciclasine, an alkaloid isolated from the bulbs of narcissus, possesses antimitotic activity against S-180 in ascites form suggesting thereby that it acts essentially as a metaphasic or preprophasic poison.

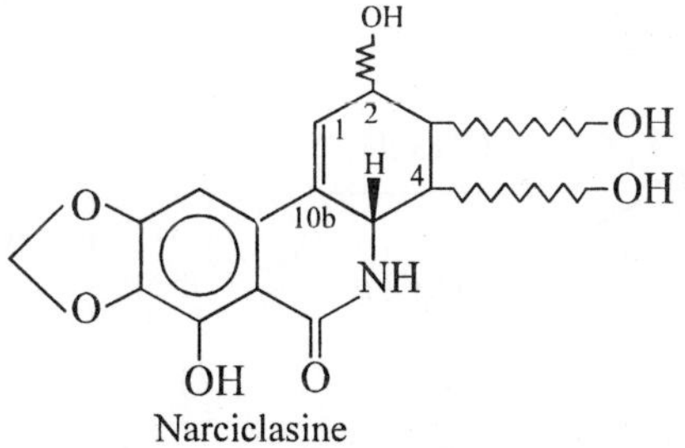

Narciclasine

Both chemical and spectral studies suggest the above structure of narciclasine, but unlike other members of the amaryllidaceae group of alkaloids it possesses no basic properties.

2.4.1.1. Mechanism of Action

The mechanism of action of colchicine and narciclasine described under section 2.4.1 will be treated as under :

2.4.1.1.1. Colchicine. The precise mechanism of action of this '*drug*' is not yet known, although it is believed to minimize appreciably leukocyte motility, phagocytosis, and also lactic acid production, thereby lowering the deposition of **urate crystals** and the **inflammatory response.** In fact, all these effects combinedly relate to the interference of colchicine upon the cellular mitotic spindles progressively.

It is found to be absorbed very well after oral administration ; and almost 31% gets bound to plasma protein. It is usually eliminated by the faecal and urinary routes.

CAUTION. *The 'drug'* **must be given with great caution particularly to debilitated and aged patients ; and also for those who have a history of cardiac, renal, hepatic, GI, or hematological problems.**

2.4.1.1.2. Narciclasine. The '*drug*' exert its action due to its inherent antimitotic agent. It also inhibits protein synthesis. It is regarded to be the most active antitumour agent of the *Amaryllidaceae* alkaloids.

2.4.2. Tertiary Amines

It includes a good number of dimeric, acyclic and phenanthro compounds and a few of them are discussed below :

(*i*) Dimeric indole alkaloids : *e.g.,* Vinblastine ; Vincristine ;

(*ii*) Dimeric tetrahydroisoquinolines : *e.g.,* Thalicapine ; Thalidasine ;

(*iii*) Acyclic tertiary amines : *e.g.,* Solapalmitine ; Solapalmitenine ;

(*vi*) Phenanthroquinilizidines : *e.g.,* cryptoleurine ;

(*v*) Phenanthroindolizidines : *e.g.,* Tylophorine ; Tylocrebrine ; Tylophorinine ; Phenanthroin-dolizidine ;

2.4.2.1. Dimeric Indole Alkaloids

So far about 72 alkaloids have been isolated from *Vinca rosea* Linn, genus *Catharanthus roseus* (Family : *Apocynaceae*). Out of these 24 dimeric alkaloids only six possess antineoplastic activity but specifically two *i.e.,* vincristine, vinblastine, are used clinically in human neoplasms. These are cell-cycle specific agents.

A. *Vinblastine* BAN

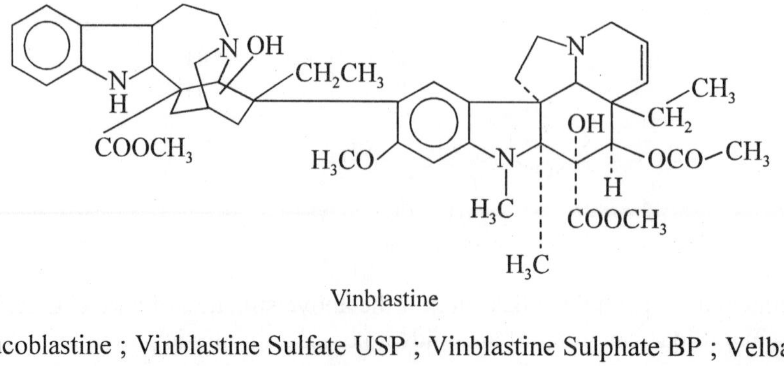

Vinblastine

Vincaleucoblastine ; Vinblastine Sulfate USP ; Vinblastine Sulphate BP ; Velban[(R)] (Lilly) ;

The alkaloid vinblastine is made up of two moieties namely : catharanthine and vindoline which is found to occur in the plant.

It is used in the treatment of Hodgkin's disease, monocytic a drug of third choice in the treatment of neuroblastoma, breast tumours and mycosis fungoides. It is combined with vincristine in the treatment of lymphocytic and myeloblastic leukemia in children.

Dose. Intravenously as a solution containing 1 mg per ml in sodium chloride injection.

B. *Vincristine* BAN ; *Vincristine Sulfate* USAN :

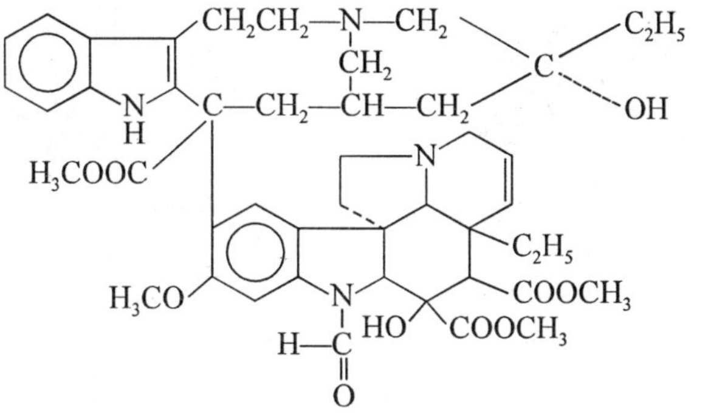

Vincristine Sulfate USP : Vincristine Sulphate BP ;

Oncovin[R] (Lilly) ; Vincasar PFS ;

It is employed for the treatment of acute leukemia in children, neuroblastoma, Wilm's tumour and rhabdomyosarcoma. It is found to induce remission in lymphosarcoma and Hodgkin's disease. It is also used in combination therapy with daunomycin and prednisone in dramatic remission of leukemia.

Dose. Intravenously as a solution containing 0.01 to 1 mg per ml in sodium chloride injection.

2.4.2.1.1. Mechanism of Action. The mechanism of action vinblastine and vincristine sulfate shall be treated as under :

2.4.2.1.1.1. Vinblastine. The '*drug*' specifically interferes with the assembly of the microtubules, by effectively combining with tubulin, thereby causing a mitotic arrest in the metaphase. Besides, there exists enough supportive evidence that vinblastine exerts its antitumour effect significantly with glutamate and aspartate metabolism. It is, however, pertinent to mention here that the extent of antineoplastic spectrum and the degree of toxicity are distinctly different in comparison to vincristine, that incidently also interacts with tubulin. It has been found that in plasma the '*drug*' is almost 75% protein bound. It usually manifests a **three-compartment kinetics,** of which the second-phase essentially exhibit a half-life ranging between 1—1.5 hours, and an elimination half-life varying between 18-40 hours. Vinblastine is metabolized extensively by the liver ; and, therefore, the dosage regimen has got to be reduced by almost 50% in such patients who have comfirmed impaired liver function.

2.4.2.1.1.2. Vincristine Sulfate. The '*drug*' progressively gets combined to the protein tubulin, and subsequently provides a check upon the assembly of microtubules, thereby causing a complete disruption of various cellular processes, including essentially mitosis and spindle formation. Besides, vincristine appreciably suppresses the strategic syntheses of proteins and RNA.

It manifests a three-compartment kinetics, having the half-lives of 0.08, 2.3 and 8.5 hour respectively.It is usually secreted directly into the bile upto 70%. Nearly 12% gets excreted in urine. As the '*drug*' cannot penetrate into the brain ; therefore, it has a little usage for the CNS leukemias.

2.4.2.2. Dimeric Tetrahydroisoquinolines

In fact only two alkaloids have been isolated from the roots of *Thalictrum dasycarpum* (Family : *Ranunculaceae*) by systematic fractionation namely : thalicarpine and thalidasine.

A. *Thalicarpine*

It has also been isolated from *Thalictrum minus* Linn., *Thalictrum revoluctum* and *Hernandia ovigera* (Family : *Hernandiaceae*).

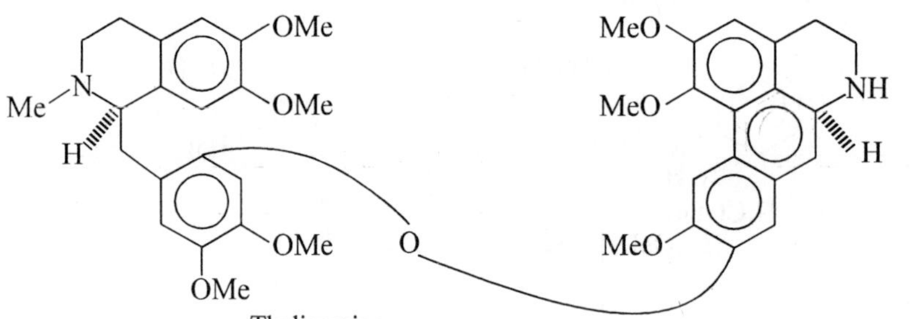

Thalicarpine

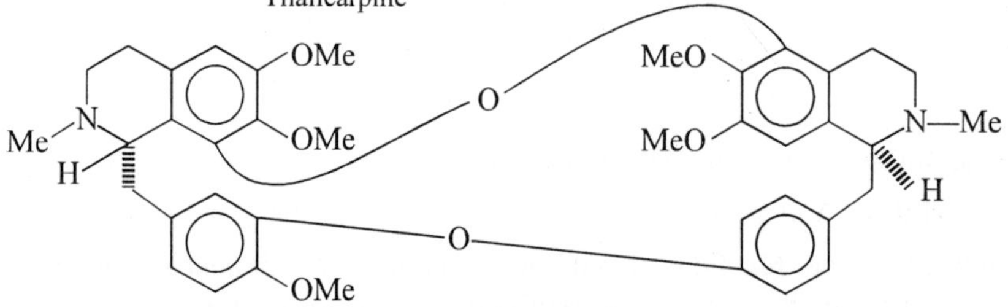

Thalidasine

Both these compounds have shown activity in mice, dog and rats against Walker 256 carcinomas.

2.4.2.3. Acyclic Tertiaryamines

The Bolivian plant *Solanum tripartitum* (Family : *Solanaceae*) gave two alkaloids, namely : solapalmitine and solapalmitenine.

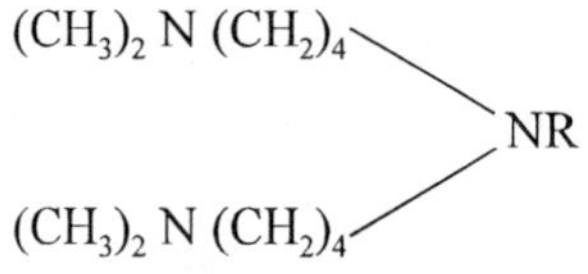

Solapalmitine

$R-CO(CH_2)_{14}CH_3$

Solapalmitenine

$R-CO \cdot CH-CH(CH_2)_{12} CH_3$

Both these alkalodis have shown *in vivo* activity against Walker 256 and their therapeutic indices do not call for further clincial studies.

2.4.2.4. Phenanthroquinolizidines

Cryptoleurine has been isolated from *Bochmeria cylindrica* (Family : *Urticaceae*) and it is found to possess highly specific cytotoxic action against Eagle's KB carcinoma but inactive against many experimental tumours. A number of its analogues have been synthesized for antineoplastic studies.

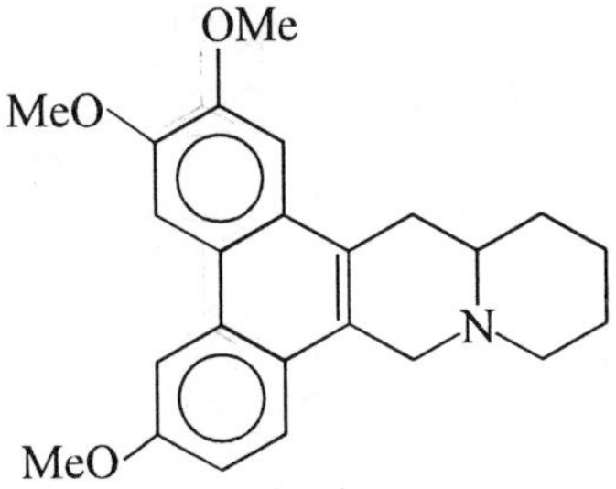

Cryptoleurine

2.4.2.5. Phenanthroindolizidines

Four alkaloids have been isolated from *Tylophora crebriflora* (Family : *Asclepiadaceae*) by systematic fractionation, namely : Tylophorine ; Tylocrebrine, Tylophorinine, and Phenanthroindolizidine.

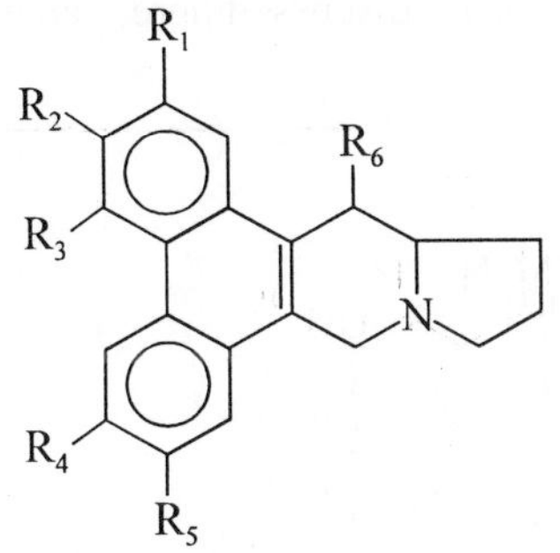

Tylophorine : $R_1 = R_2 = R_4 = R_5 = OCH_3$ & $R_3 = R_6 = H$;

Tylocrebrine : $R_1 = R_6 = H$ & $R_2 = R_3 = R_4 = R_5 = OCH_3$;

Tylophorinine : $R_1 = R_3 = H$; $R_6 = OH$ & $R_2 = R_4 = R_5 = OCH_3$;

Phenanthroindolizidine :
$R_1 = R_2 = R_3 = R_4 = R_5 = R_6 = H$;

Tylophorine is active against C-755 and W-256 and tylocrebrine against C-755, P-388, lymphocytic leukemia and L-1210.

2.4.3. Heterocyclic Amines

These alkalodis namely :camptothecin, hydroxycamptothecin and methoxy camptothecin were isolated from the Chinease tree *Camptotheca acuminata* (Family : *Nyssaceae*).

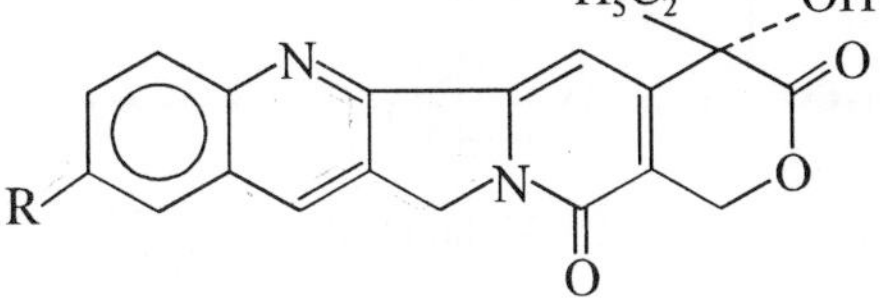

Camptothecin : R = H ;

Hydroxycamptothecin : R = OH ;

Methoxycamtothecin : R = OCH_3 ;

Both camptothecin and hydroxycamptothecin are found to be active against rodent leukemia and solid tumours.

2.4.4. Lactones

Podophyllotoxin and deoxypodophyllotoxin are the two alkaldois obtained from the Himalayan shrub *Podophyllam emodi* and the May Apple *Podophyllum peltatum* (Family : *Berberidaceae*).

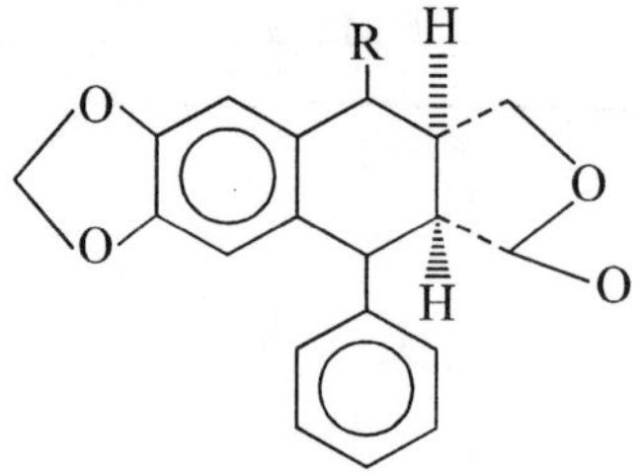

Podophyllotoxin : R = OH ;

Deoxypodophyllotoxin : R = *o*-glucosyl ;

Podophyllotoxin is an aromatic lactone that arrests the metaphase activity in the DNA synthesis.

2.4.5. Glycosides

Two glycosides that possess antineoplastic properties are discussed here, namely : Mithramycin (Aureolic Acid) and β-Solamarine.

A. *Mithramycin* BAN, USAN,

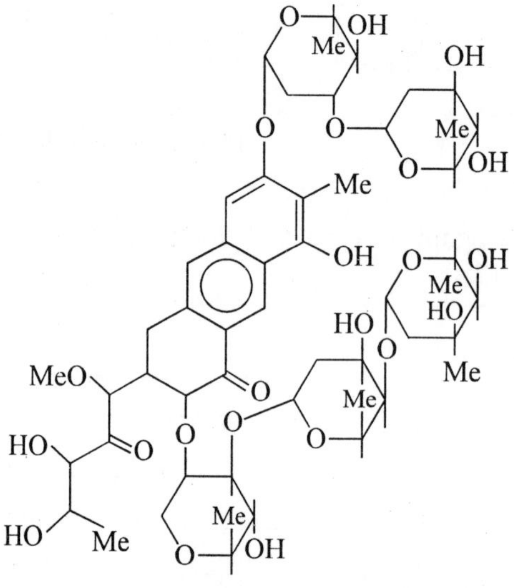

Mithramycin

Plicamycin ; Aureolic Acid ; USP ; Mithracin[R] (Pfizer-Roarig ; Dome) ;

It is isolated from *Streptomyces argillaceus*. It is employed in the treatment of breast cancer, malignant lymphomas and carcinoma of the stomach.

Dose. For hypercalcamia and hypercalcuria : usual 25 mcg per kg daily by slow iv infusion for 3 or 4 days.

B. β-*Solamarine* :

The steroidal alkaloidal glycoside β-solamarine is isolated from woody night-shade *Solanum dulcomara* Linn., (Family : *Solanaceae*).

It is found to be active against S-180, strain.

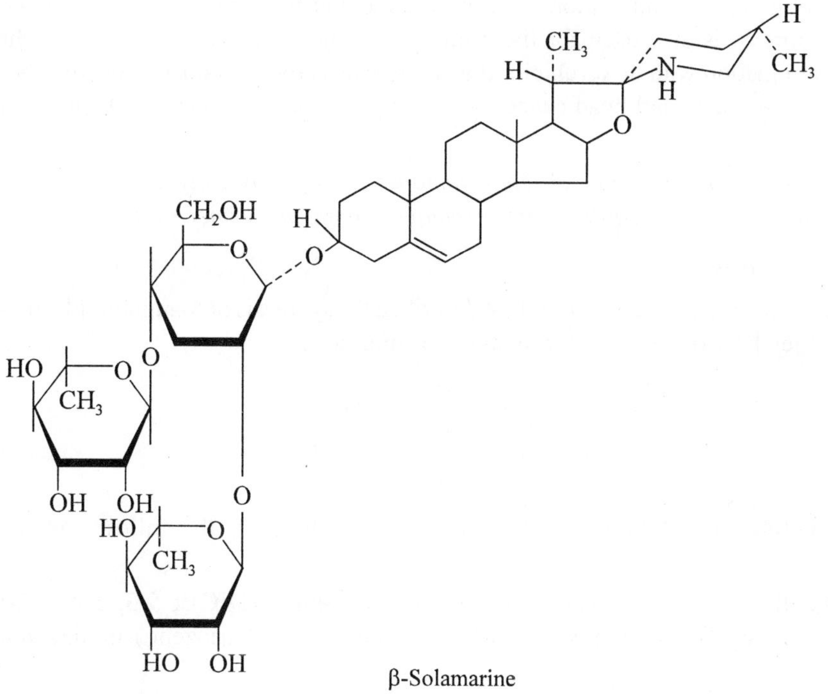

β-Solamarine

2.4.5.1. Mechanism of Action. The mechanism of action of medicinal compound discussed under section 2.4.5 shall be treated sparately in the sections that follows :

2.4.5.1.1. Mithramycin (Plicamycin) : The '*drug*' exerts its action by getting itself bound to **guanine-rich DNA** and thereby helps in inhibiting **DNA-dependent RNA polymerase**. It predominantly acts during the S-phase. As it is found to suppress *osteoclast activity**, it is invariably employed to treat *malignant hypercalcemia*** that is both unresponsive to *conventional treatment* and other severe, *refractory hypercalcemias.*

2.5. Miscellaneous Compounds

There are various compounds that exert neoplastic activity both belonging to synthetic and natural origins. A few such compounds are described below :

Examples : Cisplatin, Imidazole Triazines, Hycanthone, Pipobroman,

A. Cisplatin BAN, USAN,

$$cis—[Pt\ (NH_3)_2\ Cl_2]$$

Cisplatine ; *cis*-Dichlorodiamine platinum ;

Platinex[R] (Bristol-Meyers) ; Neoplatin[R] (Mead-Johnson) ;

The effectiveness of transition-metal complexes, particularly platinum complexes, as experimental antineoplastic agents has been reported in recent years. Cisplatin is the prototype platinum complex having antineoplastic activity.

---

* A device for fracturing bones for therapeutic purposes.

**Neoplasms which essentially cause dissolution of bone salts.

It is employed in combination with vinblastine and bleomycin for the treatment of metastatic testicular tumours. It is also used for the remission of metastaticovarian tumours when given either alone or in combination with doxorubicin. It also exhibits activity against a host of other tumours, such as :: cervical cancer, neck and head cancer, penile cancer, bladder cancer and small-cell cancer of the lung.

Dose. Usual, for metastatic testicular tumours 20 mg/m² iv daily for five days, followed every 3 weeks for 3 courses : for metastatic ovarian tumours 50 mg/m² iv once every 3 weeks.

B. *Imidazole Triazines :*

Windans and Langenbeck (1923) first described the synthesis of 5-aminoimidazole-4-carboxamide (AIC) which they later on used for the synthesis of purines :

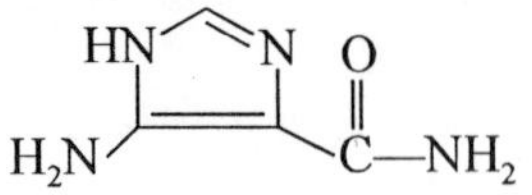

AIC

Its structural modification resulted into the synthesis of the following *three* compounds, namely :

5-(3, 3-dimethyl-1-triazeno) imidazole-4-carboxamide (DIC) ; 5-3, 3-bis (2-chloroethyl)-1-triazeno) imidazole-4-carboxamide (BIC) ; 5-(3-monomethyl-1-triazeno) imidazole-4-carboxamide (MIC) ;

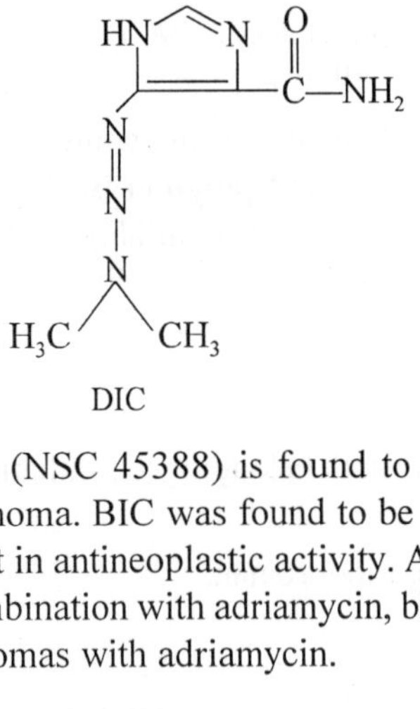

DIC 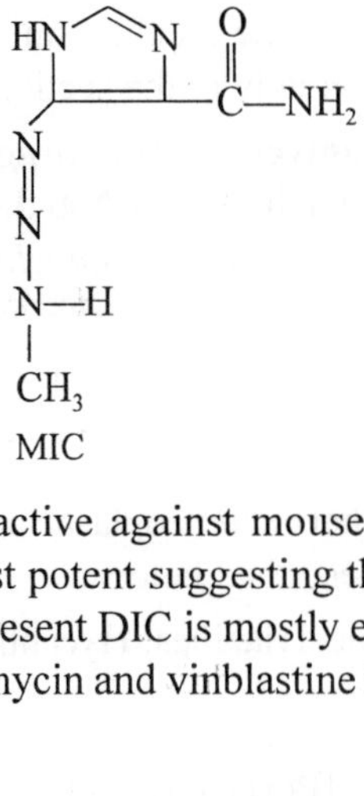   MIC 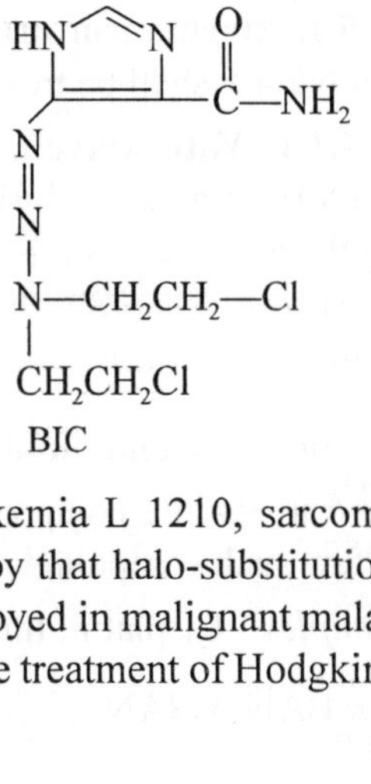   BIC

DIC (NSC 45388) is found to be active against mouse leukemia L 1210, sarcoma 180 and adenocarcinoma. BIC was found to be most potent suggesting thereby that halo-substitution are often more potent in antineoplastic activity. At present DIC is mostly employed in malignant malanoma. It is used in combination with adriamycin, bleomycin and vinblastine in the treatment of Hodgkin's diseases and in sarcomas with adriamycin.

C. *Hycanthone* USAN

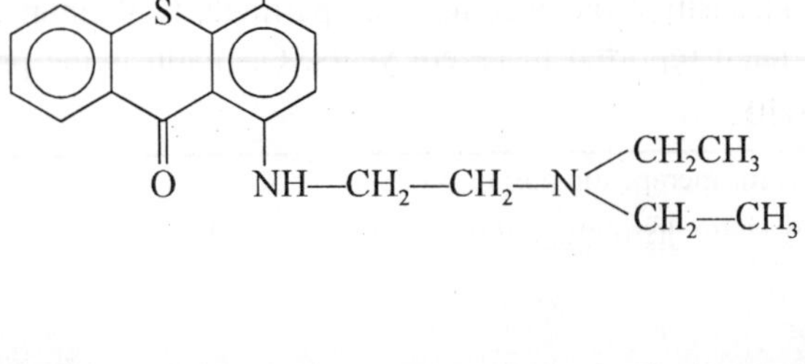

[(Diethlamino-2-ethyl) amino]-1-hydroxymethyl-4-thioxanthenone-9 ;

Etrenol[R] (Winthrop) ;

Hycanthone, which was earlier identified as an antischistosomal drug, found to possess antineoplastic activity in animals. It is comparatively non-toxic. Besides, it is an intercalating agent which inhibits both DNA and RNA synthesis.

D. Pipobroman USAN,

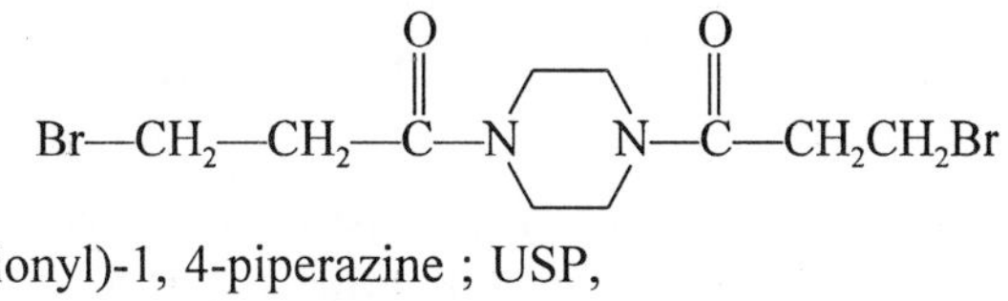

$$Br—CH_2—CH_2—\overset{\displaystyle O}{\overset{\displaystyle \|}{C}}—N\diagup\diagdown N—\overset{\displaystyle O}{\overset{\displaystyle \|}{C}}—CH_2CH_2Br$$

Bis-(bromo-3-propionyl)-1, 4-piperazine ; USP,

Vercyte[R] (Abbott) ;

Synthesis :

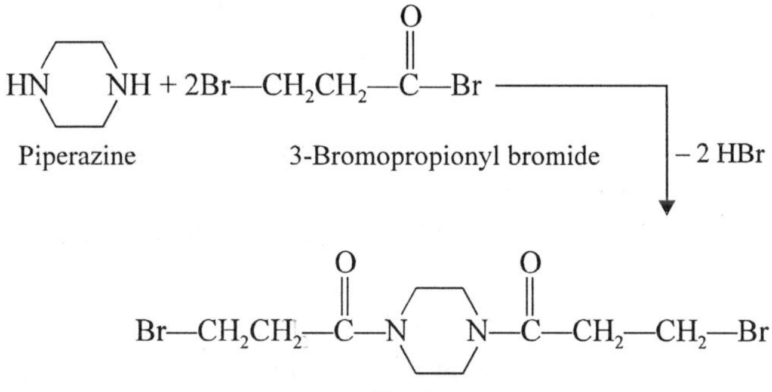

$$HN\diagup\diagdown NH + 2Br—CH_2CH_2—\overset{\displaystyle O}{\overset{\displaystyle \|}{C}}—Br \xrightarrow{\quad -2\,HBr\quad}$$

Piperazine   3-Bromopropionyl bromide

$$Br—CH_2CH_2—\overset{\displaystyle O}{\overset{\displaystyle \|}{C}}—N\diagup\diagdown N—\overset{\displaystyle O}{\overset{\displaystyle \|}{C}}—CH_2—CH_2—Br$$

Pipobroman

It is prepared by the interaction of piperazine with two moles of 3-bromopropionyl bromide.

It is used in patients with chronic granulocytic leukemia refractory to busulfan. It is also employed for the treatment of polycythemia vera.

Dose. Usual, initial : 1 to 1.5 mg per kg body weight daily.

E. Asparaginase USAN *; Colaspase* BAN ;

L-Asparaginase amidohydrolase :

Leunase[R] (May & Baker) : Elspar[R] (Merck, Sharp & Dohme) ; Crasnitin[R] (Bayer) ;

It is a preparation from *Escherichia coli* containing the enzyme L-asparaginase amidohydrolase.

It is used in patients suffering from acute lymphocytic and other leukemias.

Dose. Intravenously : 1000 units per kg body weight daily for 10 days following treatment with vincristine or prednisone.

2.5.1. Mechanism of Action. The mechanism of action of certain medicinal compounds discussed under section 2.5 shall now be dealt with individually as under :

2.5.1.1. Cisplatin. The '*drug*' essentially cross-links DNA ; and, therefore, behaves like alkylating antineoplastic agents. In general, the platinum complex acts as a potent inhibitor of DNA polymerase. Based on adequate supportive evidences it has been duly established that there exists a bondage between DNA and platinum complex, wherein the two Cl⁻ ions are duly displaced by N or O atoms of purines.

This evidence is fully substantiated by concrete experimental findings, such as : (*a*) enhanced sedimentation coefficient ; (*b*) hyperchromicity shown by the DNA-UV-spectrum ; and (*c*) selective and specific reaction occurring between the Pt-complex and guanine over other bases.*

Cisplatin is *not* well absorbed by oral administration, and hence, must be given IV. The '*drug*' gets bound to plasma proteins to the extent of 90%. It fails to cross the blood-brain barrier (BBB). It gets secreted chiefly *via* renal route, partly by tubular secretion ; however, the overall pattern is found to be '**nonlinear**' in nature. It has been observed that the prevailing distribution half-life of the unbound drug is 25-49 minutes and the elimination half-life of total Pt ranges between 58-73 hours, which may get extended upto 240 hours in *anuria*.**

Note. Sodium thiosulphate decomposes cisplatin and complxes with Pt, and in this manner affords protection against renal damage and certain other toxicity.

2.5.1.2. Pipobroman. The '*drug*' exerts its action due to its alkylating properties. Importantly, it is invariably held in reserve for usage in such patients who have virtually turned refractory to *X-irradiation* and *busulfan* in the severe case of **leukemia** and **phlebotomy**.***

CAUTION : It should not be used in pregnancy.

2.5.1.3. Asparaginase. It has been observed that the ensuing protein synthesis in a good number of normal and malignant cell types depends partially on exogenous asparagine ; and in a few cells like-*leukemic cells* and *lymphoblasts* is dependent almost completely. Consequently, the enzymatic destruction of asparagine by the enzyme asparaginase duly injected into plasma usually deprives the dependent cells of the essential asparagine thereby causing predominantly *three* vital effects, namely : (*i*) *partial cell-death ; (ii) arrest cell growth ;* and (*iii*) *tumour regression.*

Asparaginase (the enzyme) is found to protect certain tissues and malignnant tumours from some known antimetabolites, such as : methotrexate, ara-C, presumably by directly preventing DNA synthesis.

Ervinia (Porton) *asparaginase* is observed to be less sensitizing in comparison to that obtained from *E. coli*. Besides, Erwinia asparaginase is designated as an '*orphan drug*' which is virtually reserved for usage particularly in patients who are found to be allergic to **asparaginase** obtained from *E. coli*. Interestingly, both enzymes do exhibit **immunosuppressant activity.**

The '*drug*' exhibits extremely poor extravascular tissue penetration ; and, therefore, gets cleared from plasma in a quite sluggish and unpredictable manner. Nevertheless, the elimination is *biphasic*, having an inital half-life of 4-9 hours and a terminal half-life ranging betwen 1.4 to 1.8 day.****

2.6. Hormones

Hormones has the ability to supress mitosis in lymphocytes and this effect is duly utilized in the treatment of neoplastic diseases. Adrenocorticosteroids specifically are effective in the treatment of leukemia in children and in the management of hemolytic anaemia and hemorrhagic complications of thrombocytopenia that mostly occur in malignant lymphomas and chronic lymphocitic leukemia. Acute

---

* Sartorelli AC and Johns DJ (eds) : *Handbook of Experimental Pharmaeology,* Vol. 38, Pt. 2, Springer = Verlag, New York, pp. 829-838, 1975.

**Absence of urine formation.

*** The surgical opening of a vein to withdraw blood.

****Physicians' Desk Reference, Medical Economics, Oradell N.J., **33rd** edm, (p. 749) 1979.

lymphoblastic leukemias in children are better treated with corticosteroids rather than antimetabolities and remission take place more rapidly. The hormones have been beneficial in breast cancer and other carcinomas although palliative effects are of short duration.

A few important compounds are discussed here, namely : Megestrol ; Mitotane and Testolactone ;

A. *Megestrol* BAN, ; *Megestrol Acetate* USAN :

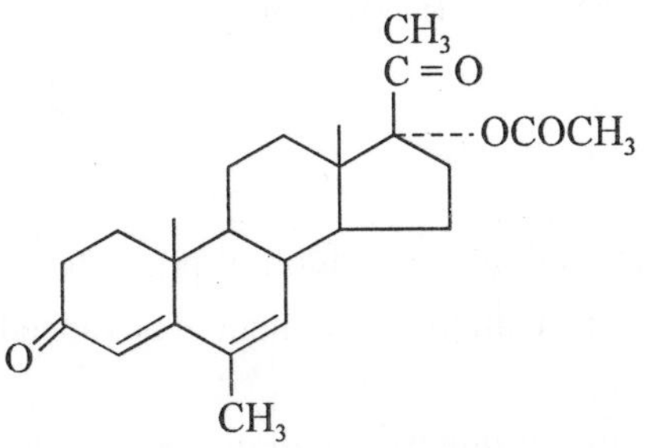

17-(Acetoxy-6-methyl-pregna-4, 6-diene-3, 20-dione, Megesterol Acetate BP (1973) :

Ovarid[R] (Glaxo) ; Megestat[R] (Bristol) ; Megace[R] (Mead-Johnson) ;

It is indicated for the palliative treatment of endometrial carcinoma and advanced breast cancer when other methods of medication are not effective.

Dose. Usual, 160 mg/day in four equal doses in breast cancer, ; 40-320 mg/day in equal divided doses in endometrial cancer.

B. *Mitotane* USAN ;

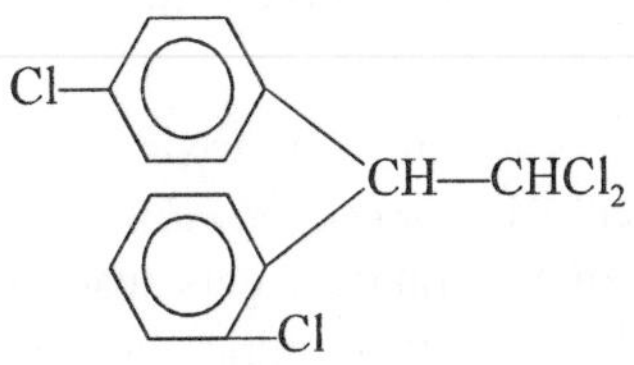

1, 1-Dichloro-2-(*o*-chlorophenyl)-2-(*p*-chlorophenyl) ethane ; *o, p*-DDD ;

USP ; Lysodren[R] (Bristol) ;

It is indicated mainly for the treatment of inoperable adrenal cortical carcinoma.

Dose. Usual, 8-10 g per day, divided into 3 or 4 equal doses.

C. *Testolactone* USAN ;

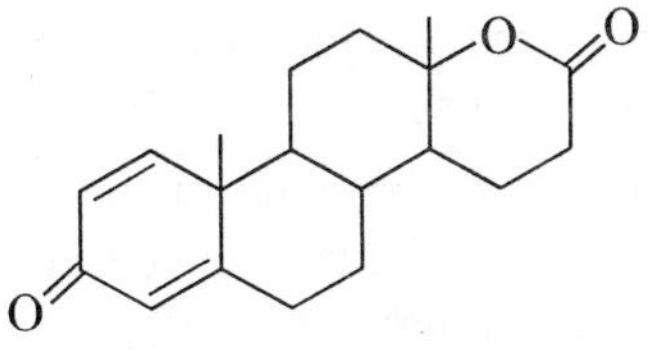

1-Dehydrotestololactone

USP ; Teslac[R] (Squibb) ;

Synthesis :

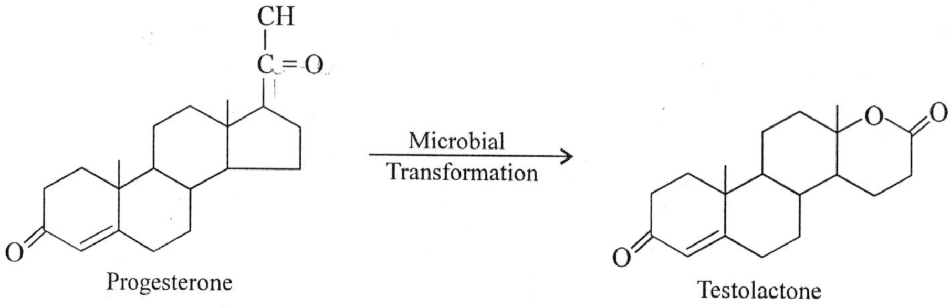

It may be prepared by microbial transformation of progesterone.

It is invariably employed in the palliative treatment of advanced breast cancer in postmenopausal women.

Dose. Usual, 250 mg 4 times per day by mouth or 100 mg intramuscularly thrice weekly.

2.6.1. Mechanism of Action. The mechanism of action of some drug substances discussed under section 2.6 shall be treated individually in the section that follows :

2.6.1.1. Magestrol Acetate. It is well established that not only normal, but also well-differentiated neoplastic target cells to possess a plethora of strategically located '*hormone receptors* ; and eventually they bank upon the hormones for stimulation.* Importantly, the rather comparatively less differentiated neoplastic cells invariably become independent of the ensuing '*hormonal control*' and thereby lose their specific receptors eventually. Evidently, a few malignant tumours are solely hormone dependent and responsive to hormone-based therapy, on the contrary others are independent and naturally altogether unresponsive. Hence, magestrol acetate exert its action on hormone dependent tumours *e.g.*, endometrial carcinoma and breast cancer.

2.6.1.2. Mitotane. The '*drug*' is found to be toxic to the adrenal cortex and, therefore, it is exclusively indicated for the treatment of *inoperable adrenal cortical carcinoma.* It is metabolized in the liver. Approximately 40% of a single oral dose gets absorbed ; whereas, the '*drug*' gets excreted in urine in the form of an *unidentified metabolite* ranging between 10-25%, and almost 60% is excreted absolutely unchanged in faeces. The remainder of the '*drug*' gets stored in the adipose tissues *in vivo.*

2.6.1.3. Testolactone. The '*drug*' obtained *via* microbial transformation of progesterone** is found to be devoid of any androgenic activity in the usual recommened dosage regimens.

2.7. Immunotherapy

It is now an established fact that the human body continually produces cells having neoplastic potential which are destroyed by our immune surveillance system. The very formation of tumours suggests that this system is impaired. In other words, suppression of body's immune system by these agents easily results to development of serious viral, bacterial and fungal infections.

Biochemical modulation of the action of some of the antineoplastic agents has more or less provided a means of improving their specificity for tumour cells. Biological modifiers specific antibodies such as interferons, interleukins and agents that might affect or arrest cancerous growth by inducing terminal differentiation have been also applied but there exists only limited evidence till date that these

---

* Fried J *et. al. J. Am. Chem. Soc.* **75** : 5764, 1953.

agents can affect widely disseminated cancers, It may, however, be ascertained confidently that treatment with biological response-modifiers amalgamated with improved form of chemotheraphy will ultimately lead to significant enhancement both in the arrest and even cure of wider spectrum of neoplasma.

One important member of this group is discussed here, namely : Interferon Alfa-2*a* recombinant.

A. *Interferon Alfa-2a, Recombinant*

It is prepared on a large scale from a strin of *E. coli* having essentially a plasmid produced by the technique of genetic engineering, otherwise known as recombinant DNA technology, consisting of an interferon alfa-2*a* gene from human leukocytes.

It is employed in subjects above the age of 18 years for the treatment of hairy cell leukemia.

Dose. In hairy cell leukemia the dose of interferon alfa-2a and alfa-nl is 3 million units daily by deep intramuscular or subcutaneous injection until there is improvement or for up to 24 weeks, then redcued to a maintenance dose of 3 million units 3 times a week.

2.7.1. Mechanism of Action. The mechanism of action of the interferon alfa-2*a*, recombinant is discussed as under :

2.7.1.1. Interferon Alfa-2a, Recombinant. *The 'drug'* enhances class I histocompatibility molecules on lymphocytes, increases the production of ILs-1 and -2,* regulates antibody responses, and above all enhances NK cell** activity. Besides, it also inhibits cancerous tumour-cell growth by virtue of its inherent ability to inhibit protein anabolism (synthesis) *in vivo*. The *'drug'* is antiproliferative ; and, therefore, may also serve as immunosuppressive. It is, however, partinent to state here that the prevalent action of the *'drug'* upon the NK cells is believed to be the most important and critical factor for its prevailing **antineoplastic profile.**

The *'drug'* also exhibits antiviral activity, specifically against the RNA viruses. Furthermore, it appreciably enhances the strategic targetting of monoclonal antibody-tethered cytotoxic drugs to the corresponding malignant cells.

The *'drug'* is not absorbed when administered through mouth. However, by IV route it exclusively disappears within a span of 4 hours, but by the IM or sub-cutaneous route disapp-earance gets prolonged to 6-7 hours.

Probable Questions for B. Pharm. Examinations

1. What is a neoplasm ? What are the causations of neoplasm ? Give the structure, name and uses of at least **three** potent drugs employed as antineoplastic agents belonging to :

   (*a*) Natural plant source

   (*b*) Synthetic drugs.

2. How would you classify the **'antineoplastic agents'** ? Give the structure, chemical name and uses of **one** important member from each category.

3. Mustards, methanesulphonates, ethylenimines and nitrosoureas constitute **four** vital categories of the **'Alkylating Agents'** employed for the treatment of neoplasms. Discuss the synthesis of the following drugs :

   (*a*) Chlorambucil

   (*b*) Busulfan

---

* Mediate most of the toxic and therapeutic effects.

**Natural killer cells.

(c) Triethylene melamine

(d) Carmustine.

4. (a) How would you classify **'Antimetabolities'** ?

   (b) Give the structure, chemical name and uses of the following :

   (i) Methotrexate

   (ii) Meracaptopurine

   (iii) Fluorouracil

   (iv) Azaserine.

   (c) Discuss the synthesis of any **one** drug stated above.

5. 'Recognition of **antibiotics'** as an important class of *antineoplastic agents* is quite recent'. Justify the statement with reference to the follwoing drugs :

   (a) Dactinomycine

   (b) Daunorubicin.

6. Classify the **'plant products'** employed in the treatment of malignant disease. Give structure, name and uses of **one** potent drug from each category.

7. Discuss the synthesis of the following :

   (a) Pipobroman

   (b) Lomustine

   (c) Cytarabine.

8. Give a comprehensive account of **'hormones'** that are potent as antineoplastic agents. Support your answer with suitable examples.

9. Give a brief account of the following :

   (a) Immunotherapy in cancer

   (b) Pharmacokinetics, pharmacodynamic and mode of action of antineoplastic agents.

10. Discuss the following with regard to antineoplastic agents :

   (a) Dimeric tetrahydroisoquinoline

   (b) Acyclic tertiaryamine

   (c) Phenanthroquinolizidine

   (d) Phenanthroindolizidine.

RECOMMENDED READINGS

1. CG Zubrod, Historical Pespective of Curative Chemotherapy. *In Oncology.* RL Clark, RW Cumely, JE Mcloy and M. Lopeland (eds)-Being the proceedings of the Tenth International Cancer Congress Year Book, Chicago (1970).

2. HT Skipper and FM Schabel (Jr.) Quantitative and Cytokinetic Studies in Experimental Tumour Models. In *Cancer Medicine,* JF Holland and E. Frei, Philadelphia, Lea and Febiger, 3rd End., (1973).

3. J. Skoda, Azapyrimidine, Nucleoside, In *Antineoplastic and Immunosuppressive Agents* (PT 11) AC Sarterelli and GD John. (eds), Berlin, Handbuch de Experimentellen Pharmakologic (1975).

4. E. Klein, H. Milgrom, HL Stoll, F. Helm, HJ Walker and OA Holtermann. Topical 5-Fluorouracil Chemotheraphy for Pre-Malignant and Malignant epidermal Neoplasma. In Cancer Chemotheraphy 11, 1. Bradsky and S.B. Kahn (eds). New York, Grune and Stratton Inc., (1972).

5. RE Handchumacher, Cancer Chemotherapy-Examples of Current Progress and Future Perspective. In IUPHAR, 9th International Congress of Pharmacology Proceedings (Pt. 2.) W. Paton. J. Mitchell and P. Turner (eds), London, Macmillan Press Ltd., (1984).

6. SC Harvey, Antineoplastic and Immunosuppressive Drugs. In Remington's The Science and Practice of Pharmacy, 19th Ed., Vol. I & II, Easton Pennsylvania, Mac Publishing Company, (1995).

7. MB Garnick, JD Griffin, MJ Sack, RH Blum, M. Israel and E. Frei, Anthracycline. In *Antibiotics in Cancer Chemotherapy*, F.M. Muggia, CW Young and SK Carter, (eds). The Hague, Martinus Nijhoff (1982).

8. Block JH and Beale JM : *Wilson and Gisvold's Textbook of Organic Medicinal and Pharmaceutical Chemistry,* Lippincott Williams & Wilkins, New York, 11th edn., 2004.

9. Martindale *The Extra Pharmacopoeia,* (31st Edn), London, The Pharmaceutical Press (1996).

10. D Lednicer and LA Mitscher The Organic Chemistry of Drug Synthesis, New York., John Wiley and Sons (1995).

11. Williams DA and Lemke TL : *Foye's Principles of Medicinal Chemistry,* Lippincott Williams & Wilkins, New York, 5th end, 2002.

12. MC Griffiths (Ed.) *USAN and the USP Dictionary of Drug Names*, Rockville, United States Pharmacopoeial Convention Inc., (1985).

13. Index Nominum, Zurich, Swiss Pharmaceutical Society, (1982).

Antipsychotics (Tranquilizers)

1. INTRODUCTION

In general, **antipsychotics** (tranquilizers) are primarily employed for the treatment of symptoms in mental diseases, their overall influence being to free the mind from passion or disturbance and thus clam the mind *i.e.,* they cause sedation without inducing sleep.

Tranquilizers are drugs essentially used in the management and, treatment of psychoses and neuroses. They specifically exert their action on the lower brain areas to produce emotional calmness and relaxation without appreciable hypnosis sedation euphoria or motor impairment. In addition many of these drugs also display clinically beneficial actions, for instance skeletal muscle relaxants, antihypertensive, antiemetic and antiepileptic properties.

One school of thought even suggested that these drugs may be divided into *two* categories, namely : *major tranquilizers* (for psychoses) and *minor tranquilizers* (for neuroses) ; however, such an arbitrary categorization stands invalid because of their overlapping characteristic features.

More recently **antipsychotics** may be defined as—*'drugs' which ameliorate mental aberrations* * *that are invariably characteristic feature of the psychoses.'*

Positive symptoms of psychoses essentially comprise of a host of disorders, such as : mild behavioural changes anxiety, delusions, hallucinations, and sclizzophrenias. Negative symptoms are usually designated by cognitive deficits, social withdrawl, apathy, and anhedonia. * *

Interestingly, *'psychoses'* may be organic which could be either trigger off or directly related to a variety of reasons, namely :

 (*i*) particular toxic chemical influence *e.g.,* **delirium**—due to central anticholinergic drugs,

 (*ii*) a N-methyl-D-aspartate (NMDA) antagonist *e.g.,* **phencylidine**.

 (*iii*) a particular disease process *eg.,* **dementia** (cognitire deficit including memory impairment, and

 (*iv*) idiopathic conditions *i.e.,* disease without clear pathiogenesis, as of spontaneous origin.

In a broader perspective the typical **'antipsychotics'** should ideally possess the following cardinal requirements, such as :

 (*a*) high lipid solubility,

 (*b*) affinity for protein-binding (92–99%),

 (*c*) large volumes of distribution ($v_d{}^{55}$) *i.e.,* greater than 7 L.kg^{-1}, (*d*) variance in oral bioavailability (25–35%), and

(*e*) short plasma half-life between 10–20 hours.

It is, however, pertinent to mention here that though these drugs have a relatively shorter plasma half-life but their duration of action is much longer ; their metabolites may be found in the urine weeks even after the last terminal dosage ; and finally a good proportion of the drug are adequately sequestered in the various tissues.

2. CLASSIFICATION

Antipsychotics may be classified under the following categories, namely :

(*a*) Reserpine and Related Alkaloids

(*b*) Alkylene Diols

(*c*) Diphenylmethane Compounds

(*d*) Phenothiazine Compounds

(*e*) Dibenzazepines

(*f*) Butyrophenones

(*g*) Azaspirodecanediones

2.1. Reserpine and Related Alkaloids

The roots of *Rauwolfia serpentina,* a climbing shrub indigenous to India, and named after the German botanist **Rauwolf,** contains and alkaloid Reserpine which was reported to possess both tranquilizing and hypotensive properties.

Examples : Reserpine and Deserpidine.

A. *Reserpine* BAN, USAN, INN,

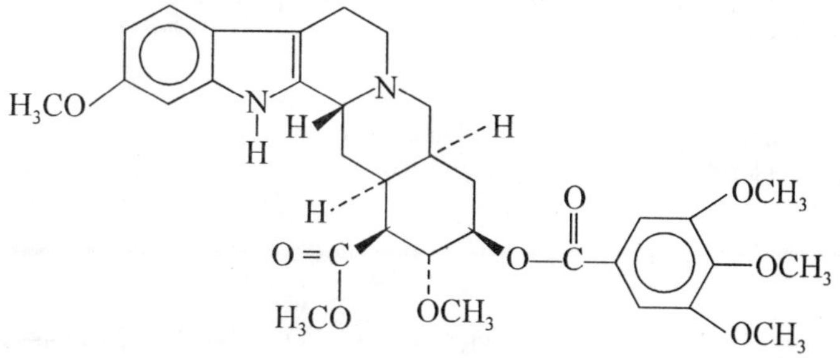

Methyl 18 β-hydroxy-11,17 a-dimethyoxy-3β, 20α-yohimban-16β-carboxylate 3,4,5-trimethoxybenzoate (ester) ;

BP ; USP ; Int. P ;

Serpasil[R] (Ciba-Geigy) ; SK-Reserpine[R] (Smith Kline & French) ; Sandril[R] (Lilly) ;

It has central depressant and sedative actions and a primarily peripheral antihypertensive effect accompanied by bradycardia. It is also used for the management and treatment of hypertensive specifically in patients with mild labile hypertension associated with tachycardia.

Dose. As sedative in anxiety states and chronic psychoses : 0.1 to 1 mg daily doses. For hypeprtension-in adults : 250 to 500 mcg daily for about 2 weeks ;

B. *Deserpidine* INN, USAN ; *Desipraminum* BAN ;

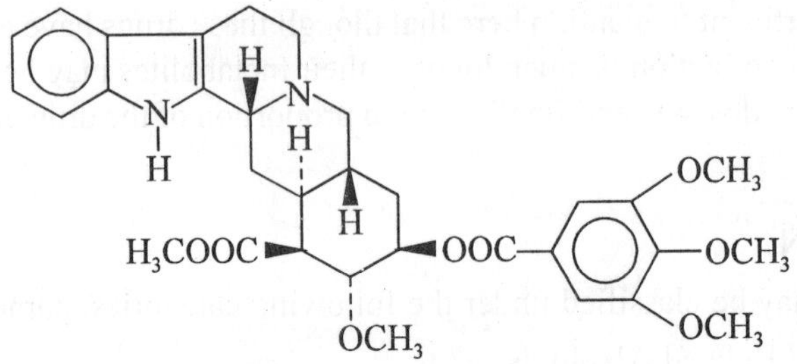

II-Desmethoxyreserpine ; BP ; USP ;

Harmonyl[R] (Abbott) ; Pertofram(R) (Geigy) ;

Its actions and uses are very much similar to those described under reserpine.

Dose. For psychiatric treatment : Average, initial, 500 mcg daily with a range of 0.1 to 1 mg ; For antihypertension ; initial 0.75 to 1 mg daily subsequently reduced to a maintenace dose of about 250 mcg per day.

Miller and Weinberg in 1956 observed that even the simple tertiary amines having the trimethoxybenzoyl group exhibits the reserpine-like activity.

Example : 3-Diethylamino propyl ester of 3, 4, 5-trimethoxy benzoic acid.

Synthesis

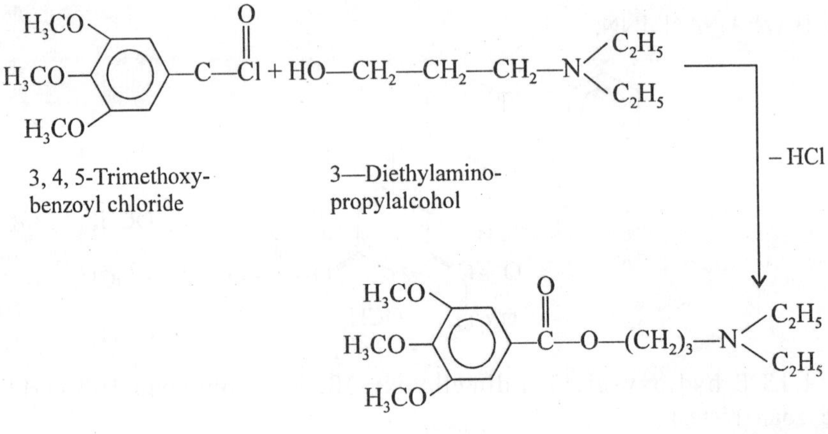

3, 4, 5-Trimethoxy-   3—Diethylamino-
benzoyl chloride   propylalcohol

It is prepared by the interaction of 3, 4, 5-trimethoxybenzoyl chloride with 3-diethylaminopropyl alcohol with the elimination of hydrochloric acid.

It is found to possess about 1/3rd the activity of reserpine.

2.1.1. Mechanism of Reaction.

The mechanism of reaction of the *'drug'* discussed under section 2.1. shall be treated separately as under :

2.1.1.1. Reserpine. The '*drug*' exerts its action to cause significant inhibition of both *neuronal* and *chromaffin granule transporters*. Consequently, the **catecholamine accumulation** gets blocked appreciably. As a net overall effect the depletion is rather slower and less complete in the adrenal medulla in comparison to other tissues. Therefore, the strategic prevention of such storage ability/capacity, the '*drug*' at its initial primary state affords a distinct catecholamine release. Subsequently, a marked and pronounced '*depletion of transmitter*' commences that usually prolongs for days through weeks. The effects of reserpine seem to be irreversible absolutely.

Reserpine exerts its antihypertensive action by virtue of its adrenergic neutronal blockade consequent to depletion of the catecholamines-containing granudles of the postganglionic sympathetic neuron. It, however, depletes both brain catecholamines and seritonins.

Phenothiazines exert their antipsychotic potency by interacting with a receptor at three marked sites X, Y, Z to produce a singificant response. The order of specific structural requirement at these sites is YZX. However, a three-carbon chain at site Y affords an optimal antipsychotic activity.

The apparant variance in the efficacy of oral administration of reserpine is due to the fact that it gets absorbed very poorly as well as erratically from the ensuing GI tract. Importantly and characteristically the '*drug*' bears a relatively longer latency of onset, and followed by a prolonged duration of action.

Note. Combinations of resperpine, with a diuretic enhances the efficacy of the former significantly.

2.1.1.2. Deserpidine. It essentially possesses almost similar phermacological activity and mechanism of action of that of 'reserpine' discussed under section 2.1.1.1. above.

2.2. Alkylene Diols

Alkylene diols are also referred to as **'propanediol carbamates'** have been used as tranquilizer. Two mportant members of this classification, namely : meprobamate and tybamate have been discussed in the chapter on '*Muscle Relaxants*' in this book.

2.3. Diphenylmethane Compounds

A number of diphenylmethane derivatives have been synthesized that exhibit antipsychotic activities.

A few such compounds are described below, namely : Pipradrol ; Captodiame ; Hydroxyzine ; Benactyzine ;

A. *Pipradrol* INN, BAN, *Pipradrol Hydrochloride* USAN,

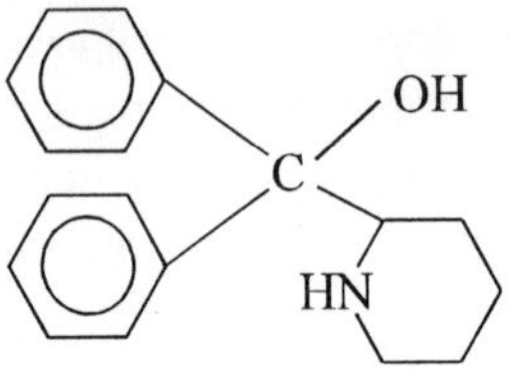

α, α-Diphenyl-2-piperidine methanol ;

Pipradrol Hydrochloride BP ;

Meratran[R] (Merrell Dow) ;

Synthesis :

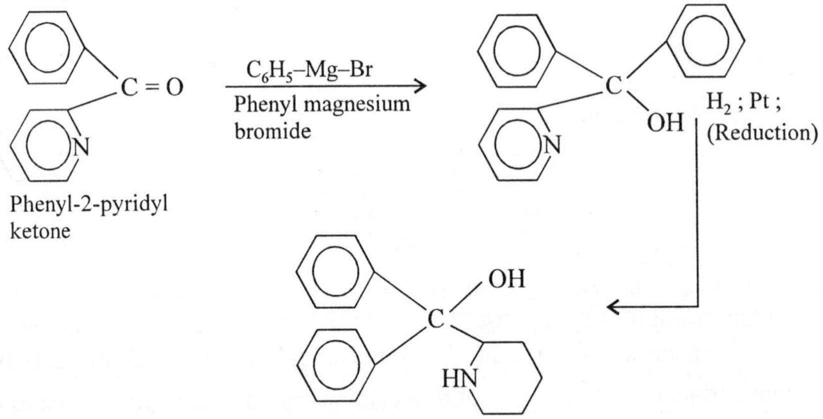

It is synthesized by Grignard reaction of phenyl-2-pyridyl ketone with phenyl magnesium bromide followed by catalytic reduction to get the official compound.

It is used for the treatment of functional fatigue and various types of depressions.

Dose. Usual, 2.5 mg twice daily.

The corresponding 4-piperidyl derivative pipradrol is also a tranquilizer used under the name of **Azacyclonal**.

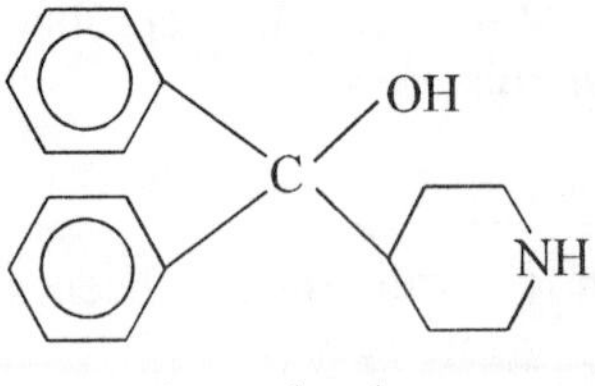

Azacyclonal

The drug is prepared from phenyl ketone exactly in the same manner as that of pipradrol. It has the same application as that of its isomer.

B. *Captodiame* INN, BAN ; *Captodiame Hydrochloride* USAN ;

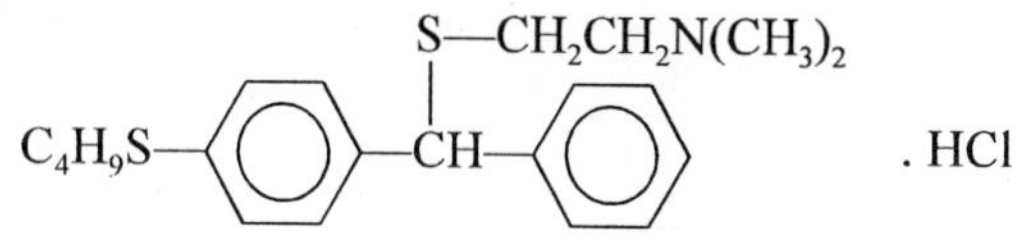

2-[*p*-(Butylthio)-α-phenylbenzylthio]-N, N-dimethylethylamine hydrochloride ; Covatin[R] (Warner-Lambert) ;

Synthesis :

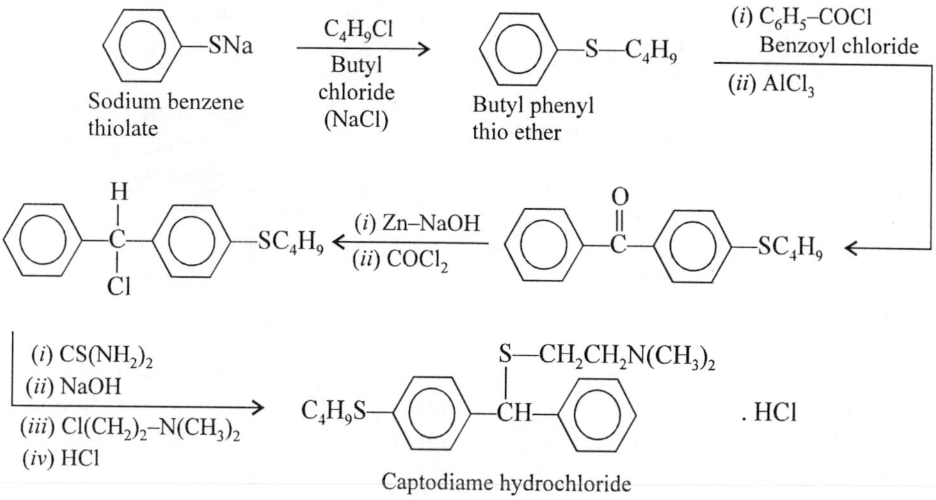

Captodiame hydrochloride

Butylphenyl thioether is first prepared by the interaction of sodium benzenethiolate and butyl chloride. The resulting product on treatment with benzoyl chloride and aluminum chloride yields butyl-p-benzoyl phenyl thioether. This on reaction with zinc and sodium hydroxide and carbonyl chloride yields an intermediate. The intermediate on treatment with thiourea, sodium hydroxide, 2-dimethyl amine ethyl chloride and hydrochloric acid gives rise to the desired compound.

It is used for the treatment of anxiety and tension. It is an excellent nonhypnotic sedative.

Dose. For anxiety and tension : 50 mg three times per day.

C. *Hydroxyzine* INN, BAN ; *Hydroxyzine Hydrochloride* USAN ;

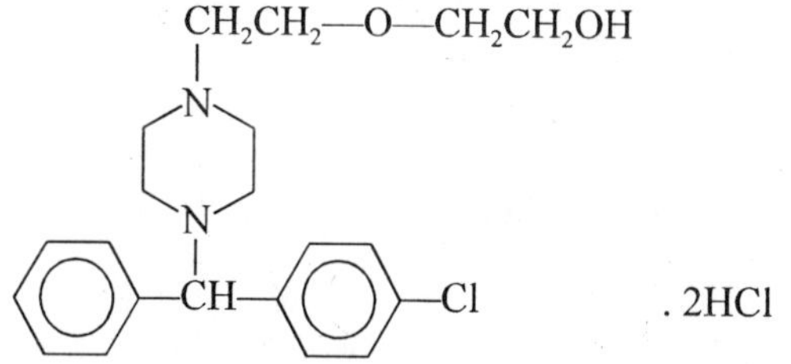

Ethanol, 2-[2-[4-chlorophenyl) phenyl methyl]-1-piperazinyl]ethoxy]-, dihydrochloride ;

Hydroxyzine Hydrochloride USP ;

Atarax$^{(R)}$ (Roerig) ; Orgatrax$^{(R)}$ (Organon) ;

Synthesis :

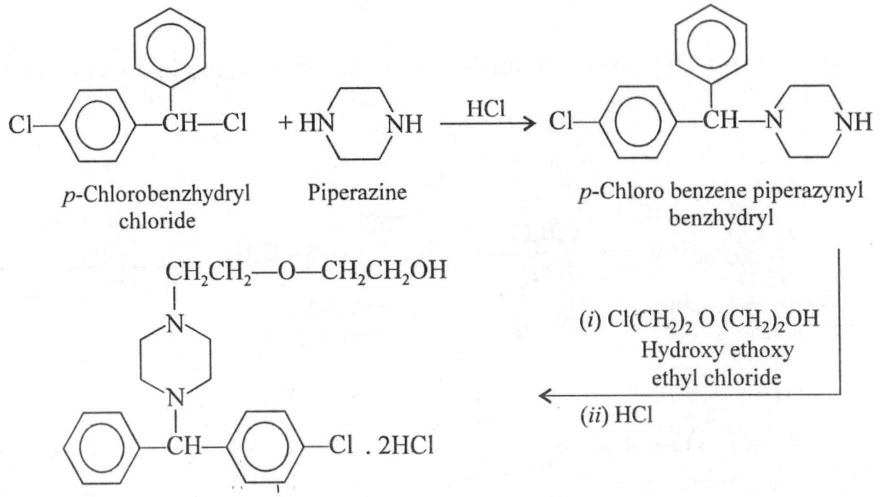

p-Chlorobenzhydryl   Piperazine   p-Chloro benzene piperazynyl
    chloride   benzhydryl

CH$_2$CH$_2$—O—CH$_2$CH$_2$OH

(i) Cl(CH$_2$)$_2$ O (CH$_2$)$_2$OH
Hydroxy ethoxy
ethyl chloride

(ii) HCl

Hydroxyzine Hydrochloride

It is prepared by interacting p-chloro benzhydryl chloride with piperazine to obtain p-chlorobenzene piperazynyl benzhydryl which is subsequently treted with β-hydroxy ethoxy ethyl chloride to give the hydroxyzine base. They official compound may be finally obtained by treating with hydrochloric acid.

It is employed for pre-and postoperative sedation. It has also been used successfully in the treatment of anxiety, tension and agitation.

Dose. Adult, IM injection as the hydrochloride in doses of 25 to 100 mg every 4 to 6 hrs ; Children : 1 mg per kg body weight im for pre-and postoperative sedation.

D. *Benactyzine* BAN ; USAN ; *Benctyzine Hydrochloride* INN ;

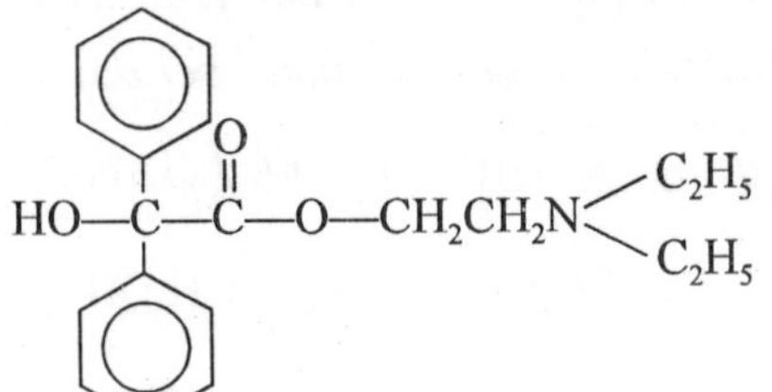

2-Diethylaminoethyl benzilate ; BPC (1959) ; Suavitil$^{(R)}$ (Dumax) ; Nutinal$^{(R)}$ (Boots) ; Cevanol$^{(R)}$ (ICI) ; Parasan$^{(R)}$ (Medix) ;

Synthesis :

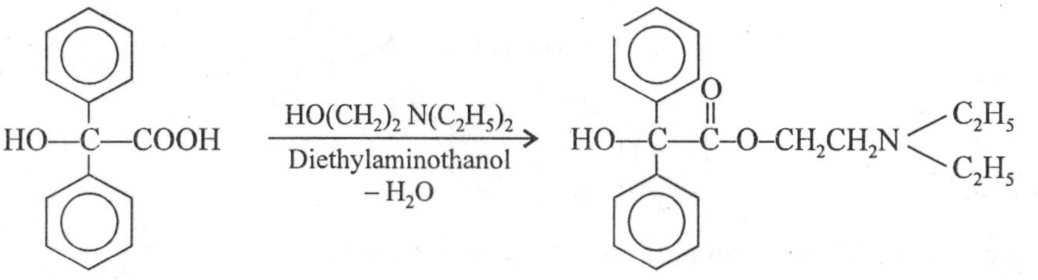

| Benzilic acid | Benactyzine |

It may be prepared by treating benzilic acid with diethylamne ethanol resulting the official compound with the elimination of a molecule of water.

2.3.1. Mechanism of Action. The mechanism of action of some of the medicinal compounds described under section 2.3. shall now be dealt with individually in the sections that follows :

2.3.1.1. Pipradrol Hydrochloride. The '*drug*' exerts its action as a stimulant of the central nervous system. Perhaps this could be the reason it is invariably included in multingredient preparations of combat antipsychotic profile to a great extent.

2.3.1.2. Captodiame Hydrochloride. The '*drug*' exerts its therapentic action to combat various types of anxiety disorders *viz.*, generalized anxiety disorders, panic attacks, phobic disorders, obsessive-compulsive disorder, post-tranmatic stress disorder, and mixed anxiety and depressive disorders. Perhaps the '*drug*' acts as a atypical antisychotic agent by virtue of its reduced tendency to produce the extrapyramidal effects.

2.3.1.3. Hydroxyzine Hydrochloride. The '*drug*' exhibits antichloinergic action ; and, therefore, its overall effects may be additive with those of *atropine* and other *belladona alkaloids*. Likewise, a host of other therapeutically potent sedative drugs it essentially requires a stringent precautionary measure with regard to its dose adjustment in such subjects who are on other CNS-depressant drugs. In the same vein, when employed as a preanaesthetic medication with other agents *e.g.*, *barbiturate(s)*, *mepreidine*, the dosage regimen need to be adjusted on an individual basis cautiously.

Note. *The potentiating effect of this drug should always be taken into consideration when it is employed in conjunction with CNS-depressants for instance : barbiturates and narcotics.*

2.3.1.4. Benactyzine Hydrochloride. The '*drug*' is found to show its action as an antidepressant as well as antimuscarinic activity. However, the antidepressant therapy has mainly been accomplished either *via* the monoamine oxidase (MAO) inhibitors or *via* the reversible inhibitors (RIMAs).

2.4. Phenothiazine Compounds

The discovery of phenothiazine as an anthelmintic dates back to 1883, however, its antithistaminic activity was revealed in 1937. The comintuous search for better drugs ultimately resulted into the synthesis of chlorpromazine in the famous Rhone-Poulenc Laboratories in France in the year 1950 which was found to possess remarkable ameliorative effect on anxiety, agitation and psychoses. This ultimately led to the synthesis of a host of structural analogues of chlorpromazine that were found to be useful as antipsychotics.

A few typical examples from this class of compounds are discussed here, namely ; Chlorpromazine, Perphenazine, Thioridazine.

A. *Chlorpromazine* INN, BAN, USAN,

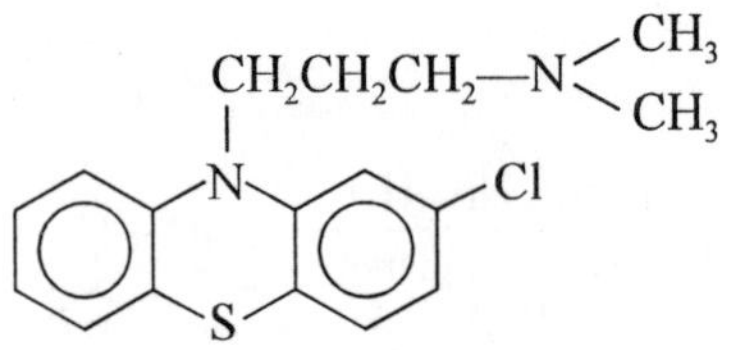

2-Chloro-10-[3-(dimethylamino) propyl] phenothiazine ;

BPC (1973) ; USP ;

Thorazine[R] (Smith Kline French) ; Promapar[R] (Parke-Davis) ; Megaphen[R]

(Bayer) ; Promacid[R] (Knoll) ;

Synthesis :

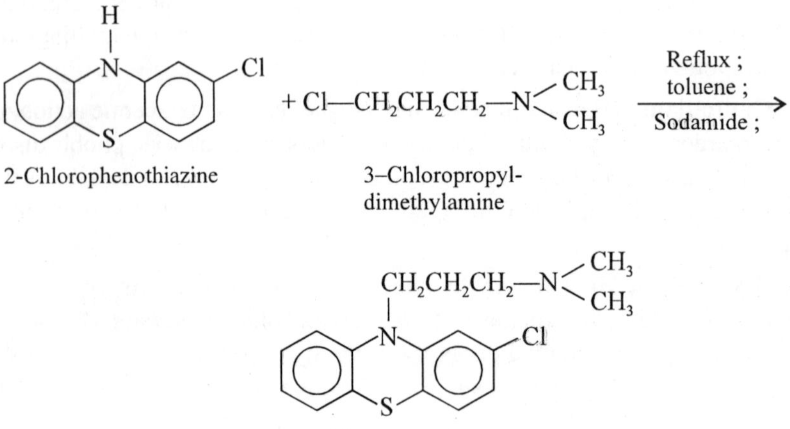

2-Chlorophenothiazine   3–Chloropropyl-
                                        dimethylamine

Chloropromazine

It is prepared by refluxing a toluene solution of 2-chlorophenothiazine and 3-chloropropyl dimethyla nine in the presence of sodamide for several hours, followed y filtration and removal of toluene under reduced pressure.

It enjoys the reputation for being the *tranquilizer* of the phenothiazine compounds. It is found to be effective in the management of manifestations of psychotic disorders and manic depressive illness (*manic pahse*), apprehension and anxiety and prior to surgery. It is also used for the treatment of moderate to severe agitation.

Dose. Tranquilizer : Adults, oral, usual, 10 to 50 mg 2 or 3 times daily to a total dose of 1 g daily when indicated ; 1m, 25 to 50 mg repreated in 1 hour upto a total dose of 1 g per day ; Children, oral, 0.55 mg/kg every 4 to 6 hours ; im, 0.55 mg/kg every 6 to 9 hours.

B. *Perphenazine* INN, BAN, USAN,

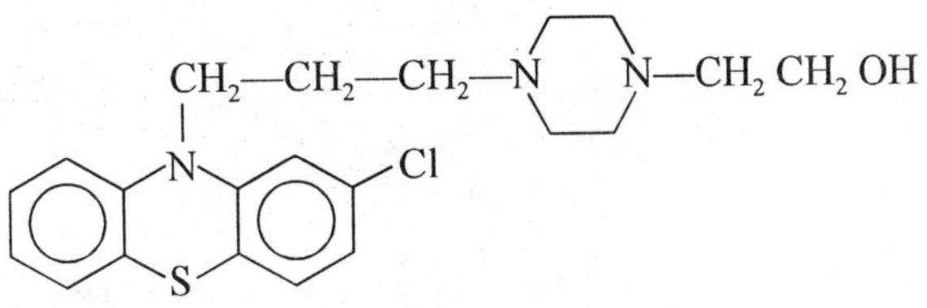

Piperazineethanol, 4-[3(2-chloro-10-phenothiazine-10 yl) propyl]- ;

BP (1973) ; USP ;

Trilafon[R] (Schering-Plough) ;

Synthesis :

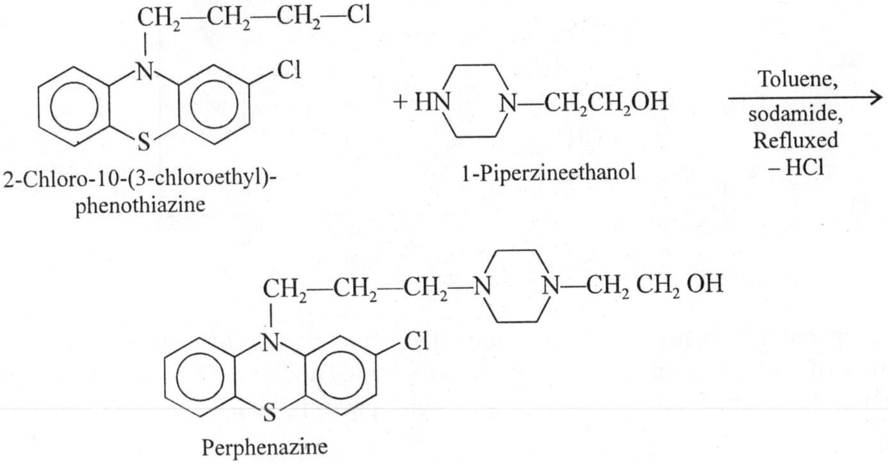

2-Chloro-10-(3-chloroethyl)-
phenothiazine

1-Piperzineethanol

Toluene,
sodamide,
Refluxed
– HCl

Perphenazine

It is prepared by refluxing a toluene solution of 2-chloro-10-(3-chloro-propyl) phenothiazine with 1-piperazineethanol in the presence of sodamide.

It is used for the mangement and treatment of neuroses.

Dose. Usual, oral, nonhospitalized patients, 2 to 8 mg thrice daily ; hospitalized patients, 8 to 16 mg 2 to 4 times a day ; im, 5 to 10 mg initially, followed by 5 mg in 6 hours.

C. *Thioridazine* INN, BAN, USAN,

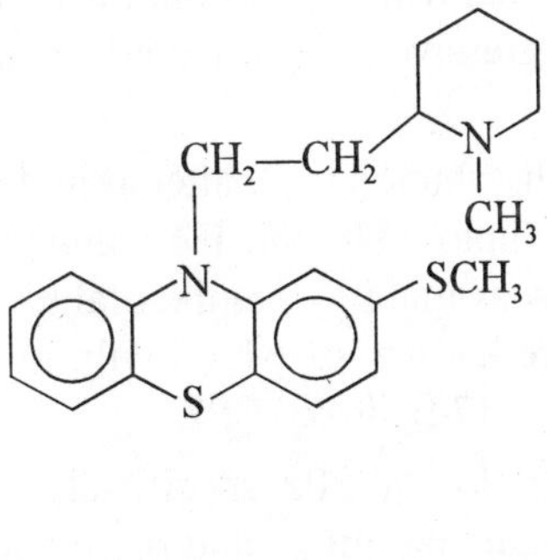

10[2-Methyl-2-piperidyl) ethyl]-2-(methylthio)-phenothiazine ;

BP (1973) ; USP ;

Mellaril$^{(R)}$ (Sandoz) ;

Synthesis :

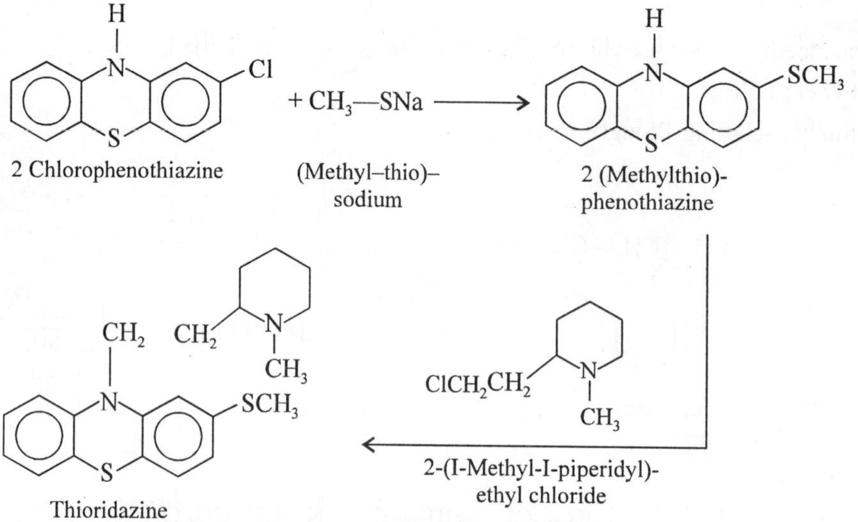

It is prepared in two steps. First step being the preparation of 2-(methylthio) phenothiazine by the interaction of 2-chlorophenothiazine with (methyl-thio) sodium. The second step involves the condensation of the resulting product with 2-(1-methyl-1-piperidyl) ethyl chloride to obtain the official compound.

Its actions and uses are very much identical to those of chlorpromazine discussed earlier.

Dose. Usual, initial, for psychoses : 50 to 100 mg 3 times daily ; for nonpsychotic emotional disturbances for instance tension and anxiety : 30 to 200 mg per day ; for children having behavioural disorders : 1 mg per kg body weight per day in divided doses.

2.4.1. Mechanism of Action. The mechanism of action of drugs described under section 2.4 shall now be dealt with as below :

2.4.1.1. Chlorpromazine Hydrochloride. The '*drug*' shows its effectiveness for the control and management of symptoms associated with mild alcohol withdrawl, moderate to acute agitation, and observed hyperactivity or apparent aggressiveness particularly in mentally disturbed children by exerting its action on the CNS.

The '*drug*' shows volume of distribution (v_d^{ss}) after a single oral administration to be 80.6 L kg^{-1}, whereas a reduction to 21.8 L kg^{-1} (upto 25%) *via* IM administration. Hence, the 4-fold difference actually reflects directly upon the low availability *via* the oral route (32%). Almost 100 metabolites of chlorpromazine (CPZ) in humans are known, of which only two are found to be **active** in humans, namely : (*i*) 11-hydroxy CPZ ; and (*ii*) 17-hydroxy-CPZ.

The effective plasma concentration of CPZ specifically in severe schizophrenic subjects have been demonstrated to vary from 30—300 ng. mL^{-1}, and plasma levels from 750-1000 ng. mL^{-1}.

CAUTION. *Levarterenol and phenylephrine are employed invariably for the control and management of hypotension.*

2.4.1.2. Perphenazine. The '*drug*' exerts its action very much similar to the one described under chlorpromazine (section 2.4.1.1).

SAR of Perphenazine. It differs chemically from *prochlorperazine* only with respect to the substitution of a hydroxyethyl (—C_2H_5OH) moiety for the methyl moiety of the latter drug as shown below :

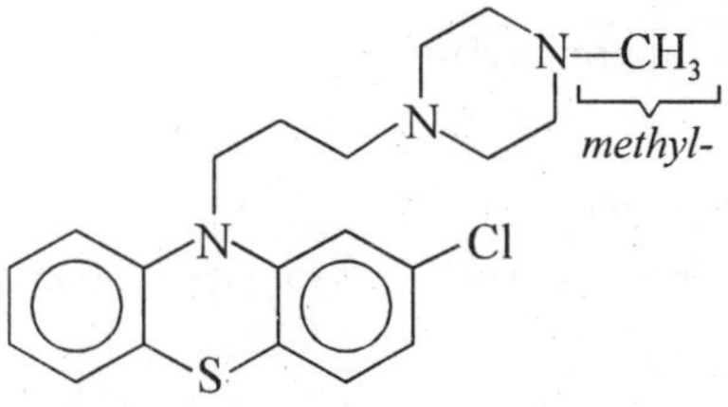

Prochlorperazine

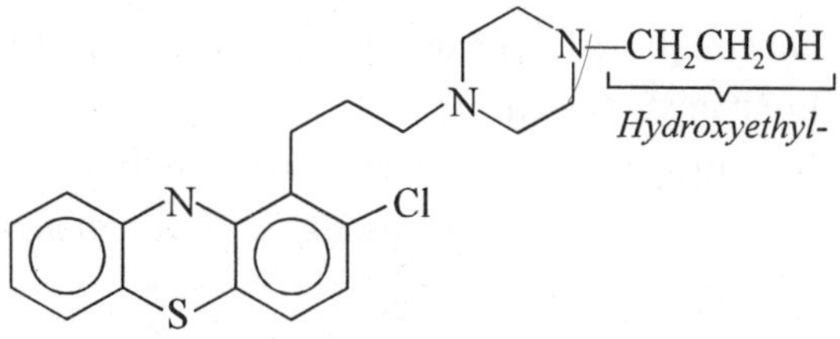

Perphenazine

2.4.1.3. Thioridazine Hydrochloride. The '*drug*' exerts its bear minimum antimetic activity and thereby gives rise to minimal **extrapyramidal stimulation** (EPS). Drowsiness and sedation are predominantly less intense in this '*drug*' in comparison to either CPZ and other similar drug substances.

The half-life seems to be particularly to a *multiphasic status i.e.,* having an *early phase* ranging between 26-36 hours ; and a definitive *late phase* varying between 26-36 hours. The '*drug*' gets bound to plasma protein to the extent of 96-99%. Importantly, thioridazine gets sulfoxidized *in vivo* into the metabolities *mesoridazine* plus a small quantum of **sulforidazine,** both of which are **active** pharmacologically.

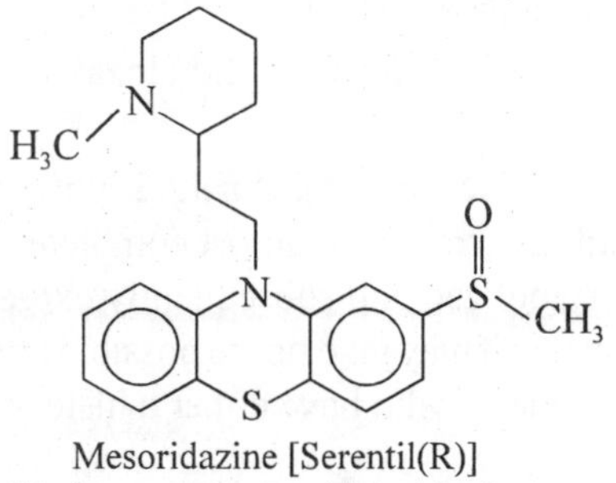

Mesoridazine [Serentil(R)]
**[Twice as potent as 'thioridazine'
in humans]**

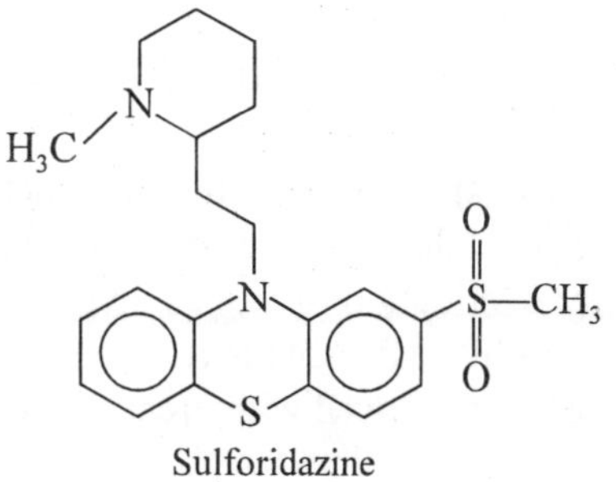

Sulforidazine
(Active)

2.5. DIBENZAZEPINES

Dibenzazepine analogous constitute another category of antisychotics which have gained recognition in late sixties.

A few important member of this class are described here, namely ; Loxapine ; Clozapine ;

A. *Loxapine* BAN, USAN, INN,

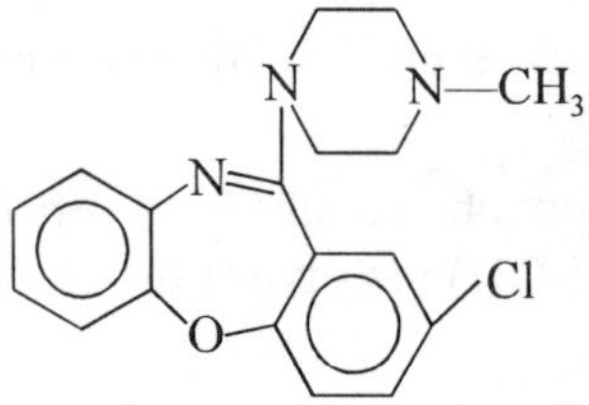

Dibenz [b, f][1, 4] oxazepine, 2-chloro-11-(4-methyl-1-piperazinyl)-;

Loxitane[R] (Lederle) ;

Its antipsychotic actions are similar to those of chlorpromazine.

Dose. Usual, oral, for psychoses : 20 to 50 mg per day initially, split in 2 to 4 divided doses.

B. *Clozapine* INN, BAN, USAN,

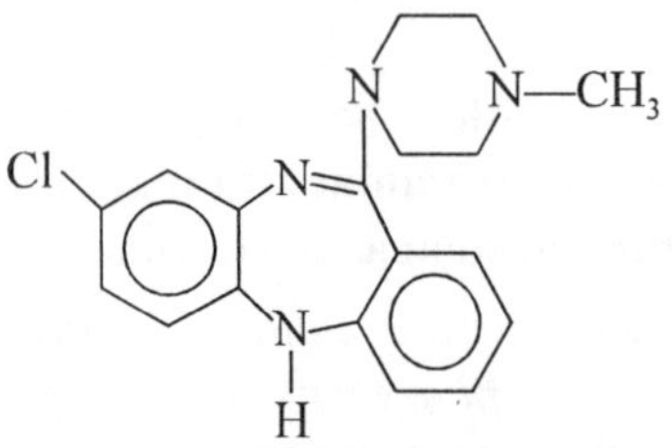

5 H-Dibenzo [*b, e*][1, 4] diazepine, 8-chloro-11-(4-methyl-1-piperazinyl)-;

Clozaril[R] (Sandoz) ;

It is a dibenzodiazepine derivative. It is an unusual antipsychotic agent which hardly produces any extrapyramidal symptoms. It possesses the ability to suppress symptoms of tardive dyskinesia.

2.5.1. Mechansim of Action. The mechanism of action of loxapine and clozapine shall now be described as under :

2.5.1.1. Loxapine Succinate. The exact mechanism of action of this '*drug*' is not yet established. It has been observed that the absorption soonafter oral administration is almost complete. Furthermore, the '*drug*' after distribution to tissues invariably gets metabolized and subsequently excreted *via* urine and faeces, largely in the first 24 hours. By virtue of the fact that it may give rise to possible **anticholinergic activity,** it must be employed with great caution in such patients who have either a history of *glaucoma* or *urinary retention* problems.

2.5.1.2. Clozapine. The '*drug*' is found to very effective and relatively rapid-acting in the treatment of schizophrenia perhaps due to several of its CNS effects that essentially differ from a host of other members of antipsychotics. It is also believed beyond any reasonable doubt that the antipsychotic actions of this '*drug*' are basically of more complex nature in comparison to other antipsychotic drugs. In addition to the above observations the '*drug*' specifically blocks dopamine D-2 and 0-1 receptors essentially in the *mesolimbic** and *mesocortical brain regions*, which may also involve covertly and overtly **cholinergic, seretonergic** and **noradrenergic** systems.

Contrary to the usual activity of antipsychotics, clozapine exhibits *regional specific anti-dopaminergic profile* having, relatively mild antagonism on the extrapyramidal dopaminergic action ;

---

* Medium sized border of a part.

and this could be responsible for its low prepensity to produce extrapyramidal side effects *e.g.,* dystonias, tardive dyskinesia.* It also causes greater blockade of dopamine D-1 receptors ; and however, it is not yet established whether such action precisely contributes to its antipsychotic therapeutic activity.

Clozapine gets absorbed rapidly from the GI tract and extensively metabolized during the first pass through the liver. The peak plasma levels usually take place in about 1.5 hour after a single oral dosage. It has been observed that there exists a **sixfold interindividual variability** in steady state plsma concenrations in subjects administered with high dosage regimen of this '*drug*'. The '*drug*' and its metabolites are excreted mostly in the urine.

2.6. BUTYROPHENONES

Janssen and coworkers synthesized a number of butyrophenones having antipsychotic potency similar to chlorpromazine :

Examples : Haloperidol ; Droperidol ;

A. *Haloperidol* INN, BAN, USAN,

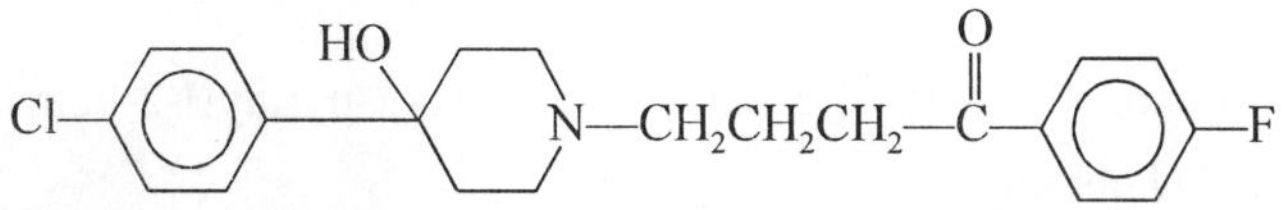

4-[4-(*p*-Chlorophenyl)-4-hudroxypiperidino]-4'-fluorobutyrophenone ;

BP (1973) ; USP ;

Haldol[R] (McNeil) ; Aloperidin[R] (Janssen) ; Serenace[R] (Searle) ;

Synthesis :

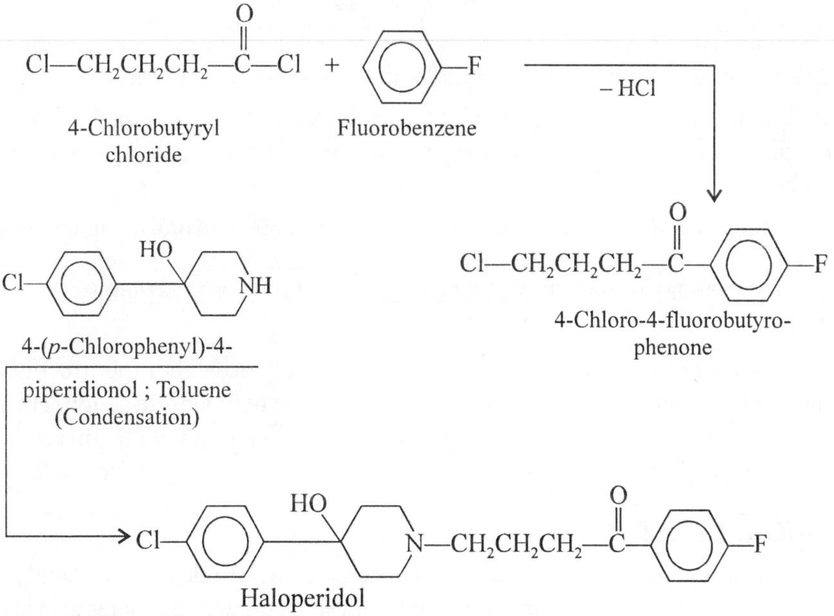

4-Chlorobutyryl chloride

Fluorobenzene

4-Chloro-4-fluorobutyro-phenone

4-(*p*-Chlorophenyl)-4-piperidionol ; Toluene (Condensation)

Haloperidol

---

*A condition of slow, rhythmical, automatic sterotyped movements, either generalized or in single muscle groups. These occur as an undesired effect of therapy with phenolliazines (psychotropic drugs).

4-Chloro-4'-fluorobutyrophenone is first prepared by the **Friedel-Craft's reaction** between 4-chlorobutyl chloride and fluorobenzene which is subsequently condensed with 4-(*p*-chloro-phenyl)-4-piperidinol in toluene to give the official compound.

It is useful in the treatment of anxiety, tension, moderate to severe agitation, hostility, and hyperactivity. It also finds its usefulness in schizophrenia, psychotic reactions related to organic brain syndromes and in *Gilles de la Tourette's disease* (unusual barking).

Dose. Usual, adult, oral, 0.5 to 5 mg 2 or 3 times daily ; Intramuscular, 3 to 5 mg ;

B. *Droperidol* INN, BAN, USAN,

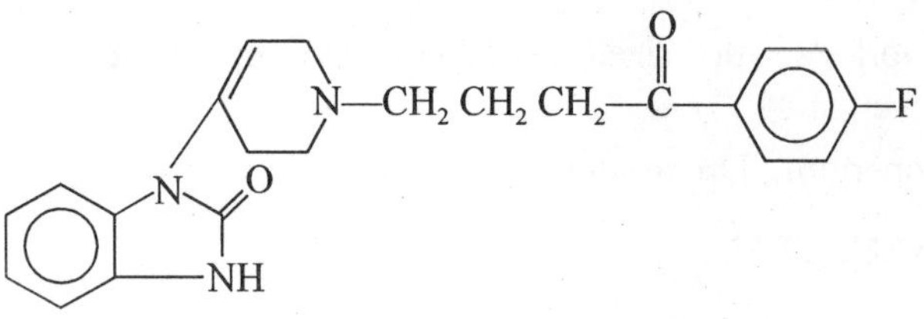

1-[1-[3-(*p*-Fluorobenzoyl) propyl]-1,2,3,6-tetrahydro-4-pyridyl]-2-benzimidazolinone ;

USP ; Inapsine[R] (Janssen) ;

It is employed for the control of agitated patients in acute psychoses. It is normally used in conjunction with an analgesic for instance fantanyl citrate or phenoperidine hydrochloride to maintain the patient in a state of neuroleptanalgesia whereby he is calm and indifferent to his surroundings and able to cooperate with the surgeon.

Dose. Premeditation : iv or im, 2.5 to 10 mg 30 to 60 minute before induction ; Induction : usual, IV, 2.5 mg per 20 to 25 lb. ; maintenance : usual, IV : 1.25 to 2.5 mg.

2.6.1. Mechanism of Action. The mechanism of action of the medicinal compounds discussed under section 2.6 shall now be treated individually in the sections that follows :

2.6.1.1. Haloperidol. The '*drug*' exhibits its activity by calming down excessive motor activity quite prevalent in '*hyperactive children*' having conduct disorders, such as : aggressivity, mood lability, impulsivity, poor frustration tolerance, and difficulty in sustaining attention.

Haloperidol shows bioavailability extending upto almost 60% *via* oral administration. Interestingly, its elimination half-life *via* oral route varies between 12-38 hours, which eventually gets lowered between 10-19 hours after IV administration. The usual therapeutic plasma concentrations ranges between 3-10 ng mL^{-1}.

2.6.1.2. Droperidol. The '*drug*' exhibits relatively low therapeutic potency, medium extrapyradimal toxicity, high sedative effect, and above all high hypotensive action. However, it is most frequently employed in the form of its combination [Innovar[R]] along with the narcotic agent *fentanyl* [Sublimaze[R]] preanaesthetically.

2.7. AZASPIRODECANEDIONES

Azaspirodecanediones have gained prominence as antipsychotic agents recently. In mid-eighties a member of this family, namely, buspirone was introduced in the United States and is discussed here.

A. *Buspirone* INN, BAN ; *Buspirone Hydrochloride* USAN,

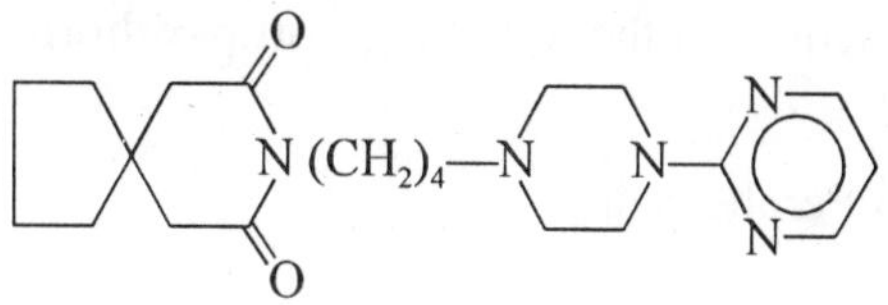

8-Azaspiro [4, 5] decane-7, 9-dione, 8-4-[4-(pyramidinyl)-1- piperazinyl]butyl]-; Buspar[(R)] (Mead Johnson) ;

It is found to be useful in the treatment of anxiety and its effectiveness is fairly comparable to that of diazepam. It alleviates anxiety without producing sedation or functional impairment. It neither promotes abuse nor physical dependence.

2.7.1. Mechanism of Action. The mechanism of action of buspirone hydrochloride will be discussed as under :

2.7.1.1. Buspirone Hydrochloride. The exact mechanism of its anxiolytic effect has not yet been established ; but however, seems to be altogether different in comparison to the **barbiturates** and the **benzodiazepines.** Most probably the '*drug*' essentially involves *multiple transmitter systems,* specifically those of the first-pass metabolism.

Buspirone attains peak plasma concentrations ranging between 1-6 mg mL^{-1}, usually take place within a span of 40-90 minutes. The '*drug*' gets bound to plasma protein to nearly 95% ; and excretion through urine varies between 29-63%, while through faeces between 18-38%. The elimination half-life of the unchanged drug is approximately between 2-3 hours.

Probable Questions for B. Pharm. Examinations

1. What are '**antipsychotics**' ? Classify them by giving examples of **one** potent compound from each category.

2. (*a*) Name the **two** major alkaloids isolated from the roots of *Rauwolfia serpentina* ued as '**antipsychotics**'.

    (*b*) Give their structure, chemical name and uses.

    (*c*) Discuss the synthesis of 3-diethylamino propyl ester of 3,4,5-trimethoxy benzoic acid given by Miller and Weinberg.

    (*d*) What is the relative potency of compound in (*c*) and that of Reserpine ?

3. Discuss the synthesis of the following **diphenylmethane analogues** as 'antipsychotics' :

    (*a*) Benactyzine

    (*b*) Captodiame

4. How would you synthesize **Perphenazine** and **Thioridazine** belonging to the phenothiazine group of '**antipsychotics**' ?

5. Describe **dibenzazepines** as potent 'antipsychotics'. Give the structure, chemical name and uses of **two** such drugs.

6. (*a*) Give the structure, chemical name of the following :

    (*i*) Haloperidol

    (*ii*) Droperidol.

(*b*) Discuss the synthesis of any **one** drug.

7. Elaborate the 'mode of action' of the following **'antipsychotics'** :

(*a*) Reserpine

(*b*) Chlorpromazine and Perphenazine.

(*c*) Loxapine

(*d*) Haloperidol

(*e*) Buspirone.

8. Give a comprehensive account of **'antipsychotics'.** Support your answer with the most potent drugs by providing structures, chemical names and uses adequately.

RECOMMENDED READINGS

1. P. Jenner and CD, Marden in *Drugs in Central Nervous System Disorders.,* DC Horwell, Ed., New York, (1985).

2. WH Oldendorf and WG Dewburst *Principles of Psychopharmacology,* 2nd Ed., WG Clark and J. Del Guidice Eds., New York Academic Press (1978).

3. Martindale *The Extra Pharmacopoeia,* 31st Edn., London, The Pharmaceutical, Press (1996).

4. D. Lednicer and LA Mitscher *The Organic Chemistry of Drug* Synthesis, New York, John Wiley and Sons (1995).

5. MC Griffith (Ed.) USAN and the USP Dictionary of Drug Names, Rockville, United States Pharmaceutical Convention Inc., (1985).

6. Index Nominum, Zurich, Swiss Pharmaceutical Society (1982).

7. PG Strange : *Antipsychotic Drugs : Importance of Dopamine Receptors for Mechanisms of Therapeutic Actions and Side Effects, pharmacol. Rev.,* **53** : 119, 2001.

8. DR Weinberger : *Anxiety at the Frontier of Molecular Medicine, N. Engl. J. Med.,* **344** : 1247, 2001.

9. AR Gennaro : *Remington the Science and Practice of Pharmacy,* Lippincott Williams & Wilkins, New York, 20th edn, Vol. II, 2000.

10. SM Roberts and BJ Price (eds) : *Medicinal Chemistry-The Role of Organic Chemistry in Drug Research*, Academic Press, New York, 1985.

11. CG Wermuth : *The Practice* of Medicinal Chemisty, Academic Press, New York, 1996.

12. FD King : *Medicinal Chemistry-Principles* and *Practice,* The Royal Society of Chemistry, London, 1994.

26   Antiviral Drugs

1. INTRODUCTION

Viruses are obligate parasites having the operational characteristics of an exogenous submicroscopic unit capable of multiplication only inside specific cells.

The size range is considerable, ranging from an approximate diameter of 200 micrometers (μm) vaccinia down to 10 (μm) for foot and mouth disease. They can be seen and identified-only with the aid of an electron microscope.

In general, viruses are essentially made up of a nucleic acid core having either deoxyribonucleic acid (DNA) or ribonucleic acid (RNA) that provides the genetic material and also forms the basis for classification of viruses.

The viruses may be conceived as particles attaching themselves to particular '*receptors*' of the succeptible cells. These receptors may be chemical configurations that combine with either viruses or allied substances of similar composition. After due attachment the viruses gain entry into the cell and subsequently multiply. Thus the newly constituted viruses are eventually realised from the cell to *paraciticize other cells of the host.* In such transformation the metabolic activity of the host cell is modified in some manner.

A number of diseases are caused by different types of viruses which are enumerated briefly as under :

S.No.	Species	Indications
1.	Herpes simplex (virus types 1 & 2)	Eye infections, skin diseases, encephalitis and genital infections
2.	Influenza A, B & C viruses	Influenza A, B and C
3.	Rabies viruses	Rabies, encephalitis
4.	Enteroviruses (polio, Coxsackle A, B echovirus)	Poliomyelitis
5.	Parainfluenza virus	Parainfluenza
6.	Variola, Vaccinia	Smallpox (variola), Cowpx (vaccinia)
7.	Variclla-zoster	Varicla (zoster), herpes zoster (shingles)
8.	Rhiniviruses	Respiratory diseases

1.1. Replication and Transformation

Viruses, in general, utilize only the *enzyme-system* of the host-cell for two purposes, namely : *first*, to synthesize DNA ; and *secondly,* to replicate virus, thereby enabling it to perform their usual metabolic activities. They may carry out either the transformation or the replication processes of the cell at the same time. By virtue of the fact that viruses are **obligate intracellular parasites,** therefore, their replication phenomenon solely depends on the *host's cellular processes.*

2. CLASSIFICATION

Antiviral drugs are broadly classified on the basis of their specific mode of action as stated below :

(*a*) Substances that inhibit early stages of viral replication

(*b*) Substances that interfere with viral nucleic acid replication

(*c*) Substances that affect translation on cell ribosomes

2.1. Substances That Inhibit Early Stages of Viral Replication

A number of antiviral drugs that block particular viruscoded enzymes produced in the host cells which are vitally required for viral replication have been included in the therapeutic armamentarium for the treatment of viral infections.

Examples : Amantadine hydrochloride ; Interferon ;

A. *Amantidine Hydrochloride* USAN ; *Amantadinum* BAN ;

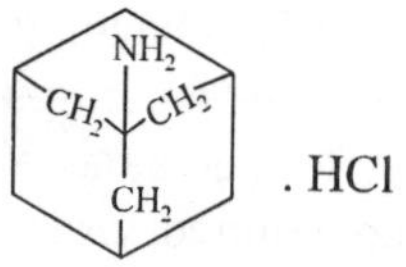

1-Admantanamine hydrochloride ;

Amantidine Hydrochloride USP ;

Symmetrel$^{(R)}$ (Geigy) ;

Synthesis :

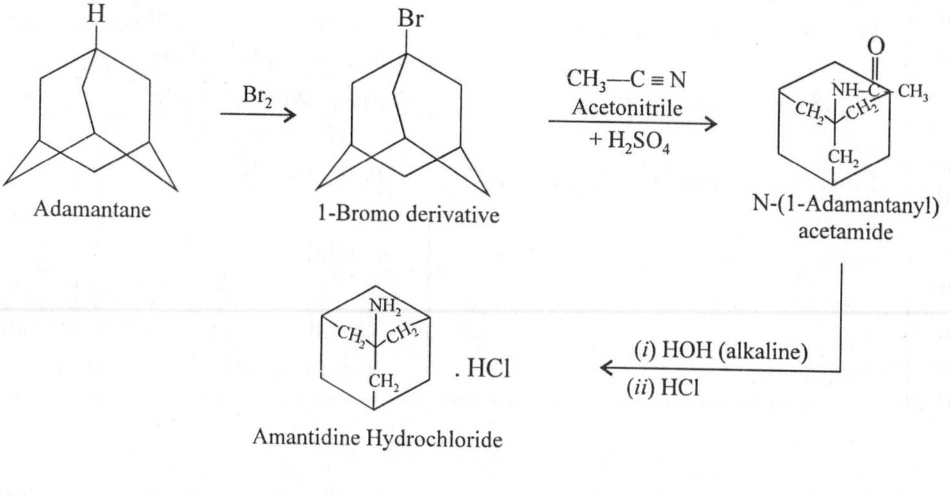

It is prepared by brominating adamantane and treating the resulting bromo derivative with acetonitrile in the presence of sulphuric acid to obtain N-(1-adamantanyl) acetamide. This on alkaline hydrolysis and treatment with hydrochloric acid affords the official compound.

It is useful only as *prophylactic against A_2 influenza virus (Asian Flu).* It broadly prevents the entry of certain viruses into the cell.

Dose. Usual, 100 mg twice daily ; For children, 1 to 9 years of age, 4-9 mg/kg and 9 to 12 years of age 100 mg twice daily.

B. *Interferon α-2B* :

Intron[R] (Parke-Davis) ; Wellferon[R] (Burroughs Wellcome) ;

Though 'interferon' was first reported in 1957 by Issacs and Lindenmann but it was recognized as an antiviral drug in 1980.

It is the protein formed by the intersection of animal cells with viruses capable of conferring on animal cells resistance to virus infection.

In general, interferons are made up of a mixture of relatively small proteins with molecular weights varying from 20,000 to 1,60,000. These are basically glycoproteins that display specific antiviral properties with species-related characteristics. So far three different categories of interferons have been isolated, characterized and studied in an elaborated manner :

These are namely :

(*a*) Alpha (α)–secreted by human leukocytes

(*b*) Beta (β)-secreted by human fibroblasts

(*c*) Gamm (γ)-secreted by lymphoid cells

It is broadly employed for the treatment of herpes zoster, herpetic–keratitis, herpes genitalis, chronic hepatitis, common cold and influenza. It also finds its usefulness in lung carcinoma, breast cancer, multiple myclomas. It is also recommended as a prophylactic agent in cytomegalovirus infection in renal transplant patients.

2.1.1.   Mechanism of Action

The mechanism of action of the various medicinal compounds discussed under section 2.1 shall be treated separately below :

2.1.1.1. Amantidine Hydrochloride

The '*drug*' is a narrow-spectrum antiviral active against almost all influenzae. A virus strains, certain C virus strains ; however, not found effective against B strains. It has been observed that the peripheral and central effects of the anticholinergic drugs are enhanced by concomitant use of amantidine.

The '*drug*' fails to undergo metabolism and almost a major portion (90%) is practically excreted unchanged in the urine. The half-life is nearly 20 hours. It attains levels in the cerebral spinal fluid (CSF) to the extent of 60% of the plasma concentration.

2.1.1.2. Interferon α-2B

The '*drug*' belongs to a family of proteins ranging between 15,000–27,000 (MW) which are found to be secreted by the lymphocytes in response to acute viral infections. It has been observed that they invariably get bound to cellular proetins. In this way, they actually exert a plethora of effects that may essentially include induction of certain enzymes, for instance : **2, 5A synthetase,** which particularly inhibits viral replication and inhibition of cell proliferation. Besides, they augment **immune-regulating activity** which includes expression of *HLA-major histocompatibility antigens* that ultimately turn out to be the targets of cytotoxic T lymphocytes.

The peak-serum level is attainable within 3–12 hour after injection. The elimination half-life ranges between 2–3 hours, and serum levels are undetectable after 16 hours. It is found that the hepatitis C replication rates are in the trillious a day, having viral half-life of nearly 5 hours. Therefore, there exists a predominant inherent mismatch between pharmacokinetic and pharmacodynamic characteristic features which ultimately ends up in enhancing the prevailing **replication rate of the virus above the baseline.**

Note. More recently interferons may also be indicated for antiinflammatory, antifibri-nogenic, and antineoplastic conditions.

2.2.   Substances That Interfere With Viral Nucleic Acid Replication

A good number of antiviral drugs exert their effect against DNA viruses either by interfering with their replication due to its similarity of structure to the nucleotide structures in natural DNA virus or by interfering with the nucleic acid replication of the virus, specifically inhibiting the early steps in DNA synthesis.

A few important compounds shall be discussed here, namely : Idoxuridine, Acyclovir, Vidarabine, Ribavirin.

A. *Idoxuridine* INN, BAN, USAN,

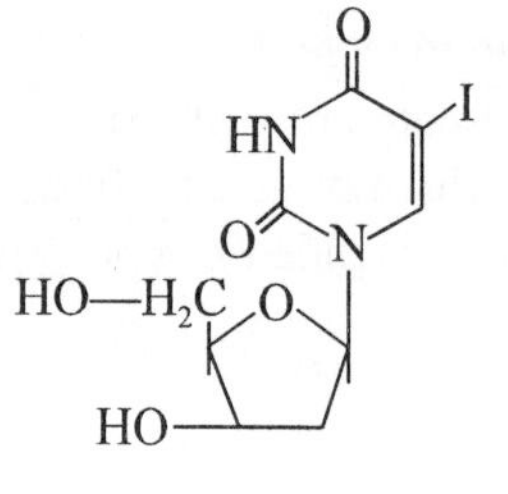

2-Deoxy-5-iodouridine ;

BP ; USP ;

Stoxil[R] (Smith, Kline & French) ; Herplex Liquifilm[R] (Allergan) ;

Synthesis :

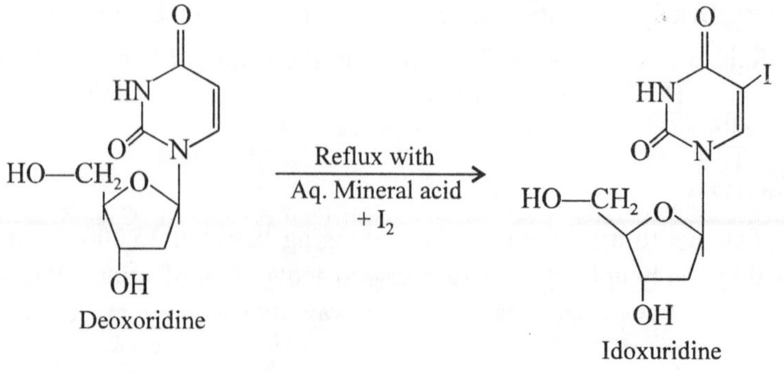

Deoxoridine   Idoxuridine

It is prepared by refluxing a solution of deoxuridine in aqueous mineral acid in the presence of iodine.

It is an antimetabolite of thymidine and, therefore, may be incorporated into deoxyribonucleic acid in place of thymidine, thus interfering with usual nuclear metabolism. It is solely employed for the topical therapy of *herpes simplex keratitis* of the eye. It has also been administered intravenously for the treatment of herpetic encephalitis.

Dose. Topical, as a 0.5% ointment 4 to 16 times a day, or 0.1 ml of a 0.1% solution every 1 to 2 hours, to the conjunctiva.

B. *Acyclovir* USAN ; *Aciclovir* INN ;

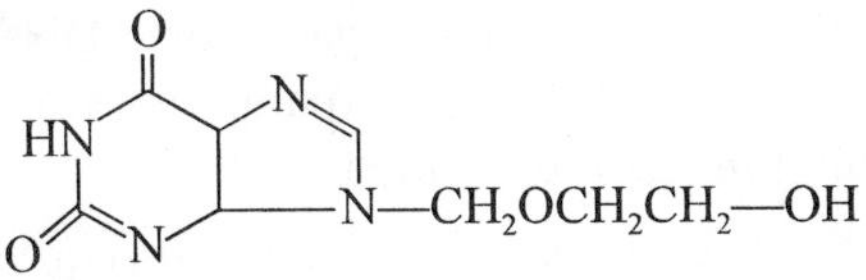

$$\text{O}$$

HN—N—CH$_2$OCH$_2$CH$_2$—OH

6H-Purin-6-one, 2-amino-1, 9 dihydro-9-[2-hydroxyethoy)-methyl]– ; Zovirax$^{(R)}$ (Burroughs Wellcome) ;

It is used for the treatment of cold sore caused by labial herpes, herpes simplex virus Type I and Type II responsible for genital herpes. It also affects the isolated of EB viruses and varicella-zoster significantly.

Herpetic keratitis and herpes genitalis are treated effectively by using an ointment containing 5% acyclovir.

Dose. For herpes virus infections in immunosupressed patients : up to 10 mg per kg body–weight every 8 hours.

C. *Vidarabine* BAN, USAN,

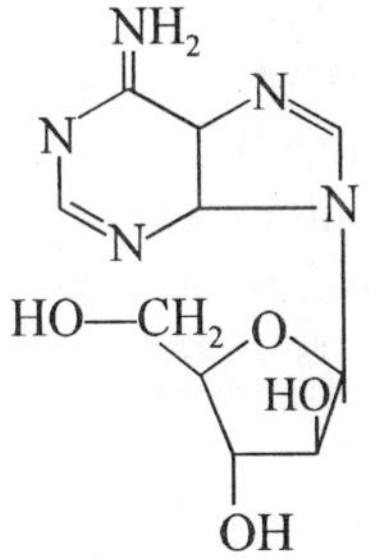

NH$_2$

HO—CH$_2$

HO

OH

β-D-Arabinofurannosyl-9-adenine ; USP ;

Vira-A$^{(R)}$ (Parke–Davis) ;

It is specifically employed for the treatment of herpes simplex virus infections belonging to Types 1 and 2 encephalitis.

It is also found to be beneficial both in varicella-zoster infections and neonatal herpes.

Dose. For encephalitis due to herpes simplex : 15 mg per kg body–weight per day for 10 days ; which is infused at a constant rate over a period of 12 to 24 hours.

D. *Ribavirin* INN, USAN ; *Trivabirin* BAN ;

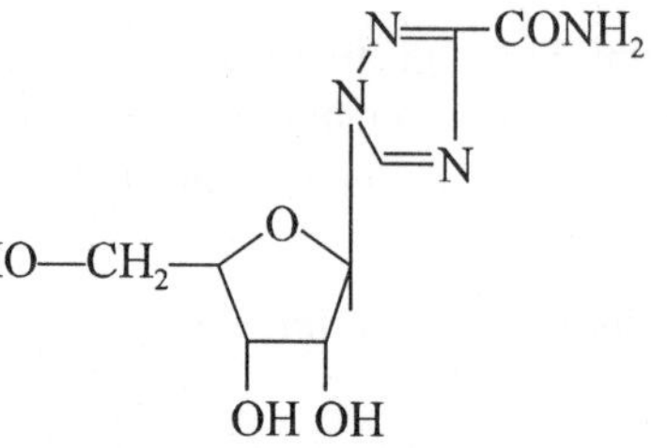

1-β-D-Ribofuranosyl-1H-1, 2, 4-triazole-3-carboxamide ; Virazole$^{(R)}$ (ICN) ;

It possesses broad–spectrum antiviral activity against both DNA and RNA viruses. It exerts its maximum activity against influenza A and B and the parainfluenza group of measles, hepatitis and viruses. It is also reported to inhibit *in vitro* replication of HTLV-III, which is concerned with AIDS.

Dose. For viral hepatitis, influenza and herpes virus infections : upto 1 g per day in divided doses.

2.2.1. Mechanism of Action

The mechanism of action of various drugs described under section 2.2 shall be dealt with separately as follows :

2.2.1.1. Idoxuridine

The '*drug*' is a *nucleic acid synthesis inhibitor*, which specifically acts as an antiviral agent against **DNA viruses** by the aid of its direct interference with their replication phenomenon given their similarity of structure. As a first and foremost step the '*drug*' undergoes phosphorylation largely by the host cell **virus-encoded enzyme thymidine kinase** into an active triphosphate form. Consequently, kinase the phosphorylated drug entity is found to inhibit **cellular DNA polymerase** surprisingly to a much lower extent in comparison to **HSV-DNA polymerase,** that is required as an absolute necessity for the ultimate synthesis of DNA. Subsequently, the triphosphate form of the drug is appropriately introduced in the course of viral nucleic acid synthesis aided by an altogether **false pairing system** which critically replaces thymidine. The overall net outcome upon *transcription* is the generation of faulty viral proteins which finally give rise to **'defective viral particles'.***

2.2.1.2. Acyclovir

In this particular instance the '*drug*' aid inside an infected cell to get converted into a '*triphosphate*', which is subsequently incorporated directly into DNA. In fact, this phenomenon terminates elongation of the DNA, and thereby prevents viral replication mechanism significantly. It fails to eradicate latent herpes. It is observed to be unpredictable to a certain extent as a '*topical prophylactic agent*' against particularly the recurrent infections caused by HSV–1 and HSV–2.

The '*drug*' gets bound to protein in plasma between 9–33%. The drug is usually excreted through urine either by oral IV administration between a range of 62–91% and 9–20% respectively. The normal usual half-life is about 2.5 hours which may get extended upto 19.5 hours in patient having renal failure.

---

*Farah A *et al.* (eds) : *Handbook of Experimental Biology,* Vol. : 38/2, Springer–Berlin, pp : 272–347, 1975.

2.2.1.3. Vidarabine

The '*drug*' gets converted by the help of cellular enzymes into corresponding mono-, di-, and tri-phosphate structural analogues which strategically interfere with viral nucleic acid replication phenomenon, particularly jeopardizing the very preliminary steps involved in DNA–synthesis. It has been observed that the antiviral effect is, in certain instances, proved to be much superior to that of *idoxuridine* and *cytarabine*.

Vidarabine is deaminated quite rapidly by the enzyme **adenine deaminase** that is usually found in serum and RBC. Interestingly, this enzyme helps in the conversion of this '*drug*', into its principal metabolite termed as **arabinosyl** hypoxanthine (ara–HX), which displays weak antiviral activity.*

2.2.1.4. Ribavirin

The '*drug*' is duly converted to metabolites which critically cause inhibition of the 5′ capping of viral *m*RNA ; so that finally the viral protein synthesis of both DNA and RNA viruses are affected directly and significantly. The **triphosphate** is believed to be the **active metabolite** that is formed invariably in lung and liver than in other tissues. Perhaps that could be a pausible explanation for its optimal activity against infections in these organs specifically. It fails to pass the blood-brain barrier (BBB). It has been found that the '*drug*' and its known emtabolites are duly excreted in the urine upto 50% and faeces upto 15%. The plasma half-life stands at 9.5 hours, whereas the half-life in erythrocytes is approximately 40 days.

Ribavirin specifically inhibits *in vitro* replication of HIV-1, which is essentially involved in AIDS.

Note. Interestingly, till date viral strains susceptible to ribavirin have not been found which may develop drug resistance, as could be observed with other antiviral agents, *e.g.*, acyclovir, idoxuridine, and bromovinyldeoxy uridine (BVDU).

2.3. Substances That Affect Translation on Cell Ribosomes

There are a few specific compounds that directly interfere with the translation of RNA message into protein synthesis on the cell ribosome thereby resulting a defect in protein inclusion into the virus. In short, virus DNA gets enhanced host-cells are mutilated and ultimately infectious virus is not generated.

Examples : Methisazone ; Arildone ;

A. *Methisazone* USAN ; *Metisazone* INN ;

1-Methyl-indole-2, 3-dione-3-(thiosemicarbazone) ;

Marboran[R] (Burroughs Wellcome) ;

---

*Chao DL and AP Kimbali, *Cancer Res.,* **32**, 1721 (1972).

Synthesis :

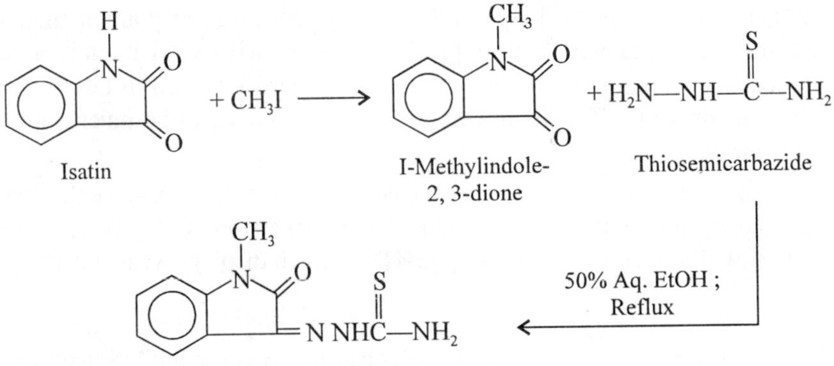

Isatin   I-Methylindole-   Thiosemicarbazide
                                2, 3-dione

Methisazone

1-Methylindole-2, 3-dione is first prepared by reacting isatin with methyl iodide, which is then treated with thiosemicarbazide in 50% aqueous ethanol and refluxed for several hours to obtain the desired compound.

It possesses prophylactic value against smallpox and alastrim. In conjunction with gamma globulin it shows its usefulness against eczema vaccinatum and vaccinia gangrenosa.

Dose. Oral, 1.5 to 3.0 g twice daily for 4 days ; as a prophylactic against smallpox-it should be administered before the 8th or 9th day of the 12-day incubation period.

B. *Arildone* INN, USAN ;

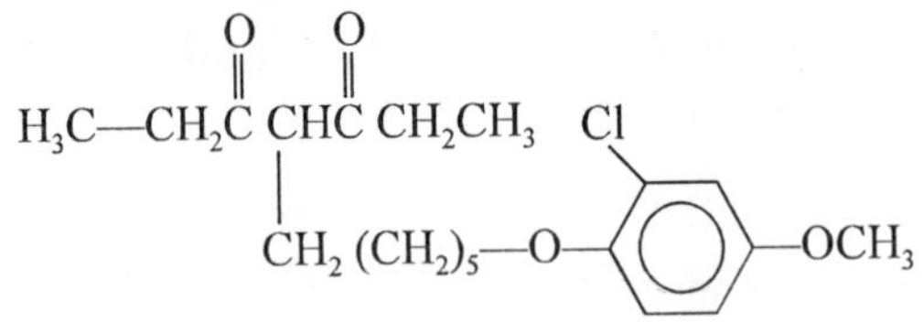

4-[6-(Chloro-4-methoxyphenoxy) hexyl]-3, 5-heptanedione ;

Win 38020[R] (Winthrop) ;

It is found to be useful against both DNA and RNA viruses, for instance herpes virus, parainfluenza virus and respiratory *syncytial viruses**. It is also used as a prophylactic in renal transport patients and to minimise herpes simplex virus infections.

2.3.1. Mechanism of Action

The mechanism of action of the medicinal compounds discussed under section 2.3 shall be treated individually in the pages that follows :

2.3.1.1. Methisazone

The '*drug*' once enjoyed the fame of being used as a prophylaxis for smallpox.

2.3.1.2. Arildone

The '*drug*' has been demonstrated to be exerting practically little effect when applied topically in the control, management and treatment of genital herpes.

---

*Viruses of the nature of a synctium *i.e.,* a multinucleated mass of protoplasm such as a striated muscle fiber.

Probable Questions for B. Pharm. Examinations

1. (*a*) Give brief account of viruses.

   (*b*) What are the various diseases caused by different types of viruses ?

   (*c*) Replication and transformation in viruses.

2. Classify the **'antiviral drugs'** on the basis of their **mode of action.** Give the structure, chemical name and uses of at least **one** potent drug from each category.

3. How would you synthesize the following :

   (*a*) Amantadine hydrochloride

   (*b*) Methisazone.

4. **Interferon** was recognized as an **'antiviral drug'** in 1980. Discuss its merits.

5. Give the names of **three** important drugs that specifically interfere with *viral nuclei acid replication.* Discuss the synthesis of **one** such drug selected by you.

6. Give the structure, chemical name and uses of two important **'antiviral drugs'** that affect translation on cell ribosomes. Discuss the synthesis of any **one** drug.

7. Discuss the following in details :

   (*a*) Important **'antiviral drugs'**.

   (*b*) Mode of action of **'antiviral drugs'.**

RECOMMENDED READINGS

1. EK Wagner and MJ Hewlett (eds) : *Basic Virology*, Blackwell Science, Malden MA, 1999.

2. Martindale *The Extra Pharmacopoeia* (31st edn), London, The Royal Pharmaceutical Society, London, 1996.

3. Remington's *The Science and Practice of Pharmacy,* Vol I & II, 20th edn., Lippincott Williams & Wilkins, New York, 2004.

4. MC Griffith (Ed) *USAN and the USP Dictionary of Drug Names,* Rockville, United States Pharmacopoeial Convention, Inc., (1985)

5. *Index Nominum,* Zurich, Swiss Pharmaceutical Society (1982).

6. D Lednicer and LA Mitscher *The Organic Chemistry of Drug Synthesia,* New York, John Wiley & Sons (1995).

7. JH Block and JM Beale (Jr.) (eds) : *Wilson and Gisvold's Textbook of Organic Medicinal and Pharmaceutical Chemistry,* 11th edn., Lippincott Williams & Wilkins, Baltimore, 2004.

8. DM Knipe and PM Howley (eds) : *Fundamental Virology,* 4th edn., Lippincott Williams & Wilkins, New York, 2001.

9. AJ Cann, '*Principles of Molecular Virology*', 3rd edn., Academic Press, New York, 2001.

27 | Newer Drugs for Newer Diseases

1. INTRODUCTION

The quest for **'better medicines for a better world'** is indeed an eternal process world wide to help the suffering mankind from dreadful and fatal ailments. The accelerated growth and spectacular advancement of science and research specifically during the past three decades, has not only contributed some really **'wonder drugs'** to the therapeutic armamentarium, but also paved the way for their availability internationally through approved and recognised government agencies.

The history of medicine reveals that some of the most potent compounds were made known to the world by the traditional healers and herbal practitioners belonging to Ancient Greek, Rome, China, Egypt, Tibet, India and Africa. The guardians of these ancient system of medicines made use of the medicinal plants, herbs and shrubs available locally to treat a variety of diseases ranging from mild fever to acute mental disorders. With the passage of time, such age–old but classical treatment started gaining popularity even outside the countries of origin by virtue of their startling therapeutic efficacy.

The advent of modern sophisticated technological techniques have helped the scientists of various disciplines to isolate, purify and characterize the medicinally active constituents present in a vast number of medicinal plants all over the world. Aside, scientists in the research laboratories have successfully synthesized tailor–made–biologically active prototype compounds possessing better therapeutic effects, lesser toxicity and fewer side effects. Scientific knowledge, thus generated, were skillfully communicated across the globe through scientific literatures, research journals and internet so that positive contributions might prove beneficial in a particular specialized field of interest.

In short, such scientific oriented investigations though may not help in the establishment of potent remedies from plant sources, yet the expository information of the various components along with the physical and chemical characteristics profusely stimulate ingenous ventures.

2. NEWER DRUGS

A host of **newer drugs** both from the natural origins and synthetic routes have been isolated/ synthesized, purified, characterized and schematically evaluated for various biological responses.

A number of newer drugs that have been found to be useful for the treatment of newer diseases are discussed here briefly :

(*i*) Prostaglandins and other Eicosanoids

(*ii*) Antilipedemic Drugs

(*iii*) Hormone Antagonists

(*iv*) Antimycobacterial Drugs

(*v*) Antithyroid Drugs

(*vi*) Cardiac Steroids and Related Inotropic Drugs

(*vii*) Heparin

(*viii*) Radiosensitizer

(*ix*) Cromakalim

(*x*) Drugs to Combat AIDS

2.1. Proistaglandins and other Eicosanoids

Earlier observation that human seminal fluid exerts direct muscle contractions of uterine tissues was thought to be caused by an acidic vasoactive substance produced in the prostate gland, that was subsequently named as *prostaglandin* (referred to as PG). It was revealed later on that the acidic substance contained a number of structurally similar prostaglandin products.

Prostaglandins (IGA through PGF) are a group of cyclopentane derivatives formed from 20-carbon polyunsaturated fatty acids. They exert a good number of physiologic properties. They are usually termed as **"local hormones"** because of the *two* cardinal facts, namely : *first*, they influence biologic processes near their point of release ; and *secondly*, they display mechanisms for their inactivation near the locus of release.

2.1.1. Nomenclature

Prostaglandin (PG) is based on the following hypothetical compound prostanoic acid :

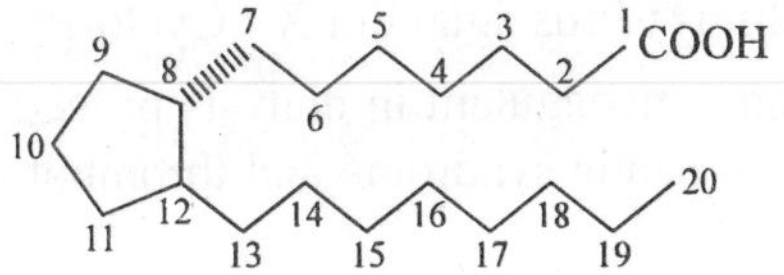

A-type compounds	:	α, β-unsaturated ketones ;
E-type compounds	:	β-hydroxyketones with *keto* moiety at C-9 and an α-OH at C-11 ;
F-type compounds	:	1, 3-diols ;
α-designation	:	Having stereochemistry of the OH at C-9, below ; and, therefore ; on the same side (*cis*) of the cyclopentane ring as the OH at C-11 ;
PGF_2	:	PG of F series (containing 2 ring OH functions) having the OH at C-9 in the α-configuration and 2 double-bonds at C-5 and C-13,
PGI_2	:	Contains a furan ring (consisting of C-6 and C-7 from the side chain and the C-9 α-oxygen) fused to the cyclopentane ring.
TXA_2	:	Origin from thrombocytes containing thromboxane A_2.
LTC_4	:	Origin from leukocytes containing leukotriene C_4.

A few typical examples are described below, namely : Aloprostadil ; Epo-prostenol ; and Misoprotol ;

A. *Alprostadil* INN, BAN, USAN,

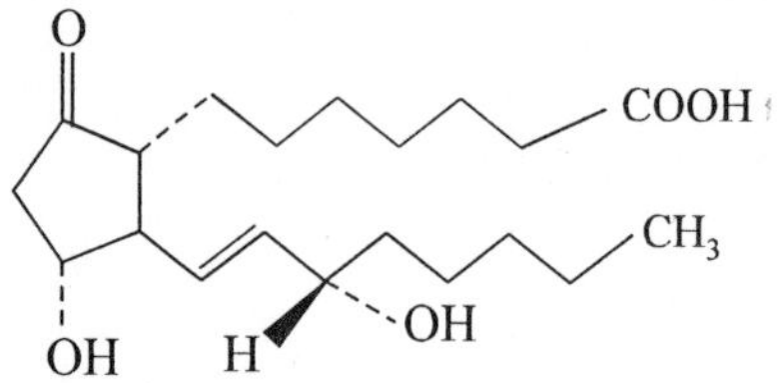

Prost-13-en-1-oic acid, 11, 15-dihydroxy-9-oxo, -(11α, 13E, 15S)- ; Prostaglandin E_1 ; PGE_1 ; Prostaglandin E_1 USP ; Prostin VR$^{(R)}$ (Upjohn) ;

It is employed temporarily to maintain potency of the *ductus arteriosus* in the management of congenital heart disease. It also finds its usefulness in peripheral vascular disease.

Dose : For congenital heart disease : by IV drip starting with doses of 100 nanograms per kg body weight per minute.

B. *Epoprostenol* INN, BAN ; *Epoprostenol Sodium* USAN ;

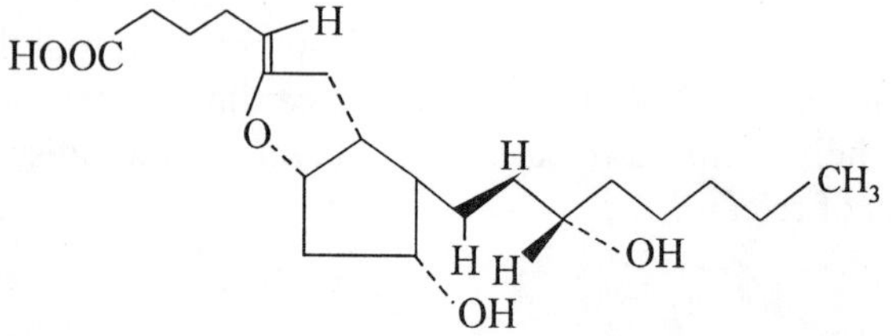

Prosta-5, 13-dien-1-oic acid, 6, 9-epoxy-11, 15-dihydroxy-, (5z, 9α, 11α, 13E, 15S)- ; PGI_2 ; PGX ; Prostacyclin ; Prostaglandin I_2 ; Prostaglandin X ; Cyclo-Prostin$^{(R)}$ (Upjohn) ;

It has been employed as an anticoagulant in dialysis procedures. It has also been used in pre-eclampsia, and in the haemolytic–uraemic syndrome and thrombotic thrombocytopenic purpura.

C. *Misoprostol* INN, BAN, USAN,

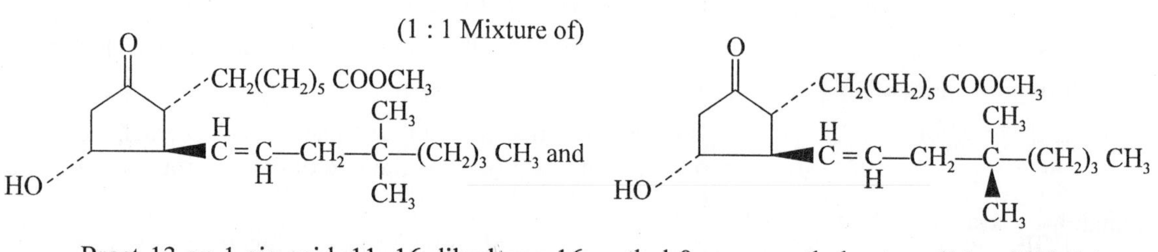

(1 : 1 Mixture of)

Prost-13-en-1-oic acid, 11, 16-dihydroxy-16-methyl-9-oxo-, methyl ester ; (11α, 13E)-(±)- ; Cytotec$^{(R)}$ (Searle) ;

It is used orally as a potent gastric antisecretory and gastroprotective agent.

Dose : Oral, 100-200 mcg 4 times daily to check gastric ulceration in susceptible individuals who are taking NSAIDS.

2.1.2. Mechanism of Action

The mechanism of action of the drugs described under section 2.1.1 shall be dealt with appropriately in the pages that follows :

2.1.2.1. Alprostadil

The '*drug*' helps to *maintain the potency of the ductus arteriosus of the faetus*. It has been observed that soon after birth, prostaglandin production falls and ductus get closed. Nevertheless, in the events involving congenital heart defects, for instance : tetralogy of Fallot, transposition of the great vessels, pulmonary atresia, pulmonary stenosis, tricuspid atresia, or imperfect artic arch, coarctation of the aorta, it is almost a primary and necessary urgent requirement that the ductus should remain patent unless and until corrective surgical measures may be carried out effectively. Therefore, in such instances, timely infusion of alprostadil (PGE) definitely helps in maintaining patency pending surgical manipulations.

The '*drug*' is also employed for treating erectile dysfunction by injection right into corpora cavernosa of the penis (Edex) or alternatively by direct insertion of a suppository into the urethra (MUSE). It exerts its action by relaxation of trabecular smooth muscle and also by reasonable dilatation of the cavernous arteries.

2.1.2.2. Epoprostenol Sodium

The '*drug*' causes vasodilation and prevents platelet aggregation. The endogenous substance is known as **prostacyclin,** which is a product of arachidonic acid metabolism having a very shot half-life. Soonafter IV infusion it gets hydrolyzed rapidly to the more stable but much less active **6-keto-prostaglandin $F_{1\alpha}$ (6-oxo-prostaglandin $F_{1\alpha}$).** Contrary to host of other (prostaglandins, this '*drug*' is not inactivated in the pulmonary circulation.

2.1.2.3. Misoprostol

The '*drug*' not only inhibits gastric acid secretion but also enhances mucosal resistance appreciably. It is believed that the '*drug*' essentially sustains as well as derives its therapeutic supremacy particularly in the GI tract by enhancing duly mucous and bicarbonate secretion by the gastric epithelium by augmenting epithelial regeneration ; besides by increasing specifically the mucosal blood flow thereby enhancing the mucosal protection to a great extent.

Misoprostol is rapidly (T_{max}, 12 minutes) and largely absorbed. It has a terminal half-life ranging between 20-40 minutes, with 80% almost excreted in the urine.

Note. It does not prevent the therapeutic benefit of NSAIDs specifically in the treatment of rheumatoid arthritis.

2.1.3. Future Developments

A plethora of prostaglandin analogoes are presently under active investigation to combat various human ailments. Some specific areas of common interest being gastroprotection in antiulcer therapy, management and control of coronary artery or cerebrovascular diseases, control of facility and above all the progress for antiasthamatics. Likewise, eiscosanoids offer a bright prospect for the treatment of immune system disorders and hypertension. It is very much sure that the application of eicosanoids as potent therapeutic drugs in the near future shall dominate the search for newer drugs for newer diseases.

2.2. Antilipemic Drugs

According to a current medical dictionary antiatherosclerotics may be defined as '*a form of simple intimal arteriosclerosis with atheromatous deposits within and beneath the intima*'.

In fact, due recognition and efforts for the management and treatment of atherosclerosis and the evolution of antilipedemic drugs dates back to a couple of centuries. Previously it was believed that fatty deposition in walls of arteries mostly take–place either in middle or in old ages, but now it has been observed that it may happen at any age whether child or youngesters.

Many scientists have laid primary stress of such abnormalities on blood serum contents, for instance : cholesterol, triglycerides, phopholipids and free fatty acids (FFA). Recent studies have revealed that patients with elevated blood-cholesterol generally have the plasma cholesterol bound in the form of β-lipoproteins which becomes higher than in 'normal individual'. These β-lipoproteins have a higher molecular weight and a higher proportion of cholesterol that the corresponding α-lipoproteins that are higher in 'normal blood'.

It is, however, pertinent to mention here that every individual suffers from atherosclerosis to a certain extent, but its incidence is more abundant in affluent countries. Perhaps the various general factors that contribute towards its frequent occurrence are, namely : emotional stress, excessive smoking, hypertension, unbalanced diet and obesity, lack of endurance type physical activity and above all sustained high–serum–cholesterol levels. More recently some specific factors are found to be responsible for this ailment, such as-immunologic and autonomic factors, coagulation and blood flow, genetic make–up and most importantly endocrinologic aberration.

Developed countries and affluent nations are footing a big chunk of their resources in their health–care–systems to combat coronary artery disease (CAD) and corinary heart disease (CHD).

A few important antilipidemic drugs which have gained recognition by virtue of their clinical significance are described here, for instance : Theofibrate ; Probucol ; and Gemfibrozil ;

A. *Theofibrate* USAN : *Etofyline Clofibrate* INN ;

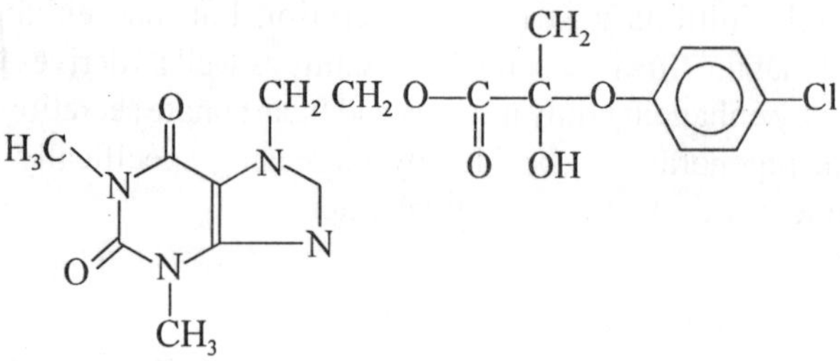

Propionic acid, 2-(4-chlorophenoxy)-2-methyl-, 2-(1, 2, 3, 6-tetrahydro-1, 3-dimethyl-2, 6-dioxo-7H-purin-7-yl) ethyl ester ;

Duolip[(R)] (L.Merckle, Germany) ;

It has been used in the treatment of hyperlipidaemias.

B. *Probucol* INN, BAN, USAN,

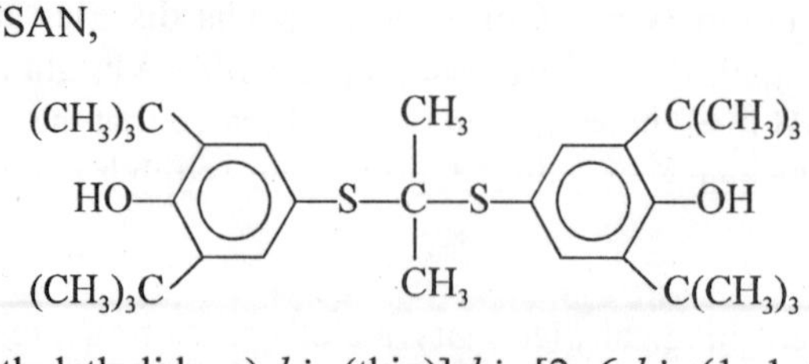

Phenol-4, 4'-[(1-methylethylidene) *bis* (thio)] *bis* [2, 6-*bis* (1, 1-dimethylethyl)]- ; Lorelco[(R)] (Merrell Dow) ;

It is employed as an adjunct to diet, to reduce elevated serum-cholesterol concentrations, particularly in type II, hyperlipoproteinamia.

Dose : Usual, 500 mg with meals, morning and evening.

C. *Gemfibrozil* INN, USAN,

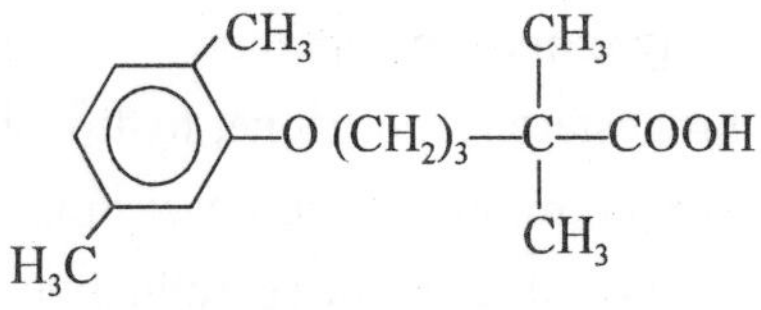

Pentanoic acid, 5-(2, 5-dimethylphenoxy)-2, 2-dimethyl)- ;

USP ; Lopid[(R)] (Parke–Davis) ;

It is used in the treatment of hyperlipidaemia.

Dose : Usual, 0.8 to 1.2 g per day in divided doses.

2.2.1. Mechanism of Action

The mechanism of action of various medicinal compounds discussed under section 2.2 shall be treated separately in the sections that follows :

2.2.1.1. Theofibrate

The '*drug*', being a structural analogue of *fibric acid*, acts on the lipoprotein metabolism. It also reduces elevated plasma concentration of cholesterol to a much lesser extent, but the overall effect is variable in nature. Its mechanism of action is not quite explicite and clear. It may also cause *regression of xanthomas.*

2.2.1.2. Probucol

The '*drug*', which is a highly lipid soluble sulphur containing *bis*-phenol, and is quite capable of minimizing LDL-cholesterol even upto 20% without causing any significant difference in the triglyceride levels. It has been amply-proved that it has the inherent double-action on enhancing bile acid secretion in one hand while on the other increasing the degradation of LDL-apo β. Besides, probucol also lowers HDL as a component of the overall cholesterol reduction.

2.2.1.3. Gemfibrozil

The '*drug*' happens to be a structural analogue of **clofibric acid.** Contrary to *clofibrate* it helps to raise HDL levels but at the same time triglyceride levels are reduced appreciably. Interestingly, the '*drug*' is invariably employed in *diet-refractory hypertriglyceridemia.* *

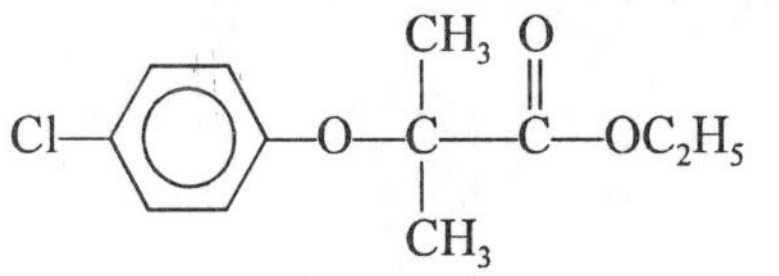

Clofibrate [Atromid-S]

---

*An increased blood triglyceride level.

2.3.   Hormone Antagonists

Hormones may be defined as—'*substances secreted by the endocrine, or ductless glands that essentially serve to integrate various metabolic processes*'.

It is interesting to observe that hormones do represent a widely diverse category of compounds, for instance ;

Amino acid derivatives *e.g.,* thyroxine, epinephnine ;

Steroids-*e.g.,* testosterone, progesterone, cortisone, hydrocortisone ;

Polypeptides/proteins-*e.g.,* corticotropin, calcitonin, insulin ;

While, hormones are solely responsible for the reproductive system, they are also the causative substances for the growth and development of cancers related to breast, prostate and uterine.

A few typical examples of hormone antagonists are discussed below.

2.3.1. Antiestrogens

Antiestrongens are such compounds which block the oestrogen activity. Compounds belonging to this category are essentially the structural analogues of the oestrogen triphenylethylene.

Example : Tamoxifen citrate ; Nitromifene citrate ;

A. *Tamoxifen Citrate* USAN ; *Tamoxifene* BAN ;

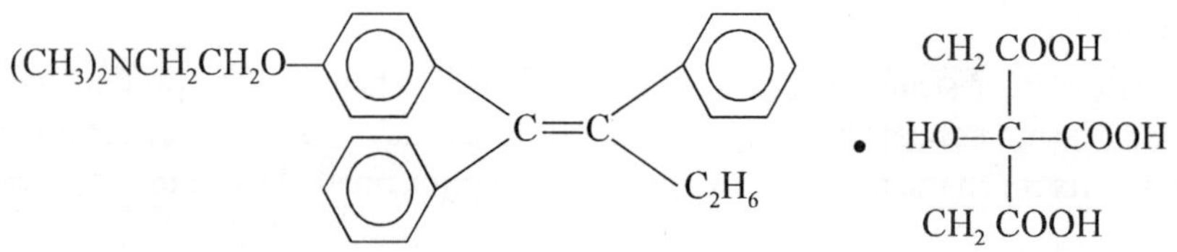

(Z)-2-[*p*-(1, 2-Diphenyl-1-butenyl) phenoxy]-N, N-dimethylethylamine citrate (1 : 1) ; Tamoxifen Citrate USP ; Nolvadex[R] (Sturat) ;

It is employed as an alternative to androgens and oestrogens in the management of breast cancer. It is also used to stimulate ovulation in infertility.

Dose : For breast cancer : oral, 10 to 20 mg of tamoxifen twice daily (or 20 to 40 mg daily) ; For stimulating ovulation : usual, 10 mg of tamoxifen two times per day on day 2, 3, 4 and 5 of the menstrual cycle, alternatively : single daily doses of 20 to 80 mg may be employed on the same days.

B. *Nitromifene Citrate* USAN

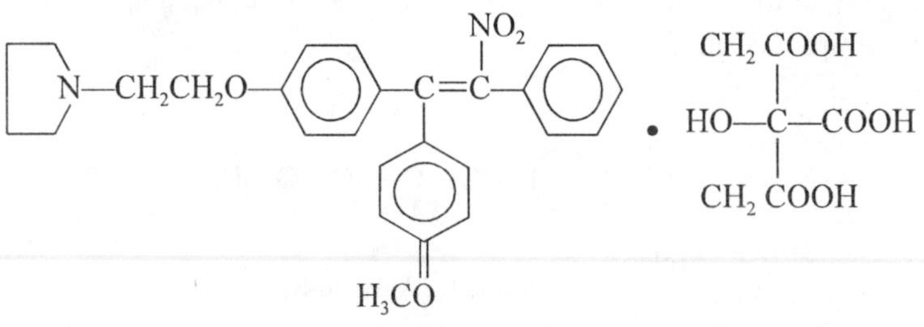

1-[2-[*p*-[α-(*p*-Methoxyphenyl)-β-nitrostyryl] phenoxyl] ethyl] pyrrolidine citrate (1 : 1) ; CN 5518[R] (Parke-Davis) ;

Nitromifene gets converted to its corresponding phenolic metabolite (methoxy moiety changes to phenolic OH group) which displays a high affinity for the oestrogen receptor.

2.3.1.1. Mechanism of Action

The mechanism of action of drugs described under section 2.3.1 shall be treated individually as follows :

2.3.1.1.1. Tamoxifen citrate

The '*drug*' is found to exert its action by competing with estrogens for the **cytosol estrogen receptors**, which eventually affords more or less complete blockade of the ensuing estrogen effect in the '*target tissue*'. However, tumours having essentially *negative receptor assays* do not respond to it.

Oncologists make use of this specific antiestrogen, tamoxifen, in depriving the malignant process of the source of these hormones, therefore, shows a much better, safer and potentially useful method of treatment.

The bioavailability *via* the oral administration of this '*drug*' ranges between 25–100%. The half-life of a single dose stands at 18 hours, but it is only 7 hours at a steady state.

2.3.1.1.2. Nitromifene Citrate

The '*drug*' exerts its action after undergoing demethylation to the corresponding phenol.

2.3.2. Antiandrogens

Antiandrogens normally prevents the binding of dihydrotestosterone to androgen receptors particularly in target tissues. They may be further classified into *two* categories, namely :

(*a*) Non-steroidal Antiandrogens

(*b*) Steroidal Antiandrogens

which shall be dealt separately as under with specific examples.

2.3.2.1. Nonsteroidal Antiandrogens

These compounds do not possess a steroidal nucleus and androgenic properties as such, but their metabolites exhibit antiandrogenic properties. Example : Flutamide ;

A. *Flutamide* INN, BAN, USAN,

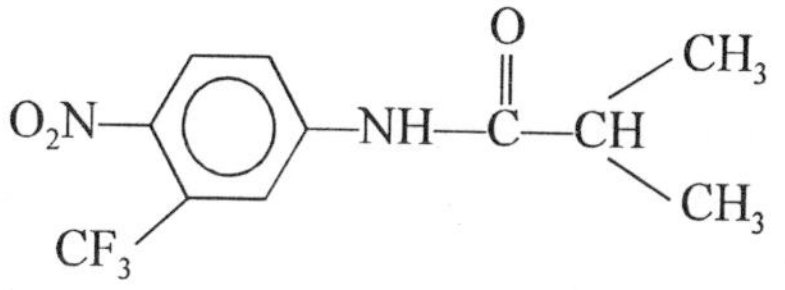

Propionamide, 2-methyl-N-[4-nitro-3-(trifluoromethyl) phenyl]- ;

Sebatrol[R] (Schering–Plough) ;

It has been used to improve urine flow in benign prostatic enlargement.

Dose : 300 mg per day.

Mechanism of Action. Flutamide being an orally active and potent competitive inhibitor of specific **nuclear androgen receptors** in target tissues *e.g.,* seminal vesicles, adrenal cortex and prostate.

Its pharmacological action is mainly on account of its major metabolite, *2-hydroxyflutamide,* as given below :

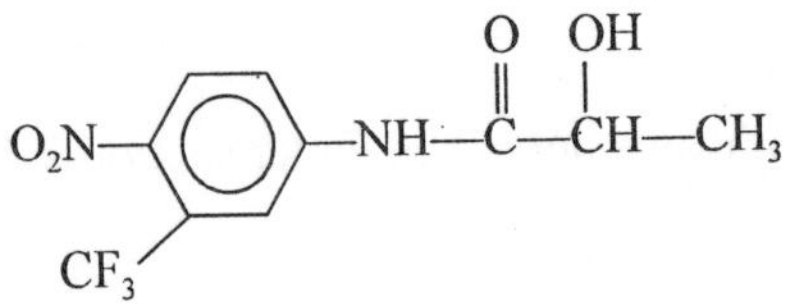

2-Hydroxyflutamide (major metabolite)

It has been observed that nearly 50% of the '*drug*' gets eliminated in the urine within a span of 3 days. Interestingly, the *hydroxylated metabolite* has a half-life which ranges between 6-22 hours depending on the dosage administered.

Caution. The 'drug' causes a high incidence of gynecomastia* and GI-discomfort to a certain extent.

2.3.2.2. Steroidal Antiandrogens

In early sixties cyproternone acetate was first discovered to possess antiandrogen activity. A few other compounds were found to exhibit similar properties.

Example. Chlormadinone Acetate ;

A. *Chlormadinone* INN, BAN ; *Chlormadinone Acetate* USAN

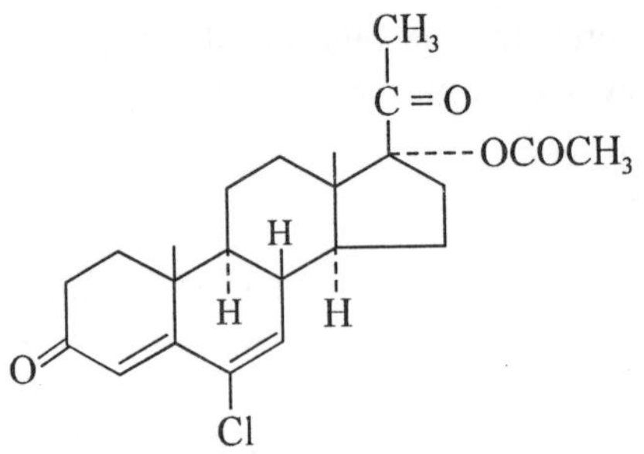

Pregna-4, 6-diene-3, 20-dione, 17-(acetyloxy)-6-chloro- ; Chlormadinone acetate BP (1968) ; Progestin(R) (Syntex) ;

It is used in the treatment of functional uterine bleeding. It also exerts very slight oestrogenic activity.

Mechanism of Action. Chlormadinone acetate is an orally active progestogen having distinct and potent antiandrogenic activity, which has been used in combination as an oral contraceptive. It serves as a hormonal antineoplastic agent.

2.3.3. Aldosterone Antagonists

The mineralocorticoid-aldosterone essentially monitors the electrolyte balance in the body by enhancing the excretion of K^+ and the retention of Na^+. Thus, aldosterone antagonists are usually employed for the effective treatment of edematous ailments and hypertension.

Example : Spironolactone ;

---

*Enlargement of breast tissue in the male.

A. *Spironolactone* INN, BAN,

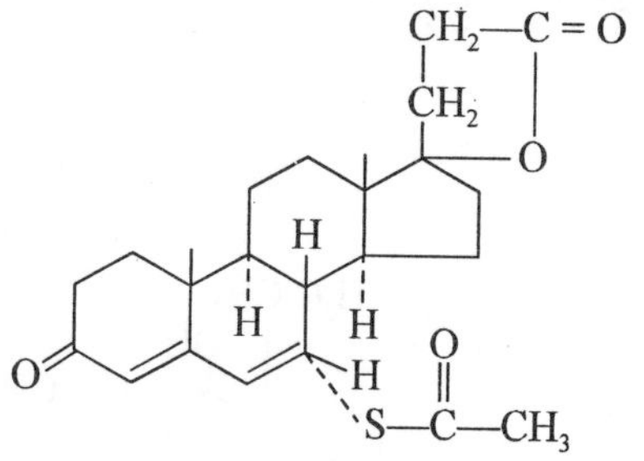

17-Hydroxy-7α-mercapto-3-oxo-17α-pregn-4-ene-21-carboxylic acid, γ-lactone acetate ; BP (1973) ; USP ; Aldactone$^{(R)}$ (Searle) ;

It is employed in the treatment of refractory oedema associated with congestive heart-failure, cirrhosis of the liver or the nephrotic syndrome. It is frequently administered along with diuretics like thiazides and frusemide whereby it adds to their natriuretic but retards thir kaliuretic effects, hence conserving potassium.

Dose : Usual, initial, 100 mg per day in divided doses ; For children : 3 mg per kg body weight per day, in divided doses.

Mechanism of Action. Spironolactone particularly competes with aldosterone at its receptor sites, that eventually triger the synthesis of the prevailing enzyme(s) which predominantly catalyze Na^+ transport when stimulated duly. Therefore, it predominantly **reverses** these electrolyte alterations by specifically causing blockade in the renal tubular action of the hormone. Thus, by initiating the inhibition of Na^+ reabsorption the drug produces significant diuresis and thereby minimizes K^+ excretion.

It has been well established that the '*drug*' exerts its action by reasonably blocking the sodium–retaining effects of aldosterone upon the distal convoluted tubule, thereby it corrects meticulously one of the most vital mechanisms solely responsible for the production of edema, but it may be noted with great emphasis that **spironolactone** is effective only in the presence of **aldosterone** specifically.

The '*drug*' gets metabolized rapidly after oral administration. However, the metabolites are usually excreted mostly in the urine and also in the bile to a certain extent. Interestingly, the most vital metabolite, **canrenone,** attains the peak plasma levels within a short span of 2–4 hours after oral administration of the '*drug*'.

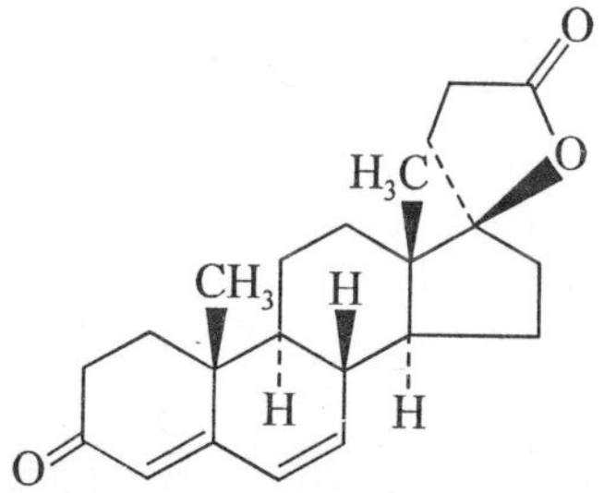

Canrenone

2.3.4. Antiprogestational Steroids

Antiprogestational steroid gained significant interest by virtue of their potential as an antifertility agent.

Example : Cestrinone ;

A. *Cestrinone* INN, USAN,

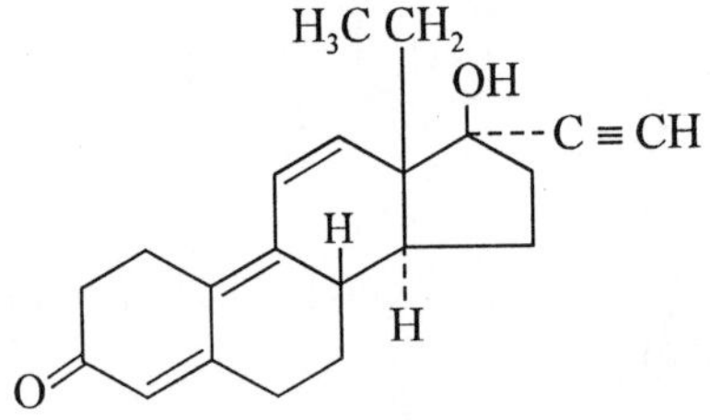

13-Ethyl-17-hydroxy-18, 19-dinor-17α-pregna-4, 9, 11-trien-20 yn-3-one ; RU-2323[R] (Roussel) ;

It has been employed as a contraceptive. It has also been used in male subjects for the suppression for spermatogenesis.

Summary

The various hormone antagonists not only help in giving a vivid picture of hormonal control mechanisms but also serve as immensely viable tool for the management and control of hormone–dependent cancer. This has, in fact, generated enough interest towards the development of newer and altogether safer hormone antagonists.

2.4. Antithyroid Drugs

In the morbid state produced by excessive secretion of the thyroid gland *i.e.,* hyperthyroidism, the only remedial measure is surgery. However, for pre-surgery treatment the patient must be administered with antithyroid drugs to abolish hyperthyroidism to a considerable extent.

Thiourea was initially found to exhibit an antithyroid activity but had to be abandoned due to their high toxicity. Later on, it was revealed that similar activity was shown by 2-thiouracil derivatives and 4-keto-2-thiopyrimidines ; more precisely 6-alkyl-2-2 thiouracils and their congeners display meaningful clinical characteristics.

A few typical examples are described here, namely, Propylthiouracil ; Methimazole ;

A. *Propylthiouracil* INN ;

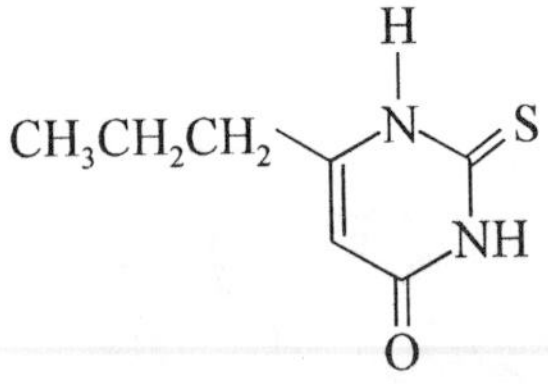

6-Propyl-2-thiouracil ; BP ; USP ; Int., IP ; Tietil[R] (Pharmacia, Sweden) ;

Synthesis

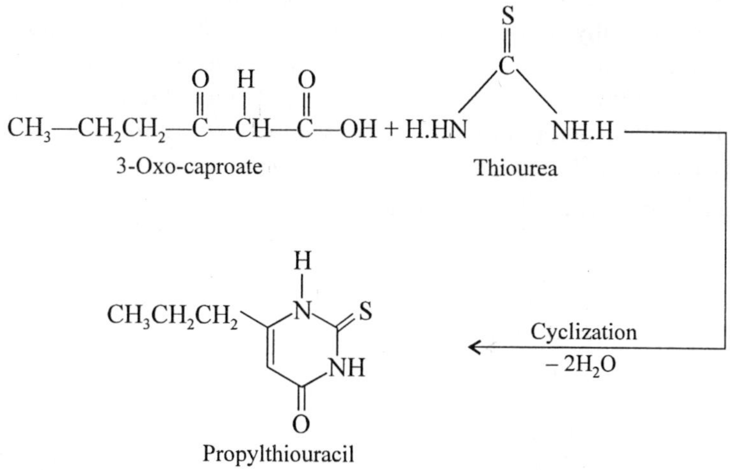

3-Oxo-caproate

Thiourea

Propylthiouracil

Cyclization
– 2H$_2$O

It is prepared by the condensation (cyclization) of 3-oxo-caproate with thiourea and elimination of two mole of water.

Polythiouracil is used in the preparation of the hyperthyroid patient for surgery. It is also employed in the complete management and treatment of hyperthyroidism spread over a period ranging from 6 months to 3 years.

Dose : For hyperthyroidism in adults, initially 200 to 300 mg per day in 3 divided doses ; when the patient attains normal basal metabolic rate (euthyroidism) the dose is usually reduced to a maintenance dose fo 50 to 75 mg daily in 2 to 3 divided doses. In children : over 10 years old, initial, 150 to 300 mg per day in 4 divided doses until the child becomes euthyroid, then usually 100 mg daily in 2 divided doses, for maintenance.

B. *Thiamazole* INN, *Methimazole* BAN, USAN ;

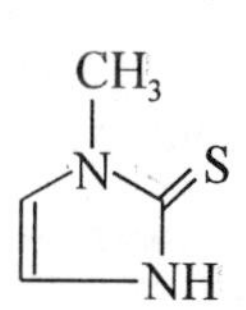

1-Methylimidazole-2-thiol ; Methimazole USP ; Tapazele$^{(R)}$ (Lilly) ;

It is an antithyroid substance that retards the formation of thyroid hormone. It acts by reducing the formation of iodotyrosines and, therefore, of thyroxine tri-iodothyroxine. It is also used in the preparation of patients for subtotal thyroidectomy.

Dose : Usual, initial, 5 to 20 mg every 8 Hrs.-when condition gets stabilized (1-2 months)-the dose is reduced to a maintenance dose of 5 to 15 mg per day ; For children : initial, 400 mcg per kg body weight per day in divided doses.

2.4.1. Mechanism of Action

The mechanism of action of the medicinal compounds discussed under section 2.4 shall now be dealt with individually as follows :

2.4.1.1. Propylthiouracil

The '*drug*' fails to cause interference with the release or usage of accumulated thyroid hormone ; and the time-gap that essentially elapses between the very initial stage of medication together with the manifestations of its antithyroid activity entirely depends upon the quantum of thyroid hormone present in the gland (thyroid). The resulting marked and pronounced *hyperplasia** of the thyroid gland which follows soonafter its administration, in fact, is a consequence of a **compensatory enhancement of thyroprotein release** as a result in the *thyroid hormone titer value of the blood.*

Propylthiouracil undergoes *tautomerism* as follows :

Propylthiouracil
(*enol*-form)

Propylthiouracil
(*keto*-form)

Because the thiol group (SH) does not invariably occur in medicinal compounds, S-glucuronide products have been duly reported for only a few drugs *e.g., propylthiouracil***, undergo conjugation with *glucuronic acid.*

The '*drug*' gets absorbed by the oral route to the extent of 75%. As the drug lowers the metabolic rate ; therefore, the dosage regimen must be regulated accordingly so as to avoid accumulation as far as possible.

2.4.1.2. Methimazole

The '*drug*' is found to be almost 10 fold as potent as propylthiouracil, besides being more prompt in eliciting an antithyroid response. It also distinctly exhibits a much more prolonged action in comparison to propylthiouracil. The plasma half-life varies between 6—8.5 hours in *hyperthyroid* patients, but 8—18 hours in *hypothyroid* ones ; and, therefore, as the '*drug*' decreases the metabolic rate, its own metabolism gets slowed appreciably which may ultimately result into **'accumulation'** unless and until the dosage is readjusted accordingly.

2.5. Antimycobacterial Drugs

There are two dreadful diseases produced by the species *Mycobacterium,* namely ; tuberculosis caused by the organisms *Mycobacterium tuberculosis* and leprosy produced by *Mycobacterium leprae.* Unfortunately both these diseases have afflicted the mankind throughout the recorded history.

In this context, the *two* categories of antimycobacterial drugs shall be discussed individually as under :

(*a*) Antitubercular Drugs

(*b*) Antileprotic Drugs

---

*Excessive proliferation of normal cells in the normal tissue arrangement of an organ. (**SYN** : *Hypergenesis*).

Lindsay RH *et al. Pharmacologit* **18 : 113, 1976.

2.5.1. Antitubercular Drugs

The first ever breakthrough in antitubercular chemotherapy took place in the year 1938 with the historical fact that sulphanilamide exhibited week bacteriostatic properties. This observation triggered off the extensive and intensive research towards the synthesis of a number of antitubercular agents which was subsequently followed by certain antitubercular antibiotics.

A few such important antitubercular drugs shall be discussed here, namely : Ethambutol ; Isoniazid ; Ethionamide ; Streptomycin ; Capreomycin Sulphate ; Rifampicin ;

A. *Ethambutel* INN, *Ethambutolum* BAN *; Ethambutol Hydrochloride* USAN ;

1-Butanol, 2, 2′-(1, 2-ethanediyldiamino) *bis*-, dihydrochloride, [S–(R*, R*)] ; Ethambutol Hydrochloride BP ; USP ; Myambutol$^{(R)}$ (Lederle) ;

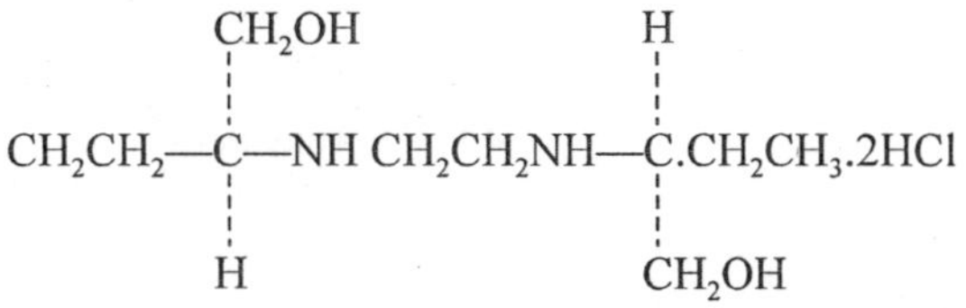

$$\begin{array}{ccc}
CH_2OH & & H \\
| & & | \\
CH_2CH_2-C-NH\ CH_2CH_2NH-C.CH_2CH_3.2HCl \\
| & & | \\
H & & CH_2OH
\end{array}$$

Synthesis

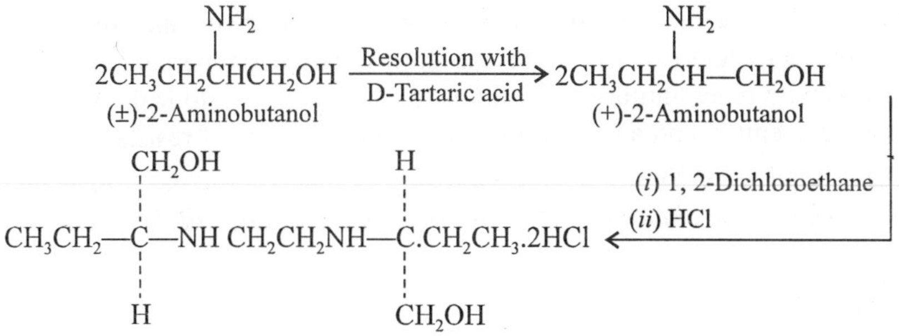

$$\begin{array}{cc}
NH_2 & NH_2 \\
| & \xrightarrow[\text{D-Tartaric acid}]{\text{Resolution with}} \quad | \\
2CH_3CH_2CHCH_2OH & 2CH_3CH_2CH-CH_2OH \\
(\pm)\text{-2-Aminobutanol} & (+)\text{-2-Aminobutanol}
\end{array}$$

$$\begin{array}{ccc}
CH_2OH & & H \\
| & & | \\
CH_3CH_2-C-NH\ CH_2CH_2NH-C.CH_2CH_3.2HCl & \xleftarrow{} \\
| & & | \\
H & & CH_2OH
\end{array}$$

(*i*) 1, 2-Dichloroethane
(*ii*) HCl

Ethambutol Hydrochloride

It is prepared by first resolving (±)-2-aminobutanol *via* its tartrate and the (+)-enantiomorph is condensed with 1, 2-dichloroethane in a suitable dehydro-chlorinating atmosphere. The resulting ethambutol is dissolved in an appropriate solvent and treated with hydrochloric acid to obtain the official compound.

It may be used alone to clear the sputum of mycobacteria within a span of 12 weeks in many cases, but bacterial resistance usually takes place in 35% of cases, and thus leading to frequent relapses. Therefore, it is employed in combination with isoniazid, pyrazinamide, cycloserine or ethionamide-such relapses are uncommon.

Dose : 15 to 25 mg/kg once a day ; low dose for new cases and high dose for use in patients that have had previous antitubercular therapy.

B. *Isoniazid* INN, USAN ; *Isoniazide* BAN ;

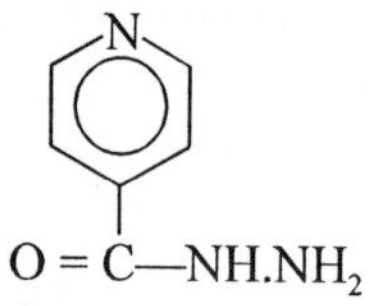

4-Pyridinecarboxylic acid, hydrazide ;

BP ; USP ; Eur. P. ; Int. P, IP ;

Continazin$^{(R)}$ (Pfizer) ; Dimacrin$^{(R)}$ (Winthrop) ; INH$^{(R)}$ (Ciba—Geigy) ; Nydrazid$^{(R)}$ (Squibb) ;

Synthesis

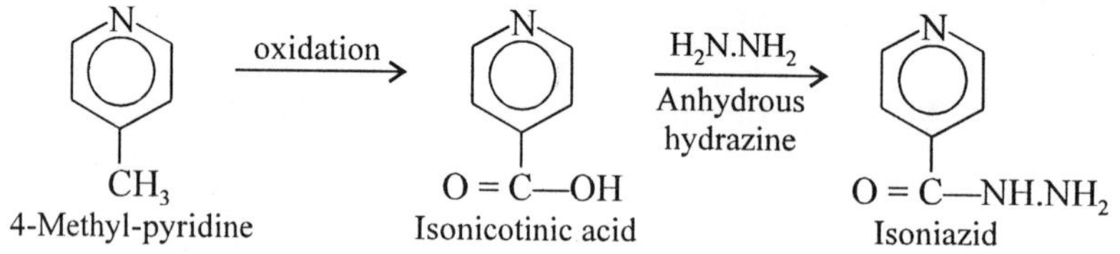

It is prepared by first carrying out the oxidation of 4-methylpyridine to obtain isonicotinic acid which upon heating with anhydrous hydrazine yields the desired compound.

It is considered to be the drug of choice for the treatment of tuberculosis. It has also been employed as a prophylactic for those who were constantly exposed to tubercular patients. It is invariably used in combination with other antitubercular drugs to achieve better clinical response, to allow lower doses of other active agent(s), and above all to retard the emergence of resistant tubercle bacilli.

Dose : Adult, oral or im, for active tuberculosis, 5 to 7 mg/kg/day singly or in divided doses ; for prophylaxis-300 mg daily ; For children : treatment, 10 to 30 mg/kg day in 2 divided doses, as prophylaxis 10 mg/kg day.

C. *Ethionamide* INN, BAN, USAN,

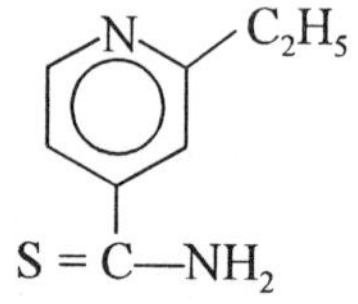

4-Pyridinecarbothioamide, 2-ethyl- ;

BP ; USP ; Eur. P ; IP ;

Tractor-SC$^{(R)}$ (Ives) ; Iridocin$^{(R)}$ (Bayer) ; Trescatyr$^{(R)}$ (May & Baker) ;

Synthesis

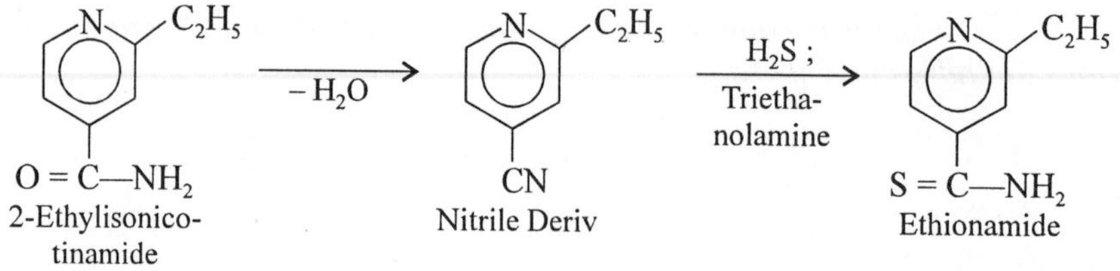

It is prepared by dehydrating 2-ethylisonicotinamide to the corresponding nitrile analogue which is then reacted with hydrogen sulfide in the presence of triethanolamine to afford the official compound.

It is used only when the usual combination of PAS, streptomycin and INH are either intolerable or ineffective.

Dose : 500 mg to 1 g per day in 3 or 4 divided doses, to be administered with meals.

Besides, these synthetic compounds discussed above, a number of *antitubercular antibiotics* have also gained significant recognition over the past few decades. A few typical members of this particular category of compounds are described below :

Example : Streptomycin ;

D. *Streptomycin* INN, *Sterptomycin Sulphate* BAN, *Sterptomycin Sulfate* USAN ;

It has been described under the chapter on **'Antibiotics'**.

E. *Capreomycin* INN, BAN ; *Capreomycin Sulfate* USAN ;

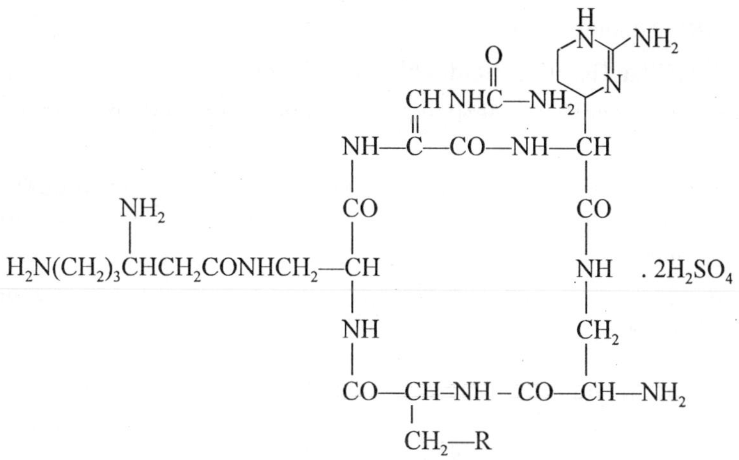

Capreomycin IA : R = OH ;
Capreomycin IB : R = H ;

Capreomycin Sulphate ;

Capreomycin Sulphate BP ; Capreomycin Sulfate USP ;

Capstat Sulfate[(R)] (Lilly) ;

It is a strongly basic antibiotic having a cyclic peptide structure isolated from *Streptomyces capreolus.* Out of the four capreomycins deignated as IA, IB, IIa and IIb, only IA and IB are found to be clinically useful.

Capreomycin resembles viomycin both chemically and pharmacologically. It is considered to be second-line antitubercular drug used in combination with other such agents. It is frequently used in place of streptomycin when either the patient is sensitive to it or the strain of *M. tuberculosis* is resistant to it.

Dose : Administered by deep IM injection, usual, daily : 1 million units, equivalent to about 1 g of capreomycin with a maximum of 20,000 units (20 mg) per kg body weight.

F. *Refampicin* INN, BAN ; *Rifampin* USAN ;

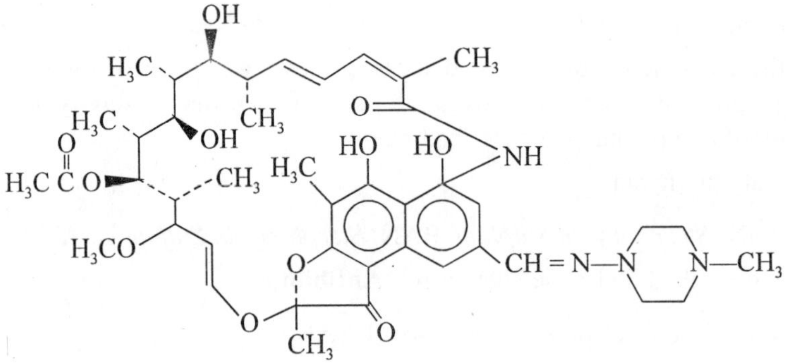

Rifamycin, 4-0-[2-(diethylamino)-2-oxoethyl]- ;

Rifampicin BP, Rifampin USP ;

Rifmactane[(R)] (Ciba–Geigy) ; Rifadin[(R)] (Marrell Dow) ;

Rifampicin-a semisynthetic derivative of naphthalene was produced by *Streptomyces mediterranei.*

It is used solely in the treatment of tuberculosis. It is interesting to observe that the rate of development of resistance of the mycobacterium is low. However, it is always employed in combination with other antitubercular drugs.

Dose : Oral, 600 mg daily m 1 to 3 divided, taken 1 hour be used in combination with at least one other antitubercular drug.

2.5.1.1. Mechanism of Action

The mechanism of action of the drug substances discussed under section 2.5.1 shall now be dealt with separately in the sections that follows :

2.5.1.1.1. Ethambutol Hydrochloride

The '*drug*' exerts its action against tubercle bacilli resistant to either isoniazid or streptomycin. It has been observed to act specifically upon the proliferating cells, evidently by causing interference with the synthesis of RNA. Importantly, when the '*drug*' is employed singly for the treatment of tuberculosis, it may help in clearing the sputum of myobacteria within a span of three months in most of the subjects ; however, bacterial resistance usually takes place in nearly 35% of cases, and relapses occur frequently. Certainly the combination with either ioniazid or other prevalent tuberculostatic drugs, the incidence of relapses occur seldomly.

The oral bioavailability ranges between 75–80%. It gets well distributed into most tissues and fluids but definitely less in CSF. The volume of distribution stands at 1.6 mL.g^{-1}. As much as 80% of the '*drug*' gets eliminated in the wine. The half-life is 3-4 hours but may extend upto 8 hours in renal failure.

2.5.1.1.2. Isoniazid

The '*drug*' exerts its action as a **tuberculocidal** particularly to the **growing organisms** (*tubercle bacilli*) and considered to be the most effective agent in the therapy of tuberculosis, and **not on the resting organisms.** It has since been well established that the '*drug*' gains an easy access to **all organs** and to **all body fluids**, including CSF, thereby rendering it of extremely special status and value in the management and treatment of tuberculosis, meningitis together with other diseases related to extrapulmonary manifestations.

Interestingly, the manner by which the '*drug*' acts as bactericidal, may be explained by the fact that it causes the bacilli to lose lipid component by a still not fully elucidated mechanism. However, the most widely accepted and reasonably justified theory suggests that the principal effect of isoniazid is accomplished *via* **inhibition of the synthesis of mycotic acids.***

It has been observed critically that a **mycobacterial-catalase-peroxidase enzyme complex** is necessarily required for the bioactivation of isoniazid.** Consequently, a reactive species, usually produced *via* the action of these enzymes on the drug is supposed invariably to attack a *very specific enzyme* needed urgently for carrying out the *mycolic acid synthesis in mycobacteria.*** Furthermore, the observed resistance to INH, believed to vary between 25—50% of the '*clinical isolate*' of the **INH-resistant strains**, is intimately associated with the apparent loss of catalase and peroxidase activities, both of which are legitimately encoded by a single gene, **kat G.**** Recently, the actual predicted target for the action of INH has been duly recognized and identified as an *enzyme* which catalyze the NADH-specific reduction of 2-*trans*-**enolyacyl** carrier protein, which is otherwise proved to be an essential step in the fatty acid elongation *i.e.,* lengthening the carbon-chain ; and subsequently the aforesaid '*enzyme*' is encoded adequately by a very specific gene, *inh A*, present in *M. tuberculosis.*****

Isoniazid is chiefly acetylated by the liver, and th rate of acetylation varies appreciably. The half-life in '*fast-acetylators*' ranges between 1—1.5 hours ; and in relatively slow ones, it varies between 2-5 hours.

Note. IM injection of INH may cause local irritation.

2.5.1.1.3. Ethionamide

The '*drug*' is regarded to be a **secondary drug** employed usually for the treatment of tuberculosis. It gets absorbed rather rapidly and completely soonafter oral administration. Importantly, it gets largely distributed throughout the body and metabolized extensively into the **inactive forms** predominantly which are ultimately excreted through the urine. It is found that almost 1% of the '*drug*' appears in the urine in an unchanged form.

SAR of Ethionamide : The *two* structural modifications to the corresponding INH-series, namely :

(*a*) isosteric replacement of the carbonyl ($-\overset{\overset{\text{O}}{\|}}{\text{C}}-$) function in INH with ($-\overset{\overset{\text{S}}{\|}}{\text{C}}-$) ; and (*b*) 2-ethyl

---

*Quomard A *et al. Antimicrob. Agents Chemother*, **35** : 1035, 1991.

Youatt J *et al. Am. Rev. Respir, Dis.,* **100, 25, 1969.

***Johnsson K *et al. J Am. Chem. Soc.,* **116** : 7425, 1994.

****Zhang Y *et al., Nature,* **358,** 591, 1992.

*****Dessen A *et al. Science,* **267,** 1638, 1995 ; Benerjee A *et al.Science,* **263,** 227, 1994.

substitution, increases antituberculostatic activity in the thioisonicotinamide series to a considerable extent.

2.5.1.1.4. Streptomycin Sulfate

It has already been discussed under **Chapter 23** section 4A.

2.5.1.1.5. Rifampin

The '*drug*' exerts its predominant activity against two vital pathogenic organisms *viz.*, *Mycobacterium tuberculosis* and *Mycobacterium leprae*. Bearing in mind the glaring fact that the rate of development of resistance of the causative organism, mycobacterium, is rather at a low ebb, it is invariably employed in combination with other antitubercular drugs.

Rifamycin is found to afford induction of the specific **hepatic drug-metabolizing enzyme system** ; and, therefore, accelerates the metabolism of digitoxin, methadone, phenytoin, β-blockers, verapamil, oral contraceptives, chloramphenicol, theophylline, besides—oral anticoagulants, estrogens, tolbutamide, barbiturates and itself.

The '*drug*' has proved to be *teratogenic* in laboratory animals ; and, hence, must be refrained particularly in pregnancy.

It is absorbed almost 100% after oral administration ; however, the food present in the stomach may delay its absorption considerably. It gets widely distributed in the body and even into the CSF. In plasma almost 98% is protein-bound. The volume of distribution (v_d^{ss}) stands at 0.9 mL.g^{-1}. Biotransformation in the liver helps in the elimination of almost 85% of the '*drug*'. An active and primary metabolite, **deacetylrifampin,** gets secreted right into the bile where it is effective therapeutically.

Cautions : (1) Risk of hepatotoxicity gets increased when used in combination with INH.

(2) Imparts distinct reddish-orange colour in urine, faeces, sweat, saliva and even tear.

(3) Soft transparent and clear '*contact lenses*' may be stained permanently.

2.6. Cardiac Steroids and Related Inotropic Drugs

A plethora of drugs are known that are found to be affecting the force of cardiac contractions. These drugs find there enormous use for the treatment of congestive heart failure by prolonging the life span of patients through pumping sufficient blood to sustain body requirements.

The inotropic drugs may be classified into the following *four* heads, namely :

(*a*) Cardiac Steroids

(*b*) Phosphodiesterase Inhibitors

(*c*) Adenylate Cyclase Stimulants

(*d*) Drugs that Enhance the Ca Sensitivity of Myocardial Contractile Proteins.

2.6.1. Cardiac Steroids

The use of the cardiac glycosides that act on the heart by causing atrioventricular conduction and vague tone has been discussed earlier. They belong to the class of *cardenolides*. Another cardiac glycoside class, the *bufadienolides,* does not warrant enough therapeutic importance. A typical example of bufadienolides is bufotalin, the aglycone portion of bufalin, a potent cardiac glycoside found in poisonous toad—skin secretions. The glycoside *k*–strophanthoside, obtained from the seeds of *Strophanthus kombe,* gives rise to the aglycone strophanthidine which also exerts cardiac contractions.

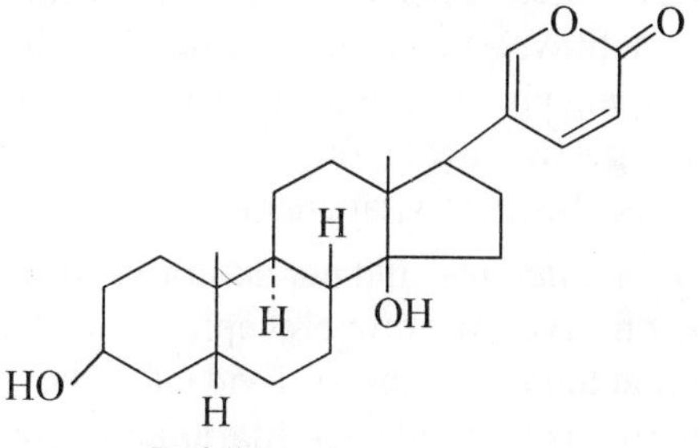

Bufalin 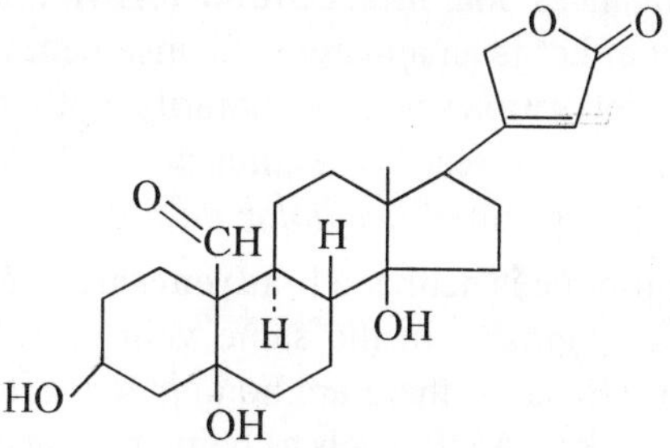   Strophanthidine

2.6.2. Phsophodiesterase Inhibitors

In the recent past a good number of 'nonglycoside inotropic agents have emerged as potentially beneficial drugs. This type of agents usually exert their action by the inhibition of a *c*AMP (cyclic adenosine monophosphate)-specific phosphodiesterase in the myocardium. The resulting inhibition ultimately leads to increased levels of *c*AMP, which through a complicated series of biochemical steps gives rise to an enhancement in intracellular Ca^{2+} and finally an elevation in muscle contractility.

A. *Amrinone* INN, BAN, USAN,

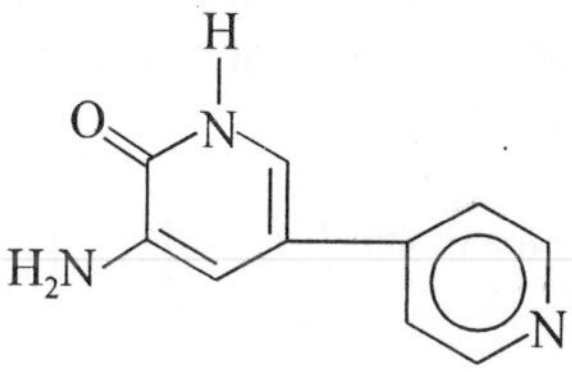

5-Amino [3, 4'-bipyridin]-6 (1H)-one ; Inocor[R] (Winthrop) ;

It is used for the treatment of heart failure. It has been given by injection and by mouth.

B. *Milrinone* INN, USAN ;

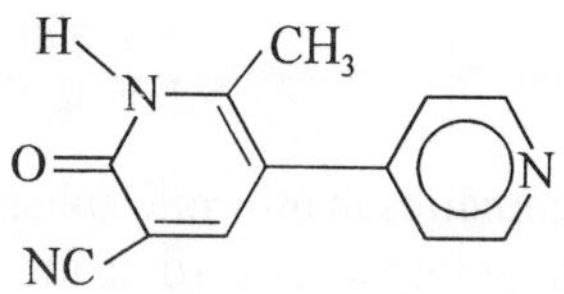

1, 6-Dihydro-2-methyl-6-oxo [3, 4'-bipyridin]-5carbonitrile ; Primacor[R] ;

It is relatively more potent than amrinone. It is found to be tolerated, having least apparent thrombocytopenia or gastrointestinal disturbance.

2.6.2.1. Mechanism of Action

The mechanism of action of the medicinal compounds described under section 2.6.2 shall now be dealt with individually as follows :

2.6.2.1.1. Amrinone

The '*drug*' categorically causes inhibition of the specific enzyme phosphodiesterase III and thereby enhances both **intarcellular cAMP** and **calcium** predominantly. Besides, in heart muscle the overall net effect is an apparent enhancement in contractility, whereas in vascular smooth muscle the outcome is relaxation solely. Sumararily, both these ensuing pharmacologic effects in unison contribute to augmentation in cardiac output particularly in **congestive heart failure** ; however, *ventricular unloading* as a result of *arteliolar dilatation* seems to be the more vital characteristic feature.

Amrinone is neither a β-**adrenergic agonist** as *dobutamine,* and nor an **inhibitor of Na⁺—K⁺–ATPase** as *digitalis.* In the same vein there is no observed α-adrenoreceptor or cholinoreceptor stimulation. Besides, there are no apparent effects on autonomic ganglia. Evidently, the '*drug*' gives rise to enhanced *c*AMP levels as being responsible for both observed **vasodilation** and **direct positive inotropy.**

Though the '*drug*' gets absorbed only to a small extent by the oral administration, it fails to show any significant therapeutic effect. Its volume of distribution (v_d^{ss}) stands at 1.2 L.kg⁻¹. It is conjugated in the liver upto 70%, while the balance gets excreted through the urine. The half-life of the '*drug*' is about 3.6 hours in normal subjects but extends upto 5–8 hours in patients having heart-failure. The duration of action ranges between 30–120 minutes.

2.6.2.1.1.2. Milrinone Acetate

The '*drug*' is found to be 20—30 folds more potent in comparison to amrinone as a positive inotropic agent, besides being somewhat more potent as an *arteriolar* and *venous dilator* probably due to the *two* basic alteration in functional moieties in it as shown below :

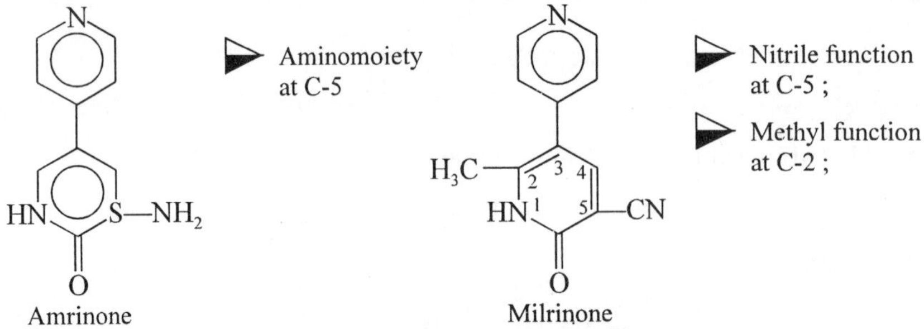

Amrinone
[Inocor(R)]

▷ Aminomoiety at C-5

Milrinone
[Primacor(R)]

▷ Nitrile function at C-5 ;
▷ Methyl function at C-2 ;

Thus, the '*nitrile*' and '*methyl*' moieties in milirone enhances its therapeutic potency significantly than amrinone. It has been observed that it improves cardiac index by 34% and lowers systemic vascular resistance by almost 31% in subjects having congestive heart failure. It is proved to be much superior to amrinone because it is not only orally active but also fails to cause either fever or *thrombocytopenia**.

The '*drug*' has a volume of distribution (v_d^{ss}) that stands at 0.4 L.kg⁻¹ and a mean half-life varying between 2—3 hours. It, however, gets secreted rapidly in the urine *via* active secretion.

2.6.3. Adenylate Cyclase Stimulants

Colforsin a diterpine directly stimulates adenylate cyclase or a structurally related protein. This causes an elevation of *c*AMP levels in the myocardium thereby activating the protein kinases and

---

*An abnormal decrease in number of the blood platelets (*Syn* : *Thrombopenia*).

enhances the intracellular Ca^{2+}. It also affects vasodilatation. It has been reported that a few physiologic effects caused by Forskolin are not mediated by cAMP.

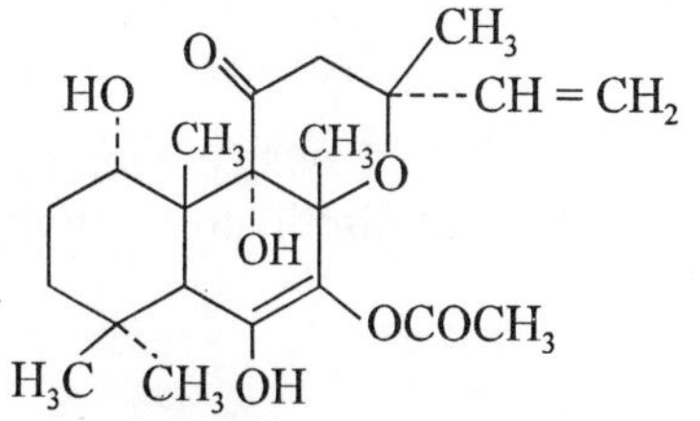

Colforsin

Colforsin [*Syn* : Boforsin ; Forskolin ; Forscolin ;] exerts its action by causing stimulation of adenylate cyclase. It has been observed that the '*drug*' possesses positive inotropic and bronchodilator effects.

2.6.4. Drugs that Enhance the Ca^{2+} Sensitivity of Myocardial Contractile Proteins

After the spectacular revelation of *amrinone*'s inotropic action a vigorous global effort was made in search of other nonsteroidal inotropic drugs. These drugs seem to enhance the effect of existing Ca^{2+} levels and subsequently cause an inotropic effect.

A few typical examples are discussed here, namely : Pimobendan ; Sulmazole ;

A. *Pimobendan* INN, USAN ;

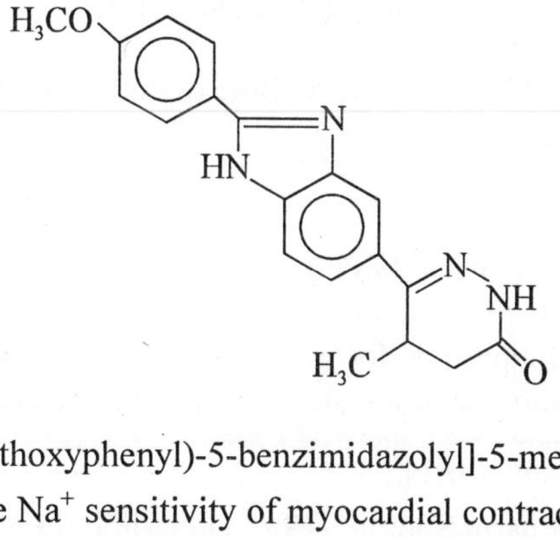

4, 5-Dihydro-6 [2-(*p*-methoxyphenyl)-5-benzimidazolyl]-5-methyl-3(2H)-pyridazinone ;

It is found to increase the Na^+ sensitivity of myocardial contractile proteins.

B. *Sulmazole* INN, USAN ;

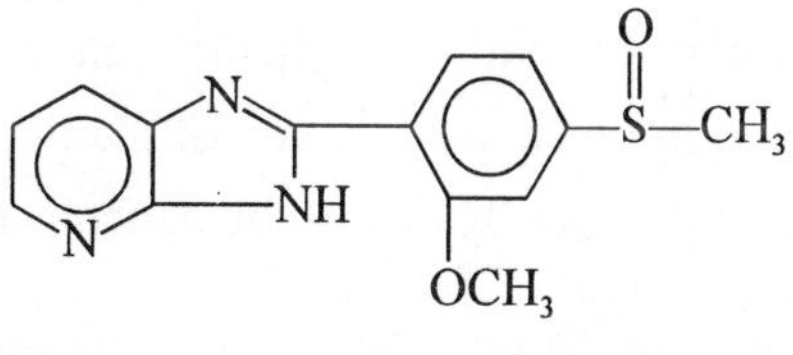

2-[2-Methoxy-4-(methylsulfinyl) phenyl]-3H-imidazo [4, 5-*b*] pyridine ;

Sulmazole, an imidazopyridine derivative also increases Na^+ sensitive of myocardial contractile proteins.

2.6.4.1. Mechanism of Action

The mechanism of action of the medicinal compounds described under section 2.6.4 shall be treated as below :

2.6.4.1.1. Pimobendan

The '*drug*' is a phosphodiesterase inhibitor having predominant calcium-sensitizing characteristic properties. It also exerts positive inotropic and vasodilatory activity ; and, therefore, has been tried in the management of heart failure.

2.6.4.1.2. Sulmazole

The cardiotonic agent (*a*) was shown to produce **'bright visions'** in some patients, which suggested that it was entering the CNS. It was indeed supported by the fact that the log P value of (*a*) was 2.5 g. In order to check the '*drug*' from entering the CNS directly, the 4-OCH$_3$ moiety was duly replaced with a 4-S(O) CH$_3$ group. This specific moiety is approximately of the same size and magnitude as that of the methoxy moiety, but certainly more hydrophilic in nature. Sulmazole has a log P value of 1.17 (about 50% less).

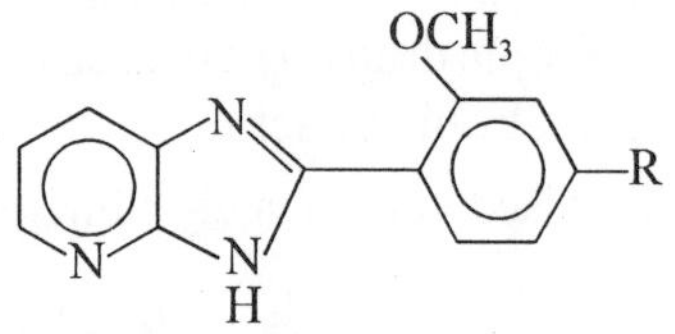

(*a*) R = OCH$_3$
(*b*) R = S(O)CH$_3$ Sulmazole

Thus, this '*drug*' turned out to be too hydrophilic in status so as to enter the CNS ; and, therefore, was devoid of CNS- side-effects.

2.7. Heparin

Heparin is a mucopolysaccharide, having a molecular weight ranging from 6,000 to 20,000, made up of several repeating units of glucuronic acid and sulphated glucosamine. It is comparatively a strong acid that forms water–soluble salts, for instance heparin sodium. The presence of a number of ionizable sulphate moieties renders the molecule a strong electronegative charge.

A. Heparin INN ; *Heparin Sodium* USAN ;

Heparin B.P. ; Int. P. ; Heparin Sodium USP ; Heparin Lock Flush[R] (Abbott) ; Lipo–Hepin[R] (Riker) ; Liquaemin Sodium[R] (Organon) ; Liquemin[R] (Roche) ;

It is prepared in large–scale from liver and lung by adopting the procedure advocated by Kuizenga and Spaulding and purifying the isolated heparin by suitable means.

Heparin may be administered intravenously by any of the two following methods, namely ;

(*a*) the continuous infusion method or by deep subcutaneous injection, and

(*b*) the intermittent injection method.

However, the continuous infusion method is generally performed as it affords a more constant anticoagulating activity besides offering lower incidence of bleeding complications. Nowadays, a constant–rate infusion pump is mostly recommended.

Dose : Usual, full-dose, parentral, the following amounts, as indicated by prothrombin–time determinations : IV, 10,000 USP heparin units initially, then 5,000 to 10,000 U every 4 to 6 hours ; infusion ; 20,000-40,000 U/L at a rate of 1000 U/ hour over a 24-hour duration ; subcutaneous ; 10,000 to 20,000 U initially, then 8,000 to 10,000 U every 8 hours or 15,000 to 20,000 U every 12 hours ; Usual, pediatric dose ; IV injection ; 50 U/kg of body weight initially, then 50-100 U/kg of body wight every 4 hours ; 50 U/kg of body weight intiially, followed by 100 U/kg, added and observed every 4 hours.

Mechanism of Action. The '*drug*' combines with **AT III***. The resulting complex then interacts with certain activating clotting factors, such as : Factors IX, X, XI and XII, in order to prevent the conversion of *prothrombin to thrombin*. It has been observed that in high concentrations the complex invariably interacts with thrombin and thereby inhibits its effects to promote conversion of *fibrinogen* into *fibrin*. In short, it inhibits the aggregation of platelets. It is indeed a fast acting drug substance which has the overwhelming plus point of being a naturally occurring substance.

2.8. Radiosensitizer

The value and importance of radioisotopes for the imaging and killing of cancerous tumours has gained cognizance for more than three decades. A significant leap forward was duly accomplished with the incorporation of monoclonal antibodies that may be employed to direct imaging or cytotoxic agents to cancer sites with abundant selectively. The most effective and successful treatment of cancer with the aid of radionucleotides is critically dependent on the attainment of a high target (*i.e.,* the tumour and secondary metastases) to non-target (*i.e.,* healthy tissue and organs) ratio. In fact, the two most important and specifically radiosenitive tissues in humans are identical as the intestinal mucosa and the bone marrow, for which the target to non-target ratios of the order of 20 : 1 must be maintained so as to avoid the lethal consequences of high radiation doses.

Very high degree of selectivity in targeting the tumour cells can only be accomplished with the aid of an antibody that binds promptly and strongly to the tumour–associated compounds (antigens) that clears swiftly from the normal tissue and which has been irreversibly radiolabelled with the appropriate radioisotoppe. Molecular biologists are effectively tackling the first two issues to constantly devising antibodies, both whole and fragments, whose rates of catabolism, clearance and uptake besides half-life in the tumour relative to other tissues may be mentioned for the desired *in vivo* duration. However, the latter problem is solely addressed by the practising chemist, who in turn is required to design, synthesize and link to the antibody an appropriate functionalized ligand whose structure is determined by the choice of radioisotopoe.

Radioisotopes may be conveniently divided into *two* categories, namely ;

(*a*) Therapeutic Radioisotopes

(*b*) Imaging Radioisotopes

2.8.1. Therapeutic Radioisotopes

The radioisotopes that are used exclusively for the treatment of various diseases are generally known as therapeutic radioisotopes. The following Table contains the different ranges of therapeutic radioisotopes :

---

*Antithrombin III.

Isotope	Half-life/h	Dose Rate/ radh^{-1} per Gg^{-1}	Total electron dose/rad from 1 Gg^{-1} at t_∞	Mean range in tissue/mm
^{90}Y	64	1.96	180	3.9
^{67}Cu	62	0.58	30	0.2
^{111}Ag	170	0.82	198	1.1
^{131}I	193	1.22	115	0.4
^{161}Tb	166	0.50	101	0.3
^{188}Re	17	1.91	44	3.3
^{199}Au	75	0.53	47	0.1

In radioimmunotherapy it is pertinent to select a specific type of isotope that should be suitable for the tumour morphology. This fact may be substantiated with the help of the following two examples :

Example-1 : *Long-range emitters :*

The densely packed lymphomas or hepatacellular carcinoma when crossfixed from the long-range emitters like ^{90}Y—it registers an appreciable contribution to the dose to the tumour. This particular aspect is very important because the radiolabelled antibody may not be in a position to penetrate densely packed tumours efficiently or may not bind uniformly to tumours where the level of surface antigen is modest.

Example-2 : *Short-range emitters :*

Interestingly the smaller tumours such as the leukemias can be more effectively treated with the short-range emitters like ^{199}Au.

2.8.2. Imaging Radioisotopes

The ideal radionucleotides or radiopharmaceuticals for organ imaging should possess the following characteristic features, namely :

(*a*) decay by γ-radiation alone and having an energy of about 200 keV ;

(*b*) possess half-life in ranging between 6 to 12 hours ;

(*c*) must be readily available ;

(*d*) should be chemically versatile *i.e.,* be easily incorporated into carrier molecules ;

(*e*) produce radiopharmaceuticals that are fairly stable both *in vitro* and *in vivo.*

Ironically enough, not a single radionucleotide is known till date which compiles all the five above mentioned criteria. There are in all about **2000 imaging radioisotopes** that are used today for organ imaging purposes.

The following table contains a few important ranges of imaging radiopharmaceuticals :

Imaging Radioisotopes :

Isotope	Half-life/h	Photon energy/keV
^{111}In	68	171
^{99m}Tc	6.02	141
^{67}Ga	80	184
^{131}I	193	364
^{123}I	13.2	159
^{64}Cu*	12.8	511

*For use in positron emission tomography (PET)

Two typical examples are described here :

2.8.2.1. Technetium-99m (^{99m}Tc)

Because it has a short-life and can be administered in relatively large doses, and because the energy of its γ-emission is readily detected, Technetium-99m is very widely employed, either as the pertechnetate or in the form of various labelled compounds, particles and colloids for scanning bone and organs. such as the brain, liver, lung, spleen and thyroid. ^{99}Tc is used in over 80% of imaging procedures.

Dose : Usual, 37 to 185 MBq (1 to 5 millicuries) in the investigation of the liver and spleen and upto 740 MBq (20 millicuries) for bone marrow). [mCi = Millicuries]

2.8.2.2. Gallium-67 (^{67}Ga)

This imaging radiopharmaceutical, in the form of its citrate, is concentrated in some tumours of the lymphatic system and other soft tissues. Localization has been reported in inflammatory lesions and Gallium-67 has been employed effectively for the diagnosis of infection.

Dose : For tumours of lymphatic system and soft tissues : 55.5 to 92.5 MBq (1.5 to 2.5 millicuries) to visualize tumour by scanning techniques.

2.9.   Cromakalim

Cardiovascular diseases is still responsible as the cardinal cause of death in the technologically advanced and developed countries and also supported by the fact that hypertension places a person on a high risk of heart attacks and strokes. About 1/5th of the population suffers from hypertension, the causes of which are not yet fully understood, but various factors for instance : genetic history, age, diet, stress and strain and smoking may be involved either fully or partially. While mild hypertension can be arrested by altering the patient's lifestyle, but non-treatment of acute hypertension may ultimately lead to enhanced risk of stroke, kidney failure and impaired vision.

Hypertension the quite and dreadful disease has been treated effectively over the past two decades with the aid of different classes of drugs, for instance ;

(*a*) β-Blockers (or β-adrenoreceptor antagonist)

(*b*) Calcium channel blockers and

(*c*) Angiotensin converting enzyme (ACE) inhibitors

More recently a new class of drugs for treating hypertension has been discovered that relaxes smooth muscle by activating potassium channels. Precisely this novel mechanism involves an enhancement in the outward movement of potassium ions through channels in the membranes of vascular smooth muscle cells, ultimately leading to relaxation of the smooth muscle. Hence, these compounds may be termed as potassium channel activators.

A. *Cromakalim*

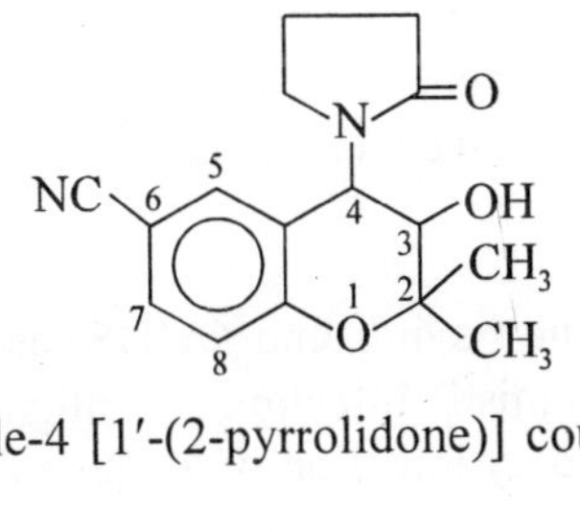

2-Dimethyl-3-hydroxy-6-nitrile-4 [1'-(2-pyrrolidone)] coumarine ;

Synthesis

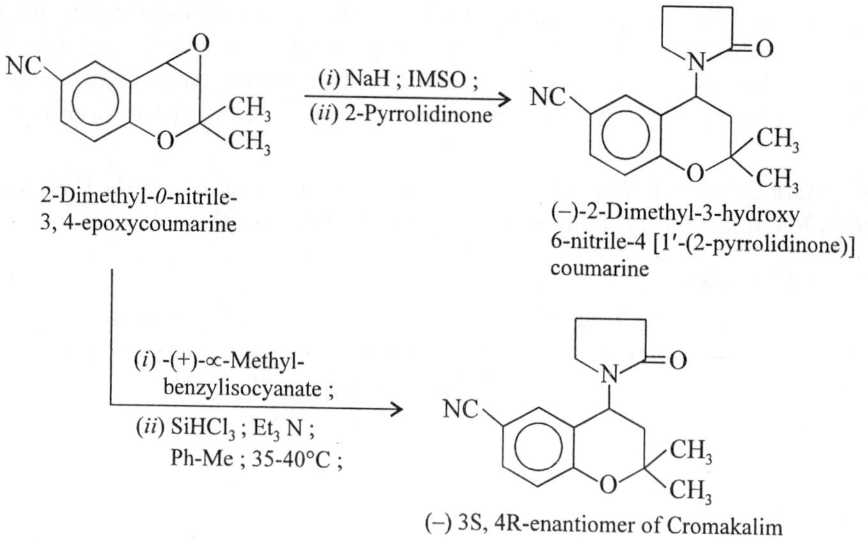

2-Dimethyl-*0*-nitrile-
3, 4-epoxycoumarine

(i) NaH ; IMSO ;
(ii) 2-Pyrrolidinone

(−)-2-Dimethyl-3-hydroxy
6-nitrile-4 [1′-(2-pyrrolidinone)]
coumarine

(i) -(+)-∝-Methyl-
benzylisocyanate ;

(ii) SiHCl₃ ; Et₃ N ;
Ph-Me ; 35-40°C ;

(−) 3S, 4R-enantiomer of Cromakalim

It is prepared by treating 2-dimethyl-6-nitrile-3, 4-epoxy coumarine *first* with sodium hydride in dimethylsulphoxide and *secondly* with 2-pyrrolidinone to obtain 2-dimethyl-3-hydroxy-6-nitrile-4 [1′ (2-pyrrolidinone)] coumarine. The resolution of the resulting cromakalim is achieved *via* the S-α-methylbenzyl carbamate to get the (-)-3S, 4R enantiomer.

It is used as an antihypertensive agent. It has also shown the ability to relax human bronchial tissue. It is an effective inhibitor of histamine-induced bronchoconstriction. The long plasma half-life of cromakalim suggests that it would be specifically beneficial in the treatment of patients suffering from nocturnal asthama.

Dose : Hypertension : oral, 1.5 mg ; Asthma ; oral, 0.5 mg at night (at the low oral dose employed, no reduction in blood pressure is observed.

Future Prospects

The future perhaps lies in the design and synthesis of selective potassium channel modulators for treating the various ailments associated with smooth muscle function. It is, however, quite evident that potassium ion channels showing diverse characteristics exist in plethora of cells in the human body and because these channels are particularly responsible in modulating cellular activity, it is very much likely that drugs which are able to influence different categories of these channels may be developed in the near future for treating many other diseases effectively. In short, cromakalim has shown to be a versatile drug having potential usage in a number of diseases. It is further anticipated that this potential will be meaningfully exploited with its active enantiomer (-) 3S, 4R cromakalim.

2.10.   Drugs to Combat AIDS

The acquired immunodeficiency syndrome (AIDS) is a condition characterized by the development of life-threatening opportunistic infection or malignancies with severe depression of the T-cell mediated immune system caused by infection with human immunodeficiency virus (HIV).

AIDS was first and foremost described as a specific entity in the US in 1981, and its frequency and mortality since have increased tremendously in geometrical proportion. In the US alone by 1991, approximately 1,20,000 cases were reported in adults and adolescents and nearly 2000 cases in children.

There is a global epidemic of AIDS perdominantly in the US, Europe, South America, Canada, Africa and South East Asea. These HIV centres rare broadly responsible for transmission of the virus to others. Unless and until serious and prompt corrective measures are not taken immediately by the respective governments in terms of sex-education, dreadfulness of this disease and meaningful precautionary measures to contain the spreading of AIDS, there is every possibility that it may even penetrate into relatively healthier countries of the world.

Various therapeutic agents in the form of specific drugs, immunoglobulins, vaccines and photochemical procedures are known to the world to combat the ever-increasing menace of AIDS. They will be discussed briefly in this chapter.

For instance : Zidovudine, Carbovir, AIDS-immunoglobulins, AIDS-vaccines, and HIV-drug under the spotlight.

A. *Zidovudine*

Thymidine, 3'-azido-3'-deoxy ; Azidothymidine ; AZT ; Retrevir$^{(R)}$ (Burroughs Wellcome) ;

Zidovudine has shown activity against human immunodeficiency virus (HIV) ; consequently, it is used for the treatment of AIDS and AIDS-related complex (ARC). It positively enhances the survival and improves the quality of life of patients with complications such as severe weight loss, fever, pneumocystosis, herpes zoster, herpes or thrush. Because it crosses the blood-brain-barrier (BBB), it has a favourable response on the neurological symptoms of AIDS.

Zidovudine is a nucleoside analogue structurally similar to thymidine and hence it is also known as **azidothymidine**.

It may be given to symptomatic patients with HIV-infection (with blood CD_4 counts of less than 200 per mm^3 or 200 to 500 per mm^3 and rapidly falling).

Its oral bioavailability ranges between 52 to 75%. The plasma-protein binding is 34 to 38%. The drug is mostly metabolized in the liver with a half-life ranging between 0.8 to 1.9 hours. Only 14% of the drug is eliminated as such through the urine.

Dose : Adult, oral, asymptomatic HIV-infection, initially 100 mg every 4 hour, while awake (500 mg a day), after 1 month dose may be reduced to 100 mg every 4 hour ; intravenous infusion, 1 to 2 mg/kg infused over 1 hr. every 4 hr. around the clock (6 times a day).

Mechanism of Action. The '*drug*' exerts its action by creeping into the retroviral DNA *via reverse transcriptase* to render a more or less nonsense sequence which essentially breaks the ensuing DNA chain synthesis. However, it has been established that the *reverse transcriptase* is nearly 100 folds more susceptible to the '*drug*' in comparison to the **mammalian DNA polymerase.** Zidovudine exhibits its therapeutic activity specifically against the human immuno deficiency virus. Consequntly, it is employed for the treatment of AIDS and AIDS-related complex (ARC). As it happens to cross the blood-chain barrier (BBB), it demonstrated a rather favourable effect on the prevailing **neurological symptoms of AIDS.** However, in the course of prolonged therapy resistance may also take place gradually.

Another school of thought suggests that *zidovudine* gains its entry into the host cells by means of diffusion, and in turn gets *phosphorylated* by the **cellular thymidine kinase.** Thus, the enzyme thymidylate kinase eventually helps in the conversion of the *monophosphate* into the *diphosphate* and the *triphosphates* respectively. It is, however, pertinent to state here that the rate determining step is actually **'the conversion to the diphosphate'** ; and, therefore, perhaps very high levels of *monophosphorylated AZT* invariably get accumulated into the cell. Besides, relatively low levels of diphoshpate and triphosphate are also present. Importantly, AZT-triphosphate affords competitive inhibition of **reverse transcriptase** specifically with regard to **thymidine triphosphate.**

Furthermore, the 3'-azido functional moiety [$—N = \overset{\oplus}{N} = \overset{\ominus}{N}$] predominantly prevents the formation of a **5′, 3′-phosphodiester bond** ; and, therefore, AZT gives rise to DNA-chain termination effectively, thereby producing an **incomplete proviral DNA** substantially.* Besides, AZT-monophosphate also executes competitive inhibition of the specific *cellular thymidylate kinase*, thereby lowering the *intracellular levels of thymidine triphosphate*. It has been reported that the '*point mutations*' at multiple sites prevailing in the reverse transcriptase invariably causes resistance that may ultimately lead to a rather lower degree of affinity for AZT.**

B. Carbovir

---

*Furman PA *et al.* Proc. Nale. Aead. Sci. USA., *873* : 8333, 1996.

Richman DD *et al. J. Infect. Dis.,* **164, 1075, 1991.

Synthesis

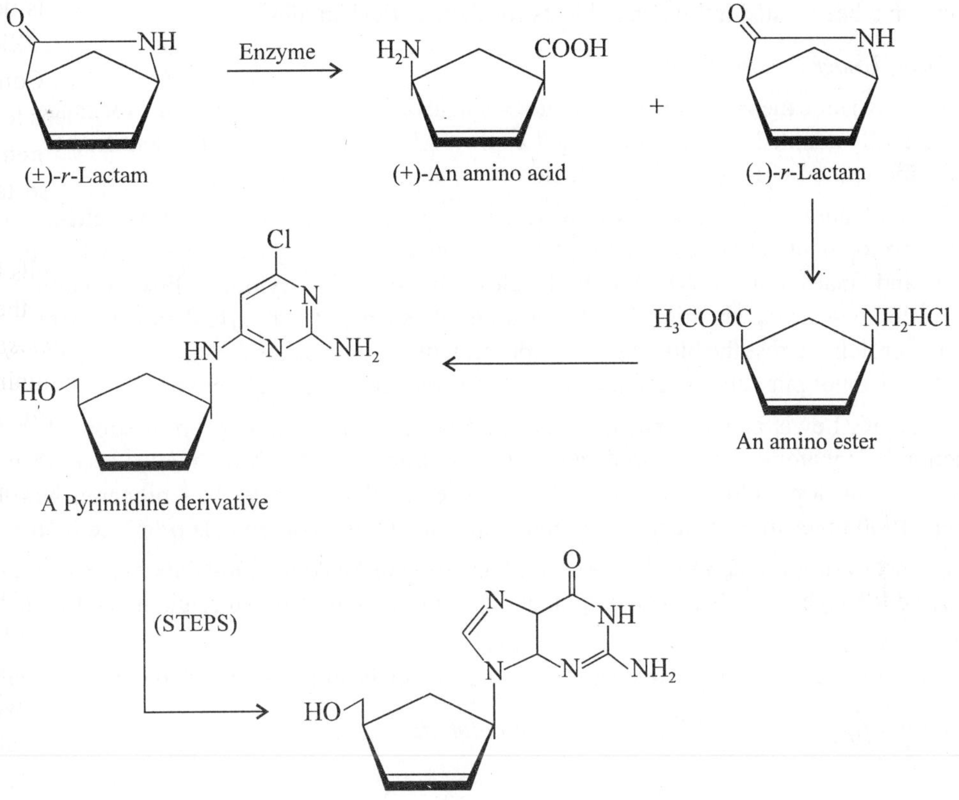

(±)-*r*-Lactam   (+)-An amino acid   (−)-*r*-Lactam

An amino ester

A Pyrimidine derivative

(STEPS)

Carbovir

The racemic mixture of the γ-lactam is hydrolyzed enantioselectively to give the corresponding amino acid in tis *dextro*-form and the lactam, (-)-γ-lactam by an enzyme from a *Pseudomonas* microorganisms. The optically pure lactam, (-)-γ-lactam is ring-opened chemically to give the corresopnding amino ester. This compound is converted *via* the pyrimidine derivative into the desired guanosine analogue *i.e.*, carbovir.

It is a cyclopentane derivative that has been prepared in optically active form by employing a chemoenzymatic total azymmetric synthesis.

Carbovir has attracted a lot of interest as a potent inhibitor of the AIDS virus *in vivo*.

Mechanism of Action. The '*drug*' is a nucleoside analogue structurally related to guanosine having distinct antiviral activity against HIV-1. Its acts as an inhibitor of viral reverse transcryptase and is under investigation in the treatment of AIDS.

C. *AIDS-Immunoglobulins* :

AIDS-immunoglobulins are also known as **HIV-immunoglobulins**. More recently, AIDS immunoglobulins preparations containing HIV-neutarlizing antibodies have been prepared from the plasma of asymptomatic HIV-positive subject. They are being tried for passive immunization in patients with AIDS or AIDS-related complex (ARC).

D. *AIDS-Vaccines :*

These are also referred to as **HIV-vaccines**. In fact, a large number of prototype vaccines against the AIDS have been tested in human subjects for their critical studies.

E. *HIV-Drug Under the Spotlight :*

Photodynamic therapy (PDT), the photochemical procedure now being used for cancer treatment, was successfully adapted to help in the fight against AIDS. A dye has been identified that can latch onto HIV and related viruses and knock them out when a light is shone on them.

The collaborative research of **David Lewis**-an organic chemist and **Ron Utecht**-a biochemist, led to the development of a series of hydrophobia compounds which get among the lipids of the viruse's envelopes and finally link to trytophan molecules. The loose dye-trytophan linkage upon irradiation gets converted to a strong chemical bond. It is an established fact that trytophan is responsible for the transport of proteins across the biological membrane, it ultimately makes the viral envelope impermeable and its DNA cannot migrate out and thus the virus is rendered non-ineffective.

Later on, Lewis and Utecht, tried to improve the single oxygen production of the dye by introduction heavy atoms onto the molecule, such as bromine. In this way, they succeeded in turning the soft-virus envelope into a concrete jacket. It revealed that the new dyes prepared with bromine atom were 1000 times more potent and effective than the earlier compounds prepared by them.

The heavy-atom containing dyes are based on 3-bromo-4-alkylamino-N-alkyl-1, 8-naphthalimide. These are yellow dyes that absorb blue light, and are similar to the optical brighteners used in washing powders.

The bromine-containing dyes were prepared by Lewis and Utecht which are described here :

A. *3-Bromo-4 (hexylamino)-N-hexyl-1, 8-naphthalimide*

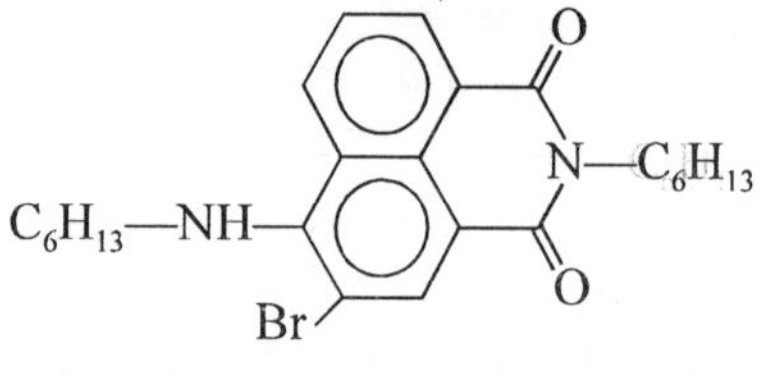

It is a monomeric naphthalimide. It acts against herpes simplex virus (HSV) at concentrations around 250 nM.

B. *Bis-amide of 3-Bromo-4 (hexylamino)-N-hexyl-1, 8-naphthalimide*

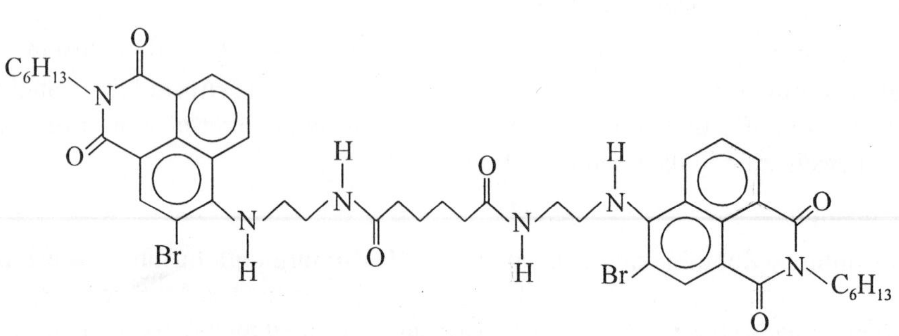

A Bis-Amide

It is a dimeric naphthalimide. It is effetive against HSV at concentrations below 100 nM.

In short, it may prove to be a good news for the potential therapy, because HSV is rather more difficult to kill than HIV.

Probable Questions for B. Pharm. Examinations

1. Discuss the role of **Prostaglandins** and **Eicosanoids** to combat the following diseases :

    (*a*) Gastric ulceration

    (*b*) Management of congenital heart disease

    (*c*) Haemolytic-uramic syndrome and anticoagulant in the dialysis procedure.

2. (*a*) What is the role of **antilipedmic drugs** ?

    (*b*) Enumerate the various factors causing **atherosclerosis.**

    (*c*) Give the structure, chemical name and uses of **three** potent **antilipedemic drugs.**

3. How would you classify **Hormonal Antagonists.** Give the structure, chemical name and uses of at least **one** potent drug from each category.

4. Write a brief note on the following :

    (*a*) Probucol

    (*b*) Nitromifene citrate

    (*c*) Flutamide

    (*d*) Chlormadinone

    (*e*) Spironolactone

    (*f*) Gestrinone.

5. Discuss the synthesis of the following :

    (*a*) Propylthiouracil from 3-oxo-caproate

    (*b*) Isoniazid

    (*c*) Cromakalim

    (*d*) Ethambutol hydrochloride.

6. Give a comprhensive account of the **antimycobacterial drugs.** Support your answer with suitable examples.

7. Describe any **two** of the following **cardiac steroids and inotropic drugs :**

    (*a*) cardiac drugs

    (*b*) phosphodiesterase inhibitors

    (*c*) adenylate cyclase stimulants

    (*d*) drugs that emulate the Ca^{2+} sensitivity of myocardial contractile proteins.

8. What are **biosensitizers ?** Classify them by citing a few appropriate examples in support of your answer.

9. Discuss **cromakalim** as an important potassium channel activator. Give its synthesis and future prospects.

10. (*a*) Give a brief account on the **global epidemic of AIDS.**

     (*b*) Discuss the synthesis of **carbovir.**

     (*c*) HIV-drug under the spotlight.

RECOMMENDED READINGS

1. JN Delgado and WA Remers '*Wilson and Gisvold's Textbook of Organic Medicinal and Pharmaceutical Chemistry',* 9th edn., Philadelphia, JB Lipincott Company (1991).

2. *Chemistry in Britain,* 27(50), 1991, p-439.

3. *Chemistry in Britain,* 27(6), 1991, p-518-520

4. *Chemistry in Britain,* 29(5), 1993, p-376.

5. *Martindale : The Extra Pharmacopoeia,* (31st Edn.), London, the Royal Pharmaceutical Society, London (1996).

6. Lemke *et al.* (Eds.) *Foye's 'Principles of Medicinal Chemistry',* 5th edn., Lippincott Williams & Wilkins, New York, 2004.

7. MC Griffiths (Ed.) *USAN and the USP Dictionary of Drug Names,* Rockville, United States Pharmacopeial Convention Inc., (1985).

8. *Index Nominum,* Zurich, Swiss Pharmaceutical Society (1982).

9. *Remington : 'The Science and Practice of Pharmacy',* 20th edn., Vol. I & II, Lippincott Williams & Wilkins, New York, 2000.

10. Gringauz, A., *Introduction to Medicinal Chemistry,* Wiley-VCH, New York, 1997.

Index

C

F

Flufenisal 222, 237, 469

Flukes 572

Fluorouracil 44, 692, 693

Flurbiprofen 458, 460, 459

Flurothyl 204, 210

Fluroxene 57, 58, 70

Folic acid 45, 514, 529

Fourth generation cephalosporins 655

Free fatty acids (FFA) 742

Friedal-Craft's reaction 401, 726

Friedel-Craft's acylation 205

Frusemide 400

FTIR-spectrophotometry 15

Full-salol principle 14

Fumarate reductase 577

Furosemide 399, 400, 402

G

G, ampicillin 466

G proteins 163, 249

γ-aminobutyric acid 192

G-protein-activated K+ channel 192

GABA 140

GABA agonist 580

GABA-transaminase 163, 165

GABA_A chloride channel 163

GABA_A receptors 141, 163

GABA_A-mediated chloride ion conductance 140

GABA_B receptors 163

Galactose 627

Gallamine 171

Gallamine triethiodide 169, 171, 189

Gallium-67 (^{67}Ga) 763

Gametocyte stage 551

Gamma-aminobutyric acid (GABA) 160, 186

Gangionic active moiety 10

Ganglionic blocking agent 9, 41, 309, 360, 361, 364

Ganglionic moiety 10

Gelsemine 360

Gemfibrozil 742, 743

General anaesthetics 55, 67, 70, 73, 75

General chracteristics of the tetracyclines 665

Genetic engineering 19

Genins 623

Genomics 13

Geographical incidence 673

Geometrical isomeric forms 616

Geometrical isomers 51

Gestogens 620

Gestogens (The Corpus Luteum Hormones) 605

Ginger 491

Gitoxigenin 623

Glimepiride 593

Glipizide 591, 593

Globin 586

Glomerular filtration 370

Glomerulonephritis 373

Glucagon 584

Glucagon secretion 592

Glucose 627

Glucuronidation 191

Glucuronides 434, 619

Glutamine antagonists 694

Glutaminergic transmission 140

Glutethimide 116, 138, 139, 145, 146

Glyburide 592, 593

Glycerol monoethers and analogues 178

Glycoamylase 597

Glycogenolysis 209

Glycopyrrolate 356, 357

Glycopyrrolate 478

Glycopyrronium bromide 354, 356, 357, 359, 360

Glycosides 697, 702

Glycylcyclines 670

Glycyrrhiza 491, 492

Glycyrrhizinic acid 492

Glypizzide 592

Gold compounds 451, 461, 463, 464

Gold sodium thiomalate 464

Gold thiomalic acid 464

Gramicidin 658

M

P